Lecture Notes in Computer Science 3726

Commenced Publication in 1973
Founding and Former Series Editors:
Gerhard Goos, Juris Hartmanis, and Jan van Leeuwen

Editorial Board

David Hutchison
  Lancaster University, UK
Takeo Kanade
  Carnegie Mellon University, Pittsburgh, PA, USA
Josef Kittler
  University of Surrey, Guildford, UK
Jon M. Kleinberg
  Cornell University, Ithaca, NY, USA
Friedemann Mattern
  ETH Zurich, Switzerland
John C. Mitchell
  Stanford University, CA, USA
Moni Naor
  Weizmann Institute of Science, Rehovot, Israel
Oscar Nierstrasz
  University of Bern, Switzerland
C. Pandu Rangan
  Indian Institute of Technology, Madras, India
Bernhard Steffen
  University of Dortmund, Germany
Madhu Sudan
  Massachusetts Institute of Technology, MA, USA
Demetri Terzopoulos
  New York University, NY, USA
Doug Tygar
  University of California, Berkeley, CA, USA
Moshe Y. Vardi
  Rice University, Houston, TX, USA
Gerhard Weikum
  Max-Planck Institute of Computer Science, Saarbruecken, Germany

Laurence T. Yang   Omer F. Rana
Beniamino Di Martino   Jack Dongarra (Eds.)

High Performance Computing and Communications

First International Conference, HPCC 2005
Sorrento, Italy, September 21-23, 2005
Proceedings

 Springer

Volume Editors

Laurence T. Yang
St. Francis Xavier University
Department of Computer Science
P.O. Box 5000, Antigonish, B2G 2W5, NS, Canada
E-mail: lyang@stfx.ca

Omer F. Rana
Cardiff University
School of Computer Science
Queen's Buildings, Newport Road, Cardiff, CF24 3AA, Wales, UK
E-mail: o.f.rana@cs.cardiff.ac.uk

Beniamino Di Martino
Seconda Università di Napoli
Dipartimento di Ingegneria dell'Informazione
Facoltà di Studi Politici e per l'Alta Formazione Europea "Jean Monnet"
Real Case dell'Annunziata via Roma 29, 81031 Aversa (CE), Italy
E-mail: beniamino.dimartino@unina.it

Jack Dongarra
University of Tennessee
Innovative Computing Laboratory
Computer Science Department
1122 Volunteer Blvd., Knoxville, TN 37996-3450, USA
E-mail: dongarra@cs.utk.edu

Library of Congress Control Number: 2005932548

CR Subject Classification (1998): D, F.1-2, C.2, G.1-2, H.4-5

ISSN   0302-9743
ISBN-10   3-540-29031-1 Springer Berlin Heidelberg New York
ISBN-13   978-3-540-29031-5 Springer Berlin Heidelberg New York

This work is subject to copyright. All rights are reserved, whether the whole or part of the material is concerned, specifically the rights of translation, reprinting, re-use of illustrations, recitation, broadcasting, reproduction on microfilms or in any other way, and storage in data banks. Duplication of this publication or parts thereof is permitted only under the provisions of the German Copyright Law of September 9, 1965, in its current version, and permission for use must always be obtained from Springer. Violations are liable to prosecution under the German Copyright Law.

Springer is a part of Springer Science+Business Media

springeronline.com

© Springer-Verlag Berlin Heidelberg 2005
Printed in Germany

Typesetting: Camera-ready by author, data conversion by Scientific Publishing Services, Chennai, India
Printed on acid-free paper   SPIN: 11557654   06/3142   5 4 3 2 1 0

Preface

Welcome to the proceedings of the 2005 International Conference on High Performance Computing and Communications (HPCC 2005) which was held in Sorrento (Naples), Italy, September 21–23, 2005.

With the rapid growth in computing and communication technology, the past decade has witnessed a proliferation of powerful parallel and distributed systems and an ever-increasing demand for the practice of high performance computing and communication (HPCC). HPCC has moved into the mainstream of computing and become a key technology in determining future research and development activities in many academic and industrial branches, especially when the solution of large and complex problems must cope with very tight timing schedules. The HPCC 2005 conference provided a forum for engineers and scientists in academia, industry, and government to address all resulting profound challenges, and to present and discuss their new ideas, research results, applications and experience on all aspects of high performance computing and communications.

There was a very large number of paper submissions (263), not only from Europe, but also from Asia and the Pacific, and North and South America. All submissions were reviewed by at least three Program or Technical Committee members or external reviewers. It was extremely difficult to select the presentations for the conference because there were so many excellent and interesting submissions. In order to allocate as many papers as possible and keep the high quality of the conference, we finally decided to accept 76 regular papers and 44 short papers for oral presentations. We believe that all of these papers and topics not only provided novel ideas, new results, work in progress and state-of-the-art techniques in this field, but also stimulated the future research activities in the area of high performance computing and communications.

The exciting program for this conference was the result of the hard and excellent work of many others, such as program vice-chairs, external reviewers, and Program and Technical Committee members. We would like to express our sincere appreciation to all authors for their valuable contributions and to all Program and Technical Committee members and external reviewers for their cooperation in completing the program under a very tight schedule. We were also grateful to the members of the Organizing Committee for supporting us in handling the many organizational tasks, and to the keynote speakers for accepting to come to the conference with enthusiasm.

Last but not least, we hope that the attendees enjoyed the conference program, and the attractions of the city of Sorrento and its peninsula, together with the social activities of the conference.

July 2005
Laurence T. Yang, Omer F. Rana
Beniamino Di Martino, Jack Dongarra

Organization

HPCC 2005 was organized by the Second University of Naples, Dept. of Information Engineering and Facoltá di Studi Politici e per l'Alta Formazione Europea e Mediterranea "Jean Monnet", in collaboration with CREATE, St. Francis Xavier University, and the University of Cardiff.

Executive Committee

General Chairs	Jack Dongarra, University of Tennessee, USA
	Beniamino Di Martino, Second University of Naples, Italy
Program Chairs	Laurence T. Yang, St. Francis Xavier University, Canada
	Omer F. Rana, Cardiff University, UK
Program Vice-Chairs	Geyong Min, University of Bradford, UK
	José E. Moreira, IBM Research, USA
	Michael Gerndt, Technical University of Munich, Germany
	Shih-Wei Liao, Intel, USA
	Jose Cunha, New University of Lisbon, Portugal
	Yang Xiao, University of Memphis, USA
	Antonio Puliafito, University of Messina, Italy
	Dieter Kranzlmueller, Johannes Kepler University Linz, Austria
	Manish Parashar, Rutgers University, USA
	Rajkumar Buyya, University of Melbourne, Australia
	Antonino Mazzeo, Università degli Studi di Napoli Federico II, Italy
	Umberto Villano, Università del Sannio, Italy
	Ian Taylor, Cardiff University, UK
	Barbara Chapman, University of Houston, USA
	Luciano Tarricone, University of Lecce, Italy
	Domenico Talia, DEIS, Italy
	Albert Zomaya, University of Sydney, Australia
Steering Committee	Beniamino Di Martino, Second University of Naples, Italy
	Laurence T. Yang, St. Francis Xavier University, Canada
Organizing Committee	Domenico Di Sivo, Second University of Naples, Italy
	Patrizia Petrillo, Second University of Naples, Italy
	Francesco Moscato, Second University of Naples, Italy

Salvatore Venticinque, Second University of
    Naples, Italy
Valentina Casola, Università degli Studi
    di Napoli Federico II, Italy
Tony Li Xu, St. Francis Xavier University, Canada

Sponsoring Institutions

Second University of Naples, Italy
Lecture Notes in Computer Science (LNCS), Springer

Program/Technical Committee

Micah Adler	University of Massachusetts, USA
Enrique Alba	University of Malaga, Spain
Sahin Albayarak	University of Berlin, Germany
Vassil Alexandrov	University of Reading, UK
Raad S. Al-Qassas	University of Glasgow, UK
Giuseppe Anastasi	University of Pisa, Italy
Irfan Awan	University of Bradford, UK
Eduard Ayguade	UPC, Spain
Rocco Aversa	Second University of Naples, Italy
David A. Bader	Georgia Institute of Technology, USA
Rosa M. Badia	UPC, Spain
Frank Ball	Bournemouth University, UK
Sameer Bataineh	Jordan University of Science and Technology, Jordan
Veeravalli Bharadwaj	National University of Singapore, Singapore
Luciano Bononi	University of Bologna, Italy
Fabian Bustamante	Northwestern University, USA
Roberto Canonico	Università degli Studi di Napoli Federico II, Italy
Denis Caromel	INRIA, France
Bruno Casali	IDS, Italy
Valentina Casola	Università degli Studi di Napoli Federico II, Italy
Luca Catarinucci	University of Lecce, Italy
Alessandro Cilardo	Università degli Studi di Napoli Federico II, Italy
Michael Charleston	University of Sydney, Australia
Hui Chen	University of Memphis, USA
Peng-sheng Chen	National Tsinghua University, Taiwan
Chen-yong Cher	IBM TJ Watson Research Center, USA
Albert Cohen	INRIA, France
Michele Colajanni	University of Modena, Italy

Program/Technical Committee (continued)

Antonio Corradi	University of Bologna, Italy
Paul Coddington	University of Adelaide, Australia
Luigi Coppolino	Università degli Studi di Napoli Federico II, Italy
Domenico Cotroneo	Università degli Studi di Napoli Federico II, Italy
Michele Colaianni	Università di Modena e Reggio Emilia, Italy
Yuanshun Dai	IUPUI, USA
Vincenzo De Florio	University of Antwerp, Belgium
Jing Deng	University of New Orleans, USA
Luiz DeRose	Cray Inc., USA
Hans De Sterck	University of Waterloo, Canada
Bronis de Supinski	Lawrence Livermore National Laboratory, USA
Beatrice Di Chiara	University of Lecce, Italy
Ewa Dielman	University of Southern California, USA
Marios Dikaiakos	University of Cyprus, Cyprus
Karim Djemame	University of Leeds, UK
Xiaojiang (James) Du	North Dakota State University, USA
Abdennour El Rhalibi	Liverpool John Moores University, UK
Alessandra Esposito	University of Lecce, Italy
Peter Excell	University of Bradford, UK
Emilio Luque Fadon	Universitat Autònoma de Barcelona, Spain
Thomas Fahringer	University of Innsbruck, Austria
Joao Pinto Ferreira	University of Porto, Portugal
Bertil Folliot	Université Pierre et Marie Curie, Paris VI, France
Domenico LaForenza	ISTI-CNR, Italy
Geoffrey Fox	Indiana University, USA
Singhoff Frank	Brest University, France
Franco Frattolillo	Università del Sannio, Italy
Chao-ying Fu	MIPS, USA
Joan Garcia Haro	University of Cartagena, Spain
Luis Javier Garcia Villalba	Complutense University of Madrid, Spain
Vladimir Getov	University of Westminster, UK
Luc Giraud	CERFACS, France
Tom Goodale	Louisiana State University, USA
John Gurd	University of Manchester, UK
Luddy Harrison	University of Illinois at Urbana-Champaign, USA
Pilar Herrero	Universidad Politecnica de Madrid, Spain
Mark Hill	University of Wisconsin-Madison, USA
Pao-Ann Hsiung	National Chung Cheng University, Taiwan
Adriana Iamnitchi	Duke University, USA
Henry Jin	NASA Ames Research Center, USA

Program/Technical Committee (continued)

Eugene John	University of Texas at San Antonio, USA
Zoltan Juhasz	University of Veszprém, Hungary
Carlos Juiz	University of the Balearic Islands, Spain
Peter Kacsuk	SZTAKI, Hungary
Hartmut Kaiser	AEI, MPG, Germany
Christos Kaklamanis	University of Patras, Greece
Helen Karatza	Aristotle University of Thessaloniki, Greece
Wolfgang Karl	University of Karlsruhe, Germany
Hironori Kasahara	Waseda University, Japan
Rainer Keller	HPC Center Stuttgart, Germany
Seon Wook Kim	Korea University, Korea
Michael Kleis	Fraunhofer-Gesellschaft, Germany
Tham Chen Khong	National University of Singapore, Singapore
Arun Krishnan	Bioinformatics Institute, Singapore
Ricky Yu-Kwong Kwok	University of Hong Kong, P.R. China
Aurelio La Corte	University of Catania, Italy
Erwin Laure	CERN, Switzerland
Mario Lauria	Ohio State University, USA
Craig Lee	Aerospace Corporation, USA
Laurent Lefevre	INRIA, France
Haizhon Li	University of Memphis, USA
Lei Li	Hosei University, Japan
Xiaolin (Andy) Li	Rutgers University, USA
Antonio Lioy	Politecnico di Torino, Italy
Tianchi Ma	University of Melbourne, Australia
Muneer Masadah	University of Glasgow, UK
Anthony Mayer	Imperial College of London, UK
Nicola Mazzocca	Università degli Studi di Napoli Federico II, Italy
Xiaoqiao Meng	University of California, Los Angeles, USA
Sam Midkiff	Purdue University, USA
Giuseppina Monti	University of Lecce, Italy
Alberto Montresor	University of Bologna, Italy
Francesco Moscato	Second University of Naples, Italy
Wolfgang E. Nagel	University of Dresden, Germany
Thu D. Nguyen	Rutgers University, USA
Salvatore Orlando	University of Venice, Italy
Djamila Ouelhadj	University of Nottingham, UK
Mohamed Ould-Khaoua	University of Glasgow, UK
Paolo Palazzari	ENEA, Italy
Stylianos Papanastasiou	University of Glasgow, UK
Symeon Papavassiliou	National Technical University of Athens, Greece

Program/Technical Committee (continued)

Jose Orlando Pereira	University of Minho, Portugal
Rubem Pereira	Liverpool John Moores University, UK
Antonio Pescape	Università degli Studi di Napoli Federico II, Italy
Antonio Picariello	Università degli Studi di Napoli Federico II, Italy
Nuno Preguica	Universidade Nova de Lisboa, Portugal
Thierry Priol	IRISA, France
Yi Qian	University of Puerto Rico, USA
Xiao Qin	New Mexico Institute of Mining and Technology, USA
Rolf Rabenseifner	HPC Center Stuttgart, Germany
Tajje-eddine Rachidi	Al Akhawayn University, Morocco
Khaled Ragab	Ain Shams University, Egypt
Massimiliano Rak	Second University of Naples, Italy
Louiqa Raschid	University of Maryland, USA
Lawrence Rauchwerger	Texas A&M University, USA
Michel Raynal	IRISA, France
Graham Riley	University of Manchester, UK
Matei Ripeanu	University of Chicago, USA
Casiano Rodriguez Leon	Universidad de La Laguna, Spain
Luigi Romano	Università degli Studi di Napoli Federico II, Italy
Stefano Russo	Università degli Studi di Napoli Federico II, Italy
Kouichi Sakurai	Kyushu University, Japan
Liria Sato	University of Sao Paulo, Brazil
Mitsuhisa Sato	University of Tsukuba, Japan
Biplab K. Sarker	University of New Brunswick, Canada
Cristina Schmidt	Rutgers University, USA
Assaf Schuster	Technion, Israel
Matthew Shields	University of Cardiff, UK
David Skillicorn	Queen's University, Canada
Alexandros Stamatakis	FORTH, Greece
Heinz Stockinger	University of Vienna, Austria
Jaspal Subhlok	University of Houston, USA
Bo Sun	Lamar University, USA
Fei Sun	Princeton University, USA
El-ghazali Talbi	University of Lille, France
Michela Taufer	University of Texas at El Paso, USA
Ian Taylor	Cardiff University, UK
Nigel A. Thomas	University of Newcastle, UK
Sabino Titomanlio	ITlink, Italy

Program/Technical Committee (continued)

Jordi Torres	UPC Barcelona, Spain
Paolo Trunfio	University of Calabria, Italy
Denis Trystram	ID-IMAG, France
Theo Ungerer	University of Augsburg, Germany
Sathish Vadhiyar	Indian Institute of Science, India
Arjan van Gemund	Delft University of Technology, The Netherlands
Srinivasa R. Vemuru	Ohio Northern University, USA
Srikumar Venugopal	University of Melbourne, Australia
Salvatore Venticinque	Second University of Naples, Italy
Lorenzo Verdoscia	ICAR, National Research Council (CNR), Italy
Paulo Verissimo	University of Lisbon, Portugal
Valeria Vittorini	Università degli Studi di Napoli Federico II, Italy
Jens Volkert	GUP, University of Linz, Austria
Ian Wang	Cardiff University, UK
Lan Wang	University of Memphis, USA
Wenye Wang	North Carolina State University, USA
Xiaofang Wang	New Jersey Institute of Technology, USA
Xin-Gang Wang	University of Bradford, UK
Xudong Wang	Kiyon Inc., USA
Yu Wang	University of North Carolina at Charlotte, USA
Shlomo Weiss	Tel Aviv University, Israel
Roland Wismuller	Siegen University, Germany
Chengyong Wu	Chinese Academy of Sciences, China
Kui Wu	University of Victoria, Canada
Peng Wu	IBM T.J. Watson Research Center, USA
Bin Xiao	Hong Kong Polytechnic University, China
Baoliu Ye	University of Aizu, Japan
Hao Yin	Tsinghua University, P.R. China
Mohammed Zaki	Rensselaer Polytechnic Institute, USA
Antonia Zhai	University of Minnesota, USA
Ning Zhong	Maebashi Institute of Technology, Japan
Bing Bing Zhou	University of Sydney, Australia
Hao Zhu	Florida International University, USA
Xukai Zou	IUPUI, USA

Reviewers

Enrique Alba
Giuseppe Anastasi
Charles Archer
Stefano Avallone
Irfan Awan
Leonardo Bachega
David A. Bader
Rosa M. Badia
Frank Ball
Paola Belardini
Luciano Bononi
Dario Bottazzi
Fabian Bustamante
Xing Cai
Roberto Canonico
Bruno Casali
Luca Catarinucci
Martine Chaudier
Hui Chen
Wei Chen
Chen-yong Cher
Alessandro Cilardo
Paul Coddington
Michele Colaianni
Luigi Coppolino
Antonio Corradi
Stefania Corsaro
Domenico Cotroneo
Jose Cunha
Pasqua D'Ambra
Yuanshun Dai
Vincenzo De Florio
Hans De Sterck
Bronis de Supinski
Filippo Maria Denaro
Jing Deng
Luiz DeRose
Beatrice Di Chiara
Xiaojiang (James) Du
Abdennour El Rhalibi
Alessandra Esposito
Que Fadon
Thomas Fahringer

Guillem Femenias
Bertil Folliot
Geoffrey Fox
Singhoff Frank
Franco Frattolillo
Joan Garcia Haro
Carlo Giannelli
Sushant Goel
Tom Goodale
Anastasios Gounaris
Mario Guarracino
John Gurd
Pilar Herrero
Xiaomeng Huang
Mauro Iacono
Adriana Iamnitchi
Juergen Jaehnert
Henry Jin
Eugene John
Zoltan Juhasz
Carlos Juiz
Hartmut Kaiser
Christos Kaklamanis
Helen Karatza
Rainer Keller
Anne-Marie Kermarrec
Seon Wook Kim
Michael Kleis
Arun Krishnan
Fred Lai
Piero Lanucara
Erwin Laure
Mario Lauria
Craig Lee
Laurent Lefevre
Haizhon Li
Xiaolin Li
Yiming Li
Ben Liang
Claudio Lucchese
Tianchi Ma
Muneer Masadah
Sam Midkiff

Armando Migliaccio
Geyong Min
Giuseppina Monti
Alberto Montresor
José E. Moreira
Mirco Nanni
Thu D. Nguyen
Salvatore Orlando
Mohamed Ould-Khaoua
Luca Paladina
Generoso Paolillo
Maurizio Paone
Stylianos Papanastasiou
Symeon Papavassiliou
Manish Parashar
Luciana Pelusi
Jose Orlando Pereira
Rubem Pereira
Antonio Pescape
Antonio Picariello
Nuno Preguica
Thierry Priol
Antonio Puliafito
Yi Qian
Xiao Qin
Rolf Rabenseifner
Tajje-eddine Rachidi
Massimiliano Rak
Omer F. Rana
Louiqa Raschid
Lawrence Rauchwerger
Graham Riley
Matei Ripeanu
Casiano Rodriguez Leon
Luigi Romano
Sergio Rovida
Stefano Russo
Kouichi Sakurai
Luis Miguel Sanchez
Pere Pau Sancho
Mitsuhisa Sato
Marco Scarpa
Cristina Schmidt

Lutz Schubert
Martin Schulz
Assaf Schuster
Matthew Shields
Claudio Silvestri
David Skillicorn
Heinz Stockinger
Jaspal Subhlok
Bo Sun
Fei Sun
El-ghazali Talbi
Kevin Tang
Michela Taufer
Ian Taylor
Nigel A. Thomas
Alessandra Toninelli
Jordi Torres
Paolo Trunfio
Yu-Chee Tseng

George Tsouloupas
Theo Ungerer
Sathish Vadhiyar
Laura Vallone
Mariangela Vallone
Arjan van Gemund
Silvia Vecchi
Alessio Vecchio
Srinivasa R. Vemuru
Srikumar Venugopal
Lorenzo Verdoscia
Valeria Vittoriani
Jens Volkert
Chuan-Sheng Wang
Ian Wang
Lan Wang
Weichao Wang
Wenye Wang
Xiaofang Wang

Xin-Gang Wang
Xudong Wang
Yu Wang
Roland Wismuller
Chengyong Wu
Kui Wu
Peng Wu
Bin Xiao
Jiadi Yu
Angelo Zaia
Mohammed Zaki
Antonia Zhai
Beichuan Zhang
Jing Zhang
Jieying Zhou
Hao Zhu
Xukai Zou

Table of Contents

Keynote Speech

Alternative Approaches to High-Performance Metacomputing
  Vaidy Sunderam 1

The EGEE Project - A Multidisciplinary, Production-Level Grid
  Dieter Kranzlmüller 2

Grid and High Performance Computing: Opportunities for Bioinformatics Research
  Albert Y. Zomaya 3

Track 1: Network Protocols, Routing, Algorithms

On Multicast Communications with Minimum Resources
  Young-Cheol Bang, SungTaek Chung, Moonseong Kim, Seong-Soon Joo 4

A New Seamless Handoff Mechanism for Wired and Wireless Coexistence Networks
  Pyung Soo Kim, Hak Goo Lee, Eung Hyuk Lee 14

Approaches to Alternate Path Routing for Short Duration Flow in MPLS Network
  Ilhyung Jung, Hyo Keun Lee, Jun Kyun Choi 24

An Elaboration on Dynamically Re-configurable Communication Protocols Using Key Identifiers
  Kaushalya Premadasa, Björn Landfeldt 33

Optimal Broadcast for Fully Connected Networks
  Jesper Larsson Träff, Andreas Ripke 45

Cost Model Based Configuration Management Policy in OBS Networks
  Hyewon Song, Sang-Il Lee, Chan-Hyun Youn 57

Analytical Modeling and Comparison of AQM-Based Congestion Control Mechanisms
  Lan Wang, Geyong Min, Irfan Awan 67

Spatial and Traffic-Aware Routing (STAR) for Vehicular Systems
  Francesco Giudici, Elena Pagani 77

Adaptive Control Architecture for Active Networks
  Mehdi Galily, Farzad Habibipour, Masoum Fardis, Ali Yazdian 87

A Study on Bandwidth Guarantee Method of Subscriber Based DiffServ in Access Network
  HeaSook Park, KapDong Kim, HaeSook Kim, Cheong Youn 93

Least Cost Multicast Loop Algorithm for Local Computer Network
  Yong-Jin Lee 99

AB-Cap: A Fast Approach to Available Bandwidth Estimation
  Changhua Zhu, Changxing Pei, Yunhui Yi, Dongxiao Quan 105

An Integrated QoS Multicast Routing Algorithm Based on Tabu Search in IP/DWDM Optical Internet
  Xingwei Wang, Jia Li, Min Huang 111

On Estimation for Reducing Multicast Delay Variation
  Moonseong Kim, Young-Cheol Bang, Hyunseung Choo 117

Track 2: Languages and Compilers for HPC

Garbage Collection in a Causal Message Logging Protocol
  Kwang Sik Chung, Heon-Chang Yu, Seongbin Park 123

Searching an Optimal History Size for History-Based Page Prefetching on Software DSM Systems
  Cristian Ruz, José M. Piquer 133

Self-optimizing MPI Applications: A Simulation-Based Approach
  Emilio P. Mancini, Massimiliano Rak, Roberto Torella, Umberto Villano 143

Track 3: Parallel/Distributed System Architectures

Efficient SIMD Numerical Interpolation
  Hossein Ahmadi, Maryam Moslemi Naeini, Hamid Sarbazi-Azad 156

A Threshold-Based Matching Algorithm for Photonic Clos Network
Switches
   Ding-Jyh Tsaur, Chi-Feng Tang, Chin-Chi Wu, Woei Lin 166

CSAR-2: A Case Study of Parallel File System Dependability Analysis
   D. Cotroneo, G. Paolillo, S. Russo, M. Lauria 180

Rotational Lease: Providing High Availability in a Shared Storage File
System
   Byung Chul Tak, Yon Dohn Chung, Sun Ja Kim,
   Myoung Ho Kim 190

A New Hybrid Architecture for Optical Burst Switching Networks
   Mohamed Mostafa A. Azim, Xiaohong Jiang, Pin-Han Ho,
   Susumu Horiguchi 196

Track 4: Embedded Systems

A Productive Duplication-Based Scheduling Algorithm for
Heterogeneous Computing Systems
   Young Choon Lee, Albert Y. Zomaya 203

Memory Subsystem Characterization in a 16-Core Snoop-Based
Chip-Multiprocessor Architecture
   Francisco J. Villa, Manuel E. Acacio, José M. García 213

Factory: An Object-Oriented Parallel Programming Substrate for Deep
Multiprocessors
   Scott Schneider, Christos D. Antonopoulos,
   Dimitrios S. Nikolopoulos 223

Track 5: Parallel/Distributed Algorithms

Convergence of the Discrete FGDLS Algorithm
   Sabin Tabirca, Tatiana Tabirca, Laurence T. Yang 233

P-CBF: A Parallel Cell-Based Filtering Scheme Using a Horizontal
Partitioning Technique
   Jae-Woo Chang, Young-Chang Kim 245

Adjusting the Cluster Size Based on the Distance from the Sink
   Sanghyun Ahn, Yujin Lim, Jaehwoon Lee 255

An Efficient Distributed Search Method
  Haitao Chen, Zhenghu Gong, Zunguo Huang 265

Practical Integer Sorting on Shared Memory
  Hazem M. Bahig, Sameh S. Daoud 271

On Algorithm for the Delay- and Delay Variation-Bounded Multicast Trees Based on Estimation
  Youngjin Ahn, Moonseong Kim, Young-Cheol Bang, Hyunseung Choo 277

Track 6: Wireless and Mobile Computing

Synchronization-Based Power-Saving Protocols Based on IEEE 802.11
  Young Man Kim 283

An Energy-Efficient Uni-scheduling Based on S-MAC in Wireless Sensor Network
  Tae-Seok Lee, Yuan Yang, Ki-Jeong Shin, Myong-Soon Park 293

Delay Threshold-Based Priority Queueing Packet Scheduling for Integrated Services in Mobile Broadband Wireless Access System
  Dong Hoi Kim, Chung Gu Kang 305

Call Admission Control Using Grouping and Differentiated Handoff Region for Next Generation Wireless Networks
  Dong Hoi Kim, Kyungkoo Jun 315

A Cluster-Based QoS Multipath Routing Protocol for Large-Scale MANET
  Hui-Yao An, Xi-Cheng Lu, Zheng-hu Gong, Wei Peng 321

A MEP (Mobile Electronic Payment) and IntCA Protocol Design
  Byung kwan Lee, Tai-Chi Lee, Seung Hae Yang 331

Enhancing Connectivity Based on the Thresholds in Mobile Ad-hoc Networks
  Wongil Park, Sangjoon Park, Yoonchul Jang, Kwanjoong Kim, Byunggi Kim 340

Call Admission Control for IEEE 802.11e WLAN Using Soft QoS
  Hee-Bong Lee, Sang Hoon Jang, Yeong Min Jang 348

Overlay Multicast Routing Architecture in Mobile Wireless Network
  Backhyun Kim, Iksoo Kim 354

Real-Time Measurement Based Bandwidth Allocation Scheme in
Cellular Networks
    Donghoi Kim, Kyungkoo Jun 360

Track 7: Web Services and Internet Computing

A Hybrid Web Server Architecture for Secure e-Business Web
Applications
    *Vicenç Beltran, David Carrera, Jordi Guitart, Jordi Torres,
    Eduard Ayguadé* 366

An Efficient Scheme for Fault-Tolerant Web Page Access in Wireless
Mobile Environment Based on Mobile Agents
    HaiYang Hu, XianPing Tao, JiDong Ge, Jian Lu 378

Class-Based Latency Assurances for Web Servers
    *Yaya Wei, Chuang Lin, Xiaowen Chu, Zhiguang Shan,
    Fengyuan Ren* 388

Workflow Pattern Analysis in Web Services Orchestration: The
BPEL4WS Example
    *Francesco Moscato, Nicola Mazzocca, Valeria Vittorini,
    Giusy Di Lorenzo, Paola Mosca, Massimo Magaldi* 395

An Effective and Dynamically Extensible DRM Web Platform
    Franco Frattolillo, Salvatore D'Onofrio 401

Track 8: Peer-to-Peer Computing

JXTPIA: A JXTA-Based P2P Network Interface and Architecture for
Grid Computing
    *Yoshihiro Saitoh, Kenichi Sumitomo, Takato Izaiku,
    Takamasa Oono, Kazuhiko Yagyu, Hui Wang, Minyi Guo* 409

A Community-Based Trust Model for P2P Networks
    Hai Jin, Xuping Tu, Zongfen Han, Xiaofei Liao 419

Enabling the P2P JXTA Platform for High-Performance Networking
Grid Infrastructures
    Gabriel Antoniu, Mathieu Jan, David A. Noblet 429

Efficient Message Flooding on DHT Network
    Ching-Wei Huang, Wuu Yang 440

An IP Routing Inspired Information Search Scheme for Semantic
Overlay Networks
  Baoliu Ye, Minyi Guo, Daoxu Chen 455

Track 9: Grid and Cluster Computing

Transactional Cluster Computing
  *Stefan Frenz, Michael Schoettner, Ralph Goeckelmann,
  Peter Schulthess* 465

CPOC: Effective Static Task Scheduling for Grid Computing
  Junghwan Kim, Jungkyu Rho, Jeong-Ook Lee, Myeong-Cheol Ko 477

Improving Scheduling Decisions by Using Knowledge About Parallel
Applications Resource Usage
  *Luciano José Senger, Rodrigo Fernandes de Mello,
  Marcos José Santana, Regina Helena Carlucci Santana,
  Laurence Tianruo Yang* 487

An Evaluation Methodology for Computational Grids
  Eduardo Huedo, Rubén S. Montero, Ignacio M. Llorente 499

SLA Negotiation Protocol for Grid-Based Workflows
  Dang Minh Quan, Odej Kao 505

Track 10: Reliability, Fault-Tolerance, and Security

Securing the MPLS Control Plane
  Francesco Palmieri, Ugo Fiore 511

A Novel Arithmetic Unit over $GF(2^m)$ for Low Cost Cryptographic
Applications
  Chang Hoon Kim, Chun Pyo Hong, Soonhak Kwon 524

A New Parity Space Approach to Fault Detection for General Systems
  Pyung Soo Kim, Eung Hyuk Lee 535

Differential Power Analysis on Block Cipher ARIA
  *JaeCheol Ha, ChangKyun Kim, SangJae Moon, IlHwan Park,
  HyungSo Yoo* 541

A CRT-Based RSA Countermeasure Against Physical Cryptanalysis
  *ChangKyun Kim, JaeCheol Ha, SangJae Moon, Sung-Ming Yen,
  Sung-Hyun Kim* 549

The Approach of Transmission Scheme in Wireless Cipher
Communication
    Jinkeun Hong, Kihong Kim 555

A New Digit-Serial Systolic Mulitplier for High Performance $GF(2^m)$
Applications
    Chang Hoon Kim, Soonhak Kwon, Chun Pyo Hong, In Gil Nam 560

A Survivability Model for Cluster System Under DoS Attacks
    *Khin Mi Mi Aung, Kiejin Park, Jong Sou Park, Howon Kim,
    Byunggil Lee* 567

Track 11: Performance Evaluation and Measurements

A Loop-Aware Search Strategy for Automated Performance Analysis
    Eli D. Collins, Barton P. Miller 573

Optimization of Nonblocking MPI-I/O to a Remote Parallel Virtual
File System Using a Circular Buffer
    Yuichi Tsujita 585

Performance Analysis of Shared-Memory Parallel Applications Using
Performance Properties
    Karl Fürlinger, Michael Gerndt 595

On the Effectiveness of IEEE 802.11e QoS Support in Wireless LAN:
A Performance Analysis
    José Villalón, Pedro Cuenca, Luis Orozco-Barbosa 605

Trace-Based Parallel Performance Overhead Compensation
    Felix Wolf, Allen D. Malony, Sameer Shende, Alan Morris 617

Reducing Memory Sharing Overheads in Distributed JVMs
    Marcelo Lobosco, Orlando Loques, Claudio L. de Amorim 629

Analysis of High Performance Communication and Computation
Solutions for Parallel and Distributed Simulation
    *Luciano Bononi, Michele Bracuto, Gabriele D'Angelo,
    Lorenzo Donatiello* 640

An Analytical Study on the Interdeparture-Time Distribution for
Different Multimedia Source Models in a Packet Switched Network
    Sergio Montagna, Riccardo Gemelli, Maurizio Pignolo 652

Performance Analysis Depend on OODB Instance Based on ebXML
  Kyeongrim Ahn, Hyuncheol Kim, Jinwook Chung 660

Performance Visualization of Web Services Using J-OCM and
SCIRun/TAU
  *Wlodzimierz Funika, Marcin Koch, Dominik Dziok, Marcin Smetek,
  Roland Wismüller* 666

Track 12: Tools and Environments for Software Development

A Metadata Model and Information System for the Management of
Resources in a Grid-Based PSE Toolkit
  Carmela Comito, Carlo Mastroianni, Domenico Talia 672

Classification and Implementations of Workflow-Oriented Grid Portals
  Gergely Sipos, Péter Kacsuk 684

YACO: A User Conducted Visualization Tool for Supporting Cache
Optimization
  Boris Quaing, Jie Tao, Wolfgang Karl 694

DEE: A Distributed Fault Tolerant Workflow Enactment Engine for
Grid Computing
  Rubing Duan, Radu Prodan, Thomas Fahringer 704

An OpenMP Skeleton for the A* Heuristic Search
  G. Miranda, C. León 717

Track 13: Distributed Systems and Applications

A Proposal and Evaluation of Multi-class Optical Path Protection
Scheme for Reliable Computing
  *Shoichiro Seno, Teruko Fujii, Motofumi Tanabe, Eiichi Horiuchi,
  Yoshimasa Baba, Tetsuo Ideguchi* 723

Lazy Home-Based Protocol: Combining Homeless and Home-Based
Distributed Shared Memory Protocols
  *Byung-Hyun Yu, Paul Werstein, Martin Purvis,
  Stephen Cranefield* 733

Grid Enablement of the Danish Eulerian Air Pollution Model
  Cihan Sahin, Ashish Thandavan, Vassil N. Alexandrov 745

Towards a Bayesian Statistical Model for the Classification of the
Causes of Data Loss
  Phillip M. Dickens, Jeffery Peden 755

Parallel Branch and Bound Algorithms on Internet Connected
Workstations
  Randi Moe, Tor Sørevik 768

Track 14: High-Performance Scientific and Engineering Computing

Parallel Divide-and-Conquer Phylogeny Reconstruction by Maximum
Likelihood
  Z. Du, A. Stamatakis, F. Lin, U. Roshan, L. Nakhleh 776

Parallel Blocked Algorithm for Solving the Algebraic Path Problem on
a Matrix Processor
  Akihito Takahashi, Stanislav Sedukhin 786

A Parallel Distance-2 Graph Coloring Algorithm for Distributed
Memory Computers
  *Doruk Bozdağ, Umit Catalyurek, Assefaw H. Gebremedhin,
  Fredrik Manne, Erik G. Boman, Füsun Özgüner* 796

Fast Sparse Matrix-Vector Multiplication by Exploiting Variable Block
Structure
  Richard W. Vuduc, Hyun-Jin Moon 807

Parallel Transferable Uniform Multi-round Algorithm for Achieving
Minimum Application Turnaround Times for Divisible Workload
  Hiroshi Yamamoto, Masato Tsuru, Yuji Oie 817

Application of Parallel Adaptive Computing Technique to Polysilicon
Thin-Film Transistor Simulation
  Yiming Li 829

A Scalable Parallel Algorithm for Global Optimization Based on
Seed-Growth Techniques
  Weitao Sun 839

Track 15: Database Applications and Data Mining

Exploiting Efficient Parallelism for Mining Rules in Time Series Data
  Biplab Kumer Sarker, Kuniaki Uehara, Laurence T. Yang 845

A Coarse Grained Parallel Algorithm for Closest Larger Ancestors in
Trees with Applications to Single Link Clustering
Albert Chan, Chunmei Gao, Andrew Rau-Chaplin 856

High Performance Subgraph Mining in Molecular Compounds
Giuseppe Di Fatta, Michael R. Berthold 866

Exploring Regression for Mining User Moving Patterns in a Mobile
Computing System
Chih-Chieh Hung, Wen-Chih Peng, Jiun-Long Huang 878

A System Supporting Nested Transactions in DRTDBSs
Majed Abdouli, Bruno Sadeg, Laurent Amanton, Adel Alimi 888

Efficient Cluster Management Software Supporting High-Availability
for Cluster DBMS
Jae-Woo Chang, Young-Chang Kim 898

Distributed Query Optimization in the Stack-Based Approach
Hanna Kozankiewicz, Krzysztof Stencel, Kazimierz Subieta 904

Track 16: Biological/Molecular Computing

Parallelization of Multiple Genome Alignment
Yiming Li, Cheng-Kai Chen 910

Track 17: Special Session on HPSRF

Detonation Structure Simulation with AMROC
Ralf Deiterding 916

Scalable Photon Monte Carlo Algorithms and Software for the Solution
of Radiative Heat Transfer Problems
Ivana Veljkovic, Paul E. Plassmann 928

A Multi-scale Computational Approach for Nanoparticle Growth in
Combustion Environments
Angela Violi, Gregory A. Voth 938

A Scalable Scientific Database for Chemistry Calculations in Reacting
Flow Simulations
Ivana Veljkovic, Paul E. Plassmann 948

The Impact of Different Stiff ODE Solvers in Parallel Simulation of
Diesel Combustion
   Paola Belardini, Claudio Bertoli, Stefania Corsaro,
   Pasqua D'Ambra 958

FAST-EVP: An Engine Simulation Tool
   Gino Bella, Alfredo Buttari, Alessandro De Maio,
   Francesco Del Citto, Salvatore Filippone, Fabiano Gasperini 969

Track 18: Pervasive Computing and Communications

A Mobile Communication Simulation System for Urban Space with
User Behavior Scenarios
   Takako Yamada, Masashi Kaneko, Ken'ichi Katou 979

Distributing Multiple Home Agents in MIPv6 Networks
   Jong-Hyouk Lee, Tai-Myung Chung 991

Dynamically Adaptable User Interface Generation for Heterogeneous
Computing Devices
   Mario Bisignano, Giuseppe Di Modica, Orazio Tomarchio 1000

A Communication Broker for Nomadic Computing Systems
   Domenico Cotroneo, Armando Migliaccio, Stefano Russo 1011

Adaptive Buffering-Based on Handoff Prediction for Wireless Internet
Continuous Services
   Paolo Bellavista, Antonio Corradi, Carlo Giannelli 1021

A Scalable Framework for the Support of Advanced Edge Services
   Michele Colajanni, Raffaella Grieco, Delfina Malandrino,
   Francesca Mazzoni, Vittorio Scarano 1033

A Novel Resource Dissemination and Discovery Model for Pervasive
Environments Using Mobile Agents
   Ebrahim Bagheri, Mahmood Naghibzadeh, Mohsen Kahani 1043

A SMS Based Ubiquitous Home Care System
   Tae-seok Lee, Yuan Yang, Myong-Soon Park 1049

A Lightweight Platform for Integration of Mobile Devices into Pervasive
Grids
   Stavros Isaiadis, Vladimir Getov 1058

High-Performance and Interoperable Security Services for Mobile Environments
  Alessandro Cilardo, Luigi Coppolino, Antonino Mazzeo, Luigi Romano 1064

Distributed Systems to Support Efficient Adaptation for Ubiquitous Web
  Claudia Canali, Sara Casolari, Riccardo Lancellotti 1070

Track 19: Special Session on LMS

Call Tracking Management Using Caching Scheme in IMT-2000 Networks
  Dong Chun Lee 1077

Correction of Building Height Effect Using LIDAR and GPS
  Hong-Gyoo Sohn, Kong-Hyun Yun, Gi-Hong Kim, Hyo Sun Park 1087

Minimum Interference Path Selection Based on Maximum Degree of Sharing
  Hyuncheol Kim, Seongjin Ahn 1096

The Effect of the QoS Satisfaction on the Handoff for Real-Time Traffic in Cellular Network
  Dong Hoi Kim, Kyungkoo Jun 1106

Author Index 1113

Alternative Approaches to High-Performance Metacomputing

Vaidy Sunderam

Department of Math & Computer Science,
Emory University, Atlanta, GA 30322, USA
vss@emory.edu
http://www.mathcs.emory.edu/dcl/

Abstract. Software frameworks for high-performance computing have long attempted to deal with the dichotomy between parallel programming on tightly-coupled platforms and the benefits of aggregating distributed heterogeneous resources. Solutions will inevitably involve compromise, but middleware architectures can alleviate the discord. The Harness project has approached high-performance distributed computing via a design philosophy that leverages dynamic reconfigurability – in terms of resources, capabilities, and communication fabrics. We provide an overview of Harness, and position its goals in the context of recent and current metacomputing systems. We then describe the salient features of the H2O framework, a core subsystem in the Harness project, and discuss its alternative approach to high-performance metacomputing. H2O is based on a "pluggable" software architecture to enable flexible and reconfigurable distributed computing. A key feature is the provisioning of customization capabilities that permit clients to tailor provider resources as appropriate to the given application, without compromising control or security. Through the use of uploadable "pluglets", users can exploit specialized features of the underlying resource, application libraries, or optimized message passing subsystems on demand. H2O is supplemented by subsystems for event handling and fault-tolerant naming services, thereby providing a comprehensive suite of software subsystems for robust and large scale metacomputing. The current status of the H2O subsystem and the overall Harness framework, recent experiences with its use, and planned enhancements are discussed.

The EGEE Project – A Multidisciplinary, Production-Level Grid

Dieter Kranzlmüller

EGEE Project, CERN,
CH-1211 Geneva 23, Switzerland
&
GUP, Joh. Kepler University Linz,
Altenbergerstr. 69, A-4040 Linz, Austria/Europe
kranzlmueller@gup.jku.at,
http://www.gup.uni-linz.ac.at/~dk

Abstract. The EGEE project[1] (Enabling Grids for E-sciencE) operates the world's largest grid infrastructure today. Combining over 15000 CPUs in more than 120 resource centers on a 24x7 basis, EGEE provides grid services to a diverse user community at institutions around the world. At the core of this infrastructure is gLite, a next generation grid middleware stack, which represents a bleeding-edge, best-of-breed framework for building grid applications. Together with gLite and a federated hierarchical structure of resource and regional operation centers, EGEE offers grid computing for a variety of different applications, including high-energy physics, biomedicine, earth sciences, computational chemistry, and astrophysics.

---

[1] EGEE is a project funded by the European Union under contract number INFSO 508833.

Grid and High Performance Computing: Opportunities for Bioinformatics Research

Albert Y. Zomaya

School of Information Technologies,
The University of Sydney,
Sydney, NSW 2006, Australia
zomaya@it.usyd.edu.au

Over the past few years the popularity of the Internet has been growing by leaps and bounds. However, there comes a time in the life of a technology, as it matures, where questions about its future need to be answered. The Internet is no exception to this case. Often called the "next big thing" in global Internet technology, Grid computing is viewed as one of the top candidates that can shape the future of the Internet. Grid Computing takes collective advantage of the vast improvements in microprocessor speeds, optical communications, raw storage capacity, World Wide Web and the Internet that have occurred over the last five years. Grid technology leverages existing resources and delays the need to purchase new infrastructure. With demand for computer power in industries like the life sciences and health informatics almost unlimited, Grids ability to deliver greater power at less cost gives the technology tremendous potential. Ultimately the Grid must be evaluated in terms of the applications, business value, and scientific results that it delivers, not its architecture. Biology provides some of the most important, as well as most complex, scientific challenges of our times. These problems include understanding the human genome, discovering the structure and functions of the proteins that the genes encode, and using this information efficiently for drug design. Most of these problems are extremely intensive from a computational perspective. One of the principal design goals for the Grid framework is the effective logical separation of the complexities of programming a massively parallel machine from the complexities of bioinformatics computations through the definition of appropriate interfaces. Encapsulation of the semantics of the bioinformatics computations methodologies means that the application can track the evolution of the machine architecture and explorations of various parallel decomposition schemes can take place with minimal intervention from the domain experts or the end users. For example, understanding the physical basis of protein function is a central objective of molecular biology. Proteins function through internal motion and interaction with their environment. An understanding of protein motion at the atomic level has been pursued since the earliest simulations of their dynamics. When simulations can connect to experimental results, the microscopic examinations of the different processes (via simulation) acquire more credibility and the simulation results can then help interpret the experimental data. Improvements in computational power and simulation methods facilitated by the Grid framework could to lead to important progress in studies of protein structure, thermodynamics, and kinetics. This talk will review the state of play and show how Grid technology can change the competitive landscape.

On Multicast Communications with Minimum Resources*

Young-Cheol Bang[1], SungTaek Chung[1], Moonseong Kim[2], and Seong-Soon Joo[3]

[1] Department of Computer Engineering,
Korea Polytechnic University,
429-793, Gyeonggi-Do, Korea
{ybang, unitaek}@kpu.ac.kr

[2] School of Information and Communication Engineering,
Sungkyunkwan University,
440-746, Suwon, Korea
moonseong@ece.skku.ac.kr

[3] Carrier Ethernet Team,
Electronics and Telecommunications Research Institute,
Daejun, Korea
ssjoo@etri.re.kr

Abstract. We have developed and evaluated an efficient heuristic algorithm to construct a multicast tree with the minimum cost. Our algorithm uses multiple candidate paths to select a path from source to each destination, and works on directed asymmetric networks. We also show that our proposed algorithm have a perform gain in terms of tree cost for real life networks over existing algorithm. The time complexity of our algorithm is $O(D(m + n\log n))$ for a m-arc n-node network with D number of members in the multicast group, and is comparable to well-known algorithm, TM [13], for a multicast tree construction in terms of the time-complexity. We have performed empirical evaluation that compares our algorithms with the others on large networks.

1 Introduction

Multicast in a network is an efficient way to send the same information from a single source to multiple destinations called the multicast group. With the deployment of the high-speed networks, new communication services involving multicast communications and real time multimedia applications are becoming widespread. These applications need very large reserved bandwidth to satisfy

---

* Research of Dr. Bang is sponsored by the Ministry of Commerce, Industry and Energy under Next-Generation Growth Engine Industry. Research of Dr. Joo is sponsored by ETRI and the Ministry of Information and Communication in Republic of Korea. Dr. Bang is the corresponding author.

their end-to-end delay requirements. Since network resources are limited in the current Internet architecture using the best-effort services, it is very critical to design multicast routing paths that use the resources, such as bandwidths, very efficiently.

An efficient solution for multicast communications includes the construction of a multicast tree that is rooted at a source and spanning to all destinations in the multicast group. Broadly, there are two types of modes to select paths that construct a multicast tree [4]; single-path mode and multiple-path mode. In the single-path mode, the tree construction mechanism consider single candidate path only. Meanwhile, multiple paths are considered as candidate paths to construct multicast tree in the multiple-path mode. Although the multiple path mode increases the chance of finding a feasible tree, the time-complexity of the mechanism is relatively very high compared to the single path mode so that special treat is needed to reduce the time-complexity when the multiple-path mode is used. Under high-speed network environments supporting the best-effort service, efficient multicast services for high-bandwidth applications require a multicast tree that should be designed to minimize the network resources as much as possible in terms of the tree cost, where the tree cost is defined by summing costs of all links in the tree.

The general problem of multicast is well-studied in the areas of computer networks and algorithmic network theory. Depending on the specific cost or criterion, the multicast problem could be of varying levels of complexity. The Steiner tree problem is to find a tree which spans a given node set S contained in a network. The minimum Steiner tree spans S with a minimum tree cost. The Steiner tree has a natural analog of the general multicast tree in computer networks [7,8,14,15]. The computation of Steiner tree is NP-complete [6], and two interesting polynomial time algorithms are KMB [9] and TM [13] algorithm. Also, an interesting polynomial time algorithm for a fixed parameter has been proposed in [10], wherein an overview of several other approximation algorithms are also provided; Distributed algorithms based on Steiner heuristic are provided in [3].

We consider a computer network represented by a graph $G = (V, E)$ with n nodes and m arcs or links, where V and E are a set of nodes and a set of arcs (links), respectively. Each link $e(i,j) \in E$ is associated with cost $C(e) \geq 0$. Consider a simple directed path $P(i_0, i_k)$ from i_0 to i_k given by $(i_0, i_1), (i_1, i_2), \ldots, (i_{k-1}, i_k)$, where $(i_j, i_{j+1}) \in E$, for $j = 0, 1, \ldots, (k-1)$, and all $i_0, i_1, \ldots, i_k$ are distinct. Subsequently, a simple directed path is referred to simply as a path. The path cost of P is given by $C(P) = \sum_{j=0}^{k-1} C(e_j)$, where $e_j = (i_j, i_{j+1})$. The tree cost of T is given by $C(T) = \sum_{e \in T} C(e)$. Let s be a node in the network, called the source, and $MC = \{d_1, d_2, \ldots, d_k\}$, where $k \leq n-1$, be the set of destination nodes. Our objective is to minimize the $C(T)$ for any network topology with asymmetric links.

In this paper, we introduce a very efficient mechanism using the multiple-path mode and apply to TM algorithm that is known to be the best-heuristic. We also show that our algorithm outperforms single-path mode algorithm with the same time-complexity. Since there may exist exponential number of paths between a source and a destination, multiple candidate paths are limited to the minimum cost paths to obtain efficient time-complexity. We strongly believe our method can be generalized to apply to any type of tree-construction algorithm that are aimed to reduce the tree costs.

The rest of the paper is organized as follows. In Section 2, we introduced previous works, and details of our algorithm will be presented in Section 3. We discuss the network model and the results of our simulation in Section 4. Conclusions are presented in Section 5.

2 Previous Works

There are two well known approaches for constructing multicast trees with minimum costs. The first approach is based on the shortest paths, and the second uses minimum spanning tree based algorithms. The algorithm TM due to Takahashi and Matsuyama [13] is a shortest path based algorithm and works on asymmetric directed networks. Also it was further studied and generalized by Ramanthan [10]. The TM algorithm is very similar to the shortest path based Prim's minimum spanning tree algorithm [12], and works as following three steps.

1. Construct a subtree, T_1, of network G, where T_1 consists a source node s only. Let $i = 1$ and $MC_i = \{s\}$
2. Find the closest node, $d_i \in (MC \setminus MC_i)$, to T_i (if tie, broken arbitrarily). Construct a new subtree, T_{i+1} by adding all links on the minimum cost path from T_i to d_i. Set $MC_i = MC_i \cup \{d_i\}$. Set $i = i + 1$.
3. If $|MC_i| < |MC|$ then go to step 2, otherwise return a final T_i

Fig. 1 (a) shows a given network topology with link costs specified on each link. Fig. 1 (b) represents the ultimate multicast tree obtained by the TM. The cost of the tree generated by the TM is 15. In the worst case, the cost of tree by TM is worse than $2(1 - 1/|MC|)C(T)$, where T is the optimal tree [13].

The algorithm KMB by Kou, Markowsky, and Berman [9] is a minimum spanning tree based algorithm. Doar and Leslie [5] report that KMB usually achiving 5% of the optimal for a large number of realistic instances. Ramanathan [10], in this comparison between parameterized TM and KMB, has shown that TM outperforms KMB in terms of the cost of the tree constructed since KMB is designed for undirected graph. To find a tree, KMB starts with constructing the complete distance network $G' = (V', E')$ induced by MC, where V' contains source node and destinations only, and E' is a set of links connecting nodes in V' for each other. In next step, a minimum spanning tree T of G' is determined. After then, a subgraph G_s is constructed by replacing each link (i, j) of T with its actual corresponding minimum cost path from i to j in G. If there exist several

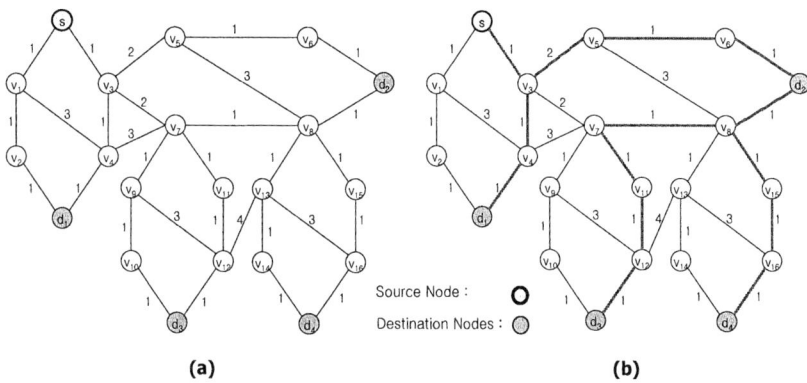

Fig. 1. Given a network (a), a multicast tree based on TM is shown in (b)

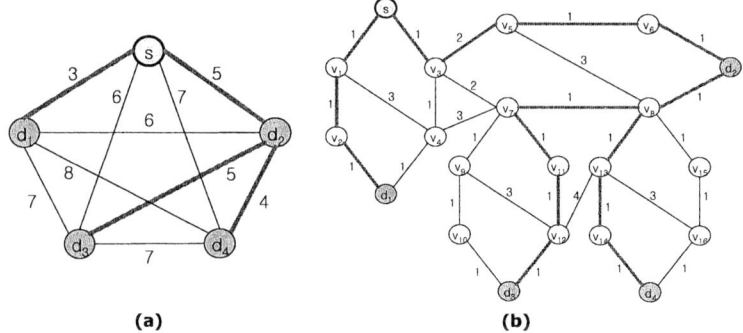

Fig. 2. (a) The complete graph and the minimal spanning tree, (b) KMB tree

minimum cost paths, pick an arbitrary one. Next step is to find the minimum spanning tree T' of G_s. In final step, delete from T' all unnecessary nodes and corresponding links. Then the resulting tree is a KMB tree. Fig. 2 (a) shows the complete graph from the given network Fig. 1 (a) and the minimal spanning tree. Fig. 2 (b) represents KMB tree by replacing each edge in the spanning tree by its corresponding shortest path in the given network. The cost of the tree generated by the KMB is 16. Even though TM outperforms KMB, this algorithm is easy to implement.

3 Algorithms

In this section, we present our proposed multiple-path mechanism that is applied to TM algorithm. Our algorithm is named the Minimum Cost Multicast Tree (MCMT). In order to define a multicast tree, MCMT is based on the modified Dijkstra's shortest path algorithm (MDSP) to select all the minimum-cost paths from source to each destination in G [1,2] and Prim's minimum spanning tree

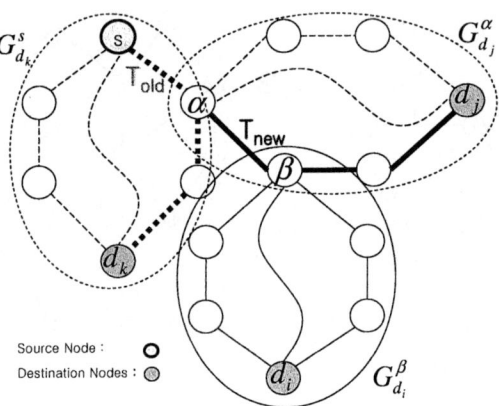

Fig. 3. The conceptual main idea of MCMT

[12]. We define a sub-graph $G_{d_i}^{\alpha}$ such that $G_{d_i}^{\alpha}$ is constructed by merging all the shortest paths from α to each d_i with $d_i \in MC$, where $G_{d_i}^{s}$ can be constructed using MDSP. Thus, any path of $G_{d_i}^{\alpha}$ is a minimum-cost path. Also note that $G_{d_i}^{\alpha}$ is an acyclic graph. Let T be a set of node that constitute a tree we want to define, and be empty initially. Let $G' = (V', E')$ be sub-graph of G with $V' \subseteq V$ and $E' \subseteq E$, where $G' = G' \cup G_{d_i}^{\alpha}$ with $G' = \emptyset$ initially. Then, the conceptual main idea of our approach is as follows; select (s, d_i) pair, called $(s, d_i)_{min}$ such that $C(P(s, d_i))$ is minimum among all (s, d_i) pairs with $d_i \in MC$, where s is a source of the multicast communication at initial step. If $G' \neq \emptyset$, find $(\alpha, d_j)_{min}$ pair with $\alpha \in V'$ and $d_j \in MC$. If α is not in T that is empty initially, we select single minimum-cost path P_{min} of $G_{d_k}^{\beta}$ that contains a node α. Once P_{min} via α is selected, nodes of P_{min} are added to T, and all other redundant nodes and links are pruned from $G_{d_k}^{\beta}$. When α is in T, then we just add $G_{d_j}^{\alpha}$ to the set G'. we repeat this process until all $G_{d_j}^{\alpha}$ with $\alpha \in V'$ and $d_{d_j} \in MC$ are considered. At the end of process, if there exist $G_{d_j}^{\alpha}$ of which P_{min} is not selected, single path from such $G_{d_j}^{\alpha}$ selected and all redundant nodes and links are removed. Then, the final sub-graph, G', is a tree, and spans all destinations.

Consider Fig.3 as an example. At initial step $G_{d_k}^{s}$ is computed. After $G_{d_j}^{\alpha}$ is constructed, P_{min} from s to d_k via α is selected and all redundant nodes and associated links are removed from $G_{d_k}^{s}$. In next step, $G_{d_i}^{\beta}$ is constructed and P_{min} of $G_{d_j}^{\alpha}$ is selected. At final step, any P_{min} is selected from $G_{d_j}^{\beta}$ and redundant nodes and links are removed. Thus, we may obtain a loop-free G' that is a multicast tree spanning destinations d_k, d_i, and d_j.

A formal algorithm and its descriptions are as follows: Let $\overleftarrow{G} = (\overleftarrow{V}, \overleftarrow{E})$ be a network with reversed link direction of $G = (V, E)$. We define $P(u, v)$ and $\overleftarrow{P}(u, v)$, where $P(u, v)$ is a path connecting u to v, and $\overleftarrow{P}(u, v)$ is a path with reversed link direction of $P(u, v)$, respectively.

Algorithm MCMT $(G, \overleftarrow{G}, s, MC)$

Begin
01. **For** each $d_i \in MC$ **Do**
02.   Compute all the minimum cost path $P(d_i, u)$ from d_i to each u with $\overleftarrow{G}$, $\forall u \in \overleftarrow{V}$
03.   Construct $G^u_{d_i}$ by merging all the minimum cost path $\overleftarrow{P}(d_i, u)$
04.   Reverse direction of all links in $G^u_{d_i}$
05. $G' = \emptyset$
06. $T = \emptyset$
07. Find d_i such that $C(P(s, d_i))$ is minimum for each $d_i \in MC$
08. $G' = G' \cup G^s_{d_i}$
09. $MC = MC \setminus \{d_i\}$
10. **For** each $d_i \in MC$ **Do**
11.   Find $P(u, d_i)$ such that $C(P)$ is minimum among all $P(u, d_i)$, $\forall u \in V'$
12.   **If** $u \notin T$ **then**
13.   Select a path $P(v, d_j)$ of $G^v_{d_j}$ that contains u
14.   Add nodes on $P(v, d_j)$ to T
15. $G' = G' \cup G^u_{d_i}$
16. $MC = MC \setminus \{d_i\}$
17. **For** each $G^u_{d_i}$ of which no path is selected yet **Do**
18.   Select a path destined to d_i
19.   Remove redundant nodes and associated links
20. Return G'
End Algorithm.

We now consider the time-complexity and correctness of the algorithm. Given a network $\overleftarrow{G} = (\overleftarrow{V}, \overleftarrow{E})$ that can be constructed by reversing link direction of G, Step 2 computes all the minimum cost path from $d_i \in MC$ to each $u \in \overleftarrow{V}$ by invoking the modified Dijkstra's shortest path algorithm with $O(m + n \log n)$ using Fibonacci heap [1,2]. Step 3 and 4 construct a directed and acyclic subgraph $G^u_{d_i}$ by merging all paths $\overleftarrow{P}(d_i, u)$ computed in Step 2. Since Step 2-4 are iterated $D = |MC|$ times and each $G^u_{d_i}$ can be constructed with $O(m)$ in worst case, the worst case time-complexity is $O(D(m + n \log n))$. Step 5-6 initialize a sub-graph G' and set of node in a tree T we want to define. Step 7-9 find an initial $G^s_{d_i}$ by comparing the cost of $P(s, d_i)$, and add $G^s_{d_i}$ to the set G'. Step 10-16 use the concept of Prim's minimum spanning tree to select a path destined to any $d_i \in MC$ for each iteration. In step 11, costs of minimum cost paths between all (u, d_i) pairs, $u \in V'$ are compared for each $d_i \in MC$, and select a path $P(u, d_i)$ such that $C(P(u, d_i))$ is minimum among all $P(u, d_i)$s with time-complexity of $O(n)$. Let $G^v_{d_j}$ be a sub-graph that contains a node u. Step 12 checks if path $P(v, d_j)$ destined to some destination d_j via node u has been already selected. If not selected yet, path $P(v, d_j)$ is selected and nodes on $P(v, d_j)$ are added to T with time-complexity of $O(n)$ in Step 13. Sub-network $G^u_{d_i}$ is added to G' with $O(m)$, and d_i is removed from MC with constant time in Step 15 and 16, respectively. The time-complexity of Step 10-16 is $O(Dm)$. After all d_i's are considered, a path $P(u, d_i)$ is selected for each $G_{d_i}^u$ of which no path is selected yet, and all redundant nodes and links are removed in Step 17-19. The time-complexities of Step 18 and 19 are clearly $O(n)$ and $O(m)$, respectively. Thus Step 13-15 can be computed in $O(m)$ in worst case. Therefore total time-complexity of algorithm MCMT is $O(D(m + n\log n))$ in worst case which is the same as that of TM.

Let us consider loop in G'. Suppose there are two minimum cost paths $P1$ and $P2$, where $P1$ is a path from u to d and $P2$ is a path from v to d, respectively. We assume that $P1$ passes through v as a relay node to reach d. Then $C(P1) = C(P(u,v)) + C(P(v,d))$ and $C(P2) = C(P(v,d))$, which means $C(P2) < C(P1)$. Thus the path $P1$ cannot be selected by Step 11. Also each $G_{d_i}^u$ that is involved

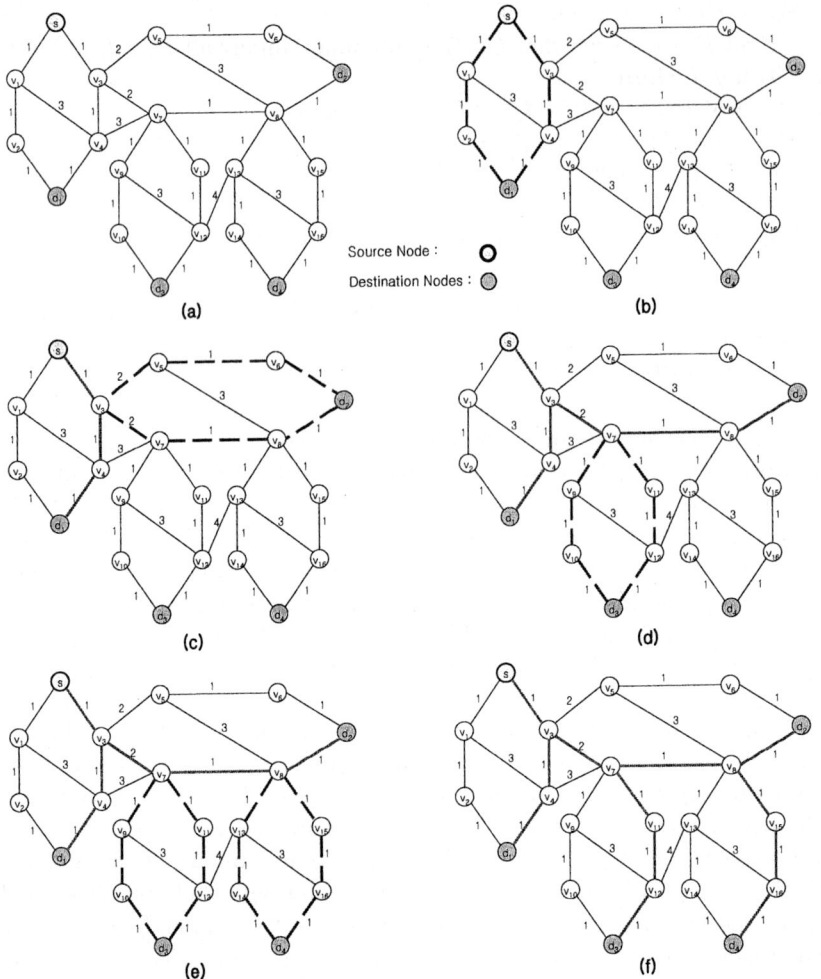

Fig. 4. Given a network (a), an illustration of the MCMT algorithm is shown in (b)–(f)

in G' has single path only so that no loop exists in a final G'. Therefore, a final G' is a tree we want to define.

Fig. 4 (a) is the same a given network in Fig. 1. Fig. 4 (b)-(f) illustrate of the procedure of creating multicasting tree. $C(T_{MCMT}) = 13$ is less than $C(T_{TM}) = 15$.

4 Performance Evaluation

As mentioned, we apply our mechanism to TM and compare with the TM algorithm that uses single-path mode. A performance measure $C(T)$ is our concern and investigated here. We now describe some numerical results with which we compare the performance for the new algorithm MCMT. The proposed one is implemented in C++. We randomly selected a source node. We generate 10 different networks for each size of given 25, 50, 100, and 200. The random networks used in our experiments are directed, asymmetric, and connected, where each node in networks has the probability of links (P_e) equal to 0.3, 0.5, and 0.7. The destination nodes are picked uniformly from the set of nodes in the network topology (excluding the nodes already selected for the destination). Moreover,

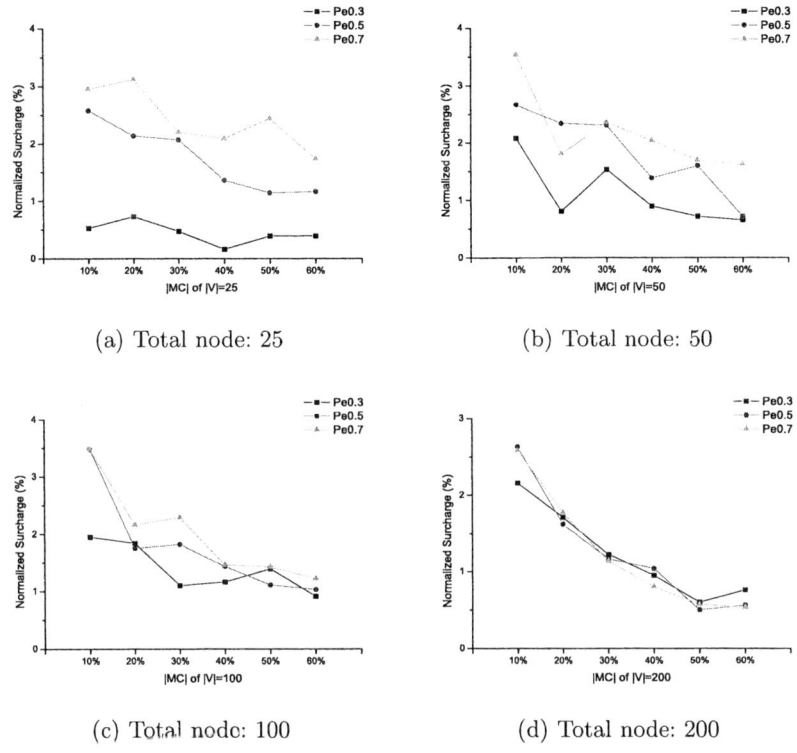

(a) Total node: 25

(b) Total node: 50

(c) Total node: 100

(d) Total node: 200

Fig. 5. Normalized surcharges versus number of nodes in networks

the destination nodes in the multicast group will occupy 10, 20, 30, 40, 50, and 60% of the overall nodes on the network, respectively [11]. We simulate 1000 times (10 different networks $\times 100 = 1000$) for each n and P_e.

For the performance comparison, we implement TM algorithm in the same simulation environment. We use the normalized surcharge, introduced in [8], of the algorithm with respect to our method defined as follows:

$$\bar{\delta}_C = \frac{C(T_{TM}) - C(T_{MCMT})}{C(T_{MCMT})}$$

In our plotting, we express this as a percentage, i.e., $\bar{\delta}_C$ is multiplied by 100. As indicated in Fig. 5, we are easily noticed that the MCMT algorithm is better than the TM algorithm. The enhancement is up to about 0.16%~3.54% in terms of normalized surcharge for the TM tree cost.

5 Conclusion

We considered the transmission of a message from a source to a set of destinations with minimum cost over a computer network. In this paper we have assumed a source routing based solution. We presented a simple algorithm that specifies a multicast tree with minimal cost using multiple-path mode. We also presented simulation results to illustrate the relative performances of algorithm. Empirical results show that the algorithm MCMT outperforms TM which is the most straightforward and efficient among algorithms known so far. For future research direction, it would be interesting to obtain an unified algorithm for any single-path mode algorithm that is used to construct a multicast tree with minimum costs.

References

1. Ravindra K. Ajuja, Thomas L. Magnanti, and James B. Orlin, Network Flows: Theory, Algorithms, and Applications, Prentice-Hall, 1993.
2. Y.-C. Bang and H. Choo, "On Multicasting with Minimum Costs for the Internet Topology," Springer-Verlag Lecture Notes in Computer Science, vol. 2400, pp. 736-744, August 2002.
3. F. Bauer and A. Varma, "Distributed algorithms for multicast path setup in data networks," IEEE/ACM Transactions on Networking, vol. 4, no. 2, pp. 181-191, 1996.
4. Shigang Chen, Klara Nahrstedt, and Yuval Shavitt, "A QoS-Aware Multicast Routing Protocol," IEEE Journal on Selected Areas in Communications, vol. 18, no. 12, pp. 2580-2592, 2000.
5. M. Doar and I. Leslie, "How Bad is Naive Multicast Routing?," Proc. IEEE INFOCOM'93, pp. 82-89, 1993.
6. M. R. Garey and D. S. Johnson, Computers and Intractability: A Guide to the Theory of NP-Completeness, W. H. Freeman and Co., San Francisco, 1979.

7. B. K. Kadaba and J. M. Jaffe, "Routing to multiple destinations in computer networks," IEEE Transactions on Communications, COM-31, no. 3, pp. 343-351, 1983.
8. V. P. Kompella, J. C. Pasquale, and G. C. Polyzoa, "Multicast routing for multimedia communications," IEEE/ACM Transactions on Networking, vol. 1, no. 3, pp. 286-292, 1993.
9. L. Kou, G. Markowsky, and L. Berman, "A fast algorithm for steiner trees," Acta Informatica, vol. 15, pp. 145-151, 1981.
10. S. Ramanathan, "Multicast tree generation in networks with asymetric links," IEEE/ACM Transactions on Networking, vol. 4, no. 4, pp. 558-568, 1996.
11. A.S. Rodionov and H. Choo, "On Generating Random Network Structures: Connected Graphs," Springer-Verlag Lecture Notes in Computer Science, vol. 3090, pp. 483-491, September 2004.
12. R.C. Prim, "Shortest Connection Networks And Some Generalizations," Bell System Techn. J. 36, pp. 1389-1401, 1957.
13. H. Takahashi and A. Matsuyama, "An Approximate Solution for the Steiner Problem in Graphs, Mathematica Japonica," vol. 24, no. 6, pp. 573-577, 1980.
14. B. M. Waxman, "Routing of multipoint connections," IEEE Journal on Selected Areas in Communications, vol. 6, no. 9, 1988.
15. Q. Zhu, M. Parsa, and J. J. Garcia-Luna-Aceves, "A source-based algorithm for delay constrained minimum-cost multicasting," Proc. IEEE INFOCOM'95, pp. 377-385, 1995.

A New Seamless Handoff Mechanism for Wired and Wireless Coexistence Networks

Pyung Soo Kim[1], Hak Goo Lee[2], and Eung Hyuk Lee[1]

[1] Dept. of Electronics Engineering, Korea Polytechnic University,
Shihung City, 429-793, Korea
pskim@kpu.ac.kr

[2] Mobile Platform Lab., Digital Media R&D Center,
Samsung Electronics Co., Ltd, Suwon City, 442-742, Korea

Abstract. This paper deals with design and implementation of seamless handoff mechanism between *wired* and *wireless* network adapters for a system with both network adapters. A unique virtual adapter is developed between different adapters and then an IP address is assigned to the virtual adapter. As a general rule, when both network adapters are available, the wired adapter is preferred due to its faster transmission speed than the wireless adapter. When wired communication via the wired adapter gets disconnected while in service, the disconnection of wired adapter is automatically detected and then wireless handoff occurs by mapping information on the wireless adapter to the virtual adapter. According to the proposed handoff mechanism, the session can be continued seamlessly even when handoff between wired and wireless network adapters occurs at lower level in a network application where both IP address and port number are used to maintain session. To evaluate the proposed handoff mechanism, actual experiments are performed for various internet applications such as FTP, HTTP, Telnet, and then their results are discussed.

1 Introduction

Most of today's computers are operating on Microsoft Windows systems which allow installations of multiple network adapters such as wired adapter and wireless adapter. In Windows systems, not only with different types of network adapters, but also with same type of network adapters, when one communication medium disconnects, all sessions of internet applications that are communicating through the corresponding adapter get disconnected automatically [1], [2]. It's because, under TCP, information on an adapter is stored in TCP control block (TCB), and once adapter disconnection is notified by TCB, TCP automatically cuts off the corresponding sessions. In other words, disconnection of an adapter means the IP address assigned to the adapter can no longer be used. Thus, when a handoff occurs from one adapter to another where different IP addresses are assigned to each network adapter, Windows systems don't support seamless handoff. It's because the session must be newly made with the

IP address of the new network adapter, since the IP address assigned to the old network adapter cannot be used any more. To solve this problem, Windows systems provide "Bridge" function that allows multiple network adapters to be combined into one virtual adapter. However, although Bridge allows multiple network adapters to share same IP address, it has some shortcomings. Even with the Bridge, once a network adapter gets disconnected, the virtual adapter notifies protocol driver about disconnection. Thus, all application sessions using TCP/IP protocol driver automatically get disconnected. In addition, this function has too long handoff latency. It is observed via the experiment that handoff took about 30 seconds, which results in the timeout of TCP applications under Windows systems.

In order to solve the problems addressed above, this paper proposes a new seamless handoff mechanism between wired (IEEE 802.3) and wireless (IEEE 802.11) network adapters for a system with both network adapters. A unique virtual adapter is developed between different adapters and then an IP address is assigned to the virtual adapter. As a general rule, when both network adapters are available, the wired adapter is preferred due to its faster transmission speed than the wireless adapter. When wired communication via the wired adapter gets disconnected while in service, the disconnection of the wired adapter is automatically detected and then wireless handoff occurs by mapping information on the wireless adapter to the virtual adapter. Through the proposed handoff mechanism, since IP address does not change, the session can be continued seamlessly even when handoff between wired and wireless network adapters occurs at lower level in a network application where both IP address and port number are used to maintain session. Finally, to evaluate the proposed handoff mechanism, actual experiments are performed for various internet applications such as FTP, HTTP, Telnet, and then their results are discussed.

In section 2, the existing network driver system is briefly shown and its limitations are discussed. In section 3, a new seamless handoff mechanism is proposed for a system with both wired and wireless network adapters. In section 4, actual experiments are performed for various internet applications. Finally, in section 5, conclusions are made.

2 Limitations of Existing Network Driver System

There is a kernel level library released by Microsoft to allow Windows systems to have networking capability. This library is called Network Driver Interface Specification (NDIS) [3]. The normal NDIS is formed with miniport drivers and protocol drivers as depicted in Fig. 1. The miniport driver is used to run network adapters and communicate with the protocol driver. The protocol driver such as TCP/IP or IPX/SPX/NETBIOS services obtains binding handle through a binding process with the miniport driver. In the binding process, the protocol driver makes bindings with all working miniport drivers. Up to the protocol driver belongs to the operating system's kernel level, and applications and others belong to user level [4]. During booting sequence of Windows systems, NDIS

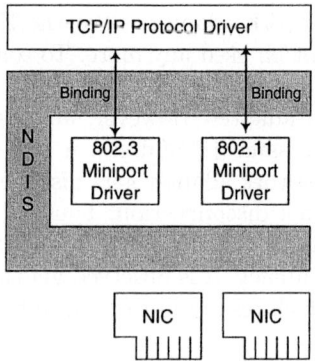

Fig. 1. Existing network driver system

initializes the miniport driver of registered adapter first. Then, as the protocol driver gets initialized, it binds with each miniport driver. Binding handle acts as a key used in transmission or reception of packets through corresponding adapter.

It is noted that the session must be newly made with the IP address of the new network adapter when a handoff occurs from the old network adapter to the new network adapter, since the IP address assigned to the old network adapter cannot be used in the new network adapter any more. Therefore, Windows systems using the existing network driver system of Fig. 1 cannot support the seamless handoff, because different IP addresses are assigned to each network adapter.

3 Proposed Seamless Handoff Mechanism

In order to solve the problems addressed in Section 2, this paper proposes a new seamless handoff mechanism between *wired* and *wireless* network adapters for a system with both network adapters.

3.1 New Network Driver System

Firstly, an intermediate driver is developed newly to modify packets sent from protocol driver then sends them to the miniport driver, and vice versa. The intermediate driver resides in between protocol driver and miniport driver, communicates with both miniport and protocol drivers as shown in Fig. 2. Note that the intermediate driver doesn't always exist, but exists only when it is needed by the developer. The intermediate drive generates the virtual protocol driver on the bottom and the virtual miniport drive on top. In other words, if we take a look at the virtual protocol drive and miniport driver on the bottom of the intermediate driver, then the intermediate driver works as the virtual protocol driver; and if we take a look at the virtual miniport driver and protocol driver on top of the intermediate driver, then the intermediate driver works as the virtual miniport driver. At this point, the virtual miniport driver is recognized by

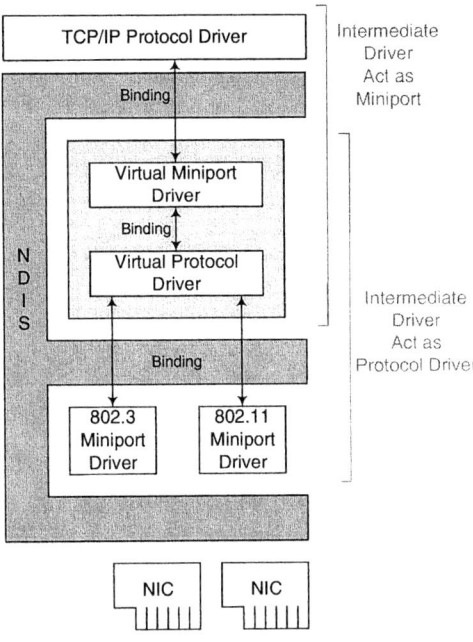

Fig. 2. New network driver system

actual upper level protocol driver as just another adapter [4]. The reason why the intermediate driver and the virtual drivers are in use is to solve some of the problems for the seamless handoff. Firstly, when there are two network adapters, each adapter must have different IP address, therefore, seamless communication is compromised during handoff. Due to this reason, a virtual adapter is used, whereas the actual protocol driver binds only with the virtual adapter to operate. This way, the layers above the protocol driver use the IP address assigned to this virtual adapter to communicate. Secondly, the intermediate drive can filter out data, which is sent to TCP/IP protocol driver in the event of a network adapter's network connection ends, to prevent session disconnection. Through this process, layers above the protocol driver do not know what is going in the lower level. Thirdly, use of the intermediate driver gives advantage of changing packet routing in real-time. By selecting the optimal adapter to be used for communication in accordance with the network connection status, packets can be transmitted and received through the optimal adapter. When this is realized, the handoff between wired and wireless network adapters can be done without disrupting the connection.

The intermediate driver in accordance with this paper can be divided into virtual miniport driver and virtual protocol driver as shown in Fig. 3. The virtual miniport driver includes virtual adapter control module. The virtual adapter control module generates a virtual adapter to control binding of an actual protocol driver to the virtual adapter. Furthermore, to set up each network adapter's property during the binding of the virtual protocol driver and a network adapter

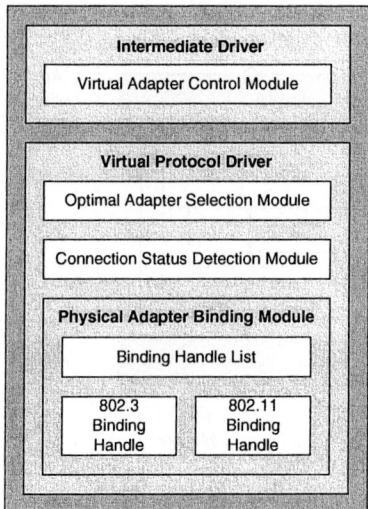

Fig. 3. Detailed diagram of the proposed driver system

at lower level, the virtual adapter control module sets all network adapters that are bound, to wireless adapter's link layer address. Here, under wired adapter, packet filtering property is set to promiscuous mode, which accepts all packets, and under wireless adapter, the property is set to direct/multicast/broadcast modes, which are typical. Hereinafter, the virtual adapter's link layer address is used as the wireless adapter's link layer address. In other words, under the wired adapter, since promiscuous mode is set, even though the virtual adapter's link layer address is set as the link layer address of the wireless adapter, corresponding packets can be transmitted and received. However, under the wireless adapter, since typical mode is used, the wireless adapter's link layer address is just used. Therefore, to upper protocol drivers, only one virtual adapter set with a link layer address is shown. Advantages from this are as follows. Firstly, by using the same link layer address, there is no need for retransmission of the address resolution protocol (ARP) packets to update ARP table in a hub [5]. Secondly, when assigning address dynamically using the dynamic host configuration protocol (DHCP), link layer address is used as one of options of DHCP protocol to identify the corresponding host. Here, by using the same link layer address again, DHCP server recognizes host as the same host and thus the same IP address gets assigned over and over [6]-[8]. Thirdly, when using link layer address in IPv6 (Internet Protocol version 6) environment assign IPv6 address through stateless auto configuration [9], [10], since the link layer address is the same, it is possible to assign an identical address.

When upper level protocol drivers request data of an adapter, the virtual adapter control module reports optimal adapter information selected by an optimal adapter selection module to an upper level protocol driver. Furthermore, the virtual adapter control module sends packet through binding handle of an optimal adapter chosen by the optimal adapter selection module. A virtual protocol driver includes an optimal adapter selection module, a connection status detection module and a network adapter binding module. The optimal adapter selection module decides to which adapter to transmit or to receive in accordance with connection information of wired and wireless network adapter which is delivered from the connection status detection module. The connection status detection module detects connection status information of the wired and wireless network adapters, then provides the information to the optimal adapter selection module in real-time. The network adapter binding module generates binding handle list through binding with all active adapters. In the binding handle list, binding handle and status information for controlling bound network adapters get stored.

3.2 Seamless Handoff Mechanism Using New Network Driver System

As illustrated in Fig. 4, when a connection event occurs while all wired and wireless network adapters are in disconnection status, the connection status detection module is used to recognize which adapter is in connection status. Based on the connection status, connection information of the corresponding adapter that is in the binding handle list of the network adapter binding module gets updated. At first, this updated information gets sent to the virtual adapter control module in real-time; then, right away the virtual adapter control module reports the connection status to the upper level protocol driver. After that, the optimal adapter selection module determines whether the connected network adapter is wired or wireless, and if it is the wired connection, then a wired adapter gets selected and sent the information to the virtual adapter control module. If a

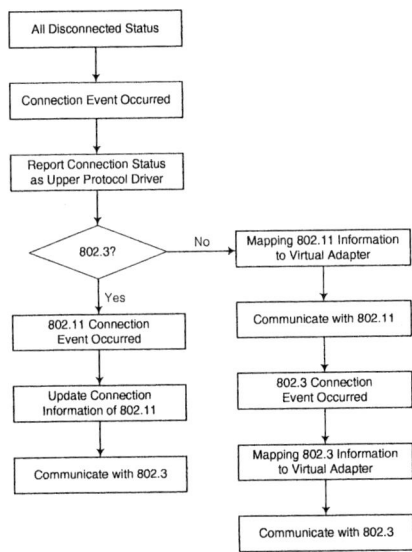

Fig. 4. Operation flow during connection event

wireless connection event occurs during wired communication, the connection status detection module recognizes connection of a wireless adapter, and this information updates only the connection information of the wireless adapter listed in the binding handle list of the network adapter binding module.

If a connection event of a wireless adapter occurs while all others are in disconnection status, the connection information obtained by the connection status detection module updates the connection status information of the corresponding adapter listed in the binding handle list of the network adapter binding module. Upon the updated information, the optimal adapter control module selects the wireless adapter as the optimal adapter. And then, the virtual adapter control module maps the information on the wireless adapter into a virtual adapter. Then, communication gets performed via the wireless adapter. If a wired connection event occurs during wireless communication, the information detected by the connection status detection module gets updated into the binding handle list and based on the updated information, the optimal adapter changes from the wireless to the wired by the optimal adapter selection module. Furthermore, the wired adapter's information gets mapped on to the virtual adapter. And at last, communication gets performed via the wired adapter. According to Fig. 5, when a disconnection event occurs during the wired communication, the connection status detection module determines whether the disconnected adapter is wired or wireless. If the adapter is the wired adapter, the corresponding adapter's connection status information gets updated to the binding handle list of the network adapter binding module. Then, the wireless adapter becomes the optimal adapter via the optimal adapter selection module, and the wireless adapter information gets mapped to the virtual adapter. Then, communication gets performed via the wireless adapter. If the disconnected adapter is not wired, the current connection status of the wired adapter must be verified. Upon the verification, if the wired adapter is still in connection, then communication is performed via the wired adapter. However, if the wired adapter is in disconnected status, the status gets reported to the upper level protocol driver, and the wired adapter's information gets mapped on to the virtual adapter.

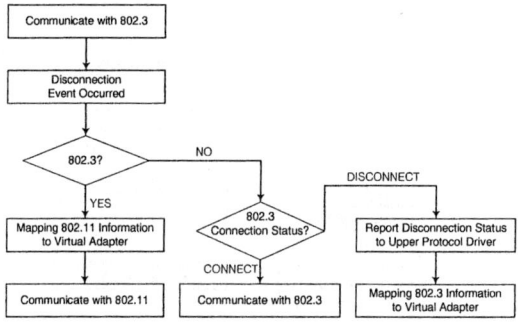

Fig. 5. Operation flow during disconnection event

4 Experiments and Results

To evaluate the proposed handoff mechanism, actual experiments are performed for various internet applications. The intermediate driver is realized on a portable PC using the driver development kit(DDK) [4] provided by Microsoft. The applications used in the experiment are FTP, HTTP, and Telnet. FTP supports large payload size and generates many packets for specific time in order to transmit large data, fast. HTTP's payload size and packet amount vary in accordance with amount of data each requested homepage provides. And for Telent, payload size and amount are relatively small since it supports interactive communication in text form.

In the experiment, the handoff latency is measured as follows. Firstly, a wired-to-wireless handoff occurs when wireless adapter takes over network connection as wired adapter loses the connection. And then, after mapping various information on the wireless adapter onto the virtual adapter, time it takes to transmit the first data packet has been measured. A wireless-to-wired handoff occurs when wired adapter connection is available during the wireless networking. And then, after mapping various information on the wired adapter onto the virtual adapter, time it takes to transmit the first data packet has been measured. Tick counter provided by Windows systems at kernel level is used as timer. Timing resolution per a tick of this tick counter is equivalent to approximately 16.3 $msec$. When testing each application, during data downloading FTP generated a handoff. For the case with HTTP, when downloading data linked to a homepage, handoff latency is measured during transmission of data via HTTP protocol. And for the case with Telnet, shell script to execute commands is made to run since packets can only be generated while sending or receiving commands through a terminal in Telnet.

Experimental results are shown in Fig. 6 and Table 1. Upon analyzing results, following two conclusions can be drawn. Firstly, wired-to-wireless handoff

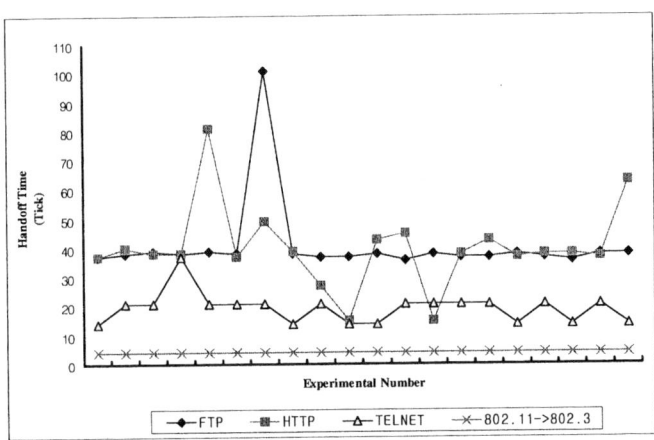

Fig. 6. Handoff Latency for Each Applications

Table 1. Mean value of handover latency

Handoff Direction	FTP	HTTP	Telnet
802.3 → 802.11	664msec	653msec	315msec
802.3 ← 802.11	25msec	25msec	25msec

latency takes as many as 10 times more than wireless-to-wired handoff latency, Secondly, a packet's payload size increases, or when packet amount per a unit time increases, handoff latency also increases. The first case is believed to happen because since 802.11 MAC protocol lacks reliability compare to 802.3 MAC protocol, some overhead had been added to 802.11 MAC protocol to overcome what it lacks [11], [12]. And the second case is believed to happen because packet size is large in different layers in accordance with Windows systems' internal scheduling rule; and in order to process many packets simultaneously, resource gets assigned to operations other than handoff. However, it is peculiar that a wireless-to-wired handoff remains constant regardless of the upper level application types. The reason for such behavior is believed to be happen because since reliability is always guaranteed for the wired adapter, and thus the impact the actual miniport driver, which drives the wired adapter, has on the actual throughput is insignificant since the driver is simple. Thus an analogy can be drawn here that the impact that the aforementioned overhead that 802.11 MAC protocol possesses is greater than that of resource sharing during a wired-to-wireless handoff. Therefore, the result, that nearly no impact is put on handoff even when packets are large, and resources being taken away to other layers due to the large number of packets, has been measured.

5 Conclusions

In this paper, the new seamless handoff mechanism between *wired* and *wireless* network adapters has been proposed for a system with both network adapters. The unique virtual adapter is developed between different adapters and then an IP address is assigned to the virtual adapter. As a general rule, when both network adapters are available, the wired adapter is preferred due to its faster transmission speed than wireless adapter. When wired communication via the wired adapter gets disconnected while in service, the disconnection of wired adapter is automatically detected and then wireless handoff occurs by mapping information on the wireless adapter to the virtual adapter. Through the proposed handoff mechanism, the session can be continued seamlessly even when handoff between wired and wireless network adapters occurs at lower level in a network application where both IP address and port number are used to maintain session. In order to evaluate the proposed handoff mechanism, actual experiments are performed for various internet applications such as FTP, HTTP, Telnet, and then their results are discussed.

In this paper, only the impacts of applications that are using TCP on handoff have been experiments. In the future, impacts that applications using transport layer other than TCP have on handoff will be analyzed, and intermediate driver that efficiently corresponds to the impacts will be developed. Furthermore, the reasons for difference in amount of time it takes to handoff from a wired adapter to a wireless adapter in accordance with the characteristics of an application will be more carefully dealt with, and plan to design and develop an intermediate driver that gets less impacts from an application.

References

1. Forouzan, B. A.: TCP/IP Protocol Suite. McGraw-Hill (1999)
2. Wright, G. R. and Stevens, W. R.: TCP/IP Illustrated Volume 2. Addison-Wesley (1995)
3. Windows Network Data and Packet Filtering [Online], Available : http://www.ndis.com (2002)
4. Microsoft: Driver Development Kit Help Documentation. Microsoft Corporation, Redmond, WA (2002)
5. Stevens, W. R.: TCP/IP Illustrated Volume 1. Addison-Wesley (1994)
6. Droms, R.: Automated configuration of TCP/IP with DHCP, IEEE Internet Computing, **3** (1999) 45–53
7. Sun Microsystems, Dynamic Host Configuration Protocol (Whitepaper), (2000) 6–10
8. Park, S.H., Lee, M.H., Kim, P.S., Kim, Y.K.: Enhanced mechanism for address configuration in wireless Internet. IEICE Trans. Commun. **E87-B** (2004) 3777–3780
9. Thomson, S., Narten, T: IPv6 stateless address autoconfiguration, IETF RFC 2462 (1998)
10. Droms, R.: Deploying IPv6, IEEE Internet Computing, **5** (2001) 79–81
11. Gast, M. S.: 802.11 Wireless Networks. O'Reilly (2002)
12. ISO/ICE.: Wireless LAN Medium Access Control (MAC) and Physical Layer (PHY) Specifications. ANSI/IEEE Std 802.11 (1999)

Approaches to Alternate Path Routing for Short Duration Flow in MPLS Network

Ilhyung Jung[1], Hyo Keun Lee[1], and Jun Kyun Choi[2]

[1] Korea Aerospace Research Institute (KARI),
45 Eoeun-dong Yuseong, Daejon, Korea
dragon@icu.ac.kr, hklee@kari.re.kr
[2] Information and Communications University (ICU),
P.O. Box 77, Yusong, Daejon, Korea
jkchoi@icu.ac.kr

Abstract. For Traffic engineering (TE) in QoS control, load balancing routing scheme of large IP backbone networks becomes a critical issue. Provisioning network resource in load balance is very difficult to ISPs because the traffic volume usually fluctuates widely over time and short duration flow has its bursty arrival process. In the paper, we propose the optimized alternate path routing schemes for short duration flow when congestion occurs. Our simulation result shows the proposed algorithm has less packet loss probability and less resource waste under heavy traffic load when we restrict the additional hop count by one or two.

1 Introduction

Many researches have been studied on the Quality of Service (QoS) to support a pre-defined performance contract between a service provider and end user. For QoS control, TE of large IP backbone networks becomes a critical issue and these TE studies have been accelerated by dynamic routing in recent years. However, provisioning network resource efficiently through dynamic routing is very difficult for ISPs because the traffic volume usually fluctuates widely over time. Recent studies show that only small amount of the flows have more than 10 packets but these flows carry the majority of the total traffic [1], [4], [8]. A short duration flow has more bursty arrival process, while a long duration flow has a less bursty arrival process. So a hybrid routing algorithm was proposed to reduce the overhead of routing complexity using these properties [1].

In this paper, we proposed an optimized alternate path selection algorithm based on the flow duration in an environment where the modified hybrid routing algorithm works. And our algorithm is simulated by the modified Routesim to measure the packet loss probability and resource waste under various link utilizations. The configuration of the paper is as follows. In Section 2, related works are reviewed and analyzed. Section 3 explains the methodologies of routing algorithms based on traffic duration time. In Section 4, we show simulation results under two types of networks. Conclusion and a summary are in presented in Section 5.

2 Previous Researches

2.1 QoS Routing in MPLS Network

For each new flow, network operator should assign a path to meet the flow's QoS requirements such as end-to-end delay or bandwidth guarantees [9]. Basically, QoS routing protocols must distribute topology information quickly and they must react to changes of network states and minimize their control traffic overhead. QoS routing also suffers from various problems such as diverse QoS specifications, dynamically changing network states in the mixture of the best-effort traffic [1], [4], [9], [6]. Those problems make the QoS routing complicated.

Moreover, the Internet traffic between particular points is unpredictable and fluctuates widely over time [8]. It is noted that most internet flows are short duration, and Link-State Update (LSU) propagation time and route computation time is relatively long to handle short duration flow [1]. A pure MPLS solution is probably too costly from the signaling point of view because the MPLS network also consists mainly of short duration flow.

2.2 Traditional Alternate Path Routing

In the traditional alternative path routing (APR), when the set up trial on the primary path fails, the call can be tried on alternative path in a connection oriented network. Simple updating of network status information may increase control traffic overhead and computational overhead. APR is a technique that can help to compensate for the routing inaccuracy and improve routing performance [4].

But alternate paths will use more resources than primary paths. Therefore, under heavy traffic load, much use of alternate routes may result in more serious congestion especially for long duration flow [3], [4], [9]. Our Algorithm uses alternate path only for the short duration flow to reduce those negative effects of the APR.

2.3 Hybrid Routing Scheme in MPLS Network

Two different schemes were proposed to allow the efficient utilization of MPLS for inter-domain traffics as well as the number of signaling operations [7], [13]. The first scheme is to aggregate traffic from several network prefixes inside a single LSP, and the second one is to utilize MPLS for high bandwidth flows and normal IP routing for low bandwidth flows. By introducing aggregation long duration flows, it is shown that performance is improved and reduced overhead via simulations [1], [4], [7], [12].

Hybrid routing algorithm in MPLS network, which is one of the second scheme, classifies arriving packets into flows and applies a trigger (e.g., arrival of some number of packets within a certain time interval) to detect long duration flow [1]. Then, the router dynamically establishes a shortcut connection that carries the remaining packets of the flow. The hybrid routing was introduced in [4], [5], [7].

3 Alternate Path Routing for Short Duration Flow in MPLS Network

This section describes the details of the proposed algorithm in modified hybrid routing. By default, router forward arriving packets onto the path selected by a traditional link-state routing protocol. When the accumulated size or duration of the flow exceeds a threshold (in terms of bytes, packets or seconds), the router in original hybrid routing scheme would select a dynamic route for the remaining packets in the flow depending on the bandwidth provisioning rule [1]. A variety of load-sensitive routing algorithms can be used in path selection for long duration flow to achieve high resource utilization and avoid congestion problems. From the insights of previous studies of load-sensitive routing, a dynamic algorithm favors short paths in order to avoid consuming extra resources in the network [6]. So, we should choose the widest-shortest path for long duration flow because long routes make it difficult to select a feasible path for subsequent long duration flows. When a link is overloaded, the router distributes some of the traffic over less-busy links by automatically "bumping" some packets in "non-optimal" directions. Packets routed in a non-optimal direction take more than the minimum required number of hops to reach their destination, however the network throughput is increased because congestion is reduced. But this load balancing routing scheme requires a number of signaling operations, link-state update messages and route computation. Therefore, simple, fast and robust algorithm is essential for bursty and unpredictable short duration flow while load-sensitive routing is used for long duration flow.

Fig. 1 show how the proposed alternate path scheme works. Each LSR checks that the link which is forwarding the packet is congested or not for short duration flow. The number of packets queued at LSRs output ports indirectly convey information about the current load. If the link is congested, we mark the congestion link infeasible and re-try the route selection. If this procedure returns "success," we check how many additional hops the path consumes. If the new route does not exceed the Max-Hopcount that is our restriction, additional hop counts, based on the original path, the new path takes the congestion test again. This procedure is important to prevent a looping problem or a resequence problem. And alternate path usually consumes excess bandwidth that would otherwise be available to other traffic. The whole network would be caused "domino" effect in that case. If the new-extracted path has fewer hops than the Max-Hopcount and sufficient bandwidth, we re-try to forward the traffic through that path. If the node can not find the re-extracted path, we mark that node as an infeasible node. Then we return the traffic to the previous node that is called crank-back node and the same procedures are repeated.

The suggested algorithm for short duration flow works on a simple distributed traditional routing scheme. The advantage of distributed routing for the short duration flow is that the routers need not keep persistent state for paths or participate in a path set-up procedure before data transfer begins. And the path set-up procedure is relatively long in delivering short duration flow.

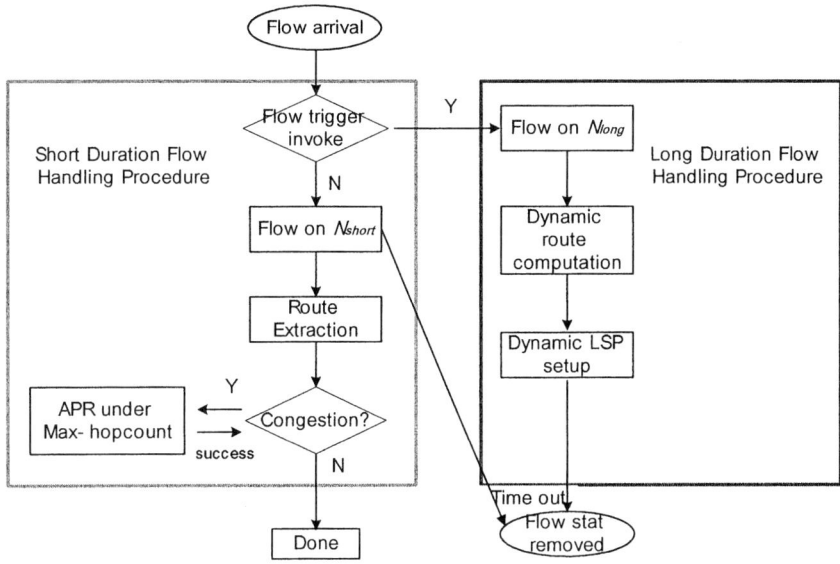

Fig. 1. Flow handling procedure of hybrid routing based on flow duration time

Since there are multiple routing paths for packets flowing between a given source-destination pair, our algorithm may cause packets to get out of sequence and may require resequencing buffers at the destination. The required buffer size in our algorithm and the length of the time-out interval can be computed using statistics of the source-to-destination delay distribution [10]. But the required resequencing buffer size can be small because the proposed algorithm limiting additional hops yields a small variance in the source-to-destination delay.

4 Simulation Result

This section evaluates our algorithm via simulation experiments. After a brief description of the simulation environment, we compare the proposed scheme to the traditional static routing algorithm at a hybrid scheme under various traffic loads. We show that, in contrast to shortest-path routing, our algorithm has lower packet loss probability and consumes minimum network resources.

4.1 Simulation Model

The simulation tool, Routesim [1] allows us to simulate the dynamics of load-sensitive routing, static shortest-path routing, and the effects of out-of-date link-state information. We adapt the alternate path routing algorithm in packet-switching level not in call-setup level. Previous studies shows that choosing a alternate path at the blocking node, that is called as Progressive Control (PC), tends to forward the traffic faster than Source Control(SC) [4]. Thus, supporting PC with or without crank-back can reduce more packet loss probability.

Implementing the algorithm has two major parts. First part is to implement a function which returns multiple alternate paths from a blocking node to the destination. Second is to implement route finding trials at crank-bank node. When a link is overutilized (for example 60% of total capacity), the link state is marked as Congestion.

4.2 Simulation Assumptions and Performance Measure Parameters

In simulations, we denote the short duration and long duration flow classes as N_{short} and N_{long}, respectively. The link capacity c_s is allocated for N_{short}, and c_l is allocated for N_{long}. For simplicity, flows are modeled as to consume a fixed bandwidth for their duration while their bandwidth changes usually occur over their session time in real networks. More studies and simulations should be conducted to get the variable data of bandwidth for flow's duration time. In choosing a traffic model, we must balance the need for accuracy in representing Internet traffic flows with practical models that are amenable to simulation of large networks. The assumptions in the simulation are as follows:

Flow Durations. In order to accurately model the heavy-tailed nature of flow durations, the flow duration in Routesim is modeled with an empirical distribution drawn from a packet trace from the AT&T World-Net ISP network [1]. The flow durations were heavy-tailed, which means that there were lots of flows with small durations and a few calls with long durations.

Flow Arrivals. We assumed flow arrivals to be a uniform traffic matrix specification with Poisson flow inter-arrival distribution. The value of λ is chosen to vary the offered network load, ρ (ρ varies from 0.6 to 0.9 in most of our experiments). This assumption slightly overstates the performance of the traditional dynamic routing scheme which would normally have to deal with more bursty arrivals of short duration flows.

Flow Bandwidth. Flow bandwidth is uniformly distributed with a 200% spread of the mean bandwidth value $\overline{b}$. The value of $\overline{b}$ is chosen to be about 1-5% (mainly 1.5 %) of the average link capacity. Bandwidth for long duration flow is assigned to be 85% while 15% for short duration flow.

Network Topology. In order to study the effects of different topologies, we used two topologies: the Random and the MCI topology. Their degrees of connectivity are quiet different, such as 7.0 and 3.37. The random graphs were generated using Waxman's model. The two topologies we described in Table 1 were widely used in other routing studies [1], [4], [9]. The 'Avg. path length' in Table 1 represents the mean distance (in the number of hops) between nodes, averaged across all source-destination pairs. Each node in the topology represents a core switch which handles traffic for one or more sources, and also carries transit traffic to and from other switches or routers.

Congestion Situation. When N_{short} reaches its capacity c_s, we call this as congestion in short duration flow. If the buffer thresholds are too small, the algorithm will overreact to normal state and too many packets will be bumped to longer routing paths.

Table 1. Topologies Used for Simulations

Topology	Nodes	Links	Degrees	Subnet	Net
Random	50	350	7.0	4	2.19
MCI	19	64	3.37	4	2.34

We simulate the algorithm under dual mode: applying a widest-shortest algorithm for long duration flow and shortest path first algorithm for short duration flow. We use the distance cost when selecting the path as "hop". To study the effect of additional hops, we vary additional hops from 1, 2 and infinite when using the algorithm. Paths are setup by on-demand routing policy for long duration flow while distributed routing for short duration flow. We evaluate the packet loss probability under various packet arrival rates in order to see the effect of additional hops for short duration flow. Average path length is checked to evaluate the proposed algorithm under heavy-traffic load. We also compare the alternate routing at the blocking node with the crank-back node. Their average path length and delay time are simulated.

4.3 Simulation Results

First, we study the performance of the proposed algorithm. The two networks with different connectivity were simulated with and without alternate routing. We considered only alternate routing at the congested node for this experiment. The simulation results are shown in Fig 2. We found from Fig 2(a) that a single additional hop in the algorithm leads to a significant improvement in the loss probability. Adding one more hop in our algorithm slightly reduces the loss probability when it is working in the range of 0.7 and 0.8 utilization. But the situation changes as the arrival rate exceeds 0.8 utilization. The packet loss probability becomes higher with two additional hops compared to that with a single alternate hop. Adding more alternate hops further degrades the packet loss probability without performance gain under any loading region.

We used the average path length (in hops) of the chosen paths as the measure of the resource usage of the paths. From Fig 2(b), as the network load increases, the number of average hops increases. The simulation results shows that the alternate path tends to be longer than the primary path for a heavy packet arrival rate, which means the alternate routing requires more network resources than the primary routing. The limit of additional hops must be carefully chosen to achieve satisfactory network performance. From the experiment, we find that less than two additional hops are sufficient to achieve the benefit of alternate routing without significantly increasing packet transfer delay in the range of 0.6 to 0.85 utilization. This result presents Braess' paradox that is an important consideration for the analysis of any network system that has alternate routes [11]. Braess' paradox says that alternate routing, if not properly controlled, can cause a reduction in overall performance.

In order to study the effect of different degrees of connectivity, we evaluate random graph with 7 degrees under the same traffic load. Dotted lines in Fig 2 show that richly connected topology significantly has low packet loss probability. Additional resources which alternate routing consumes, increases with the number of alternate hops allowed. But the bad effect of alternate routing under various packet arrival rates

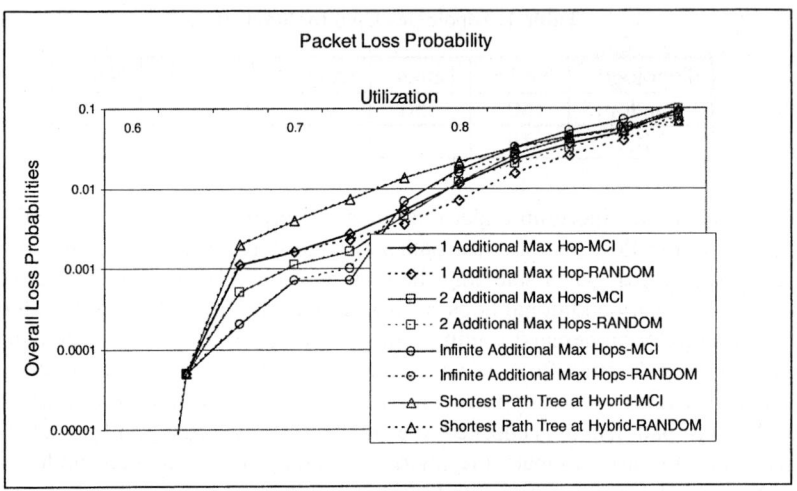

(a) Overall Packet Loss Probability

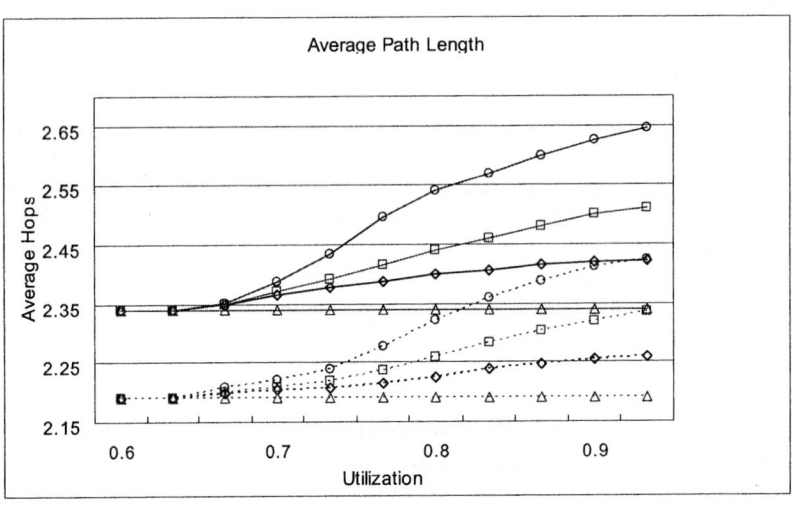

(b) Average Path Length

Fig. 2. Impact of alternate routing under MCI and Random network

is less than the one at MCI. From this experiment we find that, in Random topology, less than two additional hops are sufficient to achieve the benefit of alternate routing without significantly increasing packet transfer delay. The network with rich connectivity makes it possible to quickly disperse packets away from congested parts of the network with the proposed algorithm for short duration flow in distributed routing. So, we note the connectivity of network when we apply the alternate routing algorithm as well as note link utilization.

We also compared the algorithm with the crank-back scheme. We performed these experiments under the MCI topology and allowed only one additional hop. The packet loss probability of alternate routing at crank-back node has almost the same result with the alternate routing in at the congestion-node. However, alternate routing in our algorithm with the crank-back node yields longer paths than the proposed algorithm without crank-back.

5 Conclusions

In this paper, we introduced an efficient routing scheme that handle shortest first path algorithm combined with alternate routing for short duration traffic in hybrid routing algorithm. Our study shows that additional hops in alternate paths should be carefully restricted to avoid network resource waste under heavy load. The proposed algorithm has less packet loss probability and less resource waste because we restricted the resource by one or two. The network with rich connectivity has significantly less packet loss probability even though resource waste is the same as the network with poor connectivity. Finally, alternate routing at the congestion node and the crank-back node shows almost same packet loss probability while alternate routing at the crank-back node consumes more additional hops.

Acknowledgments

This research was supported in part by IITA (Institute of Information Technology Assessment) through MIC (Ministry of Information and Communication) and by KARI (Korea Aerospace Research Institute) and KOSEF (Korea Science and Engineering Foundation) through MOST (Ministry of Science and Technology), Korea

References

[1] A, Shaikh, Jennifer Rexford, Kang G. Shin,: 'Load-Sensitive Routing of Long-Lived IP Flows' Proceedings of ACM SIGCOMM, August 1999, Cambridge, MA, pp 215-226
[2] Keping Long, Zhongshan Zhang, Shiduan Cheng,: 'Load Balancing Algorithms in MPLS Engineering'. High Performance Switching and Routing, 2001 IEEE Workshop 2001, pp 175-179
[3] KRUPP, S 'Stabilization of alternate routing networks' Proceedings of IEEE International conference on Communications, ICC'82, Philadelphia, PA, June 1982, 3□.2.1-3□.2.5
[4] M. Sivabalan and H. T. Mouftah, 'Approaches to Link-State Alternate Path Routing in Connection-Oriented Networks' Modeling, Analysis and Simulation of Computer and Telecommunication Systems, 1998. Proceedings. Sixth International Symposium on, July 1998, pp 92-100
[5] V. Srinivasan, G. Varghese, S. Suri, and M. Waldvogel, 'Fast scalable algorithms for level four switching' in Proceedings of ACM SIGCOMM, September 1998, pp 191-202
[6] S. Chen and K. Nahrstedt, 'An overview of quality of service routing for next-generation high-speed networks: Problems and solutions' IEEE Network Magazine, November/December 1998, pp 64-79
[7] David Lloyd, Donal O'Mahony 'Smart IP Switching: A Hybrid System for Fast IP-based Network Backbones' IEEE, Jun 1998, pp 500-506

[8] A Feldmann, J. Rexford, and R. Caceres, 'Efficient policies for carrying Web traffic over flow-switched networks' IEEE/ACM Transactions on Networking, December 1998, pp 673-685
[9] Q. Ma and P. Steenkiste, 'On path selection for traffic with bandwidth guarantees' in Proceedings of IEEE International Conference on Network Protocols, Atlanta, GA, October 1997, pp 191-202
[10] Mark J. Karol and Salman Shaikh "A Simple Adaptive Routing Scheme for ShuffleNet Multihop Lightwave Networks" IEEE GLOBECOM 88, December 1988, pp 1640-1647
[11] N. G. Bean, F. P. Kelly, P.G. Taylor "Braess's Paradox in a Loss Network" Journal of Applied Probability, 1997, pp 155-159
[12] Steve Uhlig, Olivier Bonaventure "On the cost of using MPLS for interdomain traffic" Proceedings of the First COST 263 International Workshop on Quality of Furture Internet Services, 2000, pp 141-152
[13] I.F. Akyildiz, T. Anjali, L. Chen, J.C. de Oliveira, C. Scoglio, A. Sciuto, J.A. Smith, G. Uhl "A new traffic engineering manager for DiffServ/MPLS networks: design and implementation on an IP QoS Testbed" Computer Communications, 26(4), Mar 2003, pp 388-403

An Elaboration on Dynamically Re-configurable Communication Protocols Using Key Identifiers[*]

Kaushalya Premadasa and Björn Landfeldt

School of Information Technologies,
The University of Sydney,
Sydney 2006 Australia
{kpremada, bjornl}@cs.usyd.edu.au

Abstract. In this paper we elaborate on our novel concept and methodology for generating tailored communication protocols specific to an application's requirements and the operating environment for a mobile node roaming among different access networks within the global Internet. Since the scheme that we present employs a universal technique, it can be also deployed in small-scale independent networks such as sensor networks to generate application-specific lightweight transport protocols as is appropriate to its energy-constrained operating environments. Given that our proposed scheme is based on decomposing the communication protocols of the TCP/IP protocol suite, it allows sensor networks implementing the proposed scheme to easily connect to the existing Internet via a sink node consisting of a dual stack, without the loss of information in the protocol fields during the protocol translation process. We present preliminary experimental and analytical results that confirm and justify the feasibility of our method based on a practical example applicable to the sensor network environment.

1 Introduction

Networked and distributed computing is evolving at a tremendous rate. Both the applications and end hosts are becoming increasingly diversified and requirements on the behaviour and functions of the underlying communication infrastructure are following this evolution. The glue between the applications and the infrastructure is the communication protocol stack. In order to exploit the full potential of these diverse applications and nodes, the stacks should be configurable.

However, the current philosophy and design of the TCP/IP protocol stack has remained relatively intact for the past three decades and as a result, the individual protocols provide the same basic set of services to all applications regardless of individual needs.

In a networking environment such as wireless that grants the freedom of movement, the characteristics of the underlying network can change dynamically as a mobile node roams through different access networks thereby affecting the application's performance. As a result, we believe that the application requirements would also

---

[*] This research work is sponsored by National ICT Australia.

change in order to maximize resource usage and throughput given the changed conditions.

Therefore, considering that both the application requirements and the network characteristics can change dynamically in a mobile computing environment, we need to be able to dynamically re-configure the communication protocols to suit the new operating environment in order to achieve the optimum results for the new residing network.

A configurable stack also offers a distinct advantage for thin clients such as sensors operating in energy-constrained wireless environments. Since the functions required for the applications or roles of such devices are known, it is possible to streamline the implementation of the communication stack to only implement these functions. This has the distinct advantage that power consumption can be minimized since only necessary computations have to be made. This in turn will translate to longer life span for battery-powered devices.

In [1] we presented our novel concept and methodology for generating dynamic communication protocols customized to an application's requirements and the network characteristics for a mobile node roaming through different access networks within the global Internet. In this paper we elaborate on the details of our proposed method and make the following contributions:

- We describe by way of an example how the proposed method can be also deployed in energy-constrained sensor networks for generating application-specific lightweight transport protocols due to the uniformity and universality of the presented scheme.
- We show that for our proposed scheme the computational requirement to implement sequence control functionality alone on a gateway consisting of a 32-bit Intel StrongARM SA-1110 processor is 217 clock cycles compared with TCP's 1679 clock cycles, given that TCP does not allow the provision to implement only the desired functions on need basis.

2 Related Work

The related work consists of two parts. Given that our proposed scheme is based on decomposing the communication protocols of the TCP/IP protocol suite, in the first part we describe previous work related to deployment of TCP/IP within sensor networks to enable seamless connectivity with the global Internet. In the second part we describe the related work as applicable to dynamically generated protocols.

2.1 Deployment of TCP/IP for Sensor Networks

Sensor networks require the ability to connect to external networks such as the Internet through which activities such as monitoring and controlling can take place. Deploying TCP/IP directly on sensor networks would enable seamless integration of these networks with the existing Internet. However, it had been the conventional belief that TCP/IP, the de-facto standard for the wired environment is unsuitable for the wireless sensor networks consisting of nodes of limited capabilities, since its implementation needs a large resource requirement both in terms of code size and memory usage.

[2] disproved this widely accepted norm by describing two small TCP/IP implementations for micro-sensor nodes such as motes consisting of 8-bit microcontrollers, without its implementations sacrificing any of TCP's mechanisms such as urgent data or congestion control. The proposed TCP/IP implementations were written independently from the Berkeley BSD TCP/IP implementation [3], since the BSD implementation was originally written for workstation-class machines and hence not catered for the limitations of small-embedded systems. On the other hand, InterNiche NicheStack [4] is a portable BSD-derived implementation of TCP/IP that can be deployed in a more high-end sensor such as a gateway consisting of a 32-bit microcontroller such as Intel's StrongARM SA-1110 processor.

The proposed implementations of TCP/IP above for the sensor nodes have contributed to address the problem of enabling seamless connectivity of sensor networks with the global Internet. However, because TCP does not allow the facility to selectively implement functions as needed, these implementations do not allow the generation of application-tailored lightweight transport protocols for the sensor networks. For instance, an application may desire to implement the transport functionalities of error detection and sequence control but without the cost of retransmissions. By using TCP, it is not possible to satisfy such a requirement.

Hence in this paper, we elaborate on the details of our proposed scheme for dynamically generated communication protocols that has the potential and capability to address this important issue. Also, our scheme can be applied for any TCP/IP implementation including those proposed for the sensor networks since it is based on decomposing the communication protocols of the TCP/IP protocol suite. Furthermore, as a result of this latter feature, it also allows sensor networks deployed with this scheme to easily connect to the existing Internet through a gateway, with a loss-free mapping of the transport protocol information in the protocol translation process.

2.2 Dynamically Generated Protocols

Some architectural principles for the generation of new protocols were presented in [5]. These included implementation optimization techniques such as Application Level Framing and Integrated Layer Processing to reduce interlayer ordering constraints. The concept of a "protocol environment" was introduced in [6] consisting of standard communication functionalities. It proposed the ability for an application to create flexible protocol stacks using standard protocol entities thus leading to the generation of application-tailored, extensible stacks. The Xpress Transfer Protocol (XTP) was defined in [7] that consisted of a selective functionality feature allowing the selection of certain functions such as checksum processing or error control on a per packet basis.

The x-Kernel [8] provided an architecture for constructing and composing network protocols and also simplified the process of implementing protocols in the kernel. The notion of adaptable protocols was proposed in [9] and presented the generation of flexible transport systems through the reuse of functional elements termed "micro-protocols". The DRoPS project in [10] was concerned with providing the infrastructure support for the efficient implementation, operation and re-configuration of adaptable protocols and DRoPs based communication systems were composed of micro-protocols.

All these approaches have contributed to advance knowledge and demonstrate the benefits of dynamically generated protocols. However, because of the complexity involved in parsing dynamically generated headers with varying formats and compositions these approaches, unlike our proposed approach have proven too complex to realize and be widely deployed.

3 The Proposed Concept

3.1 Concept of Generating Tailored Communication Protocols

The framework for our proposed work is a modified, five-layered OSI model consisting of the Application, Transport, Network, Data Link and Physical layers.

The central idea adopted in defining tailored communication protocols is the concept that the Transport and Network layers of the proposed model can be decomposed into separate functions. For instance, the Transport layer can be decomposed into end-to-end flow control, sequence control, error detection, error recovery etc. Hence based on the application's requirements and the network characteristics, the application has the ability to select the functions it wishes to implement from layers three and four of this model. As a result, a communication protocol is generated tailored to the specific needs consisting of the functional information belonging to the requested functions.

It should be noted that details relating to how an application's requirements and network characteristics are specified are beyond the scope of this paper considering that this is an Application Program Interface (API) design issue.

3.2 Use of Key Identifiers to Differentiate Among Functional Information

In order to overcome the previous difficulties of parsing dynamically generated headers and to efficiently recover the necessary functional information from the resulting communication header, in [1] we introduced the concept of "Key Identifiers" that will be used to differentiate among the functional data belonging to the various functions. The fundamental idea that forms the basis for this concept is that each function of the Transport and Network layers of the proposed model is assigned a unique key, a method used since the 1940's [11] in communication systems, most notably in Code Division Multiple Access (CDMA) systems.

On the transmitter side, to enable the process of recovering the required functional information efficiently, each individual functional information of the Dynamically Re-Configured Communication Protocol header is firstly multiplied by the unique key of the function to which it belongs and the individual results are then summed to produce the final resulting communication header that will then be transmitted along with the Application and Data Link layer headers and the payload.

The same key used for encoding a particular functional data field is used to decode the same at the receiver. The receiver simply multiplies the received communication header with the key to recover the information.

It is worth noting that although one is generally accustomed to associating the use of Keys in the context of security, in the scheme that we present the Keys are used for the purpose none other than allowing the transmission of any combination of functional data as needed and efficient recovery of this information at a receiver. Therefore we derive a globally standardized key space for the Transport and Network layers of our proposed model such that each intermediate router and end-host maintains a table, mapping functions to keys so that the required information may be extracted.

A special key identification field is also included in the complete header that is transmitted in which each bit, relative to its position in this field, signifies whether the functional data belonging to a particular function is present as denoted by a "1" or conversely, the functional information is absent as denoted by a "0".

3.3 Process of Generating Key Identifiers

The chipping sequences that are used to encode the different functional information are the orthogonal Walsh functions. The main advantage of orthogonal codes is that they are able to completely eliminate multi-access interference as a result of their orthogonal property [12]. In the context of our work this translates to being able to correctly recover particular functional information from the received communication header, among the co-existence of many different functional data belonging to various functions in any given combination.

4 Key Distribution Approaches

Table 1 summarizes the fields that would be encoded and left un-encoded for the Transport and Network layers of the proposed model. The fields defined are derived from a composite of traditional communication protocols from the TCP/IP protocol suite.

Table 1. Summary of encoded/un-encoded fields

	Transport layer	Network layer
Encoded fields	Sequence #, Acknowledgement #, Source port #, Destination port #, Receive window, Internet checksum, Urgent data pointer, Flag Field (RST, SYN, FIN, ACK, URG, PSH, UNUSED, UNUSED)	Traffic class
Un-encoded fields		Destination address, Source address, Flow label, Hop limit

Given that the application-specific nature of sensor networks makes use of data-centric routing mechanisms as opposed to the Internet's address-centric mechanisms, only the functions applicable to transport layer will be implemented in the sensor nodes within these networks with the exception of the gateways.

There are two ways of assigning keys to fields and therefore also organizing the header information. In approach 1, each field is assigned a unique key to encode its functional information and in approach 2, the encoded fields from table 1 are grouped according to their bit lengths. Therefore, we derive three groups consisting of similar fields of 32, 16 and 8 bit lengths.

5 Mechanisms for Fast Computation

Given that computation in hardware is more efficient than in software, the multiplicative functions with respective keys would be implemented in hardware at the network layer instead of it being a full kernel implementation. Also, in order to allow fast and efficient recovery of desired functional information, an array of multipliers would be utilized to allow parallel multiplication of the received communication header by the appropriate key identifiers.

In the following section we present the initial experimental results for our proposed concept based on the mechanisms described in this section.

6 Experimental Results

6.1 Observed Results

It is of primary concern to investigate the computational overhead in using the key system compared with the standard way of sequentially extracting the header fields. We have therefore conducted a hardware simulation to determine the number of clock cycles it would take to decode a single bit that has been encoded by key lengths of 16-bits, 8-bits and 4-bits (as used by the key distribution approaches described in section 4) at an end-host, employing an array of multipliers operating in parallel as described in section 5.

The simulation was performed using Symphony EDA VHDL Simili, a software package consisting of a collection of tools that enable the design, development and verification of hardware models using VHDL, a hardware description language [13].

The experiment was conducted by encoding three bits (representing three separate bits from three different functional information) with three different keys that are generated using the key generation process described in section 3.5 above and then decoding the transmitted bits in parallel at the receiver (representing an end-host). Table 2 summarizes the results of this experiment.

Table 2. Results of the experiment

Key length (bits)	Number of clock cycles to decode each bit
16	10
8	6
4	4

Based on the results of the experiment given in table 2 above, table 3 summarizes the number of clock cycles it would take to decode a single functional information field in parallel for the proposed key distribution approaches described in section 4 above.

Table 3. Summary of clock cycles to decode a single field for the proposed key distribution approaches

	32-Bit field	16-Bit field	8-Bit field
Approach 1	32x10=320 clock cycles	16x10=160 clock cycles	8x10=80 clock cycles
Approach 2	32x4=128 clock cycles	16x6=96 clock cycles	8x4=32 clock cycles

6.2 Comparison with the Existing Method

The results of this experiment are very encouraging based on past research conducted to determine the overhead associated with TCP/IP protocol processing. In order to justify this conclusion, we firstly describe the past experiments that were conducted and present the results of those investigations. These results are then later used as a comparison point for evaluating the feasibility of our proposed method.

6.2.1 Related Work on TCP/IP Protocol Processing.

An experiment was conducted in [14] to analyze the TCP processing overhead for a modified Berkeley implementation of Unix on a 32-bit Intel 80386 processor with a clock speed of 16MHz and an instruction execution rate of 5 million instructions per second (MIPS). The TCP code was modified to give a better measure of the intrinsic costs involved. Fig.1 illustrates the methodology that was used for this analysis.

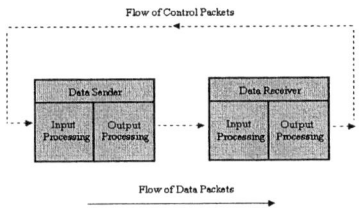

Fig. 1. Methodology used for analysis in [14]

It was discovered that for input processing at both a data sender and a data receiver the common TCP protocol-specific processing path consisted of 154 instructions of which 15 were either procedure entry and exit or initialization. In particular, the input processing for a data receiver also consisted of an additional 15 instructions for sequencing the data and calling the buffer manager, 17 instructions for processing the window field of a packet and 25 instructions for finding the Transmission Control Block (TCB) for the TCP connection. It is worthwhile mentioning that the reported instruction counts for various TCP protocol-specific tasks consisted of extracting the information from the header fields and execution of the corresponding algorithms. It was also found that the output processing of a data receiver consisted of a total of 235 instructions to send a packet in TCP. Therefore based on these results, for a data receiver, the ratio of input to output processing instructions is given by 211:235 respectively.

Although all the reported experiments conducted in [14] were on processors of a Complex Instruction Set Computer (CISC) architecture, the authors have stated that based on a separate study of packet processing code, they found little expansion of the code when converted to a Reduced Instruction Set Computer (RISC) chip. They concluded that this is because given the simplicity of the operations required for packet processing, irrespective of the processor that is used the instruction set actually utilized is a RISC set. Given that the authors have estimated the ratio of CISC to RISC instructions to be approximately 3:4 respectively, 211 and 235 instructions for an 80386 processor would be translated to 282 and 314 instructions respectively for a RISC processor.

In an independent experiment also conducted in [14] to measure the actual costs involved with a Berkeley TCP running on a Sun-3/60 workstation with a 32-bit Motorola 68020 processor with a clock speed of 20MHz and an instruction execution rate of 2 MIPS it was discovered that it took 370 instructions for TCP checksum computation which is the major overhead associated with TCP processing. This figure thus translates to 494 instructions for a RISC processor based on the above ratio of CISC to RISC instructions.

In a similar study conducted in [15] an experiment was conducted to determine the overhead associated with the Ultrix 4.2a implementation of TCP/IP software processing. The experiment consisted of one workstation sending a message to the system under test (where all measurements were taken) which then sends the same message back to the originator. All measurements were taken on a DECstation 5000/200 workstation, a 19.5 SPECint MIPS RISC machine with a clock speed of 25MHz and an instruction execution rate of 20 MIPS.

In this experiment it was discovered that for TCP messages, protocol-specific processing (that includes operations such as setting header fields and maintaining protocol state with the exception of checksum computations) consumes nearly half the total processing overhead time. It was also discovered that TCP protocol-specific processing in particular dominates the protocol-specific processing category consuming approximately 44% of the processing time. The TCP protocol-specific processing however did not include checksum computation in its classification.

In this experiment it was discovered that 3000 instructions were consumed for TCP protocol-specific processing at a data receiver. Based on the ratio of RISC processor instructions for input to output processing for a data receiver in [14], approximately 1420 instructions were thus consumed in [15] for Input processing.

6.2.2 Evaluation of the Feasibility of the Proposed Method
In order to demonstrate the viability of our proposed method we present a practical example within the sensor network environment in conjunction with the routing protocols, in which the transport layer functionality of sequence control would be very desirable. We confirm the feasibility of our approach by evaluating the number of clock cycles that would be consumed to implement this functionality alone with our

proposed method to that consumed using the traditional sequencing transport protocol TCP that does not allow the provision to implement only the desired functions on need basis.

There are no data available in relation to the processing times for the implementations of TCP/IP as described in section 2.1 for the sensor networks. However, given that InterNiche NicheStack as reported above consists of a BSD-derived implementation of TCP/IP that can be deployed in a more high-end sensor such as a gateway consisting of a 32-bit microcontroller, we therefore believe that it is fair to assume that its implementation would be very similar to those described in section 6.2.1, and we conduct our analysis based on this assumption.

In [16] a family of adaptive protocols called Sensor Protocols for Information via Negotiation (SPIN) was proposed for the Network layer for disseminating information among sensor nodes in an efficient manner. The goal of SPIN family of protocols is to conserve energy through negotiation by firstly communicating a high-level data descriptor called a meta-data that describes the data without transmitting all the data. SPIN protocols employ a simple three-way handshake mechanism for negotiation based on three types of messages: ADV, REQ and DATA. Prior to sending the DATA message, the sensor node broadcasts an ADV message containing the meta-data. Neighbours that are interested in the data then respond with a REQ message after which the DATA is sent to those sensor nodes. This process is thus repeated by the neighbour sensor nodes once they receive a copy of the interested data. Therefore, all interested nodes within the entire sensor network would eventually receive a copy of this data.

For a sensor network employing such a data transaction mechanism, it would be very desirable to be able to utilize the transport layer functionality of sequence control to allow the data receiving neighbour nodes to sequence the data, at the cost of minimum overhead. Using our proposed methodology this can be very easily fulfilled by appending to the broadcasted ADV message a dynamically generated communication header consisting of the encoded sequence number field bootstrapped with the chosen initial sequence number. As a result, the neighbours that respond with a REQ message would be all aware of the initial sequence number associated with the data transaction which is to follow. Thereafter, these nodes can use the information in the encoded sequence number field of subsequent data packets to organize the received data in their proper order.

In [14] it was discovered that it took 15 instructions for sequencing the data and calling the buffer manager that translates to 20 RISC processor instructions, based on the ratio of CISC to RISC instructions reported above. Therefore based on the ratio of RISC processor instructions for the sequence control functionality to that of input processing for a data receiver in [14], approximately 101 instructions are thus consumed in [15] for the same sequence control functionality.

Table 4 provides a summary of the instruction counts for various tasks as reported above and calculates the number of clock cycles it takes to execute these instructions on a gateway to a collection of motes consisting of a 32-bit Intel StrongARM RISC SA-1110 processor with a clock speed of 206 MHz and an instruction execution rate of 235 MIPS [17]. In this analysis we make the assumption that the instruction count for TCP checksum computation based on a RISC processor in [14] is similar for the same in [15].

Table 4. Summary of instruction counts for various tasks and the associated clock cycles for their execution

	Instruction count as per [15]	Number of clock cycles as per StrongARM SA-1110 processor
Sequencing data	101	89
Input processing at data receiver	1420	1245
TCP checksum computation	494	434

In Table 5 and in fig.2 we therefore provide a comparison of the number of clock cycles it takes on the StrongARM SA-1110 processor to implement the sequence control functionality alone with our proposed methodology to that of the traditional sequencing transport protocol TCP that does not allow the facility to implement only the desired functions on need basis. In TCP it is also not possible to disable certain functionality by simply setting a particular header field to zero since for most fields zero is a valid value.

For the results of table 5 below, it should be noted that in the case of our proposed method based on both key distribution approaches, during the time spent to decode the sequence number field, up to an additional 12 and 14 encoded functional fields can also be decoded in parallel at a node for key distribution approaches 2 and 1 respectively. Also, the value of 89 clock cycles for sequencing the data actually comprises of the cycles consumed for both extracting the information from the header field and execution of the corresponding algorithm associated with the sequence control function. Therefore, the total cycles consumed for implementing the sequence control functionality using our method would be theoretically less for both key distribution approaches than the calculated values if the cycles consumed for extracting the functional information from the header field had been subtracted.

Table 5. Comparison of clock cycles for sequencing of data for our proposed method against TCP

	Number of clk cycles for our method: key distribution approach 2	Number of clk cycles for our method: key distribution approach 1	Number of clk cycles for TCP
Decoding sequence number field	128	320	
Sequencing data	89	89	
Input processing at data receiver			1245
TCP checksum computation			434
TOTAL	217	409	1679

From the results of fig.2 below it can be clearly seen that the number of clock cycles consumed for sequencing of data using our method based on key distribution approach 2 and approach 1 are 1462 and 1270 cycles respectively less than that for the case of TCP. Also key distribution approach 2 consumes only about half the number of clock cycles compared to key distribution approach 1. Therefore through this simple but practical example we have demonstrated the feasibility and viability of our proposed method for tailoring communication protocols for the sensor network environment for which energy conservation is of utmost importance.

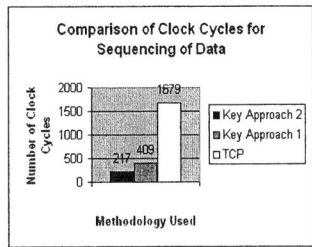

Fig. 2. Comparison of clock cycles for sequencing of data for our proposed method against TCP

7 Conclusions and Future Work

In this paper we have elaborated on the details of our novel method based on the well-known CDMA technique for dynamically generating communication protocols customized to an application's requirements and the network characteristics within the wireless environment. The advantages of being able to dynamically generate communication protocols based on a specification have been made clear, with its benefits extending to a spectrum of wireless devices with varying processing capabilities such as sensor nodes and portable laptops.

The feasibility and viability of the proposed method have been proven with initial experimental work and through comparison of these results to those obtained from past studies carried out to discover the cost of TCP/IP protocol processing overheads.

We are currently working on a header compression technique for the proposed scheme. As continuing work, we will be working toward a full system implementation of the proposed concept for the sensor networks.

References

1. Premadasa, K., Landfeldt, B.: Dynamically Re-Configurable Communication Protocols using Key Identifiers. In: to be published in the proceedings of ACM/IEEE 2nd conference on MobiQuitous 2005, San Diego, California, USA (July 2005)
2. Dunkels, A.: Full TCP/IP for 8-bit architectures. In: Poc. 1st conference on MOBISYS'03 (May 2003)
3. McKusick, M.K., Bostic, K., Karels, M.J., Quarterman, J.S.: The Design and Implementation of the 4.4 BSD Operating System. Addison Wesley, United States of America (1996)

4. InterNiche Technologies Inc NicheStack portable TCP/IP stack: www.iniche.com/products/tcpip.htm
5. Clark, D.D., Tennenhouse, D.L.: Architectural Considerations for a new Generation of Protocols. In: ACM SIGCOMM Computer Communications Review, Vol. 20, No.4 (August 1990) 200-208
6. Tschudin, C.: Flexible Protocol Stacks. In: ACM SIGCOMM Computer Communications Review, Vol. 21, No.4 (August 1991) 197-205
7. Strayer, W.T., Dempsey, B.J., Weaver, A.C.: XTP: The Xpress Transfer Protocol. Addison Wesley, United States of America (1992)
8. Hutchinson, N.C., Peterson, L.L.: The x-Kernel: An architecture for Implementing Network Protocols. IEEE Transactions on Software Engineering, Vol. 17, No.1 (January 1991) 64-76
9. Zitterbart, M., Stiller, B., Tantawy, A.: A Model for Flexible High-Performance Communication Subsystems. In: IEEE Journal on Selected Areas in Communications, Vol. 11, No.4 (May 1993) 507-518
10. Fish, R.S., Graham, J.M., Loader, R.J.: DRoPS: Kernel Support for Runtime Adaptable Protocols. In: Proc. 24th Euromicro conference '98, Vol.2, Vasteras, Sweden (1998) 1029-1036
11. Scholtz,: The Evolution of Spread-Spectrum Multiple Access Communications. In: Code Division Multiple Access Communications. Glisic, S.G., Leppanen, P.A. (eds.): Kluwer Academic Publishers (1995)
12. Rhee Y.M.: CDMA Cellular Mobile Communications & Network Security. Prentice Hall Inc., United States of America (1998)
13. Symphony EDA. URL: http://www.symphonyeda.com
14. Clark, D., Jacobson, V., Romkey, J., Salwen, H.: An Analysis of TCP Processing Overhead. In: IEEE Communications Magazine, 50th Anniversary Commemorative Issue (May 2002) 94-101
15. Kay, J., Pasquale, J.: Profiling and Reducing Processing Overheads in TCP/IP. In: IEEE/ACM Transactions on Networking, Vol. 4, No.6 (December 1996) 817-828
16. Heinzelman, W.R., Kulik, J., Balakrishnan, H.: Adaptive Protocols for Information Dissemination in Wireless Sensor Networks. In.: Proc. ACM Mobicom '99, Seattle, Washington, USA (1999) 174-185
17. Intel StrongARM SA-1110 Datasheet. URL: www.intel.com/design/strong/datashts/278241.htm

Optimal Broadcast for Fully Connected Networks

Jesper Larsson Träff and Andreas Ripke

C&C Research Laboratories, NEC Europe Ltd.,
Rathausallee 10, D-53757 Sankt Augustin, Germany
{traff, ripke}@ccrl-nece.de

Abstract. We develop and implement a new optimal broadcast algorithm for fully connected, bidirectional, one-ported networks under a linear communication cost model. For *any* number of processors p the number of communication rounds required to broadcast N blocks of data is $\lceil \log p \rceil - 1 + N$. For data of size m, assuming that sending and receiving m data units takes time $\alpha + \beta m$, the best running time that can be achieved is $(\sqrt{(\lceil \log p \rceil - 1)\alpha} + \sqrt{\beta m})^2$, meeting the lower bound under the assumption that the m units are sent as N blocks. This is better than previously known (and implemented) results, which achieve this only when p is a power of two (or other special cases), in particular, the algorithm is (theoretically) a factor two better than the commonly used, pipelined binary tree algorithm. The algorithm has a regular communication pattern based on simultaneous binomial-like trees, and when the number of blocks to be broadcast is one, degenerates into a binomial tree broadcast. Thus the same algorithm can be used for all message sizes m. The algorithm has been incorporated into a state-of-the-art MPI (*Message Passing Interface*) library. We demonstrate significant practical improvements of up to a factor 1.5 over several other, commonly used broadcast algorithms.

1 Introduction

There has recently been renewed interest in efficient, portable and easy to implement broadcast algorithms for use in *Message Passing Interface* (MPI) libraries [12], the current *de facto* standard for distributed memory parallel computers [3,9,10,13,15]. Earlier theoretical results, typically assuming a strict, one-ported communication model in which processors can either send or receive messages are summarized in [5,7]. Early implementations typically assume a homogeneous, fully connected network, and were often based on straightforward binary or binomial trees, which are inefficient for large data sizes. Better MPI libraries (for instance [6]) take the hierarchical communication system of current clusters of SMP nodes into account by broadcasting in a hierarchical fashion, and use pipelined binary trees, or algorithms based on recursive halving [13,15] that are much better as data size increases. However, these algorithms are all (at least) a factor two off from the theoretical optimum.

A theoretically better algorithm for hypercubes (later extended to incomplete hypercubes) was proposed by Johnsson and Ho [8,14]. Instead of pipelining the blocks successively through a fixed-degree tree, in this so called *edge-disjoint spanning binomial tree algorithm* the root processor sends successive blocks to its children in a round-robin fashion, each of which functions as a root in a binomial tree that is edge-disjoint from the other binomial trees. Another interesting algorithm based on so called *fractional trees* [10] provides for a smooth transition from pipelined binary tree to a linear pipeline algorithm as data size increases, and gives an improvement over both for medium data sizes. In the arguably more realistic *LogP* model [1,4] an (near) optimal algorithm was given in [11], but without implementation results. Non-pipelined broadcast algorithms in hierarchical, clustered, heterogeneous systems were discussed in [2], which proposes a quite general model for such systems.

In this paper we first give an algorithm for one-ported, fully connected networks of a pipelined broadcast algorithm that is quite similar to the algorithm of Johnsson and Ho [8] when the number of processors is a power of two. Fully connected networks are realized by crossbars as in the Earth Simulator, and the assumption is a reasonable approximation for medium sized networks for high-end clusters like Myrinet or Quadrics. Our main result extends this algorithm to arbitrary numbers of processors. An important feature of the algorithm is that it degenerates towards the binomial tree algorithm as the number of blocks to be broadcast decreases. A smooth transition from short data to long data behavior is thus possible with one and the same algorithm. The optimal algorithm has been implemented in a state-of-the-art MPI library [6], and we present an experimental comparison to other, commonly used broadcast algorithms on a 32-node AMD cluster with Myrinet interconnect, showing a bandwidth increase of more than a factor 1.5 over these algorithms.

2 Problem, Preliminaries and Previous Results

For the rest of this paper m denotes the amount of data to be broadcast, and p the number of processors which are numbered from 0 to $p-1$. Logarithms are to the base 2, and we let $n = \lceil \log p \rceil$. Without loss of generality, we assume that broadcast is from processor 0 (otherwise, processor 0 and the broadcast *root* processor exchange roles). In MPI, broadcast is a *collective operation* MPI_Bcast(buffer,count,datatype,root,comm) to be executed by all processors in the communicator comm.

We assume a fully connected, homogeneous network with one-ported, bidirectional communication, and a simple, linear cost model. A processor can simultaneously send and receive a message, possibly from two different processors, and the time to send m units of data is $\alpha + \beta m$ where α is the start-up latency and β the transfer time per unit. In the absence of network conflicts this model is somewhat accurate, although current communication networks and libraries typically exhibit a large difference in bandwidth between "short" and "long" messages. The extended *LogGP* model, for instance, attempts to capture this [1], and we discuss this problem further in Section 4. The model also does not match current clusters of SMP nodes that have a hierarchical communication system. We cater for this by broadcasting hierarchically on such systems, but do not discuss this further in this paper; see instead the companion paper [16].

2.1 Lower Bounds

In the homogeneous, linear cost model, the lower bound for broadcasting the m data is $\max(\alpha n, \beta m)$. Assuming furthermore that the m data is sent as N blocks of m/N units, the number of rounds required is $n - 1 + N$, for a time of $(n - 1 + N)(\alpha + \beta m/N) = (n - 1)\alpha + (n - 1)\beta m/N + N\alpha + \beta m$. By balancing the terms $(n - 1)\beta m/N$ and αN, the optimal number of rounds can be found as

$$N_{\text{opt}} = \sqrt{\frac{(n-1)\beta m}{\alpha}}$$

and the optimal block size as

$$B_{\text{opt}} = \sqrt{\frac{m\alpha}{(n-1)\beta}} = \sqrt{\frac{m}{n-1}}\sqrt{\frac{\alpha}{\beta}} \qquad (1)$$

for a total running time of

$$T_{\text{opt}}(m) = (n-1)\alpha + 2\sqrt{(n-1)\alpha}\sqrt{\beta m} + \beta m = (\sqrt{(n-1)\alpha} + \sqrt{\beta m})^2 \quad (2)$$

For proofs of these lower bounds, see e.g. [8,10]. Other algorithms are off from the lower bound either by a larger latency term (e.g. linear pipeline) or a larger transmission time (e.g. $2\beta m$ for a pipelined binary tree).

3 The Algorithm

In this section we give a high-level description of our new optimal broadcast algorithm. The details are filled in first for the easier case where p is a power of two, then for the general case. First, we assume that the data to be broadcast have been divided into N blocks, and that $N > n$. Note that the algorithm as presented here only achieves the $n - 1 + N$ rounds for certain combinations of p and N; in some cases up to $(n - N \bmod n - 1)$ extra rounds may be required for some processors (see Subsection 3.4).

The algorithm is pipelined in the sense all processors are both sending and receiving blocks at the same time. For sending data each processor acts as if it is a root of a(n incomplete, when p is not a power of 2) binomial tree. Each non-root processor has n different parents from which it receives blocks. To initiate the broadcast, the root (processor 0) sends the first n blocks successively to its children. The root continues in this way sending blocks successively to its children in a round robin fashion. A sequence of n blocks is called a *phase*.

The non-root processors receive their first block from the parent in the binomial tree rooted at processor 0. The non-roots pass this block on to their children in this tree. After this initial *fill phase*, each processor now has a block, and the broadcast goes into a *steady state*, in which in each round each processor (except the root) receives a new block from a parent, and sends a previously received block to a child.

A more formal description of the algorithm is given in Figure 1. The buffer containing the data being broadcast is divided into N blocks of roughly m/N units, and the ith block is denoted buffer[i] for $0 \leq i < N$.

Root processor 0:

```
/* fill */
for i ← 0, n − 1 do
    send(buffer[sendblock(i, 0)], next(i, 0))
/* steady state */
for i ← 1, N do
    j ← (i − 1) mod n
    send(buffer[sendblock(n − 1 + i, 0)], next(j, 0))
```

Non-root processor r:

```
/* fill */
i ← first(r)
recv(buffer[recvblock(i, r)], prev(i, r))
for i ← first(r) + 1, n − 1
    send(buffer[sendblock(i, r)], next(i, r))
/* first block received, steady state */
for i ← 1, N
    j ← (i − 1) mod n
    if next(j, r) ≠ 0 then /* no sending to root */
        send(buffer[sendblock(n − 1 + i, r)], next(j, r))
    ‖ /* send and receive simultaneously */
    recv(buffer[recvblock(n − 1 + i, r)], prev(j, r))
```

Fig. 1. The optimal broadcast algorithm. For the general case where p is not a power of two, small modifications of the fill phase and the last rounds are necessary.

As can be seen each processor receives N blocks of data. That indeed N different blocks are received and sent is determined by the recvblock(i, r) and sendblock(i, r) functions which specify the block to be received and sent in round i for processor r. In the next subsections we describe how to compute these functions. The functions next and prev determine the communication pattern. In each phase the same pattern is used, and the n parent and child processors of processor r are next(j, r) and prev(j, r) for $j = 0, \ldots n-1$. The parent of processor r for the fill phase is first(r), and the first round for processor r is likewise first(r). With these provisions we have:

Theorem 1. *In the fully-connected, one-ported, bidirectional, linear cost communication model, N blocks of data can be broadcast in $n-1+N$ rounds reaching the optimal running time (2).*

The algorithm is further simplified by the following observations. First, the block to send in round i is obviously

$$\mathsf{sendblock}(i, r) = \mathsf{recvblock}(i, \mathsf{next}(i, r))$$

so it will suffice to determine a suitable recvblock function. Actually, we can determine the recvblock function such that for any processor $r \neq 0$ it holds that

$$\{\mathsf{recvblock}(0, r), \mathsf{recvblock}(1, r), \ldots, \mathsf{recvblock}(n-1, r)\} = \{0, 1, \ldots, n-1\}$$

that is the recvblock for a phase consisting of rounds $0, \ldots n-1$ is a permutation of $\{0, \ldots, n-1\}$. For such functions we can, with slight modifications for the last phase, take for $i \geq n$

$$\mathsf{recvblock}(i, r) = \mathsf{recvblock}(i \bmod n, r) + n(\lfloor i/n \rfloor - 1 + \delta_{\mathsf{first}(r)}(i \bmod n))$$

where $\delta_j(i) = 1$ if $i = j$ and 0 otherwise. Thus in rounds $i+n, i+2n, i+3n, \ldots$ for $0 \leq i < n$, processor r receives blocks $\mathsf{recvblock}(i, r), \mathsf{recvblock}(i, r) + n, \mathsf{recvblock}(i, r) + 2n, \ldots$ (plus n if $i = \mathsf{first}(r)$). We call such a recvblock function a *full block schedule*. The broadcast algorithm is *correct* if the full block schedule fulfills the conditions that either

$$\mathsf{recvblock}(i, r) \in \{\mathsf{recvblock}(j, \mathsf{prev}(i, r)) \mid 0 \leq j < i\} \tag{3}$$

or

$$\mathsf{recvblock}(i, r) = \mathsf{recvblock}(\mathsf{first}(\mathsf{prev}(i, r)), \mathsf{prev}(i, r)) \tag{4}$$

for $0 \leq i < n$, i.e. the block that processor r receives in round i from processor $\mathsf{prev}(i, r)$ has been received by that processor in a previous round.

When $N = 1$ the algorithm degenerates into an ordinary binomial tree broadcast, that is optimal for small m. The number of blocks N can be chosen freely, e.g. to minimize the broadcast time under the linear cost model, or, which is relevant for some systems, to limit communication buffer space.

3.1 Powers of Two Number of Processors

For the case where p is a power of two, the communication pattern and block schedule is quite simple. For $j = 0, \ldots n-1$ processor r receives a block from processor $\mathsf{prev}(j, r) = (r - 2^j) \bmod p$ and sends a block to processor $\mathsf{next}(j, r) = (r + 2^j) \bmod p$, that is the distance to the previous and next processor doubles in each round.

The root sends n successive blocks to processors $1, 2, 4, \ldots, 2^j, \ldots 2^{n-1}$ for $j = 0, \ldots n-1$ with this pattern. The subtree of child processor $r = 2^j$ consists of the processors $(r + 2^k) \bmod p, k = j+1, \ldots, n-1$. Processors $2^j, \ldots 2^{j+1} - 1$ together form *group* j, since these processors will all receive their first block in round j. The *group start* of group j is 2^j, and the *group size* is likewise 2^j. Note that $\mathsf{first}(r) = j$ for all processors in group j. Figure 2 shows the

group:	0	1	2				3							
proc r:	1	2 3	4 5 6 7				8 9 10 11 12 13 14 15							
schedule:	0	1 0	2 0 1 0				3 0 1 0 2 0 1 0							

Fig. 2. First block schedule for $p = 16$

blocks received in rounds 0 to 3 for $p = 16$. We call the sequence of first blocks received by the processors the *first block schedule*, i.e. schedule[r] is the first block that processor r will receive (in round first(r)). It is easy to compute the schedule array: assume schedule[i] computed for groups $0, 1, \ldots, j-1$, that is for $1 \leq i < 2^j$; then set schedule[2^j] $= j$, and schedule[$2^j + i$] = schedule[i] for $i = 1, \ldots, 2^j - 1$ (incidentally this sequence is a palindrome). We need the following property of the schedule array.

Lemma 1. *Any segment of size 2^j of the first block schedule for p a power of two contains exactly $j + 1$ different blocks.*

The proof is by induction on j.

Using the first block schedule we can compute a full block schedule recvblock as follows. In round n (the first round after the fill phase) processor r receives a block from processor $r' = (r-1) \bmod p$; this can only be the block that processor r' received in its first round first(r'), so take recvblock($0, r$) = schedule[r']. For $0 < i <$ first(r) we take recvblock(i, r) to be the *unique* block in schedule[prev(i, r) $- 2^i + 1$, prev(i, r)] which is *not* in schedule[prev(i, r) $+ 1, r$] and is *not* schedule[r]. These two adjacent segments together have size 2^{i+1}, and by Lemma 1 contain exactly $i + 2$ different blocks, one of which is schedule[r] and another i of which have already been used for recvblock($i-1, r$), recvblock($i-2, r$), ..., recvblock($0, r$). The first block received by processor r is schedule[r] so recvblock(first(r), r) = schedule[r]. Finally, for $i >$ first(r) take recvblock(i, r) to be the unique block in the interval schedule[prev(i, r) $- 2^{i-1} + 1$, prev(i, r)]. By construction either recvblock(i, r) $\in$ {recvblock(j, prev(i, r)) $\mid j < i$} or recvblock(i, r) = recvblock(first(prev(i, r)), prev(i, r)). We have argued for the following proposition.

Proposition 1. *When p is a power of two the full block schedule constructed above is correct (and unique).*

The full block constructed above furthermore has the property that for all processors in group j

$$\{\text{recvblock}(0, r), \text{recvblock}(1, r), \ldots, \text{recvblock}(j, r)\} = \{0, 1, \ldots, j\} \qquad (5)$$

and recvblock(j, r) $= j$ for the first processor $r = 2^j$ in group j.

Without further proof we note that it is possible using the bit pattern of the processor numbering to compute for each processor r each block recvblock(i, r) of the full block schedule in $O(\log p)$ bit operations (and no extra space), for a total of $O(\log^2 p)$ operations per processor. This may be acceptable for a practical implementation where B_{opt} and N_{opt} are considerable.

3.2 Arbitrary Number of Processors

When p is not a power of two, the uniqueness of the first block schedule as guaranteed by Lemma 1 no longer holds (the segment of size 4 starting at processor 8 for the first block schedule for $p = 22$ in Figure 3 has $4 > 3$ different blocks). This is one obstacle for generalizing the construction to arbitrary number of processors, and solving this is the main contribution of this paper.

group:	0	1	2				3						4									
proc r:		1	2	3	4	5	6	7	8	9	10	11	12	13	14	15	16	17	18	19	20	21
schedule:	0	1	2	0	1	3	0	1	2	0	4	0	1	2	0	1	3	0	1	2	0	

group:	0	1	2				3						4									
proc r:		1	2	3	4	5	6	7	8	9	10	11	12	13	14	15	16	17	18	19	20	21
block	0	0	1	2	0	1	3	0	1	2	0	4	0	1	2	0	1	3	0	1	2	
	2	1	0	1	2	0	1	3	0	1	2	2	4	0	1	2	0	1	3	0	1	
	1	2	2	0	1	2	2	2	3	3	1	2	4	4	4	2	2	2	3	3		
	3	3	3	3	3	3	0	1	2	0	1	3	3	3	3	3	4	4	4	4	4	
	4	4	4	4	4	4	4	4	4	4	4	0	1	2	0	1	3	0	1	2	0	

Fig. 3. First (top) and full block schedules (bottom) for $p = 22$. The part of the full block schedule corresponding to the first block schedule is shown in bold.

The communication pattern must satisfy for each j that the total size of groups $0, 1, \ldots, j-1$ plus the root processor must be at least the size of group j, so that all processors in group j can receive their first block in round j. Likewise, the size of the last group $n - 1$ must be at least the size of groups $0, 1, \ldots, n - 2$ for the processors of the last group to deliver a block to all previous processors in round $n - 1$. To achieve this we define for $0 \leq j < n$

$$\text{groupsize}(j, p) = \begin{cases} \text{groupsize}(j, \lceil p/2 \rceil) & \text{if } j < \lceil \log p \rceil - 1 \\ \lfloor p/2 \rfloor & \text{if } j = \lceil \log p \rceil - 1 \end{cases}$$

and

$$\text{groupstart}(j, p) = 1 + \sum_{i=0}^{j-1} \text{groupsize}(j, p)$$

Figure 3 illustrates the definition with $p = 22$: $\text{groupsize}(0) = 1, \text{groupsize}(1) = 1, \text{groupsize}(2) = 3, \text{groupsize}(3) = 5, \text{groupsize}(4) = 11$. It obviously holds that both $\text{groupsize}(j, p) \leq 2^j$ and $\text{groupstart}(j, p) \leq 2^j$. For p a power of two $\text{groupsize}(j, p) = 2^j$, so the definition subsumes the power of two case.

Now we can define the next and prev functions analogously to (and subsuming) the powers-of-two case:

$$\text{next}(j, r) = (r + \text{groupstart}(j, p)) \bmod p$$
$$\text{prev}(j, r) = (r - \text{groupstart}(j, p)) \bmod p$$

This communication pattern leads to an exception for the fill phase of the algorithm as formalized in Figure 1. It may happen that next(j, r) = groupstart$(j+1, p) = r'$ and prev(first$(r'), r') = 0 \neq$ next(j, r). Such a **send** has no corresponding **recv**, and shall not be performed. For an example consider the full block schedule of Figure 3. Here processor 1 would attempt to send to processor 3 in fill round 2; processor 3, however, will become active in round 3 and receive from root 0.

3.3 Computing the Block Schedule

A greedy algorithm almost suffices for computing the full block schedule for the non-powers of two case. For each processor r the construction is as follows.

1. Construct the *first block schedule* schedule as described in Subsection 3.1: set schedule[groupstart(j, p)] = j, and schedule[groupstart$(j, p) + i$] = schedule[i] for $i = 1, \ldots,$ groupstart$(j, p) - 1$.
2. Scan the first block schedule in descending order $i = r-1, r-2, \ldots 0$. Record in block[j] the first block schedule[i] different from block[$j-1$], block[$j-2$], ... block[0], and in found[j] the index i at which block[j] was found.
3. If prev$(j, r) <$ found[j] either
   - if block[j] $>$ block[$j-1$] then swap the two blocks,
   - else mark block[j] as unseen,

   and continue scanning in Step 2.
4. Set block[first(r)] = schedule[r]
5. Find the remainder blocks by scanning the first block schedule in the order $i = p-1, p-2, \ldots r+1$, and swap as in Step 3.

For each r take
$$\text{recvblock}(i, r) = \text{block}[i]$$
with block as computed above.

To see that the full block schedule thus constructed satisfies the correctness conditions (3) and (5) we need the following version of Lemma 1.

Lemma 2. *Any segment of size $\sum_{i=0}^{j}$ groupstart(i, p) of the first block schedule contains at least $j + 1$ different blocks.*

Again the proof is by induction on j, but is omitted due to limited space.

When prev$(j, r) \geq$ found[j] the next block[j] is within the segment already scanned by processor prev(j, r), and taking this block as recvblock(j, r) is therefore correct. The violation prev$(j, r) <$ found[j] means that the block that has been found for processor r for round j has possibly not been seen by processor prev(j, r). To ensure correctness we could simply mark the block as unseen and continue scanning; Lemma 2 guarantees that a non-violating block can be found that has been seen by processor prev(j, r) before round j. However, since we must also guarantee Condition (5), large numbered blocks (in particular block j for processors in group j) cannot be postponed till rounds after first(r). The two alternatives for handling the violation suffice (as per this proof sketch) for the main theorem to hold.

Theorem 2. *For any number of processors p the full block schedule constructed above is correct. The full block schedule can be constructed in $O(p \log p)$ steps.*

Since the whole first block schedule has to be scanned for each r, the construction above takes $O(p)$ steps per processor and $O(p^2)$ steps for constructing the full schedule. It is relatively easy to reduce the time to $O(p \log p)$ steps for the full block schedule. The idea is to maintain for each possible block $0 \leq b < n$ a set of processors that have not yet included b as one of its found blocks block[i]. Scanning the first block schedule as above, each new processor is inserted into the block set for all blocks, and for each $b =$ schedule[r] all processors now in the set for block b are ejected and the condition of Step 3 is checked for each. Two scans of the schedule array are necessary, and since each r is in at most n queues the $O(p \log p)$ time bound follows.

Neither is, of course, useful for on-line construction at each broadcast operation, so instead the block schedule must be constructed in advance. For an MPI implementation of the broadcast algorithm this is not a problem because collective operations can only be executed by processors belonging to the same communication domain (*communicator* in MPI terminology), which must have been set up prior to the MPI_Bcast(...) call. The full block schedule can be constructed at communicator construction time and cached with the communicator for use in later broadcast operations. With a small trick to cater for the general case where the broadcast is not necessarily from root processor 0, it is possible to store the full block schedule in a distributed fashion with only $O(\log p)$ space per processor.

3.4 The Last Phase

To achieve the claimed $n - 1 + N$ number of rounds for broadcasting N blocks, modifications to the full block schedule for the last phase are necessary. Using the full block schedule defined in Section 3, after $n-1+N$ rounds each processor has received N different blocks. Some of these may, however, be larger than N (that is, recvblock(i, r) $\geq N$ for some rounds i belonging to the phase from $N-1$ to $n-1+N$), and processors for which this happens will miss some blocks $< N$. To repair this situation, a mapping of the blocks $\geq N$ to blocks $< N$ has to be found such that after $n-1+N$ rounds all processors have received all N blocks. In the rounds of the last phase where the root would have sent a block $b > N$, the block onto which b is mapped is sent (again) instead. In cases where such a mapping does not exist, the communication pattern for the last phase must also be changed. We do not describe this here.

4 Performance

The optimal broadcast algorithm has been implemented and incorporated into NEC's state-of-the-art MPI implementations [6]. We compare this implementation to implementations in the same framework of a simple binomial tree algorithm, a pipelined binary tree algorithm, and a recently developed algorithm based on recursive halving of the data [15].

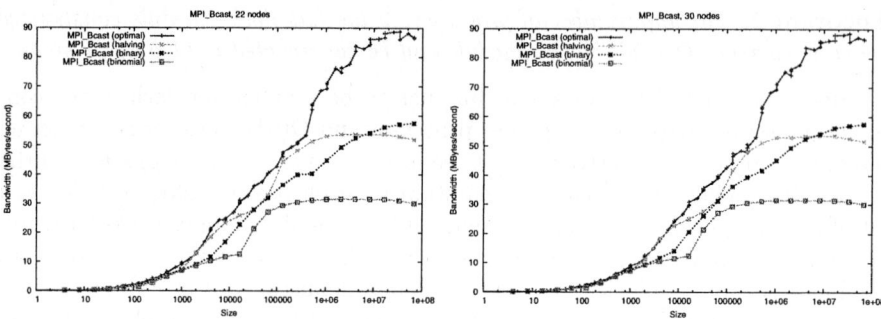

Fig. 4. Bandwidth of 4 different broadcast algorithms for fixed number of processors $p = 22$ (left) and $p = 30$ (right) with data size m up to 64MBytes

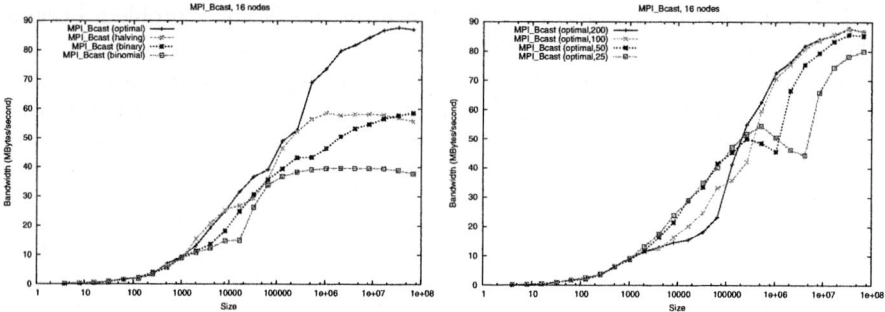

Fig. 5. Bandwidth for $p = 16$, data size m up to 64MBytes, for the piecewise linear (left) and the linear model (right) with four different values for the $\sqrt{\alpha/\beta}$ factor of equation (1)

Experiments have been performed on a 32-node, dual-processor AMD cluster with Myrinet interconnect. We consider only the homogeneous, non-SMP case here, that is the case where only one processor per node is active.

Figure 4 compares the four algorithms for fixed number of processors $p = 22$ and $p = 30$ and data size m from 0 to 64MBytes. For large data sizes the new optimal broadcast algorithm is more than a factor 1.5 faster than both the pipelined binary tree and the halving algorithm. To cater for the fact that communication bandwidth is not independent of the message size, we have, instead of using the simple, linear cost model, modeled the communication time as a piecewise linear function $t(m) = \alpha_1 + m\beta_1$ for $0 \le m < \gamma_1$, $t(m) = \alpha_2 + m\beta_2$ for $\gamma_1 \le m < \gamma_2$, ..., $t(m) = \alpha_k + m\beta_k$ for $\gamma_{k-1} \le m < \infty$ with $k = 4$ pieces for our cluster. Finding the optimum block size in this model is not much more complicated or expensive than in the linear cost model (case analysis). Figure 5 contrasts the behavior of the algorithm under the piecewise linear cost model to the behavior under the linear model with four different values for the factor $\sqrt{\alpha/\beta}$ that determines the optimum block size in equation (1). A smooth bandwidth increase

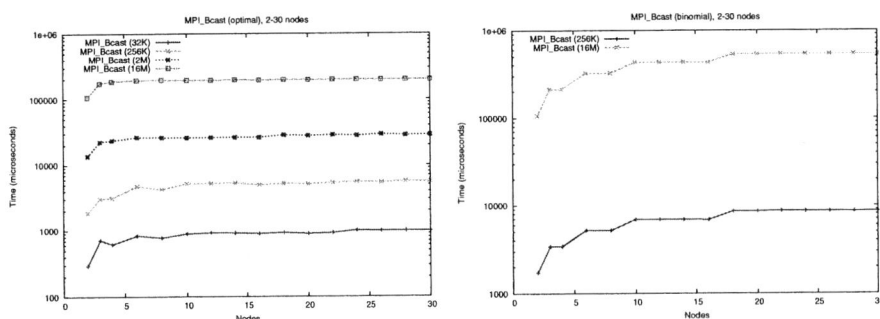

Fig. 6. Running time of the optimal broadcast algorithm for $p = 2, \ldots, 30$ processors and fixed message sizes $m = 32\mathrm{KBytes}, 256\mathrm{KBytes}, 2\mathrm{MBytes}, 16\mathrm{MBytes}$ (left). For comparison the running time for the binomial tree algorithm is given for $m = 256\mathrm{Kbytes}$ and $m = 16\mathrm{MBytes}$ (right).

can be achieved with the piecewise linear model which is not possible with a fixed, linear model.

Finally, Figure 6 shows the scaling behavior of the optimal algorithm with four fixed data sizes and varying numbers of processors. For reference, the results are compared to the binomial tree algorithm. Already beyond 3 processors the broadcast time for $m > 32\mathrm{KBytes}$ is independent of the number of processors.

5 Conclusion

We gave a new, optimal broadcast algorithm for fully connected networks for arbitrary number of processors that broadcasts N blocks over p processors in $\lceil \log p \rceil - 1 + N$ communication rounds. On a 32-node Myrinet PC-cluster the algorithm is clearly superior to other widely used broadcast algorithms, and gives close to the expected factor of two bandwidth improvement over a pipelined binary tree algorithm.

The algorithm relies on off-line construction of a communication schedule, which can be both space and time consuming. We would therefore like to be able to compute for any p and each r the $\mathsf{recvblock}(\cdot, r)$ and $\mathsf{sendblock}(\cdot, r)$ functions fast and space efficiently as is possible for the case where p is a power of two.

References

1. A. Alexandrov, M. F. Ionescu, K. E. Schauser, and C. J. Scheiman. LogGP: Incorporating long messages into the LogP model for parallel computation. *Journal of Parallel and Distributed Computing*, 44(1):71–79, 1997.
2. F. Cappello, P. Fraigniaud, B. Mans, and A. L. Rosenberg. HiHCoHP: Toward a realistic communication model for Hierarchical HyperClusters of Heterogeneous Processors. In *15th International Parallel and Distributed Processing Symposium (IPDPS01)*, pages 42–47, 2001.

3. E. W. Chan, M. F. Heimlich, A. Purkayastha, and R. A. van de Geijn. On optimizing collective communication. In *Cluster 2004*, 2004.
4. D. E. Culler, R. M. Karp, D. Patterson, A. Sahay, E. E. Santos, K. E. Schauser, R. Subramonian, and T. von Eicken. LogP: A practical model of parallel computation. *Communications of the ACM*, 39(11):78–85, 1996.
5. P. Fraigniaud and E. Lazard. Methods and problems of communication in usual networks. *Discrete Applied Mathematics*, 53(1–3):79–133, 1994.
6. M. Gołebiewski, H. Ritzdorf, J. L. Träff, and F. Zimmermann. The MPI/SX implementation of MPI for NEC's SX-6 and other NEC platforms. *NEC Research & Development*, 44(1):69–74, 2003.
7. S. M. Hedetniemi, T. Hedetniemi, and A. L. Liestman. A survey of gossiping and broadcasting in communication networks. *Networks*, 18:319–349, 1988.
8. S. L. Johnsson and C.-T. Ho. Optimum broadcasting and personalized communication in hypercubes. *IEEE Transactions on Computers*, 38(9):1249–1268, 1989.
9. S. Juhász and F. Kovács. Asynchronous distributed broadcasting in cluster environment. In *Recent Advances in Parallel Virtual Machine and Message Passing Interface. 11th European PVM/MPI Users' Group Meeting*, volume 3241 of *Lecture Notes in Computer Science*, pages 164–172, 2004.
10. P. Sanders and J. F. Sibeyn. A bandwidth latency tradeoff for broadcast and reduction. *Information Processing Letters*, 86(1):33–38, 2003.
11. E. E. Santos. Optimal and near-optimal algorithms for k-item broadcast. *Journal of Parallel and Distributed Computing*, 57(2):121–139, 1999.
12. M. Snir, S. Otto, S. Huss-Lederman, D. Walker, and J. Dongarra. *MPI – The Complete Reference*, volume 1, The MPI Core. MIT Press, second edition, 1998.
13. R. Thakur, W. D. Gropp, and R. Rabenseifner. Improving the performance of collective operations in MPICH. *International Journal on High Performance Computing Applications*, 19:49–66, 2004.
14. J.-Y. Tien, C.-T. Ho, and W.-P. Yang. Broadcasting on incomplete hypercubes. *IEEE Transactions on Computers*, 42(11):1393–1398, 1993.
15. J. L. Träff. A simple work-optimal broadcast algorithm for message passing parallel systems. In *Recent Advances in Parallel Virtual Machine and Message Passing Interface. 11th European PVM/MPI Users' Group Meeting*, volume 3241 of *Lecture Notes in Computer Science*, pages 173–180, 2004.
16. J. L. Träff and A. Ripke. An optimal broadcast algorithm adapted to SMP-clusters. In *Recent Advances in Parallel Virtual Machine and Message Passing Interface. 12th European PVM/MPI Users' Group Meeting*, volume 3666 of *Lecture Notes in Computer Science*, 2005.

Cost Model Based Configuration Management Policy in OBS Networks

Hyewon Song, Sang-Il Lee, and Chan-Hyun Youn

School of Engineering, Information and Communications University (ICU),
103-6 Munji-dong, Yooseong-gu, Daejeon, 305-714, Korea
{hwsong, vlsivlsi, chyoun}@icu.ac.kr

Abstract. The one-way reservation strategy in Optical Burst Switching (OBS) networks causes a blocking problem due to contention in resource reservation. In order to solve this problem, in this paper, we propose a configuration management policy based on the operation cost model. We develop the operation cost model based on DEB according to network status information changed by guaranteed Quality of Service (QoS) and the network status decision algorithm, and develop policy decision criteria for configuration management by providing an alternate path using bounded range of the sensitivity of this cost. Finally, throughout our theoretical and experimental analysis, we show that the proposed scheme has stable cost sensitivity and outperforms the conventional scheme in complexity.

1 Introduction

The explosive growth of Internet traffic demands huge bandwidth in optical networks. Given that fact, bandwidth has increased dramatically due to advances in wavelength-division multiplexing technology. Optical packet switching (OPS) is considered a promising solution. However, limitations in optical technology such as optical buffering have yet to be resolved. Therefore, OBS was introduced as an intermediate technology. The most important characteristic of OBS is that it uses one way reservation by which data bursts are transmitted in offset time after transmission of control packets without any acknowledgement from the destination node [1]. Due to this one way reservation, when contention occurs at an intermediate node, two or more bursts that are in contention can be drop. This is the reason why one of the critical design issues in OBS networks is finding efficient ways to minimize burst dropping resulting from resource contention.

To reduce the burst blocking probability and thus increase throughput, several viable methods are needed to solve the wavelength contention arising in OBS networks: buffering, wavelength conversion and deflection routing. In general, due to the immaturity in both optical buffering and wavelength conversion techniques, deflection routing has recently received a lot of attention. Deflection routing was first used as a contention resolution in mesh optical networks with regular topology. When a data unit arrives at an intermediate node in the network but finds that all wavelengths at the preferred port are not available, it will be switched to an alternate port. A deflection routing protocol for the OBS network has been proposed in many papers [2]-[4].

As shown in these works, applying deflection routing in an OBS network can reduce data loss and average delay compared with data retransmission from the source. However, it can not be guaranteed that the control packet will reserve all the wavelengths across the destination over the alternate path, especially when traffic load is highly congested in a wavelength routed network. In addition, most of the deflection routing schemes that have been proposed do not address implementation problems encountered in the network such as architectural issues, control and management, and others. Therefore, we study a policy based configuration management model to compensate the existing control and management schemes in an OBS network.

In this paper, we propose an operation cost model based on the Quality of Service (QoS) guaranteeing scheme by decreasing the blocking rate and complexity in OBS networks. We consider operation cost based on the Deterministic Effective Bandwidth (DEB) and the additional cost when using the alternate path. Since total operation cost varies according to network status information, we propose a configuration management policy in which the sensitivity value of the total operation cost from DEB is estimated recursively from the Configuration Information Base (CIB). Through theoretical and experimental analysis, the proposed scheme outperforms the conventional scheme in complexity.

2 Operation Cost Model Based Configuration Management Policy

2.1 Operation Cost Model

In OBS networks, by sending a control packet before forwarding a data burst, resource reservation for the data burst can be carried out. Therefore, when a contention of resource reservation, which causes blocking status and QoS degradation, occurs, a cost for QoS degradation is represented by a DEB concept [5]. The cost based on DEB in a link (i, j) between source s and destination d is defined as follows [6],

$$C_{DEB}^{ij}(t) = C_{DEB} \{ e_{D_{sd}}^{ij}(\alpha_{sd}^{ij}(t)) + \delta_{D_{sd}}^{ij}(D_{sd} - D_{sd}^{rq}) \} \qquad (1)$$

where $\alpha_{sd}^{ij}(t)$ represents the arrival curve of traffic flows in a link (i, j) between source s and destination d at time t. $e_{D_{sd}}^{ij}(\alpha)$ means DEB associated to given source s and the delay requirement D_{sd}^{rq} in a link (i, j). D_{sd} represents the actual delay of traffic that flows along to a path between source s and destination d. $\delta_{D_{sd}}^{ij}$ is a DEB sensitivity according to the delay variation, $\partial e_{D_{sd}}^{ij}(\alpha_{sd}^{ij}(t))/\partial D_{sd}$, in a link (i, j) between source s and destination d. C_{DEB} is a cost factor per unit of DEB.

Using above Eq. (1), the cost based on DEB is defined as follows:

$$C_{DEB}^{sd}(t) = \sum_{ij \in N_{sd}} C_{DEB}^{ij}(t). \qquad (2)$$

N_{sd} represents nodes that belong to the path between source s and destination d.

When a QoS constraint traffic by Service Level Agreements (SLAs) is transmitted across OBS networks, contention in the reservation process of a network resource can occur. At that time, in order to guaranteeing a required QoS for the traffic in contention within a tolerable range by the SLAs, an alternate path can be provided. In this case, the additional cost of the alternate path can be considered in two parts: the additional setup cost and the penalty resulting from transmission throughout the alternate path such as detour cost [7]. For the formulation, the variable $x_{i,j}$, which represents whether or not a link (i, j) is included in the path, is defined as

$$x_{i,j} = \begin{cases} 1, & \text{if the path includes a link } (i, j) \\ 0, & \text{otherwise} \end{cases} \quad (3)$$

Using Eq. (3), when there is an alternate path between source s and destination d, the number of passed nodes before a current node in this path, H_{sc}^{sd}, and the number of remaining nodes after a current node in this path, H_{cd}^{sd}, are represented by,

$$H_{sc}^{sd} = \sum_{\forall i,\, i+1 \in N_{sc}} x_{i,i+1} \qquad H_{cd}^{sd} = \sum_{\forall i,\, i+1 \in N_{cd}} x_{i,i+1} \quad (4)$$

N_{sc} and N_{cd} represent the number of nodes between source and a current node, and the number of nodes between a current core node and destination, respectively.

Using above Eq. (4), the cost of providing a alternate path is derived as follows:

$$C_{alt}(t) = C_{altsetup}(t) + C_{apc}(t) = C_{altsetup} \exp(\gamma \cdot H_{cd}^{sd}(t)) + C_{apc}\left\{ H_{A_{cd}}^{sd}(t) - H_{P_{cd}}^{sd}(t) \right\} \quad (5)$$

where $C_{altsetup}$ and C_{apc} are the unit cost by an additional path set up and by penalty from using an alternate path, respectively. γ is the proportional constant. $H_{cd}^{sd}(t)$ represents the number of remaining nodes after a current node in this path. $H_{A_{cd}}^{sd}(t)$ means the number of remaining nodes after a current node in the alternate path, and $H_{P_{cd}}^{sd}(t)$ means the number of remaining nodes after a current node in the primary path.

When a network provider determines that the alternate path isn't needed under the contention situation, resource reservation isn't possible, and the traffic is blocked. In this case, a penalty cost by this blocked traffic occurs, and this penalty cost is affected by the service type of traffic [7]. The penalty cost by burst drop is defined as follows:

$$C_{be}(t) = C_{be} \sum_{ij \in N_{sd}} S_{ij}(t). \quad (6)$$

C_{be} represents the penalty cost factor per unit. This cost is influenced by the service type of application. $S_{ij}(t)$ is defined as the service-specific cost function according to traffic flows on the link (i, j) between source s and destination d [7].

2.2 Cost Sensitivity Based Configuration Management Policy (CS-CMP)

Since the operation of configuration management in a network is different according to the network status, the operation cost is derived differently by the network status. In order to derive the operation cost model according to the network status for the configuration management policy, we consider the network status in an OBS network. This network status is divided into three statuses according to the guaranteed required QoS as follows: the status guaranteeing the required QoS (NS_{deb}), the status guaranteeing the tolerable QoS by providing the alternate path (NS_{alt}), and the burst drop status (NS_{be}).

In this paper, we consider burst scanning for division of the network status. Through burst scanning, the burst per channel can be measured by the number of bursts and the average burst size at a source edge node. The method for measuring the burst is to record the number of busy channels when scanning the channel periodically. The average burst size can then be obtained by dividing the amount of total traffic as the number of bursts. When the channels which the node can use are given as $L_1, L_2, L_3, \ldots, L_i$, we can expect the traffic load in the channel L_i as $T_{L_i} = B/S$ where B means the number of bursts, and S is the number of scanning [8]. In this process, we assume the burst size is larger than the period of the scanning, since when a longer period of scanning than the burst size occurs, the possibility of error in measuring traffic increases.

When the traffic load increases, the node can not assign a resource for the burst. Thus, we can expect contention situation by this measured traffic load. The network status is determined as follows:

$$NS = \begin{cases} NS_{deb} & \text{when } T_L \leq \gamma_1 \cdot C \cdot i \\ NS_{alt} & \text{when } \gamma_1 \cdot C \cdot i \leq T_L \leq \gamma_2 \cdot C \cdot i \\ NS_{be} & \text{when } T_L \geq \gamma_2 \cdot C \cdot i \end{cases} \quad (7)$$

where $0 < \gamma_1 < \gamma_2 < 1$ and $T_L = \sum_i T_{L_i}$. γ_1 and γ_2 are the utilization factor. T_{L_i} is the amount of traffic in the i th channel, L_i and T_L is the amount of traffic through a link. i is the number of channel. C is the channel capacity. If the measured traffic is under the lower boundary by the utilization, the network status is NS_{deb}. If the measured traffic is between the lower boundary and the upper boundary, the network status is NS_{alt}. If the measured traffic is over the upper boundary, the network status is NS_{be}.

Using Eq. (7) and cost functions from the previous section, the total cost function in a path between source s and destination d is derived as follows:

$$F_{sd}(t) = \begin{cases} C_{DEB}^{sd}(t), & T_L \leq \gamma_1 \cdot C \cdot i \\ C_{DEB}^{sd}(t) + C_{alt}(t), & \gamma_1 \cdot C \cdot i \leq T_L \leq \gamma_2 \cdot C \cdot i \\ C_{be}(t), & T_L \geq \gamma_2 \cdot C \cdot i \end{cases} \quad (8)$$

The total cost function means the cost in order to provide the path which guarantees QoS. When the data burst is transmitted from source s to destination d through an OBS network, if a bandwidth for guaranteeing the QoS constraint of this data burst is assigned, only the cost based on DEB is considered for the total cost function. However, if that bandwidth can't be assigned due to contention of resource or blocking status, the alternate path is needed to guarantee the QoS. In this case, the total cost function is represented by a sum of the cost based on DEB and the cost that results from providing the alternate path. Moreover, when it is no meaning that guarantees the QoS because of a continuous increment of operation cost, the total cost is represented by the penalty cost.

When the total cost from Eq. (8) is considered between source and destination, to increase this cost means that the cost for guaranteeing the required QoS increases, especially when network status changes such as the case of providing an alternate path. When the amount of traffic per each channel is expected by the burst scanning, the sensitivity of the total cost, $\zeta_F^{sd} = \partial F_{sd}(t) / \partial C_{DEB}^{sd}(t)$ from Eq. (8) and Eq. (2), means the variance of total cost according to the variance of the cost based on DEB between source s and destination d. Thus, we can derive the sensitivity according to the network status using the total cost function, F, as follows:

$$\zeta_F^{sd} = \begin{cases} 1, & T_L \leq \gamma_1 \cdot C \cdot i \\ 1 + \dfrac{\partial C_{alt}(t)}{\partial C_{DEB}^{sd}(t)}, & \gamma_1 \cdot C \cdot i \leq T_L \leq \gamma_2 \cdot C \cdot i \\ \dfrac{\partial C_{be}(t)}{\partial C_{DEB}^{sd}(t)}, & T_L \geq \gamma_2 \cdot C \cdot i \end{cases} \qquad (9)$$

When we consider the sensitivity according to $C_{DEB}^{sd}(t)$, the sensitivity value of F is dominant to the variation value of both $C_{alt}^{sd}(t)$ and $C_{be}^{sd}(t)$. In this equation, we assume that the value in the drop status NS_{be}, in which the required QoS by SLAs is not guaranteed, is not considered, since our proposed scheme relates to guaranteed QoS. Therefore, ζ_F^{sd} dominantly depends on the term, $\Delta = \partial C_{alt}(t) / \partial C_{DEB}^{sd}(t)$, which means the variation of the cost for providing the alternate path according to $C_{DEB}^{sd}(t)$.

When the alternate path is used in a contention situation in an OBS network, the cost for providing this alternate path occurs. This cost increases when the number of hops in the provided alternate path increases as shown in Eq. (5). In high channel utilization of the overall network, the selected alternate path includes many hops since high channel utilization means that most channels have a traffic load which is closer to the boundary; meaning, most nodes are under the contention situation. Therefore, as shown in Fig. 1, the value of Δ can have a positive value because the cost for the alternate path increases. However, if the utilization of channels in an overall network is closer to the boundary, it becomes more difficult to reserve the resource for the data burst. Accordingly, the selected alternate path has to include more hops. This increment of the number of hops causes an increment of the cost by Eq. (5). Thus, the value of Δ increases, so that the point in which this value exceeds the upper bound occurs. This upper bound is given by SLAs. By this upper boundary, it is determined to provide the alternate path. Therefore, the tolerable range by SLAs is represented in Fig. 1. When the sensitivity of total cost, ζ_F^{sd}, has the boundary by $\zeta_F^{sd} \leq 1+\Delta$, the value exceeding this boundary has no meaning in the network operation cost point of view, so that it need not provide an alternate path in this case.

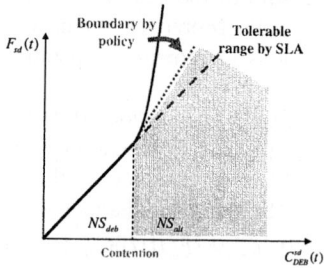

Fig. 1. Total cost F according to the DEB cost

3 Configuration Management Policy Decision Rule

In order to reflect the network status information, we make Network Status Information Base (NSIB) for collected network status information and Network Status Table (NST) as an updatable table. We assume that the value of NST, NS_{ij}, changes, and is then updated by the proposed algorithm according to the network status.

The condition factor C_h for a threshold check function is defined as follows:

$$C_h = H_{sc}^{sd} - H_{cd}^{sd}.$$ (10)

From Eq. (4), the value of this factor determines whether the current node is closer to source or destination. We have the condition factor by C_h, Q_h, as follows:

$$Q_h = \begin{cases} 1 & \text{if } C_h \geq 0 \\ 0 & \text{otherwise} \end{cases}.$$ (11)

If the current node is closer to destination d, the value of Q_h is one, otherwise, the value of Q_h is zero. Also we can obtain the other condition factor, $Q_{\delta_F^{sd}}$, from the boundary in section 2.3 and it is defined as follows:

$$Q_{\delta_F^{sd}} = \begin{cases} 1 & \text{if } \zeta_F^{sd} \leq 1+\Delta \\ 0 & \text{otherwise} \end{cases}$$ (12)

where $\Delta = \partial C_{alt}(t)/\partial C_{DEB}^{sd}(t)$. $C_{threshold}$ means the boundary of ζ_F^{sd}. If ζ_F^{sd} is within the tolerable boundary $C_{threshold}$, the value of $Q_{\delta_F^{sd}}$ is one, otherwise, the value of $Q_{\delta_F^{sd}}$ is zero. When the decision factor is represented by a sum of above two condition factors, $Q_t = w_h Q_h + w_\delta Q_\delta$ (w_h, w_δ: weighting factors), the combined threshold check function can then be stated as

$$C_{ALT} = \begin{cases} 1 & \text{if } Q_t = w_\delta + w_h \\ 0 & \text{otherwise} \end{cases}. \quad (13)$$

When the current node between source and destination is under a contention situation, if the node that is closer to destination d and the value of ζ_F^{sd}, which represents the sensitivity of the total operation cost, is within the tolerable range, the combined threshold check function C_{ALT} is one, so that the node makes a decision to deflect the alternate path. Otherwise, C_{ALT} is zero, so that the node makes a decision to drop the burst. When information is obtained from NST and NSIB, (a) of Fig. 2 shows the algorithm for decision of the threshold check function.

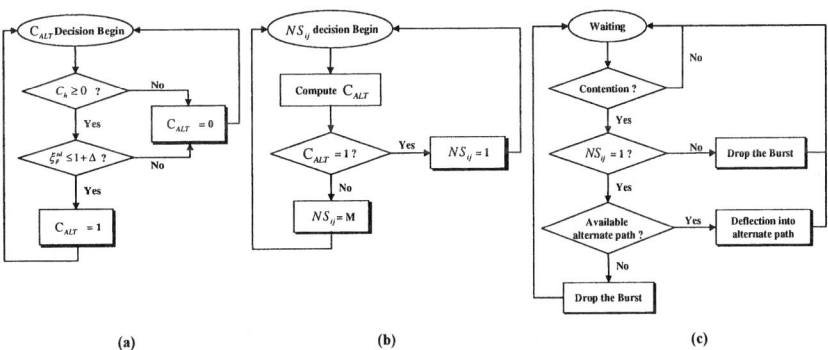

Fig. 2. The configuration management policy decision algorithm: (a) The threshold check function decision algorithm, (b) The NS_{ij} decision algorithm, (c) The CS-CMP algorithm

Next, (b) of Fig. 2 represents the algorithm for the decision of the network status on the link (i, j). We assume that the initial value of NST is zero. This algorithm is performed on the node under a contention situation. If contention occurs, the node computes C_{ALT} using the threshold check function algorithm. If C_{ALT} is one, NS_{ij} is then 1 because the network status at that time is NS_{ij}. Otherwise, NS_{ij} is M, which is bigger than the number of hops between source and destination in under NS_{be}. Finally, (c) of Fig. 2 shows the CS-CMP algorithm in order to decide the operation of configuration according to the information given by NST and NSIB. When the current node is under the contention situation, the node makes a decision whether the data burst is to be deflected to an alternate path or dropped according to the threshold check function, C_{ALT}.

4 Simulation and Results

In order to evaluate the performance of the proposed cost sensitivity based configuration management policy (CS-CMP), a simulation model is developed. We use the JET method of offset-based reservation in our simulation. The burst sources were individually simulated using the on-off model based on [2]. The simulation is carried out using a 14-node NSFNET topology. The transmission rate is 10 Gb/s, the switching time is 10 us, and the burst header processing time at each node is 2.5 us. The primary paths are computed using the shortest-path routing algorithm, while the alternate paths are the link-disjoint next shortest paths for all node pairs. Fig. 3 shows the results from this simulation.

The basic mechanism of CS-CMP is similar to CLDR. When the node is under a contention situation, an alternate path is provided by the threshold value. While CLDR uses linear programming for deflection routing, CS-CMP uses a comparison of the sensitivity of the total operation cost. As shown in Fig. 3, the blocking rate of CS-CMP increases an average of about 5.37% compared with CLDR. As well, the blocking rate of CS-CMP decreases an average of about 21.38% compared with DCR.

Moreover, in order to evaluate the configuration policy decision scheme in terms of cost, we consider the traffic source that is leaky bucket constrained with an additional constraint in the peak rate based on [9]. We assume that the NS_{be} is under a blocked situation and the service-specific cost function, $S_{ij}(t)$, is the function used in [7] according to the type of blocked service. We consider 50 different network status tables according to randomly generated traffic patterns under the given conditions. We assume an interval among incoming traffic scenarios is a monitoring interval. For each scenario, we compare the values of total operation cost function between the CLDR [2] and the proposed CS-CMP. The results are shown in Fig. 4.

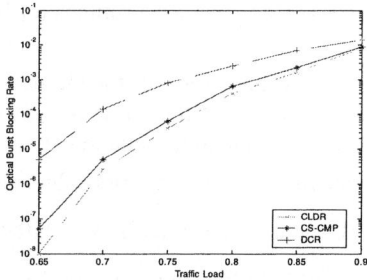

Fig. 3. Blocking rate comparison of CLDR and CS-CMP

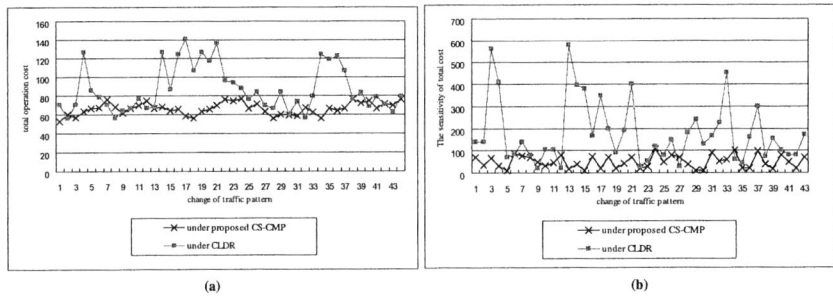

Fig. 4. (a) The total operation cost comparison, (b) The sensitivity comparison

For a comparison, the upper boundary for CS-CMP, $1+\Delta$, is assumed to be 100. In the case of CLDR, the total cost is about 4 times that of the proposed policy decision in terms of average total cost. This means that CLDR provides an alternate path in spite of high cost value. In addition, from the point of view of variation, the cost of CLDR fluctuates widely as shown in (a) of Fig. 4. Also, (b) of Fig. 4 shows that most of the sensitivity values in the case of CS-CMP are constant, at 100.

In order to compare complexity, we consider the big O function. For CLDR of [2], the complexity is represented by the iteration number of this algorithm which depends on the number of nodes. As well, each node runs linear programming in order to compute the alternate path. For this linear programming, each node has an algorithm iteration number of N^2 with the number of nodes, N. Thus, the total complexity for this algorithm can consider $O(N^3)$ with N. For DCR of [3] and CS-CMP, the complexity is computed in a similar way. Thus, the complexity for DCR depends on $O(N + N \log_2 N)$ and the complexity for CS-CMP is represented by $O(N^2)$.

5 Conclusion

In this paper, we proposed a configuration management policy for decreasing the blocking rate caused by contention as the critical issue. We also presented complexity in conventional schemes in OBS networks. For this configuration management policy, we developed an operation cost model based on DEB according to the network status information changed by guaranteed QoS. In addition, using the bounded range of the sensitivity of this cost, we proposed a network status decision algorithm, and developed policy decision criteria for configuration management by providing an alternate path. As shown in the comparison of the cost performance between our proposed scheme and conventional schemes, our scheme is performed under a stable state. As well, in comparing the blocking rate between our proposed scheme and conventional schemes, ours has good performance in terms of blocking rate. Moreover, by using the bounded range of the sensitivity of the total operation cost, our proposed scheme has a reducing effect of about 24% in terms of total operation cost. Finally, as the proposed scheme is applied to the OBS network, it is simple to implement in real networks and outperforms the conventional scheme in complexity.

Acknowledgement

This research was supported in part by the Korea Science and Engineering Foundation (KOSEF) and the Ministry of Information and Communication (MIC).

References

1. M. Yoo, C. Qiao and X. Dixit: Optical Burst Switching for Service Differentiation in the Next-Generation Optical Internet. IEEE Communications Magazine, Vol. 39, No. 2, Feb. 2001, pp. 98-104.
2. S.K. Lee, K. Sriram, H.S. Kim and J.S. Song: Contention-based Limited Deflection Routing in OBS networks. GLOBECOM '03. IEEE, Volume: 5, Dec. 2003
3. Guru P.V. Thodime, Vinod M. Vokkarane, and Jason P. Jue: Dynamic Congestion-Based Load Balanced Routing in Optical Burst-Switched Networks. GLOBECOM '03. IEEE, Volume: 5, Dec. 2003
4. Samrat Ganguly, Sudeept Bhatnagar, Rauf Izmailov and Chunmini Qiao: Multi-path Adaptive Optical Burst Forwarding. HPSR. 2004, 2004
5. J.-Y. Le Boudec: Network Calculus, Deterministic Effective Bandwidth and VBR Trunks. GLOBECOM '97, vol. 3, Nov. 1997
6. C.H. Youn, H.W. Song, J.E. Keum, L. Zhang, B.H. Lee and E.B. Shim: A Shared Buffer Constrained Topology Reconfiguration Scheme in Wavelength Routed Networks. INFORMATION 2004, Nov. 2004.
7. Xi Yang and Byrav Ramamurthy: An Analytical Model for Virtual Topology Reconfiguration in Optical Networks and A Case Study. Computer Communications and Networks, 2002. Oct. 2002
8. J.H. Yoo: Design of Dynamic Path Routing Network using Fuzzy Linear Programming. Ph. D Thesis, Yonsei University Jun. 1999
9. D. Banerjee and B. Mukherjee: Wavelength routed Optical Networks: Linear Formulation, Resource Budget Tradeoffs and a Reconfiguration Study. IEEE/ACM Transactions on Networking, Oct. 2000.

Analytical Modeling and Comparison of AQM-Based Congestion Control Mechanisms

Lan Wang, Geyong Min, and Irfan Awan

Department of Computing, University of Bradford, Bradford, BD7 1DP, U.K
{Lwang9, G.Min, I.U.Awan}@Bradford.ac.uk

Abstract. Active Queue Management (AQM) is an effective mechanism to support end-to-end traffic congestion control in modern high-speed networks. The selection of different dropping functions and threshold values required for this scheme plays a critical role on its effectiveness. This paper proposes an analytical performance model for AQM using various dropping functions. The model uses a well-known Markov-Modulated Poisson Process (MMPP) to capture traffic burstiness and correlations. Extensive analytical results have indicated that exponential dropping function is a good choice for AQM to support efficient congestion control.

1 Introduction

With the convincing evidence of traffic burstiness and correlations over modern high-speed networks, more and more powerful stochastic models have been presented to capture such traffic properties. The well-known Markov-Modulated Poisson Process (MMPP) has been widely used for this purpose owing to its ability to model the time-varying arrival rate and capture the important correlation between interarrival times while still maintaining analytical tractability. On the other hand, in very large networks with heavy traffic, sources compete for bandwidth and buffer space while being unaware of the current state of the system resources. This situation can easily lead to congestion even when the demand does not exceed the available resources [1]. Consequently, system performance degrades due to the increase of packet loss. In this context, congestion control mechanisms play important roles in effective network resource management. This study aims to develop a queueing system with an MMPP arrival process for evaluating the performance of various congestion control schemes.

End-to-End congestion control mechanisms are not sufficient to prevent congestion collapse in the Internet. Basically, there is a limit to how much control can be accomplished from the network edges. Therefore, intelligent congestion control mechanisms for FIFO-based or per-flow queue management [2] and scheduling mechanisms are needed in the routers to complement the endpoint congestion avoidance mechanisms. Scheduling mechanisms determine the sequence of sending packets while queue management algorithms control the queue length by dropping packets when necessary.

Buffer is an important resource in a router or switch. The larger buffer can absorb larger bursty arrivals of packets but tend to increase queueing delays as well. The traditional approach to buffering is to set a maximum limit on the amount of data that can be buffered. The buffer accepts each arriving packet until the queue exhausted and drops all subsequent arriving packets until some space becomes available in the queue. This mechanism is referred to as Tail Drop (TD) [3]. TD is still the most popular mechanism in IP routers today because of its robustness and simple implementation. However, "Lock-Out" and "Full Queues" [3] are the main drawbacks of TD due to dropping packets only when the congestion has occurred. The other two alternative queue disciplines, which can be applied when the queue becomes full, "Random drop on full" and "Drop front on full" can solve the "Lock-Out" problem but not "Full Queues" problem [3].

To overcome these problems and to provide low end-to-end delay along with high throughput, a widespread deployment of Active Queue Management (AQM) in routers has been recommended in the IETF publications [3]. To avoid the case that the buffer maintains a full status for a long time, AQM mechanism starts dropping packets before the queue is full in order to notify incipient stages of congestion. By keeping the average queue length small, AQM decreases the average delay and reduces the number of dropped packets, thus resulting in increased link utilisation. Two key issues in an AQM mechanism are when and how to drop packets. The first issue is mainly based on either the averaging queue length or the actual queue length. The second is based on the threshold and linear dropping function. Both have a significant impact on the average delay, throughput and probability of packet loss under bursty traffic. However, it is not clear how different dropping methods influent the performance of AQM schemes. To fill this gap, this paper proposes analytical models that can be used as powerful performance tools to investigate the effect of various dropping functions on the effectiveness of AQM.

The rest of this paper is organized as follows. Section 2 presents an overview of the existing AQM mechanisms. A two-state MMPP will be addressed briefly in Section 3. Performance results and the analysis of different dropping functions are presented in Section 4. Finally, Section 5 concludes the paper and indicates the future work.

2 Related Work

As the most popular AQM mechanism, Random Early Detection (RED) was initially described and analyzed in [4] with the anticipation to overcome the disadvantages of TD. RED monitors the status of the queue length during the past period to determine whether or not to start dropping packets. The arriving packets can be dropped probabilistically and randomly if necessary. In RED the exponentially weighted moving average $avg = (1-\omega) \times avg + \omega \times q$ is used to compare with two thresholds: $\min_{th}$ and $\max_{th}$. when $avg < \min_{th}$, no packets are dropped. When $\min_{th} \le avg < \max_{th}$, the dropping probability p_b increases linearly from 0 to $\max_p$. Once $avg \ge \max_{th}$, p_b reaches the maximum dropping probability $\max_p$ and the router drops all arriving packets.

There are five parameters in RED which can individually or cooperatively affect its performance. How to set parameters for RED was discussed by Sally in 1993 [4] and 1997 [5] separately in detail and by few current works [5]. But it is hard to choose a set of these parameter values either to balance the trade-off between various performance measures within different scenarios. As a result, most studies on RED are using the values introduced by Sally in 1997 [5].

RED is designed to minimize packet loss and queueing delay by maintaining high link utilization and avoiding a bias against bursty traffic [4]. A simple simulation scenario where only four FTP sources were considered has shown that RED performs better than TD. But against to Sally and Van's original motivation, the more scenarios are considered, the more disadvantages of RED appear. Based on the analysis of extensive experiments of aggregate traffic containing different categories of flows with different proportion, Martin *et al* [6] concluded that the harm of RED due to using the average queue size appears generally and mainly when the average value is far away from the instantaneous queue size. The interaction between the averaging queue length and the sharp edge in the dropping function results in some pathology such as increasing the drop probability of the UDP flows and the number of consecutive losses. On the other hand, Mikket *et al* [7] studied the effect of RED on the performance of Web traffic using HTTP response time (a user-centric measure of performance) and found RED can not provide a fast response time for end-user as well.

Recent researches have taken more account into how to offer a better router congestion control mechanism and compare them with each other. In [8], a modification to RED, named as Gentle-RED (GRED), was suggested to use a smoothly dropping function even when $avg \geq \max_{th}$ but not the sharp edge in the dropping function as before. In [6], an extension of GRED, named GRED-I, was suggested to use an instantaneous queue length instead of the averaging queue length, as well as one threshold and the dropping probability varying smoothly from 0 to 1 between the threshold and the queue size. The surprised results reported in [6],[8],[9] show that GRED-I performs better than RED and GRED in terms of aggregate throughput, UDP loss probability, queueing delay and the number of consecutive losses. Compared to RED, GRED appears less advantageous than GRED-I because, for RED, the averaging strategy causes more negative effects than the cooperation of the average queue length and the sharp edge in the dropping function.

3 Analytical Model

Different from the work reported in [10], we use a two-state Markov-Modulated Poisson Process (MMPP-2) to model the external traffic which can more adequately capture the properties of the inter-arrival correlation of bursty traffic than Poisson Process [11].

An m-state MMPP is a doubly stochastic Poisson Process where the arrival process is determined by an irreducible continuous-time Markov chain consisting of m different states [11]. Two different states S_1 and S_2 of MMPP-2 correspond to two different traffic arrival processes with rate λ_1 and λ_2 separately. Both δ_1 and δ_2 are the intensities of transition between S_1 to S_2. They are independent of the arrival process. MMPP-2 is generally parameterized by the infinitesimal generator $\mathbf{Q}$ of the Markov chain and the rate matrix λ, given as follows.

$$\mathbf{Q} = \begin{bmatrix} -\delta_1 & \delta_1 \\ \delta_2 & -\delta_2 \end{bmatrix} \tag{1}$$

$$\lambda = \text{diag}(\lambda_1, \lambda_2) \tag{2}$$

The duration of state i ($i=1,2$) is in accordance with an exponential distribution with mean $1/\delta_i$. The mean arrival rate of an MMPP-2 is $(\lambda_1 * \delta_2 + \lambda_2 * \delta_1)/(\delta_1 + \delta_2)$.

The rest of this section explains how to model a finite queue system with AQM under the MMPP-2 arrival process and calculate the probability of each state in the model.

A state transition diagram of the MMPP-2 model is shown in Fig. 1. The queue capacity is given as L. The average transition rate from (i, j) to $(i-1, j)$, where $(1 \leq i \leq L, 0 \leq j \leq 1)$, is μ. There is one threshold in this simple model to control the external traffic arriving into the queue. When the queue length exceeds the threshold, some packets will be dropped with a probability according to the different dropping functions. This process can be seen as a decrease of the arriving rate with some probability $(1-d_i)$, where i represents the number of customers in the system.

According to the transition equilibrium between in-coming and out-coming streams of each state and Probability Theory, the following equations are built in a general way, where the value of d_i depends on the AQM mechanism. p_{ij} is the probability that the arrival process is in state $(i, j), (1 \leq i \leq L, 0 \leq j \leq 1)$.

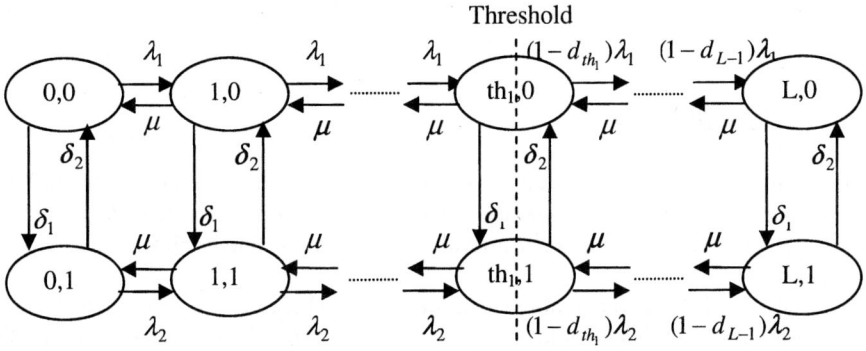

Fig. 1. A state transition diagram

$$\begin{aligned}
((1-d_0)\lambda_1 + \delta_1)p_{00} &= \delta_2 p_{01} + \mu p_{10} \\
((1-d_0)\lambda_2 + \delta_2)p_{01} &= \delta_1 p_{00} + \mu p_{11} \\
((1-d_i)\lambda_1 + \delta_1 + \mu)p_{i0} &= (1-d_{i-1})\lambda_1 p_{i-1,0} + \delta_2 p_{i1} + \mu p_{i+1,0} & 1 \le i \le L \\
((1-d_i)\lambda_2 + \delta_2 + \mu)p_{i1} &= (1-d_{i-1})\lambda_2 p_{i-1,1} + \delta_1 p_{i0} + \mu p_{i+1,1} & 1 \le i \le L \\
(\delta_1 + \mu)p_{L0} &= (1-d_{L-1})\lambda_1 p_{L-1,0} + \delta_2 p_{L1} \\
(\delta_2 + \mu)p_{L1} &= (1-d_{L-1})\lambda_2 p_{L-1,1} + \delta_1 p_{L0}
\end{aligned} \tag{3}$$

$$\sum_{i=0}^{L}\sum_{j=0}^{1} p_{ij} = 1 \tag{4}$$

Solving these equations, we can find probability p_{ij} of each state in the Markovian model as:

$$p_{ij} = a_{ij} m + b_{ij} n \tag{5}$$

with

$$a_{ij} = \begin{cases} 0 & i=0, j=0 \\ 1 & i=0, j=1 \\ \frac{(1-d_0)\lambda_1+\delta_1}{\mu} & i=1, j=0 \\ -\frac{\delta_1}{\mu} & i=1, j=1 \\ \frac{((1-d_{i-1})\lambda_1+\mu+\delta_1)a_{i-1,0}-(1-d_{i-2})\lambda_1 a_{i-2,0}-\delta_2 a_{i-1,1}}{\mu} & 2 \le i \le L, j=0 \\ \frac{((1-d_{i-1})\lambda_2+\mu+\delta_2)a_{i-1,1}-(1-d_{i-2})\lambda_2 a_{i-2,1}-\delta_1 a_{i-1,0}}{\mu} & 2 \le i \le L, j=1 \end{cases} \tag{6}$$

$$b_{ij} = \begin{cases} 0 & i=0, j=0 \\ 1 & i=0, j=1 \\ \frac{(1-d_0)\lambda_1+\delta_1}{\mu} & i=1, j=0 \\ -\frac{\delta_1}{\mu} & i=1, j=1 \\ \frac{((1-d_{i-1})\lambda_1+\mu+\delta_1)b_{i-1,0}-(1-d_{i-2})\lambda_1 b_{i-2,0}-\delta_2 b_{i-1,1}}{\mu} & 2 \le i \le L, j=0 \\ \frac{((1-d_{i-1})\lambda_2+\mu+\delta_2)b_{i-1,1}-(1-d_{i-2})\lambda_2 b_{i-2,1}-\delta_1 b_{i-1,0}}{\mu} & 2 \le i \le L, j=1 \end{cases} \tag{7}$$

$$k = \lambda_1(1-d_{l-1,0})b_{l-1,0} + \delta_2 b_{l1} - (\mu+\delta_1)b_{l0} \tag{8}$$

$$h = (\mu+\delta_1)a_{l0} - \lambda_1(1-d_{l-1,0})a_{l-1,0} - \delta_2 a_{l1} \tag{9}$$

$$n = \frac{h}{k\sum_{i=0}^{l}\sum_{j=0}^{1}a_{ij} + h\sum_{i=0}^{l}\sum_{j=0}^{1}b_{ij}} \tag{10}$$

$$m = \frac{k}{h}n \tag{11}$$

4 Performance Analysis

4.1 Dropping Functions

The linear dropping function is adopted mainly in AQM. In addition to the basic methods addressed in popular mechanisms, we also pay attention to the dropping probability belonging to an exponential distribution. Four functions of dropping probability have been shown in Fig. 2. For Functions 1 and 2, the maximum dropping probability is 0.1. For the other two functions, the maximum dropping probability is 1. Dropping probability will be increasing linearly from 0 to the maximum value in Functions 2 and 3. The increase of the dropping probability value in Function 4 follows an exponential curve. However the sharpest Function 1 keeps the maximum dropping probability from start to end.

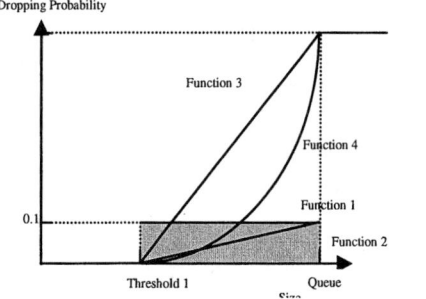
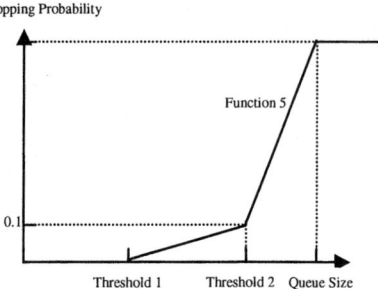

Fig. 2. Dropping Functions with One Threshold

Fig.3. Dropping Function with Two Thresholds

Function 5 is another different dropping function as shown in Fig. 3. There are two thresholds in the queue. Meanwhile two linear functions will be applied in the two intervals from threshold 1 to threshold 2 and from threshold 2 to the queue size with two maximum dropping probabilities 0.1 and 1, respectively.

4.1 Performance Results

The system performance metrics including the mean queue length (MQL), mean queuing delay (MQD), system throughput and probability of packet losses will be evaluated as follows.

We compare not only the performance variation with different threshold settings for each function, but also the corresponding performance based on different dropping functions with the same parameter settings of MMPP-2. Our comparisons consist of 3 parts which investigate the above four performance metrics.

Firstly, all dropping functions with one threshold are presented. As shown in Figs. 4-6, by increasing the threshold values, the mean queue length, mean queueing delay and throughput under each function are increasing as well. For any dropping Function, when the dropping area size (e.g. shaded area for Function 1) is decreasing, this means that the congestion control mechanism is becoming less effective, consequently these three performance measures will increase reasonably. On the other hand, the main strategy of AQM is to drop packets before congestion occurs. So loss probability indicated in Fig. 7 tends to decrease by increasing the threshold value. In all these experiences we have used e^x as Function 4. It does not give better performance as compared to other three functions. It is not like the other functions, the variation of performance caused by the exponential function with the increasing value of threshold is so small that we find there are no effects of this standard exponential dropping function on the performance.

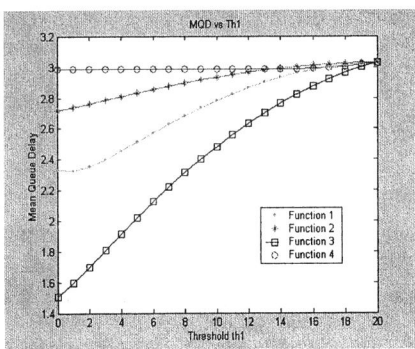

Fig. 4. MQL vs th1 for Functions 1,2,3,4

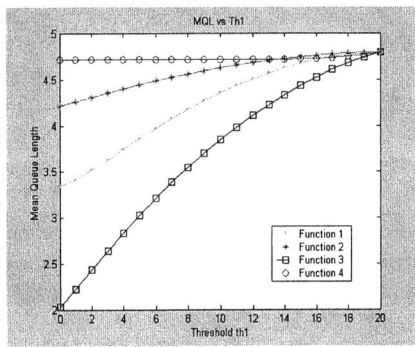

Fig. 5. MQD vs th1 for Functions 1,2,3,4

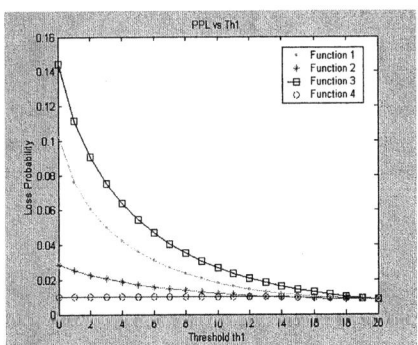

Fig. 6. Throughput vs th1 for Functions 1,2,3,4

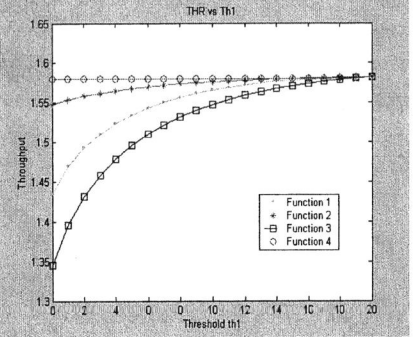

Fig. 7. PPL vs th1 for Functions 1,2,3,4

In order to find out the effects of other exponential dropping functions, we consider a more general exponential function a^x with different parameter a. We observe the effect is similar when $a \geq 2$. However along with the decrease of a from 2 to 1, the performance curves shown in Figs. 8-11 present an obvious variation. Compared with Fig. 5, especially when the threshold is small, Function 4 with $a = 1.1$ supports the mean queue delay in Fig. 9 lower than Functions 1 and 2 but higher than Function 3. Furthermore, in general, the throughput resulting from Function 4 with $a = 1.1$ is higher than that from Function 3, near to Function 1 and less than Function 2. As for the loss probability, the advantage of Function 4 is expressed in Fig. 11 compared with Fig. 7. As a result, based on different threshold values, it is possible for Function 4 to offer a better perform such as lower mean queue delay and/or higher throughput but with less packets loss probability.

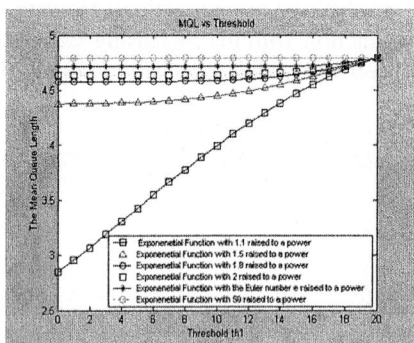

Fig. 8. MQL vs th1 for exponential functions

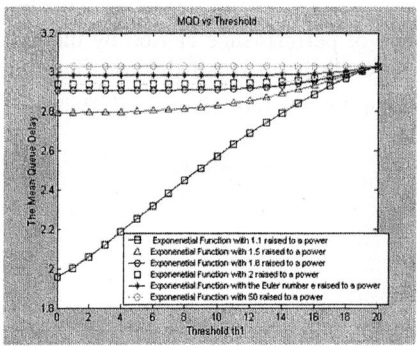

Fig. 9. MQD vs th1 for exponential functions

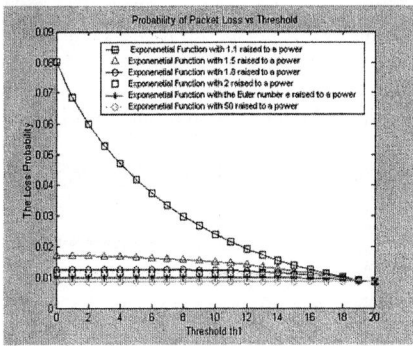

Fig. 10. Throughput vs th1 for exponential functions

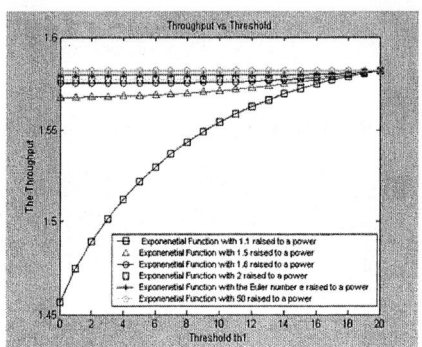

Fig. 11. PPL vs th1 for exponential functions

Finally we are also interested in the unique dropping function with two thresholds. This function can be easily changed to be Function 2 and 3 through, for example, varying the value of threshold. Figs. 12-14 depict that the mean queue length, throughput and mean queuing delay are lower when the first threshold is set to be lower. Moreover, they are increasing when enlarging the distance of two thresholds. But the corresponding results are described conversely in Fig. 15. Compared with the first group figures correspondingly, it is easy to conclude that dropping Function 5 performs between Functions 2 and 3.

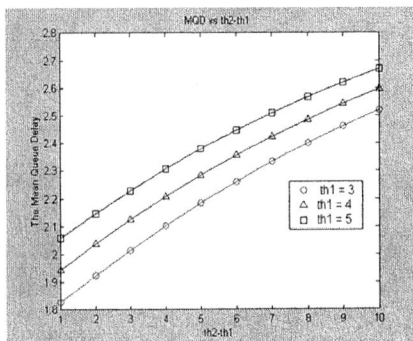

Fig. 12. MQL vs th1 for Function 5

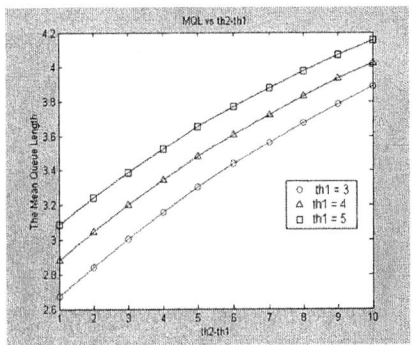

Fig. 13. MQD vs th1 for Function 5

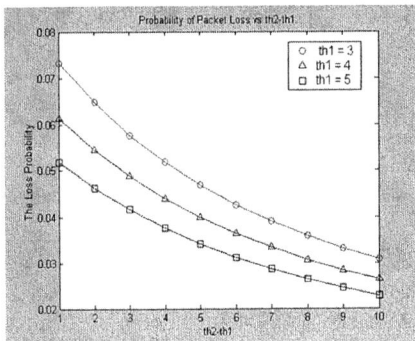

Fig. 14. Throughput vs th1 for Funciton 5

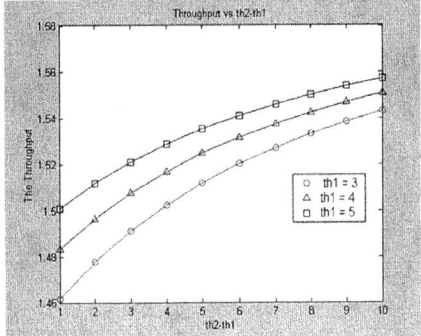

Fig. 15. PPL vs threshold for Function 5

5   Conclusions and Further Work

Traffic burstiness and correlations have considerable impacts on network performance. This paper has developed a new analytical model to calculate mean queue length, packet latency, loss probability, and throughput of a queuing system using AQM congestion control mechanisms. The distinguishing properties of this model include: (1) it can capture traffic bustiness and correlations; (2) it can generate close-form expressions to calculate the desired performance measures of the queuing systems. The analytical model has been used to investigate the performance of AQM with different dropping functions and threshold parameters. The performance results have shown that exponential dropping function is a good choice for AQM to support efficient congestion control.

References

1. McDYSAN, D. E.: QoS & Traffic Management in IP & ATM Networks. McGraw Hill. (1999)
2. ryu Seungwan, Christopher rump, Qiao, C.M.: Advances in Internet congestion Control. IEEE Communications. Vol. 5. No.1. (2003)
3. Braden, B. et al.: Recommendations on Queue Management and Congestion Avoidance in the Internet. IETF RFC2309. (1998)
4. Floyd, S., Jacobson, V.: Random Early Detection Gateways for Congestion Avoidance. IEEE/ACM Transactions on Networking. Vol. 1. (1993) 397-413
5. Floyd, S.: RED: Discussions of Setting Parameters. http://www.icir.org/floyd/REDparameters.txt. (1997)
6. May, M., Diot, C., Lyles, B., Bolot, J.: Influence of Active Queue Management parameters on aggregate Traffic Performance. INRIA.RR3995. (2000)
7. Christiansen, M., Jeffay, K., Ott, D., Smith, F.D.: Tuning RED for Web Traffic, IEEE/ACM Trans. Network. Vol 9. No. 3. (2000) 249-264
8. Floyd, S., Fall, K.: Promoting the Use of End-to-End Congestion Control in the Internet. IEEE/ACM Trans. Network. Vol 7. No.4. (1999) 485-472
9. Brandauer, C., Iannaccone, G., Diot, C., Ziegler, T.: Comparison of Tail Drop and Active Queue Management Performance for Bulk-date and Web-like Internet Traffic, Proc. ISCC (2001) 122-129
10. Bonald, T., May, M., Bolot, J.C.: Analytic Evaluation of RED Performance, Proc.IEEE INFOCOM. (2000) 1415-1424
11. Fischer, W., K.Meier-Hellstern, The Markov-modulated Poisson process (MMPP) cookbook. Performance Evaluation, 1993, 18(2), pp.149-171.
12. Min, G., Ould-Khaoua, M.: On The Performance of Circuit-switched Networks in The Presence of Correlated Traffic. Concurrency Computat. Pract.Exper. 16. (2004) 1313-1326

Spatial and Traffic-Aware Routing (STAR) for Vehicular Systems

Francesco Giudici and Elena Pagani

Information and Communication Department,
Università degli Studi di Milano, Italy
{fgiudici, pagani}@dico.unimi.it

Abstract. In this paper, we propose a novel position-based routing algorithm for vehicular ad hoc networks able to exploit both street topology information achieved from geographic information systems and information about vehicular traffic, in order to perform accurate routing decisions. The algorithm was implemented in the NS-2 simulator and was compared with three other algorithms in the literature.

1 Introduction

Progresses in wireless technologies and decreasing costs of wireless devices are leading toward an increasing, pervasive, availability of these devices. Recently, research took interest in possibilities opened by equipping vehicles with wireless devices. Vehicles carrying wireless devices are able to connect with one another in an ad hoc mobile network. These systems can be useful for several distributed applications, ranging from vehicular safety, to cooperative workgroup applications and fleet management, to service retrieval (e.g., availability of parking lots), to entertainment for passengers. Several problems are still to be solved: a main issue is to design effective and efficient routing algorithms appropriate for the characteristics of these systems. A promising approach seems to be using *position-based* routing: routing is performed basing on the current geographic position of both the data source and destination.

In this paper, the novel *Spatial and Traffic-Aware Routing* (STAR) algorithm is proposed, that overcomes drawbacks of other solutions proposed in literature.

2 System Model

In this work, we consider *Vehicular Ad Hoc Networks* (VANETs). As in Mobile Ad Hoc Networks (MANETs) [1], devices – in this case, vehicles – are equipped with wireless network interface cards. Nodes are required to have unique identifiers. The network topology dynamically changes as a consequence of vehicle movements, possibly at high speed. Each vehicle is responsible for forwarding data traffic generated from or addressed to other vehicles. The network is completely decentralized: vehicles have no information on either the network size –

in terms of number of nodes involved – or topology. Two vehicles are said to be *neighbors* if they are in communication range. Differently from MANETs, in VANETs power saving is not of concern. Each vehicle can exploit a *Global Positioning System* (GPS) [2] to determine its own position. A GPS navigation system allows to obtain information about the local road map and the vehicle's direction of movement. For the use of STAR, digital road maps are translated into graphs, with crossroads as vertexes and streets as edges.

According to all currently proposed wireless routing algorithms, vehicles periodically exchange network-layer *beacon* messages, allowing each node to discover the identities and the positions of its own neighbors.

In VANETs, nodes are addressed through their position rather than through their network address. When a vehicle has data to send, it can discover the current position of the receiver by exploiting a *location service*. Several location services have been proposed in the literature [3,4,5,6,7]. We do not make any assumption about the location service used.

In this paper a *pure* wireless ad hoc network is considered, without any access point connected to a fixed network infrastructure. Different mobility scenarios are possible. Often, in the literature, a random-waypoint model is assumed. This model is not able to capture the peculiarities of vehicular movements. In real scenarios, vehicle movements are constrained by roads. In this article, a city mobility model is assumed, along with a Manhattan street topology.

3 Spatial and Traffic-Aware Routing (STAR)

STAR approach to vehicular routing problem is quite different from other position-based routing algorithms. Other existing algorithms [10] may fail in case they try to forward a packet along streets where no vehicles are moving. Such streets should be considered as "broken links" in the topology. Moreover, a packet can be received by a node that has no neighbors nearer to the receiver than the node itself; in this case, the problem of a packet having reached a *local maximum* arises. These problems can be overcome to some extent knowing the *real* topology, that is, by trying to use for packet forwarding only streets where vehicular traffic exists. To reach this objective the STAR algorithm is organized in two layers (fig.1): a lower layer that manages the gathering and exchange of information about network status and a higher layer for the computation of paths. As network status we mean the actual distribution of vehicles along streets. Status knowledge should not concern the whole network: collecting and exchanging information about topology can be expensive, and on the other hand this information is highly volatile due to node mobility. The higher layer takes in charge the route computation on the network topology discovered by the lower layer. Some reference points (*Anchor Points*, or APs for short) on the streets traversed by the computed routes are taken; packets are forwarded from one AP to the successive. It is convenient to compute only a *partial* path to approach the destination position by determining only a subset of Anchor Points. When a packet arrives to the last AP computed for it, the node responsible for forwarding takes in charge the characterization of the next APs. Partial successive computation of the path has a threefold advantage: (*i*) the size of packet header is fixed; (*ii*) the computation of subsequent APs is done exploiting more updated information about vehicular traffic distribution; (*iii*) subsequent APs can be computed exploiting updated information about the current position of the destination. In the following subsections we explain the functionalities deployed at each layer. For more details, interestered readers may refer to [12].

3.1 Vehicular Traffic Monitoring

We are interested in detecting two extreme situations: the presence of an high number of vehicles - queues - or the total absence of them. Streets with queues of vehicles should be preferred, as they provide several alternatives for packet forwarding, thus minimizing the risk of a packet reaching a local maximum. By contrast, the routing algorithm *must* avoid streets where vehicles are not present, because packets cannot for sure be routed over them as long as their status does not change.

Monitoring and propagation of vehicular traffic conditions are performed through the exchange of network-level *beacons* (fig.1), carrying observations of node neighborhoods. The observations are maintained in data structures managed by the *traffic monitoring* module.

A node maintains the position of its neighbors in the neighbors-table. Node neighborhood is discovered via the beacons. In the presence vector (PRV) a node maintains four node counters, which represent the number of neighbors it has toward cardinal points computed dividing the node cell into sectors as

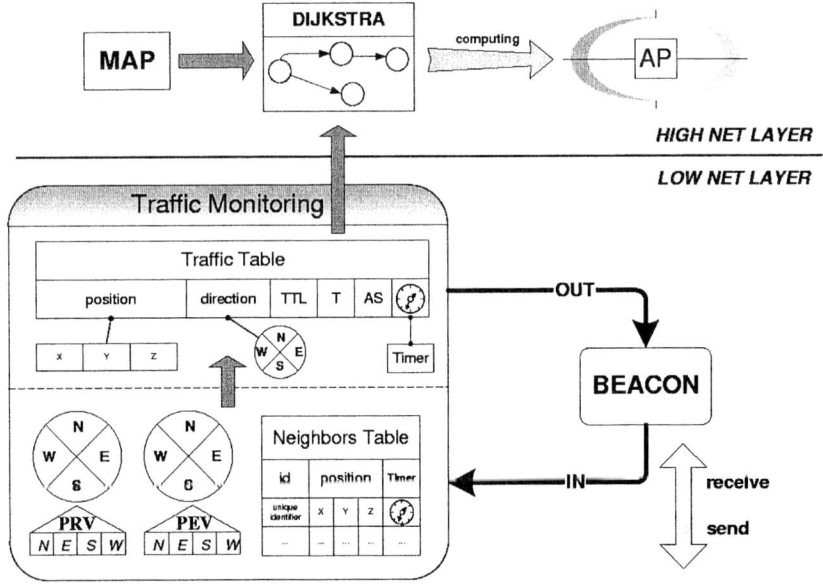

Fig. 1. Functional architecture for the STAR algorithm

shown in fig.1. Each counter is incremented when a neighbor is discovered in the corresponding direction and decremented when a neighborhood information is not refreshed for a certain time.

When a PRV counter exceeds a parameter highPR, a high concentration of vehicles exists in the corresponding direction. By contrast, an elements of PRV below parameter lowPR indicates scarce vehicular traffic along a street in the corresponding direction of the node. When one of these situations occurs, the modification of a related element in the persistence vector (PEV) is triggered. PEV has four elements as PRV. Each element can be in one of three different conditions: it could be in a *reset state* (a value equal to 0), or in a *growing state* (value > 0), or in a *shrinking state* (value < 0). In the event of a PRV counter exceeding highPR, if the corresponding PEV element is in either *reset* or *growing* state, then it is incremented; otherwise is set to *reset state*. On the other hand, in the event of a PRV element below lowPR, if the corresponding PEV element is in either *reset* or *shrinking* state then it is decremented; otherwise is set to *reset state*. PEV is used to register critical situations only when they last for a long time. When the value of an element of PEV goes out of a range [lowPE, highPE], then the information about vehicular traffic stored in the element is recorded in the traffic-table and the value of the PEV element is reset. The use of PEV is necessary to guarantee that a temporary abnormal condition is not registered, but if it lasts then it is registered in the traffic-table.

Each node has a traffic-table. Each entry in this table has five fields: position, direction, traffic bit (Tbit), already-sent bit (ASbit) and Time-To-Live (TTL). The position indicates the coordinates of the node when it first notices an anomalous vehicular traffic situation. The direction is the direction in which the traffic anomaly is taking place with respect to position. Tbit specifies the type of traffic (high or low). ASbit records whether a traffic entry has been already propagated to neighbors and TTL is the number of hops the information has to travel. Each entry has an associated traffic-timer. When the timer expires, the entry is removed from traffic-table in order to forget obsolete information about traffic anomalies.

Each node periodically broadcasts to its neighbors a beacon that contains sender identifier, sender coordinates and the vehicular traffic conditions it has in its traffic-table. Broadcasting period is determined by a beacon-timer.

When a node receives a beacon, it registers the position of the sender in its neighbors-table; PRV and PEV are possibly updated as explained before. If one of the elements of PEV becomes either lower than lowPE or higher than highPE then a new entry is added in the traffic-table. The new entry has as position the coordinates of the node, direction equal to the corresponding element of PEV ('N', 'E', 'S' or 'W'), ASbit set to zero and TTL set to a value maxTTL, which determines how far traffic information will be spread. Tbit is set to the appropriate value 'H' or 'L' according to the vehicular traffic condition detected. After updating the information about neighbors, the local traffic-table is augmented with the vehicular traffic information stored in the received beacon. Each traffic entry in the beacon is copied into traffic-table, with TTL value decremented by one. While updating the `traffic-table`, existing entries must be compared with the information carried by the beacon. In case two matching entries exist in both the beacon and the table, and they have the same `direction` and `positions` differ less than a parameter `traffic information distance` (TID), then the two entries refer to the same traffic condition. Only the one having the highest `TTL` is kept. This allows to suppress duplicate advertisements, still guaranteeing that abnormal conditions are notified. If TTLs are equal, the `traffic-timer` (determining entry expiration time) of the `traffic-table` entry is set to the initial value, and the `ASbit` is set to 0. This occurs when a traffic anomaly persists for a certain time, and is thus advertised more than once. The receiving nodes must refresh the corresponding `traffic-table` entry and re-propagate this information.

When a node's `beacon-timer` expires, a new beacon carrying node's identifier and position is created. Each entry from the `traffic-table` is copied into the beacon only if `TTL` > 0 and `ASbit` = 0. Then the `ASbit` is set to 1 to prevent diffusing multiple times a certain information. Finally the beacon is sent.

3.2 Routing and Packet Forwarding

At the higher layer (fig.1), routes are computed *on-demand* exploiting neighbors and vehicular traffic information locally owned. When a source S has a packet to send to a destination D, S builds a *weighted graph* using street map and traffic information. Edges corresponding to streets without traffic have associated a high weight. By contrast, when in a street there is high vehicular traffic, the weight of the associated edge must be decreased to privilege the choice of this street although it could characterize longer paths. As a consequence of vehicles' mobility, weights of the edges are dynamically adjusted; initial edge weights and the mechanism of weight adaptation must guarantee that weights never become negative or null.

Dijkstra's algorithm is applied to the obtained graph in order to find the shortest route. APs are computed along the streets belonging to the route. The packet header includes the destination identifier, the destination position and a limited number of APs: the packet is forwarded with GEOGRAPHIC GREEDY ROUTING [10] from one AP to the other. When the last AP is reached, the node in charge of forwarding the packet will compute other APs toward the destination, until it is reached. In case of routing failure, the *recovery procedure* adopted by STAR consists in computing a new route from the current node, exploiting updated traffic information.

3.3 Dimensioning of Parameters

STAR behavior is controlled through some parameters. The `CROSS RANGE ACCEPT` (CRA) parameter is introduced to enhance STAR traffic detection: if a vehicle is moving along a straight road it does not need to collect information about traffic in (non-existent) lateral streets. If a vehicle is far from the nearest crossroad more than `CRA` meters, then it stops detecting traffic orthogonal to the moving direction.

The Dijkstra-starting-weight (DIJSW) is the weight initially assigned to each edge when building the weighted graph, while Dijkstra-high-weight (DIJHW) and Dijkstra-low-weight (DIJLW) are respectively weight increment and decrement for an edge with associated information of low and high traffic. In a regular Manhattan street topology, to guarantee that empty streets are avoided, DIJLW must be at least three times DIJSW.[1] Moreover, to prevent an edge weight to become negative as a consequence of multiple notifications about high vehicular traffic along a certain street, the following equation should be satisfied: DIJSW > DIJHW $\times street\_length/$TID, where $street\_length$ is the length of a street between two crossroads.

The max-anchor-point (maxAP) is the maximum number of APs that are included in a packet. This parameter is related with maxTTL: if maxAP is large with respect to maxTTL, some APs are computed without relying on traffic information. By contrast, a small maxAP could waste computation time because a vehicle refrains from using the information it owns to compute a longer path. Computing more APs allows greater accuracy in choosing a path; on the other hand, this implies higher overhead in both packet header and beacon traffic, as a consequence of the higher maxTTL needed.

4 Performance Analysis

STAR performance has been analyzed using the NS-2 simulation package [11] under different parameter settings, and has been compared with results achieved by GREEDY [10] approach without any recovery procedure, GPSR [8] approach, and SAR [9] algorithm. Simulations have been performed adopting a city mobility model, applied to a Manhattan street map formed by 5 horizontal streets and 5 vertical streets. Distances between adjacent streets are equal to 400 mt., thus characterizing a regular grid. The number of nodes is 250; the communication range is 250 mt. A small street length with respect to the communication range has been chosen to advantage GREEDY and GPSR, which have not been specifically designed for vehicular networks and do not involve mechanisms to deal with mobility along streets. Nodes move along the streets and can change their direction at crossroads; maximum vehicle speed is 50 Km/h. Simulated time is 200 sec. Results are averaged on 105 packets, exchanged between 5 source-destination pairs.

Simulations have been performed in three different scenarios: with all crossroads usable by vehicles, and with 4% and 8% of crossroads without vehicular traffic.

Initial measures have been performed varying 4 parameters, namely:

- lowPR threshold assuming values 0 or 1;
- traffic entry's maxTTL assuming values 10 or 20;
- lowPE threshold assuming values -2 or -4;
- CROSS RANGE ACCEPT parameter assuming values 10, 20 or 30.

---

[1] This way, a certain point can be reached avoiding an empty street by going around a building block via three streets.

Table 1. Parameter settings used in simulations

DIJSW	4	highPR	100
DIJLW	20	maxAP	5
DIJHW	-1	TID	150

Other parameters discussed in sec.3.3 are set to values reasonable in a real environment, as shown in Table 1. Detection of dense vehicular traffic conditions has been disabled because the mobility model prevents simulating queues of vehicles. Hence, highPR has a very high value while highPE does not even need to be defined. The performance indexes measured with simulations are:

- **percentage of delivered packets**;
- **percentage of lost packets:** packets can be lost because of collisions, or because the routing algorithm fails in getting them delivered to their destinations. This index only accounts for packets lost due to failures of the routing algorithm, with STAR performing only one path re-computation before deciding to drop the packet;
- **average beacon size:** large beacons increase the probability of collisions with both other beacons and data packets; hence, this index can explain other packet losses not due to routing failures.

Initial measures allowed us to tune the values of lowPR, lowPE, maxTTL and CRA in order to achieve the best performance. This is obtained with the scenario tending to minimize the beacon size, being prudent in both detecting and signaling absence of traffic (low lowPR and lowPE thresholds) and propagating the notification only in a restricted area (low maxTTL).

The probability of delivering packets increases with higher CRA value for increasing number of empty streets, because this increases the probability that the abnormal traffic condition is detected and advertised. Effective advertisement also reduces collisions: if a packet is appropriately routed by exploiting accurate information about the traffic distribution, then the probability of an alternative route computation is reduced, thus forcing the packet to follow a shorter route and decreasing packet collision probability.

The best scenario characterized through initial simulations uses lowPR=0, maxTTL=10, lowPE=-4 and CRA=20. The measures leading to this choice are reported in [12].

We compared the best scenario with the performance obtained by GREEDY, GPSR and SAR in the same conditions. It has to be noticed that a Manhattan street map simplifies route computation. Hence, map knowledge gives both SAR and STAR less advantages over the other two policies than expected, as there are not blind alleys or street forks. Moreover, SAR implementation does not involve any recovery procedure. Analyzing simulation results, it is worth to notice that so far a realistic mobility model is still missing. The model we used does not provide the possibility of simulating queues of vehicles, which also tend to be more dense around crossroads, thus having greater probability of packet collisions.

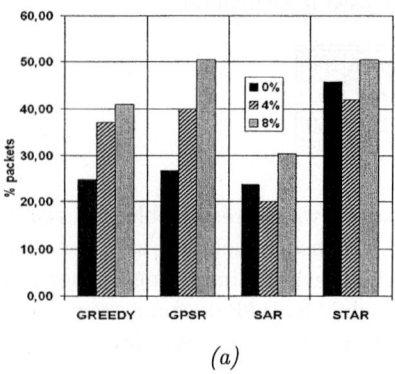 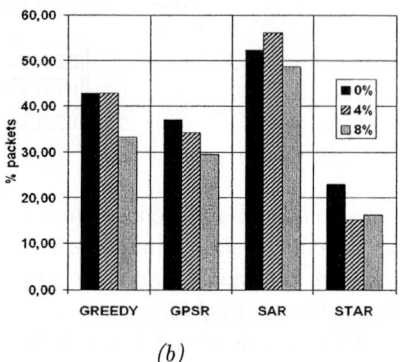

Fig. 2. Algorithm comparison: *(a)* percentage of delivered packets and *(b)* percentage of lost packets due to routing failure

Lack of queues and high collision probability worsen STAR performance with respect to a real environment. Moreover, the simulation environment does not allow to simulate radio signal attenuation due to obstacles: as a consequence, messages can be exchanged between two vehicles that actually would not be in communication range because of a building between them. This characteristic – together with streets short with respect to the communication range – advantages GREEDY and GPSR, while impacts to a lower extent on both SAR and STAR as they follow APs – and thus streets – to forward packets.

In fig.2*(a)*, the percentage of delivered packets is shown. STAR is comparable with GPSR, but it is advantaged by knowledge of street map. Both GREEDY and GPSR behave worse in the 0% scenario, because in that case vehicle density is lower than in the other scenarios as vehicles can be distributed all over the considered area. By fig.2*(b)* it can be noticed that STAR is far better than the other algorithms with respect to routing failure probability, thus confirming that STAR is effective in performing accurate routing decisions, preventing packets from reaching a local maximum. Indeed, the remaining packets not delivered to their destinations were lost because of collisions (fig.3*(b)*). On the other hand, although GPSR recovery procedure provides comparable guarantees of packet delivery in dense networks, the routes achieved with it are 1 to 2 hops longer than those computed with STAR.[2] SAR does not behave well because of both the lack of recovery mechanisms and the simplicity of the street map. If street length were higher, GREEDY and GPSR would drop many more packets because they are unable to find a longer route following streets to forward data to destination.

Distributing information about vehicular traffic drastically increases the size of network-layer beacons (fig.3*(a)*) with respect to the other considered algorithms. Because of larger beacons, collisions suffered by STAR overcome quite significantly those observed with other algorithms (fig.3*(b)*), except for scenario 0% where a lower average node density reduces collision probability.

---

[2] With path lengths of 7-8 hops on average.

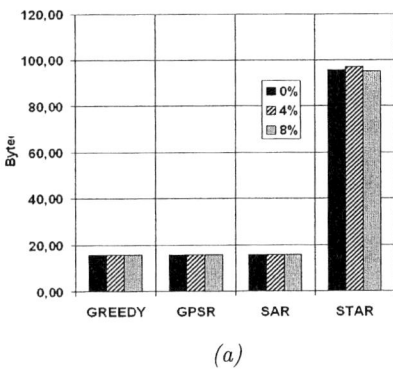 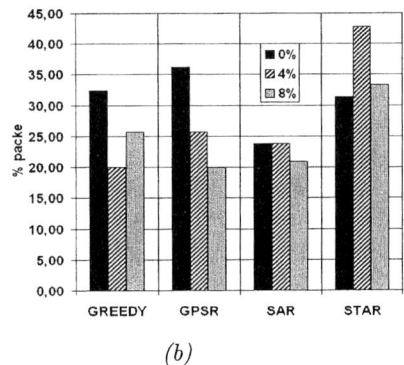

Fig. 3. Algorithm comparison: *(a)* average beacon size and *(b)* percentage of lost packets due to collisions

5 Conclusions and Future Works

In this paper we proposed a novel routing algorithm for vehicular ad hoc networks (STAR), and we measured its performance in comparison to other algorithms existing in the literature. STAR performs better than the other considered algorithms in spite of having an inaccurate mobility model that imposed disabling detection of high traffic conditions in simulations. However, STAR greatly suffers collisions. Parameters ruling traffic collection and information diffusion are the core of STAR: we are currently working on designing mechanisms for *dynamic adaptation of parameters*. A more realistic vehicular mobility model must also be developed, in order to ensure that more accurate results are obtained. Such a model would also allow to implement and evaluate the effectiveness of exchanging information about high vehicular traffic conditions. Anyway, this task is not trivial, as it involves correlating positions and speeds of different vehicles.

Acknowledgments

This work has been partially supported by the Italian Ministry of Education, University, and Research in the framework of the FIRB "Web-Minds" project.

We would like to thank Prof. Gian Paolo Rossi for useful discussions and his helpful comments in preparing the paper.

References

1. IETF MANET Working Group: *"Mobile Ad Hoc Networks"*. http://www.ietf.org/html.charters/manet-charter.html.
2. PaloWireless: *"Global Positioning System (GPS) Resource Center"*. http://palowireless.com/gps/.

3. Grossglauser, M., Vetterli, M.: *"Locating Nodes with EASE: Last Encounter Routing in Ad Hoc Networks through Mobility Diffusion"*. Proc. IEEE INFOCOM'03, Mar.2003.
4. Basagni, S., Chlamtac, I., Syrotiuk, V., Woodward, B.: *"A Distance Routing Effect Algorithm for Mobility (DREAM)"*. Proc. 4th Annual ACM/IEEE Intl. Conf. on Mobile Computing and Networking (MOBICOM'98), pp. 76-84, 1998.
5. Li, J., Jannotti, J., De Couto, D.S.J., Karger, D.R., Morris, R.: *"A Scalable Location Service for Geographic Ad Hoc Routing"*. Proc. 6th Annual ACM/IEEE Intl. Conf. on Mobile Computing and Networking (MOBICOM'00), pp. 120-130, 2000.
6. Stojmenovic, I.: *"Home Agent Based Location Update and Destination Search Schemes in Ad Hoc Wireless Networks"*. Technical Report TR-99-10, Computer Science, SITE, University of Ottawa, Sep. 1999.
7. Käsemann, M., Füssler, H., Hartenstein, H., Mauve, M.: *"A Reactive Location Service for Mobile Ad Hoc Networks"*. Technical Report TR-14-2002, Department of Computer Science, University of Mannheim, Nov. 2002.
8. Karp, B., Kung, H.T.: *"Greedy Perimeter Stateless Routing for Wireless Networks"*. Proc. 6th Annual ACM/IEEE Intl. Conf. on Mobile Computing and Networking (MOBICOM'00), Aug. 2000.
9. Tian, J., Han, L., Rothermel, K., Cseh, C.: *"Spatially Aware Packet Routing for Mobile Ad Hoc Inter-Vehicle Radio Networks"*. Proceedings IEEE 6th Intl. Conf. on Intelligent Transportation Systems (ITSC), Vol. 2, pp. 1546-1552, Oct. 2003.
10. Mauve, M., Widmer, J., Hartenstein, H.: *"A Survey on Position-Based Routing in Mobile Ad Hoc Networks"*. IEEE Network Magazine, Vol. 15, No. 6, pp. 30-39, Nov. 2001.
11. Fall, K., Varadhan, K.: *"The Network Simulator – NS-2"*. The VINT Project, http://www.isi.edu/nsnam/ns/.
12. Giudici, F., Pagani, E.: *"Spatial and Traffic-Aware Routing (STAR) for Vehicular Systems"*. Technical Report RT 07-05, Information and Communication Dept., Università degli Studi di Milano, Jun.2005.

Adaptive Control Architecture for Active Networks

Mehdi Galily, Farzad Habibipour, Masoum Fardis, and Ali Yazdian

Iran Telecom Research Center, Tehran, Iran
m.galily@gmail.com

Abstract. In this paper, the general architecture of adaptive control and management in active networks is presented. The proposed Adaptive Active Network Control and Management System (AANCMS) merges technology from network management and distributed simulation to provide a unified paradigm for assessing, controlling and designing active networks. AANCMS introduces a framework to assist in managing the substantial complexities of software reuse and scalability in active network environments. Specifically, AANCMS provides an extensible approach to the dynamic integration, management, and runtime assessment of various network protocols in live network operations.

1 Introduction

Active Networking (AN) is an emerging field which leverages the decreasing cost of processing and memory to add intelligence in network nodes (routers and switches) to provide enhanced services within the network [1,2]. The discipline of active networking can be divided into two sub- fields: Strong and Moderate AN. In Strong AN, users inject program carrying *capsules* into the network to be executed in the switches and routers. In Moderate AN, network provides provision code into the routers to be executed as needed. This code can provide new network based services, such as active caching and congestion control, serve as a mechanism for rapidly deploying new protocol versions, and provide a mechanism to monitor, control, and manage networks [3]. Active Networking is related to IN (intelligent networking), which provides intelligence and service creation mechanisms in the PSTN (public switched telephone network) [4].

The most significant trends in network architecture design are being driven by the emerging needs for global mobility, virtual networking, and active network technology. The key property common to all these efforts is *adaptability*: adaptability to redeploy network assets, to rewrite communication rules, and to make dynamic insertion of new network services a natural element in network operations. Critical to the deployment and management of these future networks is the need to provide consistency and control over dynamic changes, and to limit the impact that such changes have on performance and stability, as required for robust communication. Adaptive computing environments could benefit greatly from several ongoing research efforts. Active network research [5-11], in particular, seeks to pursue this concept of adaptive computing by providing network protocols that are more flexible and extensible. Active networking is motivated by the notion that the improvement and evolution of current networking software is greatly hindered by slow and expensive standardization processes. Active networking tries to accommodate changes to network software by facilitating the safe and efficient dynamic reconfiguration of the network [12].

In this paper, an Adaptive Active Network Control and Management System (AANCMS) is proposed. Our architecture is designed to actively control, monitor, and manage both conventional and active networks, and be incrementally deployed in existing networks. The AANCMS is focused on an active monitoring and control infrastructure that can be used to manage networks. The rest of the paper is organized as follows: in Section 2 we explain the architecture of an active network node. Section 3 presents the basic structure of AANCMS. Some comments on distributed simulation are made in Section 4. Section 5 presents some possible future works on this subject. Finally, the paper is concluded in Section 6.

2 Active Network Node Architecture

Active networking technology signals the departure form the traditional store-and-forward model of network operation to a store-compute-and-forward mode. In traditional packet switched networks, such as the Internet, packets consist of a header and data. The header contains information such as source and destination address that is used to forward the packet to the next element that is closer to the destination. The packet format is standardized and processing is limited to looking up the destination address in the routing tables and copying the packet to the appropriate network port. In active networks, packets consist not only of header and data but also of code. This code is executed on the active network element upon packet arrival. Code can be as simple as an instruction to re-send the packet to the next network element toward its destination, or perform some computation and return the result to the origination node.

Apart from obvious practical advantages such as those described above, there are several properties which make active networks attractive for the future of global networking as a form of agreement on network operation for interactions between components that are logically or physically distributed among the network elements. A number of reasons have been contributing to a very long standardization cycle, as observed in the activities of the Internet Engineering Task Force (IETF). Most importantly, the high cost of deploying a new function in the infrastructure, required extreme care and experimentation before the whole community would to agree that a standardized protocol or algorithm is good enough. Diversity, competition and other conflict creating conditions also contribute to severe delays in standardization and thus deployment of new services. In active networks, functionality can be deployed in the infrastructure dynamically, and can be easily removed or replaced. This offers more flexibility in deploying early implementations (without having to stall on the standardization process), protocol bridged (that translates between different revisions/generations of a service as in the active bridge [13]), and most importantly, services themselves: users are free to customize the network infrastructure to fit their needs, when such needs emerge. The key component enabling active networking is the *active node*, which is a router or switch containing the capabilities to perform active network processing. The architecture of an active node is shown in Fig. 1, based on the DARPA active node reference architecture [14].

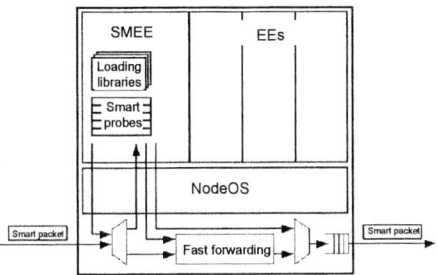

Fig. 1. Architecture of an Active Node

3 AANCMS Structure

In addition to work in the active network community, new standards are being proposed to assist in the distribution and maintenance of end-user applications [15-17]. These standards attempt to introduce more timely and cost-effective mechanisms for distributing and maintaining application software via the network, allowing users to install or update software components by simply accessing HTML-like pages. However, extending such mechanisms to include the deployment and maintenance of system-level software is more difficult. The addition of system-level networking software must be done carefully to avoid potentially costly mistakes, and must also be properly coordinated with the management infrastructure if such changes are to be properly monitored and controlled.

While the trend toward adaptable protocol and application-layer technologies continues, the control and assessment of such mechanisms leaves open broader questions. Future networks could greatly benefit from simulation services that would allow network engineers to experiment with new network technologies on live network operations, without compromising service. Live traffic-based simulation services would provide engineers insight into how a proposed alteration would affect a network, without committing the network to potentially disruptive consequences.

Finally, the management of adaptive networks would greatly benefit from sophisticated monitoring tools to help assess the effects of runtime alterations and detect when those effects result in activity outside a boundary of desired operation. AANCMS is intended to streamline and, at the same time, enrich the management and monitoring of active networks, while adding new support to the network management paradigm to assist network designers. The AANCMS is pursuing a unified paradigm for managing change in active network computing environments. Underlying this framework is a conceptual model for how elements of technology from network management, distributed simulation, and active network research can be combined under a single integrated environment. This conceptual model is illustrated Fig. 2.

AANCMS gains from *discrete* active networking the ability to dynamically deploy engineering, management, and data transport services at runtime. AANCMS leverages this capability with network management technology to (1) integrate network and system management with legacy standards (SNMP, CMIP) to provide a more flexible and scalable management framework, (2) dynamically deploy mechanisms to collect network statistics to be used as input to network engineering tools and higher-level assessment tools, and (3) assist network operators in reacting to significant changes in the network.

AANCMS targets an active network environment, where powerful design and assessment capabilities are required to coordinate the high degree of dynamism in the configuration and availability of services and protocols. To this end, we have formulated architecture of a network management and engineering system that, while inheriting some components from current NM technology, introduces distributed simulation as an additional tool for design and performance assessment. Some components of the AANCMS architecture map very well to already existing technology. Recognizing this, the architecture has been explicitly designed to accommodate other network management engineering solutions. The AANCMS architecture is divided into data, assessment, and control layers. Fig. 3 shows how the data and information flow through the layers. The data layer operates at the data packet level and offers a set of services for the manipulation of network data. The assessment layer performs analytical reviews of network behavior to extract relevant semantic information from it. The control layer performs higher-order functions based on expert knowledge.

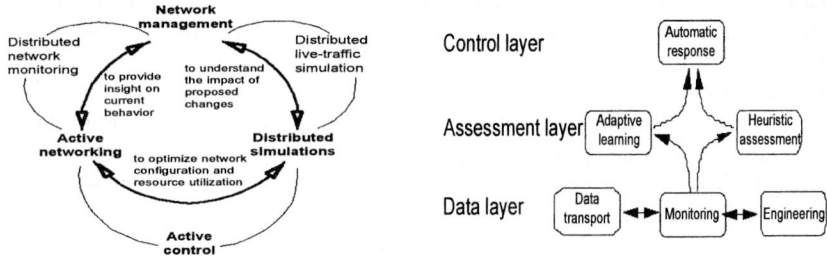

Fig. 2. Conceptual Framework of AANCMS   **Fig. 3.** AANCMS architecture

The AANCMS architecture has been designed to reuse and integrate software components derived from significant advances in network alarm correlation, fault identification, and distributed intrusion detection. In particular, the assessment and control layers of the AANCMS architecture perform tasks analogous to alarm correlation and fault analysis of the types currently proposed by network management expert systems. All the components constituting these logical layers may be independently deployed and configured on machines throughout the network using common system management support. The implementation of each of these logical layers may use (1) existing non-active technology properly fitted to be dynamically deployed (thus implementing the discrete active networking approach) or (2) new active networking technology. AANCMS may distribute data-layer services on machines across domains, but deploys assessment and control layer services in machines within the domain they manage. Depending on the amount of resource sharing resulting from the deployment of active networking services, the assessment layer may also be distributed across machines in multiple domains. Because the control layer must possess a significant amount of authority to perform changes in the network, it should be deployed only within a single domain. Several control services may then cooperate at the inter-domain level to exchange information for making better control decisions about their respective domains. The following sections describe the data layer, which embodies the most innovative features of our architecture. The assessment and control layer will not be further discussed in this paper. The foundation of the AANCMS architecture is the data layer, which is composed of engineering, monitoring, and data transport services. Although presented as a single layer, it is useful to recognize and distinguish the various modules that may populate this layer. For this reason, we decompose the data layer into three distinct data service types, all of which may benefit from dynamic deployment in the network.

4 Distributed Simulation

Adaptable and configurable networks will require code repositories to store and retrieve deployable applications. This idea has already appeared in several network management designs where deployable monitors can be dynamically inserted to key points in a network. Under AANCMS we are reusing and extending these concepts in the development of generic and reusable simulation models, which are deliverable as part of an AANCMS simulation service. In particular, we are developing simulation models that allow network engineers to compose and design experiments dynamically, which may then use traffic models derived form network traffic observed from spatially distributed points in a network. The traffic models may be (more traditionally) derived at the NM station and then re-exported to the simulation nodes or derived in the network itself through a distributed modeling approach (i.e. deploy a specialized monitoring application that creates and feeds the models to the network engineering services). The following briefly summarizes the benefits of extending simulation into the network management framework, and how issues of resource utilization can be controlled and balanced against the fidelity of simulation results.

5 Conclusion

In this paper, an adaptive control and management system for active networks (termed AANCMS) was proposed. The AANCMS is focused on an active monitoring and control infrastructure that can be used to manage networks. As the dynamic deployment of network services becomes standard technology to support user applications, network operators will require an efficient and flexible infrastructure to assist them in network design, configuration, and monitoring. The quality of future network management, monitoring, and engineering tools and standards will be crucial in determining the speed at which networking will evolve toward a more dynamic architecture.

References

1. D. L. Tennenhouse and D. J. Wetherall. Towards and active network architecture. ACM Computer Communication Review, 26(2):5–18, Apr. 1996.
2. K. L. Calvert, S. Bhattacharjee, E. Zegura, and J. P. Sterbenz. Directions in active networks. IEEE Communications Magazine, 36(10), Oct. 1998.

3. W. Jackson, J.P.G. Sterbenz, M. N. Condell, R. R. Hain. Active network monitoring and control, Proc. DARPA Active Networks Conference and Exposition (DANCE.02), 2002.
4. A.V. Vasilakos, K.G. Anagnostakis, W. Pedrycz, Application of computational intelligence techniques in active networks, Soft Computing (5), pp. 264-271, 2001.
5. D. S. Alexander, M. Shaw, S. M. Nettles, and J. M. Smith. Active bridging. Proceedings of the ACM SIGCOMM'97 Conference, Cannes, France, September 1997.
6. J. Hartman, U. Manber, L. Peterson, and T. Proebsting. Liquid software: A new paradigm for networked systems. Technical Report 96-11, University of Arizona, 1996.
7. U. Legedza, D. J. Wetherall, and J. V. Guttag. Improving the performance of distributed applications using active networks. IEEE INFOCOM'98, 1998.
8. J. Smith, D. Farber, C. A. Gunter, S. Nettle, M. Segal, W. D. Sincoskie, D. Feldmeier, and S. Alexander. Switchware: Towards a 21st century network infrastructure. http://www.cis.upenn.edu/~switchware/papers/sware.ps, 1997.
9. D. J. Wetherall, J. V. Guttag, and D. L. Tennenhouse. ANTS: A toolkit for building and dynamically deploying network protocols. Proceedings of IEEE OPENARCH'98, 1998.
10. Y. Yemini and S. da Silva. Towards programmable networks. Proceedings IFIP/IEEE International Workshop on Distributed Systems: Operations and Management, L'Aquila, Italy, October 1996.
11. L. Ricciulli. Anetd: Active NETwork Daemon. Technical Report, Computer Science Laboratory, http://www.csl.sri.com/ancors/Anetd , SRI International, 1998.
12. G. Minden, W.D. Sincoskie, J. Smith, D. Tennenhouse, D. Wetherall, "A survey of Active Network Research", IEEE Communications, Vol. 35, No. 1, pp 80-86, January 1997.
13. Alexandr D.S., Show M., Nettles S.M. and Smith J.M., Active bridging, in Proc, 1997 ACM SIGCOMM Conference, pp.231-237, 1997.
14. K. Calvert, ed. Architectural Framework for Active Networks. AN draft, AN Architecture Working Group, 1998.
15. Van Hoff, J. Giannandrea, M. Hapner, S. Carter, and M. Medin. The HTTP Distribution and Replication Protocol. http://www.marimba.com/standards/drp.html, August 1997.
16. Van Hoff, H. Partovi, and T. Thai. Specification for the Open Software Description (OSD) Format. http://www.microsoft.com/standards/osd/, August 1997.
17. S. Crane and N. Dulay and H. Fossa and J. Kramer and J. Magee and M. Sloman and K. Twidle, Configuration Management For Distributed Software Services, Integrated Network Management IV, 1995.

A Study on Bandwidth Guarantee Method of Subscriber Based DiffServ in Access Network

HeaSook Park[1], KapDong Kim[1], HaeSook Kim[1], and Cheong Youn[2]

[1] Electronics and Telecommunications Research Institute,
161 Gajeong-Dong Yuseong-gu Daejeon, 305-700, Korea
{parkhs, kdkim71, hskim}@etri.re.kr
[2] Dept. of Computer Science, Chungnam National University
cyoun@cs.cnu.ac.kr

Abstract. In this paper, we describe the structure of the access network and we propose bandwidth guarantee scheme for subscriber and service. The scheme uses two kinds of the classification table, which are called "service classification table" and "subscriber classification table." Using the classification table, we can identify the flow of the subscriber and service. Also, we compute the number of hash table entry to minimize the loss ratio of flows using the M/G/k/k queueing model. Finally, we apply to deficit round robin (DRR) scheduling through virtual queueing per subscriber instead of aggregated class.

1 Introduction

The traditional IP network offers best-effort service only. To support quality of service(QoS) in the Internet, the IETF has defined two architectures: the Integrated Services or Intserv[2], and the Differentiated Services or Diffserv[3]. They have important differences in both service definitions and implementation architectures. At the service definition level, Intserv provides end-to-end guaranteed or controlled load service on a per flow (individual or aggregate) basis, while Diffserv provides a coarser level of service differentiation. By enabling QoS, which essentially allows one user gets a better service than another. The internet is a media transmitted data. Gradually, it needs the transmission of a triple play service such as voice and video. Ethernet passive optical network (EPON) is generally used as an optical subscriber network. The main configuration of this paper is an analysis of the QoS structure in the EPON access network. In this paper, QoS of the optical line termination (OLT) system has been implemented with the conjunction network processor and provisioning management software[1]. The remaining part of the paper is as follows. The system architecture of access network using the network processor is proposed in Sect. 2. In the Sect. 3, we propose the bandwidth guarantee scheme for subscriber and service in optical sub-scriber network. Finally, the result of simulation with analysis has been described in Sect. 4 and we give our conclusion in Sect. 5.

2 System Architecture

Figure 1 show the interconnection architecture of the OLT. The OLT system is located between Internet backbone network and fiber to the home (FTTH) subscriber through EPON. The main functions of the OLT are it offers basic routing protocol, it provides the function of IP broadcasting and subscriber authentication, and it offers the function of QoS and security. EPON master is physically interconnected with line card of OLT, and a PON bridge is consisted to ONU and ONT through optical splitter, and then, subscribers are connected. Authentication server+ performs two kinds of functions. One is authentication function for subscriber and service. Another is the function setting of the QoS profile for subscriber and service.

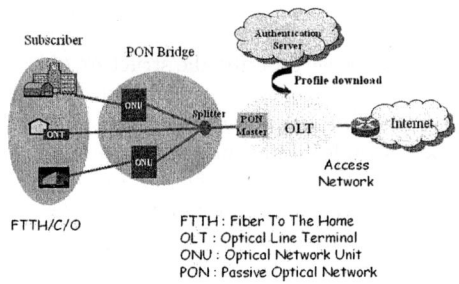

Fig. 1. The Structure Of PON Access Network

3 Bandwidth Guarantee Scheme

Packets of broadcasting channel using multicasting address are copied in egress network processor of the OLT and transmitted to the PON bridge area. Therefore, we could not distinguish the address of subscriber in OLT. So, we don't know the information of the bandwidth of a subscriber. For bandwidth guarantee of the subscriber and service in OLT system, we should be resolved the number of two problems. First, QoS control server should allocate dynamically the bandwidth and QoS profile of subscribers and services. Second, Network processor in line card of the OLT must provide the packet transmission function through classification table to distinguish subscribers and services.

3.1 Hierarchical Classification

Figure 2 shows data structure of the hierarchical classification table and the flow of dynamic bandwidth allocation. The 6-tuple classifier takes selected fields of an IPv4 header and the input port as a lookup key. The lookup result contains flow_id (identifier of a packet flow) and class_id (a relative identifier of a target QoS queue). To accommodate color-aware srTCM, the classifier may also set color_id. If a hash table lookup fails, the classifier sets default values the above metadata variables.

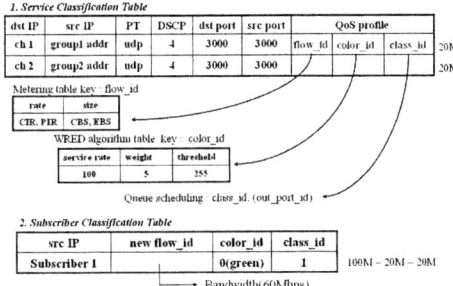

Fig. 2. Hierarchical Classification Table

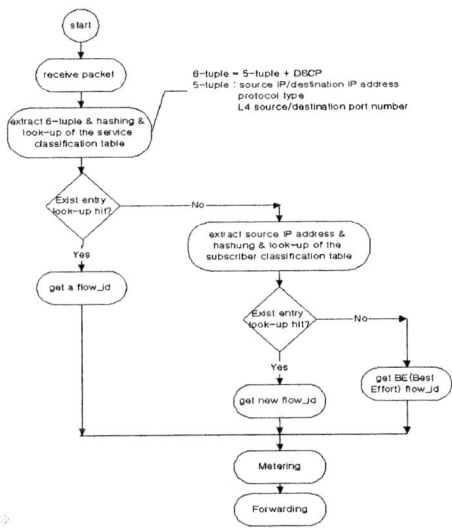

Fig. 3. Flow Chart of Packet Classification

Figure 3 shows the processing flow of packet received in ingress micro engine. When packet is received, it extracts 6-tuple of the packet, hashing and then look-up service classification table with the result value of the hashing. If look-up is success then we get a flow_id, color_id and class_id of the entry. If look-up is not success, the packet is not a packet of a registered service. So, we need one more look-up the subscriber classification table with the source IP address of the packet. If, the look-up is success, it is a packet of registered premium subscriber. We get a new flow_id, color_id, and class_id. If second look-up is failed, the packet is using best-effort service for normal subscriber. So, we get a default flow_id, color_id and class_id.

3.2 Number of Hash Table Entry

Arrival process of flows are according to a poisson process with rate λ. There are k rooms in the hash table, each of which stores a single flow information. If there is no room in the hash table at a new flow arrival, the new flow is rejected and lost. The sojourn times of flows are assumed to have distribution B(x) and density function b(x). Then the system can be modeled by the M/G/k/k queueing system. In this queueing model, the number of flows in the hash table is according to the following distribution[5]: **Expression 1**:

$$P\{n \text{ customers in the system}\} = \frac{\frac{(\lambda E[\mathbf{B}])^n}{n!}}{\sum_{i=0}^{k} \frac{(\lambda E[\mathbf{B}])^i}{i!}}, 0 \leq n \leq k \tag{1}$$

Then the loss probability due to the limitation of the hash table size is given by **Expression 2**:

$$P_{loss} = \frac{\frac{(\lambda E[\mathbf{B}])^k}{k!}}{\sum_{i=0}^{k} \frac{(\lambda E[\mathbf{B}])^i}{i!}} \tag{2}$$

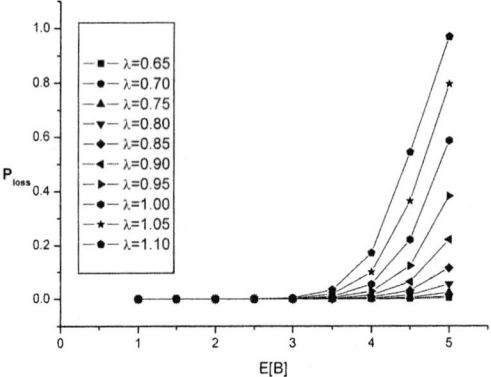

Fig. 4. Loss Probability According to E[B]

Figure 4 shows the loss probability according to E[B]. We have to regard that loss probability is increased hastily when E[B] is more than 3.

3.3 Queueing and Scheduling per Subscriber

Consider a DRR (Deficit Round Robin) scheduler with n queues having weights $\phi_i, (1 \leq i \leq n)$. Then it is known that the latency θ_i of queue i is given by[4] **Expression 3**:

$$\theta_i = \frac{1}{r}\left[(F - \phi_i)\left(1 + \frac{\overline{P_i}}{\phi_i}\right)\sum_{j=1}^{n}\overline{P_i}\right] \tag{3}$$

Where, $\overline{P_i}$ is the maximum packet size of flow i and $F = \sum_{i=1}^{n} \phi_i$. Now assume

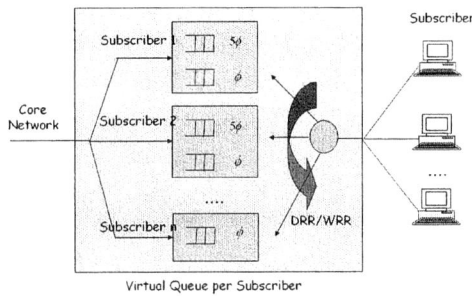

Fig. 5. The Example of DRR Scheduling

that all weights are fixed. Then, when we want to accommodate CBR type real time traffic with delay bound δ in queue i, it is necessary to satisfy $\theta_i < \delta$ from which we can get the maximum allowable number of queues. Figure 5 shows the DRR algorithm allocating virtual queue per subscriber. Subscriber n has BE (best effort) service and subscriber 1 and 2 have a internet service of best effort and premium service. Packets received from core network queued to virtual queues according to subscriber and transmitted to subscriber through DRR algorithm. In this paper, we allocates the weight ϕ to the BE service and weight 5ϕ to the premium service to simplify the computation. We assume the following conditions to induce value n satisfy the delay bound. $\phi_i = \phi, (1 \le i \le n)$, premium subscribers occupies 20% of the total subscriber, $P_i = MTU$(Maximum Transfer Unit) where, 64Kbytes with worst case, and $\gamma = 10Gbps$ (The performance of network processor).

4 Simulation and Analysis

Figure 6 shows the relation of latency θ_i and the number of subscriber n. We know there is difference between premium service and best effort service. For

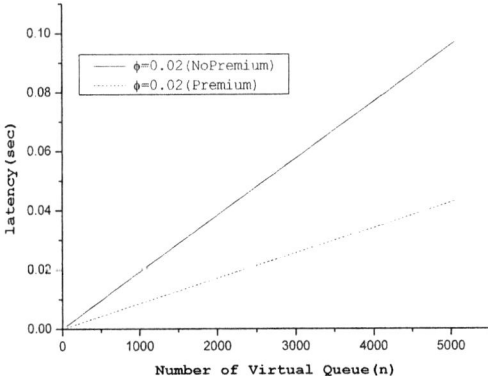

Fig. 6. Maximum Subscriber According to Latency

example, if the delay bound δ of the target system is 20ms~50ms then, the number of subscriber meets the delay bound is 1000~2700. This result derived according to worst case. Therefore, there is difference between the design of a real system network and the result of in this paper.

5 Conclusion

Based on the basic idea of DiffServ, we decouple the control plane and data plane functions in optical subscriber network. The low overhead and implementation simplicity makes a hierarchical look-up of classification table an attractive candidate for adoption in optical subscriber network based IP DiffServ.

References

1. H. Park, S. Yang, and C. Youn : The implementation of the Cos/QoS using network Proces-sor in the OLT System, OECC/COIN, (2004) 174–175
2. S. Shenker R. Braden, and D. Clark : Integrated services in the Internet architecture: an over-view, Internet RFC 1633, (1994)
3. S. Black et al. : An Architecture for Differentiated Services, IETF RFC 2475, (1998)
4. L. Lenzini, E. Mingozzi, and G. Stea : Tradeoffs Between Low Complexity, Low La-tency,and Fairness With Deficit Round-Robin Schedulers, IEEE/ACM Trans. on Network-ing, Vol. 12, No. 4, (2004) 68–80
5. D. Gross and C. Harris : Fundamentals of Queueing Theory, Third Edition, Wiley, (1998)

Least Cost Multicast Loop Algorithm for Local Computer Network

Yong-Jin Lee

Department of Technology Education,
Korea National University of Education
yjlee1026@daum.net

Abstract. Minimal cost loop problem consists of finding a set of loops to minimize the total link cost of end-user nodes while satisfying the traffic constraints. This paper presents a heuristic algorithm using trade-off criterion based on the node exchange and node transfer among loops by the initial algorithms. Simulation results show that the proposed algorithm produces about ten percent better solution than previous algorithms in short execution time. Our algorithm can be applied to find multicast loops in local computer network.

1 Introduction

Minimal cost loop problem (MCLP) [1] is to form a collection of loops that serve all user nodes with a minimal connection cost. End-user nodes are linked together by a loop that is connected to a port in the backbone node such as switch. The links connecting end-user nodes have a finite capacity and can handle a restricted amount of traffic, thus limit the number of end-user nodes that can be served by a single loop.

A similar problem is quality of service (QoS) routing problem, which is to select feasible paths that satisfy various QoS requirements of applications in a network. Multiple additively constrained QoS routing problem is referred to as multiple constrained path selection (MCPS) [2,3]. Delay constrained least cost (DCLC) path selection problem [4,5] is to find a path that has the least cost among the paths from a source to a destination for which the delay remains under a given bound. To solve the above problems, network configurations such as link connectivity, link cost, and delay should be available. On the other hand, the MCLP is to find the network configuration composed of several loops to satisfy the traffic capacity constraint. Since the problem definition of MCLP is different from those of MCPS and DCLC, solutions to MCPS and DCLC can not be applied to MCLP.

The MCLP is NP-hard [6]. Therefore, the heuristic approach is desirable for the MCLP, which generates a feasible solution in a reasonable time. In this paper, we present a heuristic algorithm that makes use of a set of loops by the existing algorithms based on traveling salesman problem (TSP) or node partition algorithm (NPA). Our algorithm improves the solution by exchanging and transferring nodes based on the suggested heuristic rules. The proposed algorithm reduces total cost in a short computation time. In addition, it can be applied to the finding of multicasting loops from the source node to several end-nodes in local computer network. The rest of the paper is organized as follows. We begin by describing modeling for the MCLP.

Section 3 describes the proposed algorithm for the MCLP. Section 4 discusses performance evaluation and section 5 concludes the paper.

2 Modeling and Initial Algorithms for the MCLP

We now consider the modeling of the MCLP. Assume that there is graph, $G=(V, E)$, where $V=\{0,1,2,...,n\}$, and link cost between node i and node j is d_{ij} ($i,j \in V$-$\{0\}$ and $(i,j) \in E$). Q represents maximum number of nodes served by single loop. Index 0 represents the source node and can be viewed as an object such as the backbone router with several ports. End-user nodes originate traffics and can be regarded as hosts or switching hubs.

The problem formulation for the MCLP is described by Eq. (1). The objective of the MCLP is to find a collection of least-cost loops rooted at node 0. L_p is the pth loop ($p=1,2,...,lcnt$: where $lcnt$ is the number of loops in the solution) whose two endpoint nodes are connected the source node. No more than Q nodes can belong to any loop of the solution because one port of source node has a finite capacity, which can process a restricted amount of traffics generated at end-nodes. In Eq. (1), x_{ij} represents link between node i and j (i,j: $i=1,2,..n$; $j=1,2,...,n$). If link (i, j) is included in any loop (L_p) of the solution, then x_{ij} is set to 1. The total number of links should be more or than equal to $\lceil n/Q \rceil (Q+1)$.

$$\text{Minimize} \sum_{i,j} d_{ij} x_{ij}$$

S.T.

$$\sum_{i,j \in L_p} x_{ij} \leq Q \quad (1)$$

$$\sum_{i,j} x_{ij} \geq \lceil n/Q \rceil (Q+1)$$

$$x_{ij} = 0 \text{ or } 1$$

We will use solutions to IOTP, QIOTP and node partition algorithm (NPA) [1] as the initial solution to our algorithm in Section 3. IOTP and QIOTP heuristics convert a given TSP tour into feasible loops satisfying that the maximum number of nodes in a single loop is less than the predefined threshold (Q). NPA divides nodes into several clusters satisfying the constraint and finds node sets included in each cluster. It uses parallel nearest neighbor rule and backtracking.

3 Minimal Cost Loop Heuristic Algorithm

Our algorithm (hereafter referred to minimal cost loop heuristic (MCLH)) uses the simple heuristic rules based on node exchange and transfer in order to improve solution starting from the initial solution by IOTP, QIOTP, and NPA.

An initial topology is obtained by applying TSP algorithm [7] to each set of nodes ($node(p)$, $p=1,2,..,lcnt$) in the initial solution. Then, MCLH finds the initial network cost (NEW_{cost}). For the node pair (i, j) $(i<j)$, which are belonged to node sets of other loops, it finds trade-off criterion. The concept of trade-off criterion is described as followings: It is assumed that $i \in node(sub1)$, $j \in node(sub2)$, $sub1 \neq sub2$, and $node(sub1)$ and $node(sub2)$ are ordered sets, $\{i-1, i, i+1\}$ and $\{j-1, j, j+1\}$ respectively on TSP tour. Here, $sub1$ and $sub2$ represent indexes of loop including node i and node j, respectively. Both loop $sub1$ and loop $sub2$ are TSP tours. Also, $node(sub1)$ and $node(sub2)$ represent sets of node indexes on loop $sub1$ and loop $sub2$, respectively. Exchange heuristic rule (ks_{ij}) is defined as $(d_{i,j-1}+d_{i,j+1}+d_{j-1,j}+d_{j,j+1}) - (d_{i-1,j}+d_{i+1,j}+d_{i,j+1}+d_{i,j-1})$. If ks_{ij} is positive, solutions can be improved by exchanging node i for node j. On the node pair (i, j) which are belonged to node set of the same loop, ks_{ij} is set to $-\infty$. Starting from the node pair (i, j) with the maximum positive value of ks_{ij}, we obtain two new loops by exchanging node i for node j. TSP algorithm is applied to these two node sets again and two new loops are found. A reason to apply TSP algorithm to only two node sets is because other loops are already TSP tours.

We next compare the newly obtained cost (D_{cost}) with the previous cost (NEW_{cost}). If D_{cost} is less than NEW_{cost}, which is substituted for D_{cost} and the corresponding ks_{ij} is set to $-\infty$. Also, we repeat the above procedure after computing ks_{ij} matrix again because the changed locations of node i and j affect the trade-off criterion. Otherwise, the corresponding ks_{ij} is set to $-\infty$ and we repeat the above procedure on node pair (i, j) with the maximum value among the ks_{ij} matrix, which is relevant to former NEW_{cost}.

MCLH starts to transfer nodes from one loop to other loops based on trade-off criterion (ps_{ij}) when it can not exchange nodes any more. The question whether each loop contains Q nodes or trade-off's are negative for all node pair (i, j) is examined. If so, the algorithm is terminated. Otherwise, the node pair (i, j) with the maximum positive value of ps_{ij} is found. By transferring node j to the loop including node i, the solution is improved. In this case, only two endpoint nodes in each loop are considered. The reason is why solutions can be improved by transferring endpoint nodes. The trade-off criterion for node transfer is defined as followings: If $sub1$ is equal to $sub2$ or $kcnt(sub1)+1$ is greater than Q, node j can not be transferred to the loop $sub1$. Here, $kcnt(sub1)$ represents the number of nodes in loop $sub1$. In this case, ps_{ij} is set to $-\infty$. Otherwise, ps_{ij} is computed. It is assumed that $i \in node(sub1)$, $j \in node(sub2)$, $sub1 \neq sub2$, and $node(sub1)$ and $node(sub2)$ are ordered sets, $\{i-2, i-1, i\}$ and $\{j, j+1, j+2\}$ respectively on the TSP tour. Possible transfer trade-off criterion (ps_{ij}) is defined as $(d_{0i} + d_{j,j+1}) - (d_{ij} + d_{0,j+1})$. Entire procedure of MCLH is described as followings:

Algorithm: Minimal cost loop heuristic
(1) Obtain $node(p)$ by using the initial algorithm and find NEW_{cost} by applying TSP algorithm to $node(p)$.
(2) If node i and j are included in the different $node(p)$, Compute the trade-off criterion (ks_{ij}) for node pair (i,j). Otherwise set $ks_{ij} = -\infty$. If all ks_{ij}'s are negative, stop.
(3) While there is (i,j) such that $ks_{ij} > 0$, exchange node i and j with the maximum positive ks_{ij}. Find D_{cost} by applying TSP algorithm to two changed node sets.
(4) If $D_{cost} \geq NEW_{cost}$, restore node (i, j) to the previous state. Set $ks_{ij} = -\infty$ and repeat step 3. Otherwise, set $NEW_{cost} = D_{cost}$, $ks_{ij} = -\infty$ and go to step 2.
(5) If $kcnt(p) = Q$ for all p $(p=1,2,..,lcnt)$, stop.

(6) If node i and j are included in the different $node(p)$ and for loop p containing node i, $kcnt(p)+1$ is less than or equal to Q, compute the trade-off criterion(ps_{ij}) for node pair (i,j). Otherwise set $ps_{ij} = -\infty$. If all ps_{ij} are negative, stop.
(7) While there is (i,j) such that $ps_{ij} > 0$, transfer node j to $node(p)$ containing node i with the maximum positive ps_{ij}. Find D_{cost} by applying TSP algorithm to two changed node sets.
(8) If $D_{cost} \geq NEW_{cost}$, restore node (i, j) to the previous state. Set $ps_{ij} = -\infty$ and repeat step 7. Otherwise, set $NEW_{cost} = D_{cost}$, $ps_{ij} = -\infty$ and go to step 6.

Lemma 1. Time complexity of the MCLH algorithm using either solution of IOTP or QIOTP as the initial algorithm is $O(Qn^2)$.

Proof) First, we consider step 1-4 of the MCLH algorithm. To obtain the initial solution of step 1, we apply the TSP algorithm to Q nodes up to the number of loops times. So, time complexity is (multiply the number of loops by time complexity of TSP algorithm) = $\lceil n/Q \rceil O(Q^2) = O(Qn)$. Here, the maximum number of links is $\lceil n/Q \rceil(Q+1)$ and time complexity is $O(Qn^2)$. In step 2, ks_{ij}'s are computed for node pair (i, j), $\forall (i, j)$ ($i \in node(sub1)$, $j \in node(sub2)$; $sub1 \neq sub2$) and the maximum number of ks_{ij}'s to be computed is $\lceil n/Q \rceil Q$. Thus, the time complexity of step 2 is $O(n)$. In the worst case, step 2 should iteratively be executed up to the number such that ks_{ij}'s are positives. In this case, the maximum number of iterations for $ks_{ij} > 0$ is $\lceil n/Q \rceil Q$. Time complexity for exchanging node i and j with the maximum positive ks_{ij} is $O(\lceil n/Q \rceil Q) = O(n)$. Since we apply the TSP algorithm to only two changed node sets, the time complexity for this is $O(Q^2)$. After all, time complexity of step 3 is $Q\lceil n/Q \rceil \cdot \text{maximum}[O(n), O(Q^2)] = O(Q^2n)$. From this result, time complexity of step 1-4 is $O(Qn^2)$. In the same manner, time complexity of step 5-8 becomes $O(Qn^2)$. Therefore, total execution time of the MCLH algorithm is maximum $[O(n^2), O(Qn^2)] = O(Qn^2)$.

4 Performance Evaluation

In order to evaluate the MCLH algorithm, computational experiments were carried out using two types of network defined according to the location of the source node. The coordinates of nodes were randomly generated in a square grid of dimensions 100 by 100. We used 20, 40, 60, and 80 as the number of nodes (n). Q was used from 2 to $n/4$. Computational experiments were carried out on an IBM-PC. Programming languages were C. The mean savings rate is defined as

$$\text{Mean savings rate (\%)} = ((A - B) / A) * 100 \qquad (2)$$

Fig. 1 represents mean savings rate. In Fig. 1, IOTP+MCLH, QIOTP+MCLH, and NPA+MCLH represent mean savings rate obtained by letting A be the solutions of IOTP algorithm, QIOTP algorithm, and NPA, and B be the solutions by applying the MCLH algorithm to the solution of IOTP algorithm, QIOTP algorithm, and NPA, respectively. For the savings rate, different results are obtained depending on using

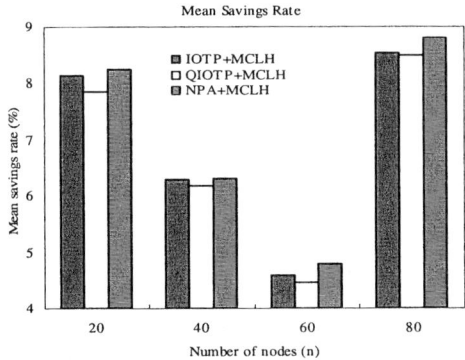

Fig. 1. Mean savings rate

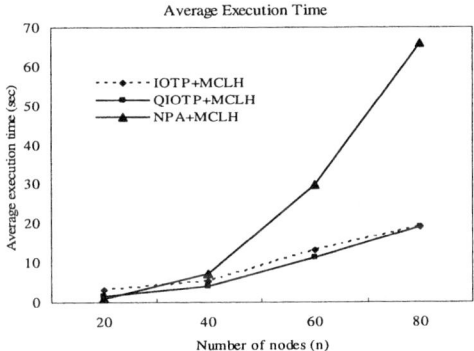

Fig. 2. Average execution time

which initial algorithm. Among initial algorithms, NPA is the most superior in average property. Increasing of Q and n doesn't affect the result. Also, it is known that traffic volume at node is not related with the savings rate.

Fig. 2 depicts average execution time (CPU time: second) in ten runs of simulation. In Fig. 2, IOTP+MCLH is the average execution time when the MCLH algorithm is applied to the solution of IOTP algorithm. QIOTP+MCLH is the average execution time when the MCLH algorithm is applied to the solution of QIOTP algorithm. NPA+MCLH is the average execution time when the MCLH algorithm is applied to the solution of NPA.

The sequence of average execution time is QIOTP+MCLH < IOTP+MCLH < NPA+MCLH. When comparing Fig. 1 with Fig. 2, the MCLH algorithm yields about ten percent improvement over the initial algorithms by adding 6 second to the exccution time of the latter. Also, when the number of nodes is 80, the execution time of the MCLH algorithm alone is about 10 second. Thus, we can conclude that our MCLH algorithm has the relatively short execution time.

5 Conclusions

In this paper, we have described a novel heuristic algorithm for the MCLP. This heuristic improves solutions by exchanging or transferring nodes between loops based on the suggested trade-off criteria. Memory property of our algorithm was derived and through simulation, it was shown that the proposed algorithm improves the existing solutions by ten percent in short execution time. Our algorithm is applied to find multicasting loops in local computer network. Future works include the more efficient algorithm in time complexity at a lower cost.

References

1. Lee, Y.: Minimal Cost Heuristic Algorithm for Delay Constrained Loop Network. International Journal of Computer Systems Science &Engineering, Vol. 19, No. 4, CRL Publishing. (2004) 209-219.
2. Cheng, G. and Ansari, N.: On Multiple Additively Constrained Path Selection. IEE Proc. Communications, Vol. 149, No. 5. (2002) 237-241.
3. Cheng, G. and Ansari, N.: Finding All Hop(s) Shortest Path. IEEE Communications Letters, Vol. 8, No. 2. (2004) 122-124.
4. Cheng, G. and Ansari, N.: Achieving 100% Success Ratio in Finding the Delay Constrained Least Cost Path. Proc. of IEEE GLOBECOM '04. (2004) 1505-1509.
5. Juttner, A., Szyiatowszki, Mecs, I., and Rajko, Z.: Lagrange relaxation based method for the QoS ruoting problem. Proc. IEEE INFOCOM '01. (2001) 859-869.
6. Lenster, J. and Kan, R.: Complexity of Vehicle Routing and Scheduling Problems. Networks, Vol. 11. (1981) 221-227.
7. Magnanti, T., Ahuja, R. and Orlin, J.: Network Flows- Theory, Algorithms, and Applications, Prentice-Hall. (1993).

AB-Cap: A Fast Approach to Available Bandwidth Estimation[*]

Changhua Zhu, Changxing Pei, Yunhui Yi, and Dongxiao Quan

State Key Lab. of Integrated Services Networks, Xidian Univ.,
Xi'an 710071, China
{chxpei, chhzhu}@xidian.edu.cn

Abstract. A new active measurement approach to Internet end-to-end available bandwidth estimation, AB-cap, is proposed, which is based on the concept of self-induced congestion. In this approach, the transmission rates of probing packets are increased step by step and a new decision criteria, NWID (Normalized Weighted Increased Delay), is defined. Another idea is that sliding-window decision is applied to determine the available bandwidth in one measurement. Simulation and test results show that the measurement precision is improved with slightly adding probing overhead compared with pathchirp.

1 Introduction

The available bandwidth of a network path is the maximum throughput that the path can provide for a new flow without reducing the throughput of the cross traffic along the path[1] (so-called maximum residual bandwidth). It is determined by the tight link of the end-to-end path[2]. Available bandwidth is one of the important performance metrics of network, which can be applied to research and development of network protocol, control scheme and network application, etc.

Since the available bandwidth changes with variation of burst cross traffic in Internet path, the method to measure it should be fast, accurate and light load on network. Available bandwidth can be obtained by accessing the interfaces information of MIB(Management Information Base) through agents in routers along the path by using the SNMP network management protocol. However, such access is typically available only to administrators but not to end users. Even network administrators sometimes need to determine the bandwidth from hosts under their control to hosts outside their infrastructures, so they must rely on end-to-end measurement. In addition, routers are implemented by various ways and this method is limited by the time of polling the routers, the obtained available bandwidth is not immediate. So the available bandwidth is obtained mainly by measurement. The current measurement methods can be classed into two categories: one is based on statistical models of cross traffic, also named Probe Gap Model(PGM), in which available bandwidth is computed by the mathematical formula of sent gap, received gap, bottleneck

---

[*] This work is supported by National Natural Science Foundation of China (NSFC, No.60132030).

bandwidth and available bandwidth, and measuring related parameters, e.g. Spruce[3], ABwE[4], ab-probe[5], etc. Another is based on the idea of self-induced congestion, also named Path Rate Model (PRM), in which the principle is as follows: if the sending rate of probe packets is faster than available bandwidth, the probe packets queue at some routers so that end-to-end delay increase gradually; if the sending rate of probe packets is slower than available bandwidth, the probe packets experience little delay(approximately no delay). While observing the delay variation in receiving host and deciding the time at which congestion begins, the available bandwidth can be obtained, e.g. pathload [1][6], TOPP [7], Pathchirp[8]. PGM can be cooperated with PRM, e.g. IGI (Initial Gap Increasing)[9], PTR(Packet Transmission Rate)[9]. In Pathload CBR(Constant Bit Rate) traffic is sent with adaptively adjusting the sending rate and the sending rate converges to the available bandwidth. Pathload is large in time cost and heavy in traffic load, in addition it may over estimate the true value[3]. IGI can not estimate accurately in high bottleneck link utilization[3]. We find the model based on PGM is far approximate, so the PRM based method is studied in this paper.

We propose a new method, AB-cap, which is also based on the principle of self-induced congestion. In AB-cap, the sending rate of probe packets is increased step by step and the result can be obtained in one measurement.

2 Algorithm Description

In AB-cap, the related parameters of probe packets are as shown in Table 1.

Table 1. The related parameters in AB-cap

Symbols	Parameters
l^n	packet size
N	the number of packets at each sending rate
$t_{s,n}$	the sending time of probe packet (refer to local clock of sender)
$t_{d,n}$	the receiving time of probe packet (refer to local clock of receiver)
$\Delta ts_{n,n+1}$	the sending time interval of contiguous probe packets
$\Delta td_{n,n+1}$	the received time interval of contiguous probe packets
d_n	the end-to-end delay of probe packet
λ_s	the frequency of sender referring to the clock of receiver whose frequency is 1
Δ_0	offset between the clock of sender and the clock of receiver

In Table 1, superscript or subscript n denotes the n-th probe packet. For the n-th and (n+1)-th probe packets, we have

$$t_{d,n} = t_{s,n}/\lambda_s + d_n + \Delta_0, \quad t_{d,n+1} = t_{s,n+1}/\lambda_s + d_{n+1} + \Delta_0$$

So

$$\Delta td_{n,n+1} = \tfrac{1}{\lambda_s}\Delta ts_{n,n+1} + (d_{n+1} - d_n) \qquad (1)$$

We observed that the total end-to-end delay, which is about a few or several hundred milliseconds[10], varies with different paths and clock skew is about 10^{-6}. So the clock skew is neglected in AB-cap, that is, $\lambda_s \approx 1$. We have

$$d_n = \sum_{l=1}^{n-1}(\Delta td_{l,l+1} - \Delta ts_{l,l+1}) + d_1 \qquad (2)$$

Let d_n' denote relative delay $d_n - d_1$.

In order to reduce the influence of the probe traffic on network and avoid the heavy load incurring by feedback-based sending rate adjustment algorithm (The sending rate means the ratio of packet size to the sending time interval of contiguous probe packets and the unit is bps). Our probing scheme is as follows: the sending rate of probe packet is increased step by step in a packet train. The number of packets at each sending rate is $N(N>1)$. Finally, available bandwidth can be obtained by analysis of the trend of relative delay variation. The detailed algorithm can be given as follows:

Firstly, we divide the measured relative delay sequence into multiple segments, each segment with the same sending rate. Then we compute the mean of each segment. Notice that the number of received packets in each segment may be less than the number of sending packets because of the existence of packet loss. We neglect the lost packet and take the packets preceded and followed this lost packet as the contiguous packets. If there is N_i packets in the i-th segment, then $N_i \leq N$. Let the mean of the relative delays at the i-th segment be y_i, then

$$y_i = \frac{1}{N_i}\sum_j d_j' \;, \text{ the } j\text{-th relative delay belongs to the } i\text{-th segment} \qquad (3)$$

Secondly, we adopt slide-window processing scheme with window size M. Furthermore, we define a new value, NWID (Normalized Weighted Increased Delay) as

$$NWID = \frac{\sum_{m=1}^{M} w_m (y_{m+1} - y_m)}{\sum_{m=1}^{M} w_m |y_{m+1} - y_m|} \qquad (4)$$

Where w_m is weight value. The weights are assigned as follows: the sum of all weights is 1 and the larger the subscript m, the larger the weight. The larger weights are assigned to the delay increasements with larger subscript in the slide window which are more sensitive to congestion and influence estimation results more significantly. In addition, from the way of measurement if the delay with larger subscript increasements are negative and the absolutes are larger, then the turning point of delay variation moves backward further. If $NWID > V_{th}$ (in which V_{th} is the decision threshold), then the turning point is determined.

Finally, If the condition is met L times continuously, that is to say, $NWID > V_{th}$ always comes into existence when window is slid L times, then the starting point of the first window which meet the condition is determined as the turning point and the corresponding sending rate is the available bandwidth.

3 Simulation and Measurement of AB-cap

3.1 Simulation Results and Analysis

AB-cap is simulated by use of ns2.26. The simulation topology is shown in Fig. 1. The link from R3 to R4 works in 10 Mb/s, and others work in 100Mb/s. The probing packet size is 1K Bytes. The sending rate of probe packet begins with 200 kb/s and increases 200 kb/s every 10 packets. There are 5 sources which generate cross traffic whose sources and destinations are CT1->CT5, CT3->CT7, CT2->CT4, CT6->CT8, CT9->CT10, respectively. They are all Pareto On/Off distributed. The parameters packet size(Bytes), ON duration(ms), OFF duration (ms) and shape factor are:1024, 5, 5, 1.5; 400, 5, 5, 1.5; 600, 7, 3, 1.2; 40, 7, 3, 1.2; 100, 7, 3, 1.2, respectively. The mean value of sending rates is set in order to obtain different throughput of cross traffic in simulation.

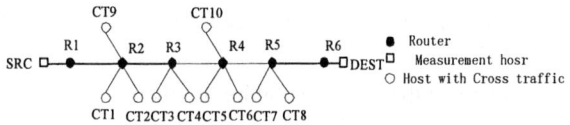

Fig. 1. Simulation topology

We simulate AB-cap in three different cross traffic scenarios. By adjusting the sending rate of cross traffic packets, the available bandwidth is designated by 3 Mb/s (scenario 1), 5 Mb/s (scenario 2), 7 Mb/s (scenario 3), respectively.

The weights are selected as follows: $w(1) = w(2) = 0.05, w(3) = 0.2$, $w(4) = 0.3, w(5) = 0.4$. The decision threshold, $V_{th} = 0.75$, length of each fragment $M = 5$ and $L = 5$. We compared the simulation results with Pathchirp,

Table 2. The simulation results: AB-cap vs. Pathchirp

Measurement method	Scenario 1		Scenario 2		Scenario 3	
	Traffic load	Measured results	Traffic load	Measured results	Traffic load	Measured results
Pathchirp	310 KB	2.1 Mbps	260 KB	3.7 Mbps	350 KB	5.4 Mbps
AB-cap	500 KB	3.3 Mbps	430 KB	5.2 Mbps	500 KB	6.8 Mbps

which parameters are selected as follows: $P = 1000 Bytes, r = 1.2, L = 5, F = 1.5$. The meant of the parameters sees [8]. We adopt the mean of the 20 measured results, shown in Table 2.

From table 2, we can see that PathChirp underestimate the available bandwidth. This is determined by the weakness of PathChirp algorithm itself[8]. AB-cap improves the measurement precision by slightly adding the traffic overhead.

3.2 Measurement in Test Bed

We also build a test bed which consists of 4 Huawei routers (2 AR4640 routers and 2 R2611 routers) and several hosts, as shown in Fig.2. The host "sender" sends probing packets to the host "receiver", they are applied to test the algorithm. The cross traffic is generated by "hostA" and received by "hostB". The cross traffic is generated to make the available bandwidth be 3 Mb/s (scenario 1), 5 Mb/s (scenario 2), 7 Mb/s (scenario 3), respectively. The test results and comparison with pathchirp are shown in Table 3. The results are nearly the same as the simulation results.

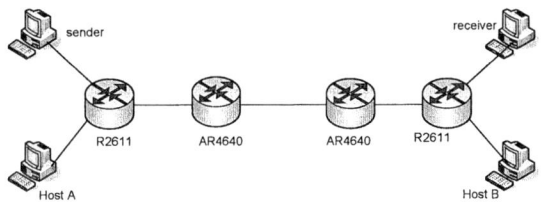

Fig. 2. Test bed for available bandwidth measurement

Table 3. The measured results: AB-cap vs. Pathchirp

Measurement method	Scenario 1		Scenario 2		Scenario 3	
	Traffic load	Measured results	Traffic load	Measured results	Traffic load	Measured results
Pathchirp	320KB	2.0Mbps	290KB	3.9MB	350KB	5.1Mbps
AB-cap	486KB	3.1Mbps	510KB	4.9MB	500KB	6.7Mbps

During the experiment we found: (1) when the path from "sender" to "receiver" is in light load, relative delay is approximately constant in the beginning segments of probe. So, the AB-cap algorithm based on formula (4) may fail when the number of probe packets is larger than M+L. (2) The probe packets may be lost because of some reasons (e.g. congestion, TTL time out, etc.). The lost packets are omitted, or their delays are estimated by the mean value of the two adjacent packets' delays in the simulation and test bed. Based on these two facts, we enhance AB-probe as follows: (1) Modify the decision rule and add a new condition---at least one of the abstract values of the slope of two consequent probing packets in the sliding window is larger than 1 during the process of determining the turning point of delay. (2) Discard the probe in which lost packets occur before the turning point.

4 Conclusions and Further Work

A new method to measure the end-to-end Internet available bandwidth, AB-cap, is proposed, which is based on the concept of self-induced congestion and in which the transmission rates of probing packets are increased step by step and sliding-window decision is adopted. In this approach, the selection of decision threshold value, which determines the measurement results, is very important. The value, 0.75, is recommended and is adopted in our simulation as empirical value. Further work will be focused on taking large mount of measurements in operational networks, discussing the end-to-end available bandwidth behaviors and investigating Internet application based on available bandwidth estimation results.

References

[1] M. Jain and C. Dovrolis. Pathload: A Measurement Tool for End-to-End Available Bandwidth. In Passive and Active Measurements Workshop, Fort Collins, CO, March 2002
[2] Ravi Prasad, Constantinos Dovrolis, Margaret Murray and kc claffy, Bandwidth Estimation: Metrics, Measurement Techniques, and Tools, IEEE Network, November/December, 2003, pp.27-35
[3] Jacob Strauss, Dina Katabi, Frans Kaashoek, A Measurement Study of Available Bandwidth Estimation Tools. In Proceedings of the ACM SIGCOMM Internet Measurement Conference (IMC03), Miami, Florida, November 2003.
[4] J. Navratil, R. L. Cottrell, ABwE: A Practical Approach to Available Bandwidth Estimation[EB/OL], Passive and Active Measurement Workshop (PAM), March 2003
[5] Manthos Kazantzidis, How to measure available bandwidth on the Internet[EB/OL], Tech. Rep. 010032, Department of Computer Science, University of California, Los Angeles, Los Angeles, CA, USA, Sept. 2001. http://www.cs.ucla.edu/~kazantz/1.html
[6] M. Jain and C. Dovrolis. End-to-End Available Bandwidth: Measurement Methodology, Dynamics, and Relation with TCP Throughput. IEEE/ACM Transactions on Networking, Vol.11, No.4, 2003, pp.537-549
[7] B. Melander, M. Bj"orkman, and P. Gunningberg, A new end-to-end probing and analysis method for estimating bandwidth bottlenecks, in Proc. IEEE Globecom, 2000, pp. 415-420
[8] V. J. Ribeiro, R. H. Riedi, R. G. Baraniuk, J. Navratil, and L. Cottrell. pathChirp: Efficient Available Bandwidth Estimation for Network Paths. In Passive and Active Measurement Workshop, 2003.
[9] N. Hu and P. Steenkiste. Evaluation and Characterization of Available Bandwidth Techniques. IEEE Journal Selected Areas in Communications. Vol.21, No.6, August, 2003. pp. 879-894
[10] Zhu Changhua, Pei Changxing, Li Jiandong, Xiao Haiyun, Linear Programming Based Estimation of Internet End-to-end Delay, Journal of Electronics and Information (in Chinese), 2004,Vol.26, No.3, pp.445-451

An Integrated QoS Multicast Routing Algorithm Based on Tabu Search in IP/DWDM Optical Internet[*]

Xingwei Wang, Jia Li, and Min Huang

College of Information Science and Engineering, Northeastern University,
Shenyang, 110004, P.R. China
wangxw@mail.neu.edu.cn

Abstract. An integrated QoS multicast routing algorithm in IP/DWDM optical Internet is proposed in this paper. Considering load balancing, given a multicast request and flexible QoS requirement, to find a QoS multicast routing tree is NP-hard. Thus, a tabu search based algorithm is introduced to construct a cost suboptimal QoS multicast routing tree, embedding the wavelength assignment procedure based on segment and wavelength graph ideas. Hence, the multicast routing and wavelength assignment is solved integratedly. Simulation results have shown that the proposed algorithm is both feasible and effective.

1 Introduction

In IP/DWDM optical Internet, the integrated QoS (Quality of Service) routing scheme supporting the peer model is necessary [1]. It has been proved NP-hard [2]. In general, most of the existing algorithms aim simply at minimizing the cost of the tree and often only cope with rigid QoS constraints, i.e. the QoS requirements have strict upper or lower bounds [3]. However, due to the difficulty in accurately describing the user QoS requirements and the inaccuracy and dynamics of the network status information, we believe flexible QoS should be supported. This motivates our work. Considering load balancing, given a multicast request and flexible QoS requirement, a tabu search [4] based algorithm is introduced to construct the multicast routing tree, embedding the wavelength assignment procedure based on segment and wavelength graph ideas. Hence, the multicast routing and wavelength assignment is solved integratedly, which could optimize both the network cost and QoS.

2 Model Description

IP/DWDM optical Internet can be considered to be composed of optical nodes (such as wavelength routers or OXCs) interconnected by optical fibers. Assume each optical node exhibits multicast capability, equipped with optical splitter at which an optical

---

[*] This work is supported by the National Natural Science Foundation of China under Grant No. 60473089 and No. 70101006; the Natural Science Foundation of Liaoning Province in China under Grant No. 20032018 and No. 20032019.

signal can be split into an arbitrary number of optical signals. In consideration of the still high cost of wavelength converter, assume only some optical nodes are equipped with full-range wavelength converters. Assume the conversion between any two different wavelengths has the same delay. The number of wavelengths that a fiber can support is finite, and it may be different from that of others.

Given a graph $G(V,E)$, where V is the set of nodes representing optical nodes and E is the set of edges representing optical fibers. If wavelength conversion happens at node $v_i \in V$, the conversion delay at v_i is $t(v_i)=t$, otherwise, $t(v_i)=0$. The set of available wavelengths, delay and cost of edge $e_{ij}=(v_i,v_j) \in E$ are denoted by $w(e_{ij}) \subseteq w_{ij} = \{\lambda_1, \lambda_2, \cdots, \lambda_{n_{ij}}\}$, $\delta(e_{ij})$ and $c(e_{ij})$ respectively, where w_{ij} is the set of supported wavelengths by e_{ij} and $n_{ij} = |w_{ij}|$.

A multicast request is represented as $R(s,D,\Delta)$, $s \in V$ is the source node, $D=\{d_1,d_2,\cdots,d_m\} \subseteq \{V-\{s\}\}$ is the destination node set, and Δ is the required end-to-end delay interval. Suppose $U=\{s\} \cup D$. The objective is to construct a multicast routing tree from the source to all the destinations, i.e. $T(X,F)$, $X \subseteq V$, $F \subseteq E$.

The total cost of T is defined as follows:

$$Cost(T) = \sum_{e_{ij} \in F} c(e_{ij}). \quad (1)$$

To balance the network load, those edges with more available wavelengths should be considered with priority. The edge cost function is defined as follows:

$$c(e_{ij}) = n - |w(e_{ij})|. \quad (2)$$

$$n = \max_{e_{ij} \in E}\{n_{ij}\}. \quad (3)$$

Let $P(s,d_i)$ denote the path from s to d_i in T. The delay between s and d_i along T, denoted by PD_{sd_i}, can be represented as follows:

$$PD_{sd_i} = \sum_{v_i \in P(s,d_i)} t(v_i) + \sum_{e_{ij} \in P(s,d_i)} \delta(e_{ij}). \quad (4)$$

The delay of T is defined as follows:

$$Delay(T) = \max\{PD_{sd_i} | \forall d_i \in D\}. \quad (5)$$

Let $\Delta = [\Delta_{low}, \Delta_{high}]$, and the user QoS satisfaction degree is defined as follows:

$$Degree(QoS) = \begin{cases} 100\% & Delay(T) \leq \Delta_{low} \\ \dfrac{\Delta_{high} - Delay(T)}{\Delta_{high} - \Delta_{low}} & \Delta_{low} < Delay(T) < \Delta_{high} \\ 0\% & Delay(T) \geq \Delta_{high} \end{cases}. \quad (6)$$

3 Algorithm Design

3.1 Solution Expression

A solution is denoted by binary coding. Each bit of the binary cluster corresponds to one node in $G(V,E)$. The graph corresponding to the solution S is $G'(V',E')$. Let the function $bit(S,i)$ denote the ith bit of S, $bit(S,k)=1$ iff $v_k \in V'$. The length of the binary cluster is $|V|$. Construct the minimum cost spanning tree $T_i'(X_i', F_i')$ of G'. T_i' spans the given nodes in U. However, G' may be unconnected, thus S corresponds to a minimum cost spanning forest, also denoted by $T_i'(X_i', F_i')$. It's necessary to prune the leaf nodes not in U and their related edges in T_i', the result is denoted by $T_i(X_i, F_i)$, and assign wavelengths to T_i.

3.2 Wavelength Assignment Algorithm

The objective is to minimize the delay of the multicast routing tree by minimizing the number of wavelength conversions, making $Degree(QoS)$ high. If T_i is a tree, assign wavelengths; otherwise, the solution is unfeasible.

(1) Constructing Auxiliary Graph AG

Locate the intermediate nodes with converters in T_i, and divide T_i into segments according to them, i.e., the edges having wavelength continuity constraint should be merged into one segment. Number each segment.

In AG, add node a_0 as the source node, and create node a_j according to segment j, where $j = 1,2,\cdots,m$, m is the number of segments. Each node in AG can be considered to be equipped with wavelength converter.

Assume a_k $(1 \le k \le m)$ corresponds to the segment that the source node in T_i belongs to. Add a directed edge (a_0, a_k) between a_0 and a_k, making the intersection of the available wavelength set on each edge in segment k as its available wavelength set. For each node pair a_{j_1} and a_{j_2} $(1 \le j_1, j_2 \le m, j_1 \ne j_2)$, if segments j_1 and j_2 are connected in T_i, add a directed edge (a_{j_1}, a_{j_2}) between a_{j_1} and a_{j_2}, making the intersection of the available wavelength set on each edge in segment j_2 as its available wavelength set.

(2) Constructing Wavelength Graph WG

Transform AG to WG. In WG, create $N*w$ nodes, N is the number of nodes in AG, and w is the number of wavelengths available at least on one edge in AG. All the nodes are arranged into a matrix with w rows and N columns. Row i represents a corresponding wavelength and column j represents a node in AG, $i = 0,1,\cdots,w-1$ and $j = 0,1,\cdots,N-1$. Create edges in WG, where a vertical edge represents a wavelength conversion at a node, assigning 1 as its weight, and a horizontal edge represents an actual edge in AG, assigning 0 as its weight. The construction method is shown in [5].

(3) Wavelength Assignment

Treat WG as an ordinary network topology graph. Find the shortest paths from the source node column to each leaf node column in WG using Dijkstra algorithm, and construct the multicast routing tree T_{WG}. Map T_{WG} back to AG, and denote the resulting subgraph in AG by T_{AG}. Since in WG all the nodes in one column correspond to the same node in AG, those shortest paths that are disjoint in T_{WG} may intersect in T_{AG}. Thus, pruning some edges is needed.

Map the paths in WG back to the paths and wavelengths in AG, and then map them back to the paths and wavelengths in T_i, thus wavelength assignment is completed.

3.3 Generating Initial Solution

A destination-node-initiated joining algorithm [6] is adopted to find an initial feasible solution, leading the algorithm to be more robust with fewer overheads.

3.4 Fitness Function

Fitness function $f(S)$ is determined by $Cost(T_i)$ and $Degree(QoS)$ together:

$$f(S) = \frac{Cost(T_i) + [count(T_i) - 1] * \rho}{Degree(QoS)}. \tag{7}$$

$count(T_i)$ is the number of trees in T_i, ρ is a positive value. If T_i has more than one tree, add a penalty value to the cost of T_i and take a smaller value for $Degree(QoS)$.

3.5 Tabu List

The method of generating neighbors is: choose one node not in U randomly, and take the reverse value for the corresponding bit in its solution. Select neighbors randomly to form the candidate set. Tabu object is 0-1 exchange of the components of solution vectors. Tabu length is constant t. If a tabued solution in the candidate set could give a better solution than the current optimum one, meaning that a new region has been reached, make it free. Whenever all the solutions in the candidate set are tabued, the object with the shortest tabu term will be freed. If the non-improved iteration times become greater than a given number, clear the tabu list, trying to drive the search into a new region and to get a better solution. Both "stopping after definite iteration times" and "controlling the change of the object value" are adopted as termination rules.

4 Simulation Research

Simulation research has been done over some actual network topologies and mesh topologies with 20 nodes to 100 nodes. Several example topologies are shown in Fig.1.

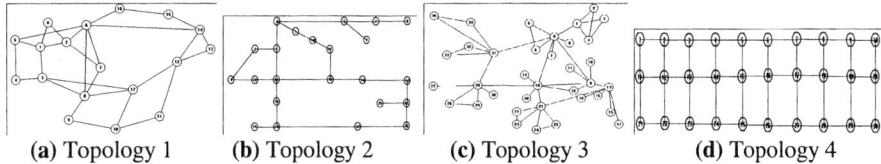

(a) Topology 1　　(b) Topology 2　　(c) Topology 3　　(d) Topology 4

Fig. 1. Example topologies

4.1 Auxiliary Graph Effect Evaluation

Compare the runtime of the proposed wavelength assignment algorithm (S&WG-based) with that of WG-based [5], the results are shown in Fig. 2. In most cases, the S&WG-based is faster than the other one.

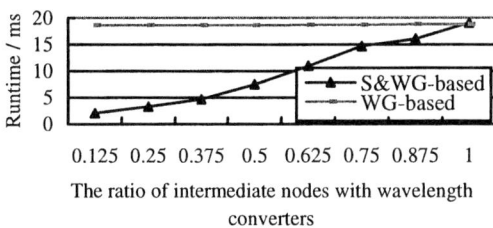

Fig. 2. Evaluation on auxiliary graph effect

4.2 Multicast Routing Tree Cost Evaluation

Comparing solutions obtained by the proposed algorithm with the optimal ones obtained by exhaustive search, the results are shown in Table 1. The solutions obtained by the proposed algorithm are rather satisfied, sometimes are even optimal.

4.3 QoS Evaluation

The delay of the tree obtained by the proposed algorithm and its counterpart obtained without considering *Degree(QoS)* are compared. Take Topology 1 of Fig. 1 (a) as example, simulation results are shown in Fig. 3. The QoS of the multicast routing tree is improved effectively and efficiently.

Table 1. Evaluation on multicast routing tree cost

Topology	Obtained tree cost vs. optimal tree cost			
	≤ 1%	≤ 5%	≤ 10%	>10%
Topology 1	0.99	0.01		
Topology 2	0.99			0.01
Topology 3	1			
Topology 4	0.78		0.22	

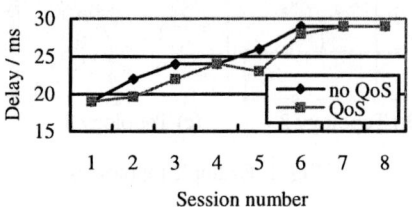

Fig. 3. Evaluation on QoS

5 Conclusions

An integrated algorithm for flexible QoS multicast routing in IP/DWDM optical Internet is proposed. Given a multicast request and flexible QoS requirement, a tabu search based algorithm is introduced to construct QoS multicast routing tree, embedding the wavelength assignment procedure based on segment and wavelength graph ideas. Hence, the multicast routing and wavelength assignment is solved integratedly. Simulation results have shown that the proposed algorithm is feasible and effective.

References

1. Rajagopalan B.: IP over Optical Networks: a Framework. IETF-RFC-3717 (2004)
2. Ramaswami R., Sivarajan K. N.: Routing and Wavelength Assignment in All-Optical Networks. IEEE/ACM Transactions on Networking. Vol. 3. No. 5 (1995) 489-500
3. Carlos A. S. O., Panos M. P.: A Survey of Combinatorial Optimization Problems in Multicast Routing. Computers & Operations Research, Vol. 32. No. 8 (2005) 1953-1981
4. George M. W., Bill S. X., Stevan Z.: Using Tabu Search with Longer-Term Memory and Relaxation to Create Examination Timetables. European Journal of Operational Research. Vol. 153. No. 1 (2004) 80-91
5. Wang X. W., Cheng H., Li J., *et al*: A Multicast Routing Algorithm in IP/DWDM Optical Internet. Journal of Northeastern University (Natural Science). Vol. 24. No. 12 (2003) 1165-1168
6. Wang X. W., Cheng H., Huang M.: A Tabu-Search-Based QoS Routing Algorithm. Journal of China Institute of Communications. Vol. 23. No. 12A (2002) 57-62

On Estimation for Reducing Multicast Delay Variation*

Moonseong Kim[1], Young-Cheol Bang[2], and Hyunseung Choo[1]

[1] School of Information and Communication Engineering,
Sungkyunkwan University,
440-746, Suwon, Korea
{moonseong, choo}@ece.skku.ac.kr
[2] Department of Computer Engineering, Korea Polytechnic University,
429-793, Gyeonggi-Do, Korea
ybang@kpu.ac.kr

Abstract. The core-based multicast routing protocol plays a significant role in many multimedia applications such as video-conferencing, replicated database updating and querying, and etc. However, existing core-based multicast routing protocols construct only the shortest paths between the core and the members in a multicast group without optimizing the quality of service requirements. In this paper, we propose an efficient algorithm for multicast delay variations and tree cost. The efficiency of our algorithm is verified through the performance evaluation and the enhancements are up to about 2.5% $\sim$ 4.5% and 3.8% $\sim$ 15.5% in terms of the multicast delay variation and the tree cost, respectively. The time complexity of our algorithm is $O(m(l + n\log n))$.

1 Introduction

The multicast tree problem can be modelled as the Steiner tree problem which is NP-complete. Several heuristics that construct low-cost multicast routing based on the problem have been proposed. Not only the tree cost as a measure of bandwidth efficiency is one of the important factors, but also there is another important factor to consider real-time application in the multicast network. During video-conferencing, the person who is talking has to be heard by all participants at the same time. Otherwise, the communication may lose the feeling of an interactive screen-to-screen discussion. They are all related to the multicast delay variation problem. Our research subject is concerned with the minimization of multicast delay variation.

The rest of the paper is organized as follows. In Section 2, we state the network model for the multicast routing, the problem formulation, and the weighted factor algorithm [3]. Section 3 presents the details of the proposed algorithm. Then, we evaluate the proposed algorithm by the simulation model, in section 4. Finally, section 5 concludes this paper.

---

* This work was supported in parts by Brain Korea 21 and the Ministry of Information and Communication in Republic of Korea. Dr. H. Choo is the corresponding author and Dr. Bang is the co-corresponding author.

2 Preliminaries

We consider that a computer network is represented by a directed graph $G = (V, E)$ with n nodes and l links or arcs, where V is a set of nodes and E is a set of links, respectively. Each link $e = (i, j) \in E$ is associated with two parameters, namely link cost $c(e) > 0$ and link delay $d(e) > 0$. The delay of a link, $d(e)$, is the sum of the perceived queueing delay, transmission delay, and propagation delay. We define a path as sequence of links such that $(u, i), (i, j), \ldots, (k, v)$, belongs to E. Let $P(u, v) = \{(u, i), (i, j), \ldots, (k, v)\}$ denote the path from node u to node v. If all nodes u, i, j, $\ldots$, k, v are distincts, then we say that it is a simple directed path. We define the length of the path $P(u, v)$, denoted by $n(P(u, v))$, as a number of links in $P(u, v)$. The path cost of P is given by $\phi_C(P) = \sum_{e \in P} c(e)$ and the path delay of P is given by $\phi_D(P) = \sum_{e \in P} d(e)$.

For the multicast communications, messages need to be delivered to all receivers in the set $M \subseteq V \setminus \{s\}$ which is called the multicast group, where $|M| = m$. The path traversed by messages from the source s to a multicast receiver, m_i, is given by $P(s, m_i)$. Thus multicast routing tree can be defined as $T(s, M) = \bigcup_{m_i \in M} P(s, m_i)$ and the messages are sent from s to M through $T(s, M)$. The tree cost of tree $T(s, M)$ is given by $\phi_C(T) = \sum_{e \in T} c(e)$ and the tree delay is $\phi_D(T)) = max\{ \phi_D(P(s, m_i)) \mid {}^\forall P(s, m_i) \subseteq T, {}^\forall m_i \in M \}$.

The multicast delay variation, δ, is the maximum difference between the end-to-end delays along the paths from the source to any two destination nodes.

$$\delta = max\{ |\phi_D(P(s, m_i)) - \phi_D(P(s, m_j))|, {}^\forall m_i, m_j \in M, i \neq j \}$$

The issue defined above is to minimize multicast delay variation.

$$min\{ \delta_\alpha \mid {}^\forall P(s, m_i) \subseteq T_\alpha, {}^\forall m_i \in M \},$$

where T_α denotes any multicast tree spanning $M \cup \{s\}$.

M. Kim, et al. have recently proposed a unicast routing algorithm [3] that is based on new factor which is probabilistic combination of cost and delay and its time complexity is $O((l + nlogn)|\{\omega_\alpha\}|)$. The unicast routing algorithm is quite likely a performance of a kth shortest path algorithm which has the high time complexity.

3 The Proposed Algorithm

The goal of this paper is to propose an algorithm which produces multicast trees with low multicast delay variation. The core placement problem [5] with optimized multicast delay variation is to find a core node. Core Based Tree (CBT) [2] is to first choose some core routers, which compose the core backbone. The CBT chooses a core node for minimizing delay variation. Contrary to the CBT, our proposed algorithm selects an weight ω. As the first step in our proposed algorithm, we assume that the function $\psi(s, m_k, \omega) = \phi_D(P_\omega(s, m_k))$ is increasing continuous function. We do not take up a reason for the assumption. For further details of the reason, see the performance evaluation in [3]. Our proposed

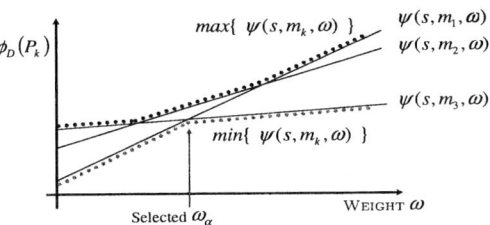

Fig. 1. The ω selection

algorithm with optimized multicast delay variation is to find an weight ω, which can satisfy the following condition:

$$min\{\ max\{\psi(s,m_k,\omega)\} - min\{\psi(s,m_k,\omega)\}\ |\ ^\forall m_k \in M,\ ^\forall \omega \in [0,1]\ \}, \quad (1)$$

where source node s and multicast group M. If we have several weights satisfied the condition (1), we take the largest weight. Though path delays are same, path costs are different. Because a path cost is decreasing as a weight is increasing [3], we choose the largest weight as ω.

Theorem 1. *Let s be a source node and M be a multicast group. Our proposed algorithm always finds ω such that satisfied the condition (1).*

Proof. Let $\xi(\omega)$ be the continuous function such that satisfied the condition (1). If all ψ are parallel curves, then the function ξ is constant function. So, the function ξ has a minimum. Otherwise, there always exists ω_γ such that $\xi'(\omega_\gamma - \epsilon) < 0 < \xi'(\omega_\gamma + \epsilon)$ for arbitrary positive real number ϵ. So, $\xi(\omega_\gamma)$ is locally minimum. Let ω_α be the weight such that satisfied $min\{\xi(\omega_\gamma)\}$ for every γ. Therefore $\xi(\omega_\alpha)$ is minimum for ξ. □

Theorem 2. *Our proposed algorithm always constructs a multicast tree.*

Proof. By theorem 1, there exist ω_α such that satisfied the condition (1). We take $\omega = max\{\omega_\alpha\}$, we always construct a multicast routing tree. □

Theorem 3. *Let $G(V,E)$ be a given network with $|V| = n$ and $|E| = l$. Suppose that M is a multicast group with $|M| = m$ and an ordered set of weights is $\{0, 0.1, 0.2, ..., 0.9, 1\}$. The expected time complexity of our proposed algorithm is $O(m(l + n\log n))$.*

Proof. We use the Fibonacci heaps implementation of the Dijkstra's algorithm [1], $O(l + n\log n)$. The cardinality of the set of weights is 11. Because we use the above the one for each ω, the time complexity is $O(11(l + n\log n)) = O(l + n\log n)$ for searching 11 paths. Since $|M| = m$, the total time complexity is $O(m(l + n\log n))$ for each destination node. □

Table 1. The manner by which the CBT selects a core node

		v_1	v_2	v_3	v_4	v_5	v_6	v_7	v_8	v_9
	v_5	4	2	5	4	0	2	3	4	6
destination	v_6	4	4	3	2	2	0	3	2	5
	v_9	6	8	8	4	6	5	3	3	0
max_i		6	8	8	4	6	5	3	4	6
min_i		4	2	3	2	0	0	3	2	0
$difference_i$		2	6	5	2	6	5	**0**	2	6

Our algorithm is described, making use of the example. Fig. 2(a) shows the given network topology G. And the link delays and the link costs are presented on each link as a pair $(delay, cost)$. From Table 1, we can know that the node v_7 has the minimum delay difference. Hence, the CBT selects the node v_7 as core node and constructs the multicast tree as Fig. 4(b). But the multicast delay variation is 6 and the tree cost is 20. The set of weights is $\{0, 0.1, 0.2, \ldots, 0.9, 1\}$. We use the unicast algorithm [3] for the source node v_1 and destination nodes $\{v_5, v_6, v_9\}$. Then we obtain the Table 2. We draw the graph for each ψ

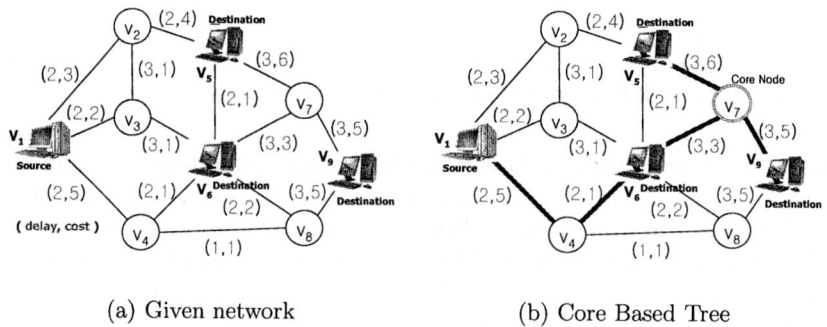

(a) Given network   (b) Core Based Tree

Fig. 2. A given network G and the core node v_7 with minimum distance between any destination nodes in the CBT

Table 2. All paths $P(v_1, m)$ for each ω and $m \in M$

ω	$P(v_1, v_5)$ for each ω	ϕ_C	ϕ_D
0.0, 0.1, 0.2, 0.3, 0.4, 0.5	$\{(v_1, v_2), (v_2, v_5)\}$	7	4
0.6, 0.7, 0.8, 0.9, 1.0	$\{(v_1, v_3), (v_3, v_6), (v_6, v_5)\}$	4	7
ω	$P(v_1, v_6)$ for each ω	ϕ_C	ϕ_D
0.0, 0.1, 0.2, 0.3, 0.4, 0.5, 0.6	$\{(v_1, v_4), (v_4, v_6)\}$	6	4
0.7, 0.8, 0.9, 1.0	$\{(v_1, v_3), (v_3, v_6)\}$	3	5
ω	$P(v_1, v_9)$ for each ω	ϕ_C	ϕ_D
0.0, 0.1, 0.2, 0.3, 0.4, 0.5	$\{(v_1, v_4), (v_4, v_8), (v_8, v_9)\}$	11	6
0.6, 0.7, 0.8, 0.9	$\{(v_1, v_3), (v_3, v_6), (v_6, v_8), (v_8, v_9)\}$	10	10
1.0	$\{(v_1, v_3), (v_3, v_6), (v_6, v_7), (v_7, v_9)\}$	11	11

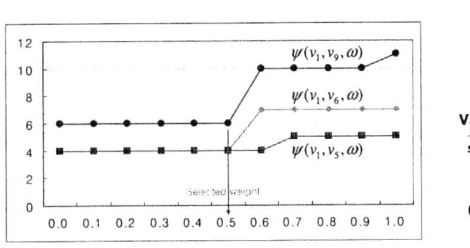
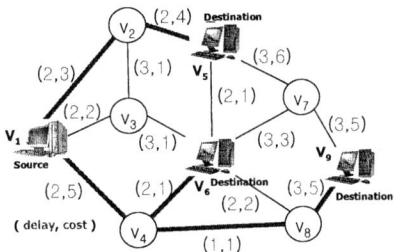

(a) The ω selection in Table 2   (b) The proposed multicast tree

Fig. 3. The multicast tree construction

and select the weight satisfied the condition (1). We take the weight 'ω: 0.5', and construct the multicast tree as Fig. 3(b). Hence, the multicast delay variation is 2 and the tree cost is 19. Since the delay variation is $2 < 6$ and the tree cost is $19 < 20$, the performance of our algorithm is better than the CBT.

4 Performance Evaluation

The proposed algorithm is implemented in C. We randomly selected a source node and destination nodes. We generate 10 different networks [4] for size 100. The random networks used in our experiments are the probability of links (P_e) equal to 0.3. The destination nodes are picked uniformly from the set of nodes in the network topology (excluding the nodes already selected for the destination). Moreover, the destination nodes in the multicast group will occupy 10, 20, 30, 40, 50, and 60% of the overall nodes on the network, respectively. We simulate

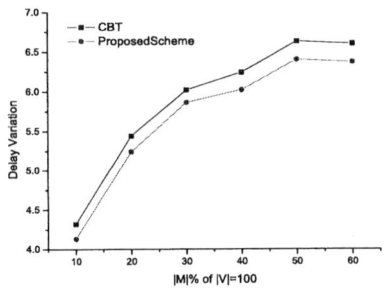
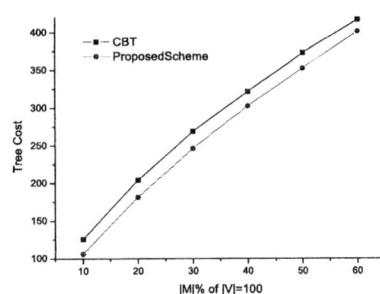

(a) Average delay variation graph for each destination nodes   (b) Average tree cost graph for each destination

Fig. 4. Simulation results

1000 times (10 different networks ×100 trials = 1000). For the performance comparison, we implement the CBT in the same simulation environment. We use the efficiency of the algorithm with respect to delay variation and tree cost. In our plotting, we express these as a percentage. As indicated in Fig. 4(a) and Fig. 4, we are easily noticed that the proposed algorithm is better than the CBT. The enhancement is up to about 2.5%~4.5% and 3.8%~15.5% in terms of the multicast delay variation and the tree cost, respectively.

5 Conclusion

We examine the problem of constructing minimum delay variation and minimum multicast tree cost from the source to the destination nodes. We proposed an algorithm using expected multiple paths. The expected multiple paths are obtained by new factor which is introduced in [3]. The new factor is efficiently combining two independent measures, the cost and the delay. The weight ω of new factor plays on important role in combining the two measures [3]. If the ω approximates 0, then the path delay is low. Otherwise the path cost is low. Our proposed algorithm performs better than the CBT in terms of delay variation and tree cost. Also, the time complexity is $O(m(l + n \log n))$.

References

1. R. K. Ajuja, T. L. Magnanti, and J. B. Orlin, *Network Flows: Theory, Algorithms and Applications*, Prentice-Hall, 1993.
2. A. Ballardie, B. Cain, and Z. Zhang, "Core Based Trees (CBT Version 3) Multicast Routing," Internet draft, 1998.
3. M. Kim, Y.-C. Bang, and H. Choo, "The Weighted Factor Algorithm with Efficiently Combining Two Independent Measures in Routing Paths," Springer-Verlag Lecture Notes in Computer Science, vol. 3483, pp. 989-998, May 2005.
4. A. S. Rodionov and H. Choo, "On Generating Random Network Structures: Connected Graphs," Springer-Verlag Lecture Notes in Computer Science, vol. 3090, pp. 483-491, August 2004.
5. M. Wang and J. Xie, "Core placement algorithm for CBT multicast routing under multiple QoS requirements," IEE Electronics Letters, vol. 38, no. 13, pp. 670-672, June 2002.

Garbage Collection in a Causal Message Logging Protocol*

Kwang Sik Chung[1], Heon-Chang Yu[2], and Seongbin Park[2]

[1] Dept. of Computer Science, Korea National Open University,
169, Dongsung-dong, Jongno-Gu, Seoul 110-791, Korea
kchung0825@knou.ac.kr
[2] Dept. of Computer Science Education, Korea University,
1, 5-ka, Anam-dong, Sungbuk-ku, Seoul, Korea
{yuhc, psb}@comedu.korea.ac.kr

Abstract. This paper presents a garbage collection protocol for message content logs and message determinant logs which are saved on a stable storage. Previous works of garbage collections in a causal message logging protocol try to solve the garbage collection of message determinant log and force additional checkpoints[5,6,7]. In order to avoid the sympathetic rollback, we classify the fault tolerance information into message determinants logs and message contents logs. Then we propose new definitions for garbage collections conditions for message determinant logs and message content logs and present a garbage collection algorithm. To represent determinants of messages, a data structure called MAG (Modified Antecedence Graph) is proposed. MAG is an extension of Antecedence Graph of Manetho system [7] and it is used for garbage collections conditions of message determinant logs and message content logs. Unlike Manetho system that needs additional messages for garbage collection of message content logs, our algorithm does not need additional messages. The proposed garbage collection algorithm makes 'the lazy garbage collection effect' because it relies on the piggybacked checkpoint information in send/receive message. 'The lazy garbage collection effect' provides the whole system with an efficient and simple recovery protocol.

1 Introduction

As a distributed computing system develops, one task is partitioned and executed on several processes. Therefore the probability of failure occurrence becomes high and many fault tolerance methods have been proposed. Among these, two fault tolerance methods are prevalent: one is the message logging method and the other is the checkpointing method. The fault tolerance information can use all of a stable storage and garbage collection becomes neccesary. To keep the fault tolerance information, the checkpointing methods save the states of processes and the message logging methods save the message contents logs and deterministic order of message sending and receiving[3,4]. Message logging methods are classified as pessimistic, optimistic, and causal methods and they rely on the *piecewise deterministic* (PWD) assumption.

---

* Seongbin Park is the corresponding author.

Under this assumption, the log-based rollback-recovery protocol can identify all of the nondeterministic events executed by each process and for each such event, log a determinant that contains the information that is necessary to replay the event should it be necessary during recovery. Pessimistic log-based rollback-recovery protocols guarantee that orphans are never created when a failure occurs. Pessimistic protocols log at a stable storage, the determinant of each nondeterministic event before the event is allowed to affect the computation. In optimistic logging protocols, processes log determinants asynchronously at a stable storage. These protocols make the optimistic assumption that logging finishes before a failure occurs[8,9,11]. Causal message logging methods save the message determinants in a volatile storage of several processes and have the advantage of optimistic message logging methods by asynchronously saving the fault tolerance information in a stable storage. Previous causal message logging methods save the message determinants as the fault tolerance information and a recovery procedure is based on message determinants logs. But for the computing performance the message contents logs are needed for the recovery and previous message logging methods save the message contents logs as the fault tolerance information[6,7,8,12]. This paper is organized as follows. In section 2, we describe related works. In section 3, we give definitions of the message contents log and message determinants log. We also explain the conditions for garbage collections. In section 4, we present a garbage collection algorithm for the message contents log and message determinants log. Section 5 concludes the paper.

2 Related Works

Fault tolerance methods assume the existence of a stable storage and a volatile storage. The garbage collection in fault tolerance methods decides the time to delete the fault tolerance information. Previous garbage collection in fault tolerance methods is based on consistent global checkpoints[3,4] or additional message exchange[7,11]. Garbage collection methods based on consistent global checkpoints can delete the fault tolerance information when the consistent global checkpoints are constructed[3,4,10]. Construction of consistent global checkpoints keeps the garbage collection of fault tolerance valid. But causal message logging methods do not guarantee the consistent global checkpoints. Thus the garbage collection in causal message logging methods is more complex than optimistic and pessimistic message logging protocol and an additional message for garbage collection of message logs is necessary. Manetho system can delete the message logs after the checkpoints of process. The Manetho system propagates the causal information in an antecedence graph which provides every process in the system with a complete history of the nondeterministic events that have causal effects on its state [7,8]. An antecedence graph has a node representing each nondeterministic event that precedes the state of a process and edges correspond to the happened-before relation [8,13]. The antecedence graph that is happened before a checkpoint can be deleted. Thus in Manetho system, nodes of an antecedence graph before a checkpoint need not be used for recovery. The unnecessary nodes in an antecedence graph can be deleted from a stable storage by exchanging the state interval information so that additional messages and forced checkpoints are needed. Since the Manetho does not roll back before the latest checkpoint, if a process q receives a message from p and does not take a checkpoint yet, p requests checkpoints of q. If all processes requested from p take checkpoints, p can delete the fault tolerance information in the state interval between checkpoints. Each process maintains the state interval index of other processes before garbage collection. The Manetho system with the antecedence graph achieves low failure-free overhead, limited rollback and fast output commit. But it has a complex recovery protocol and garbage collection overhead. In order to reduce the additional messages we define message logs as fault tolerance information and propose the conditions for garbage collection of fault tolerance information in this paper.

3 Conditions for Garbage Collections

A distributed system N is a collection of n processes. A message is denoted as m and m.source is the message sender identification and m.dest represents the message destination identification, respectively. The computation state of a distributed system is changed by the event of m [3]. Depend(m) represents the union of receiver process of m and receiver processes of m' that a process belonging to Depend(m) sends. Fault tolerance information is classified into message contents and message determinants. The former is denoted as $content_m$ and the latter $\#_m$. Fault tolerance information about a message, FTI_m is defined as follows.

$$FTI_m \stackrel{def}{=} \{content_m\} \cup \{\#_m\}$$

[Definition 1]
Message determinants information consists of <m.source, m.ssn, m.rsn> and message contents information consists of <m.contents, m.ssn, m.rsn>. (m.source : process identifier sending message m, m.contents : contents of message m, m.ssn : sending sequential number for message m, m.rsn : receiving sequential number for message m).

$Log(\#_m)$ is the set of processes that save $\#_m$ in a volatile storage and is called message determinants logs. $Log(content_m)$ is the set of processes that save $content_m$ in a volatile storage and is called message contents logs. An orphan process p is the element of $N-C$, where C is a set of faulty processes and $Log(\#_m)$ of p is subset of C. An orphan process is defined as a process with an orphan message that cannot guarantee the same message determinant regeneration of the failure-free operation during recovery[8]. But we extend the definition of an orphan process in order to avoid a sympathetic rollback.

[Definition 2]

$$orphan\ process\ p\ of\ C \stackrel{def}{=} \left\{ \begin{array}{c} (p \in N - C) \\ \wedge (\exists m : ((p \in Depend\ (m)) \\ \wedge (Log\ (\#_m) \subseteq C))) \end{array} \right\} \cup \left\{ \begin{array}{c} (p \in N - C) \\ \wedge (\exists m : ((p \in Depend\ (m)) \\ \wedge (Log\ (content_m) \subseteq C))) \end{array} \right\}$$

The condition that an orphan process does not occur even though a failure occurs is as follows.

$$\forall m : (((Log\,(\#_m) \subseteq C) \vee (Log\,(content_m) \subseteq C)) \Rightarrow (Depend\,(m) \subseteq C)) \quad (1)$$

If the set of processes that save $\#_m$ or $content_m$ of m is not included in C, the set of processes that depend on m is the subset of C.

$$\forall m : (Depend\,(m) \subseteq (Log\,(\#_m) \cup Log\,(content_m)) \quad (2)$$

Thus the set of processes depending on m is the subset of processes that saves $\#_m$ or $content_m$. Processes satisfying both of the above conditions (1) and (2) are not orphan processes in spite of a failure. In order to avoid the sympathetic rollback that occurs when messages are lost, sender-based message logging methods are necessary[9]. Thus causal message logging methods have to save the message contents logs. And message contents logging conditions have to satisfy the following condition.

[Condition 1]
In causal message logging methods, a message before time t satisfies the below two conditions if and only if the whole system keeps consistent in spite of a failure at time t. ($\Box$:temporal always operator, $\Diamond$:eventual always operator, f :the number of failure)

i) $\forall m \ \Box((|\,Log\,(\#_m)\,| \leq f\,) \Rightarrow$
$(((Depend\,(m) \subseteq (Log\,(\#_m) \wedge \neg stable\,(\#_m)))) \wedge$
$\Box((Depend\,(m) = (Log\,(\#_m) \wedge \neg stable\,(\#_m))))))$

ii) $\forall m : \Box(\neg stable(\#_m) \Rightarrow ((Log(content_m) \subseteq C) \Rightarrow \Diamond(Depend(m) \subseteq C)))$

[Proof] : i) and ii) are for the safety property of distributed algorithm. In order to keep the whole system consistent and satisfy that the number of $Log(\#_m)$ is less than that of failure f, the set of processes that depend on message m is eventually same as the intersection of the set of processes saving $\#_m$ log and the set of processes that don't save the $\#_m$ log in a stable storage. If all processes that are dependent on m save $\#_m$ log, f failures don't make rollback. i) and ii) always have to be satisfied till the termination of algorithm. If the set of processes saving $content_m$ is the subset of C and eventually the set of processes causally dependent on m would be the subset of C, there is no process that saves $\#_m$ in a stable storage. As a result, a message satisfying i) and ii) does not make a sympathetic rollback. ∎

The set of processes that save $\#_m$ in a stable storage is $stable\,(\#_m)$. The causal message logging methods satisfying [Condition 1] keep the whole system consistent. We use the checkpoints and message determinants information in order to decide the time for garbage collection. In *Manetho* system, sender-based and optimistic message logging method are used[7]. But for the garbage collection of message logs, additional messages are needed. To avoid the additional messages, we record the checkpoint time with the message orders. The message determinants logs with checkpointing time are piggybacked with messages in the form of a graph. Based on the message determinants logs with checkpoints time the message determinants logs and message contents logs can be deleted. A data structure representing message determinants information with checkpoints time is defined as follows.

[Definition 3]

> i) $MAG = (V, E)$
> ii) $V_{p_{i,k}} = \{<i, j> \cup C_{p_{i,k}} \mid i: \text{process id},$
> $\quad j:\text{event number}, k:\text{checkpoint number},$
> $\quad C_{p_{i,k}} : \text{checkpoint of process } p \}$
> iii) $E = <V_{p_{i,k}}, V_{p_{i,l}}>$
> iv) $<V_{p_{i,k}}, V_{p_{i,l}}> \neq <V_{p_{i,l}}, V_{p_{i,k}}>$

MAG consists of vertices and edges. Vertices represent the message send and receive event. There are message send events, message receiving events and checkpointing events in MAG. Edges represent the events orders.

[Condition 2]
At t, if the following condition for the message in a causal message logging method is satisfied, message determinants log can be deleted from a stable storage.

$$\exists m, \exists P_k : (deliver_{m.dest}(m) \wedge Chpt_{P_k}(\#_m)) \Rightarrow garbage(\#_m)$$

If there exists a process P_k satisfying $Chpt_{P_k}(\#_m)$, and satisfying $garbage(\#_m)$ in the above equation, the whole system keeps the states consistent.

[Proof] : Let us assume that a process P_k saves the message determinants logs of message m and $Chpt_{P_k}(\#_m)$ in a stable storage. Thus a failure that depends on m makes P_k rolled back to $Chpt_{P_k}(\#_m)$. But P_k can maintain the message determinants logs of m. A faulty process can receive message determinants logs of message m from P_k and replay the consistent executions. In [Condition 1], if $stable(\#_m)$ is satisfied, any process in $Depend(m)$ does not make an orphan message in spite of a failure in $Depend(m)$. Thus fault tolerance information $\#_m$ in MAG need not be kept in a stable storage. All processes that save the $Chpt_{P_k}(\#_m)$ of m can delete the fault tolerance information of m in MAG. ∎

[Condition 3]
At t, message contents log is deleted from a stable storage if and only if the following condition for the message in causal message logging method is satisfied.

$$\exists m : (deliver_{m.dest}(m) \wedge Chpt_{m.dest}(\#_m)) \Rightarrow garbage_{m.source}(content_m)$$

If $m.source$ satisfies $Chpt_{m.dest}(\#_m)$ and $garbage_{m.source}(content_m)$ in the above equation, the whole system keeps the states valid or consistent. ($Chpt_{m.dest}(\#_m)$ is the event of checkpoint with message determinants log of $m.dest$ and $garbage_{m.source}(content_m)$ is the event of garbage collection of message contents logs by $m.source$.)

[Proof] : Since $\#_m$ is kept by $m.dest$, $m.dest$ that takes a checkpoint after receiving a message need not be rolled back before $Chpt_{m.dest}(\#_m)$ in spite of a failure. Thus $m.source$ can delete $content_m$. If $m.dest$ takes a checkpoint, $stable(\#_m)$ is satisfied. Although a fault occurs in a process that depends on m, $m.dest$ recovers the fault process. Since the faulty process needs not be rolled back before a latest checkpoint, $content_m$ in $m.source$ can be deleted. As a result $garbage_{m.source}(content_m)$ makes no orphan process and keeps the whole system state consistent in spite of a failure. ∎

4 Garbage Collection of Fault Tolerance Information

In figure 1 when process p_i receives a message m, p_i can delete the message contents logs and message determinants logs based on MAG. In order to delete message contents logs, p_i calculates the union set of MAG in message m and MAG in p_i. Using newly calculated MAG, message contents logs satisfying [Condition 3] can be marked as target information and determinants logs satisfying [Condition 2] can be marked as target information as well. The target information can be deleted from a stable storage.

```
P_j : sending process , P_i : receiving process
procedure Garbage_content_M {
    MAG ← P_i's MAG ∪ m_{p_j}.MAG ;
    while ((P_k ∈ MAG ) ∧ (m_{p_i}^{P_k} ∉ MAG ))
    stable_content_M _log
    = garbage(stable_content_M _log, content_{m_{p_i}^{P_k}});
}
procedure Garbage _#_M {
    MAG ← P_i's MAG ∪ m_{p_j}.MAG;
    while (P_k ∈ MAG )
    MAG = garbage (MAG , MAG depended with
    P_k before P_k's Chpt );
    P_i's MAG = MAG ;
}
procedure garbage(Source Information, Target Information) {
    garbage_information=Target Information;
    remove garbage_information on stable storage;
}
```

Fig. 1. Garbage Collection Algorithm

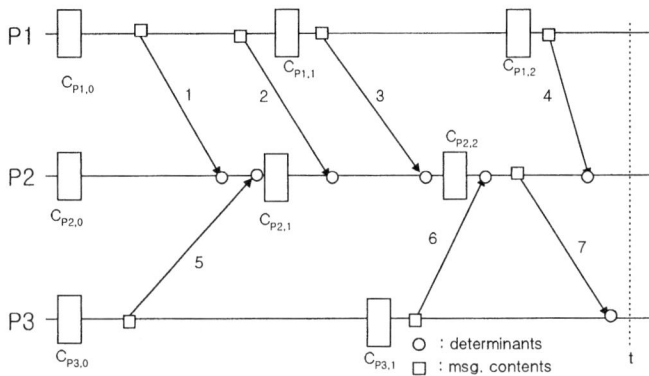

Fig. 2. System processing diagram

We use *MAG* in order to avoid additional messages for a state interval index.

Message determinants logs can be maintained by each process. And all message determinants logs satisfying [Condition 2] can be deleted from a stable storage. At time t of figure 2, each process should maintain the union of determinant information and contents information. The union of determinant information and contents information is used to recover the failed process and to trace the fault tolerance information.

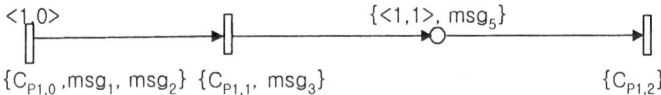

Fig. 3. Message determinant information of P_1

At time t of figure 3, since P_1 does not receive any message from other process, P_1 can not decide the deletion of message determinants logs of any message according to *MAG*. Process P_2, because of *MAG* of process P_2, delete the useless determinant information. And the useless contents information is not present in figure 4.

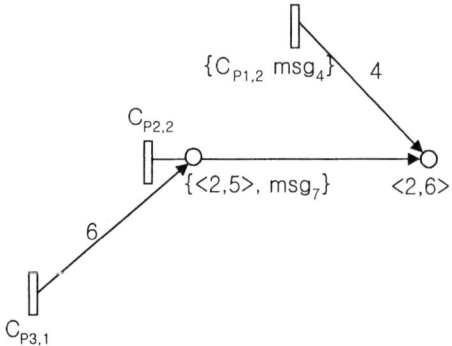

Fig. 4. *MAG* of P_2

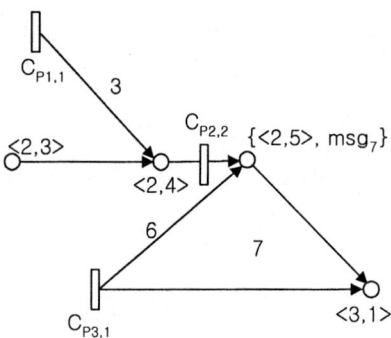

Fig. 5. *MAG* of P_3

In figure 5, if P_3 would receive a message from other process, it could do garbage collection of determinant information and contents information according to the *MAG* of the received message and [Condition 1], [Condition 2], and [Condition 3].

If the proposed garbage collection algorithm is run based on [Condition 2] and [Condition 3], the whole system has to keep the valid states or consistent states.

[Theorem 1] If there exists a process P_k satisfying $Chpt_{P_k}(\#_m)$, and satisfying $garbage(\#_m)$ in [Condition 2], the whole system keeps the states consistent.

[Proof] : Let us assume a process P_k save the message determinants logs of message m and $Chpt_{P_k}(\#_m)$ in stable storage. Thus failure dependent with m makes P_k rolled back to $Chpt_{P_k}(\#_m)$. But P_k can maintain the message determinants logs of m. A fault process can be received message determinants logs of message m from P_k and replay the consistent executions. In [Codition 1], if $stable(\#_m)$ is satisfied, any process in $Depend(m)$ does not make an orphan message in spite of failure in $Depend(m)$. Thus fault tolerance information $\#_m$ in *MAG* need not be kept in a stable storage. All processes save the $Chpt_{P_k}(\#_m)$ of m can delete the fault tolerance information of m in *MAG*. ∎

[Theorem 2] If $m.source$ satisfies $Chpt_{m.dest}(\#_m)$ and $garbage_{m.source}(content_m)$ in [Condition 3], the whole system keeps the states valid or consistent

---

[Proof] : Since $\#_m$ is kept by $m.dest$, $m.dest$ that takes a checkpoint after receiving a message need not be rolled back before $Chpt_{m.dest}(\#_m)$ in spite of a failure. Thus $m.source$ can delete $content_m$. If $m.dest$ takes a checkpoint, $stable(\#_m)$ is satisfied. Although a fault occurs in process dependent with m, $m.dest$ recovers the fault process. Since the fault process need not be rolled back before a latest checkpoint,

$content_m$ in $m.source$ can be deleted. As a result $garbage_{m.source}(content_m)$ makes no orphan process and keeps the whole system state consistent in spite of a failure. ∎

[Theorem 3] $garbage(\#_m) \supset garbage_{m.source}(content_m)$

[Proof] : i) When $m.dest$ takes a checkpoints, If $\#_m$ is recorded in checkpoint of $m.dest$, $m.dest$ would not be rolled back before the latest checkpoint. It is proved in [Theorem 2]. Thus $content_m$ need not be maintained in $m.source$, And $\#_m$ need not be maintained in any processes except for $m.dest$. Thus $garbage(\#_m) \supset garbage_{m.source}(content_m)$ is correct.

ii) A process P_k is an element of $Depend(m)$ and is not $m.dest$ takes a checkpoint.

A checkpoint of P_k saves $\#_m$. But $content_m$ in $m.dest$ is needed when $m.dest$ is rolled back before the receiving event of m. Thus $m.source$ has to maintain $\#_m$ by [Condition 2] and $garbage_{m.source}(content_m) = \phi$. $\#_m$ is not maintained by any processes except for , because is saved as checkpoint of P_k. Thus $garbage(\#_m) \supset garbage_{m.source}(content_m)$ is correct. ∎

5 Conclusions

In this paper, we proposed conditions for garbage collection of the message determinants logs and message contents logs. We classified the message logging information into message determinants logs and message contents logs, and proved that message determinants logs depend on message contents logs. For the decision of garbage collection of message contents logs and message determinants logs we defined MAG that presents the message receiving event and checkpoints order. Finally, we proposed an algorithm that deletes the message contents logs and the message determinants logs using MAG. The proposed garbage collection algorithm does not need the additional messages. As a result, the number of forced garbage collection decreases and the performance of the whole system is improved. The algorithm can not delete the fault tolerance information until the related message with MAG is received and the garbage collection is performed lazily. But lazy garbage collection does not violate the consistency of the whole system. The size of messages is bigger than size of ordinary messages. We are currently investigating various ways to reduce the size of MAG.

References

1. R. Koo, S. Toueg, "Checkpoint and Rollback-Recovery for Distributed Systems," IEEE Trans. S. E., Vol. 13, (1987) 23-31
2. Sean W. Smith, et al, "Completely Asynchronous Optimistic Recovery with Minimal Rollbacks," Proc. of IEEE FTCS-25, 1995, pp361-370.

3. M. V. Sreenivas, Subhash Bhalla, "Garbage Collection in Message Passing Distributed Systems," Proceeding of International Symposium on Parallel Algorithms/Architecture Synthesis, IEEE Computer Society Press, (1995) 213-218
4. M. V. Sreenivas, Subhash Bhalla, "Garbage Collection in Message Passing Distributed Systems," Parallel Algorithms/Architecture Synthesis, 1995. Proceedings. First Aizu International Symposiumon , (1995)15-17
5. Lorenzo Alvisi, Keith Marzullo. "Message Logging: Optimistic, Causal and Optimal," In Pro IEEE Int. Conf. Distributed Computing Systems, (1995)229-236
6. Lorenzo Alvisi, Bruce Hoppe, Keith Marzullo. "Nonblocking and orphan-free message logging protocols," In Proceedings of 23rd Fault-Tolerant Computing Symposium, (1993)145-154
7. E. L. Elnozahy, W. Zwanepoel. "Manetho: Transparent rollback-recovery with low overhead, limited rollback and fast output commit." IEEE Transactions on Computers, 41(5)(1992)526-531
8. E. N. Elnozahy, L. Alvisi, Y. Wang, and D. B. Johnson, "A Survey of Rollback-Recovery Protocols in Message-Passing Systems," ACM Computing Surveys. vol. 34, No. 3, (2002)375-408
9. Jian Xu, Robert H. B. Netzer, Milon Mackey, "Sender-based Message Logging for Reducing Rollback Propagation," Parallel and Distributed Processing, (1995)602-609
10. D. Manivannan, Mukesh Singhal, "A Low-Overhead Recovery Technique Using Quasi-Synchronous Checkpointing," Proceedings of the 16th ICDCS, (1996)100-107
11. D. B. Johnson, W. Zwaenepoel, "Sender-based Message Logging," In Digest of papers:17 Annual International Symposium on Fault-Tolerant Computing, (1987)14-19
12. K.S. Chung, Y. Lee, H Yu, and W. Lee, "Management of Fault Tolerance Information for Coordinated Checkpointing Protocol without Sympathetic Rollbakcs," Journal of Information Science and Engineering, (2004)379-390
13. LAMPORT, L. "Using time instead of timeout for fault-tolerant distributed systems," ACM Transactions on Programming Languages and Systems, (1984)254–280

Searching an Optimal History Size for History-Based Page Prefetching on Software DSM Systems

Cristian Ruz[1] and José M. Piquer[2]

[1] Escuela de Ingeniería Informática, Universidad Diego Portales
cristian.ruz@udp.cl
[2] Depto. Ciencias de la Computación, Universidad de Chile
jpiquer@dcc.uchile.cl

Abstract. This work presents a study regarding the search for an optimal value of the history size for the prediction/prefetching technique *history-based prefetching*, which collects the recent history of accesses to individual shared memory pages and uses that information to predict the next access to a page. On correct predictions, this technique allows to hide the latency generated by page faults on the remote node when the access is effectively done. Some parameters as the size of the *page history* structure that is stored and transmitted among nodes can be fine-tuned to improve the prediction efficency.
Our experiments show that small values of history size provide a better performance in the tested applications, while bigger values tend to generate more latency when the *page history* is transmitted, without improving the prediction efficiency.

Keywords: Distributed shared memory, data prefetching, distributed systems.

1 Introduction

Software distributed shared-memory (DSM) systems provide programmers with a virtual shared memory space on top of low cost message-passing hardware, and the ease of programming of a shared memory environment, running on a network of standard workstations [1]. However, in terms of performance, DSM systems suffer from high latencies when accessing remote data due to the overhead of the underlying message-passing layer and network access [2,3]. To address these issues several latency-tolerance techniques have been introduced. One of these techniques is called *prefetching*, it reduces latency by sending data to remote nodes in advance of the actual data access time.

Many *prefetching* techniques have been proposed. In this work we will focus on *history-based prefetching*, a prediction/prefetching strategy that has proved useful to reduce latency issues for a DSM system on certain applications that show a regular memory access pattern [4]. In order to reduce latency, this technique collects the recent history of accesses to individual shared memory pages, and uses that information to predict the next access to a page through the identification of memory access patterns.

Throughout the execution of an application, the number of accesses made to a certain page can be very large. Hence it could be highly inneficient to store the whole history of accesses made to each page. The solution to this problem is to store only the M most recent accesses. The parameter M therefore determines the size of the *page history*.

This work presents a study regarding the search for an optimal value of the parameter M. There are adventages and disadvantages of giving M a big value. On one hand, a big history is more likely to contain the information required to make a correct prediction, therefore reducing latency. But, on the other hand, since the *page history* must travel along with the ownership of the page from one node to another, a bigger history involves the size of the messages that must be sent to grow, causing latency to increase due to the excess of data that has to be transmitted.

A series of experiments were done with three applications running over a page-based DSM system, on a 14-nodes linux cluster. Results show that small values for M provide a better performance in the tested applications. Bigger values tend to generate more latency when *page history* is transmitted, and does not provide a major benefit in the prediction efficiency.

The rest of this paper is organized as follows. In section 2, some related work regarding other prefetching techniques is discussed. Section 3 describes the prediction technique, and the issue of the size of the *page history*. In section 4, experimental results and analysis are presented. Finally, section 5 gives conclusions and perspectives of future work.

2 Related Work

Bianchini et al. have done important work in prefetching techniques for software DSM systems. They developed the technique $B+$ [5] which issues prefetches for all the invalidated pages at synchronization points. The result is a high decrease of page faults, but at the cost of sending too many pages that will not be used and increasing bytes transfer. They also presented the *Adaptive++* technique [6] that predicts data access and issues prefetches for those data prior to actual access. Their work uses a local per-node history of page accesses that records only the last two barrier-phases and issues prefetches in two modes: repeated-phase and repeated-stride. Lee et al. [7] improved their work, using an access history per synchronization variable. *History-based prefetching* uses a distributed per-page history to guide prefetching actions, in which multiple barrier-phases can be collected leading to a more complete information about the page behavior. Prefetches are only issued at barrier synchronization events.

Karlsson et al. [8], propose a history prefetching technique that uses a per-page history, and exploits the producer-consumer access pattern, and, if the access pattern is not detected, uses a sequential prefetching. *History-based prefetching* differs in that it supports more access patterns, and the *page history* mechanism provides more flexibility to find repeated patterns that are not categorized. Also, if no pattern is detected, no prefetching action is generated, avoiding useless prefetches.

3 History-Based Prediction

History-based prediction is a technique that allows processors to prefetch shared memory pages and make them available before they are actually accessed. Using a correct prefetching strategy, page faults are avoided and the latency due to interruptions and message waiting from other nodes is hidden, improving the overall application performance.

Historical information about page behavior is collected between two consecutive barrier events. Predictions are generated inside a barrier event to make sure that no other node may generate regular consistency messages, hence avoiding overlapping of those messages and prefetching actions. When every node has reached a barrier, a *prediction phase* is executed, in which every node makes predictions using the information collected, and speculatively sends pages to other nodes. After that, the barrier is released.

3.1 Page History

History-based prediction uses a structure that stores the access history for each shared memory page. This structure is called *page history*. A *page history* is a list H, composed of M *history elements*: $H = \langle h_1, h_2, \ldots, h_M \rangle$. Each *history element* h_i represents one access to one shared memory page and contains the following information regarding the node that accessed the page: the access mode used; the number of the execution phase when the access was actually made; and a list of predictions that were made in the past when this history element was the last in the *page history*. Only the last M *history elements* are kept, reflecting the last M accesses to the page.

A *page history* is updated using a sequential-consistent model. A *page history* migrates between nodes along with the ownership of the page where it belongs, every time a node gets permission to write on that page. At any time, only the owner of the page can update its *page history*.

The *page history* of page p is updated by the owner of p when a local read fault or write fault on p is detected, and also when remote faults over p are received. After the *page history* of p has been updated, the owner may reply to the remote request according to the rules of the consistency protocol.

3.2 Prediction Strategy

Predictions are made at the end of each execution phase, when all nodes have reached a barrier, to avoid overlapping with regular consistency actions. Every node executes a predictive routine for each page that it owns.

The predictive routine attempts to find a pattern on the *page history*, by looking for the previous repetition of the last D accesses seen, and predicting

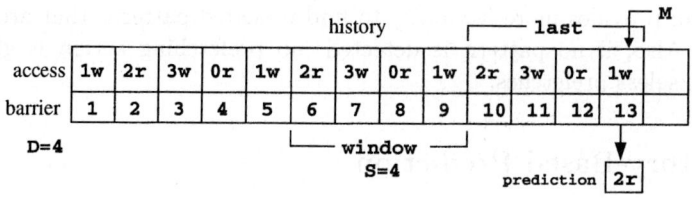

Fig. 1. Prediction of access 2r with $D = 4$. The match was found when $S = 4$.

the same behavior that was seen, in the past, after that sequence. In applications that show a regular memory access pattern, the repetition of a certain sequence of accesses can be expected, so that this prediction should be correct in most situations.

If the routine can not deduce a possible next access, then no prediction is made ; otherwise, another routine is called to actually execute the predictions.

A *page history* is analyzed using a fixed-size window W of length $D < M$, $W = \langle w_1, w_2, \ldots, w_D \rangle$ and comparing them to the last D accesses on *page history*, stored in the list $Last = \langle h_{M-(D-1)}, \ldots, h_{M-1}, h_M \rangle$. $Last$ is compared to $W_S = \langle h_{M-(D-1)-S}, h_{M-(D-2)-S}, \ldots, h_{M-1-S}, h_{M-S} \rangle$ for $S \in \{1, 2, ..., M-D\}$ in increasing order. If both lists happen to be equal, then the access h_{M-S+1} is predicted as the next access for the page, as shown in Fig. 1.

```
p.Prediction():
    if (p.status == PREFETCHED)
        return 0;
    for(i=0;i<D;i++)
        last[i] = p.history[M-(D-i-1)];
    for(s=1;s<M-D;s++) {
        for(i=0;i<D;i++)
            window[i] = p.history[M-(D-i-1)-s)];
        if(compare(last,window)) {
            p.history[M].addPrediction(p.history[M-s+1)]);
            return 1;
        }
    }
    return 0;
```

If the status of the page is PREFETCHED, it means that the page is owned by the node because of a previous prefetching action and was not accessed during the previous *execution phase*. In this situation the page was predicted but it was not used by the destination node, so it should not be predicted again.

3.3 Pattern Analysis

The page behavior may also be predicted by the identification of three predefined page access patterns and taking appropriate actions for each one. If no pattern can be detected, then the prediction routine described in section 3.2 can be used.

Pattern detection is based on the categorization described by Monnerat and Bianchini [9], where page behavior can be classified as 1PMC, MIG or MW.

1PMC is a *one producer - multiple consumers* pattern, where there is only one node that writes on the page, and many readers. This behavior has a low rate of page ownership transfers. MIG stands for *migratory* pages. It is a pattern where the page ownership is transferred to a different node in each execution phase. This behavior allows a good hit ratio to be reached using the basic prediction strategy of *history-based prefetching*. MW is a *multiple writer* pattern, in which one page is owned by different nodes inside the same execution phase. In this case it is hard to make a good prediction. Since the prediction strategy only works at the end of an execution phase, at most 1 page fault will be avoided and the $N-1$ remaining writes will still generate page faults. The case may be even worse if the nodes must compete for the access to the page, which will make the access almost random at every phase.

3.4 Page History Size

Previous work [4] showed the usefulness of *history-based prefetching* to be applied on some applications that show a regular memory access pattern. Experiments considered a fixed size M for *page history* assuming this choice was good enough for each application.

The size M of the *page history* data structure determines the amount of history that is kept to make prediction decisions, and that is sent to remote nodes along with the ownership of the related page. Theoretically, a large value of M could improve the prediction accuracy, because a greater amount of information is transmitted every time a *page history* is transferred between two nodes. On the other hand, a smaller history size is faster to transmit, but may not have enough information to deal with some situations and could lead to wrong predictions.

4 Experiments

The experiments were executed on a platform of 14 Pentium IV processors, 256MB RAM, running linux Fedora Core 1. All computers execute the applications over DSM-PEPE [10], a page-based software DSM system designed to execute parallel applications over a shared-memory environment on multicomputers and different consistency protocols. Tested applications follows.

- LIFE is an implementation of Conway's *Game of Life* [11] on a 2048 × 2048 circular matrix. The parallelization is done through stripes. On each iteration, each node computes a different stripe of the matrix. Processors wait in a barrier before computing the next iteration.
- SHEAR is an implementation of the *Shearsort* algorithm [12] to sort integers inside a 1024 × 1024 matrix. The execution goes through a fixed number of alternate row-phases and column-phases. The parallelization is done through stripes. At every iteration, each node works over a fixed set of consecutive rows, or consecutive columns.

– GRAPH is a distributed *single-source shortest-path* search on a graph of N vertexes on shared memory [13]. Each node is assigned a fixed set of vertexes to evaluate. On the kth iteration, nodes compute the shortest path from each vertex to a distance of k vertexes away, using the information from their neighbors, if necessary. After N iterations, the weight of the shortest-path to every node has been calculated.

4.1 Methodology

Each experiment was executed under a sequential-consistent protocol, using *history-based prefetching* and different values for M. Also, a normal execution without any kind of prefetching was done.

Statistics collected in each case include execution time, locally-generated read-faults and write-faults, prefetches done for *read-only* and *read-write* modes, and correct prefetches detected.

The metrics used to evaluate the effectiveness of the technique are *coverage* and *hit ratio*, as it has been used in previous related works [4,3]. *Coverage* refers to the percentage of page faults which are eliminated by prefetching pages in advance. *Hit Ratio*, or *utilization*, is defined as the percentage of valid prefetches among total prefetches. A *valid prefetch* is a prefetch that successfully avoids the generation of a *page fault*. Coverage and hit ratio are calculated as follows:

$$\text{Coverage} = \frac{\text{Valid Prefetches}}{\text{Total Page Faults}}, \quad \text{Hit Ratio} = \frac{\text{Valid Prefetches}}{\text{Total Prefetches}} \quad (1)$$

A technique that shows a high *coverage* avoids a high percentage of page faults. As an example, prefetching all shared memory pages would achieve a high coverage, but at the cost of a low hit ratio. On the other side, a technique could provide a high *hit ratio* making only correct predictions, but covering only a low percentage of all page faults. A good prefetching strategy should aim to get both a high coverage, and a high hit ratio.

4.2 Results

Results obtained are shown for each application. The parameter $D = 14$ was used as the fixed size of the prediction window. This value is based on the number of nodes where the applications were ran, based on the assumption that the window size should be at least as big as the number of active processors in order to be able to reflect at least one action of each one of them. A further study is required to validate this assumption.

Results are presented in terms of the time taken by the system to complete the execution, including the time required to execute the prediction routine and the prefetching actions; and the coverage and hit ratio obtained for each value of the *page history* structure size, M.

LIFE. LIFE is a case of an extremely regular application. The shared memory access pattern induced by the matrix division alternates between reads and

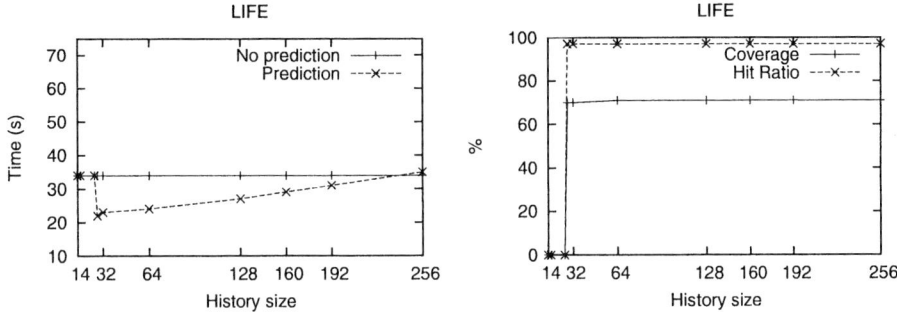

Fig. 2. Execution time, coverage and hit ratio for LIFE application

writes of different nodes in a repetitive sequence along iterations. Pages present a migratory pattern, in which by every two iterations, page ownership changes from one node to another, and then returns to the first one. This pattern is quite suitable for *history-based prefetching*, since the sequence of page accesses always follows the same cycle.

This case shows the benefit of a small size for the *page history* structure. Execution time linearly increases as the history size grows. On the other hand, the efficiency of the predictions is not affected as *coverage* and *hit ratio* remains the same. For history size values lower than 14, the techniques makes no predictions. For values greater than 150, the execution time increases over the no-prediction execution, but giving no improvement in the quality of the prediction. This is due to the greater amount of history that has to be transmitted every time the ownership of a page is transferred.

SHEAR. SHEAR is a case of a difficult application for this strategy: row-phases produce a uniform page access per node due to the row-assignment, but column-phases produce *false sharing*, since when sorting a column a node must access every page, and this page must be written by every node in the same phase. This produces a multiple-writer access pattern and an almost random order in the access sequence to a page, since nodes must compete for the access to a page every time a column is sorted.

The SHEAR application is a case where few predictions can be done, so the cost of searching for patterns through the page history in order to make predictions, and the additional page faults generated because of wrong predictions, begins to take importance. For small history size values, the execution time is almost the same as in the no-prediction execution, while for bigger values the execution time increases as a consequence of the page faults generated by wrong *read-write* predictions, a higher number of entries in the *page history* structure, and a bigger time taken to find patterns.

Coverage remains with a low value because only a little number of page faults are successfully avoided. Inside that little number, however, the *Hit Ratio* is high enough to show the accuracy of the predicting strategy and tends to stabilize as the *page history* size increases.

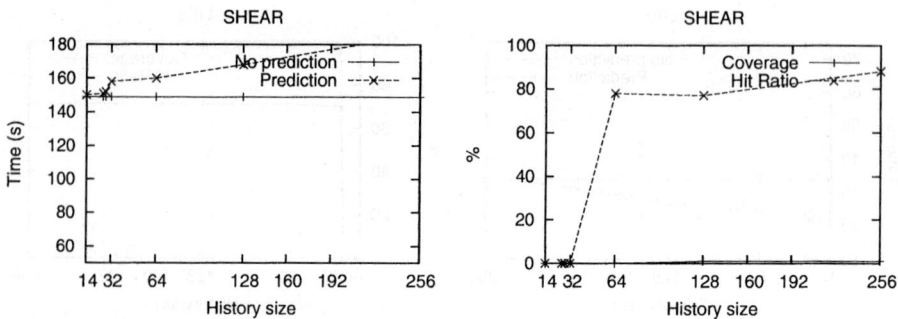

Fig. 3. Execution time, coverage and hit ratio for SHEAR application

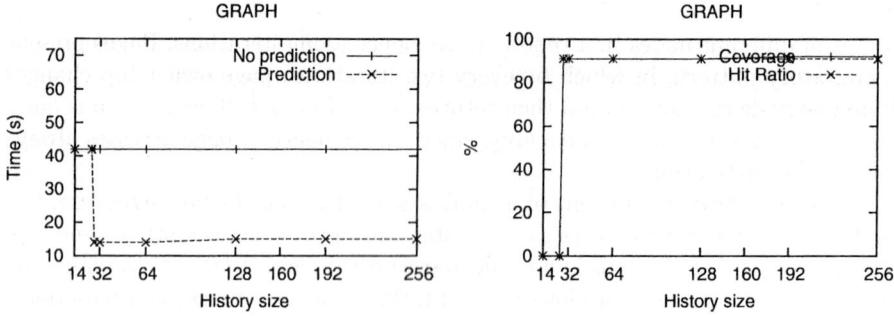

Fig. 4. Execution time, coverage and hit ratio for GRAPH application

GRAPH. GRAPH presents a case for a 1PMC pattern due to the *node-to-page* allocation. At each iteration, all nodes update the information of their vertexes regarding shortest distances to other vertexes, writing on their pages and, in the next iteration, that information is read by the other nodes to update their information on the next phase. This way, for every page there is only one node that writes on it, and every other node only reads from it, producing the *one-producer multiple-consumers* access pattern.

In this application, almost no page ownership is transferred among nodes because of the 1PMC pattern. Each node makes read access to pages owned by remote nodes, and writes only on locally owned pages. When *page history* size is increased the cost associated to *page history* transfers is not relevant, as *page history* is seldom transferred, so the execution time only varies depending on the additional time required to generate predictions.

Coverage is not affected by the *page history* size, and *hit ratio* barely increases, showing that the quality of the prediction is not affected by a bigger amount of information available.

4.3 Analysis

The results show that the size of the *page history* does not significatively increase the quality of the predictions made. *Coverage* and *hit ratio* are generally not harmed by the storage of a small list of past access, providing that it is big enough to allow a pattern to be found. Once again the fixed parameter D of the window size, reflecting the number of nodes involved in the execution, has proved to be useful. In most applications, predictions can be made with a *page history* size being at least two times the window size used. A further study should prove this conjecture.

Applications can achieve a better performance by looking only at a small list of past accesses, rather than having bigger amounts or information. In all situations, maintaining a complete history of events will lead to a poor performance, as the time taken to transmit and search through the history overcomes the prediction improvement.

5 Conclusions and Future Work

This work presented an experimental study on finding an optimal size for the M parameter used in *history-based prefetching*. This parameter represents the size of the *page history* that is stored, and that is used to make predictions about the future shared memory access behavior of the nodes on a software DSM system.

Results show that the tested applications achieve a better performance when the *page history* structure stores a small portion of the most recent shared memory accesses made to each page. Large amounts of information lead to a poor performance when the page ownership, and therefore, the *page history* is repeatedly transferred among nodes, and has to be searched.

The quality of the prediction, measured in terms of *coverage* and *hit ratio*, is, in most cases, barely improved by a bigger *page history* size. In the applications tested, a small size of M provides a prediction efficiency almost as good as that obtained with a large size.

As a future work, a similar study has to be done to measure the influence of the D parameter, that represents the size of the prediction window used to find patterns. Our conjecture, based on the fixed size used in this work and the results obtained, is that the history size should be at least two times bigger than the window size, and probably not bigger.

References

1. Li K. and Hudak P.: Memory Coherence in Shared Virtual Memory Systems. ACM Transactions on Computer Systems **7** (1989) 321–359
2. Cox A. L. and Dwarkadas S. and Keleher P. and Lu H. and Rajamony R. and Zwaenepoel W.: Software Versus Hardware Shared-Memory Implementation: A Case Study. In: Proc. of the 21th Annual Int'l Symp. on Computer Architecture (ISCA'94). (1994) 106–117
3. Pinto R. and Bianchini R. and De Amorim C. R.: Comparing Latency-Tolerance Techniques for Software DSM Systems. IEEE Transactions on Parallel and Distributed Systems **14** (2003) 1180–1190
4. Ruz C.: History-based Prefetching for Software DSM Systems. Master's thesis, Pontificia Universidad Católica de Chile (2005)

5. Bianchini R. and Kontothanassis L. I. and Pinto R. and De Maria M. and Abud M. and De Amorim C. L.: Hiding Communication Latency and Coherence Overhead in Software DSMs. In: Proc. of the 7th Symp. on Architectural Support for Programming Languages and Operating Systems (ASPLOSVII). (1996) 198–209
6. Bianchini R. and Pinto R. and De Amorim C.L.: Data Prefetching for Software DSMs. In: ICS '98: Proceedings of the 12th International Conference on Supercomputing, ACM Press (1998) 385–392
7. Lee S. and Yun H. and Lee J. and Maeng S.: Adaptive Prefetching Technique for Shared Virtual Memory. In: First IEEE International Symposium on Cluster Computing and the Grid, CCGRID, IEEE Computer Society (2001) 521–526
8. Karlsson M. and Stenström P.: Effectiveness of Dynamic Prefetching in Multiple-Writer Distributed Virtual Shared-Memory Systems. Journal of Parallel and Distributed Computing **43** (1997) 79–93
9. Monnerat L. R. and Bianchini R.: Efficiently Adapting to Sharing Patterns in Software DSMs. In: Proc. of the 4th IEEE Symp. on High-Performance Computer Architecture (HPCA-4). (1998) 289–299
10. Meza F. and Campos A. E. and Ruz C.: On the Design and Implementation of a Portable DSM System for Low-Cost Multicomputers. In: International Conference on Computational Science and Its Applications, ICCSA 2003. Number 2667 in Lecture Notes in Computer Science, Montreal, Canada, Springer-Verlag (2003) 967–976
11. Gardner M.: Mathematical games: The fantastic combinations of John Conway's new solitaire game 'Life'. Scientific American **223** (1970) 120–123 The original description of Conway's game of LIFE.
12. Sen S. and Scherson I. D. and Shamir A.: Shear Sort: A True Two-Dimensional Sorting Technique for VLSI Networks. In: ICPP. (1986) 903–908
13. Wilkinson B. and Allen M.: Parallel programming: techniques and applications using networked workstations and parallel computers. Prentice-Hall, Inc. (1999)

Self-optimizing MPI Applications: A Simulation-Based Approach

Emilio P. Mancini[1], Massimiliano Rak[2], Roberto Torella[2], and Umberto Villano[1]

[1] RCOST and Dip. di Ingegneria, Università del Sannio
{epmancini, villano}@unisannio.it
[2] Dip. di Ingegneria dell'Informazione, Seconda Università di Napoli
{massimiliano.rak, r.torella}@unina2.it

Abstract. Historically, high performance systems use schedulers and intelligent resource managers in order to optimize system usage and application performance. Most of the times, applications just issue requests of resources to the central system. This centralized approach is an unnecessary constraint for a class of potentially flexible applications, whose resource usage may be modulated as a function of the system status. In this paper we propose a tool which, in a way essentially transparent to final users, lets the application to self-tune in function of the status of the target execution environment. The approach hinges on the use of the MetaPL/HeSSE methodology, i.e., on the use of simulation to predict execution times and skeletal descriptions of the application to describe run-time resource usage.

1 Introduction

The presence of geographically distributed software systems is pervasive in current computing applications. In commercial and business environments, the majority of time-critical applications has moved from mainframe platforms to distributed systems. In academic and research fields, the advances in high-speed networks and improved microprocessor performance have made clusters or networks of workstations and Computational GRIDs an appealing vehicle for cost-effective parallel computing.

However, the systematic use of distributed systems can be frustrating, due to difficulties in resource usage optimization and development of applications with good performance. Usually high performance systems, such as clusters or NOWs (Network of Workstations), adopt resource allocation systems and schedulers (e.g., Nimrod [1], MOAB [2] or MAUI [3], or the GRAM [4] component in Globus [5]). Following this approach, applications act as resource requesters; when the application starts, they request the needed resource. GRID environments, such as Globus, offer languages to describe the resources an application needs (RSL [6]). When the resources are actually available, the application is started. The main limit of this approach is that application requests are "static", i.e., the total amount of resources needed by an application cannot vary in function of the system status. For example, an application might declare how many nodes or processor it needs, but it has no way to perform this choice taking into account the system status: the resource allocator locks it until all the requested nodes or processors are available. This makes the problem of performance portability [7] and performance engineering [8] really hard to manage.

Application developers, on the other side, try to develop codes that are as flexible as possible, in order to easily optimize them on systems which continuously change their architecture, due to new node acquisition inside a cluster, or to the distribution in a GRID environment [9, 10]. Du, Ghosh, Shankar and Sun [11] propose an alternative approach: they develop a tool that allows MPI tasks to migrate from a node to another in order to optimize resource usage. This approach, as always in code migration, has the problem of the large overhead due to migration and to the system monitoring that triggers migrations.

In this paper we propose a new approach, which involves the application in resource allocation, even if hides this process from final users. The idea is to rely on a tool that predicts the optimal application configuration (for example, in terms of number of nodes needed) using performance data collected using direct measurement and simulation to predict the system usage. The tool, MHstarter, is invoked every time users launch a new instance of the application. The tool hinges on an existing description language and simulation environment (MetaPL and HeSSE, respectively [12,13,14,15,16,17,18]). In other words, newly-developed services or applications can exploit the simulator to predict the system behavior in real, fictitious or future possible working conditions, and to use its output to help optimize themselves, anticipating the need for new configurations or tunings.

The reminder of the paper is structured as follows: the next section introduces the HeSSE/MetaPL prediction methodology. Then the tool architecture and its components are described, explaining how it can be used. After this, a simple case study is proposed. Finally, the conclusions are drawn and our future work is outlined.

2 MetaPL/HeSSE Methodology

HeSSE is a simulation tool that allows the user to simulate the performance behavior of a wide range of distributed systems for a given application, under different computing and network load conditions.

The HeSSE compositional modeling approach makes it possible to describe Distributed Heterogeneous Systems by interconnecting simple *components*. Each component reproduces the performance behavior of a section of the complete system at a given level of detail. A HeSSE component is basically an object, hard-coded with the performance behavior of a section of the whole system. More detailed, each component has to reproduce both the functional and temporal behavior of the subsystem it represents. In HeSSE the functional behavior of a component is the service set exported to the other components. So connected components can ask other components for services. The temporal behavior of a component describes the time spent servicing.

System modeling is performed primarily at the logical architecture level. For example, physical-level performance, such as the one resulting from a given processor architecture, is generally modeled with simple analytical models or by integral, and not punctual, behavioral simulation. In other words, the use of a processor to execute instructions is modeled as the total time spent computing, without considering the per-instruction behavior.

HeSSE uses traces to describe applications. A trace is a file that records all the actions of a program that are relevant for simulation. For example, the trace for an MPI application is a sequence of CPU burst and requests to the runtime environment. Each trace is the representation of a specific execution of the parallel program.

Trace files are simulation-oriented application descriptions, typically obtained through application instrumentation. When the application is not available, e.g., it is still being developed, they can be generated using prototypal languages. In the past years the HeSSE framework was provided with an XML-based metalanguage for parallel programs description, MetaPL [16,15,18]. MetaPL is language independent, and can support different programming paradigms or communication libraries. The core MetaPL notation can be extended through *Language Extensions* (XML DTDs), which introduce new constructs into the language. Starting from a MetaPL program description, a set of (extensible) filters makes it possible to produce different program *views*, among which are trace files that can be used to feed the HeSSE simulation engine. The detailed description of the MetaPL approach to program description, and of the trace generation process, is out of the scope of this paper and can be found in [15, 18].

The simulation and analysis process of a given application can be represented graphically as in Fig. 1. It is subdivided in three steps: System Description, Simulation and Results Analysis. Unsatisfactory analysis results may lead to the use of new or improved algorithms/code structures and to further analysis sessions.

The System Description phase includes:

- MetaPL metacode production (Application Description);
- system architecture model definition (System Architecture Modeling);
- evaluation of time parameters (model tuning).

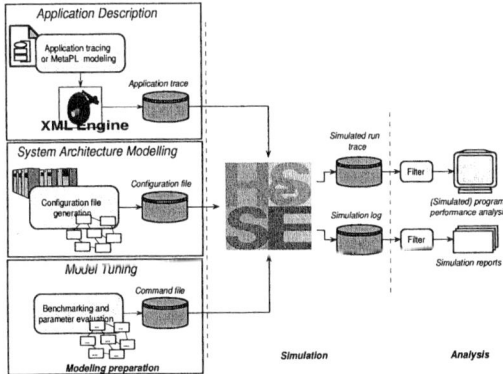

Fig. 1. HeSSE simulation session

```
<MetaPL>
  <Code>
    <Task name="test"><Block>
      <CodeBlock coderegion="Init" time="10" />
      <Loop iteration="10" variable="nstep" >
        <Block>
          <CodeBlock coderegion="Calc" time="10" />
        </Block>
      </Loop>
      ...
    </Block></Task>
  </Code>
</MetaPL>
```

Fig. 2. A simple "core" MetaPL description

```
<MetaPL>
  <Code>
    <Task name="test"><Block>
      <CodeBlock coderegion="Init" time="10" />
      <MPI_Init />
      <MPIAllGather dim="6720" />
      <MPI_Finalize />
    </Block></Task>
  </Code>
  <Mapping>
    <NumberOfProcesses num="4" />
  </Mapping>
</MetaPL>
```

Fig. 3. MetaPL description of a simple MPI code

The application description step consists essentially of the development of MetaPL prototypes, which make it possible to generate the trace files needed to drive the simulation execution. The system architecture model definition consists of the choice (or development, if needed) of the HeSSE components useful for simulation, which are successively composed through a configuration file. At the end of this step, it is possible to reproduce the system evolution. The last step consists of running benchmarks on the target system, in order to fill the simulator and the MetaPL description with parameters (typically) related to timing information.

Among the extensions currently available for MetaPL, there is one that enables the description of application developed using MPI. Figure 2 shows an example of description exploiting only the *Core* of the MetaPL language. It is a description of a simple task containing a loop. Figure 3 describes an MPI application that performs just an AllGather primitive; the Mapping element contains information useful to define the execution conditions, such as the number of processes involved.

3 Simulation-Based Autonomic Applications: MHStarter

MetaPL and HeSSE help users to predict the performance of a given system configuration for the execution of a target application. In this paper we propose their integration in a new tool, MHstarter, which automatically uses them to optimize the execution of MPI applications.

From a user perspective, the tool is a sort of application launcher, just like mpirun, even if it needs to be suitably configured before the application is started. The user should declare a new project, and provide the tool with the application executable and its MetaPL description. It should be noted that, according to the MetaPL/HeSSE methodology, application descriptions are generated in an integrated way with the final application code, thus making it possible to evaluate the performance tied to the design choices in a proactive way. MHstarter creates a project, in which it stores both the executable and the corresponding MetaPL description. Then the tool analyzes the MetaPL description, and, using a suitable MetaPL filter, it defines the configurations and the traces used to run the simulator for each configuration chosen. For any given problem dimension the user has to "prepare" the application launch, so that the tool is able to reduce the overhead for application starting. Once the project is created, the application can be started, just giving to the tool the reference to the project, whose name is the same of the executable code.

Every time that the user asks the starter to launch the application, the tool performs one or more simulations, retrieves results of previous simulations, automatically analyzes this information and chooses the optimal configurations for the application. Finally, the application is started on the chosen nodes with optimal parameters.

Figure 4 summarizes the tool architecture. The main component of the proposed solution is the starter (in the following we will call MHstarter both the complete tool and the starter component). It is the executable used by final users to launch the applications. Its performance affects the application startup time. Along with the starter, the tool provides two simple programs, create and prepare that create a new project, and prepare the program for execution, respectively. Creation and Preparation components are useful to reduce the amount of time spent in choosing the optimal configurations, at application startup.

In addition to the starter, the tool provides a data collector, which acts as a local daemon; at regular time intervals, the collector queries the available nodes, collecting such data as the computational workload on each node or network load. The collector stores all the information collected in a file representing the system status. MHstarter uses the file to update, off-line, the simulator configuration files. The collector works independently of user requests, so it does not affect the performance of the application.

On the bad side, the proposed approach has the effect to make application latency grow. In fact, the tool needs to perform several simulations to choose the optimal hardware/software configuration, and this affects the application execution time. As anticipated, in order to reduce the time spent to launch the

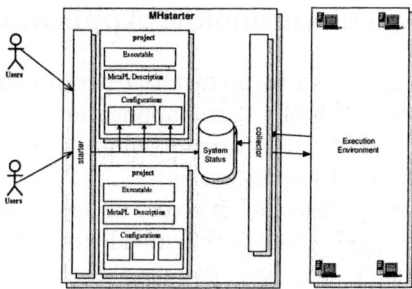

Fig. 4. Overview of the MHstarter architecture

applications, a large set of information is collected off-line (i.e. during project creation and application preparation), thus reducing the number of simulations performed when the application starts. The idea is that when the final user creates a new project and prepares it for startup (choosing the application parameters such as problem dimension), a set of simulations takes place, collecting results about possible useful configurations. When the user starts the application, most of this information is available.

3.1 Configuration Choice

In order to choose the optimal configuration, the tool has to evaluate:

- the system status, i.e., the current availability of compute nodes;
- the application-manageable parameters, i.e. the parameters available for optimization.

In this paper we use as optimization parameters the number of processes making up the application and several different allocation policies. The target system considered is a generic SMP cluster, with N nodes and m processors per node. For simplicity's sake, we will assume that the processors are not hyperthreaded.

The system status is modeled simply as the mean load on every processor of the target environment. We represent the status by means of a Workload Matrix (WM), whose rows represents nodes, and columns processors. In other words, the element WM_{ij} is the mean CPU usage of the j-th processor of the i-th node. The data collector component, acting as a daemon, updates continuously the system status. In the existing implementation, the collector simply reports the actual status of the processors. Future versions of the tool might provide an extrapolation based on CPU usage history and current requests.

Under the assumptions mentioned above, the MHstarter has to evaluate the optimal system configuration, and hence how many process to launch and on which processors. We represent a process allocation configuration by means of an Allocation Matrix (AM). Element AM_{ij} of the matrix is 1 if a process of the target application will be started on the j–th CPU of the i-th node, 0 otherwise. The MHstarter is able to convert this representation into HeSSE configuration files tuned on the target environment whenever necessary. It should be noted that the number of possible configurations grows with cluster dimension, and this may lead to an excessive time spent in simulation.

A few considerations make it possible to limit the number of configurations useful for analysis:

- A process which shares its CPU with another task *may* be a bottleneck, i.e. sometimes it is better to does not use a processor, if the application shares it with another task.
- In a SMP cluster, a node *may* be a bottleneck, even if it has a free processor, if there is CPU load external to the application on one of node CPUs. In other words, sometimes it is better to have four CPUs on two biprocessor nodes, than five CPUs on three biprocessor nodes.
- It is possible to simulate the time spent in execution on the system without load external to the application, before the user asks to start the application.

In order to reduce the application startup latency, the user has to ask off-line to the tool to prepare application executions for the given problem dimension. The tool performs a set of simulation and collects the estimated time, so that when the application will start, it is already known teh variation of the application execution times as a function of the number and distribution of available processors. Note that usually the user starts the application multiple times with the same problem dimensions (even if with different data).

So, given a system status wm, we can define two allocation matrices:

$am = P(wm, t)$ where $am_{ij} = 1 \Leftrightarrow wm_{ij} < t$, i.e., we start a process on the processors whose usage is under the given threshold.

$am = N(wm, t)$ where $am_{ij} = 1 \Leftrightarrow (\forall j, wm_{ij} < t)$, i.e., we start a process only on nodes whose CPUs has *all* usages under the given threshold.

Due to these considerations, `MHstarter` simulates at project startup all the possible application configurations with CPUs without load external to the application, and stores the results. When the application starts the tool, it:

- simulates the application behavior using all available processors **and** the actual system status;
- retrieves the results of the simulations (already done) for the allocated matrix, given by $P(wm, 0)$, $N(wm, 0)$;
- It compares the results, chooses the best configuration, generates a suitable MPI machine file, and starts the application.

Using this approach, only one simulation is performed before the application, even if simulation results of many other configurations are already available, is started, and it is extremely likely that optimal configuration is one of the proposed. The only drawback of this approach is that the user has to request explicitly a new project set-up when he is interested to a different problem dimension.

4 A Case Study

The proposed technique is able to optimize the performance behavior of a configurable application, even if it introduces an overhead due to the time spent for simulation. In order to judge the validity of the approach, we present here a description of the technique, as applied on a real system running an application developed by MetaPL/HeSSE. In particular, the case study is an implementation of the Jacobi method for resolving iteratively linear equations systems. We firstly modeled and developed the proposed algorithm. Then, we used the MHstarter to launch the application. The launcher chooses the optimal application parameters, using the simulation environment to predict the optimal execution conditions, and starts the application.

4.1 The Case Study Application

The application chosen to show the validity of the approach is an implementation of the Jacobi method for solving iteratively linear systems of equations. This numerical method is based on the computation, at each step, of the new values of the unknowns using those calculated at the previous step. At the first step, a set of random values is chosen. It works under the assumption that the coefficient matrix is a dominating-diagonal one [19]. Due to the particular method chosen, parallelization is very simple. In fact, each task calculates only a subset of all the unknowns and gathers from the other tasks the rest of the unknowns vector after a given number of steps. Figure 5 shows a partial MetaPL description of the above described application (in order to improve readability, some XML tags not useful for code understanding have been omitted).

This application is particularly suitable for our case study, because its decomposition can be easily altered, making it possible to run it on a larger/smaller number of processors, or even to choose uneven work sharing strategies. Note that in this paper the only parameter used for optimization will be the number

```
<MetaPL>
  <Code>
    <Task name="Jacobi"><Block>
      <CodeBlock coderegion="MPI_Init" time="10" />
      <CodeBlock coderegion="Init" time="10" />
      <Loop iteration="10" variable="nstep" >
        <Block>
          <CodeBlock coderegion="Calc" time="10" />
          <MPIAllGather dim="6720" />
        </Block>
      </Loop>
      ...
    </Block></Task>
  </Code>
</MetaPL>
```

Fig. 5. Jacobi MetaPL description

of processes into which the application is decomposed. Of course, this affects the duration of CPU bursts, i.e., of the times spent in Calc code regions. The study of the effect of uneven decompositions will be the object of our future research.

The execution environment is Orion, a cluster of the PARSEC laboratory at the 2nd University of Naples. Orion is an IBM Blade cluster with an X345 Front-End. Its seven nodes are SMP double-Pentium Xeon 2.3 GHz, 1GB RAM memory, over GigaEthernet. The system is managed through Rocks [20], which currently includes Red Hat ES 3 and well known administration tools, such as PBS, MAUI and Ganglia. A Pentium Xeon 2.6 GHz, 1GB RAM, is adopted as front-end.

4.2 Results

We used a synthetic workload to model different working conditions of the computing cluster. In practice, the synthetic workload was injected on a variable number of nodes, in order to model different global workload status of the target machine. Our objective was to model the presence of other running applications or services. In each status, the application was started using both the standard mpirun launcher, and the newly proposed one. Then, the performance figures were compared. The objective was to prove that MHstarter can react to the presence of an uneven additional workload, finding out automatically a decomposition of the application tuned to the CPU cycles available in each node, and taking suitably into account the difference between intra-node and extra-node communications. In any case, our expectation was that the more uneven the CPU cycles available (due to the workload injected), the more significant should be the performance increase due to the adoption of the self-optimizing launcher.

In the following, we will show for brevity's sake the system behavior in a simple working condition, evaluating the tool performance as compared to the absence of control (use of mpirun). Then, we will present a more systematic test of the effects of the synthetic workloads.

The system status first considered is the following: in two of the seven available nodes of the system is present a synthetic load. The workload is injected asymmetrically in only one of the two CPUs available per node. The collector component updates the system status file, so that, when the application starts, the workload matrix and the allocation matrices are:

$$wm = \begin{pmatrix} 0 & 0 \\ 0 & 0 \\ 0 & 100 \\ 0 & 0 \\ 0 & 99 \\ 0 & 0 \\ 0 & 0 \end{pmatrix} \quad P = \begin{pmatrix} 1 & 1 \\ 1 & 1 \\ 1 & 0 \\ 1 & 1 \\ 1 & 0 \\ 1 & 1 \\ 1 & 1 \end{pmatrix} \quad N = \begin{pmatrix} 1 & 1 \\ 1 & 1 \\ 0 & 0 \\ 1 & 1 \\ 0 & 0 \\ 1 & 1 \\ 1 & 1 \end{pmatrix}$$

Table 1 shows the application response times, both the simulated and the actually measured one, together with the resulting prediction error. In the first column of the table, W is the trivial configuration (use of all nodes and of all

Table 1. Application response time

Configuration	real	simulated	%error
W	65.7	66	0.45%
P	52.7	53	0.57%
N	60.4	61	0.99%

CPUs, independently of the presence of the workload), whereas P and N are the configurations described in section 3.

The results show that the best configuration is P, which gets an execution time 20% lower (13 sec less, in absolute time). The application execution time with MHstarter (including the starting overhead) is $t = 54.5s$, whereas with mpirun is $t = 65.7s$. This means a 16.8% performance increase due to the use of our technique.

The above-shown example is a single case in which the approach is advantageous. In order to show that it can be effective (i.e., it improves application performance) in almost all the relevant cases, we measured the application response time using MHstarter and mpirun in different system status, injecting, as mentioned before, synthetic workload in a variable number of nodes (once again, asymmetrically in only one CPU per node).

Figure 6 shows the application response time using both launchers, varying the number of nodes into which the extraneous workload was injected. mpirun uses always the trivial application decomposition (even decomposition over all the CPUs of all nodes), while MHstarter uses a different configuration each time, the one found by proactive simulation. Figure 7 shows the performance increase obtained using MHstarter in the seven proposed conditions. Note that on the right side of the diagram (where workload is injected in 5, 6 and 7 nodes) the standard launcher performs better. This is due to the number of available CPU cycles becoming more and more even as the number of nodes into which the synthetic workload is injected rises. After all, in these conditions MHstarter chooses as optimized one the "trivial" configuration, just as mpirun, but introduces a

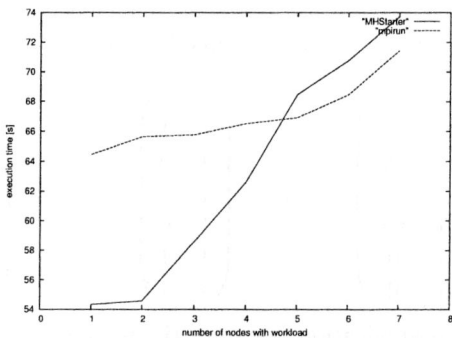

Fig. 6. Response times with MHstarter and mpirun for a variable number of nodes with injected workload

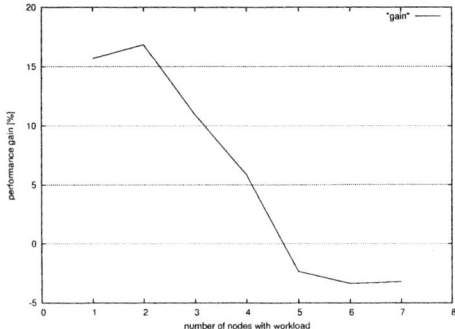

Fig. 7. Performance increase using `MHstarter`

larger overhead. It is important to point out that the overhead is almost independent of the execution time of the application. Hence, if the problem dimension grows, the effect of overhead tends to become negligible. On the other hand, Figure 7 also shows that the use of `MHstarter` is extremely advantageous when the available CPU cycles are unevenly distributed, and can lead to performance improvements up to 15%. This result confirms our expectations, and shows the validity of the approach.

5 Conclusions

In this paper we have proposed an innovative approach to application self-optimization. The proposed tool, in a way transparent to the final user, simulates a set of possible target system configurations and chooses the one that optimizes the application response time. The tool limits the number of simulations, collecting a large set of results off-line, in order to reduce the overhead linked to application starting. The approach has been tested on a real case study, an implementation of the Jacobi algorithm running on an SMP cluster. The results of the tests show the substantial validity of the technique. Future work will extend the tool, in order to optimize the prediction of the system status, by finding the expected load on the basis of the history and of current user requests. The next releases of the tool will also take into account for optimization other application-dependent parameters, automatically chosen from the MetaPL description.

References

1. Buyya, R., Abramson, D., Giddy, J.: Nimrod/G. An architecture of a resource management and scheduling system in a global computational grid. In: Proc. of HPC Asia 2000, Beijing, China (2000) 283–289
2. luster Resources Inc.: Moab Grid Scheduler (Silver) Administrator's Guide. (2005) http://www.clusterresources.com/products/mgs/docs/index.shtml.

3. Jackson, D.B., Jackson, H.L., Snell, Q.: Simulation based hpc workload analysis. In: IPDPS '01: Proceedings of the 15th International Parallel & Distributed Processing Symposium, Washington, DC, USA, IEEE Computer Society (2001) 47
4. Globus Alliance: WS GRAM: Developer's Guide. (2005) http://www-unix.globus.org/toolkit/docs/3.2/gram/ws/developer.
5. Foster, I., Kesselman, C., Tuecke, S.: The anatomy of the Grid: Enabling scalable virtual organization. In: The International Journal of High Performance Computing Applications. Volume 15, num. 3., USA, Sage Publications (2001) 200–222 http://www.globus.org/research/papers/anatomy.pdf.
6. The Globus Alliance: The Globus Resource Specification Language RSL. (2005) http://www-fp.globus.org/gram/rsl_spec1.html.
7. Foster, I., Kesselman, C., Nick, J.M., Tuecke, S.: The physiology of the Grid: An open grid services architecture for distributed systems integration. (2002) http://www.globus.org/research/papers/ogsa.pdf.
8. Smith, C., Williams, L.: Software performance engineering for object-oriented systems. In: Proc. CMG, CMG, Orlando, USA (1997)
9. Kleese van Dam, K., Sufi, S., Drinkwater, G., Blanshard, L., Manandhar, A., Tyer, R., Allan, R., O'Neill, K., Doherty, M., Williams, M., Woolf, A., Sastry, L.: An integrated e-science environment for environmental science. In: Proc. of Tenth ECMWF Workshop, Reading, England (2002) 175–188
10. Keahey, K., Fredian, T., Peng, Q., Schissel, D.P., Thompson, M., Foster, I., Greenwald, M., McCune, D.: Computational grids in action: the national fusion collaboratory. Future Gener. Comput. Syst. **18** (2002) 1005–1015
11. Du, C., Ghosh, S., Shankar, S., Sun, X.H.: A runtime system for autonomic rescheduling of mpi programs. In: ICPP '04: Proceedings of the 2004 International Conference on Parallel Processing (ICPP'04), Washington, DC, USA, IEEE Computer Society (2004) 4–11
12. Rak, M.: A Performance Evaluation Environment for Distributed Heterogeneous Computer Systems Based on a Simulation Approach. PhD thesis, Facoltà di Ingegneria, Seconda Università di Napoli, Naples, Italy (2002)
13. Mazzocca, N., Rak, M., Villano, U.: The transition from a PVM program simulator to a heterogeneous system simulator: The HeSSE project. In J. Dongarra et al., ed.: Recent Advances in Parallel Virtual Machine and Message Passing Interface, Lecture Notes in Computer Science. Volume 1908., Berlin (DE), Springer-Verlag (2000) 266–273
14. Mazzocca, N., Rak, M., Villano, U.: Predictive performance analysis of distributed heterogeneous systems with HeSSE. In: Proc. 2001 Conference Italian Society for Computer Simulation (ISCSI01), Naples, Italy, CUEN (2002) 55–60
15. Mazzocca, N., Rak, M., Villano, U.: The MetaPL approach to the performance analysis of distributed software systems. In: Proc. of 3rd International Workshop on Software and Performance (WOSP02), IEEE Press (2002) 142–149
16. Mazzocca, N., Rak, M., Villano, U.: MetaPL a notation system for parallel program description and performance analysis parallel computing technologies. In V. Malyshkin, ed.: Parallel Computing Technologies, Lecture Notes in Computer Science. Volume 2127., Berlin (DE), Springer-Verlag (2001) 80–93
17. Mancini, E., Rak, M., Torella, R., Villano, U.: Off-line performance prediction of message passing application on cluster system. In J. Dongarra et al., ed.: Recent Advances in Parallel Virtual Machine and Message Passing Interface, Lecture Notes in Computer Science. Volume 2840., Berlin (DE), Springer-Verlag (2003) 45–54

18. Mancini, E., Mazzocca, N., Rak, M., Villano, U.: Integrated tools for performance-oriented distributed software development. In: Proc. SERP'03 Conference. Volume 1., Las Vegas (NE), USA (2003) 88–94
19. Quinn, M.J.: Parallel Computing: Theory and Practice. 2 edn. McGraw-Hill (1994)
20. Papadopoulos, P.M., Katz, M.J., Bruno, G.: NPACI Rocks: tools and techniques for easily deploying manageable linux clusters. Concurrency and Computation: Practice and Experience **15** (2003) 707–725

Efficient SIMD Numerical Interpolation

Hossein Ahmadi[1], Maryam Moslemi Naeini[1], and Hamid Sarbazi-Azad[2,1]

[1] Sharif University of Technology, Tehran, Iran
[2] IPM School of Computer Science, Tehran, Iran
{h_ahmadi, moslemi}@ce.sharif.edu, azad@sharif.edu, azad@ipm.ir

Abstract. This paper reports the results of SIMD implementation of a number of interpolation algorithms on common personal computers. These methods fit a curve on some given input points for which a mathematical function form is not known. We have implemented four widely used methods using vector processing capabilities embedded in Pentium processors. By using SSE (streaming SIMD extension) we could perform all operations on four packed single-precision (32-bit) floating point values simultaneously. Therefore, the running time decreases three times or even more depending on the number of points and the interpolation method. We have implemented four interpolation methods using SSE technology then analyzed their speedup as a function of the number of points being interpolated. A comparison between characteristics of developed vector algorithms is also presented.

1 Introduction

Many interpolation algorithms for various applications have been introduced in the literature [6]. In general, the process of determining the value of a specific point using a set of given points and their corresponding values is named *interpolation*. In other words, interpolation means to fit a curve on a set of points. Interpolation on large number of points requires a great amount of computation power and memory space. Parallelism is one of the most practical approaches to increase the performance of different interpolation methods [3, 4, 5].

Since the arrival of modern microprocessors with single-instruction multiple-data stream (SIMD) processing extensions, little effort has been made to increase the performance of time-consuming algorithms using the SIMD computational model [1, 2, 10]. In the SIMD model, processors perform one instruction on multiple data operands instead of one operand as scalar processors do. In other words, in the SIMD model data operands are vectors. Therefore, to reach an appropriate speedup using the SIMD model we should focus on packing data elements to form data vectors.

In this paper, we show how various interpolation methods can adapt to gain speedup using SIMD computational model. The fact that how these methods implicitly allow parallelism greatly affects the amount of speedup that can be achieved by vectorizing them. The discussed interpolation methods are Lagrange, Newton-Gregory forward, Gauss forward, and B-Spline. For each of these methods, a generic vector algorithm is designed and then implemented in assembly and C++ (using their SIMD instruction sets) for Pentium processor.

The rest of paper is organized as follows. Section 2 introduces SIMD processing and available technology in current SIMD microprocessors. In Section 3, four common interpolation methods are briefly introduced. The vector algorithms for SIMD computation of the four interpolation methods are presented in Section 4. In Section 5, experimental evaluation of the implemented algorithms is reported. Finally, conclusions and future work in this line are suggested in Section 6.

2 SIMD Processing

The SIMD processing model allows performing a given operation on a set of data instead of a single (scalar) data. Such a computational model is now supported by many new processors in the form of SIMD extensions. Examples of such extensions are Intel's MMX, SSE, SSE2 and SSE3 technologies [7], AMD's 3DNow! technology [8], Motorola's AltiVec technology implemented in PowerPCs, and Sun's VIS technology [12]. We can assume these units as a vector processor with different vector sizes. For example 3DNow! Technology uses 8 packed 64-bit register to perform SIMD operations. When single precision floating-point data elements are used, 3DNow! extension appears as a 2-element vector processing model, while SSE technology using 128-bit registers can perform operations on 4 single precision floating-point elements simultaneously, which means a 4-element vector processing model.

We will focus our implementations on one of the most popular technologies, Intel's SSE technology. The packed single-precision floating-point instructions in SSE technology perform SIMD operations on packed single-precision floating-point operands. Each source operand contains four single-precision floating-point (32-bit) values, and the destination operand contains results of the parallel operation performed on corresponding values of the two source operands. A SIMD operation in Pentium 4 SSE technology can be performed on 4 data pairs of 32-bit floating-point numbers.

In addition to packed operations, the scalar single-precision floating-point instructions operate on the low (least significant) double words of the two source operands. The three most significant double words of the first source operand are passed through to the destination. The scalar operations are similar to the floating-point operations in the sequential form. It is important to mention that the proposed algorithms are designed to be implemented on any processor supporting vector or SIMD operations. Consequently, all algorithms are described with a parameter k, independent of the implementation, which represents vector size of the vector processor and, in a way, indicates the level of parallelism.

3 Interpolation Methods

In this paper, we consider four well-known interpolation methods: Lagrange, Newton-Gregory forward, Gauss forward and B-Spline. Lagrange interpolation is one of the most appropriate methods to be executed on a vector processor. Moreover, the simplicity of sequential implementations makes this method be used in many interpolation applications. Lagrange interpolation, for a set of given points (x_m, y_m), $0 \leq m \leq N-1$, in point x is carried out as

$$f(x) = \sum_{m=0}^{N-1} L_m(x) y_m \,. \tag{1}$$

where $L_m(x)$ is called Lagrange polynomial [6], and is given by:

$$L_m(x) = \frac{\prod_{\substack{0 \leq i \leq N-1 \\ i \neq m}} (x - x_i)}{\prod_{\substack{0 \leq i \leq N-1 \\ i \neq m}} (x_m - x_i)} \,. \tag{2}$$

It requires a long computation time to carry out the above computation in sequential form. Hence, this interpolation method is usually implemented on a number of parallel systems with different topologies, such as the k-ary n-cube [3].

Newton-Gregory forward method is based on a difference table. The table contains elements as [6]

$$\Delta^k f_n = \Delta^{k-1} f_{n+1} - \Delta^{k-1} f_n, \quad \Delta f_n = f_{n+1} - f_n \,. \tag{3}$$

and the computation for equally spaced input points can be realized as

$$f(x) = f_0 + \binom{s}{1} \Delta f_0 + \binom{s}{2} \Delta^2 f_0 + \cdots + \binom{s}{n} \Delta^n f_0, \quad s = \frac{x - x_0}{h} \,. \tag{4}$$

where h is the difference between x values.

Similarly, Gauss forward method uses a difference table; indeed, it operates on different path in the difference table. The Gauss forward interpolation for given points (x_m, y_m), $-N/2 \leq m \leq (N+1)/2$, can be formulated as [6]

$$f(x) = f_0 + \binom{s}{1} \Delta f_0 + \binom{s}{2} \Delta^2 f_{-1} + \binom{s+1}{3} \Delta^3 f_{-1} + \binom{s+1}{4} \Delta^4 f_{-2} + \cdots + \binom{s+[(n-1)/2]}{n} \Delta^n f_{-[n/2]} \tag{5}$$

B-Spline is probably one of the most important methods in surface fitting applications. It has many applications especially in computer graphics and image processing. B-Spline interpolation is carried out as follows [8]:

$$y = f_i (x - x_i)^3 + g_i (x - x_i)^2 + h_i (x - x_i) + k_i, \quad \text{for } i = 1, 2, \cdots, n-1, \tag{6}$$

By applying boundary and smoothness conditions in all points, parameters f, g, h, and k can be obtained as:

$$f_i = \frac{D_{i+1} - D_i}{6l_i}, \quad g_i = \frac{D_i}{2}, \quad h_i = \frac{y_{i+1} - y_i}{l_i} - \frac{l_i D_{i+1} + 2l_i D_i}{6}, \quad k_i = y_i \cdot \quad (7)$$

while D is the result of a tri-diagonal equations system, and $l_i = x_{i+1} - x_i$.

$$\begin{aligned}
&2(l_0 + l_1)D_1 + l_1 D_2 = m_1 \\
&l_{i-1}D_{i-1} + 2(l_{i-1} + l_i)D_i + l_i D_{i+1} = m_i, \ 2 \le i \le n-2 \\
&l_{n-2}D_{n-2} + 2(l_{n-2} + l_{n-1})D_{n-1} = m_{n-1} \\
&m_i = 6(\frac{y_{i+1} - y_i}{l_i} - \frac{y_i - y_{i-1}}{l_{i-1}})
\end{aligned} \quad (8)$$

4 Implemented Algorithms

To achieve the highest speedup, one can try to use vector operations for the interpolation process as much as possible. Using vector computing independent operations on different data elements in a k-element vector, a speedup of k is expected. Some other factors on SSE technology also help us to make our algorithm gain a speedup of more than k due to fewer accesses to memory hierarchy, faster data fetching from internal cache, and better use of pipelining. Obviously, for the methods with low data dependencies between different steps, better speedup can be obtained.

For any of the four methods, a well optimized vector algorithm running on a general vector processor is designed and implemented using Pentium 4's SSE instruction set. In what follows, we briefly explain the implementation of Lagrange, Newton-Gregory forward, Gauss, and B-Spline interpolation techniques. Next section reports experimental results for the performance evaluation of implemented algorithms.

For Lagrange interpolation, first, the common dividend in all Lagrange factors, $L_i(x)$, is computed. Each factor can be derived by dividing this common dividend with $(x - x_i)$. Next, for each factor the divisor is calculated. After computing Lagrange factors, the final value is calculated using Eq. (2). All operations involved in these steps can effectively exploit SIMD operations in SSE; thus, a noticeable speedup is expected.

The proposed algorithm for B-Spline interpolation is based on solving tri-diagonal equations system associated with D values. The method is based on simultaneous substitution in Gauss-Jordan method solving a system of equations [9]. More detailed, in the first iteration, values of l and m are initialized in a completely parallel manner. In the second iteration, values for b' and d' are the results of elimination of x_{i-1} from i-th equation which are computed for all k values. In the last iteration back-substitution is performed to obtain x_i values. $\log(k)$ operations are necessary to eliminate or back-substitute k elements. Therefore, we may expect a speedup of $k / \log(k)$, which is 2 for SSE technology with $k = 4$ (4/log(4) = 2). Figure 1 shows a pseudo code of the vector algorithm for B-Spline method.

for $i \leftarrow 0$ to (n/k)

$[l_{i \times k}, l_{i \times k+1}, \cdots, l_{i \times k+k-1}] \leftarrow [x_{i \times k+1}, x_{i \times k+2}, \cdots, x_{i \times k+k}] - [x_{i \times k}, x_{i \times k+1}, \cdots, x_{i \times k+k-1}]$

$Y \leftarrow ([y_{i \times k+1}, y_{i \times k+2}, \cdots, y_{i \times k+k}] - [y_{i \times k}, y_{i \times k+1}, \cdots, y_{i \times k+k-1}]) / [l_{i \times k}, l_{i \times k+1}, \cdots, l_{i \times k+k-1}]$

$[m_{i \times k}, m_{i \times k+1}, \cdots, m_{i \times k+k-1}] \leftarrow (Y - Y') / 6$

for $i \leftarrow 0$ to (n/k)

$V1 \leftarrow [m_{i \times k}, m_{i \times k+1}, \cdots, m_{i \times k+k-1}]$

$V2 \leftarrow [-l_{i \times k} / d'_{i \times (k-1)}, -l_{i \times k+1} / d'_{i \times (k-1)+1}, \cdots, -l_{i \times k+k-1} / d'_{i \times (k-1)+k-1}]$

$V1 \leftarrow V1 + V1(1) \times V2$

$V2 \leftarrow V2 \times V2(1)$

$[b'_{i \times k}, b'_{i \times k+1}, \cdots, b'_{i \times k+k-1}] \leftarrow V1 + V1(2) \times V2$

$V1 \leftarrow -[l_{i \times k}, l_{i \times k+1}, \cdots, l_{i \times k+k-1}] \times [l_{i \times k}, l_{i \times k+1}, \cdots, l_{i \times k+k-1}]$

$V2 \leftarrow 2 \times ([0, l_{i \times k+1}, \cdots, l_{i \times k+k-1}] + [0, l_{i \times k+2}, \cdots, l_{i \times k+k}])$

$V3 \leftarrow [1, 0, 0, \cdots, 0]$

$V4 \leftarrow [1, 1, \cdots, 1]$

$V1' \leftarrow V1 \times V3(1) + V2 \times V1(1)$

$V2' \leftarrow V1 \times V4(1) + V2 \times V2(1)$

$V3' \leftarrow V3 \times V3(1) + V4 \times V1(1)$

$V4' \leftarrow V3 \times V4(1) + V4 \times V2(1)$

$[d'_{i \times k}, d'_{i \times k+1}, \cdots, d'_{i \times k+k-1}] \leftarrow (V1 \times V3'(2) + V2 \times V1'(2)) / (V3 \times V3'(2) + V4 \times V1'(2))$

for $i \leftarrow (n/k)$ to 0

$V1 \leftarrow [b'_{i \times k} / d'_{i \times k}, b'_{i \times k+1} / d'_{i \times k+1}, \cdots, b'_{i \times k+k-1} / d'_{i \times k+k-1}]$

$V2 \leftarrow [-l_{i \times k-1} / d'_{i \times (k-1)}, -l_{i \times k} / d'_{i \times (k-1)+1}, \cdots, -l_{i \times k+k-2} / d'_{i \times (k-1)+k-1}]$

$V1 \leftarrow V1 + V1(1) \times V2$

$V2 \leftarrow V2 \times V2(1)$

$[D_{i \times k}, D_{i \times k+1}, \cdots, D_{i \times k+k-1}] \leftarrow V1 + V1(2) \times V2$

Fig. 1. Pseudo code for B-Spline method

For Newton-Gregory forward method, we begin with the computation of the difference table. Then, each Δf_i value is multiplied by $\binom{s}{i}$ and added to S, and the sum of calculated terms is computed. Adopting the same technique, we can realize the Gauss forward method as well.

Both methods, Newton-Gregory and Gauss, use a sequential summation procedure to carry out $f(x)$. Therefore, the parallelism is exposed only in computing the difference table. Because of great data dependency between successive steps in these methods, very high speedup cannot be expected.

5 Performance Analysis

As the Intel's SSE technology is the most common SIMD extension used in today's personal computers, we implemented the presented algorithms on Intel's Pentium 4 processor using single precision floating point. Thus, an approximate performance gain of 4 is expected. The speedup is computed with two different reference implementations: the SIMD C++ and assembly codes, and their equivalent sequential (non-SIMD) codes. We analyze the performance of implemented methods by speedup curves as a function of the number of points being interpolated.

The system used to perform our performance analysis had the following configuration:

CPU: *Intel Pentium 4 Mobile Processor at 1.8 GHz*
Main memory: *512 MB of DDR (266 MHz) RAM*
Cache memory: *512KB L2 Cache*
Operating system: *Microsoft Windows XP Professional Edition*
Compiler: *Microsoft Visual Studio .NET 2003*

To compute the speedup for a given number of interpolated points, execution time of the SIMD code is divided by the execution time of equivalent sequential code.

Figure 2 shows the speedup for Lagrange interpolation method. It is predictable that speedup grows as the number of input points increases because the effect of sequential parts of the program in the total running time decreases. The sequential parts of the code consist of loop control codes, initialization codes, and some serial arithmetic operations which have a high data dependency in different steps.

There are also fluctuations in the graph with peaks on multiple of 4, the vector length in SSE extension. This effect is common in register-register vector computing architecture [11]. When the number of interpolated points is not a multiple of 4, operations on the last data block has less speedup than previous data blocks; therefore, the total speedup will be reduced. By increasing the number of interpolated points, operations on the last data block get less portion of total execution time and fluctuations will diminish. Performing Lagrange interpolation on more than 50 single-precision floating point numbers will cause the result to loose the precision. That is because of large number of factors in Eq. (2). Hence, the graph of Lagrange interpolation is drawn only up to 50 points.

Note that in Newton and Gauss methods, as shown in Figure 3 and Figure 4, the ripple in the speedup curves is negligible in contrast with Lagrange method. The reason is the varying size of the difference table during computation. The table size can be multiple of 4 in the first step but not in following three steps. Consequently, the initial table size does not result in visible extra instruction execution and a speedup drop.

In many applications it is required to carry out the value of interpolated curve in several points. In this case, a parallel algorithm is designed based on Newton-Gregory forward and Gauss forward methods to interpolate at k points simultaneously. In this implementation a higher amount of speedup is gained which is presented in Figure 5.

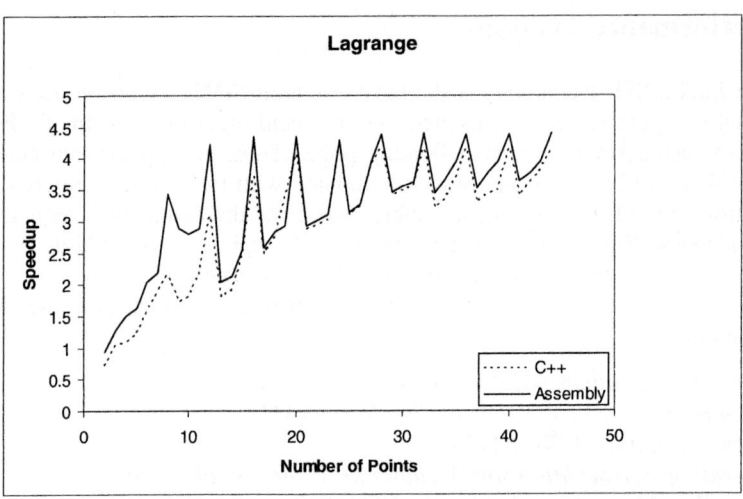

Fig. 2. Speedup of Lagrange method

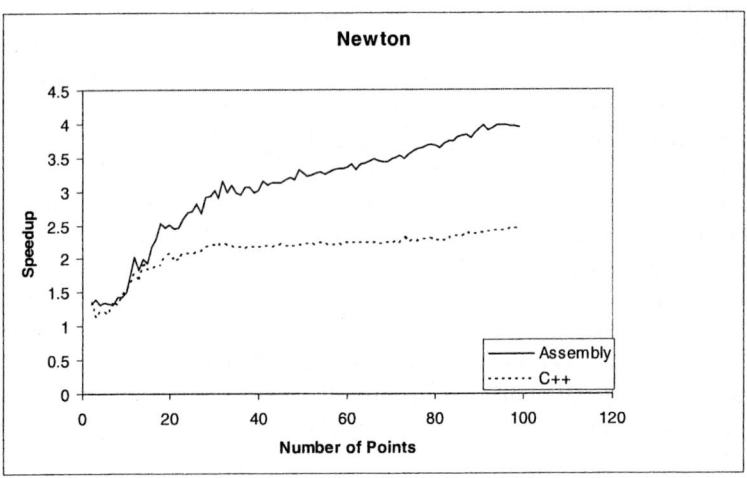

Fig. 3. Speedup of Newton-Gregory forward method

For B-Spline, as it was discussed in previous section, a speedup of about 2 is obtained. Similar to Lagrange method, fluctuations are completely visible in B-Spline method because of its noticeable extra process for the points. The interesting behavior shown in the figure 6 is the better performance achieved by the C++ code with respect to the assembly code. This is due to the low-performance code generated by the complier for sequential C++ code.

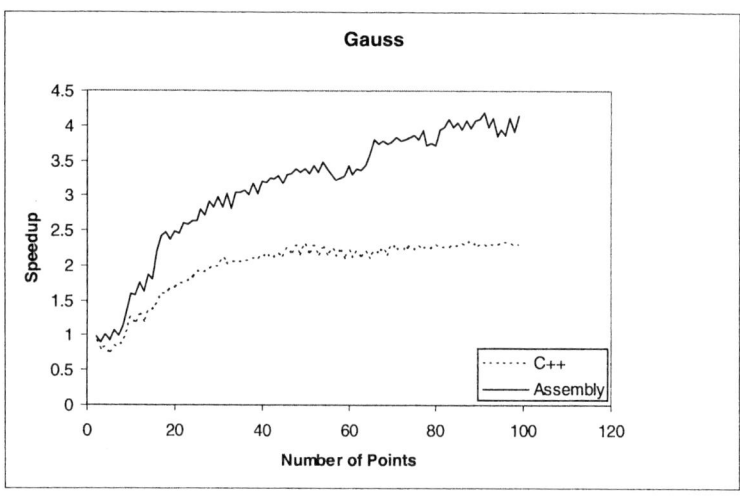

Fig. 4. Speedup of Gauss forward method

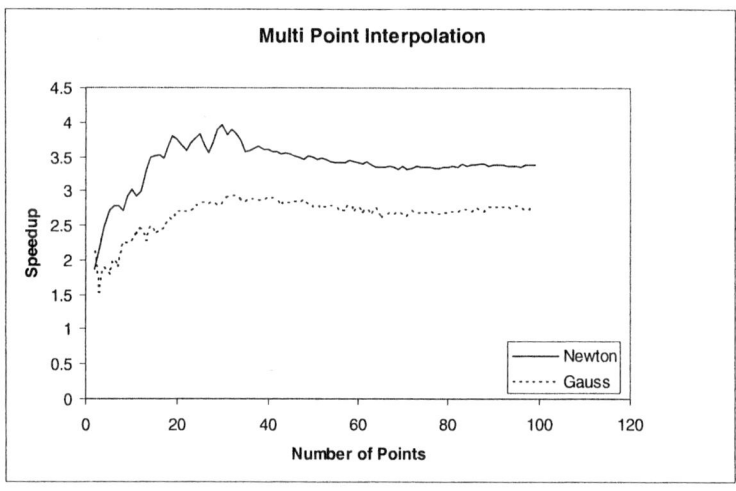

Fig. 5. Speedup of Gauss and Newton-Gregory forward methods on interpolating multi points

Finally, a comparison between the gained speedup of different methods is presented in Table 1. As it is shown in the table, Lagrange and B-Spline interpolation algorithms achieved the highest and the lowest speedup respectively.

Table 1. Speedup of four interpolation algorithms for 40 input points

	Lagrange	Newton forward	Gauss forward	B-Spline
Assembly Implementation	4.44	3.17	3.24	1.56
C++ Implementation	4.16	2.21	2.23	1.81

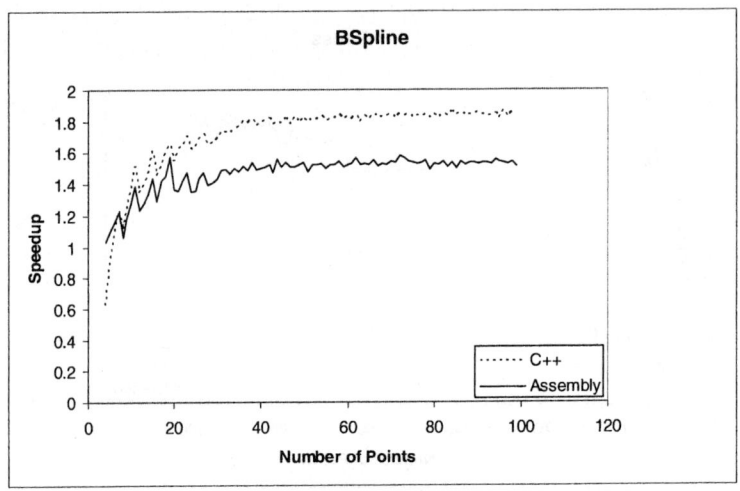

Fig. 6. Speedup of B-Spline method

6 Conclusion

We designed and implemented SIMD codes for four interpolation methods, namely Lagrange, Newton-Gregory forward, Gauss forward, and B-Spline, using Intel's SSE extension in Pentium 4 processors. Our performance analysis showed a noticeable speedup achieved for each case, ranging from 1.5 to 4.5.

Results showed that Lagrange method can achieve a high performance when executed in SIMD mode. In the case of interpolating multiple points, Newton-Gregory forward and Gauss forward methods also exhibit a good speedup. The completely dependent operations in B-Spline method make it very difficult to run in a parallel way.

The use of other SIMD extensions and comparing their effectiveness, for implementing interpolation techniques can be considered as future work in this line. Also, studying other interpolation techniques and implementing them using SIMD operations is also another potential future work.

References

1. A. Strey, On the suitability of SIMD extensions for neural network simulation, *Microprocessors and Microsystems*, Volume 27, Issue 7, pp. 341-35, 2003.
2. O. Aciicmez, Fast Hashing on Pentium SIMD Architecture, *M.S. Thesis, School of Electrical Engineering and Computer Science*, Oregon State University, May 11, 2004.
3. H. Sarbazi-Azad, L. M. Mackenzie, and M. Ould-Khaoua, Employing k-ary n-cubes for parallel Lagrange interpolation, *Parallel Algorithms and Applications*, Vol. 16, pp. 283-299, 2001.
4. B. Goertzel, Lagrange interpolation on a tree of processors with ring connections, *Journal Parallel and Distributed Computing*, Vol. 22, pp. 321-323, 1994.

5. H. Sarbazi-Azad, L. M. Mackenzie, and M. Ould-Khaoua, A parallel algorithm-architecture for Lagrange interpolation, *Proc. Parallel and Distributed Processing Techniques and Applications*, Las Vegas, pp. 1455-1458, 1998.
6. C. F. Gerald and P.O. Wheatley, Applied Numerical Analysis, 4th ed., Addison-Wesley, New York, 1989.
7. Intel Corporation, Intel IA-32 Architecture Developer's Manual, 2004.
8. AMD Corporation, 3DNow! Developer's Manual, 2003.
9. Chung, Lin, and Chen, Cost Optimal Parallel B-Spline Interpolations, *ACM SIGARCH Computer Architecture News*, Vol. 18, Issue 3, pp. 121 – 131, 1990.
10. S.Thakkar, T.Huff, Internet Streaming SIMD extensions, *Computer*, Vol.32, no.12, pp. 26 - 34, 1999.
11. K. Hwang, Advanced Computer Architecture: Parallelism, Scalability, Programmability, McGraw-Hill, 1993.
12. Sun Microsystems, VIS Instruction Set User's Manual, November 2003, http://www.sun.com/processors/vis/

A Threshold-Based Matching Algorithm for Photonic Clos Network Switches

Ding-Jyh Tsaur[1,2], Chi-Feng Tang[1], Chin-Chi Wu[1,3], and Woei Lin[1]

[1] Institute of Computer Science, National Chung-Hsing University
[2] Department of Information Management, *Chin Min Institute of Technology*
[3] Department of Information Management, Nan Kai Institute of Technology
djc@ms.chinmin.edu.tw, s9156006@cs.nchu.edu.tw,
wcc007@nkc.edu.tw, wlin@nchu.edu.tw

Abstract. This study concerns a threshold-based matching algorithm for Clos (t-MAC) network switches. The studied Clos network uses a central photonic switch fabric for transporting high-speed optical signals and electronic controllers at the input and output ports. The proposed matching algorithm is used to solve the scheduling problems associated with the Clos network. The matching algorithm incorporates cut-through transmission, variable-threshold adjustment and preemptive scheduling. The performance is evaluated using the bursty and unbalanced traffic model, and simulation results obtained by t-MAC are presented. The results demonstrate that the proposed algorithm can markedly reduce switching latency by approximately 45 to 36 % below that of Chao's c-MAC, for frame lengths from 10 to 512 cells.

Keywords: Photonic switch, Clos network, MAC, optical packet switching, Packet Scheduling.

1 Introduction

Networking has undergone rapid development during the last decade, as demonstrated for example, by the increase in the number of Internet users, their requests for more bandwidth and the appearance of optical transportation techniques such as DWDM (dense wavelength division multiplexing). More requests for bandwidth and high-tech optical transportation approaches have considerably increased the difficulties associated with designing high-speed packet routers. Current routers cannot supply sufficient switch capacity to satisfy future requests for networking bandwidth. However, the multi-stage switching structure increases the switching capacity. This study uses a three-stage Clos network [1], [2] instead of the original single-stage switch structure.

A matching algorithm for Clos, MAC [3], [4], [5], [6], [7] is executed two phases, port-to-port matching and route assignment, to transport cells smoothly though the Clos network from the input ports to the output ports. Scheduling by a single cell is replaced with scheduling by a small group of cells, removing the rigid initial time constraints. The loosened scheduling period is called a frame, and consists *r* time slots.

Figure 1 presents the system structure used by MAC, which mainly comprises a package scheduler (PS) and a three-stage photonic Clos network switch. The package scheduler includes k scheduling input modules (SIM) and k scheduling output modules (SOM), which correspond to the input and output modules, respectively, of the three-stage Clos network switch. The crosspoint switch links the scheduling input modules to the output modules. Within a frame, the ingress line card (ILC) delivers relevant information about cells in the VOQ to PS, which handles competition problems and finds the paths of a frame in the middle hierarchy (using MAC). Then, PS returns relevant information to ILC through the Clos network. The buffer of ELC is called a reassembly queue (RAQ), and restores the received cells to the states of the initial packets.

In this structure, cells are stored in electronic signals in the hardware and transported through the optical Clos network. When the networking transportation speed increases and the structure size rises, the electronic line card becomes a bottleneck. Recently, a new switch algorithm called the concurrent matching algorithm for Clos(c-MAC), proposed by Choa [4], [5], solves the problems of port-to-port matching and route assignment, as well as the heavy computation that results from the conventional concentrated packet scheduler structure, by exploiting the concept of distribution.

The threshold-based matching algorithm for Clos (t-MAC) proposed herein is to assign VOQs with heavier loads a greater probability of successful preemptive matching, enabling each input port within a frame to transmit more cells, thereby increasing transportation efficiency of the overall network.

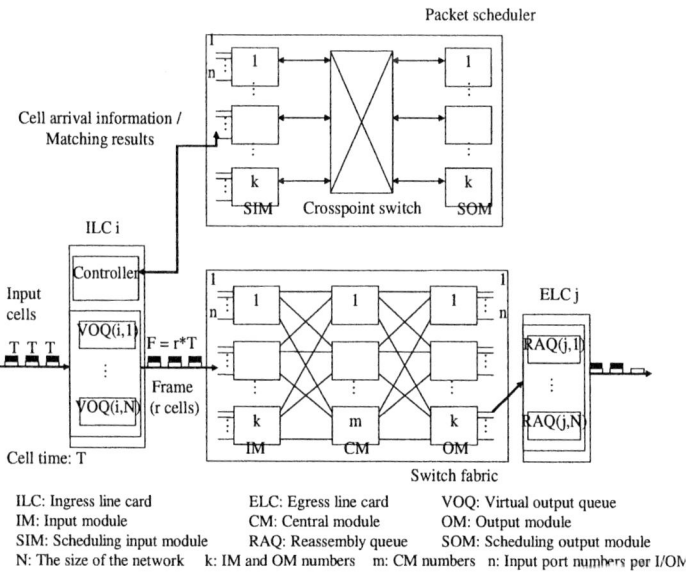

Fig. 1. The System Structure Used by MAC

The rest of this article is organized as follows. Section 2 describes the structure of packet schedulers and details manipulation using MAC. Section 3 describes the system structure and presents the new implementation of t-MAC. Section 4 presents the simulation results and the evaluations to estimate efficiency; the final section draws conclusions.

2 Structure of Packet Scheduler

Figure 2 depicts the structure of the packet scheduler, which consists k scheduling input modules (SIM) and k scheduling output modules (SOM). These match the input modules (IM) and output modules (OM) of the Clos network, respectively. Each scheduling input module includes n input port arbiters (IPA) and n virtual output port arbiters (VOPA). Each VOPA maps to n output ports of the matched output modules in the current period. Each scheduling input module has an input module arbiter (IMA) and each scheduling output module has an output module arbiter (OMA). The design seeks to solve the problems of route assignment. Crosspoint switches control links between SIM and SOM. Each SIM comprises $n \times N$ counters, each of which holds the corresponding number of cells in the virtual output queue.

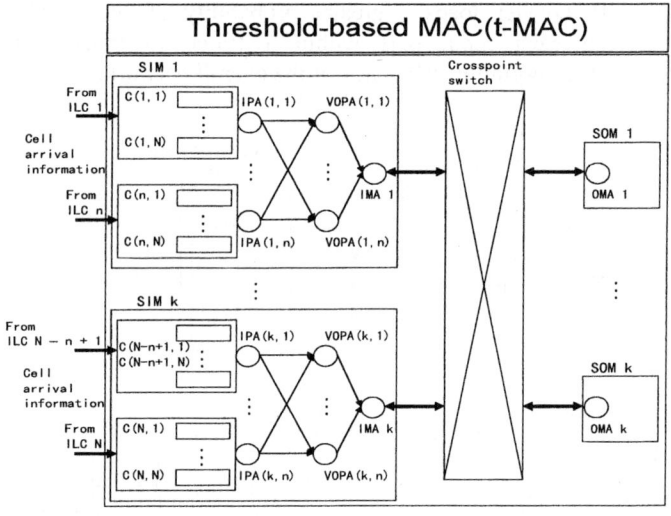

SIM: Scheduling input module   IMA: Input module arbiter   IPA: Input port arbiter   SOM: Scheduling output module
OMA: Output module arbiter   VOPA: Virtual output port arbiter $C(i,j)$: Counter for number of cells from ILCi destined to ELCj

Fig. 2. The Packet Scheduler Structure of t-MAC

The t-MAC scheme divides a frame period into k matching periods, as show in Fig. 3. In each matching period, each input module is matched to one of k output modules. Therefore, k sets of matches between input modules and output modules are generated in each matching period. In each matching period, t-MAC handles port-to-port matching and route assignment problems. In the first phase, a derivative approach exhaustive dual round-robin matching (EDRRM) [8], [9], [10], is used to solve port-to-port matching problems. Exhaustive dual round-robin matching retains the previous successful matching between input and output ports and reduces the number of inefficient tasks conducted to find re-matches in the solving of problems associated with throughput reduction. However, exhaustive dual round-robin matching causes starvation in the input ports. Hence, t-MAC adds a timer to each input port such that when the period for which the cells remain in the virtual output queue exceeds the time threshold, the cells emit high-priority signal requests to transmit cells because that indicate starvation. In the second phase, a parallel matching scheme is used to handle route assignment, and thus reduce the time required, and thus determine which matches can be transported through the Clos network. The route setup of the Clos network switches to transport successfully matched frames smoothly from input ports to output ports.

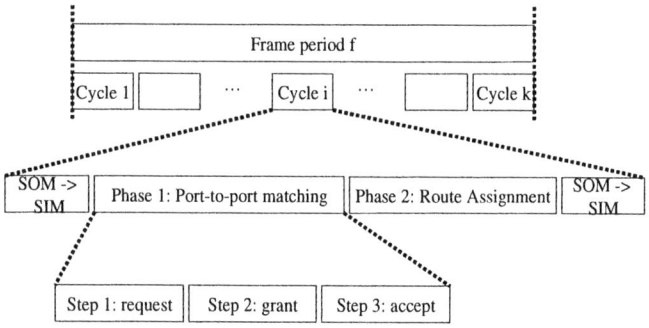

Fig. 3. The Time Frame of t-MAC

3 System Description

The system structure of t-MAC, as presented in Fig. 4, comprises a packet scheduler, n input line cards, n output line cards, a photonic switch fabric and a system clock.

3.1 System Architecture and Specification

This study proposes a threshold-based matching algorithm (t-MAC), exhibits distribution, cut-through, the variation of threshold values depended on load, and the high-priority matching can preempt low-priority matching. The strategy of t-MAC is that VOQs with heavier loads should be more likely to be successfully matched by preemption, enabling each input port within a frame to transmit more cells, improving the transportation efficiency of the whole network.

The design concepts of t-MAC are as follows.
- **Distribution**: The system design and algorithm are partially borrowed from c-MAC.
- **Pipeline Operation**: Within each frame period, the packet scheduler performs matching calculations and input ports transport cells through the Clos network according to the matching results in the preceding frame period.
- **Cut-Through**: Within a particular frame, after all permitted cells in the virtual

output queue have been transported, new cells in VOQ can be directly transported out. This design concept differs from that of c-MAC.
- **Variable Threshold Mechanism**：This mechanism is used to distinguish the priority threshold values associated with variations of system input offered load, a novel characteristic of t-MAC.
- **Preemptive Matching**：In a particular frame, a matching request with a high priority can rob a matching request with a low priority, is a new technique of t-MAC.

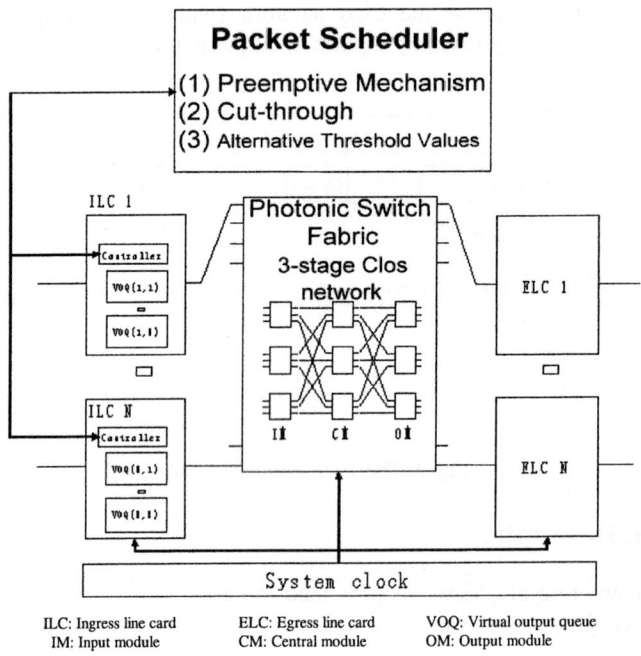

Fig. 4. The System Structure of t-MAC

3.2 Threshold-Based Matching Algorithm for Clos

The t-MAC divides a frame period into k matching periods. In the packet scheduler, k sets of matching pairs between IM and OM are formed within each matching period. At the beginning of each frame period, the initial matching sequence moves to ensure fair matching.

The threshold value is determined by system input offered load ρ and the number of time slots r required by packet schedulers, so the variation threshold value $T_w = \rho \times r$, distinguishing the priorities more accurately.

The scheduling problems involve two phases of computation - port-to-port matching and routing assignment. They are described as follows.

3.2.1 Threshold-Based Port-to-Port Matching

The algorithm associated with the first phase is described in detail below:

Step 1 : Request

(1) If the wait time of the cells in the input port exceeds the time threshold T_w, then the request become the high-priority will be transmitted.

(2) Each matched IPA in a preceding frame, is high-priority matching, and if **cell numbers ≥ threshold**, then the request signal for high-priority in the previous matching will be emitted.

(3) Unmatched input ports will transmit request signals based on the corresponding virtual output port arbiters and number of cells in the present matching period.

(4) Of all the input ports that have completed low-priority matching, if the number of cells associated with the new matches ≥ **threshold** value, then the request for the high-priority will be emitted.

Step 2 : Grant

(1) Assume that, in the preceding frame period, priority matching occurs and the **cell numbers ≥ threshold**, then the output port must grant the request for high-priority matching.

(2) Unmatched output ports return grant signals in a round-robin fashion, based on the received request signals.

(3) If the input ports that are successful matches with low-priority receive requests of high-priority, then they will pick out a request signal, using the round-robin method and the grant signals of high-priority will be re-transmitted.

Step 3 : Accept

(1) Unmatched IPA receives one or more grants, according to their priorities, and transmits acceptance signals to corresponding output ports according to a round-robin schedule.

(2) If all the input ports that have completed low-priority matching have received high-priority grant signals, then they will select high-priority one, starting at the location with a high-priority indicator, and sending out high-priority acceptance signals to corresponding output ports.

3.2.2 Concurrent Routing Assignment

The second phase handles routing assignments. t-MAC uses the parallel matching scheme to solve problems of routing assignment and thus minimize computational complexity. The set of output links of each IM_i are labeled by $IMA_i (1 \leq I \leq k)$, and the set of input links of each OM_j are labeled by OMA_j ($1 \leq j \leq k$), as shown in Fig.5. Each IMA_i and OMA_j contains exactly m elements to denote the state of each physical link. When the value of the element is zero, that link was not matched. When the value of the element is one, the link is matched. Only a vertical pair of zeros in IMA_i and OMA_j must be found to identify a routing path between IM_i and OM_j. Figure 5 reveals that two available paths can be obtained from IMA_i and OMA_j.

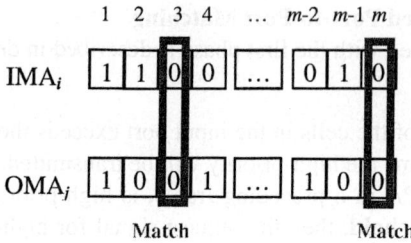

Fig. 5. Parallel matching scheme

Following matching in the first and second phases, if the match is generated by the preemption mechanism, released output ports and snatched input ports must be released so that they can compete in the following matching period. Only when both the first and second phases are successful, are the indicators of input port arbiters and virtual output port arbiters updated.

4 System Simulation Results and Evaluation

The simulator of t-MAC uses the bursty traffic model to simulate practical packet networking flow. Delay is defined as the required number of time slots during which a cell goes from the entry of ILC to the exit of the ELC.

For simplicity, the following notation is used to represent the original parameters.

N : Size of the network.
r : Required number of time slots during which packet schedulers perform a calculation; is also the number of cells in a frame.
w : Probability of unbalanced flow.
l : Mean length of burst.
p : Mean offered load of cells that enter the network.
T_w : Time threshold.
m : Number of central modules associated with Clos.
n : Number of input ports in each input module (IM).
k : Number of input and output modules.

Each simulation result displays figures formed in 1,000,000 time slots. The system begins to collect statistics after 1,000,000 time slots to yield more accurate statistical results, and thus prevent instability of the system at the beginning of the operation.

4.1 Comparison of c-MAC and t-MAC Performance Using Various Traffic Models

Given various traffic patterns, c-MAC and t-MAC associated with Clos switches are dissimilar . Three traffic models are used in the simulation to elucidate the dissimilarity.

4.1.1 Bernoulli Arrival of Traffic and On-Off Traffic

In the on-off traffic model, each input can alternate between active and idle periods with geometrically distributed durations. During an active period, cells destined for the same output, arrive continuously in consecutive time slots. The probability that an active or an idle period will end in a time slot is fixed. In Fig. 6, p and q denote the probabilities that a period is active and idle, respectively. The duration of an active or an idle period is geometrically distributed, as follows.

$$\Pr[\text{active period} = i \text{ slots}] = p(1-p)^{i-1}, i \geq 1,$$
$$\Pr[\text{idle period} = j \text{ slots}] = q(1-q)^{j}, j \geq 0.$$

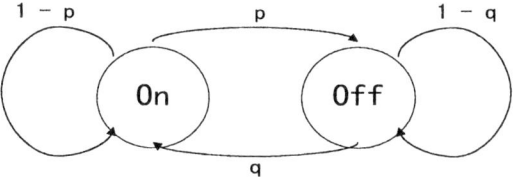

Fig. 6. On-Off model

Notably, an active period is assumed to include at least one cell. An active period is usually called a burst. The mean burst length b is given by Eq. (1), and the offered load ρ is the fraction of time for which a time slot is active, as in Eq. (2).

$$b = \sum_{i=1}^{\infty} ip(1-p)^{i-1} = \frac{1}{p} \tag{1}$$

$$\rho = \frac{\frac{1}{p}}{\frac{1}{p} + \sum_{j=0}^{\infty} jq(1-q)^{j}} = \frac{q}{q+p-pq} \tag{2}$$

Notably, the Bernoulli arrival process is in fact a special instance of the bursty geometric process, for which $p + q = 1$.

4.1.2 Unbalance Traffic

An unbalanced probability variable w is defined to measure the performance of t-MAC with an unbalanced traffic distribution. Also, the traffic load from input port s to output port d, $\rho_{s,d}$, is defined as in Eq. (3), where N is the size of network. When $w = 0$, the cells will be sent uniformly to all output ports. When $w = 1$, the network traffic is in a circuit switching. The cells from input port i will be sent to output port i.

$$\rho_{s,d} = \begin{cases} \rho(w + \frac{1-w}{N}) & , \text{if } s = d \\ \rho\frac{1-w}{N} & , \text{otherwise} \end{cases} \tag{3}$$

4.2 Comparison of Throughput and Delay

For various ρ values, t-MAC and c-MAC can both transport out of the network most of the cells that entered the network, as presented in Fig. 7. In terms of throughput, t-MAC equals c-MAC. A throughput of nearly 100% can be achieved for variously sized frames without expansion in a three-stage Clos-network photonic switch.

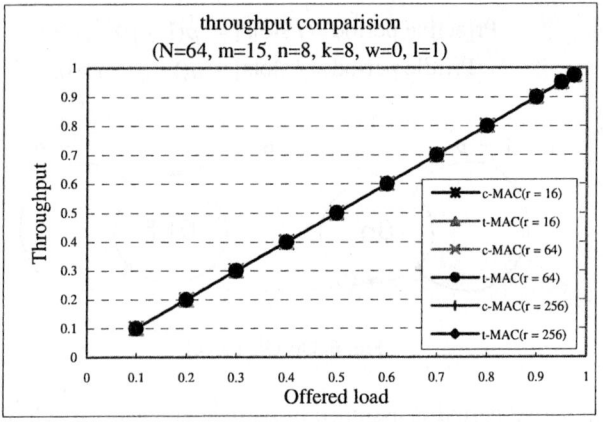

Fig. 7. Throughput Comparison of c-MAC vs. t-MAC

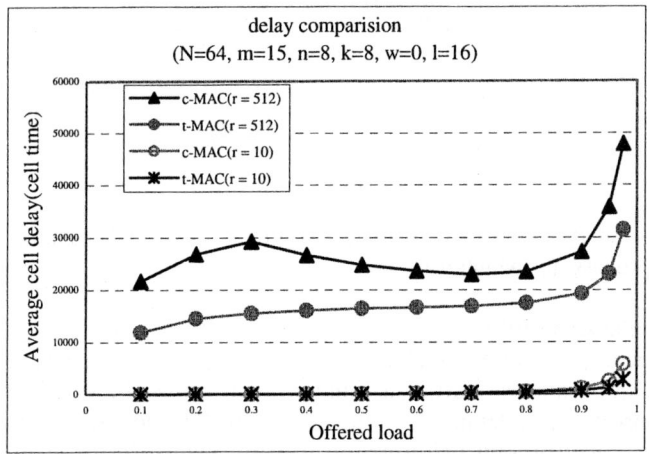

Fig. 8. Delay Comparison of c-MAC and t-MAC

As presented in Fig. 8, for $r = 10$ and $\rho < 0.8$, the average delay generated by t-MAC only slightly exceeds that generated by c-MAC. When $\rho \geq 0.8$, the average delay caused by t-MAC is less than that caused by c-MAC. The sum of average cell delays over all offered loads associated with c-MAC, is 10012 time slots, and that associated with t-MAC is 5476, or approximately 45% lower than for c-MAC. When $r = 512$, t-MAC outperforms c-MAC in terms of average delay, by approximately 36%.

As Figs. 7 and 8 reveal, t-MAC and c-MAC do not differ in throughput, t-MAC outperforms c-MAC in terms of delay. In particular as r increases, cells can be sent out earlier using t-MAC.

Figure 8 shows that the delay of c-MAC has a rising abnormality at load is about 0.3, because c-MAC was used successfully to match input ports with low-priority, which still send out requests, but follow-up handling does not occur. Most matches with low-priority are generated during low load, and in the following matching period, these matches transmit requests according to cell information in the VOQ. Therefore, these input ports with low-priority matching snatch more output ports, but they use only the first successfully matched output port. This rising abnormality apparently increases with the number of time slots required by the packet schedulers to execute computation.

4.3 Effect of the Length of Frame on Delay

As Fig. 9 reveals, the average cell delay of t-MAC increases with ρ. When ρ is set in the range 10% ~90%, the average cell delay increases slowly with ρ. When ρ exceeds 90%, the average cell delay increases rapidly. With the rise in r, the delay produced by t-MAC also rises.

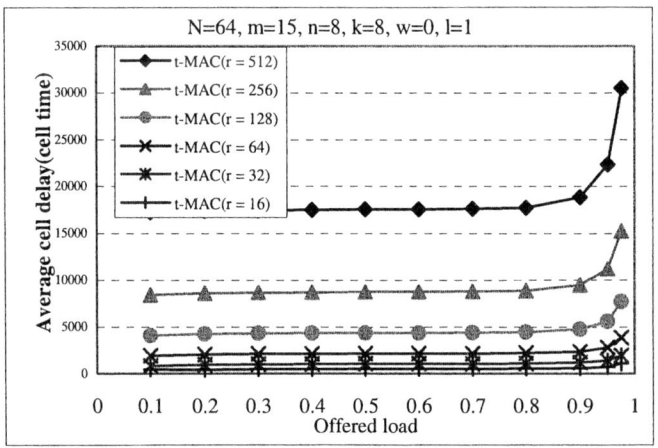

Fig. 9. Average Delay Variation of t-MAC

4.4 Effect of Unbalanced Flow on Efficiency

In Fig. 10, when $w = 0$, cells generated by each input port will be delivered uniformly to all of the output ports. When $w = 1$, the cells generated by the input port i will be sent to the output port i. When t-MAC is used, the output remains at over 85.3%

during unbalanced flow while $r = 1$. As r rises, the output of t-MAC decreases during unbalanced flow.

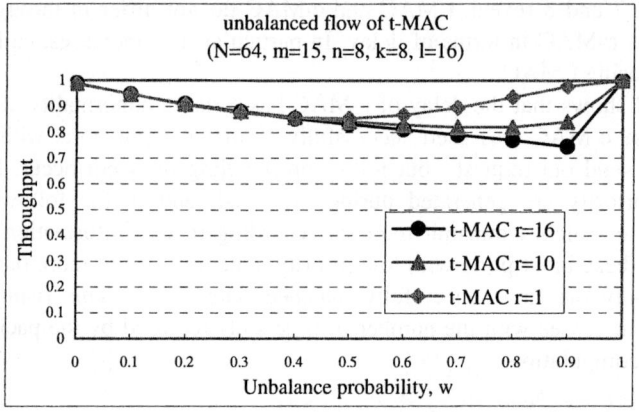

Fig. 10. Throughput Variations of t-MAC under Unbalanced Flow

4.5 Variation of Efficiency with Number of Central Modules

In Fig. 11, when $m \geq n$, the throughput of t-MAC is 97.5% (so the dispatch is almost complete), because the number of central modules can affect the matches produced in the first phase, and almost no matches are dropped during the second phase.

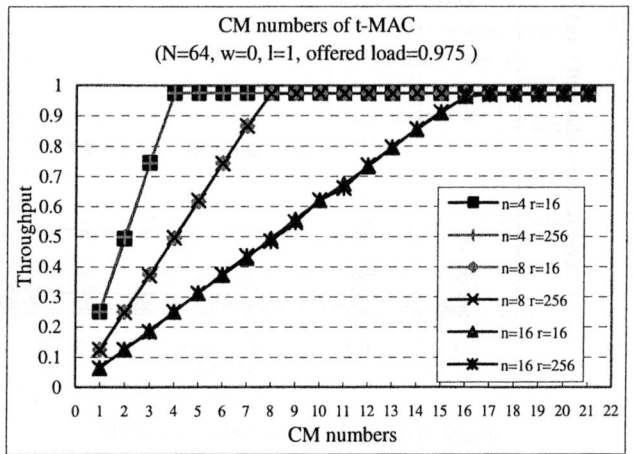

Fig. 11. Throughput Variation of t-MAC with Different CM Numbers

4.6 Comparison Between Fixed Threshold and Variable Threshold

In Fig. 12, t-MAC with a variable threshold is demonstrated to have a shorter delay than t-MAC with a fixed threshold. Under fixed threshold with heavy load, the VOQ cells more easily exceed the fixed threshold when the offered load is increased. Restated, the number of high-priority matching requests is increased, indicating that the entire frame period can be fully utilized. At the same time, the effect limited from verifying priorities via fixed threshold. However, with a high offered load, the improvement obtained with a variable threshold is limited. Therefore, when the offered load is increased, the offset between the fixed threshold and the variable threshold decreases.

In Fig. 12, t-MAC with a fixed threshold depicts a surge when the offered load is at 0.2. This abnormal phenomenon arises from that decrease in the offer load, and the fact that the number of cells of VOQ cannot exceed the threshold. Also, almost all of the requests generated have low priorities. More importantly, all requests are treated similarly when the number of cells is below the fixed threshold value. Accordingly, selecting a high-priority request that is closest to a fixed threshold for threshold-based matching is impossible, increasing the delay. However, a variable threshold allows a threshold to be set based on the offered load. This threshold mechanism can be utilized to solve the problem with respect to the increase in the delay, using a fixed threshold under low offered load.

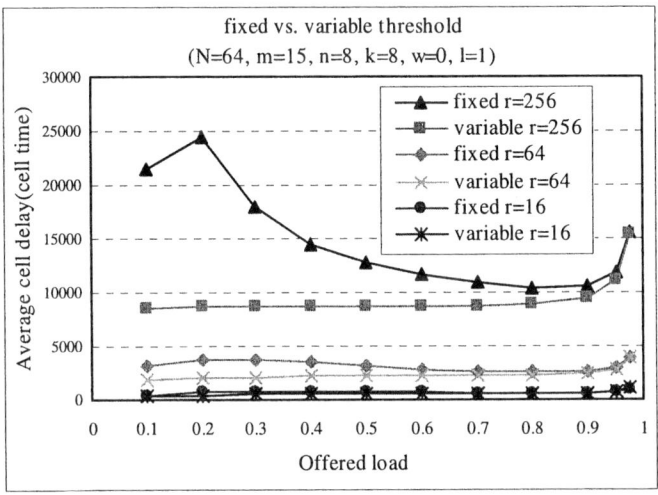

Fig. 12. Performance of t-MAC at average delay time with fixed threshold vs. variable threshold

4.7 Evaluating Performance for Various Busty Lengths

In Fig. 13, when the burst length is one, each cell arrives at each VOQ uniformly, without causing a situation in which all cells must arrive at a particular VOQ during a particular period. At this point, the number of cells of each VOQ is steadily increases.

Also, each VOQ is maintained in a situation that each VOQ has cells to send. At all times, the number of matches generated by the Packet Scheduler is held between 61 and 63 (full connection is 64). Also, each connected physical line is stable. Furthermore, the average delay increases very slowly. When the burst lengths are 16 and 32, and a cell (head) arrives at a VOQ, the remaining cells (body) are likely to arrive at that specific VOQ. Hence, numerous cells can be transmitted after t-MAC has chosen the desired VOQ. When ρ is low, the cells that are unselected by VOQ are under conditions in which VOQ either does not contain any cells or does not contain many cells. When $\rho<90\%$, the bursty traffic enables all cells that arrive at the input port to be sent out after the matching is calculated by t-MAC. Then, the delays at either $l=16$ or $l=32$ are shorter than that a $l=1$.

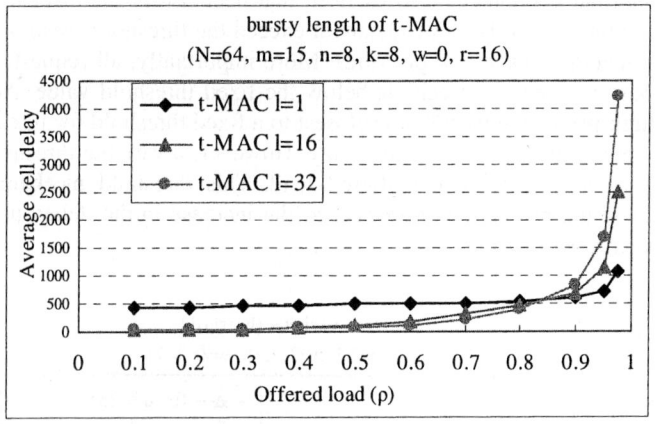

Fig. 13. Performance of observing t-MAC during an average delay time in response to different bursty length

When the load is heavy($\rho \geq 90\%$), cells can arrive at the input port continuously. Under bursty traffic, unselected VOQs can accumulate a large amount of cells. In contrast, although the same accumulation phenomenon occurs when $l=1$, cells in each VOQ do not arrive at the system continuously. Specifically, assuming that cells in VOQs cannot be transmitted until after a long period of waiting.. There will be no issue for concerning about generating a large amount of delay time. At this point, the average delay follows the order $l=1 < l=16 < l=32$.

5 Conclusions

The simulation results indicate the following three characteristics of t-MAC.

1. When the preemption mechanism is used, the VOQ with a heavier load provides a greater probability of successful matching.
2. The variation threshold mechanism improves the accuracy of the system used to determine the high/low priority when the offered load is low.

3. Cut-through reveals that the results generated by packet schedulers are fully used.

These three characteristics are such that more cells are transported from each input port within each frame period, so the average latency of t-MAC is approximately 45% to 36% lower than that of c-MAC, and the throughput is close to 100% in the three-stage Clos-network photonic switch. Finally, the t-MAC is more efficient than the earlier c-MAC.

References

[1] H. J. Chao, Cheuk H. Lam, and Eiji Oki, "Broadband Packet Switching Technologies," Wiley, 2001.
[2] H. J. Chao, K. L. Deng, and Z. Jing, "Packet scheduling scheme for a 3-stage Clos-network photonic switch," IEEE ICC 2003, Anchorage, Alaska, May 2003, pp. 1293-1298.
[3] H. J. Chao, K-L. Deng, and Z. Jing, " A Petabit Photonic Packet Switch(P3S)," *IEEE INFOCOM '03*, San Francisco, CA, Apr. 2003.
[4] H. J. Chao, S. Y. Liew, and Z. Jing, "Matching Algorithms for Three-Stage Bufferless Clos Network Switches," *IEEE Communications Magazine*, pp. 46 -54, October. 2003.
[5] H. J. Chao, K-L. Deng, and Z. Jing, "PetaStar: A Petabit Photonic Packet Switch," *IEEE JSAC* Special Issue on High-Performance Optical/Electronic Switches/Routers for High-Speed Internet, vol. 21, no. 7, Sept. 2003.
[6] H. J. Chao, S. Y. Liew, and Z. Jing, "A Dual-Level Matching Algorithm for 3-stage Clos-Network Packet Switches," *Proc. Hot Interconnects 11*,Stanford Univ., CA, Aug. 2003.
[7] H. J. Chao, K. L. Deng, and Z. Jing, "Packet scheduling scheme for a 3-stage Clos-network photonic switch," *IEEE ICC 2003*, Anchorage, Alaska, May 2003, pp. 1293-1298.
[8] E. Oki, Z. Jing, R. Rojas-Cessa, H. J. Chao, "The dual round robin matching switch with exhaustive service," *Merging Optical and IP Technologies*, pp. 58-63, May. 2002.
[9] Y. Li,S.S. Panwar,and H.J.Chao,"Performance analysis of a dual round robin matching switch with exhaustive service," *Global Telecom-munications Conference, 2002. GLOBECOM '02. IEEE*, Vol. 3, pp. 2292-2297, Nov. 2002.
[10] E. Oki, Z. Jing, R. Rojas-Cessa, H. J. Chao, "Concurrent round-robin-based dispatching schemes for Clos-network switches," *IEEE/ACM Transactions on Networking*, Vol. 10, No. 6, Dec. 2002.

CSAR-2: A Case Study of Parallel File System Dependability Analysis*

D. Cotroneo[1], G. Paolillo[1,**], S. Russo[1], and M. Lauria[2]

[1] Dipartimento di Informatica e Sistemistica, Universita' di Napoli "Federico II",
Via Claudio 21, 80125, Napoli, Italy
{cotroneo, gepaolil, sterusso}@unina.it
[2] Dept. of Computer Science and Engineering, The Ohio State University,
2015 Neil Ave., Columbus, OH 43210, USA
lauria@cse.ohio-state.edu

Abstract. Modern cluster file systems such as PVFS that stripe files across multiple nodes have shown to provide high aggregate I/O bandwidth but are prone to data loss since the failure of a single disk or server affects the whole file system. To address this problem a number of distributed data redundancy schemes have been proposed that represent different trade-offs between performance, storage efficiency and level of fault tolerance. However the actual level of dependability of an enhanced striped file system is determined by more than just the redundancy scheme adopted, depending in general on other factors such as the type of fault detection mechanism, the nature and the speed of the recovery. In this paper we address the question of how to assess the dependability of CSAR, a version of PVFS augmented with a RAID5 distributed redundancy scheme we described in a previous work.

1 Introduction

Parallel scientific applications need a fast I/O subsystem to satisfy their demand of aggregate bandwidth. In particular in clusters environment applications will benefit from a parallel file system (PFS) that can exploit the high-bandwidth and low latency of high performance interconnect such as Myrinet and Gbps Ethernet. PFS such as PVFS [1], can improve significantly the performance of I/O operations in clusters by using striping across different cluster's node. The main

---

* This work has been partially supported by the Consorzio Interuniversitario Nazionale per l'Informatica (CINI), by the Italian Ministry for Education, University, and Research (MIUR) in the framework of the FIRB Project "Middleware for advanced services over large-scale, wired-wireless distributed systems (WEB-MINDS)", by the National Partnership for Advanced Computational Infrastructure, by the Ohio Supercomputer Center through grants PAS0036 and PAS0121, and by NSF grant CNS-0403342. M.L. is partially supported by NSF DBI-0317335. Support from Hewlett-Packard is also gratefully acknowledged.
** G. Paolillo performed this work while visiting dr. M. Lauria's group at the Ohio State University.

objective in the construction of such architecture for data intensive applications continues to be the performance, but the current direction is also toward systems that provide high availability and reliability level. One of the major problems with the striping is the reliability because the striping of data across multiple server increases the likelihood of the data loss. Classical RAID approaches are usable locally to the server to provide tolerance to the disk failure, but if a server crashes, all the data on that server will be inaccessible until the server is recovered. To solve this problem many redundancy techniques across the servers have been proposed since 1990 [2]. Although the disk failures on storage node or more generally storage node failures are the most studied in such systems, simple data redundancy is not sufficient to protect the parallel file system. Due to the concurrency of the client accesses and to the dependency among the storage nodes introduced with the redundant schemes that was not present in the original striped file systems, end failure might damage the system by violating the original semantic of file system. The failure of a process running on a client node might happen during a write operation leaving the system in a inconsistent state (i.e., system error). An error resulting from a client node failure might becomes a system failure (i.e., semantic violation) subsequently to a recovery reconstruction based on corrupted data. So far, to the best of our knowledge, there are no works in the literature that address the semantic issues of parallel file systems dealing with both client and server failures. This paper analyzes dependability issues related to the striped file systems. We adopted the CSAR [3] parallel file system such as a case of study and improve its dependability characteristics in spite of a new type of fault. Quantitative assessment shows the reliability and availability improvements achieved from the enhanced CSAR.

1.1 Data Replication for Parallel File Systems

In order to deal with the system failures in the context of parallel file systems different strategies can be adopted but they all have to take in consideration the added overhead in terms of performance and architectural cost. The objective of guaranteeing an high system availability is achieved through fault tolerance. The increasing number of storage nodes involved exposes the system to failure resulting from a disk or node failure. The idea to extend the well known RAID techniques to the distributed case was explored for the first time in the 1990 by Stonebraker and Schloss [2]. Based on this idea many solutions have been proposed. All the mirroring strategies like RAID-10, such as orthogonal striping and mirroring [4] in spite of an low space-efficiency, storage cost equal to 2, provides a good level of reliability since the maximum number of disk/node failures tolerable is n/2. The RAID-5 technique provides a better space-efficiency, variable storage cost of (n+1)/n, it can tolerate only a single disk/node failure. Differently from mirroring, RAID-5 uses a redundant disk block per each stripe as the parity check for that stripe. Those RAID-like techniques theoretically have the same read performance even though practical experiments have shown that RAID-5 can exhibit slightly higher performance [4]. Instead, for parallel writes, RAID-10 requires a double number of disk accesses respect to the simple striping. In RAID-x, the large write is reduced respect to the mirroring, for more details see the reference [4]. In RAID-5, the small write has the well known problem to pre-read the parity and the old data to compute the new parity and execute finally the write. Furthermore, in a distributed striped file system, the lack of a centralized controller introduces a new problem for RAID-5 not present in disk array: each stripe write need to be executed atomically to avoid simultaneous read or update of shared parity blocks. The optimal choice among the above techniques depends on parallel read/write desired performance, the level of required fault tolerance, and the cost-effectiveness in specific I/O processing applications. We concentrate on the RAID-5 technique rather than mirroring because it achieves single fault tolerance with a much lower storage cost by using a fewer redundant disk space. Indeed, even though the storage cost is decreasing rapidly, having the double number of storage nodes also doubles the system exposure to these type of failures. Although all these works address the disk failure problem on parallel I/O architectures, none of them deal with client failure that also represents a potential source of system failure. This paper presents a strategy to achieve tolerance with respect to both disk and client failure by using RAID-5 together with a distributed reservation mechanism to make the system recoverable from those kind of faults.

2 Design of CSAR-2

The combination of the requirements of high performance and fault tolerance in the context of PFS make the use of different optimized solution possible to accomplish a specific tradeoff. The aim of this study is to preserve the good PVFS performance and its semantic in a cost-effective way while making it robust to the following types of fault: *i)* disk failure; *ii)* storage node failure; *iii)* client failure. One way to deal with disk failure is to replicate the data across different nodes of the architecture by extending the well known RAID techniques to the distributed case. However, assuming that one of redundant techniques is applied to the parallel file system, a client failure could represent a failure for the whole system since the file system semantic could be violated. Figure 1 shows an example of semantic violation due to a client failure in the case of mirroring and RAID5. This violation consist of a not deterministic reconstruction of corrupted file during the recovery procedure due to a client failure occurred in the middle of a write operation. Actually, due to the large number of components involved the parallel file system should provide data redundancy but also consider client failure to be a common case.

2.1 Fault Model

Generally, distributed file systems should be prepared to handle several types of failures: a server's disk can fail, a server can fail, communication links can be broken, and clients can fail. Usually server failures are assumed to be fail-stop in that the server does not show any degradation before the failure, and when it fails, it fails by stopping all processing completely. For each one of those faults

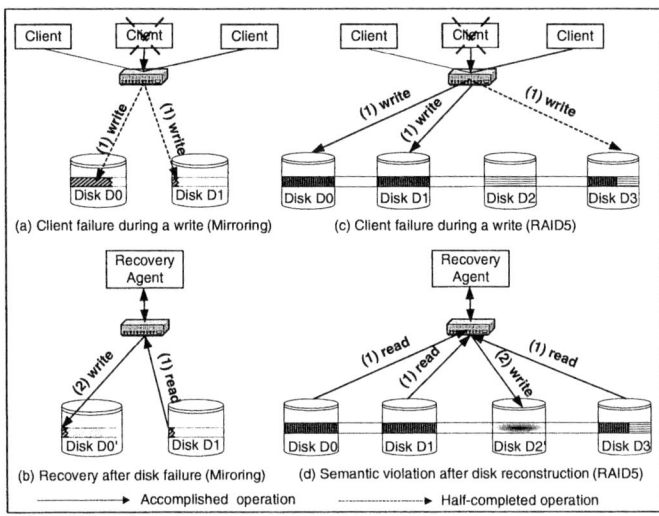

Fig. 1. Violation of File System Consistency

the file system should handle it in such a way that the consistency and semantic guarantees of the system will not be violated and when possible activate automatic recovery without human intervention. We are primarily concerned with three types of faults: *i)* storage node failure caused by hardware, operating system (OS) or application software faults in the node or by fault on the communication link; *ii)* disk failure on storage node; *iii)* client failure caused by hardware and software but the contribution of hardware errors to the client failure rate is fairly small. Most of the client failures are likely to be due to software cause [6].

2.2 Fault Detection

In RAID-5 the non-overlapping write operations that share one stripe need to be executed in a mutually exclusive way because they could update simultaneously the parity block of the shared stripe. For this reason a mechanism is needed to assure the sequential execution of the write operations that involve regions of the same file on the same stripe. The mechanism adopted in CSAR [3] makes use of queues on the servers to store the pre-read requests of the parity block that precede every small write (i.e., write of partial stripe) on the same stripe. The pre-read of the parity block is an operation preceding every small write and thus it can be used as a synchronization point: only after the update of the parity block from the current write is completed, the first pre-read request in queue will be served and so on. We modify the CSAR file system so that it can tolerate the client, server, and disk failure.

Server Failure Detection. The basic idea is to use a mutual fault detection between clients and servers by exploiting as much as possible the already existent interactions between them. This choice allows to contain the detection overhead during the normal operation of the system preserving the performance but on the other hand could increase the detection time of server node failure. In fact, if the server crashes when there is no client that is accessing the system, the detection will be postponed to the first access. Instead, if the server crashes during an access the involved client will detect suddenly the failure and will inform the recovery agent about the failed server. It is worth noting that the delay in the detection of node failure does not change the resulting file system consistency in that no operations will be performed in the while.

Client Failure Detection. We detect the client failure only when it is necessary, that is during a write operation. To enable the server nodes to detect the client failure a timeout mechanisms is used in the write phases. But the only timeout mechanism is not sufficient. In fact, a single client write involves at the same time more writes on different servers and the correct completion of a part of them represents an undetectable system error from the single server point of view. For this reason we modified the protocol concerning the write operation adding a reservation phase before each write. When a client wants to perform a write, it sends a short reservation message with the information about the data to write (i.e., file, region in the file and write size) to each server involved in the operation and starts a timer. Each server replies to the client by sending an acknowledgment message and starting a local timer. Only if the client receives all the acknowledgments from the servers by the timeout, it concludes that no server is failed before the write starts and it stops the timer and starts the write. During the write phase each client message exchanged with the servers is acknowledged and thus eventually server failures can be detected by timeout. The server can also detect client failure during the write phase in that the server knows the amount of data that it should receive before the timeout expires. Each server, on which the timeout expires, will notify the client failure to the recovery agent.

2.3 Recovery Procedures

Recovery from Client Failure. When a client failure is detected by one of the server involved in the write, it informs the recovery agent about the region affected by the failed write. The recovery agent undertakes the following steps: *i)* read the data written up to that moment from the failed client; *ii)* compute the parity blocks of the involved stripes; *iii)* write the new parity blocks. This recovery procedure can be performed during the normal system working and thus the system availability does not undergo modifications.

Recovery from Server Failure. Just after the node failure is detected, all the remaining servers are informed that the recovery procedure is in progress and so they will reject all the future requests and complete the ones in progress, already acknowledged in the reservation phase, in order to reach as soon as possible a consistent state in which to block the system. During the time interval between the server failure and the notification of failure to all the other servers no new write operation will be accepted because the faulty server will not reply to any request. After the server is recovered all his stripe units involved in write operations that were in progress at the failure time will be updated according the information of the other servers. Different failures could affect the server, for software failures (e.g., operating system, application software) it is sufficient the node reboot, while for hardware failures it is necessary the substitution of faulty component. To distinguish between the two faults usually it is performed node reboot and only if the problem persists the failure is considered hardware.

Recovery from Disk Failure. The disk failure recovery procedure differs from the hardware server failure only because after the substitution of the failed disk, requiring a human intervention, it proceeds to the reconstruction of the data on the new disk. The reconstruction operation will take a time proportional to the amount of data that was present at the failure instant. For each stripe the reconstruction consists of parallel block read operations on the *n-1* servers, computation of the XOR function on these data and finally the write of the resulting block on the new disk.

3 Dependability Evaluations

In this paper we make the following assumptions in the dependability model: *i)* all component failure events are mutually independent; *ii)* the failure and repair time are exponentially distributed; *iii)* the metadata server is assumed more dependable than the other nodes in the cluster and for this reason we do not consider the failure of this component in our model; *iv)* the faults of network host interface are comprised in the host fault.

3.1 Reliability

In PVFS in the current form, with no data redundancy mechanisms, the single disk failure brings the system to the failure because it involves the loss of data. Instead, the client failure does not represent a failure because PVFS does not have data redundancy to keep consistent. Therefore, for PVFS the reliability is $R_{PVFS}(t) = e^{-(N\lambda_{disk})t}$ where λ_{disk} is the disk failure rate and N is the number of server nodes. As for PVFS with distributed RAID-5, two events could lead the system to the failure. The first event is the second disk failure that entails the data loss and the second event is the single disk failure after a client failure during write operation. The second event decreases the theoretical reliability of the RAID-5 scheme. To evaluate the reliability loss due to client failure event we conducted some simple numerical examples by using the Markov model in Figure 2. The notations, symbols and values in the dependability models are listed as in Table 1. The client failure rate during a write operation is predicted as $\lambda_{client/write} = \lambda_{client} * \alpha * \beta$, where λ_{client} is the failure rate of a processing node in a cluster environment, α is the fraction of total execution time spent performing I/O operations and β is the fraction of time spent for write operations with respect to the whole I/O time. It is worth noting that it is hard to extract generic values that can represent scientific application workload. The

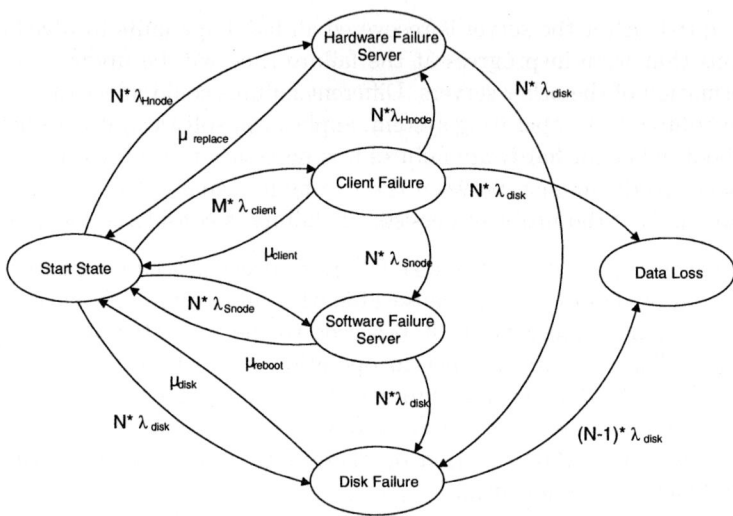

Fig. 2. Markov model

Table 1. Symbols, notations and values of dependability models

Symbol	Meaning	Value
N	Number of storage nodes	variable
M	Number of clients	variable
$1/\lambda_{disk}$	Mean time to disk failure	500,000 hours
λ_{client}	Client failure rate	0.00094 per hour
$\lambda_{client/write}$	Client failure rate during write operation	0.06278 per year
λ_{Snode}	Software server node failure rate	0.00094 per hour
λ_{Hnode}	Hardware server node failure rate	1 per year
$1/\mu_{client}$	Mean time to client repair	0.5 seconds
$1/\mu_{disk}$	Mean time to disk repair	8 hours
μ_{reboot}	Software server repair rate	20 per hour
$\mu_{replace}$	Hardware server repair rate	6 per day
α	Fraction of total execution time spent performing I/O operations	0.3177
β	Fraction of whole I/O time spent for write operations	0.024

time spent on I/O operations depends on many factors such as for instance the type of application (CPU intensive vs. I/O intensive) or the level of parallel I/O provided by the architecture. Figure 3.a shows the reliability values related to the three different schemes spread over a period of 2 years: *i)* PVFS in the current implementation; *ii)* PVFS with distributed RAID-5 without client failure treatment; *iii)* PVFS with distributed RAID-5 and client failure treatment. The reliability values concerning the schemes with client failure treatment have been obtained solving the model in Figure 2 by using the SHARPE

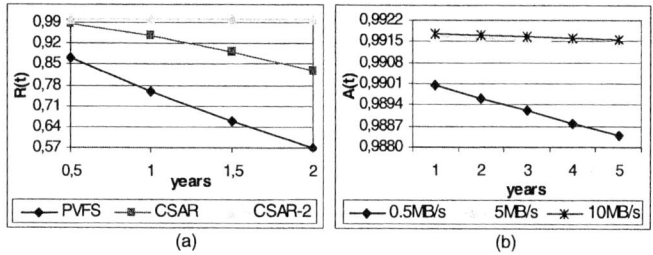

Fig. 3. (a) Reliability comparison for 16 server nodes and 8 clients; (b) Sensitivity analysis of availability

package [5]. As for the scheme without client failure treatment, it has been solved by means the same model without the transition from *Client Failure* to *Start State* that represents the recovery action. The system configuration is composed by 16 server nodes and 8 clients. The disk repair rate μ_{disk} has been obtained adding the time to replace the failed disk, estimated around 4 hours according to the mean time to repair for hardware in [8], and the time to reconstruct the entire disk at a reconstruction speed of 5 MB/second. The time to recover from the client failure $1/\mu_{client}$, instead, is composed by two terms, the time to detect the fault and the time to fix the parity. The detection time is the sum of the timeout that the server nodes use to determine the client failure plus the time to inform the recovery agent about the client failure. The time to fix the parity depends on the number of stripes involved in the failed write. All these times are measured on a prototype in which this scheme is implemented. In particular, the λ_{client} value has been chosen according the prediction of software failure rate for processing node in [6], while α and β have been extracted as mean values from the work [7] focused on the workload characterization of I/O intensive parallel applications. The numerical example, summarized in Figure 3.a, shows the remarkable improvements that can be obtained in terms of reliability by using the client failure treatment respect to the RAID-5 only technique. These improvements become more pronounced when the number of clients and server nodes increase in that the rate of the two transitions on the upper branch of the Markov model in Figure 2 increases correspondingly. The comparison with PVFS is simply resumed by the fact that, with a configuration of 16 servers, after two years it provides a probability of data loss equal to 0.43 respect to 0.00014 of CSAR-2.

3.2 Availability

To evaluate the system availability for the improved PVFS we use the Markov model in Figure 2. This model presents an absorbing state that corresponds to the system failure due to the data loss. For that reason we cannot study the steady state availability but only the instantaneous availability of the system that represents the probability that the system is properly functioning at time

t. The system becomes unavailable during the recovery procedures from failures that concern the server nodes. A way to reduce the system vulnerability is to recover it as soon as possible in spite of the system availability. We decide to stop the service just after the server failure detection to speedup the recovery action and preserve the system consistency. As for the client failure, it does not involve degraded service for the system in that after the error detection the recovery agent takes the place of the failed client fixing the parity data while all the other clients continue to use the system. Nonetheless, the client failure influences slightly also the availability in that it expose the system to the failure under single disk failure for the short time necessary for his recovery. The availability provided by PVFS can be estimated through his reliability in that in the absence of a repair, availability $A(t)$ is simple equal to the reliability $R(t)$. If we consider as system fault the only disk failure ignoring the server failure we can use the reliability values, related to 16 server nodes on Figure 3.a, as an upper bound of his availability and compare it with the availability of the enhanced PVFS for the same configuration. So, with CSAR-2 the availability is always greater than 0.9915 over a period of three years while PVFS just after six months provides an availability under 0.8692.

Sensitivity Analysis. The time to recover from the disk failure is composed by the time to replace the disk and the time to reconstruct the whole data. It represents a crucial parameter for the dependability characteristic of the system. In fact, the shorter this interval of vulnerability, the lower is the probability of a second concurrent disk failure. The reconstruction time is strongly dependent on the particular system architecture (i.e., disk access time, network bandwidth) and on the possible optimizations in the reconstruction procedure. We present a sensitivity analysis in which we vary the reconstruction speed and observe the relative changes in the corresponding dependability. The values of reconstruction speed have been chosen on the basis of the measurements performed on the prototype of CSAR-2 for a system configuration of 16 servers and 8 clients. In particular, 0.5, and 5 MB/sec are the reconstruction speed relative to the Gigabit Ethernet and Myrinet network, respectively, while 10 MB/sec is the speed of an hypothetical system with higher disk and network performance. Figure 3.b shows that the instantaneous availability of CSAR-2 relatively to 5 and 10 MB/sec are quite similar and both are substantially better than the one relative to 0.5 MB/sec. The reason is that when the reconstruction is very fast, the mean time for a human intervention, fixed to 4 hours in our experiments, become the predominant factor. Only with the reconstruction speed of 0.5 MB/sec, the bottleneck is represented by the network bandwidth and most of the recovery time is spent in the data reconstruction. This observation suggests that, in order to improve the availability, the network speed is as important as the time to replace the hardware of the cluster. Furthermore, the amount of availability improvement achieved by increasing the reconstruction speed from 0.5 to 5 MB/sec is equal to at least 0.0016 (from 0,99003 to 0,99166) which entails a reduction of the combined system outages in a year of 16 percent.

4 Conclusions and Future Work

This paper analyzes the dependability issues related to the striped file systems. We enhance CSAR, a PVFS version augmented with a distributed RAID5 scheme, by adding fault detection and recovery mechanisms and evaluate the reliability and availability improvements by means of dependability models using the parameter of CSAR-2 prototype. The dependability analysis of the system shows a 4 nine reliability while the sensitivity analysis based on the data reconstruction speed shows a reduction of the combined system outage time up to 16 percent. Future work will aim to: *i)* extend the dependability results with more system configurations; *ii)* assess the performance cost of CSAR-2 by means of standard benchmark.

Acknowledgments

We wish to thank Manoj Pillai at Ohio State University for the insightful discussions we had with him and for his assistance with the CSAR code.

References

1. P. H. Carns, W. B. Ligon III, R. B. Ross, and R. Thakur. PVFS: a parallel file system for Linux clusters. In Proc. of the 4th Annual Linux Showcase and Conference, Pages: 317-327, Atlanta, GA, 2000. Best Paper Award.
2. M. Stonebraker, G. A. Schloss, Distributed RAID-a new multiple copy algorithm, Proceedings of Sixth Int. Conf. on Data Engineering, 5-9 Feb. 1990, Pages: 430-437.
3. M. Pillai, M. Lauria, CSAR: Cluster Storage with Adaptive Redundancy, ICPP-03, pp. 223-230, October 2003, Kaohsiung, Taiwan, ROC.
4. K. Hwang, H. Jin, and R. S. C. Ho, Orthogonal Striping and Mirroring in Distributed RAID for I/O-Centric Cluster Computing, IEEE Trans. on Parallel and Distributed Systems, Vol. 13, no. 1, Jan. 2002.
5. K. S. Trivedi, SHARPE 2002: Symbolic Hierarchical Automated Reliability and Performance Evaluator, Proceedings of Int. Conf. on Dependable Systems and Networks. 23-26 June 2002 Page(s):544.
6. V. B. Mendiratta, Reliability analysis of clustered computing systems, Proceedings of The 9th Int. Symp. on Software Reliability Engineering, Nov. 1998 pp. 268-272.
7. E. Smirni, D. A. Reed, Workload Characterization of Input/Output Intensive Parallel Applications, Proceedings of the 9th Int. Conf. on Computer Performance Evaluation: Modelling Techniques and Tools, 1997. Vol.1245, Pages: 169-180
8. H. Sun , J. J. Han, H. Levendel, A generic availability model for clustered computing systems. Proceedings of Pacific Rim Int. Symposium on Dependable Computing. 17-19 Dec. 2001 Page(s):241-248.

Rotational Lease: Providing High Availability in a Shared Storage File System[*]

Byung Chul Tak[1], Yon Dohn Chung[2,**], Sun Ja Kim[1], and Myoung Ho Kim[3]

[1] Embedded S/W Div., Electronics and Telecommunications Research Institute,
161 Gajeong-dong, Yuseong-gu, Daejeon, Korea
{bctak, sunjakim}@etri.re.kr
[2] Dept. of Computer Engineering, Dongguk University,
26, 3-ga, Pil-dong, Chung-gu, Seoul, Korea
ydchung@dgu.edu
[3] Division of Computer Science, Korea Advanced Institute of Science and Technology,
373-1 Guseong-dong, Yuseong-gu, Daejeon, Korea
mhkim@dbserver.kaist.ac.kr

Abstract. Shared storage file systems consist of multiple storage devices connected by dedicated data-only network and workstations that can directly access the storage devices. In this shared-storage environment, data consistency is maintained by lock servers which use separate control network to transfer the lock information. Furthermore, lease mechanism is applied to cope with control network failures. However, when the control network fails, participating workstations can no longer make progress after the lease term expires. In this paper we address this limitation and present a method that enables network-partitioned workstations to continue file operations even after the control network is down. The proposed method works in a manner that each workstation is rotationally given a predefined lease term periodically. We also show that the proposed mechanism always preserves data consistency.

1 Introduction

Wide-spread use of Internet and increased computing power enabled applications to process more data and file systems have accordingly evolved to utilize network technologies. Recently the development of new medium such as fiber channel has enabled another form of shared storage network file system such as Storage Area Network[1,2,3,4,5]. Shared storage file systems use dedicated network for data traffic to enhance the performance. Storages are transparently shared so that workstations or servers attached to the storage area network use them as if they are local devices. This structure effectively eliminates the possible bottleneck point which was a major drawback of conventional network attached storage. Servers have local cache to store data read from the storages. However, the use of cache inevitably raises cache coherency

---

[*] This work was done as a part of Information & Communication Fundamental Technology Research Program, supported by Ministry of Information & Communication in Republic of Korea.
[**] Corresponding author.

and data consistency problems. To cope with this problem some of the servers are designated as lock managers. Lock managers use control network, which is usually an Ethernet connecting the servers, to transmit lock information.

For network file systems it is crucial that the file system be able to continuously provide reliable services. Among many possible failures in the shared storage environment, our approach deals with the failure of the control network. Conventional approach uses leases to reclaim the locks thereby forcing disconnected servers to stay idle until the problem is physically taken care of, whereas our approach allows servers to continue using the file system.

The remainder of this paper is organized as follows. Section 2 introduces some related works. Section 3 describes the proposed method in detail. Section 4 discusses experiments and analyzes characteristics of the proposed method. Finally in Section 5 we conclude with summary and future directions.

2 Related Work

The shared storage file system requires a mechanism to control access to the shared storage and this is commonly achieved by locking mechanism. But usually lock alone is not enough and the lease[6] mechanism is used together. Lock managers issue per-system lease and if the lease is not renewed within the lease term, lock managers invalidate all the locks issued to the server. StorageTank[2] uses per-system lease for data coherency and opportunistic renewal method in order to minimize the lease renewal cost.

Among many possible failures of the shared storage file system we are mainly concerned with the failure of the control network. Failure of the control network implies that servers are partitioned into several groups. In GPFS[4] only the majority partition is allowed to continue using the storage device. In order to determine the majority partition, GPFS[4] employs a kind of group membership protocol. It also uses a fencing function provided by the SAN switch to block I/O requests from the minority group. CXFS[5] adopts similar group membership protocol. This group membership technique has a drawback of only allowing certain server to survive while others stop indefinitely. And since GPFS[4] and CXFS[5] both uses locking mechanism, they are unable to reclaim the issued locks.

3 The Rotational Lease

3.1 Network Failures in the Shared Storage File System

The shared storage file system uses separate network for data traffic between servers and storage devices to gain performance enhancement. Lock information among servers or file requests from clients are transmitted through the local area network which acts as a control network as in Figure 1 (a). The failure of the control network creates network partitions and partitioned servers can no longer exchange lock information with the lock managers or renew the lease. Although some servers on the same partition as the lock managers may still be able to continue to work, the file system as a whole does not function any more.

Our proposed method tries to overcome this limitation by enabling the file system to continue working after the network partitioning occurs. The proposed method extends the lease concept and regards leases as being revived periodically after it expires. The idea is that participating servers can continue to use the file system if their lease is renewed automatically by predetermined order and time duration. This way of using extended lease concept is in effect making servers to wait for their turn to use the storage in rotation, and in this sense we call it a **Rotational Lease** method.

3.3 Periods and Active Intervals of the Rotational Lease

Automatic and periodic renewal of the lease requires that the renewal order be predetermined and servers be able to correctly find the time interval of their renewed leases. To do so we first need to introduce the concept of active intervals and periods in the Rotational Lease. Figure 1 (b) illustrates the concept.

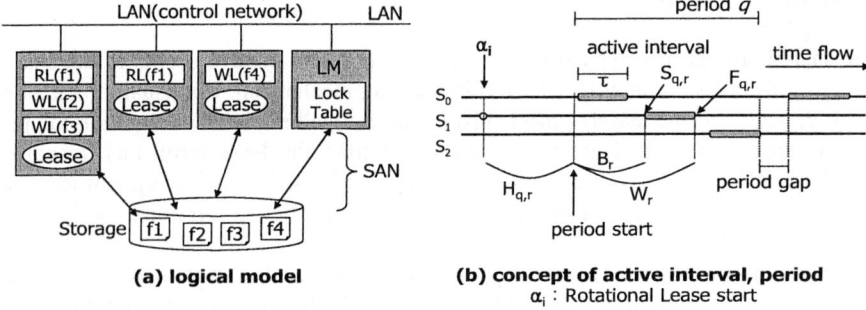

Fig. 1. Logical model of the shared storage file system and the concept of the active interval and period of the Rotational Lease

Determining the active interval involves time computation based on each server's internal clock. But some complication arises from the fact that real clocks deviate from the absolute time and the amount of deviation differs slightly from clock to clock. We define the clock drift rate as the maximum ratio of relative clock speed among the servers' internal clocks per unit time similar to the relative clock rate of StorageTank[2].

ρ : Maximum clock drift rate in range of ≥ 1. The time length t at one server falls within the $(t/\rho, t\rho)$ range at another.
τ : Time length of the active interval, or the Rotational Lease term.
S_i : i-th server.

The start time of the active interval can be found in two steps: finding the start time of the period and determining the time to wait to enter the active interval from the start of the period. According to Figure 1 (b) the value of interest is the start time of the active interval which is denoted by $S_{q,r}$ and we could establish the following equation.

$$S_{q,r} = \alpha_i + H_{q,r} + B_r \qquad (1)$$

The constant α_i of Equation 1 represents the time point where S_i enters the Rotational Lease mode. The active intervals obtained from Equation 1 must possess the following property to ensure the data consistency.

Definition 1. Non-overlapping property : Let $S_{p,r}$ and $S_{q,s}$ be two arbitrary active intervals where p, q are the period numbers and r, s the active interval numbers and $S_{p,r} < S_{q,s}$ holds. If $F_{p,r} \leq S_{q,s}$ is satisfied, then it is said to have the non-overlapping property.

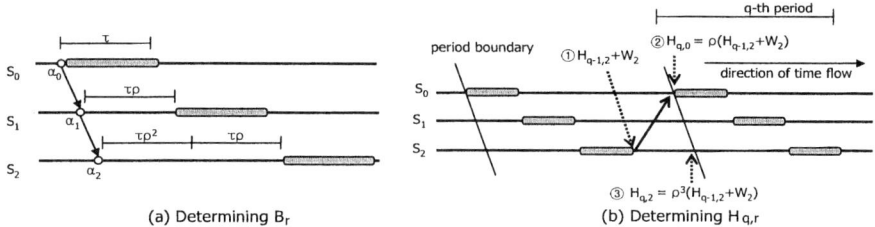

Fig. 2. Determining B_r and $H_{q,r}$

$H_{q,r}$ and B_r on the right hand side of Equation 1 needs to be rewritten in terms of constants, period numbers and active interval numbers. Figure 2 (a) illustrates the idea of transforming B_r and $H_{q,r}$. Generalizing in terms of active interval number r gives us equations for and they are used to claim the following theorem.

Theorem 1. The start time of the active interval acquired by the following equation satisfies the non-overlapping property.

$$S_{q,r} = \alpha_i + H_{q,r} + B_r \text{ where } B_r = \tau \sum_{x=1}^{r} \rho^x \text{ and } H_{q,r} = \rho^{r+1}\left(H_{q-1,m-1} + W_{m-1}\right)$$

3.4 Detection of the Failure

For detecting the moment of failure, we employ heartbeat messages that periodically visit every server once and return to the originator. CXFS[5] and GPFS[4] also use heartbeat messages for failure detection. Heartbeat messages are propagated in the order of active interval assignment. The event of network failure is not known to any servers at the very moment. It is not until the next heartbeat message is initiated that servers detect the network failure. Message originator sends out the message and waits for the return. If the message is not returned within appropriate time limit, it concludes that a network problem has occurred. Other servers detect the failure if the heartbeat message does not arrive from the predefined predecessor within the expected time limit.

3.5 Determining the Rotational Lease Parameters

Since smaller number of active intervals means better throughput, it is desirable to minimize the number of active intervals by assigning as many servers as possible to a single active interval. One way of determining whether two servers have no overlapping region of data set is using the directory structure of the file system. Servers declare their data region by designating a set of directories and it means that the server will only use the data within the subtree. Once this data region is declared, it is transformed into the parent-child relationship graph. In this graph, nodes represent servers and edges represent the parent-child relationship. From this parent-child relation graph we apply the graph coloring algorithm to find the optimal active intervals. The presence of edge means that two nodes have a common data region, and thus they must not be in the same active interval. It is equivalent to assigning a color to a node and assigning different color to adjacent nodes with minimal number of colors. It is known that the graph coloring problem is NP-complete[7]. Many heuristics have been developed, and any of these algorithms can be readily applied to find practical number of active intervals.

4 Simulation and Analysis

Table 1. Parameter used and their value range

Parameter	Actual Value Range	Value Range used
Drift rate	$1+1 \times 10^{-3} \sim 1+1 \times 10^{-6}$ sec/sec	$1+1 \times 10^{-3} \sim 1+1 \times 10^{-6}$ sec/sec
Number of Servers	$10 \sim 100$	10, 30, 50
Active interval length	Several msec ~ several sec	200, 500, 1000, 2000 msec

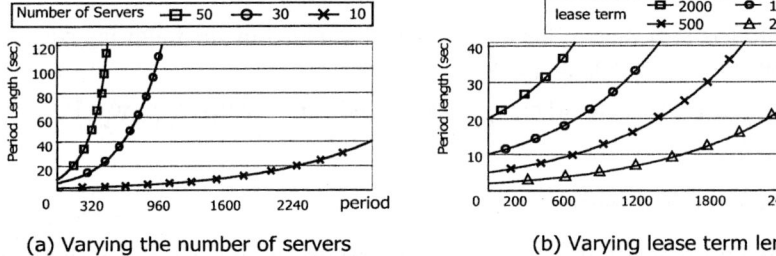

(a) Varying the number of servers
(lease term: 200ms, Drift rate:1.0001)

(b) Varying lease term length
(Num of servers: 10, Drift rate:1.0001)

Fig. 3. The effect of varying parameters

The behavior of the proposed method is examined using a software simulator. In the simulation, we examined the effect of varying parameters. The parameters used are the number of systems, Rotational Lease term, and clock drift rate. Table 1 shows the actual range of these parameters and the value range used in the experiment.

Figure 3 (a) shows the effect of varying the number of servers, and Figure 3 (b) shows the effect of varying lease terms. We see that larger number of servers and larger lease term both cause the period gap to widen faster, but the number of system affects more than the length of lease term. Note that the period length shows an exponential growth. However according to our simulation it showed that the performance degradation due to exponential increase of period length is not serious. The period length stays reasonably low long enough that it gives enough time for physically recovering from the network failure.

5 Conclusion

We proposed a method that improves availability of the shared storage file system by enabling it to continue functioning under the failure of the control network. The proposed method, which we call the Rotational Lease, extends the conventional lease concept by automatically renewing the lease in a predetermined order in the event of network failure. Each server with pre-assigned active interval number claims an access right to the storage during the renewed lease term in rotation.

In order to properly apply the Rotational Lease, other conditions must be considered. It was assumed in the paper that participating servers were able to finish up the file operations within the active interval. However, it is possible that servers may fail to do so due to many causes. In future work some measures will be added to protect data consistency in these exceptional cases.

References

1. Steven R. Soltis, Thomas M. Ruwart, Matthew T.O'Keefe, The Global File System, Proceedings of the Fifth NASA Goddard Space Flight Center Conference on Mass Storage Systems and Technologies, Sept 17-19, 1996
2. R. C. Burns, R. M. Rees, and D. D. E. Long, Safe Caching in a Distributed File System for Network Attached Storage, In Proceedings of the International Parallel and Distributed Processing Symposium (IPDPS), IEEE, 2000.
3. Chang-Soo Kim, Gyoung-Bae Kim, Bum-Joo Shin, "Volume Management for SAN environment", In Proceedings of the International Conference on Parallel and Distributed Systems, 2001
4. Frank Schmuck and Roger Haskin. GPFS: A Shared-Disk File System for Large Computing Clusters. Proceedings of the Conference on File and Storage Technologies (FAST'02), pp. 231-244, 2002
5. CXFS: A high-performance, multi-OS SAN file system from SGI. SGI White Paper. URL http://www.sgi.com/products/storage/tech/file_systems.html.
6. C. Gray and D. Cheriton, "Lease: An efficient fault-tolerant mechanism for distributed file cache consistency," Twelfth ACM Symposium on Operating Systems Principles, pp. 202 210, 1989.
7. Michael R. Garey and David S. Johnson. Computers and Intractability: A Guide to the Theory of NP-Comleteness. W.H.Freeman, 1979.

A New Hybrid Architecture for Optical Burst Switching Networks

Mohamed Mostafa A. Azim[1], Xiaohong Jiang[1], Pin–Han Ho[2], and Susumu Horiguchi[1]

[1] Faculty of Engineering, Graduate School of Information Sciences,
Tohoku University, Japan
[2] Department of Electrical and Computer Engineering, University of Waterloo,
Ontario, Canada
{mmazim, jiang, susumu}@ecei.tohoku.ac.jp,
pinhan@bbcr.uwaterloo.ca

Abstract. Optical Burst Switching (OBS) has been proposed as a new paradigm to achieving a practical balance between Optical Circuit Switching (OCS) and Optical Packet Switching (OPS). Hybrid Optical Networks (HON) that combines OCS and OBS has emerged to grant the OBS network the ability to support connection-oriented applications. Although efficient in terms of reducing the burst loss probability as well as improving the overall network performance, it provides a limited degree of Quality of Service (QoS). What we mean by QoS is the capability of a network to provide a guaranteed level of availability while satisfying strict limits on delay. In this paper, we propose new hybrid architecture for OBS networks. The proposed architecture qualifies the OBS network to support different classes of QoS.

1 Introduction

With the continuing advances of the Wavelength Division Multiplexing (WDM) technology, the amount of raw bandwidth transported through a single fiber optic has increased dramatically to an order of terabit-per-second (Tbps) of effective bandwidth. In order to efficiently harness this bandwidth in a cost-effective way, an appropriate switching architecture must be developed. This switching scheme must be able to handle different traffic patterns with various traffic characteristics as well as differentiated QoS requirements.

Optical Burst Switching (OBS) [1] has emerged as a promising balanced approach between coarse optical circuit switching and fine grain optical packet switching. In OBS, several data packets are assembled together to form a burst. A control packet is sent ahead of the burst on a control channel (separate from data channels) to setup a connection without waiting for acknowledgement for the connection establishment. The data burst follows the header after an offset time; a value is chosen to be long enough to allow the control packet to be processed at each intermediate node while the burst is buffered in the electronic domain at the source node. So, no fiber delay lines (FDLs) are necessary at intermediate nodes to buffer the burst while its header is being processed. Thus OBS is efficiently suitable for short traffic flows that are bursty in nature. However, for large and constant bit rate (CBR) flows, OBS is not appropriate for several reasons. First, due to the control packet processing overhead at intermediate nodes or, equivalently, the offset time during which the data burst must be buffered in electronic domain at the source node. Second, the burst assembly/disassembly overhead as well as the frequent switch fabric reconfiguration to switch a burst obscure the applicability of this technique for transporting such traffic flows.

Therefore, Hybrid Optical Network (HON) architectures that combine both Optical Circuit Switching (OCS) and OBS have been proposed recently [2, 3]. The current architecture of HON classifies the incoming traffic into short-lived (best-effort) and long-lived traffic. With the former, the best-effort traffic that is bursty in nature is supported by OBS, while in the latter; the real-time applications are transported using OCS. Therefore, this architecture consists of two main modules, namely the OBS module (HON-B) and the OCS module (HON-C). The HON-B module is responsible for best effort traffic in the form of small flows. While the HON-C module is responsible for large traffic flows. In [4] the authors proposed a QoS provisioning algorithm using flow-level service classification in the available HON. In particular, they proposed using OBS for transporting short-lived traffic including delay sensitive traffic while using OCS for transporting long-lived traffic including loss-sensitive traffic aiming at maximizing the network utilization while satisfying users' QoS requirements. Although the available HON architectures are efficient in terms of reducing the burst loss probability as well as improving the overall network throughput, they only provide a limited degree of QoS, in terms of the ability of the network to provide a guaranteed capacity (not best-effort) while satisfying the delay limits of the corresponding applications.

Internet traffic is a mixture of voice, real-time video and data. As the traffic load on the Internet grows, and as the variety of applications grows with it, providing differing levels of QoS to different traffic flows is imperative. More specifically, some short-lived traffic are delay sensitive (e.g., IP telephony) while others are not (e.g., FTP, email, web browsing). Similarly, not all long-lived traffic flows are delay insensitive. For example, a safety-critical application, such as *remote surgery* (the ability for a doctor to perform surgery on a patient even though they are not physically in the same location, also known as telesurgery) may require a guaranteed capacity and impose strict limits on maximum delay. Available OBS-based hybrid architectures will not be able to satisfy the versatile requirements of QoS-critical applications in foreseeable future.

In this paper we propose a framework under which an OBS network can support different classes of QoS in terms of traffic characteristics (e.g., bursty or constant bit rate) and performance requirements (e.g., delay sensitivity). In our architecture, delay sensitive traffic can be given preferential treatment as such applications do not tolerate the bottleneck of requesting a bandwidth that can not be reserved due to scarcity of resources or burst loss caused by resource contention at intermediate nodes. In this paper, in order to consider the QoS demands in an OBS network, we introduce the virtual topology concept to the available HON architecture such that the new HON structure combines OBS, OCS and Optical Virtual Circuit Switching (OVCS). With OVCS, the proper set of lightpaths is established constructing the virtual topology. Traffic with strict QoS requirement can be transported right away through an available lightpath. In this way, our proposed architecture will be able to providing an end-to-end guaranteed QoS.

The rest of this paper is organized as follows. In section 2, we review the virtual topology concept. In section 3, we present the proposed architecture for HON as well as the traffic classification model. We conclude the paper in section 4.

2 Virtual Topology Concept

The virtual topology optimization problem has been well studied in the literature [5,6]. Ideally for a network with N nodes, we would like to establish lightpaths between all $N(N-1)$ pairs. However, this is not practically feasible mainly due to the lack of resources (wavelength, transceivers) and the limited nodal degree of the network nodes. Therefore, the goal of virtual topology design problem aims at selecting a set of lightpaths that optimizes the performance (throughput, delay, packet loss,etc.) while satisfying certain constraint, which is proved to be NP-complete [7].

In this paper, the *physical topology* of the network refers to the set of all network nodes and the fiber-optic links connecting them. A *lightpath* between a node pair is an optical channel consists of a path between the node pair as well as a wavelength along all links of that path. A lightpath is established by reserving a wavelength channel on all links and configuring the switching nodes along the path. Two lightpaths that share the same link in the network must use different wavelengths (wavelength continuity constraint).

The virtual topology (also known as logical topology) consists of the set of all network nodes and lightpaths that have been established between all end nodes. A lightpath in the virtual topology is called a virtual link or virtual circuit (VC). An example physical topology is shown in Fig. 1 and its logical topology shown in Fig. 2.

In Fig. 1 and Fig.2 we assume that the network has a single wavelength per fiber. Note that, when there is no wavelength available in certain link in the physical topology, it is deleted from the virtual topology. Therefore, a physical topology and its virtual topology may not contain the same set of physical links.

The selection of the proper candidate lightpaths of the virtual topology as well as when the virtual topology can be reconfigured to be in tune with the dynamic changes in the traffic patterns have a great influence on the network performance. Therefore we dedicate an elaborate study of them, which is not in the scope of this paper.

The purpose of constructing the virtual topology in the new hybrid architecture is to overcome the following shortcomings of the current hybrid architecture.

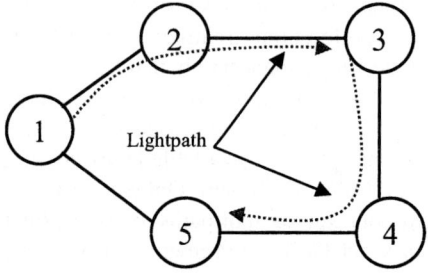

Fig. 1. Physical topology

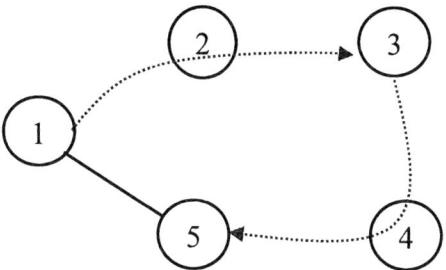

Fig. 2. Virtual topology

First, with HON-B, a connection needs to wait in electronic domain at the source node for a time interval equal to the offset time. This introduces a bottleneck when the connection is rejected several times. Consequently, it may fail to meet the delay requirements of a given request.

Second, HON-B only supports best-effort traffic. This means that, there is no guarantee of delivering this traffic. However, some short-lived traffic requires a guaranteed capacity.

Third, Fairness. The available HON architectures are sacrificing the performance of small flows for maintaining a good performance for large flows [2]. In other words, short-lived traffic will always be assigned to be transported by OBS. However, on the internet, some short-lived traffic (e.g., IP-telephony) requires strict limits on jitter and delay that makes it unsuitable for being transported by OBS. Therefore, we need to ensure that lower priority applications, like e-mail and web surfing, do not impact mission-critical applications such as ATM (Automatic Teller Machine) processing even if they are short lived traffic.

Fourth, although HON-B has been used for transporting delay-sensitive traffic [4], it can not guarantee a bounded delay due to resource contention at intermediate nodes. Even when employing a contention resolution mechanism, there is no guarantee that the burst will be delivered to its destination. It is also notable that, resource contention does not only influence the delivery delay, but also wastes the network bandwidth along the path between the ingress node and the node at which contention occurs, thus reducing the probability of accepting any upcoming traffic.

Therefore we are motivated to propose our hybrid architecture for OBS networks that overcomes the above mentioned shortcomings of the available HON architectures.

3 The New Architecture

The proposed HON architecture is shown in Fig. 3, in which the network consists of 3 main components namely, the optical burst switching (HON-B) component, the circuit switching (HON-C) component and the virtual circuit switching (HON-VC) component, responsible for transporting all classes of traffic. These components may share all wavelength channels, or we may dedicate a subset of the wavelength channels for each component. Our traffic classifier performs two types of classification. At first, we classify the traffic flow according to its traffic characteristics (short/long) then we conduct a finer classification for satisfying the QoS requirements in terms of delay. In our architecture, we provide three main traffic classes.

Traffic class 1: dynamic traffic flows (short-lived or long-lived traffic) with strict delay and/or packet-loss requirements are assigned to this supreme traffic class that uses HON-VC.

Traffic class 2: short-lived traffic that does not require an end-to-end guaranteed QoS will be served by HON-B.

Traffic class 3: long-lived traffic that does not impose a rigid requirement on delay will be suitable for HON-C.

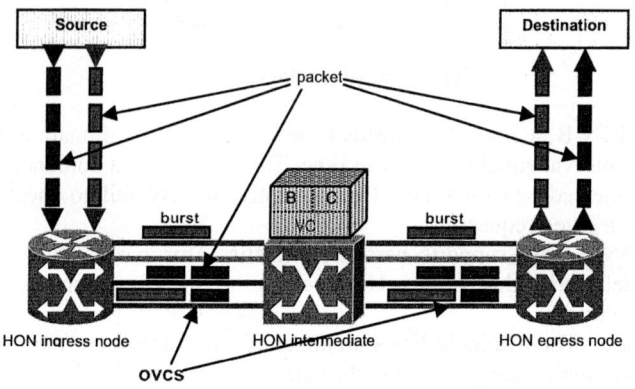

Fig. 3. The New Hybrid Switching Architecture

As mentioned in section 2, with OBS, the burst loss due to the contention degrades the performance in terms of throughput and delivery delay especially under high workloads. In our framework we make use of the HON-VC for resolving resource contention. Specifically, when contention occurs between two connection requests, instead of dropping one of the bursts, it can be rerouted to its destination through the available HON-VC. This introduces a new type of traffic we refer to as "Burst over OVCS". In this case, the burst is switched from the ingress node up to the intermediate node, at which contention occurs through burst switching. Then lightpath switching is used up to the egress node. Fig. 4 shows different traffic classes provided by the proposed architecture. Fig. 5 also shows the traffic classification flowchart at an ingress node of the new HON architecture.

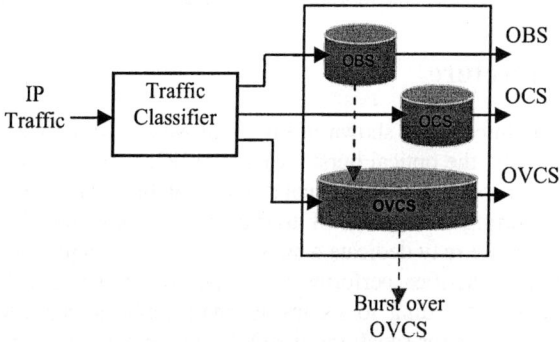

Fig. 4. Different traffic classes

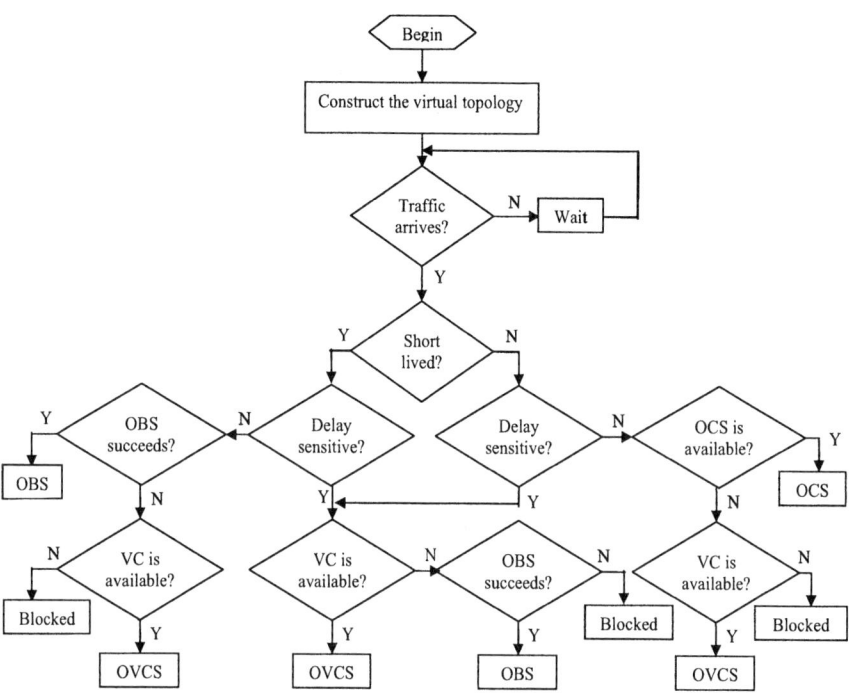

Fig. 5. Traffic classification flowchart at ingress node

The flowchart shown in Fig. 5 presents our proposed model for traffic classification. At first we establish the virtual topology, and then when a traffic flow arrives it is initially classified as short-lived or long-lived. According to the QoS requirement of each traffic flow, a finer classification is done based on the delay sensitivity of the application corresponding to the traffic flow. So, finally we have four groups of traffic and we have to select the most suitable switching technique for each group. Delay sensitive traffic best fits to be transported through OVCS. Therefore, delay-sensitive, short-lived traffic and long-lived traffic groups will be routed through HON-VC. If no VC is available for the corresponding connection, it may be routed through OBS; otherwise, it will be blocked. On the other hand, for non delay-sensitive applications their traffic flows will be routed based on their traffic characteristics. Short-lived traffic that is not sensitive to delay will be routed through HON-B. Long-lived traffic flows that can tolerate delay will be switched to HON-C. In either case, if the traffic can not be routed to the destination due to scarcity of resources, it may be routed through the available HON-VC. Otherwise it is blocked.

In summary, the new hybrid switching architecture aims at providing a guaranteed level of capacity as well as satisfying the user's QoS requirements by utilizing the available VCs. It provides the following advantages.

Traffic does not need to wait for an offset time as the case of OBS or for an acknowledgement after a round-trip propagation delay as the case of OCS. Alternatively; the traffic will be transported right away after being received at an ingress node.

Employing the virtual topology grants an end-to-end assured bandwidth (not best effort) that is suitable for loss-sensitive traffic as well as satisfying the QoS demands of such traffic.

The proposed model for traffic classification gives preferential treatment to delay sensitive traffic while not sacrificing the performance of the short-lived traffic for the sake of maintaining good performance of the long-lived traffic.

The embedded virtual topology will help resolving the resource competition at the contended nodes, thus reducing the overall burst loss probability.

4 Conclusion

In this paper, a new hybrid architecture for OBS networks that supports different classes of QoS has been proposed. Our architecture combines OBS, OCS, and OVCS for providing a finer class of QoS. In the proposed architecture, delay sensitive and packet-loss critical traffic can be given preferential treatment. We also proposed a model for traffic classification at an ingress node of a network. We expect that the framework proposed in this paper will be useful for providing guaranteed capacity as well as satisfying the QoS requirements of the real time applications of the next generation optical internet. Currently we are conducting extensive simulations to support our proposed architecture.

Acknowledgement

This research is partly supported by the Grand-In-Aid of scientific research (B) 14380138 and 16700056, Japan Science Promotion Society.

References

[1] C. Qiao and M. Yoo, "Optical burst switching (OBS)–a new paradigm for an optical Internet," *Journal of High Speed Networks*, vol. 8, no. 1, pp. 69–84, 1999.
[2] Chunsheng Xin, Chunming Qiao, Yinghua Ye, Sudhir Dixit, "A Hybrid Optical Switching Approach", *Proc. Of Globecom2003* vol. 22, no. 1, pp. 3808-3812, Dec 2003.
[3] G. M. Lee, B. Wydrowski, M. Zukerman, J. K. Choi and C. H. Foh, "Performance Evaluation of Optical Hybrid Switching System," Proceedings of IEEE GLOBECOM 2003,pp. 2508-2512, San Francisco, Dec. 2003.
[4] Gyu Myoung Lee, Jun Kyun Choi: Flow Classification for IP Differentiated Service in Optical Hybrid Switching Network. ICOIN 2005, pp. 635-642.
[5] R. Ramaswami and K. N. Sivarajan, "Design of logical topologies for wavelength-routed optical networks," *IEEE J. Select. Areas Commun.*, vol. 40, pp. 840-851, June 1996.
[6] R. Dutta and G. Rouskas, "A Survey of Virtual Topology Design Algorithms for Wavelength Routed Optical Networks," *Opt. Net. Mag.*, vol. 1, pp. 73-89, Jan. 2000.
[7] I. Chlamtac, A. Ganz, and G. Karmi, "Lightpath communications: An approach to high bandwidth optical WAN's," *IEEE Transactions on Communications*, vol. 40, pp. 1171–1182, July 1992.

A Productive Duplication-Based Scheduling Algorithm for Heterogeneous Computing Systems

Young Choon Lee and Albert Y. Zomaya

Advanced Networks Research Group, School of Information Technologies,
The University of Sydney, NSW 2006, Australia
{yclee, zomaya}@it.usyd.edu.au

Abstract. The scheduling problem has been shown to be NP-complete in general cases, and as a consequence many heuristic algorithms account for a myriad of previously proposed scheduling algorithms. Most of these algorithms are designed for homogeneous computing systems. This paper presents a novel scheduling algorithm for heterogeneous computing systems. The proposed method is known as the Productive Duplication-based Heterogeneous Earliest-Finish-Time (PDHEFT) algorithm. The PDHEFT algorithm is based on a recently proposed list-scheduling heuristic known as the Heterogeneous Earliest-Finish-Time (HEFT) algorithm which is proven to perform well with a low time complexity. However, the major performance gain of the PDHEFT algorithm is achieved through its distinctive duplication policy. The duplication policy is unique in that it takes into account the communication to computation ratio (CCR) of each task and the potential load of processors. The PDHEFT algorithm performs very competitively in terms of both resulting schedules and time complexity. In evaluating the PDHEFT algorithm a comparison is made with another two algorithms that have performed relatively well, namely, the HEFT and LDBS algorithms. It is shown that the proposed algorithm outperforms both of them with a low time complexity.

1 Introduction

Task scheduling problems have been extensively studied for many years. However, due to the NP-complete nature of the task scheduling problem in most cases [1] heuristic algorithms account for a myriad of existing scheduling algorithms. Therefore, the time complexity of a task scheduling algorithm is one of the most fundamental factors in determining its quality. In addition to time complexity, the minimization of the schedule length is another main objective of a task scheduling algorithm. Hereafter, scheduling and task scheduling are used interchangeably.

Heuristic based scheduling algorithms are normally the ones favored by a large number of researchers. Three major sub categories of heuristic based scheduling are list scheduling, clustering and task duplication. List scheduling in the heuristic based category is preferred to other scheduling techniques. This is due to the fact that list scheduling algorithms [2], [3], [4], [5], [6], [7], [8] tend to produce competitive solutions with lower time complexity compared to those of the algorithms in the other subcategories [9]. The two fundamental phases commonly found in list scheduling are task prioritization and processor selection.

This paper proposes a duplication-based scheduling algorithm known as the Productive Duplication-based Heterogeneous Earliest-Finish-Time (PDHEFT). It is based on a recent list-scheduling algorithm, HEFT. The PDHEFT algorithm schedules tasks in a task graph for heterogeneous computing systems with a distinctive duplication policy. The duplication policy is unique in that it takes the communication-to-computation cost ratio (CCR) of each task and the potential load of processors into consideration in order to avoid redundant duplications that might increase the schedule length. Despite the adoption of an additional duplication phase the time complexity of the PDHEFT algorithm still remains the same as that of the HEFT algorithm but with better quality schedules.

The target system used in this work consists of heterogeneous processors/machines that are fully interconnected. The inter-processor communications are assumed to perform with the same speed on all links without contentions. It is also assumed that a message can be transmitted from one processor to another while a task is being executed on the recipient processor which is possible in many systems.

The remainder of this paper is organized as follows. Section 2 introduces some background material on scheduling problems. The proposed algorithm is described in great detail in Section 3. In Section 4, the evaluation results are presented and explained with conclusions following in Section 5.

2 Scheduling Problem

Parallel programs, in general, can be represented by a directed acyclic graph (DAG). A DAG, $G = (V, E)$, consists of a set of v nodes, V, and e edges, E. A DAG is also called a task graph or macro-dataflow graph. In general, the nodes represent tasks partitioned from an application and the edges represent precedence constraints. An edge $(i, j) \in E$ between task n_i and task n_j also represents the inter-task communication. More specifically, the output of task n_i has to be transmitted to task n_j in order for task n_j to start its execution. A task with no predecessors is called an entry task, n_{entry}, whereas an exit task, n_{exit}, is one that does not have any successors.

A task is called a *ready task* if all of its predecessors have been completed. A *level* is associated with each task. The level of a task is defined to be:

$$level(n_i) = \begin{cases} 0, & if\ n_i = n_{entry} \\ \max_{n_j \in imed\_pred(n_i)} \{level(n_j)\} + 1, & otherwise \end{cases} \quad (1)$$

where $imed\_pred(n_i)$ is the set of immediate predecessor tasks of task n_i.

The weight on a task, n_i denoted as w_i represents the computation cost of the task. In addition, the computation cost of a task, n_i is on a processor, p_j is denoted as $w_{i,j}$.

The weight on an edge, denoted as $c_{i,j}$ represents the communication cost between two tasks, n_i and n_j. However, communication cost is only required when two tasks are assigned to different processors. In other words, the communication cost when they are assigned to the same processor can be ignored, i.e., 0. The average computation cost and average communication cost of a task, n_i are denoted as $\overline{w_i}$ and $\overline{c_i}$, respectively. The former is the average computation cost of task n_i over all the processors in a given system. The latter is the average communication cost between task n_i and its successor tasks.

Three frequently used task prioritization methods are *t-level*, *b-level* and *s-level*. The t-level of a task is defined as the summation of the computation and communication costs along the longest path of the node from an entry task in the task graph. The task itself is excluded from the computation. In contrast, the b-level of a task is computed by adding the computation and communication costs along the longest path of the task from an exit task in the task graph (including the task). The only distinction between the b-level and s-level is that the communication costs are not considered in the s-level.

The CCR is a measure that indicates whether a task graph is communication intensive, computation intensive or moderate. For a given task graph, it is computed by the average communication cost divided by the average computation cost on a target system. The CCR of a task n_i is defined by:

$$CCR(n_i) = \frac{\sum_{n_j \in imed\_succ(n_i)} c_{i,j}}{S_i} / \overline{w_i} \qquad (2)$$

where S_i corresponds to the number of the immediate successors of task n_i.

If a set P of p processors exists for scheduling, the earliest start time and the earliest finish time of a task n_i on a processor p_j are denoted as $EST(n_i, p_j)$ and $EFT(n_i, p_j)$, respectively. The earliest start time of task n_i on processor p_j is defined to be whichever is the maximum: the earliest available time of the processor or the communication completion time of the task of its predecessors. The predecessor of a task n_i from which the communication completes at the latest time is called the most influential parent (MIP) of the task denoted as $MIP(n_i)$. The communication completion time, also called data arrival time of task n_j from task n_i is denoted as $CCT(n_i, n_j)$. The schedule length (SL), also called *makespan*, is defined as $max\{EFT(n_{exit})\}$ after the scheduling of v tasks in a task graph G is completed.

3 Proposed Algorithm

3.1 Algorithm Description

The PDHEFT algorithm uses the same fundamental operations used by the HEFT algorithm, such as task prioritization and processor selection. However, the major performance gain of the PDHEFT algorithm is from the distinctive duplication policy it proposes. The decision of the duplication is made based on two factors: the CCR of a task and the potential load of the processors. These factors differentiate the PDHEFT algorithm from other duplication based scheduling algorithms. The PDHEFT algorithm consists of three main phases:

- Task Prioritization Phase – assigns priorities and levels to tasks and arranges the tasks in decreasing order by b-level value.
- Processor Selection Phase – selects the processor on which the task finishes the earliest.
- Duplication Phase – determines which task to be duplicated by taking into account the CCR of each task and the load of the processors in the target system.

The workings of the PDHEFT algorithm are given in Fig. 1.

---

1. Compute average communication and computation costs of tasks
2. Compute b-level values, levels and the number of tasks in each level of all tasks
3. Sort the tasks in a scheduling list in decreasing order by b-level value
4. **while** there are unscheduled tasks in the list **do**
5.   Select the first unscheduled task, n_i, from the list
6.   **for** each processor p_j in P **do**
7.   Compute $EFT(n_i, p_j)$
8.   Add $EFT(n_i, p_j)$ to the processor reference list, PR in increasing order by finish time
9.   **endfor**
10.   Assign task n_i to the processor, p^* that minimizes its finish time
11.   Remove p^* from PR
12.   **if** task has out-degree of two or more and the number of parallelized tasks in the same level of that of task n_i is not greater than P **then**
13.   Duplicate task n_i as many as its out-degree based on duplication criteria
14.   **endif**
15. **endwhile**

Fig. 1. The PDHEFT algorithm

The time complexity of the PDHEFT algorithm is $O(ep)$ which is identical to that of the HEFT algorithm. For a dense graph, the number of edges is proportional to $O(v^2)$. Thus, the time complexity of the PDHEFT algorithm is in $O(v^2p)$. This is far lower than the time complexities of the two versions of the LDBS algorithm, that are $O(v^3ep^3)$ and $O(v^3ep^2)$.

3.2 Duplication Policy

The duplication of a task is considered as soon as the task is scheduled for the first time using the processor selection procedure of the PDHEFT algorithm.

The duplicability of the task is then checked based on its CCR and out-degree. A task with an out-degree of two or more is regarded as a candidate for duplication. In order not to increase the time complexity of the PDHEFT algorithm, duplications apply only to the task currently being scheduled. In addition, the necessary information for duplication is obtained during processor selection phase.

As mentioned earlier, the duplication policy developed in this research introduces two distinct measures to determine if the duplication of a task is allowed. The first measure is the CCR of each task in a task graph. The CCR of a task rather than the CCR of a task graph is used because in the PDHEFT algorithm the duplication of a task only concerns the task and its successor tasks, i.e. a sub-graph. It was assumed that the CCR of this sub-graph, therefore, is a more significant factor than the CCR of the task graph. This leads to the fact that there might be a noticeable difference between the CCR of a task and that of the task graph. It is to be noted that duplicating tasks in a 'computation intensive' task graph is normally impractical.

Moreover, duplicating computation intensive tasks in a 'communication intensive' task graph may cause an increase of the output schedule. The CCR of a task is used to compute the *allowance* of the finish time of the task when duplicating. This means that the finish time of a task varies in a heterogeneous computing system depending on the processor that executes it. This allowance is an additional time that can be provided for the finish time of the task. If its finish time on the particular processor on which it is being duplicated is substantially larger than its minimal finish time the duplication of the task has a high possibility of being an unproductive. The allowance computation algorithm is presented in Fig. 2.

A task is classified into four types: computation intensive, moderate, communication intensive and extremely communication intensive. This classification is conducted based on the CCR of the task. The thresholds of the four types are shown in Fig. 2. These thresholds may need to be changed according to the target application model.

---

if CCR of task $n_i < 0.5$ **then**   /* computation intensive */
  Let *allowance* = $\overline{c_i}$
else if CCR of task $n_i < 1.0$ **then**   /* moderate */
  Let *allowance* = $\overline{w_i}$
else if CCR of task $n_i < 5.0$ **then**   /* comm. intensive */
  Let *allowance* = $\overline{c_i}$
else   /* extremely communication intensive */
  Let *allowance* = $\overline{c_i}$ / 2
endif

Fig. 2. The algorithm for computing allowance

In addition to CCRs of tasks, the potential load of the processors in a given system is used to predict whether duplicating a task becomes a source of delays or interruptions for the execution of remaining unscheduled tasks.

More precisely, if there are more tasks in a level than the number of processors the tasks in that level are not considered for duplication.

The duplication algorithm is shown in Fig. 3. The Duplicate function is called if a task satisfies the second measure of the PDHEFT's duplication policy; that is, the function is called if the number of tasks in the same level as that of the task is no greater than the number of processors.

```
1. Duplicate( )
2. Compute the allowance of task n_i
3. while task n_i is not duplicated as many as the number of its out-degree and there
   are unchecked processors do
4.    Select p_j from PR
5.    Let allowedftime = EFT(n_i, p*) + allowance
6.    if EFT(n_i, p_j) < allowedftime then
7.       Duplicate task n_i on p_j
8.    endif
9. endwhile
10.end
```

Fig. 3. The duplication algorithm of the PDHEFT algorithm

4 Performance Results and Comparison

The comparative evaluation of the PDHEFT algorithm is presented in this section. Comparisons have been conducted between two previously proposed scheduling algorithms, HEFT and LDBS [10], and the PDHEFT algorithm. The two former are chosen because they have been shown to deliver competitive output schedules. In addition, their target system configurations are the same as those used for the PDHEFT algorithm.

The two performance metrics used for comparison are the *normalized schedule length* (NSL) and time complexity. Typically, the schedule length of a task graph generated by a scheduling algorithm is used as the main performance measure of the algorithm. The normalized schedule length is defined as:

$$NSL = \frac{\text{schedule length obtained by a particular algorithm}}{\text{schedule length obtained by the PDHEFT}} \quad (3)$$

4.1 Test Parameters

The proposed algorithm and the two previously proposed algorithms, HEFT and LDBS are extensively experimented with various types of both randomly generated and well-known application task graphs. The three well-known parallel applications used for our experiments are the Laplace equation solver [11], the LU-decomposition [12] and Fast Fourier Transformation [13]. The numbers of random and well-known application task graphs are 1566 and 270, respectively. The common parameters used to populate the variations of the task graphs are:

- 9 different CCRs of 0.1, 0.2, 0.3, 0.5, 1.0, 2.0, 3.0, 5.0 and 10.0,
- 3 different processor heterogeneity values of 100, 200 and random.

The processor heterogeneity value of 100 is defined to be the percentage of the speed difference between the fastest processor and the slowest processor in a given system.

4.2 Performance Results with Random Task Graphs

The test results obtained from the random task graphs are presented in two different categories. The first category is as shown in Fig. 4 where comparisons between the three algorithms are conducted with various graph sizes on a computing system consisting of 20 heterogeneous processors. In the second category, an increasing number of processors are used as shown in Fig. 5.

Although the tests for each category are carried out with nine different CCRs as mentioned in Section 4.1, three significant test results are presented. As shown in Figs. 4 and 5, they are CCRs of 0.1, 1.0 and 10.0.

4.2.1 Comparisons with Various Graph Sizes

It is clearly shown in Fig. 4 that the PDHEFT algorithm delivers quite competitive schedule lengths irrespective of different graph sizes and CCRs. The schedule lengths obtained from communication intensive and moderate task graphs shown in Figs. 4a and 4b indicate that the PDHEFT algorithm best suits task graphs consisting of fine-grain tasks with large communication costs. This is also true for the LDBS algorithm. However, its performance drops noticeably with computation intensive task graphs as shown in Fig. 4c. The main reason for this is because LDBS does not take CCR into account. It is observed that many of the duplications tend not to contribute to shortening schedule lengths when scheduling computation intensive task graphs. The PDHEFT algorithm, however, overcomes this drastic degradation by restricting redundant duplications. It, therefore, tends to give shorter schedule lengths for computation intensive task graphs compared to those generated by LDBS.

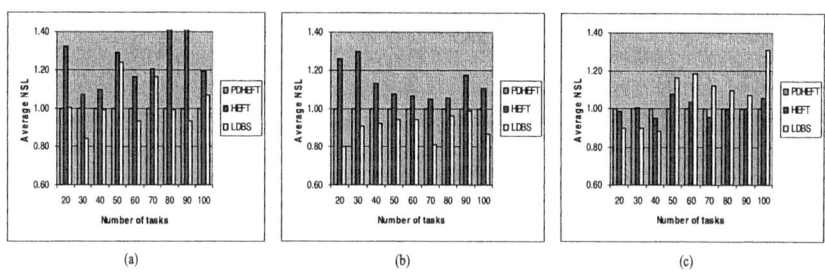

Fig. 4. Average NSL of DAGs with (a) CCR = 10/1, (b) CCR = 1/1 and (c) CCR = 1/10 under the PDHEFT, HEFT and LDBS algorithms with respect to graph size

The average schedule length of the PDHEFT algorithm computed based on the first test set shown in Fig. 4 is 11% on average and 23% at best smaller than that of the HEFT algorithm. It is observed that the LDBS algorithm delivers 6% smaller average schedule length than that of the PDHEFT algorithm for communication intensive and moderate task graphs. However, the PDHEFT generates an average schedule length that is 4% on average, 24% at best and 20% for computation intensive task graphs smaller than that of the LDBS algorithm.

4.2.2 Increasing Number of Processors

The test results shown in Fig. 5 are obtained from the second test set, 1323 task graphs. The patterns that were found in Fig. 4 are re-confirmed in Fig. 5. First, the PDHEFT algorithm performs very reliably regardless of the different characteristics of task graphs. Second, as shown in Fig. 5a the LDBS algorithm tends to deliver longer schedule lengths than that of the PDHEFT algorithm when the processors in a given system are overloaded even though task graphs are communication intensive.

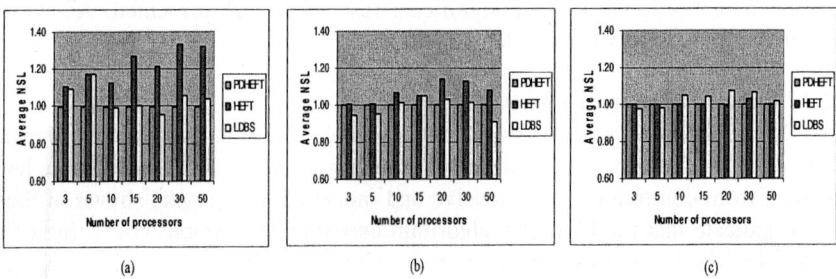

Fig. 5. Average NSL of DAGs with (a) CCR = 10/1, (b) CCR = 1/1 and (c) CCR = 1/10 under the PDHEFT, HEFT and LDBS algorithms with respect to increasing number of processors

Finally, the duplication for computation intensive task graphs does not improve the performance of the LDBS algorithm in terms of schedule length. The average schedule length of the HEFT algorithm for computation intensive task graphs as shown in Fig. 5c also proves that the LDBS algorithm performs some redundant duplications. With the second test set, the PDHEFT algorithm overall outperforms the HEFT and LDBS algorithms. The average schedule length of the PDHEFT algorithm is 8% on average and 18% at best smaller than that of the HEFT algorithm. The average schedule length of the LDBS algorithm is 2% on average longer than that of the PDHEFT algorithm.

4.3 Performance Results with Well-Known Application Task Graphs

The performance results of the PDHEFT algorithm obtained from the experiments conducted with a wide range of different task graphs of the three well-known applications once again confirm its better practicability over the other two algorithms. As shown in Fig. 6 the PDHEFT algorithm achieves a consistent performance irrespective of various types of task graphs.

Note, that the LDBS algorithm with a large number of processors tends to outperform both the PDHEFT and HEFT algorithms. This in fact indicates the impracticability of the LDBS algorithm. More specifically, when there are relatively more tasks in a task graph than the number of processors in a given system, which is quite normal in practice, the LDBS algorithm tends to generate a longer schedule length compared to that of the PDHEFT and HEFT algorithms. Moreover, the performance of the LDBS algorithm drops noticeably when a task graph contains a number of levels on each of which many tasks have the same or similar upward rank as shown in Fig. 6b,

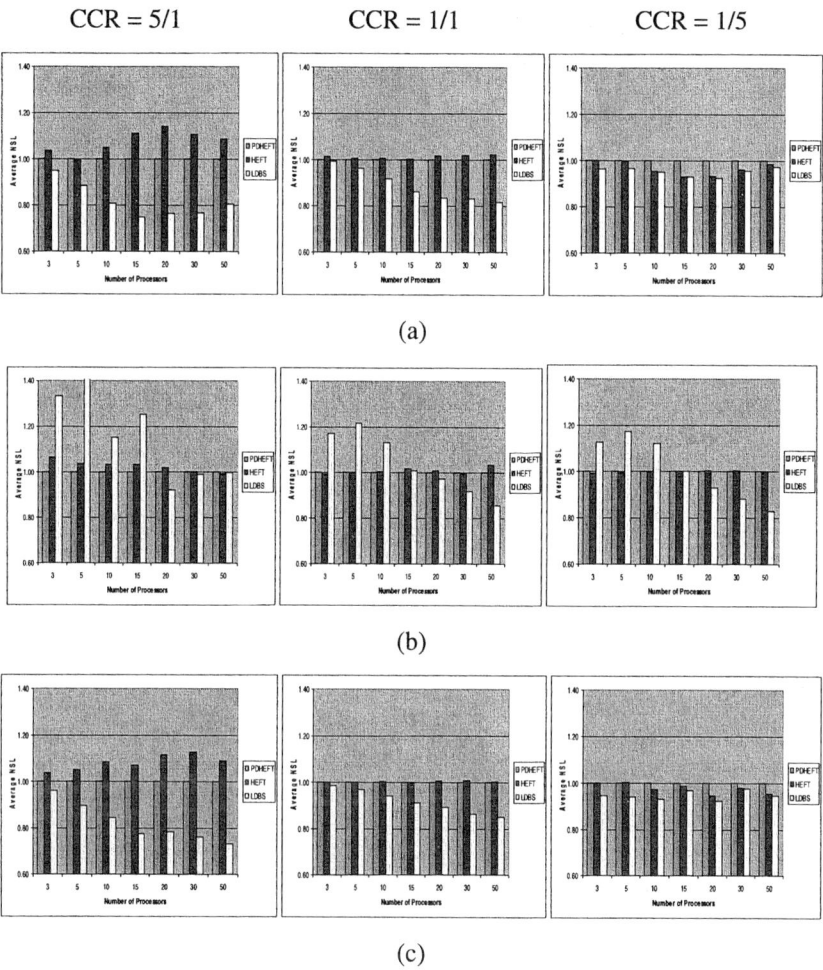

Fig. 6. Average NSL of Well-known Application DAGs (a) Laplace (b) LU (c) FFT

the performance results with the LU task graphs. This shortcoming occurs because the LDBS algorithm tries to duplicate even though the load of the processors is high. This leads to the necessity of predicting the load of the processors that is one of the main characteristics of PDHEFT's duplication policy.

5 Conclusion

In this paper, a new duplication based scheduling algorithm, called the PDHEFT algorithm was presented, for heterogeneous computing systems. The algorithm is based on a previously proposed and well-known algorithm, called the HEFT algorithm. A number of intensive experiments with various test configurations have been conducted. Based on the test results, the PDHEFT algorithm showed its practicability and mostly outperformed existing algorithms including HEFT and LDBS. Because of its robust duplication policy, it delivers competitive schedule lengths regardless of the characteristics of task graphs and the processor configuration in a given system. The robust duplication policy is achieved by taking two very influential factors of the scheduling process into account. They are the CCR of a given task graph and the load of the processors in a given system. In addition to the high quality of the output schedules the low time complexity of the proposed method should be highly attractive.

Acknowledgements

Professor Albert Y. Zomaya's work is supported by an Australian Research Council grant no. DP0452884.

References

1. Garey, M.R. and Johnson, D.S.: Computers and Intractability: A Guide to the Theory of NP-Completeness. W.H. Freeman and Co. (1979) 238–239.
2. Topcuoglu, H., Hariri, S. and Wu, M.: Performance-Effective and Low-Complexity Task Scheduling for Heterogeneous Computing. IEEE TPDS. V. 13. No. 3. (2002) 260-274.
3. Radulescu, A. and Gemund. A.J.C.: Fast and effective task scheduling in heterogeneous systems. Proc.HCW (2000) 229-238.
4. Radulescu, A. and Gemund, A.J.C.: On the complexity of list scheduling algorithms for distributed memory systems. Proc. 13th ACM Int'l Con. Supercomputing. (1999) 68–75.
5. Radulescu, A. and Gemund, A.J.C.: FLB: Fast Load Balancing for distributed-memory machines. Proc. ICPP. (1999) 534–541.
6. Sih, G.C. and Lee, E.A.: A Compile-Time Scheduling Heuristic for Interconnection-Constrained Heterogeneous Processor Architectures. IEEE TPDS. V. 4. No. 2. (1993) 175–187.
7. Kruatrachue, A. and Lewis, T.G.: Grain Size Determination for Parallel Processing. IEEE Software. (1988) 23–32.
8. Hwang, J.J., Chow, Y.C., Anger, F.D. and Lee, C.Y.: Scheduling Precedence Graphs in Systems with Interprocessor Communication Times. SIAM J. Computing. V. 18. No. 2. (1989) 244–257.
9. Kwok, Y.K. and Ahmad, I.: Benchmarking the Task Graph Scheduling Algorithms. Proc. First Merged Int'l Parallel Symp./Symp. Parallel and Distributed Processing Conf. (1998) 531–537.
10. Dogan, A. and Ozguner, R.: LDBS: A Duplication Based Scheduling Algorithm for Heterogeneous Computing Systems. Proc. ICPP. (2002) 352–359.
11. Wu, M.-Y. and Gajski, D.D.: Hypertool: A Programming Aid for Message-Passing Systems. IEEE TPDS. V. 1. No. 3. (1990) 330-343.
12. Lord, R.E., Kowalik, J.S., and Kumar, S.P.: Solving Linear Algebraic Equations on an MIMD Computer. J. ACM. V. 30. No. 1. (1983) 103-117.
13. Cormen, T.H., Leiserson, C.E., and Rivest, R.L.: Introduction to Algorithms. MIT Press. (1990).

Memory Subsystem Characterization in a 16-Core Snoop-Based Chip-Multiprocessor Architecture

Francisco J. Villa, Manuel E. Acacio, and José M. García

Departamento de Ingeniería y Tecnología de Computadores,
Universidad de Murcia, 30071 Murcia, Spain
{fj.villa, meacacio, jmgarcia}@ditec.um.es

Abstract. In this paper we present an exhaustive evaluation of the memory subsystem in a chip-multiprocessor (CMP) architecture composed of 16 cores. The characterization is performed making use of a new simulator that we have called DCMPSIM and extends the Rice Simulator for ILP Multiprocessors (RSIM) with the functionality required to model a contemporary CMP in great detail.

To better understand the behavior of the memory subsystem, we propose a taxonomy of the L1 cache misses found in CMPs which subsequently we use to determine where the hot spots of the memory hierarchy are and, thus, where computer architects have to place special emphasis to improve the performance of future dense single-chip multiprocessors, which will integrate 16 or more processor cores.

Keywords: Dense chip-multiprocessors, memory subsystem, snoop-based cache-coherence, high-performance interconnection networks.

1 Introduction

As integration scales grows, the number of transistors available on a die is increasingly becoming larger, and billion transistor chips will soon be possible. The role of computer designers is to translate all this raw potential into increased computational power. For this, efficient architectures must be designed. One of the approaches recently proposed in the literature to make efficient use of this huge number of transistors are single chip-multiprocessors [1,2]. A CMP integrates several processor cores onto a chip, as well as other resources such as the cache hierarchy and the interconnection network. As a result of this, the traditional advantages of parallel architectures are kept, but some of their drawbacks, such as wire delays or network latencies, are minimized.

The viability and importance of chip multiprocessors is further supported by a number of recently announced commercial, small-scale CMP designs [3,4]. However, nowadays the best organization of the components in this kind of architecture is not clear, and there is still much work to be done in order to improve the performance and maximize the utilization of the resources in a

CMP. Besides, state-of-the-art CMPs and recent commercial releases integrate 2, 4 or at most 8 processor cores onto the chip, but they do not deal with future dense-CMPs (or D-CMPs), in which 16 processor cores or more are expected to be integrated. D-CMPs impose new restrictions that do not appear in current chip-multiprocessors and, thus, it is necessary to evaluate the problems of this kind of architecture.

This paper presents an exhaustive evaluation of the memory subsystem in a chip-multiprocessor (CMP) architecture composed of 16 cores. To the best of our knowledge this is the first characterization for a dense-CMP architecture with a detailed execution model for each core. As we are specially interested in the behavior of the memory hierarchy, we propose a taxonomy of the private L1 cache misses in terms of how these misses are satisfied. This taxonomy provides statistics which help us to identify the bottlenecks of a future chip multiprocessor. The evaluation is accomplished making use of a new simulator which is an extension of the well-known RSIM (Rice Simulator for ILP Multiprocessors) [5] and that we have called DCMPSIM. It models accurately a CMP with a configurable number of cores, private L1 caches connected together via a split-transaction bus and a shared, multibanked L2 cache.

The key contributions of this work are:

- We perform a detailed memory subsystem evaluation of a 16-core snoop-based CMP architecture when executing parallel workloads. This contrasts with the majority of previously published works, as CMPs has been used usually to execute multiprogrammed workloads.
- We present a novel taxonomy of the L1 cache misses found in a CMP.
- We show that the main bottleneck of a 16-core snoop-based D-CMP is the shared bus. Besides, we see that the vast majority of transactions snooped by the cache controllers are not concerned with the lines in that cache, which induces too much unnecessary work.

The rest of the paper is organized as follows. Section 2 summarizes the related work. Section 3 describes briefly some implementation issues of the new simulator. We present a taxonomy of L1 cache misses found in CMPs in Section 4. In Section 5 we show a detailed performance evaluation of a CMP composed by 16 cores. Finally, Section 6 contains our main conclusions and outlines some future work.

2 Related Work

Some works have evaluated the usefulness of the coherence actions obtaining similar results than ours in the context of a 4-way SMP in [6] and in the context of a 16-core CMP [7]. There are some other recent works in the literature dealing with the performance of the memory hierarchy in CMP architectures. Beckmann and Wood present several proposals to reduce the impact of wire delays on large, shared L2 caches in future CMPs [8]. They implement a directory-based coherence protocol and a 2D-mesh interconnection network, evaluating a CMP composed of 8 processor cores.

Liu et al. [9] study several L2 cache organizations in order to increase the utilization (and in last term, the performance of the memory hierarchy) of this structure. They propose a mechanism that dynamically assigns L2 splits to each processor. In this way, they obtain better utilization of the L2 cache, taking into account the demands of each processor at every moment. However, they do not consider other possible bottlenecks, such as the presence of the shared bus. As in the previous work, they simulate a CMP composed of 8 cores. In [10] the authors propose a central coherence unit and a new cache coherence protocol in order to reduce shared-bus transaction time. They evaluate several configurations with a variable number of processors, ranging from 2 to 8. However, the integration scale and memory subsystem latencies assumed in the paper differ highly from those found nowadays, so their results are not comparable with ours.

Finally, some other authors have studied the memory hierarchy when combining chip-multiprocessing with speculative multithreading, although most of them focus on the support for thread-level memory speculation [11,12,13]. Among them, Yanagawa et al. [14] perform a complexity analysis of a cache controller designed by extending a MSI controller to support thread-level memory speculation. They use a directory-based mechanism to maintain coherence, and find that the main component of memory latency is the delay incurred when accessing the directory.

3 DCMPSIM: A Detailed CMP Simulator Based on RSIM

DCMPSIM models the architecture shown in Figure 1. In this architecture, we have an arbitrary number of processor cores, each one with its unified L1 instruction/data cache. All the cores share a unified, multibanked L2 cache through a split-transaction bus. Finally, main memory is interleaved, and each L2 bank is connected to a main memory module.

As we stated in Section 1, our tool has been derived from Rice Simulator for ILP Multiprocessors (RSIM) [5]. RSIM implements a directory protocol [15] in order to keep cache-coherence between nodes; directory protocols are commonly employed in machines composed of a large number of processors (in these cases, it is imperative that the coherence protocol scales with the number of processors, which is not possible with snoop-based protocols) or when there is an unordered interconnection network, such as a mesh or a torus, so it is not possible to rely in the network to ensure a partial order of the memory references.

However, our CMP architecture interconnects the private L1 caches via a shared, ordered, split-transaction bus. In this case, cache-coherence can be maintained by using a snoop-based protocol. As most CMPs do, we have implemented the MOESI snooping protocol [16]. MOESI protocol introduces an *owned* state which is used when one or more caches have a valid copy of the line, main memory does not hold a valid copy, and thus, one of the caches with a copy of the line is responsible for providing the line when it is requested by another processor. This mechanism is an optimization of the MESI protocol, which tries to take advantage of the shorter access time of a small cache when compared with a bigger memory structure.

When implementing a snoop based protocol, it is necessary to adopt some design decisions concerning how conflicting requests are managed. In our design, the L1 cache controllers snoop replies as well as requests (this is not the case of L2 cache controller, which only snoops requests), so read requests to the same location are optimized. When a cache controller sees that there is an outstanding read to the same line, it does not put the new request into the bus, but obtains the value through the reply to the original request. In the case of an outstanding write, subsequent reads or writes are not allowed to proceed until the previous write is completed. These design decisions are commonly taken in most SMP designs [17].

4 A Taxonomy of the L1 Cache Misses Found in CMPs

We are interested in measuring the latency of L1 misses in order to evaluate the performance of the memory subsystem. L1 miss latency can be divided into three categories, corresponding to the cycles spent at the L1 cache controller, the shared bus and those spent obtaining data. More specifically, these three components are the following:

- $T_{controller}$ is the time spent in the L1 cache controller. It includes the time spent until the request is inserted in the bus as well as the time spent processing the corresponding reply.
- T_{bus} includes the time taken to obtain the bus and to transmit the packets.
- T_{mem} is the time needed to get data from the structure which provides it (it can be another L1 cache, one of the L2 cache banks or main memory).

Furthermore, to identify more clearly where the hot spots of the memory hierarchy are, we present a taxonomy of the L1 cache misses found in a CMP

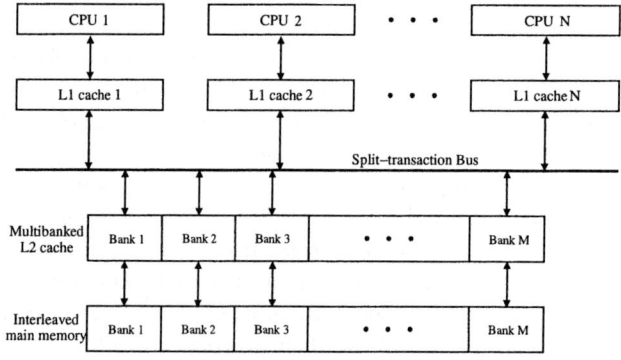

Fig. 1. The CMP architecture implemented

architecture (more generally, a taxonomy of the misses of the lower level of private caches) in terms of which memory structure provides the requested data. In this way, we can determine whether a structure is more critical than another and, even more, how future memory hierarchy optimizations will affect each memory component. The classification assumes a MOESI coherence protocol and identifies four categories:

1. Misses satisfied by another L1 cache (or $-to-$ misses): the line is in a single cache, or the line is in several caches and one of them has the line in *Owned* state.
2. Misses satisfied by the L2 cache (Hit L2 misses): no L1 cache can provide the line, and the L2 has a valid copy of it.
3. Misses satisfied by main memory (Mem misses): neither L1 caches nor the L2 cache have a copy of the line, so data must be obtained accessing main memory.
4. Invalidation or upgrade misses (Inv misses): the faulting cache has a valid copy of the line in *Shared* state but it tries to write the memory line, for which exclusive access is needed. It is necessary to place a *BusUpgrade* request in the bus to invalidate the rest of the copies of the line (if any) and gain exclusive access to it. For this kind of misses, the T_{mem} component of the latency is zero because no data is needed.

5 Experimental Results

In this section, we present a detailed performance evaluation of a CMP composed of 16 out-of-order processor cores similar to the MIPS R10000 processor with an optimized implementation of the sequential consistency memory model that includes load speculation and allows stores to graduate before completion. The architecture is similar to Piranha [1] but twice as many processors are simulated. In Table 1 we can see the configuration of the architecture we have evaluated. L1 cache sizes have been set commensurate to the total number of cores.

Table 1. Base architecture configuration

Parameter	Value
Number of cores	16
L1 size	8KB
L1 associativity	4-way
L1 latency	1 cycle tags + 1 cycle data
L2 size	2MB
Number of L2 banks	8
L2 associativity	8
L2 latency	2 cycles tags + 8 cycles data
Line size	32 bytes
Memory latency	120 cycles
Bus Arbitration	3 cycles
Bus cycle	3 cycles
Bus width	32 bytes

Table 2. Applications and input sizes used in this work

Application	Input size
BARNES-HUT	4096 bodies, 4 time steps
EM3D	38400 nodes, degree 2, 15% remote, 25 time steps
FFT	256K complex doubles
OCEAN	130x130 ocean
RADIX	1M keys, 1024 radix
UNSTRUCTURED	Mesh.2K, 5 time steps
WATER-NSQ	512 molecules, 4 time steps

Table 3. Classification of L1 accesses (on average)

Application	L1 hit	MSHR Coalesced	Bus Coalesced	$-to-$	L2 Hit	Mem.	Inv.
BARNES-HUT	75.6%	11.1%	0.25%	0.09% (0.67%)	13.17% (99.1%)	0.002% (0.02%)	0.03% (0.2%)
EM3D	74.6%	9.5%	0.001%	0.46% (1.77%)	10.05% (38.74%)	15.42% (59.47%)	0.004% (0.02%)
FFT	82.58%	11.16%	0%	0.0006% (0.01%)	2.95% (47.13%)	3.31% (52.85%)	0.0003% (0.005%)
OCEAN	75.87%	12.87%	0.06%	0.71% (6.38%)	8.5% (75.94%)	1.75% (15.59%)	0.24% (2.1%)
RADIX	89.02%	3.11%	0%	0.06% (0.71%)	5.74% (72.95%)	2.06% (26.2%)	0.01% (0.13%)
UNSTRUCT	81.42%	6.37%	0.05%	6.88% (56.62%)	4.34% (35.7%)	0.01% (0.12%)	0.91% (7.56%)
WATER-NSQ	86.56%	9.95%	0.002%	1.91% (54.66%)	1.57% (44.84%)	0.0001% (0.001%)	0.02% (0.5%)

Table 2 describes the benchmarks we have used in our experiments. This set of parallel scientific applications covers a variety of computation and sharing patterns. BARNES-HUT, FFT, OCEAN, RADIX and WATER-NSQ belong to the SPLASH-2 benchmark suite [18]. EM3D is a shared memory implementation of the Split-C benchmark [19]. UNSTRUCTURED is a computational fluid dynamics application [20]. The input sizes for the applications have been chosen taking into account the cache sizes and number of cores in the baseline architecture.

We can see, in Table 3, a classification of how the accesses to the L1 cache are solved. Most of these accesses are captured at the L1 cache, whose hit rates range from 74.6% to 89.02%. The number of occasions in which an access matches an outstanding request issued by the same processor (MSHR Coalesced) is greater than 10% for some applications (this is a consequence of using an optimized implementation of sequential consistency); however, the number of accesses matching an outstanding request issued by another processor (Bus Coalesced) is zero or near to zero for all the applications. The last four columns correspond to the taxonomy presented in Section 4. We show the percentage over the total number of accesses to the L1 caches and, in brackets, the percentage over the total number of misses.

For two applications (UNSTRUCTURED and WATER-NSQ) a significant number of misses are satisfied by another L1 cache ($-to-$ misses). For the remaining applications, most misses are satisfied by the L2 cache, except in

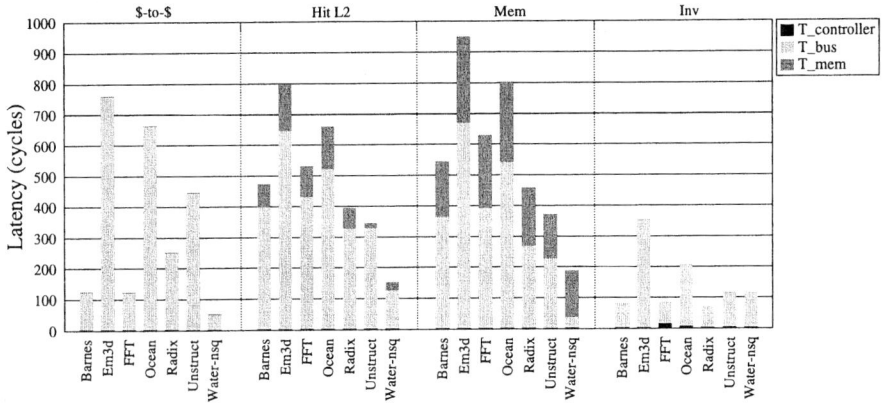

Fig. 2. Average Latency for $-to-$, Hit L2, Mem and Inv misses

the case of EM3D, for which 59.47% of the misses reach main memory. In all the applications (except in the case of UNSTRUCTURED), the number of Inv misses represents a small fraction of the total.

Once we have seen this classification of the L1 accesses, we are going to analyze the latencies suffered by each type of L1 miss. In Figure 2 we can see the average latency for $-to-$, Hit L2, Mem and Inv misses divided into the components described in Section 4 for all the applications. The main component of the overall L1 miss latency for all the miss types is the time spent at the bus. Logically, the miss type with lower latency is the Inv miss, as there is not a *memory* component and we do not have to wait for a reply. We also see that the time spent at the controller is negligible when compared with the bus latency.

We see that latencies are very variable, even for the same application if we compare different miss types. To better understand this erratic behavior, we must analyze the statistics in Table 4 (as well as those presented in Table 3). Table 4 shows the total number of requests and replies snooped by cache controllers (columns 1 and 2). Columns 3 and 4 contains the average number and percentage

Table 4. Classification of the snooped requests and replies

Application	Requests snooped	Replies snooped	Useful requests	Useful replies	L2 useful requests
BARNES-HUT	3,758,643	3,749,512	198,456 (5.27%)	251,489 (6.7%)	97,541 (2.59%)
EM3D	9,475,489	9,424,412	10,052 (0.11%)	587,125 (6.23%)	1,145,478 (12.15%)
FFT	4,912,331	4,912,085	45 (0.0009%)	307,022 (6.25%)	613,951 (12.5%)
OCEAN	4,119,636	4,033,051	19,259 (0.47%)	258,823 (6.42%)	471,302 (11.44%)
RADIX	3,292,750	3,288,383	1,495 (0.05%)	205,806 (6.26%)	408,114 (12.39%)
UNSTRUCT	20,726,502	19,168,246	831,566 (4.01%)	1,301,762 (6.79%)	928,474 (4.48%)
WATER-NSQ	2,837,894	2,823,757	97,868 (3.45%)	177,463 (6.28%)	159,081 (5.61%)

of useful requests and replies snooped by each L1 controller. By useful request we mean a request that implies some action over the local copy of the line at the L1 cache, and by useful reply we mean a reply that contains data and the cache is waiting for. Finally, column 5 shows the number of useful requests snooped by each L2 cache bank (the L2 cache does not snoop replies). The main result is the small number of useful requests (very insignificant in most applications). This implies that in the vast majority of occasions, it is not necessary that the L1 cache controllers snoop the requests that appear in the bus, as these requests are referred to a line that it is not in the cache. We can also see, comparing columns 1 and 2, how the number of snooped requests does not match exactly the number of snooped replies. This is caused by the Inv misses, which do not need a reply.

We have seen that one of the main hot spots in this architecture is the bus, so in order to evaluate the potential of future techniques aimed at alleviating this bottleneck, we simulate the same architecture but with an ideal network, in which the arbitration delay is zero cycles and the bus cycle duration is equal to the processor cycle. This network is equivalent to have one bus working at the same frequency than the processor for each private L1 cache.

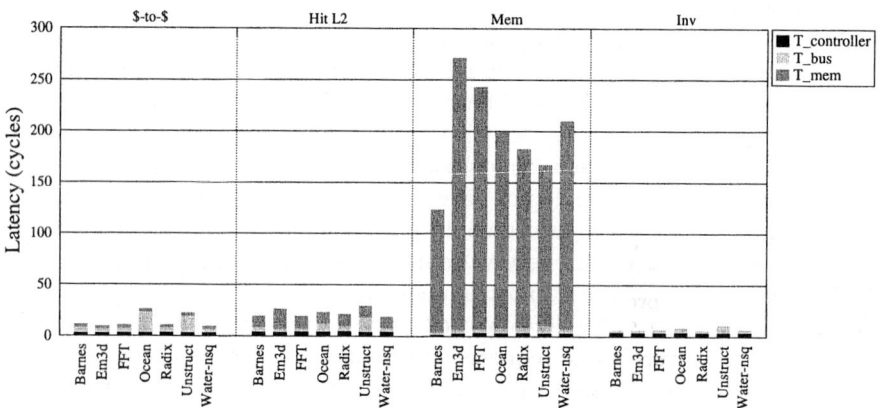

Fig. 3. Average Latency for $-to-$, Hit L2, Mem and Inv misses with an ideal network

When we use an ideal interconnection network, bus latencies are drastically reduced (see Figure 3). The overall latency for $-to-$, Hit L2 and Inv misses is reduced by a factor of 10 on average. Reductions for Mem misses are less impressive, as the *memory* component is hard limited by main memory latency. We can also see that the T_{mem} component of the latency is approximately 30% greater for Mem misses. As we have removed the bus bottleneck, requests arrive at a higher rate to memory controllers and, thus, these controllers are more loaded than in the baseline architecture. These results demonstrate the potential savings of future proposals using a high-performance point-to-point interconnection network providing more bandwidth than a bus, although different cache coherence protocols are to be implemented.

6 Conclusions

In this paper we have presented a L1 cache miss taxonomy of a snoop-based D-CMP (dense chip-multiprocessor) based on how the misses are satisfied. Our results point out that the main bottleneck of this kind of architecture is the shared bus, concluding that this type of interconnection does not provide enough bandwidth for a CMP composed of 16 cores. We have shown that a snoop-based coherence protocol induces too much unnecessary work at the cache controllers, as the majority of snooped transactions are not concerned with the lines in that cache. By simulating a perfect interconnection network, we have seen that it is possible to reduce the latencies of L1 misses by a factor up to 20 in some cases. This gives us a theoretical limit for future proposals in which the bus is replaced by a higher-performance interconnection network.

As future work, we are interested in modeling some different point-to-point interconnection networks and evaluate their performance. This may require some modifications in the coherence mechanism, so it will be necessary to design a cache-coherence protocol tailored to the particularities of a CMP architecture.

Acknowledgments

This work has been supported by the Spanish Ministry of Ciencia y Tecnología and the European Union (Feder Funds) under grant TIC2003-08154-C06-03.

References

1. Barroso, L.A., Gharachorloo, K., McNamara, R., Nowatzyk, A., Qadeer, S., Sano, B., Smith, S., Stets, R., Verghese, B.: "Piranha: A Scalable Architecture Based on Single-Chip Multiprocessing". In: Proc. of 27th Int'l Symp. on Computer Architecture. (2000) 282–293
2. Hammond, L., Hubbert, B.A., Siu, M., Prabhu, M.K., Chen, M., Olukotun, K.: "The Stanford Hydra CMP". IEEE Micro **20** (2000) 71–84
3. Kalla, R., Sinharoy, B., Tendler, J.M.: IBM Power5 Chip: A Dual-Core Multithreaded Processor. IEEE Micro **24** (2004) 40–47
4. Krewell, K.: UltraSPARC IV Mirrors Predecessor. Micro. Report, pp. 1-3 (2003)
5. Hughes, C.J., Pai, V.S.P., Ranganathan, P., Adve, S.V.: RSIM: Simulating Shared-Memory Multiprocessors with ILP Proccesors. IEEE Computer **35** (2002) 68–76
6. Moshovos, A., Memik, G., Falsafi, B., Choudhary, A.: "JETTY: Filtering Snoops for Reduced Energy Consumption in SMP Servers". In: Proc. of 7th Int'l Symp. on High-Performance Computer Architecture. (2001) 85–96
7. Ekman, M., Dahlgren, F., Stenström, P.: "Evaluation of Snoop-Energy Reduction Techniques for Chip-Multiprocessors". In: Proc. of 1st Workshop on Duplicating, Deconstructing and Debunking. (2002) 2–11
8. Beckmann, B., Wood, D.: "Managing Wire Delay in Large Chip-Multiprocessor Caches". In: Proc. of 37th Int'l Symp. on Microarchitecture. (2004) 319–330
9. Liu, C., Sivasubramaniam, A., Kandemir, M.: Organizing the Last Line of Defense before Hitting the Memory Wall for CMPs. In: Proc. of 10th Int'l Symp. on High Performance Computer Architecture. (2004) 176–185

10. Takahasi, M., Takano, H., Kaneko, E., Suzuki, S.: "A Shared-bus Control Mechanism and a Cache Coherence Protocol for a High-performance On-chip Multiprocessor". In: Proc. of 2nd Int'l Conference on High-Performance Computer Architecture. (1996) 314–322
11. Hammond, L., Willey, M., Olukotun, K.: "Data Speculation Support for a Chip Multiprocessor". In: Proc. of the 8th Int'l Symp. on Architectural Support for Parallel Languages and Operating Systems. (1998) 58–69
12. Krishnan, V., Torrellas, J.: "A Chip-Multiprocessor Architecture with Speculative Multithreading". IEEE Transactions On Computers **48** (1999) 866–880
13. Steffan, J.G., Colohan, C.B., Zhai, A., Mowry, T.C.: "A Scalable Approach to Thread-Level Speculation". In: Proc. of 27th Int'l Symp. on Computer Architecture. (2000) 1–12
14. Yanagawa, Y., Hung, L.D., Iwama, C., Barli, N.D., Sakai, S., Tanaka, H.: "Complexity Analysis of A Cache Controller for Speculative Multithreading Chip Multiprocessors". In: Proc. of 10th Int'l Conference on High Performance Computing. (2003) 393–404
15. Culler, D.E., Singh, J.P., Gupta, A.: Parallel Computer Architecture: A Hardware/Software Approach. Morgan Kaufmann Publishers, Inc. (1999)
16. Sweazey, P., Smith, A.J.: A Class of Compatible Cache Consistency Protocols and their Support by the IEEE Futurebus. In: Proc. of 13th Int'l Symp. on Computer Architecture. (1986) 414–423
17. Charlesworth, A.: "The Sun Fireplane Interconnect". In: Proc. of SC2001 High Performance Networking and Computing Conference. (2001) 1–14
18. Woo, S.C., Ohara, M., Torrie, E., Singh, J.P., Gupta, A.: "The SPLASH-2 programs: Characterization and Methodological Considerations". In: Proc. of 22nd Int'l Symp. on Computer Architecture. (1995) 24–36
19. Culler, D.E., Dusseau, A., Goldstein, S.C., Krishnamurthy, A., Lumetta, S., Luna, S., von Eicken, T., Yelick, K.: "Parallel Programming in Split-C". In: Proc. of Int'l SC1993 High Performance Networking and Computing Conference. (1993) 262–273
20. Mukherjee, S.S., Sharma, S.D., Hill, M.D., Larus, J.R., Rogers, A., Saltz, L.: "Efficient Support for Irregular Applications on Distributed-Memory Machines". In: Proc. of 5th Int'l Symp. on Principles and Practice of Parallel Programing. (1995) 68–79

Factory: An Object-Oriented Parallel Programming Substrate for Deep Multiprocessors

Scott Schneider, Christos D. Antonopoulos, and Dimitrios S. Nikolopoulos

Department of Computer Science, The College of William and Mary,
Williamsburg, VA 23187-8795
{scotts, cda, dsn}@cs.wm.edu

Abstract. Recent advances in processor technology such as Simultaneous Multithreading (SMT) and Chip Multiprocessing (CMP) enable parallel processing on a single die. These processors are used as building blocks of shared-memory multiprocessor systems, or clusters of multiprocessors. New programming languages and tools are necessary due to the complexities introduced by systems with multigrain, multilevel execution capabilities. This paper introduces Factory, an object-oriented parallel programming substrate which allows programmers to express multigrain parallelism, but alleviates them from having to manage it. Factory is written in C++ without introducing any extensions to the language. Because it leverages existing C++ constructs to express arbitrarily nested parallel computations, it is highly portable and does not require extra compiler support. Moreover, Factory offers programmability and performance comparable to already established multithreading substrates.

Keywords: Multithreading substrate, Object-oriented parallel programming, Deep parallel architectures, Multiparadigm parallelism, Portability, Programmability.

1 Introduction

Conventional processor technologies capitalize on increasing clock frequencies and on using the full transistor budget to exploit ILP. The diminishing returns of such approaches have shifted the focus of computer systems designers to clustering and parallelism. Current mainstream processors such as SMTs, CMPs and hybrid CMP/SMTs exploit coarse-grain thread-level parallelism at the microarchitectural level [1,2]. Thread-level parallelism is pervasive in high-end microprocessor designs as well [3,4].

Disparity in memory access latencies and the multiple levels of parallelism offered by emerging hardware necessitate programming languages, libraries and tools that enable users to express and control such parallelism. Furthermore, programmers need the means to map different granularities of parallelism to the different levels offered by emerging hardware. Unfortunately, current industry standards for expressing parallelism, such as MPI [5] and OpenMP [6], do not rise to these challenges, because they are designed and implemented with optimized support for a flat parallel execution model and provide little to no additional support for multilevel execution models.

In this paper, we present *Factory*, an object-oriented parallel programming substrate written entirely in C++. Factory was designed as a substrate for implementing next-generation parallel programming models that naturally incorporate multiple levels and types of parallelism controlled by an intelligent runtime system. Factory is functional as a standalone parallel programming library without requiring additional compiler or preprocessor support. The main goals of Factory are to:

- Provide a clean, object-oriented interface for writing parallel programs.
- Provide a type-safe parallel programming environment.
- Define a unified interface to multiple types of parallelism.
- Allow effective exploitation and granularity control for multilevel and multitier parallelism within the same binary.
- Provide a pure C++ runtime library which does not need external interpreter or compiler support.

We outline the design and implementation of Factory and evaluate its performance using a multi-SMT compute node as a target testbed. Our primary contribution is a concrete set of object-oriented capabilities for expressing multiple forms of parallelism in a unified manner, along with generic runtime mechanisms that enable the exploitation of multiple forms of parallelism in a single program. As such, Factory can serve as a runtime library for next-generation, object-oriented parallel programming systems that target deep, multigrain parallel architectures. Factory also makes contributions in the direction of implementing more efficient object-oriented substrates for parallel programming. Its features include lock-free synchronization, flexible scheduling algorithms that are aware of SMT/CMP processors and hierarchical parallel execution, and localized barriers for independent sets of work units.

The rest of this paper is organized as follows: Section 2 briefly discusses prior work which relates to Factory. In Section 3 we present the design of Factory. Section 4 compares Factory's performance with other multithreading programming models and substrates and shows that Factory can exploit the most commonly used forms of parallelism without compromising performance. We discuss future work and conclude the paper in Section 5. An extended version of this paper can be found in [7].

2 Related Work

There is a large body of earlier work in multithreading programming models and object-oriented frameworks for parallel programming. We focus on recent and active projects with strong relation to Factory.

Cilk [8] is an extension to C with explicit support for multithreaded programming. Cilk is designed to execute strict multithreaded computations and provides firm algorithmic bounds for the execution time and space requirements of these computations. Although Factory shares some functionality with Cilk, it has a different and broader objective, since its focal point is the exploitation of multilevel and multiparadigm parallelism, including task-level, loop-level and divide-and-conquer parallelism.

OpenMP [6] is an industry standard for programming on shared memory multiprocessors. OpenMP is particularly suitable for expressing loop based parallelism in multithreaded C, C++ and Fortran programs. Instead of explicitly extending the language, programmers use compiler directives that adhere to the OpenMP standard to express parallelism. The current OpenMP standard has limitations related to the orchestration and scheduling of multiple levels of parallelism. A limited form of static task-level parallelism is supported in OpenMP via the use of parallel sections. Dynamic task-level parallelism is not currently supported by the OpenMP standard, although some vendors, such as Intel, provide platform-specific implementations [9,10]. Factory differs from OpenMP in that it provides a generic object-oriented programming environment for expressing multiple forms of parallelism explicitly and in a unified manner, while providing the necessary runtime support for effectively scheduling all forms of parallelism.

X10 [11] is an ongoing project at IBM to develop an object-oriented parallel language for emerging architectures. The proposed language has a rich set of features, including C++ extensions to describe clustered data structures, extensions to define activities (threads) for both communication and computation and associate these activities with specific nodes, and other features. We view Factory as a complementary effort to X10, which places more emphasis on the runtime issues that pertain to the management of multigrain parallelism, without compromising expressiveness and functionality. Furthermore, Factory can be used as a supportive runtime library for extended parallel object-oriented languages such as X10.

3 Design

The design of Factory focuses on leveraging existing C++ constructs to express multiple types of parallelism at multiple levels. We find the mechanisms provided by C++ expressive enough that we do not have to resort to defining a new language or language extensions which require a separate interpreter or compiler. The combination of inheritance and a sophisticated type system allows the design of a clean, well defined, high-level interface for the development of efficient parallel code. The implementation of Factory solely in C++ and exclusively at user level makes it a multithreading substrate that is portable across different architectures and operating systems. A more detailed presentation of the design, including a small object allocator optimized for object reuse across multiple threads, can be found in [7].

3.1 Enabling Multiparadigm Parallelism with C++

C++ enables the programmer to define class hierarchies. Factory exploits this feature to define all types of parallel work as classes which inherit from a general work class. However, deeper in the hierarchy, classes are dissociated according to the type of work they represent. In the context of this paper we focus on task- and loop-parallel codes, however the Factory hierarchy is easily extensible to other forms of parallelism as well.

Inheritance allows the expression of different kinds of parallelism, with different properties, via a common interface. Factory exploits the C++ templates mechanism in order to adapt the functionality and the behavior of the multithreading runtime according to the requirements of the different forms of parallel work. As a result, Factory allows programmers to easily express different kinds of parallel work, with different properties, through a common interface. At the same time, they can efficiently execute the parallel work, transparently using the appropriate algorithms and mechanisms to manage parallelism.

Work as Objects: Objects are the natural way to represent chunks of parallel work in an object-oriented programming paradigm. Parallel work can be abstracted as an implementation of an algorithm and a set of parameters, which in turn can be directly mapped to a C++ object (represented as a work_unit). Specific chunks of the computation are consequently represented as objects of a work unit class.

The user-defined member function work() implements the computation for the specific work unit, and its member fields serve as the computation's parameters. For each type of computation the programmer defines a new class. Objects instantiated from this class represent different chunks of the computation. At runtime, Factory executes the work() member function of each work_unit object.

More details on the Factory interface can be found in [7].

Work Inheritance Hierarchy: All different kinds of Factory work units export a common API to the programmer as a way to enhance programmability. However, in order to differentiate internally between different kinds of work units and provide the required functionality in each case, Factory work units are organized in an inheritance hierarchy. The hierarchy structure facilitates the addition of new types of work, or the refinement of existing types, without interfering with unrelated types.

The work_unit base class is the root of the work inheritance hierarchy. It defines the minimal interface that a work unit must provide. Programmer defined work units do not inherit directly from work_unit, but rather from classes at the leaves of the inheritance tree, which correspond to particular types of work.

The tree_unit class derives directly from work_unit, and is used to express parallel codes that follow a dependence driven programming model. Work units which derive from tree_unit are organized as a dependence tree at runtime, which is used by Factory to enforce the correct order of work unit execution. Both task_unit and loop_unit derive from tree_unit and they are used by programmers to define task- and loop-parallel work chunks respectively. These classes provide the required support and functionality for the efficient execution of each specific type of parallel computation. A plain_unit can, in turn, be used for codes that are not dependence-driven and directly manage the execution of work chunks at the application level.

Work Execution: All the interaction of applications with the Factory runtime system occurs through an object of the `factory` class[1]. While `work_unit` classes are used to express the parallel algorithms, the `factory` class provides the necessary functionality for their creation, management and execution. The `factory` class defines member functions for starting and stopping parallel execution, as well as creating, scheduling, and synchronizing work units.

3.2 Scheduling

Factory incorporates a generic, queue-based runtime system based on local, per execution context work queues. The later are implemented using non-blocking, lock-free FIFO and LIFO queue management algorithms [12]. The queue hierarchy can be easily extended in order to map more accurately to the target parallel architecture. We have implemented several kinds of scheduling algorithms based on LIFO and FIFO execution order of work units, but programmers can also define their own, according to the specific needs of their applications. Our performance evaluation section demonstrates that Factory schedulers achieve identical or better performance than both generic and customized, application-embedded user-level schedulers.

Factory uses kernel threads as execution vehicles. Each execution vehicle is bound to a specific execution context and has its own local work queue, from which it receives work through the active scheduling algorithm. Load balancing is achieved via work stealing from remote queues. Factory provides hierarchy-conscious work stealing algorithms, which favor work stealing between execution contexts close in the architectural hierarchy.

3.3 Synchronization

Factory provides support for the efficient execution of dependence-driven parallel codes. Each work unit employs a *children* counter to keep track of the number of in-flight children work units. As a result, a dependence tree is dynamically formed and maintained at runtime. The leaves of the tree are work units without dependencies, which are either currently executing, or are ready to execute in the future. The internal nodes represent work units whose execution is blocked because they have to wait for the termination of their children before they can continue.

Correct order of execution is enforced through Factory barriers, which operate on a particular work unit. The execution is either blocked until all children work units in the dependence subtree of the calling work unit have terminated (`child_barrier()` member function of the `factory` class), or until both the children and the work unit itself have terminated (`root_barrier()` member function).

Whenever a barrier prevents further execution of a work unit, the corresponding execution vehicle is not blocked. The user-level scheduling algorithm is invoked, and the execution vehicle starts executing other work units. When

---

[1] Throughout the paper we use the notation Factory to refer to the multithreading substrate and `factory` to refer to the class.

the dependencies of the blocked work unit are satisfied, the blocked work unit is allowed to resume.

4 Performance Evaluation

We have experimentally evaluated the performance of Factory on a multilevel parallel architecture, namely an SMT-based multiprocessor. We compare Factory against other popular parallel programming models, namely OpenMP, Cilk and parallelization using POSIX threads.

Our experimental platform is a quad SMP, based on Intel Hyper-Threaded (HT) processors. Intel HT processors share most of the internal processor resources between 2 simultaneously executing threads. The system is equipped with 2 GB of main memory and runs Linux (2.6.8 kernel). We created our binaries using the Intel Compiler suite for 32-bit applications (version 8.1).

We experimented using both microbenchmarks to assess the overheads for managing parallelism and parallel applications to compare Factory against the aforementioned parallel programming models. All experiments throughout our evaluation have been executed 20 times. We report the average timings across all 20 repetitions. The 95% confidence interval for each data point has always been lower than 1.7% of the average.

4.1 Minimum Granularity of Exploitable Parallelism

The minimum granularity of parallelism that can be effectively exploited by any multithreaded substrate is directly related to the degree of overheads—both architecture-specific and software-related—associated with the creation and management of parallel jobs.

The minimum granularity experiment consists of a variable number of pause assembly instructions. The number of the instructions is reduced until a break-even point is identified, at which point the sequential execution is as fast as the parallel one. The sequential execution time of the number of instructions corresponding to the break-even point is the minimum granularity. We represent work with pause instructions because they incur as minimal interference as possible when executed simultaneously on the different execution contexts of a single HT processor. The minimum granularity is also a factor of the number of threads used for the parallel execution. We thus evaluate the minimum granularity for the parallel execution with 2, 4 and 8 threads which are either packed on as few or spread to as many physical CPUs as possible. The different binding schemes allow the evaluation of both intra- and inter-processor parallelism overheads.

Table 1 summarizes the measured minimum exploitable granularity of Factory and the other multithreading systems. We compare Factory against Cilk, which supports only strict multithreaded computations with recursive task parallelism, and OpenMP. For the latter, we distinguish between the task- and loop-minimum granularities, as the OpenMP runtime uses different mechanisms for each. For task parallelism we use Intel compiler's workqueue extensions to OpenMP [6,10]. Factory uses the same mechanisms for creating parallel work

Table 1. Comparison of the minimum granularity of effectively exploitable parallelism

	2 Threads		4 Threads		8 Threads
	1 CPU	2 CPUs	2 CPUs	4 CPUs	4 CPUs
Factory	6.2μsec	6.2μsec	10μsec	10μsec	26μsec
Cilk	121μsec	81μsec	153μsec	153μsec	222μsec
OpenMP task	20μsec	20μsec	26μsec	24μsec	202μsec
OpenMP loop	10μsec	6.2μsec	6.2μsec	4.2μsec	68μsec

units, regardless of whether these work units are used for task- or loop-parallelism. As a result, it is represented by only one entry in the table.

Factory's minimum task granularity is finer than Intel's task queue implementation in OpenMP and remains competitive with OpenMP's loop granularity even though Intel's implementation of loop- and task-level parallel execution is heavily optimized. At the same time, Factory proves able to exploit significantly finer granularity than Cilk. Although the point where Cilk achieves speedup is relatively high, the break-even point is significantly lower, close to the performance of OpenMP tasks. This behavior can be attributed to the fact that for very fine-grain parallel work, the Cilk runtime actually schedules multiple tasks to the same kernel thread. Hence, Cilk requires a relatively large work load before multiple threads are used to execute it. Both Cilk and OpenMP perform better when threads are spread to as many physical CPUs as possible. Factory overheads, on the other hand, are uncorrelated with thread placement, making it a more predictable multithreading substrate for deep, multilevel parallel systems.

4.2 Factory vs. Other Programming Models

Radiosity is an application from the Splash-2 [13] benchmark suite which computes the equilibrium distribution of light in a scene. It uses several pointer-based data structures and has an irregular memory access pattern. The code uses application-level task queues and applies work stealing for load balancing. Radiosity tests Factory's ability to handle fine grain synchronization, since it is sensitive to the efficiency of synchronization mechanisms [14]. It also allows a direct comparison of Factory with POSIX Threads as underlying substrates for the implementation of hand crafted parallel codes. Porting the original code to Factory required just the conversion of the task concept to a work unit object. Both implementations were executed with the options `-batch -largeroom`. The performance results are depicted in Figure 1.

Factory consistently performs better than the POSIX Threads, mainly due to its efficient, fine-grain synchronization mechanisms. There is a 17% performance improvement from 4 to 8 threads, which is significantly less than the 72% improvement from 1 to 4 threads. This degradation is caused by each Radiosity thread using almost all shared execution resources.

We tested Factory using both LIFO and FIFO lock-free scheduling policies. LIFO execution ordering yielded better performance due to temporal locality, since data shared between the parent and children work units are likely to be found in the processor cache if a LIFO ordering is applied.

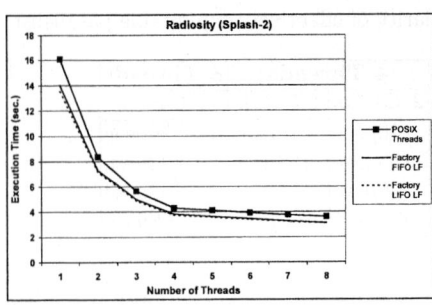 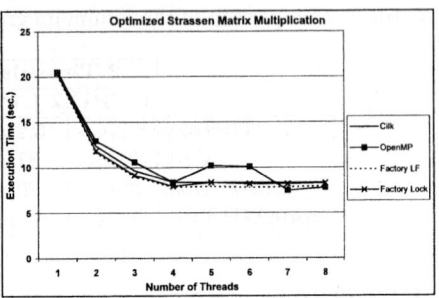

Fig. 1. Performance of Factory and POSIX Threads Radiosity implementations

Fig. 2. Performance of Factory, Cilk, and OpenMP taskq for parallel, Strassen matrix multiplication

As a next step, we experimented with an optimized parallel implementation of the Strassen algorithm from the Cilk distribution. The algorithm is applied on 2048x2048 double precision floating point matrices. The OpenMP version of the application is based on Intel's OpenMP extensions for the support of task queues. Once again, the conversion to the Factory programming model was straightforward. We replaced recursive Cilk functions with work unit classes (task_unit). The conversion to OpenMP was also simple: recursive calls to Cilk functions have just been preceded by OpenMP task directives.

As shown in Figure 2, we also experimented with lock-free and lock-based queue implementations in Factory. All four implementations attain good scalability until 4 threads. After that point, at least one processor is forced to execute threads on both SMT contexts. When more than 4 threads are used, the OpenMP implementation suffers erratic performance. Cilk is not affected by intra-processor parallelism. It should be noted that Cilk's work stealing algorithm avoids locking the queues in the common execution scenario [15]. The Factory implementation that uses a lock-based queue implementation also suffers a performance degradation at 5 and 6 threads. However, the problem is solved if lock-free queues are used.

The performance degradation at 5 and 6 threads is related to synchronization. Previous studies indicate that lock-free algorithms are more efficient than lock-based ones under high contention or multiprogramming [12]. The execution of more than one thread on the execution contexts of SMT processors often has similarities to a multiprogrammed execution on a conventional SMP. As a result of resource sharing, SMT-based multiprocessors may prove more sensitive to the efficiency of synchronization mechanisms than conventional SMPs.

5 Conclusions and Future Work

We have presented Factory, an object-oriented parallel programming substrate, which allows the exploitation of multiple types of parallelism on deep parallel architectures. Factory uses a unified and clean interface to express different, potentially nested forms of parallelism. Its implementation allows its use both as a standalone parallel programming library and as a runtime system for high-level object-oriented languages for parallel programming. The performance optimizations of Factory include lock-free synchronization for internal concurrent data structures and scheduling policies which are aware of the topology of hierarchical parallel systems. We have presented performance results that illustrate the efficiency of the central mechanisms for managing parallelism in Factory and justify our design choices for these mechanisms. We have also presented results obtained from the implementation of parallel applications with Factory which show that Factory performs competitively with other parallel programming models for shared-memory systems.

We regard Factory as a viable means for programming emerging parallel architectures and for preserving both productivity and efficiency. We plan to extend Factory in several directions, including the introduction of hierarchical scheduling algorithms that are aware of the type of parallelism they manage and the integration of precomputation into the runtime.

Acknowledgments

This material is based in part upon work supported by the National Science Foundation under Grant Numbers CAREER:CCF-0346867, and ITR: ACI-0312980. Any opinions, findings, and conclusions or recommendations expressed in this material are those of the author and do not necessarily reflect the views of the National Science Foundation.

References

1. Tullsen, D.M., Eggers, S., Levy, H.M.: Simultaneous Multithreading: Maximizing On-Chip Parallelism. In: Proceedings of the 22th Annual International Symposium on Computer Architecture. (1995)
2. Hammond, L., Hubbert, B.A., Siu, M., Prabhu, M.K., Chen, M., Olukotun, K.: The Stanford Hydra CMP. IEEE Micro **20** (2000) 71–84
3. Takayanagi, T., Shin, J., Petrick, B., Su, J., Leon, A.: A Dual-Core 64b UltraSPARC Microprocessor for Dense Server Applications. In: Proc. of the 41st Conference on Design Automation (DAC'04), San Diego, CA, U.S.A. (2004) 673–677
4. Cascaval, C., Castanos, J., Ceze, L., Dennea, M., Gupta, M., Lieber, D., Moreira, J., Strauss, K., H. S. Warren, J.: Evaluation of a Multithreaded Architecture for Cellular Computing. In: 8th International Symposium on High-Performance Computer Architecture (HPCA-8), Cambridge, MA, U.S.A. (2002) 311–321
5. Forum, M.P.I.: MPI: A Message-Passing Interface Standard. Technical Report UT-CS-94-230 (1994)
6. OpenMP Architecture Review Board: OpenMP Application Program Interface. Version 2.5 edn. (2005)
7. Scott Schneider, C.D.A., Nikolopoulos, D.S.: Factory: An Object-Oriented Parallel Programming Substrate for Deep Multiprocessors. Technical Report WM-CS-2005-06, The College of William and Mary (2005) http://www.cs.wm.edu/ scotts/papers/wm-cs-2005-06.pdf.

8. Blumofe, R., Joerg, C., Kuszmaul, B., Leiserson, C., Randall, K., Zhou, Y.: Cilk: An Efficient Multithreaded Runtime System. In: Proceedings of the 5th Symposium on Principles and Practice of Parallel Programming. (1995)
9. Shah, S., Haab, G., Petersen, P., Throop, J.: Flexible Control Structures for Parallelism in OpenMP. Concurrency: Practice and Experience **12** (2000) 1219–1239
10. Xinmin, T., Girkar, M., Shah, S., Armstrong, D., Su, E., Petersen, P.: Compiler and Runtime Support for Running OpenMP Programs on Pentium and Itanium architectures. In: Proceedings of the Eighth International Workshop on HighLevel Parallel Programming Models and Supportive Environments, Nice, France (2003) 47–55
11. Ebcioglu, K., Saraswat, V., Sarkar, V.: X10: Programming for Hierarchical Parallelism and Non-Uniform Data Access. In: 3rd International Workshop on Language Runtimes. (2004)
12. Michael, M.M., Scott, M.L.: Simple, Fast, and Practical Non-Blocking and Blocking Concurrent Queue Algorithms. In: Proceedings of the 15th annual ACM Symposium on Principles of Distributed Computing (PODC'96), Philadelphia, Pennsylvania, U.S.A. (1996) 267–275
13. Woo, S.C., Ohara, M., Torrie, E., Singh, J.P., Gupta, A.: The SPLASH-2 Programs: Characterization and Methodological Considerations. In: Proceedings of the 22th International Symposium on Computer Architecture, Santa Margherita Ligure, Italy (1995) 24–36
14. Radović, Z., Hagersten, E.: Efficient Synchronization for Non-Uniform Communication Architectures. In: Supercomputing '02: Proceedings of the 2002 ACM/IEEE conference on Supercomputing, Los Alamitos, CA, USA, IEEE Computer Society Press (2002) 1–13
15. Frigo, M., Leiserson, C.E., Randall, K.H.: The Implementation of the Cilk-5 Multithreaded Language. In: PLDI '98: Proceedings of the ACM SIGPLAN 1998 conference on Programming language design and implementation, New York, NY, USA, ACM Press (1998) 212–223

Convergence of the Discrete FGDLS Algorithm

Sabin Tabirca[1], Tatiana Tabirca[1,*], and Laurence T. Yang[2]

[1] University College Cork, Computer Science Department,
College Road, Cork, Ireland
s.tabirca@cs.man.ac.uk
[2] St. Francis Xavier University, Computer Science Department,
Antigonish, B2G 2W5, Canada
lyang@stfx.ca

Abstract. The Feedback-Guided Dynamic Loop Scheduling (FGDLS) algorithm [1] is a recent dynamic approach to the scheduling of a parallel loop within a sequential outer loop. Earlier papers have analysed convergence under the assumption that the workload is a positive, continuous, function of a continuous argument (the iteration number). However, this assumption is unrealistic since it is known that the iteration number is a discrete variable. In this paper we extend the proof of convergence of the algorithm to the case where the iteration number is treated as a discrete variable. We are able to establish convergence of the FGDLS algorithm for the case when the workload is monotonically decreasing.

1 Introduction

It is widely recognised that loops are a very important source of parallelism in many practical applications. Since a significant overhead in many parallel implementations is represented by load imbalance, a number of algorithms have been designed to schedule loop iterations to processors of a shared-memory machine in an optimal way (so-called loop scheduling algorithms).

An important class of loop scheduling algorithms is based on *Guided Self-Scheduling* (Polychronopoulos and Kuck [8]) or some variant of *Guided Self-Scheduling* (see for example Eager and Zahorjan [4], Hummel et al. [5], Lucco [6], Tzen and Ni [14]). These algorithms divide the loop iterations into a relatively large number of chunks which are assigned to processors from a central queue. One of the motivations for this approach is the assumption that each execution of a loop is independent of any previous executions of the same loop, and therefore has to be rescheduled 'from scratch'. Important overheads such as additional synchronisation, loss of data locality, and reductions in the efficiency of loop unrolling and pipelining can be caused by this approach.

The class of *Affinity Scheduling* algorithms (Markatos and LeBlanc [7], see also Subramanian and Eager [9] for variants of *Affinity Scheduling*) is an attempt to ameliorate some of this loss of performance. Rather than maintaining a single central queue these algorithms are based on per-processor work queues with exchange of work

---

* This work was supported by Boole Centre for Research in Informatics at National University of Ireland, Cork.

(chunks of the loop iteration) if required. The underlying assumption of affinity scheduling algorithms remains that each execution of a parallel loop is independent of previous executions.

Feedback Guided Dynamic Loop Scheduling (FGDLS) is a relatively recent scheduling method ([1], [2]), which deals directly with a sequence of similar or identical parallel loops (see Figure 1). This loop structure is very important since it frequently occurs in a number of theoretical [10], [11] and practical applications [2]. The convergence of the FGDLS method has been studied in [3] and [12]. In these papers the workload is assumed to be a continuous positive function of a continuous argument (iteration number). However, the approach is artificial since in reality the workload is a positive function of the discrete argument (iteration number).

1.1 The FGDLS Algorithm

The FGDLS algorithm aims to determine an optimal schedule, across p processors $P_1, P_2, ..., P_p$, for the sequence of parallel loops given in Figure 1. The (unknown) workloads of the parallel loop are assumed to be given by the values $\{w_i, i = 1, 2, ..., n\}$ (so that w_i is the workload of the call to the routine loop body(i)). The FGDLS algorithm calculates a block partitioning of the parallel (i) loop, where l_j^t and h_j^t are the lower and upper bounds of the loop block assigned to Processor j on outer iteration t. These bounds clearly should satisfy the simple equations

$$l_1^t = 1; \quad h_p^t = n; \quad l_{j+1}^t = h_j^t + 1, j = 1, 2, ..., p-1. \tag{1}$$

FGDLS starts with some initial loop bounds $\{(l_j^1, h_j^1), j = 1, 2, ..., p\}$ that are chosen arbitrarily. At the end of the outer iteration t, the new bounds $\{(l_j^{t+1}, h_j^{t+1}), j = 1, 2, ..., p\}$ are calculated from the bounds $\{(l_j^t, h_j^t), j = 1, 2, ..., p\}$ by approximately balancing the observed execution times. Assuming that the observed execution times $\{T_j^t, j = 1, 2, ..., p\}$ are given by

$$T_j^t = \sum_{i=l_j^t}^{h_j^t} w_i, \quad j = 1, 2, ..., p, \tag{2}$$

a piecewise constant approximation of the workload at the iteration t can be formed as

$$\hat{w}_i^t = \frac{T_j^t}{h_j^t - l_j^t + 1}, \quad l_j^t \leq i \leq h_j^t, \quad j = 1, 2, ..., p. \tag{3}$$

```
do sequential t = 1, nsteps
    do parallel i=1,n
        call loop_body(i)
    end do
end do
```

Fig. 1. The FGDLS loop structure

The piecewise constant workloads $\hat{w}_i^t$ can be interpreted as the mean observed workload per loop iteration index on the outer iteration t. It is this piecewise constant function that is approximately equidistributed amongst the p processors to define the new loop bounds $\{(l_j^{t+1}, h_j^{t+1}), j = 1, 2, ..., p\}$:

$$\sum_{i=l_j^{t+1}}^{h_j^{t+1}} \hat{w}_i^t \approx \frac{1}{p} \sum_{i=1}^{n} \hat{w}_i^t = \frac{1}{p} \sum_{k=1}^{p} T_k^t, \quad j = 1, 2, ..., p. \tag{4}$$

These new bounds also satisfy

$$l_1^{t+1} = 1; h_p^{t+1} = n : l_{i+1}^{t+1} = h_i^{t+1} + 1, i = 1, 2, ..., p-1.$$

In order to find expressions for these new bounds two new functions are introduced [13]. Firstly, the function f^t gives the partial sums of the piecewise constant workloads

$$f^t(i) = \sum_{k=1}^{i} \hat{w}_k^t, \quad i = 1, 2, ..., n, \tag{5}$$

($f^t(i)$ is a piecewise linear function that approximates the cumulative workload), and, secondly, the corresponding f^t–inferior part function is given by [13]

$$f_{[]}^t(x) = i \Leftrightarrow f^t(i) \leq x < f^t(i+1). \tag{6}$$

Using these functions the upper bounds $\{h_j^{t+1}, j = 1, 2, ..., p\}$ are given by

$$h_0^{t+1} = 0, \quad h_j^{t+1} = f_{[]}^t \left(f^t(h_{j-1}^{t+1}) + \overline{W} \right), \quad j = 1, 2, ..., p, \tag{7}$$

where $\overline{W} = \frac{1}{p} \sum_{i=1}^{n} \hat{w}_i$ is the target (balanced) workload for each of the p processors.
It can be shown (see [13]) that the functions f^t and $f_{[]}^t$ satisfy the following lemmas.

Lemma 1. *If $l_j^t \leq i \leq h_j^t$ then*

$$f^t(i) = \sum_{q=1}^{j-1} T_q^t + \frac{T_j^t}{h_j^t - l_j^t + 1} (i - l_j^t + 1). \tag{8}$$

Lemma 2. *If $f^t(h_{j-1}^t) < x \leq f^t(h_j^t)$ then*

$$f_{[]}^t(x) = h_{j-1}^t + \left[\left(x - \sum_{q=1}^{j-1} T_q^t \right) \frac{h_j^t - l_j^t + 1}{T_j^t} \right], \tag{9}$$

where $f_{[\ ]}$ represents the inferior part function.

2 Convergence of the FGDLS Algorithm

In this section the convergence of the FGDLS algorithm is considered. For the fixed workloads $\{w_i, i = 1, 2, ..., n\}$ we can find the optimal bounds $\{(l_j^*, h_j^*), j = 1, 2, ..., p\}$ so that

$$\sum_{i=l_j^*}^{h_j^*} w_i \approx \frac{1}{p}\sum_{i=1}^{n} w_i, \quad j=1,2,\ldots,p, \qquad (10)$$

where $l_1^* = 1$; $h_p^* = n$; $l_{i+1}^* = h_i^* + 1, i = 1,2,\ldots,p-1$.

It can be shown that the optimal bounds also satisfy the equations:

$$h_0^* = 0, \ h_j^* = f_{[]}\left(f(h_{j-1}^*) + \overline{W}\right), \ j=1,2,\ldots,p, \qquad (11)$$

where $f(i) = \sum_{k=1}^{i} w_k, i = 1,2,\ldots,n$, represent the partial sums of the workloads (the cumulative workload) and the corresponding inferior part function is given by

$$f_{[]}(x) = i \Leftrightarrow f(i) \leq x < f(i+1).$$

The problem of convergence of the FGDLS algorithm can be stated as follows:

Convergence of Discrete FGDLS Algorithm: *Given the fixed, strictly positive, workloads $\{w_i, i=1,2,\ldots,n\}$ and the initial upper bounds $\{h_j^1, j=0,1,\ldots,p\}$, find conditions such that the upper bound sequences $\{h_j^t\}, t>0$ are convergent and $\lim_{t\to\infty} h_j^t = h_j^*, j=0,1,\ldots,p$.*

Since $h_0^t = h_0^* = 0$, and $h_p^t = h_p^* = n, \forall\, t>0$, we find that the convergence holds trivially for the cases $j=0$ and $j=p$. Recall that the upper bounds are integers so that the sequence $\{h_j^t, t=1,2,\ldots\}$ is convergent to h_j^* whenever, $\exists t_0 > 0$ such that

$$h_j^t = h_j^*, \forall t \geq t_0. \qquad (12)$$

Thus, we have to establish that the upper bounds h_j^t are equal to the optimal bound h_j^* from some index t_0 onwards.

In the following we analyse the convergence of the FGDLS scheduling algorithm for the case when the workloads are monotonically decreasing; we assume that the workloads satisfy the inequalities

$$w_1 \geq w_2 \geq \ldots \geq w_n. \qquad (13)$$

We prove by induction that Equation (12) holds whenever Equation (13) is satisfied. Firstly we show that the sequence of bounds $\{h_1^t\}, t>0$, is convergent to h_1^*. Inductively, we assume that the sequences $\{h_1^t\}, t>0$, $\{h_2^t\}, t>0$, ..., $\{h_j^t\}, t>0$ are convergent (to $h_1^*, h_2^*, \ldots, h_j^*$, respectively) and prove that the sequence $\{h_{j+1}^t\}, t>0$ is convergent to h_{j+1}^*.

2.1 The Convergence of $\{h_1^t\}, t > 0$.

Recall that the upper bound h_j^{t+1} satisfies the equation $h_j^{t+1} = f_{[]}^t\left(f^t(h_{j-1}^{t+1}) + \overline{W}\right)$. Since, $h_0^t = 0$ and $f^t(0) = 0$ we find that the upper bound h_1^{t+1} satisfies

$$h_1^{t+1} = f_{[]}^t\left(f^t(h_0^{t+1}) + \overline{W}\right) = f_{[]}^t(\overline{W}) = h_{j-1}^t + \left[\left(\overline{W} - \sum_{q=1}^{j-1} T_q^t\right)\frac{h_j^t - l_j^t + 1}{T_j^t}\right], \qquad (14)$$

where j is the index that satisfies $f^t(h_{j-1}^t) < \overline{W} \leq f^t(h_j^t)$. Some simple properties of the bounds $\{h_1^t, t=1,2,\ldots\}$ are given in following lemma.

Lemma 3. *The upper bounds $\{h_1^t, t = 1, 2, ...\}$ satisfy the following inequalities:*

1.
$$f^t(h_{j-1}^t) < \overline{W} \leq f^t(h_j^t) \Rightarrow h_{j-1}^t < h_1^{t+1} \leq h_j^t. \tag{15}$$

If the workloads decrease then

2.
$$h_1^{t+1} = h_1^t \Rightarrow h_1^{t+1} = h_1^t = h_1^*. \tag{16}$$

3.
$$h_1^t \leq h_1^* \Rightarrow h_1^t \leq h_1^{t+1} \tag{17}$$

4.
$$h_1^t \geq h_1^* \Rightarrow h_1^t \geq h_1^{t+1} \tag{18}$$

Proof. From (14) we know that h_1^{t+1} satisfies

$$f^t(h_1^{t+1}) \leq \overline{W} < f^t(h_1^{t+1} + 1), \tag{19}$$

and from (11) h_1^* satisfies

$$f(h_1^*) \leq \overline{W} < f(h_1^* + 1).$$

1. When $f^t(h_{j-1}^t) < \overline{W} \leq f^t(h_j^t)$ the definition of h_1^{t+1}, together with Equations (19) and (15), directly gives

$$h_{j-1}^t \leq h_1^{t+1} \leq h_j^t.$$

2. When $h_1^{t+1} = h_1^t$ we have, from Equation (19), that

$$f^t(h_1^{t+1}) \leq \overline{W} < f^t(h_1^{t+1} + 1) \Rightarrow f^t(h_1^t) \leq \overline{W} < f^t(h_1^t + 1) \Rightarrow \tag{20}$$

$$f(h_1^t) \leq \overline{W} < f(h_1^t) + \hat{w}_{h_1^t+1}^t \Rightarrow f(h_1^t) \leq \overline{W} < f(h_1^t) + \frac{\sum_{i=l_2^t}^{h_2^t} w_i}{h_2^t - l_2^t + 1}. \tag{21}$$

Since, the workloads are monotonically decreasing we find that

$$\frac{\sum_{i=l_2^t}^{h_2^t} w_i}{h_2^t - l_2^t + 1} \leq \frac{\sum_{i=l_2^t}^{h_2^t} w_{h_1^t+1}}{h_2^t - l_2^t + 1} = w_{h_1^t+1},$$

and thus

$$f(h_1^t) \leq \overline{W} < f(h_1^t) + w_{h_1^t+1} = f(h_1^t + 1),$$

which implies that

$$h_1^t = h_1^*.$$

3. If $h_1^t \leq h_1^*$ it follows that $f^t(h_1^t) = f(h_1^t) \leq f(h_1^*) \leq \overline{W}$. If $f^t(h_1^t) = \overline{W}$ then $f^t(h_1^t) = \overline{W} < f^t(h_1^t + 1)$ and thus $h_1^{t+1} = h_1^t$. When $f^t(h_1^t) < \overline{W}$, let j be the index such that $f^t(h_{j-1}^t) < \overline{W} \leq f^t(h_j^t)$, then $h_1^t \leq h_{j-1}^t$. Thus we find $h_1^t \leq h_{j-1}^t \leq h_1^{t+1}$.
4. The case $h_1^t \geq h_1^*$ is similar to case 3. ♠

Equations (17, 18) establish that the upper bounds $\{h_1^t, t > 0\}$ behave monotonically. For example, when the upper bound h_1^t is less than, or equal to, the upper bound h_1^*, we find that the new upper bound h_1^{t+1} is greater than, or equal to, h_1^t.

Lemma 4. *If the workloads $\{w_i, i = 1, 2, ..., n\}$ are monotonically decreasing then the sequence $\left\{\frac{h}{\sum_{i=1}^{h} w_i}, h = 1, 2, ..., n\right\}$ is monotonically increasing.*

Proof. The difference between two consecutive terms of the sequence is given by:

$$\frac{h+1}{\sum_{i=1}^{h+1} w_i} - \frac{h}{\sum_{i=1}^{h} w_i} = \frac{(h+1)\sum_{i=1}^{h} w_i - h\sum_{i=1}^{h+1} w_i}{(\sum_{i=1}^{h} w_i)(\sum_{i=1}^{h+1} w_i)} = \quad (22)$$

$$= \frac{(h+1)\sum_{i=1}^{h} w_i - h\sum_{i=1}^{h} w_i - h\, w_{h+1}}{(\sum_{i=1}^{h} w_i)(\sum_{i=1}^{h+1} w_i)} = \frac{\sum_{i=1}^{h} w_i - h\, w_{h+1}}{(\sum_{i=1}^{h} w_i)(\sum_{i=1}^{h+1} w_i)}. \quad (23)$$

Since the workloads are monotonically decreasing we find that

$$\sum_{i=1}^{h} w_i - h \cdot w_{h+1} \geq 0$$

and therefore

$$\frac{h+1}{\sum_{i=1}^{h+1} w_i} - \frac{h}{\sum_{i=1}^{h} w_i} \geq 0.$$

Thus the sequence increases. ♠

Theorem 1. *If the workloads $\{w_i, i = 1, 2, ..., n\}$ decrease then the upper bounds $\{h_1^t, t > 0\}$ converge to h_1^*.*

Proof. Two cases are analysed in the following.

- **Case 1.** $h_1^t \leq h_1^*, \forall t > 0$.
  Equation (17) gives that $h_1^t \leq h_1^{t+1}, \forall t > 0$. Thus the sequence of bounds $\{h_1^t, t > 0\}$ increases and is bounded above by h_1^*, and therefore it converges.
- **Case 2.** $\exists t_0 > 0$, **such that** $h_1^{t_0} \geq h_1^*$.
  By induction, we prove that $h_1^t \geq h_1^*, \forall t \geq t_0$. Let us suppose that this holds for t so that the upper bound h_1^t satisfy $h_1^t \geq h_1^*$. Since $h_1^t \geq h_1^*$ we find $f'(h_0^t) = 0 < \overline{W} \leq f'(h_1^t)$, therefore the index j from Equation (23) is 0. Equation (23) can be re-written as:

$$h_1^{t+1} = \left\lfloor \overline{W} \frac{h_1^t}{\sum_{i=1}^{h_1^t} w_i} \right\rfloor. \quad (24)$$

Since, the sequence $\left\{\frac{h}{\sum_{i=1}^{h} w_i}, h = 1, 2, ..., n\right\}$ increases and $h_1^t \geq h_1^*$, we find that

$$\frac{h_1^t}{\sum_{i=1}^{h_1^t} w_i} \geq \frac{h_1^*}{\sum_{i=1}^{h_1^*} w_i} \Rightarrow h_1^{t+1} = \left\lfloor \overline{W} \frac{h_1^t}{\sum_{i=1}^{h_1^t} w_i} \right\rfloor \geq \left\lfloor \overline{W} \frac{h_1^*}{\sum_{i=1}^{h_1^*} w_i} \right\rfloor.$$

Since $\sum_{i=1}^{h_1^*} w_i \leq \overline{W}$, we find that $\frac{\overline{W}}{\sum_{i=1}^{h_1^*} w_i} \geq 1$ and therefore $h_1^{t+1} \geq h_1^*$. Therefore, $h_1^t \geq h_1^*, \forall t \geq t_0$ holds.
From Equation (18) we find that $h_1^t \geq h_1^{t+1}, \forall t \geq t_0$. Hence, the sequence of upper bounds $\{h_1^t, t > 0\}$ is monotonically decreasing and is bounded below by h_1^* so that it converges.

In both of the above cases we find that the sequence $\{h_1^t, t > 0\}$ converges. Therefore, we find that there exists an index $t_0 > 0$ such that the sequence is constant $h_1^t = h_1^{t+1}$, $\forall t > t_0$. Finally, we apply Equation (16) to obtain that $h_1^t = h_1^*$, $\forall t > t_0$. ♠

2.2 The Induction Step

In this subsection we present the induction step which proves that if the sequences $\{h_k^t, t > 0\}$ are convergent to h_k^* for $k = 1, 2, ..., j-1$ then the sequence $\{h_j^t, t > 0\}$ is convergent to h_j^*. Given that the sequences $\{h_k^t, t > 0\}$ are convergent we know that $\exists t_0 > 0$ such that

$$h_k^t = h_k^*, \ \forall t \geq t_0, \ k = 1, 2, ..., j-1. \tag{25}$$

Thus, for $t \geq t_0$ the upper bound satisfies

$$h_j^{t+1} = f_{[]}^t\left(f^t(h_{j-1}^{t+1}) + \overline{W}\right) = f_{[]}^t\left(f^t(h_{j-1}^t) + \overline{W}\right) = f_{[]}^t\left(f(h_{j-1}^*) + \overline{W}\right). \tag{26}$$

Let $u(j)$ be the index such that

$$f^t(h_{u(j)-1}^t) < f(h_{j-1}^*) + \overline{W} \leq f^t(h_{u(j)}^t).$$

Then the upper bounds $\{h_1^t, t = 1, 2, ...\}$ satisfy

$$h_1^{t+1} = h_{u(j)-1}^t + \left[\left(f(h_{j-1}^*) + \overline{W} - f(h_{u(j)-1}^t)\right) \frac{h_{u(j)}^t - l_{u(j)}^t + 1}{\sum_{i=l_{u(j)}^t}^{h_{u(j)}^t} w_i}\right]. \tag{27}$$

Lemma 5. *The upper bounds* $\{h_j^t, t = 1, 2, ...\}$ *satisfy:*

1.
$$f^t(h_{u(j)-1}^t) < f(h_{j-1}^*) + \overline{W} \leq f^t(h_{u(j)}^t) \Rightarrow h_{u(j)-1}^t < h_j^{t+1} \leq h_{u(j)}^t. \tag{28}$$

If the workloads decrease then

2.
$$h_j^{t+1} = h_j^t \Rightarrow h_j^{t+1} = h_j^t = h_j^*. \tag{29}$$

3.
$$h_j^t \leq h_j^* \Rightarrow h_j^t \leq h_j^{t+1}. \tag{30}$$

4.
$$h_j^t \geq h_j^* \Rightarrow h_j^t \geq h_j^{t+1}. \tag{31}$$

Proof. The proof is similar to the proof of Lemma 3.

Thus, we find the same monotonic behaviour for the upper bounds $\{h_j^t, t > 0\}$ as for $\{h_1^t, t > 0\}$.

Lemma 6. *If the workloads* $\{w_1, w_2, ..., w_n\}$ *decrease then the sequence*

$$\left\{\frac{h - h_{j-1}^*}{\sum_{i=h_{j-1}^*+1}^{h} w_i}, \ h = h_{j-1}^* + 1, ..., n\right\} \text{ is monotonically increasing.}$$

Proof. The result follows directly by applying Lemma 4 for the workloads $\{w_j, j = h^*_{j-1}+1,...,n\}$.

Theorem 2. *If the workloads $\{w_i, i = 1,2,...,n\}$ decrease monotonically then the sequence of upper bounds $\{h^t_j, t > 0\}$ converges to h^*_j.*

Proof. We again analyse two cases.

- **Case 1.** $h^t_j \leq h^*_j$, $\forall t > t_0$.
  Based on Equation (30) we find $h^t_1 \leq h^{t+1}_1$, $\forall t > t_0$, which means that the sequence of bounds $\{h^t_j, t > 0\}$ is monotonically increasing and is bounded above by h^*_j, and therefore it converges.
- **Case 2.** $\exists t_1 \geq t_0$ such that $h^{t_1}_j \geq h^*_j$.
  By induction we prove that $h^t_1 \geq h^*_j$, $\forall t \geq t_1$. Let us suppose that this holds for t so that $h^t_j \geq h^*_j$. Since $h^t_j \geq h^*_j$ we find

$$f^t(h^t_{j-1}) = f(h^*_{j-1}) \leq f(h^*_{j-1}) + \overline{W} \leq f^t(h^t_j),$$

therefore the index $u(j)$ is j so that two terms in Equation (27) reduce. Equation (27) becomes

$$h^{t+1}_j = h^t_{j-1} + \left\lceil \overline{W} \frac{h^t_j - l^t_j + 1}{\sum_{i=l^t_j}^{h^t_j} w_i} \right\rceil. \tag{32}$$

Since, the sequence $\left\{ \frac{h}{\sum_{i=h^*_{j-1}+1}^{h} w_i}, h = h^*_{j-1} + 1, ..., n \right\}$ is monotonically increasing and $h^t_j \geq h^*_j$ we find that

$$\frac{h^t_j - l^t_j + 1}{\sum_{i=l^t_j}^{h^t_j} w_i} = \frac{h^t_j - h^t_{j-1}}{\sum_{i=h^t_{j-1}+1}^{h^t_j} w_i} \geq \frac{h^*_j - h^*_{j-1}}{\sum_{i=h^*_{j-1}+1}^{h^*_j} w_i} \Rightarrow$$

$$h^{t+1}_j = h^t_{j-1} + \left\lceil \overline{W} \frac{h^t_j - h^t_{j-1}}{\sum_{i=h^t_{j-1}+1}^{h^t_j} w_i} \right\rceil \geq h^*_{j-1} + \left\lceil \overline{W} \frac{h^*_j - h^*_{j-1}}{\sum_{i=h^*_{j-1}+1}^{h^*_j} w_i} \right\rceil.$$

Based on $\sum_{i=h^*_{j-1}+1}^{h^*_j} w_i \leq \overline{W}$ we have that $\frac{\overline{W}}{\sum_{i=h^*_{j-1}+1}^{h^*_j} w_i} \geq 1$ so that $h^{t+1}_j \geq h^*_{j-1} + \left[h^*_j - h^*_{j-1}\right] = h^*_j$. Therefore, $h^t_j \geq h^*_j, \forall t \geq t_1$.

Based on Equation (18) we find that $h^t_j \geq h^{t+1}_j$, $\forall t \geq t_0$. Hence, the sequence of upper bounds $\{h^t_j, t > 0\}$ decreases and is bounded below by h^*_j so that it converges.

Equation (29) finally gives that the upper bounds are constant, $h^t_1 = h^*_1, \forall t > t_2$. ♠

In conclusion we have proved that

- The sequence $\{h_1^t, t > 0\}$ converges to h_1^*.
- If the sequences $\{h_k^t, t > 0\}$ converge to h_k^* for all $k < j$ then the sequence $\{h_j^t, t > 0\}$ converges to h_j^*.

Therefore the sequences of upper bounds $\{h_j^t, t > 0\}$ converge to h_j^* for all $j = 1, 2, ..., p$.

One might reasonably expect also to prove the convergence of the FGDLS algorithm in the case when the workload is monotonically increasing. Unfortunately, it has not been possible to establish convergence in this case and moreover we give a counter example that demonstrates convergence to a periodic solution in this case.

2.3 Numerical Results

In this section some numerical results are presented to illustrate the convergence of the FGDLS algorithm. Firstly, the workloads $\{w_i = 1001 - i, i = 1, ..., 1000\}$ are considered. Note that the workload decreases so that the FGDLS algorithm converges. The initial upper bounds are $h^1 = (250, 500, 750, 1000)$ with the corresponding lower bounds are $l^1 = (1, 251, 501, 751)$. The sequence of upper bounds $\{h_j^t, j = 1, 2, 3, 4\}$ for the first 5 iterations is given below.

```
t=1    250   500   750   1000
t=2    142   298   497   1000
t=3    134   292   498   1000
t=4    133   291   497   1000
t=5    133   291   497   1000
```

The corresponding execution times $T_j^t = \sum_{i=l_j^t}^{h_j^t}(1000 - i), j = 1, 2, 3, 4$, are given by

```
t=1  218,625  156,125   93,625   31,125
t=2  131,847  121,602  119,798  126,253
t=3  124,955  124,267  124,527  125,751
t=4  124,089  124,425  124,733  126,253
t=5  124,089  124,425  124,733  126,253
```

and are displayed in Figure 2. In this case convergence is achieved in only 5 steps.

Secondly, we investigate the case when the workloads $\{w_i = i, i = 1, 2, ..., 1000\}$ are monotonically increasing. The initial upper bounds are $h^1 = (250, 500, 750, 1000)$ with corresponding lower bounds $l^1 = (1, 251, 501, 751)$ The sequence of upper bounds $\{h_j^t, j = 1, 2, 3, 4\}$ for the first 6 iterations are

```
t=1    250   500   750   1000
t=2    499   699   856   1000
t=3    499   705   864   1000
t=4    499   706   865   1000
t=5    499   705   864   1000
t=6    499   706   865   1000
```

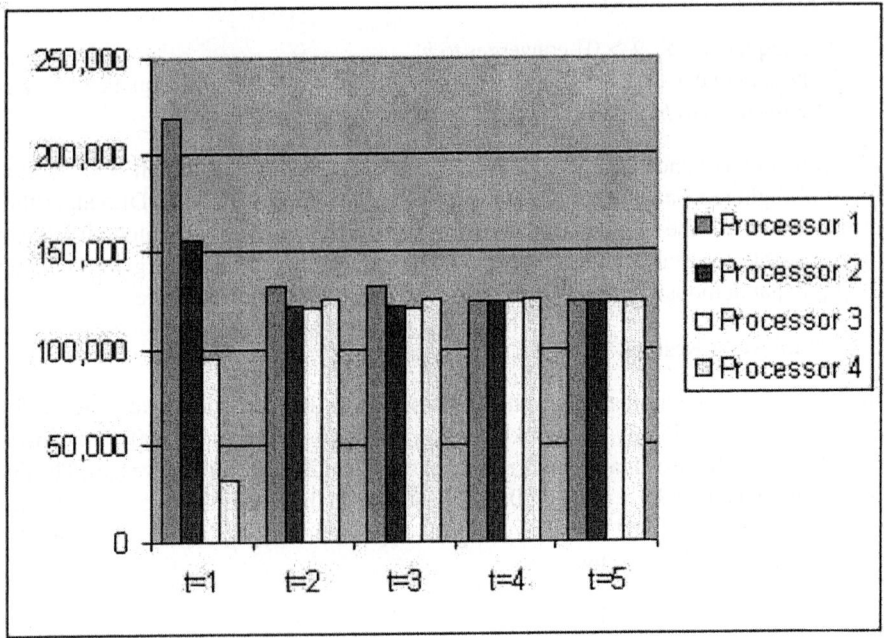

Fig. 2. The Running Times for the Workloads $w_i = 1001 - i$, $i = 1, ..., 1000$

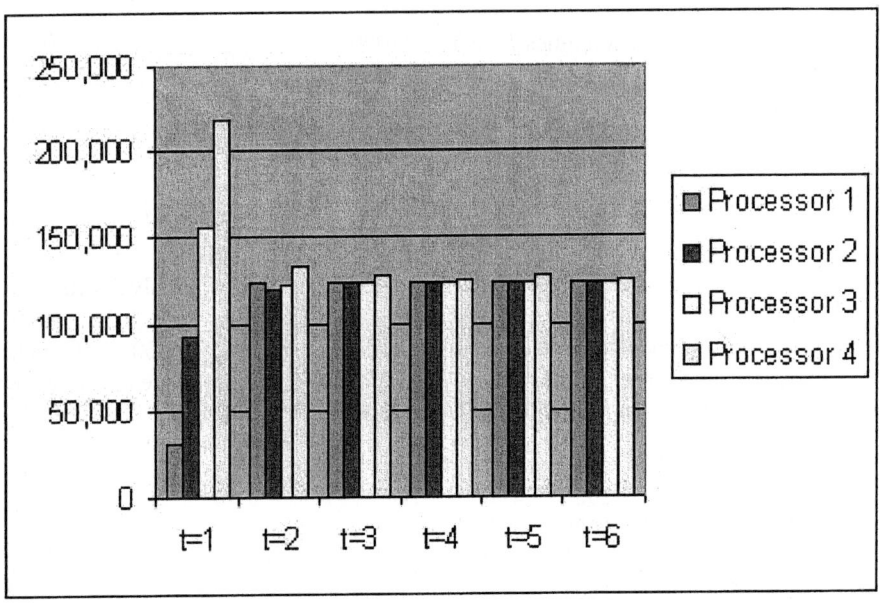

Fig. 3. The Running Times for the Workloads $w_i = i$, $i = 1, ..., 1000$

and the corresponding execution times $T_j^t = \sum_{i=l_j^t}^{h_j^t} i, j = 1,2,3,4$, are given by

```
t=1   31,375    93,875   156,375   218,875
t=2  124,750   119,900   122,146   133,704
t=3  124,750   124,115   124,815   126,820
t=4  124,750   124,821   124,974   125,955
t=5  124,750   124,115   124,815   126,820
t=6  124,750   124,821   124,974   125,955
```

and are displayed in Figure 3. In this case the convergence is not achieved since the second and third upper bounds h_2^t, h_3^t are periodic. Although the algorithm does not strictly converge, one can see that an acceptable load balance is achieved.

3 Conclusions

This paper has developed a convergence study for the FGDLS algorithm under the realistic assumption that the workloads $\{w_i, i=1,2,...,n\}$ are discrete. The convergence of the algorithm has been established in the case when the workloads are monotonically decreasing. Two numerical examples are presented; one demonstrates convergence in the case of a monotonically decreasing workload, the second illustrates failure to converge in the case of a monotonically increasing workload.

References

1. J.M. Bull, "Feedback Guided Loop Scheduling: Algorithms and Experiments", *Proceedings of Euro-Par'98*, Lecture Notes in Computer Science, Springer-Verlag, 1998.
2. J.M. Bull, R.W. Ford, and A.,Dickinson, "A Feedback Based Load Balance Algorithm for Physics Routines in NWP", *Proceedings of Seventh ECMWF Workshop on the Use of Parallel Processors in Meteorology*, World Scientific, 1996.
3. J.M. Bull, R.W.Ford, T.L.Freeman, and A. Hancock, "A Theoretical Investigation of Feedback Based Load Balance Algorithm", *Proceedings of Ninth SIAM Conference on Parallel Processing for Scientific Computing*, SIAM Press, 1999.
4. D.L. Eager and J. Zahorjan, "Adaptive Guided Self-Scheduling", *Technical Report 92-01-01*, Department of Computer Science and Engineering, University of Washington 1992.
5. S.F. Hummel, E. Schonberg, and L.E. Flynn, "Factoring: A Practical and Robust Method for Scheduling Parallel Loops", *Communications of the ACM*, vol. 35, no. **8**, pp.90-101, 1992.
6. S. Lucco, "A Dynamic Scheduling Method for Irregular Parallel Programs", *Proceedings of ACM SIGPLAN '92 Conference on Programming Language Design and Implementation*, pp.200-211, San Francisco, CA, 1992.
7. E.P. Markatos and T.J.LeBlanc, "Using Processor Affinity in in Loop Scheduling on Shared Memory Multiprocessors", *IEEE Transactions on Parallel and Distributed Systems*, vol. 5, no. **4**, pp. 379–400, 1994.
8. C.D.Polychronopoulos and D.J. Kuck, "Guided Self-Scheduling: A Practical Scheduling Scheme for Parallel Supercomputers", *IEEE Transactions on Computers*, vol. C-36, no. **12**, pp.1425-1439, 1987.
9. S. Subramanian and D.L. Eager, "Affinity Scheduling of Unbalanced Workloads", *Proceedings of Supercomputing'94*, IEEE Comp. Soc. Press, pp.214-226, 1994.

10. T.Tabirca, L.T.Freeman and S. Tabirca, "A Theoretical Application of Feedback Guided Dynamic Loop Scheduling", *Proceedings of the NATO Advanced Research Workshop on Advanced Environments, Tools and Applications for Cluster Computing*, Lecture Notes in Computer Science, Springer-Verlag, Vol. **2325**, pp.287-292, 2001.
11. T.Tabirca, L.T.Freeman, S. Tabirca and T.L.Yang, "An Application of Feedback Guided Dynamic Loop Scheduling to the Shortest Path Problem", *Proceedings of the International Conference on Parallel and Distributed Processing Techniques and Applications (PDPTA'02)*, CSREA Press, Bogart, Georgia, pp.1786-1789, 2001.
12. T.Tabirca, L.T.Freeman, S. Tabirca and T.L.Yang, "Feedback Guided Dynamic Loop Scheduling; A Theoretical Approach", *T M Pinkston Proceedings of the 2001 ICPP Workshops*, IEEE Computer Society Press, pp.115-121, 2001.
13. T.Tabirca, S.Tabirca, L.T. Freeman, T. Yang, "An $O(p+\log p)$ Algorithm for the Discrete FGDLS", Proceedings of The 2003 International Conference on Parallel Processing, ICPP-HPSECA 2003, Taiwan, pp. 164-170, 2003.
14. T.H.Tzen and L.M.Ni, "Trapezoid Self-Scheduling Scheme for Parallel Computers", *IEEE Trans. on Parallel and Distributed Systems*, vol. 4, no. **1**, pp. 87–98, 1993.

P-CBF: A Parallel Cell-Based Filtering Scheme Using a Horizontal Partitioning Technique

Jae-Woo Chang and Young-Chang Kim

Dept. of Computer Engineering, Chonbuk National University,
Chonju, Chonbuk 561-756, South Korea
{jwchang, yckim}@dblab.chonbuk.ac.kr

Abstract. To efficiently retrieve high-dimensional data in data warehousing and multimedia database applications, many high-dimensional index structures have been proposed, but they suffer from the so called 'dimensional curse' problem, i.e., the retrieval performance becomes increasingly degraded as the dimensionality is increased. To solve this problem, the cell-based filtering (CBF) scheme has been proposed, but it shows a linear decrease in performance as the dimensionality is increased. In this paper, we propose a parallel CBF scheme using a horizontal partitioning technique, which is called P-CBF, so as to cope with the linear decrease in retrieval performance. To achieve it, we construct our P-CBF scheme under an SN(Shared Nothing) cluster-based parallel architecture. In addition, we present data insertion, range query processing and k-NN query processing algorithms which are suitable for the SN architecture. Finally, we show that our P-CBF scheme achieves good retrieval performance in proportion to the number of servers in the SN architecture and that it outperforms a parallel version of the VA-File when the dimensionality is over 10.

1 Introduction

As the use of the Internet becomes more widespread with the development of network technology, it is necessary to deal with a large amount of multimedia data, such as images and videos. For the content-based retrieval of multimedia data, an object in a multimedia database can be defined as an n-dimensional feature vector. In addition, data warehousing provides for the organization of the information, so that the user can find the information that best corresponds to his or her request, by finding some patterns and tendencies among the data. For this, the information used in data warehousing can be composed of n-dimensional attribute vectors having several attributes. Therefore, it is necessary to research on high-dimensional index structures for efficiently retrieving high-dimensional data in data warehousing and multimedia database applications.

Thus, high-dimensional index structures such as K-D-B-tree [1], VAMSplit k-d-tree [2] and TV-tree[3], have been proposed. However, most of them were found to cause the so called 'dimensional curse' problem, in that the retrieval performance becomes increasingly degraded as the dimensionality is increased [5][6]. To solve this problem, the Cell-Based Filtering (CBF) scheme and the VA-file scheme were proposed. The VA-file performs filtering by using vector approximation information [7]. The CBF scheme performs filtering by using signatures, and shows good performance by redefining the maximum and minimum distances for good filtering [8]. However, both of these schemes show a linear decrease in retrieval performance as the dimensionality increases. In order to alleviate this linear decrease in retrieval performance, it is necessary to make use of a parallel processing technique. In this paper, we propose a parallel CBF scheme using a horizontal partitioning technique, which is called P-CBF. In order to maximize its retrieval performance, we construct our P-CBF scheme under an SN(Shared Nothing) cluster-based parallel architecture. For our P-CBF scheme, we also devise data insertion, range query processing, and k-NN query processing algorithms which are suitable for the SN parallel architecture.

This paper is organized as follows. In Section 2, we introduce the conventional CBF scheme. In Section 3, we propose our parallel CBF scheme using a horizontal partitioning technique. In Section 4, we provide experimental performance results. Finally, we draw our conclusions and suggest future works in Section 5.

2 Cell-Based Filtering (CBF) Scheme

The conventional high-dimensional index structures cause the 'dimensional curse' problem, in that the retrieval performance becomes even worse than that of a sequential scan when the dimensionality is high. To solve this problem, the VA-file scheme [7] and CBF scheme [8] were proposed. The VA-file minimizes the 'dimensional curse' problem by scanning a sequential file in which an approximation of each cell is stored. However, since the distance between the user's query and the cells is not the real distance between the query and the object, the VA-file scheme is affected by the error distance caused by the data distribution within a cell and the size of the cell. Meanwhile, the CBF scheme minimizes the error distance. The CBF scheme improves the retrieval performance by filtering out cells effectively using the signature of a feature vector, as well as the distance between the feature vector and the center of the cell containing it. To transform an object feature vector into a signature, the CBF scheme first divides a high-dimensional data space into cells, and it creates a signature which corresponds to a cell containing the object feature vector. The signature is constructed by concatenating the representatives of each dimension for the cell. When a user query is given, at first a query signature is generated by a signature generation algorithm. Secondly, the CBF scheme performs the filtering process, by sequentially searching each signature stored in a signature file. After obtaining the range of each cell through its signature, it regards candidate signatures as those signatures which are within the search range of the query vector. The CBF scheme is also able to improve the filtering process, by using a combination of the signature information and the cell distance. Finally, it retrieves the real feature vectors corresponding to a given query, by checking the candidate vectors corresponding to candidate signatures.

3 Parallel CBF Scheme Using a Horizontal Partitioning Technique

Though the CBF scheme alleviates the 'dimensional curse' problem, it still exhibits a linear decrease in retrieval performance as the dimensionality increases. We propose a parallel CBF scheme (P-CBF) using a horizontally-partitioned technique, which counteracts the linear performance decrease due to the increase in the dimensionality.

3.1 Overall Architecture of Our P-CBF

To respond to a user query, the CBF scheme sequentially scans signatures and feature vectors, which are stored in a signature file and a data file, respectively, on a physical disk. However, our P-CBF scheme distributes signatures and feature vectors over multiple processors by means of a declustering technique, and stores them on separate physical disks. In the existing declustering techniques, either horizontal or vertical partitioning is employed. In order to design an efficient declustering algorithm for our P-CBF scheme, it is necessary to consider the two properties of the CBF scheme.

Property 1: All signatures of the CBF scheme should be scanned sequentially in order to respond to a given query.

When a query is processed, the P-CBF scheme scans all of the signatures contained in the signature file. Thus, if the signatures are distributed evenly over multiple servers by a horizontal partitioning technique, the number of page accesses required is approximately equal to B/S, where S is the number of servers and B is the total number of blocks containing all of the signatures. If they are distributed into multiple servers by a vertical partitioning technique, however, we need approximately B/S page accesses, and additional merging time is also required for the partitioned signatures.

Property 2: The feature vectors corresponding to the candidate signatures in the CBF scheme should be partially searched in order to respond to a given query.

When a query is processed, the P-CBF scheme scans only those feature vectors corresponding to the candidate signatures contained in the data file. Thus, if the feature vectors are distributed evenly over multiple servers by a horizontal partitioning technique, the number of page accesses required is approximately equal to N/S, where S is the number of servers and N is the number of feature vectors corresponding to the candidate signatures. If the feature vectors are distributed over multiple servers by a vertical partitioning technique, however, we need approximately N/S page accesses, and additional merging time for the partitioned feature vectors is also required.

Based on the above two properties, we propose a parallel CBF scheme, called P-CBF, which distributes both signatures and feature vectors over multiple servers by means of a horizontal partitioning technique. Figure 1 shows the overall architecture

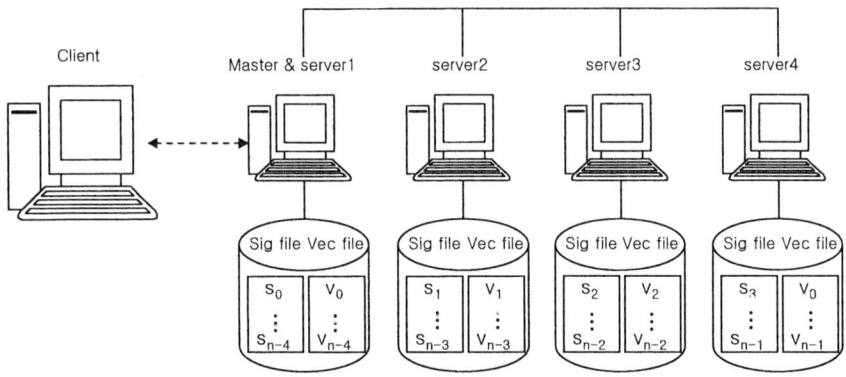

Fig. 1. Overall architecture of our P-CBF

of the parallel CBF scheme under the SN parallel architecture with four processors, i.e., servers. A special server serves as the master. The master distributes the inserted data over the multiple servers using a horizontal partitioning technique, and allows the servers to store the data on their own local disks. Meanwhile, the master also performs the parallel processing of a query received from a client, by first sending the query to multiple servers and allowing them to process it simultaneously. Then, the master computes the final answer based on the results obtained from the servers and transmits the final answer to the client.

3.2 Data Insertion

There are two files in our P-CBF scheme, i.e. the signature file and the data file. The signature file contains the signatures, as well as the distances between the feature vectors and the center of the cells in which they are contained. The data file includes the feature vectors of the objects and the object identifiers obtained from the database. To create the signature file, we first generate a signature from a given feature vector by using a signature generation algorithm. Then, the signature generated is merged with the distance between the feature vector and the center of the cell containing it. Finally, the signature, which has been merged with the distance, is stored in the signature file.

To store both the signature and data files in our P-CBF scheme, it is necessary to distribute them over multiple servers under a SN parallel architecture. In order to accomplish the uniform distribution of data over multiple servers using a horizontal partitioning technique, we devise an equation to assign an original n-dimensional vector (Vectori) and its signature (Signaturei) into a specific processor as follows.

$Pi = (int)[Signature_i >> (rand\_num(\text{the radius of Vector}_i) * Max\_SR)] \% Np$ Eq.(1)

Here, Pi means a processor to be assigned, Np means the number of processors used for parallelism, and Max_SR means the maximum number of shifting and rotating a signature. The rand-num() function generates a random number ranging from 0 to 1 and makes use of the radius of an n-dimensional vector as its seed. In addition, '>>' means a bit operator to shift and rotate a binary code, and (int)[] means an automatic type conversion from a binary code to an integer. Eq.(1) allows for the nearly uniform distribution of a set of vectors in the data file and their signatures over multiple processors (or servers) because even some n-dimensional vectors residing on the same cell can be distributed into different processors using the equation. Figure 2 shows the algorithm of data insertion.

```
Insertion_MasterNode (Vector data){
  Class server[MAXSERVER];
  // choose server id Pi to store vector using Eq(1)
  Pi = select_server(MAXSERVER, data);
  server[Pi].Insertion_ServerNode( data ); }
Insertion_ServerNode (Vector data) {
  signature = generate_signatre( data );
  write_data( signatureFile, signature );
  write_data( vectorFile, data ); }
```

Fig. 2. Data insertion algorithm

3.3 Range Query Search

A range query is expressed as a boundary which has an arbitrary distance from a given query point and it is processed as a way of searching for all of the objects that exist within the boundary. For a range query, a user inputs both a query vector and a distance value. The range query searches for all of the objects included within a circle which is centered on the query vector in data space and whose diameter is equal to the distance value. We first transform the query vector into a query signature, and then search the cells within the boundary given by the query signature. To respond to a range query in our P-CBF scheme, it is necessary to access multiple servers simultaneously, since both the signature and data files are distributed over multiple servers under a SN cluster-based parallel architecture. When a user query is processed, the master node creates as many threads as there are servers. Each thread sends the query to its own server. Each server searches its own signature and data files in parallel, thus retrieving the feature vectors that match the query.

```
Range_MasterNode ( query ) {
  transfer_query_to_server( query );
  list = Range_ServerNode( MAXSERVER );
  return integrate_result( list ); }
Range_ServerNode ( qeury) {
  signaturelist = Generate_siglist( signatureFile );
  qsig = generate_signature( query );
  candidatelist = find_sig ( signatureFile, qsig );
  for(;candidatelist!=NULL;)
  {   getdata( vectorFile, candidatelist.data );
      if(compute_range( candidatelist.data, query )
         <= query.range ) add_list( resultlist );
      candidatelist = candidatelist->next; }
  transfer_to_master ( resultlist ); }
```

Fig. 3. Range query processing algorithm

In this way, our P-CBF scheme improves the overall retrieval performance, by simultaneously searching multiple sets of signature and data files, because the same amount of data is distributed over multiple servers under a SN parallel architecture. Figure 3. shows the algorithm of range query processing.

3.4 k-Nearest Neighbor (k-NN) Search

The purpose of the k-NN search query is to find the nearest k objects to a query point in data space. The k-NN algorithm of our P-CBF consists of three filtering phases and an additional phase for integrating the result lists obtained by these three filtering phases, since the signature and data files are distributed over multiple servers under an SN parallel architecture. Figure 4 shows the algorithm of k-NN query processing.

```
KNN_MasterNode ( query ) {
 transfer_query_to_server( query );
 KNN_ServerNode( MAXSERVER );
 list = integrate_result();
 return sortlist(list); }
KNN_ServerNode ( query ) {
 sigbuf_list = find_sig ( query, signatureFile );
 // First Phase
 for( i=0; i<query.k ; i++)
 {   add ( cndlist, sigbuf_list );
     sigbuf_list = sigbuf_list->next; }
 for(;sigbuf_list!=null;)
 {   if(compare_dist(cndlist, sigbuf_list)>0)
     { delete(cndlist); add(cndlist, sigbuf_list ); }
     sigbuf_list=sigbuf_list->next; }
 // Second Phase
 for(;cndlist!=NULL;)
 {   if(compare_dist(cndlist.mindist, k_maxdist)>0)
         delete(cndlist);
     cndlist=cndlist->next; }
 // Third Phase
 for(;cndlist!=NULL;)
 {   if(compare_dist(cndlist.mindist, k_dist)>0)
     {   delete(cndlist); continue; }
     getdata( vectorFile, cndlist.data );
     if(compare_dist(cndlist.data,query)<=k_dist )
     {   add( resultlist, cndlist.data );
         delete(cndlist); } }
 transfer_to_master ( resultlist ); }
```

Fig. 4. k-NN query processing algorithm

In the first phase, we first create a thread for each server and send the k-NN query to each thread. Secondly, the CBF instance of each server generates a signature list (sigbuf_list) by searching all of the signatures in its own signature file. Thirdly, we generate the first candidate signature list by inserting signatures sequentially into the candidate list, without performing any comparison, until the number of signatures to be inserted is k. Once the number of signatures in the candidate list is k, each thread compares the next signature with the k-th candidate signature in the candidate list. If the next signature has a shorter distance from the given query point than the k-th candidate signature, then the k-th candidate signature is deleted from the candidate list and the next signature is inserted into it. Otherwise, we continue to compare the next signature with the k-th candidate signature until there are no more signatures to be compared. In the second phase, we reduce unnecessary page accesses by deleting cells whose lower bound is greater than the k-th upper bound (k_maxdist) in the candidate signature list. In the third phase, we obtain a result list by retrieving those real vectors which correspond to the candidate signatures of the candidate list. To accomplish this, we first compare the lower bound of the last candidate cell with the k-th object distance (k_dist). If k_dist is less than the lower bound of the last candidate cell, the last candidate cell is deleted from the candidate list. Otherwise, we calculate the distances between the query point and the real objects and generate a result list by obtaining the nearest k objects. Secondly, we transmit the result list to the master node. In the case where the number of servers is N, the number of nearest neighbors obtained from the multiple servers is k*N. Finally, we integrate the result lists obtained from the multiple servers and find the final k nearest objects in the master node.

4 Performance Analysis

For the performance analysis of our P-CBF scheme, Table 1 describes our experimental environment. At first, we compare our P-CBF scheme with the conventional CBF scheme in terms of the data insertion time and the search times for both range and k-NN queries. Secondly, we compare our P-CBF scheme with a parallel version of the VA-File.

Table 1. Experimental environment

System Environment	4 servers (each 450 MHz CPU, HDD 30GB, 128 MB Memory) Redhat Linux 7.0 (Kernel 2.4.5), gcc 2.96 (g++)	
Data Set	Synthetic data	2 million data (10, 20, 40, 50, 60, 80, 100-dimensional data)
	Real data	2 million data (10, 20, 50, 80-dimensional data)
Retrieval time	Range query	Retrieval time for searching 0.1 % of data
	k-NN query	Retrieval time for searching 100 nearest objects

4.1 Experimental Performance Analysis

To estimate the insertion time, we measure the time needed to insert two millions pieces of synthetic data. Table 2 shows the insertion time for both the existing CBF scheme and our P-CBF scheme. In the CBF scheme, it takes about 240, 480 and 1860 seconds to insert 10-, 20- and 100-dimensional data, respectively. In our P-CBF scheme, the same operations take about 250, 450 and 1930 seconds, respectively. It can be seen from the insertion performance result that our parallel CBF scheme is nearly the same as the existing CBF scheme. This is because our P-CBF can reduce the time required to insert data by using multiple servers, but it requires additional time due to the communication overhead.

Table 2. Insertion time for synthetic data (unit:sec)

Dimension / Scheme	10	20	40	50	60	80	100
Existing CBF	236.50	478.89	828.49	992.78	1170.40	1505.90	1862.16
P-CBF	256.12	445.42	824.02	1001.93	1188.42	1543.62	1930.10

The range query is used to search for objects within a certain distance from a given query point. For our experiment, we use a radius value designed to retrieve 0.1% of the data from the two million synthetic data. Table 3 shows the retrieval time for the range and k-NN query. The performance improvement metric of our P-CBF scheme against the existing CBF scheme can be calculated by means of Eq. (2), where PT and CT refer to their performance measurements (retrieval times) of our P-CBF scheme and that of the existing CBF scheme, respectively. In the existing CBF scheme, it takes about 13, 16, 38 and 60 seconds to respond to a range query in the case of 10-, 20-, 50- and 80-dimensional data, respectively. In our P-CBF scheme, it takes about 1.6, 2.2, 7.3 and 11.6 seconds for the same operations, respectively. When the number of dimensions is 80, the performance improvement metric of our P-CBF scheme is about 520. This is because our P-CBF scheme utilizes a large buffer under the SN parallel architecture with four servers.

$$\frac{1}{PT/CT}*100 \qquad \text{Eq.(2)}$$

The purpose of the k-nearest neighbors (k-NN) query is to search for the k objects which best match a given query. For the purpose of performance analysis, we measure the time needed to respond to a k-NN query for which k is 100. In the case of 10-dimensional data, it takes about 3.6 seconds to retrieve the data for the CBF scheme and 1.4 seconds for the P-CBF scheme. This is because in our P-CBF scheme, we retrieve objects from four servers under the SN parallel architecture simultaneously. In the case of 100-dimensional data, the CBF scheme requires about 76.4 seconds to respond to a k-NN query, while the P-CBF scheme requires about 25.6 seconds. Thus, it is shown that the performance improvement metric of the P-CBF scheme is about 300-400%, depending on the dimension of the data. The performance improvement for a k-NN query is relatively low, compared with that for a range query. This is because the overall retrieval performance for a k-NN query is very sensitive to the distribution of the data. That is, it entirely depends on the lowest retrieval performance among the four servers. When the number of servers is D, the overall performance of a k-NN query is not linearly increased in proportion to D. This is because an additional time is required to integrate the result lists obtained from the multiple servers.

Table 3. Retrieval time for range and k-NN Query using syntactic data

Scheme	Dimension	10	20	40	50	60	80	100
range	Existing CBF	13.20	16.42	31.91	37.93	44.96	59.50	74.27
range	P-CBF	1.59	2.24	5.92	7.27	7.87	11.60	14.04

Performance improvement metric		830%	733%	539%	522%	571%	513%	529%
k-NN	Existing CBF	3.59	17.80	31.92	39.28	46.72	61.60	76.39
	P-CBF	1.44	2.62	8.28	10.78	11.02	18.97	25.62
Performance improvement metric		249%	679%	386%	364%	424%	325%	298%

4.3 Comparison with Parallel VA-File

In this section, to verify the usefulness of the P-CBF scheme as a high dimensional indexing scheme, we compare P-CBF with the parallel VA-file. Table 4 describes the retrieval time for a range and k-NN query using two million pieces of real data. As dimensionality is higher, the P-CBF scheme achieves better retrieval performance than the parallel VA-file. This is because the P-CBF scheme performs good filtering by redefining the maximum and minimum distances as the dimensionality is increased.

In the case of the parallel VA-file, it takes about 1.8 seconds to respond to a k-NN query involving 20-dimensional data and about 11 seconds for 80-dimensional data. In the case of the P-CBF scheme, it takes about 1.8 seconds for 20-dimensional data and 10.5 seconds for 80-dimensional data. Thus, the performance of the P-CBF scheme is slightly better than that of the parallel VA-file in case dimensionality is high.

Table 4. Retrieval time for range and k-NN query using real data

Scheme	Dimension	10	20	50	80
range	Parallel VA-file	1.71	2.27	4.37	12.96
	Our P-CBF	1.49	1.82	2.75	10.08
k-NN	Parallel VA-file	1.51	1.82	7.17	11.32
	Our P-CBF	1.60	1.77	6.06	10.44

5 Conclusions and Future Work

Most of the conventional indexing schemes work well at low dimensionality, but perform poorly as the dimensionality of feature vectors increases. The CBF scheme was proposed to overcome the inefficiency of the conventional indexing schemes at high dimensionality. As the dimensionality is increased, the retrieval performance of the CBF scheme decreases linearly. To cope with this problem, we proposed the P-CBF scheme, which could uniformly distribute both signatures and feature vectors over multiple servers using a horizontal partitioning method under the SN parallel architecture. We showed from the performance analysis that our P-CBF scheme provided a near linear improvement in proportion to the number of servers, for both the range and k NN queries. We also showed that our P-CBF scheme outperformed the parallel VA-file when the number of dimensions is greater than 10. In the future work, it is required to apply our P-CBF scheme to real applications, such as data warehousing and multimedia databases, so as to be used as a high dimensional indexing scheme

References

1. J. T. Robinson, "The K-D-B-tree : A Search Structure for Large Multidimensional Dynamic Indexes", Proc. ACM SIGMOD Int. Conf. on Management of Data, pp. 10-18, 1981.
2. D.A. White and R. Jain, "Similarity Indexing : Algorithms and Performance", In Proc. of the SPIE : Storage and Retrieval for Image and Video Databases IV, Vol. 2670, pp.62-75, 1996.
3. H.I. Lin, H. Jagadish, and C. Faloutsos, "The TV-tree : An Index Structure for High Dimensional Data", VLDB Journal, Vol. 3, pp. 517-542, 1995.
4. S. Berchtold, D. A. Keim, H-P. Kriegel, "The X-tree : An Index Structure for High-Dimensional Data, Proceedings of the 22nd VLDB Conference, pp.28-39, 1996.
5. S. Arya, D.M. Mount, O. Narayan, "Accounting for Boundary Effects in Nearest Neighbor Searching', Proc. 11th Annaual Symp. on Computational Geometry, Vancouver, Canada, pp. 336-344, 1995.
6. Berchtold S., Bohm C., Keim D., Kriegel H. -P, "A Cost Model for Nearest Neighbor Search in High-Dimensional Data Space", ACM PODS Symposium on Principles of Databases Systems, Tucson, Arizona, 1997.
7. Roger Weber, Hans-Jorg Schek, Stephen Blott, "A Quantitative Analysis and Performance Study for Similarity-Search Methods in High-Dimensional Spaces," Proceedings of 24rd International Conference on Very Large Data Bases, pp.24-27, 1998.
8. S.-G. Han and J.-W. Chang, "A New High-Dimensional Index Structure Using a Cell-based Filtering Technique", In Lecture Notes in Computer Science 1884(Current Issues in Databases and Information Systems), Springer, pp. 79-92, 2000.
9. C. Faloutsos, "Design of a Signature File Method that Accounts for Non-Uniform Occurrence and Query Frequencies", ACM SIGMOD, 165-170, 1985.
10. J.-K. Kim and J.-W. Chang, "Horizontally-divided Signature File on a Parallel Machine Architecture," Journal of Systems Architecture, Vol. 44, No. 9-10, pp. 723-735, June 1998.
11. J.-K. Kim and J.-W. Chang, "Vertically-partitioned Parallel Signature File Method," Journal of Systems Architecture, Vol. 46, No. 8, pp. 655-673, June 2000..
12. N. Roussopoulos, S. Kelley, F. Vincent, "Nearest Neighbor Queries", Proc. ACM Int. Conf. on Management of Data(SIGMOD), pp. 71-79, 1995.

Adjusting the Cluster Size Based on the Distance from the Sink[*,**]

Sanghyun Ahn[1], Yujin Lim[2], and Jaehwoon Lee[3]

[1] School of Computer Science, University of Seoul, Seoul, Korea
ahn@venus.uos.ac.kr
[2] Department of Information Media, University of Suwon, Suwon, Korea
yujin@suwon.ac.kr
[3] Department of Information and Communications Engineering,
Dongguk University
jaehwoon@dongguk.edu

Abstract. One of the most important issues on the sensor network with resource-limited sensor nodes is prolonging the network lifetime by effectively utilizing the limited node energy. The most representative mechanism to achieve a long-lived sensor network is the clustering mechanism which can be further classified into the single-hop mode and the multi-hop mode. The single-hop mode requires that all sensor nodes in a cluster communicate directly with the cluster head (CH) via single hop and, in the multi-hop mode, sensor nodes communicate with the CH with the help of other intermediate nodes. One of the most critical factors that impact on the performance of the existing multi-hop clustering mechanism (in which the cluster size is fixed to some value, so we call this the fixed-size mechanism) is the cluster size and, without the assumption on the uniform node distribution, finding out the best cluster size is intractable. Since sensor nodes in a real sensor network are distributed non-uniformly, the fixed-size mechanism may not work best for real sensor networks. Therefore, in this paper, we propose a new dynamic-size multi-hop clustering mechanism in which the cluster size is determined according to the distance from the sink to relieve the traffic passing through the CHs near the sink. We show that our proposed scheme outperforms the existing fixed-size clustering mechanisms by carrying out numerical analysis and simulations.

Keyword: Sensor Network, Clustering, Wireless Network.

1 Introduction

The wireless sensor network is the network composed of wireless sensor nodes distributed over a specific area to monitor the current condition within that area.

---

[*] This research was supported by the MIC(Ministry of Information and Communication), Korea, under the Chung-Ang University HNRC-ITRC (Home Network Research Center) support program supervised by the IITA (Institute of Information Technology Assessment).
[**] This work was supported by grant No. R01-2004-10372-0 from the Basic Research Program of the Korea Science & Engineering Foundation.

Sensor nodes recognize and measure some requested phenomena, and send the sensed data to the sink via the wireless channel. The sink collects and analyzes data from sensor nodes. The sensor network is different from the mobile ad hoc network in the sense that sensor nodes have lower mobility and more restricted energy and denser distribution.

One of the most important issues in the sensor network is to prolong the network lifetime. In general, the network lifetime is defined as the time when for the first time any sensor node experiences energy depletion. Major part of node energy consumption comes from the radio communication.

In the sensor network, there are two categories of approaches to reducing the node energy consumption. The first approach is to turn off the radio of a node which does not need to send or receive data at the MAC and the network layers [1] [2] [3] [4]. The second approach is using the data aggregation to reduce the amount of the transmitted data for the reduction of the communication cost. The most representative mechanism belonging to this approach is the clustering mechanism. The clustering mechanism is very useful for those applications requiring scalability to efficiently handle several hundreds to thousands of sensor nodes. In the clustering mechanism, sensor nodes form a number of clusters and send their sensed data to the cluster heads (CHs) of the clusters that they belong to instead of sending them to the sink. Each CH aggregates collected data and sends the aggregated data to the sink in lieu of sensor nodes in its cluster.

The clustering mechanism can be further classified into the single-hop and the multi-hop clustering mechanisms according to the communication mode within a cluster [5]. In the single-hop mode, all the sensor nodes in a cluster communicate with the CH via single hop and, in this case, data is not relayed from sensor nodes to the CH by other intermediate sensor nodes. Because the communication between sensor nodes and the CH is direct, sensor nodes are not allowed to transmit data to the CH simultaneously and, therefore, the contention-less MAC protocol (such as TDMA) is preferred and each sensor node is required to send a join message to the corresponding CH.

On the other hand, in the multi-hop clustering, sensor nodes communicate with the CH via multiple hops and intermediate sensor nodes relay data to the CH, so there is no requirement on the contention-less MAC protocol. However, sensor nodes near the CH may suffer from extra overhead of relaying data between the CH and other sensor nodes.

Another aspect that we have to consider is the node distribution within a sensor network. Since there is no guarantee on the uniform distribution of sensor nodes, the node density within a cluster may be different from that within other clusters, so in the single-hop mode some specific CHs may get overloaded. On the other hand, the multi-hop mode can control the overhead imposed on a CH by determining the best cluster size. However, since the node distribution within a real sensor network is not uniform, it is infeasible to find out the best cluster size for a real sensor network.

Therefore, in this paper, we propose a clustering mechanism which can prolong the entire network lifetime by controlling the load on each CH with adjusting the size of each cluster according to the distance from the sink.

The rest of this paper is organized as follows: in section 2, we introduce the representative clustering mechanisms. In section 3, our proposed distance-based dynamic-size multi-hop clustering (DDMC) mechanism is described in detail. Section 4 presents the performance analysis by carrying out the numerical analysis and simulations. Section 6 concludes this paper.

2 Related Work

LEACH [6] and HEED [7] are the most representative clustering mechanisms using the single-hop mode. LEACH [6] is the mechanism whose goal is to balance the load on each CH by allowing each sensor node to become a CH in a round-robin fashion by applying Eq.1. P is the probability of a sensor node being elected as a CH (ex. P = 0.05), r indicates the number of the current round, and G is the set of nodes not elected as CHs for $\frac{1}{P}$ rounds. The node elected as a CH announces itself as a newly elected CH by broadcasting an advertisement message. Each non-CH sensor node receiving advertisement messages decides a cluster that it is going to join and sends a message notifying its join to the corresponding CH. This procedure is called the set-up stage and, once all sensor nodes join clusters, they enter into the steady-state. In the steady-state, each sensor node transmits its sensed data to the CH of the cluster that it belongs to instead of directly sending them to the sink. The CH receiving data from sensor nodes reduces the amount of the transmitted data by aggregating the collected data, and sends the aggregated data to the sink. A round is composed of the set-up and the steady-state stages and, for each round, a new CH is elected.

$$T(n) = \begin{cases} \frac{P}{1-P \times (r \bmod \frac{1}{P})} & \text{if } n \in G \\ 0 & \text{otherwise} \end{cases} \quad (1)$$

HEED [7] tries to increase the network lifetime by assigning the same probability of being a CH to each node and electing the node with the largest amount of available energy as a CH using Eq.2. $E_{residual}$ is the amount of available energy and E_{max} is the initial node energy.

$$CH_{prob} = CH_{prob} \times \frac{E_{residual}}{E_{max}} \quad (2)$$

S. Bandyopadhyay [8] has proposed a multi-hop clustering mechanism in which a sensor node elected as a CH with probability p broadcasts an advertisement message of its becoming a CH and other nodes relay this message up to k hops (which is the cluster size). In this case, the most critical factor that affects the performance is the cluster size and the best cluster size has been calculated with assuming the uniform distribution of sensor nodes. However, in the real network environment, it is almost impossible to distribute sensor nodes uniformly

```
compute Dist (distance from sink) using any message from sink
// Dist is distance from sink (e.g., Interest message from sink in Direct Diffusion)


compute  $CH_{prob} \leftarrow CH_{prob} \times \dfrac{E_{residual}}{E_{max}}$   // in case of HEED for Cluster Head election
TTL← default_TTL    // TTL means cluster size (e.g., default_TTL is 2)


if ( $CH_{prob} \geq Random(0,1)$ ) {   // I am a Cluster Head
        my_status ← Cluster Head
        if ( $\lfloor Dist/Threshold \rfloor > 0$ ) {    // e.g., Threshold is 3
           TTL ← TTL + $\lfloor Dist/Threshold \rfloor$ × Increment   // e.g., Increment is 2   }
        broadcast an advertisement message
}
```

Fig. 1. Algorithm of the Distance-based Dynamic-size Clustering (DDMC) mechanism

and, with non-uniform distribution of sensor nodes, it is not feasible to compute the best cluster size. Therefore, in this, paper, we propose a multi-hop clustering mechanism that dynamically adjusts the cluster size based on the distance from the sink.

3 Distance-Based Dynamic-Size Multi-hop Clustering Mechanism

We assume a contention-based MAC protocol, homogeneous sensor nodes (i.e., all sensor nodes with the same capability) and the multi-hop mode for our proposed mechanism. The multi-hop mode is adopted since it gives a higher probability of aggregation than the single-hop mode since usually the single-hop mode requires more CHs.

The existing multi-hop clustering mechanism assumes all clusters have the same size (we call this the fixed-size mechanism) and tries to compute the best cluster size with assuming the uniform node distribution. However, the node distribution in most real sensor networks is non-uniform (i.e., each cluster may have different node density) making the fixed-size mechanism less efficient in improving the network lifetime. Also, in a multi-hop cluster mechanism, nodes closer to the sink may have to relay more traffic than other nodes. Therefore, it may be more preferable to make clusters near to the sink smaller than others since the CH of a smaller cluster consumes less energy than that of a larger one. If the size of a cluster is determined according to the distance from the sink, the overhead imposed on a CH near the sink can be alleviated and this may extend the network lifetime.

In our proposed clustering mechanism, the distance-based dynamic-size multi-hop clustering (DDMC) mechanism, a node is elected as a CH by applying either Eq.1 or 2 (i.e., the CH election mechanism of either LEACH or HEED) (in this paper, we do not focus on how to elect a CH, but on the way of adjusting the cluster size to prolong the network lifetime) and the elected CH broadcasts an advertisement message with TTL being set to the size of the corresponding cluster. The initial cluster size is set to a default value and, after that, the cluster size is increased according to the distance from the sink. The distance (or hops) of a CH from the sink is determined from the information included in the messages sent by the sink (for instance, the interest message of the direct diffusion can provide this information). A CH increases its cluster size by using the following equation:

$$TTL \leftarrow TTL + \lfloor Dist/Threshold \rfloor \times Increment \qquad (3)$$

A sensor node receiving advertisement messages decides the cluster that it is going to join, but it does not need to send a join message to the corresponding CH since the contention-based MAC protocol is used (i.e., the CH does not have to maintain the list of nodes within the cluster). A sensor node not receiving any advertisement message within a specific time interval becomes a CH, and a sensor node receiving more than one advertisement messages selects the CH which seems to be the nearest to itself (the TTL value within an advertisement message can be used for this purpose). Once a cluster is formed, sensor nodes within the cluster send data to the CH and the CH aggregates the collected data and transmits the aggregated data to the sink.

4 Performance Evaluation

4.1 Numerical Analysis on a Sample Sensor Network

In order to evaluate the performance of the proposed DDMC mechanism, we numerically analyze the DDMC and the fixed-size multi-hop clustering mechanism for a simple sensor network shown in figure 2. To simplify the analysis, the sensor network is represented as a grid with 8×8 cells and the size of each cell is $r \times r$ and at each crosspoint a sensor node is placed. The transmission range of each node is $\sqrt{2}r$ and the sink is placed on the lower right corner of the grid. For the simplification of the analysis, we have assumed that each sensor node knows the direction to the CH of the cluster to which it belongs and that to the sink. The communication between a CH and the sink is carried out by multiple hops with the help of intermediate nodes. Also we assume that all the sensor nodes are triggered at the same time (i.e., all the sensor nodes sends sensed data to the sink simultaneously) and each monitoring cycle consists of sensing by sensors and collecting the sensed data by the sink.

For the DDMC mechanism, the size of the clusters located within the area of $4 \times \sqrt{2}r$ from the sink is set to $2r \times 2r$ and that of other clusters to $4r \times 4r$ as shown in figure 2. For the fixed-size multi-hop clustering mechanism, two cases

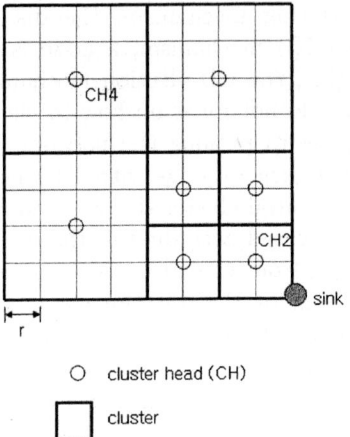

Fig. 2. The DDMC mechanism with 2r × 2r and 4r × 4r clusters

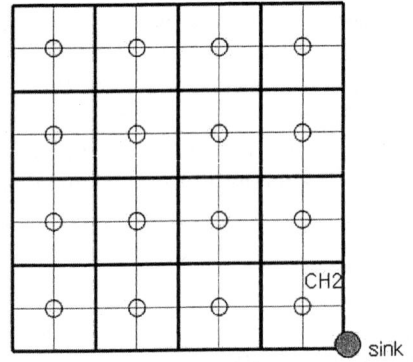

Fig. 3. A sensor network using the 2r × 2r fixed-size multi-hop clustering mechanism

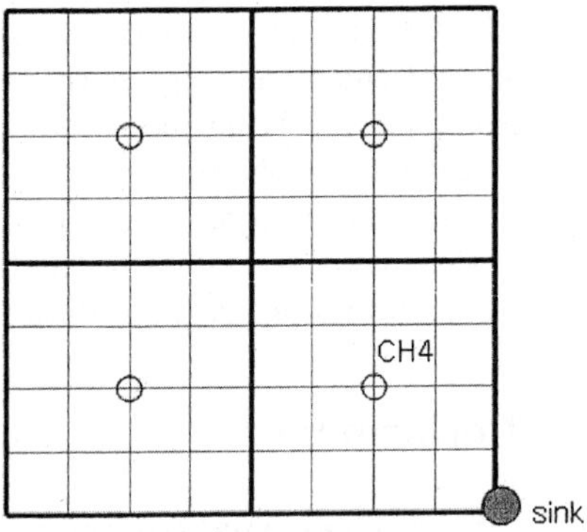

Fig. 4. A sensor network using the 4r × 4r fixed-size multi-hop clustering mechanism

with 2r × 2r and 4r × 4r cluster sizes are analyzed. From now on, 'Fixed(2)' will represent the case of 2r × 2r cluster size (figure 3), and 'Fixed(4)' that of 4r × 4r cluster size (figure 4).

We assume the same communication model of [5] which follows a simple communication model for the transceiver similar to the one in [9]. In this communication model, the amount of energy required to transmit a packet over distance x is $l + \mu x^k$, where l is the amount of energy spent in the transmitter electronics circuitry and μx^k is that spent in the RF amplifiers to counter the propagation loss. Here, μ takes into account the constant factor in the propagation loss term, as well as the antenna gains of the transmitter and the receiver. The value of the propagation loss exponent k is highly dependent on the surrounding environment. In the free space environment, the value of k is 2 and, in other environments, the value of k is high from 3 to 5. When receiving a packet, only the receiver circuitry is invoked and, as a result, the energy spent on receiving a packet becomes l. Therefore, the amount of energy spent on relaying a packet over distance x is $2l + \mu x^k$.

Under the assumption that nodes know the shortest routes to the sink, there can exist more than one routes from a CH to the sink. For the simplification of the analysis, we assume the worst case in which aggregated packets from all CHs to the sink are relayed by the CH which is the nearest to the sink and nodes on the boundary of more than one clusters belong to those clusters.

In the DDMC mechanism, the node determining the network lifetime can be either the CH which is the nearest to the sink (labeled with 'CH2' in figure 2) or the CH of the 4r × 4r cluster which is the nearest to the sink (labeled with 'CH4' in figure 2). For each cycle, the energy spent by CH2 and CH4 is

$$E_{DDMC\_CH2} = \{f(9) \times (l + \mu(\sqrt{2}r)^k) + 8l\}$$

$$+ 3 \times f(25) \times \{l + (l + \mu(\sqrt{2}r)^k)\} + 3 \times f(9) \times \frac{l + (l + \mu(\sqrt{2}r)^k)}{2} \quad (4)$$

and

$$E_{DDMC\_CH4} = f(25) \times (l + \mu(\sqrt{2}r)^k) + 24l \quad (5)$$

respectively, and here f(x) means the number of packets resulted from aggregation of x packets.

In Fixed(2), the CH which is the nearest to the sink (labeled with 'CH2' in figure 4) determines the network lifetime and the energy spent by CH2 for a cycle is

$$E_{Fixed\_CH2} = \{f(9) \times (l + \mu(\sqrt{2}r)^k) + 8l\} + 15 \times f(9) \times \{l + (l + \mu(\sqrt{2}r)^k)\} \quad (6)$$

And, in Fixed(4), the CH which is the nearest to the sink (labeled with 'CH4' in figure 3) determines the network lifetime and the energy spent by CH4 for a cycle is

$$E_{Fixed\_CH4} = E_{DDMC\_CH4} + 3 \times f(25) \times \{l + (l + \mu(\sqrt{2}r)^k)\} \quad (7)$$

If we set l, μ and k to 0.21mJ, 5.46pJ and 4 (from the same system parameter values used in [5]) and f(x) = 1 for any x and r = 100m, then E_{DDMC_CH2} = 5.9388mJ, E_{DDMC_CH4} = 5.4684mJ, E_{Fixed_CH2} = 11.59mJ and E_{Fixed_CH4} = 7.3836mJ. From this result, we can see that the DDMC mechanism outperforms the fixed-size multi-hop clustering mechanism. In the above computation, we have assumed f(x) = 1 for any x which means each CH aggregates any number of collected packets to only 1 packet (that is, only one packet is transmitted by a CH to the sink for each cycle). For a larger value of f(x) (note that f(x) ≥ 1 and f(x) is positively proportional to x), the DDMC mechanism will give much better performance than the fixed-size clustering mechanism since the fixed-size multi-hop clustering mechanism consumes more energy for a larger f(x).

4.2 Simulations

For the simulation, we have used the NS-2 simulator and the sensor network extension package of NRL [10]. Simulations are performed for the range of 1000m × 1000m with randomly distributed 120 ~ 180 nodes. The initial energy for each sensor node is set to 7 J (Joule) and each sensor node can store up to 50 packets. If the available energy of a sensor node is less than or equal to 10^{-4} J, the node is assumed to be not working. The transmission range of a sensor node is 150m and the packet length is 100 bytes. Each sensor node is triggered by an event at every second, so each sensor node has to send its sensed data to the sink at every second.

The performance of the DDMC mechanism is compared with the fixed-size multi-hop mode, and the CH selection rules of LEACH and HEED are both applied. The initial cluster size is set to 2 and the cluster size is increased by 2 whenever the distance from the sink becomes multiple of 3. The performance evaluation factors considered for the simulation are the network lifetime and the variance of the available energy of each node.

Figures 5 and 6 show the network lifetime with varying the number of nodes. Figure 5 is the result obtained by using LEACH as the CH election mechanism, and Fixed(4) and Fixed(2) represent the fixed-size multi-hop mechanism with fixing the cluster size to 4 and 2, respectively. In the multi-hop mode, if all the sensor nodes receiving an advertisement message forward the message, the overhead caused by the forwarding can become severe especially in a dense sensor network. Therefore, we have tried to reduce this overhead by allowing nodes located near the boundary of a cluster (by using the Received Signal Strength Indicator (RSSI)) to forward the advertisement message. Figure 5 shows that the network lifetime of Fixed(2) is shorter than that of Fixed(4). The reason for this is that the fixed 2-hop clustering requires more CHs than the fixed 4-hop clustering, resulting in more advertisement messages. And the DDMC mechanism shows better performance than Fixed(4) since the DDMC mechanism alleviates

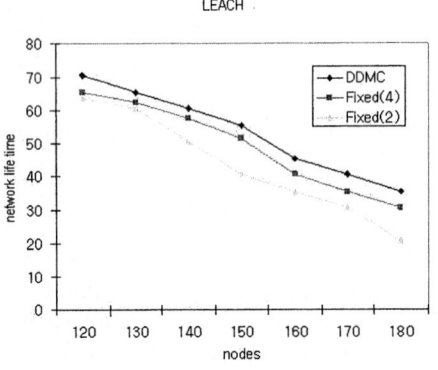

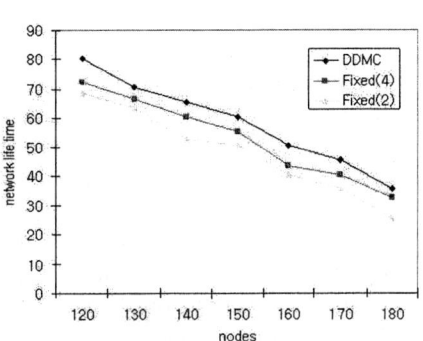

Fig. 5. Network lifetime vs. the number of nodes (using LEACH)

Fig. 6. Network lifetime vs. the number of nodes (using HEED)

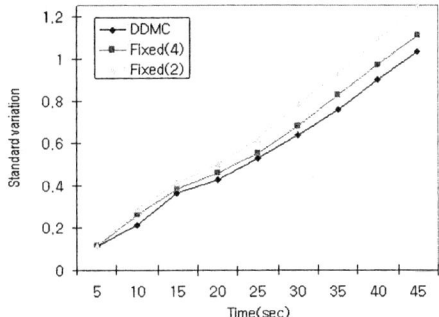

Fig. 7. Standard variation of available energy of each sensor node (using HEED)

the overhead imposed on the CHs near the sink by keeping their cluster sizes small and reduces the number of CHs by increasing the cluster size according to the distance from the sink. Figure 6 shows the network lifetime when HEED is used for the election of CHs. Overall, HEED gives longer lifetime than LEACH. This shows that HEED considering the available energy of a node is more effective in prolonging lifetime than LEACH making all nodes become CHs in a round-robin fashion. As a summary, the DDMC mechanism gives about 10 % increase of the network lifetime than Fixed(4) and 20 ∼ 25 % than Fixed (2), which is almost the same results as those obtained by the numerical analysis .

Figure 7 shows the standard variation of the available energy on each node with using HEED as the CH election mechanism. As shown in the figure, the DDMC mechanism yields smaller standard variation of the available energy, which indicates that traffic imposed on each sensor node is relatively well balanced than other schemes.

5 Conclusion

One of the most important issues on the sensor network with resource-limited sensor nodes is prolonging the network lifetime by effectively utilizing the given energy. The most representative mechanism to prolong the network lifetime is the clustering mechanism. The clustering mechanism is classified into the single-hop mode and the multi-hop mode. In the single-hop mode, all the sensor nodes in a cluster communicate with the CH via single hop and, as a result, the contention-less MAC protocol is preferred. On the other hand, the multi-hop mode does not need the contention-less MAC protocol. One of the main issues on the multi-hop mode is to determine the best cluster size for prolonging the network lifetime. However, it is almost impossible to find out the best cluster size for a real sensor network with non-uniform node distribution.

Therefore, in this paper, we have proposed a clustering mechanism to prolong the network lifetime by adjusting the cluster size according to the distance from the sink. The performance of the proposed clustering mechanism, the distance-based dynamic-size multi-hop clustering (DDMC) mechanism, is evaluated by carrying out the numerical analysis and the simulation. The results from the numerical analysis and the simulation show that the proposed DDMC mechanism outperforms the existing fixed-size clustering mechanism in terms of the network lifetime and the standard variation of the available energy on each node (i.e., load balancing).

References

1. W. Ye, J. Heidemann, and D. Estrin, "An energy-efficient mac protocol for wireless sensor networks", IEEE Infocom, pp1567-1576, June 2002.
2. A. Cerpa and D. Estrin, "ACENT: Adaptive self-configuring sensor networks topologies", IEEE Infocom, pp1278-1287, June 2002.
3. B. Chen, K. Jamieson, H. Balakrishnan, and R. Morris, "SPAN: An energy-efficient coordination algorithm for topology maintenance in ad hoc wireless networks", ACM/IEEE Mobicom, pp85-96, July 2001.
4. V. Kawadia and P. R. Kumar, "Power control and clustering in ad hoc networks", IEEE Infocom, pp459-469, April 2003.
5. V. Mhatre and C. Rosenberg, "Design guidelines for wireless sensor networks: communication, clustering and aggregation", Ad-hoc networks journal, Elsevier science, vol. 2, pp45-63, 2004.
6. W. R. Heinzelman, A. Chandrakasan, and H. Balakrishnan, "Energy-efficient communication protocol for wireless microsensor networks", IEEE Hawaii international conference on system sciences, January 2000.
7. O. Younis and S. Fahmy, "Distributed Clustering in Ad-hoc Sensor Networks: A hybrid, energy-efficient approach", IEEE Infocom, pp629-640, March 2004.
8. S. Bandyopadhyay and E. J. Coyle, "An energy efficient hierarchical clustering algorithm for wireless sensor networks", IEEE Infocom, pp1713-1723, April 2003.
9. W. Heinzelman, A. Chandrakasan and H. Balakrishnan, "An Application-Specific Protocol Architecture for Wireless Microsensor Networks", IEEE transactions on wireless communications, Vol. 2, No. 4, Oct. 2002.
10. NRL's sensor network extension to ns-2, http://nrlsensorsim.pf.itd.nrl.navy.mil/

An Efficient Distributed Search Method[*]

Haitao Chen, Zhenghu Gong, and Zunguo Huang

School of Computer Science, National University of Defense Technology,
Changsha, Hunan, China
nchrist@163.com

Abstract. The big challenge of constructing P2P applications is how to implement efficient distributed file searching in complex environment which implies huge-amount users and uncontrollable nodes. FriendSearch is introduced to improve the efficiency and scalability of distributed file searching. FriendSearch introduces a new hybrid architecture in which the storage and search of raw file is based on DHT network, but the storage and search of meta-data is based on unstructured P2P network. FriendSearch learns interest similarity between participating nodes and uses it to construct friend relations. The forwarding of queries is limited to friend nodes with similar interests. Simulation tests show that FriendSearch algorithm is both efficient and scalable.

1 Introduction

With the development of Internet applications in scope and depth, the role of ordinary nodes changes from just receiving content passively from servers to acting as a supplier of Internet content. P2P technology provides a new application pattern which enables the edge nodes participate the Internet application as both clients and servers at the same time. It offers powerful support for the construction of huge and complicated distributed network application. P2P file sharing applications which allow ordinary user to share files in local disk to others, have become one of the most popular Internet applications.

The big challenge of constructing P2P application is how to implement efficient distributed file searching in complex environment which implies decentralized and huge-amount users, uncontrollable nodes with unbalanced computing capacity and network connection. This paper presents a new distributed file search methods – FriendSearch. FriendSearch constructs friend relations between nodes based on interest similarity and limits query broadcast to nodes with similar interests.

2 Related Work

The current mainstream P2P file sharing applications can be divided into centralized search model, broadcast search model, and hierarchical search model. Napster[1] is a typical centralized search model. It can search effectively and reliably. But the

---

[*] This research is supported by the National Grand Fundamental Research 973 Program of China under Grant No.2003CB314802, also the National High-Tech Research and Development Plan of China under Grant No.2003AA142080.

directory server is a single point of failure and performance bottleneck. Gnutella[2] is a typical broadcast search model. The main disadvantage of broadcast model is high bandwidth consumption which leads to bad scalability. Furthermore, the search results of this model is uncertain which means the documents existing somewhere in the network maybe can not be located. Kazaa[3] and JXTASearch[4] are typical hierarchical search models. The model enhances the stability and scalability through super-nodes. But the super-nodes are new performance bottle-neck. Also the communication between super-nodes depends on broadcast routing, which restricts scalability.

The problems of existing distributed file search systems include high bandwidth consumption, poor search pattern, bad ranking of results and so on. According to architecture, these projects can be partitioned into search in structured P2P network and search in unstructured P2P network.

Many researches concentrate on search in unstructured P2P network, such as Freenet[5], NeuroGrid[6], APPN[7], GS[8], Alpine[9] and so on. Freenet[5] can guarantee the anonymity of publisher, reader, and storage space supplier. The requests will be routed to the most possible positions. Its performance is almost as good as DHT, but it still has some uncertainty. NeuroGrid[6] abstracts the knowledge of documents distribution from the search results and makes routing decision based on keywords distribution of other nodes. The disadvantages of NeuroGrid are: the information updating speed is slow; the result of query is uncertain; the size of route table is interrelated with the number of files and the number of nodes in the network. APPN[7] optimizes search process through the construction of associated rule. One kind of simple associated rule is possession rule which means owning some special files. The spending of establish rule and the choice strategy of rules are still problems which APPN faces. GS[8] establishes shortcuts among nodes based on principle of interest locality. The method is simple and effective, but it lacks inspection for large scale network. Alpine [9] manages the sharing information by groups, and each member of a group will evaluate other members' trust degree according to the satisfaction with their services. For this method the establishing of group depends on the user contact out of the P2P network, also the scalability of group is limited.

Structured P2P network can finish search process in several limited hops, which provide a good method for deterministic search. Some researches devote to realize fuzzy searching in DHT network. PSearch[10] makes use of LSI (Latent Semantic Indexing) to construct semantic space of files and maps the file vector space to CAN space. PSearch only supports keywords query and is a promising method for text search in distributed environment. Semplesh[11] presents a method of mapping the RDF Triples to DHT in which every item of RDF Triples will be mapped to DHT network once. There are problems for Semplesh such as low search efficiency, heavy workload for popular item and lacking support for substring matching.

3 Search Based on Friend Relations

3.1 Hybrid Architecture

This paper presents hybrid architecture- HA. HA includes file search layer and metadata layer, in which storage and search of raw documents are based on structure P2P while the storage and search of metadata are based on unstructured P2P network. This architecture combines the advantages of deterministic search in structured P2P network and the advantages of fuzzy search in the unstructured P2P network. Unstructured P2P network can express easily the complex relations among the file meta-data and meet the users' diverse query needs. DHT network can effective solve problems such as file moving, file replicating and download. It also supports for the discovery of file relations.

Each node of HA participates in two kinds of networks at the same time. The construction of unstructured network relies on the DHT network. The nodes publish raw files in DHT network. Then these raw files will establish relations according to their metadata. At last the nodes can establish friend relations according the relating of raw files. The relations of files and friend relations between nodes belong to the metadata layer. Based on the friend relations between nodes, most search requests can be restricted in a very limited scope.

3.2 Algorithm of Constructing Friend Relations

One key problem of FriendSearch is how to construct overlay network of friend relations based on file possession relations. The basic process of constructing friend relations includes three steps. 1) First each node collects all nodes that share same files as friend candidates in the bottom DHT network. 2) Then it ranks these friend candidates from high to low according the number of sharing same files. 3) At last it chooses the first k nodes as its friend nodes. If we view friend relations as directed edge, the graph that is make of nodes and friend relations is a directed graph with high clustering coefficient. The total overhead of this algorithm is linear with number of nodes. But the computation is distributed. Computing overhead of each node is only linear with its sharing files, which is obviously an acceptable overhead. At regular intervals, the algorithm has to run. Also the algorithm can run only when the search success rate is low enough. Simulation tests show that friend relations between nodes are very stable, so the algorithm needs not to run with high frequency.

3.3 Search Algorithm Based on Friend Relations

Search algorithm based on friend relations includes two steps. 1) First it makes use of two level friend relations to search. Original query nodes will broadcast query to all its direct friend nodes. If direct friend nodes fail, the query will be forwarded to the friend nodes of its direct friend nodes. Most requests can receive responses in first step. 2) Then if the friend search fails, it adopts efficient DHT-based flooding search as supplement. The search process uses cache to improve performance.

4 Simulation Test

For fully test of FriendSearch algorithm, we adopt many groups of test data and many test measures. We use simulation data and web log data to test the performance of pure multi-hops friend relations in section 4.1 and section 4.2.

We evaluate the performance of search algorithm using the follow targets. Success rate denotes the proportion of success rate of search. Search Consumption is the

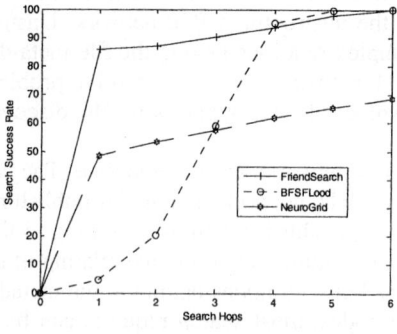 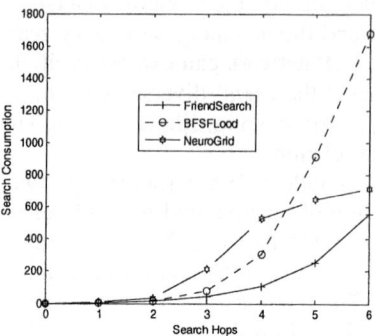

Fig. 1. Hops VS Search Success Rate **Fig. 2.** Hops VS Search Consumption

number of peers in the system involved in query processing for each query. A smaller query scope increases system scalability. Query hop stands for the average delay for reply to come back.

4.1 Simulation Test Based on Simulating Data

NeuroGrid simulator [12] is a generic P2P simulator developed by Tokyo University of Japan. It has many configurable parameters and can implement simulation of P2P file sharing system with good expansibility. We implement FriendSearch search algorithm on the base of NeuroGrid simulator.

NeuroGrid simulator has several configuration parameters and can generate different test data. We found that test results of FriendSearch were stable. So here we just use typical results. Simulation tests compare FriendSearch with BFSFLood[13] and NeuroGrid[5]. BFSFlood is an improved version of flooding algorithm, which randomly chooses fixed number of neighbor nodes to forward query. Figure 1 shows that FriendSearch gains more than 80 percent search success rate at first hop, the search success rate enhance slowly with the searching hop increased. Figure 2 shows search consumption augment rapidly with the search hop increased, and FriendSearch keep the lowest rising speed.

We can draw these conclusions from the test results:

◆ The size of route table of FriendSearch algorithm is fixed and small. NeuroGrid algorithm can get good performance at the cost of more than one hundred of items in its route table, so the cost of maintenance is high.
◆ Search success rate of FriendSearch algorithm is very high. Especially it can get more than 80 percent success rate just at the first hop.

4.2 Simulation Test Based on Web Log

Web log data and P2P access data are similar at some ways and they both obey some same laws. Collecting the web log data is much easier than P2P data, so many researches [7] [8] adopt web log data as test data of P2P search research.

We adopt three groups of wildly-used web log data to test FriendSearch algorithm.

◆ Boston [14] is the web log data of Boston University which contains 558261 records, 538 nodes and 9431 files.
◆ Berkeley [14] is the web log data of Berkeley University which contains 1703836 records, 5222 nodes and 116642 files.
◆ Boeing [14] is the web log data of Boeing Corporation which contains 4421526 records, 28895 nodes and 254240 files.

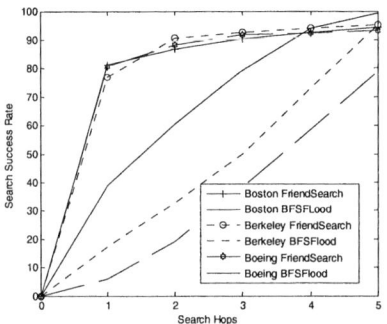 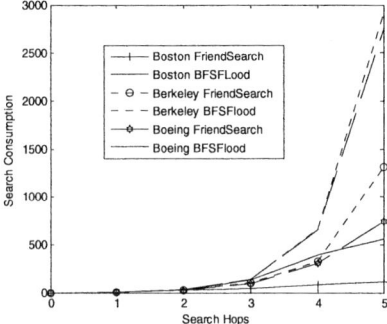

Fig. 3. Search Success Rate **Fig. 4.** Search Consumption

These three groups of web log data respectively represent access case of different scale networks. The test method is similarly with the method in [7]. The test results are shown in figure 3 and figure 4. We can draw these conclusions from the test results:

◆ Three groups of test results of FriendSearch algorithm are similar, which shows FriendSearch algorithm's stability.
◆ Simple friend relations construction can guarantee very high search success rate. Search success rates of FriendSearch algorithm in three groups of test data are very high, especially the searching success rate can exceeds 75 percent at the first hop.
◆ Search consumption of FriendSearch is low and the search efficiency of FriendSearch algorithm is high.

5 Conclusions

This paper researches on distributed file search in complex environment and presents a new method of search sharing files. FriendSearch constructs friend relations between nodes based on search interests and sharing files. The search requests firstly are forwarded to friend nodes. Only failed requests will continue to broadcast in DHT flooding pattern.

Simulation tests show that FriendSearch algorithm is efficient and stable. FriendSearch brings the performance to within an order of magnitude of improvement compared with classical algorithms such as BFSFlood[13], NeuroGrid[6] and so on. The future researches include: 1) more effective algorithm of constructing friend relations. For example it can take file rarity into count. 2) Adding semantic description on friend relations to improve the expansibility of system.

References

1. Napster. www.napster.com. 2005.
2. Gnutella. www.gnutella.com. 2005.
3. Kazaa. www.kazaa.com. 2005.
4. JxtaSearch. http://search.jxta.org/. 2005.
5. Clarke, I., Sandberg, O., Wiley, B. and Hong T. W. Freenet: A Distributed Anonymous Information Storage and Retrieval System. In Proc of the Workshop on Design Issues in Anonymity and Unobservability, Ed. Federrath H., Berkeley, CA, July 2000.
6. Joseph, S.R.H. NeuroGrid: Semantically Routing Queries in Peer-to-Peer Networks. International Workshop on Peer-to-Peer Computing, Pisa (2002).
7. Edith Cohen, Amos Fiat, Haim Kaplan. Associative Search in Peer to Peer Networks: Harnessing Latent Semantics. In Proc of INFOCOM 2003.
8. Kunwadee Sripanidkulchai, Bruce Maggs, Hui Zhang. Efficient Content Location Using Interest-Based Locality in Peer-to-Peer Systems, In Proc of INFOCOM 2003.
9. Alpine. http://www.cubicmetercrystal.com/alpine/. 2005.
10. Chunqiang Tang, Zhichen Xu, Sandhya Dwarkada. Peer-to-Peer Information Retrieval Using Self-Organizing Semantic Over-lay Networks. In Proc of SIGCOM 2003.
11. Semplesh. http://www.plesh.net/. 2005.
12. Sam Joseph, An Extendible Open Source P2P Simulator. P2P Journal. 2003.
13. V. Kalogeraki, D. Gunopulos, and D. Zeinalipour-Yazti. A Local Search Mechanism for Peer-to-Peer Networks. In Proc of CIKM, 2002.
14. webtraces. http://www.web-caching.com /traces-logs.html.

Practical Integer Sorting on Shared Memory

Hazem M. Bahig and Sameh S. Daoud

Department of Mathematics, Faculty of Science, Ain Shams University,
Cairo, Egypt
hbahig@asunet.shams.edu.eg

Abstract. Integer sorting is a special case of the sorting problem, where the elements have integer values in the range $[0, n-1]$. We evaluate the behavior and performance of two parallel integer sorting algorithms in practice. We have used Ada tasks facilities to simulate the parallel sorting algorithms on a machine consisting of 4-processors. Both algorithms are based on self-index method on EREW PRAM [3,6]. Also, we study the scalability of the two parallel algorithms and compare between them.

1 Introduction

The sorting problem is a rearrangement of the elements of an array $X = \{x_0, x_1, \cdots, x_{n-1}\}$ into an array $X' = \{x'_0, x'_1, \cdots, x'_{n-1}\}$ such that $x'_i \leq x'_{i+1}$, $\forall 0 \leq i < n-1$. Integer sorting is a special case of the sorting problem, where the elements are drawn from the restricted domain $[0, n-1]$. The parallel integer sorting is used in many fields such as simulation and combinatorics.

Many algorithms are suggested to solve the integer sorting problem by parallel processing. The difference between them are depend on the kind of parallel models, techniques, deterministic or not, and domain of elements. Some of these algorithms are based on counting (self-index) technique under the computational model PRAM(Parallel Random Access Machine PRAM). PRAM consists of p processors that can communicate via a shared memory of m cells. There are many different ways for the processors to read/write the datum from/into the shared memory. One of them when no two processors are allowed to read or write the same shared memory cell simultaneously. In this case, the model is called **Exclusive Read Exclusive Write EREW PRAM**. The counting sort algorithm is based on determine, for each input element x_i, the number of elements less than x_i.

In this paper, we studied the performance and scalability of two EREW PRAM algorithms for integer sorting. The algorithms are based on self-index technique [3,6]. Both algorithms were implemented in a high-level language, Ada95, and run on a multiprocessors machine consisting of 4-processors, Compaq Proliant 7000. Our experimential results are coincide with the theoretical analysis and the algorithm which required large storage is faster then the other.

The structure of the paper is as follows. In Section 2, we will give the definitions of some algorithms that are used as subroutines in the parallel algorithms. Also, we will give an overview of the two parallel integer sorting algorithms. The performance and scalability of the two algorithms are given in Section 3. Finally, in Section 4, we will give the conclusion of our work.

2 Overview

2.1 Fundamental Algorithms

The parallel integer sorting algorithms use the following algorithms as subroutines:

(1) **Prefix Computation:** given an array $X = \{x_0, x_1, \ldots, x_{n-1}\}$ and an associative binary operator $\oplus$. The prefix computation algorithm is to compute the n prefix operations $s_i = x_1 \oplus x_2 \oplus \ldots \oplus x_i$, $\forall 0 \leq i < n$ [2]. The running time for sequential and parallel prefix algorithms are $O(n)$ and $O(n/p + \log p)$ respectively.

(2) **Compaction:** given an array $X = \{x_0, x_1, \ldots, x_{n-1}\}$ such that only k elements of X have nonzero values, and the remaining elements are equal to zero value. The compaction algorithm is the problem of moving the nonzero elements into the first k consecutive locations [2]. The sequential and parallel compaction algorithms take $O(n)$ and $O(n/p + \log p)$ time respectively.

(3) **Self-index Sort Algorithm:** the self-index sort carries outsorting of n elements in the range $[0, n-1]$ by directly transferring the element into a relative offset based on the value of this element. This algorithm is similar to the counting algorithm [14]. The self-index sorting algorithm takes $O(n)$ sequential time.

2.2 Parallel Self-index Integer Sorting

We shall give here a quick discussion for the two algorithms [3,6].

The first algorithm, which is called parallel self-index integer sorting (PSIIS), uses $\sqrt{n}$ processors to sort n integers in the range $[0, n-1]$ in $O(\sqrt{n})$ time. The algorithm has optimal cost $O(n)$, but takes nonlinear space. The algorithm is based on the self-index method and consists of the following stages.

Stage 1 (Number of repetitions): compute the repetitions of each element in X.
Stage 2 (Compaction): compact the repetition array R_1 to R'.
Stage 3 (Reallocation): reallocate the elements of R'.

The second algorithm, which is called parallel group self-index integer sorting (PGSIIS), uses $\sqrt{n}$ processors to sort n integers in the range $[0, n-1]$ in $O(\sqrt{n})$ time. The algorithm has optimal cost $O(n)$ and takes linear space. The algorithm is based on dividing the elements into $\sqrt{n}$ groups, not necessary of equal size, then it uses the self-index method to sort them. The algorithm consists of the following stages.

Stage 1 (Partition): divide the elements of array X into $\sqrt{n}$ groups S_i, not necessarily of equal size, such that $x_j \in S_i$ if $x_j \in [i\sqrt{n}, (i+1)\sqrt{n} - 1]$.

Stage 2 (First position): determine the starting position of each group S_i in the sorted array X.
Stage 3 (Number of repetitions): compute the repetition of each element in S_i by using the self-index method using $(|S_i|\text{Div}\sqrt{n})$ processors, $\forall 0 \leq i < \sqrt{n}$, where $|S_i|$ is the number of the elements in S_i.
Stage 4 (Compaction): compact the result of stage 3.
Stage 5 (Reallocation): reallocate the elements of S_i after applying stages 3 and 4 into a sorted array X using $(|S_i|\text{Div}\sqrt{n})$ processors.

3 Implementation

Hardware: we examine the performance of implementation of PSIIS and PGSIIS on a Compaq Proliant 7000 computer. The machine consists of four Intel Pentium processors on the same processor board. Each processor has a clock speed of 200 MHz and memory size 256 MBytes. The machine working under Windows NT server operating system.

Language: we select Ada95 langauge (Aonix-compiler) to implement our algorithms. Ada95 is an example of language that deal explicitly with parallelism. The programs are executed concurrently by using tasks.

Data Sets: our data input obtained by calling the Ada library random generator **Random()**, which is a function in the generic package **Discrete Random**. A sufficiently long sequence of random numbers is obtained by successive calls of the function **Random()**. The sequence, that is generated, is approximately uniformly distributed over the range of the result subtype. The range of elements is taken from long integer instead of integer, since the range of data exceeds the integer range.

3.1 Running Time

We examine the running time for both algorithms by running them for the data inputs 100K, 144K, 196K, 256K, 324K and 400K, where K=1024. For each input data we examine the algorithms many times. For the cases 100K, 144K, 196K, 256K and 324K, we examine the algorithms 100 times. For the case 400K we examine the algorithms 60 times. We take the average of times to be the algorithm time. We measure the time (in second) of two algorithms by using the clock subroutine **clock()** in the package **Ada Calender**. The results of the running time are represent in Figure 1.

From the results in Figure 1 we observe that:
(1) The relation between the number of elements and the execution time is almost linear for the two algorithms, see Figure 1.
(2) The PSIIS algorithm is faster than the PGSIIS algorithm. The cost of speedup of PSIIS algorithm than PGSIIS algorithm is approximately equal to 9%. The reasons for the speedup of the time are:(i)the number of computations in PGSIIS algorithm is greater than PSIIS algorithm, (ii) the PGSIIS algorithm

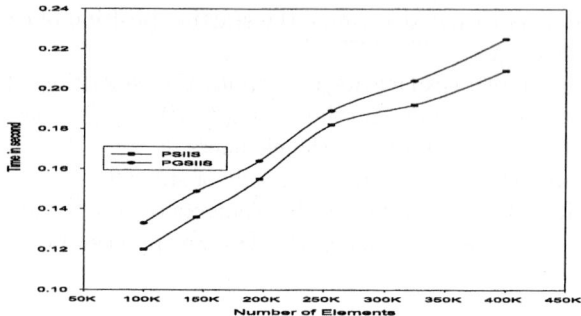

Fig. 1. Running time for both algorithms

uses a linked list which is slower in manipulation than arrays, and (iii) the classes in PGSIIS are not reallocated all at the same time.

3.2 Distribution Time

We examine the relative cost of the stages of PSIIS and PGSIIS as shown in Figures 2. For PSIIS execution time, Figure 2 (a) shows that the stages 1, 2, and 3 take approximately $50\%, 16\%$ and 34% of the execution time. The reason for large time consumed by stage 1 is that the outer loop of step 3 consists of $\sqrt{n} - 1$ sequential iterations. For PGSIIS execution time, Figure 2 (b) shows that the stages 1, 2, 3, 4 and 5 take approximately $23\%, 16\%, 33\%, 12\%$ and 16% of the execution time. The time consumed by stage 1 is not small because the creation of dynamic list takes a lot of times. Also, the time of stage 3 takes large time because some classes have number of elements greater than or equal to $2\sqrt{n}$ which required execution of the two steps of stage 1 for the PSIIS algorithm.

3.3 Scalability

Scalability of parallel system is a measure of its capability to increase speedup on proportion to the number of processors. The scalability of the parallel integer sorting algorithms are studied in two directions:

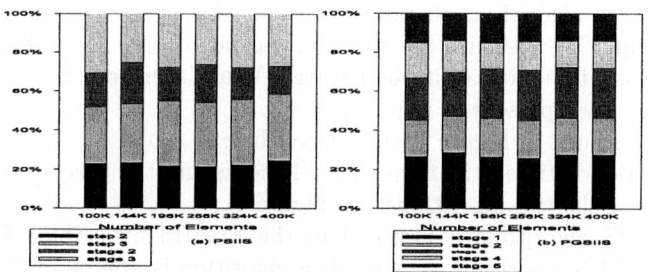

Fig. 2. Distribution time for both algorithms

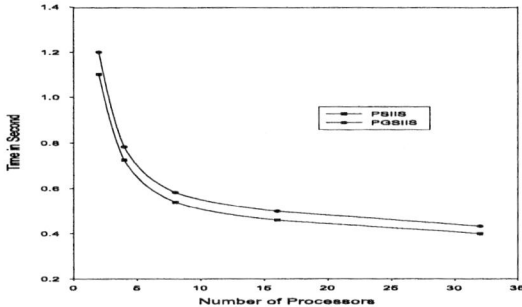

Fig. 3. Scalability of integer sorting with respect to machine size for both algorithms

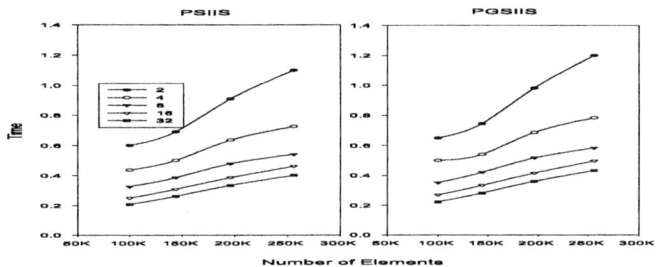

Fig. 4. Scalability in machine and problem size for both algorithms

1. Scalability of integer sorting as a function of machine size. The results in Figure 3 examine the scalability of integer sorting as a function of machine size for PSIIS and PGSIIS respectively. The number of processors that used are 2, 4, 8, 16 and 32. The results in Figure 3 indicate that for a fixed input size $N = 256K$, the running times of the two parallel algorithms decrease as the number of processors increase. i.e the relationship between them is inverse.

2. Scalability of integer sorting as a function of number of processors and problem size. We examine the scalability of integer sorting as a function of problem size with different number of processors for PSIIS and PGSIIS algorithms. The values of problem size, that tested, are 100K, 144K, 196K and 256K for the number of processors 2, 4, 8, 16, and 32. Figure 4 shows that there exists a linear (almost) dependence between the running time of parallel integer sorting and the total number of elements.

4 Conclusion

The studying parallel integer sorting practically is very important. In this work we tried to use Ada tasking technique to implement PSIIS and PGSIIS algorithms for integer sorting problem. The results of applying the two algorithms showed that tasking can simulate the parallel computations fairly. The results of these implementation agree with the excepted theoretical results. Also, we found that the PSIIS algorithm is faster than PGSIIS algorithm, but requires large space. From the results and discussions, we can ask many questions such as: can we remove the linked list from PGSIIS algorithm to speedup the time? and can we partition the array X into $\sqrt{n}$ classes of equal size?

References

1. S. Akl. Parallel Sorting Algorithms. Academic Press Inc. 1985.
2. S. Akl. Parallel Computation: Models and Methods. Prentice Hall, Upper Saddle River, New Jersey, 1997.
3. H. Bahig, S. Daoud, and M. Khairat. Parallel Self-Index Integer Sorting. J. of Supercomputing, Vol. **22**, No. 3: 269–275, 2002.
4. K. Chandy. Writing Correct Parallel Programs. In Proc. 7th International Parallel Processing Symposium, 630–634, 1993.
5. R. Karp, and V. Ramachandran. Parallel Algorithms for Shared-Memory Machines. In Handbook of Theoretical Computer Science, Vol. **A** 870–941. Elsevier Science Publisher, North Holland, Amsterdam, 1990.
6. M. Khairat, S. Daoud and H. Bahig. Optimal Parallel Integer Sorting on EREW PRAM. In Proc. of the 2000 Symposium on Performance Evaluation of Computer and Telecommunication Systems. Vancouver, British Columbia, 370–374, 2000.
7. D. Knuth. The Art of Computer Programming: Sorting and Searching. Addison-Wesley, 1973.
8. H. Mayer and S. Jahnihen. The Data-Parallel Ada Run-Time System, Simulation and Empirical Results. In Proc. 7th International Parallel Processing Symposium, 621–627, 1993.
9. None. Ada 95 Quality and Style: Guidelines for Professional Programmer. Department of Defense Ada Joint Program Office, Software Productivity Consortium, 1995.
10. M. Paprzycki and J. Zalewski. Parallel Computing in Ada: An Overview and Critique. ACM Ada Letter, Vol **XVII**, No. 2, 55–62, 1997.
11. R. Perrott. Parallel Languages. In Handbook of Parallel and Distributed Computing. Edited by Albert Zomaya, 843–864. MCGraw-Hill, 1996.
12. J. Sibeyn, F. Guillaume and T. Seidel. Practical Parallel List Ranking. In Proc. 4th Symposium on Solving Irregularly Structured Problems in Parallel. LNCS **1253**: 25–36, 1997.
13. R. Volz, R. Theriault, G. Smith and R. Waldrop. Distributed and Parallel Execution in Ada83. In third Workshop on Parallel and Distributed Real-Time System, 52–61, 1995.
14. S. Wang. A New Sort Algorithm: Self-Indexed Sort. ACM SIGPLAN Notices, Vol. **31**, No. 3: 28–36, March 1996.
15. M. Wolfe. Program Portability Across Parallel Architectures - SIMD/MIMD/SMPD/Shared/Distributed. In Proc. 8th International Parallel Processing Symposium, 658–661, 1994.

On Algorithm for the Delay- and Delay Variation-Bounded Multicast Trees Based on Estimation*

Youngjin Ahn[1], Moonseong Kim[1], Young-Cheol Bang[2], and Hyunseung Choo[1]

[1] School of Information and Communication Engineering,
Sungkyunkwan University, 440-746, Suwon, Korea
{watchman, moonseong, choo}@skku.edu
[2] Department of Computer Engineering,
Korea Polytechnic University,
429-793, Gyeonggi-Do, Korea
ybang@kpu.ac.kr

Abstract. With the multicast technology, demands for the real-time group applications through multicasting is getting more important. An essential factor of these real-time strategy is to optimize the Delay- and delay Variation-Bounded Multicast Tree (DVBMT) problem. In this paper, we propose a new algorithm for the DVBMT solution. The proposed algorithm outperforms other algorithms up to 9%~25% in terms of the delay variation.

1 Introduction

Not only the tree cost as a measure of bandwidth efficiency is one of the important factors for the QoS, but also networks supporting real-time transmission are required to receive messages from source node in a limited amount of time. Therefore, we should consider the multicast end-to-end delay and delay variation problem [3]. In this paper, we study the delay variation problem under the upper bound on the multicast end-to-end delay. We propose an efficient algorithm in comparison with the Delay and Delay Variation Constraint Algorithm (DDVCA) [4]), known as the best algorithm so far. Even if the time complexity of our algorithm is same as the one of DDVCA, the proposed algorithm would have the better performance than DDVCA has in terms of the multicast delay variation. The rest of the paper is organized as follows. In Section 2, we state the network model for the multicast routing, and the problem formulation. Section 3 presents the details of the proposed algorithm. Finally, section 4 concludes this paper.

---

* This work was supported in parts by Brain Korea 21 and the Ministry of Information and Communication in Republic of Korea. Dr. H. Choo is the corresponding author and Dr. Bang is the co-corresponding author.

2 Related Works

We consider that a computer network is represented by a directed graph $G = (V, E)$ with n nodes and l links or arcs, where V is a set of nodes and E is a set of links, respectively. Each link $e = (i, j) \in E$ is associated with delay $d(e) \geq 0$. The delay of a link, $d(e)$, is the sum of the perceived queueing delay, transmission delay, and propagation delay. We define a path as sequence of links such that $(u, i), (i, j), \ldots, (k, v)$, belongs to E.

Let $P(u, v) = \{(u, i), (i, j), \ldots, (k, v)\}$ denote the path from node u to node v. If all $u, i, j, \ldots, k, v$ are distinct, then we say that it is a simple directed path. For a given source node $s \in V$ and a destination node $d \in V$, $(2^{s \to d}, \infty)$ is the set of all possible paths from s to d.

$$(2^{s \to d}, \infty) = \{ P_k(s, d) \mid \text{all possible paths from } s \text{ to } d, \; ^\forall s, \; d \in V, \; ^\forall k \in \Lambda \},$$

where Λ is an index set. The path-delay of P_k is given by $\phi_D(P_k) = \sum_{e \in P_k} d(e)$, $^\forall P_k \in (2^{s \to d}, \infty)$. $(2^{s \to d}, \Delta)$ is the set of paths from s to d for which the end-to-end delay is bounded by Δ. Therefore $(2^{s \to d}, \Delta) \subseteq (2^{s \to d}, \infty)$.

For the multicast communications, messages need to be delivered to all receivers in the set $M \subseteq V \setminus \{s\}$ which is called the multicast group, where $|M| = m$. The path traversed by messages from the source s to a multicast receiver, m_i, is given by $P(s, m_i)$. Thus multicast routing tree can be defined as $T(s, M) = \bigcup_{m_i \in M} P(s, m_i)$ and the messages are sent from s to M through $T(s, M)$. The multicast end-to-end delay constraint, Δ, represents an upper bound on the acceptable end-to-end delay along any path from source node to a destination node. The multicast delay variation, δ, is the maximum difference between the end-to-end delays along the paths from the source to any two destination nodes.

$$\delta = max\{ \; |\phi_D(P(s, m_i)) - \phi_D(P(s, m_j))|, \; ^\forall m_i, m_j \in M, \; i \neq j \;\}$$

The issue defined and discussed in [3], initially, is to minimize multicast delay variation under multicast end-to-end delay constraint. The authors referred to this problem as Delay- and delay Variation-Bounded Multicast Tree (DVBMT) problem. The DVBMT problem is to find the tree that satisfies

$$min\{ \; \delta_\alpha \mid \; ^\forall P(s, m_i) \in (2^{s \to m_i}, \Delta), \; ^\forall P(s, m_i) \subseteq T_\alpha, \; ^\forall m_i \in M \; ^\forall \alpha \in \Lambda \;\},$$

where T_α denotes any multicast tree spanning $M \cup \{s\}$.

3 Proposed Algorithm

The algorithm gives an explicit solution about the DVBMT problem. We define the $MODE$ function, since the location of the core node influences the multicast delay variation. In addition to the $MODE$ function, we measure the delay variation for each mode, using the CMP function. The $MODE$ and CMP functions are deployed as follows.

$$MODE(c) = \begin{cases} \text{I} & if \quad c = s \\ \text{II} & if \quad \exists m \text{ in } P(s,c), \text{ where } {}^\forall m \in M \\ \text{III} & if \quad \exists s \text{ in } P(m,c), \text{ where } {}^\forall m \in M \\ \text{IV} & if \quad \text{II and III} \\ \text{V} & otherwise \end{cases}$$

where s is source node.

$$CMP(x) = \begin{cases} |(ds_{core} + max\_delay) - ds_{m_k}| & if \quad x = \text{II or III or IV} \\ dv_{core} & otherwise \end{cases},$$

where $core = MODE^{-1}(x)$, $max\_delay = max\{min\{\phi_D(P(core,m))\} \mid {}^\forall m \in M \setminus \{m^* \in M \mid \exists m^* \text{ in } P(s,core) \bigvee \exists s \text{ in } P(m^*, core)\}\}$.

As above functions $MODE$ and CMP, $MODE$ I is that the core node corresponds to the source node. In this case, CMP keeps dv_{core}. Otherwise, $MODE$s II~V are executed.

In the case of $MODE$ II [1], Fig. 1 (a) shows that there exists the destination node from the source node to the core node. CMP which can be computed stores the sum of the delay from the core node to the destination node and max_delay. The third case just satisfies $MODE$ III. In this case, the proposed algorithm initially adds the delay from the source node to the core node to max_delay and subtracts the delay from the source node to the adjacent destination node, and stores the value with CMP. If the value is negative, CMP accepts the absolute value of it. The fourth case, MODE which satisfies II and III should find the factor to determine CMP, having situation that the destination nodes exist around the source node. The measure of the comparison is determined with the minimum delay from the source to the associated destinations. We can decide this situation because the shortest delay value from the source node affects the delay variation of the created tree. Therefore, the proposed algorithm selects either $MODE$ II or III. In $MODE$ V, if the proposed algorithm doesn't satisfy $MODE$s I~IV, CMP stores dv_{core}. With respect to the time complexity, the proposed algorithm follows DDVCA so that it equals to $O(mn^2)$.

Fig. 2 shows the given network topology, and the link delays are presented on each link. It is supposed that the multicast end-to-end delay constraint Δ

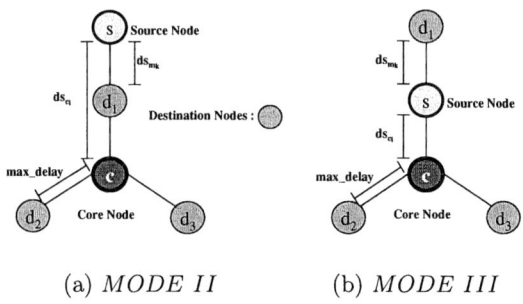

(a) MODE II   (b) MODE III

Fig. 1. Main idea of $MODE$ function

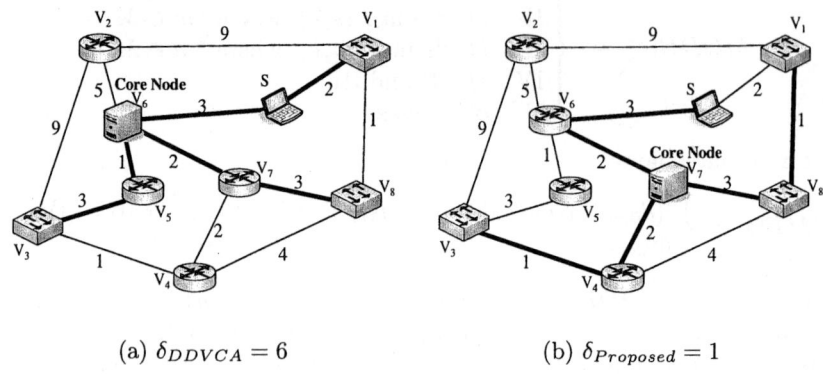

(a) $\delta_{DDVCA} = 6$ (b) $\delta_{Proposed} = 1$

Fig. 2. DDVCA and proposed algorithm

Table 1. The method by which proposed algorithm selects a core node

		s	v_1	v_2	v_3	v_4	v_5	v_6	v_7	v_8
source	s	0	2	8	7	7	4	3	5	3
	v_1	2	0	9	6	5	6	5	4	1
destination	v_3	7	6	9	0	1	3	4	3	5
	v_8	3	1	10	5	4	6	5	3	0
max_i			7	6	10	6	5	6	5	4	5
min_i			2	0	9	0	1	3	4	3	0
dv_i			5	6	1	6	4	3	1	1	5

equals 10. Fig. 2 (a) and (b) give the multicast tree using DDVCA and the proposed algorithm, respectively. From Table 1, we can know that the nodes which have the minimum multicast delay variation dv_{min} are v_2, v_6, and v_7. However, node v_2 is excluded, because the upper delay bound Δ is 10. And then, DDVCA randomly chooses node v_6 as the core node, but the proposed algorithm chooses node v_7 as the core node. The propsed algorithm selects the less value of $CMP(MODE(v_7)) = 4 - 3 = 1$ than the value of $CMP(MODE(v_6)) = |(3+5) - 2|$. In Fig. 3, when dv_{v_6} and dv_{v_7} have the same value, the proposed algorithm take the wise selection node with v_7. As a result, the multicast delay variation of DDVCA is 6, but the one of proposed algorithm is 1.

4 Performance Evaluation

We compare our proposed algorithm with the DDVCA in terms of multicast delay variation. We describe the generation of random network topologies for the evaluation and the simulation results based on the network topology generated [2]. We now describe some numerical results, comparing the performance of the

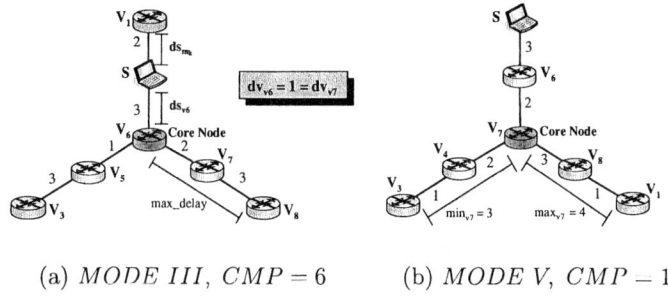

(a) $MODE\ III,\ CMP = 6$ (b) $MODE\ V,\ CMP = 1$

Fig. 3. Core selection by CMP

proposed algorithm. Our algorithm is implemented in C^{++}. The 10 different network environments are generated for each size of given 100, 200, and 400 nodes. A source node is randomly selected and destination nodes are picked uniformly from the set of nodes except the source in the network topology. Moreover, the destination nodes in the multicast group occupy 10% to 60% of the all nodes on the network, respectively. An upper bound Δ is randomly valued. We simulate 1000 times ($10 \times 100 = 1000$) for each environment and P_e=0.3. For the performance comparison, the proposed algorithm and DDVCA are implemented in the same condition. Fig. 4 shows the simulation results for the multicast delay variations. It is apparent that the multicast delay variation of the proposed algorithm outperforms that of DDVCA. Fig. 4 (d) shows inefficiency in case of the number of nodes. We define inefficiency in the following equation.

$$\overline{\delta} = \frac{\delta_{DDVCA} - \delta_{Proposed}}{\delta_{DDVCA}}$$

We present it as percentage (i.e, $\overline{\delta} \times 100\%$). The enhancement is up to about 9%~25% in terms of the multicast delay variation.

5 Conclusion

In this paper, we consider the transmission of a message that guarantees certain bounds of the end-to-end delays as well as the multicast delay variations computer network. The time complexity of our algorithm is $O(mn^2)$, which is the same as that of DDVCA. Furthermore, our algorithm results in the more efficient multicast delay variation than DDVCA.

References

1. M. Kim, Y.-C. Bang, and H. Choo, "Efficient Algorithm for Reducing Delay Variation on Bounded Multicast Trees," Springer-Verlag Lecture Notes in Computer Science, vol. 3090, pp. 440-450, September 2004.

2. A.S. Rodionov and H. Choo, "On Generating Random Network Structures: Connected Graphs," Springer-Verlag Lecture Notes in Computer Science, vol. 3090, pp. 483-491, September 2004.
3. G. N. Rouskas and I. Baldine, "Multicast routing with end-to-end delay and delay variation constraints," IEEE J-SAC, vol. 15, no. 3, pp. 346-356, April 1997.
4. P.-R. Sheu and S.-T. Chen, "A fast and efficient heuristic algorithm for the delay- and delay variation bound multicast tree problem," Information Networking, Proc. ICOIN-15, pp. 611-618, January 2001.

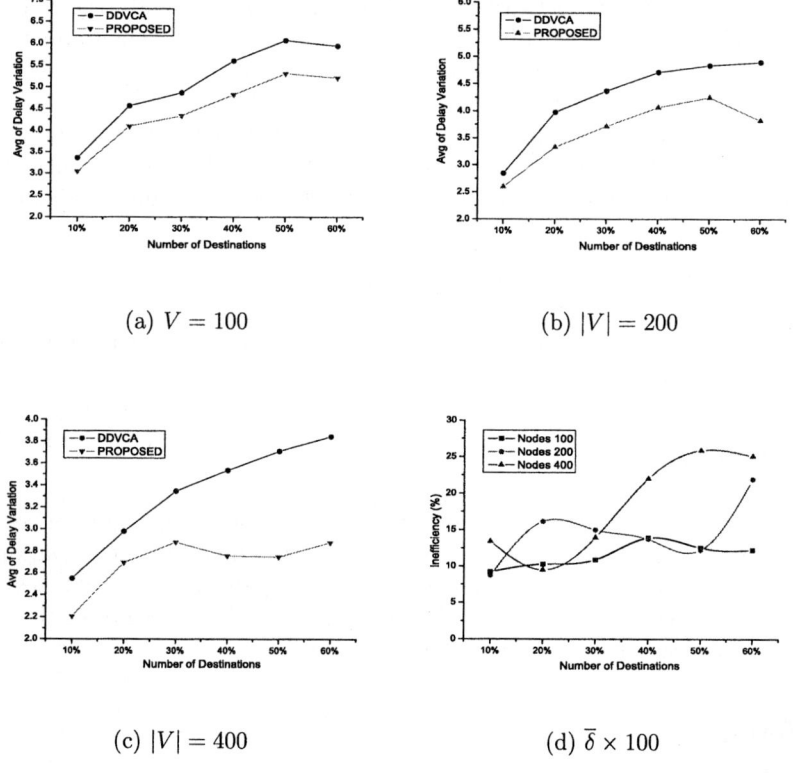

(a) $V = 100$

(b) $|V| = 200$

(c) $|V| = 400$

(d) $\bar{\delta} \times 100$

Fig. 4. The multicast delay variations and inefficiency of three different networks, $P_e = 0.3$

Synchronization-Based Power-Saving Protocols Based on IEEE 802.11

Young Man Kim

Kookmin University, School of Computer Science,
Seoul, 136-702, South Korea
ymkim@kookmin.ac.kr
http://cclab.kookmin.ac.kr/

Abstract. *Power-saving* is a critical issue for almost all kinds of portable devices. In this paper, we consider the design of power-saving protocols for *mobile ad hoc networks (MANETs)* that allow mobile hosts to switch to *low-power sleep mode*. The MANETs being considered in this paper are characterized by unpredictable mobility and multi-hop communication. In MANET, we study how the node synchronization affects the performance in addition to the protocol complexity. We propose three synchronous power management protocols, namely *Synchronous PFAI (SPFAI)*, *Efficient SPFAI (ESPFAI)*, and *Non-MTIM SPFAI (NSPFAI)* protocols, which are directly applicable to IEEE 802.11-based MANETs.

1 Introduction

Among the various network architectures, the design of *mobile ad hoc network(MANET)* has attracted a lot of attention recently. A MANET is one consisting of a set of mobile hosts which can communicate with one another and roam around at their will. No base stations are supported in such an environment, and mobile hosts may have to communicate with each other in a *multi-hop* fashion. Applications of MANETs occur in situations like battlefields, major disaster areas, and outdoor assemblies.

One critical issue for almost all kinds of portable devices supported by battery powers is *power-saving*. Without power, any mobile device will become useless. Furthermore, battery technology is not likely to progress as fast as computing and communication technologies do. Hence, how to lengthen the lifetime of batteries is an important issue, especially for MANET.

Solutions addressing the power-saving issue in MANETs can generally be categorized as follows:

1. *Transmission Power Control*: In wireless communication, transmission power has strong impact on bit error rate, transmission rate, and inter-radio interference. These are typically contradicting factors. In [1], power control is adopted to reduce interference and improve throughput on the MAC layer. How to determine transmission power of each mobile host so as to determine the best network topology, or known as *topology control*, is addressed in [2]. How to increase network throughput by power adjustment for packet radio networks is addressed in [3].

2. *Power-Aware Routing*: Power-aware routing protocols have been proposed based on various power cost functions [4], [5], [6], [7]. In [4], when a mobile hosts battery level is below a certain threshold, it will not forward packets for other hosts. In [6], five different metrics based on battery power consumption are proposed. Reference [7] considers both hosts lifetime and a distance power metric. A hybrid environment consisting of battery-powered and outlet-plugged hosts is considered in [5].
3. *Low-Power Mode*: More and more wireless devices can support low-power sleep modes. IEEE 802.11 [8] has a power-saving mode in which a radio only needs to be awake periodically. HyperLAN allows a mobile host in power-saving mode to define its own active period. An active host may save powers by turning off its equalizer according to the transmission bit rate. Comparisons are presented in [9] to study the power-saving mechanisms of IEEE 802.11 and HIPERLAN in ad hoc networks. Bluetooth [10] provides three different low-power modes: *sniff*, *hold*, and *park*.

This paper studies the management of power-saving (PS) modes for IEEE 802.11-based MANETs. As far as we know, the power-management problem for IEEE 802.11-based multi-hop MANETs has not been addressed seriously in the literature. Existing standards, such as IEEE 802.11 and HYPERLAN, do support PS modes, but assume that the MANET is fully connected. Bluetooth also has low-power modes, but is based on a master-slave architecture, so time synchronization is trivial. The works [11], [12] address the power-saving problem, but assume the existence of access points. A lot of works have focused on multi-hop MANETs on issues such as power-aware routing, topology control, and transmission power control (as classified above), but how to design a PS mode is not well studied. Recently, Tseng[13] proposes three asynchronous power-saving protocols managing PS mode in IEEE 802.11.

In this paper, we study the effect of node synchronization in the power-saving MANET protocol. We target ourselves at IEEE 802.11-based LAN. First, a representative asynchronous power-saving protocol is selected as a reference protocol. Then, we enforce node synchronization by adding a synchronization procedure into that protocol. Protocol synchronization invites two advantages; the *significantly enhanced performance* and the *reduced implementation complexity*. In particular, we propose three efficient protocols in the synchronized power-saving MANET protocol class.

The rest of this paper is organized as follows. Power-saving mode in IEEE 802.11 is reviewed and an asynchronous reference protocol, called PFAI, is presented in Section 2. In Section 3, we introduce three synchronized versions (SPFAI, ESPFAI, and NSPFAI) of the PFAI protocol. Section 4 concludes this paper.

2 Power-Saving in IEEE 802.11

In this section, the power-saving modes in IEEE 802.11 are reviewed. Since the PS mode of IEEE 802.11 is designed only for a single-hop (or fully connected) ad hoc network, the PS protocol for multi-hop ad hoc network remains as an open problem. Recently, Tseng [13] proposes three power-saving protocols that are all operating asynchronously due to the difficulty of establishing the node synchronization. However, the asynchronism in the PS protocol creates two problems: the inevitable complex preparation for broadcasting and the reduced performance. To consider the performance effect of node asynchronism in detail, we choose a representative power-saving protocol, called *Periodically-Fully-Awake-Interval(PFAI)* protocol, from [13]. For the convenience of explanation, the short description of PFAI protocol is also presented. In the next section, three synchronous variations of the PFAI protocol are introduced to clarify the defect of asynchronous PS protocol.

2.1 Power-Saving Modes in IEEE 802.11

IEEE 802.11 [8] supports two power modes: *active* and *power-saving (PS)*. The protocols for *infrastructure networks* and *ad hoc networks* are different. As for the detailed PS operation in infrastructure network, refer to [13]. In an ad hoc network, PS hosts also wake up at the beginning of each period spaced by a fixed *beacon interval(BI)*. The short interval at the beginning of BI that PS hosts wake up is called the *ATIM window*. It is assumed that hosts are fully connected and all synchronized, so the ATIM windows of all PS hosts will start at about the same time. In the beginning of each ATIM window, each mobile host will contend to send a beacon frame. Any successful beacon serves as the purpose of synchronizing mobile hosts clocks. This beacon also inhibits other hosts from sending their beacons. To avoid collisions among beacons, a host should wait a random number of slots between 0 and $2 \times CW_{min} - 1$ before sending out its beacon.

After the beacon, a host with buffered unicast packets can send a direct ATIM frame to each of its intended receivers in PS mode. ATIM frames are also transmitted by contention based on the DCF access procedure. After transmitting an ATIM frame, the mobile host shall remain awake for the entire remaining period. On reception of the ATIM frame, the PS host should reply with an ACK and remains active for the remaining period. The buffered unicast packets should be sent based on the normal DCF access procedure after the ATIM window finishes. If the sender does not receive an ACK, it should retry in the next ATIM window. As for buffered broadcast packets, the ATIM frames need not be acknowledged. Broadcast packets then can be sent based on contention after the ATIM window finishes.

The PS mode of IEEE 802.11 is designed for a single-hop (or fully connected) ad hoc network. When applied to a multi-hop ad hoc network, the two problems may arise so that they will pose a demand of redesigning the PS mode for multihop MANET. The first problem concerns *node synchronization*. Since IEEE 802.11 assumes that mobile hosts are fully connected, the transmission of a beacon frame can be used to synchronize all hosts beacon intervals. So the ATIM windows of all hosts can appear at around the same time without much difficulty. However, in a multi-hop MANET, clock synchronization is a difficult job because communication delays and mobility are all unpredictable.

Neighbor discovery makes another problem. In a wireless and mobile environment, a host can only be aware by other hosts if it transmits a signal that is heard by the others. For a host in the PS mode, not only is its chance to transmit reduced, but also the same thing occurs to the chance to hear others. As reviewed above, a PS host must compete with other hosts to transmit its beacon. A host will cancel its beacon frame once it hears others beacon frame. This may run into the detection failure of neighbors. Thus, many existing routing protocols that depend on neighbor information may be impeded.

In the next subsection, one simple power-saving protocol, called PFAI protocol, that resolves these problems by utilizing more beacons and awake intervals, is explained in short.

2.2 Asynchronous Power-Saving Protocol for MANET

In this subsection, we summarize one asynchronous power-saving protocol, called PFAI(Periodically-Fully-Awake-Interval) protocol, that allows mobile hosts to enter PS mode in a multi-hop MANET. For each PS host, it divides its time axis into a number of fixed-length intervals called *beacon interval(BI)*, as IEEE 802.11 PS modes do. In each BI, there are three windows called *active window*, *beacon window*, and *MTIM window*, as shown in Fig. 1.

During the active window, the PS host should turn on its receiver to listen to any packet and take proper actions as usual. The beacon window is for the PS host to send its beacon, while the MTIM window is for other hosts to send their MTIM frames to the PS host. The MTIM frames serve the similar purpose as ATIM frames in IEEE 802.11; here MTIM is used to emphasize that the network is a mutil-hop MANET. Excluding these three windows, a PS host with no packet to send or receive may go to the sleep mode. AW, BW, and MW

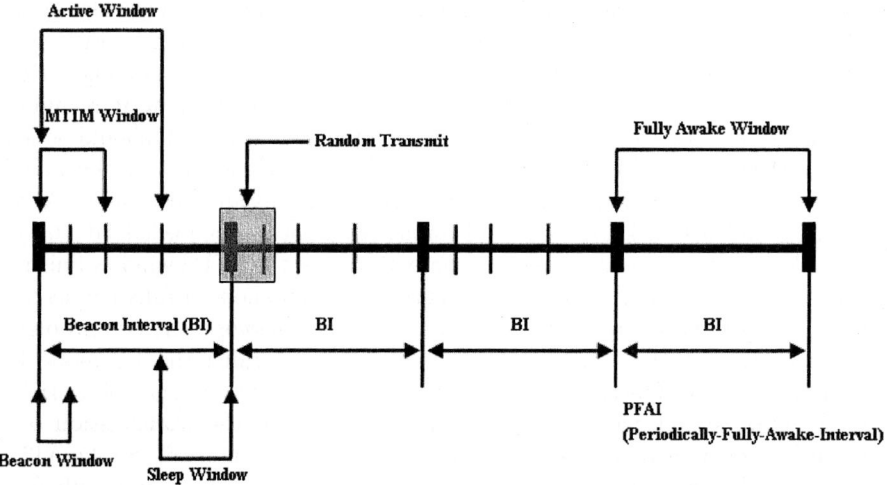

Fig. 1. Beacon interval, various types of windows, and periodically-fully-awake-interval(T=4)

are denoted as the length of an active window, a beacon window, and an MTIM window, respectively. In the beacon window (resp., MTIM window), hosts can send beacons (resp., MTIM frames) following the DCF access procedure. Each transmission must be led by a SIFS followed by a random delay ranging between 0 and $2 \times CW_{min} - 1$ slots.

To save the power and, at the same time, discover neighbors at maximum, two additional types of beacon intervals, *low-power intervals* and *PFA intervals(PFAIs)*, are introduced. Each low-power interval starts with an active window, which contains a beacon window followed by a MTIM window, such that $AW = BW + MW$. In the rest of the time, the host can go to the sleep mode. On the other hand, each PFA interval also starts with a beacon window followed by a MTIM window. However, the host must remain awake in the rest of the time, i.e., $AW = BI$. The PFAIs arrive periodically every T intervals(T is called *PFAI parameter*), and the rest of the intervals are low-power intervals.

Intuitively, the low-power intervals is for a PS host to send out its beacons to inform others its existence. The PFA intervals are for a PS host to discover who are in its neighborhood. It is not hard to see that a PFA interval always has overlapping with any hosts beacon windows, no matter how much time their clocks are asynchronous. By collecting other hosts beacons, the host can predict when its neighboring hosts will wake up. Fig. 1 shows an example with $T = 4$ intervals.

Like 802.11 PS mode, PFAI protocol utilizes MTIM-ACK sequence to make a reservation before sending a data frame, as shown in Fig. 2. During the receivers MTIM window, the sender contends to send its MTIM packet to the receiver. The receiver, on receiving the MTIM packet, will reply an ACK after SIFS and stay awake in the remaining of the beacon interval. After the MTIM window, the sender will contend to send the buffered packet to the receiver based on the DCF procedure.

The situation is more complicated for broadcasting since the sender may have to deal with multiple asynchronous neighbors. To reduce the number of transmissions, these asynchronous neighbors are divided into groups to notify them separately in multiple runs. When a source host S intends to broadcast a packet, it first checks the arrival time of the MTIM windows of all its neighbors. Then S picks the host, say Y, whose first MTIM window arrives earliest. Based on Y s first MTIM window, S further picks those neighbors whose MTIM windows have overlapping with Y s first MTIM window. These hosts, including Y, make a group together and S will try to notify them in one MTIM frame. After this notification, S considers the rest of the neighbors that have not been notified yet in the previous MTIM and repeats the same procedure again to initiate another MTIM frame. The process is repeated until all its neighbors have been notified.

A neighbor, on receiving a MTIM carrying a broadcast indication, should remain awake until a broadcast packet is received or a timeout value expires for which a timeout value of two beacon intervals is recommended. The source S, after notifying all neighbors, can contend to send its buffered broadcast packet after the last neighbors MTIM window passes. Broadcast packets should be sent based on the DCF procedure too.

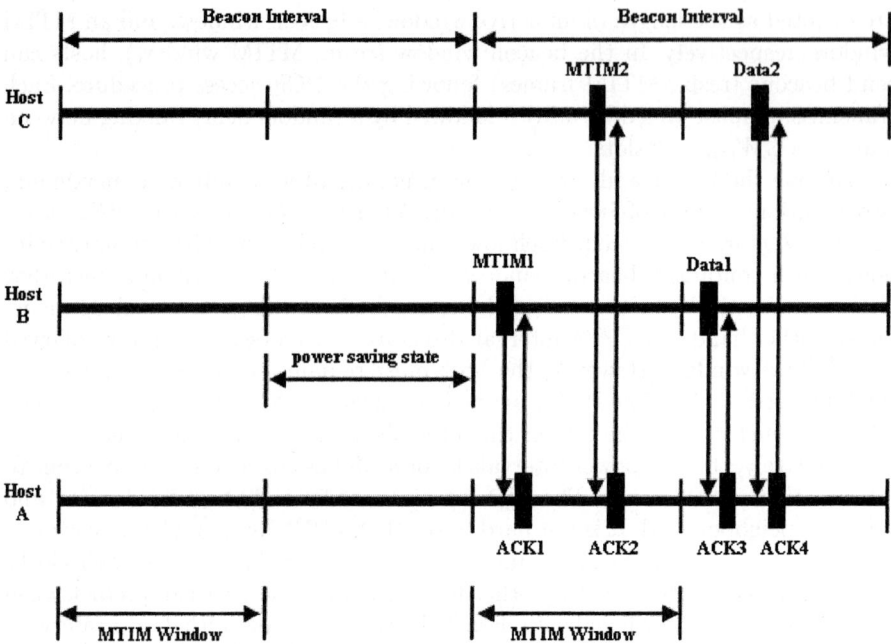

Fig. 2. MTIM frame and data frame transmission procedure

3 Synchronous Power-Saving Protocols for IEEE 802.11

In this section, we introduce the synchronous version of the PFAI protocol, called *SPFAI(Synchronous PFAI)* protocol. Then, we modify the MTIM management procedure to enhance the performance of SPFAI protocol. We denote this modified SPFAI protocol as *ESPFAI(Efficient Synchronous PFAI)* protocol. Finally, we present a reference protocol called *NSPFAI(Non-MTIM Synchronous PFAI)* in which the MTIM window and its packet exchange are removed from SPFAI protocol.

3.1 SPFAI(Synchronous PFAI) Protocol

PFAI protocol has two inherent defects in relation to the broadcast function. Note that the major practical ad hoc routing protocols, e.g. DSDV, AODV, DSR, and TORA, utilize the broadcast communication within their fundamental routing discovery stage. As described in the previous section, in PFAI protocol, a broadcast packet is realized by a sequence of point-to-point unicast reservation messages due to asynchronism. In other words, the number of transmitted control packets per broadcast packet is proportional to the number of neighbors, inevitably resulting in the performance degradation in addition to the expensive power consumption waiting in active mode until all neighbors are notified with the forthcoming broadcast packet. Since the broadcast management requires to keep track of the beginning of MTIM window for each neighbor and send the

MTIM packets during the corresponding windows, the protocol implementation becomes much more complex than 802.11 PS modes do.

On the other hand, the above two problems disappear if all nodes operate synchronously. A broadcast frame may be sent only one time during the common MTIM window. Furthermore, there is no need to manage the asynchronous MTIM window of the neighbors, reducing the implementation complexity.

SPFAI protocol, proposed in this subsection, is the same as PFAI protocol except that the former has the following node synchronization procedure and removes the complex broadcast preparation procedure necessary to make a broadcast transmission.

Node Synchronization Algorithm (at node n)

1. *(Initialization)* Node n has four local parameters concerning the clock synchronization: (i) the host ID of node n, $id(n)$, that is uniquely set up in the initialization stage, (ii) the current local clock time, $clock(n)$, that is initialized with any value including random number, (iii) the clock synchronization ID, $synid(n)$, to which the local clock is synchronized at the last time, and (iv) the clock time to live, $ttl(n)$, up to which BIs $synid(n)$ remains effective at node n. The last two parameters, $synid(n)$ and $ttl(n)$, are initialized to $id(n)$ and ttl_{max} such that ttl_{max} is larger than the maximum number of hops in any MANET configuration.

2. *(Step 1)* In the beginning of each beacon interval, node n decrements $ttl(n)$ by one if $synid(n)$ is different from $id(n)$. As the new value of $ttl(n)$ becomes zero, the effectiveness of the current clock synchronization node $synid(n)$ becomes out of date so that $synid(n)$ and $ttl(n)$ are reset to $id(n)$ and ttl_{max}. Following this adjustment, node n constructs and broadcasts a beacon message consisting of 4-tuple parameters, $(id(n), clock(n), synid(n), ttl(n))$, during MTIM window, as depicted in Fig. 3.

3. *(Step 2)* When node n receives a beacon message, $(id(m), clock(m), synid(m), ttl(m))$, from node m, it compares the local 2-tuple parameter $(synid(n), ttl(n))$ with the corresponding value in the beacon, $(synid(m), ttl(m))$. If the latter is bigger in lexicographical order than the former, the local parameters, $clock(n)$, $synid(n)$ and $ttl(n)$ are updated with the synchronization parameters in the beacon, $clock(m)$, $synid(m)$ and $ttl(m)$, respectively.

Suppose that node k is the node with the highest ID in a MANET configuration. In the above procedure, the beacon from node k will update the clocks of its neighbors with $clock(k)$. This clock synchronization event will propagate eventually to the most distant nodes in the network. Note that the synchronization clock source has a finite life time such that, as node k becomes out of power

Fig. 3. Beacon frame format

or shut down, the network will be resynchronized eventually to second-highest-priority node that had been synchronized to node k.

The broadcast packet in SPFAI protocol is transmitted directly during the MTIM window, that is different from the broadcast in the PFAI protocol where an MTIM frame is transmitted to each neighbor within the corresponding MTIM window and, finally, one broadcast packet is sent after all neighbors are remaining in the active mode. However, the SPFAI protocol omits the unnecessary MTIM frame transfers. In addition, it saves the power by transmitting the broadcast packet immediately rather than delaying it until all the neighbors are waken up to receive.

3.2 ESPFAI(Efficient Synchronous PFAI) Protocol

Both PFAI and SPFAI protocols employ the MTIM and ACK frames for each node to make an independent reservation for the actual data packet transmission. To reduce this transmission overhead, in this subsection, the MTIM management scheme is analyzed to derive a better solution.

Suppose that nodes i and j have some data to send to node k. According to the rule defined in PFAI and SPFAI, nodes i and j should send their own independent MTIM frames and node k reply with two ACK frames. The MTIM operation thus invokes four frame transmissions. In general, the frame transmission number will increase proportional to the number of sending parties. In the above example, either node i or j finishes MTIM-ACK exchange with node k before the other does. Suppose node i and k already made the frame exchange. Since node j is also the neighbor of node k, the former can hear the ACK from node k to node i, so that it is obvious to node j that node k will remain in active mode during the current beacon interval to receive the reserved data frame. Node j may succeed to send its own data frame without MTIM-ACK exchange, if the data transfer of node j is faster than that of node i. Otherwise, node k will enter the sleep mode immediately after receiving the data frame from node i. However, this undesirable behavior can be removed, if we make a slight modification to the MTIM management as follows; *even if the receiving node(e.g., node k in the above) got all reserved data frames from the corresponding nodes, it should wait until the maximum backoff delay time goes by without any transmission detection.* In the above example, node j still have the chance to send its data frame to node k, even if node k already succeeded to receive the reserved data from node i. Thus, a slight loss of power due to remaining in active mode up to the maximum backoff delay time yields better throughput performance. The SPFAI protocol upgraded with this new MTIM management scheme is called *Efficient SPFAI(ESPFAI)* protocol.

3.3 NSPFAI(Non-MTIM Synchronous PFAI) Protocol

A reference protocol called *Non-MTIM SPFAI(NSPFAI)* protocol is given in this subsection. NSPFAI protocol is an extreme version of ESPFAI protocol. In ESPFAI protocol, the MTIM-ACK sequence is shared by the neighboring nodes so that the data transfer under ESPFAI protocol occurs more efficiently than that under SPFAI. In extreme, the MTIM-ACK sequence may be omited. In other words, the MTIM window can disappear in the power-saving protocol with some additional modifications. Because reservation for data transfer is not allowed without MTIM-ACK seqnence, each node assumes that there may be some sending node(s) in the current beacon interval. After the beacon transfer, the node(say, node k) remains in active mode. If the maximum backoff delay time passes without the occurance of any frame transmission in the air, it is obvious that no other node has data to transfer to itself and thus, node k safely enters the sleep mode until the next beacon interval. Otherwise, it should wait until the current frame transmission ends and repeats the idle time check operation.

Note that node k should also wait through the sequence of unrelated data transmissions even in case that no node has data to send to it, so that the power consumption becomes more elevated than that of SPFAI. On the other hand, the algorithm complexity is greatly reduced and the data transmission interval is enhanced.

4 Conclusion

In this paper, we have addressed one main issue that the node synchronization in the 802.11 power-saving protocol is useful for the performance improvement and the reduction of the protocol complexity. First of all, the broadcast transmission is easily implemented in the synchronous power-saving protocol, as described in the proposed SPFAI protocol. On the other hand, the asynchronous protocol like PFAI must create a temporary pseudo-synchronization state among all neighbors by using multiple sequences of MTIM-ACK packets before the actual broadcast transmission. Since almost all routing methods adopt the broadcast function in the neighbor discovery, the characteristics of the broadcast function provides a significant impact toward the network performance. For example, the performance of the SPFAI protocol is much better than that of the PFAI protocol, since the former significantly simplifies the broadcast procedure of the latter. In addition, the optimized MTIM management scheme is devised and applied to SPFAI so that the resulting ESPFAI protocol improves the performance further than SPFAI does. The node synchronization also reduces the protocol complexity. We are going to make a comprehensive simulation to evaluate three proposed protocols. In the simulation, we will also analyze the performance effects of two parameters: PFAI parameter and MTIM window size.

References

1. S. L. Wu, Y. C. Tseng, and J. P. Sheu, Intelligent Medium Access for Mobile Ad Hoc Networks with BusyTones and Power Control, IEEE Journal on Selected Areas in Communications, vol. 18, pp. 1647-1657, Sep 2000.
2. L. Hu, Topology Control for Multihop Packet Radio Networks, IEEE Transactions on Communications, vol. 41, pp. 1474-1481, Oct 1993.

3. C. F. Huang, Y. C. Tseng, S. L. Wu, and J. P. Sheu, Increasing the Throughput of Multihop Packet Radio Networks with Power Adjustment, International Conference on Computer, Cummunication, and Networks, 2001.
4. J. Gomez, A. T. Campbell, M. Naghshineh, and C. Bisdikian, A Distributed Contention Control Mechanism for Power Saving in randomaccess Ad-Hoc Wireless Local Area Networks, Proc. of IEEE International Workshop on Mobile Multimedia Communications, pp. 114-123, 1999.
5. J. H. Ryu and D. H. Cho, A New Routing Scheme Concerning Power- Saving in Mobile Ad-Hoc Networks, Proc. of IEEE International Conference on Communications, vol. 3, pp. 1719-1722, 2000.
6. S. Singh, M.Woo, and C. S. Raghavendra, Power-Aware Routing in Mobile Ad Hoc Networks, Proc. of the International Conference on Mobile Computing and Networking, pp. 181-190, 1998.
7. I. Stojmenovic and X. Lin, Power-aware Localized Routing in Wireless Networks, Proc. of IEEE International Parallel and Distributed Processing Symposium, pp. 371-376, 2000.
8. LAN MAN Standards Committee of the IEEE Computer Society, IEEE Std 802.11-1999, Wireless LAN Medium Access Control (MAC) and Physical Layer (PHY) specifications, IEEE, 1999.
9. H.Woesner, J. P. Ebert,M. Schlager, and A.Wolisz, Power-Saving Mechanisms in Emerging Standards for Wireless LANs: The MAC Level Perspective, IEEE Persinal Communications, pp. 40-48, Jun 1998.
10. J. C. Haartsen, The Bluetooth Radio System, IEEE Persinal Communications, pp. 28-36, Feb 2000.
11. A. K. Salkintzis and C. Chamzas, An In-Band Power-Saving Protocol for Mobile Data Networks, IEEE Transactions on Communications, vol. 46, pp. 1194-1205, Sep 1998.
12. T. Simunic, H. Vikalo, P. Glynn, and G. D. Micheli, Energy Efficient Design of Portable Wireless Systems, Proc. of the International Symposium on Low Power Electronics and Design, pp. 49-54, 2000.
13. Y.C.Tseng, C.S.Hsu and T.Y.Hsieh, Power-Saving Protocols for IEEE 802.11-Based Multi-Hop Ad Hoc Networks,IEEE INFOCOM,2002.

An Energy-Efficient Uni-scheduling Based on S-MAC in Wireless Sensor Network

Tae-Seok Lee[1,2], Yuan Yang[1], Ki-Jeong Shin[2], and Myong-Soon Park[1,*]

[1] Internet Computing Lab,
Dept. of Computer Science and Engineering, Korea University, Korea
{tsyi, yy, myongsp}@ilab.korea.ac.kr
[2] Korea Institute of Science and Technology Information, Korea
kjshin@kisti.re.kr

Abstract. S-MAC is a MAC (Medium Access Control) protocol as specialized for the wireless sensor network in order to sacrifice transmission delay and extend the working life of the whole sensor nodes. S-MAC, which was made by modifying IEEE 802.11, reduced the energy wasted in the wireless sensor network by using a periodic listen and sleep scheduling method. The first task that the wireless sensor network using S-MAC performs is to find a neighbor node, and then choose and broadcast its own schedule. In this process, as a result of using a time selected randomly and scheduling, diversified schedules are generated. In the end, schedule clusters, which are different from one another, get to be made, and nodes on these cluster borders fail to communicate with one another due to an inconsistent listen time. In the existing paper on S-MAC, this problem is solved by placing a border node to adopt schedules which are different from one another. However, compared with other nodes, the border node consumes energy more, and its efficiency of data transmission according to the broadcast type is lower. This paper suggested a method for unifying schedules through the H-SYNC (Heartbeat-SYNC) method for solving such problem of the diversified scheduling of S-MAC, and as a result of comparing and evaluating its performance through simulation, it was found that the border node consumed energy less.

1 Introduction

Recently, many studies have been conducted on the sensor network for the purpose of a pure research or detection in the military field, or sensing in the ubiquitous computing environment [3]. However, in such wireless sensor network without any central control system or the mobile ad hoc network like Ad-hoc [5], IEEE 802.11 [2] does not show a satisfactory performance. In the wireless sensor network field where the wasted power is particularly highlighted as a big problem, the conventional standardized MAC protocol shows a poor performance falling short of one's expectations [4].

In case a great number of sensors for a sensor network are scattered in a specific area, it is considered that if a node's position is once fixed, it is almost impossible to relocate it or withdraw it. Therefore, the price of the sensor node itself should be low, and it should be able to work for a long term by using power very efficiently. As an

---

* Corresponding Author: Myong-Soon Park; E-mail: myongsp@ilab.korea.ac.kr.

approach for meeting this requirement, Sensor-MAC (S-MAC) protocol [4, 13] has been recently suggested. S-MAC has dramatically solved the existing power waste problem in the wireless sensor network which was incidental to IEEE 802.11.

As S-MAC basically adopts the listen and sleep method, it keeps a node to be in the sleep state for a long time, and thereby, it makes power not wasted unnecessarily in the idle state. In addition, by using a technique for optimizing power consumption in the functions of the sensor network, such as message passing or adaptive listening, it meets several necessary conditions.

However, even though this technique of S-MAC has the said advantages, S-MAC has a problem that it is difficult to apply it actually to the initial listen and sleep scheduling method of each node.

Since each sensor network node makes a listen and sleep schedule thereof initially on the basis of a time as determined randomly, in case neighbor schedules are completely inconsistent in the cycle thereof, a failure to communicate with one another occurs. Therefore, it transmits a SYNC packet periodically among nodes and thereby, it informs a schedule of each node and unifies some schedules, it does not unify the listen and sleep cycle completely. Even in the actual implementation thereof, all nodes are supposed to work 10 seconds per 2 minutes while they are in the listen state in order to find an independent schedule node [12].

In order to solve this problem, this paper has suggested a policy for using a unified synchronizing signal in place of multiple schedules of border nodes. It requires a simple and extendable algorithm for simplifying and defining the control of initial operation of the sensor network so that it may not have individual schedules in the wireless sensor network environment. The reliability and the performance of the suggested algorithm, which satisfied such condition, were verified through simulation, and further compared with those of the conventional S-MAC. Through such comparisonal results, its advantages and disadvantages were grasped. Finally, a direction for improvement of such algorithm was presented.

This paper is constructed as follows: Section 2 looks into assumptions and objectives of the S-MAC as reported in the reference studies; Section 3 analyzes the four problems of S-MAC; Section 4 describes an initial working algorithm of the sensor network, and verifies the reliability of the unified scheduling method through H-SYNC; Section 5 evaluates the performance of the suggested method; and Section 6 draws a conclusion.

2. Related Work

S-MAC is a Medium Access Control protocol equivalent to a link layer on the OSI 7 Layer as designed so that it may be suitably used in such sensor network environment, which was suggested by Wei Ye, John Heidemann, Deborah Estrin in 2002 [4].

Locally managed synchronizations and periodic sleep-listen schedules based on these synchronizations form the basic idea behind the Sensor-MAC (S-MAC) protocol [12]. Neighboring nodes form virtual clusters to set up a common sleep schedule. If two neighboring nodes reside in two different virtual clusters, they wake up at listen periods of both clusters. A drawback of S-MAC algorithm is this possibility of following two different schedules, which results in more energy consumption via idle listening and overhearing.

Schedule exchanges are accomplished by periodical SYNC packet broadcasts to immediate neighbors. The period for each node to send a SYNC packet is called the synchronization period. Figure 1 represents a sample sender-receiver communication. Collision avoidance is achieved by a carrier sense, which is represented as CS in the figure. Furthermore, RTS/CTS packet exchanges are used for unicast type data packets.

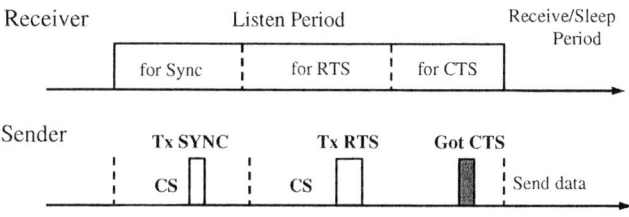

Fig. 1. S-MAC Messaging Scenario [12]

Advantages: The energy waste caused by idle listening is reduced by sleep schedules. In addition to its implementation simplicity, time synchronization overhead may be prevented with sleep schedule announcements [14].

Disadvantages: Broadcast data packets do not use RTS/CTS which increases collision probability. Adaptive listening incurs overhearing or idle listening if the packet is not destined to the listening node. Sleep and listen periods are predefined and constant, which decreases the efficiency of the algorithm under variable traffic load [14].

3 Analysis of Problems with the Diversified Scheduling Method

3.1 Problem of Delay Time in Stabilization of Schedule

What is considered in the S-MAC scheduling method is self-configuration. Each node should be able to sense a neighbor node around it and form a network topology for itself. S-MAC transmits a SYNC packet so that a randomly generated schedule may be maintained and extended. A long time is taken until scheduling of each node is stabilized. In case 10 nodes are actually available, they wait for about 100 seconds. If the number of nodes is increased, a longer time for stabilization is taken in proportion to the number of nodes.

3.2 Problem of Data Transmission and Synchronization in Border Node

The procedure for each node to have its own schedule is that in the first place, it waits for a given time, and then if it fails to receive a SYNC packet, it sets its own schedule and broadcasts a SYNC packet to neighbor nodes. As far as possible, many nodes receive the SYNC packet and use the listen and sleep policy having the same period, but the whole nodes are not unified under one schedule. Therefore, an independent schedule cluster having an independent schedule gets to be made. And, a node between heterogeneous schedules gets to receive SYNC packets which are different from each other and work as a border node [9, 10].

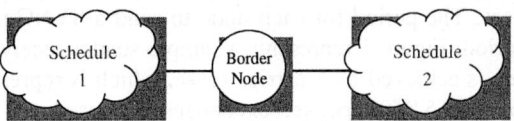

Fig. 2. Cluster and Border Node adopting Each Different Schedule

A listen time section and a sleep time section are arranged depending upon a time of the border node adopting both schedules as shown in the Figure 3, as follows:

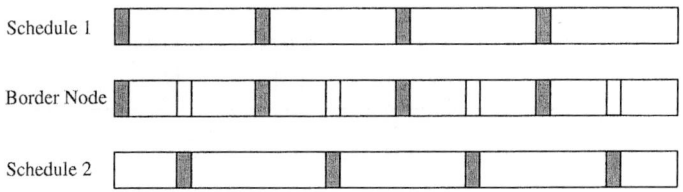

Fig. 3. Schedule in the Cluster and the Border Node

As shown in the figure, the border node adopts both kinds of schedules so that the listen section gets to be extended two times as long as other nodes. This just means that power consumption of the border node is two times as much as power consumption of a common node. If the border node gets to be dead due to extreme power consumption, communication between different schedule clusters will not be available any longer. And, since power consumption is increased in proportion to the number of different schedules adopted by the border node, a link connectivity failure of the whole network occurs problematically.

As an option for solving the border node problem in S-MAC, there is a method for the border node to adopt one schedule depending upon the SYNC packet as received first. However, also in this case, a SYNC packet for synchronization should be broadcast twice to clusters on both sides. In order to know all the schedules of neighbor nodes in this way, each and all nodes should a schedule table and in this process, a SYNC packet overhead gets to occur.

3.3 Problem of an Isolated Schedule Cluster

In case an independent schedule cluster having one neighbor node or more is overlapped with or identical with another cluster of which a period is different, it gets to be synchronized with the same schedule period by receiving a SYNC packet.

However, in case such schedule clusters are completely inconsistent in terms of time, there occurs a problem that they fail to acknowledge each other permanently. In this case, even though sensor nodes are present in the sensor network, they fail to communicate one another. S-MAC's solution to this problem is that per a given period, all nodes work in the listen state for a given time.

In an actual implementation, all nodes are embodied so that they may sense a SYNC packet and seek neighbor nodes while they stay in the listen state for 10 seconds per 2 minutes. This is an operation going against the basic purpose of S-MAC, that is to say, reducing energy for an idle listen time, which means that a lot of energy gets to be consumed in order to find a hidden schedule cluster node.

3.4 Problem of SYNC Packet Overhead

In order to solve the time error problem resulting from the long-term clock drift of the timer existing in the sensor node, it broadcasts the SYNC packet periodically and thereby informs and transmit the relative sleep time. Only by doing so, the listen and sleep period does not get to be inconsistent through adjustment of the timer.

As a result of measuring the clock drift against the sensor node used in the implementation example as described in the paper on S-MAC, it was about 0.2ms per second at maximum [12]. In case the listen time and the sleep time of S-MAC are set at 0.5 second and 1 second, respectively, the SYNC packet is transmitted one time per 10 times of the listen and sleep period in the actual implementation so that the clock drift occurs one time per about 15 seconds.

Wherein, it can be known that since the clock drift is 0.2ms per second, considering the listen time of 0.5 second, as about 2,500 seconds pass, a time error of 0.5 second at maximum occurs. If the passage is increased by 0.1 time in order to ensure that the node works stably, even though the SYNC packet is transmitted one time per 250 seconds, no problem will occur.

However, S-MAC has a problem that many SYNC packets are generated in order to broadcast the schedule time periodically per each schedule cluster. A time error resulting from a long-term clock drift is a very critical factor in the actual implementation.

4 Proposed Method: Unification of Schedule According to the H-SYNC Method

4.1 Method for Solving Problem of Delay Time in Stabilization of Schedule

The S-MAC scheduling method enables self-configuration on the basis of the use of a randomly selected time, but it has a problem that schedules, which are locally diversified, are generated. Therefore, by using the unified scheduling method, the following problems resulting from different schedule periods can be solved; delay time in stabilization of schedule, generation of a border node, generation of an isolated schedule cluster, and SYNC packet overhead as independently transmitted by each and every schedule cluster.

Several assumptions for application of the unified schedule are as follows:

- In the first place, all nodes should have a listen and sleep period which is constantly determined.
- All nodes should work according to the unified synchronizing signal (Heartbeat-SYNC).

- The *H-SYNC* (Heartbeat-SYNC) should be periodically broadcast by the sink node.
- At the beginning of communication, the *awake packet* is transmitted from the sink node.
- The sink node should have more energy than other nodes.

This method is to let each node follow the unified schedule by letting each node work according to the unified synchronizing signal (Heartbeat-SYNC). This is to broadcast a constant synchronizing signal to the whole sensor network just as the heart beats constantly in our body. Characteristically, the sensor network has a structure to collect the information sensed by each sensor node and transmit it to the sink node. If the sink node is not available, the sensor node does not need to work [16]. Therefore, the sink node transmits *H-SYNC* signals. Also, the sink node can work with it having more energy than other sensor nodes.

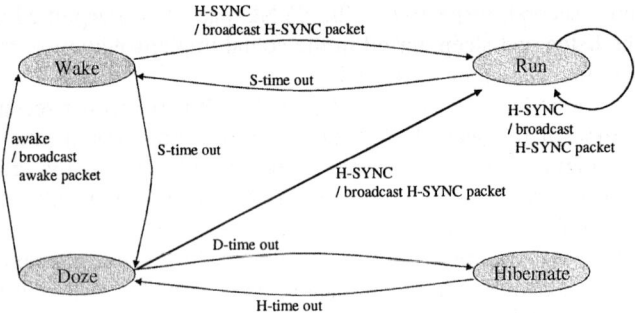

Fig. 4. State Transitions

In order to solve the problem of delay time in stabilization of schedule, it is necessary to synchronize the whole schedules of the network. This could be graphed into a state diagram as follows.

As shown in the Figure 4, the sensor node has four kinds of working states, and the working state of the sensor node gets to be changed through the *awake* packet and the *H-SYNC* packet. It works by using three kinds of timers.

- The *hibernate* state is an initial node state in which the node works in the sleep state reducing energy consumption at maximum. This state is periodically changed to the *doze* state according to the *H-time*
- The *doze* state is to sense the *awake* packet and the *H-SYNC* packet, in which the node listens a wireless signal for a time as set by the *D-timer*.
- The *wake* state is to broadcast the *awake* packet to neighbor nodes, which induces a neighbor node so that it may be shifted to the *wake* state. In this state, the node transmits the *awake* packet periodically every *D-time*.
- In the *run* state, the node works by the periodic listen and sleep method in the same manner as S-MAC, and it controls the *H-SYNC* packet by its timer and broadcasts it again to neighbor nodes.

- The *awake* packet is a message to change the node's state from the *doze* state to the *wake* state, which is a signal packet to change neighbor nodes promptly so that they may be in the *wake* state.
- The *H-SYNC* packet is a signal transmitted from the sink node and broadcast to the whole sensor nodes. According to this signal, the whole nodes reset their timer, and a listen and sleep time thereof gets to be synchronized.
- The *H-timer* is to set a time for changing the node's state periodically from the *hibernate* state to the *doze* state.
- The *D-timer* is to set a time for which the node waits until it receives the *awake* packet or the *H-SYNC* packet in the *doze* state.
- The *S-timer* is to set a time for which the node waits until it receives the subsequent *H-SYNC* packet.

Each node begins to work initially in the *hibernate* state, and its state is periodically changed to the *doze* state. If it receives the *awake* packet or the *H-SYNC* packet, its state is changed to the *awake* state or the *run* state. The *awake* state is to induce the whole sensor nodes to be promptly changed so that they may be in the working state by changing promptly neighbor nodes in the *doze* state so that they may be in the *awake* state. In the *wake* state and the *doze* state, the node can receive data with it being in the listen state, but it can transmit data only after it receives the *H-SYNC* packet and its state is changed to the *run* state. In the *hibernate* state, the node is in the sleep state so that it fails to receive data and further transmit data. In the *run* state, the node receives the *H-SYNC* packet so that the timer is synchronized, and therefore, it transmits/receives data according to the fixed listen and sleep period.

If all nodes follow such state change algorithm, in the end, the whole network gets to follow the unified schedule fit for the *H-SYNC* packet of the sink node.

4.2 Solution to Problem of Data Transmission and Synchronization in Border Node

As assumed above, all nodes have a fixed sleep and listen time of the same period, and each node's timer gets to be reset in conformity with *H-SYNC* packet transmitted from the sink node. Therefore, each neighbor node can transmit and receive data at the listen time with it being in the listen state. When a delay time in transmission of the *H-SYNC* packet is actually corrected and the clock-drift error is considered, in an

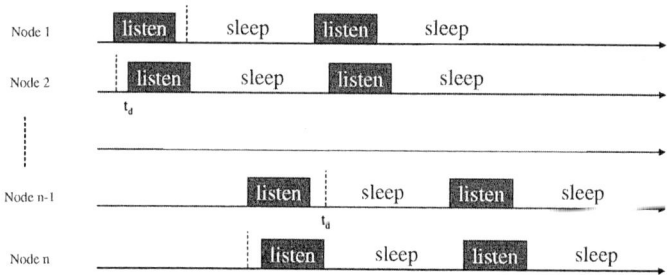

Fig. 5. Relationship of Listen and Sleep Timing in Unified Schedule

adjacent neighbor node, such as Node 1 and Node 2 or Node n-1 and Node n, it is possible to maintain td, which is short, versus a listen time. Of course, in Node 2 and Node n, a great time error will occur much depending upon a distance and the number of hops, but in the sensor network, communication is done between adjacent hops [11], and therefore, an error occurring in the process of broadcasting the *H-SYNC* packet can be disregarded.

By doing so, in the whole sensor network, the border node generated in the conventional S-MAC can be removed and the long-term connectivity of the sensor network can be assured through synchronization by the unified schedule without using diversified schedules.

4.3 Solution to Problem of Isolated Schedule Cluster

Like the case of the problem relating to data transmission and synchronization in the border node as above described, because the unified synchronizing method is used, any isolated schedule cluster is not generated. Even in case a new node is added, the node receives the *H-SYNC* packet in the doze state and promptly works in the *run* state so that it is easy to expand the sensor network. Also, since the working schedule of the neighbor node is the same, there is no need to maintain a schedule table of the neighbor node. It results in an advantage enabling a cost for manufacturing sensor nodes to be reduced. Further, all nodes do not need to be in the listen state for a given time periodically in order to find an isolate schedule cluster in the conventional S-MAC.

4.4 Solution to Problem of SYNC Packet Overhead

As shown in Figure 6, the *awake* packet and the *H-SYNC* packet are periodically transmitted from the sink node, and thereby, a clock drift time error of each node is adjusted. As a result of measuring a clock drift with respect to the sensor node as used in the implementation example described in the paper on S-MAC, it was found to be about 0.2ms per second at maximum. In case a listen time and a sleep time of S-MAC are set at 0.5 second and 1 second respectively, since the clock drift is 0.2ms per second, considering the listen time of 0.5 second, if about 2,500 seconds pass, a time error of 0.5 second at maximum occurs. If the passage is increased by 0.1 time in order to ensure that the node works stably, even though the SYNC packet is transmitted one time per 250 seconds and thereby it works in the error range of 0.05 second at maximum, no

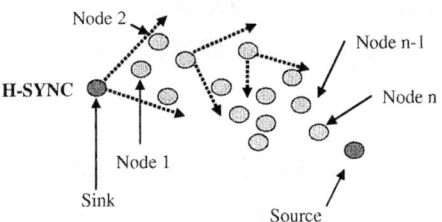

Fig. 6. Broadcasting of H-SYNC Packet

problem will occur. By doing so, transmission of the SYNC packet was improved so that the SYNC packet, which was transmitted every 15 seconds per the diversified schedule cluster in S-MAC, might be transmitted one time per 250 seconds.

As above described, synchronization was induced with the H-SYNC packet and the diversified schedules were integrated into a unified schedule as a method for solving the four problems occurring in S-MAC. Thereby, all problems were solved, and further, such improvements as reduction of a cost for manufacturing sensor nodes, simplification of the working method and enhancement of extensibility got to be available.

5 Simulation

5.1 Configuration of Simulation

Simulation was done in NS-2 [8], and the following parameters were set up.

Table 1. Configuration of Simulation

Duty Cycle = 10%	H-Time = 20sec
Listen Time = 0.5sec	D-Time = 1sec
Sync Packet Size = 9Bytes	S-Time = 250sec
RTS/CTS/ACK Size = 10Bytes	Routing Protocol = DSR
H-SYNC Packet Size = 9Bytes	Tx, Rx, Idle Power = 30, 20, 0.05
Awake Packet Size = 9Bytes	Initial Energy = 100
Data Packet Size = 50Bytes	Simulation Time = 1,000sec

Meanwhile, the adaptive listening was not applied, and it was assumed that traffic is to transmit 50 bytes every 200 seconds and all nodes should be fixed without any mobility.

5.2 Scenario 1: Result of Comparison when 2 different Schedules Are available

When 10 nodes are available, Schedule 1 is adopted by Node 0~Node 4, while Schedule 2 is adopted by Node 5~Node 9. Node 4 and Node 5, which are border nodes, adopt the other node's node each other.

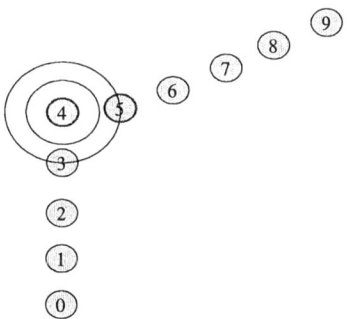

Fig. 7. Scenario 1

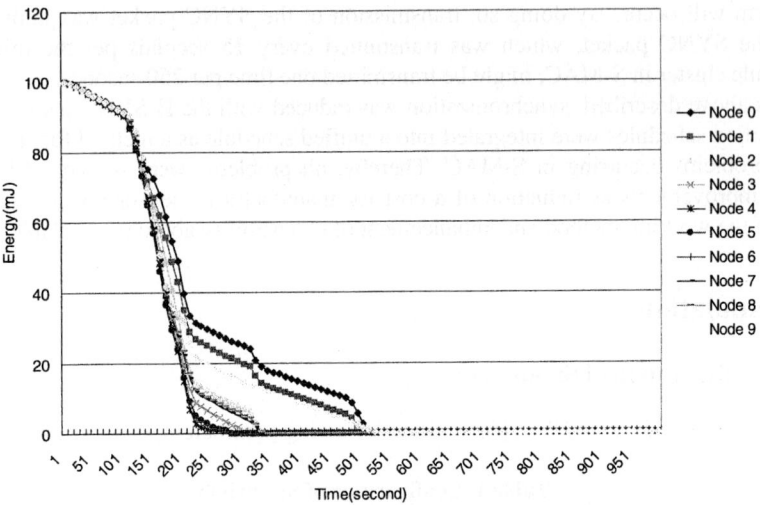

Fig. 8. When S-MAC is applied in Scenario 1

It can be known that in case Node 4 and Node 5 become border nodes by adopting a heterogeneous schedule respectively through the process of searching for a neighbor node, energy gets to be sharply reduced. And it can be known that the border node gets to be dead around about 281 seconds so that connection is cut off between the two different schedule networks.

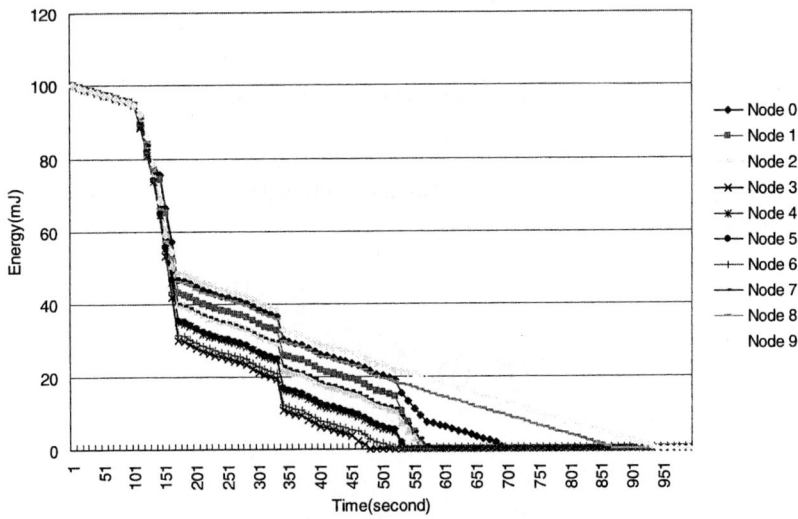

Fig. 9. When H-SYNC is applied in Scenario 1

As shown in the Figure 9, it can be known that in case H-SYNC is applied to the node, a life span of the network is extended up to about 481 seconds, as it works according to the unified schedule. This extended life span is about 1.71 times as long as the life span thereof in case H-SYNC is not applied to it.

5.3 Scenario 2: Result of Comparison when 4 Heterogeneous Schedules are Available

When 20 nodes are available, Schedule 1 is adopted by Node 0~Node 4, Schedule 2 is adopted by Node 5~Node 9, Schedule 3 is adopted by Node 10~Node14 and Schedule 4 is adopted by Node 15~Node 19. Node 4, Node 5 and Node 10, which are border nodes, adopt the other node's node mutually. It can be known that in case H-SYNC is applied to the node, a life span of the network is extended up to about 481 seconds, as it works according to the unified schedule. This extended life span is about 2.08 times as long as the life span thereof in case H-SYNC is not applied to it.

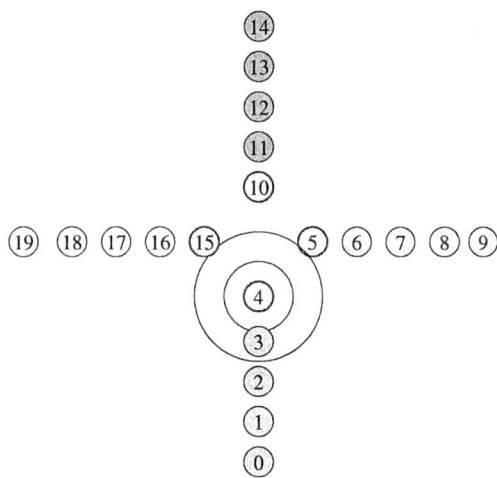

Fig. 10. Scenario 2

6 Conclusions and Future Work

The listen and sleep method may cause a synchronization problem between respective nodes. That is to say, there may occur a situation that synchronization goes amiss between network nodes so that they fail to communicate with one another. In S-MAC, this problem is solved by letting border nodes receive all the different schedules in the cluster using the heterogeneous scheduling policy. However, this solution causes power consumption of the border node to be greatly increased, and as a result, power gets to be quickly exhausted. In the end, communication gets to be unavailable between both clusters.

In this study, we intended to overcome this situation by adopting a policy for enabling the whole network to be unified by a single schedule. As a specific algorithm for removal of the border node, the H-SYNC method was suggested, and its performance and stability was verified through simulation thereof. Generally, as the number of heterogeneous schedules was greater, its performance was enhanced in proportion to it. According to the result of simulation thereof, it had an effect to extend the lifespan of the whole network in the range of about 171% to 208%.

References

1. Wei Ye and John Heidemann, "Medium Access Control in Wireless Sensor Networks", USC/ISI Technical Report, Oct. 2003.
2. The Institute of Electrical and Electronics Engineers, Inc. IEEE Std 802.11 – Wireless LAN Medium Access Control(MAC) and Physical Layer(PHY) specifications, 1999 edition.
3. James Weatherall, Alan Jones, "Ubiquitous Networks and their applications", IEEE Wireless Communications 2002.
4. Wei Ye, John Heidemann and Deborah Estrin, "An Energy-Efficient MAC Protocol for Wireless Sensor Networks", in Proc. IEEE INFOCOM, New York, pp. 1567-1576, June 2002.
5. Phil Karn, "MACA - a New Channel Access Method for Packet", In Proceedings of the 9th ARRI/CRRL Amateur Radio Computer Networking Conference, September 1992.
6. Suresh Singh and C.S. Raghavendra, "PAMAS – Power Aware Multi-Access Protocol with Signaling for Ad hoc Network", ACM Computer Communications Review, July 1998.
7. Saikat Ray, Jeffrey B. Caruthers and David Starobinski "RTS/CTS – Induced Congestion in Ad Hoc Wireless LANs", WCNCC 2003.
8. Network Simulator – 2, http://www.isi.edu/nsnam/
9. Rong Zheng, Jennifer C. How and Lui sha, "Asynchronous Wakeup for Ad Hoc Networks: Theory and Protocol Design", UIUCDCS-R-2002-2301, October, 2002.
10. Cedric Florens, Robert McElice, "Scheduling Algorithms for Wireless Ad Hoc Sensor Networks", IEEE GLOBECOM 2002.
11. Kay Romer, Eth Zurich, "Time Synchronization in Ad Hoc Networks", Mobihoc, 2001.
12. Wei Ye, John Heidemann and Deborah Estrin, "Medium Access Control With Coordinated Adaptive Sleeping for Wireless Sensor Networks", IEEE/ACM Transactions on Networking, VOL. 12, No. 3, pp. 493-506, June 2004.
13. Koen Langendoen and Gertjan Halkes, "Energy-Efficient Medium Access Control", Delft University of Technology, 2004.
14. Ilker Demirkol, Cem Ersoy and Fatih Alagoz, "MAC Protocols for Wireless Sensor Networks: a Survey", the Network Research Laboratory of the Computer Engineering Department of Bogzici University, 2004.
15. G. Lu, B. Krishnamachari and C.S. Raghavendra, "An Adaptive Energy-Efficient and Low-Latency MAC for Data Gathering in Wireless Sensor Networks", Proceedings of 18th International Parallel and Distributed Processing Symposium, Pages: 224, 26-30 April 2004.
16. Alberto Cerpa and Deborah Estrin, "ASCENT: Adaptive Self-Configuring sEnsor Networks Topologies", INFOCOM 2002.

Delay Threshold-Based Priority Queueing Packet Scheduling for Integrated Services in Mobile Broadband Wireless Access System

Dong Hoi Kim[1] and Chung Gu Kang[2]

[1] The Electronics and Telecommunications Research Institute,
Gajeong-dong, Yuseong-gu, Daejeon, Korea
donghk@etri.re.kr
[2] The college of Information and Communications,
Korea University, Seoul, Korea
ccgkang@korea.ac.kr

Abstract. In this paper, we present an opportunistic packet scheduling algorithm to support both real-time (RT) and non-real-time (NRT) services in mobile broadband wireless access (MBWA) systems. Our design objective is to determine the maximum number of RT and NRT users with respect to the overall service revenue while satisfying individual QoS requirements, e.g., the maximum allowable packet loss rate for RT traffic and the minimum reserved bit rate for NRT traffic. As opposed to a typical priority queueing-based scheduling scheme in which RT users are always served a prior to NRT users while NRT users are served with the remaining resource, the proposed scheme takes the urgency of the RT service into account only when their head-of-line (HOL) packet delays exceed a given threshold. The delay threshold-based scheduling scheme allows for leveraging the multi-user diversity of NRT users, eventually maximizing the overall system throughput. By evaluating the proposed approach in an orthogonal frequency division multiple access/frequency division duplex (OFDMA/FDD)-based mobile access system, it is shown that the overall system throughput can be significantly improved in terms of the number of users or total service revenue.

1 Introduction

In recent years, the opportunistic packet scheduling algorithms have been the increasing interests for supporting various types of data services in the emerging mobile broadband wireless access (MBWA) systems. In particular, OFDMA is considered as one of the most spectrally-efficient multiple access alternatives for these systems, as it fully leverages the multi-user diversity along with the frequency diversity inherent to the OFDM (orthogonal frequency division multiplexing) scheme. As a transmission data rate varies with a location-dependent channel condition, especially under the cellular structure with an aggressive frequency reuse, most of the packet scheduling algorithms considered in these systems deal with the non-real-time (NRT) services. One particular example is the proportional fairness (PF) packet scheduling algorithm [1]. Furthermore, there exist various types of modification to PF algorithm so as to improve the data throughput [2]. As the RT service must be also supported in these systems, packet scheduling algorithms for the RT service class are independently developed, e.g., EXP scheme [3].

In actual system to support both RT and NRT services, two different approaches can be used. One is the fixed-priority approach, which always prioritizes the RT service class over the NRT service class. In other words, NRT traffic is served with the resources remaining after all of the RT traffic has been served. This is essentially a form of priority queuing (PQ) scheme [4]. The other approach is to use two different types of priority metrics, each specified for an individual service class. The priority metric for each class is selected to take their relative urgency, throughput, and fairness among both RT and NRT service class users into account under the varying channel conditions. Once the priority metrics have been evaluated individually, a user with the highest metric value is served first, regardless of its service class. The EXP/PF scheme in [5] is an example of this particular approach.

In this paper, we propose a different type of packet scheduling scheme, called a delay threshold-based priority queueing (DTPQ) scheme, which is designed to support both RT and NRT services in an integrated manner. The maximum allowable packet loss rate and the minimum reserved bit rate are the QoS parameters under consideration for RT and NRT traffic, respectively. The design objective of the proposed packet scheduler is to determine the optimal number of RT and NRT users with respect to the overall service revenue while satisfying individual QoS requirements. As opposed to a typical priority queueing-based scheduling scheme in which RT users are always served a prior to NRT users while NRT users are served with the remaining resource, the proposed scheme takes the urgency of the RT service into account only when their head-of-line (HOL) packet delays exceed a given threshold. In other words, as long as the maximum allowable packet loss rate requirement is satisfied, RT users can be delayed so as to maximize the throughput for the NRT users, leveraging the multi-user diversity of NRT users.

By evaluating the proposed approach in an orthogonal frequency division multiple access/frequency division duplex (OFDMA/FDD)-based mobile access system, it is shown that the system capacity can be significantly improved in terms of the number of RT and NRT service users.

2 Problem Formulation

2.1 Motivation

For RT service class, there is a pre-specified packet loss rate requirement, which is governed by the maximum allowable delay, W_{max}. A HOL packet will be dropped if its delay exceeds W_{max}, the corresponding packet will be dropped and thus, the corresponding QoS requirement may not be met. In other words, as the HOL packet delay for RT service class increases towards W_{max}, a higher priority must be given to the RT service class users. However, if this prioritization is performed too early, then QoS requirement of the RT service class will be over-enforced,, which subsequently hurts the performance of NRT service class.

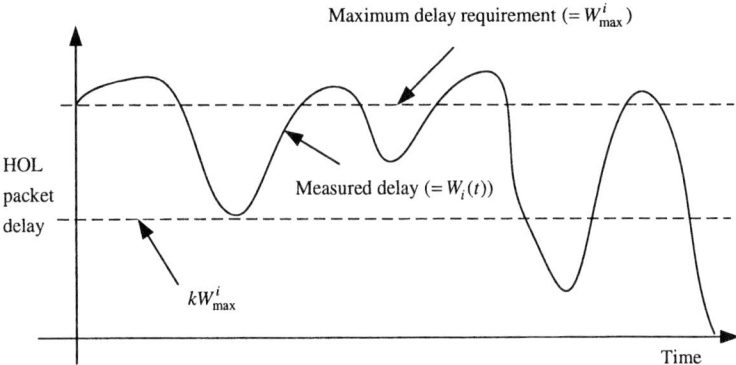

Fig. 1. Relative HOL packet delay variation subject to delay requirement

In the proposed scheme, we introduce a HOL delay threshold as a design parameter, which determines which service class to be served first. As shown in Fig. 1, the delay threshold is given by kW_{max}, where k is a control parameter that determines the priority of one service class over the other. Note that kW_{max} must be less than W_{max}, i.e., $0 \le k \le 1$. As long as HOL packet delays of all RT service class users do not exceed kW_{max}, then NRT service class users are scheduled. Otherwise, RT service class users are scheduled. Meanwhile, an appropriate priority metric for opportunistic scheduling is selected for each service class. In other words, all users in the same service class are served on the basis of their own priority metrics. In the current discussion, we consider exponential (EXP) scheduling algorithm [3] for RT service class and proportional fairness (PF) scheduling algorithm [1] for NRT service class.

The proposed approach consists of two different steps, one which deals with prioritizing the two different service classes and the other that prioritizes the users in the same service class. Here, the control parameter k must be determined to maximize a given objective function. Given the parameter k, for example, the total numbers of RT and NRT users that satisfy the target outage probability are denoted by $N_{RT}(k)$ and $N_{NRT}(k)$, respectively. Note that the optimal value of parameter k changes according to the operator's policy, e.g., depending on the service charges for the different traffic types subject to the prescribed QoS requirement. Note that delay requirement for the RT service class users is immediately violated with $k = 1$, in which case none of the RT service class users can be served unless there remains the resource after serving all RT service class users.

2.2 QoS and Performance Measures

The performance of RT service class is specified by the packet loss rate, which is defined as a ratio of the number of dropped packets to the total number of packets. Packet is dropped when the packet delay exceeds a specified delay, W_{max}. Let $\mathbf{U}_{RT}$ and $\mathbf{U}_{NRT}$ denote a set of users belonging to RT service and NRT service classes, respectively. If the numbers of dropped and transmitted packets during a given simulation interval for user i are denoted by $N_i^{(d)}$ and $N_i^{(s)}$, respectively, the packet loss rate at time t for user i is given as follows:

$$PLR_i(t) = \frac{N_i^{(d)}(t)}{N_i^{(d)}(t) + N_i^{(s)}(t)}, i \in \mathbf{U}_{RT}. \tag{1}$$

For each RT service class user, QoS requirement is imposed by the target packet loss rate, denoted by PLR_{max}, i.e., $PLR_i(t) \leq PLR_{max}$, $\forall i \in \mathbf{U}_{RT}$. Furthermore, the system outage probability can be specified in terms of the number of RT users that do not meet the packet loss rate requirement. More specifically, the system outage probability for the RT service class users at time t is defined as follows:

$$P_{out}^{(L)} = \frac{Num(PLR_i(t) > PLR_{max})}{N_{total}} \tag{2}$$

where $Num(PLR_i(t) > PLR_{max})$ denotes the number of users whose packet loss rate exceeds the target packet loss rate while N_{total} denotes the total number of RT service class users.

Meanwhile, the performance of NRT service class is specified by the average bit rate to be supported during a packet call. Assuming that there are K packet calls for a user session i, the average bit rate is defined as follows [6]:

$$\overline{R}_i = \frac{1}{K} \sum_{k=1}^{K} \frac{B_i^{(k)}}{t_d^{(k)} - t_a^{(k)}}, \quad i \in \mathbf{U}_{NRT}. \tag{3}$$

where $B_i^{(k)}$ denotes the number of bits transmitted during k-th packet call of user session i, while $t_a^{(k)}$ and $t_d^{(k)}$ denote the arrival and departure times of the k-th packet, respectively. For each NRT service class user, QoS requirement is imposed by the target minimum bit rate, denoted by R_{min}, i.e., $\overline{R}_i \leq R_{min}$, $\forall i \in \mathbf{U}_{RT}$. Furthermore, the system outage probability can be specified in terms of the number of NRT users that do not meet the minimum bit rate requirement. More specifically, the system outage probability for the NRT service class users is defined as follows:

$$P_{out}^{(R)} = \frac{Num(\overline{R}_i < R_{min})}{N_{total}} \tag{4}$$

where $Num(\overline{R}_i < R_{min})$ denotes the number of users whose minimum bit rate does not exceed the target minimum bit rate.

2.3 Problem Formulation

In general, the objective of the proposed DTPQ scheduling scheme intends to maximize the number of users in the system, increasing the total service revenue. In order to take into account the operator's policy on charging for RT and NRT service class users, we consider the weighting factors, ω_{RT} and ω_{NRT}, which represent per-user revenue for RT and NRT service, respectively. Then, overall revenue is given as follows:

$$f(k) = \omega_{RT} N_{RT}(k) + \omega_{NRT} N_{NRT}(k).. \qquad (5)$$

The control parameter k must be determined to maximize the objective function (5), while satisfying the individual QoS requirements in terms of their outage performances. Now, our design problem can be formally stated as follows:

$$\max_{0 \le k \le 1} \{\omega_{RT} N_{EXP}(k) + \omega_{NRT} N_{PF}(k)\}.. \qquad (6)$$

subject to

$$\begin{cases} P_{out}^{(L)} = \Pr\{PLR_i(t) > PLR_{\max}\} \le \tilde{P}_{out}^{(L)}, & \forall i \in \mathbf{U}_{RT} \\ P_{out}^{(R)} = \Pr\{\overline{R}_i < R_{\min}\} \le \tilde{P}_{out}^{(R)}, & \forall i \in \mathbf{U}_{NRT} \end{cases}$$

where $\tilde{P}_{out}^{(L)}$ and $\tilde{P}_{out}^{(R)}$ are the maximum allowable outage probabilities for RT and NRT users, respectively.

3 Performance Analysis

3.1 System Model

In this paper, we consider a downlink of the OFDMA/FDD system with 1,536 subcarriers derived from a total bandwidth of 20 MHz. Each frame is composed of 20 slots, each having a fixed length of 1 ms, i.e., a frame length of 20 ms. In each slot, all subcarriers are shared among all cells for a frequency reuse factor of 1. In each cell, a subset of subcarriers, defined as a subchannel, is used as the basic resource allocation unit. More specifically, we have a total of 12 subchannels in the current system, i.e., 128 subcarriers in each subchannel. Furthermore, we do not take the control sections and other overhead into account, but simply assume that all subcarriers are used for data transmission. All of the subcarriers in each subchannel are selected by following a pre-specified random pattern and thus, SIR for each user is measured by taking the average of all subcarriers assingned. The modulation order and coding rate is determined by the instantaneous SIR, following the prescribed AMC table, which specifies the minimum SIR required to meet a target frame error rate, e.g., 0.1 %.

Table 1. AMC Mode

Required SIR (dB)	Modulation Scheme	Coding Rate
1.5	BPSK	1/2
4.0	QPSK	1/2
7.0	QPSK	3/4
11.0	16 QAM	6/16
13.5	16 QAM	3/4
18.5	64 QAM	3/4

Table 2. System Parameters

Parameter	Value
System	OFDMA/FDD
Downlink Channel BW	20 MHz
OFDM symbol duration	100 μs
Total number of subcarriers	1,536
Number of subcarriers per subchannel	128
Number of subchannels	12
Frame period	20 ms
Slot period	1 ms

In our simulation, we follow the system specification described in [7], including the AMC table as shown in Table 1. All other system parameters are summarized in Table 2. Video streaming traffic and world wide web (WWW) traffic are considered as typical applications for RT and NRT services in the current application. In the sequel, their traffic models are detailed.

Near Real-Time Video Streaming Traffic Model

Video streaming consists of a sequence of frames which are emitted regularly with interval T. Every frame in the streaming is identically composed of a certain same number of slices, and each slice corresponds to a single packet. Note that the number of slices in each frame is the same, while those slices belonging to the same frame can have different sizes. In this paper, we consider the video streaming traffic with a rate of 32 kbps, a frame interval of 100 ms (T = 100), and eight slices per frame. The distribution parameters of the slice sizes and inter-arrival time are as shown in Table 3 [8][9], where α denotes the shape of the PDF for the packet size. Meanwhile, the minimum and maximum allowed packet sizes are denoted by β and γ.

NRT WWW Traffic Model

An NRT traffic model of world wide web (WWW) services, given in terms of their sessions, is defined by the time intervals of the WWW browsing usage. Each session consists of a number of packet calls, each of which also comprises a burst of packets. In other words, each packet call is made up of a burst of packets, which is a key characteristic of NRT traffic source. The specific parameters for WWW surfing, unconstrained delay data, 144 kbps traffic model are given in Table 4 [9], where the average packet size is denoted by μ_L.

Table 3. Video streaming traffic model parameters [8]

Category	Distribution	Parameters
Slice size	Truncated Pareto	α =1.2, β = 20 byte, γ=125 byte
Slice inter-arrival time	Truncated Pareto	α =1.2, β =2.5 ms, γ=12.5 ms

Table 4. WWW-application traffic model parameters

Information types	Distribution	Parameters
Number of packet calls per session	Geometric	5
Reading time between packets calls	Geometric	412 sec
Number of packets within a packet call	Geometric	25
Inter-arrival time between packets (within a packet call)	Geometric	27.7 ms
Packet size	Truncated Pareto	$\alpha = 1.1$, $\beta = 81.5$ bytes, $\gamma = 66666$ bytes, $\mu_L = 480$ bytes

3.2 Simulation Parameters

In our simulation, we consider a hexagonal cell layout with a reference cell and 6 surrounding cells in the first tier, each with an omni-directional antenna. Even though a larger number of tiers could be taken into account for improving the accuracy of the assessment of interference from other cells, it is not essential as the current analysis is merely focused on studying the effectiveness of the proposed scheme. The radius of each cell is fixed at 1 km. Mobile stations are uniformly distributed throughout each cell and move with velocities that are uniformly distributed in the range of [3,100] km/h in a random direction. The large-scale path loss [10] and log-normal shadowing [11] are taken into account for the simulation. Our path loss model and log-normal shadowing follow a typical outdoor mobile communication propagation model, which is often found in the 3GPP standardization context. More specifically, we use the following path loss model:

$$L = 128.1 + 37.6 \log_{10} R \quad (9)$$

where R is the distance from a base station (BS) to the mobile station in km. We assume that the BS has perfect channel knowledge. The transmission power of the BS is set to 12 W, which is equally distributed among all 12 subchannels, assuming that all cells are fully loaded, meaning that all subcarriers are fully allocated in each cell. We assume that there are 12 subchannels available for each time slot. Assuming that all subcarriers in each subchannel are randomly distributed in the frequency domain, it is acceptable that the same power is assigned to each subchannel in the average sense. More specifically, we assume that each subchannel is assigned a power level of 1 W, i.e., a total power of 12 W is allocated to the BS, which is equally distributed over all subchannels.

Only the payload of each packet is considered, e.g., excluding the additional bits for the header. The maximum delay allowed for RT video streaming traffic is 200 msec. For the WWW surfing unconstrained delay data (UDD) 144 kbps traffic model, the minimum required data rate is set to 50 kbps. In order to overload the system for a short duration, the reading time between packet calls is set to 2.6592 seconds. All results are obtained during the simulation time of 100 sec. Meanwhile, the maximum allowable outage probability is set to $\tilde{P}_{out}^{(L)} = \tilde{P}_{out}^{(R)} = 0.1$ for all service classes, i.e., $\Pr\{\Pr\{W_i(t) > 0.2 \text{ sec}\} > 0.01\} < 0.1$ for RT video streaming service and $\Pr\{\overline{R}_i < 50 \text{ kbps}\} < 0.1$ for NRT WWW service.

3.3 Simulation Results and Discussion

Fig. 2 shows that the number of users satisfying the outage probability of 0.1 varies with the parameter, k. Without any loss of generality, we only consider discrete values of k, i.e., $k = 0, 0.1, 0.2, \ldots, 0.9, 1.0$. When $\omega_{RT} : \omega_{NRT} = 1:1$, we find that the total number of RT and NRT users supported by DTPQ scheme is maximized at $k=1$. When $\omega_{RT} : \omega_{NRT} = 5:1$, however, the objective function is maximized at $k = 0.3$. It implies that the optimum delay threshold with respect to total revenue exists for the different per-user revenue.

In Fig. 3, outage performance of PQ and DTPQ schemes are compared for $\omega_{RT} : \omega_{NRT} = 5:1$. Note that DTPQ scheme with $k = 0$ is equivalent to the conventional PQ scheme. As expected, the outage performance of NRT service class varies with k while it is not the case for RT service class. In fact, a performance of NRT service users is significantly improved with the proposed DTPQ scheme while that of RT service is not compromised. We note that the maximum number of NRT users that can be admitted is increased almost as much as 30% by DTPQ scheme with $k = 0.3$. It confirms the optimal trade-off for opportunistic packet scheduling between RT and NRT service classes subject to their individual QoS requirement.

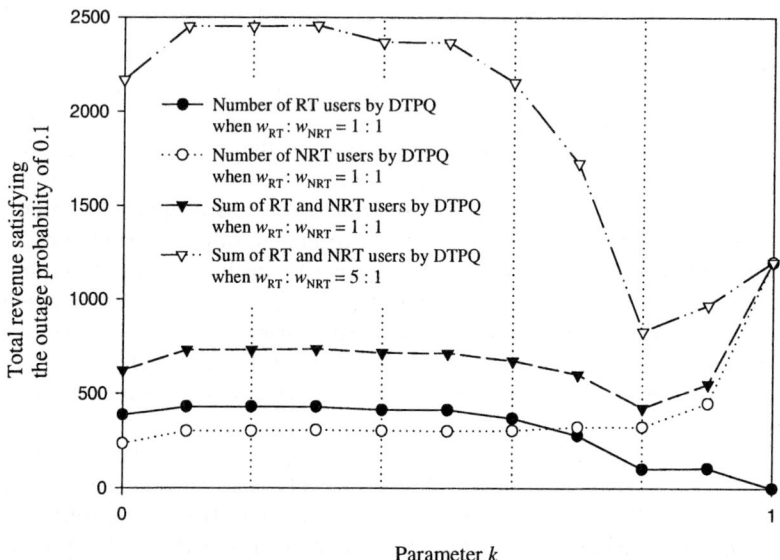

Fig. 2. Total revenue satisfying $\tilde{P}_{out}^{(L)} = \tilde{P}_{out}^{(R)} = 0.1$ as a function of the design parameter k

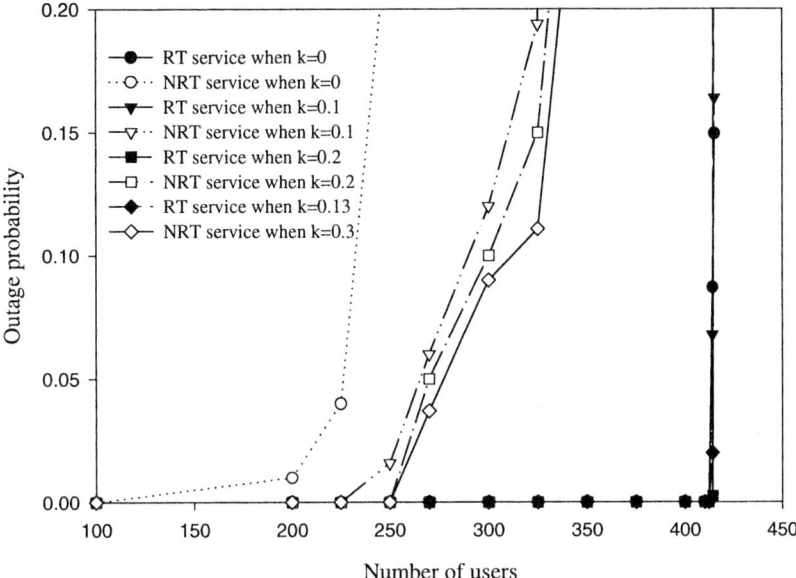

Fig. 3. Outage probability performance: PQ vs. DTPQ schemes ($\omega_{RT} : \omega_{NRT} = 5:1$)

4 Conclusion

In this paper, we proposed a novel priority queueing packet scheduling scheme to maximize total throughput of mobile broadband wireless access system, in which both RT and NRT service classes are supported at the same time. It is based on the delay threshold that trades off the packet loss rate performance of RT service class with average data throughput of NRT service class with the fixed data rate. The proposed delay threshold-based priority queueing (DTPQ) scheme provides a means of maximizing the multi-user diversity gain for NRT service class while satisfying QoS requirement for RT service class. Our simulation results have demonstrated that the number of NRT service users can be increased almost as much as 30% by DTPQ scheme. Furthermore, it has been shown that the delay threshold is a critical design parameter as it must be varies with the traffic load and channel condition. In our future work, a type of adaptive control scheme must be addressed to determine the dynamic optimal delay threshold that varies with the network condition.

References

1. A. Jalali, R. Padovani, and R. Pankaj, "Data Throughput of CDMA-HDR a High Efficiency-High Data Rate Personal Comm. Wireless System," VTC 2000-Spring, vol. 3, 2000, pp. 1854-1858.

2. Dong Hoi Kim, Byung Han Rye, and Chung Gu Kang, "Packet Scheduling Algorithm Considering a Minimum Bit Rate for Non-realtime Traffic in an OFDMA/FDD-Based Mobile Internet Access System," ETRI Journal, Vol. 26, Number 1, pp. 48~52, February 2004.
3. S. Shakkottai and A. Stolyar, "A study of Scheduling Algorithms for a Mixture of Real- and Non-Real-Time Data in HDR," 17th International Teletraffic Congress (ITC-17), Sept. 2001.
4. Geoff Huston, Internet Performance Survival Guide: QoS Strategies for Multiservice Networks, Wiley Computer Publishing, 2000.
5. Jong Hun Rhee and Dong Ku Kim, "Scheduling of Real/Non-real Time Services in an AMC/TDM System: EXP/PF Algorithm," LNCS, Vol. 2524/2003, pp. 506~514, August 2003.
6. 3GPP, "Feasibility Study for OFDM for UTRAN enhancement (Release 6)," 3G TR25.892 V0.2.0, March, 2003.
7. Qingwen Liu, Shengli Zhou, and Georgios B. Giannakis, "Cross-Layer Combining of Adaptive Modulation and Coding with Truncated ARQ over Wireless Links," IEEE Transactions on wireless communications, vol. 3, no. 5, pp. 1746~1755, Sept. 2004.
8. Draft 802.20 Permanent Document, Traffic Models for IEEE 802.20 MBWA System Simulations, July 8, 2003.
9. J. P. Castro, The UMTS Network and Radio Access Technology, John Wiley & Sons, Inc., New York, NY, 2001.
10. 3GPP, "Physical Layer Aspects of UTRA High Speed Downlink Packet Access (Release 2000)," 3G TR25.848 V4.0.0, March, 2001.
11. Recommendation ITR-R M.1225, "Guidelines for evaluation of radio transmission technologies for IMT-2000," 1997.

Call Admission Control Using Grouping and Differentiated Handoff Region for Next Generation Wireless Networks

Dong Hoi Kim[1] and Kyungkoo Jun[2]

[1] Mobile Telecommunication Research Laboratory,
Electronics and Telecommunications Research Institute, Korea
donghk@etri.re.kr
[2] Department of Multimedia System Engineering,
University of Incheon, Korea
kjun@incheon.ac.kr

Abstract. We propose a novel call admission control scheme which improves the handoff drop and the new call block probabilities of high priority services, minimizing the negative impact on low priority services, in multi–service cellular networks. The proposed scheme is a mix of the grouping and the differentiated handoff region scheme. The simulation results revealed that our hybrid scheme improved the drop and the block probabilities of the high priority services compared with the conventional reconfiguration scheme.

1 Introduction

We propose a call admission control scheme which improves the handoff drop and new call block probabilities of the high priority services, minimizing the negative impact on the low priority services, in multi–service cellular networks.

In [1], a reconfiguration–based call admission control strategy is proposed. In this scheme, services have multiple levels of required bandwidth. Depending on the available bandwidth in the cell, the call admission control can adaptively determine which level of bandwidth to be allocated to the services. In particular, to accept the admission requests of high priority services when the remaining bandwidth is insufficient, the bandwidths occupied by existing low priority services are degraded to lower level, and the resulting surplus bandwidth is reallocated to the incoming high priority service.

In [5][6], a handoff ordering–based call admission control scheme is proposed to give priority to the handoff of high priority services. In this scheme, the handoff requests are queued in a target cell and processed according to priority. The highest priority is given to the requests with high service priority, low received signal strength (RSS) from current serving cell, and large change in current RSS from the previous measured RSS.

In this paper, we propose a hybrid call admission control scheme mixing two schemes: the grouping scheme and the differentiated handoff region scheme. These two schemes are evolved from the aforementioned reconfiguration and the handoff ordering strategies, respectively.

By the grouping, the services which do not fully utilize their assigned bandwidth are grouped into a set, and then forced to share the bandwidth in round robin with others in the same group rather than being allocated separate bandwidth, thus benefiting *bandwidth-efficient* services, which are generally real-time and, at the same time, have higher priority than others. The grouping increases the total admission capacity of a cell, as a result, leading to the improved block and drop probabilities of both the low and high priority services.

In the differentiated handoff region scheme, the high priority services have larger handoff region than the low priority services. The different region sizes are determined set by employing separate RSS thresholds depending on the service priority. It targets particularly the improvement of the handoff drop probability of the high priority services.

This paper is organized as follows. Section 2 discusses our proposed call admission control strategy in detail and presents the algorithm flow chart. Section 3 shows the simulation results for the performance analysis of our proposed scheme. Section 4 concludes this paper.

2 Call Admission Control Scheme

In this section, we describe a call admission control scheme favoring high priority services in multi-service cellular networks. The proposed scheme is a mix of two strategies: *grouping* and *differentiated handoff region*.

Grouping. It reallocates the wasted bandwidth assigned to low priority services to incoming high priority services. It enables to accept the admission requests of high priority services even when there remains insufficient bandwidth in the cell.

The scheme works as follows. When high priority services request the admission to a cell, but there is insufficient bandwidth remaining, the existing low priority services hand over their bandwidth in order to accept those requests. Instead, they form a bandwidth-sharing group in which they are able to continue their services by sharing the bandwidth in round robin. The bandwidth allocated to the group is much smaller than the total sum of the bandwidth owned by each group members. Because of the intermittent presence of the think times, the actual bandwidth usage times of each service may not overlap so often, thus their services are able to proceed without noticeable service degradation even with the round robin sharing.

Differentiated Handoff Region. It is devised to improve the handoff drop probability of high priority services over low priority services by adopting different sizes of handoff region according to the service priority. We set the handoff region of high priority services larger than that of low priority services. As a result, high priority services start the handoff procedure earlier and last longer than low priority services, leading to the improved handoff drop probability.

In general, the handoff is performed by measuring the received signal strength (RSS) of the pilot channel emitted by the base stations. By setting two thresholds for the RSS, *handoff threshold* is used to indicate the start of the handoff procedure, while *receive threshold* is the last point beyond which the handoff trial fails unless the handoff to a target cell is completed yet.

The differentiated handoff region can be implemented by adopting different thresholds for services depending on their priorities. To enlarge the handoff region of high priority services, it needs to increase the handoff threshold, while decrease the receive threshold. On contrary, to shrink the handoff region for low priority services, it is accomplished by decreasing the handoff threshold and increasing the receive threshold.

In our proposed scheme, we set the different RSS thresholds according to service priorities in such a way that for high priority services, the handoff region spans entire overlapped area between the call–departing cell and the call–arriving cell, while that of low priority services locates within the narrow area of the overlaid region. In the case of the departing cell, the RSS threshold that triggers the handoff of the high priority services is set higher than that of the low priority services, thus having the effect of starting the handoff of the high priority services earlier than that of the low priority services. On contrary, the minimum RSS threshold below which the handoff is failed is set lower in the case of the high priority services than the low priority services, thus being able to lengthen the interval during which the handoff is tried. In the case of the arriving cell, the RSS threshold from which the incoming handoff can be accepted is set lower in the case of the high priority services than the low priority services, while the maximum RSS threshold beyond which the incoming handoff is failed is set higher for the high priority services than the low priority services.

The algorithm pseudo code that implements our proposed schemes is as follows. It shows how the two proposed schemes are related each other. New call request is handled by the grouping scheme first, then, when the call is about to perform the handoff, the differentiated handoff scheme takes effect.

```
 1: New Call Admission Request to Cell A
 2: if requested_BW ≤ available_BW_of_cell_A then
 3:     Accept the Request
 4: else
 5:     if The Request is High Priority Service then
 6:         Perform the Grouping at Cell A
 7:         if requested_BW ≤ available_BW_of_cell_A_after_Grouping then
 8:             Accept the Request
 9:         else
10:             Reject the Request
11:         end if
12:     else
13:         Reject the Request
14:     end if
15: end if
16: Monitor RSS
17: while RSS ≥ Handoff_Threshold_of_Cell_A do
```

18:     Keep Monitoring RSS
19: **end while**
20: Handoff Request to Cell B   ▷ Handoff Procedure
21: **if** $requested\_BW \leq available\_BW\_of\_cell\_B$ **then**
22:     Handoff Success to Cell B
23: **else**
24:     **if** The Request is High Priority Service **then**
25:         Perform the Grouping at Cell B
26:         **if** $requested\_BW \leq available\_BW\_of\_cell\_B\_after\_Grouping$ **then**
27:             Handoff Success to Cell B
28:         **else**
29:             Wait in the Handoff Queue of Cell B
30:         **end if**
31:     **else if** The Request is Low Priority Service **then**
32:         **if** Find the Group with the Available Bandwidth **then**
33:             Handoff Success to Cell B
34:         **else**
35:             Wait in the Handoff Queue of Cell B
36:         **end if**
37:     **end if**
38:     **if** $RSS \leq Receive\_Threshold\_of\_Cell\_B$ **then**
39:         Handoff Failure, Drop
40:     **else**
41:         Sort Requests in the Handoff Queue of Cell B
42:         Start the Handoff Procedure Again
43:     **end if**
44: **end if**

3 Performance Analysis

Through the simulation, we compare the performance of our proposed scheme with a reconfiguration scheme [1], which is a conventional call admission control scheme, labeled as "reconfiguration", in the following figures. The reconfiguration technique is to reassign the partial bandwidth of the low priority services to the high priority services when a new high priority service asks its admission to a cell which has less remaining bandwidth than required for the admission acceptance. As the performance metric, we measure the handoff drop probability and the new call block probabilities for the high priority service and the low priority service, respectively, and compare them with those of the reconfiguration scheme. The required bandwidth for each service class is: 16 Kbps. The ratio between the total numbers of services is set to 50:50.

The simulation results show that our hybrid scheme mixing the grouping and the differentiated handoff had the effect to improve both the handoff drop probability and the new call block probability of high priority services over those of low priority services.

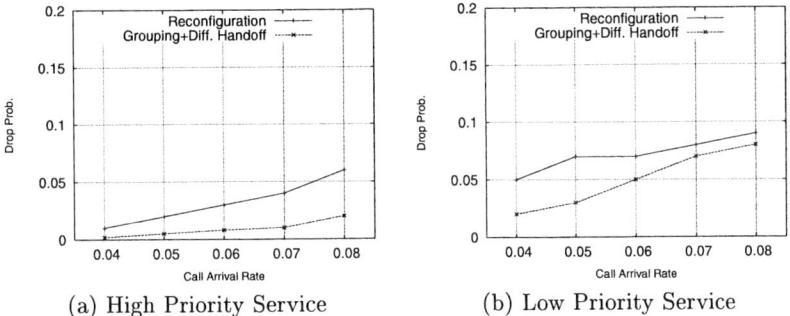

Fig. 1. The comparison of the handoff drop probabilities for each class

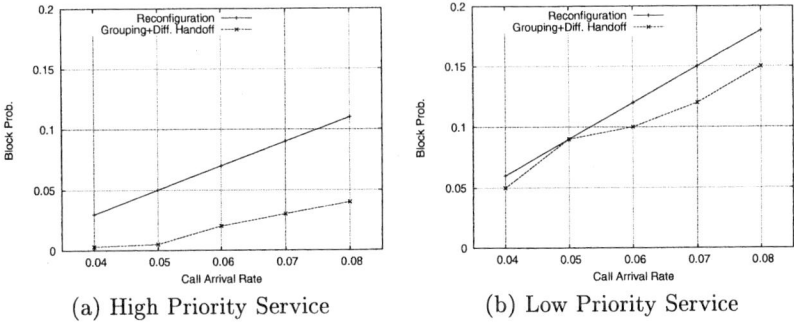

Fig. 2. The comparison of the new call block probabilities for each class

Figure 1 shows the comparison between the handoff drop probabilities measured for each class. The handoff probabilities of the schemes employing the grouping and the differentiated handoff support, labeled "grouping+diff.handoff", in the figure) are lower than the scheme without it (labeled as "reconfiguration"). As described in the earlier sections, it is because the grouping scheme increases the total admittance capacity of the cells by having the low priorities services share the bandwidth. In addition, our differentiated handoff scheme lowered the handoff drop probability of the high priority service less than that of the low priority service for all arrival rates. It is the result of enlarging the handoff region of the high priority services than that of the low priority services.

Figure 2 shows the comparison between the new call block probabilities measured for each class. First of all, for all classes, we observe that the new call block probabilities are higher than the corresponding handoff drop probabilities because both schemes, our scheme and the reconfiguration strategy, prioritize the handoff requests than the new call requests. Secondly, similar to the results shown in Figure 1, the new call block probabilities of our proposed scheme are lower than the scheme without it. It is due to the increased cell capacity by the grouping scheme as described.

4 Conclusions

We proposed the hybrid scheme which improves the handoff drop probability and the new call block probability of the high priority services, minimizing the negative impact on the low priority services, in multi–service cellular networks. The proposed scheme is a mix of the grouping and the differentiated handoff region scheme. By the grouping, the total admission capacity of a cell increases, as a result, leading to the improved block and drop probabilities of both the low and high priority services. The differentiated handoff region scheme targets particularly the drop probability of the high priority services. Through the simulation, we observed that our proposed scheme improved the drop and block probabilities compared with the reconfiguration scheme.

References

1. Ye, J., Hou, J., Papavassiliou, S. : A Comprehensive Resource Management Framework for Next Generation Wireless Networks : IEEE Transactions on Mobile Computing, vol. 1, No. 4, 2002.
2. Bettstetter, C. : Mobility Modeling in Wireless Networks : Categorization, Smooth Movement, and Border Effects : Mobile Computing and Communications Review, vol. 5, no. 3, 2001.
3. Navidi, W., Camp, T. : Stationary Distributions for the Random Waypoint Mobility Model : IEEE Transactions on Mobile Computing, vol. 3, no. 1, 2004.
4. Liu, Z., Niclausse, N., Jalpa-Villanueva : Traffic model and performance evaluation of Web servers : Performance Evaluation vol. 46, 2001.
5. Ebersman, H. and Tonguz, O. : Handoff Ordering Using Signal Prediction Priority Queuing in Personal Communication Systems: IEEE Transactions on Vehicular technology, Vol.48, No.1, January 1999.
6. Chang, R. and Leu, S. :Handoff Ordering Using Signal Strength for Multimedia Communications in Wireless Networks: IEEE Transactions on Wireless Communications, Vol.3, No.5, September 2004.

A Cluster-Based QoS Multipath Routing Protocol for Large-Scale MANET*

Hui-Yao An, Xi-Cheng Lu, Zheng-hu Gong, and Wei Peng

School of Computer, National University of Defense Technology, 410073,
Changsha, P.R. China
anthony_cs@21cn.com
{xclu, zhgong, wpeng}@nudt.edu.cn

Abstract. To support oS routing in MANET (Mobile Ad hoc Networks) is a core issue in the research of MANET. Numerous studies have shown the difficulty for provision of Quality-of-Service (QoS) guarantee in Mobile Ad hoc networks. This paper proposes a scheme refer to a cluster-based QoS multipath routing protocol (CQMRP) that provides QoS-sensitive routes in a scalable and flexible way in mobile Ad Hoc networks. In the strategy, each local node just only maintain local routing information of other clusters instead of any global ad hoc network states information. It supports multiple QoS constraints. We evaluated the performance of the protocol using the OPNET simulator, the result shows that this protocol can provide an available approach to QoS multipath routing for mobile Ad Hoc networks.

1 Introduction

An Ad Hoc Networks is a peer-to-peer mobile network consisting of large number of mobile nodes. These nodes create an instant network on demand and may communicate with each other via intermediate nodes in a multi-hop mode, i.e., every node can be a router. Ad hoc networks may be the only solutions in many situations where instant infrastructure is needed and no central backbone system and administration (like base stations and wired backbone in a cellular system) exist. However, node mobility and limited communication resources make QoS provision in MANETs routing very difficult. Mobility causes frequent topology changes and may break existing paths. The advantage and inherent nature of MANETs have led to research interest in routing.

Multipath routing is indeed a viable alternative to MANET routing approach. There have been some works on multipath routing in ad hoc networks. Multipath-DSR (M-DSR)[10] is a simple multipath extension of the popular DSR[1], in which alternate routes are maintained so that they can be utilized when the primary one fails. In AODV-BR[3], an extension of AODV[2], multiple routes are maintained and utilized only when the

---

* This research was supported by the National Grand Fundamental Research 973 Program of China under Grant No. 2003CB314802 and the National Natural Science Foundation of China under Grant No. 90104001.

primary route fails. However, traffic is not distributed to more than one path. Multiple Source Routing protocol (MSR)[4] proposes a weighted round-robin heuristic-based scheduling strategy among multiple paths in order to distribute load, but provides no analytical modeling of its performance. In [5], the positive effect of alternate path routing (APR) on load balancing and end-to-end delay in mobile ad hoc networks has been explored. Split multi-path routing (SMR), proposed in [6], focuses on building and maintaining maximally disjoint paths, however, the load is distributed in two routes per session. A framework for multi-path routing and its analytical model in mobile ad hoc network was proposed in [7]. This scheme, utilizing M-for-N diversity coding technique, solved the inherent unreliability of the network by adding extra information overheads to each packet. CHAMP[9] uses cooperative packet caching and shortest multipath routing to reduce packet loss due to frequent route breakdowns.

From the research survey of literature for multi-path routing strategy, there are still many issues in applying multi-path routing techniques into mobile ad hoc networks that are to be covered. On the one hand, in most of the routing protocols, the traffic is distributed mainly on the primary route [3-6, 9]. It is only when this route is broken that the traffic is diverted to alternate routes. Clearly, load-balancing is not achieved by using these routing mechanisms. Although there are some routing protocols which distribute traffic simultaneously on multiple paths [7,8], there has not been a routing protocol which could dynamically cope with the changes of topology in ad hoc network. On the other hand, all the routing don't takes into consideration that the routing control overhead will increase quickly with the number of the networks node increasing, These lead to scalability problem and reliability problem. As a result, there is a demand for a multi-path routing strategy that can not only cope with the dynamics of the network to balance efficiently the load on the network but also can provide QoS to satisfy the application of requirement.

Utilizing clustering algorithm to construct hierarchical topology may be a good method to solve these problems. Clustering management [12-17] has five outstanding advantages over other protocols. First, it uses multiple channels effectively and improves system capacity greatly. Second, it reduces the exchange overhead of control messages and strengthens node management. Third, it is very easy to implement the local synchronization of network. Fourth, it provides quality of service (QoS) routing for multimedia services efficiently. Finally, it can support the wireless networks with a large number of nodes.

This paper presents a hierarchical QoS multipath routing protocol for MANET (CQMRP) which uses clustering's hierarchical structure management to search effectively for multiple paths and distributes traffic among diverse multiple paths. It not only ensures fast convergence but also provides multiple guarantees for satisfying multiple QoS constraints. CQMRP also allows that an Ad Hoc group member can join/leave the cluster dynamically. The rest of the paper is organized as follows. Section 2 introduces Cluster Architecture and the Modeling, Section 3 introduces CQMRP algorithm in detail. We describe the simulation model and present our performance results in Section 4, and conclude the paper in Section 5.

2 Cluster Architecture and the Modeling

2.1 Cluster Architecture

The CQMRP is based on a multi-level hierarchical scheme, which is given Fig.1.In general, the clustering problem of MANET depends on the network topology, geographical location of nodes (or routers), connectivity, signal range, mobility as well as the relativity between nodes. In the viewpoint of clustering hierarchical networks, each node of MANET can be considered as 0th-level. A region that consists of such several nodes can be called first-level clusters are combined to form second-level clusters. Several first-level clusters are combined to form second-level cluster. Similarly, k^{th}-level clusters can be defined. The proposed mobility-based hierarchical clustering algorithm[11] is used, which can result in variable-size clusters depending on the mobility characteristics of the nodes. Each first-level cluster contains at least one node and does not overlap with any other first-level clusters. Second-level cluster contains only first-level clusters and they don't overlap. All nodes that are within the same first-level cluster are called local nodes. The node that has links connecting the nodes in other clusters in called bridge node (or domain border router). The local nodes are called 0th-level bridge nodes. The nodes that connect two first-level clusters are called first-level bridge nodes, and so on. As far as multipath routing is concerned, a network is usually represented as weighted digraph G= (N, L), where N denotes the nodes and L denotes the set of communication links connecting the nodes. |N| and |L| denote the number of nodes and links in the MANET, respectively. Without loss of generality, only digraphs are considered in which there exists at most one link between a pair of ordered nodes. Associated with each link are parameters that describe the current status of the link.

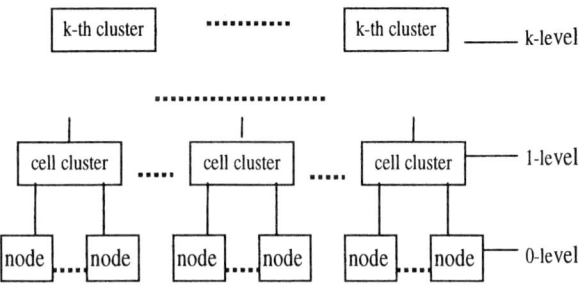

Fig. 1. Cluster Architecture

2.2 QoS Model

A node is assumed to keep the up-to-date local state about all outgoing link. The state information of link l_i^j includes 1) $DT_{l_i^j}$, the delay of link l_i^j including the radio propagation delay, the queue delay, and the protocol-processing time; 2) $BW_{l_i^j}$, the residual (unused) bandwidth of the link; and 3) $CO_{l_i^j}$, which can be simply one as a hop count or a function of the link utilization. In order to make a preference of stationary links over transient links, the cost of a transient link should be set much higher than that of a stationary link. Let $s \in N$ be source node of a MANET, and $d \in \{V - \{s\}\}$ be a set of destination nodes. Similarly, for any node $i \in N$, one can also define some metrics: delay function DT_{n_i}, cost function CO_{n_i}. The delay, bandwidth, and cost of a path $p_k = \{s, i, j, ..., m, t\}$ are defined as follows:

$$DT(p_k) = \sum_{l \in p_k} DT_{l_i^j} + \sum_{i \in p_k} DT_{n_i} \; ; \quad BW(p_k) = \min\{BW_{l_s^i}, BW_{l_i^j}, ..., BW_{l_m^d}\} ;$$

$$CO(p_k) = \sum_{l \in p_k} CO_{l_i^j} + \sum_{i \in p_k} CO_{n_i}$$

The QoS-based multipath routing problem is to find that satisfies some QoS constraints:

Delay constraint: $DT(p_k) \leq DT$ (1)

Bandwidth constraint: $BW(p_k) \geq B$ (2)

Cost constraint: $CO(p_k) \leq CO$ (3)

Where DT is delay constraint, B is bandwidth constraint and CO is cost constraint. In the above QoS constraints, the bandwidth is concave metric, and the delay and cost are additive metrics. For simplicity, we assume that all nodes have enough resource, i.e., they can satisfy the above QoS constraints. Therefore, we only consider the link's or edges' QoS constraints, because the links and the nodes have equifinality to the routing issue in question. The characteristics of edge can be described by a three-tuple (DT, B, CO), where DT, B and CO denote delay, bandwidth and cost, respectively. For simplicity, we also mainly consider the former two QoS constraints of the above QoS constraints (Equation 1-3).

3 CQMRP

3.1 Virtual Route Discovery

The protocol CQMRP is N-stage routing decision process, as explained below. A cluster is denoted by $C_i = \{N_i^j\}$, where N_i^j is the member of cluster C_i. Let CH_i be the cluster head of C_i.

When a source node s ($s \in C_i$) seeks to set up a connection to a destination d, s sends a route request message (RREQ) to its cluster head CH_i. The RREQ message includes the following fields {*source-address, destination-address, session-id, P_{lower} (DT, B, CO), path-quality, virtul-route*}. The procedure of virtual route is detailed as follow (Fig.2).

```
Set VirtualRouteSet ∈ { }
Set CandidateRouteSet ∈ { }
int VirtualRouteDiscovery(id, CandidateRouteSet){
if ( s ∈ $C_i$ and d ∈ $C_i$ ){
    setup multiple path $p_k = \{s, N_i^1, N_i^2, \cdots, d\}$ ;
    insert path $p_k$ into CandidateRouteSet;
    VirtualRouteSelection(id, VirtualRouteSet); }
if ( s ∈ $C_i$ and d ∉ $C_i$ ){
    search for a stable and optimal route as a directional guideline{ $s, C_2, \cdots, C_{n-1}, d$ };
    setup multiple path $p_k = \{s, N_i^j, \cdots, d\}$ ;
    insert path $p_k$ into CandidateRouteSet;
    VirtualRouteSelection(id, VirtualRouteSet); }
    return failure; /* Unable to find a set so far */   }
int VirtualRouteSelection(id, VirtualRouteSet){
    For each path $p_k$ ∈ CandidateRouteSet
        Compute path-quality $P_{pk}(DT(p_k), BW(p_k), CO(p_k))$ ;
        if (path-quality( $P_{pk}(DT(p_k), BW(p_k), CO(p_k))$ ) ≥ $P_{lower}$ ) {
            insert path $p_k$ into VirtualRouteSet(VR);
        }
    }
}
```

Fig. 2. Virtual route discovery procedure

Finally, it will choose all maximal disjoint, loop-freedom reliable paths from virtual route set. We call the above paths selected virtual route because they just are possible routes.

3.2 Reverse Link Labeling and Traffic Distribute

The reverse link labeling algorithm tries to find as many as possible real routes that are along the virtual path with loop-freedom and satisfy the QoS requirement for this particular session as well. The destination d generates a one-hop broadcast, sending the reverse labeling message which includes the following fields: (*Source Address(s), Labeling Source Address(l), Session ID, QoS Requirements ($P_{lower}(DT, B, CO)$, Virtual Route (VR), Hop (H), Pathquality($P_{pk}(DT(p_k), BW(p_k), CO(p_k))$)*).

Before starting the reverse-link labeling phase, d sets L as its IP address, H as 0 and $DT(p_k)$ as 0 while other fields are the same with those in the route request message. Every node that receives the reverse labeling message checks whether it meets the following conditions in order to broadcast the packet again after:

- increasing H by 1;
- adding its delay to $DT(p_k)$;
- recording l, H and $DT(p_k)$ into its routing table;

- replacing l with its IP address, l must meet the following requirement:
  It belongs to a cluster head that is in the virtual route VR.
  It has enough bandwidth: $BW(p_k) \geq B$.

  The accumulated delay $DT(p_k)$ does not exceed the delay requirement in QoS: $DT(p_k) \leq DT$.

  The hop number H doesn't exceed the maximum hop H_{max}.

Thus, more than one route will be discovered between s and d that comprise of links labeled by session ID.

We classify these paths into optimal path, shortest path and so on. For some particular requirement application, we classify all data packets (or users) into different service levels. Source node can select the proper path for the different service level applications. For the generally applications, it will calculate the path weight value, according path-quality message included in the paths messages and utilize M-for-N[9] diversity coding technique to fragment the data packet is into smaller blocks. Then according to the weight value of the path, distribute different number of blocks over the available paths. The larger the weight value of the path is, the more the blocks is distributed over the path.

3.3 Dynamic Route Repairing and Maintaining

When a cluster member node does not receive three HELLO packets continuously from its cluster head, it considers that the wireless link between them is broken. Thus, it must find a new cluster head, which is one hop from it, or becomes itself a cluster head if it cannot hear any existing cluster head.

If the route used to forward packets is broken due to node mobility or some link can't meet the QoS requirement, the node deletes the entry of this link from its routing table and selects another redundant labeled links that meet the requirement to forward information. The session traffic, QoS requirement and the link label of the link are switched to the new link.

4 Simulation

We use OPNET modeler to simulate our proposed algorithm. In our simulation, the channel capacity of mobile hosts is set to the same value: 2Mbps. We use the distributed coordination function (DCF) of IEEE 802.11 for wireless LANs as the MAC layer protocol. It has the functionality to notify the network layer about link breakage. In our simulation, 300 mobile nodes move in a 1500 meter x 500 meter rectangular region for 900 seconds simulation time. A pause time of 0 seconds presents continuous motion, and a pause time of 900 seconds corresponds to no motion. We change node number from 50 to 500 to investigate the performance influence of node number increase. The simulated traffic is 20 Constant Bit Rate (CBR). The interval time to send packets is 250ms. The size of all data packets is set to 512 bytes. Simulation time is 4 hours for every session. For each scenario, 10 runs with different random seeds were conducted and the results were averaged.

We use two different ways to study CQMRP algorithm. In one method, we compare CQMRP and a multipath algorithm named MSR with plane structure. The other method is to compare CQMRP and unipath routing (CBRP)[18]. We evaluate mainly the performance according to the following metrics:

Control overhead: The control overhead is defined as the total number of routing control packets normalized by the total number of received data packets.

Average end-to-end delay: The end-to-end-delay is averaged over all surviving data packets from the sources to the destinations.

Load balancing: In network graph G = (N, L), We define a state function f: N → I where I is the set of positive integers. f(n) represents the number of data packets forwarded at node v. Let CoV (f) = standard variance of f / mean of f. We use CoV (f) as a metric to evaluate the load balancing. The smaller the CoV (f) is, the better the load balancing is

Success Delivery Rate (SDR): $SDR = \dfrac{\text{Number of Data Received}}{\text{Number of Data Originated}}$

Figure 3 and 4 shows that the control overhead for unipath routing is less than multipath routing. The control overhead of CQMRP is lower than that of MSR, especially when the node number increases large enough. The bigger the size of the network is, the lower the cost of CQMRP is relative to MSR.

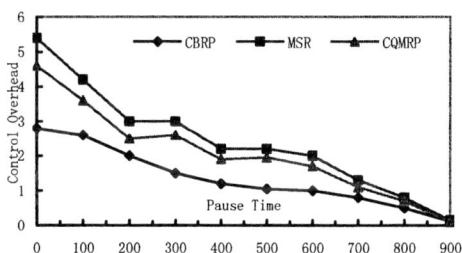

Fig. 3. Control overhead with varying speed

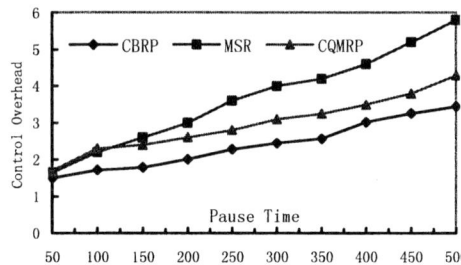

Fig. 4. Control overhead with varying network nodes

Figure 5 shows the results of average end-to-end delay. From Figure 5, the unipath routing has slightly higher average end-to-end delay compared to multipath routing and the average end-to-end delay of CQMRP is slightly higher than that of MSR.

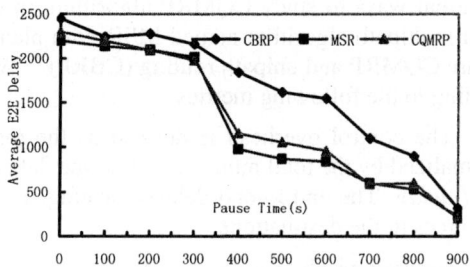

Fig. 5. Average end-to-end delay with varying speed

Figure 6 gives the results of load balancing. The CoV of network load for the uni-path routing is higher than that for the multipath routing. The CoV of network load for CQMRP is lower than that for SMR, this is because that the load of SMR is distributed in two routes per session and the load of CQMRP is distributed in all the available routes per session (we assume the number of the paths is 4).

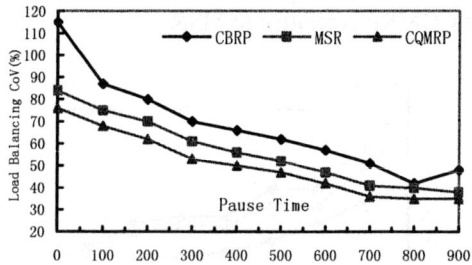

Fig. 6. CoV of the network load with varying speed

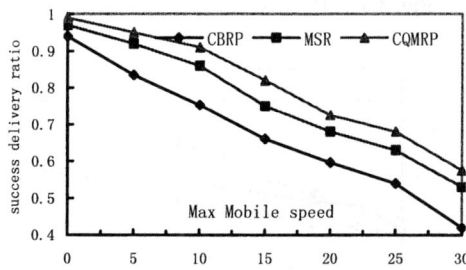

Fig. 7. Success delivery rate with varying max mobility speed

Figure 7 shows the success delivery ratio for CQMRP, CBRP and MSR. It illustrates that our proposed CQMRP outperforms CBRP and MSR at any mobility speed ranging from 1 to 30 meters/second. In addition, the simulation results demonstrate the ability of multipath routing to obtain consistent success delivery ratio regardless of the change in node mobility speed.

5 Conclusion and Future Work

CQMRP distributes traffic among diverse multiple paths to the sharing rate of channel. It not only ensures fast convergence but also provides multiple guarantees for satisfying multiple QoS constraints. It uses clustering's hierarchical structure diverse to decrease routing control overhead and improve the networks scalability. It can balance the network load, dynamically deal with the changes of network topology and improve reliability. These benefits make it appear to be an ideal routing approach for MANETs. However, these benefits are not easily explored because the data packet that is fragmented into smaller blocks must be reassembled at the destination node, it maybe lead to error and increase control overhead. In the future, we will do some work on the dynamically distribute traffic into multiple paths algorithm and error correction packet segmentation algorithm to improve the performance of CQMRP.

References

1. D. Johnson and D.Maltz. Dynamic source routing in ad hoc wireless networks. Mobile Computing, chapter 5, pages 153–181. Kluwer Academic Publishers, 1996.
2. C. Perkins, and E. Royer. Ad-hoc On-Demand Distance Vector Routing. Proc. of the 2nd IEEE Workshop on Mobile Computing Systems and Applications, pp. 90-100, February, 1999.
3. S.J. Lee and M. Gerla, AODV-BR: Backup Routing in Ad Hoc Network, IEEE WCNC 2000, pp. 1311-16.
4. L.Wang et al, Multipath Source Routing in Wireless Ad Hoc Networks, Canadian Conf. Elec. Comp. Eng., Vol. 1, 2000, pp. 479-83.
5. M.R. Pearlman et al, On the Impact of Alternate Path Routing for Load Balancing in Mobile Ad Hoc Network works, MobilHOC 2000, p.150.
6. S.J. Lee and M. Gerla, Split Multi-path Routing with Maximally Disjoint Paths in Ad Hoc Networks, ICC'01.
7. A. Tsirigos, Z. J. Haas, Multi-path Routing in the Present of Frequent Topological Changes, IEEE Communications Magazine, Nov, 2001.
8. Huiyao An, Xicheng Lu, Wei Peng, "A Cluster-Based Multipath Routing for MANET" Proc. of Med-Hoc-Net 2004, Bordum, June 2004 pp.405-413
9. Alvin Valera, Winston K.G. Seah and SV Rao, Cooperative Packet Caching and Shortest Multipath Routing in Mobile Ad hoc Networks, IEEE INFOCOM 2003
10. A. Nasipuri and S.R. Das, On-Demand Multi-path Routing for Mobile Ad Hoc Networks, IEEE ICCCN'99, pp. 64-70
11. M. Gerla and T. C. Tsai, "Multicluster, mobile, multimedia radio network," ACM-Baltzer J. Wireless Networks, vol. 1, no. 3, 1995, pp. 255-65.
12. A. Bhatnagar and T. G. Robertazzi, "Layer Net: a new self-organizing network protocols," Proc. IEEE MILCOM '90, pp. 845-849.
13. M. Gerla and T. C. Tsai, "Multicluster, mobile, multimedia radio network," ACM-Baltzer J. Wireless Networks, vol. 1, no. 3, 1995, pp. 255-65.
14. A. Alwan, R. Bagrodia, N. Bambos et al.,"Adaptive mobile multimedia networks," IEEE Personal Commun., Apr. 1996, pp. 34-51.
15. A. B. McDonald and T. F. Znati, "A mobilitybased framework for adaptive clustering in wireless ad hoc networks," IEEE J. Select. Areas Commun., vol. 17, no. 8, Aug. 1999, pp. 1466-1487.

16. C. R. Lin and M. Gerla, "Adaptive clustering for mobile wireless networks," IEEE J. Select. Areas Commun., vol.15, no. 7, Sep. 1997, pp. 1265-1275.
17. W. Chen, N. Jain and S. Singh, "ANMP: ad hoc network management protocol," IEEE J. Select. Areas Commun., vol. 17, no. 8, Aug. 1999, pp. 1506-1531.
18. Y. T. Mingliang Jiang, Jinyang Li, "Cluster based routing protocol," IETF Internet Draft, draft-ietf-manet-cbrp-spec-01.txt, July 1999. [Online]. Available: http://www.comp.nus.edu.sg/ tayyc/cbrp/

A MEP (Mobile Electronic Payment) and IntCA Protocol Design[*]

Byung kwan Lee[1], Tai-Chi Lee[2], and Seung Hae Yang[1]

[1] Department of Computer Science & Engineering, Kwandong University
bklee@kwangdong.ac.kr, yang7177@chollian.net
[2] Dept. of Computer Science, Saginaw Valley State University, University Center, MI 48710
lee@svsu.edu

Abstract. This paper proposes an MEP (Mobile Secure Electronic Payment) protocol, which uses ECC (Elliptic Curve Cryptosystem with F2m not Fp) [1, 2, 3], SHA (Secure Hash Algorithm) and BSB instead of RSA and DES. To improve the strength of encryption and the speed of processing under mobile environment, the public key and the private key of ECC are used for BSB [5, 6] algorithm, which generates session keys for the data encryption. In particular, when ECC is combined with BSB, the strength of security is improved significantly. As the process of the digital envelope used in the existing SET protocol is removed by the BSB algorithm in this paper, the processing time is substantially reduced. In particular, when authenticating each other, instead of using external CA, MEP uses IntCA (IntraCA), which is more suitable for mobile environment. In addition, the use of multiple signatures has some advantages of reducing the size of transmission data as an intermediate payment agent and avoiding the danger of eavesdropping of private keys.

1 Introduction

SSL (Security Socket Layer) and SET (Secure Electronic Transaction) protocol based on electronic payment has improved message integrity, authentication, and non-repudiation. Such a protocol is related directly to cryptography for security and consists of an asymmetric key algorithm, RSA for authentication and non-repudiation, DES for the message confidentiality, Hash algorithm and SHA for message integrity. But the disadvantage of this protocol is that the speed of processing is slow because of long key size. For mobile environment, ECC (Elliptic Curve Cryptosystem) technique is very important to Cryptography. This paper proposes an MEP(Mobile Electronic Payment) protocol, which uses ECC instead of RSA. To improve the strength of encryption and the speed of processing, the public key and the private key of ECC are used for BSB algorithm, which generates session keys for the data encryption. Therefore, the digital envelope used in the existing SET protocol can be removed by the BSB algorithm, which makes MEP protocol better than SET by simplifying the complexity of dual signature.

---

[*] This research was supported by the program for the training of Graduate Students in Regional Innovation, which was conducted by the Ministry of Commerce, Industry and Energy of the Korean Government.

2 Basic Concepts

2.1 Encryption and Decryption Algorithm

As shown in Figure 1, the user A computes a new key $k_A(k_BP)$ by multiplying the user B's public key by the user A's private key k_A. The user A encodes the message by using this key and then transmits this cipher text to user B. After receiving this cipher text, The user B decodes with the key $k_B(k_AP)$, which is obtained by multiplying the user A's public key, k_AP by the user B's private key, k_B. Therefore, as $k_A(k_BP) = k_B(k_AP)$, we may use these keys for the encryption and the decryption.

2.1 SET (Secure Electronic Transaction) Protocol

With digital signatures, encryption for message, and digital envelope, SET offers confidentiality, data integrity, authentication, and non-repudiation over an open network. But the public key cipher algorithm, which is used in digital signature and digital envelope of SET is slow in processing data. In general, SET uses SHA and RSA for a digital signature and envelope, and DES for encryption and decryption of message. However, RSA has a possibility to be destroyed by factorization, and thus SET can be weak. The procedure of encoding for SET consists of digital signature, encryption for message, and digital envelope. The generation process of digital signature is that it produces a message digest from plaintext by one-way hash function and then signs it with sender's private key. The encryption of message means that original messages, digital signature, and certificate are encoded with session keys.

The decryption of SET is as follows. First of all, the receiver decodes the digital envelope with receiver's private key and then acquires the transmitted session key. Second, the cipher text is decoded and the message of plaintext, digital signature, and certificate are generated. Third, the receiver separately generates a message digest from the message by using a hash function. In addition, in the digital signature there is a message digest that is decoded with sender's public key.

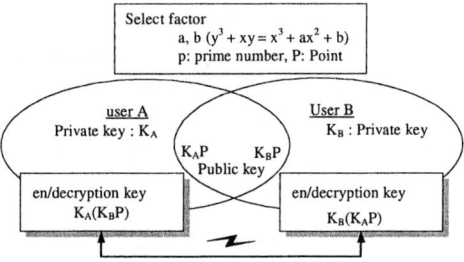

Fig. 1. Concept of en/decryption of ECC

3 The Proposed MEP (Mobile Electronic Payment) Protocol

The existing SET uses RSA in digital signature and DES in message encryption. Our proposed MEP protocol uses ECC instead of RSA. Because of this, the strength of encryption and the speed of processing are improved. Besides, in message encryption, MSEP utilizes BSB algorithm to generate session keys and cipher text. The encryption and decryption processes are shown in Figures 2 and 3 respectively. First, the public key and private key from ECC are put into BSB algorithm and then generates the session keys. Second, the BSB algorithm encodes the message by applying these keys. Since the receiver has his own private key, MEP can remove digital envelope, which enhances the speed for processing a message, and strengthens the security for information. In addition, if digital envelope is removed, MEP doesn't need to transfer a session key to the other, which may avoid the possibility of being intercepted and it can reduce part of the encryption procedure. Therefore, it simplifies a dual signature and decreases a communicative traffic over a network as compared with the existing SET.

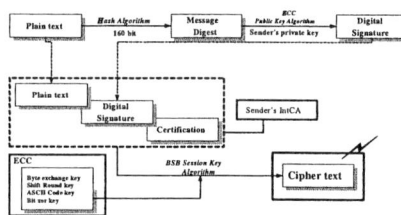

Fig. 2. Encryption of MEP   **Fig. 3.** Decryption of MSEP

3.1 F_2^m ECC (Elliptic Curve Cryptosystem)

3.1.1 Elliptic Curves Over F_2^m [7]

A non-supersingular elliptic curve $E(F_2^m)$ over F_2^m defined by the parameters a, b $\in$ F_2^m, b • 0, is the set of solutions (x, y), x $\in$ F_2^m, y $\in$ F_2^m, to the equation

$$y^2 + xy = x^3 + ax^2 + b$$

together with an extra point 0, the point at infinity. The number of points in $E(F_2^m)$ is denoted by #$E(F_2^m)$. It follows from the Hasse theorem that

$$q + 1 - 2\sqrt{q} \leq \#E(F_2^m) \leq q + 1 + 2\sqrt{q},$$

where $q = 2^m$. Furthermore, #$E(F_2^m)$ is even.

The set of points $E(F_2^m)$ is a group with respect to the following addition rules:

i) $0 + 0 = 0$.
ii) $(x, y) + 0 = (x, y)$ for all $(x, y) \in E(F_2^m)$.
iii) $(x, y) + (x, x + y) = 0$ for all $(x, y) \in E(F_2^m)$ (i.e., the inverse of the point (x, y) is the point $(x, x + y)$).
iv) Rule for adding two distinct points that are not inverses of each other :
     Let $(x_1, y_1) \in E(F_2^m)$ and $(x_2, y_2) \in E(F_2^m)$ be two points such that $x_1 \neq x_2$.
     Then $(x_1, y_1) + (x_2, y_2) = (x_3, y_3)$, where
     $x_3 = L + L + x_1 + x_2 + a$,
     $y_3 = L(x_1 + x_3) + x_3 + y_1$,

$$\text{and } L = \frac{y_1+y_2}{x_1+x_2}.$$

v) Rule for doubling a point :
Let $(x_1, y_1) \in E(F_2^m)$ be a point with $x_1 \neq 0$. Then $2(x_1, y_1) = (x_3, y_3)$, where
$x_3 = L^2 + L + a,$
$y_3 = x_1^2 + (L + 1) x_3,$

$$\text{and } L = x_1 + \frac{y_1}{x_1}.$$

The group $E(F_2^m)$ is abelian, which means that $P + Q = Q + P$ for all points P and Q in $E(F_2^m)$.

3.1.2 The Encryption/Decryption Process of Elliptic Curve Algorithm

The encryption/decryption process of elliptic curve algorithm is as follows:

Step 1] User A : Select an irreducible polynomial $f(x)$ and generate a field F_2^m.
Step 2] user A : Choose elliptic curve of the following form and vector values a, b.
$$E: y^2+xy=x^3+ax^2+b, a,b \in F_2^m$$
Find an initial point P on elliptic curve.
Step 3] user A : Compute $K_A P$ after selecting integer K_A.
Step 4-1] user A : Register $f(x)$, E, a, b, P and $K_A P$ to the open list.
Step 4-2] user B : After selecting random integer K_B as a secret key, register a public key $K_B P$ of user B, $f(x)$, E, a, b, and P in the open list.
Step 5] user B : Compute $K_B(K_A P) = (c_1, c_2)$ using public key $K_A P$ of user A in the open list.
Step 6] user B : Encrypt message m by $K_B(K_A P) = (c_1, c_2)$ and send to user A.
Step 7] user A, B : Change the result to integer and bit string to create shared secret key.

3.2 BSB Algorithm

The MEP protocol uses the same hash function, SHA as SET protocol. The session keys are generated by using the shared private key of ECC and data is encrypted by using the keys within the BSB algorithm. The proposed BSB algorithm consists of session key generation and data encryption. And the data encryption is divided into three phases, which are inputting plaintext into data blocks, byte-exchange between blocks, and Shift Round, ASCII Code, bit-wise XOR operation between data and session key.

3.2.1 Key Generation

The proposed BSB algorithm uses a 100-bit session key to perform the encryption and decryption. Given the sender's private key $X = X_1 X_2 ... X_m$ and the receiver's public key, $Y = Y_1 Y_2 ... Y_n$, we concatenate X and Y to form a key N (i.e. $N = X_1 X_2 ... X_m Y_1 Y_2 ... Y_n$), and then compute the session keys as follows:

1. If the length (number of digits) of X or Y exceed five, then the extra digits on the left are truncated. And if the length of X or Y less than five, then they are

padded with 0's on the right. This creates a number N' = X_1' X_2' X_3' X_4' X_5' Y_1' Y_2' Y_3' Y_4' Y_5'. Then a new number N'' is generated by taking the modulus of each digit in N' with 10

2. The first session key E1 is computed by taking bit-wise OR operation on N'' with the reverse string of N''.
3. The second session key E2 is generated by taking a circular right shift of E1 by one bit. And repeat this operation to generate all the subsequent session keys needed until the encryption is completed

3.2.2 Data Encryption

The process of data encryption proceeds as follows:

[step1] key generation using column
The first generated byte exchange keys is computed by using:
$\quad$ mat_{a1}(row, col), mat_{a2}(row, col), ..., mat_{a10}(row, col), mat_{b1}(row, col), ...,
$\quad$ mat_{c1}(row, col), ..., mat_{d1}(row, col), ..., mat_{d10}(row, col)
and by using these keys, the bytes of the two randomly selected blocks are exchanged by using these keys in step1.

[step 2] key generation using row
The second generated byte exchange keys is computed by using :
$\quad$ mat_{a1}(row, col), mat_{a2}(row, col), ..., mat_{a10}(row, col), mat_{b1}(row, col), ...,
$\quad$ mat_{c1}(row, col), ..., mat_{d1}(row, col), ..., mat_{d10}(row, col)
and by using these keys, the bytes of the block generated in step1 are exchanged once more by using these keys generated in step 2 to improve the strength of security.

[Step 3] Shift Round Key is generated by using the result of step 2
$\quad$ (N=prototype key, T_num=table sequence number, n=byte exchange pair number, k=1∈10 number, t=10∈1 number)

[Step 4] index table is generated by using ASCII code.

[Step 5] The byte exchange of the result of the step 4 can be obtained by using the keys previously generated in the same way as step4. Therefore the process is omitted here.

[Step 6] Bit-wise XOR between data and session key

After the byte-exchange is done, the encryption proceeds with a bit-wise XOR operation on the first 10 byte data with the session E1 and repeats the operation on every 10 bytes of the remaining data with the subsequent session keys until the data block is finished. During the process, the bit-wise XOR operation transforms a character into a meaningless one, which adds another level of complexity to the attackers.

3.2.3 Decryption

Decryption procedure is given as follows. First, a receiver generates a byte exchange block key E1 and a bit-wise XOR key E4 by using the sender's public key and the receiver's private key. Second, the receiver decrypts it in the reverse of encryption process with a block in the input data receiving sequence. The receiver does bit-wise XOR operation bit by bit, and then, a receiver decodes cipher text by using a byte-exchange block key E1 and moves the exchanged bytes back to their original positions. We reconstruct data blocks in sequence by using the decoded data block number.

4 IntCA

The certificate of IntCA proposed in this paper consists of header and body part. If a receiver's valid period of header part which is granted from CA(Certificate Authority) is valid, A sender fetches the receiver's public key and encrypts his own message with the public key. Figure 4 shows the structure of IntCA.

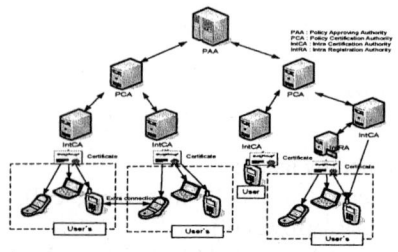

Fig. 4. The proposed IntCA authentication   **Fig. 5.** IntCA structure

4.1 IntCA Design

Figure 5 shows the structure of IntCA different from exisiting CA. IntCA which grants certification policy from PCA consists of wired/wireless CA, RA, IntServer, IntRA Server, and WTLS IntCA Server. The registration process through each server is done in WPKI Server or Web Server. Directory Server for storing certificates and CRLs(Certificate Revocation Lists) and OCSP Server for updating certificates exists in the interior of IntCA.

4.2 IntCA Certificate

Figure 6 shows the format of certificate of IntCA which is simpler than external CA.

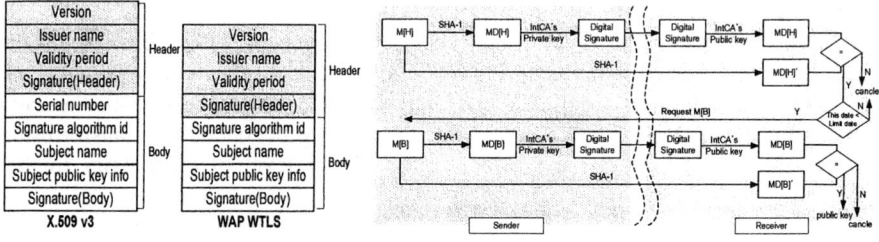

Fig. 6. Certificate format   **Fig. 7.** Message integrity of IntCA

4.3 IntCA Operation

A transaction is done between a sender and a receiver by using Figure 7. The process to transfer this certificate is shown in Figure 7.

4.4 Multiple Signature

In the proposed ASEP protocol, the multiple signature is used instead of Dual Signature of existing SET.

(1) User A generates message digests of OI (order information) and PI (payment information) separately by using hash algorithm, concatenates these two message digests; produces $MD_B MD_C$; and hash it to generates MD (message digest). Then the user A encrypts this MD by using an encryption key, which is obtained by multiplying the private key of user A to the public key of the receiver. The PI to be transmitted to user C is encrypted by using BSB algorithm. The encrypted PI is named CPI.

(2) User B generates message digest MD_B' with the transmitted OI from user A. After having substituted MD_B' for the MD_B of $MD_B MD_C$, the message digest MD is generated by using hash algorithm. User B decrypts a transmitted DS_B, and extracts MD from it. User B compares this with MD generated by user B, certificates user A and confirms the integrity of message. Finally, user B transmits the rest of data, $MD_B MD_C$, CPI, DS_C to user C.

(3) User C decrypts the CPI transmitted from user B, extracts PI, and generates message digest (MD_C) from this by using hash algorithm; substitutes this for MD_B of $MD_B MD_C$ transmitted from user B, and produces message digest (MD) by using Hash algorithm. Then the user C decrypts the DS_C transmitted from user B and extracts message digest (MD). Again, the user C compares this with the MD extracted by user C, verifies the certificate from the user A, and confirms the integrity of the message. Finally, the user C returns an authentication to the user B.

5 Performance Evaluation

5.1 The Comparison of ECC with RSA

In this paper, the proposed MEP protocol uses ECC instead of RSA. In comparison with RSA, the results of the encryption and decryption times are shown in Tables 1 and 2 respectively, which indicate that encryption and decryption time of ECC are much less than those of RSA.

Table 1. A comparison for encryption time (unit: μs)

Key size (byte) / Method of encryption	RSA	$ECC(F_p)$	$ECC(F_2^m)$
5	0.05	0.05	0.03
10	0.54	0.20	0.03
15	1.54	0.29	0.03
20	2.55	0.38	0.04
25	4.33	0.42	0.04
50	5.53	0.85	0.04
100	7.28	1.30	0.03

Table 2. A comparison for decryption time (unit: μs)

Key size (byte) / Method of encryption	RSA	ECC(F_p)	ECC(F_2^m)
5	0.11	0.10	0.03
10	0.55	0.50	0.04
15	1.20	0.80	0.04
20	3.08	1.10	0.04
25	6.21	1.15	0.05
50	8.06	2.12	0.04
100	9.95	3.21	0.03

5.2 BSB and DES

Tables 3 and 4 show the mean value of encryption time of BSB and DES by executing every number of block about message twenty times. According to Table 4, we can conclude that BSB is faster than the existing DES in encryption time. In addition, the security of BSB is enhanced by using Byte-exchange, Shift Round, ASCII Code and Bit-wise XOR Therefore, the strength of the encryption is improved and more time is saved for encryption and decryption than DES.

Table 3. A comparison for encryption time (unit: μs)

Method of encryption / Number of block	1	2	3	4	5
BSB	0.016	0.027	0.036	0.042	0.053
DES	0.120	0.134	0.144	0.154	0.174

Table 4. A comparison for decryption time (unit: μs)

Method of encryption / Number of block	1	2	3	4	5
BSB	0.015	0.019	0.030	0.034	0.041
DES	0.120	0.135	0.144	0.165	0.166

6 Conclusion

The proposed MEP protocol employees ECC and BSB algorithm instead of RSA and DES used in the existing SET. As a result, it speeds up the encryption process by reducing communication traffic for transmission, simplifying dual signature. In addition, the security for information is strengthened which prevents session keys from being intercepted from attackers on the network. The proposed BSB, which uses byte-exchange, Shift Round, ASCII Code and the bit operation increases data encryption speed. Because during the encryption process, the BSB algorithm performs byte exchange between blocks, and then the plaintext is encoded through bit-wise XOR operation, it rarely has a possibility for cipher text to be decoded and has no problem to preserve a private key.

Moreover, the proposed MEP protocol has a simple structure, which can improve the performance with the length of session keys. From the standpoint of the supply for key, the CA (Certificate authority) has only to certify any elliptic curve and any prime number for modulo operation, the anonymity and security for information can be guaranteed over communication network.

References

[1] N. Koblitz, Elliptic Curve Cryptosystems. Math. Comp. **48** 203-209 (1987).
[2] V.S. Miller, *Use of elliptic curve in cryptography.* Advances in Cryptology-Proceedings of Crypto '85, Lecture Notes in Computer Science **218,** pp. 417-426, Springer-Verlag, (1986).
[3] G. Harper, A. Menezes, and S. Vanstone, *Public-key Cryptosystem with very small key lengths.* Advances in Cryptology-Proceedings of Eurocrypt '92, Lecture Notes in Computer Science **658,** pp. 163-173, Springer-Verlag, (1993).
[4] Ecommercenet, <http://www.ezyhealthmie.com/Service/Editorial/set.htm>.
[5] I.S. Cho, D.W. Shin, T.C. Lee, and B.K. Lee, *SSEP (Simple Secure Electronic Payment) Protocol Design.* Journal of Electronic and Computer Science, pp. 81-88, Vol. 4, No. 1, Fall 2002.
[6] .S. Cho and B.K. Lee, ASEP (Advanced Secure Electronic Payment) Protocol Design. Proceedings of International Conference of Information System, pp. 366-372, Aug. 2002.
[7] IEEE P1363 Working Draft, Appendices, pp.8, February 6, 1997.

Enhancing Connectivity Based on the Thresholds in Mobile Ad-Hoc Networks

Wongil Park[1], Sangjoon Park[2], Yoonchul Jang[3],
Kwanjoong Kim[4], and Byunggi Kim[1]

[1] Department of Computer, Soongsil University, Korea
[2] Information & Media Technology Institute, Soongsil University, Korea
[3] System Operation & Management Consulting, Sysgate, Korea
[4] Department of Computer and Information, Hanseo University, Korea
bgkim@ssu.ac.kr

Abstract. In general mobile nodes in mobile Ad-hoc networks have a limited power capacity. Therefore power management is an important issue. Some protocols are proposed considering node power consumption, such as MTPR, MBCR, MMBCR, CMMBCR. But they have no measures on link breakdown from power exhaustion of relay nodes. We propose three algorithms to prevent link breakdown and to extend the connectivity of the routes in mobile Ad-hoc networks. In these ways, the lifetime of routing nodes can be extended and the link connectivity can be enhanced. Moreover the delay due to reacquisition of the route is reduced and the throughput degradation from link breakdown is avoided.

1 Introduction

Ad-hoc network is a multi-hop wireless network with no fixed infrastructure. Ad hoc networks can be usefully deployed in applications such as disaster relief, tether les¹s classrooms, and battle field situations. In ad hoc networks, the power supply of the individual nodes is limited, the wireless bandwidth is limited, and the channel condition can vary greatly. Moreover, since the nodes can be mobile, routes may constantly change. Therefore, many groups such as a MANET (Mobile Ad hoc Networks) have tried to solve these problems. Since the existing routing protocols don't consider a limited battery of the mobile node, they can not support a reliable transmission efficiently. The MPTR [1], MBCR [2], MMBCR [3][4], CMMBCR [2][5] are suggested to solve these battery consumption problems. These protocols transmit data through a power management by a power consumption equation. But, since the node with any frequent movement should continuously maintain a path for a stable data transmission, it has many overheads. We propose three algorithms to prevent link breakdown and to extend the connectivity of the routes in mobile Ad-hoc networks. When the received signal strength decreases below a threshold, the

---

* This work was supported by the Korea Research Foundation Grant (KRF-2004-005-D00147).

proactive arrangement of an alternate route is performed. In case that residual power goes below threshold, the node broadcasts a negative signal. On receiving this signal, the neighbor nodes establish alternate routes except for the signaling node. A relay node, operating in multi-channel mode selectively rejects any link establish request when the power requirement of the new link is larger than the threshold. The threshold is computed based on the residual power of the node. In these ways, the lifetime of the relay nodes is extended and the link connectivity is enhanced. Moreover the delay resulted from the reacquisition of the route is reduced and the throughput degradation from any link breakdown is avoided.

The rest of the paper is organized as follows. In Section 2, we propose three schemes for routing connectivity enforcement. Section 3 shows the performance results for the proposed scheme, and conclusions are given in Section 4.

2 Routing Connectivity Enforcement

The nodes in the ad-hoc network have many path changes because there is frequent movement [7][9]. Therefore, the link maintenance problem between the nodes is very important in the ad-hoc network. In general, if a source node doesn't receive ACK, transmitted from a destination node; it uses the retransmission scheme or operates the link recovery by recognizing a connection failure. This paper proposes three mechanisms to enhance the path connectivity by extending the path lifetime in an Ad-hoc network.

2.1 Detour Routing Based on Signal Strength

On the asymmetric channel assigned dynamically, the up-link node sends data to a destination by a downward channel. The down-link node sends a response message to a source by an upward channel. The Signal attenuation occurs because of the environment variables on link of middle node or sending and receiving node. The power status of the nodes is assumed to be environment variables in this paper. Signal attenuation occurs according to the power consumption of node; these lead to the instability of the connection. Hence, the reconfiguration process of the path between the sending and the receiving node is an executed in the existing method due to the close of the link connection. These lead to a delay or decrease in the processing rate.

Fig. 1 show the acquisition process of the detour routing before the disconnection of the link for continuity of the link when signal strength for up-link node is decreased in an asymmetric channel. When the signal strength that the node *b* receives from the node a, is below the threshold value, the node *b* transmits the control message (C_Msg), that the signal strength is weak, to the node *a* before the disconnection of the link in Fig. 1(a). After first searching for the nearest node, the node *a* that received the control message transmits the path request message of which node can receive the data at normal strength so as to configure a detour routing the new destination node before the link is disconnected.

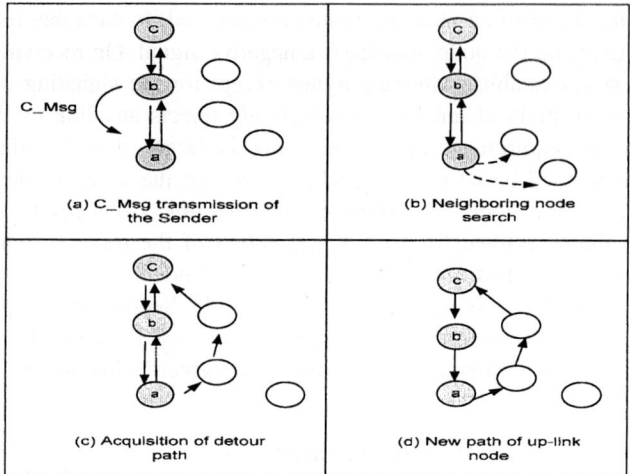

Fig. 1. Detour routing by signal strength

$RSSI_{Nn}$: the signal strength received;

T_{RSSI} : the threshold of signal strength set at the initial stage

c_msg(N_n, N_m) : Control message sent to N_m by N_n;

if $RSSI_{Na} \leq T_{RSSI}$ then

send c_msg(N_b, N_a);

repeat

R_n = shortest_path(N_a, N_c);

until($N_b \notin R_N$)

endif

Fig. 2. Detour routing algorithm

Fig. 1(c) shows the path to the destination node c using the shortest path algorithm before the link disconnection. However, the routing information of the node b is excluded from the routing table. In Fig.1(d), it shows the detour routing acquired. It is substituted by a new path in case of the disconnection a link by signal attenuation. Fig. 2 presents the detour routing algorithm. When a distance between two nodes is d, the strength of the received signal RSSI(d) is as follows [6].

$$RSSI(d) = RSSI(d_0) - 10 n \log(d / d_0) \quad (1)$$

where $RSSI(d_0)$ is the basic signal strength about standard distance d_0 between two nodes. N is a parameter of path loss and has the value 2.76[3][8].

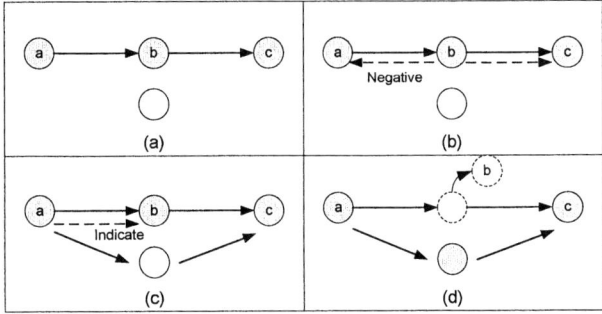

Fig. 3. Link avoidance based on the threshold

2.2 Link Avoidance Based on the Threshold

When the intermediate node that relays data among the other nodes consumes all the power, the link including the intermediate node is broken. We propose a new mechanism to solve this problem. All the intermediate nodes compare their residual power capacity with the threshold set in advance. If the power is below the threshold, the nodes should establish an alternate path in advance. Fig 3 shows a link avoidance procedure based on the threshold. The source node a sends data to the node c via the intermediate node b [Fig. 3-(a)]. The intermediate node b recognizes that the residual power reaches to the threshold and sends a negative signal to the neighboring nodes [Fig. 3(b)]. After the source node a which receives a negative signal, re-establishes on other path to the destination node c, it sends a message to notify the intermediate node b of an establishment of an alternate path [Fig. 3(c)]. The routing including the node b is excluded in this procedure. When the intermediate node b leaves, the existing link is broken and the alternate path becomes a new path [Fig. 3(d)].

P_{Nn} : Power capacity of Nn

T_{power} : Threshold value of power capacity established in an early stage;

n_msg(Nn) : Negative message, that Nn transmits to neighbors

i_msg(Nn , Nm) : Indicate message, that Nn transmits to node Nm

if $P_{Nn} <= T_{power}$ then

  send n_msg(N_b);

  repeat

    R_N = shortest_path(Na , Nc);

  until($N_b \notin R_N$)

  send i_msg(Na , N_b);

endif

Fig. 4. Link avoidance algorithm

Fig. 4 presents the pseudo-code to the link avoidance algorithm. We know that the highest power consumption is to the transmission section if the operation of node is classified in five sections (transmit, receive, signaling, silence and idle) [3]. In Fig. 4, if the remaining power of a node is lower than the threshold T_{POWER}, the node sends the negative message, and the link avoidance algorithm will be implemented. Therefore, dead nodes can be decreased and the continuity of the path can be increased as the threshold value applies to the link avoid algorithm of Fig. 4. If a detour routing cannot be found, the node that is below the threshold value will be used in the existing method since the connection cannot be maintained if that be low powered node isn't used.

2.3 Restrictive Relay

The intermediate node can relay many links in the multi-channel environment. The node which has a multi-channel should have a minimum residual power. In this section, we propose a new way to extend the lifetime of an intermediate node. When the intermediate node receives a path request message from the other node, it compares the node's residual power capacity with the power threshold. If the power capacity is below the threshold, the intermediate node rejects a link request. In this way, this restrictive relay mechanism can save the power of an intermediate node. The transmission cost $C_{i,j}$ from the node i to the node j is presented in [4].

$$C_{i,j} = \frac{B_i}{E_{i,j}} \qquad (2)$$

where B_i is the residual power capacity, E_{ij} is the power capacity required to send a packet from the node i to the node j. The transmission cost per a link is not the same since all nodes have a different distance and power capacity.

Fig. 5 presents the restrictive relay by the power threshold T_{POWER}. Fig. 5(a) shows the node b receives the routing request from the node a. In Fig. 5(b), if the node b power is below the power threshold, it sends the negative signal to neighbor nodes. In Fig. 5(c), the node a selects a new routing using neighbor nodes without the node b.

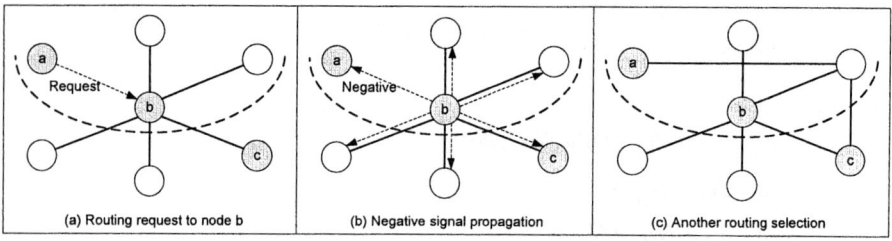

Fig. 5. Restrictive relay by the power threshold

Fig. 6 shows pseudo codes of a restrictive relay. Since the proposed mechanism can make the lifetime of an intermediate node extend using the threshold T_{POWER}, this one can increase the connectivity of an existing routing.

```
r_msg( N_n, N_m ) : routing request message
PN_n : residual power capacity;
T_POWER : Threshold for a power capacity set at the initial stage n_msg( N_n ) :
negative message sent to the neighboring nodes by N_n ;
if (r_msg( N_a, N_b )) then  /* when the node a requests a path to the node b */
    if PN_b <= T_POWER then
        reject r_msg( N_a , N_b );
        send n_msg( N_b );
        repeat
            R_N = shortest_path( N_a, N_c );
        until( N_b ∉ R_N )
    else
        accept r_msg( N_a , N_b );
    endif
endif
```

Fig. 6. Restrictive relay algorithm

3 Performance Evaluation

This section describes the simulation conducted to evaluate the performance of the proposed mechanism. In the simulation, we compare our scheme with the CMMBCR. The following are the parameters for the simulation.

Table 1. Simulation parameters

Area(m×m)	500, 1000, 2000, 3000
Number of nodes	100
Coverage of node	70m
Mobility	Random way point
Speed	1m/s
Packet size	100byte/s
Default RSSI	75dB
Threshold	25dB

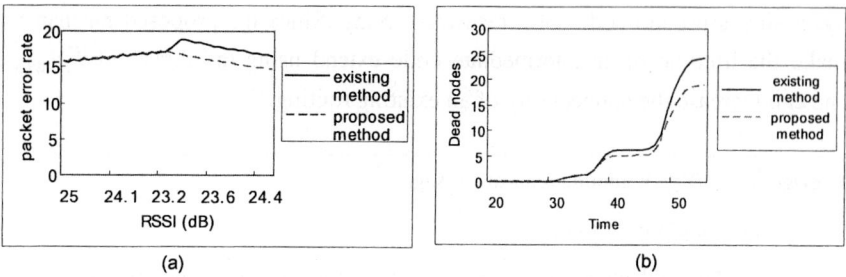

Fig. 7. Packet loss rate by signal attenuation

Fig. 7(a) shows the packet loss rate by the environmental signal attenuation. Our scheme and the proposed one have the same packet loss rate until RSSI 2.3dB. But, our scheme has less of a packet loss rate than the CMMBCR as RSSI 2.3dB indicates. The CMMBCR restarts the data transmission by re-establishing a path when the mobile node can not transmit data by an attenuation of signal any more. On the contrary, the proposed mechanism re-establishes a routing when the signal level reaches the threshold. Also when the mobile node does not have an enough power capacity or the link on the path is broken, the re-established path is used. In the next simulation, we compare the existing mechanism with the proposed link avoidance mechanism based on the threshold.

Fig. 7(b) shows a dead node number from the power consumption. The simulation is conducted by using 100 nodes in the 100×100 region. When the intermediate node reaches the threshold defined in advance, it leaves the path but the peripheral nodes can send and receive a data. As a result, since the proposed mechanism can extend the lifetime of the intermediate node, it has less numbers of dead nodes than the existing mechanism. In the next simulation, we show the performance of the proposed restrictive relay mechanism based on the threshold. The R value is the minimum residual power to maintain a path requested in Fig. 7 (a). Fig. 8 shows that the proposed mechanism has the lower acceptance rate in the higher threshold value. But, the lifetime of the intermediate node can be extended by rejecting a link request and the breakdown of the many links, which provide a relay service, can be reduced.

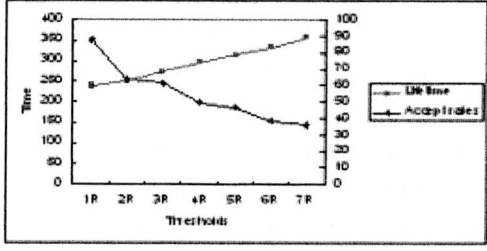

Fig. 8. Node lifetime and acceptance rate of routing request

4 Conclusion

In this paper, we propose three mechanisms to enhance the path connectivity in an ad hoc network. First, when the signal strength is weakened by environmental reasons or the signal attenuation, the receiving node compares the signal strength with a threshold and sends a control message to establish an alternate path. Second, when the power capacity of the intermediate node reaches to the threshold, the intermediate node sends a negative message to the nodes, which receive a relay service, and the sending node establishes an alternate path. After that, the intermediate node leaves the path but the peripheral nodes can send and receive a data continuously. Third, when the intermediate node receives a path request message from the other node, it compares the node's residual power capacity with the threshold. If the power capacity is below the threshold, the intermediate node rejects the link request. From the proposed schemes, the lifetime of the relay nodes can be extended and the link connectivity can be enhanced. Moreover the delay due to the reacquisition of the route is reduced and any throughput degradation from the link breakdown is avoided.

References

[1] Keith Scott, Nicholas Bambos "Routing and Channel Assignment for Low Power Transmission in PCS," In Proc. of IEEE ICUPC'96, pp.498-502, Oct. 1996.
[2] C.-K. Toh, "Maximum Battery Life Routing to Support Ubiquitous Mobile Computing in Wireless Ad Hoc Networks," IEEE Communications Magazine , vol.39, no.6, pp.138-147, June. 2001.
[3] Mark Stemm, Paul Gauthier, Daishi Harada, Randy H. Katz ,"Reducing Power Consumption of Network Interfaces in Hand-Held Devices," In Proc. of MoMuc, Sept. 1996.
[4] Archan Misra, Suman Banerjee, "MRPC: Maximizing Network Lifetime for Reliable Routing in Wireless Environments," In Proc. of IEEE WCNC'00, pp.800-806, Mar. 2002.
[5] Dongkyun Kim, J.J Garcia-Luna-Aceves and Katia Obraczka,"Power-aware routing based on the energy drain rate for mobile ad hoc networks," In Proc. of Computer Communications and Networks, pp.565-569, Oct. 2002
[6] Scott Y. Seidel and Theodore S. Rappaport, "914Mhz Path Loss Prediction Models for Indoor Wireless Communications in Multifloored Buildings," IEEE Trans. on Antennas and Propagation, vol.40. no.2, pp.207-217, Feb. 1992.
[7] Paramvir Bahl and Venkata N. Padmanabhan, "RADAR: An In-Building RF-based User Location and Tracking System," IEEE INFOCOM'00, pp.775-784, Mar. 2000.
[8] T.K. Philips, S.S. Panwar, A.N. Tantawi, "Connectivity properties of a packet radio network model," IEEE Trans. on Information Theory, vol.35, no.5, pp.1044-1047, Sept. 1989.
[9] Suresh Singh, Mike Woo, and C. S. Raghavendra. "Power-Aware Routing in Mobile Ad Hoc Networks," In Proc. of ACM/IEEE International Conference on Mobile Computing and Networking, pp.181-190, Oct. 1998.

Call Admission Control for IEEE 802.11e WLAN Using Soft QoS

Hee-Bong Lee[1], Sang Hoon Jang[2], and Yeong Min Jang[2]

[1] Mobile Communication Company of LG Electronics Inc.,
533, Hongye-1dong, Dongan-gu, Anyang-shi, Kyoungki-do, Korea
kari96@lge.com
[2] School of Electrical Engineering, Kookmin University,
861-1, Jeongneung-dong, Songbuk-gu, Seoul, Korea
{jangsang, yjang}@kookmin.ac.kr

Abstract. Although the MAC protocol provides service differentiation, it does not guarantee that the QoS requirement of each service will be fully satisfied. We propose a soft QoS based call admission control scheme to support the QoS requirements of multimedia traffic in IEEE 802.11e WLANs. In order to evaluate the performance of the proposed scheme, we developed the proposed algorithm in NS-2, and compared the proposed scheme with hard QoS scheme. Simulation results show that our proposed scheme outperforms hard QoS scheme in terms of throughput and the maximum number of accepted calls.

1 Introduction

The past few years have seen an explosion in the deployment of Wireless LANs (WLANs) conforming to the IEEE 802.11 standard. Hence, future WLANs such as IEEE 802.11n will be expected to support a wide variety of multimedia services. It is desired to provide a service differentiation mechanism in the IEEE 802.11 standard since these services require different QoS. The IEEE 802.11 specifies two mechanisms for transmission which are the DCF(Distributed Coordination Function) and the PCF(Point Coordination Function). DCF has advantages of simplicity and robustness. However, DCF does not provide any guaranteed QoS. The PCF provides some limited QoS supports for real-time traffic.

Recently, the IEEE 802.11e group is developing mechanisms to enhance the QoS of the original MAC standard including EDCA(Enhanced Distributed Channel Access) and HCCA(HCF Controlled Channel Access)[1]. It ensures that the packets sent by each STA(STAtion) can be differentiated by assigning different access parameters. Although the MAC protocol provides service differentiation, it does not guarantee that the QoS requirement of each service will be fully satisfied[2]. This limits the use of EDCA for many multimedia applications and then resource management becomes a significant issue.

Several call admission control(CAC) algorithms have been proposed in the literature of 802.11 networks[2][3][4]. These algorithms attempt to support the hard QoS requirements of each traffic class. Ideally, the service provided to the call should not be affected by the traffic dynamics of other calls sharing the common links. Such arrangements are usually referred to as hard QoS guarantees. Many multimedia applications, however, do not require hard QoS constraints, since the applications may work even if QoS requirements are not fully satisfied. The service requests of such type of applications are sometime referred to as soft QoS guarantees. We can improve system performance and guarantee QoS for multimedia services using soft QoS[4].

In this paper, we propose an admission control algorithm based on soft QoS to guarantee the QoS requirements of each traffic class. Our proposed algorithm provides the soft QoS requirements of each traffic class when there is unavailable bandwidth in the network. To borrow bandwidth from ongoing calls, we introduce critical bandwidth ratio to our proposed scheme. Critical bandwidth ratio is one of soft QoS parameters. The minimum amount of released bandwidth depends on critical bandwidth ratio and the priority of each traffic class.

2 Soft QoS-Based CAC

The admission control function decides whether a new call can be granted or not, depending on the status of the network resources and the level of service called for by the new request. The purpose of any admission control is to ensure that admittance of a new data flow into a resource-limited network does not degrade QoS committed by the network while optimizing the network resource usage.

In this paper, we presents a soft QoS based CAC algorithm supporting various multimedia services. The main purpose of our proposed algorithm is to admit the maximum number of new calls by soft QoS guarantees. An idea of soft QoS guarantees is to ensure reasonable QoS level of both new calls and ongoing calls in despite of borrowing bandwidth from existing calls. So, we introduce the critical bandwidth ratio to our proposed algorithm in order to support the reasonable QoS level of each call. The critical bandwidth ratio defines as the relation between the QoS requirement of users and the allocated bandwidth. This ratio can be obtained by soft QoS controller[5][6] and are summarized in Table 1. We assumed that soft QoS controller is implemented in 802.11e AP.

Table 1. Critical bandwidth ratio

	Applications	Critical bandwidth ratio
RT	Video on demand	0.7~0.8
NRT	Web traffic	0.6~0.8
	Background	0.2~0.6

Fig. 1 shows a flow chart of our proposed CAC algorithm. Upon a new call request, our algorithm calculates the total amount of bandwidth allocated to existing users. Let M, N_m be the number of traffic classes and the number of calls of mth class. Then C_{mn} is the allocated bandwidth to n^{th} call with class m, C_{total} is the total amount of available bandwidth, and C_{req} is the bandwidth requested by a new call. The total amount of bandwidth allocated to existing users is given by

$$C_{occupied} = \sum_{m=1}^{M} \sum_{n=1}^{N_m} C_{mn}. \tag{1}$$

Our algorithm checks out whether admission criterion is met or not,

$$C_{occupied} + C_{req} \leq C_{toatal}. \tag{2}$$

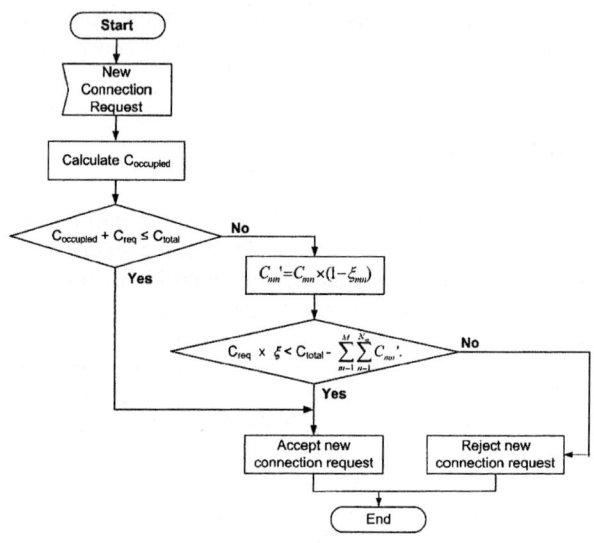

Fig. 1. Proposed CAC algorithm using soft QoS concept

If admission criterion is met, a new call is accepted in the network. When rejecting a new call, our proposed algorithm attempts to borrow the bandwidth from existing calls. The bandwidth borrowed from each existing calls is given by

$$C_{mn}(1 - \xi_{mn}) \tag{3}$$

where ξ_{mn} is the critical bandwidth ratio of nth call with class m. When borrowing bandwidth, it important to decide to the amount of bandwidth borrowed from each ongoing call fairly. The utilization of bandwidth decreases in the network if the borrowed bandwidth exceeds the bandwidth allocated to a new call. So, we update $C_{occupied}$ and then updated $C_{occupied,new}$ is given by

$$C_{occupied,new} = C_{occupied} - \sum_{m=1}^{M}\sum_{n=1}^{N_m} \frac{C_{mn}(1-\xi_{mn})}{\sum_{m=1}^{M}\sum_{n=1}^{N_m} C_{mn}(1-\xi_{mn})} \times C_{m(N_m+1)} \tag{4}$$

The proposed scheme can admit a new call even though the network does not allocated bandwidth.

3 Simulations Results

We developed a soft QoS based call admission control module in NS-2. In these simulations, no hidden stations were present and the channel was assumed to be error free. All stations with EDCA mechanism operated at 11Mbps and were configured into infrastructure mode. Also all stations have three queues. The values of parameters used in these simulations are summarized in Table 2.

Table 2. EDCA parameter

Type	Voice	Video	Data
Priority	7	5	0
AC	3	2	0
AIFS	34	43	50
CWmin	7	15	31
CWmax	15	31	1023
TXOP (msec)	3	6	0
Traffic (Kbps)	64	1024	128

Fig. 2 shows the average delay for four traffics in our proposed algorithm and hard QoS based algorithm. In this simulation result, we find out the difference of delay according to AIFS or back-off value of 802.11e QoS MAC. The delay of data and Voice traffic is same in comparison with hard QoS. In case of video traffic, our proposed algorithm shows lower delay than hard QoS.

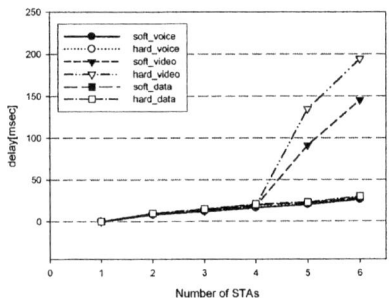

Fig. 2. Delay vs. number of STAs

Fig. 3 illustrates the traffic per throughput in our proposed algorithm. Fig. 4 shows throughput of each STA. In this simulation, we assumed that video traffic is a new call, and voice and data traffics are existing calls. Both soft QoS and hard QoS support the same throughput to voice traffic because of high priority. And the throughput of data decreases rapidly because of low priority. But in case of video traffic, our proposed algorithm reallocates resource to video traffic class by borrowing resource from voice and data traffic according to Eq. (4). So our proposed scheme can accept more STAs. Hence we can see that there is an improvement in IEEE 802.11e throughput between soft QoS scheme and hard QoS scheme. Furthermore soft QoS can guarantee reasonable QoS level for voice traffic class in despite of releasing own resource.

Fig. 4 shows the average throughput of our proposed and traditional hard QoS algorithm. We found that soft QoS scheme outperform hard QoS scheme in terms of throughput. Also, in saturation situation, the throughput of soft QoS is higher than throughput of hard QoS because each STA can share more resources.

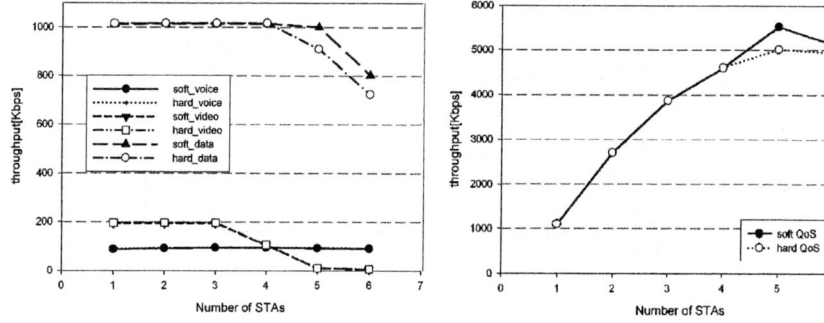

Fig. 3. Throughput comparisons according to various traffics

Fig. 4. Throughput comparison between soft QoS and hard QoS

4 Conclusion

In this paper we introduced a new framework to support QoS in the WLAN. We proposed a CAC algorithm using soft QoS in IEEE 802.11e EDCA and compared the proposed CAC algorithm with hard QoS. We evaluated the performance of CAC in IEEE 802.11e EDCA in terms of throughput, delay, and the maximum number of users. We find out that soft QoS-based CAC scheme is suitable for multimedia services. We expect that our CAC scheme could be applied in other access networks such as UWB and WCDMA.

Acknowledgment

This work was supported by the KOSEF through the grant No. R08-2003-000-10922-0 and University IT Research Center(INHA UWB-ITRC), Korea.

References

[1] IEEE 802.11 WG, Draft Supplement to Part 11: Wireless Medium Access Control (MAC) and Physical layer (PHY) specifications: Medium Access Control (MAC) Enhancements for Quality of Service (QoS), IEEE 802.11e/D5.0, July 2003.

[2] D. Pong and T. Moors, "Call Admission Control for IEEE 802.11 Contention Access Mechanism," *IEEE GLOBECOM,* Dec. 2003.
[3] Y.L.Kuo, C.H.Lu, E.H.K.Wu and G.H.Chen, "An admission control strategy for differentiated services in IEEE 802.11," *IEEE GLOBECOM,* Dec. 2003
[4] D. Gu and J. Zhang, "A New Measurement-Based Admission Control Method for IEEE802.11 Wireless Local Area Networks," *IEEE PIMRC*, Sept. 2003.
[5] D. Reininger and R. Izmailov, "Soft quality-of-service control for multimedia traffic on ATM networks," *IEEE ATM Workshop*, May 1998.
[6] D. Reininger, R. Izmailov, B. Rajagopalam, M.Ott and D. raychaudhuri, "Soft QoS control in the WATMnet broadband wireless system," *IEEE Personal Communications*, Feb. 1999.
[7] Yeong Min Jang, "Soft QoS-based Vertical Handover between cdma2000 and WLAN Using Transient Fluid Flow Model", *ICOIN2005,* Jan. 2005.

Overlay Multicast Routing Architecture in Mobile Wireless Network

Backhyun Kim and Iksoo Kim

Department of Information and Telecommunication Engineering University of Incheon,
Incheon, Korea
{hidesky24, iskim}@incheon.ac.kr

Abstract. In this paper we propose overlay multicast (Mcast) routing architecture that uses basic trees and their neighboring mobile nodes (NMN) in mobile wireless network. The basic trees are generated according to hop-counts from mobile nodes (MN) toward base node. NMNs are 1-hop away from a MN and exist in other basic trees. NMNs identify only whether they have Mcast group MNs in mapping table, and never broadcast MNs' Mcast request. But MNs on an identical basic tree must broadcast Mcast request to parent or child MNs when they don't have Mcast group in mapping table. MNs don't have Mcast routing table but have only mapping table to convert logical address into physical IP address. The proposed Mcast routing architecture has effectiveness to solve shortcomings in mobile wireless network and sensor networks since it can decrease effectively flooding traffic restricted to MNs on a basic tree.

1 Introduction

Mobile wireless networks consist of mobile hosts interconnected through wireless links and its topology may change frequently due to the nodes' movements. Mcast routing protocols for wireless networks can generally be divided into two categories; proactive and reactive. The former maintains an up-to-date route to each reachable destination MN by sending their routing information periodically. The latter is on-demand routing protocol to establish routing path when a specific MN wants to send data to a destination MN [1, 2]. Thus reactive routing protocol discovers a new path when it does not know a route to the destination MN. The basic concept of these protocols is flooding that overwhelming amount of packet transmission. This can quickly exhaust the battery of hosts and may be the source of the severe packet contention and collision. Though numerous techniques have been published, they are inefficient in reducing network traffic and/or require the significant control overhead and intensive computation on hosts [3, 4].

This paper presents a new overlay multicast routing architecture in mobile wireless network; it uses basic trees toward arbitrary or fixed base node (BN), and on-demand tree to establish a new path for Mcast or peer-to-peer (P2P) communication. For establishing basic tree, MNs send initially request packet (REQ) toward BN, and then some basic trees based on hop-count toward BN are established. On-demand tree for Mcast or P2P is established through basic trees with their NMNs. Thus MN to join in Mcast on the basic trees broadcasts Mcast join (JOIN) packet to their parent/child

MNs and NMNs. If the MNs received JOIN packets are already members of Mcast group, they support Mcast tree to the MN with the shortest path. If not, MNs on the same basic tree broadcast it to connected MNs. These operations progress until REQ packet arrives to BN. But NMNs check only their mapping tables for identifying a specific Mcast group and they never rebroadcast it. If any NMN has a list for the Mcast group, it supports Mcast tree to MN by joining Mcast group itself. Otherwise, it discards REQ packet. The shortest on-demand Mcast path is established if the tables have a list of a specific Mcast group.

The rest of this paper is as follows: Section 2 describes the structure of mobile wireless network and the role of mobile nodes, and section 3 addresses on-demand Mcast routing algorithm in mobile wireless network. Section 4 deals with simulation and analysis of the results. Finally, we discuss our conclusion in Section 5.

2 The Structure and Operation of Mobile Wireless Network

The structure of mobile wireless network for supporting multicast or P2P consists of base node (BN) and a number of mobile nodes (MNs). BN controls and manages all basic trees and tables that contain information for controlled MNs and their position. It generates basic trees and assigns the location-field of the logical address to the group of MNs that 1-hop counted towards BN ($N_{HC}=1$) as the hop count of the nearest MNs $N_{HC}=1$, where N_{HC} is the number of hop count. In this case the logical address is only location address and MNs on the same basic tree have an identical location address. Also, BN has responsibility to support P2P on-demand routing tree when a basic tree does not identify routing path from source MN to destination MN in another basic tree. And BN may provide Mcast tree using two basic trees when MNs cannot join Mcast group through their basic tree and/or NMNs.

Mobile node $N_{HC}=n$ is managed by the parent MN $N_{HC}=n-1$ toward BN. MNs accesses BN or the parent MN to establish a basic tree and always listen to the other MNs' broadcast packets. The parent MN notifies that the child MNs may be a member of its basic tree, and assigns logical address with an identical location field after receiving ACK response from the child MNs. MNs that receive the parent MN's reply packet (REP) with hop-count $N_{HC}=n$ are set their hop-count to $N_{HC}=n+1$.

Creating a basic tree is initiated by broadcasting REQ packet toward BN. Since some of MNs receive directly REP packet from BN, they become the group of the nearest MNs ($N_{HC}=1$). As the others receive REP from the nearest, 1-hoped, MNs directly or indirectly, MNs which receive it directly become 2-hop MNs. Therefore, MNs received from a MN $N_{HC}=n-1$ become n-hop MN $N_{HC}=n$. Address allocation [4,5] is initiated by BN after receiving REQ packet to establish some basic trees. When BN receives REQ, it generates location number sequentially and sends it each 1-hop MN through unicast. And the rest MNs' address allocation is assigned by the parent MN on its own tree due to hop-count of MNs during establishing a basic tree. Thus 1-hop MN that receives REP from BN identifies the child MNs and sends the same location number with peer classifying field to the child MNs. The peer classifying field discriminates MNs because the parent MN may connect with a few MNs that have the identical hop count on its basic tree. This procedure progresses until finishing the construction of a basic tree. With this logical address allocation, each MN has a tag-bit that indicates whether it is NMN on other basic trees or not, and Mcast-bit indicates whether MN has joined a specific Mcast group or not. Hence logical address consisted with n-bits is divided by 4-fields (Mcast, tag, peer classifying field, location field). In this paper, the hop count N_{HC} is restricted 5 because the probabilistic of error increases according to hop count, and each parent MN can control only 4 child MNs directly. Table. 1 shows a part of mapping table for *MN j* that has *NMNs a, b* and *c*. *MN-x* and *MN-k* on the same basic tree represented LSB 4 bits (tree xxD) have already joined Mcast group 01, and *MN-a* and *MN-c* are the members of the same Mcast group on another basic tree.

Table 1. Part of mapping table with NMNs and the status of joining Mcast for MN j

Mcast-bit	Tag	Logical Address	Physical IP Address
00	0	1C0 (MN i)	12AB0CE7
01	0	12D (MN x)	2357ABCD
01	1	12A (MN a)	15EC03425
10	0	14D (MN q)	34521257
11	0	20E (MN p)	111ACEF1
00	1	2AC (MN b)	5678ADCB
11	0	3DB (MN t)	5777234F
01	0	3BD((MN k)	2A257B0A
01	1	5C3 (MN c)	67AF7A10

The positions of MNs on a specific basic tree are found from location field of logical address. Thus, a basic tree for *MN j* includes *MN k*, *MN q* and *MN x*, and NMNs for *MN j* become *MN a*, *MN b* and *MN c* since they have an identical location field (xxD). Let the logical address of *NM j* be 23D. *MN a*, *b* and *j* are peer NMNs that have an identical hop-count $N_{HC}=4$, and N_{HC} of *MN c* is 5. The method for assigning logical address is no relation to BN except hop-1 MNs, but it depends on upper MN only. Also, Table. 1 shows *MN x*, *k* and *c* have already join Mcast group 1, *MN q* joins Mcast group 2 and *MN-p* and *t* have already join Mcast group 3 but *MN i* and *b* have not joined any Mcast group. And specific MNs' information in mapping table is modified only when the relationship among NMNs is changed according to some MNs moving to other basic trees and/or joining or pruning of Mcast group. Only information about the changed MNs is send.

3 Overlay Multicast Routing for Mobile Wireless Network

Prior to create a multicast routing tree in mobile wireless network, this paper generates basic trees that all MNs broadcast REQ packet to BN and they receive REP packet from BN or the parent MN with logical address including location field. After establishing its basic tree, when the source *MN a* wants to join a specific Mcast group, at first, the *MN a* investigates its mapping table to identify MNs on its basic tree and its NMNs on other basic trees whether they manage the Mcast group or not. If the table of *MN a* contains Mcast group, then the Mcast path will be established via its basic tree or NMNs. Otherwise, *MN a* broadcasts JOIN packet for Mcast group to its tree and NMNs. The parent and the child MNs on the basic tree and its NMNs received JOIN packet examine their mapping tables. If any table has Mcast group

MNs on basic tree or NMNs, simply it joins the Mcast group through them, otherwise only MNs on its tree broadcast Mcast JOIN packet and NMNs check only whether their table manage Mcast group. NMNs don't manage it and discard JOIN packet automatically. This procedure progresses until Mcast it arrives to BN.

When the *MN a*, the parent and the child MNs broadcast JOIN packet for joining Mcast group, some NMNs may have mapping table for requesting Mcast group. In this case, *MN a* may receive some routing information from MNs on its basic tree and/or NMNs, then *MN a* selects an on-demand Mcast routing path with minimum hop count. Otherwise, when JOIN packet to Mcast group is arrived BN, it supports a path using *MN a*'s basic tree and another basic tree managing Mcast group. Logical address on MNs is shown in Fig. 1, the 1st and the 2nd parts of it indicate location of its basic tree in mobile wireless network, the rest of it indicate to classify peer MNs with the same hop count. Thus logical address 1-2-1-1-2 means that location is the 1st basic tree from BN, the 2nd MN among 2-hop MNs, the 1st MNs among 3-hop MNs and 4-hop MNs, and the 2nd MN among 5-hop MNs on a basic tree.

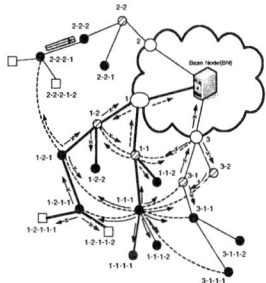

Fig. 1. The structure for generating on-demand tree using broadcast REQ packet only source and MNs on its basic tree

Let the size of REQ packet and REP packet be S_{REQ} and S_{REP}, respectively. The total traffic to establish basic trees and on-demand Mcast group tree in the whole mobile wireless network is as follows,

$$\sum_{i=1}^{h} B \cdot R_i \cdot S_{REQ} + n \cdot S_{REP} \qquad (1)$$

where, h, B, R_i, and n is the number of maximum hop count, the number of basic trees on each MN, the number of REQ packet of i'th hop MNs and the number of MNs in mobile wireless network, respectively. Equation 2 indicates the overhead for establishing on-demand Mcast tree via two basic trees from BN when NMNs on the basic tree of the source MN have not the list for a specific Mcast group in mapping table. Equation 3 is derived from NMNs on source MN's basic tree. The sum of REQ and REP is overhead for on-demand Mcast path between NMN and destination MN. Let the size of JOIN and ACK packet be S_{JOIN} and S_{ACK}, respectively.

$$n(h_{ms} + h_{BS})(S_{JOIN} + S_{ACK}) \qquad (2)$$

$$n(|h_s - h_u| + 1)(S_{JOIN} + S_{ACK}) \qquad (3)$$

where n, h_{ms}, h_{BS}, h_s and h_u is the total number of MNs requesting path, the number of hop from source MN to 1'st hop MN on its basic tree, the number of hop count toward BN, the number of hop count of source MN, and the number of hop count of MN issuing ACK, respectively. The distribution delay is the height of the tree, $O(\log_k n-1)d_{avg}$, where n is the total number of MNs to establish routing path between source and destination, and k is the number of MNs that have 1-hop toward BN in mobile wireless network, and d_{avg} is the average geometric distance of a tree edge, respectively [5].

4 Simulation and Analysis Performance of Wireless Network

Our simulation network is created within a 1000m x 1000m space with 256 MNs and a random mobility model with maximum speed of 20 m/sec and pause time of 5 seconds. The transmission range of both BS and MNs is selected from a uniform distribution 200 m, and the number of 1'st hop MNs toward BN is selected 1 to 5 and maximum hop-count is 5 and the rate of joining Mcast group is 10 to 50%. The Fig. 2 shows the percentage of MN connected Mcast tree is higher than 90% and the right side axis indicate the number of forwarding MNs by passing that don't join Mcast group. The percentage of MNs connected Mcast group tree is more than 90% and the number of MNs that don't join the Mcast group is slightly more than 30 when the percentage of MNs for joining Mcast group is 30% in mobile wireless network. The proposed Mcast routing technique supports that more than 90% of MNs for request Mcast may join Mcast tree when more than 20% MNs request joining Mcast group.

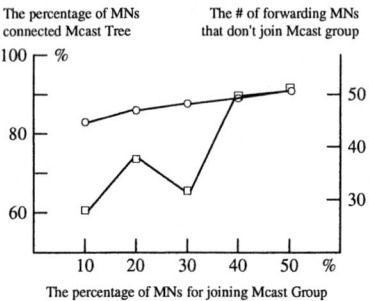
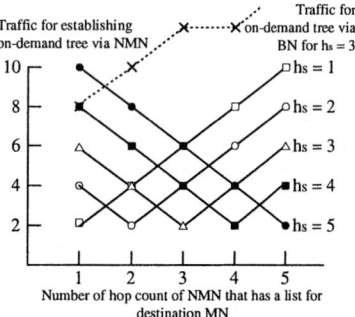

Fig. 2. The comparison of performance according to the percentage of MNs for joining Mcast group

Fig. 3. Traffic overhead for establishing on-demand tree via NMN of the source MN's basic tree according to the number of hops between source MN and NMN

The Fig. 3 shows the amount of traffic to establish on-demand Mcast tree via NMNs on source MN's basic tree according to the varying the hop-count of source MN and parent/child MN managing NMN that has a list for a specific Mcast group. In Fig. 3, the dashed line indicates traffic for on-demand multicast tree via BN and two basic trees in the case of no list for a specific Mcast group in the NMNs' mapping table. As shown in Fig. 3, the traffics are not increase or decrease uniformly according to the number of MNs and their NMNs. The reason is that hop-count is different between parent/child MN managing NMN that has a list for Mcast group and source MN.

5 Conclusion

This paper presents Mcast routing architecture in mobile wireless network using the combination of basic trees and on-demand tree. Basic trees are established easily based on hop-count toward BN from all MNs and three types of NMNs. These basic trees with NMNs may contribute to establish on-demand Mcast tree or P2P path with the shortest path, but it is established by 2-basic trees via BN if NMNs' tables don't have a list for a specific Mcast group or destination MN. The proposed routing architecture can reduce effectively flooding traffic because flooding is restricted to only MNs on a basic tree. And it is very simple for establishing Mcast routing tree or P2P path with on-demand and has robustness for MNs' mobility because MNs have only mapping table with a tag-bit and Mcast-bit that indicates MNs joined Mcast group or a specific MN's NMNs. But the logical address of mapping table may increase when MNs are concentrated in small area.

Acknowledgement

This work was supported (in part) by the Korea Science and Engineering Foundation (KOSEF) through the multimedia Research Center at University of Incheon and by University of Incheon for funding of project 2003

References

[1] D. Johnson, D. Maltz and Y. Hu, "Dynamic Source Routing Protocol for Mobile Ad-hoc Networks", Internet Draft, draft-ietf-manet-dsr-09.txt, April 2003
[2] P. Johansson, T. Larsson, N. Hedman, B. Mielczar and M. Degermark, "Scenario based Performance Analysis of Routing Protocols for Mobile Ad-hoc Networks", In Proc. of the 5th Annual Int'l Conf. On Mobile Computing and Networking (MobiCom 1999), pp195-206, August 1998
[3] B. Chen, K. Jamieson, H. Balakrishnan, and R. Morris, "Span: An Energy-Efficient Coordination Algorithm for Topology Maintenance in Ad Hoc Wireless Networks", In Proc. of MOBICOM'01, pages 85–96, July 2001.
[4] Santashil PalChaudhuri, Shu Du, Amit K. Saha, and David B. Johnson, "TreeCast: A Stateless Addressing and Routing Architecture for Sensor Networks", In Proceedings of the 4th IPDPS International Workshop on Algorithms for Wireless, Mobile, Ad Hoc and Sensor Networks (WMAN 2004), pp. 221a, IEEE, Santa Fe, NM, April 2004
[5] Kai Chen and Klara Nahrstedt, "Effective Location-Guided Overlay Multicast in Mobile Ad Hoc Networks," International Journal of Wireless and Mobile Computing(IJWMC), Special Issue on Group Communications in Ad Hoc Networks, vol. 3, 2005

Real–Time Measurement Based Bandwidth Allocation Scheme in Cellular Networks

Donghoi Kim[1] and Kyungkoo Jun[2]

[1] Mobile Telecommunication Research Laboratory,
Electronics and Telecommunications Research Institute, Korea
donghk@etri.re.kr
[2] Department of Multimedia System Engineering, University of Incheon, Korea
kjun@incheon.ac.kr

Abstract. We propose a Real–time measurement based bandwidth allocation scheme for the downlink real–time video streaming in cellular networks. Our scheme is able to maximize the bandwidth utilization, while satisfying QoS constraints, e.g., the packet loss probability. The principle of our scheme is to determine dynamically the bandwidth to be allocated at each unit time interval by measuring the queue length and the packet loss probability. The simulation results reveal that our scheme achieves the same level of performance as what can be accomplished with the pre-calculated effective bandwidth in terms of the bandwidth utilization and the packet loss probability.

1 Introduction

Considering that the video streams have the VBR-originated bursty nature, it is the non-trivial task to determine the proper amount of bandwidth to be allocated for the service of the video streams in terms of the utilization and the QoS satisfaction. For instance, allocating the bandwidth equal to the mean throughput of the stream may not satisfy the required QoS constraints such as packet loss probability. On the other hand, allocating the peak rate bandwidth is not only impractical but also undesirable in the sense of the bandwidth utilization, while the QoS requirements are met nevertheless.

To cope with the aforementioned problem, a statistical concept is devised, i.e. *effective bandwidth* [1]. It lies between the peak rate and the mean rate, thus is able to satisfy the requirements of both the utilization and the QoS. The exact effective bandwidth for streams can be calculated only by processing the actual traces of the data streaming rate. However it is impractical to obtain the exact traffic traces in a real cellular network environment. Therefore, to complement these shortcomings of the effective bandwidth, several schemes have been proposed to estimate the effective bandwidth in an on-line measurement-based way without using the trace files [2] [3] [4] [5].

In this paper, we propose a real–time measurement based bandwidth allocation scheme for the downlink video streaming. Without the traffic traces, it is still feasible to achieve the same level of performance as what can be accomplished with the trace-based effective bandwidth in terms of the bandwidth utilization and the QoS satisfaction. Our scheme can be also used to enhance the call admission control and the packet scheduling of the cellular networks.

This paper is organized as follows. The second section describes the system model we assume in this paper. The third section presents the real–time measurement based bandwidth allocation scheme in detail as well as the system modification required to integrate our proposed scheme. The fourth section compares the utilization and the packet loss probability of our scheme with those of the pre-calculated effective bandwidth scheme. Finally, the fifth section concludes this paper.

2 System Model

We assume the cellular network which consists of Mobile Station (MS) and Base Station (BS). MS makes a call connection request to BS, which then determines the admission of the request depending on the resource availability. Note that we consider only the downlink in our proposed scheme.

In our model, BS consists of three layers, each of which is named as layer 3, layer 2, and layer 1. The layer 3 is the IP layer. The layer 2 employs a Call Admission Control (CAC) module, a packet classifier, a set of queues, and a packet scheduler. The layer 1 is the physical radio interface. The CAC module of the layer 2 is to decide whether to accept or deny the call connection requests. The decision is made based on the resource availability. The queues, another component of the layer 2, are the buffers storing the packets from external networks before being transmitted to the MSs. Separate queues are allocated for each flow.

The packet classifier distinguishes the packets based on their belonging flows and puts the packets into corresponding queues. Finally, the packet scheduler decides which queue to occupy the radio interface to transmit its packets at each unit time. We assume that the packet scheduler employs the Modified Largest Weighted Delay First (M-LWDF) [7] algorithm for this scheduling task.

3 Real–Time Measurement Based Bandwidth Allocation Model and Algorithm

The principle of our proposed allocation scheme is to dynamically adjust the bandwidth per time interval in order to minimize the unused portion of the allocated bandwidth, while meeting the required QoS constraints. To determine the bandwidth for next timer interval, our scheme checks on the measured bandwidth utilization and the QoS satisfaction level during current interval.

A prominent advantage of our suggested scheme is that the effective bandwidth can be calculated on the fly in real time without a priori knowledge about the target traffic. Thus, our scheme can be employed for the real time video conference. In such real time streaming service, the effective bandwidth is rarely known in advance, and the pre-calculation of the effective bandwidth is impractical.

Our real-time measurement based scheme requires two additional components to the system model described in the previous section. : Effective Bandwidth Allocator (EBA) and a set of token buckets. The EBA regularly fills the token buckets attached to the queues with the tokens at the beginning of every unit time interval. The loaded amount of the tokens represents the allocated bandwidth for next period. Only the queues with available tokens are able to transmit the packets when they are selected by the packet scheduler for transmission. The unused tokens are flushed out at the end of the interval before the EBA reloads the buckets, otherwise the remaining tokens from a previous period could disturb the bandwidth allocation scheme. The total amount of tokens allocated to the token buckets should not exceed the total bandwidth available at the cell. Thus, in addition to the bandwidth allocation role, the token buckets controlled by the EBA are able to guarantee the minimum throughput of the associated flows.

By measuring the queue length and the experienced packet loss probability, the EBA determines the utilization of the allocated bandwidth and the provided QoS level, then based on these statistics, decides the bandwidth to be allocated for next interval by adjusting the amount of tokens to be added. The details of the bandwidth allocation procedure is discussed shortly.

Figure 1 shows the flow diagram of our proposed algorithm. The bandwidth amount for next unit time is determined by measuring the queue length at the end of every unit time interval. In our proposed scheme, we use the queue length as the metric to represent the utilization of the allocated bandwidth. For example, the queue length shorter than threshold is understood as the sign of the waste of the allocated bandwidth, and vice versa. According to the queue length comparison, either the increase or the decrease of the amount of bandwidth for next unit interval is determined compared with that of current interval.

To determine to decrease the allocated bandwidth, over two consecutive unit time periods, the measured queue length should be shorter than the threshold, TH_{short}. In the case of the increase, only one unit time period during which the queue length is longer than the threshold, TH_{long}, is required. The difference in the required number of unit time periods between when deciding the increase and the decrease stems from that fact that our proposed scheme is skewed in the direction of conservative QoS provision.

Once the increase or the decrease of the bandwidth is determined, next step is to determine the range of change by considering the accumulated packet loss probability. Depending on the measured packet loss probability, the magnitude of the increase or the decrease factor is determined, and then the factor is multiplied to the current amount of allocated bandwidth as in Equation 1.

$$BW_{next} = BW_{cur} * factor_{change} \qquad (1)$$

In the case of the increase, if the measured packet loss probability is bigger than a threshold, $TH_{pktLoss}$, relatively large value of the increase factor is employed in order to prevent further packet loss by raising the allocated bandwidth rapidly. On the other hand, if the packet loss is smaller than $TH_{pktLoss}$,

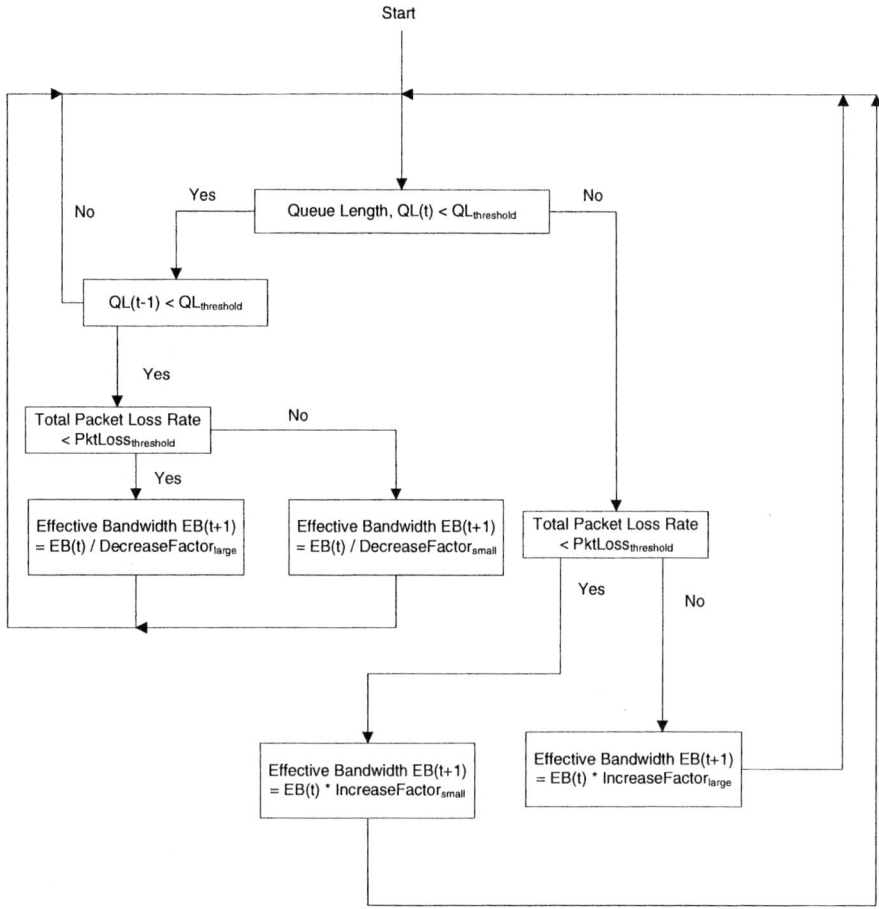

Fig. 1. The flow of the real–time measurement based bandwidth allocation algorithm

comparably small factor is used to increase the bandwidth slowly since there still exists extra room for lost packets before reaching the threshold.

In the case of the decrease, we utilize also the packet loss probability to determine the magnitude of the decrease factor. However, the determination logic is contrary to that of the increase case. For example, if the packet loss probability is bigger than $TH_{pktLoss}$, relatively small factor is used to prevent packet loss, and vice versa. It contributes to the conservative bandwidth decrease in which the chance of further packet loss is reduced gradually as the measured packet loss probability reaches near $TH_{pktLoss}$.

4 Performance Evaluation

We evaluate our proposed scheme in terms of bandwidth utilization and the satisfaction of required packet loss probability through the simulation, and compares the results with those of the peak rate allocation case and the pre–calculated effective bandwidth case. For simulation, we use the simulated video streaming traffic as described in [6]. As the algorithm parameters, TH_{short} and TH_{long} are set to 50 and 100 respectively, while $TH_{pktLoss}$ is 0.00098.

For the methods to be compared with our scheme, in the peak rate allocation case, 32 Kbps is statically allocated during the simulation, while in the pre–calculated effective bandwidth, we have computed the effective bandwidth, 27.05 Kbps, and assigned it as a fixed bandwidth during the simulation.

Figure 2 shows the simulation results of the allocated bandwidth utilization and the measured packet loss probability. In the case of the peak rate allocation, its measured packet loss probability is 0 while its utilization is below 80%, leading to the conclusion that the peak rate allocation is impractical method in terms of the utilization. On the contrary, the pre–calculated effective bandwidth yields the packet loss probability, 0.00098, very close to the required probability 0.001, while maintaining the utilization as high as 94.4%. Our proposed scheme produces the bandwidth utilization, 93.9%, and the packet loss probability, 0.0008, which are similar to those of the pre–calculated effective bandwidth case. The reason that the utilization of our scheme is slightly lower than that of the pre–calculated case is due to the conservative QoS provision of our scheme.

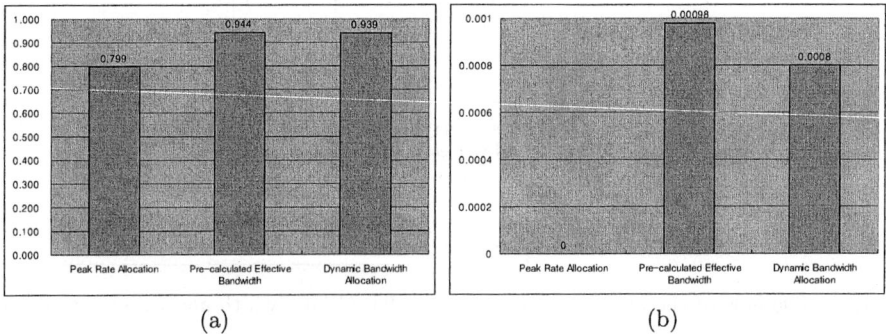

Fig. 2. The comparison of (a) the utilization and (b) the packet loss probability among the peak-rate allocation, the pre–calculated effective bandwidth and our proposed scheme

5 Conclusions

In this paper, we proposed a real–time measurement based bandwidth allocation scheme for the downlink real–time video streaming in the next generation cellular networks. Our scheme is able to maximize the utilization of the allocated bandwidth, while satisfying the required packet loss probability by dynamically determining the amount of bandwidth to be allocated by measuring the queue length and the packet loss probability. By the use of the simulation, it has been shown that our scheme produces similar results to those of the pre–calculated effective bandwidth case. The superiority of our scheme is that our approach does not need a priori knowledge about the video traffic.

References

1. F. Kelly, "Notes on Effective Bandwidths", Stochastic Networks : Theory and Applications, Oxford University Press, 1996.
2. F. Ramos, P. Luan, and L. Lee, "An Effective Bandwidth Allocation Approach for Self-Similar Traffic in a Single ATM Connection," Proceedings of the GlobeCom 1999
3. J. Evans and D. Everitt, "Effective Bandwidth–based Admission Control for Multi-service CDMA Cellular Networks," IEEE Transaction on Vehicular Technology, vol. 48, no. 1, January, 1999.
4. S. Valaee and J. Gregoire, "Resource Allocation for Video Streaming in Wireless Environment," Proceedings of IEEE International Symposium on Wireless Personal Multimedia Communications.
5. H. Li, Changcheng Huang, M. Devetsikiotis, and G. Damm, "Extending the Concept of Effective Bandwidths to Diffserv Networks", Proceedings of CCECE 2004.
6. F. Khan, "Traffic Models for IEEE 802.20 MBWA System Simulations (Ver 02)", IEEE 802.20 Working Group on Mobile Broadband Wireless Access.
7. M. Andrews, K. Kumaran, K. Ramana, A. Stolyar, and P. Whiting, "Providing Quality of Service over a Shared Wireless Link", IEEE Communication Magazine, Feb. 2001.

A Hybrid Web Server Architecture for Secure e-Business Web Applications

Vicenç Beltran, David Carrera, Jordi Guitart, Jordi Torres, and Eduard Ayguadé

European Center for Parallelism of Barcelona (CEPBA),
Computer Architecture Department, Technical University of Catalonia (UPC),
C/ Jordi Girona 1-3, Campus Nord UPC, Mòdul C6, E-08034,
Barcelona, Spain
{vbeltran, dcarrera, jguitart, torres, eduard}@ac.upc.es
http://www.bsc.es/eDragon

Abstract. Nowadays the success of many e-commerce applications, such as on-line banking, depends on their reliability, robustness and security. Designing a web server architecture that keeps these properties under high loads is a challenging task because they are the opposite to performance. The industry standard way to provide security on web applications is the use the Secure Socket Layer (SSL) protocol to create a secure communication channel between the clients and the server. Traditionally, the use of data encryption has introduced a negative performance impact over web application servers because it is an extremely CPU consuming task, reducing the throughput achieved by the server as well as increasing its average response time. As far as the revenue obtained by a commercial web application is directly related to the amount of clients that complete business transactions, the performance of such secure applications becomes a mission critical objective for most companies. In this paper we evaluate a novel hybrid web server architecture (implemented over Tomcat 5.5) that combines the best aspects of the two most extended server architectures, the multithreaded and the event-driven, to provide an excellent trade-off between reliability, robustness, security and performance. The obtained results demonstrate the feasibility of the proposed hybrid architecture as well as the performance benefits that this model introduces for secure web applications, providing the same security level than the original Tomcat 5.5 and improved reliability, robustness and performance, according to both technical and business metrics.

1 Introduction

Many e-commerce applications must offer a secure communication channel to their clients in order to achieve the level of security and privacy required to carry out most commercial transactions. But the cost of introducing security mechanisms in on-line transactions is not negligible. The industry standard for secure web communications is the Secure Socket Layer (SSL) protocol, which is generally used as a complement to the Hypertext Transport Protocol (HTTP) to create the Secure HTTP protocol (HTTPS). The primary goal of the SSL protocol is to provide privacy and reliability between two applications in communication over the Internet. This is achieved using a combination of public and private cryptography to encode the communication between the peers.

The use of public key cryptography introduces an important computational cost to the web containers. Each time a new connection attempt is accepted by a server, a cryptographic negotiation takes place. This initial handshake is required by the peers in order to exchange the encryption keys that will be used during the communication. The cost of the initial handshake is high enough to limit enormously the maximum number of new connections than can be accepted by the server in a period of time, as well as degrading the performance of the server to unacceptable levels, as it can be seen in [5].

The architectural design of most existing web servers is based on the multithreaded paradigm (Apache[11] and Tomcat [7] are widely extended examples), which assigns one thread to each client connected to the server. These threads, commonly known as worker threads, are in charge of attending all the requests issued by their corresponding client until it gets disconnected. The problem associated to this model is that the maximum number of concurrent clients accepted by the server is limited to the number of threads created. In front of this situation, the solution adopted by most web servers is to impose an inactivity timeout to each connection client, forcing the connection to get closed if the client does not produce any work activity before the timeout expires.

The effect of closing client connections is relatively irrelevant when working with plain connections, but it becomes tremendously negative when dealing with secure workloads. Closing connections, especially when the server is overloaded, increases the number of cryptographic handshakes be performed and reduces remarkably the capacity of the server, which results in an important impact over the maximum throughput achieved by the server.

On the other hand, an alternative architectural design for web servers is the event-driven model, already used in Flash[9] and in the SEDA[12] architecture. This model comes to solve the problems associated to the multithreading paradigm, especially in client-server environments. But this model lacks of the innate ease of programming associated to the multithreading model, making the task of developing web servers remarkably more complex.

In this paper we evaluate a hybrid web server architecture oriented to the use of secure communication protocol that exploits the best of each one of the discussed server architectures. With this hybrid architecture, an event-driven model is applied to receive the incoming client requests. When a request is received, it is serviced following a multithreaded programming model, with the resulting simplification of the web container development associated to the multithreading paradigm. When the request processing is completed, the event-driven model is applied again to wait for the client to issue new requests. With this, the best of each model is combined and, as it is discussed in following sections, the performance of the server is remarkably increased, especially in presence of secure communication protocols.

The rest of the paper is structured as follows: section 2 describes the HTTP/s protocol, section 3 discusses the characteristics of the multithreaded, event-driven and hybrid architectures; later, in section 4, we present the execution environment where the experimental results presented in this work were obtained and, finally, section 5 presents the experimental results obtained in the evaluation of the hybrid web server and section 6 gives some concluding remarks and discusses some of the future work lines derived from this work.

2 HTTP/S and SSL

HTTP/S (HTTP over SSL) is a secure Web protocol developed by Netscape. HTTPS is really just the use of Secure Socket Layer (SSL) as a sublayer under its regular HTTP application layering.

The SSL protocol provides communications privacy over the Internet. The protocol allows client/server applications to communicate in a way that is designed to prevent eavesdropping, tampering, or message forgery. To obtain these objectives it uses a combination of public-key and private-key cryptography algorithm and digital certificates (X.509).

The SSL protocol does not introduce a new degree of complexity in web applications structure because it works almost transparently on top of the socket layer. However, SSL increases the computation time necessary to serve a connection remarkably, due to the use of cryptography to achieve their objectives. This increment has a noticeable impact on server performance, which has been evaluated in [5]. This study concludes that the maximum throughput obtained when using SSL connections is 7 times lower than when using normal connections. The study also notices that when the server is attending non-secure connections and saturates, it can maintain the throughput if new clients arrive, while if attending SSL connections, the saturation of the server provokes the degradation of the throughput.

The SSL protocol fundamentally has two phases of operation: SSL handshake and SSL record protocol. We will do an overview of the SSL handshake phase, which is the responsible of most of the computation time required when using SSL. The detailed description of the whole protocol can be found in RFC 2246 [10]. The SSL handshake allows the server to authenticate itself to the client using public-key techniques like RSA, and then allows the client and the server to cooperate in the creation of symmetric keys used for rapid encryption, decryption, and tamper detection during the session that follows. Optionally, the handshake also allows the client to authenticate itself to the server. Two different SSL handshake types can be distinguished: The full SSL handshake and the resumed SSL handshake. The full SSL handshake is negotiated when a client establishes a new SSL connection with the server, and requires the complete negotiation of the SSL handshake. This negotiation includes parts that spend a lot of computation time to be accomplished. We have measured the computational demand of a full SSL handshake in a 1.4 GHz Xeon machine to be around 175 ms.

The SSL resumed handshake is negotiated when a client establishes a new HTTP connection with the server but using an existing SSL connection. As the SSL session ID is reused, part of the SSL handshake negotiation can be avoided, reducing considerably the computation time for performing a resumed SSL handshake. We have measured the computational demand of a resumed SSL handshake in a 1.4 GHz Xeon machine to be around 2 ms. Notice the big difference between negotiate a full SSL handshake respect to negotiate a resumed SSL handshake (175 ms versus 2 ms).

Based on these two handshake types, two types of SSL connections can be distinguished: the new SSL connections and the resumed SSL connections. The new SSL connections try to establish a new SSL session and must negotiate a full SSL handshake. The resumed SSL connections can negotiate a resumed SSL handshake because they provide a reusable SSL session ID (they resume an existing SSL session).

3   Web Server Architectures

There are multiple architectural options for a web server design, depending on the concurrency programming model chosen for the implementation. The two major alternatives are the multithreaded model and the event-driven model. In both models, the work tasks to be performed by the server are divided into work assignments that are assumed each one by a thread (a worker thread). If a multithreaded model is chosen, the unit of work that can be assigned to a worker thread is a client connection, which is achieved by creating a virtual association between the thread and the client connection that is not broken until the connection is closed. Alternatively, in an event driven model the work assignment unit is a client request, so there is no real association between a server thread and a client.

Multithreaded Architecture with Blocking I/O

The multithreaded programming model leads to a very easy and natural way of programming a web server. The association of each thread with a client connection results in a comprehensive thread life-cycle, started with the arrival of a client connection request and finished with the connection close. This model is especially appropriate for short-lived client connections and with low inactivity periods, which is the scenario created by the use of non persistent HTTP/1.0 connections. A pure multithreaded web server architecture is generally composed by an acceptor thread and a pool of worker threads. The acceptor thread is in charge of accepting new incoming connections, after what each established connection is assigned to one thread of the workers pool, which will be responsible of processing all the requests issued by the corresponding web client.

The introduction of connection persistence in the HTTP protocol, already in the 1.0 version of the protocol but mainly with the arrival of HTTP/1.1, resulted in a dramatic performance impact for the existing multithreaded web servers. Persistent connections, which means connections that are kept alive by the client between two successive HTTP requests that in turn can be separated in time by several seconds of inactivity (think times), cause that many server threads can be retained by clients even when no requests are being issued and the thread keeps in idle state. The use of blocking I/O operations on the sockets is the cause of this performance degradation scenario. The situation can be solved increasing the number of threads available (which in turn results in a contention increase in the shared resources of the server that require exclusive access) or introducing an inactivity timeout for the established connections, that can be reduced as the server load is increased. When a server is put under a severe load, the effect of applying a shortened inactivity timeout to the clients lead to a virtual conversion of the HTTP/1.1 protocol into the older HTTP/1.0, with the consequent loss of the performance effects of the connection persistence. If we use HTTP/S, open a new connection for each request can decrease dramatically the throughput of the server, specially when the server are in a overloaded state. In this situation the number of initial SSL handshakes increase quickly and saturates the web server.

In this model, the effect of closing client connections to free worker threads reduces the probability for a client to complete a session to nearly zero. It is especially important when the server is under overload conditions, where the inactivity timeout is dynamically decreased to the minimum possible in order to free worker threads as quickly as possible, which provokes that all the established connections are closed during think times. This causes a higher competition among clients trying to establish a connection with the server. If we extend it to the length of a user session, we obtain that the probability of finishing it successfully under this architecture is still much lower than the probability of establishing each one of the connections it is composed of, driving the server to obtain a really low performance in terms of session completions. This situation can be alleviated increasing the number of worker threads available in the server, but this measure also produces an important increase in the internal web container contention with the corresponding performance slowdown.

Event-Driven Architecture with Non-blocking I/O

The event-driven architecture completely eliminates the use of blocking I/O operations for the worker threads, reducing their idle times to the minimum because no I/O operations are performed for a socket if no data is already available on it to be read. With this model, maintaining a big amount of clients connected to the server does not represent a problem because one thread will never be blocked waiting a client request. With this, the model detaches threads from client connections, and only associates threads to client requests, considering them as an independent work units. An example of web server based on this model is described in [9], and a general evaluation of the architecture can be found in [2].

In an event driven architecture, one thread is in charge of accepting new incoming connections. When the connection is accepted, the corresponding socket channel is registered in a channel selector where another thread (the request dispatcher) will wait for socket activity. Worker threads are only awakened when a client request is already available in the socket. When the request is completely processed and the reply has been successfully issued, the worker thread registers again the socket channel in the selector and gets free to be assigned to new received client requests. This operation model avoids worker threads to keep blocked in socket read operations during client think times and eliminates the need of introducing connection inactivity timeouts and their associated problems. This architecture is specially efficient to deal with HTTP/S protocol because it never close connections unnecessarily. With this approach the server can recover the computation spent to establish the connection serving a number of request for each connection.

A remarkable characteristic of the event-driven architectures is that the number of active clients connected to the server is unbounded, so an admission control[4] policy must be implemented. Additionally, as the number of worker threads can be very low (one should be enough) the contention inside the web container can be reduce to the minimum.

Hybrid Architecture

In this paper we evaluate a hybrid architecture that can take benefit of the strong points of both discussed architectures, the multithreaded and the event driven. In this hybrid architecture, the operation model of the event-driven architecture is used for the assignment of client requests to worker threads (instead of client connections) and the multithreaded model is used for the processing and service of client requests, where the worker threads will perform blocking I/O operations when required. This architecture can be used to decouple the management of active connections from the request processing and servicing activity of the worker threads. With this, the web container logic can be implemented following the multithreaded natural programming model and the management of connections can be done with the highest possible performance, without blocking I/O operations and reaching a maximum overlapping of the client think times with the processing of requests.

In this architecture the acceptor thread role is maintained as well as the request dispatcher role from the pure event-driven model and the worker thread pool (performing blocking I/O operations when necessary) from the pure multithreaded design. This makes possible for the hybrid model to avoid the need of closing connections to free worker threads without renouncing to the multithreading paradigm. In consequence, the hybrid architecture makes a better use of the characteristics introduced to the HTTP protocol in the 1.1 version, such as connection persistence, with the corresponding reduction in the number of client re-connections (and the corresponding bandwidth save). When the server is overloaded and the clients use the HTTP/S protocol the fact of keep connections open improve the throughput of the server because the computation spent in each connection established is recovered with the subsequents requests.

4 Testing Environment

The hardware platform for the experiments presented in this paper is composed of a 4-way Intel Xeon 1.4 GHz with 2GB RAM to run the web servers and a 2-way Intel XEON 2.4 GHz with 2 GB RAM to run the benchmark clients. For the benchmark applications that require the use of a database server, a 2-way Intel XEON 2.4 GHz with 2 GB RAM was used to run MySQL v4.0.18, with the MM.MySQL v3.0.8 JDBC driver. All the machines were running a Linux 2.6 kernel, and were connected through a switched Gbit network. The SDK 1.5 from Sun was used to develop and run the web servers.

All the tests are performed with the common RSA-3DES-SHA cipher suit. Handshake is performed with 1024 bit RSA key. Record protocol uses triple DES to encrypt all application data. Finally, SHA digest algorithm is used to provide the Message Authentication Code (MAC).

The workload for the experiments was generated using a workload generator and web performance measurement tool called Httperf [8]. This tool, which support both HTTP and HTTPS protocols, allows the creation of a continuous flow of HTTP/S requests issued from one or more client machines and processed by one server machine: the SUT (System Under Test). The configuration parameters of the benchmarking tool used for the experiments presented in this paper were set to create a realistic workload, with non-uniform reply sizes, and to sustain a continuous load on the server. One of the parameters of the tool represents the number of new clients per second initiating an interaction with the server. Each emulated client opens a session with the server. The session remains alive for a period of time, called session time, at the end of which the connection is closed. Each session is a persistent HTTP/S connection with the server, used by the client to repeatedly send requests, some of them pipelined. The requests issued by httperf were extracted from the RUBiS [1] application. A secure dynamic content application is characterized by the long length of the user sessions as well as by the high computational cost of the first connection (initial SSL handshake) and subsequents request to be serviced (including embedded requests to external servers, such as databases).

5 Experimental Results

In this section we will evaluate and compare the performance of the proposed hybrid server architecture in front of the out-of-the-box Tomcat architecture, under a secure workload as well as under a plain workload.

The workload generator used for the experiments, Httperf, was configured using a client timeout value of 10 seconds and according to the configuration described in section 4. Each individual benchmark execution had a fixed duration of 30 minutes.

5.1 Secure Dynamic Content

Looking at figure 1, it can be seen that the throughput of the hybrid server is always better than the original Tomcat, moreover the difference between them grows up as the load is increased. When the load is low and the original Tomcat architecture is below saturation, the benefits of the hybrid architecture are already exposed. The use of the non-blocking I/O API (NIO), which offers a higher performance than the standard stream-based Java I/O, leads the hybrid architecture to offer a higher performance than the original Tomcat server. By when the saturation point of the Tomcat server is reached, the hybrid architecture is exposing a throughput that is a 50% higher than the original multithreaded architecture. When the load is increased beyond the saturation

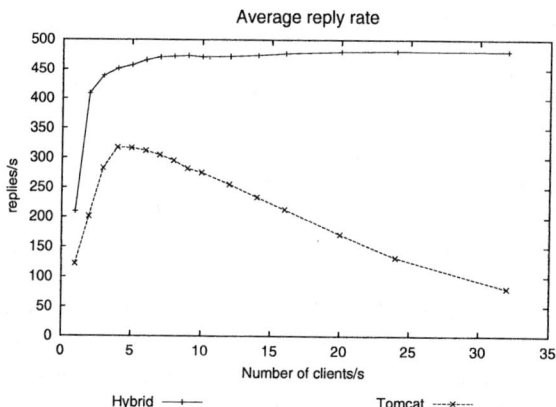

Fig. 1. Reply throughput comparison between Tomcat server and hybrid server

point, the performance of the hybrid server keeps nearly constant while the out-of-the-box Tomcat server starts reducing its output level linearly with the load increase. The benefits of the hybrid architecture beyond the saturation point are explained by the higher use of the connection persistence characteristics of the HTTP/1.1 protocol that the hybrid architecture makes in front of the original multithreaded approach. As more TCP connections are reused, more clients can keep their connections established and consequently less connection re-establishments will be required. When the considered workload uses data encryption, reducing the number of connection establishments is synonymous of less SSL handshakes negotiations and in consequence an important reduction of the processing requirements for the server system.

Usually, the workload produced over secure e-business applications is session-based. At the beginning, the client gets connected. After that, the session requests are issued and correspondingly processed by the server. Finally, when the session is finished, the

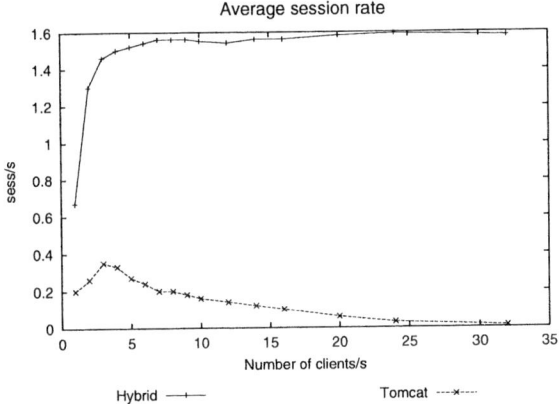

Fig. 2. Session throughput comparison between Tomcat server and hybrid server

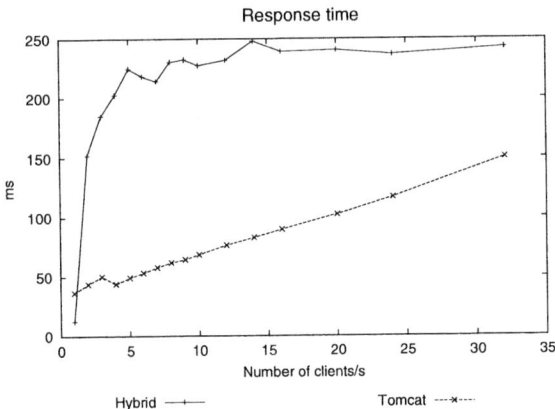

Fig. 3. Response time comparison between Tomcat server and hybrid server

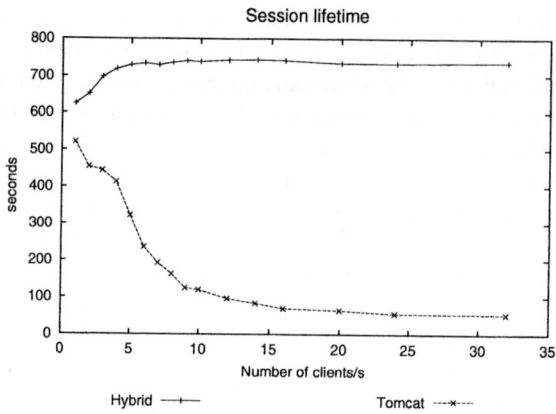

Fig. 4. Lifetime comparison for the sessions completed successfully

client gets disconnected and the user session is considered completed. The need of client re-connections introduced by the limitations of the multi-threaded server model produces an increasing difficulty to complete user sessions successfully. This effect is shown in figure 4. As it can be seen, when the saturation point is reached, the average length of the user sessions successfully completed for the original Tomcat server starts decreasing. The reason of this phenomena is the higher number of re-connections needed by the longest sessions when the server is saturated under the multi-threaded architecture. In the hybrid architecture, instead, the independence between the number of connected clients and the number of processing threads in the server allows this architecture to avoid the need for re-connections and their associated SSL handshakes, which allows long sessions to be completed without the difficulties introduced by the multi-threaded architecture.

Additionally to giving a higher chance to the long-lived user sessions to be completed, the hybrid architecture also increases the global average of sessions completed successfully, as it can be seen in figure 2. The original multithreaded Tomcat server starts reducing the number of sessions completed successfully beyond its saturation point. This is caused by the higher number of sessions that are aborted by the clients when numerous re-connections are rejected by the server when it is getting increasingly overloaded. As the load increases in the server, the chance for a connection attempt to be accepted is decreased. As user sessions, for the multithreaded model server, require a number of re-connections each one, driving the server to higher loads results in a higher probability for the clients to consider the server unavailable if several connection attempts are rejected, which leads to a higher number of clients aborting their navigation sessions under these conditions. On the other hand, the hybrid architecture avoids the need for client re-connections for a user session to be completed, which allows a higher number of them to finish successfully, independently of the system load.

The benefits of the hybrid architecture presented above are not directly translated to the response time offered by it. As it can be seen in figure 3, the average response time obtained for the out-of-the-box Tomcat server is better than the obtained for the

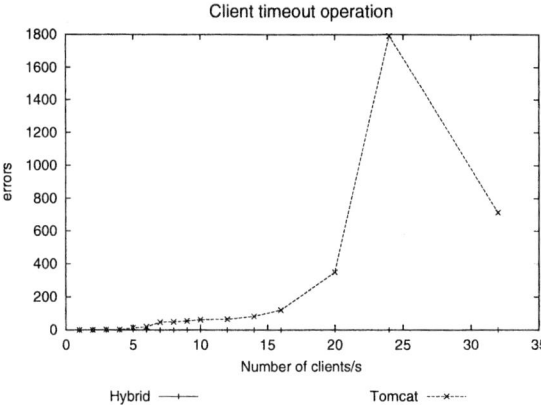

Fig. 5. Number of client timeouts under a secure dynamic content workload

hybrid architecture, although it is not unacceptable. A response time bounded to less than 250ms is more than acceptable and, moreover, it remains constant with the load. In our opinion a slightly increase in the average response time obtained is more than tolerable considering the benefits presented for the server throughput obtained for the hybrid architecture.

Finally, another benefit of the hybrid architecture can be seen in figure 5, where it can be observed the number of requests sent to the server that not produce a response after an acceptable period of time. In the benchmark application this time is expressed as a timeout assimilated to the amount of time a human client would expect a reply from its web browser before considering a page request failed. For the experiments, this timeout value was set to 10 seconds. As it can be observed, the hybrid architecture produces no client timeouts while the number of errors generated by the pure multithreaded Tomcat server grows with the system load. This situation is produced, for the hybrid architecture, by the use of an overload control mechanism as well as by the independence between the number of connected clients to the server and the amount of concurrent server threads processing requests, being it possible to keep the latest in a low value and reducing in this way the contention caused by the multithreaded architecture as well as increasing the obtained performance for the server. It can also be seen that the number of timeout operation errors observed for the original Tomcat server decreases when the load level is driven beyond a certain point. This can be explained because under that level of overload, the server starts rejecting connections because no server capacity remains available to process the volume of incoming connection attempts and less clients remain connected concurrently.

5.2 Secure vs Plain Dynamic Content

Once it was proved that for secure content workloads the proposed hybrid architecture provides a higher performance than the original pure multithreaded Tomcat server in several ways, we wanted to measure the performance gap observed for both servers when subject to a plain workload and secure workload. This was achieved by rerunning

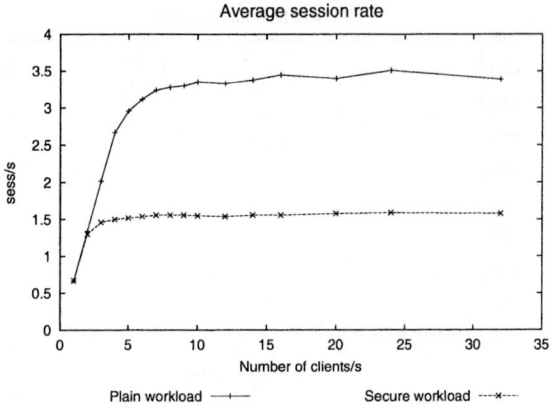

Fig. 6. Session throughput for Hybrid server

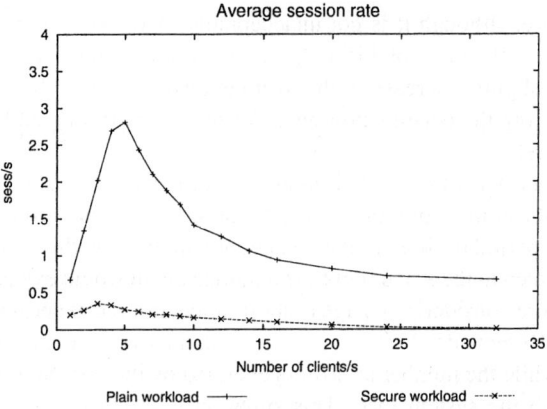

Fig. 7. Session throughput for Tomcat server

the benchmark application with the HTTP protocol instead of HTTPS protocol used in the previous experiments. More results with plain workload are discussed in [3].

It can be seen in figures 6 and 7 that another remarkable effect of the hybrid architecture is that it reduces the impact on the performance caused by the introduction of security in the workload from a factor of 7 in the original Tomcat server to only 2 for the hybrid architecture when measuring the throughput in successfully completed user sessions.

6 Conclusions and Future Work

In this paper we have proved that the use of a hybrid web server architecture that takes the best of the two more extended server architectures, the multithreaded and the event-driven, can boost the performance of the server, especially under session-based workloads and even more when the workloads use encryption techniques. As it has been demonstrated, the benefits of this architecture are especially noticeable in the throughput of the server, in terms of individual requests as well as for user sessions, particularly when the server is overloaded. The modified Tomcat server beats the original multithreaded Tomcat in nearly all the performance paramenters studied and minimizes the impact of incorporating secure connections to a web application with respect to the out-of-the-box Tomcat.

This work is a first step toward the creation of an autonomic web container, which will be deployed alone or as a web tier of an application server platform. The hybrid architecture reduces the complexity of the tuning of a web container (the size of the worker treads pool is only related to the server capacity and is not a limiting factor for the amount of connected clients as well as the server timeouts disappear), which is an important step toward the implementation of autonomic servers (see [6] for more details on the autonomic computing topic). Another interesting topic can be the integration and evaluation of this hybrid architecture in load balancers and clustering configurations.

References

1. C. Amza, A. Chanda, E. Cecchet, A. Cox, S. Elnikety, R. Gil, J. Marguerite, K. Rajamani, and W. Zwaenepoel. Specification and implementation of dynamic web site benchmarks, 2002.
2. V. Beltran, D. Carrera, J. Torres, and E. Ayguadé. Evaluating the scalability of java event-driven web servers. In *2004 International Conference on Parallel Processing (ICPP'04)*, pages 134–142, 2004.
3. V. Beltran, D. Carrera, J. Torres, and E. Ayguadé. A hybrid web server architecture for e-commerce applications. In *The 11th International Conference on Parallel and Distributed Systems (ICPADS 2005) July 20 - 22, 2005, Fukuoka, Japan*, 2005.
4. H. Chen and P. Mohapatra. Session-based overload control in qos-aware web servers. In *INFOCOM*, 2002.
5. J. Guitart, V. Beltran, D. Carrera, J. Torres, and E. Ayguadé. Characterizing secure dynamic web applications scalability. In *19th International Parallel and Distributed Processing Symposium, Denver, Colorado (USA). April 4-8, 2005*, 2005.
6. IBM Research. *Autonomic computing*. See *http://www.research.ibm.com/autonomic*.
7. Jakarta Project. Apache Software Foundation. *Tomcat*. See *http://jakarta.apache.org/tomcat*.
8. D. Mosberger and T. Jin. httperf: A tool for measuring web server performance. In *First Workshop on Internet Server Performance*, pages 59—67. ACM, June 1998.
9. V. S. Pai, P. Druschel, and W. Zwaenepoel. Flash: An efficient and portable Web server. In *Proceedings of the USENIX 1999 Annual Technical Conference*, 1999.
10. T.Dierks and C. Allen. *The TLS Protocol, Version 1.0. RFC 2246. January 1999*.
11. The Apache Software Foundation. *Apache HTTP Server Project*. See *http://httpd.apache.org*.
12. M. Welsh, D. E. Culler, and E. A. Brewer. SEDA: An architecture for well-conditioned, scalable internet services. In *Symposium on Operating Systems Principles*, pages 230–243, 2001.

An Efficient Scheme for Fault-Tolerant Web Page Access in Wireless Mobile Environment Based on Mobile Agents[*]

HaiYang Hu, XianPing Tao, JiDong Ge, and Jian Lu

State Key Laboratory for Novel Software Technology,
Nanjing University, 210093, China
{hhy, txp, gjd}@ics.nju.edu.cn, lj@nju.edu.cn

Abstract. Web Page access in mobile wireless environments suffers more challenges for the limited memory of mobile devices and the unreliable wireless network environments. It requires some facilities in the wireless environments so that to overcome the difficulties of unreliable wireless link, mobility and mobile devices capability limitation. This paper presents a fault-tolerant Web page access scheme, which uses mobile agents to facilitate seamless logging of access activities for recovery from failure. Compared with other schemes by performance analysis, our scheme keeps its efficiency with the additional fault-tolerant characteristic.

1 Introduction

Wireless access to Internet enables a user equipped with a wireless capable device to access the Internet anywhere and perform his desired operation [1]. However, for the limited hardware capability, when performing wireless Web page access, the user usually prefetches some frequently used web pages into the local cache of his mobile unit (MU) and then disconnects MU from the web server to save its battery energy and the wireless communication cost. During disconnection, the user uses the Web pages cached in MU to access the Web and the write operations to these Web pages performed by the user are recorded into a log. When the MU reconnects to networks again, the log is sent to the Web server for reintegration and the conflicts with updates performed by other users need to be resolved. And these three phrases are termed as hoarding, disconnection, and reintegration.

Until now, various mechanisms proposed in [2], [4], [5], [6], [7] support disconnected Web page access in wireless environments in the above three phases. However, fewer of these works have concerned on how to keep the wireless Web page access process fault-tolerant, when the MU suffers a handoff or even a failure without losing its efficiency. This article presents a new scheme naming MAWA based on mobile agents to implement a fault-tolerant Web page access progress, which uses mobile agents to facilitate seamless logging of access activities for the future recovery from failure. The motivation of our scheme is inspired by [17], [19], [20] and [21].

---

[*] Funded by NNSFC (60233010, 60273034, 60403014), 973 Program of China (2002CB312002), NSFC of Jiangsu Province (BK2002203, BK2002409), 863 Program of China.

The reminder of this paper is organized as follows. Section 2 overviews the related works. Section 3 presents the system framework. Section 4 discusses the fault-tolerant propagation scheme. Performance analysis of the scheme is given in section 5. Finally, section 6 concludes the paper.

2 Related Works

Until now, most existing schemes for wireless Web page access just support read-only Web page access when MU disconnects from Internet. However, with the rapidly increasing of distributed web authoring and form-based electronic commerce web applications [7], supporting to write operations is also needed.

Work reported in [11] proposes the concepts of locking and versioning. In [12], the Caubweb system designs a HTTP client proxy running on MU to cache staging updates during the disconnection phrase. In [5], the system uses a cache manager named Venus on the client side. Upon reintegration, the Venus resynchronizes its local cache with the server. If Venus detects a divergence, an application specific resolver (ASR) is invoked to resolve the difference. In [14], the authors integrate the concept of coherency interval as proposed in [9] with the concepts of versioning and locking as proposed in [11] to support disconnected writing operations for wireless Web access, and presents three update propagation algorithms.

However, fewer of these works concern on how to cope with the scenario when the MU suffers a handoff or even a failure. Recently, using mobile agents in mobile computing environments becomes a hot research area. The systems such as [19], [20], [21] give us a new insight into the wireless Web page access. Inspired by these works, we present a new Web page access mechanism based on mobile agents to resolve the above problems. It uses mobile agent between the Web server and MU to support fault-tolerant wireless Web access while without losing its efficiency.

3 System Framework

A mobile computing system (see fig.1) is composed of a static network and a dynamic wireless network. The static network comprises the fixed hosts and the communication network. Some of the fixed hosts, namely, base stations, are augmented with a wireless network and they provide gateways for communication between wireless and static network. A mobile unit (MU) can connect with static network with the supporting of base stations. When the mobile unit moves from one physical cell to another, the base station responsible for it will be changed. This process called handoff is transparent to the mobile unit itself.

There are two kinds of agents in MAWA as follows:

Base station agent (BsAg). This static agent resides on base station with responsibility for creating, storing, sending and receiving an instance of MuAg according to the request from MU or other BsAgs. It maintains the local MuAg queue, and transfers the message or data of Web pages between MU and its MuAg respectively.

MU agent (MuAg). Created by BsAg, this agent is a mobile agent. It is responsible for a certain MU. In MAWA, MU doesn't communicate with Web server directly. All the data of Web pages and messages sent from one part are first transmitted to

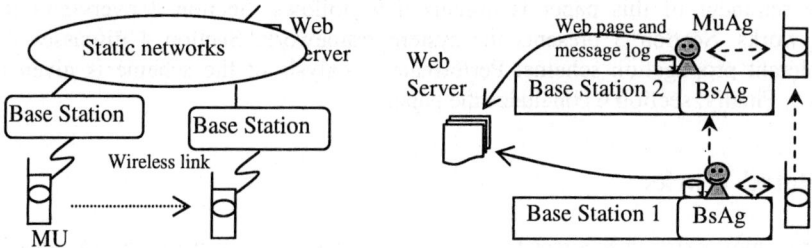

Fig. 1. Mobile computing system   **Fig. 2.** The system framework of MAWA

MuAg, and it sends them to the other. MuAg makes records of these Web pages and messages in a log on the base station. This log also plays an important role in our scheme. When MuAg gets some Web pages of new versions from Web server, it will send them to MU. When MU suffers a failure, MuAg uses the log to help it recover back. When MU moves to another base station, the correlated MuAg and its log are also sent to that base station with the help of BsAg.

4 Fault-Tolerant Web Page Access Scheme

In our scheme, we integrate the concept of versioning and locking proposed in [11], and MU disconnects from network for a coherency interval as proposed in [9]. To analyze the fault-tolerance scheme quantitatively, also as proposed in [14], we need to define the following parameters. For a Web page i cached by MU, the update rate done to it by MU is denoted as λ_i^m and the update rate done by all other users is denoted as λ_i^o. Thus, for read-only cached Web pages, its λ_i =0. There are two general cost parameters, C_m and C_w. C_m is the average one-way communication cost of transmitting a simple message over the wired network; C_w is the average one-way communication cost of transmitting a data packet carrying a Web page over the wired network. α is denoted as the ratio of the bandwidth of wired network to the bandwidth of wireless network. Then the average cost of transmitting a message from base station to MU over wireless link is denoted as αC_m, and the average cost of transmitting a data packet over wireless network is αC_w. C_{rm} is the cost of resolving Web page conflicts performed by MU. When a MuAg migrates from one base station to another, the cost is denoted as C_{Ag}. The rate of MU suffering a handoff is denoted as λ_h. The failure rate of MU is denoted as λ_f. We suppose that both handoff and failure arrive at the system as an exponential distribution.

4.1 Three Kinds of Web page Update Propagation

Suppose MU is in disconnection phrase with the length of disconnection period Lx. And after the period L_x, MU reconnects to network to perform update propagation. Thus, at the time of $L_x - mnC_w$ (here, n is the number of Web pages cached by MU, m

is the number of static nodes between the base station and Web server), MuAg inquires Web server about the up-to-date versions of the Web pages which MU has prefetched before disconnection and applies to Web server for locking all these Web pages[1]. If such a Web page has already been updated by other users, Web server sends the new version Web page to MuAg. In reintegration phrase, MU connects to network again and sends base station an inquiry message[2] indicating which cached pages it have already updated in disconnection phrase. After receiving the message, BsAg in the base station searches local MuAg queue to transfer the message to the corresponding MuAg. Then there are several possible cases as follows (here we suppose MU has prefetched only one Web page i):

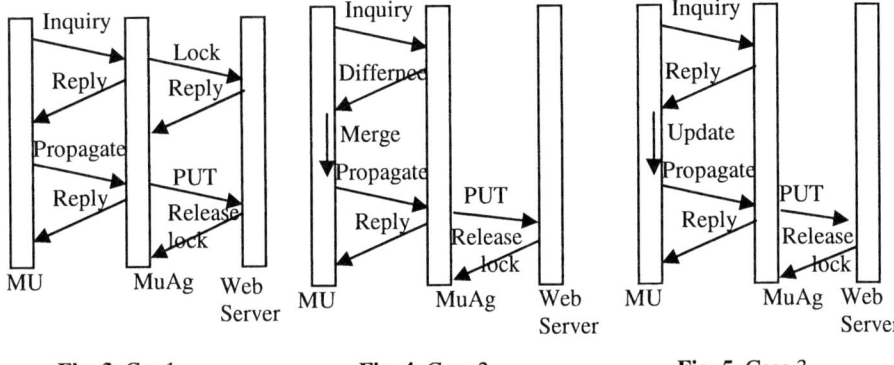

Fig. 3. Cas 1   **Fig .4.** Case 2   **Fig. 5.** Case 3

1. If the Web page i has been updated by MU (see fig.3), and not been updated by other users, then MuAg sends MU a message informing MU to propagate the modified Web page to Web server. The overall communication cost is $3\alpha C_m + \alpha C_w$.

2. Shown in fig.4, for the Web page i cached by MU, if some other user has already updated it when MU is in disconnection phrase, then MuAg sends the data of the new version Web page to MU. Based on the new version received, MU applies a merge algorithm to resolve the update conflict and sends a data packet carrying the differences of the updated Web page to MuAg. MuAg propagates the updates to Web server and releases the lock. The overall communication cost is $2\alpha C_m + 2\alpha C_w + C_{rm}$.

3. For the Web pages i cached by MU, if there are no updates to it during disconnection phrase, MU must perform a *forced update* on the page (because of the real-time requirements of online Web applications) when it reconnects with network again (see figure 5). Thus, after receiving a reply message indicating *forced update*

---

[1] We will see late that all the Web pages cached by MU will have to be updated by it. So, MuAg has to lock them all to ensure that there are no updates from other users when MU is updating them.
[2] For a certain Web page cached, MU sets in the inquiry message the identifier "U" meaning that this page has not been updated, "A" meaning that this page has been updated; MuAg sets in the message sent to MU the identifier "P" meaning that the updates to this page can be propagated, "F" meaning that this page needs force updating, "R" meaning that this page needs to resolve conflicts.

from MuAg, MU performs forced updates (the cost is also the same as C_{rm}) and propagates the updates to MuAg. After MuAg propagates the updated Web page sent by MU to Web server, MuAg releases the lock. So the overall communication cost in this case is $3\alpha C_m + \alpha C_w + C_{rm}$.

Supposing that updates to Web page i arrive at the system as an exponential distribution, then P_i, the probability of updates to page i performed by other users during MU's disconnection phrase L_x, is given as

$$p_i = 1 - e^{-\lambda_i^o L_x} \qquad (1)$$

Also, q_i, the probability that Web page i has been updated by MU during the disconnection period of L_x, is as follows

$$q_i = 1 - e^{-\lambda_i^m (L_x - mnC_w)} \qquad (2)$$

Based on the above analysis, it's easy to deduce that when MU prefetches a set of Web pages in local cache (the number is n) in hoarding phrase and updates them during disconnection period in batch. The overall communication cost is as follows:

$$C_{s1} = 3\alpha C_m + \sum_{i=1}^{n} p_i \alpha C_w + \sum_{i=1}^{n} q_i(1-p_i)\alpha C_w + \sum_{i=1}^{n}(1-q_i)(1-p_i)(\alpha C_w + C_{rm})$$

$$\sum_{i=1}^{n} q_i p_i (\alpha C_w + C_{rm}) + \sum_{i=1}^{n}(1-q_i) p_i (\alpha C_w + C_{rm}) \qquad (3)$$

The minimized value of C_{s1} is obtained as a solution of

$$\frac{\partial C_s}{\partial L_x} = 0 \quad \text{and} \quad \frac{\partial^2 C_s}{(\partial L_x)^2} > 0 \qquad (4)$$

4.3 Web Page Update Propagation Under Handoff and Failure

For the brittleness of MU's hardware, MU suffers a failure occasionally [15], [16], [17], [20]. There are three situations of MU suffering a failure: a) MU fails in disconnection phrase; b) MU fails in reintegration phrase; c) MU suffers a handoff and a failure simultaneously.

An algorithm given in MAWA to dealing with these three situations (also see fig.6):

1. After a failure, MU reconnects to network and sends a message indicating "failure/recovery" to the base station since it is aware of the failure itself.
2. BsAg on the base station receives the message and searches local MuAg queue to find the corresponding MuAg responsible for the MU.
    a) If such a MuAg can be found, BsAg transfers the message to it.
    b) If such a MuAg can not be found, BsAg knows that MU has moved from some other base station here. With the help of VLR and HLR (the cost is denoted as C_f), BsAg finds the previous base station which MU had connected with and sends a message to it. The BsAg' on that base station receives this message and finds out that MuAg. BsAg' sends the MuAg and its log to BsAg. BsAg adds the MuAg into its local MuAg queue and transfers the message indicating " failure/recovery" sent by MU to it.

3. MuAg receives the message and checks its log:

a) If MU has not propagated any updated Web pages, MuAg sends MU the follow things: a message indicating which pages MU hasn't propagated the updates, and the Web pages of up-to-date version that MU had prefetched.

b) If MU has already propagated some updated Web pages, MuAg sends MU three following things: a message indicating which pages MU hasn't propagated updates, the updated pages that MU has already propagated, and the Web pages of up-to-date version that need forced updating.

4. MU receives them

a) If the failure occurs when MU is in disconnected phrase, MU will be in disconnection phrase again.

b) If the failure occurs when MU is in reintegration phrase, MU proceeds with its update propagation as presented in section 4.1.

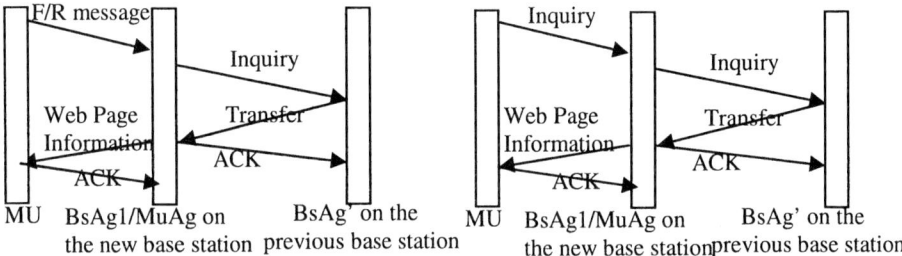

Fig. 6. Fault-tolerance Web Pageupdate propagation process

Fig. 7. Update propagation process under a handoff

Since the failures occurring in disconnected phrase doesn't influence the overall cost of update propagation in reintegration phrase, we only take the case that failure occurs in reintegration phrase into accounts. Summarizing 1-4 above, the average cost of propagating the updates of n pages prefetched when MU suffers a failure in reintegration phrase is:

$$C_{s3} = (1 - P_f)C_{s2} + P_f((1 - P_h)(C_{s1} + \alpha C_m + \alpha n C_w) + P_h(C_{s1} + \alpha C_m + \alpha n C_w + C_m + nC_w + C_f + C_{Ag})) \quad (5)$$

In (11), P_f, the probability that MU suffers a failure in reintegration phrase, is as follows

$$P_f\{L_x < X < L_x + C_s \mid X > L_x\} = \int_{L_x}^{L_x + C_s} \lambda_f e^{-\lambda_f t} dt / (1 - F(L_x)) = 1 - e^{-\lambda_f C_s}$$

P_h, the probability of MU suffering a handoff in reintegration phrase, is as follows

$$P_h\{L_x < X < L_x + C_s \mid X > L_x\} = \int_{L_x}^{L_x + C_s} \lambda_h e^{-\lambda_h t} dt / (1 - F(L_x)) = 1 - e^{-\lambda_h C_s}$$

C_{s2}, the average cost of Web page update propagation when MU suffering a handoff in reintegration phrase is shown below. The process of update propagation when

MU suffering a handoff is analogous to the fault-tolerance algorithms when MU suffering a failure, and the detailed process is given in fig.7. The cost of C_{s2} is as follows

$$C_{s2}= C_{s1}(1-P_h)+(C_{s1}+\alpha C_m +C_f +C_m +C_{Ag} +nC_w)P_h$$

5 Performance Study

Table 1 lists the values of input parameters needed. And the values of these parameters come from [14], [16] and [20].

Table 1. Input parameters

Input Parameter	Value
Failure rate λ_f	(0.005, 0.2)
Handoff rate λ_h	(0.003125, 0.2)
Wireless network factor α	10
Update rate for web page i λ_i ($=\lambda_i^o + \lambda_i^m$)	(0,15) updates/hour
Average cost of transferring a message over wired network C_m	0.01 second
Average cost of transferring a web page over wired network C_w	(0.5, 5) second
Average cost of resolving update conflicts C_{rm}	(30, 180) second
Average cost of transferring a MuAg C_{Ag}	0.1 second
Bandwidth of a wireless channel	9.6 K/S
Wired link factor for Intra-MSC BSs message transfer	2 seconds
Wired link factor for Inter-MSC BSs message transfer	3 seconds

5.1 Performance Analysis

Figs.8-13 shows the influence of handoff rate λ_h and failure rate λ_f on the recovery cost of MU in MAWA. We fix C_w=1.5 second, C_{rm}=100 seconds, C_m=0.01 second, λ_i=10 updates/hour, λ_i^m / λ_i varies in the set of {0.2, 0.6, 1}, λ_h varies in the set of {0.003125, 0.00625, 0.0125, 0.025, 0.05, 0.1, 0.2}, and λ_f varies in the set of {0.005, 0.02, 0.1}. Observed from the figures 8-10, both for single-page update propagation (in this scenario, MU prefetches and update one single Web page each time) and for multiple-page update propagation (in this scenario, MU prefetches and update ten Web pages and the λ_i of each Web page is selected randomly), the recovery cost increases with λ_f increased and this means that MU suffers more failures during a fixed period. For a fixed λ_f, the recovery cost also increases with λ_h increased and this means that when MU suffers a failure, the probability of MU suffering a handoff simultaneously increases too. Noticed that in the case of single-page update propagation, when λ_i^m / λ_i =1, all the updates to Web page are only done by MU, and all the updates propagated by MU will be accepted by Web server. So MU can prolong L_x

freely without worrying about performing forced updates. And the value of C_s is not large so the probability of suffering a handoff and a failure is distinctly increased with the increased value of λ_h and λ_f. However in the case of multiple-page update propagation, though the value of λ_f and λ_h vary in a large range, both $P_f \approx 1.0$ and $P_h \approx 1.0$ for a large value of C_s. So, though with different values of λ_f and λ_h, the recovery cost changes little.

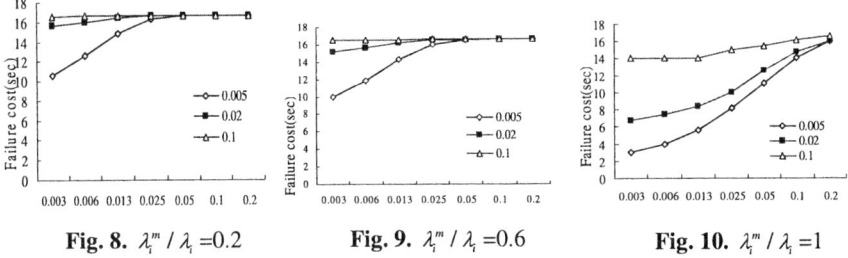

Fig. 8. $\lambda_i^m / \lambda_i = 0.2$ **Fig. 9.** $\lambda_i^m / \lambda_i = 0.6$ **Fig. 10.** $\lambda_i^m / \lambda_i = 1$

Fig. 8-10. Effect of λ_f and λ_h on the recovery cost of single-page update propagation

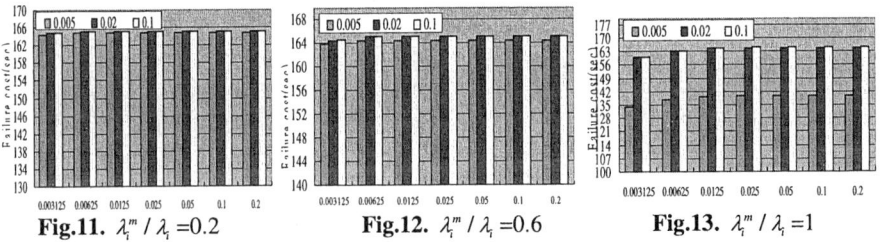

Fig.11. $\lambda_i^m / \lambda_i = 0.2$ **Fig.12.** $\lambda_i^m / \lambda_i = 0.6$ **Fig.13.** $\lambda_i^m / \lambda_i = 1$

Fig. 11-13. Effect of λ_f and λ_h on the recovery cost of multiple-page update propagation

Our fault-tolerant scheme also improves the efficiency of reintegration process. To demonstrate this, we compare it with the SPUPA (single page update propagation algorithm), MPUPA (multiple-page update propagation algorithm) proposed in [14]. In single Web page update propagation, We fix $\lambda_i^m / \lambda_i = 0.6$, $\lambda_i = 10$ updates/hour, $C_w = 1.5$ second, $C_m = 0.01$ second, and $C_{rm} = 30$ seconds in the range of [30, 180]. Seen from the figure 14, when $C_{rm} = 30$, the average cost of MAWA is about 26.7% lower than the cost of SPUPA. In multiple Web pages update propagation, MU are assumed to update ten Web pages in each reintegration phrase and the updated rate λ_i of each Web page i is selected in the range of [0,15] randomly. We fix $C_w = 1.5$ second, $C_{rm} = 30$ seconds, $C_m = 0.01$ second. See from figure 16, MAWA shows its obviously efficiency, for the cost of MPUPA is about 24.6% higher than the cost of MPUPA. The reason is that, in MAWA, MU first sends MuAg a message to inquire the state of the cached Web pages and then decides to take the following actions. In this way, much wireless communication cost on unnecessary page update propagation is avoided. And as a result, it saves the wireless communication cost greatly. Also in MAWA, all the Web page data is sent to MU by MuAg not by Web server, so the communication cost on wired network is saved in some way.

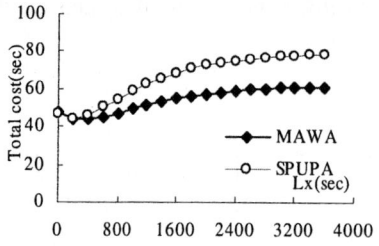

Fig. 14. MAWA compared with SPUPA in single-page update propagation

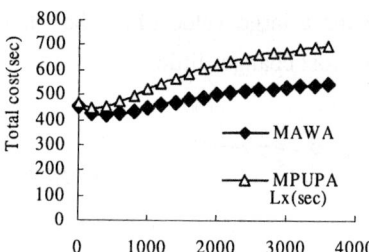

Fig. 15. MAWA compared with MPUPA in multiple-page update propagation

6 Conclusion

A fault-tolerant Web page access mechanism called MAWA is presented in this paper for MU to perform Web page update propagation in wireless mobile environments. This scheme uses mobile agents between Web server to facilitate seamless logging of MU's access activities, thus to help MU recover back when it suffers a handoff or even a failure. Compared with other schemes by performance tests, though having the additional fault-tolerant characteristic, our scheme can keep its efficiency.

References

1. Chander Dhawan.: Mobile Computing: A systems Integrator's Handbook.Mc Graw Hill, USA(1998)
2. J. Jing, A.S. Helal.: Client-Server Computing in Mobile Environments. *ACM Computin Survey*, vol. 31(2).(1999) 117-157
3. E. Pitoura and G. Samaras.: Data Management for Mobile Computing. Kluwer Ac-ademic Publishers(1998)
4. H. Chang.: Web Browsing in a Wireless Environment:Disconnected and Asynchronous Operation in ARTour Web Express. *Proc. Third ACM/IEEE Conf. Mobile Computing and Networking*, (1997) 260-269
5. J.J. Kistler, Satyanarayanan.:Disconnected Operation in the Coda File System *ACM Trans. Computer Systems*, vol. 10(1). (1992)3-25
6. Z. Jiang, L. Kleinrock.:Web Prefetching in a Mobile Environment, *IEEE Personal Comm.* vol. 5(5).(1998) 25-34
7. M.S. Mazer, C.L. Brooks.:Writing the Web while Disconnected, *IEEE Personal Comm.* vol. 5(5),(1998) 35-41.
8. A. Joshi, S. Weerawarana, E. Houstis.:On Disconnected Browsing of Distributed Information. *Proceeding of the 7th IEEE Workshop on Research Issues in Data Engineering RIDE*, (1997) 101-107

9. R. Floyd, R. Housel, C. Tait.:Mobile web access using eNetwork Web Express.*IEEE Personal Communications*, Vol. 5(5). (1998) 47-52
10. M. Liljeberg, T. Alanko, M. Kojo, H. Laamanen, K. Raatikainen. Optimizing World Wide Web for weakly connected mobile workstations: An indirect approach. *Proceeing of the 2nd International Workshop on Services in Distributed and Networked Environments*, (1995) 153-161
11. E.J. Whitehead Jr, M.Wiggins. WEBDAV: IEIF Standard for Collaborative Authoring on the Web. *IEEE Internet Computing*, Vol.2(5) (1998) 34-40
12. M.S. Mazer,C.L. Brooks. Writing the web while disconnected. *IEEE Personal Communications*, Vol. 5(5), (1998) 35-41
13. M.F. Kaashoek, T. Pinckney, J.A. Tauber. Dynamic documents: mobile wireless access to the WWW. *IEEE Workshop on Mobile Computing Systems and applications*,Santa Cruz, CA, Dec. (1994) 179-184
14. Ing-Ray Chen, Ngoc Anh Phan, I-Ling Yen. Algorithms for Supporting Disconnected Write Operations for wireless Web Access in Mobile Client-Server Environments. *IEEE Transactions on mobile computing*, VoL.1(1) (2002) 46-58
15. Debra VanderMeer, Anindya Datata, Kaushik Dutta. Mobile User Recovery in the context of Internet Transactions. *IEEE Transactions on Mobile Computing*, Vol.2(2). (2003) 132-146
16. Taesoon Park, Namyoon Woo, Heon Y. Yeon. An Efficient Recovery Scheme for Mobile Computing Environments. *Proceedings of the Eighth International Conference on Parallel and Distributed Systems*. (2001) 53-60
17. Cris Pedregal-Martin, Krithi Ramamritham. "Support for recovery in mobile System. *IEEE Transactions on Computers*. Vol.51(10).(2002) 1219-1224
18. Z.jiang and L.Kleinrock. Web Prefetching in a Mobile Environment. *IEEE Personal Communication*, Vol.5(5). (1998) 25-34
19. Mohammad A.H., Mitsuji M. MAMI: Mobile Agent based System for mobile Internet. *Proceedings of the IEEE/WIC/ACM International Conference on Web Intelligence(WI'04)*.(2004) 241-250
20. Sashidhar G., Vijay, K. Recovery in the Mobile Wireless Environment Using Mobile Agents. *IEEE Transactions on Mobile Computing*, Vol.3(2). (2004) 180-191
21. Paolo B., Antonio Corradi., Cesare S. Mobile Agent middleware for mobile computing. *IEEE Computer*. Vol.34(3). (2001) 73-81

Class-Based Latency Assurances for Web Servers

Yaya Wei[1], Chuang Lin[1], Xiaowen Chu[2], Zhiguang Shan[3], and Fengyuan Ren[1]

[1] Department of Computer Science and Technology, Tsinghua University, China, 100084
[2] Department of Computer Science, Hong Kong Baptist University, Hong Kong, China
[3] State Information Center, China
{yywei, clin}@csnet1.cs.tsinghua.edu.cn; chxw@comp.hkbu.edu.hk
Phone: 86-10-62772487

Abstract. This paper presents a fuzzy control approach for process assignment in order to guarantee absolute percentile delay in web servers. Comparing with mean delay used in previous works, percentile delay introduces strong nonlinearities to the plant; it is therefore difficult to achieve satisfactory performance by a linear controller, which is commonly used to guarantee mean delay performance. In this paper we propose a fuzzy controller. Its key feature is independence of the plant and robust to model uncertainty, which is very suitable to the strong nonlinear web servers. The experiments have demonstrated the efficiency of the proposed fuzzy controller in handing the variety of requests.

Keywords: Web server, Quality of service (QoS), Resource assignment, Fuzzy control.

1 Introduction

Recent research has highlighted the importance of QoS differentiation in terms of delay guarantee for web servers. Delay can be classified into two categories. One is the *mean delay*, which is defined as averaging all values of the monitoring variable within a sampling time period [1][2]. Another more important delay metric is the *percentile delay*. Percentile value, such as 90th percentile delay, is obtained by sorting all samples (called *Tosample*) ascendingly and getting the (0.9* *Tosample*)[th] delay value. Recent researches have focused on guaranteeing percentile delays. For example, the authors in [4] try to provide 95th percentile delay guarantee. Welsh and Culler [5] describe an adaptive approach to overload control in the context of the SEDA Web server. SEDA decomposes Internet services into multiple stages. The control variable of each stage is a targeted 90th percentile response time. In fact, comparing with the mean delay, percentile delay is a more reasonable metric considered by web server operators [3]. The reason is that mean delay cannot guarantee the response time of each user; instead it only guarantees the average

---

[1] This work is supported by the National Natural Science Foundation of China (No.60429202, 90412012 and 60373013), NSFC and RGC (No. 60218003), the National Grand Fundamental Research 973 Program of China (No.2003CB314804), and a Faculty Research Grant of Hong Kong Baptist University (FRG/03-04/II-22).

The model of the fuzzy system, comprising the control rules and the term sets of the variables with their related fuzzy sets, was obtained through a tuning process that starts from a set of initial insight consideration. In Fig. 3, both scaled inputs (e and Δe) of the controller are defined on the common interval [-1, 1]. The numerical inputs (e_N and Δe_N) are mapped onto [-1, 1] by the input scaling factors G_e and $G_{\Delta e}$. Then the relationships between the numerical inputs and the scaling factors are as follows:

$$e = e_N \cdot G_e \qquad \Delta e = \Delta e_N \cdot G_{\Delta e}$$

G_e and $G_{\Delta e}$ are set $1/DD_0$ and $1/1.25DD_0$ separately.

From Fig. 3, it can be seen that five fuzzy term sets are chosen for both E and ΔE, which are negative big (*NB*), negative small (*NS*), zero (*Z*), positive small (*PS*) and positive big (*PB*). For the output *U*, we define four more fuzzy term sets, i.e., negative huge (*NH*), negative medium (*NM*), positive medium (*PM*), and positive huge (*PH*), since more term sets can increase flexibility and enable us to make the controller react accurately such as neither too small nor large.

The control rules are described in Table 1. To design these rules, we add the expert heuristics about our specific system.

- To make the controller produce a lower overshoot and reduce the settling time, the controller output is set at a small value when the error is big (it may be PB or NB), but E and ΔE are of the opposite signs. For example, if E is PB and ΔE is NB, then U is PS.
- To make the controller respond rapidly, the controller is set at a large value when the error is big, and E and ΔE are of the same sign (i.e. the process is now not only far away from the set point but also it is moving farther away from it), then the output should be made large to prevent from further worsening the situation. For example, if E is PB and ΔE is PB, then U is PH.

4 Experimentation

We have modified the source code of Apache 1.3.9 web server on a Linux platform to implement our proposed adaptive architecture and fuzzy control algorithm. All experiments were conducted on a test-bed of PCs. All PCs had a 600MHz Celeron processor and 256 MB RAM and ran Linux-2.4.18. The servers and the clients were connected by a 10Mbps Ethernet hub. The clients generate web traffic using the Surge workload generator [6].

4.1 Latency Guarantee at Overload

The experimental setup is as follows.

- Server: The total number of processes was configured to 128. And the sampling period was set to 30 sec.
- Client: Two client machines generate requests for class 0 and one client machine for class 1. In the first 1000 sec, one client machine simulated 84 class 0 clients and one client machine simulated 180 class 1 clients. At 1300 sec, a third client machine was started simulating 26 class 0 clients.

The experiment results is shown in Fig. 4~Fig. 5. The 90th percentile delay in the open-loop (without any controller) is shown in Fig. 4. It can be seen that the 90th percentile delay of class 0 can not be guaranteed all the time. The proposed fuzzy controller was activated on 240 sec in order to avoid warming up phrase. Fig. 5 shows

Table 1. Fuzzy control rules

E \ ΔE	NB	NS	Z	PS	PB
NB	U:NH	U:NB	U:NB	U:NM	U:NS
NS	U:NB	U:NS	U:NS	U:Z	U:PS
Z	U:NS	U:Z	U:Z	U:Z	U:PS
PS	U:Z	U:Z	U:PS	U:PS	U:PM
PB	U:PS	U:PM	U:PM	U:PM	U:PH

the behavior of class 0 with the proposed fuzzy controller. It can be seen that the proposed fuzzy controller reallocated processes to class 0 to reconverge the 90th percentile delay of class 0 to the set point successfully.

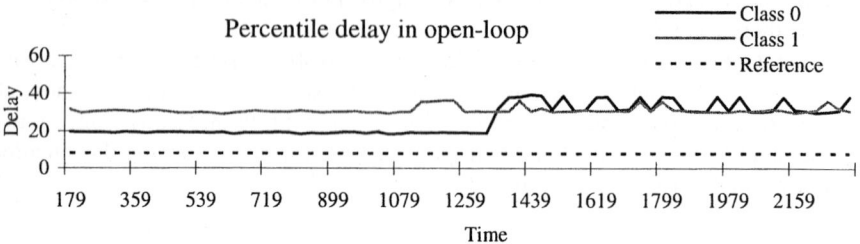

Fig. 4. Percentile Delays for Class 0 and Class 1 in Open-Loop

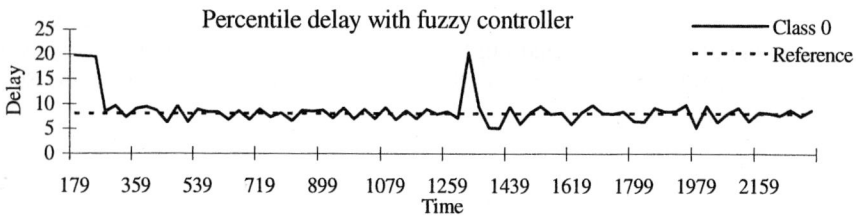

Fig. 5. Percentile Delay for Class 0 with the Fuzzy Controller

4.2 Adaptation to Various Plants

The design of fuzzy controllers does not require the exact models of the plants. It provides a general feedback control method and is not specific to the details of the system. To investigate the robustness of the proposed controller, we placed our web server on another machine with a different configuration from the previous one.

The configuration of the original machine is called server 0. Another machine (called server 1), whose model is assumed the same as sever 0, had a 1.66GHz AMD Athlon processor and 480MB RAM, which has a much higher speed than server 0. Both machines run the same modified Apache software. The workload is set up as follows. The class 0 has 90 users and the class 1 has 180 users. To avoid the starting phase, the controller was activated at 240 sec. It can be seen from Fig. 6 that the delays in server 0 and server 1 are much similar. It shows that the fuzzy controller designed for server 0 can also work well for server 1.

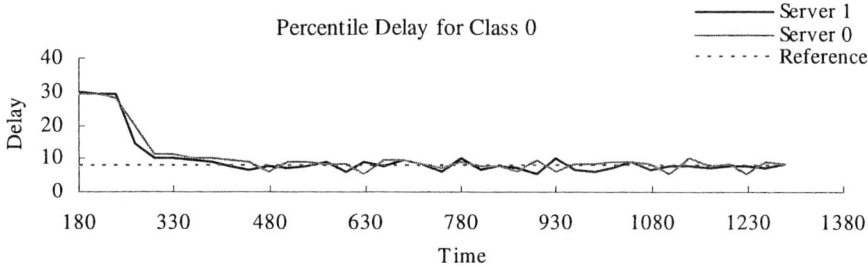

Fig. 6. Percentile Delays with the Fuzzy Controller in Server 0 and Server 1

5 Conclusions

In this paper, we have proposed a novel fuzzy controller to guarantee percentile delay for web server. The key characteristic of the proposed controller are (1) it is independent of the model of the plant; (2) it provides a non-linear control action; and (3) it adapts to various web servers with different speeds. The experimental results show that the proposed controller is efficient to guarantee the percentile delay of the premium class.

References

1. S. C.M. Lee, J. C.S. Lui, and D.K.Y. Yau. A proportional-delay DiffServ-Enabled web server: Admission control and dynamic adaption. IEEE Trans. On Parallel and distributed system, Vol. 15, No. 5, May 2004.
2. M. Crovella, R.Frangioso, and M. Harchol-Balter. Connection scheduling in web servers. Proc. USENIX Symp. Internet Technologies and System, Oct. 1999.
3. Abhinav Kamra, Vishal Misra, Erich Nahum Yaksha. A Controller for Managing the Performance of 3-Tiered Websites. International Workshop on Quality of Service (IWQOS), 2004.
4. Vikram Kanodia, and Edward W. Knightly. Ensuring Latency Targets In Multi-Class Web Servers. IEEE Trans. on Parallel and Distributed Systems, Vol. 14, No. 1, January 2003.

5. M. Welsh and D. Culler. Adaptive overload control for busy Internet servers. Proc. USENIX Symp. Internet Technologies and Systems, San Francisco, CA, March 2003.
6. P. Barford and M. E. Crovella. Generating representative web workloads for network and server performance evaluation. ACM SIGMETRICS'98, Madison WI, 1998.
7. K. M. Passino and S. Yurkovich. Fuzzy control. Menlo Park, CA: Addison Welsey Longman, 1998.

Workflow Pattern Analysis in Web Services Orchestration: The BPEL4WS Example

Francesco Moscato[1], Nicola Mazzocca[2], Valeria Vittorini[2], Giusy Di Lorenzo[2], Paola Mosca[3], and Massimo Magaldi[3]

[1] Seconda Università di Napoli, Dipartimento di Ingegneria dell'Informazione,
Real Casa dell'Annunziata, Via Roma 29, 81031 Aversa (CE), Italy
francesco.moscato@unina2.it

[2] Università degli studi di Napoli Federico II,
Dipartimento di Informatica e Sistemistica,
via Claudio 21, 80121 Napoli, Italy
{nicola.mazzocca, valeria.vittorini}@unina.it

[3] Atos Origin, Pozzuoli (Na) - Italy
{paola.mosca, massimo.magaldi}@atosorigin.com

Abstract. Web Services are becoming the prominent paradigm for distributed computing and electronic business. Web services composition is an emerging paradigm for enabling application integration within and across organization boundaries and for building complex Value Added Services (VAS). Different languages are emerging to describe web services composition, but no effort has been dedicated to systematically evaluating the capabilities and limitations of these languages and to formally define their constructs semantics, in order to allow a well defined execution of pattern describing web services interactions. The work of this paper intends to analyze in-depth the Business Process Execution Language for Web Services (BPEL4WS), presenting a new methodology used to state in a formal way which workflow patterns can be executed by using the BPEL4WS constructs.

Keywords: Web Services, Orchestration, Choreography, Workflow language, Workflow patterns, pattern analysis, semantics, BPEL4WS.

1 Introduction

The Web has become the means for organizations to deliver goods and services and for customers to search and retrieve services that match their needs. Web services are self-contained, internet-enabled applications capable not only of performing business activities on their own, but also possessing the ability to engage other web services in order to complete higher-order business transactions building *composite services* combining existing elementary services. Usually the services used in the context of a composite service are called its component services. Actually two main approaches are used to build *composite services*. They are called *orchestration* and *choreography* [7]. Business interactions require long-tuning interactions that are driven by an explicit process model. This raises the need for web services composition languages such as BPEL4WS [1], WSCI [2], BPML [3] or XPDL [4]. Except for the Web Services Choreography Interface (WSCI) Language, all other languages are based on workflow [6] concepts. The main problem in analyzing patterns is that orchestration languages lack of formal definitions of their constructs semantics. In the following a new approach to analyze workflow patterns will be described. This approach forces the formal definition of constructs semantics of the language to analyze, avoiding misleading in the composite web services execution.

2 Related Works

The workflow pattern approach to analyze languages expressiveness was introduced by V.M.P. van der Aalst in [5]. Patterns are a three-part rule, which expresses a relation between a certain context, a problem, and a solution. In the case of workflow patterns, they represent the solution to the problem of composing web services in order to build a VAS through the use of workflow models and languages. The main problem that rises with the pattern analysis of workflow languages, is that (if it is possible) a model of the workflow language or system has to be available in order to remove all ambiguities during the analysis processes. This is not possible for all standards. For these reasons, some efforts were made to define *ex-novo* new workflow languages based on formal definitions, like Yet Another Workflow Language (YAWL) [8], which constructs definitions are based on Petri-Nets formalism. Anyway, at the best of our knowledge, the most used language used to describe composite web services is BPEL4WS and for this reason we will focus our attention on this language. The Pattern analysis for BPEL4WS language introduced in [5], has one problem: the analyst have to know how to realize the pattern in the given language. It is an analysis performed *by example* and cannot be automated in any way. In order to solve this problem in the following sections we will propose an automatic, reproducible methodology that allows to analyze patterns, allowing the identification of all the way they can be implemented in a given workflow language.

3 Pattern Analysis Methodology

Web services composition and workflow management are related in the sense that both are concerned with executable processes. Therefore, much of the functionality in workflow management systems [5], is also relevant for web services composition languages like BPEL4WS, XLANG and WSFL. Thus workflow patterns (WPs) defined in [5] have been investigated to study their expressiveness. The WPs have been compiled from an analysis of workflow languages. They capture typical control flow dependencies encountered in workflow modeling. For this reason, this kind of analysis excludes data manipulation and resources allocation perspectives.

The proposed pattern analysis methodology is founded on the concept of constructs *operational semantics*[9] of a given language.

The definition of operational semantics of the language to analyze in order to state if a given workflow pattern can be described by the given language or not. As introduced in the previous sections, the main problem of workflow languages for orchestration is that they lack a formal definition of their semantics. So the proposed methodology has two advantages:

- It forces a formal description of semantics of a given language, defining once for all which kind of steps are needed to execute language constructs and not leaving its definition to workflow engine vendors.
- It allows a fully automatic way to investigate if a given pattern (or even a given workflow process) can be executed by a given language.

Our approach is different from the approach described in [5] where the analyst has to know in which way a pattern can be implemented by using language constructs. It proceeds by example and without a fixed methodology.

The steps needed to perform the pattern analysis are:

1. A definition of the operational semantics of a given language must be provided.
2. The semantics rules must be translated into a *prolog* rule based system.
3. A description of the patterns to analyze must be provided as rules to prove in the *prolog* system.
4. The prolog system evaluates if a given pattern can be realized by composing the language constructs. In the case of affirmative results, the inferential engine also provides all the possible construct combinations to realize the pattern.

In the following we apply our methodology to the BPEL4WS language and to the patterns previously described. For brevity sake only an example of the defined rules, describing one of the behavior of the *sequence* construct, is reported:

$$\frac{\sigma_S = exec, L_A \stackrel{first}{\to} a_T \cdot L_{A_C}, \sigma_{S_{a_T}} \stackrel{\mu}{\to} \sigma_{S_{a_T}}^{exec}}{\langle L_A, \sigma_S \rangle \stackrel{sequence}{\to} \langle L_{A_C}, \sigma'_S \rangle}$$

This rule applies when the sequence is already started. This means that some activities composing the sequence were previously terminated. In such case, the *first* activity to process is the activity a_T and the remaining activities in the list (L_{A_C}) to be processed are pruned of this activity that is executed by applying the μ transition. The new state of the sequence is the same of the previous one, except for the state of the the the activity a_T that evolved from the *noexec* to the *exec* state.

As result, for brevity's, some implementations of the *sequence* pattern derived from the operational rules are shown in the following. Many solutions to implement this pattern with BPEL4WS were discovered. In Fig.1 three different implementations of this pattern are shown. In Fig.1 a) simply the *sequence*

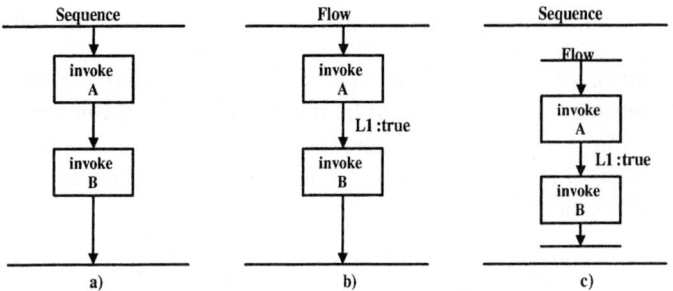

Fig. 1. Sequence Pattern Implementations

activity is used. Inside the sequence, to execute the **A** and **B** web services, two **invoke** activities are used.

In Fig.1 b) and c) alternative implementations for the pattern are shown. The first one is realized by using the *flow* activity and **links** introducing a precedence relation between the **A** and **B** invoke activities. The *invoke A* activity is the SourceLink for the link **L1** (with the transition condition true) while the *invoke B* is the TargetLink for the same link. In this way the activity *invoke B* can be executed only after the termination of the *invoke A* activity. The implementation of Fig.1 c) is simply a sequence activity with only one component activity: the flow depicted in Fig.1 b). This is not an efficient implementation but it is reported to show how our approach is able to find all implementations in a given language of a workflow pattern.

4 Case Study

In this section, we show how the methodology is be able to prove if a real business process can be implemented in the BPEL4WS language. Since *prolog* proof also provides different implementation suggestions, we will use them to implement the whole composite web service. The service to realize is a VAS having to provide consulting suggestions about insurance agencies and their policies. It has to communicate with web services of insurance agencies in order to request contracts or to stipulate policies and it has also to assure a way to permit secure transfers by using trusted Credit Card payment web services. Since all the patterns needed for the VAS are supported by the BPEL4W language, we use some of the solutions to the needed pattern implementation presented in the previous section to implement the VAS service. In Fig.2 the first part of the implemented VAS is shown. A flow activity is used to invoke all agencies web services (**invoke 1..4**). In order to collect results, an assign activity is used. This activity is the merge point and a link is defined (L1) to assure precedences relations. Then the web services used to manage contracts and to build the policy list to submit to the user is invoked (**invoke F**). Results are then submitted to the user (**Reply 1**). Also this activity must be executed after the previous invoke and another link is used (L2).

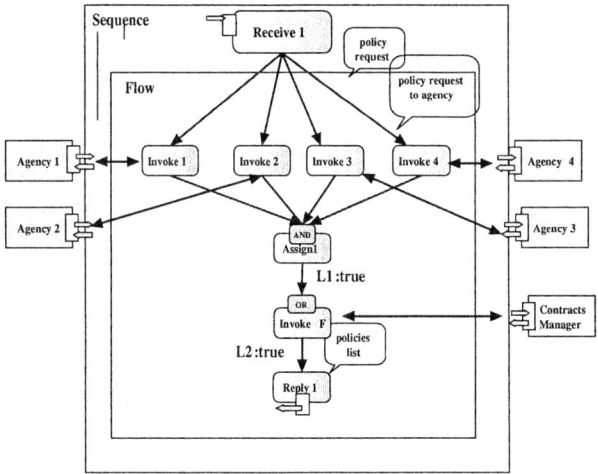

Fig. 2. CS Parallel Split and Synchronized Merge implementation

This implementation was derived directly from the application of the methodology described in the previous sections once the needed patterns to describe the VAS were identified.

5 Conclusion and Future works

In this paper we have shown an automatic methodology to analyze workflow languages for web service orchestration. The methodology is based on the definition of operational semantics of the language to analyze. This definition is useful because usually emerging standards of languages for orchestration lack of a formal definition of the semantics of their constructs, leaving to orchestration engine the task of defining them. This threaten the portability of the language and must be avoided defining once for all in a formal way the languages semantics.

The methodology is based on traducing the operational semantics rules into a prolog program and the patterns are analyzed by defining appropriate rules to prove in the prolog system. The proof is totally automatic and provides also information on how the patterns can be implemented.

Future works aim to analyze also communication patterns and to extend the methodology also to other orchestration languages.

References

1. *Business Process Execution Language for Web Services (BPEL)*. Available on-line at: http://www-128.ibm.com/developerworks/library/specification/ws-bpel, May 2003
2. *Web Services Choreography Interface (WSCI)*. Available on-line at: http://www.w3.org/TR/wsci, 2002

3. *Business Process Management Language (BPML)*. Available on-line at: http://www.bpmi.org/bpml-spec.htm
4. WfMC. Workflow process definition interface - XML Process Definition Language. Available on-line at: http://www.wfmc.org/standards/docs/XPDL_version2_draft_2005-05-09.zip, May 2005.
5. P. Wohed, W.M.P. van der Aalst, M. Dumas, and A.H.M. ter Hofstede. Analysis of Web Services Composition Languages: The Case of BPEL4WS. In Proc. of 22nd International Conference on Conceptual Modeling (ER 2003), vol. 2813 of Lecture Notes in Computer Science, pages 200-215. Springer-Verlag, Berlin, 2003.
6. R. Allen. *Workflow: An Introduction*. Extracted from the Workflow Handbook 2001, Workflow Management Coalition, Available at: http://www.wfmc.org/standards/docs.htm
7. C. Peltz. *Web Services Orchestration and Choreography*. IEEE Computer, Vol.36, n.10, 2003, pp. 46-52.
8. W.M.P. van der Aalst and A.H.M. ter Hofstede. *Workflow Patterns: On the Expressive Power of (Petri-net-based) Workflow Languages*. In K. Jensen, editor, Proceedings of the Fourth Workshop on the Practical Use of Coloured Petri Nets and CPN Tools (CPN 2002), volume 560 of DAIMI, pages 120, Aarhus, Denmark, August 2002. University of Aarhus.
9. G. Winskel. *The Formal Semantics of Programming Languages: An Introduction (Foundations of Computing)*. The MIT Press, February 5, 1993

An Effective and Dynamically Extensible DRM Web Platform

Franco Frattolillo and Salvatore D'Onofrio

Research Centre on Software Technology,
Department of Engineering, University of Sannio, Italy
frattolillo@unisannio.it

Abstract. The digital rights management platforms (DRMps) are the web platforms employed by web content providers (CPs) to ensure the copyright protection of the digital contents they distribute on the Internet. However, the current DRMps that employ watermarking technologies to protect multimedia digital contents implement "centralized" service models in which the protection process is directly implemented by the managers of the DRMps, i.e. CPs, which are also the copyright owners. This has given rise to documented problems that nowadays affect the DRMps and make their use difficult in a web context. This paper presents a DRMp characterized by a distributed architecture that enables CPs to exploit copyright protection services supplied by web service providers (SPs) according to protection processes controlled by trusted third parties (TTPs). The proposed approach exploits web-oriented programming technologies to enable CPs and SPs to dynamically cooperate in the protection process without imposing a tight coupling among them.

1 Introduction and Motivations

DRMps have gained popularity as the software platforms developed to implement the copyright protection of digital contents distributed on the Internet [1]. Among them, the platforms devoted to protect multimedia digital contents usually exploit watermarking or fingerprinting procedures [2] to enable the copyright owner to insert an invisible watermark identifying the buyer within any copy of content that is distributed. The main aim is to make it possible to establish if a user is illegally in possession of a content distributed by a CP as well as who has initially bought and then illegally shared it on the Internet [1].

However, the current DRMps are essentially based on "centralized" service models in which the overall protection process is directly implemented and under the control of the managers of the DRMps, i.e. CPs, which are also the copyright owners. As a consequence, a content protection directly applied by CPs might not take into account the buyers' rights, since the watermark is autonomously inserted by the copyright owner, i.e. the seller, without any control. This has given rise to two main documented problems which prevent current DRMps from implementing correct protection processes: the "customer's right problem" [3,4] and the "unbinding problem" [5].

A possible solution to the above problems consists in designing DRMps that enable CPs to resort to independent TTPs able to implement watermarking and dispute resolution protocols [4,5,6] by which CPs can both obtain an adequate protection of their contents distributed on the Internet and determine the identity of guilty buyers with undeniable evidence. However, this means that DRMps have to be characterized by new "distributed" software architectures that enable web entities playing different roles to dynamically cooperate in the content protection process without imposing a tight coupling among them [7,8]. In particular, in order to enhance the flexibility and modularity of the new distributed DRMps, the involved TTPs should behave solely as the guarantors of the protection process, whereas such process should be implemented by specific SPs whose services should be dynamically integrable into the DRMps. Thus, it should be possible to differentiate the generation of the fingerprinting codes and the control activity of the protection process from the watermark insertion. As a consequence, a new watermarking procedure could be dynamically chosen and applied "on the fly" without forcing the involved TTPs to directly implement it.

The idea of adopting distributed approaches in designing DRMps involving distinct web entities, such as CPs, TTPs and SPs, is nowadays considered a clever way to address the problems reported above [4,5,6]. In fact, CPs often have neither the technical competence nor the economical advantage to directly apply complex or not certificated watermarking procedures to their distributed contents. They appear to be more involved in improving their consolidated web consumer- or business-focused applications, rather than implementing new services based on advanced technologies that are not part of their original core business. On the other hand, SPs are web entities that have knowledge and expertise in the use of web programming technologies, and their core business is just to supply specialized and certificated software services to CPs. On the contrary, TTPs are basically watermark certification authorities (WCAs). Therefore, they usually take charge of implementing watermarking and dispute resolution protocols, whereas they should not be specialized in implementing watermarking procedures. In fact, this model has already proven highly successful in the Internet, where SPs enable the building of web applications for CPs with good ideas but little time for technology. Therefore, SPs can be considered well suited to deploy certificated copyright protection services on behalf of CPs, whereas TTPs can limit their action to the sole role of guarantors of the protection process.

This paper presents the architecture of a distributed DRMp that enables SPs to supply copyright protection services based on watermarking technologies on behalf of CPs in a secure and flexible context. Thus, CPs that exploit the proposed platform can take advantage of a copyright protection system provided by SPs and governed by WCAs without having to directly implement it. On the other hand, SPs can follow the proposed approach to implement certificated protection procedures, which can be easily and dynamically exploited by CPs.

The outline of the paper is as follows. Section 2 sketches the architecture of the proposed DRMp. Section 3 describes the service framework by which SPs can supply their services to the platform. In Section 4 a brief conclusion is available.

2 The Architecture of the Proposed DRM Platform

The proposed platform, whose architecture is depicted in Figure 1, consists of three main parts. The first part groups the web servers of a CP and represents the "front-end" tier of the platform seen by user clients. The second part is represented by the WCA and is the "middle" tier of the platform. The third part represents the "back-end" tier of the platform and is basically composed of the web services [9] implemented by SPs and made available to the platform by means of a purposely designed "service framework".

The platform has been developed according both to the WS-Security (Web Services Security) specifications, which define a set of SOAP header extensions for end-to-end SOAP messaging security, and to a "federated model" for the identity management of the web services' operators [10].

The front-end tier of the platform is represented by the CP's web applications, whose behavioral logic is assumed to be adapted to the distributed architecture of the platform. This means that a CP wanting to exploit the proposed DRMp has to modify its application in order to integrate it in the new interaction scheme involving the WCA.

WCA acts as a TTP and so it provides the DRMp with the security services needed to execute the watermarking and dispute resolution protocols. As a consequence, its software architecture has a static internal characterization with respect to such assumptions. On the contrary, the back-end tier of the platform has been purposely designed as a service framework in order to make the architecture modularly extensible by web services dynamically loaded and supplied by distinct and external SPs. In particular, the service framework does not directly expose the web services that it groups, but it hides them behind a *unified access point*, which acts as a unique interface toward the external web entities for

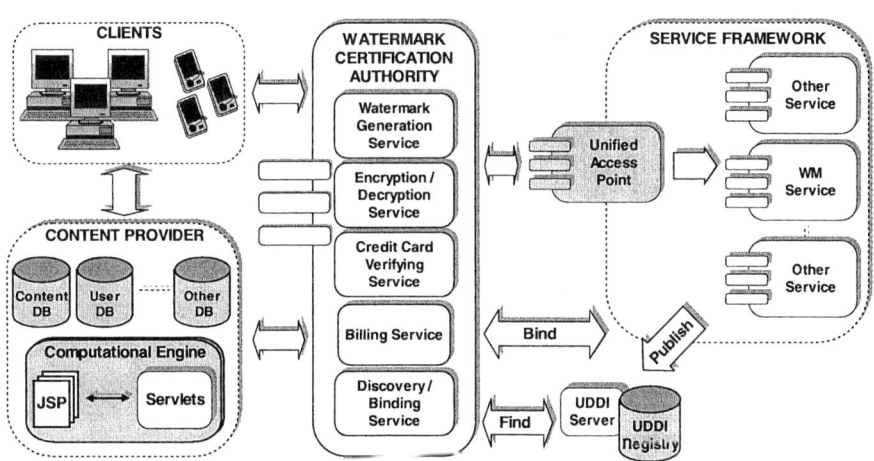

Fig. 1. The architecture of the proposed DRMp

all the web services internally loaded. Such an interface, designed itself as a web service, takes charge of receiving the service requests specified according to what published in the UDDI registries and of dispatching them to the web services internally loaded according to the strategies implemented by the framework.

The choice of implementing the service framework as a "wrapper" for a set of web services hidden from external users and exposed through a *unified access point* enables the services that have to be supplied by the framework to be abstractly expressed. In fact, the wrapper allows the service framework to publish in the UDDI registries simplified and standard versions for the interfaces of the web services that are then dynamically loaded in the framework, and this makes such interfaces independent of the possible implementations developed by SPs.

3 The Service Framework

This section focuses on the main implementation details concerning with the back-end tier of the proposed DRMp, which has been designed as a service framework (see Figure 2). In particular, to support flexibility and to make the service integration easy and dynamical, the service framework has been developed in Java. It consists of two main sections: the former comprises the internal services of the framework, whereas the latter comprises the web services supplied by SPs and dynamically loaded in the framework.

The internal services have been implemented in several software components whose interfaces have been defined according to a "component framework"

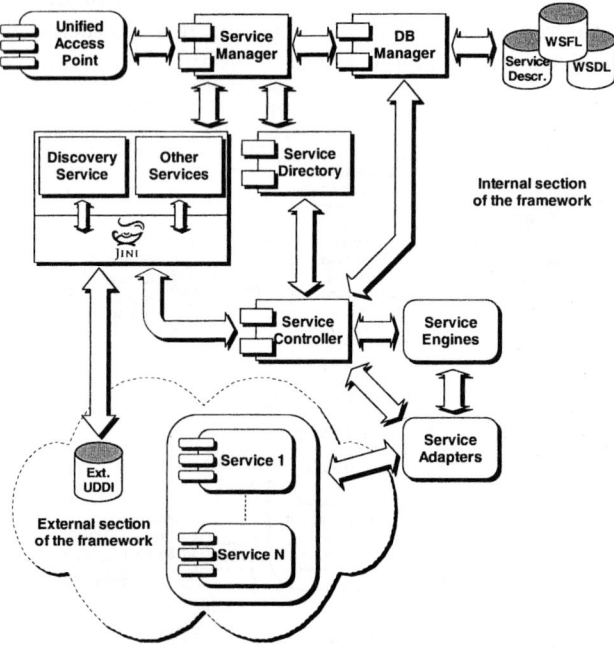

Fig. 2. The architecture of the "service framework"

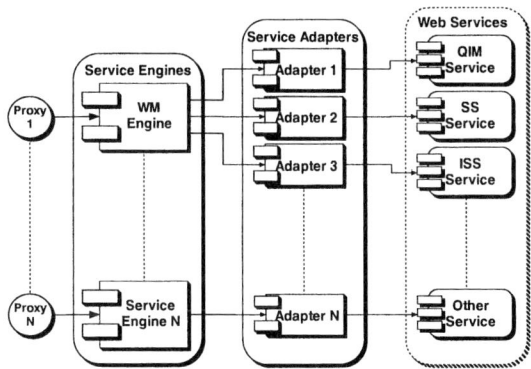

Fig. 3. The invocation scheme implemented within the "service framework"

approach [11]. In particular, this approach can simplify the development of complex services by defining a set of rules and contracts governing the interaction of a targeted set of components in a constrained domain [12]. Therefore, each component of the framework has been designed so as to implement the set of behavior rules that define the framework execution environment. Thus, the internal structure of the service framework is made up by a set of cooperating components that define the skeleton of the framework, i.e. the set of minimal services needed to enable SPs to dynamically integrate their web services in the DRMp. To this end, to facilitate the integration among the framework components and the external web services, four specific design patterns [13] have been used: "inversion of control", "separation of concerns", "proxy" and "adapter".

The first pattern suggests that the execution of components has to be controlled by the framework. Thus, the flow of control is determined by the framework that coordinates all the events produced by transactions. The second pattern allows a problem to be analyzed from several points of view, each of which is addressed independently of the others. Therefore, the internal services have been implemented as collections of well-defined, reusable, easily understandable and independent components. The third pattern is exploited to generate surrogate objects for the external web services grouped in the framework in order to control access to them. Finally, the fourth pattern is used to convert the specific interfaces of the external web services into simplified interfaces that can be invoked by the internal components of the framework. In fact, the proxy and adapter patterns make it possible to dynamically load the external web services into the framework without imposing constraints on their public interfaces.

Figure 4 shows the scheme of the interactions, indicated by numbers in the following, taking place when a service request is issued to the framework. In particular, the request is received by the *unified access point*, which is a web service that acts as a service dispatcher toward all the external web services dynamically loaded into the framework and made available to CPs. Therefore, when the *unified access point* receives a service request from outside (1), it contacts the *service manager* (2), which has the main task of managing the request.

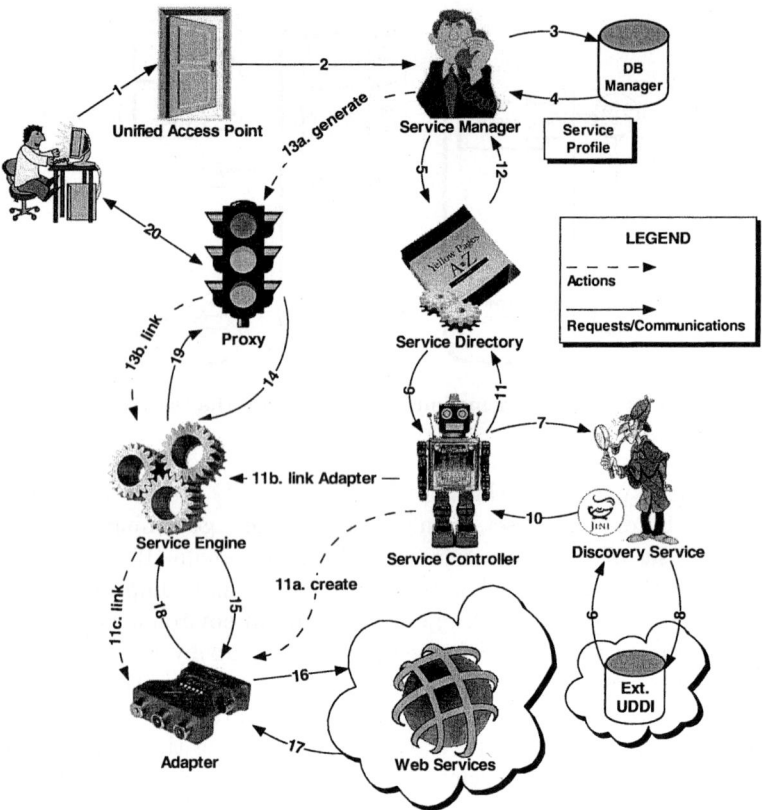

Fig. 4. The interaction scheme implemented within the "service framework"

The *service manager* accesses the *database manager* (3), which makes it possible to determine the characteristics, i.e. the "profile" (4), that a web service provided by the framework should have in order to match the requirements derived from the external request issued. Then, the *service manager* searches the *service directory* (5) for a web service having the profile thus determined. If the service results in being already loaded and available within the framework, the *service directory* returns the information about the service (12) received from the *service controller* (6,11) to the *service manager*, which thus can generate a specific *proxy* object (13a,13b) that has the task of routing the service request coming from outside (20) to the corresponding *service engine* according to the invocation scheme shown in Figure 3 (14,19) . Such an engine is an intermediate software layer that makes it possible to translate the service invocations performed on the *proxy* objects to invocations on *adapter* objects (15,18). In fact, a *service engine* receives service requests coming from the *proxy* objects and specified according to the simplified web interfaces published by UDDI registries, and takes charge of invoking the corresponding and selected web services according to their particular interfaces (15,16,17,18). To this end, in order to enable a

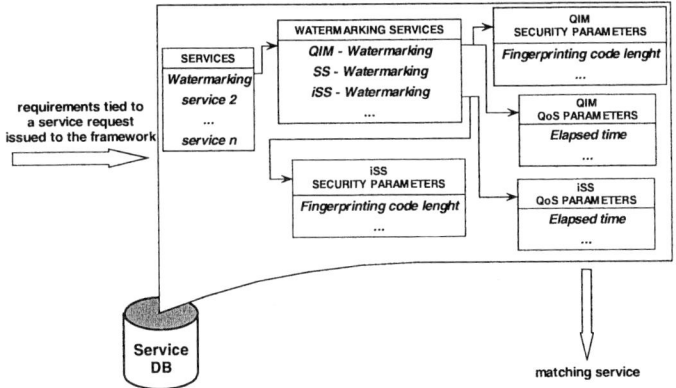

Fig. 5. The service description database

dynamical loading of external web services into the framework without imposing constraints on the web services' interfaces, the *service engines* exploit the *adapter* objects, which are dynamically generated and loaded whenever an external web service is discovered and loaded into the framework (11a,11b,11c) [14,15].

If the required service is not available within the framework, the *service directory* has to start the *discovery service* supplied by specific framework components implemented by exploiting the JINI framework [16]. The *discovery service* is activated via the *service controller* (6,7). It can search UDDI registries to discover the required service (8) and can return the result to the *service directory* (9,10,11), which communicates it to the *service manager* (12). When this phase ends, the new external web service results in being loaded into the framework, and its *adapter* object is generated and linked to the corresponding *service engine* by the *service controller* (11a,11b,11c). Then, the *service manager* can generate the *proxy* object (13a,13b) that will receive the service requests issued to the framework and coming from outside (20).

Finally, to enable the service framework to determine the "profile" that a web service should have in order to match an external service request, the *database manager* manages a repository (see Figure 5) which specifies some main parameters able to characterize the web services' implementations. Such parameters can specify, for example, the protection level achievable by a given service implementation as well as the implementation behavior with respect to web transactions. Therefore, the available repository makes it possible to determine a particular service profile depending on both protection and QoS (quality of service) requirements specified in the service requests originally issued to the framework.

4 Conclusions

The paper has presented the architecture of a distributed DRM platform implemented by using web services and object-oriented programming technologies, and by which SPs can dynamically supply copyright protection services on behalf of

CPs according to a protection process governed by WCAs acting as TTPs. CPs that exploit the proposed platform can take advantage of a copyright protection system based on TTPs without having to directly implement it. On the other hand, SPs can follow the proposed approach to implement certificated copyright protection procedures which can be easily exploited by CPs independently of their particular interfaces. Furthermore, the implementation approach exploited for the service framework and based on design patterns makes the DRMp modular and easily extensible. Finally, the capability of managing some parameters that synthetically specify the behavior of each service implementation can improve the flexibility as well as the overall performance of the platform.

References

1. Barni, M., Bartolini, F.: Data hiding for fighting piracy. IEEE Signal Processing Magazine **21** (2004) 28–39
2. Cox, I., Bloom, J., Miller, M.: Digital Watermarking: Principles & Practice. Morgan Kaufman (2001)
3. Memon, N., Wong, P.W.: A buyer-seller watermarking protocol. IEEE Trans. on Image Processing **10** (2001) 643–649
4. Qiao, L., Nahrstedt, K.: Watermarking schemes and protocols for protecting rightful ownership and customers rights. Journal on Vis. Commun. Image Representation **9** (1998) 194–210
5. Lei, C.L., Yu, P.L., et al.: An efficient and anonymous buyer-seller watermarking protocol. IEEE Trans. on Image Processing **13** (2004) 1618–1626
6. Katzenbeisser, S.: On the design of copyright protection protocols for multimedia distribution using symmetric and public-keywatermarking. In: Procs of the 12th Int'l Workshop on Database and Expert Systems Applications. (2001) 815–819
7. Frattolillo, F., D'Onofrio, S.: Applying web oriented technologies to implement an adaptive spread spectrum watermarking procedure and a flexible DRM platform. In Montague, P., et al., eds.: Procs of the 3rd Australasian Information Security Workshop. Volume 44 of Conferences in Research and Practice in Information Technology., Newcastle, Australia (2005) 159–167
8. Frattolillo, F., D'Onofrio, S.: Implementing a simple but flexible DRM web platform. In: Procs of the 3rd Int'l Conference on Computing, Communications and Control Technologies, Austin, Texas, USA (2005)
9. Brunner, R., , et al.: Java Web Services Unleashed. SAMS Publishing (2001)
10. Shin, S.: Secure Web Services, http://www.javaworld.com. (2005)
11. Szyperski, C.: Component Software. Beyond Object-Oriented Programming. Addison Wesley (1997)
12. Johnson, R., Foote, B.: Designing reusable classes. Journal of Object-Oriented Programming **1** (1988) 22–35
13. Gamma, E., Helm, R., et al.: Design Patterns. Addison Wesley (1995)
14. Gannod, G.C., Zhu, H., Mudiam, S.V.: On-the-fly wrapping of web services to support dynamic integration. In: Procs of the 10th Working Conference on Reverse Engineering, Victoria, British Columbia, Canada (2003)
15. Lampe, M., Althammer, E., Pree, W.: Generic adaptation of Jini services. In: Procs of the Ubiquitous Computing Workshop, Philadelphia, PA, USA (2000)
16. Richards, W.K.: Core JINI. Prentice-Hall (1999)

JXTPIA: A JXTA-Based P2P Network Interface and Architecture for Grid Computing

Yoshihiro Saitoh, Kenichi Sumitomo, Takato Izaiku, Takamasa Oono, Kazuhiko Yagyu, Hui Wang, and Minyi Guo

Department of Computer Software, University of Aizu,
Aizu-Wakamatsu, Fukushima 965-8580, Japan
hwang@u-aizu.ac.jp

Abstract. Peer-to-Peer (or P2P) systems are distributed systems that provide high quality information sharing services to many users distributed over the interconnection network. To extend a P2P network system with the mechanisms of distributed computing, we developed a flexible JXTA-based P2P network interface and architecture, called JXTPIA. We have implemented JXTPIA network, Homework Distribution module, Trigger Distribution module, and Data Sharing module. The JXTPIA system provides the basic functionalities for distributed/Grid computing, such as resources allocation, task scheduling, task assignment, network structure constructing and maintenance, and data sharing, etc. Experimental results show that it executes the large applications efficiently. Further performance evaluation and experimental data analysis will be conducted.

Keywords: JXTPIA system, P2P computing, Distributed/Grid computing, JXTA, Network structures, Task distribution.

1 Introduction

A Peer-to-Peer system is a collection of host computers, called peers, connected within a physical network such as the Internet, where peers attend P2P network with same capabilities. P2P is also a set of technologies and services, such as file sharing, content distribution, way of collaborated work, and data-mining on the network, by which users can directly exchange information possibly without using any server. There are many techniques to develop and construct P2P networks and applications, such as SETI@HOME [1], Napster [2], and Gnutella [3], etc. However, these systems adopt individual implementations of P2P network. It is clear that their network structures are optimized for their own purposes. Later on, there are some structured p2p systems, such as Pastry [4], Tapestry [5], and Chord [6], etc. These approaches have demonstrated the ability of P2P network to serve as a robust, scalable substrate for a variety of applications.

To extend an open P2P network system with the mechanisms of distributed computing, we developed a flexible JXTA-based P2P network interface and architecture, called JXTPIA. The goal of this work is to develop a simple, flexible, and robust technique for distributed computing using the p2p technology. It can also be extended into Grid computing. The JXTPIA system provides the basic functionalities for distributed/Grid computing, such as task scheduling, network structure constructing and maintenance, and data sharing, etc. The JXTPIA system can release programmers who work on distributed computing but lack a flexible distributed computing platform since it provides a P2P network interface and architecture for Grid computing in details.

This paper is organized as follows. Section 2 describes related work, while Section 3 present the JXTPIA system design. In Section 4, we implement JXTPIA system components. Section 5 illustrates an experiment on JXTPIA, and conclusions are given finally.

2 Related Work

JXTA [7][8][9] constructs a virtual P2P network on the existing network, which implements a P2P network and provides the API to use the network for the user. JXTA only supplies the basic low level protocols for P2P networks, but not for distributed systems.

Globus [10][11] architecture is the *de facto* standard for computational grid [12]. Globus Toolkit provides the basic system to achieve Grid computing, and it is very close to satisfy the demand of Grid computing. However, Globus concentrates on high-end resources, such as supercomputers, large clusters, etc. for distributing high-end application loads on them. In our experience, using Globus Toolkit is only a small part of cost on developing a Grid computing program. Globus requires programmers an additional overload and much stress.

One of the early efforts is Java Market project [13], which tries to utilize wasted computational power using Java applet and web technologies. User submits Java applications into a Java Market and the Java Market transforms it to a contributing machine and executes it using the web browser.

Nimrod [14] is an economic or market based resource management and scheduling system that manages the execution of parametric studies across distributed computers. It takes responsibility for the overall management, as well as resource trading and quality of service based scheduling. Nimrod/G is utilizing the resources in grid, instead of normal generic computer as in SETI@HOME [1].

Compute Power Market (CPM) [15] is a market-based resource management on Internet-wide computational resources. It transforms computers connected across the Internet into a computational market by renting the processing power, storage and special devices from idle resources.

JNGI [16] is a distributed computing framework based on JXTA that users can use to submit jobs to JXTA groups. These jobs can be split and distributed among several peers. Redundancy within JXTA peer groups ensure that failures do not affect job completion. P3 [17] is a middleware on pure P2P facilities provided by JXTA for distributed computing using volatile personal computers.

3 JXTPIA System Design

In this section, we present the JXTPIA system design with dividing into modules, such as JXTPIA network, Homework Distribution system (JHD), Trigger Distribution system (JTD), and Data Sharing system (JDS). The relationships of these modules are illustrated in Figure 1.

Fig. 1. The layer model of the JXTPIA system

In Figure 1, the bottom layer of JXTPIA system, JXTPIA network layer, is built upon JXTA P2P network. JXTPIA network layer creates network structures and provides some interfaces to connect JXTA P2P network with the upper layers. The top layer is JXTPIA User Interface, which connects user with the JXTPIA system. The core of JXTPIA system includes JHD module, JTD module, and JDS module, etc. JHD provides users the interfaces to distribute Homework system to JXTPIA network. JTD distributes task information to peers and gets feedbacks from peers. JDS keeps the data communication among these system layers.

3.1 Ring-Based Network Structure

In JXTPIA network, peers play with each other with one of the three roles: superleader peer, leader peer, and worker peer. As shown in Figure 2, superleader peer supervises the shape of JXTPIA network. Superleader peers hold and manage all information of JXTPIA leader peers. A *leader* peer acts as a group leader of *worker* peers in a worker ring. Each leader peer attends a worker ring to share some information with worker peers.

All peers in the JXTPIA system are members of JXTPIA world group. Also there are leader peer group (all of leader peers), superleader peer group (all of superleader peers), and worker peer groups. These groups are piled, and then peers construct a network connection as rings.

JXTPIA ring-based structure, shown in Figure 2, is built for its scalability and solidity. Structured P2P overlay networks such as Chord [6], Pastry [4], and Tapestry [5] effectively implement scalable distributed hash tables (DHTs). We also implement DHTs into the JXTPIA ring-based structure, where each node in the network has a unique node identifier and each data item has a unique key. As one of P2P network's characters, there is no particular server in the JXTPIA system. This character may cause serious problems since all peers can

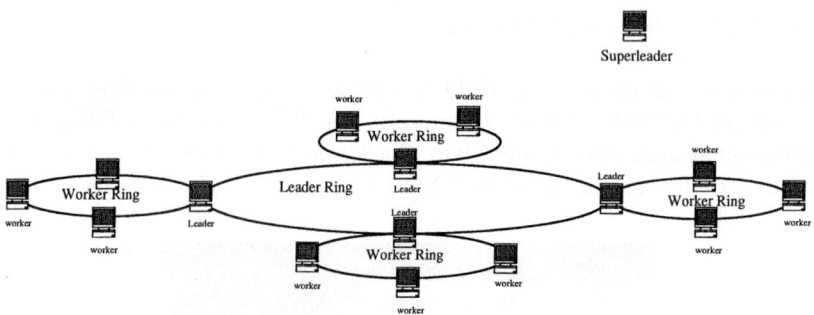

Fig. 2. The outline of JXTPIA rings

attend or leave the P2P network freely. No one can guarantee the existence of all peers at any time. To avoid the damage of peer and/or data loss, a system called "Triangle backup system" was introduced into JXTPIA. In this idea, a three-peer set works on JXTPIA system as just one node.

3.2 Message-Listener Model

To get the system scalability, a *message-listener* model is adopted in the system. The message-listener model can handle large numbers of clients/connections, spread across multiple machines. It can minimize what needs to be stored in memory, and make it possible for multiple classes to re-use instances of event handlers and events. The message-listener model is similar to the event-listener model in JAVA. In the event-listener model, the class which needs the feedback from GUI components implements the event listener, and then the instance of the class is resided in the GUI components. When some events occur on the component, it calls all of the listeners from their listener table. A JAVA programmer who wants to add a new function to JXTPIA has to write a message class which implements an interface named *JxtpiaMessageBody* and a message listener which implements a *JxtpiaMessageListener* interface.

There is a little difference between event-listener model and message-listener model. The former calls all listeners registered on a component, but the latter makes the filtering by the message type. The message listener on JXTPIA, which is called *JxtpiaListener*, has a *getTarget()* method to bind the listener to a specific message. This method returns the message type of the target message. JXTPIA network registers the listener to an extended hash table using the value returned from this method as a key. When a message arrives at a peer, the peer selects a listener with the key included in the message as a hash-key in the extended hash table, then calls the entire listener with this key.

3.3 JXTPIA Network Constructing Procedures

At the beginning, there is no peer on the JXTPIA network. With respect to our solution, the peer which wants to attend the network tries to bind the *JXTABiDiPipe* with the superleader peer pipe advertisement. If the binding succeeds, there exists a superleader peer, and then the peer applies to the superleader peer for its role in the network. Oppositely, failure of this binding means that there is no superleader peer, so the peer starts to work on this network as a superleader peer.

After connecting to a superleader peer, a peer sends a *Role-Request-Message* to fetch a role. The superleader peer decides the peer as a leader peer or a worker peer, and then it sends the *Role-Response-Message* to the peer. If all of worker rings are full, the superleader peer assigns the leadership to the peer. If there is a worker ring which is not filled by workers, the superleader peer assigns the peer as a worker.

A *Role-Response-Message* includes some important information. When a leader role is assigned, the leader peer creates two pipe advertisements in order to connect with the lower leader peer and the upper leader peer. Peer can begin to work as a leader on the leader ring just according to this information. On the other hand, there is just a pipe advertisement to connect the leader peer supervising the worker ring that the peer belongs to. The peer has to get pipe advertisements to connect to the lower and upper peers with sending a *Worker-Connection-Request-Message*.

3.4 Self-recovering Mechanism

It is obvious that the network system based on P2P network has a danger of losing a peer from its network structure, so the system must prepare for a self-recovering system to avoid the collapse. The JXTPIA system has a self-recovering mechanism as a countermeasure of the collapse.

Normally, every peer sends an *acknowledgment-message* to check the existence of the lower peer periodically with incrementing an *acknowledgment counter*. The length of period can be decided by the property file. Peer which receives an *acknowledgment-message* returns an *acknowledgment-response* message to the upper peer. When the peer receives the message, the peer authorizes the existence of the lower peer, and resets the *acknowledgment counter*, which is also written in a property file. When the value of *acknowledgment counter* becomes over a constant (say, 4), the peer authorizes the loss of its lower peer, whereupon the peer starts to recover the ring. There are two cases of peer loss: *worker peer loss* and *leader peer loss*. The self-recovering mechanism keeps peer information table updated and assigns a new leader peer in case of leader peer loss.

4 JXTPIA System Implementation

The ways to use JXTPIA are assumed below.

1. Distributing HWS tasks.
2. Distributing arguments for HWS tasks.
3. Getting results of HWS tasks.

To distribute its program into JXTPIA, peer must prepare a Java Archive (JAR) file, *Homework Set (HWS)*. An HWS task includes a JAVA program *HWS*

Abstract, which extends JXTPIA abstract class. The JXTPIA system distributes an HWS task into the JXTPIA system by calling JHD module.

4.1 JHD Module

Each HWS task is piled on the peer which distributes it. Normally, an HWS task has mainly three types of information: distribution, searching, and transportation information. HWS distribution information includes its *name*, its *digest*, and its *size* for identifying the HWS task. HWS searching information involves an *owner peer ID*, an *owner leader ID*, a *destination peer ID*, and a *destination leader ID* for searching for the HWS task. We require its *session ID*, its *tag number*, and its *HWS data* information of an HWS task for transportation. It is especially important that the HWS task is not distinguished by just its name, but also its MD5 value since its name is not unique. A hash function such as MD5 is a one-way operation that transforms a data string of any length into a shorter, fixed-length value. Not any two strings will produce the same hash value. So it is easy to distinguish two different HWSs with MD5.

JHD permits peers to get the HWSs from these peers which use them very often to keep the load-balancing on network traffics related to the transportation of HWSs. To distribute an HWS task, we conduct it with the trigger distribution module in the JXTPIA system.

4.2 JTD Module

JTD provides an interface to distribute and search a task in JXTPIA network. When distributing a task to another peer, JTD makes a copy of the task to that peer. When searching for a task is needed, JTD searches for the task within JXTPIA network, and/or transfers the task to the requested client peer. JTD also provides an interface to execute tasks. After transferring the tasks and arguments to the peer, JTD executes the task and receives the results.

In short, the system executes same codes with different arguments simultaneously. JTD is the module to supervise the procedures of distributing the arguments to the designated peers and the trigger of starting to execute the assigned HWS task. As mentioned above, HWSs distributed to other peers are piled on the local disk space of the peer.

JTD works based on two priority principles:

- Peers within the same worker ring have the higher priority to be triggered,
- Fairness has the higher priority.

Based on the principle of *"peer within the same worker ring has the higher priority to be triggered,"* the task-requesting peer gathers the information of working peers within the same worker ring first. After all peers on the worker ring are assigned the trigger, the task-requesting message is sent to the leader peer on the neighbor worker ring to gather the information of worker peers within the neighbor worker ring. This way causes the unfair situations, because peers in same worker ring may request more HWS tasks than peers int other worker rings.

Meanwhile, the task-requesting scheduler based on *"fairness has the higher priority"* behaves like following.

1. The leader peer getting one peer from its own group.
2. The leader peer sending a request to the lower leader peer to get one peer from the lower leader group.
3. The lower leader peer applying for sending another request to the next lower leader peer.

JTD repeats these three actions to fill the requested tasks. Based on the these actions, JTD keeps the JXTPIA system within a relatively small margin of load imbalance.

4.3 JDS Module

Two principal elements in the JXTPIA system are *task* and *resources (or data)*. One task commonly has a large number of resources as its targets. For example, in the case of matrix multiplication, the task is the program which describes the algorithm of matrix multiplication, and the resources are two matrices being targets of the matrix multiplication.

Resources can be any types of files such as a plain text file and/or a binary file storing serialized objects. If a peer wants to issue a task, all it has to do is to outfit an HWS and the resources associated with it. The JXTPIA Data Sharing module searches for the resources within JXTPIA Network, and/or transfers the resources to the client peers. To share data or resources efficiently in a P2P network, JXTPIA network needs efficient search and distribution algorithms. In section, we present a searching algorithm and two distribution algorithms.

The searching algorithm's mechanism is that a peer simply keeps asking its next leader peer until the resource is found. Since a leader peer's DHT has the information of worker peers in its group, resources can be found quickly.

A ring distribution algorithm (RDA) distributes a resource to the next worker peer in same worker ring. Simultaneously, the resource is distributed to the next leader peer. When receiving a resource, a worker peer informs its leader peer to updates its resource list, and the leader peer transfers the resource to both the next leader peer and the next worker peer. When no resources need to be distributed, the distribution process is complete. RDA can quickly distribute a resource because the resource is simultaneously distributed in each ring.

With the wide distribution algorithm (WDA), resources are distributed to two destination peers each time. Hence, the number of distribution destinations exponentially increases like a tree structure. The receiving peer transfers the resources again until no more resources need to be distributed.

From the experimental simulations [20], we found that RDA is suitable for distributing small resources and/or a large amount of resources, but poor at searching for a resource in a large network. On the contrary, WDA is suitable for distributing large resources and/or searching few resources in a large network, but poor at distributing a large amount of resources.

5 Experimental Results and Discussion

We have tested the JXTPIA system by conducting experiments on JXTPIA: searching concealed key generated by Data Encryption Standard (DES) [22], and simulating heat diffusion in the air [25], etc. The environments for these experiments are the same as in Table 1.

Table 1. Experimental environments of the JXTPIA system

Machine	SunBlade150
CPU	UltraSPARC-IIe 550MHz
Memory	512 MB
OS	Solaris8
Java Virtual Machine	Sun Java 1.4.05
JXTA	JXTA ver 2.2

The key searching problem is selected as a benchmark of JXTPIA system. Since searching for 56-bit DES key takes too much time in the JXTPIA system, we conduct a 24-bit sample. Originally, the key searching problem is not a good parallel problem. But we modify it here to be a good sample for parallelizing. Below are the steps of this experiment.

1. Preparing for original sentence "Hello JXTPIA" and creating a cryptogram with DES.
2. Sending the program which can make cryptogram based on DES with original sentence and an optional key to worker peer in JXTPIA network. Simultaneously, sending the checking range which will be explained in the following paragraph.
3. The peers which undertake the tasks create a cryptogram with an original sentence "Hello JXTPIA" and an optional key within the checking range, and check whether the generated cryptogram is the same as an original one sent with arguments or not.

The difficulty of key searching may depends on the value of the key. For example, if the key is 000001x0, there is just one step to find the key if we start from the 000000x0. So to hold an equal opportunity for all keys, we set finishing all bits checking as the condition of this problem. In our key searching experiment, we divide the key into two parts: the first 6 bits and remainder bits. The first 6 bits are used for a mark of each task. That is, every version of this key searching are divided into $2^6 = 64$ tasks from 000000x0 to 111111x0. The left bits of the key is used for expressing the checking range, for example in 8-bit version of this program, peer which undertakes the task with the 001100 as a prefix must check the key from 00110000 to 00110011. The results of this experiment are indicated in Figure 3. We can see the speed-up close to the ideal one.

Due to the spatial limitation of this paper, we will not describe some another experiments on JXTPIA, such as Heat diffusion experiment. Further performance evaluation and experimental data analysis will be conducted in another paper.

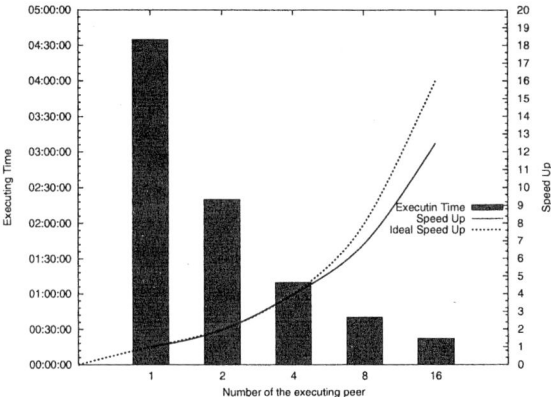

Fig. 3. Result of key searching (24-bit). Execution times are shown in histograms, while ideal speedup and actual speedup are shown in dash line and solid line, respectively.

6 Conclusions

In this paper, we present a flexible JXTA-based P2P network interface and architecture for Grid computing, JXTPIA. We have implemented main JXTPIA system components, such as network module, Homework Distribution module, Trigger Distribution module, and Data Sharing module. Experimental results show that it executes the large applications efficiently.

Further research on JXTPIA will be studied. JXTPIA allows the user execute JAVA code on the computer which attends to the JXTPIA network. Thus no existence of security system is a fatal defect, because the ill-intentioned programmer can execute any code on the PCs in JXTPIA network. The implementation of a security system checking the legal/illegal users is the most important things for JXTPIA.

The current JXTPIA system doesn't support JAVA Micro Edition (J2ME). In the next phase of development, we will support for J2ME. User can get computation power from JXTPIA network through mobile devices, such as cellular phones, PDA, and so on.

References

1. SETI@HOME. Search for Extraterrestrial Intelligence at home, http:// setiathome.ssl.berkeley.edu/.
2. Napster. http://www.napster.com, 2004.
3. Gnutella. http://www.gnutella.com/.
4. Pastry Project. http://research.microsoft.com/ antr/Pastry/.
5. Ben Y. Zhao, *et al.* Tapestry: A resilient global-scale overlay for service deployment. IEEE Journal on Selected Areas in Communications, 2003.
6. I. Stoica, *et al.* Chord: A scalable peer-to-peer lookup service for Internet applications. Technical Report TR-819, MIT, March 2001.

7. JXTA Project. http://platform.jxta.org.
8. Bernard Traversat. Project JXTA 2.0 Super-Peer Virtual Network, http://www.jxta.org/ project/www/docs/JXTA2.0protocols1.pdf.
9. Li Gong. JXTA: A Network Programming Environment. IEEE Internet Computing, 5(3):88–95, May/June 2001.
10. Globus 2.4. http://www.globus.org/gt2.4/, 2004.
11. Ian Foster, et al. The Anatomy of the Grid: Enabling Scalable Virtual Organizations, International J. Supercomputer Applications, 15(3), 2001;
    I. Foster, et al. A Metacomputing Infrastructure Toolkit. International J. Supercomputer Applications, 11(2):115-128, 1997.
12. Ian Foster, Carl Kesselman. The Grid: Blueprint for a New Computing Infrastructure. Morgan Kaufmann Publisher, 1998.
13. Yair Amir, Baruch Awerbuch and Ryan S. Borgstrom. The Java Market: Transforming the Internet into a Metacomputer. Technical Report CNDS-98-1, Johns Hopkins University, 1998.
14. Nimrod Project. http://www.csse.monash.edu.au/ davida/nimrod/.
15. Compute Power Market Project. http://grid.cs.usm.my/cpm.htm.
16. JNGI Project. http://jngi.jxta.org/.
17. Personal Power Plant (P3) Project. http://p-three.sourceforge.net/.
18. Winny Archive. http://nynode.info/, 2005.
19. Condor Project. http://www.cs.wisc.edu/condor/, 2004.
20. Kenichi Sumitomo, Takato Izaiku, Yoshihiro Saitoh, Hui Wang, and Minyi Guo. Effective Resource Allocation and Task Assignment in a JXTA-Based Grid Computing Platform JXTPIA, Third International Symposium on Parallel and Distributed Processing and Applications (ISPA'2005), Nanjing, China, Nov. 2005.
21. Naoya Henmi. High Performance Peer-to-Peer Matrix Multiplication using by JXTA. Graduation thesis, University of Aizu, 2003.
22. Data Encryption Standard (DES). http://www.itl.nist.gov/fipspubs/fip46-2.htm.
23. The Java Community Process Program. http://www.jcp.org/en/home/, 2004.
24. J2SE 5.0. http://java.sun.com/j2se/, 2004.
25. Heat Diffusion Problem. http://www.netlib.org/pvm3/book/node56.html.

A Community-Based Trust Model for P2P Networks*

Hai Jin, Xuping Tu, Zongfen Han, and Xiaofei Liao

Cluster and Grid Computing Lab, Huazhong,
University of Science and Technology, Wuhan, 430074, China
hjin@hust.edu.cn

Abstract. Trust management is a key issue for P2P networks. Previous works build peer's reputation just based on ratings of other individual peers. In this paper we present a novel trust construction approach, called *CommunityTrust*. The approach evaluates peers' reputation taking other peers' ratings into account as well as recommendations of some related communities and peers. Our preliminary simulation results show that *CommunityTrust* model significantly improves the trust computation accuracy of P2P e-Commerce systems.

1 Introduction

Trust management is a key issue for peer-to-peer networks, especially in e-Commerce or file sharing environments. These systems [4][7][10], obey that a peer's reputation only depends on its history transactions. Each peer needs to have many transactions to achieve steady-state quality levels. Moreover, the whole system needs much more time to reach a steady-state, and it is unfair that any newcomer with potential good quality is only assigned with an initial low reputation value. This serious limits the performance gains of introducing "trust" into systems.

This paper proposes another trust model, called *CommunityTrust*. It solves this problem discussed above through evaluating reputation based on others' ratings as well as recommendations of some related communities and peers. Experiments shows that this model can remarkably increase reputation value of the newcomer with good quality, which improves trust computation accuracy of the whole system. It mends the sources of trust value of most existing reputation-based trust models only taking into account history transaction records. We demonstrate that the community-based trust model will reflect a peer's trustworthiness more accurately compared with existing trust models.

The paper is structured as the following: in section 2, some previous works will be surveyed. Then a new trust model, named *CommunityTrust*, as well as some implementation issues will be proposed in section 3. Some simulations will be presented in section 4. Finally, we conclude the paper in section 5.

---

* This work is supported by National Science Foundation under grants No.60433040.

2 Related Works

Trust management has attracted many research attentions [8][9][12][13]. Dellarocas [2] has analyzed the economic efficiency of eBay-like online reputation reporting mechanisms. This mechanism relies exclusively on a binary reputation mechanism, where raters are given the opportunity to rate past transactions using one of two values, commonly interpreted as "positive" and "negative". Yu and Singh [1] present an idea on building a social network among agents supporting participants' reputation. Pujol et al. [5] propose a generalized algorithm that extracts the reputation in a general class of social networks. A main characteristic of these approaches is that the data management is completely centralized.

Aberer and Despotovic propose a reputation based management system for P2P systems [6]. However, their trust metric simply summarizes the complaints a peer receives and is very sensitive to the skewed distribution of the community and misbehaviors of peers. P2PRep [3] is a P2P protocol based on Gnutella, where servants can keep track of information about the reputation of other peers and share them with others. Some incentives for combating free riding on P2P networks are proposed [11]. Kung et al. [4] propose an eigenvector-based method to compute the trust of other nodes. PeerTrust [7] introduces three basic trust parameters and two adaptive factors in computing trustworthiness of peers. It also defines a general trust metric to combine these parameters.

EigenTrust [10] focuses on a Gnutella like P2P file sharing network. Their approach is based on the notion of transitive trust and address the collusion problem by assuming there are peers in the network that can be pretrusted. While the algorithm shows promising results against a variety of threat models, we argue that the pretrusted peers may not be available in all cases and a more general approach is needed. Another shortcoming of their approach is that the implementation of the algorithm is very complex and requires strong coordination and synchronization of peers. Furthermore, all of those P2P trust systems do not consider the factor of community.

3 Principle of Community Trust Model

In this paper we propose a dynamic P2P trust model for managing the trustworthiness of peers in P2P network, called *CommunityTrust*. In this section, we give the description of *CommunityTrust* model.

3.1 Create a Community

If a peer wants to create a community on his behalf, it can create a self-signed X.509 certificate, consisting information about the community distinguished name, community type, community description. If a certificate is only used to sign a recommendation for others instead of enrolling members, the certificate is *Private Certificate*, and the signer of Private Certificate is called *Private Recommender*; otherwise if a certificate is used to sign other certificates to enroll community members, it is called *Community Certificate*, and the signer of community certificate is called *Community Recommender*. A certificate, signed by a community certificate, is still called *Community Certificate* and acts as a child CA of the root CA. It inherits the right to sign other community certificate. Both the root and child CA control the depth of the CA hierarchy. The community certificate prevented from signing other community certificate is called *End Community Certificate*.

3.2 Join and Leave a Community

To join a community, a peer just needs to get a community certificate from a member in that community. The community certificate can specify the lifetime and the type. Usually the lifetime of certificate is short for security consideration. When it is expired, the peer can apply for another one after presenting the old certificate to any other members in that community. When a peer wants to leave a community, it sends a request to its signer who introduces it to that community. Then the signer will update the information from the peer which stores the requester's trust value.

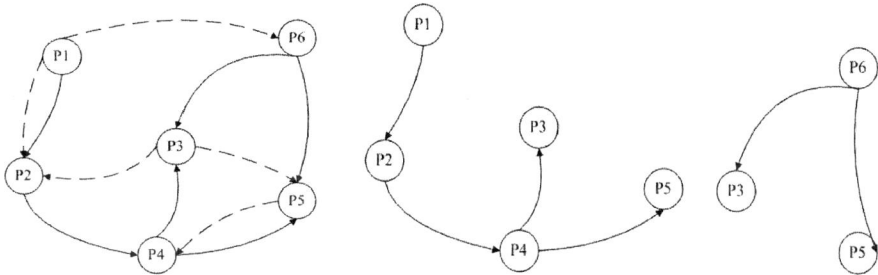

Fig. 1. Several peers construct a pseudo graph

Fig. 2. Two trees generated from the graph in Fig.1

A graph is denoted by $G = (V, E)$. Solid directed edge $(P_i, P_j)_r$ means that Peer P_i signs a community certificate to P_j, and a dotted directed edge $(P_i, P_j)_v$ means that P_i signs private recommendation in his own opinion to P_j. In Fig.1, the vertex set is $V = \{P_1, P_2, P_3, P_4, P_5, P_6\}$ and edge set is $E = E_{virtual} \bigcup E_{real}$, where $E_{virtual}$ denotes private recommendations, and E_{real} denotes community recommendations. $E_{virtual}=\{(P_1,P_2)_v, (P_1,P_6)_v, (P_3,P_2)_v, (P_5,P_4)_v, (P_3,P_5)_v\}$, and $E_{real} = \{(P_1,P_2)_r, (P_2,P_4)_r, (P_4,P_5)_r, (P_6,P_5)_r, (P_4,P_3)_r, (P_6,P_3)_r\}$. Because the solid directed edge denotes community recommendation, a node can register as more than one community member. The two trees which denote community hierarchy in Fig.1 are shown in Fig.2.

3.3 General Trust Evaluation Metrics

Before we present how to evaluate a peer's trust value, we give some definitions of metrics.

Definition-1: For any two peers, P_m, P_i, $SatNum(P_m, P_i)$ denotes times that P_m has satisfied P_i in its history, $FailNum(P_m, P_i)$ means times that P_m does not satisfy P_i. Total number of transactions between P_m and P_i is denoted by $TotalTransNum(P_m, P_i)$. So $TotalTransNum(P_m, P_i) = SatNum(P_m, P_i) + FailNum(P_m, P_i)$.

Definition-2: For any two peers, P_m, P_i, Local Trust, $t(P_m, P_i)$, denotes P_m's local trust rating about P_i, $t(P_m, P_i) \in [-1, 1]$, $t(C_j, P_n)$ means community C_j's trust rating about P_n. In this model, it will be set to 1, namely, $t(C_j, P_n) \equiv 1$.

Definition-3: Global Trust, $T(P_n)$, denotes the global trust value about P_n. $T(C_j)$ denotes the global trust value about C_j where C_j denotes the i-th Community.

Definition-4: $PeerNum(P_n)$ denotes number of peers which have had transactions with P_n, $CommNum(P_n)$ denotes number of communities that have signed certificates to P_n, and $TotalPeerNum(C_j)$ denotes the total peer number in C_j community.

In this model, several factors affect a peer's reputation value, they are: 1) other peer's ratings; 2) other peer's private recommendations; 3) the community identities the peer belongs to. A peer's trust value can be computed like:

$$T(P_n) = \alpha Private\_recomm + \beta Commmunity\_recomm. \quad (1)$$

In equation (1), α and β are weights and subject to $\alpha + \beta = 1$. $Private\_recomm$ denotes private recommendation and ratings from peers, $Commmunity\_recomm$ denotes recommendations from communities. A peer can give another peer $Private\_recomm$ for two reasons. If P_m has several transactions with P_i, it gives a feedback $t(P_m, P_i)$ to rate P_i. If two peers P_m, P_i are good friends in real world, P_m trusts P_i and gives a recommendation $t(P_m, P_i)$, in this case $t(P_m, P_i) \equiv 1$.

As each peer's trust value is different, a peer with high trust value is expected to take higher weights, so

$$Private\_commen = \sum_{i=1}^{PeerNum(P_n)} \frac{T(P_i)}{\sum_{i=1}^{PeerNum(P_n)} T(P_i)} t(P_i, P_n) \quad (2)$$

where $t(P_m, P_i)$ can be computed as:

$$t(P_m, P_i) = \begin{cases} \alpha' \times \frac{SatNum(P_m,P_i) - FailNum(P_m,P_i)}{TotalTransNum(P_m,P_i)} \\ +\beta' \times \frac{TotalTransNum(P_m,P_i)}{\sum_{j=1}^{PeerNum(P_i)} TotalTransNum(P_j,P_i)} \\ 1 \end{cases} \quad (3)$$

In equation (3), $\frac{TotalTransNum(P_m,P_i)}{\sum_{j=1}^{PeerNum(P_i)} TotalTransNum(P_j,P_i)}$ reflects the total transactions having a minor effect on the reputation value, α' and β' are subject to $\alpha' + \beta' = 1$ and $\beta' < \alpha'$. How to determine their value is an open question. Simply, α' and β'

are assumed to 0.9 and 0.1 respectively. If P_m signs a private recommendation instead of submitting a feedback for P_i, then $t(P_m, P_i)$ is set to 1.

Community_recomm is a comprehensive effect of a peer's community membership. Communities with better reputation will take higher weights. From definition 4, assume each community completely trusts peer P_n, namely $t(C_j, P_n) \equiv 1$, then *Community_recomm* can be computed like this:

$$Community\_recomm = \sum_{j=1}^{CommNum(P_n)} \frac{T(C_j)}{CommNum(P_n)} t(C_j, P_n)$$

$$= \sum_{j=1}^{CommNum(P_n)} \frac{T(C_j)}{CommNum(P_n)} \quad (4)$$

$T(C_j)$ denotes the trust value of a community. A community's trust depends on its members' trust, in our model it is defined as the average of its member's trust value. $T(C_j)$ can be computed as follows:

$$T(C_j) = \frac{\sum_{k=1}^{TotalPeerNum(C_j)} T(P_k^{C_j})}{TotalPeerNum(C_j)} \quad (5)$$

where $P_k^{C_j}$ denotes that the k-th peer in community C_j.

3.4 Responsibility of a Recommender

From equation (1), it can be seen that *Community_recomm* is mostly larger than 0, a peer likely makes profit from its community membership. But from equation (4), a community's trust value is the average of its members' trust value. Suppose x percents of peers are malicious, from equation (1)~(5), the good peers will be undervalued with βx percent at most. To protect a community's reputation, every member has a responsibility to assure every peer he recommend is a good one. In this section we propose a punishment policy to prevent a careless recommender from enrolling a malicious peer.

We assume that those having a close recommending relationship with a malicious peer will be suspicious. That is, if a community member cheats others, or does not make anyone satisfied, its private recommenders, its community recommenders, and even the ones he recommended either in private style or in community style will have a higher probability of cheating than others.

In Fig.3, there are 3 dotted-line circles, they represent 3 peers sets. The peers in the smallest circle connected to P_0 by a dotted-line denoted by $S_{Private\_recomm}$, $S_{Private\_recomm} = \{P_2, P_5\}$. The peers in the middle circle are straightly connected to P_0, and the peers between the bigger circle and the middle circle have a path length 2 connecting to P_0. The closer a peer to a malicious peer, the more he has potential threats. Additionally, we differ a peer's child node from its parent node. In this model, the parent introduces a peer as a community member and the child is introduced as a community member by the parent peer. For example, suppose P_0 provide bad service to others, and then P_0 will certainly

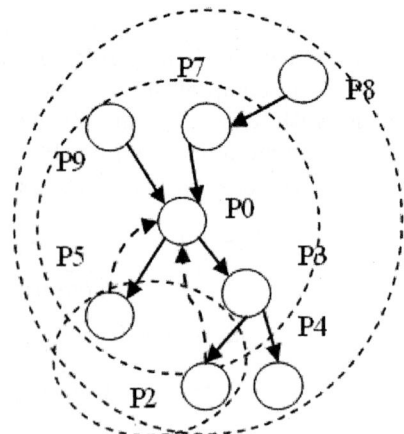

Fig. 3. Punishment to an unsatisfying peer and its potential confederates

get a negative feedback, so do its parents and children. $T(P) = (1 - \delta)T(P)$, where $P \in S_{Private\_recomm}, 1 \geq \delta \geq 0$. $T(P) = (1 - \Delta_n)T(P)$, where $P \in S_{Surround}(n)$ denotes a set in which peers have a path length n to the malicious peer. $S_{Surround(i)} \cap S_{Surround(j)} = \emptyset$. Δ_n denotes a factor that will affect the n-th surrounding peers, certainly, $n = 1, 2, 3, \cdots$, and $0 \leq \Delta_n \leq \Delta_{n-1} \cdots \leq \Delta_1 \leq 1$. In Fig.3, $S(2) = \{P_4, P_2, P_8\}$ and $S(1) = \{P_3, P_5, P_9, P_7\}$.

We only consider the responsibility of a peer which introduces a malicious peer to community or sign a private recommendation to a malicious peer. Since its introducing a malicious peer directly destroys its community member's reputation, one should be more careful using its community membership introducing right than its private recommendation right. So $0 \leq \Delta_1 \leq 1$, $\Delta_2 = \Delta_3 \cdots \Delta_n = 0$, $1 > \delta > 0$, $\Delta_1 \geq \delta$. In order to simplify process, we punish P_{child}'s private recommender P_{pri} using the following equation:

$$FailNum(P_{pri}, P_i) = FailNum(P_{pri}, P_i) + 1 \quad (6)$$

and punish P_{child}'s community recommender by using equation (7):

$$FailNum(P_{comm}, P_i) = FailNum(P_{comm}, P_i) + 2 \quad (7)$$

As shown in equations (6) and (7), once a peer sends a private recommendation or a community recommendation to a malicious peer, it will be punished for undermining interest of the community. If a group of peers construct a malicious community, the mistake of any member will be magnified which will decay the community's reputation quickly. It helps others know their bad intent earlier. Thus, bad peers will not make any profit through building a community.

3.5 Implementation Issues

Suppose each peer has an average of *commNum* of community memberships, and the average of community size is *commSize*, then CommunityTrust will increase an average of $commNum \times commSize$ times of communications compared with traditional trust model (which does not consider communities' trust). To alleviate the heavy collecting operation, this paper proposes an approximate way to compute trust value.

Before computing $T(C_j)$, the system will check whether it already exists in cache. A cache miss will certainly cause a new computing. But if the $T(C_j)$ is found in cache, and the period since last computing has not exceeded an specified $Inter$, the cached value will be used instead of collecting a new one. The cached value $T_{old}(C_j)$ is likely affect the current community value, assuming the effect of $T_{old}(C_j)$ decays exponentially. Let $T_{new}(C_j)$ denote the new fetched trust value from the peer responsible for providing the trust value of C_j. $T(C_j)$ is defined as a function of $T_{old}(C_j)$ and $T_{new}(C_j)$ in equation (8).

$$T(C_j) = \begin{cases} T_{old}(C_j) & : t \leq Inter \\ T_{old}(C_j)e^{(Inter-t)} + (1 - e^{(Inter-t)})T_{new}(C_j) & : t > Inter \end{cases} \quad (8)$$

In equation (8), if $T_{new}(C_j)$ is 0, it means that the community may not exist any more, then this community's effect will decay slowly as expected. A peer, responsible for managing a community's trust value, applies similar cache scheme to the management of the trust value of the member of the community given in equation (9).

$$T(P_n) = \begin{cases} T_{old}(P_n) & : t \leq Inter \\ T_{old}(P_n)e^{(Inter-t)} + (1 - e^{(Inter-t)})T_{new}(P_n) & : t > Inter \end{cases} \quad (9)$$

In equation (9), $T_{new}(P_n)$ denotes the new fetched trust value of P_n and $T_{old}(P_n)$ denotes the cached trust value of P_n.

4 Performance Evaluation

The number of members in p2p CommunityTrust society is $PeersNum$, and the number of transactions each peer executed is denoted by $TransactionNum$. $MaliciousRate$ is used to model the rate of the total malicious peers, and the number of the community in this society is denoted by $CommunityNum$. The size of a community is set to be $CommunitySize$. If a peer sends its transaction request, $RateOfResponsedPeers$ of the total peers will respond to it. Some of the above variations are assumed to be set to constants while others remain varying. Those constants are summarized in Table 1.

All experiment results are run over 20 times averagely. The first experiment compares CommunityTrust model with another two models. One is the random model, in which a peer randomly selects one among the responders. Those responders are randomly generated among all the peers to perform its transaction. The other is the traditional trust model, in which a peer's trust value is independent of the community. Actually, this model is a special case of CommunityTrust model, since if the coefficient of private-recommendation is set to 1, CommunityTrust model is the same as the traditional trust model. As for CommunityTrust model, it has two modes: CommunityTrust without punish and CommunityTrust with punish.

Table 1. Summary of simulation parameters

Parameter	Value
$TransactionNum$	1024
$PeersNum$	1024
$CommunityRate$	0.02
$CommunitySize$	10000
$RateOfResponsedPeers$	0.8(CT-Punish mode) Randomly(CT-No-Punish)

In the former mode, peers could carelessly enroll members or sign a private recommendation since it will not be punished for enrolling malicious peers. However, in the latter one, a peer will be serious when executing the same rights since it will be punished for making a mistake. The probability of good peers to enroll a good peer as their community member, denoted by *probabilityOfGood*, is different in such two modes. In CommunityTrust without punish mode, it is set to the rate of the good peers. That is, a good peer will randomly select one. In CommunityTrust with punish mode, it is set to a constant 0.8. The malicious peer is defined as the one who never executes a transaction successfully and the good peer is defined as the one who completes transactions at a successful-rate of its trust value.

Fig.4 shows that CommunityTrust trust model can remarkably improves the accuracy of trust computation if the rate of malicious peers is larger than 0.3 compared to the traditional reputation model. Considering punishment scheme, the accuracy of CommunityTrust with punish model is a little more than CommunityTrust without punish mode.

Fig. 5(a) and Fig. 5(b) give the accuracy of trust computation depending on α, the coefficient of the *private_recomm*. In both figures, $\alpha = 1$ represents the traditional trust model. When $\alpha = 0.8$, the CommunityTrust without punish mode gets the highest accuracy of computation, and $\alpha = 0.7$ makes CommunityTrust with punish mode get the highest accuracy of computation.

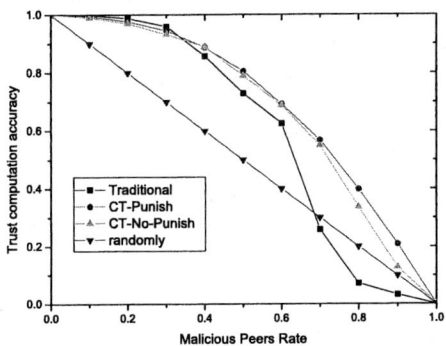

Fig. 4. Accuracy of trust computation

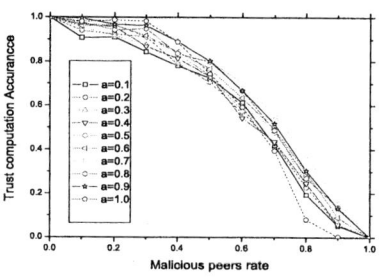 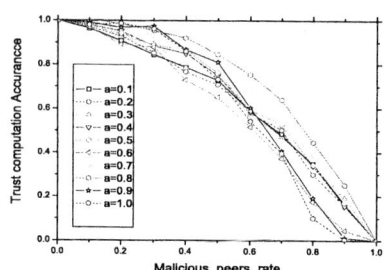

(a) CommunityTrust without punish   (b) CommunityTrust with punish

Fig. 5. Accuracy of trust computation when α varies

5 Conclusions

In this paper, we have designed CommunityTrust, a reputation-based trust supporting framework, to evaluate the trustworthiness of peers. Based on our approach, a newcomer can achieve higher probability of successful transaction and the whole system can reach a steady-state with less time. Initial simulation experiments have demonstrated that our approach remarkably improves the accuracy of trust computation.

References

1. B. Yu and M. P. Singh, "A Social Mechanism of Reputation Management in Electronic Communities", *Proc. of the 4th International Workshop on Cooperative Information Agents*, Lecture Notes in Computer Science, vol.1860, Springer, 2000.
2. C. Dellarocas, "Analyzing the Economic Efficiency of Ebay-Like Online Reputation Reporting Mechanisms", *Proc. of 3rd ACM Conf. Electronic Commerce*, 2001.
3. F. Cornelli, E. Damiani, S. D. C. di Vimercati, S. Paraboschi, and P. Samarati, "Choosing Reputable Servents in a P2P Network", *Proc. 11th Intl World Wide Web Conf.*, 2002.
4. H. T. Kung and C. H. Wu, "Differentiated admission for peer-to-peer systems: incentivizing peers to contribute their resources ", *Proc. of Workshop on Economics of Peer-to-Peer Systems*, 2003.
5. J. M. Pujol, R. Sanguesa, and J. Delgado, "Extracting Reputation in Multi-Agent Systems by Means of Social Network Topology", *Proc. of 1st Intl Joint Conf. Autonomous Agents and Multiagent Systems*, 2002.
6. K. Aberer and Z. Despotovic, "Managing Trust in a Peer-to-Peer Information System", *Proc. of ACM Conf. Information and Knowledge Management (CIKM)*, 2001.
7. L. Xiong and L. Liu, "PeerTrust: supporting reputation-based trust for peer-to-peer electronic communities", *IEEE Transactions on Knowledge and Data Engineering*, vol:16, July 2004, Pages:843 - 857.

8. M. Chen and J. P. Singh, "Computing and Using Reputations for Internet Ratings", *Proc. of 3rd ACM Conf. Electronic Commerce*, 2001.
9. P. Resnick, R. Zeckhauser, E. Friedman, and K. Kuwabara, "Reputation systems: Facilitating trust in Internet interactions", *Communications of the ACM*, 43(12): 45 - 48, 2000.
10. S. D. Kamvar, M. T. Scholsser, and H. Garcia-Molina, "The EigenTrust Algorithm for Reputation Management in P2P Networks", *Proc. 12th Intl World Wide Web Conf.*, 2003.
11. S. D. Kamvar, M. T. Schlosser, and H. Garcia-Molina, "Incentives for Combatting Freeriding on P2P Networks", *Technical report*, Stanford University, 2003.
12. Y. Atif, "Building Trust in E-Commerce", *IEEE Internet Computing*, vol.6, no.1, 2002.
13. Z. Despotovic and K. Aberer, "Maximum Likelihood Estimation of Peers' Performance in P2P Networks", *Proc. of Workshop on Economics of Peer-to-Peer Systems*, 2004.

Enabling the P2P JXTA Platform for High-Performance Networking Grid Infrastructures

Gabriel Antoniu[1], Mathieu Jan[1], and David A. Noblet[2]

[1] IRISA/INRIA, Campus de Beaulieu, 35042 Rennes cedex, France
Mathieu.Jan@irisa.fr
[2] University of New Hampshire, Department of Computer Science,
Durham, New Hampshire 03824-3591, USA

Abstract. As grid sizes increase, the need for self-organization and dynamic reconfigurations is becoming more and more important, and therefore the convergence of grid computing and Peer-to-Peer (P2P) computing seems natural. Grid infrastructures are generally available as a federation of SAN-based clusters interconnected by high-bandwidth WANs. However, P2P systems are usually running on the Internet, with a non hierarchical network topology, which may raise the issue of the adequacy of the P2P communication mechanisms on grid infrastructures. This paper evaluates the communication performance of the JXTA P2P platform over high-performance SANs and WANs, for both J2SE and C bindings. We analyze these results and we evaluate solutions able to improve the performance of JXTA on such networking grid infrastructures.

Keywords: High performance networking, grid computing, P2P, JXTA.

1 Using P2P Techniques to Build Grids

Nowadays, scientific applications require more and more resources, such as processors, storage devices, network links, etc. Grid computing provides an answer to this growing demand by aggregating resources made available by various institutions. As their sizes are growing, grids express an increasing need for flexible distributed mechanisms allowing them to be efficiently managed. Such properties are exhibited by Peer-to-Peer (P2P) systems, which have proven their ability to efficiently handle millions of interconnected resources in a decentralized way. Moreover, these systems support a high degree of resource volatility. The idea of using P2P approaches for grid resource management has therefore emerged quite naturally [1,2].

The convergence of P2P and grid computing can be approached in several ways. For instance, P2P services can be implemented on top of building blocks based on current grid technology (e.g., by using grid services as a communication layer [3]). Conversely, P2P libraries can be used on physical grid infrastructures, for example, as an underlying layer for higher-level grid services [4]. This provides a way to leverage scalable P2P mechanisms for resource discovery, resource replication and fault tolerance. In this paper, we focus on this second approach.

Using P2P software on physical grid infrastructures is a challenging problem. Grid applications often have important performance constraints that are generally not a usual requirement for P2P systems. One crucial issue in this context is the efficiency of data transfers. Using P2P libraries as building blocks for grid services, for instance, requires the efficient use of the capacities of the networks available on grid infrastructures: System-Area Networks (SANs) and Wide-Area Networks (WANs). Often, a grid is built as a cluster federation. SANs, such as Giga Ethernet or Myrinet (which typically provide Gb/s bandwidth and a few microseconds latency), are used for connecting nodes inside a given high-performance cluster; whereas WANs, with a typical bandwidth of 1 Gb/s but higher latency (typically of the order of 10-20 ms), are used between clusters. This is clearly an unusual deployment scenario for P2P systems, which generally target the edges of the Internet (those with low-bandwidth and high-latency links, such as Digital Subscriber Line, or DSL, connections). Therefore, it is important to ask: are P2P communication mechanisms adequate for a usage in such a context? Is it possible to adapt P2P communication systems in order to benefit from the high potential offered by these high-performance networks?

As an example, JXTA [5] is the open-source project on which, to the best of our knowledge, most of the few attempts for realizing P2P-grid convergence have been based (see the related work below). In its 2.0 version, JXTA consists of a specification of six language- and platform-independent, XML-based protocols that provide basic services common to most P2P applications, such as peer group organization, resource discovery, and inter-peer communication. To our knowledge, this paper is the first attempt to discuss the appropriateness of using the JXTA P2P platform for high-performance computing on grid infrastructures, by evaluating to what extent its communication layers are able to leverage high-performance (i.e. Gigabit/s) networks. A detailed description of the communications layers of JXTA can be found in [6]. This paper focuses on the evaluation of JXTA-J2SE and JXTA-C[1] over both SANs and WANs. It also discusses ways to integrate some existing solutions for improving the raw performance.

The remainder of the paper is organized as follows. Section 2 introduces the related work: we discuss some JXTA-based attempts for using P2P mechanisms to build grid services and we mention some performance evaluations of JXTA. Section 3 describes in detail the experimental setup used for both SAN and WAN benchmarks. Sections 4 and 5 present the benchmark results of JXTA over these two types of networks. Finally, Section 6 concludes the paper and discusses some possible future directions.

2 Related Work

Several projects have focused on the use of JXTA as a substrate for grid services. The Cog Kit JXTA Project [7] and the JXTA-Grid [8] project are two examples. However, none of these projects are being actively developed and not one has released any prototypes. The Service-oriented Peer-to-Peer Architecture [4] (SP2A) project aims at using P2P routing algorithms for publishing and discovering grid services. SP2A is based on two specifications: the Open Grid Service Infrastructure (OGSI) and JXTA. None of the projects above has published performance evaluations so far. Finally, JUXMEM [9] proposes to use JXTA in order to build a grid data-sharing service. All of these projects mentioned above share the idea of using JXTA as a low-level interaction substrate over

---

[1] The only two bindings compliant to JXTA's specifications version 2.0.

a grid infrastructure. Such an approach brings forth the importance of JXTA's communications performance.

JXTA's communication layers have so far only been evaluated at a cluster level, or over the Internet via DSL connections, but not over grid architectures with high-performance clusters interconnected with high-bandwidth WANs. The performance of JXTA-J2SE communication layers has been the subject of many papers [10,11,12,13,14,15] and has served as reference for comparisons with other P2P systems [16,17,18]. The most recent evaluation of JXTA-J2SE is [6], which also provides an evaluation of JXTA-C, but only over Fast Ethernet LAN networks. However, it gives hints on how to use JXTA in order to get good performance on this kind of networks.

3 Description of the Experimental Setup

For all reported measurements we use a *bidirectional bandwidth* benchmark (between two peers), based on five subsequent time measurements of an exchange of 100 consecutive message-acknowledgment pairs sampled at the application level. We chose this test as it is a well-established metric for benchmarking networking protocols, and because of its ability to yield information about important performance characteristics such as bandwidth and latency. Benchmarks were executed using versions 2.2.1 and 2.3.2 of the J2SE binding of JXTA. For the C binding, the CVS head of JXTA-C from the 18th of January 2005 was used. Both bindings were configured to use TCP as the underlying transport protocol. When benchmarks are performed using JXTA-J2SE, the Sun Microsystems Java Virtual Machine (JVM) 1.4.2 is used and executed with |-server -Xms256M -Xmx256M| options. The use of other JVMs is explicitly noted and executed with equivalent options. Also note that when the Java binding of JXTA is benchmarked, an additional warm-up phase based on 1000 consecutive message-acknowledgment pairs is performed. Finally, the JXTA-C benchmarks are compiled using |gcc| 3.3.3 with the |O2| level of optimization. Tests were performed on the following two types of networks, typically used for building grid infrastructures.

SAN Benchmarks. The networks used for the SAN benchmarks are Giga Ethernet and Myrinet (GM driver, version 2.0.11). When the network layer is Myrinet, nodes consist of machines using 2.4 GHz Intel Pentium IV processors, outfitted with 1 GB of RAM each, and running a 2.4 version Linux kernel. For Giga Ethernet, nodes consist of machines using dual 2.2 GHz AMD Opteron processors, also outfitted with 1 GB of RAM each, and running a 2.6 version Linux kernel. Since direct communication amongst nodes of a SAN-based cluster is available, direct communications between peers has also been configured. Note that this is allowed by JXTA specifications and therefore does not require the use of additional peers.

WAN Benchmarks. The platform used for the WAN benchmarks is the Grid'5000 French national grid platform [19]. Tests were performed between two of the Grid'5000 clusters located in Rennes and Toulouse. On each side, nodes consist of machines using dual 2.2 GHz AMD Opteron processors, outfitted with 1 GB of RAM each, and running a 2.6 version Linux kernel. The two sites are interconnected through a 1 Gb/s link, with an average measured latency of 11.2 ms. Note that as direct communication between nodes is possible within the Grid'5000 testbed, we configured JXTA peers to enable direct exchanges. As for the SAN benchmarks, this allowed by JXTA specification, even if this is clearly an unusual deployment scenario for P2P systems, where direct communication between peers is the exception rather than the rule (because of firewalls, etc.). However, let us stress that on some grids direct communication is only available between cluster front-ends. In that case, additional evaluations would be necessary.

Communication Protocols. JXTA communication layers provide three basic mechanisms for inter-peer communication, with different levels of abstraction. The *endpoint service* is JXTA's lowest, point-to-point communication layer which provides an abstraction for available underlying transport protocols. Messages sent by this layer are comprised of a series of named and typed *message elements* [20]. These elements may be required by higher communication layers or added by the application (e.g. the message payload). The *pipe service*, built on top of the endpoint layer, provides virtual communication channels (or *pipes*), which are dynamically bound to peer endpoints at runtime, thus allowing developers to abstract themselves from dynamic, runtime changes of physical network addresses. In this paper, we focus on point-to-point pipes, called *unicast pipes*. Finally, on top of pipes, the *JXTA sockets* add a data-stream interface, and implement reliability guarantees. JXTA sockets extend the BSD socket API, while still preserving the main feature of pipes: independence from the physical network. However, it should be noted that this layer is not part of the core specifications of JXTA. It is currently only available in JXTA-J2SE.

4 Performance Evaluation of JXTA over System-Area Networks

This section analyzes the performance of JXTA's communications layers on SANs. Note that for Myrinet, the *Ethernet emulation* mode of GM 2.0.11 is used and configured with jumbo frames. This mode allows Myrinet to carry any packet traffic and protocols that can be transported by Ethernet, including TCP/IP. Although this capability is bought at the cost of losing the main advantage of a Myrinet network (e.g. OS-bypass mode), it allows the same socket-based benchmarks to be run unmodified. On this configuration, the bandwidth and latency of plain sockets is around 155 MB/s and 60 μs respectively, whereas on Giga Ethernet it is around 115 MB/s for the bandwidth and 45 μs for the latency (average values between C and Java sockets). These values are used as a reference performance bound.

4.1 Analysis of JXTA-J2SE's Performance

JXTA-J2SE Endpoint Service. Figure 1 shows that the endpoint service of JXTA 2.2.1 nearly reaches the bandwidth of plain sockets over SAN networks: 145 MB/s over Myrinet and 101 MB/s over Giga Ethernet. However, Figures 2 also shows that the bandwidth of the JXTA 2.3.2 endpoint layer has decreased: drops of 32 MB/s over Myrinet and 20 MB/s over Giga Ethernet are observed. These lower bandwidths affect all versions of JXTA above its release 2.2.1 and are explained by a new implementation of the endpoint layer that shipped with JXTA 2.3. The profiling of JXTA has pointed

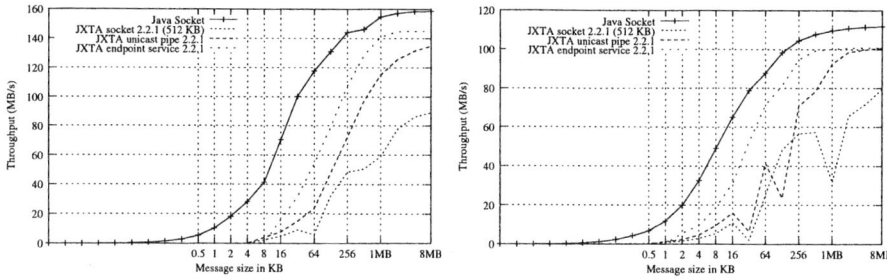

Fig. 1. Bandwidth of each layer of JXTA 2.2.1 as compared to Java sockets over a Myrinet network (left) and a Giga Ethernet network (right)

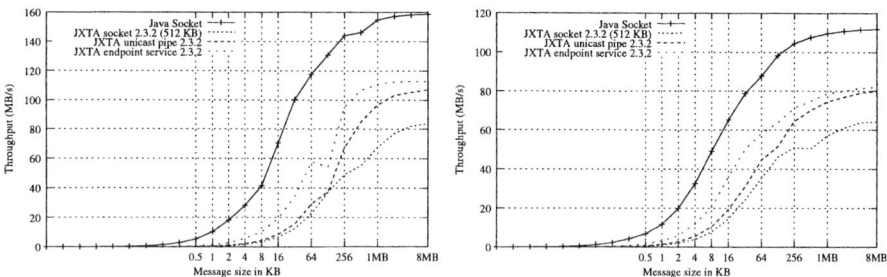

Fig. 2. Bandwidth of each layer of JXTA 2.3.2 as compared to Java sockets over a Myrinet network (left) and a Giga Ethernet network (right)

Table 1. Latency results for JXTA-J2SE and JXTA-C

Version of JXTA	JXTA-J2SE 2.2.1		JXTA-J2SE 2.3.2		JXTA-C	
Network	Myrinet	Giga Ethernet	Myrinet	Giga Ethernet	Myrinet	Giga Ethernet
Endpoint service	890 μs	357 μs	624 μs	294 μs	635 μs	322 μs
Unicast pipe	1.9 ms	834 μs	1.7 ms	711 μs	1.7 ms	727 μs
JXTA socket	3.3 ms	1.3 ms	2.4 ms	977 μs		

out that this drop of performance is due to the mechanism used for limiting the size of messages sent by the endpoint layer. Moreover, since JXTA 2.3, the limit has been lowered to 128 KB of application-level payload (larger messages are dropped). This limitation was introduced into JXTA to in order to promote some fairness in resource sharing among peers on the network, for instance when messages must be stored on relay peers (the type of peer required to exchange messages through firewalls). However, as no relay peers are needed when using SANs, we removed this limit. Table 1 shows that latency results of JXTA-J2SE have improved since version 2.2.1. The latency of the JXTA 2.3.2 endpoint service over Giga Ethernet reaches a value under 300 μs. Moreover, it goes down even further to 268 μs and 229 μs when using the SUN 1.5 and IBM 1.4.1 JVMs, respectively. The difference between Myrinet and Giga Ethernet results is due to the hardware employed, as the Ethernet emulation mode is used for Myrinet.

JXTA-J2SE Unicast Pipe. In addition, Figure 1 and 2 demonstrate a bandwidth degradation for JXTA-J2SE. For example, while JXTA 2.2.1 unicast pipe attains a good peak bandwidth of 136.8 MB/s over Myrinet, its 2.3.2 counterpart reaches a bandwidth of only 106.5 MB/s. A similar performance degradation can be observed on Giga Ethernet. However, the shape of the curve of unicast pipes 2.2.1 on Giga Ethernet has not been explained so far. We suspect the first drop is due to a scheduling problem. On the other hand, the reason of the drop at 128 KB of application payload is still unknown. At the same payload size, a smaller drop for JXTA unicast pipes 2.3.2 over Giga Ethernet can be observed, but no link have been established with the previously mentioned drop, as this drop also occurs at the endpoint level. Overall, the small performance degradation as compared to the endpoint layer is explained by the composition of a pipe message: the presence of an XML message element requiring a costly parsing prevents this layer from reaching the performance of the endpoint layer. Moreover, as shown on table 1, this extra parsing required for each pipe message also affects latency results: compared to the endpoint layer, latencies increase by more than 400 μs. However, unicast pipes are still able to achieve latencies in the sub-millisecond range, at least on Giga Ethernet.

JXTA-J2SE Sockets. As opposed to previous layers, JXTA sockets are far from reaching the performance of plain Java sockets. Indeed, in their default configuration (e.g. with an output buffer size of 16 KB), JXTA sockets 2.2.1, for instance, attain a peak bandwidth of 12 MB/s over a Myrinet network. As for unicast pipes, a similar low bandwidth result is reported on Giga Ethernet. We were able to significantly improve the bandwidth and achieve 92 MB/s by increasing the size of the output buffer to 512 KB, as shown on Figures 1 and 2. As for the unicast pipes, the irregular shape of JXTA sockets 2.2.1 curves has not been explained so far. Again, we suspect the first drop is due to a scheduling problem. The next drop may be due to some message losses when the message size is around the size of the output buffer, since many reliability issues have been fixed up to JXTA 2.3.2. Table 1 highlights the progress being made by JXTA on latency, as only JXTA Sockets 2.3.2 on Giga Ethernet is able to reach a latency under one millisecond.

Discussion. In conclusion, JXTA-J2SE 2.2.1 communication layers are able to nearly saturate SANs, but only at the endpoint and pipe levels. The measurements revealed that the bandwidth of JXTA 2.2.1 is higher than JXTA 2.3.x. Latency results have largely improved since JXTA 2.2.1, but without reaching reasonably good performance for SANs. Finally, this evaluation has also highlighted that, in their default configuration, JXTA sockets achieve a very poor bandwidth. However, this result can significantly be improved by increasing the output buffer size. This requires the JXTA socket programmer to explicitly set this parameter in the user code. Based on these results, we can conclude that JXTA-J2SE can be adapted in order to benefit from the potential offered by SANs, at least on the bandwidth side.

4.2 Analysis of JXTA-C's Performance

Figure 3 shows the bandwidth measurements of all the communications layers of JXTA-C over SANs. Note that, as in the previous section, C sockets are used as an upper reference bound. The peak bandwidths of the endpoint service over Myrinet and Giga

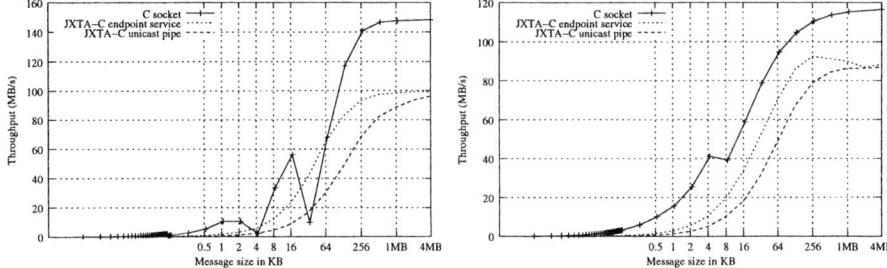

Fig. 3. Bandwidth of each layer of JXTA-C as compared to C sockets over a Myrinet network (left) and a Giga Ethernet network (right)

Ethernet are 100 MB/s and 92.5 MB/s, respectively. The upper layer (unicast pipe) reaches bandwidths of 97.6 MB/s and 87.7 MB/s over Myrinet and Giga Ethernet, respectively. These unsatisfactory results are due to memory copies being used in the implementation of the endpoint layer of JXTA-C. Table 1 highlights reduced latencies, especially on Giga Ethernet, as compared to results published in [6]: 820 μs for the endpoint layer and 1.99 ms for the pipe layer. To achieve this improvement, we modified the implementation of the endpoint layer of JXTA-C by disabling TCP packet aggregation mechanism, as it adds significantly to the latency. Consequently, the buffering mechanism is now performed within the endpoint layer allowing one TCP packet to be sent for a single JXTA message with a minimal latency. These modifications have been committed into the CVS of JXTA-C and are publicly available.

Based on this evaluation, we can conclude that, in their current implementation, the communication layers of JXTA-C are not able to saturate SANs. The non-zero copy implementation of the endpoint layer prevents JXTA-C from approaching Gb/s bandwidths available over SANs. Note, however, that JXTA-C is in the process of being revived; so we believe that the performance of JXTA-C will increase in the near future such that it will be able to efficiently use SANs.

4.3 Fully Exploiting SAN Capacities

In all previously reported evaluations based on Myrinet, the Ethernet emulation mode of GM is used. However, this removes the ability to by-pass the IP stack of the OS and introduces unneeded overhead. Consequently, communication layers are unable to fully exploit the capacities offered by Myrinet: full-duplex bandwidths of nearly 2 Gb/s and latencies of less than 7 μs thanks to zero-copy communication protocols. PadicoTM [21] is a high-performance framework for networking and multi-threading which allows middleware systems to transparently take advantage of such features. In this section, we focus on the *virtual sockets* feature offered by PadicoTM, which provides a way to directly access GM network interfaces. This is achieved by dynamically mapping, at runtime, standard sockets functions on top of GM API functions, without going through the TCP/IP stack. Zero copy is therefore possible and allows, for example, plain sockets to transparently reach a bandwidth of more than 230 MB/s and latency of 15 μs on Myrinet, compared to 160 MB/s and 51 μs without PadicoTM.

We have successfully ported JXTA-C to PadicoTM, without changing one line of code of JXTA-C. We only performed some minor modifications inside the OS-independent layer used by JXTA-C: the Apache Portable Runtime (APR). We changed from the default Posix thread library on which APR is based, to the Marcel [22] thread library used by PadicoTM. However, these modifications could be automatically achieved by a single |sed| command. An improvement of 32 MB/s for the bandwidth of JXTA-C's endpoint layer has been measured resulting in a peak bandwidth of 140 MB/s, thus reaching over 1 Gb/s. On the latency side, no significant improvements have been observed, as the non-zero copy communication layers prevents JXTA-C from fully benefiting from the OS-bypass feature of Myrinet. Note that we did not use PadicoTM with JXTA-J2SE, since PadicoTM currently supports only the open-source Kaffe JVM. Unfortunately, this JVM is not compliant with Java specification version 1.4 and, therefore, is unable to run the Java binding of JXTA.

In conclusion, our experiments with PadicoTM show that JXTA could fully benefit from the potential performance of SAN networks if: 1) the implementation of all JXTA communication layers should respect a zero-copy policy, 2) PadicoTM adds support for JXTA-compatible JVMs (e.g. compliant to version 1.4 of Java's specifications). However, we believe that these issues will be solved in the near future.

5 Performance Evaluation of JXTA over Wide-Area Networks

We performed the same type of measurements on WAN networks. Note that we had to tune network settings of nodes used for this benchmark. Our default maximum TCP buffer size initially set to 131072 bytes was limiting the bandwidth to only 7 MB/s. Based on the *bandwidth* ∗ *delay* law, we computed a theoretical maximum size of 1507328 bytes and increased this value by an arbitrary factor of 1.2. Therefore, we set the maximum TCP buffer sizes on each node to 1959526 bytes; |ttcp| configured with this value measured a raw TCP bandwidth of 107 MB/s, a reasonable level of performance.

JXTA-J2SE's Performances. As for SAN Giga Ethernet benchmarks, Figure 4 shows that the endpoint layer and unicast pipes of JXTA-J2SE are able to perform similarly to plain sockets over a high-bandwidth WAN of 1 Gb/s. This level of performance was reached by modifying JXTA-J2SE's code in order to properly set TCP buffer sizes to 1959526 bytes before binding sockets on both sides. Using the default setting, a bandwidth of only 6 MB/s was reached for JXTA 2.2.1 and less than 1 MB/s for JXTA 2.3.2. As opposed to SAN benchmarks, both versions of JXTA-J2SE achieve the same performance. This can be explained by the the fact that the higher latency of WANs hides the cost of the mechanism implemented for limiting the size of JXTA messages. Figures 4 also points out the same performance degradation for JXTA sockets as for SAN benchmarks. However, JXTA socket 2.3.2 achieves a higher bandwidth compared to its 2.2.1 counterpart. Performance drops of unicast pipes and JXTA sockets for JXTA-J2SE 2.2.1 for message size of 4 MB have not been explained so far.

JXTA-C's Performances. Figure 5 shows similar results for the communication layers of JXTA-C over WANs compared to the SAN benchmarks: both layers reach a

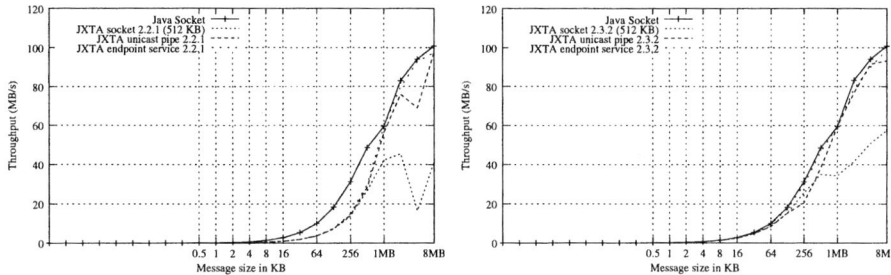

Fig. 4. Bandwidth of each layer for JXTA-J2SE 2.2.1 (left) and 2.3.2 (right) compared to Java sockets over a high-bandwidth WAN

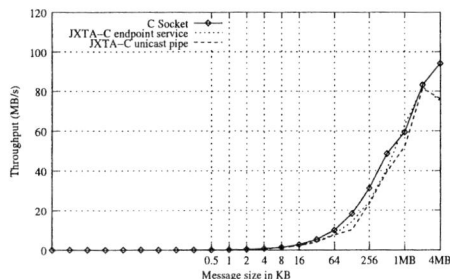

Fig. 5. Bandwidth of each layer for JXTA-C compared to C sockets over a high-bandwidth WAN

bandwidth slightly above 80 MB/s, for a message size of 2 MB. The observed performance degradation after this message size has not been explained.

Discussion. Based on this evaluation, we can conclude that JXTA's communication layers, when used on high-bandwidth WANs, are able to reach the same bandwidths as for SAN benchmarks. Both versions of JXTA-J2SE (and not only JXTA-J2SE 2.2.1) are able to efficiently use the bandwidth available on links used for interconnecting sites of a grid, whereas the non-zero copy communication layers prevents JXTA-C from saturating this type of links.

6 Conclusion

In the context of the current efforts for building grid services on top of P2P libraries, an important question is: to what extent is it reasonable to rely on P2P mechanisms to support the dynamic character of the grid? Are P2P techniques only useful for resource discovery? Or is there a way to take one step further, and efficiently exploit P2P data communication mechanisms? The question of the adequacy of P2P communication mechanisms for performance-constrained usages is therefore important in this context.

In this paper we show that, if it is correctly tuned, the JXTA platform can deliver adequate performance (e.g., over 1 Gbit/s bandwidth) and, furthermore, we explain how this performance can be improved thanks to specialized external libraries. First, we evaluated the basic performance of the communication layers of the JXTA generic P2P framework on the Grid5000 testbed. Giga-Ethernet and Myrinet links are used as SANs within each site, while high-bandwidth WANs interconnect the various sites. We then provided an analysis of the performance of the J2SE and C bindings of JXTA over these two types of networks. We show that the JXTA-J2SE 2.2.1 communication layers are able to nearly saturate SANs, at the endpoint and pipe levels, whereas the bandwidth is poor at the JXTA socket level. We also explain how to improve this bandwidth by tuning the TCP output buffer size. Finally, we show that performance can further be improved on SANs by using the PadicoTM environment, which provides a direct access to the API of the Myrinet driver by by-passing the OS and does not require any modifications of JXTA. The overall conclusion of these experiments is that JXTA may provide adequate communication performance that are required by grid computing applications.

However, these evaluations have revealed some weaknesses of JXTA in both the SAN and WAN areas. JXTA-J2SE bandwidths have degraded since JXTA 2.3, hindering JXTA from saturating SAN links. Moreover, the communication layers of JXTA-C do not follow a zero-copy policy, therefore limiting the bandwidth and latency results. Therefore, in spite of our initial efforts, JXTA-C still needs some improvements in order to be able to fully benefit of the available bandwidth provided by SANs. When these issues are solved a bandwidth of over 200 MB/s should be reached through the use of PadicoTM. On the WAN side, we plan to use parallel streams for both bindings of JXTA, in order to allow an efficient use of high-bandwidth WANs. Again thanks to PadicoTM, this functionality will be transparently available to JXTA-C, whereas for JXTA-J2SE this would require implementing this functionality. Finally, it would also be interesting to measure the impact of on-the-fly compression techniques available in PadicoTM for WAN transfers.

References

1. Foster, I., Iamnitchi, A.: On Death, Taxes, and the Convergence on Peer-to-Peer and Grid Computing. In: 2nd International Workshop on Peer-to-Peer Systems (IPTPS '03). Number 2735 in Lect. Notes in Comp. Science, Berkeley, CA, Springer-Verlag (2003)
2. Talia, D., Trunfio, P.: Toward a Synergy Between P2P and Grids. IEEE Internet Computing 7 (2003) 94–96
3. Talia, D., Trunfio, P.: A P2P Grid Services-Based Protocol: Design and Evaluation. In: Euro-Par 2004: Parallel Processing. Number 3149 in Lect. Notes in Comp. Science, Pisa, Italy, Springer-Verlag (2004) 1022–1031
4. Amoretti, M., Conte, G., Reggiani, M., Zanichelli, F.: Service Discovery in a Grid-based Peer-to-Peer Architecture. In: International Workshop on e-Business and Model Based IT Systems Design, Saint Petersburg, Russia (2004)
5. The JXTA project. http://www.jxta.org/
6. Antoniu, G., Hatcher, P., Jan, M., Noblet, D.A.: Performance Evaluation of JXTA Communication Layers. In: 5th International Workshop on Global and Peer-to-Peer Computing (GP2PC '05), Cardiff, UK (2005) Held in conjunction with the 5th IEEE/ACM International Symposium on Cluster Computing and the Grid (CCGrid '2005).
7. Cog Kit JXTA project. http://www-unix.globus.org/cog/projects/jxta/
8. JXTA-Grid project. http://jxta-grid.jxta.org/

9. Antoniu, G., Bougé, L., Jan, M.: JuxMem: Weaving together the P2P and DSM paradigms to enable a Grid Data-sharing Service. Kluwer Journal of Supercomputing (2005) To appear.
10. Halepovic, E., Deters, R.: The Cost of Using JXTA. In: 3rd International Conference on Peer-to-Peer Computing (P2P '03), Linköping, Sweden, IEEE Computer Society (2003) 160–167
11. Halepovic, E., Deters, R.: JXTA Performance Study. In: IEEE Pacific Rim Conference on Communications, Computers and Signal Processing (PACRIM '03), Victoria, B.C., Canada, IEEE Computer Society (2003) 149–154
12. Seigneur, J.M.: Jxta Pipes Performance. (2002)
13. Seigneur, J.M., Biegel, G., Jensen, C.D.: P2P with JXTA-Java pipes. In: 2nd international Conference on Principles and Practice of Programming in Java (PPPJ '03), Kilkenny City, Ireland, Computer Science Press, Inc. (2003) 207–212
14. Halepovic, E., Deters, R.: JXTA Messaging: Analysis of Feature-Performance Tradeoffs. Submitted for publication (2005)
15. Shudo, K., Tanaka, Y., Sekiguchi, S.: P3: Personal Power Plant. GGF10: Open Grid Service Architecture - Peer-to-Peer Research Group (OGSA-P2P RG) (2004)
16. Baehni, S., Eugster, P.T., Guerraoui, R.: OS Support for P2P Programming: a Case for TPS. In: 22nd International Conference on Distributed Computing Systems (ICDCS '02), Vienna, Austria, IEEE Computer Society (2002) 355–362
17. Junginger, M., Lee, Y.: The Multi-Ring Topology - High-Performance Group Communication in Peer-to-Peer Networks. In: 2nd International Conference on Peer-to-Peer Computing (P2P '02), Linköping, Sweden, IEEE Computer Society (2002) 49–56
18. Tran, P., Gosper, J., Yu, A.: JXTA and TIBCO Rendezvous - An Architectural and Performance Comparison. (2003)
19. Grid'5000 project. http://www.grid5000.org/
20. JXTA specification project. http://spec.jxta.org/
21. Denis, A., Pérez, C., Priol, T.: PadicoTM: An Open Integration Framework for Communication Middleware and Runtimes. Future Generation Computer Systems **19** (2003) 575–585
22. Danjean, V., Namyst, R., Russell, R.: Integrating Kernel Activations in a Multithreaded Runtime System on Linux. In: Parallel and Distributed Processing. Proc. 4th Workshop on Runtime Systems for Parallel Programming (RTSPP '00). Volume 1800 of Lect. Notes in Comp. Science., Cancun, Mexico, In conjunction with IPDPS 2000. IEEE TCPP and ACM, Springer-Verlag (2000) 1160–1167

Efficient Message Flooding on DHT Network

Ching-Wei Huang and Wuu Yang

Department of Computer and Information Science,
National Chiao-Tung University, HsinChu, Taiwan, R.O.C
{rollaned, wuuyang}@sp.cis.nctu.edu.tw

Abstract. For the high scalability, DHT network becomes popular in P2P development in these few years. In comparing to flooding-based searching in unstructured P2P network, DHT network provides an efficient lookup. However, flooding still plays an important role in P2P systems. Some fundamental functions such as information collection, dissemination, or keyword searching can benefit from an efficient flooding mechanism. In this paper, we present a DHT network in which one flooding request generates $O(N)$ messages where N is the system size. Moreover, our method considers message locality. Message forwarding across different autonomous systems are reduced significantly. Base on our flooding mechanism, information broadcasting and aggregation are fulfilled without much effort.

Keywords: Peer-to-Peer System, DHT, Flooding, Broadcast, Aggregation.

1 Introduction

In the last few years, unstructured P2P networks are commonly regarded as inefficient and poor in scalability. DHT networks such as Pastry [2] and Chord [3] gradually become mainstream for fast and efficient lookup. Although message flooding is unnecessary in DHT network, it should not be abandoned so soon. We believe that flooding-based mechanisms such as information collection and propagation are becoming important. For example, P2P systems such as Viceroy [4] and Symphony [5] need to know network size to determine some global variables. These variables have to be broadcasted to all nodes for tuning routing performance. Such information collection and aggregation can benefit from an efficient flooding mechanism. On the other hand, keyword search is a fundamental function in peer-to-peer file sharing systems. Up to today, message flooding is still the most simple and effective (but inefficient) solution.

The cost of message flooding is another consideration. The Internet is a collection of autonomous systems connected by routers. An *autonomous system* [14, 15] is a set of routers under a single technical administration that uses an interior gateway protocol and common metrics to route packets within the autonomous system and an exterior gateway protocol to route packets to other autonomous systems. If the messages travel across different autonomous systems frequently, the communication will be prohibitively expensive. The research [16] shows that only 2 to 5 percent of connections in Gnutella are within a single autonomous system.

In this paper, we present a DHT network called *Distributed Search Environment (DSE)*. DSE is composed of two mechanisms : *Overlay Network Control* and *Smart Message Routing*. The former maintains routing tables and provides lookup service. The later implements an efficient message flooding. Our major consideration is to reduce redundant forwardings and traffic across autonomous systems. The simulation shows that one flooding request generates $2.93 * N$ messages, where N is the system size. Based on the efficient message flooding, we developed an tree-based method to broadcast and aggregate information on DHT network without redundant forwardings.

The remainder of this paper is organized as follows. Section 2 and 3 presents NC and SMR mechanism, respectively. Section 4 presents the application. Section 5 gives related work and Section 6 concludes this paper.

2 Overlay Network Control

DSE is a server-less overlay network. Every newly joining node initially connects to some default bootstrap nodes. Once any connection is built, it joins the system and updates neighbors automatically. These default bootstrap nodes act exactly as the other nodes except that they are initial neighbors of every new node.

We first define some terms here for later discussions. A *lookup* is a request which is initialized by a node to search a given key. The request eventually reaches a node which is closest to the key globally. A *flooding* request is initialized by a node, and forwarded to every node in the system. Given a message M which is forwarded from node H_1 to node H_2 in one hop, H_1 is called the *source* of M, and H_2 is called the *destination* of M.

2.1 ID Space

Like other DHT networks, each node and key in DSE has a unique ID, which is a hexadecimal digits string of length $DepthMax$, where $0 < DepthMax < 32$. We define the *depth* of each digit as its position in the ID (indexed from 0). The prefix of ID of size ℓ, denoted by $Prefix(ID, \ell)$, is defined as the substring of the first ℓ digits. For well definition, $Prefix(ID, \ell)$ is an empty string for all $\ell < 0$. The $\ell'th$ digit of ID is denoted by $depthChar(ID, \ell)$. Figure 1(a) shows an ID and its prefix of size 6.

For distinguishing a node ID from a key, the former is denoted by NID. The NID is composed of two parts. The first part is called ISPID, which is a string of two hexadecimal digits. ISPID identifies the ISP (Internet Service Provider) the node is using. The second part is a $DepthMax - 2$ hexadecimal digits string. When a new node joins the system for the first time, the application on the node hashes user-related data such as login time, IP address, and personal information, to generate the second part of NID. Notice that the NID can be treated as permanent because most users seldom change their ISP.

For a given digit C in the circle of hexadecimal digits, we define the left side and right side of C to be the half part in clockwise and anti-clockwise direction, respectively. The digit at just the half position is in the left side if both C and

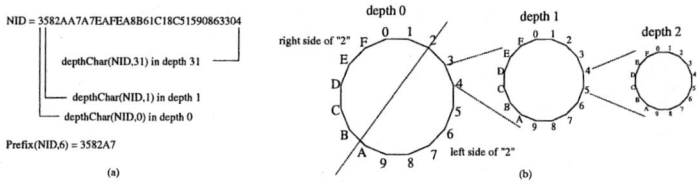

Fig. 1. ID prefix and ID space

the digit are odd or even, or in the left side otherwise. Figure 1(b) shows the circle and both sides of digit "2" in depth 0.

The distribution space of the ID can be regarded as a ring structure in Figure 1(b). The hashing function promises that all IDs distribute over the entire ID space uniformly. Given a set of ID $\{D_i\}$ in which each ID shares a common prefix of length ℓ but has different $depthChar(D_i, \ell)$, each ID falls in the same segment of first $\ell - 1$ depths but different segments of depth ℓ.

NID Distribution: Although hashing function generates an ID of up to 32 digits, it is not necessary to use all of them. Let us consider a system of N nodes in which all IDs distribute over ID space uniformly. Given a NID X and ℓ is the number of digits of a ID, the set of IDs which shares a common $Prefix(X, \ell)$ contains at most $N/16^\ell$ elements. To make each ID distinguishable, ℓ should be large enough so that there is only one ID (which is X itself) to have $Prefix(X, \ell)$. Therefore, the lower bound of ℓ is $\lceil log_{16} N \rceil$. In other words, for a system of N nodes, *DepthMax* should be set to at least $\lceil log_{16} N \rceil$ to make each NID unique. In the following sections, all NIDs are represented in 8 digits.

ID Distance: Given two IDs $A = a_0 a_1 ... a_7$ and $B = b_0 b_1 ... b_7$, we define the distance between A and B, denoted by $IDD(A, B)$, as the length of the shortest path between them in ID space ring. More precisely, let $D = \left| \Sigma_{i=0}^{7} (a_i - b_i) * 16^{7-i} \right|$, $IDD(A, B)$ is defined as D if $D < 16^7$, or $16^8 - D$ otherwise. Given three nodes A, B, and C, we say that B is closer to A than C if $IDD(A, B) < IDD(A, C)$.

Autonomous Systems: To understand the whole autonomous systems is a big challenge, DSE considers ISPs instead. In order to determine the correct ISP to which a node belongs, each node maintains a table of mappings of IP address ranges to their ISPIDs. Since the mappings may changes as time goes on, the table must be able to be updated. DSE implements the update mechanism by message flooding. The system administrator maintains and distributes the most correct mapping table on a regular time schedule. Usually, address ranges of ISPs are available to the public. For example, the ISP address ranges to be used in our simulation are collected from the Internet supervisor bureau [17].

2.2 Routing Table

Like Chord and Pastry, every node in DSE maintains a routing table. The node in the routing table is called the *neighbor*. The routing table contains two sets of neighbors: far neighbor set and near neighbor set. The far neighbor set contains a

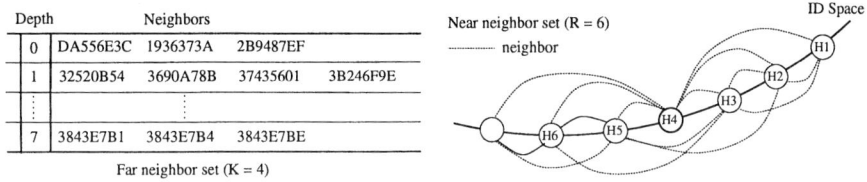

Fig. 2. Neighbor sets of node H_4 (whose NID is $3843E7B9$)

number of neighbors for each depth of NID. Given a node H_x, a node H_y is a far neighbor of H_x in depth ℓ implies that $Prefix(NID_x, \ell-1) = Prefix(NID_y, \ell-1)$. The near neighbor set contains a number of closest neighbors. Notice that the near neighbor set contains two special neighbors called predecessor and successor, which are the closest neighbors in the left and right side in the ID space ring. Suppose each node contains at most K near neighbors where $K > 0$, then for two given nodes H_1 and H_2, the symmetry of ID distance implies that if H_1 is a near neighbor of H_2, then H_2 is a near neighbor of H_1. The example in Figure 2 shows the far neighbor set and near neighbor set of a node.

Given a system of N nodes, the number of neighbors of each node can be estimated as follow. From previous discussion of NID distribution, there are $\lceil log_{16} N \rceil$ depths in which neighbors may exist. Suppose each node has a near neighbor set of size R, and there are at most K neighbors in each depth, then each node has $R + K * \lceil log_{16} N \rceil$ neighbors.

Sharing: The neighbor relation in DSE is not always symmetric (which means that A is a near neighbor of B implies that B is a near neighbor of A). For near neighbors, the symmetry of ID distance promises that the neighbor relation is always symmetric. But for far neighbors, however, it is not true. In order to perform an efficient filtering for message floodings, each node stores a copy of each neighbor's routing table. For near neighbors, to synchronize routing tables each other in minimal cost, each node only sends changed parts of its routing table to every near neighbor. For far neighbors, to obtain their routing tables, the local node queries them periodically.

Connections: In most DHT networks, a node connects to a neighbor only when a message is forwarded to it. DSE follows the same design with an exception that each node builds persistent connections to its predecessor and successor. With these persistent connections, node leaving can be detected and routing tables of other nodes can be updated for it. Section 2.4 gives a more detail discussion for it.

2.3 Lookup

In DHT networks, there are two models of lookup forwarding: recursive and iterative. A recursive lookup is forwarded from it's originator to a closest neighbor. The receiver also forwards it to a closest neighbor. The process repeats until the request is forwarded to the most closest node in the system. For an iterative lookup, the originator does not forward it. Instead, it queries a closest neighbor (to the key of the lookup) for a closest neighbor. The originator iteratively queries nodes for a closest neighbor. The process repeats until a node is queried but found no more closer neighbor to be return. The research [19] shows that the recursive lookup is more efficient than the iterative lookup, but it suffers a problem of lost requests. Additional effort is necessary to detect and recover lost lookups. Besides the lookup, the nature of message flooding is more similar to the recursive lookup. For efficiency and consistency, DSE adopts the recursive lookup model.

Lost Request Detection: To solve the problem of lost lookups, each node monitors every forwarding request. Given a node H_0 which forwards a request R to a neighbor H_1. The connection between them is held temporarily. When H_1 received R and made a decision that R will be forwarded or finalized (which means H_1 is the final destination), an acknowledgement of R is sent back to H_0. The connection is held as long as H_0 has not received the acknowledgement. If H_0 detects the connection is broken but not yet receives the acknowledgement, H_1 will be removed from the routing table and H_0 forwards R to a new closest neighbor. If the waiting for the acknowledgement exceeds a reasonable period, H_0 does not remove H_1 from its routing table, but just forwards R to the second closest neighbor. Obviously, lost requests bring a problem of redundant forwardings. DSE solves this by a simple solution. Each request has a unique tag which is composed of the requester's NID, address, and a timestamp. Each node stores tags of recently received requests. Each time a new request arrives, it will be discarded if its tag was stored before.

2.4 Neighbor Update

When a new node joins or a node leaves the system, the routing table should be initialized (for the case of join), and other nodes update their routing tables for it (for both cases). We discuss the maintenance of routing tables for each case.

Node Join: When a new node H_0 joins the system, it search neighbors in three steps: predecessor and successor, other near neighbors, and finally far neighbors.

Predecessor and Successor. H_0 initializes a lookup with a key of its NID. The lookup will reach a node, say H_1, which is closest to H_0. Let H_p and H_s be the predecessor and successor of H_1, respectively, then H_0 must locate in the segment (H_p, H_1) or (H_1, H_s) in ID space. Therefore, H_0 simply queries H_1 to acquire its predecessor and successor. When H_0 builds persistent connections to its predecessor and successor, they also update routing tables for H_0.

Near Neighbors. Suppose a node contains at most R near neighbors, to obtain them can be done by the following steps. After the predecessor and successor are connected, H_0 assigns a special request with a counter of $R/2$, and forwards it to the predecessor. The predecessor receives the request, decreases the counter, and forwards it to its predecessor. The forwarding repeats until the counter becomes zero. Each node that received the request informs H_0 and become a

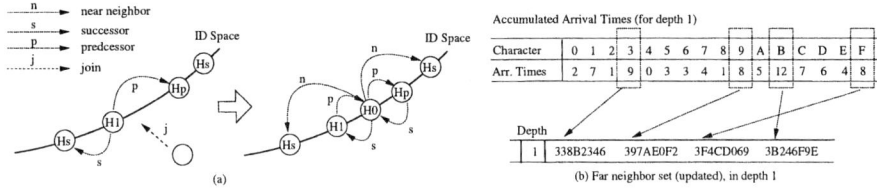

Fig. 3. (a) A node H_0 joins the system. (b) far neighbor selection based on a dynamic condition.

near neighbor of H_0. Similarly, H_0 forwards another request to its successor to obtains another half near neighbors. Figure 3(a) shows the process to obtain predecessor, successor, and near neighbors.

Far Neighbors. Most DHT networks consider static routing table, which means that neighbors are selected based on a static condition. In Chord, for example, node j exists in the routing table of node i if there are 2^k nodes between them in ID space ring. We believe that such static condition based selection of neighbors is not necessary. We propose a dynamic selection rule for far neighbors. Each node checks the key of each arrived lookup, and stores the accumulated arrival times for each character (of the key), and for each depth. With more characters of keys are accumulated, we can adjust far neighbors such that their $depthChars$ have highest frequency for each depth. The adjustment is achieved by parsing the tag of each arrived request, and inserts the originator to the far neighbor set if some $depthChar$ of its NID for some depth is higher than the $depthChar$ of the current neighbor for that depth.

On the other hand, to increase the opportunity of accepting better far neighbors, each new node initiates a number of lookups with random keys when it joins the system. When these lookup requests are forwarded from a node to another, the far neighbor selection is performed. Obviously, these keys are randomly generated so that each node in the system has an equal chance to receive them. Figure 3(b) shows that the far neighbor set in Figure 2 has been updated: three new neighbors (in depth 1) join the set, for their higher accumulated arrival times.

Node Leave: To leave the P2P system, a node can either gracefully leave by notifying other nodes or leave without any notification (e.g. node failures). DSE is designed to handle frequent ungraceful leaving of nodes while remain the robustness and performance of the system. Given a node H_0 and its predecessor H_1 and successor H_2. When H_1 detects H_0 leaves, H_1 sets H_2 to be the new successor. Similarly, H_2 also detects H_0 leaves and sets H_1 to be the new predecessor. Then H_1 and H_2 negotiate each other to let only one of them handles the leaving of H_0. Be default, the node with a bigger NID takes the role: H_2 simply initiates lookup requests, and each with a key of NID of one neighbor of H_0 (remember that H_2 stores a copy of H_0's routing table). Every node that received the request removes H_0 from the routing table if it exists.

3 Smart Message Routing

In this section, we discuss the efficiency of flooding requests. Given a message M which is forwarded from one node to another, we say that M is redundant if it arrived at the same destination more than once. Message flooding inevitably generates redundant messages. SMR provides two methods to reduce them: *redundant forwarding prevention* and *duplicated forwarding avoidance*. The difference between them is the timing to perform. The former works before the message is forwarded (it is not yet, but may be redundant later), while the later works after a message arrived at a node (it is already redundant). For the part of duplicated forwarding avoidance, Section 2.3 already provided an implementation.

3.1 Redundant Forwarding Prevention

The redundant forwarding prevention filters forwardings carefully so that a flooding request generates $O(N)$ messages, where N is the system size. To fulfill such filtering, three prerequisites should be held: (1) each node stores a copy of each neighbor's routing table. Section 2.2 presents an implementation of it. (2) the clustering coefficient[1] of the system should be as high as possible. (3) each flooding request contains a list called BCTL (means *broadcast traverse list*) which stores a number of latest visited nodes. The capacity of BCTL is set as multiples of the number of near neighbors. We use the term *"BCTL of multiple K"* to denote that the capacity is K multiplies the number of near neighbors. BCTL of multiple 0 means that redundant forwarding prevention is disabled. For easier discussion, the notation $BCTL(M, H, 0)$ means the BCTL before a request M is forwarded to its destination H, and $BCTL(M, H, 1)$ means the BCTL after M is forwarded out from H. On the other hand, the replacement of BCTL items adopts first-in-first-out policy. BCTL always keeps a number of latest visited nodes.

To prevent redundant message forwardings, we consider two geometric structures in the overlay network topology: triangles and quadrangles. The redundant forwarding prevention includes three stages. The first stage filters visited nodes. The second stage detects redundant forwardings on triangles. The third stage detects redundant forwardings on quadrangles.

First Stage: Given a node, the first stage prevention focuses on the nodes to which a request will be forwarded. Before a request is forwarded by the local node, each neighbor in the BCTL will be filtered out because these nodes are visited. Then, the rest of neighbors are pushed into the BCTL before the request is forwarded to them. In the example in Figure 4(a), there are two groups $\{H_1, H_2, H_3, H_4, H_5\}$ and $\{H_6, H_7, H_8, H_9, H_{10}\}$. Node H_2 originates a flooding request M. The request M is first sent to neighbors H_1, H_3, and H_5. Hence,

---

[1] Suppose that a node V has K neighbours; then at most $K * (K - 1)/2$ connections can exist between them. Let C_v denote the fraction of these allowable connections that actually exist. Define the clustering coefficient C as the average of C_v over all V [11].

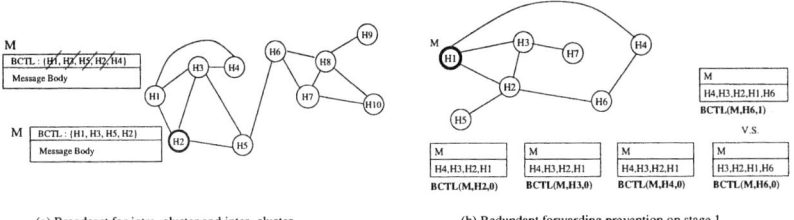

Fig. 4. (a) Broadcasting across groups and (b) redundant forwarding prevention on first stage

$BCTL(M, H_2, 1) = \{H_1, H_2, H_3, H_5\}$. After H_1 received M, it will not be forwarded to H_3 because H_3 is in $BCTL(M, H_1, 0)$.

The capacity of BCTL also affects the performance of prevention. In the example in Figure 4(b), suppose each node has four near neighbors and we setup BCTL of multiple 1. Node H_1 initiates a flooding request M to its neighbors H_2, H_3, and H_4. With the first stage prevention, the message forwarding $H_2 \to H_3$ and $H_3 \to H_2$ are filtered out by node H_2 and H_3, respectively. When M is forwarded to H_6 by H_2, $BCTL(M, H_6, 0)$ becomes $\{H_3, H_2, H_1, H_6\}$ and H_4 is removed from the set because of the BCTL capacity. When M is forwarded to H_4 by H_6, a redundant forwarding occurs. If we setup BCTL of multiple 2, such redundant forwarding can be prevented because $BCTL(M, H_6, 0)$ still contains H_4.

Second Stage: The second stage prevention focuses on forwardings by different sources to the same destination. Before a node forwards a request, if it knows that one of its neighbors will also forward it to the same destination, then only one of them should forward it. The choice of the forwarder depends on their NIDs. When several nodes prepare to forward the same request to the same destination, only the node with largest NID forwards it.

Given a flooding request M and a node H, $NBRQ(H)$ is the set of neighbors of H, $TS(M, H)$ is $NBRQ(H) - BCTL(M, H, 0)$ and $NS(M, H)$ is $NBRQ(H) \cap BCTL(M, H, 0)$ (the notation "-" means set difference). $TS(M, H)$ represents the remaining nodes to which M will be forwarded by H after the first stage. $NS(M, H)$ represents the neighbors of H that are already in $BCTL(M, H, 0)$. The concept of the second stage prevention is: for each node Y in $TS(M, H)$, if there is a node X in $NS(M, H)$ such that Y is a neighbor of X and the NID of X is larger than the NID of H, then H discards the forwarding of M to Y.

Consider an example in Figure 5(a). The string inside each angle bracket, e.g. <ID461>, denotes the NID. After H_1 forwarded the request M to neighbors H_2, H_3, H_4 and H_5, the forwarding of $H_2 \leftrightarrow H_3$ and $H_3 \leftrightarrow H_4$ are filtered by the first stage prevention. Now consider the forwarding $H_2 \to H_6$, $H_3 \to H_6$ and $H_4 \to H_6$. With the second stage prevention, H_3 discovers two facts: (1) the NID of H_2 (which is "ID2192") is greater than that of H_3 (which is "ID47"), (2) $NBRQ(H_2)$ and $NBRQ(H_3)$ contain a common destination H_6. Therefore,

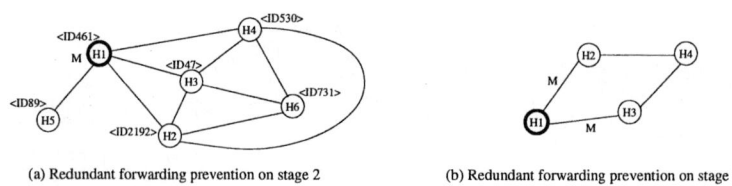

(a) Redundant forwarding prevention on stage 2   (b) Redundant forwarding prevention on stage 3

Fig. 5. Redundant forwarding prevention on second and third stage

H_3 discards the forwarding of M to H_6. The similar fact will also be discovered by H_4. As a consequence, M is forwarded to H_6 only by H_2. For the view of geometric structures, redundant forwardings on triangles $\{H_1, H_2, H_3\}$ and $\{H_1, H_3, H_4\}$ are considered to be detected.

Third Stage: The third stage prevention addresses on redundant forwarding on quadrangles. Consider an example in Figure 5(b) where node H_1 forwards a request M to neighbors H_2 and H_3. Since there is no connection between H_2 and H_3, their common neighbor H_4 will receive M twice. Such duplicated forwarding can be detected in advance by H_1: by checking the intersect of $NBRQ(H_2)$ and $NBRQ(H_3)$, H_1 found an element H_4. The third stage prevention reduces duplicated forwardings by inserting $BCTL(M, H_2, 0)$ with additional element H_4 before M is forwarded to H_2. As for the forwarding of M to H_3, no additional insertion is required. In a general form, the third stage prevention (handled by H_0 for the message M) searches each pair (H_x, H_y) of its neighbors such that $H_x \bigcap H_y$ is not empty. Each element of $H_x \bigcap H_y$ will be inserted into $BCTL(M, H_x, 0)$. Figure 6 lists the complete algorithm of redundant forwarding prevention, including the first, second, and third stage.

Performance: The performance of redundant forwarding prevention is highly related to two factors: the topology of the system and the capacity of BCTL. For a system with higher clustering coefficient, more redundant forwardings are

```
function broadcast(msg M) {
  // + : set union
  // * : set intersect
  // - : set difference
  // self : local node
  TS:=self.NBRQ - M.BCTL;
  NS:=self.NBRQ * M.BCTL;
  SQ:=[];
  foreach (X in TS) {
    if (exist G in NS such that
        G.NID>self.NID && X in G.NBRQ)
      continue;
    SQ:=SQ + {X};
  }
```
```
  M.BCTL:=M.BCTL+SQ;
  foreach (Y in SQ) {
    A:=[];
    foreach (Z in SQ) {
      if (Y.NID>Z.NID || Z in Y.NBRQ))
        continue;
      A:=A+(Y.NBRQ * Z.NBRQ);
    }
    M.BCTL:=M.BCTL+A;
    checkBCTL(M.BCTL);   // reduce size if exceed
    forward(M,Y);   // forward M to node Y
  }
}
```

Fig. 6. Redundant forwarding prevention algorithm

prevented because BCTL of each message contains more nodes which are neighbors each other. As for the capacity of BCTL, larger BCTL always prevents more redundant forwardings because requests contain more information of visited nodes. The next section will show how large the BCTL should be set to prevent most redundant forwarding.

On the other hand, the path length of a flooding request determines how fast it can reach all nodes in the system. Since each node broadcasts a flooding request to its neighbors. It can be induced that the average length of flooding requests is $O(log_R N)$ where N is the system size and R is the maximal number of near neighbors for each node.

3.2 Simulations

To verify the performance of DSE, we adopt a strategy of developing both a real system and a simulator. The real system, with a project name *apia* [18], is publicly available. DHT-based lookup and flooding-based search are both provided. User-related data such as NIDs of nodes, search requests, online period, ..., etc, are collected for our simulation. The first part of SMR, duplicated forwarding avoidance, is always enabled so that each request will never be forwarded infinitely. In the simulation, each node has at most 20 near neighbors, and each node initiates lookup requests and flooding requests periodically. Our major concern is the average ratio of the number of messages generated by one flooding request to the system size. The ratio is denoted as MRR (means message redundant ratio). We also use $SW(S_1, S_2, S_3)$ to represent the switch to enable or disable redundant message preventions for the first, second, and third stage.

The first observation is the relation between MRR and $BCTL$ of multiple K under $SW(1,1,0)$. We record MRR for different system size. The result in Figure 7(a) shows that larger BCTL capacity always reduces more redundant messages, but the reduction slows down. MRR under $BCTL$ of multiple K remains at a lower bound of 3.26 for $K \geq 4$. $BCTL$ of multiple 5 should be large enough to prevent redundant forwardings as much as possible. In the next, we observe MRR under $SW(0,0,0)$, $SW(1,0,0)$, $SW(1,1,0)$, and $SW(1,1,1)$. The

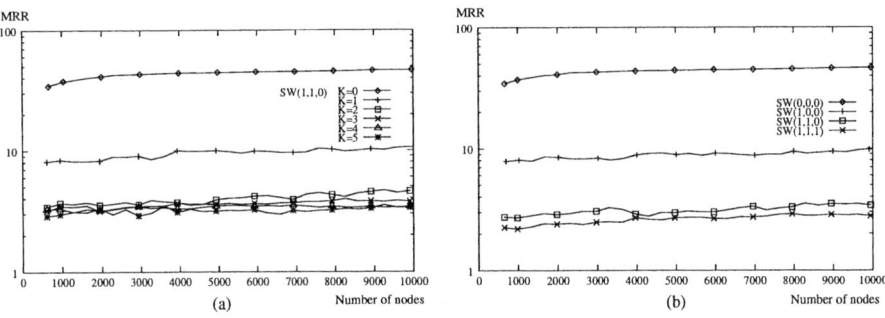

Fig. 7. (a) MRR under BCTL of multiple K where K=0, 1, 2, 3, 4, 5. (b) MRR under different combinations of prevention stages.

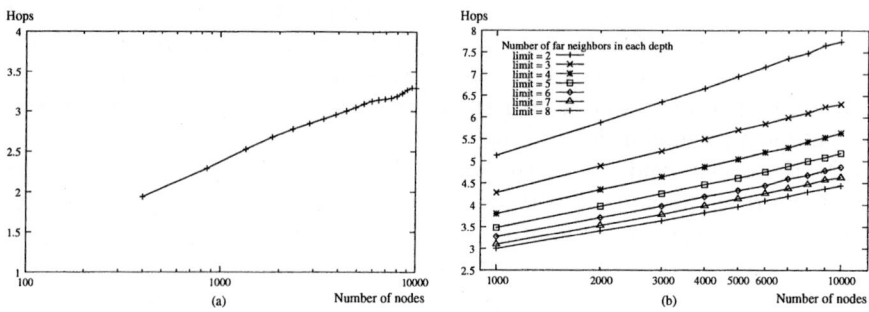

Fig. 8. (a) Average routing path length of a flooding request. (b) average number of hops of a lookup request, under different limit of the number of far neighbors in each depth.

Table 1. Five major ISPs and their respective message localities

ISP name	HINET	TANET	SEED	APOL	GIGA	Others	Across ISPs
Number of nodes	3,580	2,813	1,263	895	548	901	N/A
ISPID omitted	9.71%	5.65%	1.97%	1.08%	0.52%	N/A	80.56%
ISPID considered	28.27%	21.79%	11.18%	8.75%	4.11%	N/A	18.14%

result in Figure 7(b) shows that the first stage prevention drops MRR from 45 to 9.87. The second stage prevention further lowers MRR to 3.26. The third stage prevention reduces MRR down to 2.93. As we can see, MRR under $SW(1,1,1)$ is closer to the lower bound 1.

The third observation focuses on the average hops of a flooding request to reach every node in the system. The result in Figure 8(a) shows that the the average number of hops is a little bit higher than $log_{20} N$, the theoretical value. The next observation we concern is the average hops of a lookup request to reach its final destination. The number of far neighbors in each depth is an important factor to the performance of lookup. The simulation runs under different size limits of far neighbors, and different system size. The result in Figure 8(b) shows that the more far neighbors in each depth, the fewer hops a lookup takes to reach the final destination. But the reduction of hops slows down.

In the next simulation, the system contains 10,000 nodes whose IP addresses and ISPIDs come from the collected data of the real system. We trace each forwarding in flooding requests, including source and destination. The simulation is performed under two cases: ISPID is omitted and ISPID is considered. The result in Table 1 shows that DSE effectively promotes the message locality. In all messages whose sources belong to the biggest ISP, 28.27% of them move to their destinations within the same ISP. In comparison to the message locality of Gnutella, which is 2% to 5% in average. Our method reduces traffic across different autonomous systems significantly.

4 An Application

Based on the smart message routing, we developed an application to fulfill information broadcast and aggregation by building a spanning tree on a P2P network. The application assumes there is an administrator who wants to broadcast information to all nodes and aggregate informations from all nodes. Our method is described as follow. Each node maintains two lists called Q_{up} and Q_{down}. For an arrived flooding request R. Q_{down} stores destinations of R and Q_{up} stores the latest 7 visited sources of R if they are available (obtained from the BCTL of R). Notice that Q_{down} excludes those neighbors at which duplicated forwarding of R occurred. The administrator H_0 performs the task in two phases: spanning tree building, and information delivery on the tree.

Spanning Tree Building: To build a spanning tree, H_0 just initiates a flooding request R. Once every node received R, the set of Q_{down} of each node forms a top-down spanning tree, and the set of Q_{up} (only last one visited source of R is considered) of each node forms a bottom-up spanning tree. Now H_0 can perform information broadcast on the top-down spanning tree, and information aggregation on the bottom-up spanning tree. The cost to build such a spanning tree is the cost to flood a flooding request, which is $2.93 * N$ where N is the system size. The cost to perform an information broadcast or aggregation is just N.

Tree Maintenance: Since the leaving of nodes may break the spanning tree, a bottom-up recovery mechanism is used to reconnect broken trees. Consider an example in Figure 9 where node H_3 is leaving the system. When H_4 detects neighbor H_3 left, it searches Q_{up} for the node before H_3 (which is H_2). Then, H_4 connects to H_2 and query it to update its own Q_{up}. Meanwhile, H_2 also updates its Q_{down} to include H_3 and remove H_2. Similarly, H_5 connects to H_2 when it detects neighbor H_3 left. Since Q_{up} stores the latest seven visited nodes before the request arrives at the local node, according to research [1] where the average availability of each node is assumed to be 0.5, the probability that there is at least one alive node in Q_{up} is 99.99%. In other words, the capacity of Q_{up} guarantees the spanning tree can always be reconnected if it is broken. On the other hand, a new node which joins the system can easily joins the spanning tree. Given a new node H_x which just joins the system. After H_x connected it predecessor and successor, and initialized its routing table, it queries a neighbor for its Q_{up}. Meanwhile, the neighbor also updates its Q_{down} to include H_x.

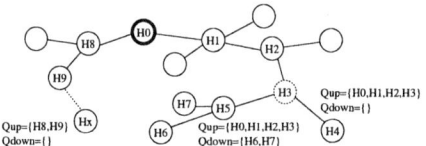

Fig. 9. Node joining and leaving on a spanning tree

Performance: The time to broadcast and aggregate information is in proportional to the height of the spanning tree. Since the tree is built by flooding a flooding request, the height of the tree is the largest path length of the request. The result in Figure 8(a) shows that the average number of hops of a flooding request is $O(logN)$ where N is the system size. Therefore, we induce that the average time to complete a broadcast or aggregate is $O(logN)$.

5 Related Work

Research [6] proposed a method to broadcast in DHT networks efficiently. The research considers DHT networks as a form of distributed $k - ary$ search and perform broadcast based on building a spanning tree. Although the research proposed an efficient broadcast, it highly depends on the correct structure of DHT networks. Leavings of nodes in the middle of the hierarchy hurt the performance of broadcasting seriously. The research does not address on recovery capability to handle leavings or failures of nodes. Research [7] suffers the same problem, either. Research [13] also adopts tree building method for information aggregation and broadcast. The method utilizes a predefined mapping function to make a connection from each node to its parent. The property of the mapping function makes the resulting graph a directed tree converging at a unique root. The work provides a simple and fast method to construct a spanning tree, however the maintain cost is considerable. Branch balancing is necessary to reduce the height of the tree.

Research [8] adopts a gossip based protocol to fulfill efficient flooding on unstructured network such as Gnutella. Their work reduces redundant messages by controlling (1) the number of neighbors to which a message is forwarded to. (2) the number of forwardings of a message. The work also considers the degree of each node. The nodes with lowest degree are chosen first to forward messages. Although the work claims that 60% redundant messages can be reduced while 96% nodes can still be reached, the cost is still significant. Moreover, the work trades off reduction of redundant messages against reduction of reachability of nodes and latency to complete a flooding.

Since the problem that our work tries to solve can be regarded as the multicast over the entire network, some related works are listed as follow. Research [12] proposed an application-level multicast scheme based on CAN network. Instead of constructing source-rooted distribution tree which is widely adopted in many works such as [9]. The work builds a small CAN for each multicast group. The multicast is fulfilled by careful flooding on the small CAN. The simulation result shows that over 97% of nodes receive no duplicate messages. Although the work provides a considerable improvement over the naive flooding, the solution is strictly based on the coordinate space structure of the CAN. Research [10] gave a detail evaluation of application-level multicast on peer-to-peer networks. The work focuses two main classes of P2P routing algorithms: Cartesian hyperspace based overlay such as CAN, and ring-based overlay such as Chord and Pastry. The multicast can be accomplished by either flooding or tree building.

The evaluation result shows that tree building based multicast (on ring overlay) is more efficient than flooding based multicast (on hyper-space overlay). The biggest disadvantage of the later is the cost of overlay construction for each group.

6 Conclusion

We developed a DHT network which fulfills both lookup and flooding. Closer nodes are connected each other so that the clustering coefficient of the system is maintained at high value. Based on this structure, efficient flooding can be fulfilled through careful message filtering on geometric structures of triangles and quadrangles. For a system of size N, a flooding request generates $2.93 * N$ messages. A lookup request, on the other hand, takes $O(log_{16} N)$ hops to reach its final destination. We propose a neighbor selection rule based on dynamic conditions. According to the frequency of each character in each depth, of lookup keys. Far neighbors can be replaced for better performance of lookups.

Smart message routing considers message locality. Message forwardings across different autonomous systems are reduced significantly. Based on the efficient flooding, we developed an application to fulfill information broadcast and aggregation in much less effort. By a flooding request, a spanning tree is built, and information broadcast and aggregation are achieved by message deliveries on the tree, without redundant traffic. Moreover, the tree has an ability of recovery. Broken trees can be detected and reconnected automatically.

References

[1] Ranjita Bhagwan, David Moore, Stefan Savage, and Geoffrey M. Voelker. Replication Strategies for Highly Available Peer-to-Peer Storage. Proceedings of FuDiCo: Future directions in Distributed Computing, June 2002.
[2] A. Rowstron and P. Druschel. Pastry: Scalable, decentralized object location, and routing for large-scale peer-to-peer systems. Lecture Notes in Computer Science, 2218:329–350, 2001.
[3] Ion Stoica, Robert Morris, David Karger, M. Frans Kaashoek, and Hari Balakrishnan. Chord: A Scalable Peer-to-peer Lookup Service for Internet Applications. Proceedings of ACM SIGCOMM 2001, San Deigo, CA, August 2001.
[4] MALKHI, D., NAOR, M., AND RATAJCZAK, D. Viceroy: A scalable and dynamic emulation of the butterfly. In Proceedings of the 21st annual ACM symposium on Principles of distributed computing, 2002.
[5] G. S. Manku, M. Bawa, and P. Raghavan. Symphony: Distributed hashing in a small world. Proc. 4th USENIX Symposium on Internet Technologies and Systems, 2003.
[6] Sameh El-Ansary, Luc Onana Alima, Per Brand, Seif Haridi, Efficient Broadcast in Structured P2P Networks, In The 2nd International Workshop on Peer-to-Peer Systems (IPTPS'03), March 2003.
[7] Y. Chawathe and M. Seshadri. Broadcast Federation: an application layer broadcast internetwork. In Proceedings of NOSSDAV 2002.

[8] Marius Portmann, Aruna Seneviratne. Cost-effective broadcast for fully decentralized peer-to-peer networks. Computer Communications 26(11). 1159-1167, 2003.
[9] A Rowstron, AM Kermarrec, M Castro, P Druschel. SCRIBE: The design of a large-scale event notification infrastructure. Networked Group Communication, 2001.
[10] Miguel Castro, Michael B. Jones, Anne-Marie Kermarrec, Antony Rowstron, Marvin Theimer, Helen Wang, Alec Wolman. An Evaluation of Scalable Application-Level Multicast Built Using Peer-to-Peer Overlays. IEEE INFOCOM, 2003.
[11] D. J. Watts and S. H. Strogatz. Collective dynamics of small-world networks. Nature, 393:440, 1998.
[12] S Ratnasamy, M Handley, R Karp, S Shenker. Application-level multicast using content-addressable networks. Proceedings of Third International Workshop on Networked Group Communication, 2001.
[13] Ji Li, Karen Sollins, Dah-Yoh Lim. Implementing aggregation and broadcast over Distributed Hash Tables. ACM SIGCOMM Computer Communication Review, Vol 35, 2005.
[14] Matei Ripeanu, Ian Foster and Adriana Iamnitchi. Mapping the Gnutella Network: Properties of Large-Scale Peer-to-Peer Systems and Implications for System Design. IEEE Internet Computing Journal, 6(1), 2002.
[15] RFC 1772. Autonomous System. http://www.faqs.org/rfcs/rfc1772.html
[16] Karl Aberer, Magdalena Punceva, Manfred Hauswirth and Roman Schmidt. Improving Data Access in P2P. IEEE Internet Computing, 6(1), 2002.
[17] TWNIC IP Address Distribution List. http://rms.twnic.net.tw/twnic/User/Member/Search/main7.jsp?Order=ORG.ID
[18] apia project. http://apia.peerlab.net/
[19] F. Dabek, J. Li, E. Sit, J. Robertson, M. Frans Kaashoek, and R. Morris. Designing a DHT for low latency and high throughput. In Proceedings of the USENIX Symposium on Networked Systems Design and Implementation, March 2004.

An IP Routing Inspired Information Search Scheme for Semantic Overlay Networks

Baoliu Ye[1,2], Minyi Guo[1], and Daoxu Chen[2]

[1] Department of Computer Software, the University of Aizu,
Aizu-Wakamatsu City, Fukushima 965-8580, Japan
{yebl, minyi}@u-aizu.ac.jp
[2] State Key Laboratory for Novel Software Technology, Nanjing University,
Nanjing 210093, P. R. China
cdx@nju.edu.cn

Abstract. How to efficiently locate desired information over the self-organized decentralized Peer-to-Peer (P2P) networks is one of the main challenges that greatly affect the realization of its potential advantages. In this paper, we present an IP routing inspired solution to address this problem in a distributed and scalable manner. In our scheme, we first build a hierarchical P2P Semantic Overlay Network (SON) via taking advantages of the hierarchical nature of document semantics, then implement an IP routing inspired routing algorithm to perform information search over the SON. Simulation results demonstrate that our proposal could significantly improve the system performance versus random overlay network with low cost.

1 Introduction

P2P computing is increasingly attracting enormous attention in research community. It introduces an interesting paradigm for pooling a large number of autonomous computing nodes decentralized over the Internet together to cooperatively share information and services. However, the new features of P2P systems make traditional multi-server oriented distributed search techniques inapplicable. The recent flurry of research also shows that it is difficult to efficiently locate desired information over such kind of self-organized decentralized networks.

We deem that the essence of P2P information search is closely similar to the IP routing problem. Both of them aim at finding the target object (location) through a shortest path without maintaining global topology information at each node. The success of the Internet shows that traditional IP routing techniques could be a good source for exploring P2P information searching schemes. In this paper, we focus on studying the feasibility of extending traditional IP routing algorithms to address the P2P information search problem in a scalable manner while without infringing node autonomy. This paper describes the details of the proposed IP routing inspired P2P information search scheme and gives simulation results. The rest of this paper is organized as follows: in the next session, we discuss the related work. Section 3 gives the concept of Semantic Classification Tree (SCT). Section 4 describes the construction of a hierarchical P2P SON. Section 5 presents the IP routing inspired information search algorithm. Section 6 shows our initial experiment result. Finally we summarize the work of this paper in section 7.

2 Related Work

In response to the problem of how to make P2P networks scalable and effective, the research community proposed many schemes via making different tradeoffs between information search efficiency and topology maintenance overhead. Although centralized indexing directory based P2P applications such as Napster [1] are very popular in today's Internet, these systems always suffer from single point of failure and performance bottleneck at the index server. Loosely coupled unstructured P2P networks, such as Gnutella [2], are extremely robust, but the search cost is high as the total number of messages is exponentially increased over the number of hops traversed per search. On the other hand, Distributed Hash Table (DHT) technique based tightly structured P2P networks [3~5] always employ rigid structures to precisely regulate information location and efficiently access distributed resources. Although these networks only require $O(log\ n)$ steps to complete a search (n is the total number of nodes), its ability to operate with extremely unreliable nodes has not yet been demonstrated. More recently, there is another class of semi-structured P2P networks [6~9] which tries to cluster nodes into different communities through capturing and exploiting the inherent interest proximity and/or behavior similarity among nodes so as to improve search efficiency. These semi-structured P2P network based techniques are very scalable to large dynamic P2P systems. Unfortunately, it is difficult to classify and model the interest topics precisely. Therefore, how to locate the desired information over the self-organized decentralized P2P system in a scalable manner without infringing node autonomy is still a hot issue to be addressed.

3 Semantic Classification Tree

Suppose a P2P system consists a set of nodes P, where each node p maintains a set of documents D_p. We use D denote the set of all documents in the P2P system and define a SON associated with semantics s (denoted as SON_s) as follows:

$$SON_s = (P_s,\ E_s)$$

where

$$P_s = \{p_i \in P \mid \exists link(p_i, p_j, s)\}$$
$$E_s = \{< p_i, p_j > \mid \exists link(p_i, p_j, s) \land (p_i, p_j \in P_s)\}$$

If there exists a triple *link (p_i, p_j, s)*, then node p_i and p_j are neighbors associated with semantics *s*. Since the documents in a P2P system could be divided into different semantic dimensions, a typical P2P SON always consists of several different sub-SONs where each node may be logically linked to a relatively small set of other nodes.

In practice, there are many options in choosing a feasible semantic classifier for organizing a P2P SON due to multi-semantics within a document. On one hand, we need a precise semantic classifier which produces sub-SONs with a small number of nodes, so as to reduce the search cost incurred. On the other hand, we need a coarse semantic classifier which makes each node only belong to a small number of sub-SONs, so as to constrain the topology maintenance overhead involved. To balance the above incompatible requirements, we propose a SCT by employing a style/substyle based classification hierarchy. The SCT conceptually is a layered semantic structure of the document information and presents a reference for classifying documents according to semantics. Each node in a SCT represents a specific semantics and could be identified by its path to the root. For simplicity, we first give several concepts related to a SCT. Here N denotes the semantic node set of a SCT.

Definition 1 Node Set of Root Path. A node's node set of root path consists all the nodes along its path to the root. We use $NOP(n, r)$ to denote node n's node set of root path in a SCT rooted at node r.

Definition 2 Parent-SON. If $n_i, n_j \in N$ and n_i is the parent node of n_j, then we call SON_{n_i} the Parent-SON of SON_{n_j}.

Definition 3 Least Common Ancestor. If node n ($n \in N$) satisfies the following conditions, then we call n the least common ancestor of nodes n_i and n_j ($n_i, n_j \in N$).
  1) $n \in NOP(n_i, r)$ and $n \in NOP(n_j, r)$;
  2) node n has the shortest distance to both nodes over the SCT.

Let $LCA(n_i, n_j)$ represent the least common ancestor of node n_i and n_j.

Definition 4 Semantic Distance. The semantic distance is the shortest path between two semantic nodes. Let $SD(n_i, n_j)$ ($n_i, n_j \in N$) denote the semantic distance between node n_i and n_j. We have $SD(n_i, n_j) = SD(n_i, LCA(n_i, n_j)) + SD(n_j, LCA(n_i, n_j))$.

In essence, a SCT is an information classification scheme which depicts the semantic relationship. All the son nodes together form a semantic partition about their parent node. In order to achieve a good tradeoff between information search efficiency and topology maintenance overhead, we define the rules for constructing a SCT, i.e., the resulted SCT must meet the following properties.

Property 1. Each document should be able to be classified as a member of document set of root node.

Property 2. If node n_i and n_j ($n_i, n_j \in N$) are two son nodes of node $n (n \in N)$, Let D_{n_i} and D_{n_j} represent the document set of node n_i and n_j, respectively. Then we have $D_{n_i} \cap D_{n_j} = \Phi$ and $D_n = \Sigma D_{child(n)}$.

Property 3. If $n_j \in NOP(n_i, r)$ ($n_i, n_j \in N$), then $D_{n_i} \subset D_{n_j}$.

4 Hierarchical P2P Semantic Overlay Network

Usually there are two different kinds of strategy for assigning nodes to a P2P SON with the SCT. In a conservative strategy, a node joins all sub-SONs as long as it has a document associated with corresponding semantics. This strategy could guarantee to get the desired documents so far as they really exist. But it increases both the size of each sub-SON and the number of connections to be maintained at each node. With an optimistic strategy, a node will join a sub-SON only if it has "significant" number of documents belongs to that semantics. The optimistic strategy reduces the size of a sub-SON as well as the number of connections of a node. However, under this strategy, some documents may be neglected and cannot be found due to its minor proportion.

To control the maintenance overhead and improve document availability, we present a threshold based node-to-SON assignment algorithm. A node joins a SON_c only if the percentage of its documents associated with semantics c exceeds a threshold α ($0<\alpha<1$). Assume S_p is the semantic set of node p, N_s ($\sum N_s=1$) is the percentage of documents corresponding to semantics s ($s \in S_p$) at this node. Algorithm 1 gives the pseudo code of the node-to-SON assignment algorithm.

Algorithm 1: Node-to-SON assignment algorithm

```
NodeAssign (Node p)
BEGIN
FOR each semantics s belongs to S_p
  IF N_s ≥ α THEN
    Let node p join SON_s
    Remove s from S_p
  END IF
END FOR
FOR any two semantics s_i, s_j in S_p
  s=LCA(s_i, s_j) //Compute their least common ancestor
  S_lca=S_lca+{s} //Add s to the least common ancestor set S_lca
END FOR
WHILE S_p ∪ S_lca is not empty DO
  IF root semantics r is the only element of S_p ∪ S_lca THEN
    Let node p join SON_r
    End the algorithm
  END IF
  Select a s from S_lca
  Let S_off be the semantic set made up of all the offspring semantics of s,

  IF s=LCA(s_i, s_j) is true for any two semantics s_i, s_j (s_i,
  s_j ∈ S_off) THEN
    N_s = ∑ N_s' , (s' ∈ S_off)
    IF N_s ≥ α THEN
      Let node p join SON_s
      Remove s from S_lca
    END IF
    Remove each node s' (s' ∈ S_off) from either S_p or S_lca
  END IF
END WHILE
END
```

Although there already exists a hierarchical semantic relationship among the sub-SONs, the P2P SON generated by the node-to-SON assignment algorithm still is a flat connected graphics. To regulate a IP network alike hierarchical P2P SON and implement an IP routing inspired information search scheme, each node must select one of the assigned sub-SONs as its host SON and only appear as a member of its host SON in the hierarchical P2P SON. We also select a semantic gateway node from each SON. Once a node cannot find any available route to the target, it will forward a request to the semantic gateway. In order to learn more route information, the semantic gateway must join the parent-SON of its host SON and exchange route information with other sibling semantic gateways. Figure 1 shows a hierarchical P2P SON.

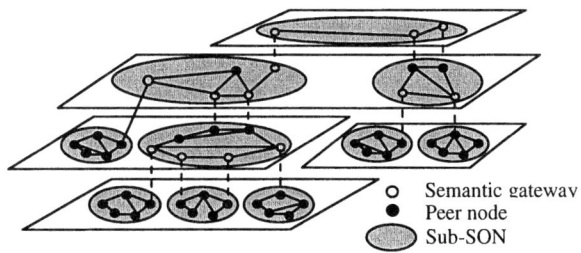

Fig. 1. A Hierarchical P2P SON

5 Information Routing Scheme

5.1 Routing Information Management

The semantic gateway provides a default route for forwarding requests, but it is unwise to let semantic gateways process all the requests. In fact, it is possible and helpful to exchange route information among node by employing a Distance-Vector routing algorithm, so as to reduce the load on semantic gateways and improve the search efficiency. We implement a distance vector $\mathbf{D}_p$ and a nexthop vector $\mathbf{R}_p$ at each node p to record all the reachable routes within H hops.

$$\mathbf{D}_p = \begin{bmatrix} D(p,s_1) \\ D(p,s_2) \\ \cdot \\ \cdot \\ \cdot \\ D(p,s_N) \end{bmatrix}, \quad \mathbf{R}_p = \begin{bmatrix} R(p,s_1) \\ R(p,s_2) \\ \cdot \\ \cdot \\ \cdot \\ R(p,s_N) \end{bmatrix}$$

Where $D(p,S_i)$ denotes the shortest distance from node p to semantic overlay network S_i; N is the total number of nodes in a P2P system; $R(p, S_i)$ represents the next forwarding node to semantic overlay network S_i at node p. For a new node, the initial value of each component in $\mathbf{R}_p$ is null, the initial value of $D(p,S_i)$ is computed with formula (1).

$$D(p,S_i) = \begin{cases} 0, & p \text{ belongs to } S_i \\ \infty, & \text{otherwise} \end{cases} \quad (1)$$

It is important to update routing information immediately in response to the network variation of the hierarchical P2P SON. Usually, routing maintenance should deal with node join and node leave, respectively.

Node Join

1) Routing Learning. The routing learning is to acquire all the available routes to other sub-SONs within H hops. Since the neighbors of node p have all the routes within H hops started from themselves, node p sends a route acquiring request to its neighbors. Once received all the distance vectors and nexthop vectors returned by these neighbor nodes, node p computes available routes as follows: If $D(p,S_i) \neq 0$, then select a neighbor node with the shortest distance to S_i according to formula (2):

$$d = \underset{y \in A}{Min}(D(y,S_i)+1) \qquad (2)$$

Where A is the neighbor list of node p. If $d \leq H$, then update corresponding components in both $\mathbf{D}_p$ and $\mathbf{R}_p$ with $D(p,S_i)=d$, $R(p, S_i)=y$.

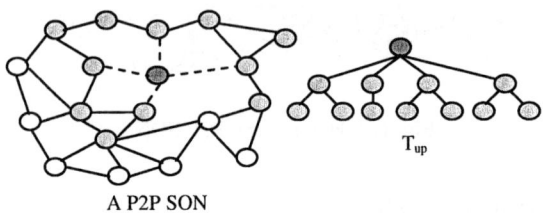

Fig. 2. A shortest spanning tree for routing update ($H=2$)

2) Routing Update. In essence, the routing update process is to publish available routing information along a shortest spanning tree T_{up} rooted at node p, which covers all the surrounding nodes (see Figure 2). Here *the surrounding nodes* of node p include all the nodes within a radius of H hops of p. For simplicity, we use P^i to denote all the nodes at level i ($i \leq H$). After receiving the distance vector and nexthop vector from a parent node, each son node updates its own vector according to algorithm 2. Here we assume node y is the parent of node x.

Algorithm 2: Routing update algorithm

Step 1: If there has a fresh route entry to a new SON S_i in $\mathbf{D}_x$ and we have $R(y, S_i) \neq x$, $D(y,S_i) < H$, then update $\mathbf{D}_x$, $\mathbf{R}_x$ according to the following:

$$D(x,S_i) = D(y,S_i)+1$$
$$R(x,S_i) = y$$

Step 2: If there has a shorter route to SON S_i in $\mathbf{D}_y$, then update $\mathbf{D}_x$, $\mathbf{R}_x$ with the above formula.

Step 3: For each component in $\mathbf{R}_x$ that satisfies with $R(x, S_i)=y$, if $D(y,S_i)<H$ is true, then update corresponding component in $\mathbf{D}_x$ with the above formula; otherwise update it as follows:

$$D(x,S_i) = \infty$$
$$R(x,S_i) = \Phi$$

Step 4: Node x forwards $\mathbf{D}_x$, $\mathbf{R}_x$ to all its son nodes.
Step 5: Repeat the above steps until all the nodes in T_{up} are updated.

An IP Routing Inspired Information Search Scheme for Semantic Overlay Networks

Node Leave

1) Friendly leaving. In this case, a node notifies its neighbors before leaving. Once a node p leave a P2P SON, all the routes via p will be disabled. Thus we must delete these unreachable routes. In this case, node p must notify its surrounding nodes before leaving and build a shortest spanning tree T_{up} rooted at itself. Then modify the route records along the tree by performing algorithm 3:

Algorithm 3: Routing update algorithm for friendly leaving

Step 1: Modify the component value in $\mathbf{D}_p$ and $\mathbf{R}_p$ with the following formula. And then forward the updated $\mathbf{D}_p$ and $\mathbf{R}_p$ to its son nodes.
$$D(p, S_i) = \infty$$
$$R(p, S_i) = \Phi$$

Step 2: After receiving the updated vectors from the parent node, current node x updates the value of $\mathbf{D}_x$ and $\mathbf{R}_x$ satisfying with $R(x, S_i)=y$ and $D(x, S_i) \neq \infty$ as follows:
$$D(x, S_i) = \infty$$
$$R(x, S_i) = \Phi$$

Step 3: Node x forwards the updated $\mathbf{D}_x$ and $\mathbf{R}_x$ to all its son nodes.

Step 4: Repeat the above steps until all the nodes in T_{up} finish its route updating.

---

2) Abnormal leaving. An abrupt failure is an exception even caused by some unpredictable reasons. We employ an event-driven based route updating mechanism to deal with this problem in the case of detected an unreachable route. Suppose node p leaves the system exceptionally. In our scheme, this event won't be found until a neighbor a tries to forward a request to node p. Here we also construct a shortest spanning tree T_{up} rooted at node a that covers all of its surrounding nodes. And then correct the route records along the tree with algorithm 4.

Algorithm 4: Routing update algorithm for abnormal leaving

Step 1: Modify all the components of $\mathbf{R}_a$ that meet $R(a, S_i)=p$ with the following formula, then forward the updated $\mathbf{D}_p$ and $\mathbf{R}_p$ to its son nodes.
$$D(a, S_i) = \infty$$
$$R(a, S_i) = \Phi$$

Step 2: After receiving the updated vectors from the parent node, current node x updates the value of $\mathbf{D}_x$ and $\mathbf{R}_x$ that satisfy with $R(x, S_i)=y$ and $D(x, S_i) \neq \infty$ according to the above formula.

Step 3: Node x forwards the updated $\mathbf{D}_x$ and $\mathbf{R}_x$ to all its son nodes.

Step 4: Repeat the above steps until all the nodes in T_{up} finish its route updating.

5.2 Routing Implementation

When a message is received, a peer first checks whether this request falls within one of its SONs. If that is the case, it starts search-within-SON by flooding the message to peers in its neighbor List associated with this SON. Otherwise, it will perform search-across SON (inter-domain search) by navigating over the hierarchical P2P SON.

At the inter-domain search stage, if current node p has no available route record to the destination SON S_d (i.e., $D(p,S_d) = \infty$), it forwards the request to its default semantic gateway directly. Otherwise, node p computes the semantic distance between the host SON S_h and destination SON S_d. If $DOP(S_h, S_d)-1<D(p,S_d)$, then forward the request to the default semantic gateway; elsewise, forward it to $R(p, S_d)$. If a peer finds the next forwarding node is unreachable, it starts the event-driven based route updating mechanism and forwards the request to the semantic gateway. Once the request arrives in the desired SON, it switches to intra-domain search.

There are two scenarios that may cause our algorithm failure in finding a desired document within the destination SON. One scenario is that we can't fall into the right SON due to an inaccurate classification about the query. The other scenario is that some documents may not be associated to the right SON owing to low proportion. Our policy to this problem is to forward the request to its parent-SON and start another intra-domain search scheme in the case of we can't find the target document within the desired SON.

6 Performance Evaluation

We evaluate the performance of the IP routing-inspired P2P information search scheme through simulation. Our simulation environment includes 2,000 nodes and 3,000 documents. These documents could be classified into 40 kinds of semantics and are uniformly distributed over the system. Each node exchanges routing information with all the nodes within 3 hops (i.e., $H=3$). The default document aggregation threshold is 15% (i.e., $\alpha = 15\%$). Each node randomly issues a document search request.

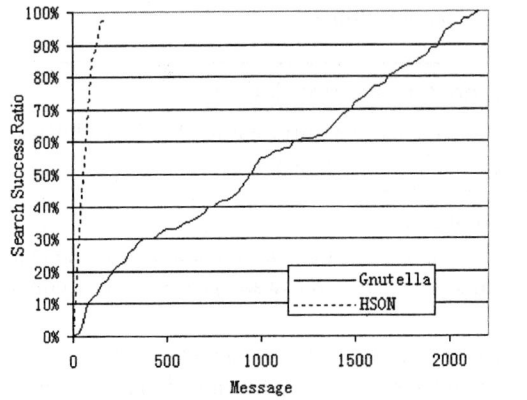
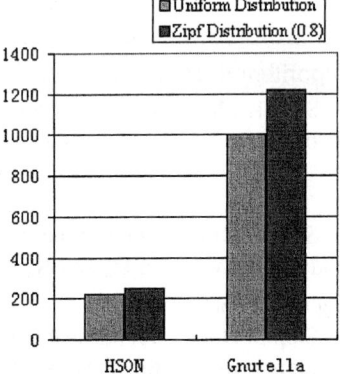

Fig. 3. Search success ratio *vs.* Message

Fig. 4. Search efficiency under different document popularity

Figure 3 shows the evolution of the search success ratio over the number of messages transmitted. From this figure we can see that the success ratio increases along with the number of messages. Figure 3 also indicates that the success ratio of our Hierarchical SON (HSON) is significantly higher than that of Gnutella. Gnutella directly employs a blind search method to flood request messages to all the neighbor nodes, so the number of transmitted messages exponentially increases over the number of hops traversed. Our IP routing inspired information search scheme will first navigate the request to the target sub-SON according to the semantic routing information before performing the flood operation. Thus effectively constrain the flooding space.

Figure 4 analyses the search efficiency with different document popularity. In this experiment we compare the performance result over uniform distribution and *Zipf's* distribution. From this figure we could see that the averaged message of a successful search under *Zipf's* distribution is a little higher than that of uniform distribution for both HSON and Gnutella. The reason is that most of the pop documents are distributed over a few nodes when the document popularity meets *Zipf's* distribution. Therefore most of the nodes would have to traverse many intermediate nodes before finding the desired pop document.

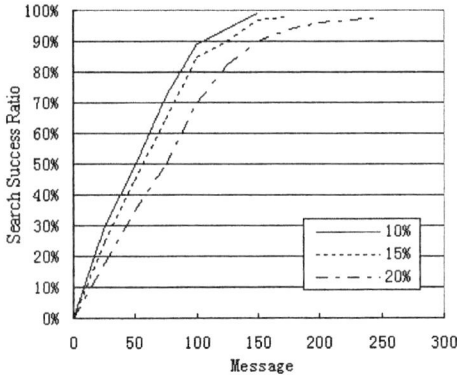

Fig. 5. Search success ratio *vs.* document aggregation threshold

Figure 5 demonstrates the search success ratio versus the number of messages transmitted under different document aggregation threshold. From this graph, we can see that a higher document aggregation threshold would result in a lower search success ratio under a specific searching space. In fact, when the document aggregation threshold is small, each node will join more sub-SONs. And it is possible to learn more routing information and potentially reduce the inter-domain search distance. On the other hand, if we enlarge the document aggregation threshold, then the more documents would have to be merged with a coarser semantics so as to assign the node to a sub-SON. Consequently, we may not be able to get these documents by executing an intra-domain search within the sub-SON with respect to their right semantics. In this case, we will have to forward the request to the parent-SON and start another intra-domain search so as to get success, thereby introducing more messages.

7 Conclusion

In this paper, we propose an IP routing inspired distributed information search scheme to solve the information search problem over self-organized decentralized P2P systems. In our solution, the SCT provides a flexible model to control the tradeoff between information search efficiency and topology maintenance overhead. The hierarchical P2P SON is a typical loosely coupled distributed P2P overlay network, which can guarantee the autonomy of nodes as well as the scalability of the system. The IP routing inspired information search scheme is a non-deterministic distributed routing algorithm. Actually, it can support both partial-match and exact search. The simulation results demonstrate that our proposal could significantly improve the system performance versus random overlay network with low cost.

In the future, we will conduct more simulations and experiments to evaluate and revise our scheme, with the purpose to implementing an efficient and practical solution to support information search over large-scale dynamic P2P systems.

Acknowledgement. This work is partially supported by the National Basic Research Program of China (973) under Grant No. 2002CB312002; the National Natural Science Foundation of China under Grant No. 60402027; the National High-Tech Research and Development Program of China (863) under Grant No. 2004AA112090.

References

1. Napster, http://www.napster.com.
2. http://www.clips2.com.
3. S. Ratnasamy, M. Handley, R. Karp, "Application-level Multicast using Content-Addressable Networks," In Proceedings of 3rd International Workshop on Networked Group Communication, November 2001
4. I. Stoica, R. Morris, D. Karger, M. F. Kaashoek, and H. Balakrishnan, "Chord: A scalable peer-to-peer lookup service for internet applications," In Proceedings of ACM SIGCOMM, San Diego, California, August 2001.
5. P. Druschel, A. Rowstron, "Pastry: Scalable, distributed object location and routing for large-scale peer-to-peer systems," ACM SIGCOMM, 2001.
6. C. Tang, Z. Xu, and S. Dwarkadas, "Peer-to-peer information retrieval using self-organizing semantic overlay networks," In Proceedings of the 2003 Conference on Applications, Technologies, Architectures, and Protocols for Computer Communications (SIGCOMM), pages 175–186. ACM, ACM Press, 2003.
7. A. Loser, F. Naumann, W. Siberski, W. Nejdl, and U. Thaden, "Semantic overlay clusters within super-peer networks," In Proceedings of the International Workshop on Databases, Information Systems and Peer-to-Peer Computing, 2003.
8. M. Bawa, G. S. Manku, P. Raghavan, "SETS: search enhanced by topic segmentation," in Proceedings of the 26th annual international ACM SIGIR conference on Research and development in information retrieval, Toronto, Canada, July 2003.
9. M. Khanbatti, K. D. Ryu, P. Dasgupta, "Structuring peer-to-peer networks using interest-based communities," Int'l Work. on Databases, Info. Sys. and Peer-to-Peer Computing, 2003.
10. A. Crespo, H. Garcia-Molina, "Semantic Overlay Networks for P2P systems," In Proceedings of ICDCS'02, Vienna, Austria, July 2002.

Transactional Cluster Computing

Stefan Frenz, Michael Schoettner, Ralph Goeckelmann, and Peter Schulthess

frenz@vs.informatik.uni-ulm.de

Abstract. A lot of sophisticated techniques and platforms have been proposed to build distributed object systems. Remote method invocation and explicit message passing on top of traditional operating systems are complex and difficult to program. As an alternative the distributed memory idea simplifies and unifies memory access, but performance drawbacks caused by expensive distributed locking mechanisms obviated this approach for cluster computing. To avoid these slowdowns more and more weak consistency models try to avoid these slowdowns by burdening the programmer with explicit synchronization. Our research group has developed a fast transaction system for a distributed heap that guarantees semantically correct access to shared memory.

Keywords: distribution, consistency, transactions, shared memory, operating system.

1 Introduction

Traditional distributed shared memory systems offer implicit replication giving rise to the question about consistency. Many strategies have been proposed to deal with the trade-off between strict consistency models, which enforce transparent replication, and relaxed models, which try to improve performance by accepting reads to outdated data and burdening concurrency control to the application. Relaxed consistency with its contingently appearance of inconsistent data values is unacceptable for sharing pointers or heap structures. On the other hand, guarding each access to shared memory lessens the benefit of cluster computing because of its immense overhead and latency.

This paper describes an alternative approach to a cluster network protocol, that supports strong consistency with implicit communication for a distributed heap or object storage with equated cluster participants. The symmetric cluster design enables fully concurrent execution without a dedicated coordinating node and the transactional consistency model provides a natural application behavior without explicit communication or synchronization. Semantical groups are executed as in atomic transactions, and the speculative memory access successfully hides the latencies normally inherent to network. High performance for parallelized algorithms is combined with reliable values allowing even pointer and heap management without explicit synchronization by the programmer. We developed a standalone operating system for a lean testing environment, which is described in section 2.1 together with the semantics of optimistic transactions in section 2.2. The design and implementation algorithms of a protocol for transaction support in a equated cluster environment are presented in section 3.1, which is followed by evaluation and comparison in section 4.

2 Testing Environment

2.1 Plurix

Plurix was created at the University of Ulm as standalone operating system for research and educational purposes mainly considering distributed algorithms like matrix operations or ray tracing and distributed applications like cooperative working. Main aspects are the development of protocols and the investigation of programming concepts for distributed systems. The central event loop and the cooperative multitasking are adopted from the innovative Oberon system developed by Wirth and Gutknecht [Wir92], combined with comfortable data sharing done with distributed shared memory. Using the Java language with some extensions for operating system's requirements offers type-safe and easy programming and access to a wealth of already written software. Plurix uses a specially developed compiler [Sch02], which compiles Java sources directly into Intel machine code and abandons the Java Virtual Machine.

Currently, Plurix communicates via a Fast Ethernet in a local area network, but is ready to switch to Gigabit Ethernet. The distributed shared memory uses Transactional Consistency [Wen03] to enable a full featured heap avoiding dangling references. The concept is proven to correctly execute parallel transactions [Wen02]. The current implementation is optimized in respect of page-sized granularity to enable utilization of hardware accelerated memory management of current x86 computers. However, the protocol design is not bound to this approach.

2.2 Speculative Transactions

As a cooperative multitasking operating system Plurix avoids the overhead typically encountered in preemptive environments. Programs running in the Plurix environment are divided into small execution blocks, which are executed atomically and usually integrate few semantically complete units. Each atomically executed block is called transaction and conforms to the well known ACID paradigm [Hae83], [Dad96]:

- Each transaction either succeeds or is completely rolled back (atomicity), which is guaranteed by the memory management [Goe03]. One reason for a roll-back may be a self-abort of the transaction, for example when the transaction has detected an error condition and wishes to leave the heap unchanged. The second reason for a roll back is a forced abort in case of an access collision between two transactions.
- A transaction starts with consistent state of memory and after committing leaves it in a consistent state (consistency). All data and code [Goe03] resides in the distributed heap, and the heap may be changed only from within a transaction. This is provided by the kernel, which manages the heap with only few local state variables, which can not be accessed by a user transaction.
- The results of a transaction are not propagated until its commit (isolation), which means: before a successful commit the changes done by a transaction are only visible to this transaction. In case of an abort this isolation prevents cascading

aborts and inconsistency of memory. The programmer can define semantic blocks, which require atomic calculation or should not interfere with others [Fre04].
- The results of a transaction survive the transaction even in case of a subsequent system error (durability). In databases, durability is ensured by saving each transaction's results to disc, but for cluster operations the term "durability" is more flexible and may be specified for each environment and application [Fre02]. The realization of durability was tested in two ways. First, we replicated within the cluster. Then, we made the same test with a pageserver that holds and buffers committed pages. Both cases provided an adequate measure of durability, but each proved to be advantageous for different applications.

At the end of each transaction the addresses of pages modified by this transaction are published in a semantically atomic way to avoid interleaving between multiple ending transactions. The usual way is just to send the addresses, which results in an invalidate-semantic: outdated data is simply removed and will be requested on demand. Another strategy is to send the data of each modified page along the address, which corresponds to update-semantic: outdated data is replaced by fresh data and immediately accessible. In both cases every concurrently running transaction must check the received writeset against its own read- and writeset to detect possible memory access conflicts. A collision will abort the local transaction, which does not require any network communication because of the isolation property of transactions.

3 Transactional Consistency Protocol

3.1 Design

The protocol must provide at least basic exchange and control mechanisms for communication between nodes. For transactional consistency this includes but is not limited to management of cluster-time, transfer of page data, reliable and atomic delivery of writesets and algorithms for error detection and recovery.

As far as possible the protocol should handle requests without additional state machines and should answer all requests directly in the interrupt service routine. This leads on the one hand to a efficient and fast response behavior and on the other hand to less indirections in the code and therewith better readability. The complexity seems to grow for the protocol caller if the protocol does not handle timeout errors itself, but as the caller can not proceed without completely fulfilled requests, there has to be a wait anyway, which can be easily equipped with timeout functionality.

LAN communication hardware typically offers non-reliable multicasts and unicasts. Because of packet loss or packet overtaking inside switches and network cards, the protocol must validate each packet. Therefore each packet contains a timestamp from the sending machine, which lets the receiving machine decide whether the packet is valid, is outdated, or indicates a critical packet miss. The transition from one time-slot to the next is implicitly done with the first and last packet of a commit, because the writeset, which is delivered inside the commit-packets, is critical to the system consistency and has to be processed atomically. During the commit phase no other packet type is allowed, as data is in a transient state.

3.2 Algorithms for Implementation

The packet types required for consistent communication will be discussed in this section. Each packet contains information about the current logical time of the sending node, the membership to a cluster and a packet type. Packets may contain additional data of fixed or variable length, but in order to remain compatible with Fast Ethernet, the overall size may not exceed 1536 bytes.

General Packet Layout
Fast Ethernet is able to transmit specially tailored packet frames, so there would be no need of IP headers or similar. But in order to communicate across routers and to coexist with other nodes in a heterogeneous network, all packets contain an IP header. The Fast Ethernet header consists of destination and source MAC address (each 6 bytes) and a frame type (2 bytes) which is fixed to IP. The standard IP-header with 20 bytes includes source and destination IP and the protocol type, which is fixed to an otherwise unused value. As mentioned above, each packet carries a time stamp from the sender (commit number, 8 bytes) and an identifier for the particular heap or cluster, which is in our implementation a 8-byte-value. Furthermore there is a type with subtype (2 bytes) and a parameter for this packet type (4 bytes), which results in a header of 22 bytes. All headers together are 54 bytes. Depending on the packet type there may be additional data bytes up to 1444 bytes yielding a maximum packet size of 1498 bytes.

Validation of a Received Packet
Each packet contains the logical time of the sender, which is compared to the local logical time with several results possible: If the received time stamp matches the local time stamp, the packet is valid. Otherwise the packet is invalid, and the higher stamp indicates the current time. If the received time stamp is lower, the sender may have missed a writeset and must be notified. A received time stamp greater than the local time indicates that the local machine has lost a writeset, which must then be requested explicitly. Figure 1 at the end of this chapter shows the corresponding flowchart.

Alive Request and Alive Acknowledge
To determine the presence of other nodes (e.g. at start-up), a packet with type "alive request" is sent as multicast. Each running node in the cluster with matching heap number will answer this request with "alive acknowledge". A starting node has no information about the current time in the cluster, so the sender-time is cleared. As the acknowledge packet contains the current time, the requesting node can update its time.

Page Request and Page Data
Page-based distributed shared memory as implemented in Plurix typically uses paging across the network. During execution a transaction may access a not present page at any time resulting in a page fault. The page fault handler verifies the address of the accessed page and requests the page from the cluster by sending a packet "page request" with the corresponding page address. For object based distributed memory systems the page number is replaced by an object id correspondingly.

The node owning this page will answer the page request with one or more packets of type "page data", which all contain a CRC for the complete data to ensure correct transmission and reassembly of the transferred page. As the page size is 4096 bytes and the maximum packet size of Fast Ethernet frames is 1536 bytes, the data can not be sent forthright in a single frame and would require three packets without additional algorithms. Because statistics showed that many pages are completely cleared, there is a special message "page empty". Checking of page emptiness is done during calculation of the CRC. In object-based systems the variable size of packets has to be considered similarly.

Since the transmission of a large number of packets consumes the bandwidth of the network, optional compression should be supported. Even though compression may spare bandwidth, it consumes CPU-time on the sending and receiving host and should only be used with sufficient advantage. In our implementation the page to send is compressed in temporary memory and the size of the compressed block is checked. If compression does not save at least one packet, the page is sent uncompressed even if compression is enabled because the network bandwidth saved is not significant.

Instead of retransmitting a single frame in case of uncommon packet loss, the requesting node simply re-requests the page after a timeout, so there is no additional bookkeeping or buffering inside the answering node.

Token Request and Token Granted
A terminating transaction attempts to commit by sending its writeset to all participating nodes. Since a transaction may conflict with another, one commit operation has to complete before the next commit can be started. To guarantee this Plurix implements a token mechanism using the packet type "token request" and the corresponding answer "token granted".

To avoid token loss in case of a packet loss, a node sending the token stores the recipient of the token in a private local memory location, so the token can be resent to this node if this node requests the token again.

Writeset Frame and Recovery with Writeset Request
The writeset which is broadcast at the end of a transaction contains the addresses of all modified pages. The committing node becomes owner of all modified pages and all outdated copies on other nodes must be discarded. Because of the limitations of the maximum packet length, only up to 370 addresses are sent in a single Fast Ethernet packet. If the transaction modified more than 370 pages, the writeset must be split into multiple frames. Although there may be several packets, the writeset must be evaluated atomically to ensure consistency of memory.

If a writeset frame is lost by a node or the network, the affected node has to request the writeset again. Therefore each node must store previously sent writesets for a short time, until most likely all other nodes have received them.

3.3 Fairness

The basic implementation does not support fairness for repeatedly aborted transactions. But the forward validation scheme allows sophisticated commit election strategies instead of simple "first wins" strategy. A protocol for improved fairness will need to take into account detailed information about the conflicting transactions to choose the optimal set of surviving transactions. There are many decision strategies about which set of transactions should survive and which set should be aborted. The

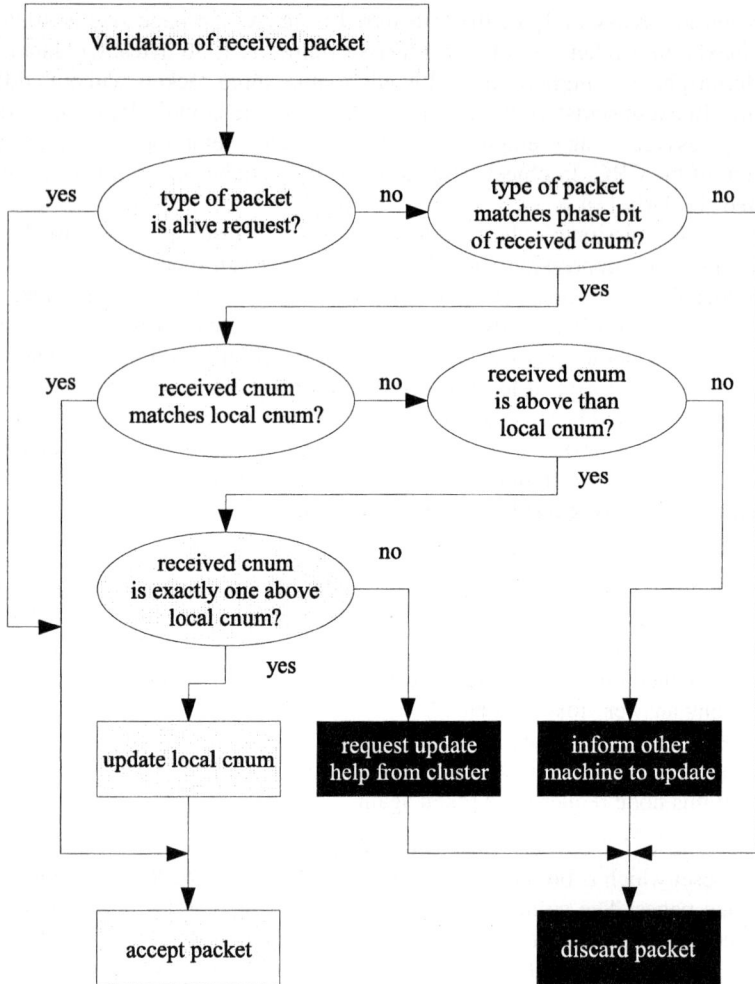

Fig. 1. Packet Validation

metrics consider completely different application's behavior, which comprise different amount of parameters. Although the fairness of the decision generally increases by adding more parameters, it depends on the type of running applications which algorithm performs best.

3.4 Checkpointing

Transactional Consistency offers a natural opportunity to checkpoint a cluster: The system architecture is prepared to discard modifications and to request current versions from the cluster, and the isolation property of transactions will let the cluster computation proceed during collection of data for a checkpoint. Of course only the distributable part of memory is checkpointed by the cluster, so each node has to make sure that local state can be restored from the distributed memory or can be regenerated by the kernel or by the application. To save also the local state in the distributed memory is neither trivial, because kernel and device states usually are not distributed, nor fast enough, because the local kernel and device drivers change their state frequently and quickly, which would make the cluster coordination to a bottleneck. Fortunately many of these states are not critical to the application and can be generated from small portions of basic information.

For example much information about an open file is kept in the kernel, but really critical to the application is only the filename, the position inside the opened file and the access rights. Usually the application only gets a handle to the file and all other information is kept in the kernel memory, which would require special attention during checkpointing. By moving the required information to the shared heap enables the application to access an opened file even after a fallback, because the kernel is able to restore the information from filename, position and access rights.

4 Evaluation and Comparison

All tests are done with a cluster of twelve single CPU machines with Fast Ethernet connection over an Allied Telesyn AT-8024 switch. The nodes are Athlon XP2500+ with 512 MB DDR-RAM equipped with 3Com 905 card on an Asus A7V8X-X.

4.1 Network Measurements

The usable bandwidth is important to cluster systems and depends on the operating systems architecture, the protocol stack and the network interface. Therefore it is at first interesting to know the overhead for transferred data and on the other hand to know the maximum achievable throughput for page data transfer. The page data transfer packets are in the current implementation split up into three packets with 1400, 1400 and 1296 bytes of payload, if the page is not empty. Otherwise a single frame with no (extra) payload is enough. Additionally for each packet a header of 54 bytes is required, containing the information for the Ethernet, IP and the DSM. The Plurix-Protocol contains 22 bytes and is treated as payload because of the information about current time and packet type (like "requested page is empty"). The behavior of our implementation shows that Fast Ethernet works well for packets with more than 300 bytes and allows a bandwidth above 11 megabytes per second. The theoretical maximum transfer rate of Fast Ethernet with 100 megabits per second is 12500 kilobytes per second, which is assumed to be 100% for utilization in the following table. As for all packets there is an additional preamble and CRC on the Ethernet medium, the overhead increases with smaller packets.

Data pages are transferred at about 11.7 MB/s, which means about 2900 pages with data can be transmitted per second on the current infrastructure. Empty pages are transferred at about 2.4 MB/s resulting in a maximum of about 37200 empty pages per second. Network switches aggregate the bandwidth, so the measured values are the limit of direct one-to-one connections of which many can take place at the same time.

Ethernet			Payload		
Packetsize bytes/packet	Throughput kB/sec	Utilization of Ethernet	Data bytes/packet	Throughput kB/sec	Utilization of Packet
1500	11970	95,76%	1468	11714	97,87%
1000	11870	94,96%	968	11490	96,80%
576	11663	93,30%	544	11015	94,44%
272	11145	89,16%	240	9834	88,24%
160	6291	50,33%	128	5033	80,00%
96	3513	28,10%	64	2342	66,66%
64	2381	19,05%	32	1191	50,00%

4.1 Protocol and Transaction Measurements

A critical time for a transactional system is the overhead at the beginning and at the end (both aborting and committing) of a transaction. As we have implemented a first wins strategy with a token, the token latency becomes an interesting value, too. For a page-based system, the minimum time between page request and complete reception of a page is important as well as the maximum throughput (see section above).

	minimum	normal	upper
begin of transaction	15 µs	17 µs	25 µs
end of transaction, abort / commit	6 / 16 µs	12 / 18 µs	24 / 26 µs
latency for empty / filled page	97 / 780 µs	99 / 783 µs	102 / 797 µs
token latency	66 µs	71 µs	94 µs

The begin of transaction is independent from the cluster and has an overhead depending on the previous transaction, because all page table entries have to be reset and initialized. Most transactions are short and have only few page table entries to be modified, so this usually takes about 16 microseconds but may take up to 25 microseconds after a a data intensive transaction.

The end of a transaction may be either an abort with resetting all information or a commit with publication of all modifications. The measured times give both values if the token is already at the committing node. Otherwise there is an additional call to get the token, which is around 70 microseconds.

As mentioned the page latency is very important to page based distributed systems. Each access to a not present page requires a page request with appropriate answer from the cluster. In the measurements the time taken to request and answer one page is given for both a filled and an empty page. The filled page needs three packets with 4258 bytes (4096+3*54), the empty page one packet with 54 bytes (see section 4.1).

To verify these values we made an additional test with 1000 empty transactions committing on a single node that already possesses the token. The time needed was 32110 ms, straight below the normal value of n*(BOT+EOT) coming to 31-35 ms.

4.2 Application Measurements

Currently a raytracer and the well-known successive-over-relaxation algorithm were implemented to test the programming model and the behavior of our system. The raytracer as fully parallelized application performs very well with nearly linear speed-up. With larger image sizes the speed-up grows, as the initial distribution of the scene and matrix is less relevant compared to the computation.

	Raytracer Execution Time in Milliseconds					
Nodes	*1024x768*	*1448x1086*	*2048x1536*	*2896x2172*	*4096x3072*	*5792x4344*
1	78067	156004	312121	623460	1247697	2494528
2	39780	79136	157716	314178	625691	1249590
4	19988	39624	79088	157293	314168	626804
8	10231	19961	39842	78883	157073	313449
12	7381	13910	27448	53328	105430	210028

	Raytracer Speedup					
Nodes	*1024x768*	*1448x1086*	*2048x1536*	*2896x2172*	*4096x3072*	*5792x4344*
1	1,000	1,000	1,000	1,000	1,000	1,000
2	1,962	1,971	1,979	1,984	1,994	1,996
4	3,906	3,937	3,947	3,964	3,971	3,980
8	7,630	7,815	7,834	7,904	7,943	7,958
12	10,577	11,215	11,371	11,691	11,834	11,877

The successive-over-relaxation needs much more communications in comparison to the raytracer and has a single-calculator-multi-wait behavior, so linear speed-up is not reached. To provide comparability to other systems, we made measurements under Linux with JPVM [jpvm] and Aleph [Her99], [aleph] in the same cluster, too. Both other systems use the Sun Java Environment. The following tables show these results, where for each matrix size the relative speed up to the single node time and to the Plurix single node time is given to provide both internal speed-up and comparison.

SOR time	2048x2048			4096x4096		
Nodes	Plurix	Aleph	JPVM	Plurix	Aleph	JPVM
1	121,48	533,02	676,78	458,49	2122,37	2721,14
2	70,10	281,06	367,39	250,70	1084,08	1414,04
4	45,22	155,92	191,27	142,57	566,66	737,63
8	36,66	155,66	118,65	98,31	330,80	413,52
12	37,52	120,52	100,20	87,35	270,01	326,92

SOR sdup	2048x2048			4096x4096		
Nodes	Plurix	Aleph	JPVM	Plurix	Aleph	JPVM
1	1,00	0,23/1,00	0,18/1,00	1,00	0,22/1,00	0,17/1,00
2	1,73	0,43/1,90	0,33/1,84	1,83	0,42/1,96	0,32/1,92
4	2,69	0,78/3,41	0,64/3,54	3,22	0,81/3,75	0,62/3,69
8	3,31	0,78/3,42	1,02/5,70	4,66	1,39/6,42	1,11/6,58
12	3,24	1,01/4,42	1,21/6,75	5,25	1,70/7,86	1,40/8,32

The JPVM shows very good internal speed up as Aleph does, but the overall performance is about five times slower than Plurix. Therefore the speed-up can be better easily, because the time to communicate is much less than the time to calculate. This results in a faster execution on two nodes running Plurix than twelve nodes running JPVM. For Aleph and JPVM the communication is done explicitly, whereas the programmer does not have to manage the data exchange in Plurix. For the small matrices the Plurix system has the maximum speed-up at 8 nodes, whereas the maximum speed-up for the bigger matrix is not reached with even twelve nodes. Both JPVM and Aleph never show this point of return at those measurements, but as the overall execution time is much higher than with Plurix, the communication is not expected to be the bottleneck.

5 Related Work

Transactions are well known in the database world for more than two decades and are implemented for example in PostgreSQL [Sto87], MySQL [mysql] and Oracle [oracle]. The transaction paradigm was overwhelmingly successful in this context, and A.P. Black [Bla90] advocated its transfer to operating systems. Plurix supports transactions both for single station and cluster set-ups.

The idea of distributed shared memory systems was presented by L. Keedy in 1985 [Kee85]. There are several page-based systems with distributed shared memory like IVY [Li88], Mirage [Fle89] and TreadMarks [Kel94] with different consistency models. Mungi [Hei94] uses 64 bit processors for a distributed single address space secured with password capabilities [Hei98], the consistency is guaranteed with single-owner and write-invalidate semantic. As an alternative Plurix integrates semantically grouped accesses into a transaction and validates them in a forward validation scheme with optimistic synchronization.

Systems like Eiffel** [Hug93], Thor [Lis99] or PerDIS [Fer99] provide transactions with distribution, but do not integrate the transaction concept into the operating system and partially use a server/client architecture. In contrast Plurix provides a flat distribution model on the kernel level.

6 Conclusion and Future Research

Transactional Consistency helps programmers to write distributed applications by providing implicit and automatic communication or synchronization. The memory and transaction management in the kernel guarantees consistent and race-free parallel execution of applications with only small overhead. The implementation of the protocol is lean and efficient, it lends itself to high performance cluster computing with low latency and near maximum throughput as shown in the measurements. Future work will be done on optimizing the protocol for Gigabit Ethernet and to include different fairness strategies.

References

[aleph] http://www.cs.brown.edu/people/mph/aleph/
[Bla90] A.P. Black: "Understanding Transactions in the Operating System Context". Proc. of the 4th workshop on ACM SIGOPS, p1-4, 1990.
[Dad96] P. Dadam: "Verteilte Datenbanken und Client/Server-Systeme". Springer-Verlag, Heidelberg, 1996.
[Fer99] P. Ferreira et al.: "PerDiS: design, implementation, and use of a PERsistent DIstributed Store". INRIA RR 3525, Rapport de Recherche, Institut National de Recherche en Informatique et Automatique, Rocquencourt, France und Lecture Notes In Computer Science, vol 1752, p427-452, 1999
[Fle89] B.D. Fleisch et al.: "Mirage: A Coherent Distributed Shared Memory Design". Proc. of 14th ACM Symp. on Operating Systems Principles, 1989.
[Fre02] S. Frenz: "Persistenz eines transaktionsbasierten verteilten Speichers". Diploma-Thesis at the University of Ulm, Distributed Systems, 2002.
[Fre04] S. Frenz, M. Schoettner, R. Goeckelmann, P. Schulthess: "Performance Evaluation of Transactional DSM". Proc. of the 4th IEEE/ACM Intern. Symposium on Cluster Computing and the Grid, Chicago, 2004.
[Goe03] R. Goeckelmann, M. Schoettner, S. Frenz, P. Schulthess: "A Kernel Running in a DSM - Design Aspects of a Distributed Operating System". Proc. of the IEEE Intern. Conf. on Cluster Computing, Hong Kong, 2003.

[Goe04]   R. Goeckelmann, M. Schoettner, S. Frenz, P. Schulthess: "Plurix, a Distributed Operating System extending the Single System Image Concept". Proc. of the IEEE Canadian Conference on Electrical and Computer Engineering, Niagara Falls, Canada, 2004.
[Hae83]   T. Haerder, A. Reuter: „Principles of Transaction-Oriented Database Recovery", Computing Surveys 15, 4, p287-317, December 1983.
[Hei94]   G. Heiser, K. Elphinstone, S. Russell, J. Vochteloo: "Mungi: A Distributed Single Address-Space Operating System". Proc. of the 17th Annual Computer Science Conference, ACSC-17, 1994.
[Hei98]   G. Heiser et al.: "The Mungi single-address-space operating system". Software: Practice and Experience, vol 28, p901-928, August 1998.
[Her99]   Maurice Herlihy: "The Aleph Toolkit: Support for Scalable Distributed Shared Objects". Proc. of the 3rd Intern. Workshop on Network-Based Parallel Computing: Communication, Architecture, and Applications, 1999.
[Hug93]   C. McHugh, V. Cahill: "Eiffel**: An Implementation of Eiffel on Amadeus, a Persistent, Distributed Applications Support Environment". TOOLS Europe '93 Conference Proceedings, p47-62, 1993.
[jpvm]   http://www.cs.virginia.edu/~ajf2j/jpvm.html
[Kee85]   J.L. Keedy, D.A. Abramson: "Implementing a large virtual memory in a Distributed Computing System". Proc. of the 18th Annual Hawaii International Conference on System Sciences, 1985.
[Kel94]   P. Keleher, A.L. Cox, S. Dwarkadas, W. Zwaenepoel: "TreadMarks: Distributed Shared Memory on Standard Workstations and Operating Systems". Proc. of the Winter 1994 USENIX Conference, 1994.
[Li88]   K. Li.: "IVY: A Shared Virtual Memory System for Parallel Computing". In Proceedings of International Conference on Parallel Processing, 1988.
[Lis99]   B. Liskov, M. Castro: "Providing Persistent Objects in Distributed Systems". Proc. of ECOOP'99, 1999.
[mysql]   http://www.mysql.com/
[oracle]   http://www.oracle.com/technology/documentation/index.html
[Sch02]   M. Schoettner: "Persistente Typen und Laufzeitstrukturen in einem Betriebssystem mit verteiltem virtuellen Speicher". PhD-Thesis at the University of Ulm, Distributed Systems, 2002.
[Sto87]   M. Stonebraker: "The Design of the POSTGRES Storage System". Proc. of the 1987 VLDB Conference, Brighton, September 1987.
[Wen02]   M. Wende et al.: "Optimistic Synchronization and Transactional Consistency". Proc. of the 4th Intern. Workshop on Software Distributed Shared Memory, Berlin, 2002.
[Wen03]   M. Wende: "Kommunikationsmodell eines verteilten virtuellen Speichers". PhD-Thesis at the University of Ulm, Distributed Systems, 2003.
[Wir92]   N. Wirth, J. Gutknecht: "Project Oberon". ACM Press, New-York, 1992.

CPOC: Effective Static Task Scheduling for Grid Computing[*]

Junghwan Kim[1], Jungkyu Rho[2], Jeong-Ook Lee[1], and Myeong-Cheol Ko[1]

[1] Department of Computer Science, Konkuk University,
Danwol-dong 322, Chungju si, Chungbuk 380-701, Korea
{jhkim, ljo, cheol}@kku.ac.kr
[2] Department of Computer Science, Seokyeong University,
Jungneung-dong 16-1, Sungbuk-gu, Seoul, Korea
jkrho@skuniv.ac.kr

Abstract. Effective task scheduling is crucial for achieving good performance in high performance computing. Many scheduling algorithms have been devised for heterogeneous computing and CPOP is one of the scheduling algorithms. In this paper we present new scheduling algorithms, CPOC and CPOC_E by modifying the CPOP. We use a cluster of processors for critical-path tasks while a single processor is used in the CPOP. This heuristic is useful for realistic Grid computing environments in which communication costs are not arbitrarily heterogeneous. In an additional heuristic the critical-path tasks are considered to finish (or start) as early as possible when non critical-path tasks are scheduled. For performance study we developed a task graph generator and a tool which would support more realistic network configuration. The experimental results show our scheduling algorithm outperforms the CPOP as well as the HEFT.

1 Introduction

Grid computing is an emerging technology for high-performance computing, which could use world-widely dispersed computing resources that may be heterogeneous and possibly owned by multiple organizations[1, 2, 3]. Effective task scheduling is crucial for achieving good performance in grid computing and many scheduling techniques have been researched[4, 5, 6].

The task scheduling problem includes assigning the tasks of an application to processors and determining the execution order of tasks on the processors. The goal of task scheduling is usually to minimize the schedule length (makespan). Most scheduling problems are known to be NP-complete except a few restricted cases[7, 8]. Many sub-optimal algorithms have been devised to reduce the schedule length because of its importance on performance.

The scheduling algorithms could be classified into *static* or *dynamic* scheduling according to whether it is done at compile time (static scheduling) or on-the-fly (dynamic scheduling)[9]. In static scheduling an application is usually represented by DAG(Directed Acyclic Graph).

---

[*] This work was supported by Konkuk University (2003).

Our research focuses on static task scheduling for heterogeneous environments with task graphs represented by DAG. Especially we started our work by considering two heuristics, HEFT(Heterogeneous Earliest Finish Time) and CPOP(Critical-Path-On-a-Processor), which had been proposed by Topcuoglu et al.[10]. The literature remarks the CPOP has lower performance than the HEFT and the HEFT gives very good quality of schedules with low cost. However, we would note two points to enhance the algorithms.

First, they experimented with an assumption that the execution time of a task is arbitrarily heterogeneous. Second, the communication cost among processors is arbitrarily heterogeneous whether the processors are close together or not.

We introduce more realistic assumptions for grid computing. First, we generate networks which range from WAN to LAN. It is more realistic for grid environments that the cost is expensive in WAN than in LAN instead of arbitrary heterogeneity. Second, we separate architectural heterogeneity from speed heterogeneity of processors. Processors may have not only diverse clock speed but also diverse architectural characteristics.

The rest of the paper is organized as follows. In section 2 we describe related work about task scheduling and in section 3 our scheduling heuristics are proposed and compared with the previous heuristics. In section 4 the experimental results are presented and analyzed. Finally section 5 gives concluding remarks and future work.

2 Related Work

The heuristic-based algorithms can be classified into a few categories: TDB(Task-Duplication Based)[11], UNC(Unbounded Number of Clusters)[12, 13, 14], and other scheduling algorithms. The other groups are further classified into BNP(Bounded Number of Processors)[15] and APN(Arbitrary Processors Network)[16, 17] scheduling algorithms. While the BNP assumes contention-free network and no routing strategies, the links are not contention-free in the APN. The idea of TDB algorithms is to reduce the communication overhead by duplicating tasks and allocating them on multiple processors redundantly. In each step of the UNC algorithms some clusters (or tasks) are merged to reduce the completion time. The UNC needs an additional step for mapping the clusters onto the available processors.

Most popular scheduling technique is *list scheduling*. The common idea of list scheduling heuristics is to make a scheduling list and schedule tasks from the front of the list. So it consists of two phases: a *task prioritizing phase* and a *processor selection phase*. On the task prioritizing phase priority of each task is computed to make a ready list, and the most appropriate processor is selected for the current highest-priority task on the processor selection phase. In the following sections we describe two previous heuristics for heterogeneous computing, HEFT and CPOP.

2.1 HEFT (Heterogeneous Earliest Finish Time)

The HEFT consists of two phases: a task prioritizing phase and a processor selection phase. On the task prioritizing phase the priority of each task is computed with upward rank, $rank_u$, which is based on average computation time and average communication time. The upward rank of a task i is defined by

$$rank_u(i) = \overline{w}_i + \max_{j \in succ(i)} (\overline{c}_{ij} + rank_u(j)) \ . \tag{1}$$

$succ(i)$ is the set of immediate successors of task i, $\overline{c}_{ij}$ is the average communication cost of edge (i, j) and $\overline{w}_i$ is the average computation cost of task i. On the processor selection phase the task with the highest priority is picked for allocation and a processor which ensures the earliest-finish-time is selected. The earliest-finish-time is considered using insertion-based allocation. When we use the insertion-based, a task can be inserted before previously allocated task as long as there is free time slot.

The HEFT scheduling algorithm reflects two heuristics. First, a task which has higher upward rank is more important, and preferred for allocation to other tasks. Intuitively, the upward rank of a task reflects the average remaining cost to finish all tasks after that task starts up. Second, simple but effective idea is earliest-finish-time approach. However, as it pursues the earliest-finish-time of the current task, it may fall into local optima like a greedy method.

2.2 CPOP (Critical-Path-On-a-Processor)

The CPOP scheduling algorithm was introduced in the same literature as the HEFT. It differs from the HEFT in that not only the priority of a task differs from that of the HEFT, but the tasks with the highest priority (critical-path tasks) are handled specially. The priority of a task i is given by $rank_u(i) + rank_d(i)$. The *downward rank* of task i, $rank_d(i)$ is defined by

$$rank_d(i) = \max_{j \in pred(i)} (rank_d(j) + \overline{w}_j + \overline{c}_{ji}) \ . \tag{2}$$

$pred(i)$ is the set of immediate predecessors of task i. The task with the highest priority is thought of as being on a critical path, and at that time the priority is thought of as critical path length, |CP|. Intuitively, the upward rank of a task is the expected time to finish the last task after the task starts, and the downward rank of a task is the expected elapsed time before the task starts from the entry task. So if the summation of the upward rank and the downward rank of a task is the highest, the task is thought of as being on a critical path.

For the set of tasks which are on a critical path, CP, the algorithm finds a processor, p_{CP}, which minimizes the sum of computation time of all CP tasks. On the processor selection phase each task of a ready queue is selected for allocation and if the selected task is on a critical path, the processor p_{CP} is used, otherwise a processor which minimizes the earliest-finish-time of the task is used.

3 Proposed Scheduling Heuristic

In this section we present our scheduling heuristics. First we propose CPOC(Critical-Path-On-a-Cluster) algorithm by modifying the CPOP algorithm. Then, two enhanced versions with respect to the CPOP and the CPOC will be presented.

3.1 CPOC (Critical-Path-On-a-Cluster)

In the HEFT and the CPOP algorithms it is assumed that processor-to-processor communication costs are arbitrarily heterogeneous. Practically there may be a bias in communication costs as grid computing environment ranges from very fast local area network to wide area network having long latency. The communication latency is possibly very small if the two peers are located in the same local area network, while the latency is very high in wide area network.

The CPOP algorithm schedules all CP tasks on a processor which minimizes overall computation cost of the critical path. Since only one processor is used for the critical path, there is no communication cost among CP tasks. We assume that communication cost among processors which are in the same local area network are negligible. So we use a cluster of processors to execute CP tasks instead of a single processor. This heuristic is profitable since it is possible to choose any processor within the cluster for each CP task. It gives more chance to minimize the sum of computation time of CP tasks.

<task prioritizing phase>
1 Compute $rank_u(i)$ and $rank_d(i)$ for each task i.
2 Assign $rank_u(i) + rank_d(i)$ to each task i as priority.

<processor selection phase>
3 Find a set CP of tasks having the largest value of $rank_u(i) + rank_d(i)$.
4 c_{CP} = find_critical_path_cluster(CP)
5 ReadyQ.insert(k) where k is the entry task.
6 while (not ReadyQ.empty()) {
7   i = ReadyQ.delete()
8   if (task i is in CP)
9   schedule i on a processor within cluster c_{CP} to minimize EFT(i).
10   else
11   schedule i on a processor to minimize EFT(i).
12   for each immediate successor k of task i,
13   if (every immediate predecessor of task k has been already scheduled)
14   ReadyQ.insert(k)
15 }

Fig. 1. CPOC scheduling algorithm

Like CPOP algorithm the CPOC algorithm consists of a task prioritizing phase and a processor selection phase(see Fig. 1). The task prioritizing phase of the CPOC is exactly the same as that of the CPOP. However, on the processor selection phase, for the CP tasks the algorithm tries to find a cluster, c_{CP}, which minimizes the sum of computation time of all CP tasks. Each task of a ready queue is considered in turn for allocation. If the selected task is on a critical path, the cluster c_{CP} is used and a processor is selected within the cluster to minimize the earliest-finish-time of the task. Otherwise a processor is selected among all processors.

3.2 Enhancement of Algorithms: Earliest Start Time for CP tasks

The processor allocation scheme based upon earliest-finish-time reflects the heuristic that the current task is most important now and should finish as early as possible. In the HEFT scheduling algorithm the task with the highest priority is always selected for allocation, so the earliest-finish-time approach matches with it well. In the CPOP or the CPOC algorithms, however, tasks are not selected in order of priority because only ready tasks can be selected and the tasks with higher priority may not be ready. In HEFT scheduling algorithm any task with the currently highest priority is always ready.

Since the current ready task is not of highest priority among unscheduled tasks, our goal should not be to complete the current task as early as possible in the CPOP or the CPOC. Instead, we should try to shorten the CP length, in other words, make the CP tasks finished as early as possible. So we could pursue the earliest start (or finish) time for the CP tasks when we schedule other than CP tasks. The goal is to make the CP tasks start as early as possible when we schedule the immediate predecessors of the CP tasks.

The enhanced scheduling algorithm CPOC_E (or CPOP_E) differs from original CPOC (or CPOP) in allocating the immediate predecessors of CP tasks. When a ready task is scheduled, it selects a processor which minimizes the earliest start time of a successor task if the successor is a CP task. Just three lines are modified for the CPOC_E compared to line 10 of Fig. 1 (see Fig. 2).

```
10-1   else if (let j = succ(i) and j is in CP)
10-2       schedule i on a processor to minimize EST(j).
10-3   else
```

Fig. 2. CPOC_E scheduling algorithm

4 Experiments

In this section we would describe experimental results over randomly generated task graphs and network graphs. As described above, the five scheduling algorithms will be compared with two performance metrics.

4.1 Task Graph Generation

Task graphs are randomly generated by a task graph generator which has been developed in this work. Most of input parameters for the task generator are similar to those of the HEFT. The followings are the list of the parameters:

- d: Out-degree.
- α: Shape of a task graph. $\sqrt{v}/\alpha$ is the mean of height of the task graph where v is the number of tasks.
- β: Heterogeneity of processor speed. If $\beta=0$, it is exactly homogeneous.
- γ: Heterogeneity of processor architecture. If $\gamma=0$, it is exactly homogeneous.

- *ccr*: Communication to computation ratio. If *ccr*>>1, it represents communication cost is very high compared to computation cost.
- π : Ratio of non-terminal tasks.

Among the above parameters γ and π are newly introduced in this work. Also, β has somewhat different meaning compared to Topcuoglu's. While all heterogeneity is represented only by β in Topcuoglu's, we use both β and γ. We can generate more various task graphs with both parameters. If $\gamma=0$, there would be no architectural difference among processors which mean there can be only heterogeneity of processor speed.

To evaluate and compare the five heuristics, we generated the set of task graphs with $\pi=0.95$. The total number of task graphs used in our experiments was 183,750.

4.2 Network Graph Generation

In grid computing the computational nodes can reside in wide area network as well as local area network. So communication time is not arbitrarily heterogeneous and the latencies among some nodes are very low while the latencies among other nodes are very high. To reflect this situation and model realistic networks a network generator is used[18].

Table 1. Bandwidth and delay parameters

	bandwidth (Mbytes/s)	fixed delay (ms)	delay / unit distance (ms/unit)
WAN-to-WAN	1000	15	0.1
WAN-to-MAN	100	100	0.01
MAN-to-MAN	100	10	0.1
MAN-to-LAN	100	5	0.01
LAN-to-LAN	1	1	0.01

Since the output of the generator contains just node-to-node distances, we convert it into latency matrix by multiplying the distance by the unit delay and adding the fixed delay. The final node-to-node latency is determined as the smallest value among the values of all possible paths using the Floyd-Warshall shortest path algorithm. The bandwidth matrix can be obtained from the smallest value along the path. Table 1 shows parameters which are used for computation of the latency and the bandwidth matrices.

4.3 Performance Results

To compare the schedule of each algorithm, we would use two performance metrics. Though performance goal is to minimize schedule length, we need a normalized metric since each task graph has various schedule length. The first metric, SLR(Schedule Length Ratio) is defined by the following[10].

$$SLR = \frac{schedule\_length}{\sum_{i \in CP_{MIN}} \min\{w_{ik}\}} \quad . \tag{3}$$

min{ w_{ik} } is the minimum value among computation costs of task i when it is assigned to processor k, and CP_{MIN} is the set of tasks on a critical path assuming every task has the minimum computation cost. Since the denominator would be lower bound of schedule length, SLR cannot be less than 1.

The other metric is speedup which is given by dividing the upper bound of schedule length by the actual schedule length. The upper bound is computed by assigning all tasks to a single processor which minimizes the sum of computation time, *i.e.*, it means the best sequential execution time.

Performance comparison is presented by Fig. 3. In most cases it shows the order of performance: CPOC_E, CPOP_E, HEFT, CPOC and CPOP. The smaller SLR we have the better performance we get and also the higher speedup the better performance. Generally, as the SLR decreases the speedup increases, and vice versa. However, for out-degree, α, and γ, the results are not accordance with the general tendency.

As out-degree of a task graph increases, the lower bound of schedule (a denominator of SLR) would rapidly increase. Since the schedule length slowly increases, SLR decreases as shown in the Fig. 3. Even though out-degree of a task graph increases, the upper bound of schedule hardly changes. Because all edges are zeroed and there is no communication time when the upper bound is computed. The out-degree of a task graph does not affect the upper bound while the schedule length increases, so the speedup decreases.

As α increases the lower bound rapidly decreases compared to the actual schedule length. So SLR would increase as α increases. Similarly to the case of out-degree, the upper bound hardly change for α. Undoubtedly the shape of a task graph does not have any correlation to the upper bound. Since the schedule length can be reduced for increment of α, speedup may increases.

γ represents heterogeneity of architecture, so the higher γ gives a chance to obtain less computation time for each task. It means very rapid decrement of the lower bound which results in rapid increment of SLR. Though it is obvious that architectural

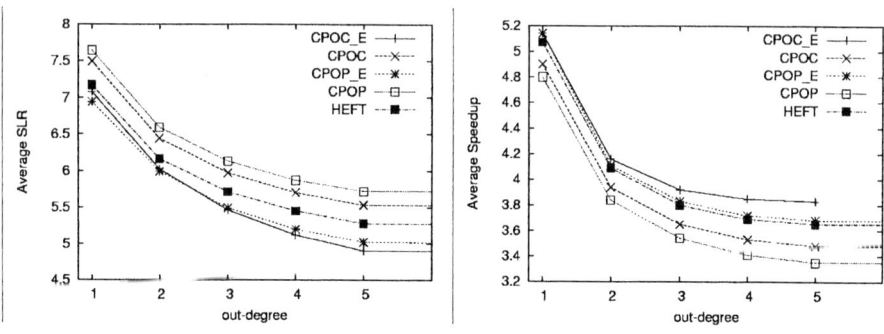

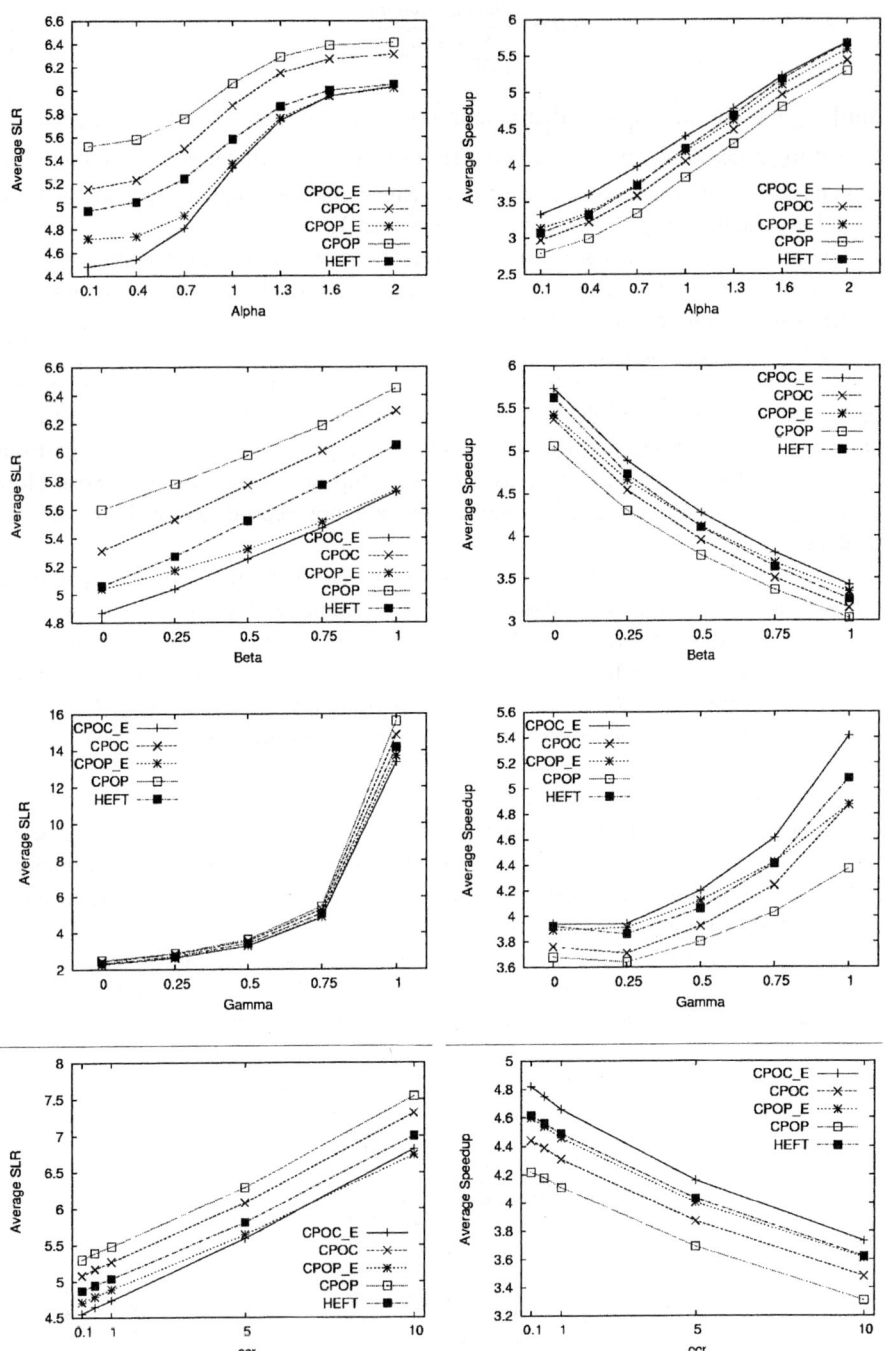

Fig. 3. Performance comparison of HEFT, CPOP, CPOC, CPOP_E and CPOC_E

heterogeneity gives more chance to shorten the length of each task, when we use a single-processor assignment it affects the upper bound in a very little amount. So speedup quickly increases.

Generally speaking, if the communication time increases, the performance would decreases. As *ccr* increases, SLR would increases and speedup would decreases.

5 Conclusions

We proposed new static scheduling algorithms which are profitable for Grid environments. We started from CPOP scheduling algorithm and modified it in two points.

First, we focused on that there could be locality of communication costs in the real world. Especially the communication latency cannot be arbitrarily heterogeneous when we assume grid computing environments. We used a cluster instead of a single processor for the CP tasks. The experimental results show that the CPOC outperforms the CPOP.

Second, we modified the processor selection heuristic of the CPOP which is to complete each task as early as possible. Our heuristic is based on the belief that some tasks are more important than others on even processor selection phase. When we schedule non-CP tasks we pursue the earliest-start-time of CP tasks instead of its own earliest-finish-time. The experimental results show that the CPOP_E outperforms the CPOP and the CPOC_E outperforms the CPOC.

We developed a task graph generator for comparison of the scheduling heuristics. The task graph generator uses not only the parameters which have been presented in the HEFT but also our own parameters such as architectural heterogeneity. For experiments we developed a tool to convert the output of a network generator into latency and bandwidth data.

The experimental results show our heuristic CPOC_E generates better schedules than CPOP and HEFT. The results show even the CPOP_E which is enhanced by our heuristic almost outperforms the HEFT.

Remaining works are to develop the various metrics for comparison of scheduling algorithms and to find out strength of each scheduling algorithm by experimenting on the larger and finer-grained input set. The experiments on real-world applications also remain in the future work.

References

1. I. Foster and C. Kesselman (ed.), *The Grid: Blueprint for a New Computing Infrastructure*, Morgan Kaufmann, 1999.
2. I. Foster, C. Kesselman and S. Tuecke, "The Anatomy of the Grid: Enabling Scalable Virtual Organizations," In Proc. of 1st International Symposium on Cluster Computing and the Grid (CCGRID01), 2001.
3. Fran Berman, Geoffrey C. Fox, and Anthony J. G. Hey, *Grid Computing: Making the Global Infrastructure a Reality*, John Wiley & Sons, 2003.
4. C. Boeres and A. Lima, "Hybrid Task Scheduling: Integrating Static and Dynamic Heuristics," In Proc. of the 15th Symposium on Computer Architecture and High Performance Computing (SBAC-PAD'03), 2003.

5. H. Chen and M. Maheswaran, "Distributed Dynamic Scheduling of Composite Tasks on Grid Computing Systems," In Proc. of the International Parallel and Distributed Processing Symposium (IPDPS'02), 2002.
6. N. Fujimoto and K. Hagihara, "Near-Optimal Dynamic Task Scheduling of Precedence Constrained Coarse-Grained Tasks onto a Computational Grid," In Proc. of the 2nd International Symposium on Parallel and Distributed Computing (ISPDC'03), 2003.
7. H. El-Rewini and H. H. Ali, "Task Scheduling in Multiprocessing Systems", IEEE Computer, pp.27-37, Dec. 1995.
8. Y. Kwok and I. Ahmad, "Static Scheduling Algorithms for Allocating Directed Task Graphs," ACM Computing Surveys, Vol. 31, No. 4, pp.407-471, Dec. 1999.
9. T. L. Casavant and J. G. Kuhl, "A Taxonomy of Scheduling in General-Purpose Distributed Computing Systems," IEEE Trans. on Software Engineering, Vol. 14, No. 2, pp.141-154, Feb. 1988.
10. H. Topcuoglu, S. Hariri and M. Wu, "Performance-Effective and Low-Complexity Task Scheduling for Heterogeneous Computing," IEEE Trans. on Parallel and Distributed Systems, Vol. 13, No. 3, pp.260-274, March 2002.
11. I. Ahmad and Y. Kwok, "On Exploiting Task Duplication in Parallel Program Scheduling," IEEE Trans. on Parallel and Distributed Systems, Vol. 9, No. 9, pp.872-892, 1998.
12. M. Wu and D. D. Gajski, "Hypertool: A Programming Aid for Message-passing Systems," IEEE Trans. on Parallel and Distributed Systems, Vol. 1, No. 3, pp.330-343, 1990.
13. T. Yang and A. Gerasoulis, "DSC: Scheduling Parallel Tasks on an Unbounded Number of Processors," IEEE Trans. on Parallel and Distributed Systems, Vol. 5, No. 9, pp.951-967, 1994.
14. Y. Kwok and I. Ahmad, "Dynamic Critical-path Scheduling: An Effective Technique for Allocating Task Graphs to Multiprocessor," IEEE Trans. on Parallel and Distributed Systems, Vol. 7, No. 5, pp.506-521, 1996.
15. J. Hwang, Y. Chow, F. D. Anger and C. Lee, "Scheduling Precedence Graphs in Systems with Interprocessor Communication Times," SIAM J. Comput., Vol. 18, No. 2, pp.244-257, April 1989.
16. G. C. Sih and E. A. Lee, "A Compile-time Scheduling Heuristic for Interconnection-constrained Heterogeneous Processor Architectures," IEEE Trans. on Parallel and Distributed Systems, Vol. 4, No. 2, pp.75-87, Feb. 1993.
17. H. El-Rewini and H. H. Ali, "Scheduling Parallel Program Tasks onto Arbitrary Target Machines," Journal of Parallel and Distributed Computing, Vol. 9, No. 2, pp.138-153, 1990.
18. M. B. Doar, "A Better Model for Generating Test Networks," IEEE Global Telecommunications Conference, Nov. 1996.

Improving Scheduling Decisions by Using Knowledge About Parallel Applications Resource Usage

Luciano José Senger[1], Rodrigo Fernandes de Mello[2], Marcos José Santana[2], Regina Helena Carlucci Santana[2], and Laurence Tianruo Yang[3]

[1] Universidade Estadual de Ponta Grossa,
Departamento de Informática, Av. Carlos Cavalcanti,
4748, 84030-900, Ponta Grossa, PR
ljsenger@uepg.br

[2] Universidade de São Paulo,
Instituto de Ciências Matemáticas e de,
Computação, Caixa Postal 668, 13560-970 São Carlos, SP
{mello, mjs, rcs}@icmc.usp.br

[3] St. Francis Xavier University,
Department of Computer Science,
Antigonish, NS, B2G 2W5, Canada
lyang@stfx.ca

Abstract. This paper presents a process scheduling algorithm that uses information about the capacity of the processing elements over the communication network and parallel applications in order to allocate resources on heterogeneous and distributed environments. The information about the applications is composed by the resources usage behavior (percentage values related to CPU's utilization, network send and network receive) and by the prediction of the execution time of tasks that make up a parallel distribution. The knowledge about the resources usage is obtained by means of the Art2A self-organizing artificial neural network and by a specific labeling algorithm; the knowledge about the execution time is obtained through the learning techniques based on instances. The knowledge about the application execution features, combined with the information about the computing capacity of the resources available in the environment, are used as an entry to improve the decisions of the proposed scheduling algorithm. Such algorithm uses genetic algorithm techniques to find out the most appropriate computing resources subset to support the applications. The proposed algorithm is evaluated through simulation by using a model parameterized with the features obtained from a real distributed scenario. The results obtained by the evaluation show that the scheduling that uses the genetic search allows a better allocation of computing resources on environments composed of tens of computers on which the parallel applications are composed by tens of tasks.

Keywords: Scheduling, high performance computing, genetic algorithms.

1 Introduction

Advances in hardware and software technology continuously improve performance of parallel systems based on heterogeneous distributed computing. This technological evolution allows heterogeneous systems, which are a group of diverse autonomous computers linked by distinct networks, to support a variety of workloads, including parallel, sequential and interactive tasks [2]. When used to support parallel applications (i.e. computational applications composed of a group of communicating tasks), distributed computers are treated as distributed processing elements (PEs) and compose a MIMD distributed memory (*Multiple Instruction Multiple Data*) parallel architecture [11]. Heterogeneity on distributed computers are derived from three main sources: the *architectural heterogeneity*, where PEs have distinct architectures (e.g., a personal computer and either a high-end workstation or a special-purpose computer); the *configuration heterogeneity*, where PEs have the same architecture, but distinct computing power (different amounts of memory, clock speeds and disk storage capacities); and the *network heterogeneity*, ranging from slow communication links to high-speed networks.

One of the challenges in such systems is to develop scheduling algorithms that assign tasks of parallel applications to heterogeneous machines. Scheduling algorithms may be implemented either by the user application or by distributed software and must deal with decisions: (i) which parallel application (among all submitted to the system) should be executed; (ii) which amount of computing resources should be reserved to each application; (iii) which resources best support the application requirements; and (iv) administrative questions, such as workstation reservation for interactive support. These decisions are influenced by different factors such as typical system workload, diversity of applications, different user requirements and conflicting scheduling goals [5].

In order to improve the decisions taken by the scheduling algorithms, several researchers have adopted techniques that use the knowledge about the parallel applications [13][7][17][24]. There are commonly three main sources to obtain knowledge: the description of application requirements provided by the user (or programmer) who submits the parallel application to the system; historical traces of all applications executed in a specific system over certain period of time, and runtime measurements from parallel applications. Among these knowledge sources, historical traces and runtime measurements have demonstrated a great potential to provide information aiming at classifying parallel applications and obtaining knowledge [12][25].

Such knowledge sources have motivated Senger *et. al* [22][21] to develop models for the execution time prediction of parallel applications and the evaluation of the resources behavior on such applications. In order to accomplish the prediction of the execution time of parallel distributions it has been proposed a model that applies the learning technique based on the memory (*Instance Based Learning*) [1]. In this model, the historical information of execution traces of parallel applications previously submitted to the system is used. Such historical database is used to predict the execution time of a new application initiated on the environment. This prediction is carried out through queries that use parameters with user identification, a command line used to initiate the parallel application and others. These queries bring back points to Cartesian plan, on which each point represents the execution of each one of the applications contained in the historical base. Based on the same parameters of the application initiated on the environment, it is made a local weighted regression among the points, which allows the estimation of attributes of the new application, such as the CPU usage, memory and execution time. The evaluation of the behavior of the resources usage on parallel applications is made with the objective of obtaining the percentage of CPU, hard disk, network and memory occupations, which vary due to time, i.e., during the applications execution. This variation caused by time has motivated the authors to propose a diagram of states to represent the probabilities of behavior changing of an application being executed.

The evaluation of the application behavior is made through the instrumentation of the Linux operating system [6]. Such instrumentation allows the obtainment of parameters such as number of bytes transferred by using hard disk, network, CPU's usage time and quantity of memory usage. These parameters are introduced as entries of a knowledge acquisition model that uses the Art2A self-organizing artificial neural architecture for categorization and an algorithm based on the significance of the attributes for labeling the categories. The model accomplishes the classification of entries by generating behavioral patterns of the application during the execution of the parallel application, thus allowing the obtainment of on-line behaviors for the applications. During its execution, an application may assume behaviors such as CPU-bound, Memory-bound, Disc-bound and Network-bound.

The knowledge sources studied by Senger et al. [19][20][23] have motivated the development of the scheduling algorithm proposed in this study. This algorithm uses as entry the prediction knowledge of the execution time and applications behavior to find out the most adequate set of computing resources to support a parallel application on a distributed environment composed of heterogeneous capacity computers. Such algorithm, named GAS (Genetic Algorithm for Scheduling), uses crossover and genetic algorithm mutation techniques. Genetic algorithms use biology concepts to improve the efficiency of the random search [9]. These algorithms are a particular case of the evolutive algorithms, which have been inspired on Darwin's theory.d Such theory proposes that the stronger individuals of a certain population interact among themselves, propagating their genes and generating descendants more adapted to their life conditions.

2 Scheduling Algorithm

The work of Senger et al. to predict the execution time of parallel applications [21] and to classify the behavior of the resources usage on such applications [18] have motivated the development of the GAS (Genetic Algorithm for Scheduling) scheduling algorithm. Such algorithm uses the information about time prediction, CPU's usage behavior and communication system to distribute tasks among the processing elements of a heterogeneous distributed environment. On such environment are considered the computers with distinct capacities of processing, memory, network and hard disk.

This algorithm uses the crossover and genetic algorithm mutation techniques to optimize the probalistic searches that aim to find out the best solutions for a problem. Such algorithms uses biology concepts to improve the efficiency of the random search [9]. They are a particular case of the evolutive algorithms, which have been inspired on Darwin's theory. This theory proposes that the stronger individuals of a certain population interact among themselves, propagating their genes among themselves and generating descendants more adapted to the life conditions.

In order to solve a scheduling problem, GAS generates an initial population composed of individuals, on which each individual represents a possible scheduling solution.

Therefore, each individual is represented by a vector v_i, where i varies from 1 up to the maximum number of individuals. Each vector v_i has the number of elements equal to the number of tasks of the parallel application. Each element of the vector v_i with index k represents the computer where the task k will be allocated. The number of computers in the environment is variable and pre-defined so that GAS algorithm may choose the available resources allocations. After generating an initial population of individuals, GAS submit it to a fitness function. This function accesses the capacity parameters (cap_i, in millions of instructions per second) of each processing element of the system, the transmission time of messages of size m among each one of the computers represented by a matrix $r_{w,q}$ of communication costs between two PEs w and q; and the slowdown factor ($MS(i, t, u)$) imposed by the memory usage of each processing element i at instants t (instants on which scheduling operations are requested due to the arrival of new applications to the system), such factor varies according to the occupation of the main and secondary memories (u). These three parameters are obtained through the model by Mello and Senger [15] that uses a suite of benchmarks and regressions to obtain the behavior equations related to the memory occupation.

In addition, the fitness function receives the execution time prediction of each task of the parallel application represented by a vector t_k, where $k \in [1, s]$ and s represent the number of tasks that makes up the parallel application. Each element of this vector provides information on the processing load of a taks k, in millions of instructions. This parameter is obtained through the application of the IBL algorithm by Senger et al. [21]. The prediction error obtained through the authors' experiments is taken into account by the GAS algorithm. The fitness function also receives the behavior of each task of the parallel application, expressed on the percentage of its processing time dedicated to the network processing and operations, such as messages sending and receiving. The function also receives a vector l_t that represents the load of each processing element at the instant t. Such vector is obtained through the load readings of the processing elements at the scheduling instant. By using such parameters, the fitness function calculates the ability of each individual to solve the scheduling.

This ability quantifies how good is the resource allocation proposed by the individual. The ability function defines which individuals are the most capable ones. They are going to be kept and combined through the crossover and altered by mutations to generate new individuals. The crossover technique is used to cross individuals. On such technique, each individual, represented by a vector v_i, is bi-parted at a certain position p. The partitions are combined two by two in order to form new individuals. Each vector element may be considered as a gene and this process is considered a crossing of features. Each new generated individual should be submitted to the fitness function so that its ability has to be evaluated.

On the gene mutation technique of an individual, that is, a vector element v_i is randomly chosen and its value is altered, in order to evaluate a new task distribution situation on the distributed environment. This altered individual is submitted to the fitness function for evaluation of its ability. Each stage on which the fitness function receives a population for evaluation is denominated generation.

After choosing the best individuals from a certain generation, crossover and mutation operations may happen. The new obtained individuals are once again submitted to the fitness function. After a certain number of generations, the quality of results obtained by the generated individuals gets stabilized. On this stage, the best individual has to be chosen to represent the process scheduling on the heterogeneous distributed environment.

2.1 Genetic Search

The fitness function used by the GAS scheduling algorithm attempts to evaluate the quality of the solution obtained on each genetic evolution of the individuals of a certain population. This function receives a set of vectors v_i that contains possible solutions for task attributions of a certain parallel application to different processing elements. Each vector of this set represents an individual of the genetic algorithm. The set of vectors represents the population of possible solutions. Once created an initial population, the fitness function goes through each individual, i.e., each vector v_i, and verifies which processing elements are used to solve a scheduling. Before going through each vector v_i, the fitness function creates a copy $lcopy_t$ of the current load vector l_t of the processing elements. When going through a vector v_i, the fitness function updates the $lcopy_t$ with new loads that each element will receive by using the schedule proposed by the individual v_i. The equation 1 is used to update the load of each processing element where: $v_{i,k}$ represents each index k of the individual v_i, k should vary from 1 to the number of tasks of the parallel distribution. Each index element k of v_i presents a value, which indicates the processing element that should receive the load of the task k of the parallel application initiated on the system. $v_{i,k}$ may be used as the index $lcopy_t$ to define the processor that will receive the load.

$$lcopy_{v_{i,k}} = lcopy_{v_{i,k}} + \frac{c_j}{cap_{v_{i,k}}} \qquad (1)$$

After updating the load vector copy $lcopy_t$, it is possible to know if the system has adopted a solution proposed by a individual v_i. When this stage is completed, the fitness function evaluates the communication costs among the tasks of the parallel application. Such costs involve the delays caused in the tasks during the message sending and receiving operations. This stage uses a matrix $r_{w,q}$ of communication costs between two PEs w and q with the objective of estimating the cost related to the exchange of variable size messages among tasks located in distinct PEs. In addition to this matrix, it is also used the time percentage on which the tasks of the parallel application have been under communication status. This percentage is obtained through the application of the model by Senger et al. [18] that uses the Art2A self-organizing neural network architecture. The product achieved on this stage is the total communication cost generated by the allocation of tasks of a same parallel application on distinct PEs. This cost $C_{net}(i)$ caused by the individual v_i is given by the equation 2, where t_w and t_q are tasks that communicate to each other.

$$C_{net}(i) = \sum_{i=0}^{n} \sum_{w=i+1}^{n} r_{t_w, t_q} \qquad (2)$$

Moreover, the usage features of the CPU and network of the parallel distribution are used to generate coefficients α and β that represent, respectively, the time percentage on which the application visits the processing and communication states. An application with coefficients $\alpha = 0.1$ and $\beta = 0.9$ is performing a 10% processing of its total execution time and the 90% remainder ones are performing the communication among its tasks. Such coefficients are used by the fitness function to weight the communication and processing costs among the tasks. The fitness function applied on a individual v_i, given as $f(i)$, is presented on the equation 3, where: $C_{ucp}(i)$ is the processing cost added by the individual v_i to the environment, $C_{net}(i)$ is the communication cost caused by the scheduling solution proposed by the individual v_i, e is a positive value close to zero that eliminates the possibility of division by zero, in case one of the costs is null.

$$f(i) = \alpha \times \frac{1}{(C_{ucp}(i) + e)} + \beta \times \frac{1}{(C_{net}(i) + e)} \qquad (3)$$

3 System Modeling and Results

In order to evaluate the GAS scheduling algorithm, it was adopted an object oriented simulator [1] which implements the UniMPP (*Unified Modeling for Predicting Performance*) model proposed by Mello and Senger [15]. Experiments have been carried out over a real heterogeneous distributed environment by using the suite of benchmarks, proposed by Mello and Senger [15], which was used to parameterize the simulator. The first suite tool applied was the *mips*, which brings back the total computing capacity (pc_i) of each processing element of the

---

[1] SchedSim – available at http://www.icmc.usp.br/~mello/outr.html.

system in millions of instructions per second. The results obtained in 10 evaluated computers are 1135.31, 1145.86, 1148.65, 187.54, 313.38, 151.50, 1052.87, 1052.87, 350.00 and 350.00 $mips$.

The 10 evaluated computers are located in 4 distinct networks. Three of these networks have Fast Ethernet technology and the other one has Gigabit Ethernet technology. In order to obtain a matrix of communication costs $r_{w,q}$ between two tasks located on the processing elements w and q, the net tool of the suite of benchmarks has been applied. This tool has allowed the generation of the communication delay among each computer of the system. Using this tool, a matrix of communication costs for messages of 32 bytes is obtained. Such matrix is used by the GAS algorithm to obtain the exchange cost of a message of size m bytes through the equation 4.

$$T(m) = (r_{w,q}/32) * m \qquad (4)$$

In order to obtain the slowdown caused by the usage of the main and virtual memories, the $memo$ tool of the suite of benchmarks has been executed. Such tool has allowed the obtainment of the times when the system has gone through delays due to the occupation of the main and virtual memories. Confirming the experiments by Mello and Senger [15], it may be evidenced that the allocation of the main memories causes linear delays and the allocation of the virtual memory causes exponential delays. The computers' delay have been rounded through the equation presented on table 1, which represent the quantity of time spent, in seconds, for the allocation of x memory megabytes.

In order to carry out simulations and performance comparisons among different scheduling algorithms, the simulator of the UniMPP model has required the performance evaluation of the accesses to the hard disk for reading and writing operations.

The results obtained by the tools of the suite of benchmarks have allowed the parameterization of the distributed environment features and the utilization of the Simulator of the UniMPP model to evaluate the behavior of the GAS scheduling algorithm and compare them to the performance of other algorithms. In order to parameterize the load of the parallel applications to be submitted to the system, the workload model proposed by Feitelson [16][10] has been used. This model follows a hyperexponential of average 1500 to define the arrival times of the processes. It is used considering two ranges for the maximum quantity of PEs requested by the 8 and 64 applications. Such values allow the observation of the GAS algorithm cost under different configurations of the requested PEs.

In order to evaluate the GAS algorithm, experiments have been carried out with different quantities of applications submitted to the environment. This algorithm performance is compared to 7 scheduling and load balancing

Table 1. Memory slowdown functions

Memory Occupy	Equations
Main	$0.0069x - 0.089$
Virtual	$1.1273 * exp(0.0045x)$

algorithms adopted in the literature: DPWP, DPWP-modified, Random, Disted, Lowest, Global and TLBA.

The DPWP (*Dynamic Policy Without Preemption*) algorithm performs the parallel applications scheduling taking into account a distributed and heterogeneous execution scenario [3][4].

A modified version of the DPWP algorithm, named DPWP-modified, is also defined and implemented on the simulator. Instead of using the size of the process queue of the EP as the load index, this version uses the tasks occupation to find out the less overloaded EPs in the environment 1.

The Disted, Global, Lowest and Random algorithms are defined on the work by [26]. On this study, such algorithms have the aim of balancing the loads and they are defined through two main components: the LIM (*Load information manager*) and the LBM (*Load balance manager*). The first one is responsible for the information policy and monitors the computers' load with the objective of calculating the load indexes. The latter one defines the information policy and aims to find out the most appropriate computer in the system to which the processes should be attributed. The manner such components perform their tasks allows the definition of the distinct algorithms. Such algorithms differ from the scheduling algorithms as they are oriented to perform the load balancing, thus there is no global scheduling software to which the applications are submitted. In fact, each EP should locally manage the application tasks that reach it, initiating them locally or defining the way through which another EP will be found out to execute the tasks.

The TLBA (*Tree Load Balancing Algorithm*) algorithm aims to balance loads on heterogeneous distributed systems with feasibility of scaling growth [8][14]. Such algorithm creates a logical interconnection topology among the EPs, in a tree format, and performs the migration of tasks in order to improve the system load balance.

The figure 1 illustrates the results of the GAS, Random, Disted, Global, Lowest, DPWP and DPWP-modified scheduling algorithms for an application set that varies from 50 to 1000, with a maximum number of 8 tasks. It may be observed that the GAS scheduling algorithm shows the best performance, evidenced by the lowest response times. This highlights the importance of considering the knowledge about the applications on the task attribution of distributed environments.

On this case, the worst performance is observed on the Random algorithm. This is because this algorithm does not use information on the EPs load and on the submitted parallel applications. The DPWP, Lowest, Disted, Global and TLBA algorithms that use only the features of the environment load generate average response times higher than the ones generated by the Random algorithm, but worse than the ones generated by the GAS algorithm. The DPWP-modified algorithm shows worse results than the ones showed by the DPWPalgorithm. This indicates that, although the load index of the DPWP-modified algorithm uses the knowledge about the applications execution time, the strategy for task attribution with no distinction from the most appropriate resources jeopardizes

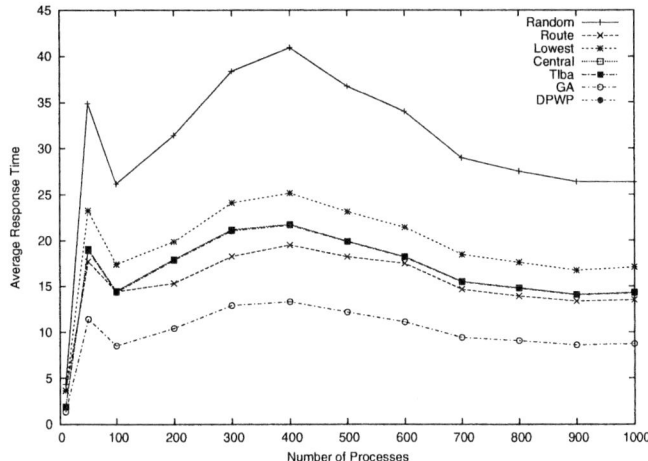

Fig. 1. Simulation results for maximum application size equal to 8 tasks

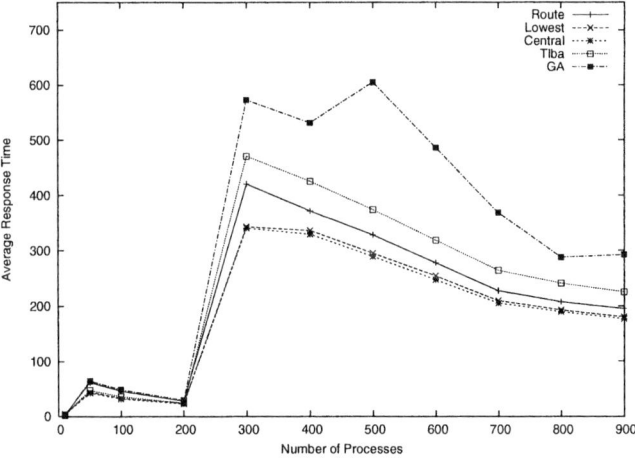

Fig. 2. Simulation results for maximum application size equal to 64 tasks

the response times experienced by the applications. The figure 2 illustrates the performance of the scheduling algorithms taking into account the maximum size of 64 to the application. The figure 3 shows the results for the DPWP and Random algorithms. In addition to these figures, it may be observed that the DPWP and Random algorithms have shown the worst results. Moreover, the GAS algorithm shows the worst results when compared to the Tlba, Disted, Global and Lowest algorithms. The reason for that is that the larger is the maximum size of the application the larger is the time for the genetic search for solution. The results are expressed by the average response time of the algorithms, considering this number of applications and maximum applications size equal to 8 and 64 tasks.

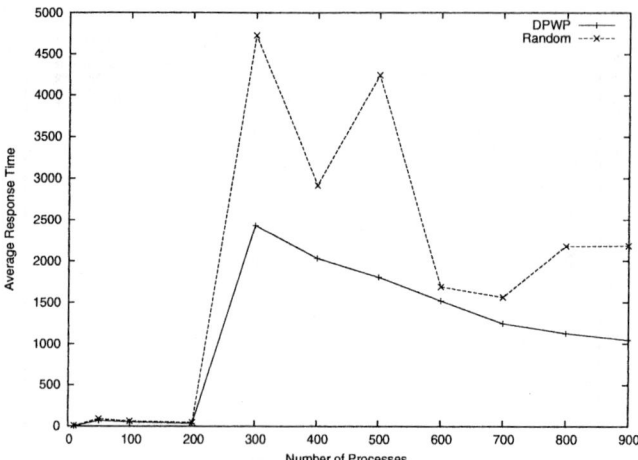

Fig. 3. Simulation results for maximum application size equal to 64 tasks (DPWP and Random only)

4 Concluding Remarks

This paper has presented a process scheduling algorithm that uses the knowledge extracted from the parallel applications to allocate computing resources on heterogeneous distributed environments. Such algorithm uses genetic algorithm techniques to find out the best resources allocation for each parallel applications received by the system.

The obtained results have allowed the conclusion that the GAS scheduling algorithm shows the best scheduling solutions for environments on which parallel applications do not have a large number of tasks. Another limitation showed by this algorithm is that, as the environment size increases, that is, as the number of available computer increases, the computing cost to obtain a better solution also increases, what may turn prohibitive the use of the algorithm.

Acknowledgment

The authors thank to Capes and Fapesp Brazilian Foundations – Fundação de Amparo à Pesquisa do Estado de São Paulo – under the process number 04/02411-9.

References

1. D. W. Aha, D. Kibler, and M. K. Albert. Instance-based learning algorithms. *Machine Learning*, 1(6):37–66, 1991.
2. T. Anderson, D. Culler, and D. Patterson. A Case for NOW (Networks of Workstations). *IEEE Micro*, 15(1):54–64, 1995.

3. A. P. F. Araújo, M. J. Santana, R. H. C. Santana, and P. S. L. Souza. DPWP: A new load balancing algorithm. In *5th International Conference on Information Systems Analysis and Synthesis - ISAS'99*, Orlando, U.S.A., 1999.
4. A. P. F. Araújo, M. J. Santana, R. H. C Santana, and P. S. L Souza. A new dynamical scheduling algorithm, international conference on parallel and distributed processing techniques and applications. In *Proceedings of PDPTA'99*, Las Vegas,Nevada, U.S.A., 1999.
5. Andrea Carol Arpaci-Dusseau. Implicit coscheduling: coordinated scheduling with implicit information in distributed systems. *ACM Transactions on Computer Systems*, 19(3):283–331, 2001.
6. D. P. Bovet and M. Cesati. *Understanding the Linux Kernel*. O'Reilly, Outubro 2000.
7. T. Brecht and K. Guha. Using Parallel Program Characteristics in Dynamic Processor Allocation Policies. *Performance Evaluation*, 28(4):519–539, 1996.
8. R. F. de Mello. *Proposta e avaliação de desempenho de um algoritmo de balanceamento de carga para ambientes distribuídos heterogêneos e escaláveis*. PhD thesis, Escola de engenharia de São Carlos (EESC), São Carlos, São Paulo, Brasil, Novembro 2003.
9. A. de P. Braga, T. B. Ludermir, and A. C. P. L. F. Carvalho. *Redes Neurais Artificiais: Teoria e Aplicações*. LTC, 2000.
10. Dror G. Feitelson. Metrics for parallel job scheduling and their convergence. In Dror G. Feitelson and Larry Rudolph, editors, *Job Scheduling Strategies for Parallel Processing*, pages 188–205. LNCS 2221, 2001.
11. M. J. Flynn and Kevin W. Rudd. Parallel architectures. *ACM Computing Surveys*, 28(1):68–70, 1996.
12. R. Gibbons. A Historical Application Profiler for Use by Parallel Schedulers. In *Job Scheduling Strategies for Parallel Processing*, pages 58–77. LNCS, 1997.
13. M. Harchol-Balter and A. B. Downey. Exploiting Process Lifetimes Distributions for Dynamic Load Balancing. *ACM Transactions on Computer Systems*, 15(3):253–285, August 1997.
14. R. F. Mello, L. C. Trevelin, M. S. Paiva, and L. T. Yang. Comparative study of the server-initiated lowest algorithm using a load balancing index based on the process behavior for heterogeneous environment. In *Networks, Software Tools and Application, ISSN 1386-7857*. Kluwer, 2003.
15. Rodrigo F. D. Mello and Luciano J. Senger. A new migration model based on the evaluation of processes load and lifetime on heterogeneous computing environments. In *16th Symposium on Computer Architecture and High Performance Computing*, Foz do Iguaçu - PR - Brazil, 2004.
16. Ahuva W. Mu'alem and Dror G. Feitelson. Utilization, predictability, workloads, and user runtime estimates in scheduling the IBM SP2 with backfilling. *IEEE Trans. Parallel & Distributed Syst.*, 12(6):529–543, Jun 2001.
17. V. K. Naik, S. K. Setia, and M. S. Squillante. Processor Allocation in Multiprogrammed Distributed-memory Parallel Computer Systems. *Journal of Parallel and Distributed Computing*, 47(1):28–47, 1997.
18. L. J. Senger, R.F.de Mello, M. J. Santana, and R. H. C. Santana. An on-line approach for classifying and extracting application behavior on linux. In *High Performance Computing: Paradigm and Infrastructure*. John Wiley and Sons Inc. (a ser publicado), 2005.

19. L. J. Senger, M. J. Santana, and R. H. C. Santana. Uma nova abordagem para a aquisiçao de conhecimento sobre o comportamento de aplicaçoes paralelas na utilizaçao de recursos. In *Proceedings of WORKCOMP*, pages 79–85, Outubro 2002.
20. L. J. Senger, M. J. Santana, and R. H. C. Santana. A new approach fo acquiring knowledge of resource usage in parallel applications. In *Proceedings of International Symposium on Performance Evaluation of Computer and Telecommunication Systems (SPECTS'2003)*, pages 607–614, 2003.
21. L. J. Senger, M. J. Santana, and R. H. C. Santana. An instance-based approach for predicting parallel application execution times. In *Proceedings of I2TS*, 2004.
22. L. J. Senger, M. J. Santana, and R. H. C. Santana. Using runtime measurements and historical traces for acquiring knowledge in parallel applications. In *International Conference on Computational Science (ICCS) (to appear)*. LNCS 3036, 2004.
23. L. J. Senger, M. J. Santana, and R. H. C. Santana. Using runtime measurements and historical traces for acquiring knowledge in parallel applications. In *International Conference on Computational Science (ICCS) (to appear)*. LNCS 3036, 2004.
24. K. C. Sevcik. Characterizations of Parallelism in Applications and their use in Scheduling. *Performance Evaluation Review*, 17(1):171–180, Maio 1989.
25. W. Smith, I. T. Foster, and V. E. Taylor. Predicting Application Run Times Using Historical Information. In *JSSPP*, pages 122–142, 1998.
26. S. Zhou and D. Ferrari. An experimental study of load balancing performance. Technical Report UCB/CSD 87/336, Universidade da Califórnia, Berkeley, Janeiro 1987.

An Evaluation Methodology for Computational Grids[*]

Eduardo Huedo[1], Rubén S. Montero[2], and Ignacio M. Llorente[1,2]

[1] Laboratorio de Computación Avanzada, Simulación y Aplicaciones Telemáticas, Centro de Astrobiología (CSIC-INTA), 28850 Torrejón de Ardoz, Spain
[2] Departamento de Arquitectura de Computadores y Automática, Universidad Complutense, 28040 Madrid, Spain

Abstract. The efficient usage of current emerging Grid infrastructures can only be attained by defining a standard methodology for its evaluation. This methodology should include an appropriate set of criteria and metrics, and a suitable family of Grid benchmarks, reflecting representative workloads, to evaluate such criteria and metrics. The establishment of this methodology would be useful to validate the middleware, to adjust its components and to estimate the achieved quality of service.

1 Introduction

Benchmarking is a widely accepted method to evaluate the performance of computer architectures. Traditionally, benchmarking of computing platforms has been successfully performed through low level probes that measure the performance of specific aspects of the system when performing basic operations, e.g. LAPACK [1], as well as representative applications of the typical workload, e.g. SPEC [2] or NPB [3]. In this sense, benchmarking has been proved helpful for investigating the performance properties of a given system, either for prediction or comparison purposes.

Grid benchmarks can be also grouped in the two aforementioned categories: low level probes that provide information of specific aspects of system's performance; and benchmarks that are representative of a class of applications. In this first category, the Network Weather Service [4] provides accurate forecast of dynamically changing performance characteristics from a distributed set of computing resources. Also, a set of benchmark probes for Grid assessment have been proposed [5]. These probes exercise basic Grid operations with the goal of measuring the performance and the performance variability of basic Grid operations, as well as the failure rates of these operations. Finally, the GridBench tool [6] is a benchmark suite for characterizing individual Grid nodes and collections of Grid resources. GridBench includes micro-benchmarks and application kernels to measure computational power, inter-process communication bandwidth,

---

[*] This research was supported by Ministerio de Ciencia y Tecnología, through the research grants TIC 2003-01321 and 2002-12422-E, and by Instituto Nacional de Técnica Aeroespacial "Esteban Terradas" - Centro de Astrobiología.

and I/O performance. In the second category, the *Grid Benchmarking Research Group*, within the *Global Grid Forum*, proposes to create a set of representative Grid benchmarks [7], which will embody challenging usage scenarios with special emphasis on large data usage. The *NAS Grid Benchmarks* (NGB) [8] was the first Grid benchmark specification available.

The aim of this paper is, firstly, to propose a set of criteria and metrics which allow evaluating the capabilities of a computational Grid environment from a user's point of view; and, secondly, to apply these criteria and metrics in the evaluation of a Grid environment based on Globus basic services using GridWay [9] as metascheduler and NGB as test programs. As an initial phase of this work, the paper-and-pencil specification of this benchmark suite for computational grids has been implemented by using the *Distributed Resource Management Application API* (DRMAA) supported by GridWay [10].

In Section 2, we describe the criteria and evaluation metrics used in this work. Then, the main characteristics of the NGB suite are detailed in Section 3. In Section 4, we describe the evaluation process of a Grid infrastructure. Finally, Section 5 presents the main conclusions of our work.

2 Criteria for Grid Evaluation

We propose functionality, reliability and performance as general criteria to evaluate a Grid environment from a user's point of view, and to guarantee the extension of its use. We have tried to keep the evaluation criteria simple and objective. In this sense, each metric is easy to measure and provides a boolean or numeric value, which is therefore also easy to compare.

In spite of the great research effort made over the last decade, application development and execution on grids require a high level of expertise due to its complex nature. Therefore, functionality should be considered as a valuable criterion, and a Grid evaluation methodology should reflect the ability of the environment to execute unattended distributed *communicating* applications. The NGB suite falls in this category, and the capability to execute it constitutes a suitable metric to test the functionality of the environment. Moreover, benchmarks should be expressed by using standard high level interfaces, like DRMAA.

Grid environments are difficult to efficiently harness due to their heterogeneous nature and unpredictable changing conditions. Adaptive scheduling and execution are some of the techniques proposed in the literature [11,12] to achieve a reasonable degree of application performance and fault tolerance. Therefore, a suitable methodology for Grid evaluation should also help to determine the reliability and dynamic adaptation capabilities of the Grid environment. As simple metrics, we propose that a job could, transparently to the user, continue its execution (at least from the beginning) in other resource when some of the following failure or loss of quality of service conditions take place [13]:

- Job cancellation (failure) or suspension (QoS loss)
- System crash (failure) or saturation (QoS loss)
- Network disconnection (failure) or saturation (QoS loss)

The coordinated performance of all the involved resources and services should be considered when analyzing the performance of a Grid infrastructure. An evaluation methodology for grids should provide tools and metrics to measure and adjust its performance. As user-level performance metrics, we propose:

- Turnaround time (T): It is the waiting time from the job execution request until the results are available.
- Productivity (P): As usual, productivity is defined as the number of completed tasks or benchmark instances per unit of time.

Moreover, it is very important to quantify the overheads of the involved components and to analyze their influence in the global performance. Therefore, we should also consider other more appropriate metrics for diagnostic and tuning purposes, interesting for application and middleware developers, and Grid architects:

- Response time (T_r): It is the time between submitting a job and the starting of the stage-in phase on the execution host. It provides information about the overhead induced by the scheduler and the Grid middleware.
- Transfer and execution time (T_{xfr} and T_{exe}): These metrics (total or averaged) are useful to evaluate the impact of data movement strategies, individual resource performance or the influence of the interconnection network.
- Resource usage (U): Represents the usage of resources throughout the benchmark execution and the achieved level of parallelism. It is defined as follows:

$$U = \frac{T_{exe}}{T}.$$

3 The NAS Grid Benchmarks

The *NAS Grid Benchmarks* [8] are presented as a data flow graph encapsulating an instance of a *NAS Parallel Benchmarks* (NPB) code in each graph node, which *communicates* with other nodes by sending/receiving initialization data.

Figure 1 shows the four families defined in NGB. Each benchmark comprises the execution of several NPB codes that symbolize scientific computation (flow solvers SP, BT and LU), post-processing (data smoother MG) and visualization (spectral analyzer FT). Like NPB, NGB specifies several different classes (problem sizes) in terms of number of tasks, mesh size and number of iterations.

4 Results

The proposed methodology does not attempt to measure the performance of the underlying Grid hardware, but the functionality, reliability and performance of the Grid environment. However, a clear understanding of the hardware configuration of the Grid resources will aid the analysis of the subsequent experiments. Table 1 shows the characteristics of the machines in our small research testbed,

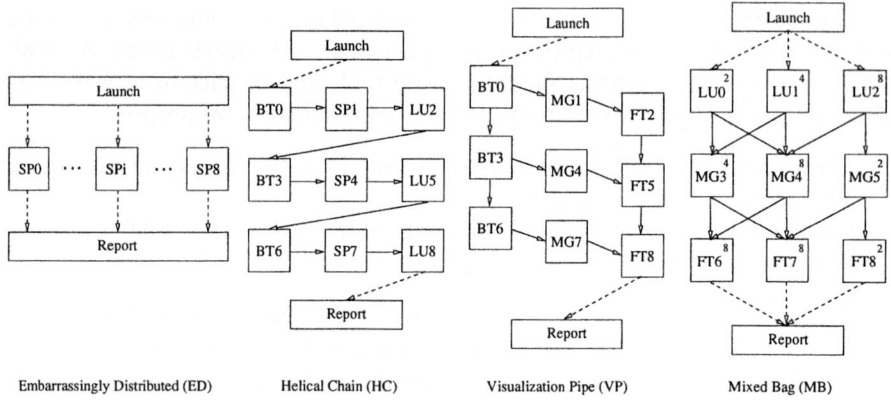

Fig. 1. The four families of the NAS Grid Benchmarks [8]

Table 1. Characteristics of the machines in the research testbed

Name	Site	Processors	Speed	Mem.	OS	DRMS
pegasus	UCM	Intel P4	2.4GHz	1GB	Linux 2.4	fork
hydrus	UCM	Intel P4	2.5GHz	512MB	Linux 2.4	fork
cygnus	UCM	Intel P4	2.5GHz	512MB	Linux 2.4	fork
cepheus	UCM	Intel PIII	600MHz	256MB	Linux 2.4	fork

based on the Globus toolkit 2.X. In the following experiments, cepheus is used as client and stores the executable and input files, and receives the output files.

Regarding the functionality criterion, the ED benchmark is a typical case of a parameter sweep application, directly supported by GridWay [14], the HC benchmark has been easily implemented using the GridWay DRMAA interface, and the VP and MB benchmarks have been programmatically implemented, through DRMAA, as workflow applications.

The HC benchmark constitutes an excellent probe to evaluate the reliability criterion of grids since output files of each task can be used as checkpoints for the next task. The metrics proposed to evaluate the reliability of the Grid environment have been implemented in the following way:

– Job cancellation or suspension: During the execution of a HC task, the job is cancelled and suspended. GridWay is able to detect the job cancellation when the task exit code is not specified and, in such case, to reschedule this task on other resource from the last saved checkpoint. Job suspension is detected when the task remains suspended longer than a given threshold.
– System crash or saturation: During the execution of a HC task, the resource where it is executing is saturated. GridWay is able to detect the performance degradation through a performance profile and, in such case, reschedule the job on a new resource from the last saved checkpoint. Resource failures are managed on an equal basis as network failures, as it is described below.

Table 2. Results obtained for each benchmark family

Metric	Description	Units	Benchmark ED.A	HC.A	VP.A	MB.A
T	Turnaround time	minutes	18.88	17.57	21.67	16.80
P	Productivity	jobs/hour	28.60	30.73	24.92	32.14
T_r	Response time	minutes	-	3.09	-	-
T_{xfr}	Transfer time	minutes	5.50	7.10	8.10	9.70
T_{exe}	Execution time	minutes	38.30	7.38	22.93	23.03
U	Resource usage	-	2.02	0.42	1.06	1.37

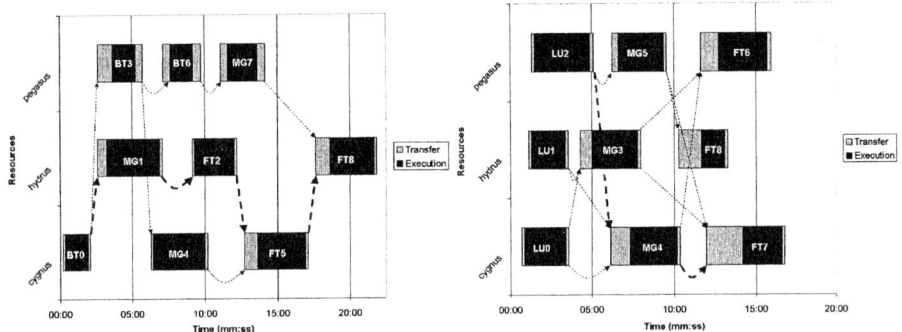

Fig. 2. Execution profile of the VP.A and MB.A benchmarks

– Network disconnection or saturation: During the execution of a HC task, the resource where it is executing is disconnected. GridWay is able to detect the disconnection by periodically probing the GRAM *job manager* in the remote resource. GridWay does not consider network saturation as a failure condition, instead, it uses network status information to rank resources [15].

Table 2 shows the values for the metrics proposed to evaluate the performance of the Grid environment. GridWay doesn't provide a measure for the response time, but can be calculated for HC.

The resource usage gives an idea of the characteristics of each benchmark family. ED is fully parallel, but the structural dependencies in the testbed (only three candidate resources) prevents a value of U close to 9. On the contrary, HC is fully sequential so it can not use the resources efficiently.

Figure 2 shows the execution profile of benchmarks VP and MB. Both exhibit some degree of parallelism ($U > 1$), that could be increased by widening the pipe (limited to three jobs) and by reducing the Grid overhead. The parallelism obtained by VP is also limited by the stages of filling and draining the pipe. MB has a wider pipe width from the beginning, which enables a better use of the resources. Both benchmarks are of great help to adjust the services of a grid, and even to compare different strategies to schedule workflows.

5 Conclusions

The presented proposal of criteria and metrics, along with the NGB suite, could be considered the first step to reach an agreed evaluation methodology for grids. The methodology provides diagnostic information of interest for middleware developers and Grid architects, that can be used to explore the behaviour and adjust the performance offered by each layer in a Grid environment. The use of standard interfaces allows the comparison between different Grid implementations, since neither NGB nor DRMAA are tied to any specific Grid middleware.

References

1. Anderson, E., Bai, Z., Bischof, C., et al.: LAPACK Users' Guide. SIAM (1999)
2. : The Standard Performance Evaluation Corporation. (http://www.spec.org)
3. Bailey, D.H., Barszcz, E., Barton, J.T.: The NAS Parallel Benchmarks. Intl. J. Supercomputer Applications **5** (1991) 63–73
4. Wolski, R., Spring, N., Hayes, J.: The Network Weather Service: A Distributed Resource Performance Forecasting Service for Metacomputing. Future Generation Computing Systems **15** (1999) 757–768
5. Chun, G., Dail, H., Casanova, H., Snavely, A.: Benchmark Probes for Grid Assessment. In: Proc. 18th Intl. Parallel and Distributed Processing Symposium (IPDPS). (2004)
6. Tsouloupas, G., Dikaiakos, M.D.: Gridbench: A Tool for Benchmarking Grids. In: Proc. 4th Intl. Workshop on Grid Computing (GRID 2003), IEEE CS (2003) 60–67
7. Snavely, A., Chun, G., Casanova, H., et al.: Benchmarks for Grid Computing: A Review of Ongoing Efforts and Future Directions. ACM Sigmetrics PER **30** (2003)
8. Frumkin, M.A., Van der Wijngaart, R.F.: NAS Grid Benchmarks: A Tool for Grid Space Exploration. J. Cluster Computing **5** (2002) 247–255
9. Huedo, E., Montero, R.S., Llorente, I.M.: A Framework for Adaptive Execution on Grids. Software – Practice and Experience (SPE) **34** (2004) 631–651
10. Herrera, J., Huedo, E., Montero, R.S., Llorente, I.M.: Developing Grid-Aware Applications with DRMAA. In: Proc. 10th Intl. Conf. Parallel and Distributed Processing (Euro-Par 2004). Volume 3149 of LNCS. (2004) 429–435
11. Berman, F., Wolski, R., Casanova, H., et al.: Adaptive Computing on the Grid Using AppLeS. IEEE Trans. Parallel and Distributed Systems **14** (2003) 369–382
12. Vadhiyar, S., Dongarra, J.: A Performance Oriented Migration Framework for the Grid. In: Proc. 3rd Intl. Symp. Cluster Computing and the Grid (CCGrid). (2003)
13. Lee, H.M., Chung, K.S., Jin, S.H., et al.: A Fault Tolerance Service for QoS in Grid Computing. In: Proc. Intl. Conf. Computational Science (ICCS). LNCS (2003)
14. Huedo, E., Montero, R.S., Llorente, I.M.: Experiences on Adaptive Grid Scheduling of Parameter Sweep Applications. In: Proc. 12th Euromicro Conf. Parallel, Distributed and Network-based Processing (PDP2004), IEEE CS (2004) 28–33
15. Montero, R.S., Huedo, E., Llorente, I.M.: Grid Resource Selection for Opportunistic Job Migration. In: Proc. 9th Intl. Conf. Parallel and Distributed Computing (Euro-Par 2003). Volume 2790 of LNCS. (2003) 366–373

SLA Negotiation Protocol for Grid-Based Workflows

Dang Minh Quan and Odej Kao

Department of Computer Science, University of Paderborn, Germany

Abstract. Service Level Agreements (SLAs) are currently one of the major research topics in Grid Computing. SLA negotiation protocol defines a business relationship between consumer and provider and is thus a necessary part of any commercial Grid application. In the Grid environment, an SLA negotiation protocol for a workflow becomes complicated because of its structure which includes many dependent sub-jobs. The complex workflow model leads to several constraints compared to single jobs and requires advanced solutions. As a continuous effort in building a full system supporting SLAs for Grid workflows, this paper presents an SLA negotiation protocol for workflows and an initial system implementation.

1 Introduction

Service Level Agreements (SLAs) [8] are currently one of the major research topics in Grid Computing, as they serve as foundation for a reliable and predictable job execution at remote Grid sites. The process of SLA composition between consumer and provider is implemented by an SLA negotiation protocol. Most of the existing SLA negotiation protocols [1,2,3,10] apply solely for Grid jobs defined as monolithic entities, where the user sends the input data to a service, computes the data – without dependencies – on this site and receives the results. The complexity of the SLA negotiation protocol grows significantly in case of workflows consisting of multiple, dependent sub-jobs. Figure 1 depicts a common scenario where a workflow in a Grid environment is executed. The execution of such a workflow differs from a single job execution in several aspects. A workflow consists of many dependent sub-jobs, each of them with individual requirements. Some sub-jobs have to run in parallel in order to fulfil the SLA conditions, so certain resources are required simultaneously. Thus, it is necessary to re-distribute the sub-jobs to different sites. Those differences lead to further challenges for systems supporting SLAs for a Grid workflow compared to the single job scenario. The challenges are found in the architecture, the language description [4], the mapping algorithm [5], etc.

This paper presents the next step of building a full system supporting SLAs for Grid workflows, which is an SLA negotiation protocol for workflows. It is organized as follows: Section 2 describes the related work, Section 3 presents the protocol and Section 4 describes the implementation architecture as well as the deployment. Finally, Section 5 concludes the paper with a short summary.

2 Related Work

Czajkowski et al. introduced a general negotiation model called Service Negotiation and Acquisition Protocol (SNAP) [3]. SNAP defines three types of SLAs to manage resources across different administrative domains, which are Task SLA (TSLA), Resource SLA (RSLA) and Bind SLA (BSLA). Those types of SLAs can be used together in order to describe complex service requirements in a distributed environment. The SNAP protocol is general and valid for many cases. However, the application of the protocol to a specific problem needs further extension to satisfy the requirements. A research contribution to express a detailed SLA negotiation and implementation is presented in [1]. The GGF Grid Resource Agreement and Allocation Protocol Working Group proposed a draft on Agreement-based Grid Service Management (OGSI-Agreement), which defines a set of OGSI-compatible portTypes through which management applications and services can negotiate.

Nassif et al. presented an agent-based SLA negotiation in Grids [10]. The negotiation module includes different forms of negotiation called bilateral, multi-issue and chaining negotiation. A bilateral negotiation occurs when only two parts are involved. In a multi-issue negotiation, different aspects are negotiated such as price, quality and time schedule. The negotiation between user_agent and network_agent is called chaining negotiation. A chaining negotiation occurs after the negotiation between user_agent and server_agent finished.

An initial work to design and implement a system supporting SLAs for Grid jobs is described in [6,7,2]. However, solely monolithic jobs are considered and supported. Shortcomings are mainly the underlying cost model and the missing consideration of SLOs (Service Level Objectives).

3 SLA Negotiation for Workflow

As shown in Figure 1, an implementation of SLAs for workflows needs the participation of three entities: a consumer, a broker and one or more providers. From consumer's point of view, the broker is responsible for satisfying the published

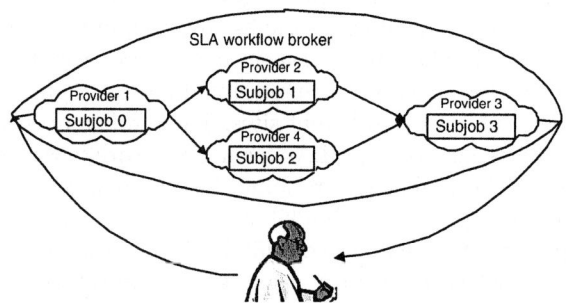

Fig. 1. Scenario of running a workflow in the grid environment

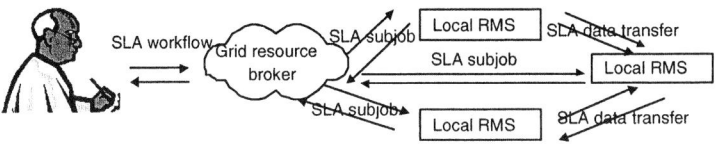

Fig. 2. Interaction among participants in SLA for workflow negotiation process

requirements. However, the broker does not execute the workflow but manages the workflow execution process. Components that run sub-jobs of the workflow are computing services with their local Resource Management Systems (RMS). The interaction among them in the negotiation process is depicted in Figure 2.

Figure 2 presents three different types of sub SLA negotiations. The User - Broker negotiation focuses on the definition of the submitted SLA. Broker - Provider negotiation considers the workflow's sub-jobs and uses the sub-job SLAs. Provider - Provider negotiation deals with the data transfer between sub-jobs (and also between providers) so the SLA part for data transfer is used. In following, the each SLA part as well as the SLA negotiation procedures are described in detail.

3.1 SLA Document Description

The SLA document defines the data transfers between two participants in the negotiation process. All three types of the SLA text - including SLA workflow, SLA sub-job, SLA data transfer - are compiled with the same SLA workflow language [4]. The structure of the SLA text is presented in Figure 3, Figure 4, and Figure 5.

3.2 Basic Negotiation Procedures

Although there are three types of SLA negotiation, the procedure remains the same, only the service attributes differ. Figure 6 describes this basic procedure as a Client/Server model. In the first step, the client creates a template SLA with some preliminary service attributes and sends those to the server. The server parses the text and checks the client requirements. In case of conflicts, a new SLA version is compiled and sent back. Modified information can be start/stop time, cost, maximal available CPUs etc. Additional information can be a new SLO, FTP address, path etc. The client verifies the new SLA version and sends it to server again. The process is repeated until the SLA is definitely accepted or rejected. After acceptance, the signing phase is completed leading to an accepted SLA, which cannot be modified.

3.3 An SLA for a Workflow Negotiation Process

Figure 7 describes the SLA negotiation for a Grid-based workflow as a time sequence. The customer sending an SLA template for workflows to the broker starts the process. The broker determines a solution and negotiates with the customer.

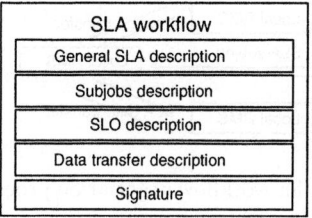
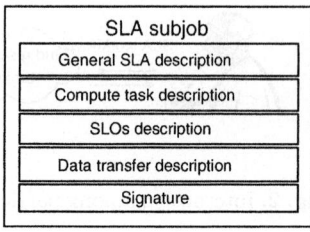

Fig. 3. SLA workflow structure

Fig. 4. SLA sub-job structure

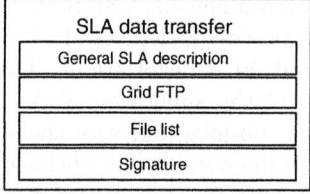

Fig. 5. SLA data transfer structure

Fig. 6. Basic SLA negotiation procedure

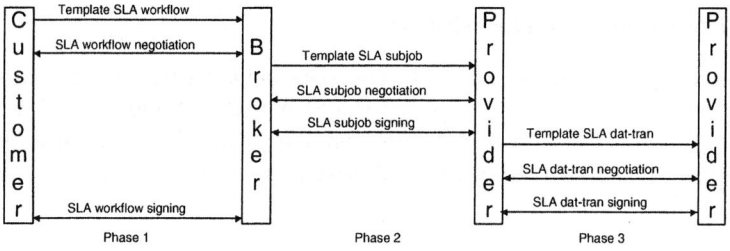

Fig. 7. SLA for workflow negotiation process

Thereafter, the broker performs an SLA sub-job negotiation with providers for all sub-jobs in the workflow. The SLA data transfer can be initiated after all related SLA sub-job negotiations are completed. In case of conflicts, the broker will find another solution and repeat phase 2 and phase 3. Following parts describe the negotiation steps with focus on the operation of each participant as well as on the modified and additional information in the SLA text.

After the preliminary mapping decision (phase 1), the broker gains detailed information on running the workflow, which is used for the negotiation (start/stop time for each sub-job, data transfer SLA, etc.). Additional information can be related for example to the SLO for workflows. However, in the SLA context, consumers can also cause a Grid job failure for example by underestimating the processing time. This will lead to job cancellation or job checkpointing after the reserved time slot expired. Thus, it is necessary for the broker to add more SLOs which impose the customer responsibility for the SLA text.

After receiving the SLA for a particular sub-job, the provider parses the document and checks if it can fulfil the requirements (Phase 2). After successful evaluation, the provider adds the FTP address and the storage path to the SLA part defining the data transfer.

Phase 3 with provider - provider negotiation process increases further the complexity. It can be done only when the two related SLA sub-job negotiations finished. If this is not the case, the destination provider will conclude that the submitted SLA data transfer belongs to none of the jobs and will discard it. In order to solve this problem, the broker monitoring the state of each SLA sub-job, plays the role of SLA transition. The template SLA from source provider is sent to the broker and then to the destination provider when appropriate.

4 System Implementation

An implemented prototype fulfils the above constraints and provides basic features, which allow any client to negotiate SLAs, monitor running workflows and to receive the result. Based on standard components such as Globus Toolkit 3.2, MySQL database, Maui ME for the local RMS, the system is compatible with the existing Grid infrastructure. The overall architecture schema is depicted in Figure 8. The online demonstration of the negotiation process as well as the execution of a workflow consisting of seven sub-jobs in cooperation of three local RMS can be found at http://pc-kao3.upb.de:9035/manual/test.html

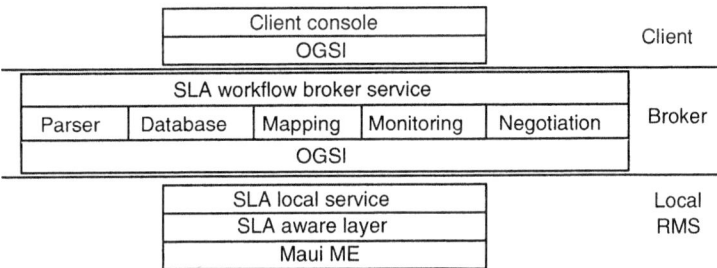

Fig. 8. System implementation architecture

5 Conclusion

An SLA negotiation process for workflow in the Grid environments is more sophisticated than the mapping of monolithic jobs contained in the workflow. This paper aims at defining that process and distinguishing the role as well as the interaction of the components participating in the negotiation. The main contribution of the paper is the definition of the components, the structure of the SLA text and of the protocol which merges all components in the SLA negotiation process. The process starts with negotiation on workflow between the consumer and the broker using the SLA workflow text. Thereafter, a series of negotiations between the workflow broker and the provider about each sub-job is performed. Finally, the negotiation on data transfer between sub-jobs is executed between providers according to the data transfer part in the SLA. The process is implemented in a prototype using existing Grid middleware.

References

1. K. Czajkowski, A. Dan, J. Rofrano, S. Tuecke, and M. Xu. "Agreement-based Grid service management (WS-Agreement)," Global Grid Forum, 2003.
2. L. Burchard, M. Hovestadt, O. Kao, A. Keller, and B. Linnert. The Virtual Resource Manager: An Architecture for SLA-aware Resource Management. Proc. IEEE CCGrid 2004, Chicago, US, 2004
3. K. Czajkowski and I. Foster and C. Kesselman and V. Sander and S. Tuecke. SNAP: A Protocol for Negotiating Service Level Agreements and Coordinating Resource Management in Distributed Systems. Proc. 8th Workshop on Job Scheduling Strategies for Parallel Processing, 2002.
4. D.M. Quan, O. Kao, "On Architecture for an SLA-aware Job Flows in Grid Environments," *Proc. 19th IEEE International Conference on Advanced Information Networking and Applications (AINA 2005).*, IEEE Press, 2005, pp. 287–292.
5. D.M. Quan, O. Kao, "Mapping Grid job flows to Grid resources within SLA context," *Proc. European Grid Conference,(EGC 2005), LNCS.*, Springer Verlag, 2005, pp. 1107–1116.
6. R. Al-Ali, O. Rana, D. Walker, S. Jha, and S. Sohail. G-QoSM: Grid Service Discovery using QoS Properties. Computing and Informatics Journal, Special Issue on Grid Computing, 21(4):363-382, 2002
7. I. Foster, C. Kesselman, C. Lee, R. Lindell, K. Nahrstedt, and A. Roy. A distributed resource management architecture that supports advance reservation and coallocation. In Proceedings of the International Workshop on Quality of Service, pages 27-36, 1999.
8. A. Sahai and S. Graupner and V. Machiraju and A. Moorsel. Specifying and Monitoring G)uarantees in Commercial Grids through SLA. Proceedings of the 3rd IEEE/ACM CCGrid2003, 2003.
9. I. Foster, C. Kesselman, and S. Tuecke. The Anatomy of the Grid: Enabling Scalable Virtual Organizations. International Journal of Supercomputing Applications, 15(3), 2002.
10. L. Nassif, J. M. Nogueira, M. Ahmed, R. Impey, A. Karmouch. Agent-based Negotiation for Resource Allocation in Grid. 3rd Workshop on computational Grids and applications, Summer program LNCC - 2005

Securing the MPLS Control Plane

Francesco Palmieri[1] and Ugo Fiore[1]

[1] Federico II University, Centro Servizi Didattico Scientifico, Via Cinthia 45,
80126 Napoli, Italy
{fpalmieri, ufiore}@unina.it

Abstract. As the Internet continues to grow, it faces an increasingly hostile environment and consequently, the need for security in network infrastructure is stronger than ever. In this scenario the Multi-Protocol Label Switching (MPLS) emerging paradigm, seems to be the cornerstone for developing most of the next generation network infrastructure-level services in the Internet. Unfortunately, due to the lack of a scalable means of verifying the authenticity and legitimacy of the control plane traffic in an MPLS domain, almost all the existing MPLS control and signaling protocols are extremely vulnerable to a variety of malicious attacks both in theory and in practice and communication between peer routers speaking the above common protocols is subject to active and passive forgery, hijacking and wiretapping activities. In this paper, we propose a robust framework for MPLS-based network survivability against security threats, by making the MPLS control and signaling protocols more secure. Our design goals include integrity safeguarding, protection against replay attacks, and gradual deployment, with routers not supporting authentication breaking the trust chain but operating undisturbed under any other respect.

Keywords: MPLS, strong authentication, integrity, label distribution, signaling.

1 Introduction

Multi-Protocol Label Switching (MPLS) is one of the emerging paradigms in Internet, and seems to be the cornerstone for developing more and more network infrastructure-level services. Almost all the new MPLS-based network services, such as Traffic Engineering, Differentiated Services QoS and layer-2/layer-3 VPN facilities need complex, reliable control and signaling protocols for information exchange between the nodes participating in an MPLS domain. In spite of this, there are many security issues that the current MPLS protocol framework does not address. The lack of strong protection based on cryptographic techniques in the original network control and signaling protocols allows malicious users to exploit the network in numerous ways by intercepting basic control information in the MPLS domain and sending fake data to subvert the network behavior. Thus, it is possible that an attacker can wiretap on the transmission link to inject anything he wants to hijack the network routers and then breach the overall infrastructure as desired. As a consequence, the need for security in the MPLS network infrastructure is stronger than ever. The main requirements are data origin authentication and data integrity for all the transactions involved in the MPLS label exchange process. Furthermore, the importance has also emerged of making the protocols transporting vital information for the network behavior, sufficiently robust. Our first goal is to identify the infrastructure-level vulnerabilities of existing MPLS domains and then develop the necessary security enhancements to be deployed for a survivable and secure MPLS framework. Accordingly, we propose a general method, implementable as a common practice or framework, to fight against these attacks based on digital signature: the originator of an MPLS control plane message, such as any label distribution or signaling protocol handler, signs the message it sends, and the corresponding recipients verify the signature so that the authenticity and integrity of the message can be protected. Our assessment shows that the framework is capable of preventing many currently unchecked security threats against the MPLS control plane without unduly overloading network elements. Even if we don't address protection from denial of service threats such as overwhelming the routers by generating spurious excess traffic, our framework may also provide an early warning of the presence of a compromised MPLS core router.

2 Related Work

The idea of signing vital protocol information in not new. Several efforts in this direction have been done for routing protocols. Foremost, of course, there is the design that Radia Perlman reported in her book [1] for signing link state information and for distribution of the public keys used in signing. However MPLS security is still an almost totally open issue. Behringer [2] and Senevirathne et al. [3] discuss two approaches to securing MPLS and analyzed the security of the MPLS framework. Senevirathne proposes an encryption approach using a modified version of IPsec working on single MPLS packets that adds significant processing delay in label forwarding nodes and overhead in packet size, and is clearly unacceptable in actual MPLS production networks. Behringer [2] analyzed MPLS security from the VPN service side making the assumption that the core MPLS network is trusted and provided in a secure manner. We make no such assumption in our work. We assume that the MPLS nodes themselves are secure, but the physical links connecting them are not – thus we will protect all the exchange of control plane messages between them using our enhanced security framework. A simple authentication scheme has been proposed for RSVP-TE [18]. However, that scheme does not cover LDP. We are proposing an unified authentication framework whose scope is the entire MPLS control plane.

3 Vulnerabilities in an MPLS Domain: The Most Common Threats

MPLS supports a variety of signaling mechanisms for label distribution and LSP set up. To date, the mechanisms the IETF has defined for basic label distribution is the *Label Distribution Protocol (LDP)* [5]. The IETF has also defined enhancements to LDP for *Constraint Route signaling (CR-LDP)* [6] and extensions to basic *RSVP for Traffic Engineering (RSVP-TE)* [7]. Both of these signaling protocols build on existing protocols to allow explicit set-up and traffic engineering of LSPs in an MPLS network. Broadly speaking, threats to MPLS control and signaling protocols come from two sources, external and internal. External threats come from outside intruders who are non-participants in the protocol. Internal threats come from compromised protocol participants such as regular MPLS nodes belonging to the domain. To perform an attack, the attacker must be able to inject arbitrary packets into the network. The goal may be to attract or redirect packets destined to other nodes for further analysis or fraudulent handling of the available network resources or just to disable the network or portions of it. An outside intruder could attack an MPLS domain in various ways:

- *Label Spoofing*: Similar to IP spoofing attacks, where an attacker fakes the source IP address of a packet, it is also theoretically possible to spoof the label of an MPLS packet to force passing of packets through disjoint VPNs, and break the VPN isolation mechanism.
- *Breaking the neighbor relationship*: This is a typical attack on *availability* in which an intelligent filter placed by an intruder on a link between two nodes could modify or change information in the label binding updates (advertisements or notifications) or even intercept traffic belonging to any data session.
- *Replay attack*: An intruder could passively collect label advertisement information. Later, the intruder could retransmit "obsolete" of fake information messages attacking the *integrity* of the MPLS domain. If obsolete information is accepted and disseminated, a normal MPLS node could make incorrect label binding decisions and send wrong information to its LDP neighbors.
- *Masquerading*: During the neighbor acquisition process, a outside intruder could masquerade a nonexistent or existing node by attaching to communication link and illegally joining in the LDP protocol domain by compromising the neighborship negotiation system. The threat of masquerading, which is typically an attack on *authenticity*, is almost the same as that of a compromised internal MPLS node.
- *Passive Listening and traffic analysis*: The intruder could passively gather exposed label binding or LSP setup information, or gain router access information such as remote terminal or SNMP access credentials . Such an attack, on information *confidentiality*, cannot directly affect the operation of the MPLS domain, but the obtained information can be used successively for hostile activity. Thus, sensitive router access, label bindings or LSP status information should be protected and their confidentiality should be the responsibility of the remote access, label distribution or signaling protocol.

If an internal MPLS node has been subverted, all information inside the node is exposed and at risk. The label information base (LIB) or the label forwarding information base (LFIB), can be directly manipulated via system commands or control interfaces. By seizing control of an MPLS node, an intruder could add a label binding entry into LFIB which will divert data traffic to a particular destination. An intruder could also randomly modify the above information to make the router advertise wrong labels or tear down existing LSPs or in the worst case creating some for of traffic blackholing. The compromised node problem has not received much attention to date, for several reasons. First, there are usually much fewer MPLS routers in a network administrative domain than hosts, and they are usually under tight control and monitoring. Second, control plane signaling is distributed and cooperative in nature (i.e. routers in an MPLS domain must coordinate their actions and cooperate to meet their protocol requirement) and thus there is a tradition of trust in all these protocols. MPLS networks must provide at least the same level of protection against all these forms of attack. There are two fundamental ways to protect the network: firstly, hardening protocols that could be abused, secondly making the network itself as inaccessible as possible. This can be achieved by a combination of packet filtering / firewalling and address hiding. In more detail, the MPLS network's internal structure information which may be useful to attackers must be totally hidden from outsiders. For example, denial-of-service (DoS) attacks against core routers will be hard if attackers do not know the targeted router's address. Thus, MPLS networks must not reveal the internal structure to customers. However, the MPLS core can be compromised by attacking the exposed edge routers. Since MPLS internal structure can be easily hidden from the outside, attackers do not know the address of any router in the core thus cannot attack them directly. To protect the edge routers whose addresses are known to the outside, we can use Access Control Lists to accept any communication only from customers' routers. On the other side, hardening MPLS control protocols is a more complex matter and several extensions have to be made to the protocols, by introducing strong authentication, no-replay and cryptography facilities, as will be detailed in the following section.

4 Hardening MPLS Control Protocols: The Basic Requirements

Generally speaking, making MPLS signaling protocol sufficiently secure requires that important label distribution and LSP setup information will be strongly authenticated between neighboring routers, since almost all the actually known attacks take advantage of the lack of authentication and integrity. Accordingly, we need to introduce some new encryption and authentication features into the above protocols. Encryption schemes keep the content of information only available to those who are authorized to have it [8]. Authentication schemes assure no-repudiation and data integrity, which protects messages against forgery or unauthorized alteration. The essential security mechanism that have to be implemented into the MPLS control protocols are described below.

4.1 Introducing Sequence Information

To prevent the replay of old information, sequence data, which can be a sequence number or a timestamp, has to be added to each update. The message sender generates packets with monotonically increasing sequence numbers. In turn, the receiver drops messages that have a lower sequence number than the previously received packet from the same source. The primary challenge posed by this requirement on long-lived MPLS control information is how to prevent sequence information from wrapping around. A sequence information must be valid for the life of a given MPLS router id. The primary advantage of sequence numbers compared to timestamps is their significantly longer life, because they will only increase when messages are sent. A sequence number can be relatively small and still provide reasonable assurance of not cycling. If timestamps were to be used, one would have to make sure that clocks were properly synchronized, e.g. via NTP. The main benefit of a timestamp would be the ease of administration provided by the well defined external reference for resetting the state. Anyway, in order to be able to check the validity of any type of sequence number, both parties must handshake and agree on some common initial information. Thus, whenever a router first establishes an LDP or RSVP relationship with a neighbor, e.g. at reboot, it must generate an initial sequence number (ISN) for that neighbor, and communicate it to the neighbor, for example transported in an LDP or RSVP HELLO message. The sequence number can then be incremented as usual, and it can be transmitted to the neighbor with each message. Another approach would be that the ingress LSR were to associate a nonce value (analogously to what is described in [9], [10], [11]) to each LDP request. That nonce would be then transmitted to the LDP neighbor, who would include it, unaltered, in the reply. Such a mechanism provides a way of matching requests and replies but offers a relatively weak protection against replay attacks, since an attacker could reuse a previously captured packet to force a long-lived LSP to close. If the nonce were to be associated to a single LSP at the time of the initial setup, with any operation involving that LSP deemed valid only if it carried that particular nonce, the nonce in the first LABEL REQUEST message should be trusted, still leaving room for attacks. We believe that a solution to this problem would be the use of symmetric encryption techniques and a trusted mutual key exchange, but this is out of the scope of this paper. We will investigate these issues in a future work.

4.2 Digitally Signing Update Information

To ensure the authenticity and integrity of the information exchanged between different routers, the originating router can digitally sign each update it generates. In addition, to allow receiving routers to validate the signature, a strong verification against a known key associated to the originating router must be performed on each update. These signatures can be used to validate control information flowing between the network elements in an MPLS domain such as a candidate label or path to a destination change before that label or path is selected for use. Asymmetric cryptography can perform digital signature, thus offering data origin authentication and data integrity, secure transmission or exchange of secret keys and data privacy (encryption and decryption). To accomplish the above functionalities on our protocol messages, each message has to be signed using a private key d, to produce an authenticated message $(m; t)$, where t is the tag on m, and this authenticated message can be verified by the sender's public key e. Hash functions can be used for providing data integrity in digital signature schemes, where the message is typically hashed first, then the hash-value is signed in place of the original message.

4.4 The Trust Model: Key Management and Distribution

The only disadvantage of asymmetric cryptography is the fact that the public keys must somehow be introduced and trusted by all the participating parties.. The best solution is given by the third trusted party (TTP) model. In this model a third party trust or Certification Authority (CA) - commonly agreed to be trusted by all interested parties – authenticates users to each other. The X.509 ITU-T Recommendation defines a feasible (and widely adopted) TTP model/framework to provide and support data origin authentication and peer entity authentication services, including formats for public-key certificates, and certificate revocation lists (CRL). The distribution of certificates can be achieved in at least three ways: the users can directly share their certificates with each other, or the certificates can be distributed via HTTP or, better, via LDAP [14]. The CAs also have the duty to publish at certain time intervals the CRLs, the black lists on which the revoked certificates are enlisted, together with the date and reason of revocation. The Online Certificate Status Protocol (OCSP) [10] enables applications to efficiently determine the (revocation) state of an identified certificate. A public key infrastructure (PKI) is usually defined as a set of cooperating CAs that issue, distribute and revoke certificates according through a hierarchical trusting policy. In a typical PKI, an authorized third party issues certificates according to a well defined hierarchal structure. A certificate binds an entity with its public key and is signed with the authorized third party's own private key. The recipient of a certificate use the authorized third party's public key to authenticate it. Using digital signature for trusting MPLS control information implicitly assumes that there should exists a Trust Authority or PKI to which the MPLS domain refers. The Trust Authority has its own private/public key pair and all the bindings between the LSRs and their public keys. Each router will be configured with its own pair of private/public key and the public key of the Trust Authority that they use to verify the trustiness of all the information obtained from the authority.

4.4 The Overall Security Scheme

We propose a digital signature scheme relying on Public Key Cryptography that takes its foundations on the work done by Murphy and Badger in [15] to secure the OSPF protocol, but is significantly modified and extended both in the basic paradigm and in key management/distribution, to cope with the features and security requirements of the MPLS control protocols. The basic idea is to add digital signature to each protocol message, and use an hash function such as message digest (like keyed MD5) to protect all exchanged message data. The originator will sign each message before sending it and the signature will stay within the message body during all the message life. This will protect the message integrity and provide authentication for all the transported data, ensuring that the data really does come from their legitimate source and has not been modified in transit. Furthermore, in the case where incorrect data is distributed by a faulty router, the signature facility provides a way to trace the problem to its source. MPLS signaling protocols have to be extended, to support on each of their vital messages, a new optional object containing the above signature and all the useful information on its size, digital signing and hashing algorithm. Eventually this object can be also used for transporting the sequence number information whose necessity has been described in the previous section. Routers providing this digital signature facility must behave as normal MPLS routers for all the protocol functionalities and can be straightforwardly mixed with router not implementing this feature. All the signature and integrity-enforcing mechanism has to be implemented, according to the same paradigm, for each signaling protocol (namely LDP and RSVP) as an optional attribute that will be transported in any message, used if recognized and otherwise silently discarded. When a router receives a signed message, it first verifies the signature using the known public key of the originator identified by the IP address associated to its LSR-id. If the signature verifies, the router accepts the message for usual processing, otherwise if the router does not knows the public key of the originator, it asks the trust authority for it, performing a standard LDAP query, and if it is available obtains and verifies the certification by using the Trust Authority's known public key and then stores the public key in its memory for further usage. For faster retrieval, public keys should be stored into the router memory by using an hash table indexed by originator LSR identifier where each public key will be kept until it expires or is revoked. If the originator public key is not available, the signed message is immediately discarded, otherwise if the originator's public key is available but verification fails, before discarding the message the public key has to be verified on-line via the OCSP protocol against the Trust authority, to cope with the case of key revocation or change. If necessary, when the known key is no more valid, the new public key is retrieved via LDAP and stored into the router's memory and a re-verification, against the new public key take place. There are still a few vulnerabilities with this security paradigm, coming from internal attackers. A subverted router is still able to forge protocol information, delete label mapping updates, and disclose LSP status information.

5 Securing the Label Distribution Protocol

LDP security is still a largely open issue since until now the LDP protocol provides no scalable standard-based mechanism for protecting the authentication, privacy and integrity of label distribution messages. LDP can actually use the TCP MD5 Signature Option to provide for the authenticity and integrity of its TCP session messages but MD5 authentication, as asserted by [16] is now considered to be too weak for this application. Recent research has in fact shown the MD5 hash algorithm to be vulnerable to collision attacks [17]. However the security requirements of label distribution protocols are essentially identical to those of the protocols which distribute routing information. In fact, with the LDP protocol, each router processes the information received from its neighbors and sends back the aggregate information. The result is that it is hard to validate the received information since the real originator of the information is obscured. Thus we need to set up an hop-by-hop trust chain in which message authentication and integrity checking has to be performed independently at each traversed node, from origin to destination. This implies a neighborship-based checking with message resigning and re-verification on each transaction. LDP has two different types of neighbors: directly connected neighbors - having a Layer 2 direct connection, hence a single IP hop, between them and non-directly connected neighbors - who are several IP hops away and connected to each other by MPLS traffic engineering tunnels that have LDP enabled on them. Signature verification, performed on each message exchanged on directly and non-directly connected neighbors, increases the level of trust in the whole MPLS domain, to the prejudice of a slight (if related to message relative frequency) overhead in message processing. In more detail, we propose the introduction of a new optional Type-Length-Value attribute, named *LDP-Security* (with section title "LDP security extensions") identified by the new type 0x8701 and structured as in fig 1 (a) at the end of the section. Our new LDP TLV is encoded, according to the standard LDP scheme, as a 2 octet field that uses 14 bits to specify a

Type and 2 bits to specify behavior when an LSR doesn't recognize the Type, followed by a 2 octet Length Field, and by a variable length Value field. The Value field itself may contain TLV encodings. That is, TLVs may be nested. Furthermore, since the value corresponding to our new TLV type has been chosen greater than 0x8000 (the high order "U" bit is set), according to the LDP specification, upon reception of this TLV in the optional parameters section of an LDP message, if it is not recognized by the accepting LDP neighbor as a known LDP type, is silently ignored and the rest of the message is processed as if the unknown TLV did not exist, without any error notification returned to the originator. So our scheme guarantees the full functionality of the label distribution mechanisms toward routers not supporting this new TLV. The first 32-bits field contains the message sequence number, calculated as explained in the previous section. The variable length signature follows, built on 32-bit words, followed by two byte codes, respectively identifying the signature algorithm (0x01 for RSA and 0x02 for DSA) and the hashing algorithm (0x01 for MD5 and 0x02 for SHA). This provides great flexibility and extensibility if other asymmetric cryptography or hashing algorithms become available. This TLV can be transported, and hence enhanced security can be provided, on almost all the LDP protocol messages, except for the Notification message, whose contents are however only informational, which doesn't allow the transport of optional parameters. To avoid label spoofing attacks, it is necessary to ensure that labeled data packets are labeled by trusted, hence authenticated, LSRs and that the labels placed on the packets are properly learned by the labeling LSRs only if the sending LSRs are trusted. Our hop-by-hop trusting model, providing implicit authentication and checking at each hop and hence natively avoids any label spoofing activity.

6 Securing The RSVP-TE Protocol

To ensure the integrity and authentication of its control messages, RSVP requires the ability to protect its them against corruption and spoofing. Corrupted or spoofed RSVP reservation requests could lead to theft of service by unauthorized parties or to denial of service caused by locking up network resources. RSVP protects against such attacks with a hop-by-hop authentication mechanism using an encrypted hash function [18]. The mechanism, providing protection against forgery or message modification, is supported by the so-called INTEGRITY objects that may appear in any RSVP message. The INTEGRITY object of each RSVP message is tagged with a one-time-use sequence number, allowing the message receiver to identify playbacks and hence to thwart replay attacks, but it does not provide confidentiality, since messages stay in the clear. Reliable and survivable use of an RSVP hop-by-hop authentication mechanism in an MPLS domain requires the availability of a stronger security model based on asymmetric cryptography, providing adequate key management and distribution infrastructure for participating routers. To implement such a strong authentication and integrity mechanism, based on the already explained neighbor based trust chain model implemented by hop-by-hop message signature and verification, as it has been done for LDP, we introduced a new optional RSVP object named STRONG-INTEGRITY defined by a Class value of 0x84 and a C-Type of 1 that can be transported in any RSVP message to be secured. Each sender supporting our enhanced security paradigm must ensure that all RSVP messages sent include a STRONG-INTEGRITY object, with the signature generated using its private key. Receivers, if supporting the above feature, have to verify whether RSVP messages, contain the new object and in this case perform signature verification according to the scheme presented in section 4.4. The Class value has been properly chosen equal to that of the already implemented INTEGRITY class, but with the high order bit set, since according to the RSVP specification when a node receives an object with an unknown class, if the high order bit in the class-num is set the node should ignore the object neither forwarding it nor sending an error message. This allows employing such new objects within an MPLS domain where some nodes do not recognize them. The object structure is very similar to that used for the new "LDP Security" TLV introduced in the previous section, as it can be seen in fig 2 (b) below. The first 32 bits field contains the usual RSVP object header, and more precisely the 16 bits object length and the two Classnum and C-type octets. Next, we find the 32 bit message sequence number, calculated as explained in the previous section. The variable length signature follows, built on 32-bit words, followed by two byte codes, respectively identifying the signature algorithm (0x01 for RSA and 0x02 for DSA) and the hashing algorithm (0x01 for MD5, 0x02 for SHA).

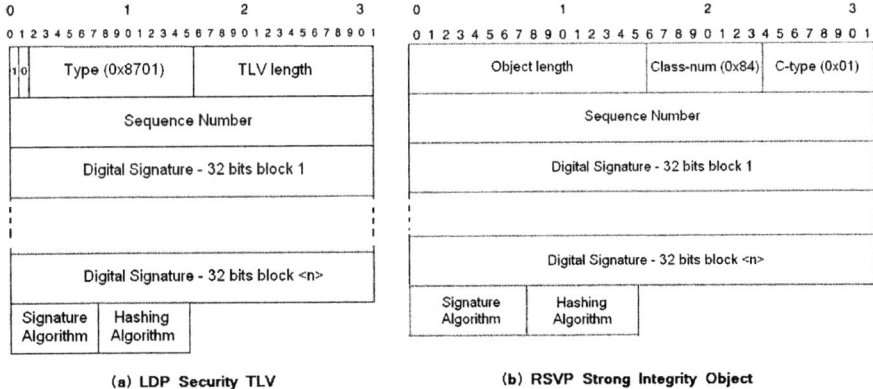

Fig. 1. The LDP and RSVP-TE security extensions

7 Implementation

To prove the concept of using digital signatures to authenticate control plane messages, we set up a minimal MPLS network using five Intel-based machines with 800 MhZ CPU (which is a common speed for actual MPLS routers), as shown in Figure 2, equipped with the Linux operating system, the mpls-linux v1.935 implementation, and, in particular, the ldp-portable v0.800 package, on which most of the implementation efforts take place. We only implemented the modification described in the previous sections on the LDP protocol. The GNU Zebra 0.94 routing software provided the necessary underlying routing functions through the supported OSPF implementation. We set up two LSPs from LER A to LER B, LSP1 and LSP2, passing respectively through LSR1 and LSR2-LSR3 and we sent packets from LER A to LER B by using both the LSPs, checking all the LDP traffic between the nodes

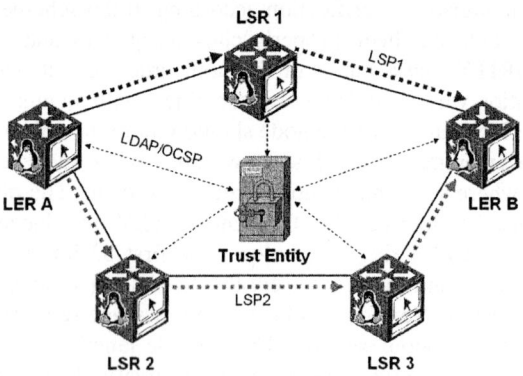

Fig. 2. The simple testbed network

Although this implementation it is only a *"proof of concept"*, it wants to emulate an active MPLS network in which hop-by-hop authentication takes place. For the LDAP and OCSP protocols, OpenLDAP v2.2.15 and OpenSSL v0.9.7e were used. Since the OCSP protocol is faster than LDAP we used it to quickly check if the certificate copy held in a router's cache is still valid. The local Certification Authority used for certificate distribution in the MPLS domain was set up based on OpenCA 0.9.2.1, with the OCSP responder daemon 1.0.2, on a Linux INTEL-powered server. The certificate cache in each router was implemented using an hash table addressed by the 32-bit LSR-id IP address and containing the corresponding public key certificates and their expiration date. For the sake of simplicity, the Linux routers were synchronized via NTP and the sequence number has been implemented as a 32-bit timestamp directly got from the Unix *time()* function. Our simple testbed demonstrated the correct operation of the proposed security framework that does not affect in any way the basic LDP functionalities of the MPLS nodes. The correct behavior of the MPLS LDP messages, has been also checked with monitoring/sniffing tools like *TcpDump* and *Ethereal*, and the traffic observations confirmed the impossibility for an internal or external attacker to eavesdrop, intercept, or modify these messages or even inject harmful messages into routing infrastructure. Finally, the LSR 2 Linux-based router has been replaced with a Cisco 3640 router, to test the backward compatibility feature through which routers with or without the above enhanced LDP security support can be mixed in the same MPLS domain. We observed that the LSP2 label switched path was successfully established, obviously without security in the last two hops in the path. The above evidence is sufficient to prove the interoperability and backward compatibility of our schema.

7.1 Storage Requirements

The storage requirement for the public keys is moderate, i.e., one key per LDP neighbor router, plus the keys for any CA involved. Only the keys pertaining to frequent talkers are kept in memory. In a production environment, the cache size will be a tradeoff between performance considerations and the overall memory requirements of the LSR processes. Note that our algorithm (see fig. 4) only allow a cache entry to be replaced after positive verification. A flooding attack, where an the router is swamped with a large number of forged messages with the sole intent of consuming resources, will not have an impact on memory. However, such attack will affect the router CPU.

7.2 Performance Considerations

In our performance analysis we first looked at the performance impact of signatures and verifications, then examined the certificate retrieval and validation cost on top of that. All the results, reported in table 1, were obtained by adding monitoring code to the various processing modules used in our security framework. One concern about the use of digital signatures to protect the integrity of control plane messages is that signature producing and verification will adversely affect the performance of the involved routers. The time it takes to produce and verify a signature can vary widely, depending on algorithm, software implementation, key length, and platform. From twenty sample functional tests ran on our testbed, using RSA and DSA algorithms with 1024-bit keys and respectively 128 and 40 bytes signature length, we observed an average verification time lower than 2ms for RSA and slightly greater that 24 ms for DSA. The Signature operation, on the opposite, was faster for DSA (16.5 ms) and significantly slower for RSA (37 ms). This result, clearly implies that the verification overheads are a minor factor compared to signing operations. But, because our model entails hop-by-hop authentication, the number of signature operations will equal the number of verifications and thus in average, considering the whole protocol transactions, both the algorithms exhibit almost the same performance. Consequently, in a real implementation there would be no significant advantage in choosing a scheme (such as RSA) where verification is much simpler to compute than signature. However, since the number of the protocol messages that must be signed (and hence the number of verifications), is very small compared to the overall traffic flowing trough the MPLS domain, because signatures are produced at relatively infrequent intervals (baring changes in the MPLS network), and since each signature can be computed at any time during the interval, we do not believe that signing will significantly damper performance. We estimated, however, that cryptography alone increases the cost of processing a LABEL REQUEST message by 70% to 110%, depending on the algorithms and parameters used. It should be taken into account, however, that the topology of our test network is quite simple. For certificate retrieval and validation we measured ad compared the average LDAP and OCSP transaction (query/response) times. Out of the total LDAP measured response latency of 8.3 ms, about 5 ms comes from the processing latency, 36% of which is contributed by back-end processing (entry retrieval from the database), and 64% by front-end processing from the client and server side (building the search filter, matching it against an entry, ASN.1 encoding/decoding of the query/result entry, sending the search result and status). The remaining 3 ms are due to connect and network latency. On the other side, the OCSP transaction time measured for one round is about 0.5–1.1 ms, the majority of which is from network latency. Consequently, we tried to keep the LDAP queries at minimum, both by caching keys locally and by using OCSP to ensure that a certificate has not been revoked.

Table 1. Performance measurement results

Algorithm	RSA	DSA	Protocol	LDAP	OCSP
Verify Time	1,8 ms	24,5 ms	Network latency	0.4 ms	0.4 ms
Sign Time	37 ms	16,5 ms	Connection latency	2.9 ms	-
Signature	128 bits	40 bits	Processing latency	5 ms	-
Key size	1024 bits	1024 bits	Total	8.3 ms	0.7 ms

Finally, it should be noted the relatively low frequency of LDP messages with respect to payload-carrying MPLS traffic can dilute the overall effect on performance. Several methods could be used to reduce the impact of the required verifications. In all of them, the verification effort could be offloaded from the processor performing the normal router functions. One offload method would be to use PCMCIA or better FPGA cards. These cards have an added benefit in that they provide increased security for the crucial private key. Another offload method would be to employ a separate processor in a multi-processor router architecture. Multi-processor architectures are now available among many router vendors, as mechanisms to separate routing computations from forwarding functions.

8 Conclusions

In this paper, we propose a robust framework, based on digital signatures, making the MPLS control and signaling protocol more secure against most of the common security threats. We designed and developed a cooperative security model, easily applicable in all the control plane protocols commonly used in the MPLS domain, based on a hop-by-hop trust chain in which message strong authentication and integrity checking has to be performed independently at each traversed node, from origin to destination. On the other side, the introduction of the above mechanisms will require the availability of a complex CA or PKI-based key management and distribution infrastructure for routers participating in a trusted MPLS domain. Anyway, our proposed cryptographic system for the protection of MPLS control plane infrastructure proved to be successful in its primary task – ensuring enhanced survivability against malicious service disruption by an intruder with exterior or interior access to an ISP's MPLS network.

References

1. R. Perlman, *Interconnections: Bridges and Routers*, Addison-Wesley, Reading Mass, 1992
2. Behringer, M., *Analysis of the Security of the MPLS Architecture*, Internet Draft < draft-behringer-mpls-security-10.txt>, IETF Network Working Group, Feb 2001.
3. T. Senevirathne, O. Paridaens, *Secure MPLS – Encryption and Authentication of MPLS Payloads*, Internet Draft, IETF Network Working Group, February 2001.

4. E. Rosen, A. Viswanathan, R. Callon, *Multiprotocol Label Switching Architecture*, IETF RFC 3031, Jan 2001
5. L. Andersson, P. Doolan, N. Feldman, A. Fredette, B. Thomas,*LDP Specification*, IETF RFC 3036, January 2001
6. B. Jamoussi, L. Andersson, R. Callon, et al. *Constraint-Based LSP Setup using LDP*, IETF RFC 3212, Jan 2002
7. D. Awduche, L. Berger et al., *RSVP-TE: Extensions to RSVP for LSP Tunnels*, IETF RFC 3209, Dec 2001
8. A. J. Menezes, P. C. van Oorschot, and S. A. Vanstone, *Handbook of Applied Cryptography*, CRC Press, Boca Raton, New York, 1997.
9. D. Maughan, M. Schertler, M. Schneider, J. Turner, *Internet Security Association and Key Management Protocol (ISAKMP)*, IETF RFC 2408, 1998
10. M. Myers, R. Ankney, A. Malpani, S. Galperin, C. Adams, *X.509 Internet Public Key Infrastructure Online Certificate Status Protocol - OCSP*, IETF RFC 2560, 1999.
11. C. Adams, P. Sylvester, M. Zolotarev, R. Zuccherato, *Internet X.509 Public Key Infrastructure Data Validation and Certification Server Protocols*, IETF RFC3029, 2001.
12. D. R. Stinson, *Cryptography Theory and Practice*, CRC Press, 1995.
13. R. Housley, W. Ford, W. Polk, D. Solo, *Internet X.509 Public Key Infrastructure Certificate and CRL Profile*, IETF RFC 2459, 1999.
14. M. Wahl, T. Howes, S. Kille, *Lightweight Directory Access Protocol (v3)*, IETF RFC 2251, Dec 1997
15. S. Murphy and M. Badger, *Digital signature protection of the OSPF routing protocol*, In Proceedings of the Symposium on Network and Distributed System Security (SNDSS'96), Feb 1996.
16. A. Heffernan, *Protection of BGP Sessions via the TCP MD5 Signature Option*, IETF RFC 2385, Aug 1998
17. Klima, V., *Finding MD5 Collisions - a Toy For a Notebook*, March 2005, (http://cryptography.hyperlink.cz/md5/MD5_collisions.pdf).
18. F. Baker, B. Lindell, M. Talwar, *RSVP Cryptographic Authentication*, IETF RFC 2747, Jan 2000

A Novel Arithmetic Unit over $GF(2^m)$ for Low Cost Cryptographic Applications

Chang Hoon Kim[1], Chun Pyo Hong[2], and Soonhak Kwon[3]

[1] Dept. of Computer and Information Engineering,
Daegu University, Jinryang, Kyungsan,
712-714, Korea
[2] Dept. of Computer and Communication Engineering,
Daegu University, Jinryang, Kyungsan,
712-714, Korea
[3] Dept. of Mathematics and Institute of Basic Science,
Sungkyunkwan University, Suwon,
440-746, Korea
chkim@dsp.daegu.ac.kr, cphong@daegu.ac.kr,
shkwon@skku.edu

Abstract. We present a novel VLSI architecture for division and multiplication in $GF(2^m)$, aimed at applications in low cost elliptic curve cryptographic processors. A compact and fast arithmetic unit (AU) was designed which uses substructure sharing between a modified version of the binary extended greatest common divisor (GCD) and the most significant bit first (MSB-first) multiplication algorithms. This AU produces division results at a rate of one per $2m - 1$ clock cycles and multiplication results at a rate of one per m clock cycles. Analysis shows that the computational delay time of the proposed architecture for division is significantly less than previously proposed bit-serial dividers and has the advantage of reduced chip area requirements. Furthermore, since this novel architecture does not restrict the choice of irreducible polynomials and has the features of regularity and modularity, it provides a high degree of flexibility and scalability with respect to the field size m.

Keywords: Cryptography, Finite Field, Multiplication, Division, VLSI.

1 Introduction

Information security has recently gained great importance due to the explosive growth of the Internet, mobile computing, and electronic commerce. To achieve information security, cryptographic systems (cryptosystems) must be employed. Among the various forms of practical cryptosystems, elliptic curve cryptosystems (ECC) have recently gained much attention in industry and academia. The main reason is that for a properly chosen elliptic curve, no known sub-exponential algorithm can be used to break the encryption through the solution of the discrete logarithm problem [1-3]. Compared to other systems such as RSA and ElGamal, this means that significantly smaller parameters/key sizes can be used in ECC for equivalent levels of security [2-3]. The benefits of having smaller key sizes include faster computation times and reduced processing power, storage, and bandwidth requirements. Another significant advantage of ECC is that even if all users employ an identical underlying finite field, each can select a different elliptic curve to use. In practice, this feature of ECC means that end users can periodically change the elliptic curve they use for encryption while using the same hardware to perform the field arithmetic, thus gaining an additional level of security [2].

Computing kP (a point or scalar multiplication) is the most important arithmetic operation in ECC, where k is an integer and P is a point on an elliptic curve. This operation can be computed by point addition and doubling. In affine coordinates, point addition and doubling can be implemented using one division, one multiplication, one squaring, and several addition operations over $GF(2^m)$ [3]. In $GF(2^m)$, addition is a bit independent XOR operation so it can be implemented in fast and inexpensive ways. Furthermore, squaring can be performed using multiplication. Therefore, it is important to design an efficient division and multiplication architecture in $GF(2^m)$ to perform the kP computations.

Three schemes have been used for computing inversion or division operations over $GF(2^m)$: 1) Repeated squarings and multiplications in $GF(2^m)$ [4], 2) Solution of a system of linear equations over $GF(2)$ [5], and 3) Use of the extended Euclid's or binary GCD algorithm over $GF(2)$ [6, 7, 8]. The first method uses successive squaring and multiplication such as $A/B = AB^{-1} = AB^{2^m-2} = A(B(B\cdots(B(B)^2)^2\cdots)^2)^2$. This method requires $m - 1$ times squarings and $\lfloor \log_2(m-1) \rfloor$ times multiplications respectively [9]. The second method finds an inverse element in $GF(2^m)$ by solving a system of $2m - 1$ linear equations with $2m-1$ unknowns over $GF(2)$. The last method uses the fact $\text{GCD}(G(x), B(x)) = 1$, where $B(x)$ is a nonzero element in $GF(2^m)$ and $G(x)$ is an irreducible polynomial defining the field, that is, $GF(2^m) \cong GF(2)[x]/G(x)$. The extended Euclid's or binary GCD algorithm is used to find $W(x)$ and $U(x)$ satisfying the relationship $W(x) \cdot G(x) + U(x) \cdot B(x) = 1$. Therefore we have $U(x) \cdot B(x) \equiv 1 \bmod G(x)$ and $U(x)$ is the multiplicative inverse of $B(x)$ in $GF(2^m)$. It is noted that the last method can be directly used to compute division in $GF(2^m)$ [7] and has area-time product of $O(m^2)$ while the first and the second schemes have $O(m^3)$.

The bit-level multiplication algorithms can be classified as either least significant bit first (LSB-first) or MSB-first schemes [9]. The LSB-first scheme processes the least significant bit of the second operand first, while the MSB-first scheme processes its most significant bit first. These two schemes have the same area-time complexity. However, while the LSB-first scheme uses three shifting registers, the MSB-first scheme only requires two shifting registers. Therefore, when power consumption is an issue, the MSB-first scheme is superior to the LSB-first scheme [9].

In this paper, we propose a novel bit-serial AU over $GF(2^m)$ for low cost ECC processors. The proposed AU produces division results at a rate of one per

$2m-1$ clock cycles in division mode and multiplication results at a rate of one per m clock cycles in multiplication mode. Analysis shows that the proposed architecture requires significantly small chip area compared to previously proposed bit-serial dividers. Therefore, the proposed AU is especially well suited to low area applications in $GF(2^m)$ such as smart cards and hand held devices.

2 Bit-Serial Divider and Multiplier for $GF(2^m)$

2.1 Bit-Serial Divider for $GF(2^m)$

Division Algorithm in $GF(2^m)$. Let $A(x)=\sum_{i=0}^{m-1} a_i x^i$ and $B(x)=\sum_{i=0}^{m-1} b_i x^i$ be two elements in $GF(2^m)$, $G(x) = \sum_{i=0}^{m} g_i x^i$ be the irreducible polynomial used to generate the field $GF(2^m) \cong GF(2)[x]/G(x)$, and $P(x) = \sum_{i=0}^{m-1} p_i x^i$ be the result of the division $A(x)/B(x) \bmod G(x)$. Then we can perform the division by using the following Algorithm I [8].

[**Algorithm I**] Binary Extended GCD Algorithm for Division in $GF(2^m)$
Input: $G(x), A(x), B(x)$
Output: V has $P(x) = A(x)/B(x) \bmod G(x)$
Initialize: $R = B(x), S = G = G(x), U = A(x), V = 0, count = 0, state = 0$
1.   for $i = 1$ to $2m - 1$ do
2.   if $state == 0$ then
3. $count = count + 1$;
4.   if $r_0 == 1$ then
5. $(R, S) = (R + S, R); (U, V) = (U + V, U)$;
6. $state = 1$;
7.   end if
8.   else
9. $count = count - 1$;
10.   if $r_0 == 1$ then
11. $(R, S) = (R + S, S); (U, V) = (U + V, V)$;
12.   end if
13.   if $count == 0$ then
14. $state = 0$;
15.   end if
16.   end if
17. $R = R/x$;
18. $U = U/x$;
19.   end for

Main Operations and Control Functions. In Algorithm I, $R = \sum_{i=0}^{m} r_i x^i$ is a polynomial with degree of m at most and $S = \sum_{i=0}^{m} s_i x^i$ is a polynomial with degree m, and $U = \sum_{i=0}^{m-1} u_i x^i$ and $V = \sum_{i=0}^{m-1} v_i x^i$ are polynomials with degree of $m - 1$ at most. As described in Algorithm I, S and V are a simple exchange operation with R and U respectively, depending on the value of $state$ and r_0.

On the other hand, R and U have two operation parts respectively. First, we consider the operations of R. Depending on the value of r_0, (R/x) or $((R+S)/x)$ is executed. Therefore, we can get the intermediate result of R as follows:

Let
$$R' = r'_m x^m + r'_{m-1} x^{m-1} + \cdots + r'_1 x + r'_0 = (r_0 S + R)/x \tag{1}$$

We can derive the following equations:
$$r'_m = 0 \tag{2}$$
$$r'_{m-1} = r_0 s_m = r_0 s_m + 0 \tag{3}$$
$$r'_i = r_0 s_{i+1} + r_{i+1}, 0 \leq i \leq m-2 \tag{4}$$

Second, we consider the operations of U. To get the intermediate result of U, we must compute the two operations of $U = (U + V)$ and $U = U/x$.

Let
$$U'' = u''_{m-1} x^{m-1} + \cdots + u''_1 x + u''_0 = U + V \tag{5}$$

We have
$$u''_i = r_0 v_i + u_i, 0 \leq i \leq m-1 \tag{6}$$

Since g_0 is always 1, we can rewrite $G(x)$ as given in (7)
$$1 = (x^{m-1} + g_{m-1} x^{m-2} + g_{m-2} x^{m-3} + \cdots + g_2 x + g_1) x \tag{7}$$

From (7), we have
$$x^{-1} = x^{m-1} + g_{m-1} x^{m-2} + g_{m-2} x^{m-3} + \cdots + g_2 x + g_1 \tag{8}$$

Let
$$U''' = u'''_{m-1} x^{m-1} + \cdots + u'''_1 x + u'''_0 = U/x \tag{9}$$

The following equations can be derived:
$$u'''_{m-1} = u_0 \tag{10}$$
$$u'''_i = u_{i+1} + u_0 g_{i+1}, 0 \leq i \leq m-2 \tag{11}$$

Let
$$U' = u'_{m-1} x^{m-1} + \cdots + u'_1 x + u'_0 = U''/x \tag{12}$$

We can derive the following (13) and (14).
$$u'_{m-1} = r_0 v_0 + u_0 = (r_0 v_0 + u_0) g_m + r_0 0 + 0 \tag{13}$$
$$u'_i = (r_0 v_{i+1} + u_{i+1}) + (r_0 v_0 + u_0) g_{i+1}, 0 \leq i \leq m-2 \tag{14}$$

In addition, the corresponding control functions of the Algorithm I are given as follows:

$$\text{Ctrl1} = r_0 \tag{15}$$

$$\text{Ctrl2} = u_0 \text{ XOR } (v_0 \text{ \& } r_0) \tag{16}$$

$$count = \begin{cases} count + 1, & if\ state == 0 \\ count - 1, & if\ state == 1 \end{cases} \tag{17}$$

$$state = \overline{state}, if \begin{cases} ((r_0 == 1)\ \&\ (state == 0))\ or \\ ((count == 0)\ \&\ (state == 1)) \end{cases} \tag{18}$$

Bit-Serial Divider for $GF(2^m)$. Based on the above main operations and control functions, we can derive a bit-serial divider for $GF(2^m)$ as shown in Fig. 1. The divider in Fig. 1 consists of a control logic, an RS-block, an m-bit bi-directional shift register block (SR-block), and a UV-block. Each functional block in Fig. 1 is described in Fig. 2, Fig. 3, and Fig. 4 respectively. As shown in Fig. 3, we remove s_0 and r^m because they are always 1 and 0, respectively. To trace the value of $count$, we use an m-bit bi-directional shift register instead of a $\log_2(m+1)$-bit up/down counter to achieve a higher clock rate. The m-bit bi-directional shift register is also used for multiplication, as will be explained in section 3. The control logic generates the control signals Ctrl1, Ctrl2, and Ctrl3 for the present iteration and updates the values of state and the $c-flag$ register for the next iteration. The 1-bit $c-flag$ register, which initially has a value of 1, is used to cooperate with the m-bit bi-directional shift register. The RS-cell computes the value of R and S of the Algorithm I, and propagates the r_0 signal to the control logic. As shown in Fig. 4, the m-bit bi-directional shift register is shifted left or right according to the value of state. When cnt_i is 1, it indicates that the value of count becomes i ($1 \leq i \leq m$). In addition, when the value of count reduces to 0, the $z-flag$ becomes 1. As a result, all the cnt_i become 0, the $c-flag$ has value 1, and state is updated to 0. This is the same condition of the first computation. The UV-cell computes the value of U and V of Algorithm I, and propagates the u_0 and v_0 signal to the control logic.

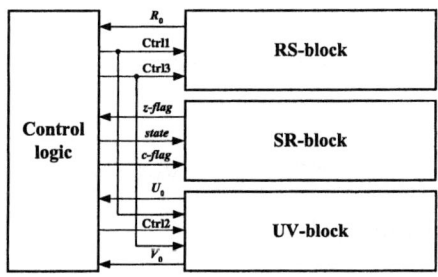

Fig. 1. Bit-serial divider for $GF(2^m)$

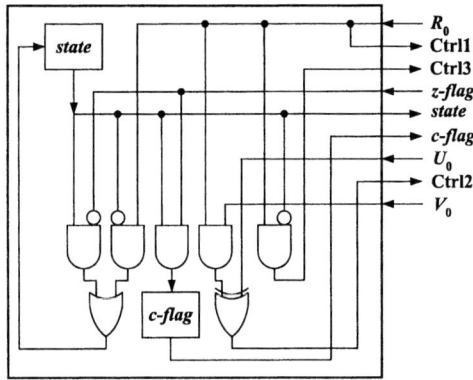

Fig. 2. The circuit of Control Logic in Fig. 1

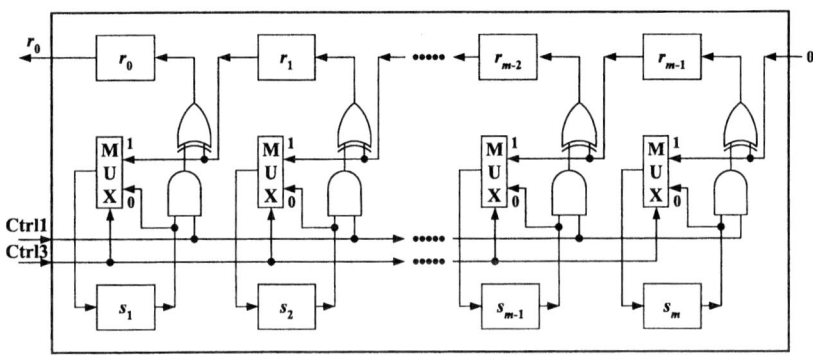

Fig. 3. The circuit of RS-block in Fig. 1

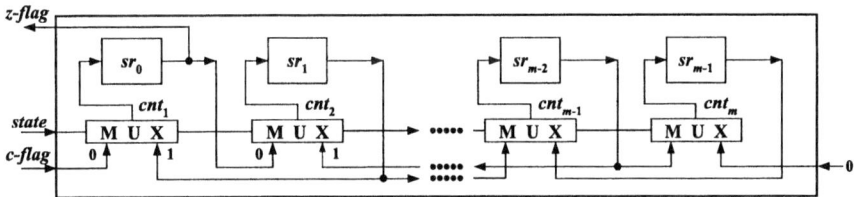

Fig. 4. The circuit of SR-block in Fig. 1

2.2 Bit-Serial Multiplier for $GF(2^m)$ Based on the MSB-First Multiplication Scheme

A bit-serial multiplier is briefly considered in this subsection. Detailed descriptions are covered in [9]. Let $A(x)$ and $B(x)$ be two elements in $GF(2^m)$, and $P(x)$ be the result of the product $A(x)B(x) \bmod G(x)$. The multiplication can

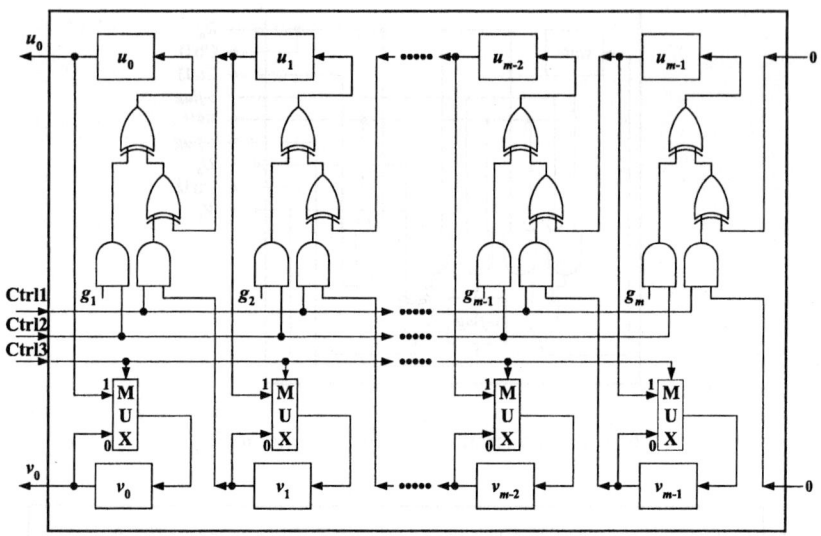

Fig. 5. The circuit of UV-block in Fig. 1

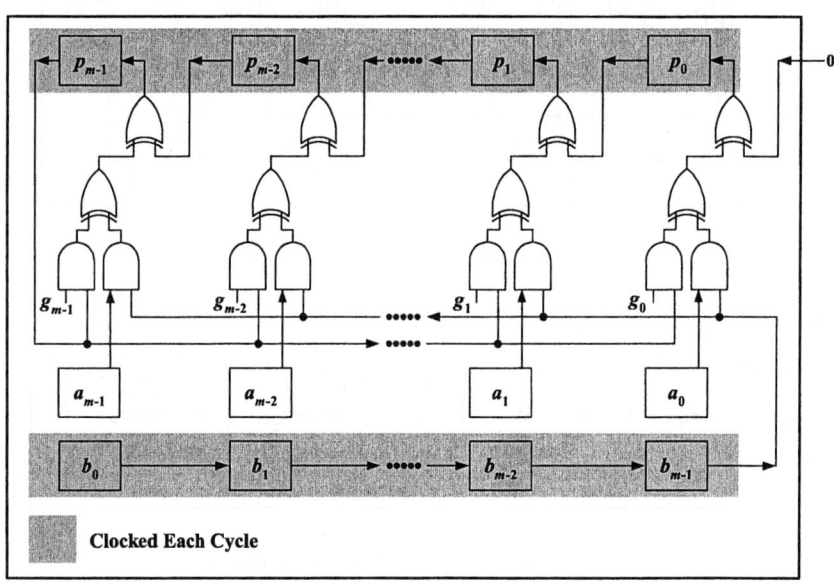

Fig. 6. MSB-first Multiplier for $GF(2^m)$

be performed by using the following MSB-first scheme:

$$P(x) = A(x)B(x) \mod G(x)$$
$$= \{\cdots [A(x)b_{m-1}x \mod G(x) + A(x)b_{m-2}]x \mod G(x) \quad (19)$$
$$+ \cdots + A(x)b_1]\}x \mod G(x) + A(x)b_0$$

The MSB-first multiplication based on (19) can be implemented using the architecture of Fig. 6.

3 A New VLSI Architecture for Both Division and Multiplication in $GF(2^m)$

Comparing the divider in Fig. 1 with the MSB-first multiplier in Fig. 6, it can be seen that: 1) The U operation of the divider is identical with the P operation of the MSB-first multiplier, except for the input values, and 2) The SR register in Fig. 4 is shifted bi-directionally, while the B register in Fig.6 is shifted unidirectionally.

To perform multiplication using the divider in Fig. 1, we modify the SR-block of Fig. 4 and the UV-block of Fig. 5. The modified SR- and UV-blocks are shown in Fig. 7 and Fig. 8, respectively. The modifications are summarized as follows:

(i) SR-block: As described in Fig. 8, we add one 2-to-1 OR gate and a $mult/div$ signal into the SR-block of Fig. 4 to shift the SR register only to the left direction when it performs multiplication. $mult/div$ is 0 for multiplication and 1 for division. Therefore, the SR register is shifted bi-directionally depending on *state* in division mode and shifted only to the left in multiplication mode.

(ii) UV-block: we add two 2-to-1 multiplexers, two 2-to-1 AND gates, and a $mult/div$ signal into the UV-block of Fig. 5. As shown in Fig. 8, in multiplication mode, by adding two 2-to-1 multiplexers, p_{m-1} and b_i are selected instead of Ctrl2 and $z-flag$, respectively. In addition, since AND gate number 1 generates 0 in multiplication mode, each a_i/b_i register selects its own value and AND gate number 2 generates 0 in division mode. As a reference, the A/V register in multiplication mode can not be clocked to reduce power consumption.

As described in Fig. 7 and Fig. 8, we can perform the MSB-first multiplication with the circuit of Fig. 7 and Fig. 8. In division mode, we clear the B/SR register in Fig. 7 the A/V register in Fig. 8, load the P/U register with U in Fig. 8, and set the $mult/div$ signal to 1. After $2m-1$ iterations, the A/V register contains the division result. In multiplication mode, we clear the P/U register in Fig. 8, load the A/V register with A in Fig. 8, load the B/SR register with B in Fig. 7, and set the $mult/div$ signal to 0. After m iterations, the P/U register contains the multiplication result.

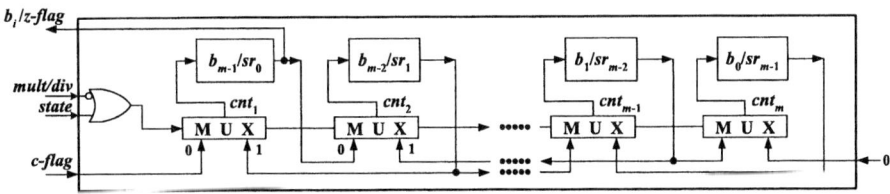

Fig. 7. Modified SR-block of Fig. 4

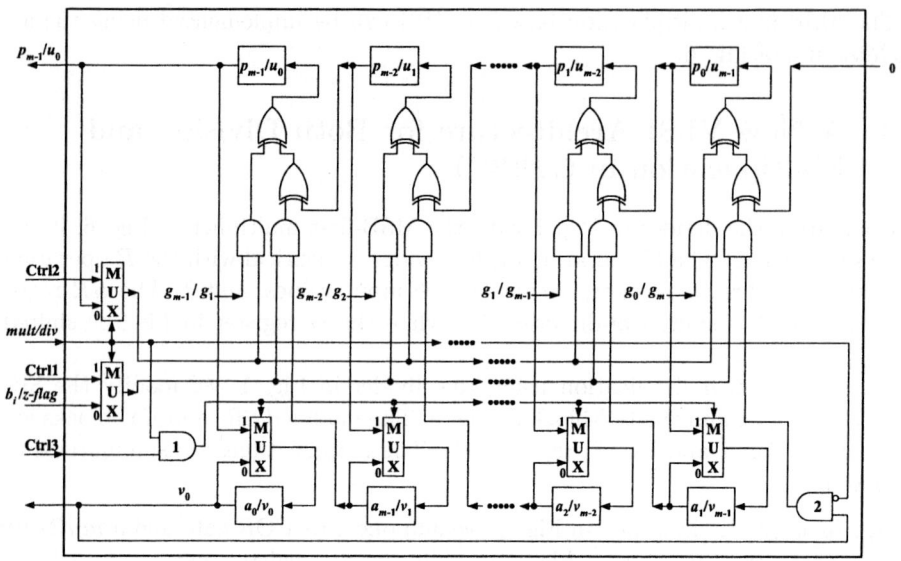

Fig. 8. Modified UV-block of Fig. 4

4 Performance Analysis

To verify the functionality of the proposed AU in $GF(2^8)$, we developed it in VHDL and synthesized it with Mentor's LeonardoSpectrum (version2002c.15), in which Altera's FPGA EP1S80F1508C6 was used as the target device. After synthesizing the design successfully, we extracted the net-list file from LeonardoSpectrum. With the net-list file, after placing and routing the synthesized design, we analyzed the timing characteristics and verified its functionality using Altera's Quartus II (version 2.0). From the timing analysis, it was estimated that the AU can run at a clock rate up to 217.56 MHz. After verifying the proposed AU's functionality, we compared the performance of the AU with previously proposed dividers. Table 1 shows the comparison results. In Table 1, a 3-input XOR gate is constructed using two 2-input XOR gates and the number of transistor (TR) estimation is based on the following assumptions: a 2-input AND gate, a 2-input XOR gate, a 2-to-1 multiplexer, a 2-input OR gate, and a 1-bit latch consist of 4, 6, 6, 6, and 8 transistors, respectively [10].

From Table 1, we can see that the computational delay time of the proposed AU is significantly less than the divider in [6], and it has the smallest number of transistor (TR). In addition, since the proposed AU can also perform multiplication with the same processing rate compared to the multiplier in Fig. 6, we can save the hardware used in the MSB-first multiplier. For reference, the MSB-first multiplier require $3m$ latches, $2m$ 2-input AND gates, and $2m$ 2-input XOR gates, i.e, it uses $44m$ TRs.

Table 1. Comparison with previously proposed dividers for $GF(2^m)$

	Brunner et al. [6]	Guo et al. [7]	Proposed AU
Throughput (1/cycles)	$1/2m$	$1/m$	Multiplication: $1/m$ Division: $1/2m-1$
Latency (cycles)	$2m$	$5m-4$	Multiplication: m Division: $2m-1$
Critical Path Delay	$T_{zero-detector} + 2T_{AND2}$ $+2T_{XOR2} + 2T_{MUX2}$	$T_{AND2} + 3T_{XOR2}$ $+T_{MUX2}$	$2T_{AND2} + 2T_{XOR2}$ $+T_{MUX2}$
Basic Components and Their Numbers	$AND_2 : 3m + 2\log_2(m+1)$ $XOR_2 : 3m$ $OR_2 : \log_2(m+1)$ Latch $: 4m + \log_2(m+1)$ $MUX_2 : 8m$	$AND_2 : 16m - 16$ $XOR_2 : 10m - 10$ Latch $: 44m - 43$ $MUX_2 : 22m - 22$	$AND_2 : 3m + 7$ $XOR_2 : 3m + 1$ $OR_2 : 2$ Latch $: 5m + 2$ $MUX_2 : 3m + 2$
# of TR	$110m + 24\log_2(m+1)$	$608m - 432$	$88m + 74$

AND_2: 2-input AND gate
XOR_2: 2-input XOR gate
OR_2: 2-input OR gate
MUX_2: 2-to-1 multiplexer
T_{AND2}: the propagation delay through one AND_2 gate
T_{XOR2}: the propagation delay through one XOR_2 gate
T_{MUX2}: the propagation delay through one MUX_2 gate
$T_{zero-detector}$: the propagation delay of $\log_2(m+1)$-bit zero-detector

5 Conclusions

In this paper, we have proposed a novel bit-serial AU over $GF(2^m)$ for low cost ECC processor applications. Unlike previously proposed architectures, the AU proposed in this paper can perform both division and multiplication in $GF(2^m)$. In addition, the AU requires significantly small chip area compared to previously proposed bit-serial dividers, and does not require additional hardware for multiplication. As a result, the AU proposed in this paper is well suited for both the division and multiplication circuitry of ECC processors. It is especially well suited to low area applications such as smart cards and hand held devices. Furthermore, since the proposed architecture does not restrict the choice of irreducible polynomials, and has the features of regularity and modularity, it provides a high flexibility and scalability with respect to the field size m.

Acknowledgement. This work was supported by the Daegu University under a research grant 2004.

References

1. IEEE 1363, *Standard Specifications for Publickey Cryptography*, 2000.
2. A. Menezes, *Elliptic Curve Public Key Cryptosystems*, Kluwer Academic Publishers, 1993.

3. I. F. Blake, G. Seroussi, and N. P. Smart, *Elliptic Curves in Cryptography*, Cambridge University Press, 1999.
4. S.-W. Wei, "VLSI Architectures for Computing exponentiations, Multiplicative Inverses, and Divisions in $GF(2^m)$," *IEEE Trans. Circuits Syst. II*, vol 44, no. 10, pp. 847-855, Oct. 1997.
5. M.A. Hasan and V.K. Bhargava, "Bit-Level Systolic Divider and Multiplier for Finite Fields $GF(2^m)$," *IEEE Trans. Computers*, vol. 41, no. 8, pp. 972-980, Aug. 1992.
6. H. Brunner, A. Curiger and M. Hofstetter, "On Computing Multiplicative Inverses in $GF(2^m)$," *IEEE Trans. Computers*, vol. 42, no. 8, pp. 1010-1015, Aug. 1993.
7. J.-H. Guo and C.-L. Wang, "Bit-serial Systolic Array Implementation of Euclid's Algorithm for Inversion and Division in $GF(2^m)$," *Proc. 1997 Int. Symp. VLSI Tech., Systems and Applications*, pp. 113-117, 1997.
8. C. H. Kim, S. Kwon, J. J. Kim, and C. P. Hong, "A Compact and Fast Division Architecture for a Finite Field $GF(2^m)$," *Lecture Notes in Computer Science*, vol. 2667, pp. 855-864, May 2003.
9. J.R. Goodman, "Energy Scalable Reconfigurable Cryptographic Hardware for Portable Applications," PhD thesis, MIT, 2000.
10. N. Weste and K. Eshraghian, *Principles of CMOS VLSI Design: A System Perspective*, 2nd ed. Reading, MA: Addison-Wesley, 1993.

A New Parity Space Approach to Fault Detection for General Systems

Pyung Soo Kim and Eung Hyuk Lee

Department of Electronics Engineering,
Korea Polytechnic University, Shihung City, 429-793, Korea
pskim@kpu.ac.kr

Abstract. This paper proposes a new parity space approach to a fault detection for general systems with noises, actuator faults and sensor faults. The proposed parity space approach could be more systematic than existing approaches since an efficient numerical algorithm is utilized. In addition, since the system and measurement noises are accounted in the proposed approach, the parity space residual will be noise-suppressed. When there are no noises and faults are constant on the interval of parity space's order, the residual for each fault is shown to be exactly equal to the corresponding fault. The proposed approach is specified to the digital filter structure for the amenability to hardware implementation.

1 Introduction

For the fault detection to enhance the reliability of systems, many authors have investigated a parity space approach by checking the consistency of the mathematical equations of the system [1]-[5]. However, in existing works, the projection matrix for a parity check is chosen arbitrarily. Therefore, as mentioned in [5], existing parity space approaches could be somewhat non-systematic. In addition, existing works in [1]-[3], might be sensitive to wide-band noise since the projection matrix is chosen without the account of noise effect. Although the existing work in [4] considers the measurement noise, it doesn't handle the actuator fault as well as the system noise. Therefore, none of the existing works listed above can handle the general system with both system and measurement noises, and both actuator and sensor faults, simultaneously.

In the current paper, a new parity space approach is proposed. For realistic situations, the proposed approach considers the general system with both system and measurement noises and both actuator and sensor faults, simultaneously. Unlike existing approaches in [1]-[4], the projection matrix is obtained from the coefficient matrix of the FIR structure filter developed in [6], [7]. Therefore, the proposed parity space approach could be more systematic than existing approaches since an efficient numerical algorithm is utilized. In addition, since the system and measurement noises are accounted in the proposed approach, the parity space residual will be noise-suppressed. Therefore, the proposed approach will work well even for systems with wide-band noises, while existing approaches doesn't not. Moreover, the proposed approach also cover both actuator and sensor faults. As an inherent property, when there is no noises and faults are constant on the interval of parity space's order, the residual for each fault is shown to be exactly equal to the corresponding fault, which cannot be obtained from existing approaches in [1]-[4].

In practice, it should be required that a fault detection algorithm can be implemented with discrete-time analog or digital hardware. In this case, the fault detection algorithm should be specified to an algorithm or structure that can be realized in the desired technology. Therefore, The proposed parity space residual algorithm is specified to the well known digital filter structure in [8] for the amenability to hardware implementation.

2 Existing Parity Space Approach

The following discrete-time system with noises and unknown faults is considered:

$$x(i+1) = Ax(i) + Bu(i) + Df(i) + Gw(i), \quad (1)$$
$$y(i) = Cx(i) + Ef(i) + v(i) \quad (2)$$

where $x(i) \in \Re^n$ is the state vector, $u(i) \in \Re^l$ and $y(i) \in \Re^q$ are the known input vector and the measured output vector. The covariances of the system noise $w(i) \in \Re^p$ and the measurement noise $v(i) \in \Re^q$ are Q and R, respectively. The fault vector $f(i) \in \Re^q$ in the system under consideration are to be represented by random-walk processes as

$$f(i+1) = f(i) + \delta(i) \quad (3)$$

where $f(i) \triangleq [f_1(i) \ f_2(i) \ \cdots \ f_q(i)]^T$, $\delta(i) \triangleq [\delta_1(i) \ \delta_2(i) \ \cdots \ \delta_q(i)]^T$ and $\delta(i)$ is a zero-mean white Gaussian random process with covariance Q_δ. It is noted that the random-walk process provides a general and useful tool for the analysis of unknown time-varying parameters and has been widely used in the detection and estimation area [9], [10].

On the finite interval $[i-N, i]$, the system (1) and (2) can be represented by the vector regression form as follows:

$$Y_N(i) - B_N U_N(i) = C_N x(i-N) + \Gamma_N F_N(i) + G_N W_N(i) + V_N(i) \quad (4)$$

where $Y_N(i) \triangleq [y(i-N)^T \ y(i-N+1)^T \ \cdots \ y(i)^T]^T$ and B_N, C_N, Γ_N, G_N are defined as follows:

$$C_N \triangleq \begin{bmatrix} C \\ CA \\ \vdots \\ CA^N \end{bmatrix}, \quad B_N \triangleq \begin{bmatrix} 0 & 0 & \cdots & 0 & 0 \\ CB & 0 & \cdots & 0 & 0 \\ \vdots & \vdots & \vdots & \vdots & \vdots \\ CA^{N-1}B & CA^{N-2}B & \cdots & CB & 0 \end{bmatrix},$$

$$\Gamma_N \triangleq \begin{bmatrix} E & 0 & \cdots & 0 & 0 \\ CD & E & \cdots & 0 & 0 \\ \vdots & \vdots & \vdots & \vdots & \vdots \\ CA^{N-1}D & CA^{N-2}D & \cdots & CD & E \end{bmatrix}, G_N \triangleq \begin{bmatrix} 0 & 0 & \cdots & 0 & 0 \\ CG & 0 & \cdots & 0 & 0 \\ \vdots & \vdots & \vdots & \vdots & \vdots \\ CA^{N-1}G & CA^{N-2}G & \cdots & CG & 0 \end{bmatrix}$$

and $U_N(i)$, $F_N(i)$, $W_N(i)$, $V_N(i)$ have the same form as $Y_N(i)$ for $u(i)$, $f(i)$, $w(i)$, $v(i)$, respectively.

The key idea of the parity space approach eliminates the unknown system state $x(i-N)$ from the equation (4) by the projection matrix for a parity check. However, in existing works, the projection matrix for a parity check is chosen arbitrarily. Therefore, as mentioned in [5], existing parity space approaches could be somewhat non-systematic. In addition, existing works in [1]-[3], might be sensitive to wide-band noise since the projection matrix is chosen without the account of noise effect. Although the existing work in [4] considers the measurement noise, it doesn't handle the actuator fault as well as the system noise. Therefore, none of the existing works listed above can handle the general system with both system and measurement noises, and both actuator and sensor faults, simultaneously.

3 New Parity Space Approach

The faults in (1) and (2) can be treated as auxiliary states as shown in the following augmented system:

$$x_a(i+1) = A_a x_a(i) + B_a u(i) + G_a w_a(i), \quad (5)$$
$$y(i) = C_a x_a(i) + v(i) \quad (6)$$

where $x_a(i) \triangleq [x(i)^T \ f(i)^T]^T$, $w_a(i) \triangleq [w(i)^T \ \delta(i)^T]^T$ and

$$A_a \triangleq \begin{bmatrix} A & D \\ 0 & I \end{bmatrix}, \quad B_a \triangleq \begin{bmatrix} B \\ 0 \end{bmatrix}, \quad G_a \triangleq \begin{bmatrix} G & 0 \\ 0 & I \end{bmatrix}, \quad C_a \triangleq [C \ E]$$

and the covariance matrix of $w_a(i)$ is the diagonal matrix with Q and Q_δ.

To obtain the projection matrix for a parity check, the filter coefficient matrix H of the FIR structure filter is introduced. When $\{A_a, C_a\}$ is observable, A_a is nonsingular, and $N \geq n+q-1$ for the system (5) and (6), the coefficient matrix H can be obtained from [6], [7] with the consideration of system and measurement noises. The coefficient matrix H has the following form:

$$H = \begin{bmatrix} H_x \\ H_f \end{bmatrix} \quad (7)$$

where H_x and H_f are the coefficient matrices for the system state $x(i)$ and the fault $f(i)$, and have dimensions of $n(qN+q)$ and $q(qN+q)$, respectively. It is noted that the coefficient matrix H has the following matrix equality:

$$H \bar{C}_N = A_a^N \quad (8)$$

where $\bar{C}_N \triangleq [(C_a)^T \ (C_a A_a)^T \ \cdots \ (C_a A_a^N)^T]^T$.

Using the equation (7) and the matrix equality (8), it will be shown that the coefficient matrix H_f in (7) can be the projection matrix at any time i for a parity check in the following theorem.

Theorem 1. *The projection matrix for a parity check can be obtained from the coefficient matrix of the FIR structure filter.*

Proof. The matrix equality (8) can be written by

$$H\begin{bmatrix}C_N & \Lambda_N\end{bmatrix} = \begin{bmatrix}H_x \\ H_f\end{bmatrix}\begin{bmatrix}C_N & \Lambda_N\end{bmatrix} = \begin{bmatrix}A^N & \sum_{l=0}^{N-1} A^l D \\ 0 & I\end{bmatrix} \quad (9)$$

where

$$\Lambda_N = \begin{bmatrix} E \\ CD + E \\ \vdots \\ \sum_{l=0}^{N-1} CA^l D + E \end{bmatrix}.$$

Therefore, (9) gives the following matrix equality:

$$H_f C_N = 0. \quad (10)$$

Pre-multiplying both sides of (4) by H_f and using the matrix equality (10) yield

$$H_f[Y_N(i) - B_N U_N(i)] = H_f[\Gamma_N F_N(i) + G_N W_N(i) + V_N(i)]. \quad (11)$$

Thus, using (10), the unknown system state term $x(i - N)$ in (11) is eliminated like existing parity space approach. Therefore, H_f is the projection matrix for parity space of order N. This completes the proof. ∎

Therefore, the parity space residual vector can be defined as

$$\mathbf{r}(i) = H_f[Y_N(i) - B_N U_N(i)]. \quad (12)$$

The equation (12) can be qualified to be a residual since it is unaffected by the unknown system state. Unlike the projection matrix in existing approaches [1]-[4], the projection matrix H_f is obtained from the coefficient matrix of optimal filter. Therefore, the proposed parity space approach may be more systematic than existing ones since an efficient numerical algorithm is utilized. Since the system and measurement noises are accounted in the proposed approach, the residual $r(i)$ will be noise-suppressed. Therefore, the proposed approach will work well even for systems with wide-band noises, while existing approaches didn't. It is also noted that the projection matrix H_f requires computation only on the interval $[0, N]$ once and is time-invariant for all windows. This means that the projection matrix H_f for a parity check can be obtained from off-line computation and thus only the residual (12) is needed in on-line computation.

However, each fault cannot be detected individually when there are multiple faults using only the residual vector (12). Thus, some manipulations are required to detect each fault individually. Therefore, each parity space residual is defined for the sth fault as

$$r_s(i) = H_f^s[Y_N(i) - B_N U_N(i)], \quad 1 \leq s \leq q \quad (13)$$

where H_f^s is the sth row of the H_f. Note that, from (10), satisfies the following equality:

$$H_f^s C_N = 0. \tag{14}$$

In addition, from $H_f \Lambda_N = I$ in (9), H_f^s satisfies the following equality:

$$H_f^s \Lambda_N^s = 1, \quad H_f^s \Lambda_N^j = 0 \ (1 \leq j \leq q, j \neq s) \tag{15}$$

where Λ_N^j is the jth column of the matrix Λ_N.

In the following theorem, when there are no noises as $w(\cdot) = 0$, $v(\cdot) = 0$, and faults are constant as $f(\cdot) = \bar{f} = [\bar{f}_1 \ \bar{f}_2 \cdots \bar{f}_q]^T$ on the interval of parity space's interval, each parity space residual $r_s(i)$ is shown to be exactly equal to the corresponding fault $\bar{f}_s$.

Theorem 2. *When there are no noises and faults are constant on the interval $[i - N, i]$, each parity space residual $r_s(i)$ is exactly equal to the corresponding fault.*

Proof. On the interval $[i - N, i]$, when there are no noises as $w(\cdot) = 0$, $v(\cdot) = 0$, and faults are constant as $f(\cdot) = \bar{f} = [\bar{f}_1 \ \bar{f}_2 \cdots \bar{f}_q]^T$, the following representation is valid:

$$Y_N(i) - B_N U_N(i) = C_N x(i - N) + \Lambda_N \bar{f}.$$

Therefore, using matrix equalities (14) and (15), the parity space residual of the sth fault $f_s(i)$ satisfies the following:

$$\begin{aligned} r_s(i) &= H_f^s [Y_N(i) - B_N U_N(i)] \\ &= H_f^s \left[C_N x(i - N) + \Lambda_N^s \bar{f}_s + \sum_{j \neq s} \Lambda_N^j \bar{f}_j \right] = \bar{f}_s. \end{aligned}$$

This completes the proof. ∎

Note that the remarkable property in Theorem 2 cannot be obtained from existing approaches in [1]-[4].

In practice, it should be required that a fault detection algorithm can be implemented with discrete-time analog or digital hardware. In this case, the fault detection algorithm should be specified to an algorithm or structure that can be realized in the desired technology. Therefore, the proposed parity space residual algorithm is specified to the well known digital filter structure in [8] for the amenability to hardware implementation. Since the projection matrix H_f^s in (13) is defined by

$$H_f^s \triangleq [h_f^s(N) \ h_f^s(N-1) \ \cdots \ h_f^s(0)],$$

the parity space residual (13) for each fault can be represented in

$$r_s(i) = \sum_{j=0}^{N} h_f^s(j) y(i-j) - \sum_{j=0}^{N} (H_f^s B_N)_j u(i-j) \tag{16}$$

where $(H_f^s B_N)_j$ is the $(j+1)$th l elements of $H_f^s B_N$. Applying the z-transformation to the parity space residual (16) yields the following digital filter structure:

$$r_s(z) = \sum_{j=0}^{N} h_f^s(j) z^{-j} y(z) + \sum_{j=0}^{N} -(H_f^s B_N)_j z^{-j} u(z) \qquad (17)$$

where $h_f^s(j)$ and $-(H_f^s B_N)_j$ become filter coefficients. It is noted that the digital filter structure (17) is a well known moving average process whose functional relation between input $u(z)$, $y(z)$ and output $r_s(z)$ is nonrecursive.

4 Conclusions

In the current paper, the new parity space approach to a fault detection has been proposed for general systems with noises, actuator faults and sensor faults. The proposed parity space approach could be more systematic than existing approaches since an efficient numerical algorithm is utilized. In addition, since the system and measurement noises are accounted in the proposed approach, the parity space residual will be noise-suppressed. When there are no noises and faults are constant on the interval of parity space's order, the residual for each fault is shown to be exactly equal to the corresponding fault. The proposed approach is specified to the digital filter structure for the amenability to hardware implementation.

References

1. Gertler, J., Monajemy, R.: Generating directional residual with dynamic parity relations. Automatica **31** (1995) 627–635
2. Filaretov, V.F., Vukobratovic, M.K., Zhirabok, A.N.: Parity relation approach to fault diagnosis in manipulation robots. Mechatronics **13** (2000) 142–152
3. Conatser, R., Wagner, J., Ganta, S., Walker, I.: Diagnosis of automotive electronic throttle control systems. Control Engineering Practice **12** (2004) 23–30
4. Jin, H., Chang, H.Y.: Optimal parity vector sensitive to designated sensor fault. IEEE Trans. Aerosp. Electron. Syst. **35** (1999) 1122–1128
5. Betta, G., Pietrosanto, A.: Instrumentation fault detection and isolation: state of the art and new research result. IEEE Trans. Instrum. Meas. **49** (2000) 100–107
6. Kwon, W.H., Kim, P.S., Han, S.H.: A receding horizon unbiased FIR filter for discrete-time state space models. Automatica. **38** (2002) 545–551
7. Kim, P.S.: Maximum likelihood FIR filter for state space signal models. IEICE Trans. Commun. **E85-B** (2002) 1604–1607
8. Oppenheim, A., Schafer, R.: Discrete-Time Signal Processing. Englewood Cliffs, NJ:Prentice-Hall (1989)
9. Ljung, L.: System Identification, Theory for The User. Englewood Cliffs, NJ:Prentice-Hall (1999)
10. Alouani, A.T., Xia, P., Rice, T.R., Blair, W.D.: On the optimality of two-stage state estimation in the presence of random bias. IEEE Trans. Automat. Contr. **38** (1993) 1279–1282

Differential Power Analysis on Block Cipher ARIA*

JaeCheol Ha[1],[**], ChangKyun Kim[2], SangJae Moon[3], IlHwan Park[2], and HyungSo Yoo[3]

[1] Dept. of Information and Communication, Korea Nazarene Univ., Korea
jcha@kornu.ac.kr
[2] National Security Research Institute, Daejeon, Korea
{kimck, ilhpark}@etri.re.kr
[3] Dept. of Electrical Engineering, Kyungpook National Univ., Korea
{sjmoon, hsyoo}@ee.knu.ac.kr

Abstract. ARIA is a 128-bit symmetric block cipher having 128-bit, 192-bit, or 256-bit key lengths. The cipher is a substitution-permutation encryption network (SPN) that uses an involutional binary matrix. This paper shows that a careless implementation of ARIA on smartcards is vulnerable to a differential power analysis attack. This attack is realistic because we can measure power consumption signals at two kinds of S-boxes and two types of substitution layers. By analyzing the power traces, we can find all round keys and also extract a master key from only two round keys using circular rotation, XOR, and involutional operations for two types of layers.

1 Introduction

In 1998, Kocher et al. first introduced power attacks including simple and differential power analysis(referred to as SPA and DPA, respectively) [1]. The power analysis attack extracts secret information by measuring the power consumption of cryptographic devices during processing. In SPA, we measure a single power trace of a cryptographic execution and analyze it to classify operations which are related to secret information. The DPA, a more advanced technique, allows observation of the effects correlated to the data values being manipulated with power consumption. Although DPA requires a more complex analysis phase, it is generally more powerful than SPA.

In fact, many papers have reported that secret key cryptosystems (AES and DES) as well as public key cryptosystems (RSA and ECC) are vulnerable to DPA [1,2,3,4]. In particular, DPA has been well studied for block ciphers that incorporate a nonlinear S-Box in their algorithm such as DES. It has been pointed out in [4] that a careless implementation of a block cipher is vulnerable to power analysis attack.

---
* This research has been supported by University IT Research Center Project.
** The first author has been supported by Korea Nazarene University research fund.

This paper will consider the ARIA (Academy, Research Institute, and Agency) block cipher, which has been developed as a national standard algorithm in Korea [5]. We will show that a raw implementation of a block cipher such as this can be broken by DPA. Our attack is targeted against an ARIA implementation on real smartcards. This paper is divided in two parts. In the first part, we show that ARIA is vulnerable to DPA. Because ARIA has two types of S-box in each round, we can apply the basic idea of DPA used in [2]. In the second part, we extract the master key (MK) using only two round key pairs. This paper also shows our experimental results of DPA on ARIA.

2 ARIA Algorithm

The ARIA 128-bit symmetric block cipher uses a 128-bit, 192-bit, or 256-bit secret key, where the number of rounds is then 12, 14 and 16, respectively. The cipher is an involution substitution and permutation encryption network. This algorithm uses a 128-bit input-output data block. The following notation is used to describe ARIA.

$S_i(x)$: The output of S-box $S_i(i = 1, 2)$ for an input x
$A(x)$: The output of the diffusion layer for an input x
$\oplus$: Bitwise exclusive OR (XOR)
$<<< n$: Left circular rotation by n bits
$>>> n$: Right circular rotation by n bits
$||$: Concatenation of two operands

2.1 Structure of ARIA

A 128-bit plaintext is XORed with the 1-round key ek_1. Each round of the cipher consists of the following three parts.

1. Round key addition: XORing with the 128-bit round key.
2. Substitution layer: Two types of substitution layers.
3. Diffusion layer: A simple involutional binary 16×16 matrix.

The more detail description can be referred in [5]

2.2 Key Scheduling

The key scheduling of ARIA consists two parts, initialization and round key generation.

The initialization part uses a 3-round Feistel cipher. Note that the master key (MK) can be of 128, 192, or 256 bits long. An input MK is separated into two 128-bit blocks (KL and KR). We first fill out KL with bits from the master key. Then, we use the remaining bits from the MK key in KR. The space remaining in KR is filled with all zeros. Then four 128-bit values W_0, W_1, W_2, and W_3 are generated from the MK as follows.

$$W_0 = KL, \quad W_1 = F_o(W_0, CK_1) \oplus KR$$
$$W_2 = F_e(W_1, CK_2) \oplus W_0, \quad W_3 = F_o(W_2, CK_3) \oplus W_1$$

In the round key generation phase, we obtain the encryption round key ek_i and decryption round key dk_i by combining the four W_i computed in the initialization phase. Recall that the number of rounds is 12, 14, or 16, depending on the size of the MK. Therefore, we need the 128-bit round keys in the 13th, 15th, or 17th round, the MK is 128, 192, or 256 bits in length respectively. The round keys are generated using the circular rotation and XOR operations.

3 DPA Attack on ARIA

3.1 Round Key Attack Using DPA

We have only considered the 12 round ARIA implementation as opposed to 14 or 16 round. In each of the 12 rounds, ARIA performs sixteen S-box lookup operations after the XOR operation with each round key. To begin with, we define a partitioning function, $D(P, b, rk_8)$. The b is a value after the S-box lookup operation and P is a plaintext. The 8 round key bits, rk_8, entering into the S-box generating output bit b are represented by $0 \leq rk_8 < 2^8$.

We have chosen N random plaintexts and measured the power consumption traces in the first round, for N ARIA encryptions where N is the number of measurements. We express these N input values as $P_1, ..., P_N$. Also let $T_{1t}, ..., T_{Nt}$ denote the N power consumption traces measured during the processing. The index t corresponds to the time of the sample. Our attack focuses on only one bit among the eight output bits of the first S-box during the processing of the first round. Let b be the value of that first bit. As shown in Figure 1, b depends on only 8 bits of the 1-round key. We make a hypothesis concerning these 8 round key bits and compute the expected values for b using these 8 bits and input values P_i. This enables us to separate the N inputs into two categories. One is a group where $b = 0$, the other where $b = 1$. And we split the data into two sets according to b as follows.

$$T_0 = \{T_{it} \mid D(P, b, rk_8) = 0\}, \quad T_1 = \{T_{it} \mid D(P, b, rk_8) = 1\} \quad (1)$$

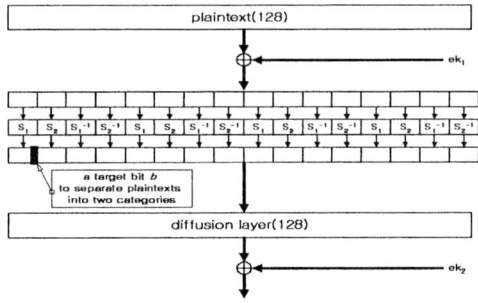

Fig. 1. The DPA attack in 1-round of ARIA

The next step is to compute the average of the power traces for each set as follows.

$$A_0[t] = \frac{1}{|T_0|} \sum_{T_{it} \in T_0} T_{it}, \quad A_1[t] = \frac{1}{|T_1|} \sum_{T_{it} \in T_1} T_{it} \qquad (2)$$

Here, the number of measurements in a trace, $N = T_0 + T_1$, depends on the sampling rate and noise influence according to measurement setup. The differential power trace of $A_0[t]$ and $A_1[t]$ is defined for $t = 1, ..., m$ as follows.

$$\Delta P[t] = A_1[t] - A_0[t]. \qquad (3)$$

If rk_8 is incorrect, the bit computed using D will differ from the actual target bit b. The partitioning function is uncorrelated to what was actually computed by the target device. Therefore, the difference $\Delta P[t]$ should approach zero as the number of measurements N approach infinity. On the other hand, if rk_8 is correct, the partitioning function will work well because the computed value for D will be equal to the actual value of the target bit b. For the other round keys, we repeat the procedure with a new target bit b in the second S-box, in the third S-box, and so on, until the final sixteenth S-box. As a result, we can find the 128 bits of the first round key. After obtaining the first round keys, we can find the second round keys in a similar way. This is reasonable because we know the outputs of the 1-round, that is, the inputs of 2-round. To find the 2-round keys, we measure the power consumption during the 2-round operations. By using this analysis method for each round, we can obtain all of the round keys.

3.2 Master Key Attack Using Partial Round Keys

As mentioned in the previous section, an adversary can find all the round keys during encryption processing by using power analysis. Now, let us examine how to search the MK given round keys. The round keys are generated using the circular rotation and XOR operations as follows.

$$\begin{aligned}
ek_1 &= (W_0) \oplus (W_1^{>>>19}), & ek_2 &= (W_1) \oplus (W_2^{>>>19}) \\
ek_3 &= (W_2) \oplus (W_3^{>>>19}), & ek_4 &= (W_0^{>>>19}) \oplus (W_3) \\
ek_5 &= (W_0) \oplus (W_1^{>>>31}), & ek_6 &= (W_1) \oplus (W_2^{>>>31}) \\
ek_7 &= (W_2) \oplus (W_3^{>>>31}), & ek_8 &= (W_0^{>>>31}) \oplus (W_3) \\
ek_9 &= (W_0) \oplus (W_1^{>>>61}), & ek_{10} &= (W_1) \oplus (W_2^{>>>61}) \\
ek_{11} &= (W_2) \oplus (W_3^{>>>61}), & ek_{12} &= (W_0^{>>>61}) \oplus (W_3) \\
ek_{13} &= (W_0) \oplus (W_1^{<<<31}), & ek_{14} &= (W_1) \oplus (W_2^{<<<31}) \\
ek_{15} &= (W_2) \oplus (W_3^{<<<31}), & ek_{16} &= (W_0^{<<<31}) \oplus (W_3) \\
ek_{17} &= (W_0) \oplus (W_1^{<<<19})
\end{aligned}$$

If a W_i is known then we can clearly extract other W_i using these key generation equations. For example, given W_0, we can compute $W_1 = (ek_1 \oplus W_0)^{<<<19}$, $W_2 = (ek_2 \oplus W_1)^{<<<19}$, and $W_3 = (ek_3 \oplus W_2)^{<<<19}$.

Now, we try to extract W_i from any of the round keys. In our attack, it should be pointed out that the above equations are just a set of linear equations in the bits of W_0, W_1, W_2 and W_3. Most of these equations are independent, hence measuring 4 ek_j values allows us to obtain all these W_i by simple elimination.

The another idea of this paper is to show that we can exploit the simple structure in these equations to break the system using 2 ek_j values rather than 4. Due to the particular structure, we can easily eliminate one of the variables to obtain the sums $W_1 \oplus W_3$ or $W_0 \oplus W_2$. We try to extract $W_1 \oplus W_3$ or $W_0 \oplus W_2$ instead of W_i. As shown in Figure 2, if we can compute $W_1 \oplus W_3$ then we can extract W_2 because of the involutional property of the two round functions F_o and F_e. In addition, if we can compute $W_0 \oplus W_2$ then extract W_1.

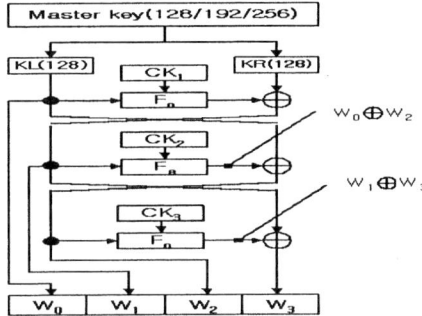

Fig. 2. Two computational points to find W_i

As an example, we try to compute $W_1 \oplus W_3$ or $W_0 \oplus W_2$ using two round keys, respectively. Each round key ek_{13}, ek_{15}, or ek_{17} is firstly rotated by 31 or 19 bits, and XORed with another round key to eliminate partial common information in two round keys. In similar way, the round key ek_{14} or ek_{16} is rotated by 31 bits and XORed with another round key. Furthermore, ARIA should use at least ek_8 and ek_{13} in encryption processing for the minimum MK size. Therefore we can extract the MK from all round keys using simple computations such as rotation, XOR and involutional operation for two layers.

Case 1 : Elimination of W_0 to find $W_1 \oplus W_3$.

$$ek_{13}^{>>>31} = W_0^{>>>31} \oplus W_1 = W_1 \oplus W_3$$

Case 2 : Elimination of W_2 to find $W_1 \oplus W_3$.

$$ek_{15}^{>>>31} = W_2^{>>>31} \oplus W_3 = W_1 \oplus W_3$$

Case 3 : Elimination of W_0 to find $W_1 \oplus W_3$.

$$ek_{17}^{>>>19} = W_0^{>>>19} \oplus W_1 = W_1 \oplus W_3$$

Case 4 : Elimination of W_1 to find $W_0 \oplus W_2$.

$$ek_{14}^{>>>31} = (W_1)^{>>>31} \oplus W_2 = W_0 \oplus W_2$$

Case 5 : Elimination of W_3 to find $W_0 \oplus W_2$.

$$ek_{16}{}^{>>>31} = W_0 \oplus W_3{}^{>>>31} = W_0 \oplus W_2$$

3.3 Master Key Attack Using Two Round Keys

After power analysis at 1-round for encryption of a plaintext, we can calculate the 1-round full keys. Additionally, after power analysis at 1-round for decryption of a ciphertext, we can calculate the final round full keys. Measuring the power differences is a simple and realistic assumption because an adversary needs only the first round keys for encryption and decryption. Note that it is not necessary to use a genuine ciphertext as an input because an adversary is only interested in the decryption operation.

Now, let us examine how to find the master key given the round keys ek_1 and ek_{13} of round 1 and 13, respectively.

$$ek_1 = W_0 \oplus W_1{}^{>>>19}, \quad ek_{13} = W_0 \oplus W_1{}^{<<<31} = W_0 \oplus W_1{}^{>>>97}$$

Therefore, if we know two round keys, then we can know a temporary value T as follows.

$$T = ek_1 \oplus ek_{13} = W_1{}^{>>>19} \oplus W_1{}^{>>>97}, \quad W_1 = T^{<<<19} \oplus W_1{}^{>>>78}$$

Figure 3 describes the bit position of the above three values. As shown in Figure 3, we assume that the first bit of W_1 is a fixed bit 0 or 1. Then we can decide $W_{1,79}$ because $T^{<<<19}$ is known. If $W_{1,79}$ is decided then we can also compute $W_{1,29}$. Here, $W_{1,i}$ means the ith bit of the W_1 and $(T^{<<<19})_i$ is the ith bit of $T^{<<<19}$. Consecutively, we can compute all odd position bits. Similarly, if the second bit of W_1 is a fixed bit 0 or 1, then we can decide all even bits. Finally, we can extract 4 values of W_1 by setting two bits $W_{1,1}$ and $W_{1,2}$. As you see, if W_i is known then we can clearly extract other W_i using key generation equations. Furthermore we can easily extract a MK from W_i as shown above description. This property used by this attack is applied for two round keys which are XORed with a value of W_i or its rotated value. For another example, we can compute the MK using $ek_1 = W_0 \oplus W_1{}^{>>>19}$ and $ek_{16} = W_0 \oplus W_1{}^{<<<19}$ because they can eliminate a value by XOR operation.

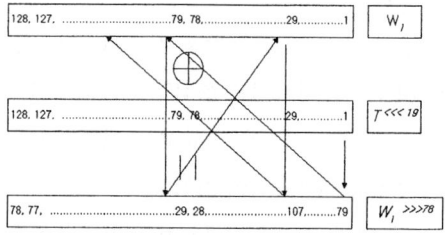

Fig. 3. Consecutive bit setting from the first bit

4 Experimental Results

In this section, we experiment with a DPA attack on a smartcard implementing ARIA. The experimental results show that ARIA is vulnerable to our DPA attack. The target smartcard of our attack has an ARM architecture based on a 32-bit CPU core and hardware crypto coprocessor. We need several experimental equipments, e.g. power supply, oscilloscope, function generator, computer simulator and so on. In the naive implementation on this smartcard, ARIA uses 12 rounds for a 128-bit key.

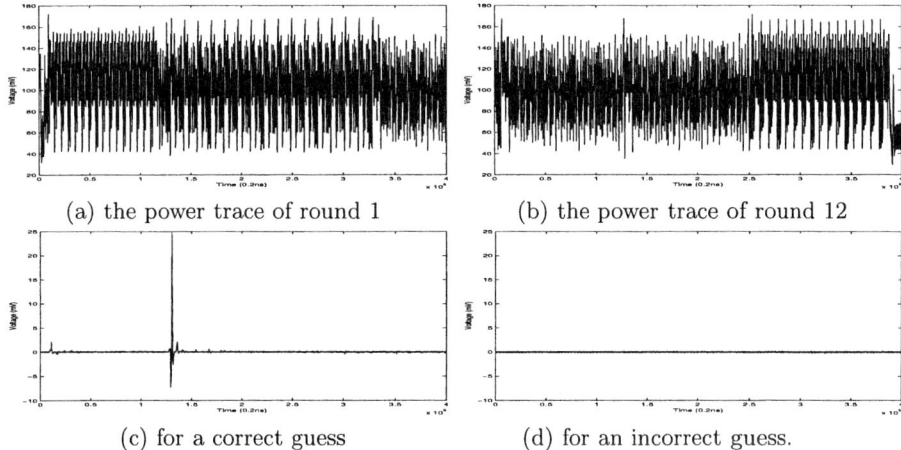

(a) the power trace of round 1   (b) the power trace of round 12

(c) for a correct guess   (d) for an incorrect guess

Fig. 4. Single and differential power traces

Figure 4(a) and 4(b) shows sample power traces of 1-round and 12-round while executing the above implementations. In Figure 4(a), it is easy to identify the portion of different operations, including the round key addition of a plaintext and a round key (approximately $0 \sim 1.2 \times 10^5$ns), the substitution by S-box (approximately 1.2×10^5ns $\sim 3.3 \times 10^5$ns), and the diffusion process (approximately 3.3×10^5ns $\sim 4.0 \times 10^5$ns). The total processing time for an encryption is about 4.8ms.

Our attack is developed by exploiting the power trace during the substitution operation prior to the diffusion operation. To generate a higher peak, we check all 8-bit outputs of the S-Box and separate the power traces into two groups based on the Hamming weight of the register after the S-box. Figure 4(c) and 4(d) show the differential power traces for a correct guess and an incorrect guess, respectively. The two power traces in Figure 4(c) and 4(d) were obtained to find an 128-bit round key by averaging 2000 random power traces

5 Conclusion

This paper has demonstrated that unprotected implementations of ARIA are vulnerable to a differential power attack. The basic concept of our power analysis is that we can separate the power traces according to the output bit of S-box used in ARIA. Furthermore, the power analysis experiment described in section 4 demonstrated the feasibility of a real attack. The useful power traces to find an 128-bit round key were obtained by averaging 2000 random power traces. Unfortunately, a master secret key can be obtained from an ARIA block cipher by combining only two round keys. Since the round key generation phase of ARIA uses only XOR and simple rotation operations, this allows the MK to be extracted from only two round keys. Additionally, Measuring the power differences is a simple and realistic because an adversary needs only the first round keys for encryption and decryption.

References

1. P. Kocher, J. Jaffe, and B. Jun, "Differential power analysis," *CRYPTO '99*, LNCS 1666, pp. 388–397, Springer-Verlag, 1999.
2. L. Goubin and J. Patarin, "DES and differential power analysis," *CHES '99*, LNCS 1717, pp. 158–172, Springer-Verlag, 1999.
3. T. Messerges, "Securing the AES finalists against power analysis attacks," *FSE '00*, LNCS 1978, pp. 150–164, Springer-Verlag, 2000.
4. S.B. Ors, F. Gurkaynak, E. Oswald, and B. Preneel, "Power-analysis attack on an ASIC AES implementation," *ITCC '04*, Volume II, pp. 546–552, 2004.
5. D. Kwon et al., "New Block Cipher : ARIA," *ICISC '03*, LNCS 2971, pp. 432–445, Springer-Verlag, 2003.
6. K. Tiri, M. Aknal, and I. Verbauwhede, "A dynamic and differential CMOS logic with signal independent power consumption to withstand differential power analysis on smartcards," *28th European Solid-State Circuits Conference*, pp. 403–406, 2002.
7. K. Tiri and I. Verbauwhede, "A logic level design methodology for a secure DPA resistant ASIC or FPGA implementation," *DATE '04*, pp. 246–251, 2004.
8. M. Akkar and C. Giraud, "An implementation of DES and AES, secure against some attacks," *CHES '01*, LNCS 2162, pp. 144–157, Springer-Verlag, 2001.
9. J.S. Coron and L. Goubin, "On boolean and arithmetic masking against differential power analysis," *CHES '00*, LNCS 1965, pp. 231–237, Springer-Verlag, 2000

A CRT-Based RSA Countermeasure Against Physical Cryptanalysis*

ChangKyun Kim[1], JaeCheol Ha[2,**], SangJae Moon[3], Sung-Ming Yen[4], and Sung-Hyun Kim[5]

[1] National Security Research Institute, Daejeon, Korea
kimck@etri.re.kr
[2] Korea Nazarene Univ., Cheonan, Choongnam, 330-718, Korea
jcha@kornu.ac.kr
[3] Kyungpook National Univ., Daegu, 702-701, Korea
sjmoon@knu.ac.kr
[4] National Central Univ., Chung-Li, Taiwan 320, R.O.C
yensm@csie.ncu.edu.tw
[5] System LSI Division, Samsung Electronics Co., Ltd., Korea
teri_kim@samsung.com

Abstract. This paper considers a secure and practical CRT-based RSA signature implementation against both side channel attacks (including power analysis attack, timing attack, and most specially the recent MRED attack) as well as the various CRT-based fault attacks. Moreover, the proposed countermeasure can resist C safe-error attack which can be mounted in many existing good countermeasures. To resist side-channel attack, a special design of random message blinding is employed. On the other hand, a countermeasure based on the idea of fault diffusion is developed to protect the implementation against the powerful CRT-based fault attacks.

1 Introduction

In this paper, we focus our attention on the CRT-based RSA private computation. The CRT-based implementation of RSA private computation can achieve a four times speedup and has already become very popular. However, a naive implementation of RSA with CRT may be vulnerable to both the CRT-based fault attack [1,3,4,5,11], the power attack [2,6,7], and the timing attack [8].

To prevent from side-channel attacks and CRT-based fault attack, some countermeasures have been reported [9,10]. However, these countermeasures for RSA with CRT suffer from either possibility of undetectable error, or become less compatible with existing standards or systems. Moreover, some papers have been repreted that they can be broken by a power attack [2,6,7].

---
* This research has been supported by University IT Research Center Project.
** The second author has been supported by Korea Nazarene University research fund.

The main contribution of this paper is that we propose a new countermeasure against CRT-based fault attacks and conventional side-channel attacks. This is more efficient and becomes totally compatible with existing RSA standards when compared with the previous countermeasures [9,10].

2 Preliminary

2.1 Vulnerability to CRT-Based Fault Attack

In the RSA with CRT (Chinese Remainder Theorem), we first compute $S_p = m^{d_p} \bmod p$ and $S_q = m^{d_q} \bmod q$, where $d_p = d \bmod (p-1)$ and $d_q = d \bmod (q-1)$. Then, S is computed by the following Gauss's recombination algorithm

$$S = (S_p \cdot q \cdot (q^{-1} \bmod p) + S_q \cdot p \cdot (p^{-1} \bmod q)) \bmod N \tag{1}$$

where both $q^{-1} \bmod p$ and $p^{-1} \bmod q$ can be pre-computed to reduce the computational load.

The CRT-based fault attack has attracted much attention for its practicality with only reasonable assumption. The CRT-based fault attack was firstly pointed out by Boneh et al. [3]. More precisely, from the faulty $\widehat{S}_p$ and the correct S_q, the faulty signature $\widehat{S}$ is computed as

$$\widehat{S} = (\widehat{S}_p \cdot q \cdot (q^{-1} \bmod p) + S_q \cdot p \cdot (p^{-1} \bmod q)) \bmod N. \tag{2}$$

Then, the following simple manipulation

$$gcd((S - \widehat{S}), N) = q \tag{3}$$

will give the secret prime q, so N is factorized. We call the above scenario the attack I.

An enhanced attack (we call it the attack II) proposed by Lenstra enables a CRT-based fault attack by providing the adversary only one faulty signature [5] under the same fault model. This enhanced attack factorize N by computing

$$gcd((\widehat{S}^e - m) \bmod N, N) = q \tag{4}$$

where e is the public exponent used to verify the signature S. This attack was later generalized in [4].

2.2 Shamir's Countermeasure

Shamir proposed a simple countermeasure against this fault attack [9]. In this countermeasure, a random prime r is selected then both $p' = p \cdot r$ and $q' = q \cdot r$ are computed. The following two partial signatures, modulo p' and q', are evaluated

$$S_p' = m^{d_p'} \bmod p', \quad S_q' = m^{d_q'} \bmod q'$$

where $d_p' = d \bmod ((p-1) \cdot (r-1))$ and $d_q' = d \bmod ((q-1) \cdot (r-1))$. Then, we check whether $S_p' \equiv S_q' \pmod r$. If the verification is correct, then it is assumed that no fault has occurred during the computation of both S_p' and S_q'. In this case the signature is obtained by computing

$$S = (S_p \cdot q \cdot (q^{-1} \bmod p) + S_q \cdot p \cdot (p^{-1} \bmod q)) \bmod N \qquad (5)$$

where $S_p = S_p' \bmod p$ and $S_q = S_q' \bmod q$.

However, the above method by Shamir has the following disadvantages. First, the probability that a fault cannot be detected is $1/r$. Second and most serious, since the size of moduli are extended from $|p|$ to $|p \cdot r|$ and also from $|q|$ to $|q \cdot r|$, respectively, the implementation of RSA with this countermeasure becomes incompatible with existing systems or platforms. Finally, Aumüller et al. described concrete results and practically approved countermeasures concerning fault attacks on RSA with CRT by using spikes attacks [1].

3 New Countermeasure Against CRT-Based Fault

3.1 The Proposed Countermeasure

Recall the fact that CRT-based fault attack is possible due to a fault occurred in either S_p or S_q but not both. This observation motivates us to develop the following *fault-diffusion* based countermeasure. The basic idea is that when a fault occurred only in $\widehat{S}_p$, then the fault will be extended into the CRT recombination algorithm and eventually makes both $\widehat{S} \neq S_p \pmod{p}$ and also $\widehat{S} \neq S_q \pmod{q}$ which disables the CRT-based fault attack. The proposed fault-diffusion based countermeasure against the CRT-based fault attack is provided in Fig. 1.

In Fig. 1, it is assumed that T_p, T_q, and T will be initialized to be random values to prevent a potential attack to skip the operation in step 2 and 3. A zero value in T will indicate error free in both computation of S_p and S_q. This assumption is reasonable since any memory or register content before being used is usually a random value. Therefore, a random value or any nonzero value can be explicitly assigned into T, T_p, and T_q initially.

Input: m, d, p, q, N
Output: S.

0. Assuming that T_p, T_q, and T are initialized to be random values.
1. $S_p = m^{d_p} \bmod p$ and $S_q = m^{d_q} \bmod q$
   where $d_p = d \bmod (p-1)$ and $d_q = d \bmod (q-1)$
2. $T_p = (m - S_p^{e_p}) \bmod p$ and $T_q = (m - S_q^{e_q}) \bmod q$
   where $e_p = e \bmod (p-1)$ and $e_q = e \bmod (q-1)$
3. $T = T_p \oplus T_q$ using XOR operation
4. $S = (S_p \cdot (q \oplus T) \cdot (q^{-1} \bmod p) + S_q \cdot (p \oplus T) \cdot (p^{-1} \bmod q)) \bmod N$
5. Check both $S \equiv S_p \pmod{p}$ and $S \equiv S_q \pmod{q}$
   If these verifications are correct, then output S.

Fig. 1. Proposed RSA implementation against CRT-based fault attack

Within the countermeasure in Fig. 1, if a faulty $\widehat{S}_p$ and a correct S_q are computed, then the proposed implementation will produce an incorrect signature S. Most importantly, in the proposed countermeasure, the generated erroneous result S cannot be exploited by the adversary to mount a CRT-based attack due to the result provided in the Theorem 1. The situation is the same if a faulty $\widehat{S}_q$ and a correct S_p are computed.

Theorem 1. *The proposed countermeasure in Fig. 1 can resist the CRT-based fault attack if the computation within the step 4 is error free.*

Proof. If either S_p or S_q is incorrect (but not both), then T_p or T_q will be a large random integer, respectively. As a result, $T = T_p \oplus T_q$ will also be a large random integer. Let $p' = p \oplus T$ and $q' = q \oplus T$. In this case, both $p' \neq p$ and also $q' \neq q$ and are both random integers. Suppose a faulty $\widehat{S}_p$ and a correct S_q are computed, then we have

$$\widehat{S} = (\widehat{S}_p \cdot q' \cdot (q^{-1} \bmod p) + S_q \cdot p' \cdot (p^{-1} \bmod q)) \bmod N. \tag{6}$$

Notice on all the following relationships that $q' \cdot (q^{-1} \bmod p) \not\equiv 1 \pmod{p}$, $p' \cdot (p^{-1} \bmod q) \not\equiv 1 \pmod{q}$, $p' \cdot (p^{-1} \bmod q) \not\equiv 0 \pmod{p}$, and $q' \cdot (q^{-1} \bmod p) \not\equiv 0 \pmod{q}$. Suppose the computation within the above step 4 is error free, then it leads to both $\widehat{S} \not\equiv S_p \pmod{p}$ and also $\widehat{S} \not\equiv S_q \pmod{q}$. This proves the claim.

3.2 The Enhancement on Exponentiation Against Side-Channel Attacks

In [2], a DPA on RSA with CRT was reported by exploiting a special category of input data called the equidistant data when performing modular reduction (abbreviated as MRED). Moreover, the computation of both S_p and S_q suffer to potential SPA if no further protection is considered.

In order to develop an implementation secure against conventional side-channel attacks and MRED attack, a countermeasure is proposed in Fig. 2. In this proposed countermeasure, random blinding technique on message m is

---

Input: m, d_p, p (or m, d_q, q)
Output: S_p (or S_q).
1. Select a random integer r and compute $r^{-1} \bmod p$
2. $C = m \cdot r \bmod p$
3. $Temp[0] = r^{-1} \bmod p$, $Temp[1] = m \cdot r^{-1} \bmod p$
4. for i from $n-1$ downto 0 do
4.1 $\quad C = C^2 \bmod p$
4.2 $\quad C = C \cdot Temp[d_{p_i}] \bmod p$
5. $C = C \cdot Temp[0] \bmod p$
6. Return($S_p = C$)

Fig. 2. The proposed enhanced exponentiation algorithm

suggested against MRED and potential timing attacks, and the algorithm also performs highly regular in each iteration in order to be secure against SPA.

For more detail on security and performance analysis, please refer to a full version of this paper.

4 Experimental Results

The target of attack is a 32-bit card chip with a crypto coprocessor of Montgomery multiplier to implement RSA with CRT speedup. The MRED attack was developed by exploiting the power trace during the period of processing the remainder prior to the CRT exponentiation. In the MRED attack, significant peaks within the differential power trace during the above mentioned period will be proportional to the Hamming distance of its related input values. Fig. 3 (a) and Fig. 3 (b) show the differential power traces for Hamming distance of one and eight, respectively. Based on the above observation of power trace and the method provided by MRED attack, RSA private prime p (or q) can be obtained. $\gcd(m, N) = p$.

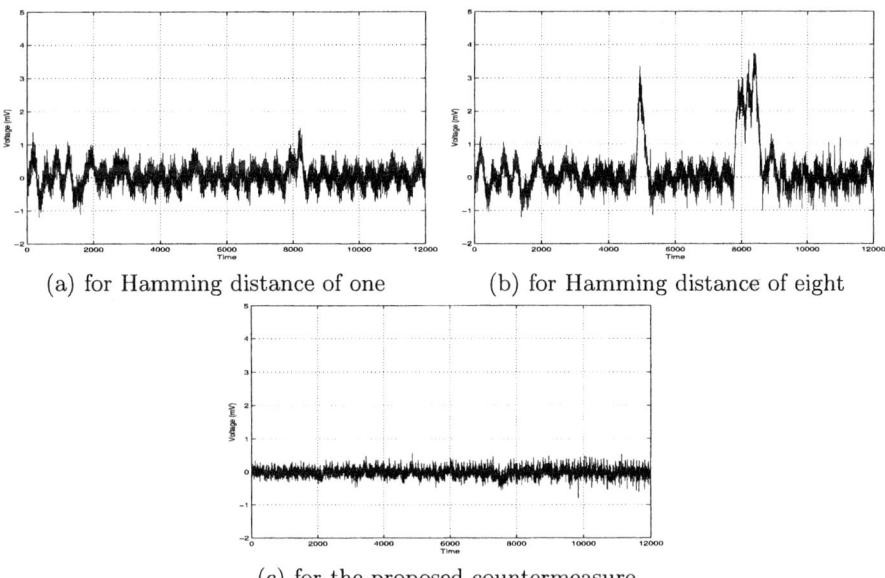

(a) for Hamming distance of one (b) for Hamming distance of eight

(c) for the proposed countermeasure

Fig. 3. Differential power traces during modular reduction

Experimental result provided in Fig. 3 (c) shows that the proposed countermeasure in Fig. 2 with random message blinding prior to modular reduction is secure against the MRED attack. In the differential power trace, no peak proportional to Hamming distance of data will be available to the attacker.

5 Concluding Remarks

Many attacks have been reported and some possible countermeasures have also been published. However, several countermeasures are not appropriate since some of them are incompatible with existing standards or hardware platform, and some others are not secure methods against side channel attacks, specially MRED and Novak's attack.

In this paper, a secure and practical countermeasure against well known side-channel attacks, the C safe-error attack, and the powerful CRT-based fault attack is proposed. This new countermeasure has low computational overhead when compared with existing ones. Besides enhanced security, it can be compatible with all hardware or software implementations on RSA with CRT.

References

1. C. Aumüller, P. Bier, W. Fischer, P. Hofreiter, and J.P. Seifert, "Fault attacks on RSA with CRT: concrete results and practical countermeasures," *CHES '02*, LNCS 2523, pp. 260–275, Springer-Verlag, 2002.
2. Bert den Boer, K. Lemke, and G. Wicke, "A DPA attack against the modular reduction within a CRT implementation of RSA," *CHES '02*, LNCS 2523, pp. 228–243, Springer-Verlag, 2002.
3. D. Boneh, R.A. DeMillo, and R.J. Lipton, "One the importance of checking cryptographic protocols for faults," *EUROCRYPT '97*, LNCS 1233, pp. 37–51, Springer-Verlag, 1997.
4. M. Joye, A.K. Lenstra, and J.-J. Quisquater, "Chinese remaindering based cryptosystems in the presence of faults," *Journal of Cryptology*, vol. 12, no. 4, pp. 241–245, 1999.
5. A.K. Lenstra, "Memo on RSA signature generation in the presence of faults," September 1996.
6. R. Novak, "SPA-based adaptive chosen-ciphertext attack on RSA implementation," *PKC '02*, LNCS 2274, pp. 252–262, Springer-Verlag, 2002.
7. K. Okeya and T. Takagi, "Security analysis of CRT-based cryptosystems," *ACNS '04*, LNCS 3089, pp. 383–397, Springer-Verlag, 2004.
8. W. Schindler, "A timing attack against RSA with the Chinese remainder theorem," *CHES '99*, LNCS 1717, pp. 292–302, Springer-Verlag, 1999.
9. A. Shamir, "How to check modular exponentiation," presented at the rump session of *EUROCRYPT '97*, Konstanz, Germany, May 1997.
10. S.M. Yen, S.J. Kim, S.G. Lim, and S.J. Moon, "RSA speedup with residue number system immune against hardware fault cryptanalysis," *ICISC '01*, LNCS 2288, pp. 397–413, Springer-Verlag, 2001.
11. S.M Yen, S.J. Moon, and J.C. Ha, "Permanent fault attack on the parameters of RSA with CRT," *ACISP '03*, LNCS 2727, pp. 285–296, Springer-Verlag, 2003.

The Approach of Transmission Scheme in Wireless Cipher Communication

Jinkeun Hong[1] and Kihong Kim[2]

[1] Division of Information and Communication, Cheonan University,
115 Anse-dong, Cheonan-si, Chungnam, 330-740, South Korea
jkhong@cheonan.ac.kr
[2] Graduate School of Information Security, Korea University,
1, 5-Ka, Anam-dong, Sungbuk-ku, Seoul, 136-701, South Korea
hong0612@hanmir.com

Abstract. This paper examines a cipher system for security in tactical network, plus an interleaving scheme is applied to the ciphered information to enhance the transmission performance over a fading channel. Experimental results showed that the BER performance of the proposed efficient interleaving scheme is higher than that of the fixed interleaving depth scheme. Of particular note is that the dynamic allocation algorithm (DAA, non-fixed interleaving depth) reduces degraded error bits by up to 53%, compared with static allocation algorithm (SAA, fixed interleaved depth) of depth 120 in 420MHz.

1 Introduction

Aviation industries are undergoing a major paradigm shift in the introduction of new network technologies [1]. Enhanced position location reporting system(EPLRS) will be fielded down to the Marine Corps infantry company level and is intended to be the primary means of secure, real time data distribution for sensor to shooter links. EPLRS uses synchronous time division multiple access (TDMA), frequency hopping, error correction coding, and embedded encryption to provide a secure transmission channel [2,3]. In previous studies about tactical networks, B. F. Donald [3] introduced digital messaging on the Comanche helicopter, the area of tactical data links, air traffic management, and software programmable radios has been investigated by B. E. White [4].

About the area of interleaving research, the subject of multiple access over fading multi-path channels employing chip interleaving code division direct sequence spread spectrum has investigated by Y. N. Link, et al. [5], the research of interleaving depth in Rayleigh fading channels has been proposed by I. C. King, et al. [6].

This paper presents a cipher system for security in tactical communication, plus an effective interleaving scheme is applied to the ciphered information to enhance the transmission performance over a fading channel. As such, a frame of ciphered information is lost if the synchronization pattern and session key for the frame are lost. Therefore, applying an interleaving method to reduce the frame loss and thereby enhance the transmission performance would seem to be an effective option that can be evaluated using the non-fixed efficient interleaving scheme.

The remainder of this paper is organized as follows. Section 2 reviews the nature of a fading channel and the cipher system. Thereafter, interleaving scheme based on a variable depth of interleaving using a non-fixed interleaving depth allocation algorithm is explained and simulation results presented in Section 3. Finally, section 4 summarizes the results of this study.

2 Characteristics of Wireless Mobile Environment

Wireless fading channel modeling is used to perform a statistical analysis based on defining the relational functions, such as the probability density function (PDF), cumulative probability distribution (CPD), level crossing rate (LCR), average duration of fades (ADF), and bit error rate (BER).

The equation of $CPD(F(L))$ for Rician fading is used as follows :

$$BER(\rho, K) = \frac{1+K}{2(\rho+1+K)} exp(\frac{-K\rho}{\rho+1+K}) \qquad (1)$$

In the above equation ρ is the C/N ratio and K is the power ratio of the direct wave and reflected waves. The crossings of the positive slopes are counted at level L. The total number of crossings N over a T second length of data divided by T seconds then becomes the level crossing rate :

$$n(L) = \frac{N}{T} \qquad (2)$$

As such, the level crossing rate of a typical fading signal can be calculated. The average duration of fades is defined as the sum of N fades at level L divided by N :

$$t(L) = \frac{\sum_{i=1}^{N} \tau_i}{N} \qquad (3)$$

Where τ_i is the individual fade.

$$F(L) = P(\gamma \leq L) = \frac{\sum_{i=1}^{N} \tau_i}{T} \qquad (4)$$

$$mbl = B \times t(L) \qquad (5)$$

Where mbl is mean burst length, B is date rate.

Now, the product of Eq. (2) and (3) becomes the CPD. From Eq. (2), (3) and (4), it can be derived mean burst length from Eq. (5) as in Table 1.

Where the variation of power deviations is down to -25dB, and the velocity of mobile device is 24Km/h. The structure of the cipher system with synchronization information includes ciphered information, consisting of a synchronization

Table 1. Mean burst length according to transmission rate

Transmission rate	Mean burst length			
	420MHz	430MHz	440MHz	450MHz
14.4Kbps	28.1bits	27.4bits	26.8bits	26.2bits
28.8Kbps	56.1bits	54.8bits	53.6bits	52.4bits
57.6Kbps	112bits	110bits	107bits	105bits

pattern, session key, and ciphered data within a period of synchronization. If the received synchronization pattern is detected normally, the error-corrected coded session key bit-stream is received and the ciphered data is deciphered.

3 Performance of DAA and Experimental Results

When ciphered information is transmitted over a Rician fading channel in which the received signal level is time variant, some of the ciphered information is lost due to burst errors, resulting a loss of the synchronization pattern and error in the session key in a period of synchronization.

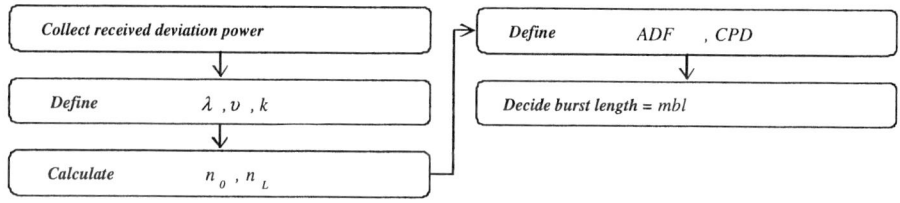

Fig. 1. Flow of dynamic allocation algorithm

Interleaving is an effective way of randomizing burst errors, plus, burst errors can not be corrected without the application of interleaving and de-interleaving. Where λ is wave length, v is mobile velocity, k is the power ratio of the direct wave and reflected waves. Also, n_0 is normalization factor $(= 2.5 \times v/c/f)$, f is frequency, c is wave speed, n_L is function of received power.

Table 2. Delayed time relative to depth of SAA

Transmission rate	Consumed time			
	30	60	120	240
14.4Kbps	1.000sec	2.000sec	4.000sec	8.000sec
28.8Kbps	0.500sec	1.000sec	2.000sec	4.000sec
57.6Kbps	0.250sec	0.500sec	1.000sec	2.000sec
115.2Kbps	0.125sec	0.250sec	0.500sec	1.000sec

Table 3. Low data error bits to depth of SAA (BER : 10^{-3})

Interleaving depth	Consumed time			
	420MHz	430MHz	440MHz	450MHz
SAA = 30	1.69×10^5	1.65×10^5	1.61×10^5	1.57×10^5
SAA = 60	1.44×10^5	1.41×10^5	1.38×10^5	1.35×10^5
SAA = 120	9.63×10^4	9.41×10^4	9.20×10^4	8.89×10^4

These interleaving schemes is evaluated in a simulation environment, the date rate is between 14.4Kbps and 115.2Kbps, the cipher communication time is 30minutes using bulk encryption, the BER is 10^{-3}, the moving velocity is 24Km/h. The performance of the DAA and SAA interleaving depth algorithms is then evaluated though simulations. Since the structure of interleaving basically depends on the interleaving depth(d), four types of DAA structure is used: $depth(d) \times span(S) = 30 \times 480, 60 \times 480, 120 \times 480, 240 \times 480$. It is explained delayed time relative to depth of SAA in Table 2. In case of the depth 60 in 28.8Kbps, the consumed time is about 1sec If the interleaving depth applied is smaller than the required depth, the resulting performance will be even worse than without interleaving. In condition of 10^{-3} bit error channel, the performance of the SAA is shown in Table 3 and Table 4, respectively. The error bits of the deciphered data without low data channel are degraded 47% at a SAA depth of 120, the error bits of the deciphered are degraded 20% at a SAA depth of 60 in 420 MHz. When the depth of the DAA is 30, 60, 120, 240, as shown in Table 4, the corrected bit rate in the DAA applied is higher than that of the other types.

Meanwhile, Table 3 presents a comparison of DAA and SAA with 430 iterations. When the delayed time when using DAA is about 802sec, however, the delayed time by the SAA depth of 30 is about 215sec, the SAA depth of 60 is 430sec, the SAA depth of 120 is 860sec. Therefore, when increasing the depth, the corrected bit rate and delayed time are enhanced. With regard to the delayed time and corrected bit rate, the performance of the proposed method is superior to that of SAA when applied to allow the delayed time of DAA.

Table 4. Comparison of DAA and SAA with 430 iterations (BER : 10^{-3}, 28.8Kbps)

Depth	Performance of corrected bits	Delay
DAA	100%	802sec
Depth = 30	7%	215sec
Depth = 60	20%	430sec
Depth = 120	47%	860sec
Depth = 240	100%	1,720sec

4 Conclusions

This paper examines a cipher system for security in tactical network, plus an interleaving scheme is applied to the ciphered information to enhance the transmission performance over a fading channel. A cipher system is proposed using an efficient interleaving scheme for the interleaving depth to enhance the transmission performance of the ciphered information.

Experimental results showed that the BER performance of the proposed efficient interleaving scheme is higher than that of the fixed interleaving depth scheme. Of particular note is that the DAA reduces degraded error bits by up to 53%, compared with SAA of depth 120 in 420MHz.

References

1. EUROCONTROL. Feasibility Study for Civil Aviation Data Link for ADS-B Based on MIDS/LINK 16. *TRS/157/02*, 2000.
2. R. Echevarria and L. L. Taylor. Co-site Interference Tests of JTIDS, EPLRS, SINCGARS, and MSE (MSRT). *Tactical Communications Conference*,1992.
3. B. F. Donald. Digital Messaging on the Comanche Helicopter. *DASC'00*, 2000.
4. B. E. White. Tactical Data Links, Air Traffic Management, and Software Programmable Radios. *DASC'99*, 1999.
5. Y. N. Link and D. W. Lin. Multiple Access Over Fading Multi-Path Channels Employing Chip Interleaving Code Division Direct Sequence Spread Spectrum. *IEICE Trans. Commun.*, 2001.
6. I. C. King and C-I C. Justin. Required Interleaving Depth in Rayleigh Fading Channels. *Globecom'96*, 1996.

A New Digit-Serial Systolic Mulitplier for High Performance $GF(2^m)$ Applications

Chang Hoon Kim[1], Soonhak Kwon[2], Chun Pyo Hong[3], and In Gil Nam[1]

[1] Dept. of Computer and Information Engineering, Daegu University,
Jinryang, Kyungsan, 712-714, Korea
[2] Dept. of Mathematics and Institute of Basic Science, Sungkyunkwan University,
Suwon, 440-746, Korea
[3] Dept. of Computer and Communication Engineering, Daegu University,
Jinryang, Kyungsan, 712-714, Korea
netsecurity@naver.com, shkwon@skku.edu, cphong@daegu.ac.kr

Abstract. This paper presents a new digit-serial systolic multiplier over $GF(2^m)$ for cryptographic applications. The proposed array is based on the most significant digit first (MSD-first) multiplication algorithm. Since the inner structure of the proposed multiplier is tree-type, critical path increases logarithmically proportional to D, where D is the selected digit size. Therefore, the computation delay of the proposed architecture is significantly less than previously proposed digit-serial systolic multipliers whose critical path increases proportional to D. Furthermore, since the new architecture has the features of regularity, modularity, and unidirectional data flow, it is well suited to VLSI implementations.

Keywords: Cryptography, Finite Field, Digit-Serial Multiplier, VLSI.

1 Introduction

In recent years, finite field $GF(2^m)$ has been widely used in various applications such as error-correcting code and cryptography [1, 2]. Important operations in $GF(2^m)$ are addition, multiplication, exponentiation, and division. Since addition in $GF(2^m)$ is bit independent XOR operation, it can be implemented in fast and inexpensive ways. The other operations are much more complex and expensive. This paper focuses on the hardware implementation of fast and low-complexity digit-serial multiplier over $GF(2^m)$, since computing exponentiation and division can be performed by repeated multiplications.

In this paper, we propose a new digit-serial systolic multiplier for high performance $GF(2^m)$ applications using a special irreducible polynomial (a polynomial of the form $G(x) = x^m + g_k x^k + \sum_{i=0}^{k-1} g_i x^i$, where $D \leq m - k$, D is the selected digit size). When input data come in continuously, the proposed array produces multiplication results at a rate of one every $N+2$ clock cycles, where $N = \lceil m/D \rceil$. Since the inner structure of the proposed array is tree-type, critical path increases logarithmically proportional to D. Therefore, the computation delay of the proposed architecture is significantly less than previously proposed digit-serial systolic multipliers whose critical path increases proportional to D. Furthermore, since the proposed architecture has the features of regularity, modularity, and unidirectional data flow, it is well suited to VLSI implementations.

2 A New Digit-Serial Systolic Array for Multiplication in $GF(2^m)$

2.1 MSD-first Multiplication Algorithm in $GF(2^m)$

Let $A = \sum_{i=0}^{m-1} a_i x^i$ and $B = \sum_{i=0}^{m-1} b_i x^i$ be two elements in $GF(2^m)$, $G = \sum_{i=0}^{m} g_i x^i$ be the irreducible polynomial used to generate the field $GF(2^m) \cong GF(2)[x]/G(x)$, and $P = \sum_{i=0}^{m-1} p_i x^i$ be the result of the multiplication $A \cdot B$ mod G, where the coefficients of each polynomial are binary digits 0 or 1. In addition, we define D and N such that D is a digit-size and N denotes the total number of digits with $N = \lceil m/D \rceil$. Then, we can derive the following MSD-first multiplication algorithm.

[Algorithm I] MSD-first Multiplication Algorithm in $GF(2^m)$
Input: $G = x^m + \sum_{i=0}^{m-1} g_i x^i, A = \sum_{i=0}^{m-1} a_i x^i, B = \sum_{i=0}^{N-1} B_i x^{Di}$,
$T = \sum_{i=0}^{m+D-1} t_i x^i = 0$
Output: $P = AB$ mod G
1.   for $i = 1$ to N do
2. $T^{(i)} = (T^{(i-1)} \bmod G)x^D + AB_{N-i}$
3.   end for
4. $P = T^{(N)}$ mod G

2.2 New Dependence Graph for Digit-Serial Multiplication in $GF(2^m)$

From Algorithm I, the main operations are $T^{(i)}=T^{(i-1)}x^D$ mod $G+AB_{N-i}$ and $P = T^{(N)}$ mod G. The critical path delay of the operations depend on selected irreducible polynomial. If we compute these operations using general irreducible polynomial, the critical path delay of the operation increases proportional to the selected digit-size D because of data dependency. On the other hand, we can reduce its critical path delay from proportional to D to logarithmically proportional to D using special irreducible polynomial (a polynomial of the form $G = x^m + g_k x^k + \sum_{i=0}^{k-1} g_i x^i$, where $D \leq m - k$). The detailed descriptions are founded in [3].

Let T'_i be the coefficients of $T^{(i)}$, then we can compute the T'_i using the following (1):

$$T'_i = \sum_{i=0}^{N} \sum_{j=0}^{D-1} (\sum_{k=1}^{D} (b_{D-k} a_{D(i-1)+j+k} + t^{(i-1)}_{m+D-k} g_{D(i-2)+j+k}) + t^{(i-1)}_{D(i-1)+j}) \quad (1)$$

where $A_N = G_{-1} = G_{-2} = 0$.

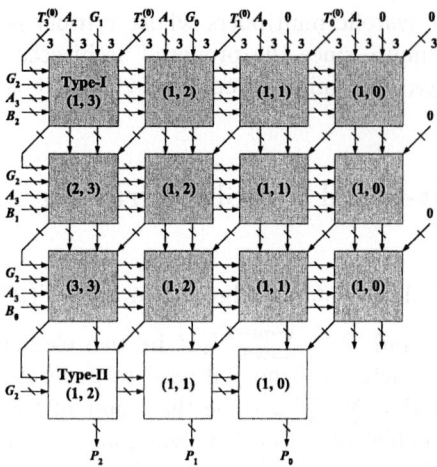

Fig. 1. A New digit-level DG for Multiplication in $GF(2^9)$ with $D=3$

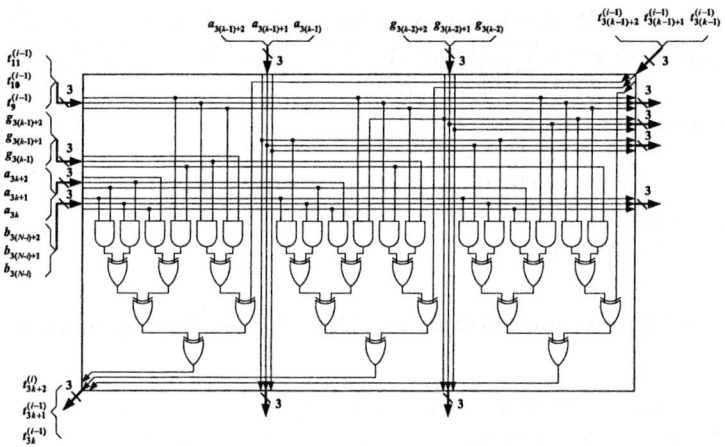

Fig. 2. Structure of Type-1 cell in Fig. 1

Let P' be the coefficients of P, then we can derive the following (2) for computing the $P = T^{(N)} \mod G$ operation.

$$P' = \sum_{i=1}^{N} \sum_{j=0}^{D-1} (\sum_{k=1}^{D} (t_{m-k}^{(N)} g_{D(i-1)+j+k}) + t_{D(i-1)+j}^{(N)}) \tag{2}$$

Based on (1) and (2), we can derive a new DG for multiplication in $GF(2^m)$ as shown in Fig. 1. The DG consists of $N \times (N+1)$ Type-1 cells and N Type-2 cells of digit-level. In Fig. 1, we assumed $m = 9$ and $D = 3$. The Type-1 cells in the i-th row of the array compute $T^{(i)} = T_{i-1} x^D \mod G + AB_{N-i}$ and the Type-2 cells perform $P = T^{(N)} \mod G$ operation respectively. The digit coefficients P_i

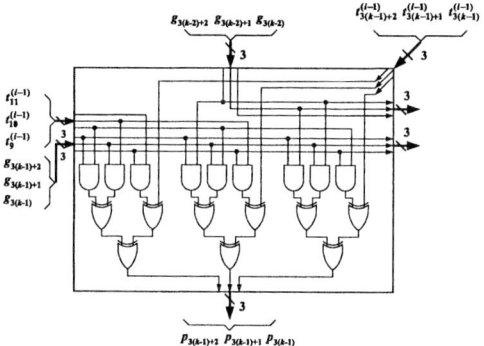

Fig. 3. Structure of Type-2 cell in Fig. 1

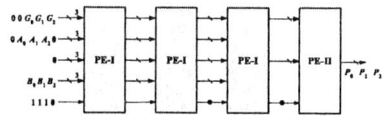

Fig. 4. One-dimensional SFG array corresponding to the DG in Fig. 1

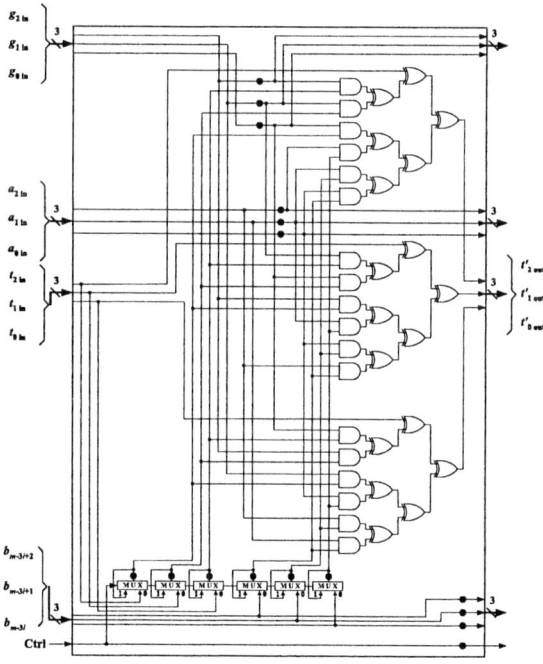

Fig. 5. Structure of PE-I in Fig. 4

of P emerge from the bottom row of the array after $(N+1)$ iterations. The structure of Type-1 and Type-2 cells are shown in Fig. 2 and Fig. 3 respectively.

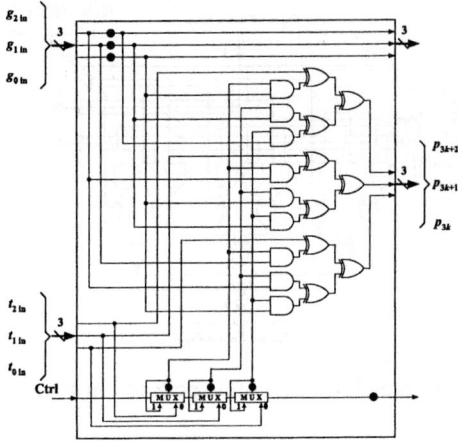

Fig. 6. Structure of PE-II in Fig. 4

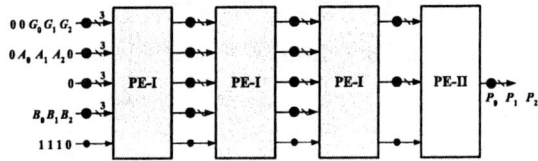

Fig. 7. A new digit-serial multiplier for $GF(2^m)$ with $m = 9$ and $D = 3$

2.3 A New Digit-Serial Systolic Array for Multiplication in $GF(2^m)$

Fig. 4 represents one dimensional signal flow graph (SFG) array for computing multiplication in $GF(2^m)$, where $m = 9$ and $D = 3$. As shown in Fig. 4, it consists of N units of processing element-I (PE-I) and one PE-II, and is controlled by a control sequence of $011\cdots 1$ with length $N+1$. The circuit of each PE is depicted in Fig. 5 and Fig. 6 respectively, where '•' denotes 1 cycle delay element. As shown in Fig. 1, since the coefficients of $T_{N-1}^{(i-1)}$ and B_{N-i} must be broadcasted to all basic cells in the i-th row of the DG in Fig. 1, we add extra $2D$ multiplexers and $2D$ one-bit latches into each PE-I, and D multiplexers and D one-bit latches into PE-II. When the control signal is in logic 0, these data are latched. By applying the cut-set systolisation techniques [6], we can obtain a new digit-serial systolic array for multiplication in $GF(2^m)$ depicted in Fig. 7. If input data come in continuously, this array produces multiplication results at a rate of one every N clock cycles after an initial delay of $3N + 2$ cycles.

3 Results and Conclusions

To verify the functionality of the proposed array in Fig 7, it was developed in VHDL and was synthesized using the Synopsis' FPGA-Express (version 2000,11-

FE3.5), in which Altera's EP2A70F1508C-7 was used as the target device. After synthesizing the circuits successfully, we extract net-list files from the FPGA-Express and simulated with the net-list files using Mento graphics' design view (VHDL-ChipSim). After verifying the functionality of the proposed array in Fig 7, we compared our architecture with some related systolic arrays having the same I/O format.

Table 1. Comparison with previously proposed digit-serial systolic multipliers for $GF(2^m)$

	Guo et al. [4]	Kim et al. [5]	Fig. 11
Throughput (1/cycles)	$1/N$	$1/N$	$1/N$
Latency (cycles)	$3N$	$3N$	$3N+2$
Circuit Requirement	$AND_2 : N(2D^2+D)$ $XOR_2 : 2ND^2$ Latch : $10ND$ $MUX_2 : 2ND$	$AND_2 : 2ND^2$ $XOR_2 : 2ND^2$ Latch : $8ND+4D$ $MUX_2 : 2ND$	$AND_2 : N(2D^2+D)$ $XOR_2 : 2ND^2$ Latch : $10ND+2D$ $MUX_2 : 2ND$
Critical Path	$T_{AND2}+3T_{XOR2}+$ $(D\text{-}1)(T_{AND2}+$ $2T_{XOR2}+T_{MUX2})$	$T_{AND2}+T_{XOR2}+$ $(D\text{-}1)(T_{AND2}+$ $T_{XOR2}+T_{MUX2})$	$T_{AND2}+\log_2(2D+1)T_{XOR2}$
Control Signal	1	1	1

Table 1 summarizes the performance comparison results, which assumed that 3-input XOR gate and 4-input XOR gate are constructed using two and three 2-input XOR gates respectively. As described in Table 1, since the inner structure of the proposed array is tree-type, critical path increases logarithmically proportional to D. Therefore, the computation delay of the proposed architecture is significantly less than previously proposed digit-serial systolic multipliers whose critical path increases proportional to D. As a result, the proposed architecture has two major advantages: 1) it has significantly less computation delay than previous architectures, and 2) if the proposed architecture is applied to ECC, which require large field size, we can select various digit size. Thus, by choosing the digit size appropriately, we can meet the throughput requirement with minimum hardware complexity. Furthermore, since the multiplier has the features of regularity, modularity, and unidirectional data flow, it is well suited to VLSI implementations.

Acknowledgement. This work was supported by the Daegu University Brain Korea IT Division in 2005.

References

1. R. E. Blahut, *Theory and Practice of Error Control Codes*, Reading, MA: Addison-Wesley, 1983.
2. I. F. Blake, G. Seroussi, and N. P. Smart, *Elliptic Curves in Cryptography*, Cambridge University Press, 1999.

3. L. Song and K. K. Parhi, "Low Energy Digit-Serial/Parallel Finite Field Multipliers," *J. VLSI Signal Processing*, vol. 19, no. 2, pp. 149-166, June 1998.
4. J. H. Guo and C. L. Wang, "Digit-Serial Systolic Multiplier for Finite Field $GF(2^m)$," *IEE Proc. Comput. Digit. Tech.*, vol. 145, no. 2, pp. 143-148, Mar. 1998.
5. C.H. Kim, S.D. Han and C.P. Hong, "An Efficient Digit-Serial Systolic Multiplier for Finite Fields $GF(2^m)$", *Proc. on 14th Annual IEEE International Conference of ASIC/SOC*, pp. 361-365, 2001.
6. S. Y. Kung, *VLSI Array Processors*, Englewood Cliffs, NJ: Prentice Hall, 1988.

A Survivability Model for Cluster System Under DoS Attacks

Khin Mi Mi Aung[1], Kiejin Park[2], Jong Sou Park[1],
Howon Kim[3], and Byunggil Lee[3]

[1] Computer Engineering Dept., Hankuk Aviation University
{maung, jspark}@hau.ac.kr
[2] Division of Industrial and Information System Engineering, Ajou University
kiejin@ajou.ac.kr
[3] ETRI (Electronics and Telecommunications Research Institute)
{khw, bglee}@etri.re.kr

Abstract. Denial of service attacks (DoS) don't necessarily damage data directly, or permanently but intentionally compromise the functionality. Even in an intrusion tolerant system, the resources will be fatigued if the intrusion is long lasting because of compromising iteratively or incrementally. In due course, the system will not provide even the minimum critical functionality. Thus we propose a model to increase the cluster system survivability level by maintaining the essential functionality. In this paper, we present the cluster recovery model with a software rejuvenation methodology, which is applicable in security field and also less expensive. The basic idea is - investigate the consequences for the exact responses in face of attacks and rejuvenate the running software/service, or/and reconfigure it. It shows that the system operates through intrusions and provides continued the critical functions, and gracefully degrades non-critical system functionality in the face of intrusions.

1 Introduction

DoS attacks don't necessarily damage data directly, or permanently but intentionally compromise the functionality. If the systems could not keep functioning, productivity can be degraded, though nothing has been damaged. Now operating systems provide few protections from DoS attacks. The system patches are also available. But there are lots of vulnerable systems with known vulnerabilities even the patches are already posted. From slammer worm attack, we have learned that even the known vulnerability and there is a patch, it infected at least 75,000 hosts, perhaps considerably more.

The nature of attacks is very dynamic because attackers have the specific intention to attack and well prepare their steps in advance. So far no respond technique able to cope with all types of attacks has been found. In most attacks, attackers overwhelm the target system with a continuous flood of traffic designed to consume all system resources, such as CPU cycles, memory, network bandwidth, and packet buffers. These attacks degrade service and can eventually lead to a complete shutdown. In this work, we address attacks mainly related to CPU usage, physical memory and swap space usage, running processes, network flows and packets. It will automatically detect potential weaknesses and reconfigure with attack patterns, which are characterizing an individual type of attack and attack profiles. We had analyzed the attack datasets and injected the attacks events into a system, and learned the prior knowledge. The next step is to restore the system to a healthy state within a set time following the predictive alerts.

In the current literature, there are significant numbers of researches, which are mainly concerned with survivability analysis but with different models [1],[2],[3], [4],[5]. In this paper, we present a novel model, using a software fault tolerance technique, software rejuvenation, to ensure an acceptable level of availability under DoS attacks.

2 Proposed Model

Significant features of various system resources may differ between specific attacks. And the response and restore methods would differ as well. In this work, the system has divided into three stages; healthy stage, restoration stage and failure stage. The model consists of five states: healthy state (H), infected state (I), rejuvenation state (R_j), reconfiguration state (R_c) and failure state (F). The healthy state represents the functioning and service providing phases. In the healthy stages, the systems aware to resist by various policies and offer proactive managements.

At the rejuvenation performing state, we need to be able to trade off between the risk of further damage against the shutting the system in an emergency stage. To this end, software rejuvenation methodologies are performed by the policies. The main strategies are occasionally stopping the executing software, cleaning the internal state and restarting by means of effectiveness of proactive managements, degrading mechanism, service stop, service restart, reboot and halt.

At the restoration stage, they may be decomposed into three types according to their specific attacks such as performing rejuvenation only, performing reconfiguration only and performing both rejuvenation and reconfiguration. For example, if an attacker carries out attack by overloading processes, causing resources to become unavailable, we will perform a rejuvenation process by gracefully terminating processes causing the resource overload and immediately restarting them in a clean state. But for the other kinds of attacks, we have to reconfigure the system according per their impact. In this case we have considered the reconfiguration state with various reconfiguration mechanisms, such as patching (operating system patch, application patch), version control (operating system version, application version), anti virus (vaccine), access control (IP blocking, port blocking, session drop, contents filtering), and Traffic control (bandwidth limit). As an example, performing rejuvenation only could deter the attacks, which cause the process degradation such as spawn multiple processes, fork bombs, CPU overload etc.

For the cases of process shutdown and system shutdown attacks, the attackers intend to halt a process or all processing on a system. Normally it happens by exploiting a software bug that causes the system to halt could cause system shutdown. In this case, just as with software bugs that are used to penetrate, so until the software bug is reconfigured, all systems of a certain type would be vulnerable. An example of attacks called mail bombardment or mail spam, the attacker accomplishes this attack by flooding the user with huge message or with very big attachments. Depending on how the system is configured, this could be counteracted by performing both reconfiguration and rejuvenation processes. To perform the various reconfiguration mechanisms, we have implemented the event manager, which contains the various strategies with respect to the various impact levels of the specific infected cases. Each type of event has its own routine, to be run when the attack takes place.

3 Steady-State Analysis with Two Nodes Through Semi-Markov Process

When only one of the states in the diagram violates the memoryless property, which means that soj0urn time in a state does not follow exponential distribution, the diagram is classified as a semi-Markov process. Semi-Markov models contain a Markov chain, which describes the stochastic transitions from state to state, and transition or 'sojourn' times, which describe the duration that the process takes to transition from state to state. We address the survivability

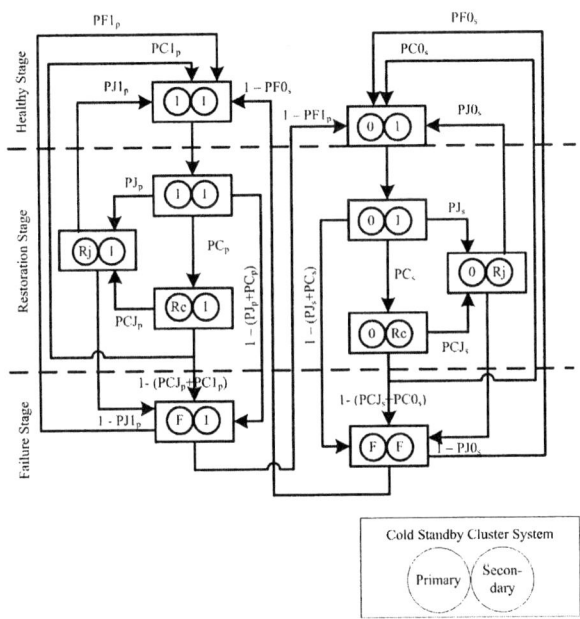

Fig. 1. Proposed Model with cold standby cluster

model with semi-Markov process. We consider a cold standby cluster with two nodes through Semi-Markov process. One node is as an active (primary) and other as a standby (secondary) unit. The failure rate of the primary node and secondary node are different, and also the effect of failure of the primary node is different from that of secondary node. The state transition diagram is shown in Figure 1. Initially the system is in state (1,1). In the infected state, the system has to figure out whether rejuvenate or reconfigure to recover or limit the damage that may happen by an attack. If the primary node has to reconfigure, the system enters state $(R_c, 1)$ otherwise enters rejuvenation state $(R_j, 1)$. If both strategies fail then the primary system enters the fail state. When the primary is infected by active attacks, the system enters state (I, 1). When the primary node fails a protection switch successfully restores service by switching in the secondary unit, and the system enters state (0,1). If the node failure occurs when the system is in one of the states : (0,I) or $(0, R_c)$, the system fails and enters state (F,F). To calculate the steady-state availability of the proposed model, the stochastic process of equation 1 was defined.

$$X(t) : t > 0 \qquad (1)$$

$XS = \{(1,1), (I,1), (R_j, 1), (R_c, 1), (F, 1), (0, 1), (0, R_j), (0, R_c), (F, F)\}$ Through SMP (Semi-Markov Process) analysis applying M/G/1, whose service time is general distribution; we have calculated the steady-state probability in each state. Healthy Stage: $\pi_{1,1} + \pi_{0,1}$ Restoration Stage: $\pi_{R_j,1} + \pi_{R_c,1} + \pi_{0,R_j} + \pi_{0,R_c}$ Failure Stage: $\pi_{F,1} + \pi_{F,F}$ As all the states shown in Figure 1 are attainable to each other, they are irreducible. Additionally, as they do not have a cycle and can return to a certain state, they satisfy the ergodicity (Aperiodic, Recurrent, and Nonnull) characteristics. Therefore, there is a probability in the steady-state of SMP for each state and each corresponding SMP can be induced by embedded DTMC (Discrete-time Markov Chain) using transition probability in each state. If we define the mean sojourn times in each state of SMP as $h'_i s$ and define DTMC steady-state probability as $d'_i s$, the steady-state probability in each state of SMP π_i can be calculated by equation 2 [6].

$$\pi_i = \frac{d_i h_i}{\sum_j d_j h_j}, i, j \in XS \qquad (2)$$

The system availability in the steady-state is defined as equation 3, which is the same as the exclusion of the probability of being in (F, 1) and (F, F) in the state transition diagram.

$$A = 1 - (\pi_{F,1} + \pi_{F,F}) \qquad (3)$$

The cluster systems are not survive in all of the rejuvenation process in the normal state (1), all of the switchover states, and the failure state (0). The survivability of cold standby cluster systems is defined as follows:

$$S = A - ((1 - \pi_{R_j,1}) + (1 - \pi_{R_c,1}) + (1 - \pi_{0,R_j}) + (1 - \pi_{0,R_c})) \qquad (4)$$

4 Numerical Results

We perform the experiments using the same system operating parameters with [7]. Failure rate of the server is one time per year and repair time is fifteen hours. Rejuvenation is scheduled at every month. The rejuvenation and switchover time are 10 and 3 minutes, respectively. The expected downtime cost per unit is 100 times greater than that of the scheduled rejuvenation cost. The number of servers is varied from simplex to multiplex (n=4), at the same time we perform software rejuvenation with the interval from 10 days (rate = 3) to infinity (rate=0: no rejuvenation). The results obviously show that the amount of survivability level increment from simplex to duplex is significant but from duplex to multiplex very little is shown. As infected states are removed frequently with high rejuvenation rates, the survivability of the cluster systems with simplex configuration increases. However, as the degree of redundancy is larger than or equal to 3, the improvement of availability is not significant. According to the required survivability level, the decision making of a rejuvenation rate is possible under consideration of various evaluation criteria such as state probabilities and downtime cost.

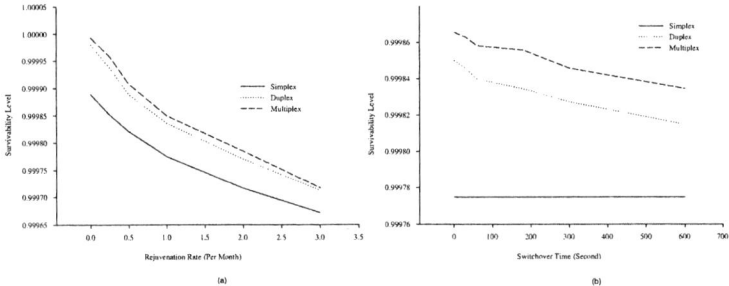

Fig. 2. (a) Survivability level changes due to rejuvenation (b) Survivability level changes due to switch over time

In the duplex configuration, failure rates are less sensitive to rejuvenation rate for availability. These results suggest that software reliability is more important than hardware reliability in improvement of the survivability of cluster systems. Figure 2 (a) and (b) show the relationships between rejuvenation rate and switch over time with survivability level. When switchover time is less than 15 minutes, a high rejuvenation rate is beneficial for improving survivability level. However when switchover time exceeds 15 minutes, frequent rejuvenation is not beneficial. Due to this fact, switch over time must be considered carefully when determining the rejuvenation policy.

5 Conclusion

We propose a model to increase the cluster system survivability level by maintaining the essential functionality. The result shows that the system operates through intrusions and provides continued the critical functions, and gracefully degrades non-critical system functionality in the face of intrusions. We have demonstrated the model can be used to analyze and proactively manage the effects of cluster network faults and attacks, and recover accordingly. We have analyzed the survivability level of cluster systems built with loosely coupled commercially available personal computers. According to the system operating parameters, we have modeled and analyzed steady state probability and survivability level of cluster systems under DoS attacks by adopting a software rejuvenation technique. We have validated the closed-form solutions of the mathematical model with experiments based on the above parameters. We have also found that software rejuvenation can be used as a preventive fault-tolerant technique and it improves the survivability of cluster system. As an ongoing work, we are performing our model with the real sojourn times of specific attacks in order to generalize it with various attacks. The integration of response time and throughput with downtime cost will provide a more accurate evaluation measure.

Acknowledgment

This work is supported by ETRI(Electronics and Telecommunications Research Institute) research fund.

References

1. S. Jha and J. Wing: Survivability Analysis of Networked Systems, Proc. of the 23rd International Conference on Software Engineering, IEEE, pp.872-874, 2001.
2. S. Nikolopoulos, A. Pitsillides and D. Tipper: Addressing Network Survivability Issues by Finding the Kbest Paths through a Trellis Graph, 16th Annual Joint Conference of the IEEE Computer and Communications Societies, Vol. 1, pp.370-377, 1997.
3. S. Liew and K. Lu: A Framework for Network Survivability Characterization, IEEE International Conference on Communications, pp.441-451, 1992.
4. K. Newport: Incorporating Survivability Considerations Directly into the Network Design Process, 9th Annual Joint Conference of the IEEE Computer and Communication Societies, pp.1963-1970, 1990.
5. Moitra, D. Soumyo and S. Konda: A Simulation Model for Managing Survivability of Networked Information Systems, SEI, Dec 2002.
6. K. Trivedi: Probability and Statistics with Reliability Queueing and Computer Science Applications, John Wiley and Sons, Inc. 2003.
7. S. Garge, "Analysis of Preventive Maintenance in Transactions Based Software Systems," IEEE Transactions on Computers Vol. 47(1), pp.96-107, 1998.

A Loop-Aware Search Strategy for Automated Performance Analysis

Eli D. Collins and Barton P. Miller

Computer Sciences Department, University of Wisconsin,
Madison, WI 53706-1685, USA
{eli, bart}@cs.wisc.edu

Abstract. Automated online search is a powerful technique for performance diagnosis. Such a search can change the types of experiments it performs while the program is running, making decisions based on live performance data. Previous research has addressed search speed and scaling searches to large codes and many nodes. This paper explores using a finer granularity for the bottlenecks that we locate in an automated online search, i.e., refining the search to bottlenecks localized to loops. The ability to insert and remove instrumentation on-the-fly means an online search can utilize fine-grain program structure in ways that are infeasible using other performance diagnosis techniques. We automatically detect loops in a program's binary control flow graph and use this information to efficiently instrument loops. We implemented our new strategy in an existing automated online performance tool, Paradyn. Results for several sequential and parallel applications show that a loop-aware search strategy can increase bottleneck precision without compromising search time or cost.

1 Introduction

Performance analysis tools aid in the understanding of application behavior. Automating the search for performance problems enables non-experts to use these tools and provides a fast diagnostic for experienced performance analysts [9,4,17,7]. A performance tool that uses dynamic instrumentation [10] can search for and identify performance problems in an unmodified program while the program runs. Our previous research in online automated search has addressed scaling with large codes, lowering instrumentation cost, and locating bottlenecks quickly [2,19,11].

This paper describes an online automated search strategy that increases the precision (granularity at which bottlenecks are identified) of an automated performance search. To demonstrate the efficacy of our strategy, we have implemented it in an existing automated performance tool, the Paradyn Parallel Performance tool [15].

An online search strategy manages its search space by focusing on functions that are currently being executed or functions that are about to be executed. To this end the Paradyn performance tool's automated search component, the Performance Consultant, uses a program's callgraph to locate bottlenecks [2].

This topdown search of program behavior matches the process an experienced programmer might use. This helps limit the amount of instrumentation inserted into a program, which improves search scalability (the ability to operate efficiently on programs with large code sizes). Searching a program's call graph identifies performance problems at function granularity.

While identifying bottlenecks at function granularity is useful in practice, the user of a performance tool would often like to know where inside the function the bottleneck is located. A large function may contain multiple distinct bottlenecks. A small function may be a bottleneck because it is called repeatedly in a loop. To better localize a bottleneck, an automated search strategy must search inside a function. This requires introducing a new level in the callgraph that improves search precision. This level must not inhibit search scalability; just adding new levels to the callgraph increases the size of the search space. The level should represent a program structure that logically decomposes a function for the user, more precisely locates bottlenecks, and partitions functions for searching. Augmenting a program's callgraph with nested loops meets these requirements. When searching code, after functions, loops are the the next natural program decomposition. Loops increase precision in several ways: (1) they may be bottlenecks themselves, especially in long running programs and scientific applications, (2) they help identify which callsites in a function are bottlenecks, and (3) they logically decompose a function for a user.

Applications often contain loops that execute for many iterations, and loops are natural sources of parallelism that both compilers and hardware exploit. For example, the OpenMP `parallel do` directive [22] allows a compiler to automatically parallelize a loop. Loop-level performance data provides valuable feedback for these optimizations.

Even if a loop is not a bottleneck, it can help to more precisely locate a bottleneck. Suppose function A contains multiple calls to bottleneck function B. If the calls to B occur at different levels in A's loop hierarchy then we might infer which calls to B within A are responsible for the bottleneck from the performance data collected for A's loops. Figure 1 depicts the callgraph for these functions with and without loop information. Without loops, we can determine if A contains a bottleneck and if that bottleneck is B. With loops, we can determine if the bottleneck in A is due to a particular loop, or a particular call to B.

Loops naturally decompose functions for a user. A search that specifies bottlenecks at loop precision is appropriate because the user can easily examine

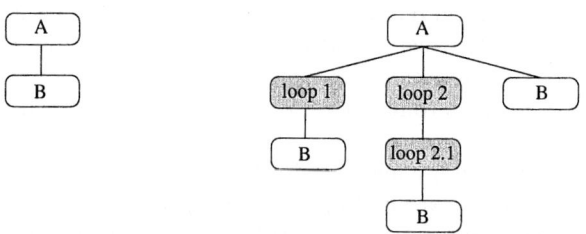

Fig. 1. A callgraph, and the same callgraph augmented with nested loops

and change the loop structure of a program. Loops also naturally decompose functions for searching. A loop may contain multiple nested loops and callsites. If a search uses an inclusive performance metric and determines that a loop is not a bottleneck then it does not have to instrument the loop's nested loops and callsites. Improving bottleneck precision can lower instrumentation cost, as we show in our experiments.

The contributions of this paper are (1) an automated search strategy that increases precision, (2) the definition of points in a function's control flow graph that correspond to loop execution, and (3) new dynamic instrumentation techniques that enable the efficient instrumentation of these points.

2 Related Work

Traditional profilers report performance data at function or statement granularity using sampling. The prof and gprof [6] profilers record flat function and callgraph profiles, respectively. Intel's VTune^{TM} Performance Analyzer [21] provides callgraph, statement, and instruction level profiles. Sampling enables data collection at a fine granularity, but only provides CPU time. For example, it is difficult to collect inclusive CPU time, elapsed time, and synchronization or I/O blocking time using sampling [19].

Some tools automate the search through a problem space similar to Paradyn's Performance Consultant. For example, Helm et al [7] use heuristic classification to control an automated search for performance problems. Finesse [17] diagnoses performance problems using search refinement across multiple runs.

Several performance tools report performance data at loop granularity. Sv-Pablo [18] allows users to instrument source code and browse runtime performance data. SvPablo correlates performance data with program source code at the level of statements, loops, and functions. Unlike Paradyn, which instruments a program's binary, SvPablo inserts instrumentation into the program source code using a preprocessor that can instrument functions and loops. Instrumenting the program's source reflects the code as the programmer wrote it, but may not fully reflect the code that was generated by the compiler.

MTOOL [5] is a tool for isolating memory performance bottlenecks. It uses a program's basic block count profile to identify frequently executed blocks to instrument. MTOOL instruments basic blocks with explicit timer calls, and aggregates basic block data to report loop, function, and whole program overheads.

HPCView [14] is a toolkit for combining multiple sets of profile data, and correlating this data with program source code. Aggregate performance data and derived metrics are presented at both function and loop levels. HPCView uses binary analysis to correlate performance data from program structures resulting from compiler transformations like loop fusion and distribution. Unlike Paradyn, which performs online automated performance analysis, HPCView is a post-mortem tool that combines the results of several program runs.

The DPOMP tool [3] uses dynamic instrumentation to collect performance data for OpenMP constructs, including `parallel do` loops. The compiler transforms OpenMP directives into function calls, which DPOMP instruments. Due to compiler optimizations, they can not always collect performance data for loop begin and end iteration events. Dynamic binary loop instrumentation enables the collection of these events by identifying loops through control flow rather than function calls inserted by the compiler.

The bursty tracing profiler [8] statically instruments a program to capture temporal profiles. To limit the overhead of their instrumentation they use counter-based sampling to switch between instrumented and un-instrumented copies of the binary. They can eliminate many checks (to switch between copies of the binary) by analyzing the program's callgraph and binary. They further limit overhead by not instrumenting "k-boring" loops (loops with no calls and at most k profiling events of interest). The bursty tracing profiler collects performance data online; performance analysis is handled offline. Our approach performs automated online performance analysis: we insert and remove instrumentation at run time based on the current performance data. We limit the overhead of fine-grain instrumentation by activating it only when necessary and using performance data to decide which parts of the program to instrument.

3 The Performance Consultant

The Performance Consultant is Paradyn's automated bottleneck detection component. It searches for application bottlenecks by performing experiments that measure application performance and try to locate where performance problems occur. Initially, the experiments measure the performance of the entire application. If the Performance Consultant discovers bottlenecks it refines the experiments to be more precise about the type and location of each bottleneck.

The experiments test *hypotheses* about why an application may suffer from performance problems. For example, one hypothesis tested by the Performance Consultant is whether the application is spending an excessive amount of time blocking on I/O operations. To activate an experiment the Performance Consultant inserts instrumentation code into the application to collect performance data. Instances when the measured value exceeds a predefined threshold for the experiment are termed *bottlenecks*.

Paradyn represents programs as collections of discrete resources. Resources include program code (modules, functions, and loops), processes, threads, machines, and synchronization objects. Paradyn organizes these program resources into trees called *resource hierarchies*. The root of each resource hierarchy is labeled with the hierarchy's name. As we move down from the root node, each level of the hierarchy represents a finer-grained description of the program. To form a *resource name* we concatenate the labels along the unique path from the root to the node representing the resource. For example, the resource name for function fact in module math.C is ⟨Code/math.C/fact⟩.

We may wish to constrain measurements to particular parts of a program. For example, we may want to measure CPU time for the entire execution of the program or for a single function or loop. An experiment's *focus* determines where the instrumentation code is inserted. Selecting a node in the resource hierarchy narrows the view to include only those nodes that descend from the selected node. For example, the focus ⟨/Code/math.C/fact, /Machine/toaster7.cs.wisc.edu, /SyncObject⟩ denotes function `fact` from module `math.C` executing on host `toaster7.cs.wisc.edu`. This focus specifies the top level SyncObject resource so it does not constrain the resources it names to any particular synchronization object or type of synchronization object.

The performance consultant *refines* its search for bottlenecks from a true experiment (an experiment whose hypothesis is true at its focus) by generating more experiments that have a more specific focus. To refine a focus, the Performance Consultant generates new foci. For example, the focus ⟨/Code/math.C/fact/loop_1, /Machine/toaster7.cs.wisc.edu, /SyncObject⟩ is refined to a particular loop in `fact` and the focus ⟨/Code/math.C/fact, /Machine/toaster7.cs.wisc.edu/8791, /SyncObject⟩ is refined to a particular process (with ID 8791) on host `toaster7.cs.wisc.edu`.

The *search history graph* records the cumulative refinements of a search. Each node represents a ⟨hypothesis, focus⟩ pair. Paradyn provides a visual representation of this graph that is dynamically updated as the Performance Consultant refines its search. This display provides both a visual history of the search and information about individual experiments such as its hypothesis, focus, whether it is active, its current measured value, and whether the experiment's hypothesis has yet to test true or false.

The cost of the instrumentation enabled by the Performance Consultant is continually monitored and limited to a user-selected threshold. New experiments generate new instrumentation requests and, when existing hypotheses test false, their instrumentation is removed.

4 Binary Instrumentation of Loops

To collect performance data at loop granularity, Paradyn must be able to instrument individual loops in a program binary. In this section we describe our implementation of loop instrumentation.

When Paradyn parses application binaries, it builds a flow graph for each function. Dominator information is calculated using the Lengauer-Tarjan algorithm [13] and is used to identify *natural loops*. We use standard definitions for basic blocks, dominators, back edges, and natural loops [1]. A *basic block* is a maximal sequence of instructions that can be entered only at the first instruction and exited only from the last instruction. A basic block M *dominates* block N, if every possible execution path in the flow graph from entry to N includes M. A *back edge* in a flow graph is an edge whose target dominates its source. The *natural loop* of a back edge M → N is the subgraph consisting of the set of nodes containing N and all the blocks from which M can be reached without passing through N. Block N is the *loop header*.

The decision to use natural loops (as opposed to any cycle in the flow graph) is reasonable given that irreducible loops are rare–you cannot create them in a structured language without using gotos. Certain compiler code replication techniques may transform reducible loops in the program's source code into irreducible loops in the program's binary, though this case is rare in our experience. The type of loop identified is not an issue of correctness; irreducible loops in a program binary are ignored, and do not hinder the instrumentation of natural loops.

If two natural loops share a header we cannot distinguish their nesting relationship, or if they are derived from a single loop in the source. In this case we combine the two loops into a single natural loop as is common in compilers [16].

To instrument loops in a flow graph, we define four *instrumentation points* in the flow graph that correspond to loop execution semantics:

1. *Loop entry* instrumentation executes when control enters a loop. We instrument the set of edges M → N such that N is the loop header and M is not a member of the loop. If M is the loop's *preheader* then one such edge exists and we may instrument M.
2. *Loop exit* instrumentation executes when control exits a loop. We instrument the set of edges M → N such that M is a member of the loop and N is not.
3. *Loop begin iteration* instrumentation executes at the beginning of each loop iteration. We instrument the loop header.
4. *Loop end iteration* instrumentation executes at the end of each loop iteration. We instrument the loop's back edge and loop exit instrumentation points.

Figure 2 (a) illustrates the location of these points for a simple loop. Loop entry and exit points are balanced. For example, when instrumenting loop entry with a start timer operation and loop exit with a stop timer operation, execution of the start timer will always be eventually followed by the execution of the stop timer. Loop begin and end iteration points are balanced as well.

Previous versions of our dynamic instrumentation [10] could instrument functions, basic blocks, and instructions. In this work, we add edge instrumentation. Edge instrumentation is not new, having been used in static binary editors, such as OM [20] and EEL [12]. Edges created by unconditional jumps can be instrumented simply by instrumenting the last instruction of the edge's source block. We do not instrument edges created by indirect jumps because they are not used for control transfer that create loops (they are typically used for jump tables and dynamic call sites). To instrument edges created by conditional jumps we create *edge trampolines*.

Figure 2 (b) illustrates how we instrument conditional jumps using edge trampolines. An edge trampoline is a code fragment that contains two new basic blocks, one that corresponds to execution of the fall-through edge and one for the taken edge (the shaded regions). We can instrument either edge by instrumenting its new basic block using our existing technique for creating instrumentation points. The conditional jump is overwritten with an unconditional jump to the edge trampoline. The conditional jump is relocated to the trampoline but is given a new target address. This simulates the execution of the original conditional jump but with our two new blocks as targets. These new blocks end with jumps to the original conditional jump targets.

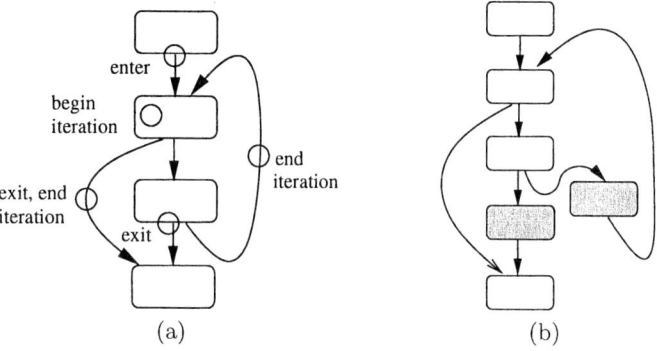

Fig. 2. Loop instrumentation points in a flow graph, indicated with circles, in (a). Instrumenting the conditional jump that creates the loop's back edge is shown in (b). An edge trampoline is used to create two new basic blocks (shaded) that correspond to the execution of the taken and fall-through edges of the conditional jump.

We use an absolute jump instruction to ensure that the edge trampoline can be reached from the application code we are instrumenting. On CISC architectures, such as the IA32, an absolute jump instruction may be larger than a PC relative jump instruction. This means the absolute jump we use to transfer control to the edge trampoline may be larger than the conditional jump that it replaces. In this case, we relocate enough instructions before the conditional jump to the head of the edge trampoline to make room. We may safely relocate all instructions in the basic block terminated by the conditional jump.

Though rare, the size of the entire basic block may be smaller than the size of the unconditional jump instruction. To handle this case, we can use a short trap instruction. The trap is caught by a handler that sets the application's PC to be the start address of the edge trampoline. The disadvantage of this approach is that performance suffers due to the cost of handling the trap.

A better strategy uses *function relocation*. When the Performance Consultant determines that the basic blocks relevant to loop instrumentation in a function are not large enough to be efficiently instrumented, it rewrites the function in a new location in the application's address space. When rewriting the function, nop instructions are inserted along with the original code to ensure that the relevant basic blocks are large enough. The Performance Consultant overwrites the beginning of the original function with a jump to the newly created copy.

5 A Loop-Aware Search Strategy

Loops form a natural extension to our callgraph-based search (see Figure 1). As such, conceptually the Performance Consultant can treat them as additional steps in its refinement. We have also defined loop instrumentation points similar to those for functions (Section 4). As a result, loops are not just conceptual steps but actual steps in the Performance Consultant's search.

The Performance Consultant performs a breadth-first search of the application's callgraph. Function entry and exit points are instrumented to collect inclusive time-based metrics. If a function is not a bottleneck then it is pruned from the search space. Otherwise, the search continues by instrumenting the functions that it calls. If a function is a bottleneck, then the search continues by instrumenting the functions that it calls that are *not* nested under any loops, as well as the function's outermost loops. If a loop is a bottleneck then we instrument its children: the functions that are called directly within this loop, and the loop's directly nested loops. If a loop is not a bottleneck then the loop, and its descendants are pruned from the search.

The addition of loops suggests that more experiments will be run. However, pruning a loop in the callgraph means its descendants are also pruned. Instrumenting a non-bottleneck loop which contains multiple function calls reduces the number of experiments run because the called functions do not have to be instrumented. Depending on the structure of the program's code, adding loops to the callgraph can cause more or fewer experiments to be run.

We found our top-down approach successful in practice, though other approaches may work as well. A search strategy can decide whether or not to instrument a function's loops based on static information like the function's depth in the call graph or the structure of the function's flow graph. Dynamic information like the function's current performance data can also be used to influence the decision. Unlike techniques that statically instrument a binary, dynamic instrumentation allows the search to use fine-grained instrumentation only when necessary, and to make the decision of which loops to instrument at run time. This enables our loop search strategy to compliment other strategies, such as Deep Start [19], that are able to quickly locate functions that are performing poorly. Once the problematic functions are found, loop instrumentation can be used to more precisely locate bottlenecks within these functions. A search strategy that dynamically evaluates the tradeoff between the low overhead of function-level instrumentation and the increased precision of loop-level instrumentation can reap the benefits of both techniques.

6 Experimental Results

To evaluate our loop-aware search strategy, we compared it to the Performance Consultant's current search strategy. We performed experiments on two sequential and two parallel (MPI and OpenMP) scientific applications (see Table 1). Table 2 lists more detailed application characteristics, including loop information. We used MPICH version 1.2.5 for our MPI implementation, version 5.2 of the Portland Group Compilers for the Fortran and C applications, and gcc version 3.3.3 for the C++ application.

For all experiments, we used 3GHz Pentium 4 PCs with Hyperthreading enabled, 2 GB RAM, running Tao Linux (a repackaged version of Red Hat Enterprise Linux 3), connected using 100 Mb Ethernet. Both MPI applications were run on identically configured PCs. The Paradyn front-end process was run

Table 1. Application characteristics

Name	Version	Type	Nodes	Language	Domain
ALARA	2.7.1	Sequential	1	C++	Induced radioactivity analysis
DRACO	6.0	Sequential	1	Fortran 90	Hydrodynamic simulation
OM3	1.5	MPI	8	C	Global ocean circulation
SPhot	1.0	MPI/OpenMP	8	Fortran 90	Monte Carlo transport code

Table 2. Application characteristics, including the number of loops, as a percent of the total number of loops, for 6 nesting levels

Name	Lines	KB	Funcs	Loops	Loops/Func	% Loops at nesting 1 2 3 4 5 6
ALARA	21,099	6,382	718	598	0.8	76 20 3 0.8 0.5 0.1
DRACO	72,305	2,516	898	5,477	6.1	47 32 16 4 1 0.1
OM3	2,673	88	28	202	7.2	40 32 21 7 0 0
SPhot	2,932	895	31	106	3.4	53 22 5 18 2 0

Table 3. Types of bottlenecks and rate of experimentation. "PC" is the default Performance Consultant search, "Loop PC" is our loop-aware search strategy.

Name	Total Bottlenecks Function Loop	Leaf Bottlenecks Function Loop	Experiments/second PC Loop PC
ALARA	11	8	4	3	0.9	1.0
DRACO	6	8	2	3	0.4	1.5
OM3	3	10	1	4	1.6	1.7
SPhot	8	10	3	5	2.9	1.9

on a different machine than the applications. We performed multiple runs of each application. During each run, we began the Performance Consultant search once the application reached steady-state behavior.

Our experiments indicated that loops were frequently bottlenecks (Table 3), which was expected since we examined scientific applications. In total, the applications contained 10 function bottlenecks that were leaf nodes in the callgraph. Of these 10, 7 contained loop bottlenecks. Loops significantly increase bottleneck precision. For example, OM3 contains a single function bottleneck, time_step, that is a leaf node in the callgraph and consumes 85% of the application's CPU time. While this information is useful, time_step is a large function that contains 90 loops. The Performance Consultant reports that 8 of these loops are bottlenecks, and that 4 of these bottleneck loops are leaf nodes in the augmented callgraph. Loop-level data indicates that a third of the time spent in time_step can be attributed to loop 12.

Figure 3 compares the rate at which bottlenecks are found using the Performance Consultant's default search strategy and our loop-aware strategy. Results are shown both considering and ignoring loop bottlenecks. Both search strategies find bottlenecks at similar rates. The loop-aware strategy finds more

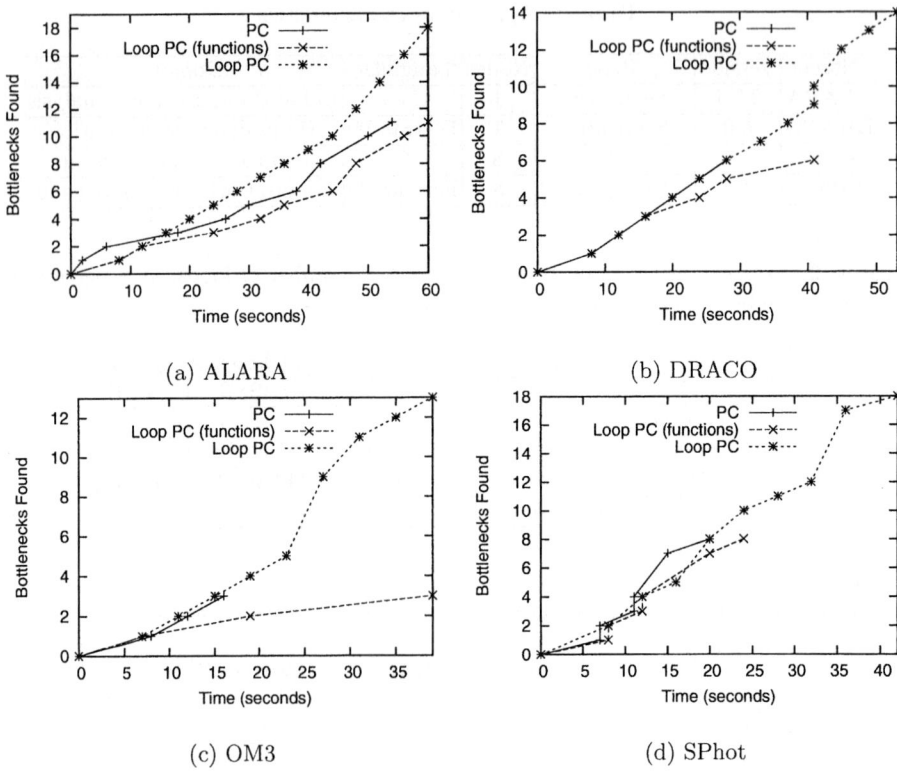

Fig. 3. Search profiles for the sequential (ALARA and DRACO) and parallel (OM3 and SPhot) applications. "PC" indicates the default Performance Consultant search, "Loop PC" indicates our loop-aware search strategy. "Loop PC (functions)" indicates that our loop-aware search strategy was used but only functions were counted as bottlenecks.

total bottlenecks (due to loops) but takes longer to identify function bottlenecks (due to the increased height of the augmented callgraph).

We did not observe significant differences in the rate of experimentation required to search loops (Table 3). For two of the applications the rate of experimentation was almost the same for both search strategies. The loop-aware Performance Consultant required a faster rate of experimentation for one application and a slower rate for the other. In general, we observed more precise results without a major change in search time or the rate of experimentation. Since we use an automated online search strategy that only instruments the loops of the functions currently in question, we only pay the cost of fine-grain instrumentation when necessary.

7 Conclusion

Searching loops proved to be a natural extension of the Performance Consultant's callgraph-based online automated search strategy. Since loops are frequently bottlenecks our strategy often provides a more precise performance diagnosis. Loops partition functions for searching without dramatically increasing the number of necessary experiments. We have defined points in a flow graph that correspond to loop execution, and presented a technique for the efficient instrumentation of these points. This mechanism enables our loop-aware search strategy to more precisely locate bottlenecks without large changes in search time or cost.

References

1. Aho, A., Sethi, R., Ullman, J., Compilers: Principles, Techniques and Tools, Addison-Wesley, 1985.
2. Cain, H. W., Miller, B. P., Wylie, B. J.N.: A Callgraph-Based Search Strategy for Automated Performance Diagnosis. *Concurrency and Computation: Practice & Experience* **14**, 3, 203-217 March 2002. Also appears as Euro-Par 2000, Munich, Germany, August 2000.
3. DeRose, L., Mohr, B., Seelam, S.: Profiling and Tracing OpenMP Applications With POMP Based Monitoring Libraries. *Euro-Par*, Pisa, Italy, August 2004, pp. 39-46
4. Gerndt, H.M., Krumme, A.: A Rule-Based Approach for Automatic Bottleneck Detection in Programs on Shared Virtual Memory Systems. *2nd Intl. Workshop on High-Level Programming Models and Supportive Environments*, Geneva, Switzerland, April 1997.
5. Goldberg, A.J., Hennessy, J.: MTOOL: A Method for Isolating Memory Bottlenecks in Shared Memory Multiprocessor Programs. *International Conference on Parallel Processing*, Augsust 1991, pp. 251-257.
6. Graham, S., Kessler, P., McKusick, M.: An Execution Profiler for Modular Programs. *Software Practice & Experience* **13** No. 8, August 1983, pp. 671-686.
7. Helm, B.R., Malony, A.D., Fickas, S.F.: Capturing and Automating Performance Diagnosis: the Poirot Approach. *Intl. Parallel Processing Symposium*, Santa Barbara, California, April 1995.
8. Hirzel, M., Chilimbi, T.: Bursty tracing: A framework for low-overhead temporal profiling. *4th ACM Workshop on Feedback-Directed and Dynamic Optimization*, Austin, Texas, December 2001.
9. Hollingsworth, J.K., Miller, B.P.: Dynamic Control of Performance Monitoring on Large Scale Parallel Systems. *International Conference on Supercomputing*, Tokyo, July 1993.
10. Hollingsworth, J.K., Miller, B.P., Cargille, J.: Dynamic Program Instrumentation for Scalable Performance Tools. *Scalable High Performance Computing*, Knoxville, Tennessee, May 1994.
11. Karavanic, K.L., Miller, B.P.: Improving Online Performance Diagnosis by the Use of Historical Performance Data. *SC99*, Portland, Oregon, November 1999.
12. Larus, J.R., Schnarr, E.: EEL: Machine-Independent Executable Editing. *Programming Language Design and Implementation* (1995), pp. 291-300.
13. Lengauer, T., Tarjan, R.E.: A fast algorithm for finding dominators in a flowgraph. *ACM Transactions on Programming Languages and Systems (TOPLAS)* **1**, 1, July 1979, pp. 121-141.
14. Mellor-Crummey, J., Fowler, R., Marin, G.: HPCView: A tool for top-down analysis of node performance. *Los Alamos Computer Science Institute Second Annual Symposium*, Santa Fe, NM, October 2001.

15. Miller, B.P., Callaghan, M.D., Cargille, J.M., Hollingsworth, J.K., Irvin, R.B., Karavanic, K.L., Kunchithapadam, K., Newhall, T.: The Paradyn Parallel Performance Measurement Tools, *IEEE Computer* **28**, 11, November 1995.
16. Muchnick, S., Advanced Compiler Design and Implementation, Morgan Kaufmann, 1997
17. Mukerjee, N., Riley, G.D., Gurd, J.R.: FINESSE: A Prototype Feedback-Guided Performance Enhancement System. *8th Euromicro Workshop on Parallel and Distributed Processing*, Rhodos, Greece, January 2000.
18. Reed, D. A., Aydt, R. A., Noe, R. J., Roth, P.C., Shields, K.A., Schwartz, B., Tavera, L.F.: Scalable Performance Analysis: The Pablo Performance Analysis Environment. Anthony Skjellum(ed). *Scalable Parallel Libraries Conference*, October 1993, pp. 104-113.
19. Roth, P.C., Miller, B.P.: Deep Start: A Hybrid Strategy for Automated Performance Problem Searches. *Concurrency and Computation: Practice and Experience* **15** 11-12, September 2003, John Wiley & Sons, pp. 1027-1046. Also appeared in shorter form in Euro-Par 2002, Paderborn, Germany, August 2002, LNCS 2400, Springer Verlag.
20. Srivastava, A., Eustace, A.: ATOM: A system for building customized program analysis tools. *Programming Language Design and Implementation* **11**, June 1994, pp. 196-205.
21. Intel®VTune™ Performance Analyzer.
    http://www.intel.com/software/products/vtune/
22. Official OpenMP Fortran Version 2.0 Specification.
    http://www.openmp.org/drupal/mp-documents/fspec20.pdf

Optimization of Nonblocking MPI-I/O to a Remote Parallel Virtual File System Using a Circular Buffer

Yuichi Tsujita

Department of Electronic Engineering and Computer Science,
Faculty of Engineering, Kinki University,
1 Umenobe, Takaya, Higashi-Hiroshima, Hiroshima 739-2116, Japan
tsujita@hiro.kindai.ac.jp

Abstract. Parallel computation applications output intermediate data periodically, and typically the outputs are moved to a remote computer for visualization. A flexible intermediate library named Stampi realizes seamless MPI-I/O operations both inside a computer and among computers. MPI-I/O operations to a remote computer are realized by its MPI-I/O processes which are invoked on a remote computer. To realize data-intensive I/O operations, a Parallel Virtual File System (PVFS) was supported in the MPI-I/O mechanism. MPI-I/O operations to a PVFS file system on a remote computer are available with seamless interfaces of the Stampi library. Among many kinds of I/O functions, nonblocking MPI-I/O functions provide overlap of computation with I/O operations, and visible I/O times can be minimized with them. Due to its architectural constraints and slow network, visible I/O times of them became long with an increase in the number of user processes and message data size. To minimize the times, a circular buffer system has been implemented in the mechanism. With the help of the circular buffer, the visible I/O times have been minimized effectively.

Keywords: MPI, MPI-I/O, Stampi, MPI-I/O process, PVFS, circular buffer.

1 Introduction

MPI [1,2] is the de facto standard in parallel computation, and almost all computer vendors have provided their own MPI libraries. But they do not support MPI communications among different MPI libraries. To realize such mechanism, Stampi [3] was developed.

Typical large-scale parallel computation periodically outputs its intermediate results as snapshots and checkpoints. A parallel I/O system is required to manage such outputs effectively, and a parallel I/O interface named MPI-I/O was proposed in the MPI-2 standard [2].

Although MPI-I/O has been implemented in several kinds of MPI libraries for I/O operations inside a computer (local MPI-I/O), MPI-I/O operations to a remote computer which has a different MPI library (remote MPI-I/O) have not been supported. To realize this mechanism, Stampi-I/O [4] was developed as a part of the Stampi library. Users can execute remote MPI-I/O operations with the help of its MPI-I/O processes which are invoked on a remote computer.

PVFS [5] has been developed and available on a Linux PC cluster. It realizes parallel I/O operations by distributing striped data among I/O nodes and gives users a virtual single file system. To realize remote MPI-I/O operations for huge amount of data by Stampi, the PVFS file system was supported in its remote MPI-I/O mechanism.

Overlap of computation with the I/O operations provided by nonblocking MPI-I/O functions is a key to minimize visible I/O cost. Although Stampi's nonblocking ones provide such overlap, overlap ratio becomes small with an increase in the number of user processes and message data size due to its architectural constraints and slow network. To improve the ratio, a circular buffer system has been implemented in the remote MPI-I/O mechanism. By caching data in the buffer using available memory on a remote computer where an MPI-I/O process is running, visible I/O cost has been minimized. In addition, optimizing configuration of the buffer has minimized the cost effectively. Since the buffer system hides I/O cost, its use is advantageous with the remote MPI-I/O.

In the following sections, outline, architecture, and preliminary performance results of the MPI-I/O mechanism are described.

2 Remote MPI-I/O Mechanism in Stampi

In this section, architecture and execution mechanism of the remote MPI-I/O operations are described.

2.1 Architecture of an MPI-I/O Mechanism

Architectural view of the MPI-I/O mechanism in Stampi is depicted in Figure 1. In an interface layer to user processes, intermediate interfaces which have MPI APIs (a part of a Stampi library) are implemented to relay messages between user processes and underlying communication and I/O systems.

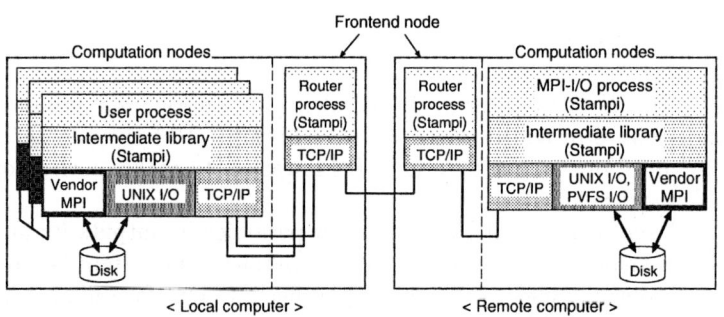

Fig. 1. Architecture of an MPI-I/O mechanism in Stampi

Stampi supports both local and remote MPI-I/O operations with the same MPI-I/O APIs. In local MPI-I/O operations, a vendor-supplied MPI-I/O library is used. If the library is not available, UNIX I/O functions are used. While remote MPI-I/O operations are carried out with the help of MPI-I/O processes which are invoked on a remote computer. I/O requests from the user processes are translated into message data, and they are transfered to the MPI-I/O processes. Bulk data are also transfered via the same communication path. The MPI-I/O processes play MPI-I/O operations on the remote computer using a vendor-supplied MPI-I/O library. If the vendor-supplied one is not available, UNIX I/O is used. To realize data-intensive remote MPI-I/O operations on a PVFS file system, PVFS I/O functions are supported in the MPI-I/O process in addition to the UNIX I/O.

2.2 Execution Mechanism

Stampi supports both interactive and batch modes in executing an MPI program. Here, execution method of remote MPI-I/O operations with an interactive system which is illustrated in Figure 2 is explained. Firstly, an MPI start-up process (MPI starter) is initiated by a Stampi start-up command (Stampi starter). Besides, a router process is invoked if computation nodes are not able to communicate outside. Then the MPI starter initiates user processes. When they call MPI_File_open(), either the MPI starter or the router process kicks off another Stampi starter process on a remote computer with the help of a remote shell command (rsh or ssh). Secondly, the starter kicks off an MPI-I/O process, and it opens a specified file. Besides, a router process is invoked on an IP-reachable node if computation nodes are not able to communicate outside directly. Remote MPI-I/O operations are available via the communication path established among the user processes and MPI-I/O process in this strategy.

Inside the MPI-I/O process, a circular buffer system is provided for UNIX I/O functions to achieve higher throughput in remote MPI-I/O operations. Each buffer size and the number of buffers are selectable by Stampi's key values, io_bufsize and io_bufcnt, respectively through an info object in an MPI program. Default values for the io_bufsize and io_bufcnt are 64 KByte and

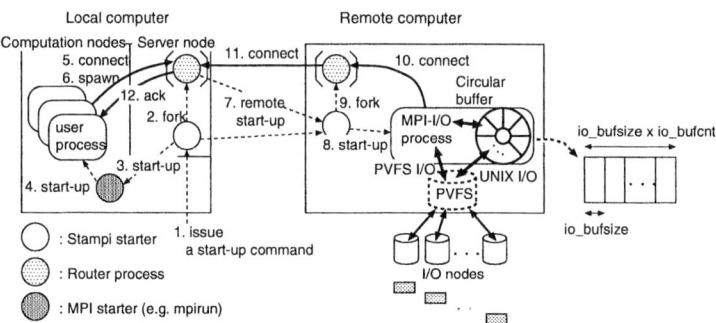

Fig. 2. Execution mechanism of remote MPI-I/O operations

32, respectively. A circular buffer whose size is io_bufsize × io_bufcnt is prepared by dynamic memory allocation when an MPI-I/O process is created. Huge amount of data is divided into pieces and they are stored once in the buffer. Later, the stored data are extracted in inter-machine data transfer (for read operations) or an I/O phase (for write operations). While PVFS I/O functions play direct I/O operations to a PVFS file system without the buffer system in the current implementation.

After the I/O operations, the file is closed and the MPI-I/O process is terminated when MPI_File_close() is called by the user processes.

2.3 Mechanism of MPI-I/O Functions

Next, mechanisms of remote nonblocking MPI-I/O operations to a PVFS file system are explained. As an example, mechanisms of MPI File read at all begin() and MPI_File_read_at_all_end() are illustrated in Figures 3 (a) and (b), respectively. When user processes call the read function depicted in Figure 3 (a), synchronization among the user processes is carried out by calling MPI_Barrier() at first. Then I/O request and associated parameters are packed in a user buffer using MPI_Pack() on each user process. The packed data is transfered to an MPI-I/O process using MPI_Send() and MPI_Recv() of the Stampi library. Inside the functions, Stampi's underlying communication functions such as JMPI_Isend() are used for nonblocking TCP socket communications. After the transfer, the received parameters are stored in an I/O request table which is created on the MPI-I/O process. Besides, a ticket number which is used to identify each I/O request is issued and stored in the table. After this operation, the ticket number and several values which are associated with the I/O operation are sent to

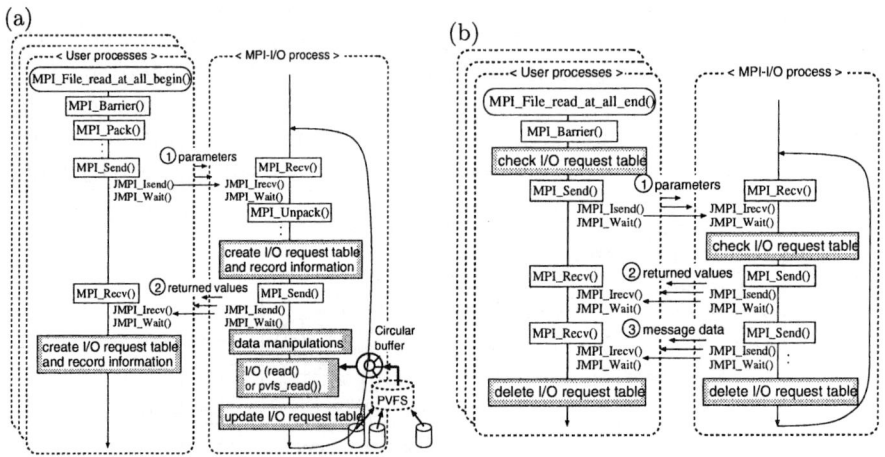

Fig. 3. Mechanisms of (a) MPI_File_read_at_all_begin() and (b) MPI_File_read_at_all_end() in remote MPI-I/O operations. MPI functions in rectangles are MPI interfaces of Stampi. Internally, Stampi-supplied functions such as JMPI_Isend() are called by them.

the corresponding user process immediately. Once the process receives them, it creates its I/O status table and stores them in it. Later, the process is able to do next computation without waiting completion of the I/O operation by the MPI-I/O process. While the MPI-I/O process retrieves an I/O request from the table and reads data from a PVFS file system using a UNIX I/O function, read(), with a circular buffer or a PVFS I/O function, pvfs_read(). In the UNIX case, the MPI-I/O process is able to store read data in the buffer and do next operation for another user process. While the MPI-I/O process waits until whole data is stored in a single user buffer in the PVFS case. After the I/O operation, the information in the I/O status table is updated. This sequence except the synchronization is repeated until all the user processes complete their I/O requests.

To detect completion of the I/O operation, MPI_File_read_at_all_end() is used to detect completion of the previous I/O function. Once this function is called on each user process, the I/O request, ticket number, and associated parameters in the I/O request table of each user process are retrieved and sent to the MPI-I/O process after synchronization among the user processes by MPI_Barrier(). After the MPI-I/O process receives them, it finds a corresponding I/O request table which has the same ticket number and reads the stored information in it. Then, several parameters and read data are sent to the corresponding user process. Finally both I/O request tables are deleted. This sequence except the synchronization is repeated until all the user processes complete the I/O request.

3 Performance Evaluation

Performance of the MPI-I/O mechanism was measured on interconnected PC clusters using an SCore cluster system [6] to evaluate effect of the proposed circular buffer. Specifications of the clusters are summarized in Table 1. A PC cluster I consisted of a server node which had a single Intel Xeon 2.4 GHz CPU and four computation nodes which had a single Intel Pentium 4 2.4 GHz CPU each. While a PC cluster II consisted of one server node and four computation nodes, which had two Intel Xeon 2.4 GHz CPUs each. Network connections among PC nodes of the clusters I and II were established on 1 Gbps bandwidth network with full duplex mode via Gigabit Ethernet switches, NETGEAR GS108 and 3Com SuperStack3 Switch 4900, respectively. Interconnection between those switches was also made on 1 Gbps bandwidth network with full duplex mode.

In the both clusters, an MPICH-SCore library [7] which was based on an MPICH [8] version 1.2.5 was available, and it was used in an MPI program on the cluster I.

In the cluster II, PVFS (version 1.6.3) was available on its server node. All the four computation nodes were used as I/O nodes for the PVFS file system. Size of dedicated disk of each I/O node was 73 GByte, and thus the PVFS system with 292 GByte (4 × 73 GByte) was available. During this test, default stripe size (64 KByte) of it was selected.

Table 1. Specifications of PC clusters. **server** and **comp** with bold font denote server and computation nodes of each cluster, respectively.

		PC cluster (I)	PC cluster (II)
Server node		\multicolumn{2}{c}{DELL PowerEdge 1600SC × 1}
Computation nodes		DELL PowerEdge 600SC × 4	DELL PowerEdge 1600SC × 4
CPU	**server**	Intel Xeon 2.4 GHz × 1	Intel Xeon 2.4 GHz × 2
	comp	Intel Pentium-4 2.4 GHz × 1	Intel Xeon 2.4 GHz × 2
Chipset		\multicolumn{2}{c}{ServerWorks GC-SL}
Memory		1 GByte DDR SDRAM	2 GByte DDR SDRAM
Local disk	**server**	\multicolumn{2}{c}{73 GByte (Ultra320 SCSI) × 1}
	comp	40 GByte (ATA-100 IDE) × 1	73 GByte (Ultra320 SCSI) × 2
Ethernet interface	**server**	Intel PRO/1000-MT (on-board)	Intel PRO/1000-XT (PCI-X)
	comp	Intel PRO/1000-MT (PCI-X)	Intel PRO/1000-XT (PCI-X)
Linux kernel	**server**	\multicolumn{2}{c}{2.4.20-28.7smp}
	comp	2.4.21-2SCORE	2.4.21-2SCOREsmp
Network driver	**server**	Intel e1000 version 5.4.11	Intel e1000 version 5.5.4
	comp	Intel e1000 version 5.4.11	Intel e1000 version 5.5.4
MPI library		\multicolumn{2}{c}{MPICH-SCore based on MPICH version 1.2.5}
Ethernet switch		NETGEAR GS108	3Com SuperStack3 Switch 4900

A router process was not invoked in this test because each computation node was able to communicate outside directly. Message data size was denoted as the size of whole message data to be transfered. A message data was split evenly among the user processes.

In performance measurement of remote MPI-I/O operations from the cluster I to the cluster II, split-collective MPI-I/O functions (MPI_File_read_at_all_begin() and MPI_File_write_at_all_begin()) with two and four user processes were evaluated. In this test, visible I/O times of them using UNIX I/O with the circular buffer and those using PVFS I/O (UNIX and PVFS methods, respectively) were measured. To optimize data transfer among the clusters, TCP_NODELAY option was activated for TCP socket connections by the Stampi start-up command.

3.1 Performance with Respect to io_bufsize

Firstly, performance was measured with respect to io_bufsize. In this test, io_bufcnt was set as a default value (32). Visible I/O times of the read and write operations are shown in Figures 4 (a) and (b), respectively. For comparison, the times using PVFS I/O are also shown in each figure. In the read operations, the times for 2 MByte in io_bufsize are the shortest in the UNIX case. The times are almost the same with those of the PVFS case for two user processes. While the times are longer than those of the PVFS case for four user processes (e.g. The time is 27 % longer than that of the PVFS case with a 256 MByte message data).

Although the write operations in Figure 4 (b) required longer time than the read operations, the times in the case of 2 MByte in io_bufsize are the shortest

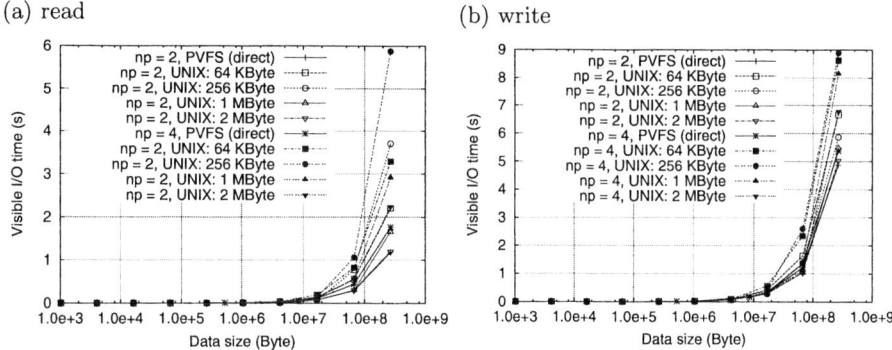

Fig. 4. Visible I/O times of (a) MPI_File_read_at_all_begin() and (b) MPI_File_write_at_all_begin() with applying default value (32) in io_bufcnt, where np denotes the number of user processes. *PVFS* and *UNIX* denote I/O operations by PVFS and UNIX I/O functions, respectively. The data sizes in the UNIX case denote io_bufsize of a circular buffer.

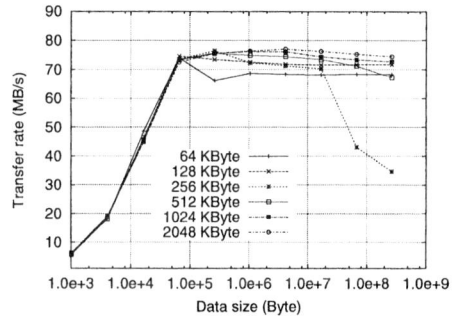

Fig. 5. Throughput of local read operations using UNIX I/O on a PVFS file system. Numbers in this figure denote chunk size for each I/O call.

in the UNIX case. Besides, the times are also shorter than those of the PVFS case with up to 64 MByte message data.

In the read operations, the times for 256 KByte in io_bufsize are the longest in the UNIX case. To evaluate this effect, performance of local read operations using UNIX I/O on a PVFS file system was measured with respect to chunk size for each I/O call. Whole I/O operation is carried out with multiple I/O calls for divided data with specified chunk size. Measured values are shown in Figure 5. As shown in this figure, throughput in the case of 256 KByte for chunk size is drastically degraded with an increase in message data size. Until now, no definitive reason for the degradation has been found yet. Anyway, the increase in the visible I/O times for remote MPI-I/O is due to the drastic performance degradation observed in reading 256 KByte each on a PVFS file system.

Through this test, it is considered that applying large number in io_bufsize made the times much short. Thus a circular buffer whose size was sufficiently

large to store almost all the data improved effect of hiding bottleneck in local
I/O operations on a PVFS file system by effective caching. But the times became
long with an increase in message data size. To minimize the times, it is preferable
to optimize an another parameter, io_bufcnt.

3.2 Performance with Respect to io_bufcnt

Next, visible I/O times were measured with applying 2 MByte in io_bufsize
with respect to io_bufcnt. The times for the read and write operations are
shown in Figures 6 (a) and (b), respectively. The read operations with applying
128 in io_bufcnt provided the most shortest times in the UNIX case. Besides,
the times are 29 % of those in the PVFS case with a 256 MByte message data
for both two and four user processes. Thus, a circular buffer with applying 128
in io_bufcnt had 256 MByte (= 2 MByte × 128) memory space in total, and
it was able to store whole data up to 256 MByte in the I/O operations. As a
result, the times were minimized by caching in the buffer.

The times of the write operations provided the similar results with the read
operations except that the times are longer than those of the read operations.
The times in the UNIX case with applying 128 in io_bufcnt are 72 % and 63 %
of those in the PVFS case with a 256 MByte message data for two and four user
processes, respectively.

To evaluate effect of the total buffer size, the visible I/O times for the I/O
operations with applying 64 KByte in io_bufsize were measured with respect
to io_bufcnt. In this test, visible I/O times with 4096 in io_bufcnt (64 KByte
× 4096 = 256 MByte) are the shortest in the UNIX case. Besides, the times are
shorter than those of the PVFS case.

Through the test results, it is confirmed that a circular buffer with a sufficient
size to store a message data has made visible I/O times short. Typically, the effect
was big in the read operations.

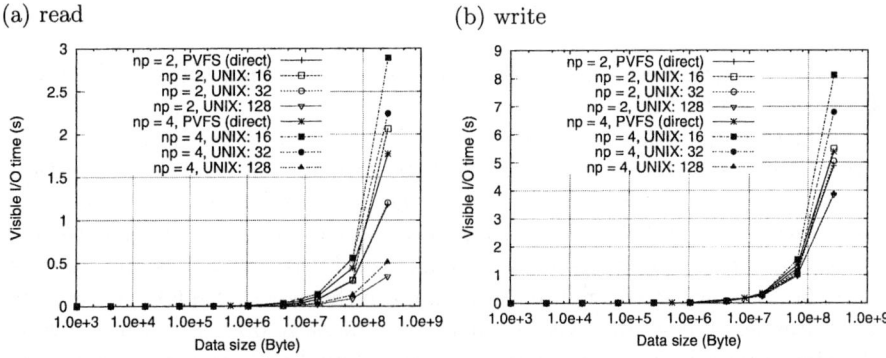

Fig. 6. Visible I/O times of (a) MPI_File_read_at_all_begin() and (b)
MPI_File_write_at_all_begin() with applying 2 MByte in io_bufsize, where
np denotes the number of user processes. *PVFS* and *UNIX* denote I/O operations by
PVFS and UNIX I/O, respectively. Numbers in the UNIX case denote io_bufcnt of a
circular buffer.

4 Related Work

ROMIO [9] provides seamless MPI-I/O interfaces to many kinds of file systems with the help of ADIO [10]. MPI-I/O operations to a remote computer where MPICH is available has been realized with the help of RFS [11]. On the other hand, Stampi provides seamless MPI-I/O operations to a remote computer which has a different MPI library with the help of its intermediate communication library and an MPI-I/O process.

Improvement in throughput of remote MPI-I/O operations has been done in ROMIO [11]. An invoked thread manages buffering of data while a main thread continues its computation, then this makes visible I/O cost short. On the other hand, Stampi has adopted a circular buffer system inside an MPI-I/O process on a remote computer. The process manages caching of data in it to minimize visible I/O times in remote MPI-I/O operations on a slow network connection.

5 Summary

A remote MPI-I/O using a circular buffer system has been realized in the Stampi library for data-intensive computation. Configuration of the buffer inside an MPI-I/O process is tunable by Stampi's key values through an info object. In performance measurement which was carried out on interconnected PC clusters, visible I/O times for both read and write functions with two and four user processes were minimized in the UNIX case with the help of the buffer compared with those for direct use of PVFS I/O. Typically, the effect was big in the read function because only an I/O request and associated parameters were transfered in issuing the function and user processes were able to do next operation without receiving a read data. Since the data was stored in the buffer once, MPI-I/O process on a remote computer was able to do next I/O operation for another waiting user process without sending the data to the corresponding user process. It is concluded that implementation of the circular buffer system is effective in the remote MPI-I/O operations. The same implementation using PVFS I/O is considered in future works. Besides, adoption in real applications is also considered.

Acknowledgments

The author would like to thank Genki Yagawa, director of Center for Promotion of Computational Science and Engineering (CCSE), Japan Atomic Energy Research Institute (JAERI), for his continuous encouragement. The author would like to thank the staff at CCSE, JAERI, especially Toshio Hirayama, Norihiro Nakajima, Kenji Higuchi, and Nobuhiro Yamagishi for providing a Stampi library and giving useful information.

This research was partially supported by the Ministry of Education, Culture, Sports, Science and Technology (MEXT), Grant-in-Aid for Young Scientists (B), 15700079.

References

1. Message Passing Interface Forum: MPI: A message-passing interface standard. (1995)
2. Message Passing Interface Forum: MPI-2: Extensions to the Message-Passing Interface. (1997)
3. Imamura, T., Tsujita, Y., Koide, H., Takemiya, H.: An architecture of Stampi: MPI library on a cluster of parallel computers. Recent Advances in Parallel Virtual Machine and Message Passing Interface, Volume 1908 of Lecture Notes in Computer Science., Springer (2000) 200–207
4. Tsujita, Y., Imamura, T., Takemiya, H., Yamagishi, N.: Stampi-I/O: A flexible parallel-I/O library for heterogeneous computing environment. Recent Advances in Parallel Virtual Machine and Message Passing Interface, Volume 2474 of Lecture Notes in Computer Science., Springer (2002) 288–295
5. Carns, P.H., Ligon III, W.B., Ross, R.B., Thakur, R.: PVFS: A parallel file system for Linux clusters. In: Proceedings of the 4th Annual Linux Showcase and Conference, USENIX Association (2000) 317–327
6. PC Cluster Consortium: (http://www.pccluster.org/)
7. Matsuda, M., Kudoh, T., Ishikawa, Y.: Evaluation of MPI implementations on grid-connected clusters using an emulated WAN environment. In: Proceedings of the 3rd IEEE/ACM International Symposium on Cluster Computing and the Grid (CCGrid 2003), 12-15 May 2003, Tokyo, Japan, IEEE Computer Society (2003) 10–17
8. Gropp, W., Lusk, E., Doss, N., Skjellum, A.: A high-performance, portable implementation of the MPI Message-Passing Interface standard. Parallel Computing **22** (1996) 789–828
9. Thakur, R., Gropp, W., Lusk, E.: On implementing MPI-IO portably and with high performance. In: Proceedings of the Sixth Workshop on Input/Output in Parallel and Distributed Systems. (1999) 23–32
10. Thakur, R., Gropp, W., Lusk, E.: An abstract-device interface for implementing portable parallel-I/O interfaces. In: Proceedings of the Sixth Symposium on the Frontiers of Massively Parallel Computation. (1996) 180–187
11. Lee, J., Ma, X., Ross, R., Thakur, R., Winslett, M.: RFS: Efficient and flexible remote file access for MPI-IO. In: Proceedings of the 6th IEEE International Conference on Cluster Computing (CLUSTER 2004), IEEE Computer Society (2004) 71–81

Performance Analysis of Shared-Memory Parallel Applications Using Performance Properties

Karl Fürlinger and Michael Gerndt

Technische Universität München,
Institut für Informatik,
Lehrstuhl für Rechnertechnik und Rechnerorganisation,
{Karl.Fuerlinger, Michael.Gerndt}@in.tum.de

Abstract. Tuning parallel code can be a time-consuming and difficult task. We present our approach to automate the performance analysis of OpenMP applications that is based on the notion of performance properties. Properties are formally specified in the APART specification language (ASL) with respect to a specific data model. We describe a data model for summary (profiling) data of OpenMP applications and present performance properties based on this data model. We evaluate the usability of the properties on several example codes using our OpenMP profiler ompP to acquire the profiling data.

1 Introduction

Tuning parallel code can often be a time-consuming and complex task. Nonetheless, it is generally important for application developers that their code uses the available resources efficiently and delivers close-to-optimal performance, since improving performance is commonly the reason exploiting parallelism in the first place.

Many tools have been devised that application developers employ to analyze the performance of their code. First it is necessary to decide whether performance problems actually exist in the code. For the list of potential tuning targets it is then also necessary to identify the *reason* for the inefficiency. Knowing the reason then allows the developer to modify the code in order to remedy the performance problem.

The process of code tuning outlined above, also sometimes referred to as the measure–analyze–modify cycle [1,13], requires the application developer to have detailed knowledge of how the application's code finally corresponds to performance delivered on some parallel machine. Various levels of abstraction added by operating system, communications middle-ware and parallel programming languages complicate the analysis of this correspondence. Furthermore, programming languages, communications middle-ware and most notably the parallel computing systems themselves evolve and change their characteristics over time and application developers need to keep track of these changes. Tradeoffs and rules-of-thumb that might be applicable for one generation of a machine can be inadequate for the next.

Performance tool builders therefore seek to support application developers by *automating* the process of performance analysis and tuning. Automation can be employed at every stage of the measure–analyze–modify cycle. While supportive tools for performance data measurement are well-understood and in widespread use (e.g., tracing and profiling tools), this is less so for approaches automating the analysis step, let alone the tuning step.

In this work we describe our advances towards automating the analysis phase of the tuning cycle, specifically for shared-memory parallel programs. Our work is based on the notion of performance *properties* that formally describe situations of inefficiency. By specifying what constitutes a performance property, a performance specialist can encapsulate domain- and platform specific knowledge that an application developer often neither has, nor wants to care about. Performance properties can also identify the reason for the inefficiency and convey hints on successful tuning strategies.

The rest of this paper is organized as follows: in Sect. 2 we give a general overview of our performance analysis approach based on performance properties. Then, in Sect. 3 we present the properties for shared-memory parallel programs and describe the data model on which the properties rely. In Sect. 4 we evaluate our approach on application examples from a test suite, designed for testing performance analysis tools (the ATS [12]) and from the OpenMP source code repository [2]. We present related work in Sect. 5 and summarize and discuss directions for future work in Sect. 6.

2 Specification of Performance Properties

Performance properties describe situations of inefficient execution. Within the APART (Automated Performance Analysis: Real Tools) working group, a language for the formal specification of performance properties was developed (the APART specification language, ASL [3,7]).

A typical property specification is shown in Fig. 1. The specification has three parts:

- A specification of the *condition* that needs to be fulfilled in order for the property to hold.
- An expression that gives the *confidence* that the property holds.
- An expression that gives the *severity* of the property, i.e., that quantifies the impact on the performance that a particular property represents.

Condition, confidence and severity are expressed with respect to entities of a specific data model. The data model contains abstractions like regions, performance summary data structures and events for a particular programming environment (i.e., there might be a different data model for OpenMP and MPI, but a combined model for mixed-parallel code is possible as well). In Fig. 1 the data model is represented by the `SeqPerf` data structure, which contains summary

```
property MpiOvhdInSeqRegInProc (SeqPerf pD)
  {
    condition  : pD.mpiT > 0;
    confidence : 1.0;
    severity   : pD.mpiT/RB(pD.exp);
  }
```

Fig. 1. The ASL specification of the `MpiOvhdInSeqRegInProc` property

(i.e., profiling) data of sequential (not thread-parallel) regions. The `condition` and `severity` specifications refer to the `mpiT` data member that gives the total time spent in MPI calls in a particular region. `RB()` refers to the ranking basis of the experiment, which allows the determination of the severity based on the time lost in a particular construct. In our case the ranking basis refers to the total execution time of the application. For online-performance analysis where the application execution is assumed to proceed in repeating phases, it could, however, also refer to the duration of a single phase.

ASL also contains mechanisms that allow the easy formulation of new properties based on existing ones (meta-properties and property templates). This allows for a compact and flexible specification of properties by a performance specialist. For details please refer to [3,7].

3 Properties for Shared-Memory Parallel Programs

A number of properties have previously been devised for shared-memory parallel programs [4]. Here we take up that work and analyze what performance data can practically be derived from the execution of OpenMP programs and which properties can be based on that data.

In our work, we rely on instrumentation of OpenMP programs that utilizes the work of Mohr et al. [10]. Since OpenMP lacks a standard performance measurement interface, Mohr et al. designed such an interface (called POMP) that exposes OpenMP program execution events to performance analysis tools. In the POMP proposal functions are called when OpenMP regions are being entered or exited, for example `POMP_Parallel_fork` and `POMP_Parallel_join` are called immediately before and after a `parallel` construct.

The `POMP_*` calls are inserted by the source-to-source instrumenter Opari [11]. A performance tool implements the `POMP_*` functions and is thus able to observe the program's execution and record performance characteristics as needed. We implemented our own POMP-based profiler called `ompP` [6] to derive the necessary profiling data for the data model presented below.

3.1 Data Model

Our ASL data model used for OpenMP shared-memory parallel programs is represented by the `ParPerf` structure holding summary data for program regions

```
ParPerf {
  Region *reg             // The region for which the summary is collected
  Experiment *exp         // The experiment where this data belongs to
  int threadC             // Number of threads that executed the region
  double execT[]          // Total execution time per thread
  double execC[]          // Total execution count per thread
  double exitBarT[]       // Time spent in the implicit exit barrier
  double singleBodyT[]    // Time spent inside a single construct
  int singleBodyC[]       // Execution count in a single construct
  double enterT[]         // Time spent waiting to enter a construct
  int enterC[]            // Number of times a threads enters a construct
  double exitT[]          // Time spent to exit a construct
  int exitC[]             // Number of times a thread exits a construct
  double sectionT[]       // Time spent inside a section construct
  int sectionC[]          // Number of times a section construct is entered
  double startupT[]       // Time required to create a parallel region
  double shutdownT[]      // Time required to destroy a parallel region
}
```

Fig. 2. The `ParPerf` structure contains summary (profiling) data for OpenMP constructs

corresponding to OpenMP constructs in the target application. The `ParPerf` structure is shown in Fig. 2 in C/C++/Java-like syntax, it has the following entries

- `exp` points to a data structure that gives general information about the conducted experiment such as when the experiment started and when it ended.
- `reg` points to a `Region` structure that holds static information about the regions of the program (such as begin and end line numbers and the type of the region, e.g., PARALLEL, CRITICAL, SINGLE, ...).
- `threadC` gives the number of threads that executed the OpenMP construct.
- The other members of `ParPerf` hold dynamic performance data in the form of timings and counts. Not all data members are defined for all region types. For example `singleBodyT` is defined only for SINGLE regions. It holds the time spent *inside* a single construct. `exitBarT` is defined for parallel regions and OpenMP worksharing regions. It represents the time spent inside the implicit exit barrier added by Opari to measure the load-imbalance in these constructs[1]. `enterT` and `exitT` are only defined for CRITICAL regions and measure the time required to enter and exit the critical section, respectively. `startupT` and `shutdownT` are only meaningful for PARALLEL regions, these timings allow the measurement of the time lost due to thread creation and teardown for parallel regions.

---

[1] To measure load-imbalance in a worksharing construct, Opari adds a `nowait` clause to the construct and inserts a `barrier` at the end of the construct.

3.2 Property Specification

The ASL performance properties for OpenMP code are defined with respect to the `ParPerf` data structure. We describe some of the more important properties below, other properties that have been defined but are not shown here are AllThreadsLockContention, FrequentAtomic, InsufficientWorkInParallel, UnparallelizedInSingleRegion, UnparallelizedInMasterRegion, ImbalanceDueToUnevenSectionDistribution and LimitedParallelismInSections.

ImbalanceAtBarrier. This property refers to an explicit OpenMP barrier directive added by the programmer. The property measures the difference in arrival time of the individual threads at the barrier. This is usually related to a situation of load imbalance the threads encounter before arriving at the barrier. Time waited by the threads at the barrier is lost and is the basis for computing the severity.

```
property ImbalanceAtBarrier(ParPerf pd) {
  let
     min = min(pd.execT[0],...,pd.execT[pd.threadC-1]);
     max = max(pd.execT[0],...,pd.execT[pd.threadC-1]);
     imbal = max-min;

  condition  : (pd.reg.type==BARRIER) && (imbal > 0);
  confidence : 1.0;
  severity   : imbal / RB(pd.exp);
}
```

ImbalanceInParallelRegion. This property (like the very similar ImbalanceInParallelLoop and ImbalanceInParallelSections) measures imbalances of the `parallel` construct (or the respective OpenMP work-sharing constructs). Opari adds an *implicit* barrier at the end of these constructs, time spent in the this barrier is accessible via `exitBarT`.

```
property ImbalanceInParallelRegion(ParPerf pd) {
  let
     min = min(pd.exitBarT[0],...,pd.exitBarT[pd.threadC-1]);
     max = max(pd.exitBarT[0],...,pd.exitBarT[pd.threadC-1]);
     imbal = max-min;

  condition  : (pd.reg.type==PARALLEL) && (imbal > 0);
  confidence : 1.0;
  severity   : imbal / RB(pd.exp);
}
```

CriticalSectionContention. This property indicates that threads contend for a critical section. Waiting time for entering or exiting the critical section is summed-up in `enterT` and `exitT`, respectively.

```
property CriticalSectionContention {
  let
     enter = sum(pd.enterT[0],...,pd.enterT[pd.threadC-1]);
     exit  = sum(pd.exitT[0],...,pd.exitT[pd.threadC-1]);

  condition  : (pd.reg.type==CRITICAL) && ((enter+exit) > 0);
  confidence : 1.0;
  severity   : (enter+exit) / RB(pd.exp);
}
```

4 Application Examples

In this section we test our approach to automate the performance analysis of OpenMP applications based on the ATS properties defined in Sect. 3. The experiments were run on a single 4-way Itanium-2 SMP system (1.3 GHz, 3 MB third level cache and 8 GB main memory), the Intel Compiler version 8.0 was used.

4.1 APART Test Suite (ATS)

The ATS [12] is a set of test applications (MPI and OpenMP) developed within the APART working group. The framework is based on functions that generate a sequential amount of work for a process or thread and on a specification of the distribution of work among processes or threads. Building on this basis, individual programs are created that exhibit a certain pattern of inefficient behavior, for example "imbalance in parallel region".

The ompP [6] output in Fig. 3 is from a profiling run of the ATS program that demonstrates the "imbalance in parallel loop" performance problem. Notice the exitBarT column and the uneven distribution of time with respect to threads {0,1} and {2,3}.

This inefficiency is easily identified by checking the ImbalanceInParallel-Loop property. The imbalance in exitBarT (difference between maximum and minimum time) amounts to 2.03 seconds, the total runtime of the program was 6.33 seconds. Therefore the ImbalanceInParallelLoop property is assigned a severity of 0.32.

```
R00003    LOOP pattern.omp.imbalance_in_parallel_loop.c (15--18)
 TID       execT      execC     exitBarT
  0         6.32        1         2.03
  1         6.32        1         2.02
  2         6.32        1         0.00
  3         6.32        1         0.00
  *        25.29        4         4.05
```

Fig. 3. The ompP profiling data for the loop region in the ATS program that demonstrates the ImbalanceInParallelLoop property

This example is typical for a number of load imbalance problems that are easily identified by the corresponding ASL performance properties. Other problems related to synchronization are also easily identified.

4.2 Quicksort

Towards a more real-world application example, we present an evaluation in the context of work performed by Süß and Leopold on comparing several parallel implementations of the Quicksort algorithm [16]. The code is now part of the OpenMP source code repository [2] and we have analyzed a version with a global work stack (called sort_omp_1.0 in [16]). In this version there is a single stack of work elements (sub-vectors of the vector to be sorted) that are placed on and taken form the stack by threads concurrently. Access to the stack is protected by two critical sections. The ompP output below shows the profiling data for the two critical sections.

```
R00002    CRITICAL         cpp_qsomp1.cpp (156--177)
TID       execT    execC    enterT    enterC    exitT    exitC
0         1.61     251780   0.87      251780    0.31     251780
1         2.79     404056   1.54      404056    0.54     404056
2         2.57     388107   1.38      388107    0.51     388107
3         2.56     362630   1.39      362630    0.49     362630
*         9.53     1406573  5.17      1406573   1.84     1406573

R00003    CRITICAL         cpp_qsomp1.cpp (211--215)
TID       execT    execC    enterT    enterC    exitT    exitC
0         1.60     251863   0.85      251863    0.32     251863
1         1.57     247820   0.83      247820    0.31     247820
2         1.55     229011   0.81      229011    0.31     229011
3         1.56     242587   0.81      242587    0.31     242587
*         6.27     971281   3.31      971281    1.25     971281
```

Checking for the CriticalSectionContention property immediately reveals the access to the stacks as the major source of inefficiency of the program. Threads content for the critical section, the program spends a total of 7.01 seconds entering and exiting the first and 4.45 seconds for the second section. Considering a total runtime of 61.02 seconds, this corresponds to a severity of 0.12 and 0.07, respectively.

Süß and Leopold also recognized the single global stack as the major source of overhead and implemented a second version with thread-local stacks.

Profiling data for the second version appears below. In this version the overhead with respect to critical sections is clearly smaller than the first one (enterT and exitT have been improved by about 25 percent) The overall summed runtime reduces to 53.44 seconds, an improvement of about 12 percent, which is in line with the results reported in [16]. While this result demonstrates a nice performance gain with relatively little effort, our analysis clearly indicates room for further improvement.

```
R00002     CRITICAL              cpp_qsomp2.cpp (175--196)
TID        execT      execC      enterT     enterC     exitT      exitC
0          0.67       122296     0.34       122296     0.16       122296
1          2.47       360702     1.36       360702     0.54       360702
2          2.41       369585     1.31       369585     0.53       369585
3          1.68       246299     0.93       246299     0.37       246299
*          7.23       1098882    3.94       1098882    1.61       1098882

R00003     CRITICAL              cpp_qsomp2.cpp (233--243)
TID        execT      execC      enterT     enterC     exitT      exitC
0          1.22       255371     0.55       255371     0.31       255371
1          1.16       242924     0.53       242924     0.30       242924
2          1.32       278241     0.59       278241     0.34       278241
3          0.98       194745     0.45       194745     0.24       194745
*          4.67       971281     2.13       971281     1.19       971281
```

5 Related Work

Several approaches for automating the process of performance analysis have been developed.

Paradyn's [9] Performance Consultant automatically searches for performance bottlenecks in a running application by using a dynamic instrumentation approach. Based on hypotheses about potential performance problems, measurement probes are inserted into the running program. Recently MRNet [15] has been developed for the efficient collection of distributed performance data. However, the search process for performance data is still centralized.

The Expert [17] performs an automated post-mortem search for patterns of inefficient program execution in event traces. As in our approach, data collection for OpenMP code is based on POMP interface. However, Expert performs tracing which often results in large data-sets and potentially long analysis time, while we only collect summary data in the form of profiles.

Aksum [8,5], developed at the University of Vienna, is based on a source code instrumentation to capture profile-based performance data which is stored in a relational database. The data is then analyzed by a tool implemented in Java that performs an automatic search for performance problems based on JavaPSL, a Java version of ASL.

6 Summary and Future Work

In this paper we demonstrated the viability of automated performance analysis based on performance property specifications. We described the general idea and explained the structure of the the Apart Specification Language (ASL). In ASL, performance properties are specified with respect to a specific data model.

We presented our data model for shared-memory parallel code that allows the representation of performance critical data such as time waited to enter a construct. The data model is designed according to restrictions on what can actually be measured from the execution of OpenMP program. We rely on the POMP specification [10] and source-to-source instrumentation added by Opari [11] to expose OpenMP execution events, since OpenMP still lacks a standard profiling interface.

We tested the efficacy of our property specification on some test applications. To acquire the performance data needed for the data model, we relied on our own OpenMP profiler ompP [6]. The examples show that the data required for checking the properties can be derived from the execution of an OpenMP program and that the performance properties given in examples in Sect. 3 are able determine the major cause of inefficiency in the presented example programs.

Tuning hints can be associated with performance properties. Currently only the name of the property (e.g., CriticalSectionContention) conveys information on the reason of the detected inefficiency. However, it is easy to augment the property specification with a more elaborate explanation of the detected inefficiency and to give advice with respect to tuning options to the application developer.

Future work is planned along several directions. First, the set of performance properties can be extended. For example, data coming from hardware performance counters are not yet included, especially cache miss counts and instruction rates can give rise to interesting and important properties.

Secondly, we are working on a larger and more automatic environment for performance analysis in the PERISCOPE project [14]. The goal is to have an ASL compiler that takes the specification of performance properties and translates them into C++ code. Compiling the code creates loadable modules used by the PERISCOPE system to detect performance inefficiencies at runtime. PERISCOPE is designed for MPI and OpenMP and supports large scale systems by virtue of a distributed analysis system consisting of *agents* distributed over the nodes large SMP-based cluster systems.

References

1. Mark E. Crovella and Thomas J. LeBlanc. Parallel performance prediction using lost cycles analysis. In *Proceedings of the 1994 Conference on Supercomputing (SC 1994)*, pages 600–609. ACM Press, 1994.
2. Antonio J. Dorta, Casiano Rodríguez, Francisco de Sande, and Arturo Gonzáles-Escribano. The OpenMP source code repository. In *Proceedings of the 13th Euromicro Conference on Parallel, Distributed and Network-Based Processing (PDP 2005)*, pages 244–250, February 2005.
3. Thomas Fahringer, Michael Gerndt, Bernd Mohr, Felix Wolf, Graham Riley, and Jesper Larsson Träff. Knowledge specification for automatic performance analysis. APART technical report, revised edition. Technical Report FZJ-ZAM-IB-2001-08, Forschungszentrum Jülich, 2001.
4. Thomas Fahringer, Michael Gerndt, Graham Riley, and Jesper Larsson Träff. Formalizing OpenMP performance properties with ASL. In *Proceedings of the 2000 International Symposium on High Performance Computing (ISHPC 2000), Workshop on OpenMP: Experience and Implementation (WOMPEI)*, pages 428–439. Springer-Verlag, 2000.

5. Thomas Fahringer and Clóvis Seragiotto Júnior. Automatic search for performance problems in parallel and distributed programs by using multi-experiment analysis. In *Proceedings of the 9th International Conference On High Performance Computing (HiPC 2002)*, pages 151–162. Springer-Verlag, 2002.
6. Karl Fürlinger and Michael Gerndt. ompP: A profiling tool for OpenMP. In *Proceedings of the First International Workshop on OpenMP (IWOMP 2005)*, 2005. Accepted for publication.
7. Michael Gerndt. Specification of performance properties of hybrid programs on hitachi SR8000. Technical report, Lehrstuhl für Rechnertechnik und Rechnerorganisation, Institut für Informatik, Technische Universität München, 2002.
8. Clóvis Seragiotto Júnior, Thomas Fahringer, Michael Geissler, Georg Madsen, and Hans Moritsch. On using aksum for semi-automatically searching of performance problems in parallel and distributed programs. In *Proceedings of the 11th Euromicro Conference on Parallel, Distributed and Network-Based Processing (PDP 2003)*, pages 385–392. IEEE Computer Society, February 2003.
9. Barton P. Miller, Mark D. Callaghan, Jonathan M. Cargille, Jeffrey K. Hollingsworth, R. Bruce Irvin, Karen L. Karavanic, Krishna Kunchithapadam, and Tia Newhall. The Paradyn parallel performance measurement tool. *IEEE Computer*, 28(11):37–46, 1995.
10. Bernd Mohr, Allen D. Malony, Hans-Christian Hoppe, Frank Schlimbach, Grant Haab, Jay Hoeflinger, and Sanjiv Shah. A performance monitoring interface for OpenMP. In *Proceedings of the Fourth Workshop on OpenMP (EWOMP 2002)*, September 2002.
11. Bernd Mohr, Allen D. Malony, Sameer S. Shende, and Felix Wolf. Towards a performance tool interface for OpenMP: An approach based on directive rewriting. In *Proceedings of the Third Workshop on OpenMP (EWOMP'01)*, September 2001.
12. Bernd Mohr and Jesper Larsson Träff. Initial design of a test suite for automatic performance analysis tools. In *Proc. HIPS*, pages 77–86, 2003.
13. Anna Morajko, Oleg Morajko, Josep Jorba, and Tomàs Margalef. Automatic performance analysis and dynamic tuning of distributed applications. *Parallel Processing Letters*, 13(2):169–187, 2003.
14. Periscope project homepage http://wwwbode.cs.tum.edu/~gerndt/home/Research/PERISCOPE/Periscope.htm.
15. Philip C. Roth, Dorian C. Arnold, and Barton P. Miller. MRNet: A software-based multicast/reduction network for scalable tools. In *Proceedings of the 2003 Conference on Supercomputing (SC 2003)*, November 2003.
16. Michael Süß and Claudia Leopold. A user's experience with parallel sorting and openmp. In *Proceedings of the Sixth Workshop on OpenMP (EWOMP'04)*, October 2004.
17. Felix Wolf and Bernd Mohr. Automatic performance analysis of hybrid MPI/OpenMP applications. In *Proceedings of the 11th Euromicro Conference on Parallel, Distributed and Network-Based Processing (PDP 2003)*, pages 13–22. IEEE Computer Society, February 2003.

On the Effectiveness of IEEE 802.11e QoS Support in Wireless LAN: A Performance Analysis*

José Villalón, Pedro Cuenca, and Luis Orozco-Barbosa

Instituto de Investigación en Informática de Albacete,
Universidad de Castilla-La Mancha,
02071 Albacete, Spain
{josemvillalon, pcuenca, lorozco}@info-ab.uclm.es

Abstract. The IEEE 802.11e draft is a proposal defining the mechanisms for wireless LANs aiming to provide QoS support to time-sensitive applications, such as, voice and video communications. The 802.11e group is currently working hardly, but the ratification of the standard has a long way to go. In this paper we carry out a performance analysis on the effectiveness of the IEEE 802.11e (EDCA) upcoming standard. We show that the defaults parameters setting recommended in the EDCA draft standard by 802.11e group do not fulfill the requirements of time-sensitive services, such as, voice and video. We show that the performance of EDCA can be improved by properly tuning its parameters.

Keywords: WLAN, IEEE 802.11, IEEE 802.11e, EDCA, QoS, Multimedia Communications and Performance Evaluation.

1 Introduction

The IEEE 802.11 Working Group is in the process of defining the IEEE 802.11e Standard: the QoS-aware standard for IEEE 802.11 WLANs [1][2]. In this paper, we carry out a performance analysis on the effectiveness of the IEEE 802.11e (EDCA) to provide QoS guarantees. Our main results show that the default parameters setting recommended by the standard do not meet the QoS requirements of time-sensitive applications. We show that the performance of EDCA can be improved by properly tuning its system parameters. Previous studies reported in the literature have evaluated the performance of the IEEE 802.11 standard [3][4][5]. However, they have not undertaken an in-depth analysis when delay bounds to the time-sensitive applications. Furthermore, we consider a multiservice scenario, i.e., a WLAN supporting four different services: voice, video, best-effort and background traffic applications.

This paper is organized as follows. Section 2 describes the upcoming IEEE 802.11e QoS enhancement standard. In Section 3, we carry out a performance analysis on the effectiveness of the IEEE 802.11e (EDCA) upcoming standard, when

---

* This work was supported by the Ministry of Science and Technology of Spain under CICYT project TIC2003-08154-C06-02, the Council of Science and Technology of Castilla-La Mancha under project PBC-03-001 and FEDER.

supporting different services, such as voice, video, best-effort and background traffic. Finally, Section 4 concludes the paper.

2 The Upcomming IEEE 802.11e Standard

The IEEE 802.11e draft standard [2] is a proposal defining the QoS mechanisms for wireless LANs for to supporting time-sensitive applications such as voice and video communications. In the IEEE 802.11e standard, distinction is made among those stations not requiring QoS support, known as nQSTA, and those requiring it, QSTA. In order to support both Intserv and DiffServ QoS approaches in 802.11 WLAN, a third coordination function is added: the *Hybrid Coordination Function* (HCF). The use of this new coordination function is mandatory for the QSTAs. HCF incorporates two new access mechanisms: the contention-based *Enhanced Distributed Channel Access* (EDCA), known in the previous drafts as the *Enhanced DCF* (EDCF) and the *HCF Controlled Channel Access* (HCCA).

One main feature of HCF is the definition of four *Access Categories* (AC) queues and eight *Traffic Stream* (TS) queues at MAC layer. When a frame arrives at the MAC layer, it is tagged with a *Traffic Priority Identifier* (TID) according to its QoS requirements, which can take the values from 0 to 15. The frames with TID values from 0 to 7 are mapped into four AC queues using the EDCA access rules. On the other hand, frames with TID values from 8 to 15 are mapped into the eight TS queues using the HCF controlled channel access rules. The TS queues provide a strict parameterized QoS control while the AC queues enable the provisioning of multiple priorities. Another main feature of the HCF is the concept of *Transmission Opportunity* (TXOP), which defines the transmission holding time for each station.

2.1 Enhanced Distributed Channel Access (EDCA)

EDCA has been designed to be used with the contention-based prioritized QoS support mechanisms. In EDCA, two main methods are introduced to support service differentiation. The first one is to use different IFS values for different ACs. The

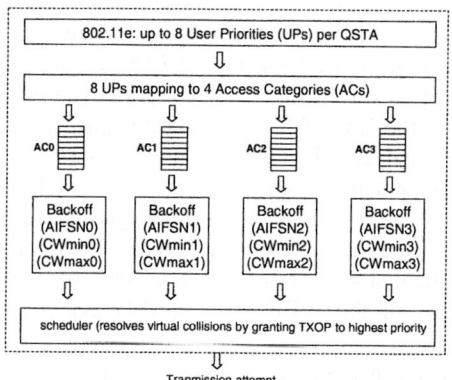

Fig. 1. EDCA

second method consists in allocating different CW sizes to the different ACs. Each AC forms an EDCA independent entity with its own queue and its own access mechanism based on DCF with its own *Arbitration Inter-Frame Space* ($AIFS\,[AC] = SIFS + AIFSN\,[AC] \times SlotTime$) and its own CW[AC] (CWmin[AC] ≤ CW[AC] ≤ CWmax[AC]) (Fig. 1). If an internal collision arises among the queues within the same QSTA, the one having higher priority obtains the right to transmit. It is said that the queue that is able to gain access to the channel obtains a transmission opportunity. Each TXOP has a limited duration (*TXOPLimit*) during which an AC can send all the frames it wants.

3 Performance Evaluation

In this section, we carry out a performance analysis on the effectiveness of the IEEE 802.11e (EDCA) upcoming standard. We demonstrate that the defaults parameters setting recommended in the EDCA draft standard [2] by IEEE 802.11e group are not the best, when the system is supporting different services, such as voice, video, best-effort and background traffic applications. We show that the performance of EDCA can considerable be improved by properly tuning its parameters.

3.1 Scenario

In our simulations, we model an IEEE 802.11b wireless LAN (using OPNET Modeler tool 10.0 [6]) supporting four types of services: Voice(Vo), Video(Vi), Best-effort(BE) and Background(BK). This classification is on line with the IEEE802.1D standard specifications. We assume the use of a wireless LAN consisting of several wireless stations and an access point connected to a wired node that serves as sink for the flows from the wireless domain. All the stations are located within a *Basic Service Set* (BSS), i.e., every station is able to detect a transmission from any other station. The parameters for the wired link were chosen to ensure that the bandwidth bottleneck of the system is within the wireless LAN.

Each wireless station operates at 11 Mbit/s IEEE 802.11b mode and transmits a single traffic type (Vo, Vi, BE or BK) to the access point. We assume the use of constant bit-rate voice sources encoded at a rate of 16 kbits/s according to the G.728 standard[7]. The voice packet size is equal to 168 bytes including the RTP/UDP/IP headers. The voice sources are randomly activated within of the interval [1,1.5] seconds from the starting time of simulation. For the video applications, we have made use of the traces generated from a variable bit-rate H.264 video encoder[8]. We have used the sequence mobile calendar encoded on CIF format at a video frame rate of 25 frames/sec. It is clear that these types of sources exhibit a high degree of burstiness characterized by a periodic traffic pattern and a high variance bit rates. The average video transmission rate is around 480 kbits/s with a packet size equal to 1064 bytes (including RTP/UDP/IP headers). Each video application begins transmitting within a random period given by t = uniform(1; 1+12/f) being *f* the frame rate. In this way, the peak periods of the source rates are randomly distributed along a GOP (*Group of Pictures*) period, a situation most likely to arise in an actual system setup. The transmission of a video frame is uniformly distributed along the time interval of a frame

(1/f). The best-effort and background traffics have been created using a *Pareto* distribution traffic model. The average sending rate of best-effort traffic is 128 kbit/s, using a 552 bytes packet size (including TCP/IP headers). The average sending rate of background traffic is 256 kbit/s, using a 552 bytes packet size (including TCP/IP headers). The traffic sources of these two latter traffic types are randomly activated within of the interval [1,1.5] seconds from the start of the simulation. Throughout our study, we have simulated the two minutes of operation of each particular scenario.

Table 1. Parameter settings evaluated

		Vo	Vi	BE	BK
AIFSN		2	2	3	7
		2	2	4	7
		2	2	5	7
		2	3	5	7
		1	2	3	7
CW_{min}		7	15	31	31
		7	31	31	31
		7	31	63	63
		15	31	31	31
		15	31	63	63
CW_{max}		15	31	1023	1023
		15	63	1023	1023
		15	127	1023	1023
		31	63	1023	1023
		31	127	1023	1023
TXOP Limit		0	0	-	-
		3	4	-	-
		3	5	-	-
		3	6	-	-
		3	7	-	-

For all the scenarios, we have assumed that one fourth of the stations support one of the four kinds of services: voice, video, BE and BK applications. We start by simulating a WLAN consisting of four wireless stations (each one supporting a different type of traffic). We then gradually increase the *Total Offered Load* of the wireless LAN by increasing the number of stations by four. In this way, the stations are always incorporated into the system in a ratio of 1:1:1:1 for voice, video, BE and BK, respectively. We increase the number of stations 4 by 4 starting from 4 and up to 36. In this way, the normalized offered load is increased from 0.12 up to 1.12. We have preferred to evaluate a normalized offered load, rather than the absolute value. The normalized offered load is determined with respect to the theoretical maximum capacity of the 11 Mbit/s IEEE 802.11b mode, i.e. 7.1 Mbit/s (corresponding to the use of the maximum packet size used by the MAC layer and in the presence of a single active station).

We start our study by setting up the parameters to the values recommended by the standards (see Table I, boldface values). This will allow us to set up a base point for comparison purposes as well as to tune up the system parameters.

3.2 Metrics

For the purpose of our performance study, the four metrics of interest are: throughput, collision rate, delay distribution and packet loss rate. To be able to compare the graphs from different levels of load (traffic patterns of different applications), we have preferred plotting the normalized throughput rather than the absolute throughput. The normalized throughput is calculated as the percentage of the offered data that is actually delivered to the destination. In order to limit the delay experienced by the video and voice applications, the maximum time that video packet and voice packet may remain in the transmission buffer has been set to 100ms and 10ms, respectively. These time limits are on-line with the values specified by the standards and in the literature. Whenever a video or voice packet exceeds these upper bounds, it is dropped. The loss rate due to this mechanism is given by the *packet loss rate due to deadline*. Our measurements started after a warm-up period allowing us to collect the statistics under steady-state conditions.

3.3 Results

EDCA makes use of different waiting intervals as a mean to provide various priority levels. These time intervals are defined by the system parameters AIFS, CWmin and CWmax. Furthermore, the use of the extra parameter TXOP can further enhance the priority-handling scheme. We start our study by setting up the different system parameters under study (see Table I) to set up a base point for comparison purposes with respect to the system parameters recommended in the draft standard [2].

Figure 2 shows the performance results as a function of the network load and for various combinations of the waiting interval, AIFS. The metrics reported in this figure are throughput, number of collisions and the number of discarded voice and video packets as well as the overall network throughput and number of packet retransmissions. The AIFS's used by the various types are denoted by BK-BE-Vi-Vo, corresponding to the AIFS used for the background, best-effort, voice and video traffic, respectively. Figure 2.a shows the results obtained for the voice traffic. The voice performance starts to degrade for loads as low as 0.4. The worst results are obtained for the combination 7-3-2-2, i.e., the recommended value in the draft standard. The best results correspond to the combinations assigning a different numerical value to each one of the AIFS's. By assigning different values; the various traffic streams do not compete simultaneously for the channel access. This is clearly demonstrated by the fact that the number of collisions reduces significantly when different values of AIFS's are used. Figure 2.b depicts the results for the video traffic. Contrary to the results for the voice traffic, the performance results obtained for the video traffic are similar for all the AIFS settings being considered. Figure 2.c shows the overall network throughput and number of packet retransmissions for all traffic types. The figure clearly shows similar results for all AIFS combinations under study. It is therefore possible to provide a better QoS to the real-time traffic without penalizing the overall network performance. This is confirmed by the fact that the setting 7 (BK) - 5 (BE) - 3 (Vi) - 2 (Vo) has provided the best results.

Figure 3 shows the performance for the voice and video traffic as well as for the overall network as a function of the network loads and for various values of the CWmin parameter. Recall that this parameter defines the initial (minimum) Backoff window size. The window size is increased after each collision. Following the same convention as above, the CWmin used for each traffic type is denoted as BK-BE-Vi-Vo. Similar to the results shown in Figure 2.a, the performance of the voice traffic

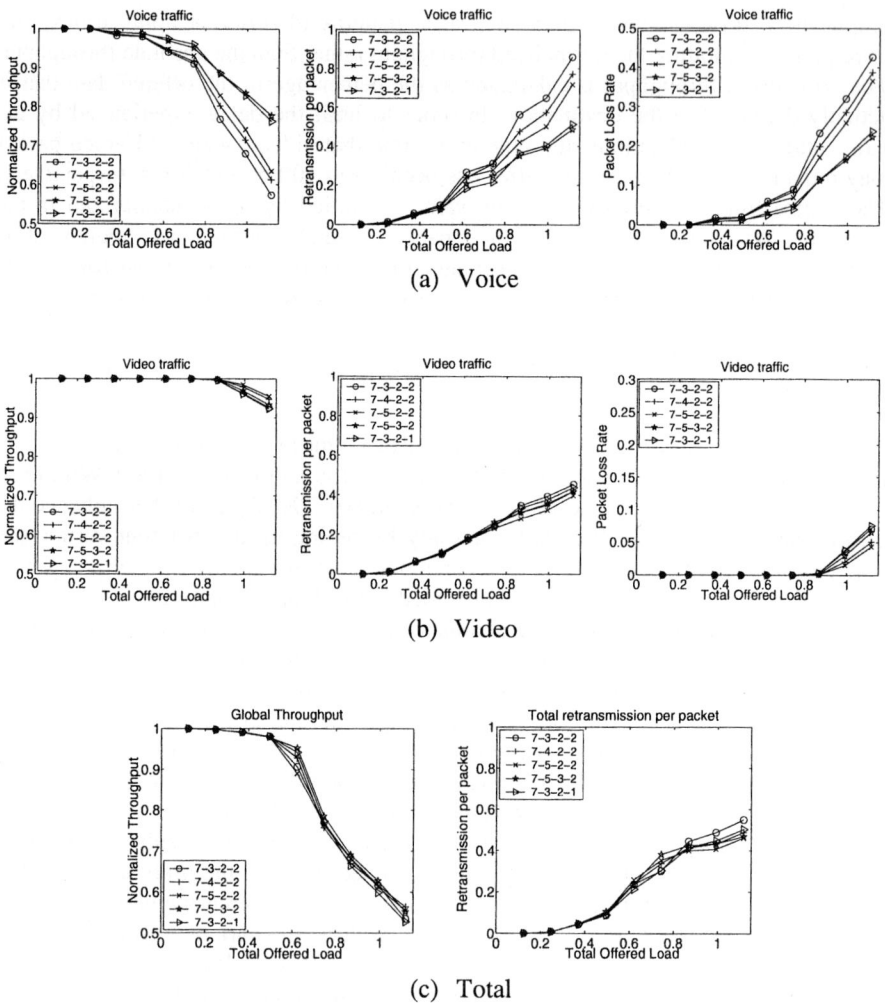

Fig. 2. Performance evaluation of EDCA using different AIFS values

heavily depends on the parameter settings. The use of a larger CWmin improves the throughput of the voice traffic as well as a significant reduction on the number of collisions experienced by the voice traffic. The results also show that it is better to use small values for the voice traffic. The results for the video traffic are depicted in Figure 3.b. The performance results for this type of traffic are very similar for all settings under consideration. It is clear from the results that the video throughput could be improved by penalizing the background and best-effort traffic, i.e., by using larger values for the CWmin used for these two other types of traffic. In this way, the collision probability for the video traffic can be reduced. The worst results are obtained for the values recommended by the standard. From the results, it is clear that it is better to spread out the values for the CWmin for each type of traffic. This is confirmed by the fact that the setting 63-63-31-7 has provided the best results.

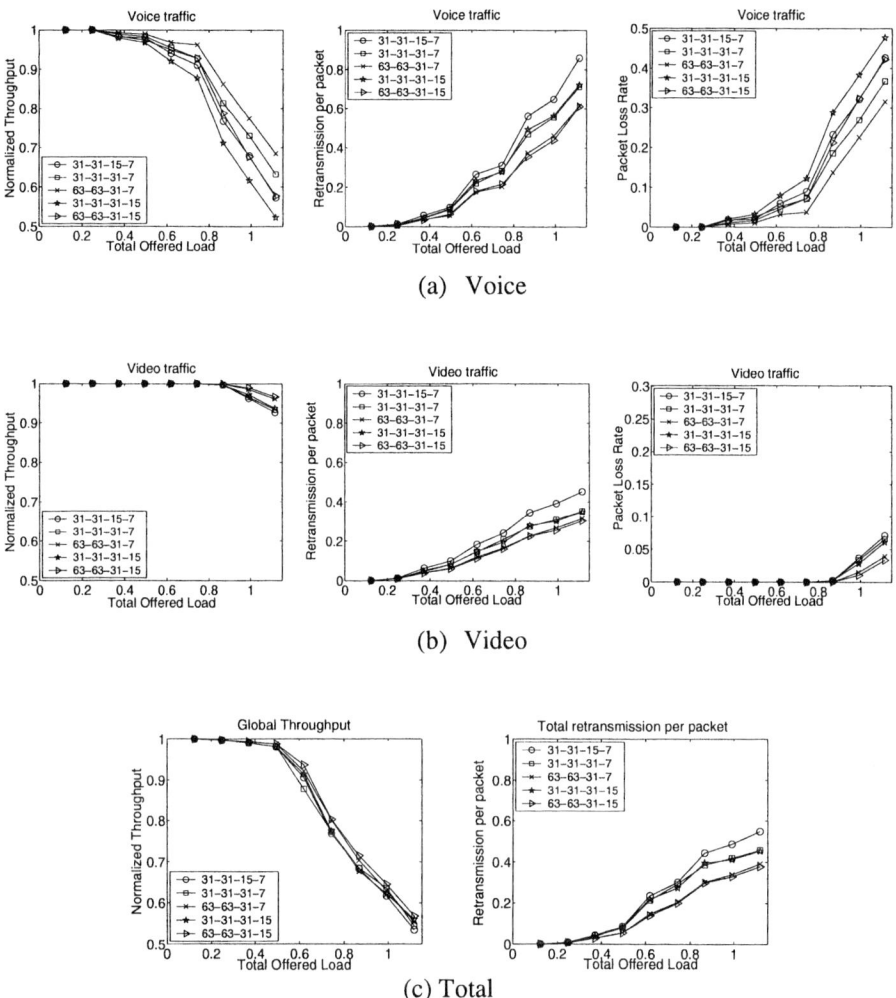

Fig. 3. Performance evaluation of EDCA using different CWmin values

Figure 4 shows the results when varying the system parameter: CWmax. The results are shown as a function of the network load and for various CWmax setting denoted as (BK y BE)-Vi-Vo. Given that this value is only used when a packet requires to be retransmitted several times, the results obtained under low loads are very similar for all combinations. It is at loads of 80% that this parameter plays an important role over the network performance. However, Figure 4.a shows that this parameter does not affect the performance of the voice traffic. This is due to the deadline defined for the voice traffic, i.e., the voice packets are discarded before reaching the CWmax. The results for the video traffic are given in Figure 4.b. Even though that by increasing this parameter, the number of video packet collisions is reduced, the number of video packets discarded increases resulting on a reduction on the video throughput. The overall network performance reported in Figure 4.c shows similar trend to the results obtained when the AIFS and CWmin have been varied.

Figure 5 shows the results for various values of the TXOPLimit parameter. The use of this performance parameter has only been activated for the real-time traffic, i.e., voice and video traffic. However, the voice traffic can not benefit of this scheme. Recall that the voice packets are dropped as soon as they exceed the prescribed deadline, i.e., 10 ms. Moreover, the packetized scheme under consideration generates a voice packet every 80 ms. Therefore, no more than two voice packets will ever be ready to be transmitted. In other words, as soon as a station has sent a packet, the station will switch to the idle state. Figure 5.a shows this situation. Even more, it is clear from the results that the best results are obtained when the TXOPLimit is not used. The figure also shows that the number of collisions encountered by the voice

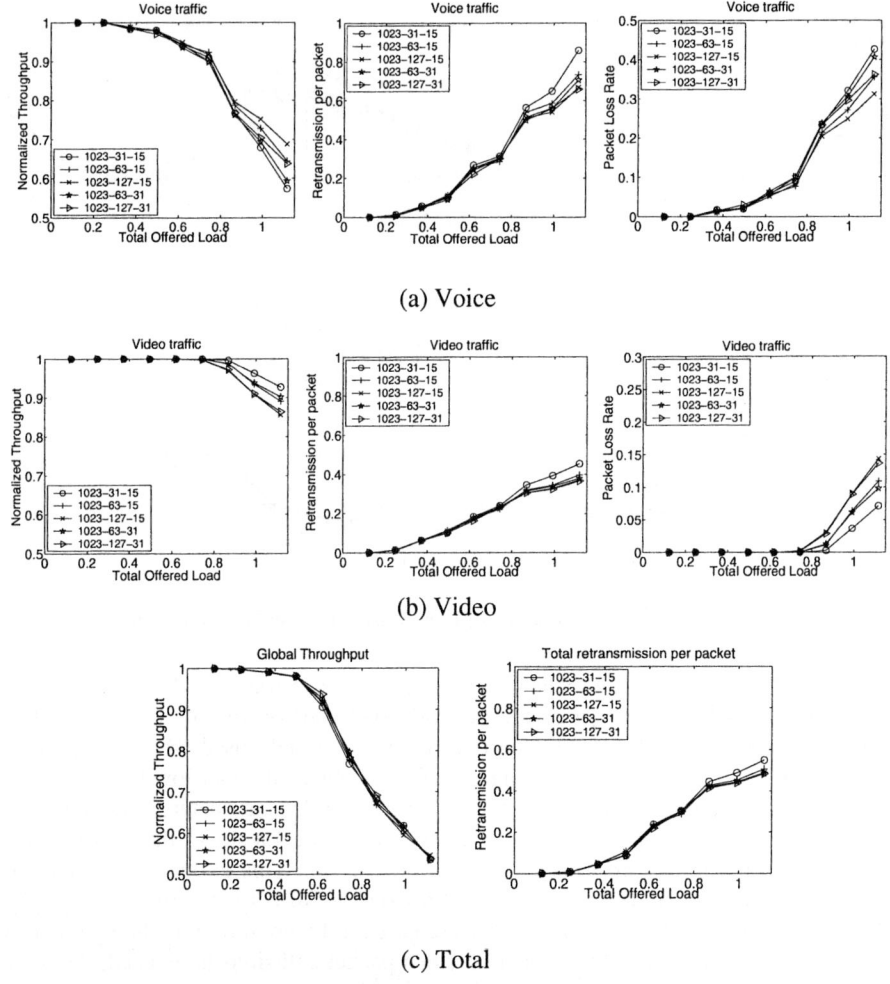

Fig. 4. Performance evaluation of EDCA using different CWmax values

packets is independent of the TXOPLimit being used. In the case of the video traffic, the use of this facility clearly improves the performance of the video applications. This is particularly useful when transmitting video frames of the I type. The results depicted in Figure 5.b clearly show that the use of the TXOPLimit parameter reduces the number of collisions encountered by the video packets. From the results, it is also clear that the value of TXOPLimit should not be set higher than 6 ms. Figure 5.c shows that the network performance is affected by the use of the TXOPLimit. It proves a useful mechanism in managing the channel access mechanism.

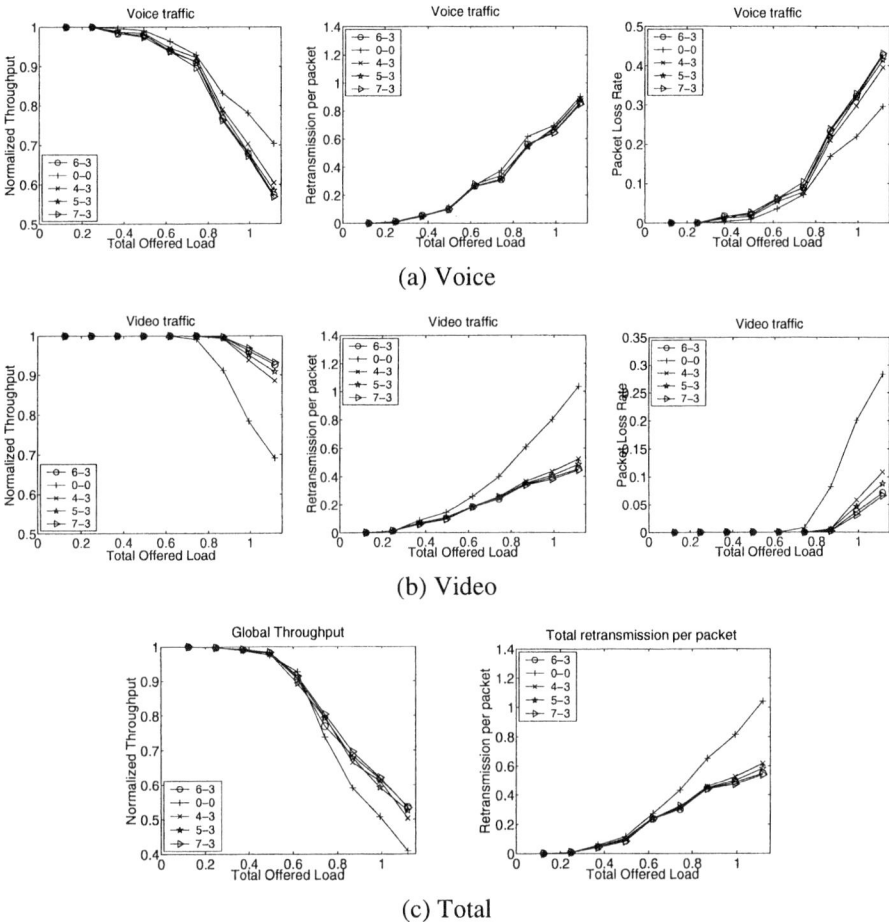

(a) Voice

(b) Video

(c) Total

Fig. 5. Performance evaluation of EDCA using different TXOPLimit values

Another important metric to be reported is the cumulative distribution function for the delay experienced by the real-time applications. Figure 6 depicts this important metric for both real-time services for a network load of 0.75. From the results obtained by varying the IFS parameter, both services, voice and video, exhibit similar

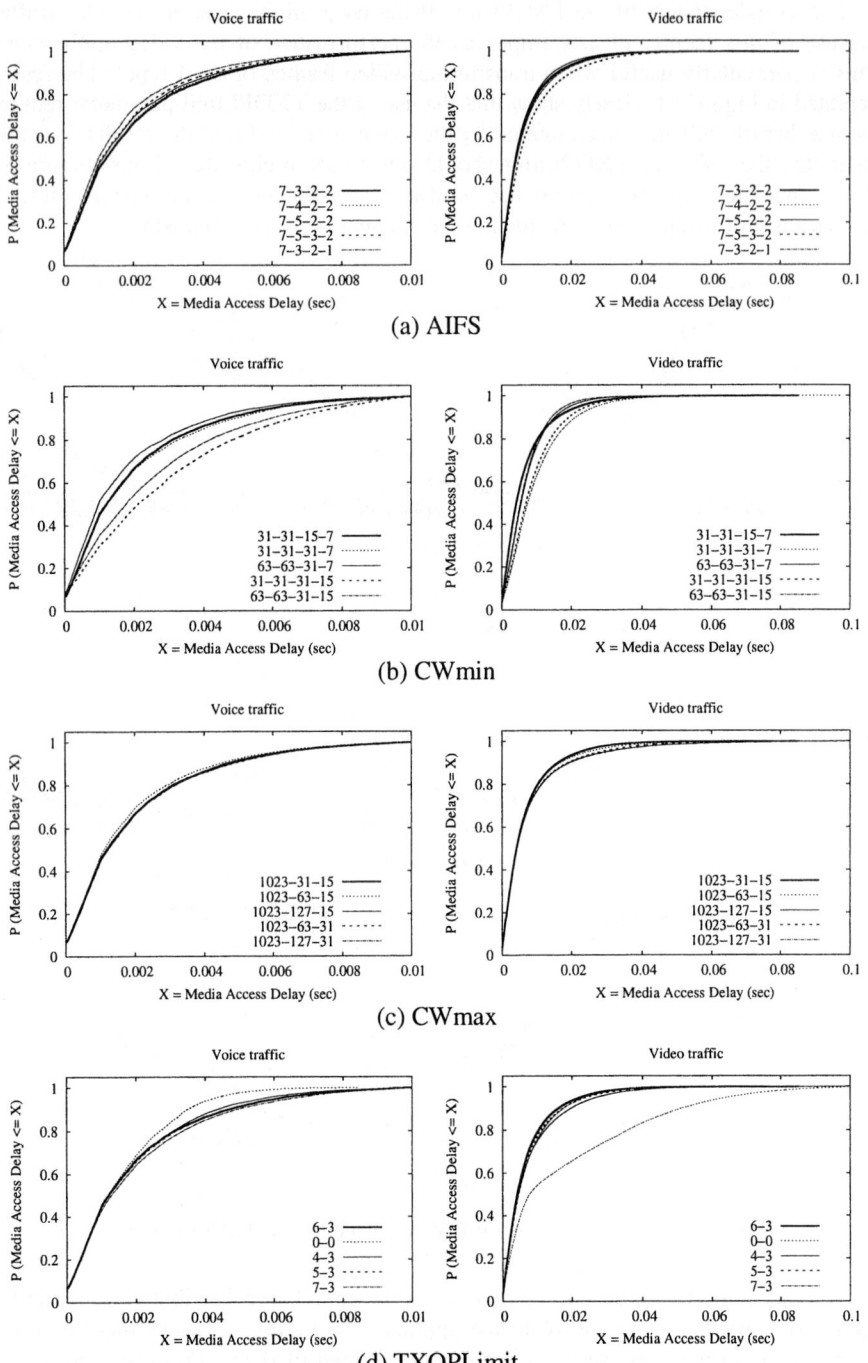

Fig. 6. Performance evaluation of EDCA: CDF of Mean Access Delay

results. For the case of the voice traffic, the delays encountered are lower when different values for the AIFS are used. The setting of the CWmin has a more significant impact over the voice performance in particular when the values used for the CWmin parameter are significantly different from one another. This is expected, since the use of a shorter AIFS allows the voice traffic to promptly access the channel. Similar results are obtained when increasing the CWmin used by the BE and BK traffics. Figure 6.c confirms once again that the CWmax does not have a clear impact over the waiting time. Finally, Figure 6.d shows the results for various values of the TXOP-Limit parameter. In the case of the video traffic, it is clear that the use of this parameter can effectively reduce its waiting time. It is also clear from the figure, that the system performance is not very sensitive to the actual setting of this parameter.

4 Conclusions

In this paper we have evaluated the IEEE802.11e. Our results show that by limiting the number of collisions, the network performance and QoS provisioning can be effectively achieved. The EDCA is unable to guarantee a good performance for loads beyond 0.75. In this latter scheme, the steeply performance drop is mainly due to the excessive number of collisions. The collisions are in turn mainly due to the fact that the AIFS parameter has been fixed to the same value for the video and voice services. Furthermore, the values used for CWmax are too short, 15 and 31 contributing to a higher collision probability. From our results, we can conclude that the values recommended by the standard do not provide the best possible results under heavy load conditions. The performance of EDCA can considerable improved by properly tuning its parameters.

We have also shown that the AIFS plays an important role for differentiating the various traffic types. Our results suggest that it is possible to provide a better service to the voice traffic by using a different value for the video traffic. In the same way, the video traffic can benefit from using a longer AIFS period for the BE traffic. We should point out that the overall network performance remains unchanged. It is therefore recommended to make use of the following setup: 7 (BK) - 5 (BE) - 3 (Vi) - 2 (Vo). Regarding the CWmin parameter, our results also show that the network performance can be greatly improved by properly setting this parameter. The voice traffic can benefit by increasing the length of this parameter for the other traffic types. It has further been shown that the video traffic can also benefit from the proper setting of this parameter. From our overall results, we recommend the use of the following set of values: 63 (BK) - 63 (BE) - 31 (Vi) - 7 (Vo). Regarding the CWmax parameter, this parameter has little effect over the voice and video performance. This is mainly due to the deadlines set up for these two traffic types, i.e., the voice and video packets are not kept for long on the buffer of the sending stations. Finally, we have examined the system sensitivity to the TXOPLimit parameter. Similar to the results obtained for the CWmax, the voice traffic does not benefit from this facility: the voice packets are discarded before the source generates the following packet. In the case of the video, it has been found that this parameter should be set to 5 or 6 ms.

References

1. LAN MAN Standards Committee of the IEEE Computer Society, ANSI/IEEE Std 802.11, "Part 11: Wireless LAN Medium Access Control (MAC) and Physical Layer (PHY) Specifications, 1999 Edition.
2. IEEE 802 Committee of the IEEE Computer Society, IEEE P802.11e/D13.0 Draft Amendment to IEEE Std 802.11, "Part 11: Wireless LAN Medium Access Control (MAC) and Physical Layer (PHY) Specifications: Medium Access Control (MAC) Quality of Service (QoS) Enhancements", April 2005.
3. A. Lindgren, A. Almquist and O. Schelén, "Quality of Service Schemes for IEEE 802.11 Wireless LANs - An Evaluation", Special Issue of the Journal of Special Topics in Mobile Networking and Applications (MONET) on Performance Evaluation of QoS Architectures in Mobile Networks, vol. 8, num. 3, pp. 223-235, June 2003.
4. W. Pattara-Atikom, P. Krishnamurthy and S. Banerjee, "Distributed Mechanisms for Quality of Service in Wireless LANs", IEEE Wireless Communications, vol. 10, num. 3, pp. 26-34, June 2003.
5. Y.Xiao, "IEEE 802.11e: QoS Provisioning at the MAC Layer". IEEE Wireless Communications. pp.72-79. June 2004.
6. Opnet.Technologies.Inc. OPNET Modeler 10.0 ©1987-2004. http://www.opnet.com
7. Coding of Speech at 16 kbit/s using Low-Delay Code Excited Linear Prediction, Std. ITU-T Recommendation G.728, September 1992.
8. ITU-T Recommendation H.264. Advanced Video Coding For Generic Audiovisual Services. May 2003.

Trace-Based Parallel Performance Overhead Compensation

Felix Wolf[1], Allen D. Malony[2], Sameer Shende[2], and Alan Morris[2]

[1] Innovative Computing Laboratory, University of Tennessee
fwolf@cs.utk.edu
[2] Department of Computer and Information Science, University of Oregon
{malony, morris, sameer}@cs.uoregon.edu

Abstract. Tracing parallel programs to observe their performance introduces intrusion as the result of trace measurement overhead. If post-mortem trace analysis does not compensate for the overhead, the intrusion will lead to errors in the performance results. We show that measurement overhead can be accounted for during trace analysis and intrusion modeled and removed. Algorithms developed in our earlier work [5] are reimplemented in a more robust and modern tool, KOJAK [12], allowing them to be applied in large-scale parallel programs. The ability to reduce trace measurement error is demonstrated for a Monte-Carlo simulation based on a master/worker scheme. As an additional result, we visualize how local perturbation propagates across process boundaries and alters the behavioral characteristics of non-local processes.

Keywords: Performance measurement, analysis, parallel computing, tracing, message passing, overhead compensation.

1 Introduction

Trace-based measurement is used to observe the performance of a parallel program when one wants to see the interoperation of multiple threads or processes of execution, as it is recorded in a time-sequence trace of events. Any performance measurement, tracing included, will introduce *overhead* during program execution due to extra code being executed and hardware resources (processor, memory, network) consumed. When performance overhead affects the program execution, we speak of *performance (measurement) intrusion*. Performance intrusion, no matter how small, can result in *performance perturbation* [6] where the program's measured performance behavior is "different" from its unmeasured performance. Whereas performance perturbation is difficult to assess, performance intrusion can be quantified by several metrics, the most important of which is dilation in program execution time. This type of intrusion is often reported as a percentage slowdown of total execution time, but the intrusion effects themselves will be distributed throughout the performance results. In the case of tracing, we will also see performance error due to intrusion (i.e., performance perturbation) in the timings of the interdependent events between the processes.

Of course, we cannot compare the measured parallel execution with the "real" parallel execution to determine the intrusion error because we do not have any information about what the execution would be like without instrumentation. All we know is that there is measurement overhead included in the trace data and that this overhead may have intruded on the parallel performance in such a way as to cause misleading performance effects. For tracing, the overhead introduced with each event measurement includes the creation of an event record and writing it to a trace buffer. If we can determine the overhead size, it may be possible to subtract this overhead for each event individually and generate a second "overhead-free" trace file. We must be careful in doing so not to violate the *happened-before* relation [1] that exists between interdependent process events.

While performance intrusion can alter program execution and, thus, perceived performance, parallel performance tools rarely attempt to adjust the performance measurements to compensate for the intrusion. Recently, we have shown overhead compensation is possible to do in parallel profiling [7,8]. However, profiling summarizes performance data and, thus, loses the performance detail captured in traces. Also, because overhead compensation must be done online, certain forms of performance perturbation cannot be resolved during profiling. With trace-based overhead compensation, we have the opportunity of preserving performance detail while dealing with more complex intrusion effects. In our earlier work, we designed the performance models necessary for trace perturbation analysis [5,10]. However, our implementation of these were for research purposes only. In this paper, we update our algorithms and build them into a robust trace measurement and analysis system.

Section 2 provides a brief background on performance intrusion and perturbation analysis. Our algorithms for overhead compensation are presented in Section 3. Section 4 outlines the implementation of these algorithms in the trace analysis tool. We demonstrate the techniques on a set of validation experiments. Section 5 describes the experimentation environment, the testcases, and the trace analysis results. Conclusions and future work are given in Section 6.

2 Tracing, Intrusion, and Perturbation

Events are actions that occur during program execution. Typical events include *interval events* that are characterized by a pair of actions marking *entry* and *exit* points, and *atomic events* that occur at a single place or time. Tools insert measurement code to track the performance of a parallel program as made visible by the instrumented events. Tracing collects event records in an *event trace*. Each record describes an event, when it occurred, and any associated data. From this information, we can see the patterns of execution behavior that contribute to the performance. The measurement intrusion in event traces displays itself as alterations of event timing and order. The goal of overhead compensation in trace analysis is to remove the time intrusion due to measurement overhead and fix its effects on event ordering, in hopes of recovering the actual performance behavior.

For discussion purposes, let us consider a message passing program. If we look at the impact of overhead on events local to a process, there is a time dilation (slowing down) of when events occur, resulting in later event time stamps compared to an actual event trace. Because the events occur locally, this dilation can be directly determined and corrected. However, every message communication links the process event streams and the evolution of process times become dependent at these points. The result of measurement overhead affects the interdependent ordering and timing of these events compared to actual. The parallel execution semantics, as reflected in the message communication operations and how the message data is used, determines process dependencies and message event ordering relationships, but only partially. Non-deterministic execution allows for alternative message event orderings, different from observed. It is important to understand that the only information we have about process interdependencies are the message communication events and when they occur in the *measured* execution. If our trace analysis does not have enough information to determine if a different (reconstructed) event order is valid, it must enforce the *same* ordering of message communication events in the "approximated" execution as in the measured execution.

In contrast with our techniques for parallel profile overhead compensation, trace overhead analysis can be thought of as a trace-driven replay of the execution where we can apply both event and timing models to correct intrusion effects. There is also opportunity in trace analysis to utilize measurements of interprocess events to improve the accuracy of the approximated execution, such as with computed message communication times. For the work reported here, it is important to remember that the goal of tracing is to observe detailed temporal performance. Thus, we hope that the overhead analysis will result in more accurate performance characterization both as it has to do with overall performance (e.g., more accurate total execution times) as well as local performance details (e.g., waiting times for individual message communications).

Other research work has sought to characterize measurement overhead as a way to bound intrusion effects or to control the degree of intrusion during execution. For instance, the work by Kranzlmüller [4] quantifies the overhead of MPI monitors using the benchmarking suite SKaMPI, and Fagot's work [2] assesses systematically the overhead of parallel program tracing. The work by Hollingsworth and Miller [3] demonstrates the use of measurement cost models, both predicted and observed, to control intrusion at runtime via measurement throttling and instrumentation disabling. Their work primarily deals with profiling-based measurement.

3 Overhead Compensation Algorithms

The trace of a parallel program's execution provides time-sequenced information about the events that occurred and when they occurred. To compensate for the overhead during measurement, we want to characterize the amount of overhead (O) for each event measurement, and subtract that overhead from the event timings. For events that are *local* to a process (we will call these *independent events*), this overhead compensation can be done directly. For events that are involved in dependent execution, we must take care not to violate happened-before time order relationships [1].

The algorithms we present below are based on our earlier work [5,10], as targeted here to MPI message passing parallel programs. We make several assumptions in these algorithms:

- Only MPI_Send() and MPI_Recv() are used for point-to-point communication.
- MPI_Send() is always non-blocking.

- Only *n-to-n* and *1-to-n* collective operations are considered.
- The per-event measurement overhead (O) is constant.
- The buffer copy time is a function of message size $C(size(msg))$.

Clearly, the values C and O are platform specific and must be measured. For C, we run experiments with different message sizes and build a table of per byte copy times to be queried during analysis. For O, we run an experiment where N trace events are generated immediately following each other. The amount of time consumed is then divided by N to give O.

Based on the assumptions above, all dependent execution is due to message communication. Point-to-point (P2P) communication involves only the sender and receiver. The dependencies in collective communication are more interesting. While it is possible to logically reduce collective communication to P2P communication, doing so may restrict the analysis from applying what is known about the collective execution semantics.

3.1 Independent Events

Independent events are events that do not directly dependent on communication. In other words, in our environment, they are not communication events. Consider the ith event on a process, $event^i$. We use the notation $event^i_m$ to denote the *measured time stamp* of the event and $event^i_a$ to denote the *approximated time stamp* of the event after trace analysis. The trace analysis moves forward in the trace for each process, computing the approximated time stamp of the next event. Thus, at any point in time during the analysis, we look to see which immediate next event on each process can be processed next.

Let us assume $event^{i-1}_a$ has been computed and the next event, $event^i$ is not a communication event. The trace analysis can determine $event^i_a$ by:

$$event^i_a = event^{i-1}_a + (event^i_m - event^{i-1}_m) - O \qquad (1)$$

Effectively, we keep the execution time between the two events, $event^i_m - event^{i-1}_m$, and subtract the overhead. To this value we then add the approximated time stamp of the predecessor event, $event^{i-1}_a$.

3.2 Dependent Events

What happens if the event is a communicaton event? Here is where things get interesting. Let us focus on P2P communication first. There are six events to consider: *enter.send*, *send*, *exit.send*, *enter.recv*, *recv*, and *exit.recv*. Among these, only *recv* directly depends on communication. Often instrumentation of message communication is done using an interposition library, such as in the MPI profiling interface, PMPI [9]. Here, the time stamps of the *send* and *enter.send* events will likely differ only by a very small amount. The same is true for *recv* and *exit.recv*.

There are two cases in the measured execution to consider for a particular P2P communication:

(m.1) $enter.recv_m \leq exit.send_m$
(m.2) $enter.recv_m > exit.send_m$

It should be understood that the send and receive events are taking place on two different processes. Condition (m.1) means that there is a temporal overlap between the MPI_Send() and the MPI_Recv() operation. Condition (m.2), in contrast, means that there is a gap between the two.

(m.1) Communication Time can be Measured. If $enter.recv_m$ occurs before $exit.send_m$, we assume that the receiver can begin processing the message as soon as it is delivered. As a result, we can calculate the actual communication time from the measured trace:

$$Comm_m = recv_m - send_m \tag{2}$$

This is important since the measured communication time is the most accurate representation of communication performance. Two cases result for the approximated execution:

(a.1) $send_a + Comm_m > enter.recv_a$

(a.2) $send_a + Comm_m \leq enter.recv_a$

In the first approximation case, the entry to the receive occurs before the communication completes, meaning that the receiver has to wait. Thus, the approximated receive time can be determined by:

$$recv_a = send_a + Comm_m \tag{3}$$

In the second case, the receive occurs after the message has already been delivered and supposedly is present at the receiving process. All that is left to do is for the receiver to copy the message into the receive buffer:

$$recv_a = enter.recv_a + C(size(msg)) * size(msg) \tag{4}$$

Again, in our experiments, we use a lookup table to determine C for difference message sizes.

(m.2) Communication Time Cannot be Measured. If the receive operation begins after the send operation has finished, we cannot use the trace measurement directly to compute the message communication time. An upper bound approximation on the communication time still comes from the communication measurement:

$$Comm_a^{upper} = recv_m - send_m \tag{5}$$

However, this time may include time a message spends sitting at the receiver before the receive begins. A lower bound approximation effectively assumes that the transmission time through the communications network is zero. In this case, we need to only account for the message copy time both at the sender and at the receiver:

$$Comm_a^{lower} = 2 * C(size(msg)) * size(msg) \tag{6}$$

Since the start times of the send and the receive operation might be significantly pulled apart in the approximated execution, there is no guarantee that the lower-bound and the upper-bound communication times together give a valid time interval for the approximated receive event, that is, the following condition might be violated:

$$recv_a > enter.recv_a \qquad (7)$$

This consistency requirement leads to the stipulation of a minimum communication time that has to be observed in both cases.

$$Comm^{min} = (enter.recv_a - send_a) + (C(size(msg)) * size(msg)) \qquad (8)$$

Now, both the lower-bound and the upper bound communication time need to be modified not to fall below the minimum, similar to what we did earlier in case (m.1 / a.2).

$$Comm_a^{lower} = max(2 * C(size(msg)) * size(msg), Comm^{min}) \qquad (9)$$

$$Comm_a^{upper} = max(recv_m - send_m, Comm^{min}) \qquad (10)$$

Finally, the bounds for the approximated receive time are:

$$send_a + Comm_a^{lower} \leq recv_a \leq send_a + Comm_a^{upper} \qquad (11)$$

3.3 Collective Communication

For our work in this paper, we also consider *n-to-n* and *1-to-n* collective communication operations. For any collective communication, we can transform the operation to point-to-point communication and then apply the formulas above to perform overhead compensation. However, it is not so easy to translate collective communication operations into their P2P equivalents. Also, collective communication has additional semantics that must still be enforced when processing the collective events. These semantics can be used to build overhead compensation algorithms specific to collective operations.

n-to-n. Consider *n-to-n* collective communication. There are two collective events of interest for each process: *enter* and *exit*. The approximated *enter* time stamp, $enter_a^i$, is determined for each process i based on the algorithm for a independent event. However, $exit_a^i$ is dependent on when the collective synchronization occurs. Let j be the process with the latest measured entry event, $enter_m^j$. Let k be the process with the latest approximated entry event, $enter_a^k$. Let l be the process with the earliest measured exit event, $exit_m^l$. Because *n-to-n* collective operations enforce collective synchronization, we could assume $exit_a$ is computed to be the same for all processes:

$$exit_a^i = enter_a^k + (exit_m^l - enter_m^j) \qquad (12)$$

Here, $exit_m^l - enter_m^j$ is a measurement (from the trace) of the time to synchronize, once all processes have entered the collective communication. However, we can also measure this synchronization time for each process individually:

$$exit_a^i = enter_a^k + (exit_m^i - enter_m^j) \qquad (13)$$

1-to-n. In the case of *1-to-n* collective communication, the translation to a P2P equivalent form will work fine for approximation purposes. The one sender (*root*) process has the events *enter.send* and *exit.send*. The multiple receivers each have events *enter.recv* and *exit.recv*. For each send-receiver pair, we translate the events as follows:

send := *enter.send*
recv := *exit.recv*

We compute *size*(*msg*) as the amount of data received by receiver from root.

Note that the actual communication time cannot be measured because it is not known which fraction of the send operation was performed on behalf of a particular receiver. Therefore, we essentially give an upper-bound approximation. Plus, as collective communication is often implemented in a tree-like fashion, receivers may be senders as well. We do not take this into consideration.

4 Implementation

We validated our model using a prototype implementation within the KOJAK performance evaluation system [12]. KOJAK is a suite of performance tools that collect and analyze trace data from parallel programs including MPI applications. Event traces are generated automatically at runtime using a combination of source code annotations or compiler-supported instrumentation and hardware counters. The analysis component uses pattern recognition to convert the traces into information about performance bottlenecks relevant to developers.

Before executing the application, it is linked to a tracing library responsible for generating the trace file at runtime. The trace files are written in the EPILOG format. Overhead-intensive activities performed by the library to support trace-file generation, such as offset measurements among local clocks for a later time synchronization or file IO operations to flush the memory buffer upon overflow, are enclosed by special records so that the compensation filter can account for these overheads. After the application has terminated, an off-line analyzer scans the resulting trace files for execution patterns, classifies them by behavior type, and quantifies their impact on the overall performance. The results can be viewed in a GUI that shows performance problems broken down by call path and process or thread (see Figure 2).

At the core of the analyzer is a library called EARL [13] reading EPILOG traces and providing abstractions that simplify the task of detecting patterns in the event stream. These abstractions include execution state information, such as the progress of collective operations at the time of a given event, and links between related events that allow, for example, the analyzer to find the send event to a given receive event. Another important feature is EARL's ability to access events randomly by their relative position in the trace file.

Because of its convenient model to access and process trace information, EARL was chosen to implement the compensation filter. The filter consists of three parts: (i) a component to measure platform-specific quantities, such as the average per-event overhead and the memory bandwidth needed to compute the buffer copy time, (ii) a queuing system to reorder the events in accordance with the approximated time stamps, and (iii) the actual compensation engine to execute the algorithms described earlier in Section 3. To accommodate cache effects, the memory bandwidth is measured for buffers of varying size. To improve scalability, the queuing systems maintains only a finite window of time stamps, events whose final position within the approximated trace can be determined and whose time stamps are no longer needed to approximate subsequent events are written to the approximated trace.

5 Experimental Results

To illustrate the effectiveness of our strategy, we examine a parallel MPI application that computes the value of π using a Monte-Carlo integration algorithm, which calculates the area under the circle function curve from 0 to 1. The program comprises of a master (or server) task that generates work packets with a set of random numbers. The master task waits for a request from any worker and sends a chunk of randomly generated numbers to it. For each pair of numbers that is given to a particular worker, it finds out whether the pair of Cartesian coordinates represented by the number is inside or outside the circle. Thus, the workers collectively estimate the value of π iteratively until it is within a given error range.

We executed the application in two modes: uninstrumented and instrumented. The instrumentation was applied to all user functions and MPI functions to generate an EPILOG trace file during execution. The number of processors was varied between 2 and 32 on a 4-way Intel Pentium III Xeon 550 MHz Linux cluster with 8 nodes. We ran the application 5 times under each configuration and took the shortest run as our representative. The time stamps in the trace files were initially collected using the RDTSC timer and later off-line synchronized with process 0 using linear interpolation between two offset measurements to compensate for different clock drifts among local clocks. For each measured trace file, two approximated trace files were generated, the first one using lower-bound approximation, the second one using upper-bound approximation. The results are shown in Figure 1.

The main fraction of the instrumentation overhead results from a small function named get_coords() that is frequently called by worker processes. Since number of invocations is proportional to the amount of work, the amount of intrusion declines as the number of worker increases. It can be seen that the execution times generated by both approximations come close to the uninstrumented time. When the dilation of execution time introduced by the instrumentation is high, our approximation proves to be most effective in terms of the percentage of overhead removed. Unfortunately, however, the lower-bound compensation is still too pessimistic in that it remains consistently above the uninstrumented execution. Reasons for this might be found in inaccurate measurements of platform constants and in simplified assumptions made by the model itself.

In Figure 1, we only investigate execution time *slow down*. The actual strength of our approach, however, is that the approximated event traces allow us to study the perturbation, that is, the qualitative change in program behavior caused by the overhead. Since interprocess communication can propagate instrumentation overhead across

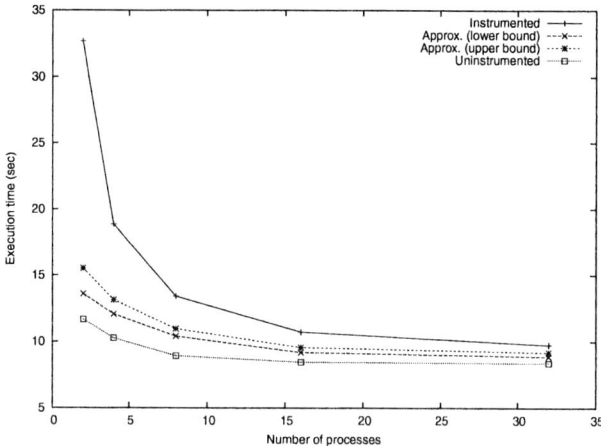

Fig. 1. Measured and approximated execution times for Monte-Carlo application

different processes, perturbation effects may be observed at a process that actually does not produce significant overhead itself. To examine qualitative perturbation effects in more detail, we applied KOJAK's pattern analyzer EXPERT to the measured and the approximated trace file. The two results have been subtracted using KOJAK'S performance algebra utility [11] to study the overhead composition in detail. The results are shown in Figure 2 displayed in the KOJAK GUI.

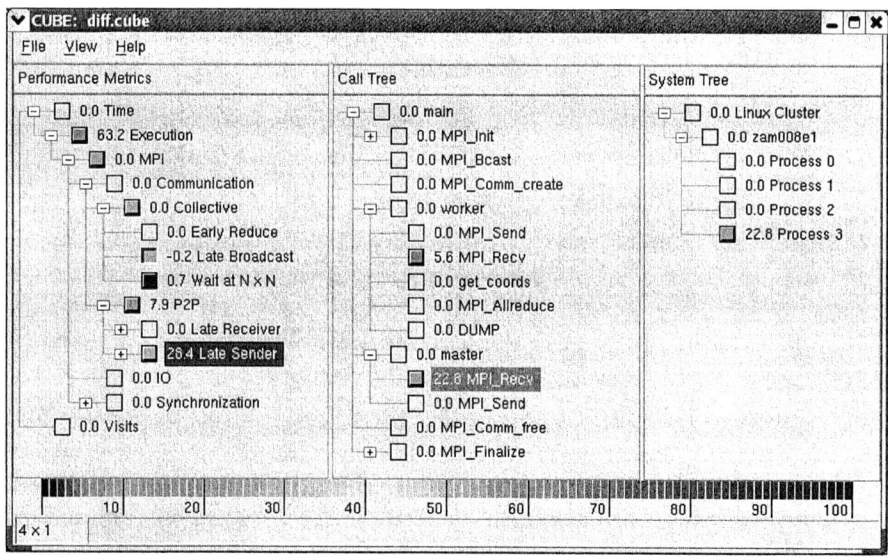

Fig. 2. Composition of overhead for 4 processors

All numbers represent percentages of the difference in the execution time between measured and approximated execution. The left pane shows the overall difference broken down by behavior type for the entire program and all processes. It can be seen that the majority (63 %) of the overhead is non-communication (i.e., metric *Execution* expanded). The non-communication overhead has been found to be a direct effect of the tracing library's operation (i.e., per-event overhead and flushing the memory buffer) and is almost exclusively caused by workers. On the other hand, a significant fraction (28 %) of the total overhead is waiting time within the *Late Sender* pattern, which describes a receiver waiting for a message that has not been sent yet. In contrast to our previous finding, this indirect effect of perturbation can be nearly exclusively (80 %) attributed to the server.

Figure 3(a) shows the measured execution of one iteration in our Monte-Carlo example with four processes in a time-line diagram. The finely striped sections represent workers processing a chunk of random numbers. The black stripes indicate calls to get_coords(), the function mainly responsible for intrusion. At the bottom, the master process (process 3) performs three send operations - one for each of the three workers, after which it starts waiting in a receive call for response. The wait state lasts until the end of the collective call performed by all workers to determine the progress of the computation.

Figure 3(b), in contrast, shows about the same amount of execution time of the lower-bound approximation. The iteration on the left finishes significantly earlier. Most notable is, however, that as a result of the reduced overhead visible on the worker time lines, the wait state found on the master time line became significantly smaller, whereas the preceding send operations remained about the same. Thus, our example demonstrates that local perturbation effects can propagate across process boundaries and significantly distort the performance behavior of non-local processes.

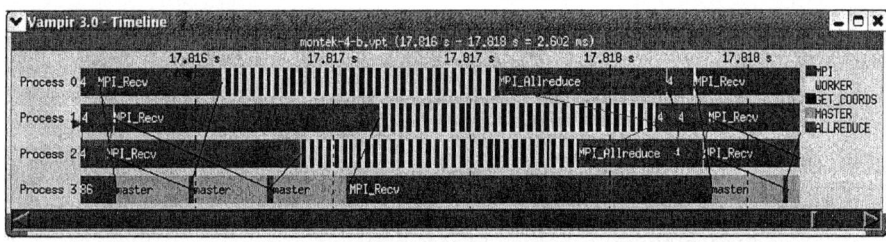

(a) Measured execution

(b) Approximated execution

Fig. 3. VAMPIR time-line diagrams of measured and approximated execution

6 Conclusion

Most parallel performance measurement tools ignore the overhead incurred by their use. Tool developers attempt to build the measurement system as efficiently as possible, but do not attempt to quantify the intrusion other than as a percentage slowdown in execution time. Our earlier work on overhead compensation in parallel profiling showed that the intrusion effects on the performance of events local to a process can be corrected [7] and it also modeled how local overheads affected performance delay across the computation [8]. However, parallel profiling only provides performance summary statistics. In order to see execution detail, tracing measurements must be used. This papers concerns the compensation of overhead during trace analysis.

The goal of trace analysis is to detect performance patterns and identify performance problems associated with certain patterns. It is important then that the timing properties of the trace data be as accurate as possible. Overhead can introduce intrusion that alters event timing structure and order, causing trace analysis to report performance problems where none are present in the "actual" execution, or to even mask performance problems that might otherwise appear.

The algorithms we designed and re-engineered in KOJAK can remove measurement overhead from a parallel trace. In doing so, we contend the performance properties [14] captured in the transformed trace data will be more representative of the performance behavior in a uninstrumented execution. The experiments presented here give powerful evidence to this conclusion. The Kojak analysis of the master-worker test case shows clearly the better performance problem identification as a result of overhead compensation. Figure 3 gives visual evidence to the improvement in detailed event time relations.

It is important to understand that we do not claim the compensated trace resulting from trace analysis is exactly the same as the trace of an uninstrumented execution, if that trace could be obtained without measurement overhead. It is even difficult to make quantitative statements about the bounds on analysis error. Indeed, the *performance uncertainty principle* [6] implies that the accuracy of performance data is inversely correlated with the degree of performance instrumentation. Our goal is to improve the tradeoff, that is, to improve the accuracy of the performance data through more intelligent trace analysis. What we are saying in this paper is that the performance results produced by applying our algorithms for trace-based overhead compensation will be more accurate that performance results produced without compensation. In addition, we are providing this overhead compensation capability in state-of-the-art tracing tools that are being distributed to the parallel computing community.

Acknowledgements

This research is supported at the University of Oregon by the U.S. Department of Energy (DOE), Office of Science contract DE-FG02-05ER25680. At the University of Tennessee, this research is supported by the U.S. Department of Energy under Grants DE-FG02-01ER25510 and DE-FC02-01ER25490.

References

1. L. Lamport, "Time, Clocks and the Ordering of Events in a Distributed System," *CACM*, **21**(7), pp. 558-565 (July 1978).
2. A. Fagot and J. de Kergommeaux, "Systems Assessment of the Overhead of Tracing Parallel Programs," *Euromicro Workshop on Parallel and Distributed Processing*, pp. 179–186, 1996.
3. J. Hollingsworth and B. Miller, "An Adaptive Cost System for Parallel Program Instrumentation," *Euro-Par Conference*, Volume I, pp. 88–97, August 1996.
4. D. Kranzlmüller, R. Reussner, and C. Schaubschläger, "Monitor Overhead Measurement with SKaMPI," *EuroPVM/MPI Conference*, LNCS 1697, pp. 43–50, 1999.
5. A. Malony, "Event Based Performance Perturbation: A Case Study," *Principles and Practices of Parallel Programming (PPoPP)*, pp. 201–212, April 1991.
6. A. Malony, "Performance Observability," Ph.D. thesis, University of Illinois, Urbana-Champaign, 1991.
7. A. Malony and S. Shende, "Overhead Compensation in Performance Profiling," *Euro-Par Conference*, LNCS 3149, Springer, pp. 119–132, 2004.
8. A. Malony and S. Shende, "Overhead Compensation in Parallel Performance Profiling," *Parallel Processing Letters*, to be pubished, 2005.
9. Message Passing Interface Forum. MPI: A Message Passing Interface Standard, Chapter 8, Profiling Interface, Juni 1995. http://www.mpi-forum.org.
10. S. Sarukkai and A. Malony, "Perturbation Analysis of High-Level Instrumentation for SPMD Programs," *Principles and Practices of Parallel Programming (PPoPP)*, pp. 44–53, May 1993.
11. F. Song, F. Wolf, N. Bhatia, J. Dongarra, and S. Moore. An Algebra for Cross-Experiment Performance Analysis. In *Proc. of the International Conference on Parallel Processing (ICPP)*, Montreal, Canada, August 2004.
12. F. Wolf and B. Mohr. Automatic performance analysis of hybrid MPI/OpenMP applications. *Journal of Systems Architecture*, 49(10-11):421–439, 2003. Special Issue "Evolutions in parallel distributed and network-based processing".
13. F. Wolf. EARL - API Documentation. Technical Report ICL-UT-04-03, University of Tennessee, Innovative Computing Laboratory, October 2004.
14. F. Wolf and B. Mohr. Specifying Performance Properties of Parallel Applications Using Compund Events. *Parallel and Distributed Computing Practices*, 4(3), September 2001. Special Issue on Monitoring Systems and Tool Interoperability.

Reducing Memory Sharing Overheads in Distributed JVMs

Marcelo Lobosco[1], Orlando Loques[2], and Claudio L. de Amorim[1]

[1] Laboratório de Computação Paralela, PESC, COPPE, UFRJ,
Bloco I-2000, Centro de Tecnologia, Cidade Universitária, Rio de Janeiro, Brazil
{lobosco, amorim}@cos.ufrj.br
[2] Instituto de Computação, Universidade Federal Fluminense,
Rua Passo da Pátria, 156, Bloco E, 3º Andar, Boa Viagem, Niterói, Brazil
loques@ic.uff.br

Abstract. Distributed JVM systems by supporting Java's shared-memory model enable concurrent Java applications to run transparently on clusters of computers. Aiming to reduce the overheads associated to memory coherence enforcement mechanisms required in such distributed JVMs, we propose two new techniques, selective dynamic diffing and lazy home allocation. To evaluate their potential benefits, both techniques were implemented in CoJVM, a distributed JVM system that we developed in a previous work. In the sequel, several experiments based on five representative concurrent Java applications were carried out using the original and modified CoJVM versions. The analysis of the experimental results showed that dynamic diffing and lazy home allocation, either in isolation or in combination, can reduce significantly memory sharing overheads due to message traffic, extra memory space, and high latency of remote memory accesses. Specifically, the application of these techniques resulted in considerable gains in performance, ranging from 9% up to 20% in four out of five applications, with speedups varying from 6.5 up to 8.1 for an 8-node cluster.

Keywords: JVM, distributed shared memory, Java, cluster computing, concurrent Java applications, high-performance computing.

1 Introduction

In previous work [1], we introduced a new Java [2] environment for high-performance computing, namely the **Co**operative **J**ava **V**irtual **M**achine (**CoJVM**). We reported performance of CoJVM ranging from 5.4 to 7.7 on 8-node cluster for several parallel Java benchmarks. The motivation behind the development of CoJVM was the fact that the use of Java's concurrency model for developing parallel applications was limited to costly shared-memory computers although clusters of low-cost computers offered an even more cost-effective computing platform for high-performance computing. Therefore, CoJVM represented a first step towards providing a distributed shared-memory (DSM) Java platform that could execute efficiently and transparently standard Java abstractions for concurrent programming. Despite the good speedups CoJVM achieved we noticed that there was considerable room for further performance improvements. Specifically, we measured a large imbalance in page distribution among the processing nodes for most applications we tested. It is wellknown that uneven page distribution impacts severely the locality of reference, which in our case increased substantially the coherence protocol overheads, including the amount of control and data messages, page access faults, and barrier times, which hurt CoJVM performance significantly.

In this work, we further address CoJVM performance issues with two new simple but effective techniques, namely selective dynamic *diffing* and lazy home allocation, which take advantage of the information CoJVM extracts at run-time about the application behavior. Selective dynamic *diffing* technique implements a bit vector to track shared memory positions that are modified at run-time and uses the bit vector to create *diffs* in an effective way. Lazy home allocation postpones the association between pages and homes until the time at which the computation phase really takes place and pages will be definitely used. Therefore, lazy home allocation will generally prevent an uneven number of pages to be assigned to any single node, as it often happens under the popular first-touch home assignment policy. Five representative kernels, MM, SOR, LU, FFT, and Radix, were used as benchmarks to evaluate the impact of our two proposed techniques on CoJVM performance. Our experimental results showed that four out five benchmarks achieved superior speedups when they were executed in optimized CoJVM versions. The only exception was MM, which performed 1% better in original CoJVM. Overall these results demonstrate the effectiveness of the proposed techniques on CoJVM performance.

This paper presents three main contributions: a) we propose and evaluate two new techniques for distributed JVMs: selective dynamic *diffing* and lazy home allocation; b) we demonstrate the effectiveness of the bit vector mechanism and associate technique we created to reduce both the amount of message traffic and extra memory space that distributed JVMs require to support a DSM model; and c) we present detailed performance analysis of the two techniques for five parallel Java benchmarks. The remainder of this paper is organized as follows. Section 2 overviews DSM systems and the CoJVM. Section 3 introduces the proposed optimization techniques. In section 4 we evaluate the impact of the optimization techniques on CoJVM. In section 5 we draw our conclusions and outline future works. Due to space limitations, we refer the reader to [1,3] where we discuss related works [4, 5, 6, 7].

2 Background

DSM. Software DSM systems provide the shared memory abstraction on a cluster of physically distributed computers. This illusion is often achieved through the use of the virtual memory protection mechanism [8]. However, using the virtual memory mechanism has a main shortcoming: the occurrence of false sharing due to the use of a large virtual page as the unit of coherence, which leads to unnecessary communication traffic. Several relaxed memory models, such as LRC [9], have been proposed to alleviate false sharing. In LRC, shared pages are write-protected so that when a processor attempts to write to a shared page an interrupt will occur and a clean copy of the page, called the twin, is built and then the page is released to write operations. In this way, changes to the page, called *diffs*, can be obtained at any time by comparing the current copy with its twin. LRC requires the programmer to use two explicit synchronization primitives to access shared data: acquire and release. Coherence-related messages are delayed until an acquire is performed by a processor. When an acquire operation is executed the acquirer receives from the last acquirer all the write-notices, which correspond to changes made to the pages that the acquirer has not seen according to the happen-before-1 partial order [10]. HLRC [11] introduced the concept of home node, in which each node is responsible for maintaining an up-to-date copy of its owned pages; then, the acquirer can request copies of modified pages from their home nodes. Pages are associated with home nodes through a first touch policy: a page is associated with the node that first referenced it. At release points, *diffs* are computed and sent to the page's home node, which reduces memory requirements in home-based DSM protocols and contributes to improve the scalability of the HLRC protocol. It is worth to point out that the extra pages required to keep the twin copies constitute a memory overhead, which can impair the scalability of memory bound parallel programs.

CoJVM. The Cooperative Java Virtual Machine (CoJVM) [1], is our DSM implementation of the standard Java Virtual Machine (JVM) for cluster computing. A distinguishing feature of CoJVM is that it does not impose changes either to the syntax or to the semantics of multithread Java programs, enabling computer clusters to become an accessible and scalable platform for concurrent Java applications. CoJVM implements the illusion of a single multiprocessing system as an interface between Java's multithreaded programming model and the cluster computing infrastructure. The heap, a Java memory area that is shared by the JVM threads [12], is allocated in the DSM space, allowing threads running in distinct cluster nodes to share data. Remote signaling and global monitors are used to implement transparently the support required to provide concurrency and synchronization to the application threads.

3 Optimization Techniques

This section presents selective dynamic *diffing* and lazy home allocation, two new techniques we devised to reduce overheads of the coherence protocol we used to implement memory sharing in CoJVM. Selective dynamic *diffing* monitors accesses to the shared memory so that it is able to dynamically create *diffs* selectively, i.e., the *diffs* only include those memory words that changed their values within a certain synchronization interval. Lazy home allocation delays the assignment of pages to home nodes until the application effectively starts its computation phase. Next, we describe the implementation of both techniques in CoJVM.

Selective Dynamic *Diffing*. At the *bytecode* level, Java distinguishes shared memory accesses from local ones. In particular, the getfield and putfield *bytecodes* are used to read and write data, respectively, that are exclusively located in shared memory. A simple version of our dynamic *diffing* technique (DDT) uses information on memory access to monitor all shared memory accesses over the execution time of an application, so that DDT enables CoJVM to avoid the use of the standard *diff* creation mechanism that most software DSM systems adopt. To do so, DDT represents the shared memory region as positions in a special bit vector and ensures that every time

DDT detects a write access to the shared memory region, it will set the corresponding bit in the bit vector. DDT will generate *diffs* simply by inspecting the bit vector for marked bits while checking the contents of their associated shared-memory positions. In this way, DDT can reduce the extra memory space that typical software DSM protocols require for twinning. However, DDT cannot eliminate all the twining operations since CoJVM's internal shared data structures are kept coherent through traditional twinning and *diffing* techniques, otherwise CoJVM's write accesses to internal shared data structures should also be instrumented, which clearly would hurt CoJVM performance. Nevertheless, the simple DDT will perform worse than the traditional *diff* creation technique as used by software DSM systems, in the following two cases: a) the elapsed time to mark *n* positions in the bit vector for one page is greater than the elapsed time to generate the *diff* for the same page; or b) an application rewrites the same value to a memory position. In this case, the traditional *diff* creation mechanism will not create a *diff* for that memory position since its value was unchanged. In contrast, the simple DDT will generate the *diff* for that memory position and thus will increase the size of the resulting *diff* for the associate page (Fig. 1a).

The selective dynamic *diffing* technique extended the simple DDT to take into account the new value being written to memory position, in that, if a new value is equal to the actual value then the bit vector is not set and also the value is not written to memory, as Figure 1b shows. Selective dynamic *diffing* can also reduce the amount of data the coherence protocol transfers. The reason is that CoJVM uses the HLRC, a well-known software DSM protocol to keep the heap consistent. However, some information stored in the heap are private to the local threads and thus are not really shared. By further distinguishing those private thread areas CoJVM avoids the HLRC protocol to unnecessarily transfer the associate data across the network.

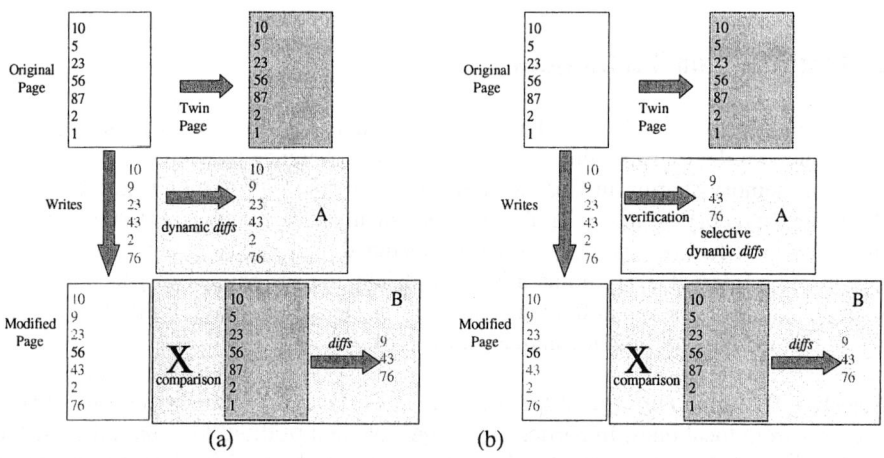

Fig. 1. a) Traditional *diffing* versus simple dynamic *diffing*. Box A (in yellow, left) shows that simple dynamic *diffing* operates while writes to the shared memory are being performed. Box B (in blue, left) shows that *diffs* are traditionally generated by comparison of the original page (the page's twin) with the modified page. b) Selective dynamic *diffing* assures that the *diffs* correspond to modified memory values.

Lazy Home Allocation. HLRC adopts a first-touch policy, where a node that first touches a data unit becomes its home node. However, we notice that it is often difficult or even impossible to the programmer to enforce at the beginning of the execution that each node touches the memory pages it will use at run time. The CoJVM implementation offers a concrete example. For instance, the Java Virtual Machine Specification [12] defines that all program variables must be initialized with the values the program specifies prior their use. Frequently, the initialization is carried out by the node where the computation starts. This means that this node will become the home node for the pages it initializes. As we demonstrated in previous work [1], this is likely to lead to a great imbalance in the distribution of pages to home nodes, which ultimately will impact negatively on the amount of data transfers across the network due to the increasing amount of node requests for remote pages. We propose the lazy home allocation technique to tackle the imbalance problem caused by HLRC's first-touch home assignment policy. In lazy home allocation the association between pages and homes is postponed until the time at which the computation phase of an application really takes place and pages will be definitely used. A naive analysis may suggest that pages be pre-initialized with the initial values (zeros) as defined by the language specification, so that the initialization phase will be avoided, and the page where data reside would not be associated with the node. The problem is that the initialization process of an object is not restricted to set its initial field values only. Other control data structures the JVM uses and which are associated with an object must also be initialized, so pages associated with the residence node of the object will be touched, as well. Lazy home allocation avoids that situation, since the association will be made only when the execution of an application effectively starts. In the case of a Java parallel application, we chose the exact moment at which the first run method is invoked by the thread the application created.

4 Experimental Methodology

Our experiments were performed on a 8-node Linux (2.2.14-5.0) cluster of 650 MHz Pentium III PCs, each of which with 512 MB RAM and 256 KB L2 cache. Each node was connected to a Giganet cLAN 5300 switch and used the VIA communication protocol [13]. We measured Giganet switch's point-to-point bandwidth at 101 MB/s to send 32 KB messages, 7.9 µs latency for sending 1-byte message, and 1.25 Gbps for aggregate bandwidth throughput.

We collected performance results of CoJVM versions for five parallel Java kernel benchmarks: Matrix Multiply (MM) Java kernel, which we developed; SOR, which we ported from Treadmarks's C benchmark suite [14]; LU, FFT, and Radix, which were ported from SPLASH-2 C benchmark suite [15]. The selected concurrent Java kernels offer distinct communication/computation ratios that make them fairly representative of regular and irregular applications that are found in typical scientific applications. MM represents an application that achieves linear speedup in clusters due to its embarrassingly parallel nature, which in turn allowed us to measure communication and synchronization overheads that CoJVM adds to maintain coherent the internal distributed JVM states. LU, FFT and SOR are concurrent Java kernels to factorize a dense matrix, calculate the Fast Fourier Transform, and solving partial differential equations using the red-black successive over-relaxation method, respectively. LU, FFT, and SOR represent regular applications with medium or coarse grain access to shared memory but with different sharing patterns that cause fragmentation. In contrast, Radix represents an irregular application that combines fine grain access to shared memory with high synchronization rates. Usually, Radix's performance scales poorly in software DSM systems. The sequential execution time for each of the five benchmarks is shown in Table 1. Note that the execution times excluded the initialization time to create remote threads or initialize array elements. We submitted the benchmarks to CoJVM five times, and reported the average execution times we obtained for each benchmarks in table 1. The standard deviation was less than 0.1% for each benchmark. All the kernels but Radix presented sequential execution times greater than 3 minutes. For Radix, the processing node's memory size (512 MB) limited Radix's largest input size to 8-Million keys and execution time to 25.80 seconds. Overall, the input data sizes we used can be considered reasonably adequate for our performance analysis of CoJVM systems.

To quantify the contribution of each run-time technique in isolation and in combination, we submitted the benchmarks to execute in six different versions of the CoJVM system: a) *Basic* version that uses none of run-time techniques; b) *Simple* dynamic *diffing* version that creates and marks the bit vector with all the write accesses the application makes to shared memory without verifying whether memory values have changed or not; c) *Selective* dynamic *diffing* version that creates the bit vector, but only marks the write accesses to shared memory that actually changed the memory values; d) *Lazy* version that allocates homes to nodes in a lazy fashion; e) *Simple+Lazy* version that combines the *Simple* with *Lazy* versions; and f) *Selective+Lazy* version that combines *Selective* with *Lazy* version.

Table 1. Program size, serial execution time in secs, and number of memory 4KB pages that the Java benchmarks required at runtime

Concurrent Java benchmark	Program Size	Serial Time - CoJVM	No. Pages
MM	1000x1000	491.94	3,230
LU	2048x2048	2,259.10	8,583
Radix	8 Million Keys	25.80	16,723
FFT	4 M Cmplx Dbls	225.94	49,524
SOR	2500x2500	306.37	5,305

To understand the experimental results better, we broke down execution time into six distinct components: 1) **Computation** indicates the time spent in useful computation; 2) **Page** indicates the amount of time spent in fetching remote pages from home nodes on page misses; 3) **Lock** is the time spent in acquiring a lock from its current owner; 4) **Barrier** is the time spent at a barrier, waiting for messages from other nodes; 5) **Overhead** time is the time spent performing protocol actions; and 6) **Handler** time is the time spent inside the remote handler, a separate thread that is used to service requests from remote nodes. In addition, we measured the amount of messages and data traffic due to memory coherence actions that were generated by both application and run-time CoJVM version. Due to space limits, those detailed figures can be found in [3].

Table 2. Speedup figures: *Basic* version, best speedup we measured, best CoJVM version, and resulting performance gain on 8 nodes

Benchmark	*Basic* Speedup	Best Speedup	Best Version	Performance Gain
MM	7,7	7,6	*Lazy*	-1%
LU	6,9	8,1	*Selective+Lazy*	17%
Radix	6,1	6,7	*Selective+Lazy*	9%
FFT	5,4	6,5	*Lazy*	20%
SOR	6,7	7,8	*Lazy*	16%

In Table 2, we present speedups for each of the CoJVM versions. For each benchmark, the table shows speedup of the *Basic* CoJVM version, the best speedup we obtained, which CoJVM version that produced it, and resulting performance gain in comparison with the *Basic* speedup. Table 2 shows that our run-time techniques were very effective in improving application performance for all the applications, which resulted in speedups between 6.5 and 8.1 on 8 nodes, except for the embarrassingly parallel MM kernel. Note that the performance gains were even more expressive if we consider that the *Basic* version speedups were already respectable, ranging from 5.4 to 6.9 for those four benchmarks.

As shown in figure 2a-left, MM benefited little from using our techniques. The reason is that it spent over 97% of the time performing local computations. In this situation, the addition of any extra code has direct impact on computation time. Indeed, despite of the benefits accrued to MM from using lazy home allocation, such as reducing the number of control messages and data traffic (Fig. 2a-right), the *Lazy* version's performance became 1% lower than that of the *Basic* version (Fig. 2a-left).

As can be seen in figure 2b-left, both *Lazy* and *Selective+Lazy* versions reduced 17% LU's execution time, whereas the *Selective* version decreased LU's execution time by 6%. Lazy home allocation and selective dynamic *diffing* were responsible for the dramatic reductions in page and barrier times (refer to [3]). These time reductions had distinct causes depending on the particular CoJVM version it used. The *Lazy* and *Selective+Lazy* versions obtained effective page distribution among the 8 nodes (Fig. 3), which eliminated most of remote page access and *diff* transfer overheads. As a result, applications spent less time in barrier waiting for *diffs* and bins (data structures that store identifiers of pages that were modified in a certain synchronization interval) transfers. For the *Selective* version, *diffs* were generated only for memory regions that were both shared and modified within a certain time interval, thus reducing network traffic (Fig. 2b-right) and decreasing barrier waiting time.

Table 3. Total number of twin pages that each application pre-allocated (including those used by JVM), when running on 2 nodes, and memory savings (% and in MB). 1 K = 1,024 pages

Benchmark	*Basic*	*Selective*	Memory Savings
MM	7 K	3 K	43% - 16 MB
LU	34 K	9 K	26% - 100 MB
Radix	25 K	5 K	20% - 80 MB
FFT	101 K	17 K	17% - 336 MB
SOR	9 K	2 K	22% - 28 MB

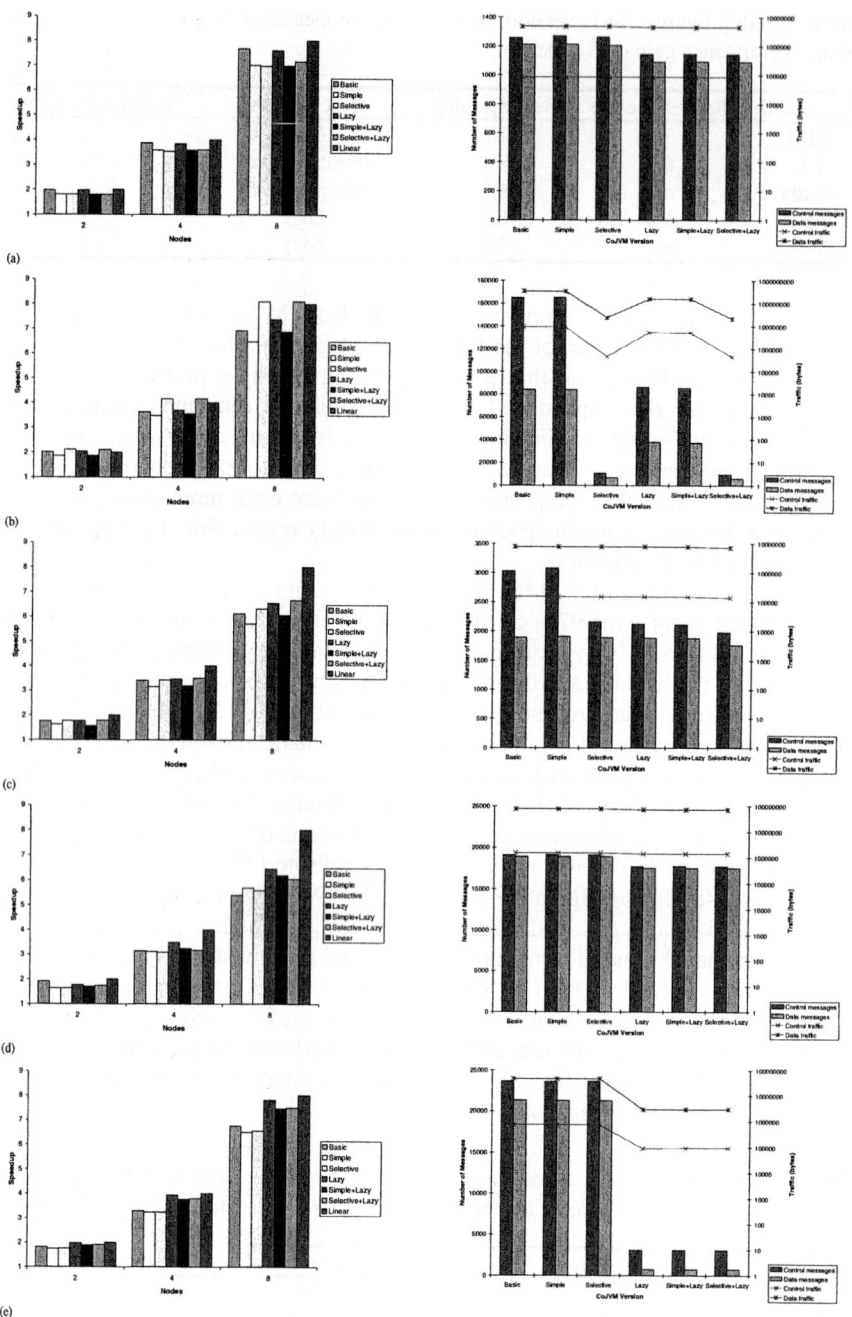

Fig. 2. *Left*: speedups on 2, 4 and 8 nodes *Right*: amount of messages, data and control traffic for on 8 nodes (right Y-axe is logarithmic) for (a) MM, (b) LU, (c) Radix, (d) FFT and (f) SOR

Figure 2c-left presents performance results for Radix under each CoJVM version. Specifically, Radix speedup improvements were 3%, 7%, and 9% for *Selective*, *Lazy*, and *Selective+Lazy*, respectively. Again, both run-time techniques contributed to reduce page and barrier times, the amount of messages and led to a small reduction in network traffic (Fig. 2c-right). We observed that the *Selective* version eliminated up to 95% of *diff* creation overheads in Radix, which surprisingly resulted in no benefit to *Selective* version performance. This was explained by the fact that even reducing overheads of *diff* creation the *Selective* version generated as much network traffic as the *Basic* version, which ultimately offset the gains due to *diff* elimination in the *Selective* version. Specifically, the *Selective* version requested 1,852 remote pages and generated 7.25 MB of data traffic, while the *Basic* version produced 1,856 remote page requests and 7.23 MB of data traffic.

Single and combined run-time techniques were most beneficial to FFT as shown in Figure 2d-left. All new CoJVM versions performed better than the *Basic* version. The *Lazy* version was the best one, as it performed as much 20% faster than the *Basic* version. This result is explained by effective page distribution among nodes (Fig. 3) and precise identification of data structures that were effectively shared by the application, which contributed to reduce considerably the page time, but without a equivalent decrease in the message and the communication traffic, which reduced slightly (Fig. 2d-right). FFT was the application that demanded the largest memory size (50,000 pages of 4KB) to execute in the *Basic* CoJVM. However, the use of dynamic *diffing* in the new CoJVM versions reduced greatly FFT's memory requirements as shown in Table 3. Most importantly, the main effect of large memory savings in FFT was to eliminate disc swap operations, which explains the superior performance of all CoJVM versions that use the dynamic *diffing* technique. Actually, FFT was the only application where the *Simple* version performed better than the *Basic* version. Recall that memory savings occurred due to elimination of twins.

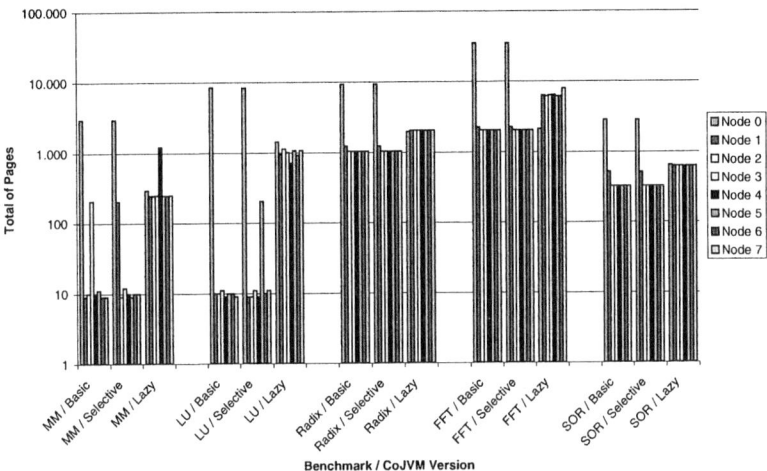

Fig. 3. Page distribution for applications in 8 nodes using three versions: *Basic*, *Selective*, and *Lazy*. The Y-axe is logarithmic.

In figure 2e-left, we present speedups that each CoJVM version produced for the SOR benchmark. All *Lazy*-based versions outperformed the *Basic* version, due to better page distribution among nodes (Fig. 3). The *Lazy*-based versions obtained an expressive decrease in the amount of *diffs* they created (from 20,727 in the *Basic* version to 182 in the *Lazy* version and 169 in both the *Simple+Lazy* and *Selective+Lazy* versions). This reduction explains their lower overheads in barrier time and less communication traffic among nodes (Fig. 2e-right). It is important to notice that the total amount of computation time was reduced in both SOR and LU when submitted to the *Selective* version. The reason was that every time a memory location had to be rewritten with the same value, selective dynamic *diffing* did not set the bit vector and also eliminated unnecessary access to memory.

5 Conclusions and Future Work

In this work, we described two new optimization techniques namely selective dynamic *diffing* and lazy home allocation that used the information the Cooperative Java Virtual Machine extracts at run-time about the application behavior. Five representative kernels, MM, SOR, LU, FFT, and Radix, were used as benchmarks to evaluate the impact of our proposed techniques on CoJVM performance. Our experimental results show that four out five benchmarks achieved superior speedups ranging from 9% to 20% when they were executed in optimized CoJVM versions, in comparison with original CoJVM version without using our techniques. The only exception was MM, which performed 1% better under original CoJVM than under the optimized versions. Overall, our techniques contributed to reduce overheads related to memory usage, number of messages, the amount of data transferred, and even computation time of CoJVM. Current CoJVM implementation is based on Sun's JDK 1.2. We plan to migrate our environment to the newest version of the Sun's released JVM, and to investigate how the techniques we proposed would perform on DSM environments that adopt languages that do not provide run-time information support.

References

1. Lobosco, M, et alli. A New Distributed JVM for Cluster Computing. 9th International Euro-Par Conference, pp. 1207-1215, Aug. 2003.
2. Arnold, K; Gosling, J. The Java Programming Language. AddisonWesley, 1996.
3. Lobosco, M, Amorim, C, Loques, O. Reducing Memory Sharing Overheads in Distributed JVMs, technical report ES-682/05, PESC/COPPE/UFRJ, May 2005.
4. Aridor, Y. et alli. A High Performance Cluster JVM Presenting a Pure Single System Image, JavaGrande 2000, pp. 168-177.
5. Fang, W, et alli. Efficient Global Object Space Support for Distributed JVM on Cluster. Int. Conf. on Parallel Processing, Aug 2002.
6. Hatcher, P. et alli, Cluster computing with Java. IEEE Computing in Science and Engineering, vol. 7, n. 2, pp 34-39, March/April 2005.
7. Veldema, R., et alli. Runtime optimizations for a Java DSM implementation. Conc. and Computation: Practice and Experience, Special Issue: ACM 2001 Java Grande-ISCOPE (JGI2001) Conference,.vol. 15, issue 3-5, pp 299-316, February 2003.

8. Li, K, Hudak, P. Memory Coherence in Shared Virtual Memory Systems. ACM Transactions on Computer Systems, 7(4):321-359, Nov 1989.
9. Keleher, P, Cox, A, Zwaenepoel, W. Lazy Release Consistency for Software Distributed Shared Memory. Int. Symp. on Computer Architecture, pp. 13-21, May 1992.
10. S. Adve and M. Hill. Weak ordering: A new definition. In Proceedings of the 17th Annual International Symposium on Computer Architecture, pp. 2-14, May 1990.
11. Zhou, Y, et alli. Performance Evaluation of Two Homebased Lazy Release Consistency Protocols for Shared Virtual Memory Systems. OSDI, Oct 1996.
12. Lindholm, T, Yellin, F. The Java Virtual Machine Specification. Addison-Wesley, 1999.
13. VIA. *VIA Specification, Version 1.0*. http://www.viarch.org. Accessed on Jan, 29.
14. Keleher, P. et. alli, Treadmarks: Distributed Shared Memory on Standard Workstations and Operating Systems. *Proceedings of the Winter 1994 USENIX Conference*, pp. 115-131, January 1994
15. Woo, S, et alli. The SPLASH-2 Programs: Characterization and Methodological Considerations. Int. Symp. on Computer Architecture, pp. 24-36, Jun 1995.

Analysis of High Performance Communication and Computation Solutions for Parallel and Distributed Simulation*

Luciano Bononi, Michele Bracuto,
Gabriele D'Angelo, and Lorenzo Donatiello

Dipartimento di Scienze dell'Informazione, Università degli Studi di Bologna,
Via Mura Anteo Zamboni 7, 40126, Bologna, Italy
{bononi, bracuto, gdangelo, donat}@cs.unibo.it

Abstract. This paper illustrates the definition and analysis of a collection of solutions adopted to increase the performance of communication and computation activities required by the implementation and execution of parallel and distributed simulation processes. Parallel and distributed simulation has been defined, and a real testbed simulation scenario has been illustrated, based on the ARTÌS simulation framework. Three classes of solutions have been proposed to improve the performance of simulations executed over commodity off-the-shelf computation and communication architectures: multi-threaded software and Hyper-Threading support by the processor architectures, data marshalling solutions for shared-memory and network-based communications, and data structure optimization for simulation events' management. All the proposed solutions have been evaluated on a testbed evaluation scenario, under variable configurations. Results obtained demonstrate that a performance improvement can be obtained by adopting and tuning the proposed solutions.

1 Introduction

"A simulation is a system that represents or emulates the behavior of another system over time. In a computer simulation the system doing the emulating is a computer program" [13]. The computer simulation is a widely adopted technique to obtain "a priori" insights of behavioral issues, and performance evaluation of theoretic, complex and dynamic systems and system architectures. At an abstract level, a simulation can be seen as a process execution managing a huge set of state variables: each variable update is activated by a simulated event. Every update may require a complex state computation, and would represent a step in the behavior of a portion of the simulated system. The simulation can be implemented by one (single or monolithic) process, or more than one (parallel or distributed) processes. Monolithic simulations may suffer the bottleneck limitation of memory and computation, being executed over single CPUs. On the

---

* This work is supported by MIUR FIRB funds, under the project: "Performance Evaluation of Complex Systems: Techniques, Methodologies and Tools".

other hand, parallel and distributed simulations could exploit aggregate memory and computation resources. Unfortunately, parallel and distributed simulations may suffer the bottleneck limitation of the communication system required to support the huge amount of synchronization and communication messages between multiple synchronized processes. The aim of this paper is to introduce main motivations and new dimensions for the research about scalability and efficiency issues of simulation frameworks executed over general purpose system architectures. Specifically, in this paper some recently introduced solutions for the processor architectures and communication network technologies have been preliminary investigated. The results presented involve the performance analysis and guidelines derived about the mixed adoption of Hyper-Threading processors, single-threaded and multi-threaded software architectures, data marshalling solutions for communications and data structure optimizations. Results obtained confirm that performance speedup can be obtained by considering and exploiting the proposed features, by offering experimental evidence to practical guidelines and limitations.

The paper structure is the following: section II illustrates some general concepts about the execution of simulation processes, that will be useful to define the assumptions and guidelines for subsequent work, section III will illustrate the state of the art in the field of parallel and distributed simulation and will introduce the simulation framework that we adopt as a testbed for our analysis (ARTÌS), section IV will illustrate the analysis of the Hyper-Threading features of processors in the context of parallel and distributed simulations, section V will illustrate the Data Marshalling concept and analysis to improve the communication efficiency, section VI will illustrate the simulation Data Structure optimization concepts and related analysis, and section VII will conclude the paper.

2 Simulation: Assumptions, Systems and Optimization

The aim of a simulation is to create a synthetic evolution of the system model, in order to capture data about the behavior of the model counterpart, that is, a real system behavior. The evolution of model entities could be defined as the history of state updates as a function of the simulated time. The system entities' evolution is emulated by a computer program that mimics their causal sequence of fine-grained state transitions, that is, the system events.

In the legacy approach for a computer simulation, the simulation software is executed on a single processing unit, by obtaining a monolithic simulation. Memory and computation bottlenecks may limit the model scalability and often require huge amount of time to complete the analysis. The need to evaluate complex systems with thousands or even millions simulated entities is often impossible to satisfy due to resources' limitations [12]. An alternative approach is based on the exploitation of parallel and distributed communication and computation systems [13]. The advantage of Parallel and Distributed Simulation (PADS) techniques is given by the exploitation of aggregate resources (memory and computation power of a collection of Physical Execution Units, PEUs) and by the potential exploitation of the model concurrency under the model-update and state-computation viewpoint. This may translate in a reduction of the computation time required to complete the analysis of the model evolution in a given scenario. A PADS framework is composed by a set of cooperating computation units managing the model state evolution in distributed way. The simulation is partitioned in a set of Logical Processes (LPs), each one managing the evolution of a subset of simulated model entities.

3 Parallel and Distributed Simulation: State of the Art

The parallel and distributed simulation (PADS) field has been widely studied in the past, resulting in the design and implementation of many tools to assist the simulation processes. In recent years, a good number of PADS tools have been proposed: Maisie [6], PDNS [7], DaSSF [1],TeD [2,15], Glomosim/Qualnet [8]. Unfortunately, weak performances and the lack of a standard have limited the adoption and interoperability of tools, and the model reuse, with low potential impact of PADS technology on real world applications. After the year 2000, under the auspices of the US Department of Military Simulation Office (DMSO), the IEEE 1516 standard for distributed simulation (High Level Architecture, HLA) has been approved [5]. ARTÌS (Advanced RTI System) is a recently proposed middleware, designed to support parallel and distributed simulations of complex system models [10,9]. A number of existing runtimes compliant to the HLA IEEE 1516 standard are available. Some runtimes suffer of implementation problems, the source code is often not available, and they miss interesting features (as entity migration support, and security). These observations stimulated the design of the ARTÌS middleware. In ARTÌS, many design optimizations have been realized for the synchronization and communication protocols' adaptation over heterogeneous execution systems (basically, Local Area Network and Shared Memory mono- and multi-processor architectures). The communication and synchronization middleware in charge of the adaptation is completely user-transparent. The middleware is able to select in adaptive way the best communication module with respect to the dynamic allocation of LPs over the execution environment. The ARTÌS runtime (RTI) kernel is realized on the top of the communication modules, and it is composed by a set of management modules, whose structure and roles have been inherited by a typical HLA-based simulation middleware.

In next sections, by following a bottom-up approach, we will introduce some details about additional optimizations that can be applied to ARTÌS (and other PADS architectures), by exploiting high computation and communication performance solutions.

4 Exploiting Advanced Processor Features: HT

The Hyper-Threading technology (HT) is a new processor architecture recently introduced by Intel [14,4]. HT technology makes a single physical processor appearing as two logical processors at the user level. To achieve best performances, the operating system should natively support the HT technology. In general, one physical execution resource (CPU) is shared between two logical processors. To obtain this effect, with low overheads introduced, the HT technology duplicates the high level portion of the architecture state on each logical processor, while logical processors share a subset of the physical processor execution resources. Some experimental results from Intel [14,4] have shown an improvement of CPU resources' utilization, together with higher processing throughput, for multi-threaded applications with respect to single-threaded executions. Under optimal assumptions and conditions, the performances shown a 30% increase. To the best of our knowledge, the influence of HT technology on PADS architectures and frameworks has not been investigated in detail. Thanks to simple heuristics, HT-enabled OSes should be able to adapt the process scheduling to the HT architecture, with the aim of optimizing the overall execution of processes. On the application side, it is quite common for parallel and distributed simulation frameworks to allocate a single LP for each available processor. This is based on the assumption that one single LP will be the only running process and would not cause context switches and other relevant overheads. On the other hand, each time the LP is blocked due to communication and synchronization delays, the CPU time would be wasted [11]. The effects of HT technology could significantly change the assumptions related to current implementation choices. Under the software architecture viewpoint, most of the modern PADS middlewares are based on multi-threaded implementation, and basically should take advantage of the HT support. It would be interesting to evaluate if the increasing in the number of logical processes could be exploited as a new dimension for PADS optimization: to concurrently run more LPs than the number of physical processors, under HT processor architectures.

4.1 The Experimental Testbed

To give answers to the above questions about HT technology and PADS assumptions, we evaluated the performances of the real ARTÌS simulation framework on a real experimental testbed, instead of relying on synthetic CPU benchmarks. First, we defined a scalable model of a complex and dynamic system, whose definition contains many of the worst model assumptions that has been identified as stressing conditions under the PADS framework optimization and simulation execution performances viewpoints: the wireless mobile ad hoc network model. The model is composed by a high number of simulated wireless mobile hosts (SMHs), each one following a Random Mobility Motion model (RMM) with a maximum speed of 10 m/s. This mobility model is far from being real, but it is characterized by the completely unpredictable and uncorrelated mobility pattern of SMHs. The system area is modeled as a torus-shaped bi-dimensional grid-topology, 10.000x10.000 space units. The torus area, indeed unrealistic, allows to simulate a closed system, populated by a constant number of SMHs. The torus space assumption is commonly used by modelers to prevent non-uniform SMHs concentration in any sub-area. The simulated space is flat and open, without obstacles. The modeled communication pattern between SMHs is a constant flow of ping messages (i.e. constant bit rate), transmitted by every SMH in broadcast to all neighbors within a wireless communication range of 250 spaceunits.

4.2 The Experimental Results

All the experiments and the analysis results shown in this paper are based on the parallel and distributed simulation of the wireless ad hoc model, under the control, optimization and management of the ARTÌS runtime (Section 3). We performed multiple runs for each experiment, and the confidence intervals obtained with a 95% confidence level (not shown in the figures) are lower than 5% the average value of the performance index. The experiments collected in this section have been executed over a PEU equipped by Dual Xeon Pentium IV 2800 Mhz, 3 GB RAM. The experiments have been divided in two groups: the first group is based on Hyper-Threading support enabled for the PEU (HT-ON), and the second one is based on Hyper-Threading support disabled directly by BIOS settings (HT-OFF). The ARTÌS implementation adopted in this analysis is itself multi-threaded, and takes advantage of the multi-threading support to manage the execution of LPs: each LP is composed by at least 3 threads (main, shared memory and network management). The standard ARTÌS implementation implements a timed wait synchronization mechanism between the threads that compose each LP. Alternative solutions for implementing communication between the threads could be based on busy waiting, or signalling-based implementations.

Figure 1 shows the wall-clock time (WCT) required to complete one simulation run, taken as a reference. The reference run is defined as the evolution of 1000 time-steps of simulated time for the wireless ad hoc model with 6000 wireless SMHs. The X coordinate (LPs) shows the number of concurrent LPs implementing the set of model entities for the reference scenario. When $LP = 1$, the simulation is strictly sequential and monolithic, that is, only one processor executes the single LP incarnating the execution of all the model entities of the simulated model. In the $LP = 2$ scenario, 2 LPs incarnate the set of model entities, and each LP is allocated on a different physical processor of the execution architecture. When $LP = [3..8]$ the ARTÌS framework introduces a load-sharing of model entities over LPs. In addition, ARTÌS supports communication layer adaptation, resulting in low latency communication between LPs. Thanks to load-sharing capability of ARTÌS, the time required for completing the simulation run (Wall Clock Time, WCT) for $LP = 2$ is better than the one obtained with one LP ($LP = 1$). When the number of LPs grows, this fact introduces overheads under the synchronization and data distribution management (DDM) viewpoints, while the concurrency at the CPU hardware level is stable. For this reason, the WCT increases when $LP \geq 2$ and shows additional overheads with HT-ON. On the other hand, results in Figure 1 show that the HT-enabled PEU (HT-ON) for $LP = 2$ gives slightly better results than the HT-disabled PEU (HT-OFF). For this experimental scenario, the minimum WCT for both HT-ON and HT-OFF curves is obtained for 2 LPs. The activa-

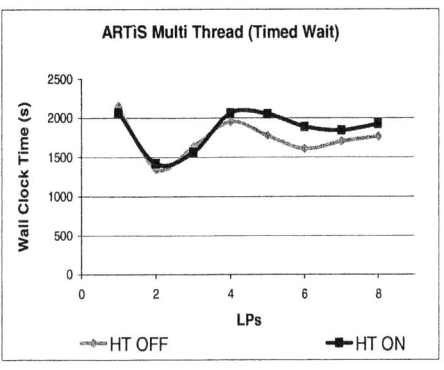 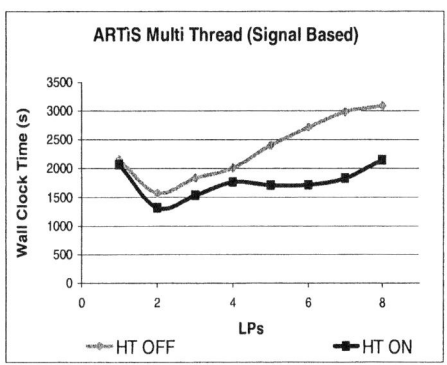

Fig. 1. PEU=1, SMH=6000, ARTÌS (Timed Wait)

Fig. 2. PEU=1, SMH=6000, ARTÌS (Signal Based)

tion of HT does not change the optimal number of LPs, but has effects on the overall performance of the simulation processes. The reason for additional overheads with HT-ON could be due to the timed wait implementation of the thread synchronization implemented by ARTÌS. For this reason, in the following Figure 2, we performed the same investigation shown in Figure 1, with a modified version of ARTÌS. The latter version of ARTÌS is still multi-threaded, but the synchronization among threads is implemented with a signal based approach. In Figure 2 results show that the HT-enabled PEU (HT-ON) performs always better than the HT-disabled (HT-OFF) version. In addition, the HT-ON scenario shows an increase in the simulation scalability, with respect to HT-OFF. In Figure 3 the HT support is evaluated with respect to a mono-threaded version of the ARTÌS runtime architecture. This means that the ARTÌS runtime is based on single-thread, which is responsible to manage all the model entities' executions and to manage the communications. This test is interesting to evaluate the behavior and performances of HT architectures when executing mono-threaded software. Figure 3 shows that the HT support may slow down single threaded

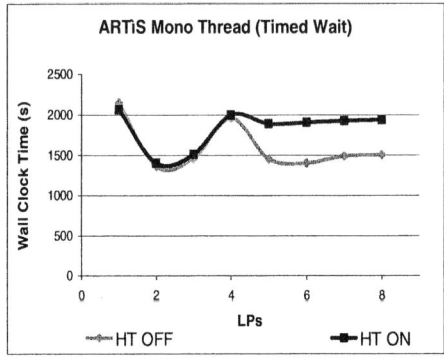

Fig. 3. PEU=1, SMH=6000, ARTÌS Monothread implementation

applications, by resulting in additional overhead. A quite strange behavior appears when $LP = 4$, showing a simulation slowdown (found also in Figure 2). The reason for this behavior requires further investigation. To summarize, in ARTÌS the HT support gives better results with signal based synchronization among threads, with respect to timed wait synchronization. On the other hand, HT support does not change the optimal number of LPs with respect to the underlying PEU architecture. When the LPs are mono-threaded, the HT support appears as not influent up to a given number of LPs (up to 4 LPs in the figures), and as an overhead when more than 4 LPs are executed (that is, when the model entities are load-shared among more than 4 LPs) in the considered PEU architecture.

5 Communication Optimization: Data Marshalling

The communication efficiency is one of the main factors determining the efficiency of a parallel or distributed simulation. The current optimization scheme in ARTÌS is based on a straightforward incremental policy: given a set of LPs on the same physical host (that is, with shared memory), such processes always communicate and synchronize via read and write operations performed within the address space of LPs, in the shared memory. Two or more LPs located on different hosts (i.e. no shared memory available) on the same local area network (or even on the Internet) would rely on standard TCP/IP connections. In ARTÌS, every interaction between LPs for synchronizing and distributing event messages is immediately performed over shared memory or network infrastructures. This kind of implementation generates much more transmissions on the communication channel, and replicates the message overheads (headers and trailers) and the channel accesses. A reduction of the overheads and channel accesses could result in increased channel utilization and reduction of the communication bottlenecks. In the following we will investigate if and how a message marshalling approach could reduce the simulation WCT. The data marshalling approach consists in the concatenation of more than one logical message in the same communication messages. In order to control the inverse trade-off degradation in the average communication latency, the data marshalling process is controlled by a timer: once every a maximum time limit the messages buffered on the LP are sent in a data marshalling packet (or frame). The proposed optimization has been applied both to shared memory and TCP/IP communications. Figure 4 shows the results for the optimization applied to a distributed simulation architecture. The hardware architecture is composed by 2 homogeneous PEUs equipped by Dual Xeon Pentium IV 2800 Mhz, 3 GB RAM, with HT-enabled, interconnected by a Gigabit Ethernet (1 Gb/sec). The simulated model for tests is the wireless ad hoc network model described in the previous section. Different scenarios have been modeled by varying the number of simulated entities (SHMs) in the model: from 3000 up to 9000 simulated mobile hosts. For each experiment, the data shown include the WCT time obtained with data marshalling ON and OFF, respectively. The data marshalling applied to this simulation testbed increased the

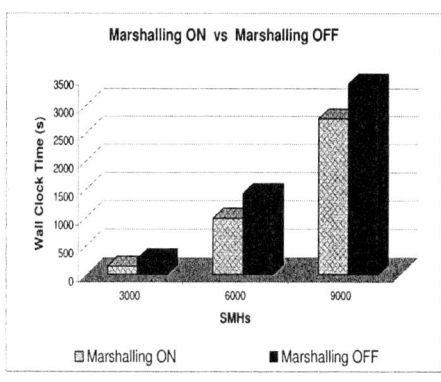 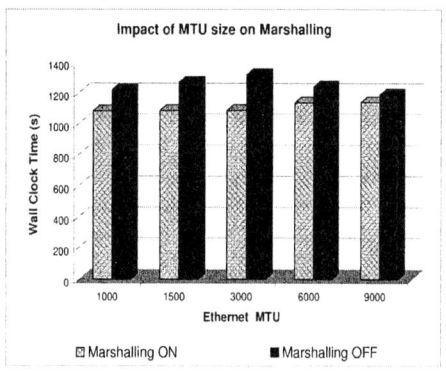

Fig. 4. PEU=2, Wall Clock Time with Marshalling ON and OFF

Fig. 5. PEU=2, SMH=6000, Wall Clock Time with Marshalling ON and OFF

WCT simulation performances: 48% for 3000 SMHs, 30% for 6000 SMHs, and 18% for 9000 SMHs (see Figure 4). When the number of SMHs increases, the percentage gain reduces: this happens because the computation required for updating and managing the states of many more SMHs becomes the predominant simulation bottleneck in this system (in the place of communication bottleneck).

The Maximum Transmission Unit (MTU) of local area network communications is another factor that could be managed under the data marshalling viewpoint. The Ethernet standard frame size has been "de facto" limited to 1500 bytes. In recent years the Ethernet bitrate has greatly increased, but the MTU size is substantially the same. Very large MTU size could reduce communication overheads (i.e. the percentage effect of headers) and could increase the network utilization. This approach is usually referred to as the adoption of "jumbo frames". For this test we have interconnected two homogeneous PEUs (defined above) by a cross-linked Gigabit Ethernet network cable. The simulation model is the wireless ad hoc model with 6000 SMHs. In Figure 5 results show that the data marshalling ON can reduce the WCT with respect to data marshalling OFF. On the other hand, results are only slightly influenced by the variation of the MTU size (from 1000 up to 9000 Bytes). The experiments shown that the adoption of jumbo frames slightly increased performances (up to 3000 Bytes) when data marshalling is OFF. When marshalling is ON, the simulation performance was almost constant up to 3000 bytes, and slightly increasing with more than 3000 Bytes.

6 Simulation Data Structures Optimization

One of the most important data structures in a computer simulation is the repository of event descriptors. Both monolithic and distributed simulations, require that future events are collected and executed in timestamp order. Every simulation may include the management of millions of events. For this reason, it is important to find a really efficient data structure at least in the support of a subset of management operations. The most frequent operations in a simulation process are: insertion of a new event descriptor (*insert()* operation), and extraction of the event with the minimum time stamp (*extract_min()* operation). Both operations should have a really low computational complexity and should be easy to implement. Some useful data structures can be adopted to assist in the implementation of the event repository definition and management: lists, hash tables, calendars and balanced trees. In most cases, the better solution is to adopt the heap data structure. "A heap is a specialized tree-based data structure. Let A and B be nodes of a heap, such that B is a child of A. The heap must then satisfy the following condition (heap property): $key(A) \leq key(B)$. This is the only restriction of general heaps. It implies that the greatest (or the smallest, depending on the semantics of a comparison) element is always in the root node. Due to this fact, heaps are used to implement priority queues" [3].

Given a binary heap, the worst case computational complexity for the *insert_heap()* and *extract_min_heap()* is $O(log_2 n)$ (because after the extraction the heap requires a heap re-organization (heapify) algorithm execution). We call a classical binary heap as the "base heap" data structure. The Base Heap (BH) demonstrates good performances in general, to implement the event repository for a simulation process. On the other hand, by considering common assumptions related to the event management and event characteristics in the simulation field, we could design even more efficient heap-based data structures. Event descriptors are usually organized as heap elements ordered by time-stamp (key). The set of events generated during a conservative simulation is usually characterized by sequential time-stamp values. Moreover, the time management of a simulation process could be time-stepped, which means that all the time-stamps of events located in the same timestep are equal. By exploiting these common properties of simulation processes it would be possible to implement an enhanced version of the heap data structure. Each node of the Enhanced Heap (EH) is now composed by a pointer to a linked list of events, including all the descriptors of events with the same time-stamp value (see Figure 6). Thanks to the principle of time-locality in the references to event descriptors in a simulation process, the access to event descriptor with time-stamp value t is followed with high probability by the accesses to event descriptors with the same time-stamp value. For this reason, by caching the pointer to the linked list of simultaneous events in the simulation, the management of the EH data structure is much more efficient than replicating search operations on the BH. This optimization allows to avoid frequent heapify operations, by working on the cached linked lists associated with heap elements. Hence, by calling a hit the insertion or extraction of one event descriptor to/from the cached list associated to current simulated time t, the *insert_heap()* and *extract_min_heap()* operations can be performed in the majority of cases with $O(1)$ complexity (given the time-locality simulation assumption, resulting in high hit ratio) in the linked list. The complexity is $O(log(k))$ in the worst case (that is, cache miss), being k the number of different event timestamps (keys) inserted in the EH. In general, with EHs, the number

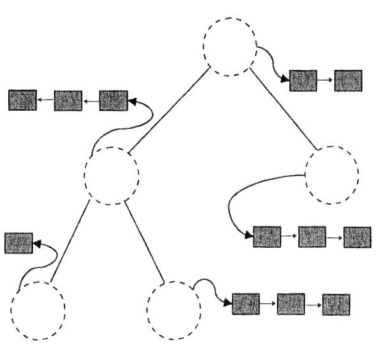 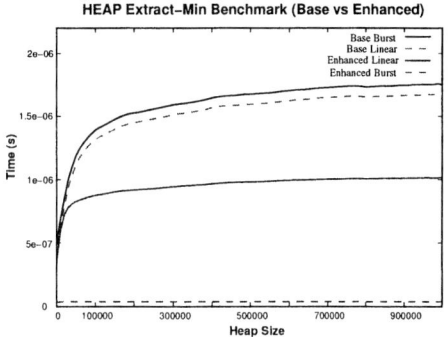

Fig. 6. An Enhanced Heap (EH) data structure

Fig. 7. Benchmarks: synthetic environment, base vs. enhanced versions

of three nodes can be reduced, by adopting only one placeholder node for the set of events with the same timestamp (key). The data structure size can be reduced by eliminating the time-stamp informations from all the event descriptors in the same list (whose time-stamp is implicitly defined by the corresponding EH node).

The Figure 7 shows the results obtained by a benchmark application (that is, not a simulation) that has been defined to test the efficiency of the EH data structure management. The curves show the average time required to insert (and to extract) a group of 1000 heap nodes in base (BH) and enhanced (EH) heap data structures, when the initial heap size has the value indicated on the X coordinate. Two kinds of heap insertion operations have been tested: the Burst insertion is defined as the insertion of a group of 1000 heap nodes with the same key (time-stamp value), while the Linear insertion is defined as the insertion of a linear sequence of 1000 heap nodes with incremental key (time-stamp value). The data shows that the management of BH with bursts insertions (Base Burst curve in the figure) obtains the worst performance, as expected. The BH with linear insertions (Base Linear curve in the figure) performs a little better than Base Burst. A great improvement in the performance is obtained with the Enhanced Heap (EH). The EH with linear insertions (Enhanced Linear curve in the figure) performs better than Base Burst (50% time reduction). The EH with burst insertions (Enhanced Burst curve in the figure) obtains almost constant performances, independent from the heap size, and results in a very good performance index.

By testing the Enhanced Heap data structure on the simulation testbed, we performed the simulation of the wireless ad hoc simulation model with variable number of SMHs over the multi-threaded version of the ARTÌS framework. The experiments have been executed over a PEU equipped by Dual Xeon Pentium IV 2800 Mhz, 3 GB RAM, with Hyper-Threading support activated. The Figure 8 shows the impact of the Heap type (Base vs. Enhanced) on the WCT performance for the simulations, with variable number of simulated entities (SMHs). As expected the WCT increases, but the effect of the Heap type is marginal.

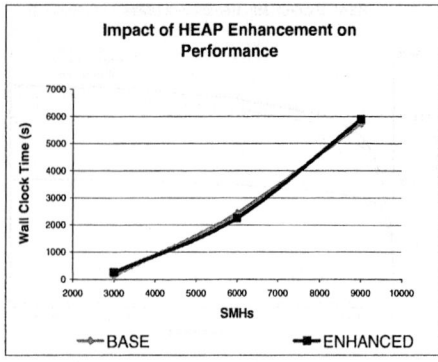
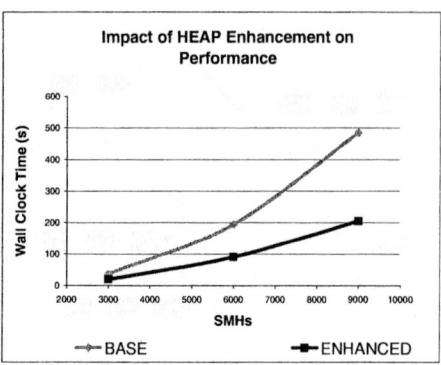

Fig. 8. Standard wireless model (computation intensive)

Fig. 9. Modified wireless model (low computation)

Only a little advantage is shown with 6000 SMHs. The reason for this fact is the high degree of computation that is required in this model for managing each event involving a subset of SMHs. Without entering in details, this simulation model is defined in order to be computation intensive. For this reason the advantages in the management of event descriptors obtained with Enhanced Heaps is hidden by the overwhelming amount of computation that follows every new event extraction or insertion. To confirm this fact, we implemented a light computation version of the same simulation model, where the computation caused by each event is reduced more or less of a factor 10. The results shown in Figure 9 confirms our expectations. In this figure, the advantage of adopting the Enhanced Heap is clear. The increase in the model complexity (number of simulated mobile hosts, SMHs) results in the increasing advantage of adopting the Enhanced Heap support for event management. The low computation required in this model, for each event, has emphasized the effect of the event management complexity in the simulation process evolution.

7 Conclusions and Future Work

In this paper we illustrated the definition and analysis of a collection of solutions adopted to increase the performance of communication and computation activities required by the implementation and execution of parallel and distributed simulation processes. Parallel and distributed simulation has been defined, and a real testbed simulation scenario has been illustrated, based on the ARTÌS parallel and distributed simulation framework. Three classes of solutions have been proposed to improve the performance of simulations executed over commodity off-the-shelf computation and communication architectures: multi-threaded software and Hyper-Threading support by the processor architectures, data marshalling solutions for shared-memory and network-based communications, and data structure optimization for simulation events' management. All the proposed solutions have been evaluated on real testbed evaluation scenarios, and under variable configurations. Results obtained demonstrate that a performance improvement, summarized by the Wall Clock Time (WCT) required to complete the simulation processes, can be obtained by adopting and tuning the proposed solutions in opportune way. The experimental analysis has provided some interesting guidelines about the way to adopt and to compose the proposed solutions, under the considered simulation testbed. Some guidelines indicate that the system bottleneck could change depending on the model and system assumptions. Most of the guidelines have been commented under the general context assumptions, and could be considered generally extensible to other simulation frameworks, models and execution scenarios. Future works include the analysis of more wide simulation scenarios, and the detailed analysis of resource utilization metrics.

References

1. Dartmouth SSF (DaSSF). http://www.cs.dartmouth.edu/research/DaSSF/.
2. GTW/TeD/PNNI. http://www.cc.gatech.edu/computing/pads/teddoc.html.
3. Heap, From Wikipedia, the free encyclopedia. http://en.wikipedia.org/wiki/Heap.
4. Hyper-Threading Technology on the Intel Xeon Processor Family for Servers. http://www.intel.com/business/bss/products/hyperthreading/server/ht_server.pdf.
5. IEEE Std 1516-2000: IEEE standard for modeling and simulation (M&S) high level architecture (HLA) - framework and rules, - federate interface specification, - object model template (OMT) specification, - IEEE recommended practice for high level architecture (HLA) federation development and execution process (FEDEP).
6. Maisie Programming Language. http://may.cs.ucla.edu/projects/maisie/.
7. Parallel / Distributed ns. http://www.cc.gatech.edu/computing/compass/pdns/.
8. SNT: QualNet. http://www.qualnet.com.
9. PADS: Parallel and Distributed Simulation group, Department of Computer Science, University of Bologna, Italy. http://pads.cs.unibo.it, 2005.
10. L. Bononi, M. Bracuto, G. D'Angelo, and L. Donatiello. ARTÌS: a parallel and distributed simulation middleware for performance evaluation. In *Proceedings of the 19-th International Symposium on Computer and Information Sciences (ISCIS 2004)*, 2004.
11. L. Bononi, G. D'Angelo, M. Bracuto, and L. Donatiello. Concurrent replication of parallel and distributed simulation. In *Proceedings of the nineteenth workshop on Principles of Advanced and Distributed Simulation*. IEEE Computer Society, 2005.
12. L. Bononi, G. D'Angelo, and L. Donatiello. HLA-based Adaptive Distributed Simulation of Wireless Mobile Systems. In *Proceedings of the seventeenth workshop on Parallel and distributed simulation*. IEEE Computer Society, 2003.
13. R. Fujimoto. *Parallel and Distributed Simulation Systems*. John Wiley & Sons, Inc., first edition, 2000.
14. D. Marr, F. Binns, D. Hill, G. Hinton, D. Koufaty, J. Miller, and M. Upton. Hyperthreading technology architecture and microarchitecture: A hypertext history. *Intel Technology Journal*, 2002.
15. J. Panchal, O Kelly, J. Lai, et al. Parallel simulations of wireless networks with TED: radio propagation, mobility and protocols. *SIGMETRICS Perform. Eval. Rev.*, 25(4):30–39, 1998.

An Analytical Study on the Interdeparture-Time Distribution for Different Multimedia Source Models in a Packet Switched Network

Sergio Montagna, Riccardo Gemelli, and Maurizio Pignolo

Italtel SpA, R&D Labs, Castelletto di Settimo Milanese, 20019 Milan, Italy
{sergio.montagna, riccardo.gemelli,
maurizio.pignolo}@italtel.it

Abstract. In this paper we studied the interdeparture time distribution of multimedia RTP traffic mixed to disturbing background traffic across a single router. Both periodic and On-Off binary Markov sources have been considered. An analytical approach for the calculation of this interdeparture time distribution has been developed. Results of these studies are the basis for estimating distribution of the interdeparture jitter in a chain of routers.

1 Introduction

The massive introduction of multimedia over IP is highlighting the importance to evaluate the effect of network elements on the distortion of the packet stream. A number of works [1-4] focused on the evaluation of the time sequence distortion of tagged traffic flowing trough a network. The delay jitter will be adopted as a measure of the distortion, the delay jitter being the difference between the interdeparture and interarrival times of packets in a stream. Results of these studies are the starting point to estimate the distribution of the interdeparture jitter in a chain of routers. In works reported in bibliography the analysis is performed on deterministic periodic and On-Off binary Markov sources.

In this paper we will focus on defining the interdeparture time distribution for a tagged traffic profile. In particular the source is a CODEC channel generated from a DSP, the rate of samples depending on the particular coding. Moreover in our work, from the point of view of analyzed sources, we extend the analysis on On-Off sources too. In section 2 the node modeling studied in this paper is defined and an exhaustive distribution of the interdeparture time is derived; results are presented in section 3.

2 Node Modeling

The model considered in this paper is illustrated in Fig. 1, where a single node is considered. The router is modeled as a buffer plus an output link managed by a scheduler. The system is fed with two different traffic sources: the Probe and the background. The probe source is the one on which we want to study the interdeparture distribution introduced from the queuing in the node. The background is

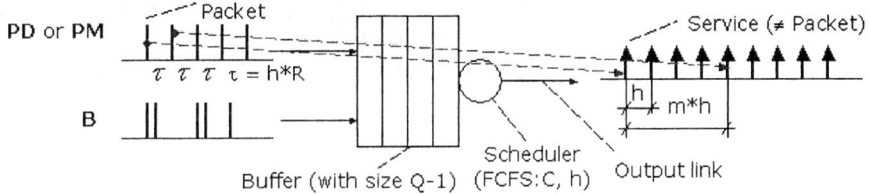

Fig. 1. Single node modeling

introduced to create a realistic environment and represents the effect of the superimposition of a wide set of RTP streams. The Probe source is modeled into two different cases: a) a discrete process with deterministic arrival times with inter-arrival time τ: in this case the probe source is labeled PD; b) a discrete process modeled by simple two state markov chain model adopted in the literature for the description of On-Off sources: in this case the probe source is labeled PM. In both a) and b) cases the rate of the source takes the constant value S. All the sources are supposed to produce RTP packets with size L. The overall system is modeled as M+G/D/1/Q. The source is M with reference to the B source, G with reference to the probe source that on its turn can become D in the case a) and an On-Off in case b), Deterministic FIFO service with service time h, single server with Q rooms in the system, Q-1 places in the buffer plus 1 in the server. All the sources PD, PM, B RTP are supposed to be scheduled with the same priority. The link capacity is C. The service time h is equal to the ratio L/C, inter-arrival time τ of the probe flow is bound to S and L from the expression τ * S = L. Denote the (integer) speed ratios by R:

$$R = \tau/h = C/S. \quad (1)$$

The system is clocked (packets are processed from scheduler at discrete instants corresponding to clock positive edge) with period h. Let the time unit h be the length of the interval required to transmit a packet on the output link. We will refer to this intervals as a time slot. Thus, packets from the source of class PD or PM can arrive at most every interval of length R, and are transmitted in one time slot, once at the head of the buffer. A further hypothesis, applicable since the system is clocked, is that deterministic events like arrivals for the PD source and services (one each h time units) are in phase, that is when a packet of the PD source arrives and the server is idle the packet is serviced immediately. On the contrary, the packet from the class B comes into the buffer and always wait the end of the time slot period before being served, also in the case the buffer is idle.

2.1 Node Modeling: PD Probe Source + B Background Source

The behavior of such a system can be studied by means of a discrete-time, finite-state Markov chain model. The chain is defined over the set T={0,1,2,.....} of the positive integers, corresponding to beginning instants of possible link transmissions. The system state is defined by the random variable (j, $0 \leq j \leq Q$) where j is equal to the total number of packets in the system. We choose to formulate the state probabilities at one time slot in terms of their values at the previous time slot. We use a subscript to specify the time dependence of the state probabilities when needed. We use standard set theoretic notation, and denote by I(E) indicator function of event E, namely $I(E) \begin{cases} 0 \text{ if } E \text{ does not occur} \\ 1 \text{ if } E \text{ occurs} \end{cases}$. Let T_{PD} the instants where an arrival of class PD occurs.

Obviously T_{PD} are subsets of T. Let us also define the following events, corresponding to the generic time t:

E_0 : t does not belong to T_{PD}; E_1 : t belongs to T_{PD}. In event E_0, no arrivals are possible for packets of class PD. In event E_1, there are arrivals for packets of class PD.

Let's introduce the r.v. (random variable) AR following a Poisson distribution and representing the number of arrivals during the time slot h from source B. Let's also introduce AR[j] that means j arrivals from source B during a time slot h. We assume a number of arrivals limited to N. The choice of the value of N guarantees a very small value of the probability. Typically AR[N] is less than 10^{-50}.

The system status is thus defined using the r.v. $P_t(x)$ representing the probability of having x packets in the system at time t. The transition equations for our model are:

$$P_{t+1}(i) = I(E_0) \cdot \{(P_t(0) + P_t(1)) \cdot AR(i)\}, i = 0. \quad (2)$$

$$P_{t+1}(i) = I(E_0) \cdot \begin{Bmatrix} \sum_{j=1}^{Q} P_t(j) \cdot \sum_{l=Q+1-j}^{N} AR(l) + \\ P_t(0) \cdot \sum_{l=Q}^{N} AR(l) \end{Bmatrix} + I(E_1) \cdot \begin{Bmatrix} \sum_{j=1}^{Q} P_t(j) \cdot \sum_{l=Q-j}^{N} AR(l) + P_t(0) \cdot \sum_{l=Q-1}^{N} AR(l) \end{Bmatrix}, i = Q. \quad (3)$$

$$P_{t+1}(i) = I(E_0) \cdot \begin{Bmatrix} \sum_{j=1}^{i+1} P_t(j) \cdot AR(i+1-j) + \\ P_t(0) \cdot AR(i) \end{Bmatrix} + I(E_1) \cdot \begin{Bmatrix} \sum_{j=1}^{i} P_t(j) \cdot AR(i-j) + \\ P_t(0) \cdot AR(j-1) \end{Bmatrix}, i = 1,2,\ldots,Q-1 \quad (4)$$

By using these transition equations we are able to monitor the transient behavior of the system starting from a given initial state distribution. This provides insights into the system's response to various perturbations, and eventually yields the steady-state probabilities, which are obtained by iterating the process until the desired accuracy is reached. More specifically, for each state of the Markov chain the percent difference between probabilities corresponding to successive interval of length τ are evaluated, and only when this difference is found smaller than some ε (usually $\varepsilon = 10^{-7}$) the procedure is stopped. This procedure is stopped according to the following rule

$$|P_t(i) - P_{t-1}(i)| / P_t(i) \leq \varepsilon, \text{ with } \varepsilon \leq 10^{-7} \text{ and } i \in \{1,\ldots,Q\}. \quad (5)$$

After convergence, the system is in steady state and $P_t(x)$ is no more function of t, thus $P(x)$. Next, we will describe the derivation of the distribution B(m) of the inter-departure between two consecutive packets belonging to PD. To derive B(m) we introduce the conditional distribution G where G(m | j) means the probability to arrive at the state m after R h-unit service conditioned to j as the initial state. j means that when a packet of PD arrives in the system it takes the j-n[th] place in the system, after R*h-unit services (R*h = τ), the next packet sent from the PD source takes the mth place in the system. This probability is calculated by the procedure outlined in (2, 3, 4) assuming equal to 1 the probability of having j (j=1,2,..Q) packets in the system when the packet belonging to the class PD arrives. The procedure evolves starting from this state and finishes after R time slots. In our analysis we suppose that Q is very large so the loss probability of packets is negligible, that B(m) is meaningful as a measure of the inter-departure for PD source. B(m) is expressed from Equation (6).

$$B(m) = \sum_{j=1}^{Q} G(m \mid j) \cdot P(j), \; m = 1,2,\dots,Q+R-1. \tag{6}$$

P is the corresponding steady state probability vector of the system, when a packet of the source PD arrives into the system. At this point the inter-departure distribution B(m) can be derived and expressed through the following formula

$$B(m) = \sum_{v=\max(1,R+1-m)}^{\min(R-m+Q,R-1,Q)} G(m+v-R \mid v) * P(v) + I(E) * G(m \mid R) * P(R) + \sum_{v=R+1}^{J} I(F) \cdot G(v+m-R \mid v) * P(v). \tag{7}$$

where:

$$J = \begin{cases} Q+R-m & \text{if } R<m+1 \\ Q & \text{if } R \geq m+1 \end{cases}, \; I(E) = \begin{cases} 1 & \text{if } (m \leq Q) \text{ and } (R \leq Q) \\ 0 & \text{otherwise} \end{cases}, \; I(F) = \begin{cases} 1 & \text{if } (m \leq Q+R-v) \\ 0 & \text{otherwise} \end{cases}.$$

2.2 Node Modeling: PM Probe Source + B Background Source

In this scenario the Probe source is modeled by two state Markov chain model. Transitions in the source model states occur at the end of a packet generation period. The sojourn times in the idle and active states are geometrically distributed with average values denoted by I and L and are measured in packets generation periods. When in the active state, the source produces a packet with probability one. The activity of the source and its offered load are respectively given by:

$$\alpha = \frac{L}{L+I}, \; \rho = \frac{\alpha}{R} \tag{8}$$

In the following, traffic class is characterized by the triplet:
(Peak bit rate, Activity, Average burst length) = (S, α, L).

The behavior of such a system can be studied by using a discrete-time, finite-state Markov Chain model. The matrix of the transition between on and idle states is C. The probability to go from state on (labeled as 1) to state idle (labeled as 0) and vice versa are:

$$C(1,0) = \frac{1}{L}, C(1,1) = 1 - C(1,0), \; C(0,1) = \frac{\alpha}{L*(1-\alpha)}, C(0,0) = 1 - C(0,1) \tag{9}$$

The chain is defined over the set T = {0,1,2,.....} of the positive integers, corresponding to beginning instants of possible link transmissions. The system state is defined by the couple of values:

$$(j,i), 0 \leq j \leq Q, 0 \leq i \leq 1 \tag{10}$$

with j equal to total number of packets in the buffer; i equal to 1 meaning that the state of the source PM is active, otherwise is in idle state. We choose to formulate the state probabilities at one time slot in terms of their values at the previous time slot. We use a subscript to specify the time dependence of the state probabilities when needed. We use standard set theoretic notation, and denote by I(E) indicator function of event E, namely:

$$I(E) \begin{cases} 0 \text{ if } E \text{ does not occur} \\ 1 \text{ if } E \text{ occurs} \end{cases}$$

Let T_1 the instants where an arrival of PM source can occur. Obviously T_1 are subsets of T. Let us also define the following events, corresponding to the generic time t:

E_0 t is not in T_1 (*No arrival possible for packet of PM source*)
E_1 t is in T_1 (*arrival for packet of PM source*)

The transition equations for our model are:

$$P_{t+1}(0,0) = I(E_0) \cdot \begin{cases} (P_t(0,0) + \\ P_t(1,0)) * A(0) \end{cases} + I(E_1) \cdot \begin{cases} P_t(0,0) * C(0,0) + P_t(1,0) * C(0,0) * A(0) + \\ (P_t(0,1) + P_t(1,1) * C(1,0) * A(0) \end{cases} \quad (11)$$

$$P_{t+1}(0,1) = I(E_0) \cdot \{(P_t(0,1) + P_t(1,1)).AR(0)\} \quad (12)$$

$P_{t+1}(Q,1) =$

$$I(E_0) \cdot \begin{cases} \sum_{j=1}^{Q} P_t(j,1) \cdot \sum_{l=Q+1-j}^{N} AR(l) + \\ P_t(0,1) \cdot \sum_{l=Q}^{N} AR(l) \end{cases} + I(E_1) \cdot \begin{cases} \sum_{j=1}^{Q} (P_t(j,0) \cdot C(0,1) + P_t(j,1) \cdot C(1,1)) \cdot \sum_{l=Q-j}^{N} AR(l) + \\ (P_t(0,1) \cdot C(1,1) + P_t(0,0) \cdot C(0,1)) \cdot \sum_{l=Q-1}^{N} AR(l) \end{cases} \quad (13)$$

$P_{t+1}(Q,0) =$

$$I(E_0) \cdot \begin{cases} \sum_{j=1}^{Q} P_t(j,0) \cdot \sum_{l=Q+1-j}^{N} AR(l) + \\ P_t(0,0) \cdot \sum_{l=Q}^{N} AR(l) \end{cases} + I(E_1) \cdot \begin{cases} \sum_{j=1}^{Q} (P_t(j,0) \cdot C(0,0) + P_t(j,1) \cdot C(1,0)) \cdot \sum_{l=Q+1-j}^{N} AR(l) + \\ (P_t(0,1) \cdot C(1,0) + P_t(0,0) \cdot C(0,0)) \cdot \sum_{l=Q}^{N} AR(l) \end{cases} \quad (14)$$

$P_{t+1}(l,0) =$

$$I(E_0) \cdot \left\{ \sum_{j=1}^{l+1} \binom{P_t(j,0) \cdot AR(l+1-j) +}{P_t(0,0) \cdot AR(l)} \right\} + I(E_1) \cdot \begin{cases} \sum_{j=1}^{l+1} (P_t(j,0) \cdot C(0,0) + P_t(j,1) \cdot C(1,0) \cdot AR(l-j+1) + \\ (P_t(0,1) \cdot C(1,0) + P_t(0,0) \cdot C(0,0)) \cdot AR(j) \end{cases} \quad (15)$$

$P_{t+1}(l,1) =$

$$I(E_0) \cdot \left\{ \sum_{j=1}^{l+1} \binom{P_t(j,0) \cdot AR(l+1-j) +}{P_t(0,0) \cdot AR(l)} \right\} + I(E_1) \cdot \begin{cases} \sum_{j=1}^{l} (P_t(j,0) \cdot C(0,1) + P_t(j,1) \cdot C(1,1) \cdot AR(l-j) + \\ (P_t(0,1) \cdot C(1,1) + P_t(0,0) \cdot C(0,1)) \cdot AR(l-1) \end{cases} \quad (16)$$

where l = 1,2,....Q-1.

By using these transition equations, the steady-state probabilities may be obtained by using the same method of the deterministic case.

Next, we will describe the derivation of the distribution B (m) of the inter-departure between two consecutive packets belonging to the same burst of the PM source. In our analysis the probability having the system full is less than 10^{-15}. With such probability values, the obtained results are not challenged from losses and are closed to the infinite buffer case:

$$B(m) = \sum_{v=\max(1,R+1-m)}^{\min(R-m+Q,R-1,Q)} \frac{G\{(m+v-R,1) \mid (v,1)\}}{\sum_{j=1}^{Q} G\{(j,1) \mid (v,1)\}} * \frac{P(v,1)}{\sum_{t=1}^{Q} P(t,1)} + I(E) * \frac{G\{(m,1) \mid (R,1)\}}{\sum_{j=1}^{Q} G\{(j,1) \mid (R,1)\}} * \frac{P(R,1)}{\sum_{t=1}^{Q} P(t,1)} + \sum_{v=R+1}^{J} \frac{G\{(v+m-R,1) \mid (v,1)\}}{\sum_{j=1}^{Q} G\{(j,1) \mid (v,1)\}} * \frac{P(v,1)}{\sum_{t=1}^{Q} P(t,1)}. \quad (17)$$

Where $j = \begin{cases} Q+R-m \text{ if } R < m+1 \\ Q \text{ if } R \geq m+1 \end{cases}$, $I(E) = \begin{cases} 1 \text{ if } (m \leq Q) \text{ and } (R \leq Q) \\ 0 \text{ otherwise} \end{cases}$.

G((s,1) | (r,1)) means the probability to arrive at the state (s,1) after R h-unit service conditioned to (r,1) as the initial state. This probability is calculated with the same method used in the deterministic case.

3 Results

In this section we will present results aiming at identifying the main characteristics of the output traffic of the source used throughout our research. We begin to analyze the case of the PD source. Fig. 2 shows: mean value, 95th percentile, 5th percentile of the interdeparture time for PD source for different values of the offered load. In this case we assume τ equal to 10 msec. and R equal to 100. Fig. 3 shows the variance of the interdeparture time distribution B. We assume two values of the time τ respectively equal to 10 and 20 msec. The value of R is chosen to guarantee that the value R/τ is equal for both cases. The offered load is equal 0.8. The results show that:

- The values of 95th and 5th percentile start nearly symmetric at low values of the offered load and then get very polarized towards the 95th percentile when the offered load increases (Fig. 2).
- The values of the variance are piecewise linear (with a knee) with R/τ (see Fig. 3).
- If the value R/τ is kept constant, the variance is independent of the τ value; in particular this is true when R/τ is greater than 1 (see Fig. 3).
- If the τ value is kept constant, the distribution B is mainly concentrated on the mean value τ for values of R increasing (variance becomes very small as R/τ increases, see Fig. 3).

To summarize: the distribution of the interdeparture is highly concentrated around the mean value but the right end tail is very long even despite of the small subtended area.

Lastly, the behavior of the interdeparture time distribution B, when the source is PM, is studied. In Table 1 we change the value of L, which is the mean number of packets sent in the ON state, keeping constant other parameters: offered load (0,8), τ (10 msec), R (100) and α (0,4). The results show that the value of L doesn't influence the value of the variance of the interdeparture-time distribution. In Fig. 4, we report the variance of the distribution B for PD and PM sources as a function of source activity assuming the following values for the parameters: τ=10 msec, R=100, offered load=0,8 and L=100. The results show that:

- The variance of the distribution B when the PM source is selected is independent of the L value (see Table 1).
- The value of the variance of the distribution B is higher in the case using PM source respect to deterministic source PD, having the same offered load. These differences are high for low values of the source activity (see Fig. 4).

Table 1. Independence of Mean and Variance from Burstiness L

L	Mean	Variance
10	10	0,11
50	10	0,11
100	10	0,11
200	10	0,11

Table 2. Comparison of variance of the jitter distribution for the PD source in the clocked and in the not clocked case

Offered Load	Variance when the system is clocked	Variance when the system is not clocked
0,2	0,004	0,002
0,4	0,0112	0,006
0,6	0,027	0,02
0,8	0,104	0,096
0,9	0,304	0,302

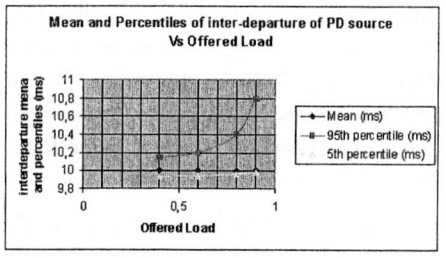

Fig. 2. Mean and Percentiles of interdeparture for PD source as a function of the offered load

Fig. 3. Variance of interdeparture for PD source as a function of R/τ

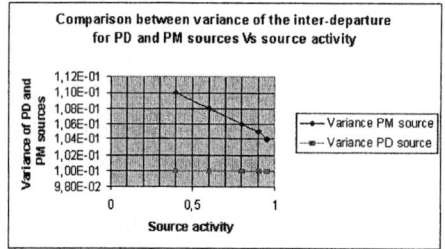

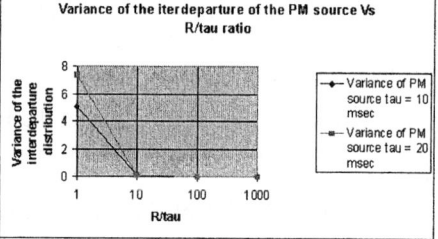

Fig. 4. Comparison between the variance of the interdeparture for PD and PM sources as a function of the source activity

Fig. 5. Variance of the interdeparture for the PM source as a function of the R/τ ratio

To interpret the results we can observe that an On-Off source (PM) with high activity is very similar to a deterministic source (PD). This is true because we only study the interdeparture time between packets belonging to the same burst (during the same On period). In Fig. 5, we compare the variance of the distribution B for two different values of the parameter τ but choosing the value of parameter R so that R/τ is equal in both cases. The offered load is equal 0.8, the activity of the source is equal 0.4 and the parameter L takes the value 100. For the sake of clarity:

- When the value R/τ is greater than 1 the variance of the distribution B does not depend on the values of the parameter τ
- If we compare these results to corresponding ones obtained with the deterministic source PD (compare Fig. 3 with Fig. 5), we get the same value of the variance when R/ τ is greater than 1. Moreover, when R/τ equals to 1 the variance in the case of PM source is greater than the one of the PD. This difference does not depend on the parameter τ but on the value L.

In Table 2 we compare the variance of the interdeparture-time distribution by the PD source when the system is clocked, as in our paper, or when the system is not clocked; in this case the packet flows into the buffer, being processed at once in case the server is idle. The results in the last case are obtained by simulation, and show that:

- An increase of the offered load has a small effect on the variance
- when the system is clocked, the variance is always higher vs. the unclocked case.

4 Conclusion

In this paper the distribution of the interdeparture time of a voice source in a node of the network was studied. The input traffic offered to the system is composed by two different sources of traffic: a first one generates traffic with exponential negative distribution, while the other source provides either a deterministic traffic (case labeled as PD source) or an On-Off source (labeled as PM source). Our results suggest:

- When the packets are produced at a rate τ greater than the unit service of the packet, the variance of the interdeparture-time distribution is very similar for both cases: PD and PM source.
- When the packets are produced at a rate equal to the unit service, the value of the variance between PD and PM source is different. Typically the On-Off source shows an higher variance figure with respect to the deterministic source; particularly when the value of the source activity is low.
- Effects on the variance of the interdeparture-time distribution B for different values of the average burst length L are negligible (see Table 1).

References

1. C. A. Fulton and San-qi Li, "Delay Jitter First-Order And Second Order Statistical Functions Of General Traffic On High Speed Multimedia Networks", *IEEE/ACM Transaction on Networking*, Volume 6, No. 2, April 1998, pp. 150-163.
2. R. Landry and I. Stavrakakis, "Study of Delay Jitter With and Without Peak Rate Enforcement", *IEEE/ACM Transaction on Networking*, Volume 5, No. 4, August 1997, pp. 543-553.
3. M. Conti, E. Gregori and I. Stavrakakis, "Large impact of temporal/spatial correlations on per-session performance measures: single and multiple node cases", *Performance Evaluation*, pp. 83-116, Vol. 41, No. 2-3, July 2000.
4. F. Guilleman and J. W. Roberts, "Jitter and bandwidth enforcement," *in Proc. IEEE Globecom'91*, Phoenix, AZ, 1991, pp. 261-265.

Performance Analysis Depend on OODB Instance Based on ebXML

Kyeongrim Ahn, Hyuncheol Kim, and Jinwook Chung

Dept. of Electrical and Computer Engineering, Sungkyunkwan University,
300 Chunchun-Dong, Jangan-Gu,Suwon, Korea, 440-746
krahn@paran.com,
{hckim, jwchung}@songgang.skku.ac.kr

Abstract. New technologies have been continuously introduced in the e-Business environment since the emergence of e-commerce. For example, ebXML (Electronic Business eXtensible Markup Language) has been adopted instead of EDI (Electronic Data Interchange) for exchanging electronic documents, or RDB (Relational Database) is being replaced by OODB (Object Oriented DataBase) for storing data. In view of the fact that ebXML is based on the object-oriented concept, OODB is more efficient than RDB at XML-based e-business system. In this paper, we design document format and Database schema with ebXML concept. We designs the XML structure (instance) of OODB instead of table or field of RDBMS(Relational Database Management System). The important thing in Database designing is the classification rule to choose element or attribute among meaningful values. It is also important to define Database configuration rules or archiving rules. That is a key point to determine system performance and efficiency. By numerical analysis, we demonstrate that the proposed DB modelling actually achieves operational efficiency and improve resource utilization.

1 Introduction

The key expectation of e-business is that it will provide standardized framework, and versatile operational functions [1][2][3]. Due to the fact that XML possesses the object-oriented characteristics inherently, we propose ebXML schema based on CCTS (Core Component Technical Specification) and the XML NDR (Naming and Design Rule) as a exchange format [4][5][6][7][8][9].

It also employs object-oriented database as a storage. Unlike RDBMS (Relational Database Management systems), store method operates by the object-unit and index (binder) definition method set frequently referenced item at retrieval [10][11][12]. The size or performance of database is governed by the number of item set by the index. To store messages defined by ebXML schema, two different DB instances, such as element-based and element-attribute structure, are also proposed and search results of each instance are analyzed respectively. The response time for request at each DB instance structure is also analyzed [13][14].

By numerical analysis, we demonstrate that the proposed DB scheme (element-attribute structure) actually achieves operational efficiency and improve resource utilization.

The rest of this paper is organized as follows. We present the background relevant to CCTS and XML NDR in section 2. In section 3, proposed DB structure is explained. In section 4, we analyze the retrieval results of the proposed DB structure. The comparative numerical analysis is also explained in section 4. Finally the paper concludes in Section 5.

2 Related Technologies

The CCTS has proposed to satisfy the interoperability of information among business applications. The CCTS is focused on the representation of business information with human-readability and machine-dependent functionality [4]. UN/CEFACT XML NDR provides a method to maximize the business information reusability and is represented as an XML schema component to support or to improve interoperability. UN/CEFACT XML NDR defines business information as XML schema language and is used whenever business information is shared or exchanged [5].

XPath have proposed to provide common structure and semantics to shared functionality between XSLT(XML Stylesheet Language Transformation) and XPointer. The fundamental purpose of XPath is moving or referencing from one position to specific position within same XML document or different XML document. To do this, XPath is modelling XML document with node tree. XQuery is notation be addressing and filtering to text. Its concept is expanding to XSL patten structure. XQuery is query language to support single structure usable at patten.

3 Proposed System and DB Architecture

3.1 Document Architecture

In this paper, we have applied to ebXML CCTS 2.01 in order to define XML document. Also, we adopted the approach that it designs the structure about an overall data and it develops individual electronic document after developing common component. Firstly, we analyzed off-line document needed to exchange data between trading partners. Next, the extracted items after analyzing are enumerated and mapped into suitable XML element of XML ACC (Aggregate Core Component) library. Hereby, the structure of business information about extracted items must be suited to ebXML CCTS 2.01 specification. [1][2].

If suitable XML element does not exist at XML ACC library, it must generate new ACC or BIE (Business Information Entity). Important thing is preferably not to modify basic CCT (Core Component Type). After BIE or ABIE (Aggregate Business Information Entity) definitions are classified, it should generate XML schema. The most important point at XML-based document definition is reusability of CC. If similar items exists already, it is desirable to re-use as CC expansion or modification.

3.2 DB Instance Architecture

We design DB structure to store XML instance defined at above section, element-based structure and element-attribute structure. We divide XML information into management information and data information. And, we store it separately due to performance and load balancing. Management information is smaller than data, then it is stored at OODB with XML format. Data(also, XML format) is stored at file system. With this store approach, DB size becomes small and performance has improved.

- Element-based Structure: It extracts characteristic value from source file and most of values are element and the others values are attribute. When the DB instance of the generated instance exponentially increases as the save times passes. Next, as instance become lager, search and response time comes to be slow. However, element-based structure is efficient in case the instance size is small.
- Element-attribute Structure: Element-attribute structure has mixed format with element and attribute of XML. it extracts characteristic value from source file and groups it into the meaningful values. Especially, we configure attribute as items having similar property or relationship. In XML characteristic, the size of XML constructed with element-attribute is smaller then constructed with element only. Hereby, the important thing is that we determine which value is element or attribute. That is a key point to judge DB sizing and performance.

4 Analysis

To evaluate the proposed DB architecture, we use OODB as a database system and we store data to OODB for following period: one day, one month and 6 months. We construct two types of DB instance, the existed structure(element-based structure) and the proposed structure(element-attribute structure). Test operation includes insert, select, delete, update and select condition is also various. With above test environment, we construct three types of DB structure. Case 1 is element-attribute structure and no binding between instances, only indexing about one XML instance. Case 2 is element-attribute structure and has whole binding and indexing between instances. Finally, case 3 is element-based structure and has whole binding and indexing same as case 2. Also, we use one or more select condition and date period. Table 1 shows the result about DB (select) operations. Fig. 2 shows the charts for result of select I operation as Table 1.

Select II operation is used to measure the response time per time when user access through WEB browser and request some data in case 1 and 3. Select

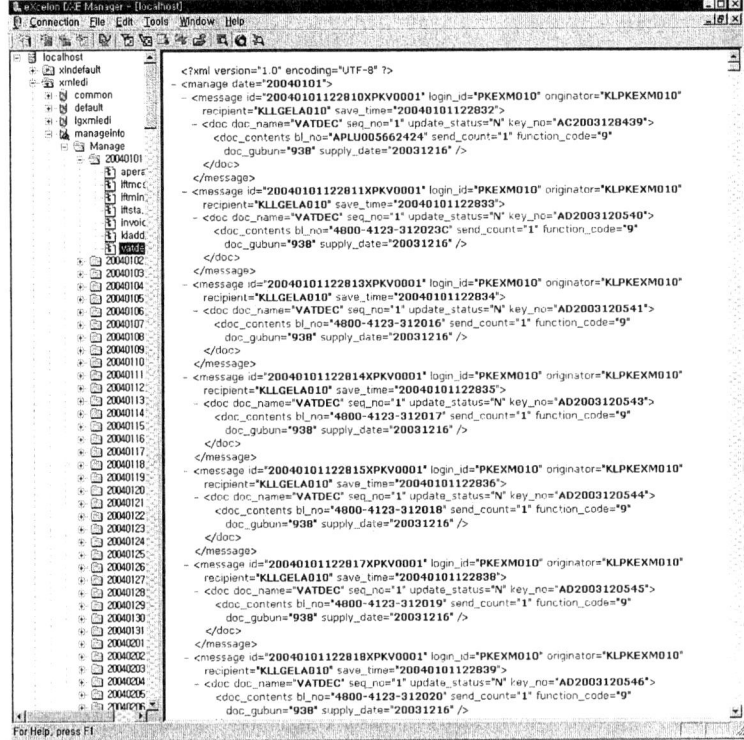

Fig. 1. Element-attribute Structure

Table 1. Select I Operation (Time units: milliseconds)

Item	Condition 1		Condition 2		Condition 3		Condition 4	
	1 Day	1 Month	1 Day	1 Month	1 Day	1 Month	1 Day	1 Month
case 1	20	3695	20	1903	230	1392	270	2284
case 2	120	2884	140	3475	271	2013	260	1523
case 3	170	5247	180	3715	892	3806	831	2794

condition is time interval(one day) and related item. Table 2 shows the result of select operation.

As above mentioned, we construct DB instance with 2 case, the existed structure(element-based structure) and the proposed structure(element-attribute structure). The DB size with the proposed element-attribute structure have decreasing 4 times less then the existed element-based structure. The average retrieval time have also decreasing from minimum 3 times to maximum 6 times.

We have found that result of case 1, 2 is similar as so far as testing. However, in case 2, as DB size grows bigger and the store period gets longer, the binding overhead grows heavier. Also, performance is not good when DB data

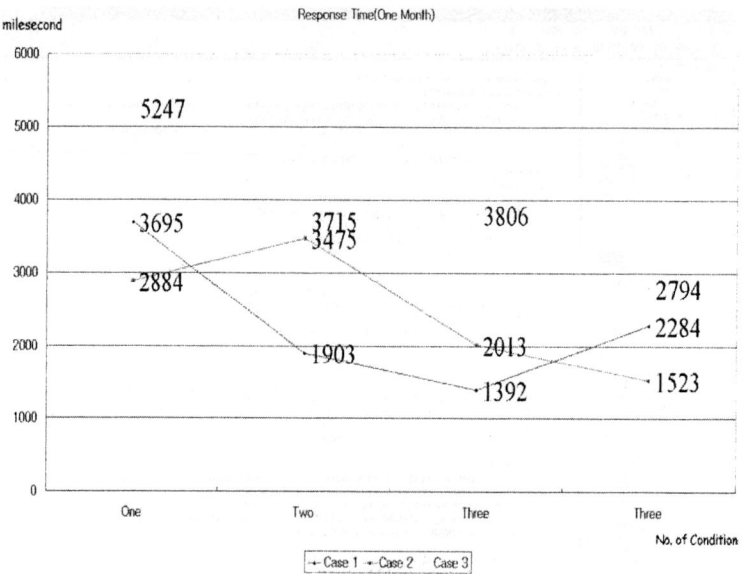

Fig. 2. Result Charts for Select I Operation

Table 2. Select II Operation (Time units: milliseconds)

Item		Case 1		Case 3	
		Generate Result	Apply Stylesheet	Generate Result	Apply Stylesheet
08:00	09:00	2953	718	859	610
09:00	10:00	3093	718	938	562
10:00	11:00	3391	501	1157	656
11:00	12:00	5125	742	1203	562
12:00	13:00	4422	282	1297	141
13:00	14:00	4109	281	953	750
14:00	15:00	4422	657	766	578
15:00	16:00	2813	266	406	125
16:00	17:00	3266	515	390	127
17:00	18:00	7093	360	406	188
18:00	19:00	4906	281	438	172
19:00	20:00	3734	594	453	203

backup (archive, delete) operation. Therefore, we recommend case 1 (element-attribute structure and no binding between instances, only indexing about one XML instance) as DB (XML) instance structure.

5 Concluding Remarks

In numerous e-business environments, the environment is rapidly changed to meet the user requirements and business status. New technology is continuously introduced into the e-business environment. For examples, for the purpose of exchanging electronic documents, ebXML has been adopted instead of EDI, and for the purpose of a database, OODB has been preferably used instead of RDB. In this paper, we introduced an electronic document and XML structures (instances) of object-oriented Database instead of tables or fields in RDBMS.

In beginning stages, we have designed DB and XML instance with element-only structure. But, as the number of transaction becomes increasing and stored data becomes large, performance and response time becomes slow. So, we decide to change XML instance structure into element-attribute structure. Then, we simulate and analyze test result and recommend element-attribute structure. Key point to judge DB sizing and performance is to determine which value is element or attribute.

Future Study is expanding current architecture to WEB. Currently, we provide presentation service with method of parsing values from XML file and viewing when user request. So, if it changes into view XML schema file with XSL (XML Stylesheet Language) without additional operation, then total processing time maybe is decreasing because of reducing some processing step. With this, response time and performance will be advanced.

References

1. MOCIE(Minister of Commerce, Industry and Energy), KIEC : 2004 e-Biz Standardization WhitePaper, KIEC-063 (2004. Jan.) 18-27, 74-197
2. MOCIE(Minister of Commerce, Industry and Energy), KIEC : 2004 e-Business WhitePaper, KIEC-068 (2004. Mar.) 32-103
3. MOCIE(Minister of Commerce, Industry and Energy), KIEC : 2002 ebXML WhitePaper, KIEC-006 (2002. Jan.) 14-23, 55-76, 208-269
4. UN/CEFACT : "Core Component Technical Specification Version 2.01, Part 8 of the ebXML Framework", (2003. Nov.)
5. UN/CEFACT : "XML Naming and Design Rules Draft 1.0", (2004. Aug.)
6. ebXML site http : //www.ebxml.org
7. Korea ebXML forum : http://ebxmlkorea.org/ebxml/index.html
8. Korea Institute of Electronic Commerce : http://www.keb.or.kr
9. ebXML Next generation EC standardization site : http://www.ebxml.or.kr
10. Dan Chang, Dan Harkey : Client/Server Data access with Java and XML, Wiley & Sons Inc., Canada
11. Sean McGrath : XML Processing with Python, Prentice-Hall Inc.
12. http://www.xmledi-group.org/xmledigroup/guide.htm - "Guidelines for using XML for Electronic Data Interchange"
13. K.R, Ahn, J.H. Ahn, B.K. Moon , WY. Lee, JW, Chung: The Converting/Transfer Agent for E-Commerce, APIS'04, (2004. Aug.)
14. K.R.Ahn, J.C.Choi, J.H.Yu, B.K.Moon, W.Y.Lee, J.W.Chung: A Study on Format Conversion and Data Transfer using FTP, The 19th spring Conference on KIPS, Vol. 10, No. 1, (2003. May.)

Performance Visualization of Web Services Using J-OCM and SCIRun/TAU

Wlodzimierz Funika[1,*], Marcin Koch[1], Dominik Dziok[1], Marcin Smetek[1], and Roland Wismüller[2]

[1] Inst. Comp. Sci., AGH, al. Mickiewicza 30, 30-059 Krakow, Poland
funika@uci.agh.edu.pl
[2] University of Siegen, Hölderlinstr. 3, D-57068 Siegen, Germany

Abstract. This paper presents a performance visualization tool, developed to provide advanced 3-D performance visualization of distributed applications, based on the integration of the J-OCM monitoring system into the SCIRun/TAU visualization environment. Following the changes in the J-OCM monitoring system, we have done extensions to SCIRun to support Web Services performance visualization.

Keywords: Performance visualization, monitoring tools, J-OCM, TAU, SCIRun, web service.

1 Introduction

Performance analysis support is a prerequisite for carrying out optimizations of distributed applications which become more and more complex. A very important role in performance evaluation is played by performance visualization [7]. Recently, Web Services became one of the fundamental elements in distributed computing on the Internet due to open standards and the capability to integrate applications coming from different platforms. As the distributed applications based on Web Service technologies become increasingly popular, our focus is on evaluating the performance of Web Service-based applications.

The overall performance of a distributed application based on Web services depends on its logic, network, but what we are interested mostly are the problems associated directly with Web services. Most such problems come from the underlying messaging and transport protocols such as SOAP and HTTP. The communication efficiency issues are associated with the throughput of connection and protocol, but on the other hand an extremely important issue is the size and complexity level of messages.

Thus accessing the monitoring data on a Web service at real time can provide information valuable for developing a distributed application. Based on this data, performance bottlenecks can be identified. Some values which can characterize the performance of Web services are: *throughput* - the number of Web service requests processed within a given time period, *latency* - the time elapsed between

---

* Corresponding author.

sending a request and receiving a response, *transaction time* - the time that passes while the Web service is finishing a complete transaction.

Our work is based on the integration of two systems: the J-OCM monitoring system [2] as a data supplier component and SCIRun/TAU/ParaVis [9] as a user interface/visualization component to form a comfortable tool for comprehensive performance analysis of Web Service applications. The work comprises extensions to the J-OCM monitoring system and the development of SCIRun-compliant modules linking the J-OCM and TAU/ParaVis.

The paper is organized in the following way: Section 2 introduces the design of the system. Section 3 gives some details on the additional components we have developed and the advantages of the presented approach. In Section 4 we are summing up the results achieved.

2 Overall Design

The system under discussion in the paper has been designed to support the performance analysis of Web Services. It is intended to provide the following features:

- ability to measure an extendible number of performance metrics (e.g. response time, latency, processing time, messages size)
- on-line monitoring and visualization
- no need in source code modification
- ability to access and manage distributed resources (Web Services)
- extendability of functionality
- easy-to-use and -configure
- low level of overhead induced on the performance of the application.

Fig. 1 presents an overview of the system architecture. Generally, the performance monitoring environment comprises two components: the first one being

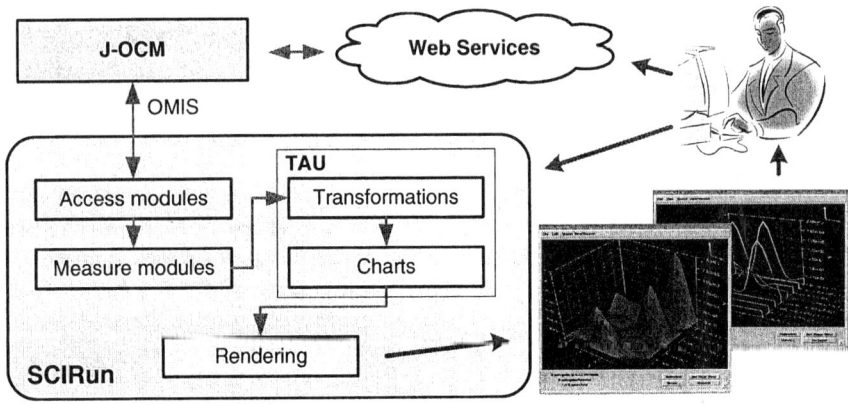

Fig. 1. System architecture

a monitoring system and the second - a visualization environment. These parts are integrated and collaborate with each other via the J-OMIS interface, a Java oriented extension to OMIS [8]. J-OMIS defines a set of services provided by the monitoring systems [1]. There are services for listing performance objects, capturing events, and executing actions on their occurrence. The monitoring system (J-OCM [2]) has been developed as an extension to the OCM monitoring system [12]. The most important feature of J-OCM is the ability to perform on-line monitoring of distributed applications, while not requiring source code instrumentation. J-OCM provides convenient access to performance objects like classes, methods, threads, or web services. Within J-OCM, these objects are identified by tokens. A *token* is an unique character string (e.g. token "jvm_j_2_n_1" identifies JVM "j_2" running on a physical machine "n_1"). All performance object types observable by the monitor form an *objects hierarchy*. On top there are "nodes" (or "hosts"). Under nodes there are JVMs, threads, classes, and methods.

J-OCM is targeted at Java applications monitoring but it is also possible to extend it for other platforms, e.g. .NET. In its original version, J-OCM did not support Web Service monitoring. Owing to the extendability of the monitoring system, J-OCM is being extended for this purpose.

J-OCM has been integrated into a visualization tool developed as a package for SCIRun [6,10]. SCIRun is a powerful Problem Solving Environment (PSE) providing a framework for building complex systems, developed at the University of Utah (USA). A system built with SCIRun consists of a number of connected modules. A single *module* is responsible for some functionality, e.g. data processing. There can be a GUI associated with each module. The module usually has one or more input and output ports. A *port* defines the data type accepted and returned by the module. The input and output ports of the modules can be connected in order to create a "*net*". Modules can be placed anywhere and easily connected; there is no limit of modules in the work area. However, since each module is a separate SCIRun thread, the performance of a great number of modules can be an issue.

SCIRun allows the developer to create additional modules (as a C++ class) and packages of modules. A GUI should be written with TCL language. In addition to SCIRun, at the University of Oregon in Eugene (USA) there has been developed the TAU/Paravis package [11,9] containing modules for the analysis and 3-D visualization of monitoring data coming from a source.

Our monitoring environment uses J-OCM as a *monitoring data source* and TAU/Paravis as a *visualization engine*. We have developed some additional modules for accessing J-OCM performance objects and for measuring performance indicators. There are also created Web Service-related modules intended for accessing performance objects like ports or operations.

3 Additional Modules

In our work we focused on the development of additional modules. The role of these modules is to enable the collaboration of J-OCM and TAU/Paravis but also to simplify configuring J-OCM. Our primary goal was to create an easily extendible architecture for J-OMIS compliant monitoring systems. For this purpose we created an additional SCIRun package of modules called "JOCM". Within the package the following categories of modules can be found:

- **Access modules.** This category of modules is responsible for accessing monitoring data directly from the J-OCM monitoring system. They provide routines for listing available performance objects (like threads or methods), attaching and detaching them from the monitor. The role of *access modules* is also to make J-OCM to enable the monitoring of events. After executing the monitoring process, *access modules* are responsible for capturing monitor events, filtering and forwarding them to the *measurement modules*.
- **Measurement modules.** This category comprises modules responsible for transforming incoming event notifications into the performance data. There is a specified metric associated with each measurement module (e.g. execution time, message size). *Measurement modules* store the performance data and transform it into the TAU/Paravis compliant format (3D matrix).

Each *access module* is associated with some performance object type and can attach performance objects of this type only. We have developed *access modules* intended to access performance objects related to Java applications (JVM, thread, class, method) and Web services (Web service, port, operation). Following our main goal, the category of *access modules* can be freely extended in order to enable handling the new kinds of performance objects. Because the structure of *access modules* is related to the hierarchy of the performance objects and tokens inside the monitoring system, it is important to maintain the relations between modules while implementing new ones. The performance objects can only be attached by *access modules* according to the hierarchy so the way the corresponding modules are to be connected is extremely important.

Measurement modules obtain raw monitoring data through *performance port*. All *access modules* have one *performance output port* and all *measurement modules* have one *performance input port*. For additional modules, it is enough to provide a *performance input/output port* to allow for collaboration with the rest of modules. Every *measurement module* composes a *performance data set*. This is an object containing all performance data related to some metric (eg. processing time or number of method invocations) for all performance objects obtained by connected *access modules*. At present the JOCM package contains *measurement modules* which can collect performance data related to: execution/processing time (summary, momentary, and average), message size (requests and responses), number of performance event occurrences, thread state and also metrics related to *garbage collector* - time of activity and amount of released memory.

The package of modules we have developed contains general classes for *access* and *measurement modules*. There is also a general class for *performance data sets* which simplifies the implementation of new metrics.

4 Concluding Remarks

Efficient performance analysis requires advanced performance visualization tools. The tool under discussion in this paper has been created to help the user investigate performance problems of distributed applications using Web Services, with a large on-line monitoring and visualization functionality. The tool is based on the integration of the J-OCM monitoring system into the TAU/SCIRun visualization environment. Since these components use different data models, the integration work involved the development of additional modules compliant with SCIRun. The monitoring part and visualization part of the tool communicate via the J-OMIS monitoring interface.

The performance data is displayed as 3-D charts: bars, histogram, scatter plot, waterfall, and terrain chart. The tool provides metrics associated with Java applications (method execution time, thread states, garbage collector activity) and Web Services (request/response processing time, message size, number of operation invocations and others).

The issues of performance evaluation of Web Services are addressed by many monitoring and testing systems. Commercial tools like AmberPoint Express [5] for .NET platform or open source ones like Apache jMeter [3] for JAVA give the developer a lot of useful features. They support advanced visualization and can easily point the application bottlenecks. However, each of them runs as a client application. They test Web Services by sending requests and counting the response time or number of fails. They can also inspect SOAP packages. The most important disadvantage is that they do not allow the developer to get insight into what really happens inside the Web service. This is the main difference between available solutions and our system. In effect our system is able to point what part of the Web Service is responsible for the performance problem (*initialization, request processing, operation invocation* or *response*).

While the J-OCM/TAU tool follows the general trends in building monitoring facilities for Web Services, our goal is to overcome the above mentioned constraints and to provide that our system can detect *which part* of the Web service (initialization, request processing, operation invocation or response) causes the performance problems encountered.

The new version of J-OCM is compliant with Java 1.5 which provides more efficient monitoring interfaces [4] than the older versions of Java. This feature causes that the monitoring system affects the performance of the monitored application to a small extent. We have performed tests in which we compared the reaction time of the Web services with and without monitoring enabled. In all test cases the overhead was quite low (ca. a few percents). The visualization tool itself doesn't slow the application down at all since it is usually running on a different computer.

Our further research is aimed at the tool's scalability, i.e. when a large number of Web services are intensively used. Finally, it is worth mentioning the performance issue of the tool: SCIRun uses an amount of physical memory, efficient CPUs and graphics device. However, this problem is expected to be overcome in the nearest future.

The first release of the J-OCM/TAU performance visualization tool is expected in the autumn of 2005.

Acknowledgements. We owe our gratitude to prof. Allen D. Malony and his team. This research was partially supported by the KBN grant 4 T11C 032 23.

References

1. M. Bubak, W. Funika, R. Wismüller, P. Metel, and R. Orlowski. Monitoring of distributed java applications. *Future Generation Computer Systems*, 19:651–663, 2003.
2. W. Funika, M. Bubak, M. Smętek, and R. Wismüller. An omis-based approach to monitoring distributed java applications. In Yuen Chung Kwong, editor, *Annual Review of Scalable Computing*, chapter 1, pages 1–29. World Scientific Publishing Co. and Singapore University Press, 2004.
3. Apache JMeter:: http://jakarta.apache.org/jmeter.
4. JVM Tool Interface:: http://java.sun.com/j2se/1.5.0/docs/guide/jvmti/jvmti.html.
5. AmberPoint Express:: http://www.amberpoint.com/solutions/express.shtml.
6. C. Johnson and S. Parker. The scirun parallel scientific computing problem solving environment. In *Proc. Ninth SIAM Conference on Parallel Processing for Scientific Computing*, 1999.
7. P. Kacsuk, G. Dózsa, J. Kovács, R. Lovas, N. Podhorszki, Z. Balaton, and G. Gombás. P-grade: A grid programing environment. *Journal of Grid Computing*, 1(2):171–197, 2004.
8. T. Ludwig, R. Wismüller, V. Sunderam, and A. Bode. *OMIS – On-line Monitoring Interface Specification (Version 2.0)*, volume 9 of *LRR-TUM Research Report Series*. Shaker Verlag, Aachen, 1997.
9. A.D. Malony, S. Shende, and R. Bell. Online performance observation of large-scale parallel applications. In *Proc. Parco 2003 Symposium*, pages 173–180. Elsevier B.V., Sept. 2003.
10. Scientific Computing SCIRun, University of Utah and Imaging Institute:: http://software.sci.utah.edu/scirun.html.
11. Computer TAU's 3-D Profile Visualizer ParaVis, University of Oregon and Information Science:: http://www.cs.uoregon.edu/research/paracomp/tau/tauprofile/dist/paravis.
12. R. Wismüller, J. Trinitis, and T. Ludwig. Infrastructure for the run-time monitoring of parallel and distributed applications. In *Proc. Euro-Par'98, Parallel Processing*, volume 1470 of *Lecture Notes in Computer Science*, pages 173–180, Southampton, UK, 1998. Springer-Verlag.

A Metadata Model and Information System for the Management of Resources in a Grid-Based PSE Toolkit

Carmela Comito[1], Carlo Mastroianni[2], and Domenico Talia[1,2]

[1] DEIS, University of Calabria, Via P. Bucci 41 c,
87036 Rende, Italy
{ccomito, talia}@deis.unical.it
[2] ICAR-CNR, Via P. Bucci 41 c, 87036 Rende, Italy
mastroianni@icar.cnr.it

Abstract. A PSE toolkit is a group of technologies within a software architecture through which multiple PSEs can be built for different application domains. This paper presents a metadata model for Grid-based PSE toolkits and the architecture of an information system based on the proposed metadata model. These two components contribute to define a general model of metadata management for supporting the design and implementation of problem solving environments on Grid.

1 Introduction

A problem solving environment (PSE) is a computer system that provides the computational features necessary to solve a target class of problems. An advancement of the PSE concept is the *PSE toolkit* concept, a group of technologies through which multiple PSEs can be built for different application domains. PSEs can benefit from advancements in hardware/software solutions achieved in parallel and distributed systems and tools. A very interesting computing model in the area of parallel and distributed computing is the Grid.

Distributed implementations of PSE toolkits can be envisioned through the exploitation of Grid features and functionalities so obtaining *Grid-based PSE toolkits*. The effective use of a Grid-based PSE requires the definition of an approach to manage the heterogeneity of the involved resources that can include computers, data, network facilities, sensors, and software tools provided by different organizations [2]. The management of such resources requires the use of metadata that, through an accurate categorization of resources, provides useful information about the features of resources and their usage modalities. As opposed to a single domain PSE, in a multi-domain PSE toolkit the structure of metadata information is not uniform: it depends on the type of the resource (i.e. software, hardware, data etc.), and on the application domain in which the resource is used. The information system of a PSE toolkit should use an efficient and flexible approach to manage metadata documents having different structures. Furthermore, such a toolkit should exploit the information services provided by the underlying Grid framework, e.g. the Index Services of the Open Grid Services Architecture (OGSA) [6].

We designed a metadata model and an information system that can be used in a Grid-based PSE toolkit to manage metadata and provide resource discovery features. The paper is organized as follows. Section 2 discusses related work. Section 3 describes the metadata model and section 4 presents an information system architecture based on the proposed metadata model. Section 5 concludes the paper.

2 Related Work

The key role of metadata management for the effective design of distributed information systems is widely recognized in the literature [8, 12]. The adoption of the service oriented model in novel Grid systems, based on the OGSA architecture, has a noteworthy impact on the management of metadata and the architecture of information services, since in such systems services and resources are provided as Grid services. The information model of service-oriented Grid frameworks is essentially based on two features:

1. Metadata describing Grid services instances is stored into XML-encoded documents, called *Service Data Elements* (SDE) in OGSA.
2. Information is collected and indexed by means of hierarchical information services (Index Services in OGSA) that collect the metadata stored in Grid services, aggregate it and provide enriched metadata information to high level browsing and querying services.

In the Grid computing community there is an effort to define the so called Semantic Grid [16], whose approach is based on the systematic description of resources through metadata and ontologies. In [8] the role of metadata in the context of the Semantic Grid is discussed. Here metadata is used to assist a three level programming model: the lowest level includes traditional code to implement a service; the next level uses agent technology and metadata to choose which services to use; the third level (workflow) links the chosen services to solve domain specific problems. When exploiting the OGSA architecture, based on the Grid Service technology [7], it is essential to integrate metadata embedded in services (SDEs) and metadata external to Grid Services, which can be stored in distributed databases and repositories with very variable scope and completeness.

Metadata management models have been proposed to address the requirements of problem solving environments. Examples of significant PSEs that use XML-based metadata models for the representation of heterogeneous resources are WebFlow and the Common Portal Application [10].

Reference [1] describes a metadata management approach based on Semantic Web technologies. An ontology system is used to produce metadata documents in three steps. The first step aims to create a hierarchy of resource classes; then, for each class, meaningful properties are defined to characterize the resources belonging to that class. Finally, the description of classes and properties, and metadata instances, are written in semantic languages such as RDF and OWL. This approach permits to exploit the richness of semantic languages, but does not take full advantage of the information services provided by service-oriented Grid frameworks.

3 A Metadata Model for a Grid-Based PSE Toolkit

In a Grid-based PSE toolkit, metadata must be used to manage the heterogeneity that comes from the large variety of resources available within each resource class [2]. As compared to a PSE designed for a single application domain, a PSE toolkit covering multiple domains must tackle a further difficulty: the structure of metadata information is not uniform but depends on the characteristics of the resource under consideration. Resources can be distinguished according to their *type* (e.g., software, data source, hardware etc.) and the *application domain* in which they are used (e.g. bioinformatics, earth observation, physics etc.). In the following, the combination of a resource type and an application domain will be referred to as a resource *category*. In other words, a resource *category* is a set of resources of a given type which can be used in a given application domain.

The following types of resources can be identified:

- Data-related resources, such as data sources and data management components.
- Software resources, among which Web and Grid services are gaining a major role.
- Hosts and hardware devices (computers, storage facilities, network connections).
- Applications modeled as workflows.

The metadata model we propose takes into account the specific characteristics of different resource types and application domains. In particular, we propose to associate a metadata document to each resource offered by the PSE toolkit and distinguish three sections within that document. More specifically, a metadata document associated to a resource is composed of the following three sections:

1. *Ontological* metadata used to identify, for each resource, the categories to which the resource belongs. Ontological metadata is generated and managed by an ontology system, which also specifies the *structure* (expressed as an XML schema) of the remaining two sections of the metadata document.
2. *Semantic* metadata used to describe and characterize the resources belonging to a given category. To this aim, for each resource category, a set of classification parameters are defined by means of an XML schema.
3. *Resource* metadata supplies specific information about a resource in order to facilitate its access and usage.

The rationale of such a distinction comes from the consideration that, in a PSE toolkit, resources must be annotated with metadata information at different levels and at different times. First of all, when a resource is published it is necessary to specify the category to which the resource belongs, in order to determine the set of users that could be interested in that resource: category specification is performed through the first section of metadata information. Furthermore, a resource should be semantically classified within its category to facilitate key services such as resource discovery and workflow composition: the second metadata section is used for this purpose. The third metadata section contains information that, once a resource has been discovered and selected, can be used to facilitate its access and use.

The proposed approach allows for taking advantage of the benefits offered by the Grid technology. In fact, the OGSA technology permits to store XML metadata information within a Grid service, on condition that such information complies with a

Service Data Description, i.e. an XML schema. This way, it is possible to exploit Grid information services to discover and access resources by examining associated metadata.

3.1 Ontological Metadata

The ontological metadata section specifies the categories to which a resource belongs (a resource belongs to multiple categories if it can be used in multiple domains) and, indirectly, the XML schemas to which semantic and resource metadata must conform. For example, TribeMCL [5] is a software tool used in the bioinformatics domain to perform data mining analysis. Ontological metadata should specify the type of the resource (it is a service-oriented software) and the application domains in which it can be used (bioinformatics and data mining). The ontological section of the metadata document related to TribeMCL is as follows:

```
<OntologicalMetadata>
  <ResourceType type="service">software</ResourceType>
  <ApplDomain>data mining</ApplDomain>
  <ApplDomain>bioinformatics</ApplDomain>
</OntologicalMetadata>
```

The element <ResourceType> specifies that the resource is a software tool, and it is offered as a service. Consequently, the *resource* metadata section must comply with the XML schema ServiceSoftware.xsd, which specifies the structure of resource metadata describing a generic service-oriented software.

Furthermore, the <ApplDomain> elements permit to establish that the software can be used under the data mining and bioinformatics domains. Therefore, the *semantic* metadata section must comply with the XML schemas used to categorize software in those two domains: SciDataMiningTools.xsd, and BioinformaticsSoftware.xsd. Such schemas are discussed in the next section.

3.2 Semantic Metadata

Semantic metadata characterizes a resource within a given category in order to facilitate the discovery and browsing of resources. As a category identifies a couple <*resource type, application domain*>, semantic metadata describes the semantics of a resource in a specific domain. For example, if we define the *scientific data mining tools* category (<*software, scientific data mining*>), an ontology system permits to determine the parameters and values that can be used to characterize an instance of that resource category. Such an ontology system produces the XML schema SciDataMiningTools.xsd (the name of the schema is generated by the ontology system in a semi-automatic way), with which the semantic metadata section must comply. Moreover, the ontology system specifies the values of ontological metadata parameters used to identify the resource category, as shown in Section 3.1.

An extract from the XML schema SciDataMiningTools.xsd is reported in Figure 1. The schema defines five elements that are used to categorize a data mining software tool and specifies, through the definition of the corresponding XML schema types, the values that can be assigned to those elements. In particular, semantic metadata elements permit to specify the type of input data sources, the type of knowledge that can be discovered, the type of technique adopted, the algorithm implemented and the driving method exploited by the mining process.

```
<schema targetNamespace="http://domain/path/SciDataMiningTools"
  xmlns="http://www.w3.org/2001/XMLSchema" …>
<simpleType name="KindOfKnowledge_value">
  <restriction base="string">
    <enumeration value="association rules"/>
    <enumeration value="clusters"/>
    <enumeration value="characteristics rules"/> …
  </restriction>
</simpleType> …
<element name="SemanticMetadata">
  <complexType>
  <sequence>
    <element name="KindOfKnowledge" type="KindOfKnowledge_value"/>
    <element name="KindOfData" type="KindOfData_value"/>
    <element name="KindOfTechnique" type="KindOfTechnique_value"/>
    <element name="Algorithm" type="Algorithm_value"/>
    <element name="DrivingMethod" type=" DrivingMethod_value"/>
  </sequence>
  </complexType>
</element>
</schema>
```

Fig. 1. An extract from the XML schema `SciDataMiningTools.xsd`

Analogously, the schema `BioinformaticSoftware.xsd` describes the structure of semantic metadata for a software tool used in the bioinformatics domain.

If a resource belongs to several resource categories (i.e., it can be used in multiple domains), the semantic metadata section is composed of as many subsections as the specified resource categories. In this case each subsection must comply with the XML schema associated to the corresponding resource category.

```
<SemanticMetadata xmlns="http://domain/path/SciDataMiningTools" …>
  <KindOfData>relational database</KindOfData>
  <KindOfKnowledge>clusters</KindOfKnowledge>
  <KindOfTechnique>statistics</KindOfTechnique>
  <Algorithm>MCL algorithm</Algorithm>
  <DrivingMethod>autonomous knowledge miner</DrivingMethod>
</SemanticMetadata>
<SemanticMetadata xmlns="http://domain/path/BionformaticsSoftware" …>
  <BiologicalFunction> protein function prediction
  </BiologicalFunction>
  <BiologicalElement>protein</BiologicalElement>
  <HasInput>BLAST protein sequence</HasInput>
  <ProducedOutput>TribeMCL protein families</ProducedOutput>
</SemanticMetadata>
```

Fig. 2. Semantic metadata section of the software `TribeMCL`

For example, the semantic metadata section associated to the software `TribeMCL` (see Section 3.1), is validated against the XML schemas `SciDataMiningTools.xsd` and `BioinformaticsSoftware.xsd`. Semantic metadata, reported in Figure 2, specifies that the software analyzes BLAST protein sequences extracted from a relational database in order to predict the protein function, uses a statistical method based on the Markov Clustering algorithm (MCL), produces clusters in the form of `TribeMCL` protein families, and is executed through an automatic process.

A similar approach is adopted for modelling the other kind of resources such as data-related resources and workflows.

3.3 Resource Metadata

The structure of resource metadata is defined through an XML schema generated by an ontology system. Resource metadata is divided into *Description* and *Usage* metadata.

Description metadata provides a concise description of a resource. It contains: a functional description of a resource, expressed in terms of the capabilities and functionalities offered by a resource; information about the quality rating of such a resource; information about the past usage of a resource.

Usage metadata gives information that specifies details on how to access and use a resource. Even if it would be preferable that all or most of the resources were offered as services (Web or Grid services), a PSE toolkit should also support non service-oriented resources. The structure of usage metadata is different for service-oriented and non service-oriented resources. Accordingly, for each type of resource (e.g. software, workflow etc.), two different XML schemas are defined. The usage metadata section of a service-oriented resource contains a reference to the WSDL document which specifies the service interface, along with the URL of the service and other information. Usage metadata related to a non service-oriented resource provides detailed XML information about the resource interface, e.g. about the correct usage of a command line interface or an API, for example in the format specified by the GGF SAGA Working Group (Simple API for Grid Applications [15]). The SAGA group aims to lower the barrier for scientific application developers to make use of the Grid by providing a small, consistent API for the operations of interest.

In the following, for three important types of resources (i.e. software components, data resources and workflows), the structure of resource metadata is briefly outlined. We extensively exploit standards that are commonly used for such resources, and in the cases in which those standards are not sufficient, we propose additional formalisms: see reference [11] for more details.

3.3.1 Software Resources

The resource metadata section of a software resource must be validated against the XML schema `ServiceSoftware.xsd`, or against the schema `GenericSoftware.xsd`, depending on whether the software is offered as a service or not. The two cases are discussed separately in the following.

Service-oriented software. The *usage* metadata section must contain at least a reference to the WSDL document describing the service. However, the WSDL language does not give semantic information about a Web service due to the limited expressive power of the XML schema formalism. The Semantic Web Services arm of the DAML program developed an OWL-based Web Service Ontology, namely OWL-

S [14], to enable automation of services on the Semantic Web. If a description of a service is furnished through the OWL-S language, the *description* subsection should contain a reference to that description. WSDL and OWL-S documents can also reference each other.

Non service-oriented software. Resource metadata should provide the same type of information which, in the case of service-oriented resources, is provided by OWL-S and WSDL documents: structure of input and output, information about the software provider, functional description etc. Such information is contained in an XML document. Details are given in [11].

3.3.2 Data Resources
Data-related resources can be classified as follows:

1. *Data resource managers*, i.e. systems designed to manage data. Examples are a file system or a DBMS.
2. *Data sources* can be files, relational databases, XML databases, transaction databases, etc.
3. *Data sets* are collections of data that are not explicitly managed by a resource manager. For example, data generated by an application or the result set of a query evaluated over a relational database.

Resource metadata for data-related resources is validated against the XML schema ServiceData.xsd, or against the schema GenericData.xsd, depending on whether the resource is offered as a service or not. The two cases are separately discussed below.

Non service-oriented data resources. Description metadata includes:

- *Structure metadata*. It contains information about both the logical/physical structure of a data source (e.g. database schemas) and the data model.
- *Capability metadata* specifies the capabilities of a data resource manager. For a DBMS such metadata specifies: language capabilities, supported operations, connection options.

Service-oriented data resources. We adopt the *Open Grid Services Architecture Data Access and Integration* (OGSA-DAI) [13] standard. It builds upon OGSA data access components to manage both relational and XML databases wrapped as Grid Data Services (GDSs). Metadata is handled through several types of XML documents including: (i) a data resource configuration document specifying the activities that a GDS can support, information on the database management system, on the connection to data resources, etc; (ii) a RoleMap file containing data sources access permissions; (iii) a registry containing information about a set of GDSs; (iv) a gridDataServicePerform document used by clients to send query and update operations to a GDS. (v) a gridDataServiceResponse document, returned by a GDS, which contains the results of query and update operations.

3.3.3 Workflows
A main purpose of a Grid-based PSE toolkit is to facilitate users in the specification of complex applications and in the construction of workflows composed by multiple tasks that must be executed sequentially or in parallel.

Several Grid-based workflow systems, such as Pegasus [4], adopt a two layer approach to build and execute a workflow: an abstract workflow is designed at a high level, and then is mapped to the set of available Grid resources, thus generating an executable workflow. In our system, a *concrete workflow* contains only well defined resources (e.g. particular software resources to be executed on specified hosts), whereas an *abstract workflow* contains at least an *abstract resource*, that is a resource defined by means of constraints on metadata properties (e.g., a software that extracts clusters from bioinformatics data). The instantiation of an abstract workflow resolves each abstract resource into a concrete resource available on the Grid.

The document that describes a concrete workflow is placed in the *resource* section of the metadata document describing the application. Details about the specification of abstract and concrete workflows with an XML formalism, and about workflow instantiation, are given in [11].

If an application is composed of Web or Grid services, a concrete workflow can also be expressed with one of the formalisms that are emerging for this purpose, such as OWL-S [14] and BPEL [3].

4 Architecture of the PSE Toolkit Information System

To properly manage metadata in a Grid-based PSE toolkit we model metadata on the basis of the above described approach and propose an information system that accomplishes two main tasks: managing metadata and supporting high-level discovery services. Figure 3 depicts the architecture of the information system.

The information system is integrated with the Globus Toolkit 3 (GT3 [9]), based on OGSA, in order to take advantage of the services offered by that framework (browsing and indexing services, information providers etc.). The information system is composed both by fully distributed components and hierarchical components. The schema repository, that stores the XML schemas generated by the ontology system, and the metadata repository, that stores the XML metadata documents, are both distributed XML databases. The ontology system and the components that are used to index, browse and search resources on the Grid are organized in a hierarchical configuration that reflects the structure of Grid Virtual Organizations. In Figure 3, components that are inherently distributed are replicated.

As mentioned in Section 3, the ontology system produces the *ontological* metadata that allows for identifying the categories to which a resource belongs by specifying the type of the resource and the application domains in which it is used. For each category, the ontology system generates the *structure* (expressed with XML schemas) of the *semantic* and *resource* metadata sections. The ontology is maintained as an OWL file in a centralized/hierarchical repository.

The rest of the section is organized as follows. Section 4.1 explains the approach used to store metadata documents and justifies the opportunity of storing metadata both in the metadata repository and within Grid services. Section 4.2 describes the main characteristics of GT3 Index Services. Section 4.3 illustrates the publishing and discovering functionalities offered by the proposed information system.

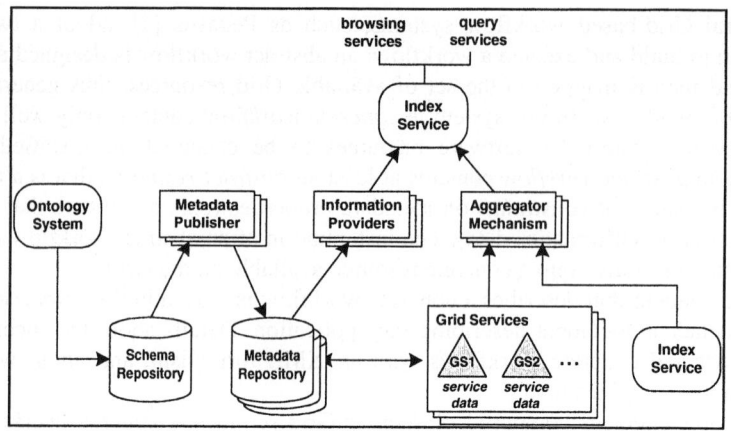

Fig. 3. Architecture of the PSE toolkit information system

4.1 Metadata Repository and Grid Services

The metadata repository stores the metadata documents related to the components/services provided by the PSE toolkit. As mentioned in Section 3, the choice of using XML schemas to define the structure of metadata allows for an efficient integration with the GT3 framework. In GT3 metadata is stored within Grid services as SDEs whose structure is defined by means of enriched XML schemas (Service Data Descriptions). As a consequence, a metadata document associated to a service-oriented resource, or part of such a document, can be retrieved from the metadata repository and stored within a Grid service.

The advantage of storing metadata both in the metadata repository and within a service is motivated as follows. The publication of metadata within a service is useful if we want to take advantage of the Grid information services offered by the Globus Toolkit. On the other hand, storing metadata in the metadata repository is useful for two reasons: (i) to give persistency and high availability to metadata; (ii) to provide a uniform point of access to metadata, including metadata describing non service-oriented resources.

However, consistency problems could arise. To tackle this issue, the metadata repository is chosen as the primary source of information. Metadata associated to a new resource is generated by the metadata publisher and stored in the metadata repository. If the new resource is a Grid service, metadata is retrieved by an *information provider* and published as SDEs. To avoid inconsistency problems, an attempt to modify an SDE requires an access to the metadata document stored in the metadata repository. If access is authorized, a lock is executed on the database, the requested modification is performed on the metadata document with a synchronous operation and finally the SDE is modified as requested.

To facilitate the searching and browsing of resources, metadata can also be aggregated and published by GT3 Index Services, as described in the next subsection.

4.2 Index Services

The GT3 information system produces, aggregates and indexes metadata related to the resources provided by a set of Grid hosts belonging to a Virtual Organization (VO). Such a system exploits the functionalities of the OGSA Index Services [6]. Metadata describing the resources of the PSE toolkit is aggregated and published on Index Services with two mechanisms, depending on the kind of resource:

1. *Non service-oriented resources.* A set of *information providers* retrieve the XML metadata documents stored in the metadata repositories of a VO, and publish them in the Index Service of that VO.
2. *Service-oriented resources.* The VO Index Service subscribes to the Grid services SDEs that have to be aggregated and indexed, in order to be notified of changes. The GT3 *aggregators* retrieve the SDEs from the Grid services and publish them in the Index Service. If the Index Services of a VO are organized in two or more levels, the *aggregators* can retrieve metadata from lower level Index Services, and publish it in higher level Index Services, as depicted in Figure 3.

Since Index Services are fed with data retrieved both from Grid services and metadata repositories, the deployed architecture provides a uniform and flexible mechanism to query and browse metadata related to all kinds of resources, including non service-oriented ones. Browsing and querying can be performed by means of the Globus Toolkit services (e.g. the Service Data Browser) or high level services offered by domain specific PSEs.

4.3 Publishing and Discovery of Resources

On top of GT3 Index Services, the proposed architecture provides high level services through which users can publish and discovery resources on the PSE toolkit.

The *publishing* functionality enables to create, modify, and delete XML metadata documents stored in the metadata repository and within Grid services. The information system offers an assisted publishing procedure that guarantees the consistency of a metadata document with the XML schemas associated to a given resource category. When a user publishes a new resource, the following steps are performed:

1. The user verifies if the new resource belongs to one of the resource categories defined by the ontology system. It can occur that:
   (a) the resource category under consideration has already been defined. In this case, the user can exploit the ontology system to fill the ontological section of the resource metadata document (see Section 3.1) and retrieve from the schema repository the structures – i.e the XML schemas - of semantic and resource metadata sections (see Sections 3.2 and 3.3).
   (b) the new resource does not belong to any defined resource category. In this case the user should use the ontology system to refine the classification of application domains, and possibly create a new resource category and the corresponding XML schemas that will be stored in the schema repository. Afterwards the user will be able to use such schemas and produce the new resource metadata document.

2. The user creates/updates the semantic and resource metadata sections by using the metadata publisher, which allows for the semi-automatically editing of metadata documents on the basis of related XML schemas.
3. At the end of the editing process, the metadata document is stored in the metadata repository.
4. If the new resource is offered as a service, its metadata document, or part of it, is also published within the service itself as a set of SDEs.
5. Metadata stored within the service can be aggregated and published by the GT3 Index Service.

The *Discovery* functionality allows users to search, locate and select PSE components and resources by examining the metadata information contained in each node of the Grid. When a user needs to discover resources having given characteristics, she/he executes the following actions:

1. To construct a query, the user must know the XML schema that defines the structure of the semantic metadata section for the resource category under consideration. If such a schema is not known, the user can browse or query the ontology system in order to retrieve it from the schema repository.
2. The user builds the query on the basis of the parameters and possible values specified in the XML schema.
3. The user submits the query to the PSE toolkit information system to discover the needed resources.

5 Conclusions

A Grid–based PSE toolkit is a group of technologies that allows for building PSEs for different application domains by exploiting the features and functionalities of a Grid infrastructure. Grid-based PSE toolkits require an efficient approach to manage the heterogeneity of the involved resources. The paper proposes a metadata model that permits to classify and describe resources needed for different domains. A metadata document, associated to each resource, includes an ontological metadata section that identifies the resource category, a semantic metadata section that characterizes resources in different application domains and assists discovery services, and a resource metadata section that gives details about how to use and access a resource. Moreover, the paper describes the architecture of an information system that allows for a uniform and flexible management of metadata. The information system exploits an ontology system, and the basic information services of a service-oriented Grid framework to aggregate and index metadata.

Acknowledgements

This work has been partially supported by the Italian MIUR FIRB Grid.it project RBNE01KNFP on High Performance Grid Platforms and Tools, by the FP6 Network of Excellence CoreGRID (Contract IST-2002-004265), and by the project KMS-Plus funded by the Italian Ministry of Productive Activities.

References

1. Aktas, M. S., Pierce, M., Fox, G. F.: Designing Ontologies and Distributed Resource Discovery Services for an Earthquake Simulation Grid, Proc. of the GGF11 Semantic Grid Applications Workshop, Honolulu, USA (2004) 1-6
2. Cannataro, M., Comito, C., Congiusta, A., Folino, G., Mastroianni, C., Pugliese, A., Spezzano, G., Talia, D., Veltri, P.: A General Architecture for Grid-Based PSE Toolkits, Workshop on State-of-the-Art in Scientific Computing PARA 04, Copenhagen, Denmark (2004)
3. Curbera, F., Goland, Y., Klein, J., Leymann, F., Roller, D., Thatte, S., Weerawarana, S.: Business Process Execution Language for WS, http://www-128.ibm.com/developerworks/webservices/library/ws-bpel/index.html
4. Deelman, E., Blythe, J., Gil, Y., Kesselman, C., Mehta, G., Vahi, K., Blackburn, K., Lazzarini, A., Arbree, A., Cavanaugh, R., Koranda, S.: Mapping Abstract Complex Workflows onto Grid Environments, Journal of Grid Computing, Kluwer Academic Publishers, Netherlands, Vol. 1, No. 1 (2003) 25-39
5. Enright A.J., Van Dongen S., Ouzounis C.A.: TribeMCL: An efficient algorithm for large scale detection of protein families, http://www.ebi.ac.uk/research/cgg/tribe/
6. Foster, I., Kesselman, C., Nick, J., Tuecke, S.: Grid services for distributed system integration, IEEE Computer, 35(6) (2002) 37-46
7. Foster, I, Kesselman, C., Nick, J., Tuecke, S.: The Physiology of the Grid: An Open Grid Services Architecture for Distributed Systems Integration, Globus Project, www.globus.org/research/papers/ogsa.pdf (2002)
8. Fox, G., Data and Metadata on the Semantic Grid, Computing in Science and Engineering, Volume 5, Issue 5, (2003)
9. The Globus Alliance: The Globus Toolkit, http://www.globus.org
10. Houstis, E., Catlin, A., Dhanjani, N., Rice, J., Dongarra, J., Casanova, H., Arnold, D., Fox, G., Problem-Solving environments, The Parallel Computing Sourcebook, M. Kaufmann Publishers (2002)
11. Mastroianni, C., Talia, D., Trunfio, P.: Managing Heterogeneous Resources in Data Mining Applications on Grids Using XML-Based Metadata, Proceedings IPDPS 2003, IEEE Computer Society Press (2003)
12. The MyGrid project. http://mygrid.man.ac.uk/myGrid/
13. The OGSA-DAI Project: Open Grid Services Architecture Data Access and Integration, http://www.ogsadai.org.uk/
14. The OWL Services Coalition: OWL-S: Semantic Markup for Web Services, http://www.daml.org/services/owl-s/1.0/owl-s.html
15. The GGF Simple API for Grid Applications (SAGA) Working Group: https://forge.gridforum.org/projects/saga-rg/
16. The Semantic Grid project: http://www.semanticgrid.org

Classification and Implementations of Workflow-Oriented Grid Portals[*]

Gergely Sipos and Péter Kacsuk

MTA SZTAKI Computer and Automation Research Institute,
H-1518 Budapest, P.O. Box 63, Hungary
{sipos, kacsuk}@sztaki.hu

Abstract. Although several grid projects have the aim to develop portals that hide the complexity of distributed computing infrastructures, only few of these portals support workflow-based applications. The number of portals that enable collaborative workflow development or collaborative resource access is even lower. The paper introduces a model to categorize workflow-oriented grid portals based on two general features: ability to access multiple grids simultaneously and support for collaborative problem solving. The generic features of the different categories are discussed and the Globus-based implementations of the various categories are demonstrated by the P-GRADE Portal.

Keywords: grid portal, grid workflow, collaborative environment, Virtual Organizations, Globus Toolkit

1 Introduction

The number of production grids has been dynamically growing for almost a decade now. The tendency raises new issues to be solved. Grid program developers would like to port their applications between grids, while end-users would like to run their applications on multiple grid systems sometimes even simultaneously. One option is that the users should learn all the details of the various grids in order to develop and/or run their applications on the different infrastructures. Obviously, this approach would require intolerably much effort from the users. Similar problems have risen in the past when the applications had to be ported among different kinds of computers. The solution was the introduction of high-level programming languages and their compilers.

In the world of grid systems portals are used to hide the low-level details of the various underlying infrastructures [14]. High-level approaches, like graphically represented technology-neutral workflows enable the porting of applications among grids [1]. However, most of the current portals do not support direct access to several grids simultaneously. If portals being connected to several grids simultaneously would be available, then users could dynamically select the best performing grid and

---

[*] The work described in this paper is supported by the Hungarian grid project (IHM 4671/1/2003) and by the Hungarian OTKA project (No. T042459).

run their workflows or parts of their workflows in multiple grids simultaneously. (Of course, the same can be achieved if meta-brokers would be connected to the portal.) Moreover, if these portals would enable the collaborative design and development of workflows by multiple users simultaneously, then researchers and developers could work together in order to integrate knowledge, data and resources into complex grid applications. Such possibilities could dramatically increase the research productivity of scientists.

There are several steps towards a completely generic multi-grid and collaborative user portal. Section 2 presents both the intermediate solutions and the most developed portal type by introducing a model based on the above discussed two properties. Section 3 gives an overview on the P-GRADE Portal, the environment used in Section 4 to present the implementation details of the different portal classes in Globus environments. Section 5 concludes by classifying current grid portals and Problem Solving Environments (PSE) with the proposed model.

2 Classification of Grid Portals

Most of current grid portals can be used to visualise grid information (like resources in the grid, job status, scientific data visualisation, etc.) and to facilitate the submission of individual jobs. Recent grid portals and PSEs support the submission of groups of jobs either as components of a parameter study or a workflow [1-9]. In the current paper we focus on portals that support the most complex (and advanced) form of job group submission, i.e. workflow submission, since we believe that very soon this feature will be a standard part of any grid portal. Such portals realise workflow developer environments and workflow enactment services on the top of the job-oriented grid middlewares.

As Flynn classified parallel computers [11] according to the number of applied processors and to the number of data units processed in parallel, we propose a classification for workflow-oriented grid portals. Our classification is based on two parameters. First, on the number of grids a given portal is connected to and can be used to execute jobs. Second, on the number of users that can access (edit, execute, control) the same workflow through the portal simultaneously. Three types of portals can be distinguished by the first property: single isolated grid portals, multiple isolated grid portals and multiple collaborative grid portals. Based on the second property two kinds of portals can be classified: isolated user portals and collaborative user portals. According to these statements Table 1 can be created.

The first row of the table represents the current grid portal approach, where a portal supports only a single grid system, i.e., users of the portal can access resources that belong to one rewarded grid. This restriction limits the usability of grids and portals in many ways. For example, just to mention one of the many limits, if a grid is overloaded, the users' jobs cannot be redirected to another, less overloaded grid. One remedy for this problem could be the access of several grids by the same portal. The second and third rows of the table represent portals that can exploit such a multi-Grid connection. These portals can be further classified whether the grids connected by them are isolated from each other or they can collaborate. Collaborative grids are able to solve the jobs of the same workflow together, executing the different workflow branches in parallel. For example, if a workflow has two parallel branches, then jobs belonging to branch *1* can be executed in grid *A*, meanwhile jobs belonging to branch *2* can be simultaneously executed in grid *B*. If such a simultaneous execution of the workflow co-ordinated by grid portal is not possible, then we say that the connected grids are isolated. In such case the whole workflow is executed either in grid *A* or in grid B^1. We refer to portals that can connect several grids in such a restricted way as xxMI portals. This class is represented in the second row of Table 1. xxMC portals, represented by the last row, are able to support the simultaneous, collaborative execution of a workflow among several connected grids.

Table 1. Classification of workflow-oriented grid portals

	Multiple Isolated users (MIxx)	Multiple Collaborative users (MCxx)
Single Isolated grid (xxSI)	MISI portals	MCSI portals
Multiple Isolated grids (xxMI)	MIMI portals	MCMI portals
Multiple Collaborative grids (xxMC)	MIMC portals	MCMC portals

Current portals support only isolated users, i.e., the grid users cannot collaborate via the portal either in order to develop complex grid applications or to collaboratively run existing applications. In our terminology collaborative workflow-oriented portals must provide the facility of controlled and concurrent access to a single grid application for multiple users during both the development and the execution phases. In our classification these workflow-oriented collaborative portals are denoted as MCxx portals and are presented in the last column of Table 1, while every other portal implementation belongs to the MIxx column.

The classification can be justified by the structure of workflow-oriented grid portals as well. A workflow-oriented grid portal is typically built from two main parts: workflow GUI and workflow manager. While the workflow GUI is the interface provided to the end-users to develop and manage workflows and to visualise results, the workflow manager is responsible for the execution of the submitted workflow graphs.

If the workflow GUI is able to serve multiple users, but each of them should work on a separate workflow graph, then the portal is a MIxx type. If the workflow GUI is able to serve multiple users and several users can work simultaneously on the same graph and can control the execution of the workflow, then the portal is a MCxx one.

Similarly, if the portal workflow manager is able to serve only a single grid the portal is an xxSI type. If the workflow manager is able to serve multiple grids but these grids should work on separate workflows, then the portal is an xxMI type. If the workflow manager is able to serve multiple grids and these grids can work simultaneously on the same workflow, then the portal is an xxMC type.

---

[1] Despite every component of a workflow is executed in the same grid the different components can run on different resources parallel.

3 The P-GRADE Portal

The P-GRADE Portal [4] is a workflow-oriented grid portal with the main goal to cover the whole lifecycle of workflow-oriented computational grid applications. It enables the graphical development of workflows consisting of various types of executable components (sequential, MPI or PVM programs), executing these workflows in Globus-based grids relying on user credentials, and finally analyzing the correctness and performance of applications by the built-in visualization facilities.

3.1 Grid Workflows in the P-GRADE Portal

Workflow applications can be developed in the P-GRADE Portal by the graphical Workflow Editor. The Editor is implemented as a Java Web-Start application that can be downloaded and executed on the client machines on the fly. The Editor communicates only with the Portal Server and it is completely independent from the grid middleware the server application is built on.

A P-GRADE Portal workflow is an acyclic dependency graph that connects sequential and parallel programs into an interoperating set of jobs. The nodes of such a graph are jobs, while the arc connections define the execution order of the jobs and the data dependencies between them that must be resolved by the workflow manager during the execution. An ultra-short range weather forecast (nowcast) grid application [12] is shown in Fig. 1 as an example for a P-GRADE Portal workflow.

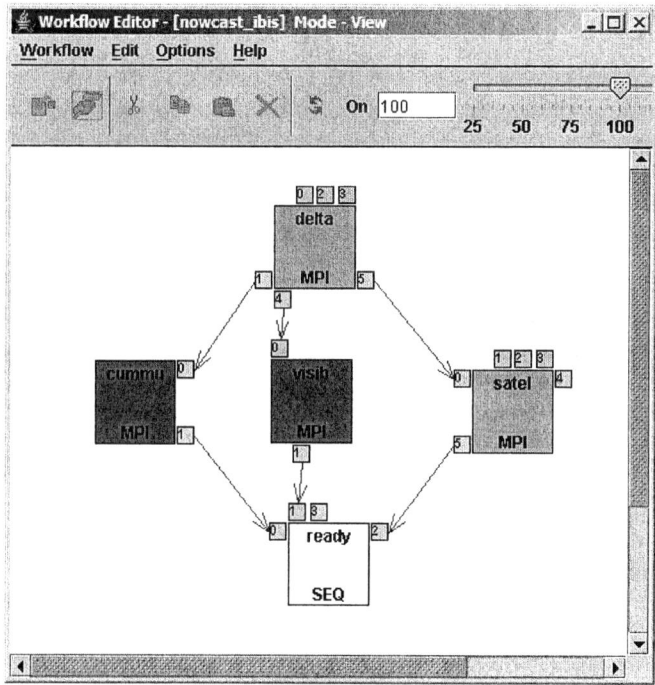

Fig. 1. The MEANDER nowcast meteorology application

Nodes (labelled as *delta, cummu, visib, satel* and *ready* in Fig. 1) represent jobs while rectangles (labelled by numbers) around the nodes are called ports and represent data files that the corresponding jobs expect or produce. Directed arcs interconnect pairs of input and output files if an output file of a job serves as an input file for another job. The semantics of the workflow execution means that a job (a node of the workflow) can be executed, if and only if all of its input files are available, i.e. all the jobs that produce input files for the job have successfully terminated, and all the user-defined input files are available on the portal server and at the pre-defined storage resources. For example, job *satel* can be launched when file *0* was produced by job *delta*, files *1* and *2* were uploaded to the portal server, and files *3* and *4* were available at the storage providers that have been specified by the user. Therefore, the workflow describes both the control-flow and the data-flow of the application. If all the necessary input files are available for a job, then the portal workflow manager transfers these files – together with the binary executable – to the site where the job has been allocated by the developer for execution. Managing the transfer of files and recognition of the availability of the necessary files is the task of the workflow manager subsystem of the Portal Server.

3.2 The Grid-Concept of the P-GRADE Portal

To achieve high portability among the different grids, the P-GRADE Portal has been built onto the Globus middleware, and particularly those tools of the Globus Toolkit 2 (GT-2) that are generally accepted and widely used in production grids today. GridFTP, GRAM, MDS and GSI [13] have been chosen as the basic underlying toolset for the P-GRADE Portal.

GridFTP services are used by the workflow manager subsystem to transfer input, output and executable files among computational resources, among computational and storage resources and between the portal server and the different grid sites. GRAM is applied by the workflow manager to start jobs on computational resources. An optional element of the Portal, the information system portlet queries MDS servers to help developers map the workflow components (executable jobs) onto computational resources. GSI is the security architecture that guarantees authentication, authorization and message-level encryption facilities for GridFTP, GRAM and MDS sites.

The choice of this infrastructure has been justified by connecting the P-GRADE Portal to several GT-2 based grid systems like the GridLab test-bed, the UK National Grid Service, and two VOs of the LCG-2 Grid (See-Grid and HunGrid VOs) [16]. Notice, that most of these grid systems use some extended versions of the GT-2 middleware. The point is that if the compulsory GRAM, GridFTP and GSI middleware set is available in a VO, then the P-GRADE Portal can be immediately connected to that particular system.

4 The Different Types of Workflow-Oriented Portals

4.1 Multiple Isolated Users, Single Isolated Grid (MISI) Portals

A MISI portal can be used by several users but every user must work on different workflow graphs individually, using some text editor or some graphical tool. After the workflow has been created it can be passed to the portal workflow manager for execution. A MISI workflow manager is connected only to a single grid and its task is to assign the jobs of the workflow to different resources of the connected grid. The resource allocation procedure can be either automatic or user-driven. It does not affect the model.

The 1.0 version of the P-GRADE Portal [4] gives one implementation for the MISI concept on the top of the GT-2 middleware. With the P-GRADE Portal 1.0 users can develop job workflows using the graphical Workflow Editor, can map the different workflow components onto computational resources of the connected single Globus VO, and finally, they can pass the workflow definition to the workflow manager for execution. The workflow manager splits the workflow up into individual jobs and submits these jobs to the specified GT-2 GRAM sites. At the same time it applies GridFTP servers to transfer input, output and executable files between the Portal Server, storage resources and computational resources according to the workflow definition.

4.2 Multiple Isolated Users, Multiple Isolated Grids (MIMI) Portals

The workflow manager subsystems of a MIMI portal must be able to cooperate with job executor services contained by multiple grids. The workflow manager subsystem must decide which grid to use to execute the different workflow components. (The decision can be made by the end-user too. This is a marginal question for the model.) Notice, that the user probably needs different certificates to access the different grids that are connected to the portal. After the grid selection is made the workflow manager must communicate with resources of one particular grid. For this the workflow manager must be provided either by the end user or some other software component with an appropriate certificate that is accepted by the selected grid.

The 2.1 version of the P-GRADE Portal [16] is one possible implementation of the MIMI concept. Users of the P-GRADE Portal 2.1 can associate proxy certificates with Globus VOs as part of the proxy download procedure. After the proxies have been mapped onto VOs, and the workflow components have been mapped onto computational resources, the workflow manager subsystem of the Portal is capable to choose the appropriate proxy to transfer files and to start up jobs on the resources of the selected VO.

4.3 Multiple Isolated Users, Multiple Collaborative Grids (MIMC) Portals

The next step enables the connection of the portal to several grids and these grids will be able to simultaneously work on different branches of the same workflow. It means that the user can assign workflow nodes to resources both from grid A and grid B and, as a result, the workflow will be collaboratively executed in these grids. If parallel branches are assigned to different grids then the workflow is executed even simultaneously on these grids. The workflow manager subsystem of an xxMC portal must be able to associate different certificates with the different components of a workflow. Although all of these certificates identify the same person, they are accepted in different grids.

Because resources from different grids (different trust domains) are involved in the execution of an xxMC workflow, data transfer between different security domains necessarily occurs. In this case none of the certificates is sufficient to perform direct file transfer between the interested resources, thus the portal server has to be used as a temporary storage for indirect file transfer. The workflow manager can use the certificate of the first workflow branch to copy the output files from the first grid onto the portal server and it can use the proxy of the second branch to copy these files onto the resources of the second grid. After the successful termination of the workflow the certificates that are accepted by the resources where the final results have been produced must be used to collect these files for the user.

The 2.1 version of the P-GRADE Portal [16] fulfils the requirements of MIMC environments. The workflow manager of the Portal is capable to associate different Globus proxies with different components of a workflow, and execute these components in different Globus VOs. Moreover, the workflow manager performs indirect file transfer through the Portal Server if it is necessary. Just like previously, the proxy selection process is based on the "proxy certificate – Globus VO" mapping table specified by the users during the proxy download procedure.

4.4 Multiple Collaborative Users, Single Isolated Grid (MCSI) Portals

MCxx denotes collaborative workflow-oriented grid portals. Contrarily to xxMC portals – where the aim is to utilise resources from multiple grids to execute a single workflow – the goal here is to integrate multiple persons' jobs, data, knowledge and certificates into a single workflow [10].

During the development phase MCSI portals must act as design environments specialised for the concurrent engineering of workflow-based grid applications. These portals enable multiple persons to contribute to a common workflow graph with their own jobs, data and certificates. In the execution phase the collaborative users can observe and control the execution process. Because the components of a MCxx workflow are defined by different persons, the workflow manager subsystems of MCxx portals must be prepared to associate different users' certificates with different parts of a single graph. The problem is similar to the one that has been discussed in Section 5.3, however, in case of an MCSI workflow the different certificates identify different persons and are valid in the same grid.

The prototype version of the Collaborative P-GRADE Portal is the only available implementation today for the MCSI concept [10]. The Collaborative P-GRADE Portal applies hard-lock based mutual exclusion mechanisms to support the concurrent development of workflow applications. The collaborative workflow manager of the portal is capable to associate the developer's proxy with each workflow component.

4.5 Multiple Collaborative Users, Multiple Isolated Grids (MCMI) Portals

MCMI portals extend the collaborative workflow development and execution support with isolated multi-grid access facility. Users of an MCMI portal can collaboratively construct a workflow graph that utilises resources from one of the grids that are connected to the portal.

The concept supposes that every component of a collaborative workflow is mapped onto resources from the same grid. If the resource selection procedure is performed by the workflow manager automatically, then this is an appropriate assumption. However, if the mapping procedure is performed by the collaborative users manually, then it is quite unlikely that resources from the same grid would be chosen. Because the Collaborative P-GRADE Portal does not give any support for the collaborative mapping procedure, it can be hardly used as an MCMI environment.

4.6 Multiple Collaborative Users, Multiple Collaborative Grids (MCMC) Portals

The most advanced grid portals will be the MCMC portals where multiple users can collaborate to create a workflow application and then to run the application in multiple collaborative grids simultaneously. Such portals require both the concurrent engineering support has been introduced in Section 5.4 and the workflow manager subsystem that has been discussed in Section 5.3. However, in the MCMC case the certificates used during the execution procedure identify different users in different grids.

The Collaborative P-GRADE Portal provides an implementation for the MCMC concept on the top of the GT-2 middleware. After the collaborative workflow development process has been finished, the short-term proxies have been downloaded and the proxies have been linked to Globus VOs, the workflow manager subsystem associates one proxy with every workflow component. The manager selects the proxy that identifies the developer of the component in the VO in which the component has been allocated for execution. After every component has been associated with a proxy the manager executes the workflow by transferring files and starting up jobs on Globus resources.

5 Related Works and Conclusions

The paper introduced a model to classify workflow-oriented computational grid portals based on two parameters: the number of clients that can collaboratively participate in the development and execution of workflow applications and the number of grids that can be applied to perform the execution procedure. The "multiple grids" class has been divided into two subclasses, so altogether six different portal types have been distinguished. Although Yu and Buyya gives a more complex classification taxonomy for workflow management systems in [17], their model does not take the discussed two dimensions into consideration. We believe that as the number of production grids and grid users rise, these two parameters will be more and more important.

Despite the current paper focused onto grid portals, the presented model can be applied for non-portal based workflow-oriented PSEs as well. Unicore [2] and Triana [1] two of the most well-known workflow-oriented PSEs belong to the simplest, to the MISI class. They provide neither multi-grid access, nor collaborative user support. Although Triana is built onto the top of the grid-independent GAT API it does not support the utilisation of multiple grids.

Pegasus [5], a Web-based grid portal is a MISI environment as well. Based on a special configuration file – filled up with GRAM and GridFTP site addresses by the administrator – Pegasus is able to map abstract workflows onto physical resources. However, consequence of the central resource list and the fact that the workflow manager cannot associate different proxies with the different workflow components, that Pegasus can be regarded only as an xxSI portal.

The workflow manager of the GridFlow portal [3] handles workflows at two levels. It manages workflows at a global grid level and schedules them at the level of different local grids. Since it does not provide collaborative development environment for end-users, according to our model the GridFlow portal belongs to the MIMC group.

Although language-based workflow definitions – such as those used by the ICENI [6], the Taverna [7], the GridAnt [8] or the Gridbus [9] workflow managers – could be concurrently engineered using a CVS server [18], none of these PSEs are capable to distinguish collaboratively engineered applications from single-user applications during the execution period. Consequently, none of these environments are capable to associate multiple users' certificate to execute a collaboratively developed workflow.

The workflow manager subsystem of the Collaborative P-GRADE Portal is based on the Condor DAGMan scheduler [15]. DAGMan executes one P-GRADE Portal specific "pre" and "post" script before and after every workflow component. These scripts perform the proxy, the job execution and the file transfer procedures. The Portal Server – in cooperation with the client side Workflow Editor – provide the concurrent workflow development and execution management support for the users.

The most advanced grid portals will be the MCMC portals where multiple users can collaborate to create a grid application and then to run the application in multiple collaborative grids simultaneously. By executing workflows on several grids simultaneously and by providing workflow-oriented collaborative environments for the end-users, these portals will revolutionise interdisciplinary research.

References

[1] Taylor, et al.: grid Enabling Applications Using Triana, Workshop on grid Applications and Programming Tools, June 25, 2003, Seattle.
[2] D. W. Erwin and D. F. Snelling. UNICORE: A grid Computing Environment. In Lecture Notes in Computer Science, volume 2150, pages 825-834. Springer, 2001.
[3] J. Cao, S. A. Jarvis, S. Saini, and G. R. Nudd:. GridFlow: WorkFlow Management for Grid Computing. In Proceedings of the 3rd IEEE/ACM International Symposium on Cluster Computing and the grid (CCGRID'03), pp. 198-205, 2003.
[4] Cs. Németh et al.: The P-GRADE grid portal In: Computational Science and Its Applications (ICCSA 2004) International Conference, Assisi, Italy, LNCS 3044, pp. 10-19
[5] G. Singh et al: The Pegasus Portal: Web Based grid Computing. To appear in Proc. Of the 20th Annual ACM Symposium on Applied Computing, Santa Fe, New Mexico, March 13-17, 2005
[6] S. McGough et al: Workflow Enactment in ICENI. In UK e-Science All Hands Meeting, Nottingham, UK, IOP Publishing Ltd, Bristol, UK, Sep. 2004; 894-900.

[7] T. Oinn, M. Addis, J. Ferris, D. Marvin, M. Senger, M. Greenwood, T. Carver and K. Glover, M.R. Pocock, A. Wipat, and P. Li. Taverna: a tool for the composition and enactment of bioinformatics workflows. Bioinformatics, 20(17):3045-3054, Oxford University Press, London, UK, 2004.
[8] G. von Laszewski et al: GridAnt: A Client-Controllable Grid Workflow System. In 37th Annual Hawaii International Conference on System Sciences (HICSS'04), Big Island, Hawaii: IEEE Computer Society Press, Los Alamitos, CA, USA, January 5-8, 2004.
[9] J. Yu and R. Buyya: A Novel Architecture for Realizing Grid Workflow using Tuple Spaces. In 5th IEEE/ACM International Workshop on Grid Computing (GRID 2004), Pittsburgh, USA, IEEE Computer Society Press, Los Alamitos, CA, USA, Nov. 8, 2004.
[10] G. Sipos, G. J. Lewis, P. Kacsuk and V. N. Alexandrov: Workflow-Oriented Collaborative grid Portals. To appear in Proc. Of European grid Conference, Amsterdam, February 14-16, 2005.
[11] M. J. Flynn: Some computer organizations and their effectiveness. IEEE Transactions on Computers. C-21 (Sept.1972). 948-960.
[12] R. Lovas, et al.: Application of P-GRADE Development Environment in Meteorology. Proc. of DAPSYS'2002, Linz,, pp. 30-37, 2002.
[13] Foster, C. Kesselman: Globus: A Toolkit-Based Grid Architecture, In I. Foster, C. Kesselmann (eds.) „The Grid: Blueprint for a New Computing Infrastructure", Morgan Kaufmann, 1999, pp. 259-278.
[14] G. von Laszewski et al.: Designing grid-based Problem Solving Environments and Portals, in 34th Hawaiian International Conference on System Science, Maui, Hawaii, 2001. Available: http://www.mcs.anl.gov/_laszewsk/bib/papers/vonLaszewski--cog-pse-final.pdf
[15] T. Tannenbaum, D. Wright, K. Miller, and M. Livny: Condor - A Distributed Job Scheduler. Beowulf Cluster Computing with Linux, The MIT Press, MA, USA, 2002.
[16] P-GRADE Portal 2.1: http://www.lpds.sztaki.hu/pgportal/
[17] Jia Yu and Rajkumar Buyya: A Taxonomy of Workflow Management Systems for Grid Computing, Technical Report, GRIDS-TR-2005-1, Grid Computing and Distributed Systems Laboratory, University of Melbourne, Australia, March 10, 2005. http://www.gridbus.org/reports/GridWorkflowTaxonomy.pdf
[18] B. Berliner: CVS II: ParalMizing software development, Proc. of Winter USENIX Technical Conference, 1990.

YACO: A User Conducted Visualization Tool for Supporting Cache Optimization

Boris Quaing, Jie Tao, and Wolfgang Karl

Institut für Technische Informatik, Universität Karlsruhe (TH),
76128 Karlsruhe, Germany
{boris, tao, karl}@ira.uka.de

Abstract. To enhance the overall performance of an application it is necessary to improve the cache access behavior. In this case, a cache visualizer is usually needed for fully understanding the runtime cache activities and the access pattern of applications. However, it does not suffice if only visualizing what happened. More importantly, a visualizer has to provide users with the knowledge about the reason for cache misses and to illustrate how the cache behaves at the runtime. This is also the goal of YACO (Yet Another Cache-visualizer for Optimization). Different from existing tools, YACO uses a top-down approach to direct the user step-by-step to detect the problem and the solution.

1 Motivation

On modern computer systems there exists a large distinction between processor and memory speed, and programs with large data sets suffer considerably from the long access latency of main memory. The cache memory, on the other hand, serves as a bridge; however, caches are usually not efficiently used. For tackling cache misses a variety of approaches have been proposed, including hardware-based cache reconfiguration [6] and user-level or compiler-based code optimization [7,2,3]. The latter is more common due to its straight-forward manner.

This kind of optimization, nevertheless, requires the knowledge about the access pattern of applications and also the runtime cache access behavior. As a direct analysis of the source code is rather tedious and often can not enable a comprehensive optimization, users usually rely on visualization tools for acquiring this knowledge. As a consequence, a set of cache visualizer has been developed [9,4,12,1]. However, these tools show only what happened and do not suffice for an exact understanding of the runtime cache and program activities.

In this case we developed YACO with the goal of showing how the application and cache behave, rather than only presenting the statistical data. For this YACO not only visualizes the changing of cache contents but also, and with more endeavor, depicts all necessary information at high level enabling the detection of cache miss reason and optimization schemes.

The resulted visualizer is optimization oriented and user conducted. This means, as illustrated in figure 1, users first acquire an overview about the cache access behavior shown by the chosen program. Based on this overview, the user can determine whether

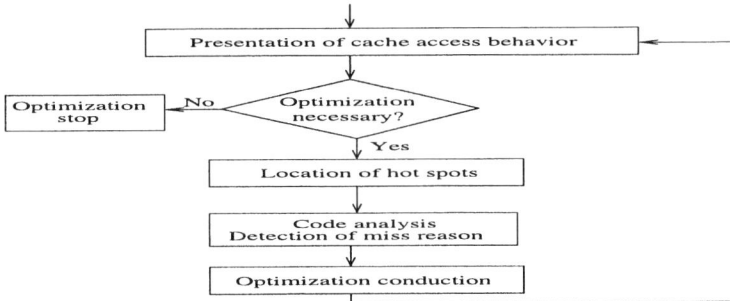

Fig. 1. Procedure of code optimization with YACO

an optimization is essential. In the next step, the access hot spots, which are responsible for poor cache performance, can be located. After that, the reasons and interrelations between memory references can be detected using YACO. This information also allows the user to select appropriate optimization scheme and related parameters, or to design novel algorithms to eliminate the detected cache problem. The impact of the optimization can be observed with YACO after running the optimized code. This process can be repeated until an acceptable cache performance is achieved.

The goal and main contribution of YACO lies in:

- providing the possibility of detailed analysis of program execution behavior with respect to different cache hierarchies
- enabling the investigation of reasons for poor cache performance
- supporting compiler developers in the task of implementing automatic optimization mechanisms
- helping programmers to directly apply adequate optimization techniques with the source code
- aiding hardware designers to evaluate the influence of various cache organization on execution behavior of applications

The rest of this paper is organized as follows. Section 2 gives a short introduction to related work in this research area. This is followed by a description of the visualization infrastructure and the source of performance data in Section 3. Then in Section 4 we demonstrate how YACO supports the optimization process with various graphical representations. After that the paper concludes in Section 5 with a short summary.

2 Related Work

For supporting code optimization and for helping users to understand the runtime cache access behavior, several visualization tools have been developed. Examples include CVT [9], CACHEVIZ [12], and Vtune [4].

Cache Visualization Tool (CVT): CVT [9] is a graphical visualization program. It is composed of a main window, which shows the active content of the whole cache memory, and a second window presenting statistics on cache misses in terms of arrays. It relies on cache profilers to achieve cache performance data.

We examine now the appropriateness of CVT for investigating the poor cache performance of an application. CVT provides information about the operations in cache during a program run and supplies different statistics. However, both cache content and statistics address only the current status. For a user it is more important to understand the regular structure of cache misses and for this not only the active but also the past status of the cache has to be presented in a single view. CVT shows only the former.

A further shortcoming with CVT lies in that it merely targets on a single cache. Although each cache level can be separately observed but the interaction between different caches in the cache hierarchy is lost. In addition, CVT shows the whole cache making it difficult to focus on a specific region where especially critical problem exists. This difficulty increases with the size of the observed cache.

CACHE Behavior VIsualiZer (CACHEVIZ): CACHEVIZ [12] intends to show the cache access behavior of a complete program in a single picture. Totally it provides three views: density view, reuse distance view, and histogram view. Within the first two views, each memory reference is presented with a pixel. Horizontal pixels indicate consecutive memory accesses. Character of a reference, i.e. hit or a kind of miss, is distinguished with color. Similarly, within the reuse distance view accesses with long reuse distance are also highlighted with specific color. The third view aims at enabling an analysis of access regularity. For this it shows different values of the chosen metric, e.g. reuse distance, in a single diagram. For the reuse distance histogram, for example, it lists the various reuse distances on the x-axis, while corresponding to each reuse distance it shows on the y-axis the number of accesses with this reuse distance.

VTune Performance Analyzer: The Intel VTune Performance Analyzer [4] is regarded as a useful tool for performance analysis. It provides several views, like "Sampling" and "Call Graph", to help identify bottlenecks. For cache performance it only shows number of cache misses. Even though this information can be viewed by process, thread, module, function, or instruction address, it does not suffice for optimizations with respect to cache behavior.

In summary, existing cache visualization tools show limited information about the cache operation and this information is mostly highly combined, allowing only to aquire an overview of the complete access behavior or, a little further, the detection of access hot spots. For an efficient, comprehensive optimization with cache locality, a more feasible cache visualizer is needed. Therefore, we implemented this user conducted and optimization oriented visualization system in order to show the various aspects of the runtime cache behavior. Using a top-down visualization procedure, it directs the users step-by-step to locate and detect the concrete object and technique for optimization.

3 Performance Data and YACO's Software Infrastructure

Performance visualizers usually heavily rely on data acquisition systems which deliver performance data for visualization. For instance, due to the limited data hardware counters can provide, Vtune only presents the number of cache misses. Besides hardware counters, profilers and simulation systems are also often deployed for acquiring performance data. However, existing systems [5,11] restrict to statistical overview data and detailed information, especially that capable of showing miss reason, is missing. In this case, we implemented CacheIn [8], a monitoring and simulation framework, to acquire comprehensive cache performance data.

CacheIn, at the top level, is comprised of a cache simulator, a cache monitor, and a software infrastructure. The former takes memory references as input and then models the complete process of data searching in the whole memory hierarchy with multilevel caches. According to the search result, it generates events for the cache monitor. The events describe both the access type and the cache activity, varying between hit, cold-miss, conflict-miss, capacity-miss, load, and replacement.

The cache monitor defines a generic monitoring facility and provides mechanisms to filter and analyze the events. More specifically, it offers a set of working modes allowing the acquisition of different kind of cache performance data, e.g. access series and cache operation sequence. This data is further delivered to a software infrastructure which produces higher level information in statistical form like histograms and total numbers. This forms the basis for YACO to exhibit the cache access behavior in different aspects.

For visualization, YACO deploys a client-server architecture to coordinate the interaction between its graphical user interface, the target system, the data source, the server, the visualization component, and the user.

To use YACO, the user first initiates a simulation run of the examined program via the graphical user interface. After the simulated execution, cache performance data collected by CacheIn is delivered to the server and processed. Also through the graphical user interface requests for specific views of user interest can be specified and further issued to the server. As the core of this infrastructure, the server is responsible for building communication, storing and processing performance data, and manipulating the visualization. For communication it establishes pipelines for connecting both the cache simulator and the graphical user interface. While the former aims at acquiring performance data, the latter is used to transfer users' visualization requirements. For control of the visualization it initiates the corresponding function for the selected view and provides the required data.

4 Visualization to Support Optimization

In Figure 1 we demonstrated an optimization process towards efficient cache performance. This section describes how the steps shown in Figure 1 are supported by YACO using a variety of graphical representations.

4.1 Performance Overview

First, the cache access behavior of the examined program must be observed in a high aggregated form, allowing the user to predict the potential performance gain when an optimization would be conducted. For this YACO provides an overview diagram which is illustrated in the left side of Figure 2.

As can be observed, this view contains two kinds of diagrams for each cache location on the whole cache hierarchy: a column diagram and a circle diagram. In the

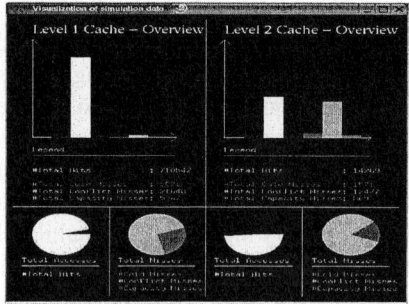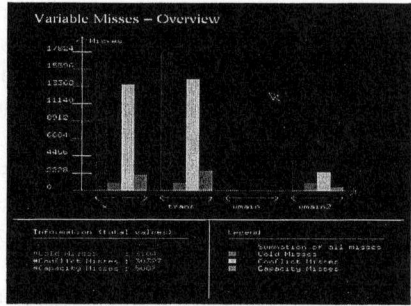

Fig. 2. YACO graphical view: Total Overview (left) and Variable Miss Overview (right)

column diagram statistics on cache hits and misses are presented. Besides the total number, cache miss is further distinguished between cold, conflict, and capacity miss. This means the height of the first colum (total access) is the sum of that with the following two columns (total hit and total miss), and the summary of the last three columns forms the total misses (the third column). Using colored columns this diagram allows the user to exactly examine the overall cache performance and further to decide if an optimization has to be performed. Similar to the column diagram, the circles also aim at showing a contrast between various metrics concerning memory references. While the left circle allows to detect how the cache overall behaves with only total hits and misses presented, the right circle shows the proportion of each miss type enabling a deeper insight in cache misses and the highlight of the most critical miss reason.

4.2 Access Hot Spots

In the next step of optimization, the access hot spots have to be located. Theses are individual data structures, functions, loops, or specific code regions with related data structures, which are responsible for poor cache performance. For this YOCO provides a set of views.

Variable Miss Overview: First, the Variable Miss Overview allows to detect the data structures which cause a significant amount of cache misses. As illustrated in the right side of Figure 2, this view presents the miss behavior of all data structures delivered by the cache simulator. The sample view in Figure 2 depicts the Fast Fourier Transformation (FFT) code, an application in the SPLASH-II Benchmarks suite [10]. For each data structure, four columns show the statistics on total misses and each miss category. Absolute numbers of the misses are depicted on the left bottom diagram, while the meaning of each column is explained on the right bottom diagram. Overall, the Variable Miss Overview allows to detect data structures that have to be optimized, e.g. x and *trans* in the FFT code.

Data Structure Information: With help of the Variable Miss Overview, data structures with large cache misses are identified. However, users still need the knowledge about the access behavior of e.g. elements and data blocks within a data structure. This helps

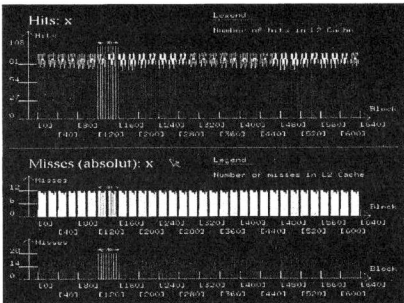 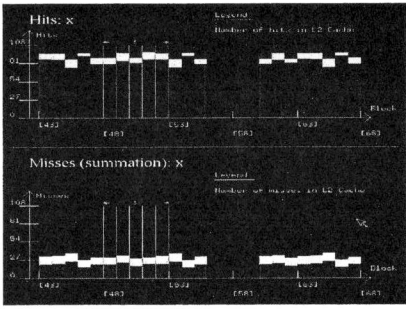

Fig. 3. YACO graphical view: Data Structure Information (left) and its Zoom-in (right)

to detect regular access structure and also to make the hot spots more concrete. For this goal the view "Data Structure Information" is implemented.

As shown in Figure 3, Data Structure Information contains two diagrams. While the top one covers the whole memory system and presents hits, the bottom one only targets on caches and shows the miss behavior. Figure 3 shows array x in the FFT code.

Within the upper diagram (left picture in Figure 3), the x-axis shows all data blocks (size of cache line) in the selected data structure with their numbers given in a stride. Corresponding to each data block, the y-axis gives the number of access hits on a location, which can be L1 cache, L2 cache, or the main memory. The sum of these hits on all locations forms the total runtime accesses to the corresponding data block. The locations are identified with different colors enabling an easy detection of regions, where most accesses are addressed in the main memory. The lower diagram presents the information about misses in a similar way. The concrete view in Figure 3 uses separate diagrams for each cache location: L1 bottom and L2 top. This allows to better observe the misses on each single cache.

For large data structures it is difficult to detect the hot spots if all data blocks are presented in the same diagram. Hence YOCO implements a zoom function for this view, where a range of blocks can be specified via a user interface and only the selected blocks are illustrated. Figure 3 (right) gives a sample view showing the misses with block 43 to 67 of array x.

3D Phase Information: The views described above cover the whole program. For data structures that are used only one time, for example within a single subroutine, the code region containing accesses to this data is clear. However, often the same data is accessed in different program stages (initialization, computation, etc.) and functions or code regions (we call them phases), and the access behavior usually changes in different phases of an application. In order to enable a further focus of the access hot spots on a code region, YACO provides the view "3D Phase Information".

The sample view in Figure 4 shows the combined misses (all caches) in different program phases with respect to the selected data structure, in this case array x in the FFT program. Each phase in this example corresponds to a single function in the program. The phases are distributed across the z-axis and each with a 2D diagram showing the miss distribution at granularity of blocks in the data structure, where blocks are

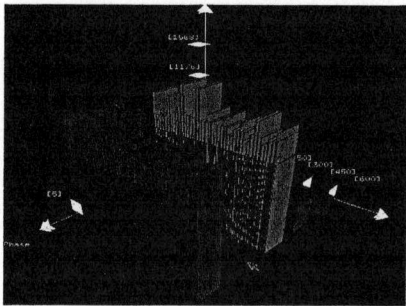 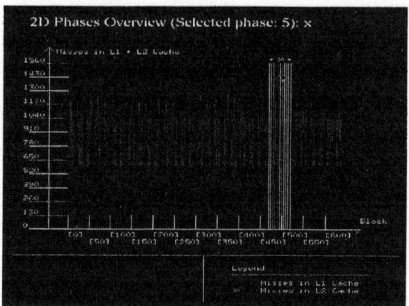

Fig. 4. YACO graphical view: 3D Phase Information (left) and its 2D zoom-in (right)

illustrated on the x-axis and number of misses to the corresponding block on the y-axis. In addition, phases are identified using different colors. Similar to the previous view, the visualization can be focused on a few of blocks enabling a better observation of the chosen data areas. In addition, this 3D diagram can be rotated in any direction allowing a deeper observation from all sides. Besides, YACO enables a zoom-in in a single phase in order to better observe a critical code region. Figure 4 (right) demonstrates the fifth phase using the same color as in the 3D view.

4.3 Access Pattern and Miss Reason

With the knowledge about the data structures and more detailed the data blocks that are responsible for poor cache performance in combination with the code region shown by the phase information, the optimization process, as illustrated in Figure 1, goes into the third step: to detect the access pattern of applications and the miss reason. For this, YACO provides another two graphical views.

Variable Trace: The first view is Variable Trace which shows the references to data blocks/elements in a data structure in the order as they are accessed. Figure 5 (left) demonstrates the array x in the FFT code. As can be seen, we use twenty fields to show the references to a data structure. Theses fields are first empty and are filled after begin of the visualization. The first reference is presented initially on the first field and then moved to the next field after the second reference to the same data structure occurs. Corresponding to each field Variable Trace shows the accessed block/element of the data structure and the type of the access, which can be a load operation or a replacement. While the name of the array and the block/element are explicitly printed in the fields, the operation type is indicated by the color of a field. As the references are filled in the diagram according to the accessed sequence, this view allows to detect e.g. the access stride between array elements and further to chose an appropriate prefetching strategy to reduce the load operations and thereby the cold misses.

Cache Set: The second view aims to exhibit the runtime activities of a cache memory. A sample view is shown in Figure 5 (right), where set 12 of the L2 cache is taken as an example. For this example, L2 is configured as a 4-way cache. Within the diagram, all lines within the selected set are illustrated in horizontal direction, with each recording

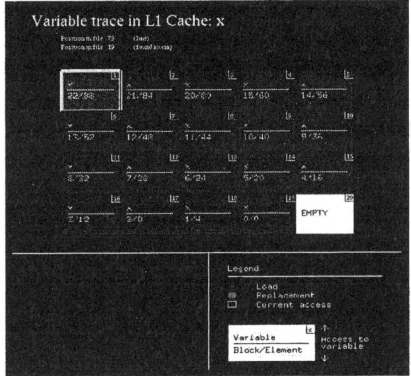 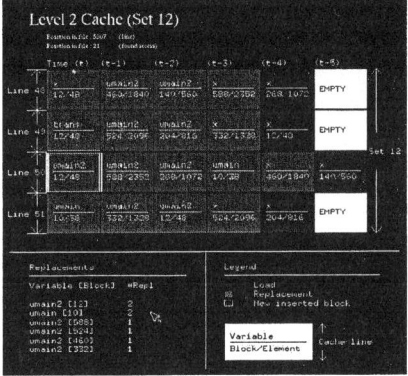

Fig. 5. YACO graphical view: Variable Trace (left) and Cache Set (right)

the most recent operations, from left to right, in the corresponding cache line. For each operation, the operation type, which can be a load, a replacement, or a hit, and the operation target, i.e. the variable or block/element in an array, are presented. Operations are inserted in the diagram according to their occurrences. The most currently occurring operation is identified with a colored frame, in this case umain2 (12/48). In addition, statistics on replacement are shown on the left bottom allowing to find data blocks that frequently enter and leave the cache.

4.4 Sample Optimization

To demonstrate the feasibility of YACO, we simulated a small code performing initialization of a two-dimensional array. In order to highlight the problem in L1 cache, we use a small size with only four cache lines. The Overview of YACO shows a 100% L1 miss with this code.

We then examine the Variable Trace view of this code. As shown in the left diagram of Figure 6, all accesses to the array is a cache miss: either load or replacement; even

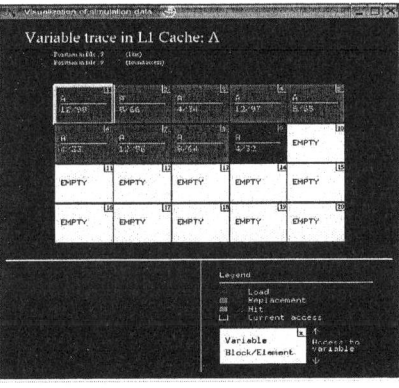 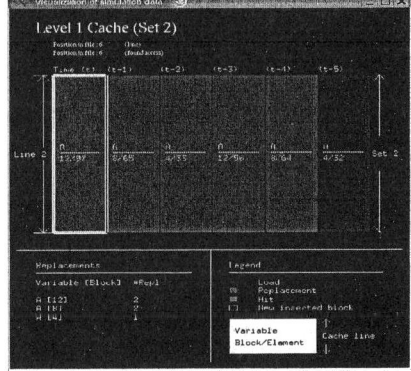

Fig. 6. Variable Trace (left) and Cache Set (right) with the sample code

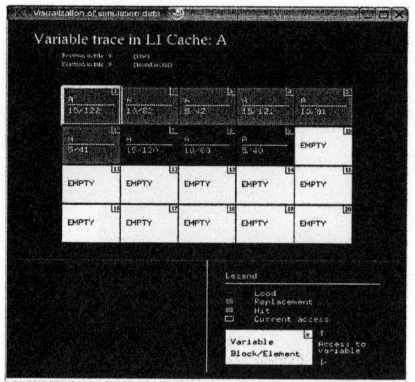 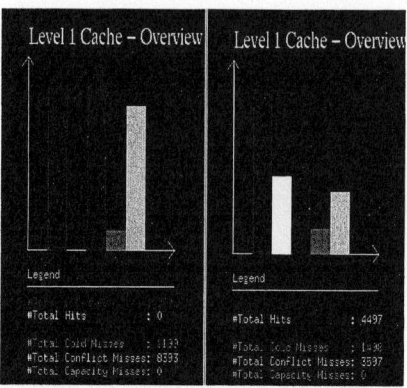

Fig. 7. Variable Trace after optimization (left) and performance comparison (right)

the access to elements in the same data block such as the fourth and seventh access to element 33 and 34 which are stored in block 4. The reason is, as depicted by the Cache Set view in the right diagram of this figure, all data blocks needed for current computation, in this case block 4, 8, and 12, are mapped in the same cache line, although other three cache lines are also available.

For optimization we insert pads with size of one cache line to the second dimension of this array, in order to change the mapping behavior of the data in the cache. Due to this padding, most accesses can be achieved in the cache, as shown in the left diagram of Figure 7. This results in a 47% improvement in cache hit ratio which is shown in the right side. This figure is a combination of two Overview views of YACO, each for the original code and the optimized version individually. It can be clearly observed that the optimization reduces nearly a half of the misses (third column).

5 Conclusion

The performance of applications significantly depends on efficient use of the cache memory. This efficiency can be enhanced through various optimizations, and for this often a visualization tool is needed for showing the runtime cache access behavior. However, existing cache visualization systems show only what happened, with either statistics on cache miss or more detailed information about the changing cache content. This does not suffice for an efficient, comprehensive optimization for which more information, especially that about cache miss reason, is required.

In this case, we introduce YACO, an optimization oriented and user conducted cache visualizer. The goal is to help the user through the whole optimization process, from deciding if an optimization is necessary, to finding the exact data structures and code regions to optimize, and up to the detection of miss reason, access pattern, and further the optimization strategies and related parameters. For this, YACO provides a variety of views showing overall performance, access behavior with specific data structures and code regions, access trace of variables, and runtime cache activities. The example, showing how to use YACO for cache optimization, has proven its feasibility.

References

1. R. Bosch, C. Stolte, D. Tang, J. Gerth, M. Rosenblum, and P. Hanrahan. Rivet: A Flexible Environment for Computer Systems Visualization. *Computer Graphics*, 34(1), February 2000.
2. Somnath Ghosh, Margaret Martonosi, and Sharad Malik. Precise Miss Analysis for Program Transformations with Caches of Arbitrary Associativity. *ACM SIGPLAN Notices*, 33(11):228–239, November 1998.
3. Somnath Ghosh, Margaret Martonosi, and Sharad Malik. Automated Cache Optimizations using CME Driven Diagnosis. In *Proceedings of the 2000 International Conference on Supercomputing*, pages 316–326, May 2000.
4. Intel Corporation. Intel VTune Performance Analyzer. available at http://www.cts.com.au/vt.html.
5. M. Martonosi, A. Gupta, and T. Anderson. Tuning Memory Performance of Sequential and Parallel Programs. *Computer*, 28(4):32–40, April 1995.
6. P. Ranganathan, S. Adve, and N. P. Jouppi. Reconfigurable Caches and their Application to Media Processing. In *Proceedings of the 27th Annual International Symposium on Computer Architecture*, pages 214–224, June 2000.
7. G. Rivera and C. W. Tseng. Data Transformations for Eliminating Conflict Misses. In *Proceedings of the ACM SIGPLAN Conference on Programming Language Design and Implementation*, pages 38–49, Montreal, Canada, June 1998.
8. J. Tao and W. Karl. CacheIn: A Toolset for Comprehensive Cache Inspection. In *Proceedings of ICCS 2005*, volume 3515 of Lecture Notes in Computer Science, pages 182–190, May 2005.
9. E. van der Deijl, G. Kanbier, O. Temam, and E. D. Granston. A Cache Visualization Tool. *IEEE Computer*, 30(7):71–78, July 1997.
10. S. C. Woo, M. Ohara, E. Torrie, J. P. Singh, and A. Gupta. The SPLASH-2 Programs: Characterization and Methodological Considerations. In *Proceedings of the 22nd Annual International Symposium on Computer Architecture*, pages 24–36, June 1995.
11. WWW. Cachegrind: a cache-miss profiler. available at http://developer.kde.org/šewardj/docs-2.2.0/cg_main.html#cg-top.
12. Y. Yu, K. Beyls, and E.H. D'Hollander. Visualizing the Impact of the Cache on Program Execution. In *Proceedings of the 5th International Conference on Information Visualization (IV'01)*, pages 336–341, July 2001.

DEE: A Distributed Fault Tolerant Workflow Enactment Engine for Grid Computing*

Rubing Duan, Radu Prodan, and Thomas Fahringer

Institute for Computer Science, University of Innsbruck,
Technikerstrasse 21a, A-6020 Innsbruck, Austria
{rubing, radu, tf}@dps.uibk.ac.at

Abstract. It is a complex task to design and implement a workflow management system that supports scalable executions of large-scale scientific workflows for dynamic and heterogeneous Grid environments. In this paper we describe the Distributed workflow Enactment Engine (DEE) of the ASKALON Grid application development environment for Grid computing. DEE proposes a de-centralized architecture that simplifies and reduces the overhead for managing large workflows through partitioning, improved data locality, and reduced workflow-level checkpointing overhead. We report experimental results for a real-world material science workflow application.

Keywords: Grid computing, checkpointing, dependence analysis, distributed enactment engine, fault tolerance, overhead analysis.

1 Introduction

Scientific workflows emerge as one of the most popular paradigm for programming Grid applications. A scientific workflow application can be seen as a (usually large) collection of activities processed in a well-defined order to achieve a specific goal. These activities may be executed on a broad and dynamic set of geographically distributed heterogeneous resources with no central ownership or control authority, that is called a *computational Grid*.

In this paper we propose a novel Distributed Enactment Engine (DEE) to reliably drive the execution of large-scale scientific workflows in dynamic and unstable Grid environments. A distributed infrastructure provides enhanced fault tolerance on its own crashes and decreases the overhead for controlling and managing the faults of various workflow parts. We propose two checkpointing mechanisms that enable Grid workflow executions to recover and resume from serious faults. In addition, we integrate existing fault tolerance techniques like task replication, retry, migration, redundancy, or exception handing. We describe

---

* This research is partially supported by the Austrian Science Fund as part of the Aurora project under contract SFBF1104 and the Austrian Federal Ministry for Education, Science and Culture as part of the Austrian Grid project under contract GZ 4003/2-VI/4c/2004.

several scenarios how our design and development is driven by a real-world material science scientific workflow. Due to space limitations we limit the scope of this paper to a set of key distinguishing features of ASKALON. Descriptions of other parts can be found in [7] that is the umbrella project of our work.

This paper is organized as follows. Section 2 describes the architecture of the DEE in detail. In Section 3 we present the two approaches taken by DEE to checkpoint and recover the execution of workflows in unstable Grid environments. In Section 4 we report experimental results that show a scalable execution control and recovery through checkpointing of a real world material science workflow application, executed in the Austrian Grid environment [4]. We provide an overview of the related work in Section 5 and conclude the paper in Section 6.

2 Distributed Enactment Engine Architecture for ASKALON

In the ASKALON application development and execution environment for Grid computing [7], the user specifies workflow applications at a high level of abstraction using the Abstract Grid Workflow Language (AGWL - [10]). AGWL is a high-level XML-based language that, with the support of the enactment engine, entirely shields the user form the technology details that implement the underlying Grid infrastructure.

DEE proposes a distributed master-slave enactment engine architecture depicted in Fig 1. The *master enactment engine* parses first the AGWL description to an internal (Java) workflow representation and sends it to scheduler for appropriate mapping onto the Grid. After the workflow has been mapped onto the Grid, the master enactment engine partitions the workflow into several sub-workflows according to the sites where the activities have been scheduled and the type of compound activities. We will present the patition mechanism in a future paper. Usually, the large scale parallel activities (i.e., hundreds of sequential activities) are sent to *slave enactment engines*. The master enactment engine monitors the execution of the entire workflow as well as the state of slave enactment engines. The slave enactment engines monitor the execution of the sub-workflows and report to the master whenever activities change their state or when the sub-workflows produce some intermediate output data relevant to other sub-workflows.

If one of the slave enactment engines crashes and cannot recover, the entire sub-workflow is marked as failed and the master enactment engine asks for re-scheduling, re-partitions the workflow, and migrates it to other Grid sites. After the initial scheduling, the master enactment engine also chooses one slaves a a backup engine (see the enactment engine on site B in Fig 1). If the master crashes, the backup becomes the master and chooses another backup slave immediately. Usually we choose the master and the backup enactment engines on machines with high CPU frequency, large memory, and good wide-area network interconnections.

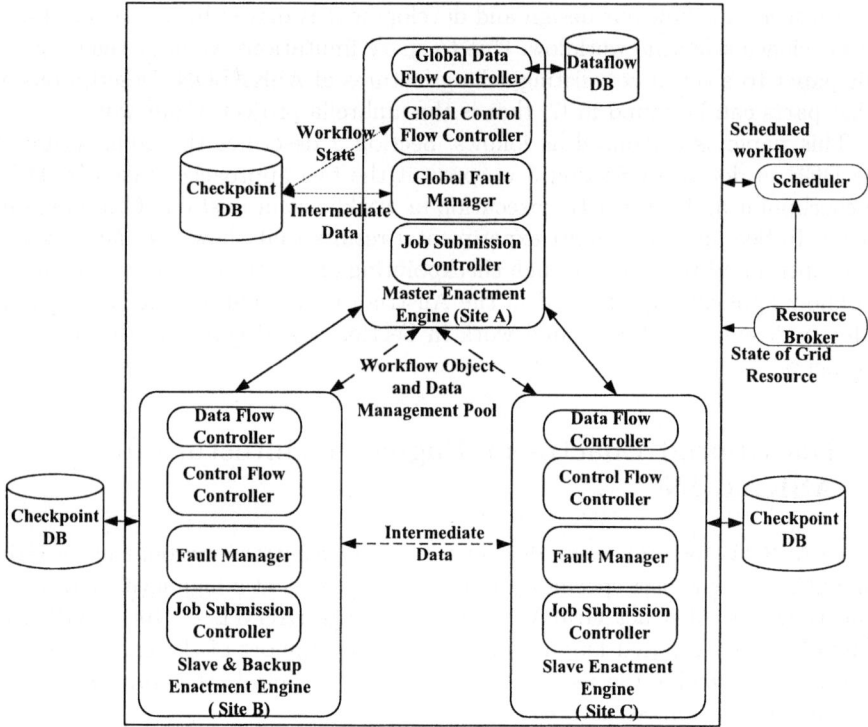

Fig. 1. The Distributed Enactment Engine (DEE) architecture

Every enactment engine consists of four main components, as follows: Control Flow Controller, Data Flow Controller, Fault Manager, Job Submission Controller. Due to the page limitation, we remove the details about them.

2.1 Distributed Data Management

An important task of the enactment engine is to automatically track data dependencies between activities. A data dependency is specified in AGWL by connecting input and output ports of different activities. At run-time, data ports may map to either data files identified using (GASS-enabled) URLs, or to objects corresponding to abstract AGWL data types. Such data ports, representing input and output data of activities, are distributed in the Grid and are statically unknown. DEE tracks the data dependencies at runtime by dynamically locating the intermediate data, as presented in the remainder of this section.

Run-time Intermediate Data Location. The enactment engine gets from the application developer the AGWL specification of the workflow. For example(see Fig 2), one workflow contains one `parallel for` loop and each loop iteration includes two serial activities A and B. After scheduling the workflow, the `parallel for` loop will be translated into a composite `parallel` activity

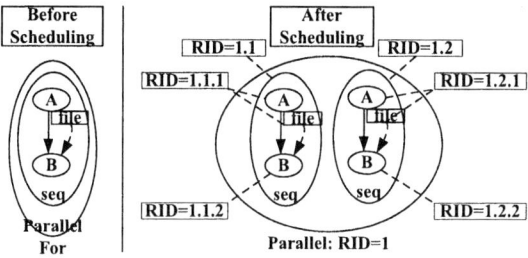

Fig. 2. Run-time intermediate data location

that unrolls the parallel loop iterations. We notice that the activities A and B, the data input ports, and data output ports (i.e., file file) inside the composite parallel activity that unrolls the parallel for loop, are the same. To solve this run-time name mismatch problem, the enactment engine associates a *run-time identification number (RID)* with each activity, generated as follows: $RID = RID_{parent} + "." + new\ RID$.

For example, the parallel (RID = 1) activity in Fig 2 is the parent of the seq (RID = 1.1) activity, which is the parent of the activity A (RID = 1.1.1). We, therefore, use the RID as an attribute of each activity, including the data input and data output ports, to retrieve the data from the correct predecessor, rather than the sibling's predecessor. For instance, in the case of activity B (RID = 1.1.2) that needs the input file file, we compare the RIDs of the two existing possibilities 1.1.2 and 1.2.2 and choose the first one since it has the correct parent identifier 1.

Transfer of Data Collections. AGWL introduces the notion of *collection*, which is a set of output data produced by one activity or (most likely) a parallel activity. A collection contains zero or more elements, where each element has an arbitrary well-defined data type, an index, and an optional name. The name and the index can be used to directly access specific elements. A collection can be used like a regular data type which is produced by almost all real-world workflow applications that we studied WIEN2k.

During our research on modelling and executing real-world applications, we identified six types of data collection transfers which we did not see in other existing workflow management systems [6,8,9,11]: a collection produced by one activity is consumed by another activity (see Fig 3(a)); a collection is produced by one activity and is consumed by every activity from a parallel activity (see Fig 3(b)); a collection is produced by one parallel activity and is entirely consumed by another activity (see Fig 3(c)); a collection is produced by one activity and each ith collection element is consumed by every ith activity from a parallel activity (see Fig 3(d)); a collection produced by one parallel activity and the entire collection is consumed by every individual activity from another parallel activity (see Fig 3(e)); a collection is produced by one parallel activity and every ith collection element is consumed by the ith activity from another parallel activity (see Fig 3(f)).

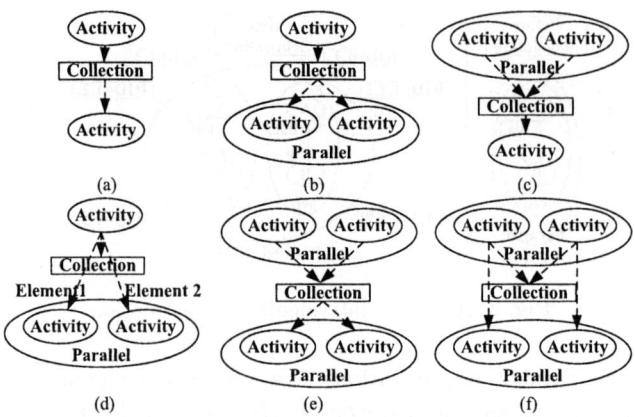

Fig. 3. Collection transfer cases

The management of the data collection transfer is the hardest problem of the DEE data dependence analysis. The problem is further complicated if several types of a collection transfer are mixed within the same workflow application, for example, Fig 3(a), 3(c), 3(e), 3(f) occur in the real-world material science application presented in Section 4.

DEE has a *run-time data transfer merge* mechanism to avoid that the same data is transferred twice, for instance in the case when a large collection is required by multiple activities scheduled on the same Grid site (see Fig 3(d)). In the future, we will also implement a *run-time activity merge* mechanism to decrease the Grid job submission overhead for multiple activities scheduled on the same site (usually of the order of 20 seconds in our GRAM configuration).

3 Checkpointing

Checkpointing and recovery are fundamental techniques for saving the application state during normal execution and restoring of a saved state after a failure to reduce the amount of lost work. There are two traditional approaches to checkpointing:

System-level checkpointing saves to the disk the image of an entire process, including registers, stack, code and data segments. This is obviously not applicable for large-scale Grid applications;

Application-level checkpointing is usually implemented within the application source code by programmers, or is automatically added to the application using compiler-based tools.

A typical checkpoint file contains the data and the stack segments of the process, as well as information about open files, pending signals, and CPU state.

We concentrate our approach on application-level checkpointing. Since it is not always possible to checkpoint everything that can affect the program behavior, it is essential to identify what is included in a checkpoint to guarantee a successful recovery. For the Grid workflow applications, a checkpoint consists of: the state of the workflow activities, and the state of the data dependencies.

DEE checkpoints a workflow application upon precise checkpointing events defined, for instance, through AGWL property and constraint constructs. Typical checkpointing events occur when an activity fails, after the completion of an important number of activities (e.g., workflow phases, parallel or sequential loops), or after a user defined deadline (e.g., percentage of the overall expected or predicted execution time). Other checkpointing events happen upon rescheduling certain workflow parts due to the dynamic availability of Grid resources or due to variable or statically unknown number of activities in workflow parallel regions. Upon a checkpointing event, the Control Flow Controller invokes the Fault Manager that stops the execution of the workflow and saves the status and the intermediate data into a *checkpoint database*.

We classify the checkpointing mechanisms in DEE as follows(see Fig 4):

Activity-level checkpointing saves the register, stack, and memory for every individual activity running on a certain processor. The advantage of the activity level checkpoint is that an individual activity can recover. At the moment we do not support activity-level checkpointing but we plan to integrate traditional system-level checkpointers like Condor [9];

Light-weight workflow checkpointing saves the workflow state and URLs to intermediate data (together with additional semantical information that characterizes the physical GASS URLs). The control-flow checkpoint is very fast because it does not back-up the intermediate data. The disadvantage is that the intermediate data remains stored on possible unsecured and volatile file systems;

Workflow checkpointing saves the workflow state and the intermediate data at the point when the checkpoint is taken. The advantage of the workflow checkpointing is that it completely backups the intermediate data such that the

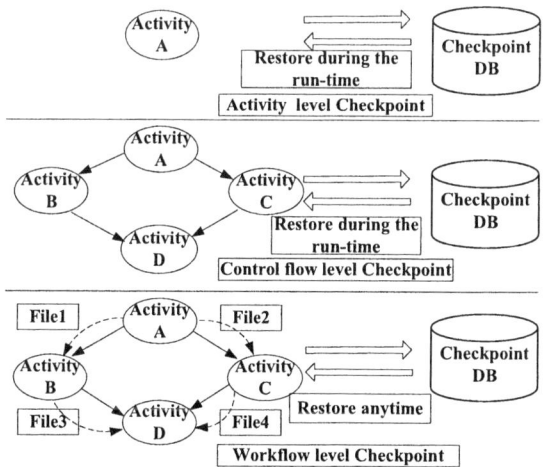

Fig. 4. The classification of the checkpoint

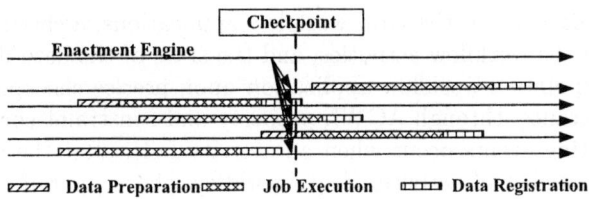

Fig. 5. Activity execution stages

execution can be restored anytime from any Grid location. The disadvantage is that the checkpointing overhead grows significantly for large intermediate data.

From the checkpointing perspective, the execution of one activity traverses three distinct stages, as shown in Fig 5: data preparation, job execution, and data register to the Data Flow Controller. Therefore, when the enactment engine calls for a checkpoint there are five job state possibilities. The job execution state counts for all the states returned by GRAM, which is our job submission interface to the Grid (SUBMITTED, QUEUED, ACTIVE, COMPLETED, and FAILED).

Definition 1 (Workflow Checkpoint). *Let $W = (AS, CFD, DFD)$ denote a workflow application, where AS is the set of the activities, $CFD = \{(A_{from}, A_{to}) \mid A_{from}, A_{to} \in AS\}$ the set of control flow dependencies and $DFD = \{(A_{from}, A_{to}, Data) \mid A_{from}, A_{to} \in AS\}$ the data flow dependencies, where Data denotes the workflow intermediate data, usually instantiated by a set of files and parameters. Let*

$$State : AS \to \{Executed, Unexecuted\}$$

denote the execution state function of an activity $A \in AS$. The workflow checkpoint *is defined by the following set of tuples:*

$$CKPT_W = \{(A, State(A), Data) \mid \forall A, A_{from} \in AS \land State(A) = Unexecuted \\ \land State(A_{from}) = Executed \land (A_{from}, A, Data) \in DFD\}.$$

As we can notice, there are two possible options for the checkpointed state of an executing activity (i.e., job execution stage). We propose three solutions to this problem:

First, We let the job run and consider the activity as *Unexecuted*; Second, We wait for the activity to terminate and set the state to *Executed*, if the execution was successful. Otherwise, we set the state to *Unexecuted*. Both solutions are not obviously perfect and therefore, we propose a third option that uses the predicted execution time of the job, as follows: Third, Delay the checkpoint for a significantly shorter amount of time, based on the following parameters: *Predicted execution time (PET)* is the time that the activity is expected to execute. DEE gets the predicted execution time from the prediction service using regression functions based on historical execution data; *Checkpoint deadline (CD)* is a predefined maximum time the checkpoint can be delayed; *Job elapsed time (JET)* is the job execution time from the beginning up to now.

We compute the state of an activity A using the following formula:

$$State(A) = \begin{cases} Unexecuted, & PET - CD \geq JET; \\ Executed, & PET - CD < JET. \end{cases}$$

This solution saves the checkpoint overhead, and let the checkpoint complete within a shorter time frame.

Another factor that affects the overhead of the workflow checkpoint is the size of the intermediate data to be checkpointed. We propose two solutions to this problem: *Output data checkpointing* stores the all the output files of the executed activities that were not previously checkpointed; *Input data checkpointing* stores all the input files of the unexecuted activities that will be used later in the execution.

For a centralized enactment engine, the input checkpoint is obviously the better choice because it ignores all the intermediate data that will not be used, which saves the file transfer (backup) overhead. In the case of DEE, the slave enactment engines do not know which intermediate data will be used later and, therefore, must use the output checkpointing mechanism. However, the solution is still efficient since the checkpoint is done locally by each slave enactment engine which saves important network file transfer overhead.

4 Experiments

In this section we show experimental results that evaluate our approach on a real-world material science workflow application. WIEN2k [2] is a program package for performing electronic structure calculations of solids using density functional theory, based on the full-potential (linearized) augmented plane-wave ((L)APW) and local orbital (lo) method. We have ported the WIEN2k application onto the Grid by splitting the monolithic code into several course-grain activities coordinated in a workflow, as illustrated in Fig 6. The lapw1 and lapw2 tasks can be solved in parallel by a fixed number of so-called *k-points*. A final activity converged applied on several output files tests whether the problem convergence criterion is fulfilled. The number of recursive loops is statically unknown.

Table 1 summarizes the Austrian Grid testbed that we used for this experiments. The Grid sites are ranked according to their individual speed in executing the WIEN2k application.

We use a WIEN2k problem size of 100 parallel k-points, which means a total of over 200 workflow activities. We started by executing the workflow on the fastest Grid site available (in Linz) and then we incrementally added new sites to the execution environment, as presented in Table 1. Fig 7(a) shows that WIEN2k considerably benefits from a distributed Grid execution until three sites. The improvement comes from the parallel execution of WIEN2k on multiple Grid sites that significantly decreases the computation of the lapw1 and lapw2 parallel sections. Beyond four Grid sites we did not obtain further improvements due to a temporary slow interconnection network of 1 Mbit per second to the

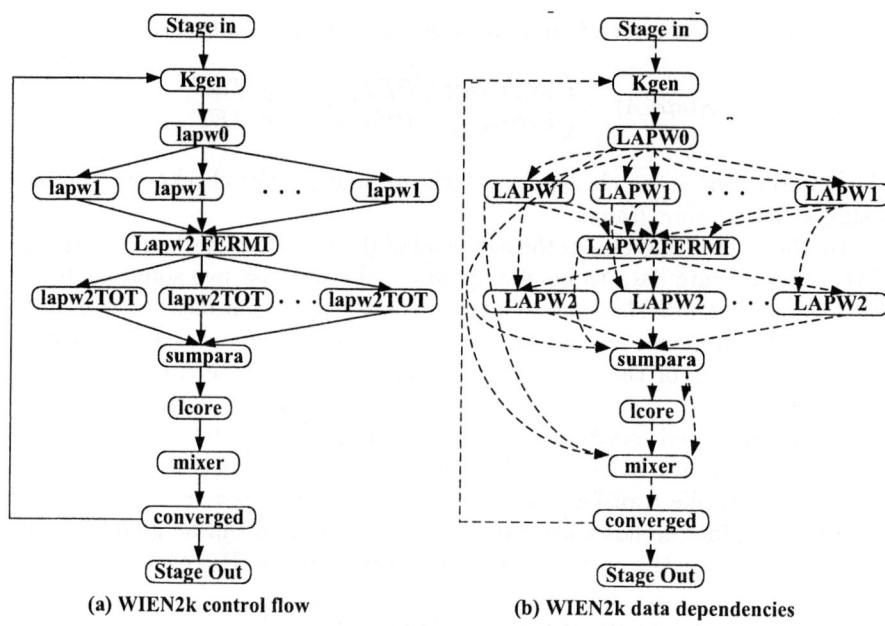

Fig. 6. The WIEN2k workflow

Table 1. The Austrian Grid testbed

Rank	Site	Architecture	#	CPU, GHz	Job Manager	Location
1	altix1.jku	NUMA, SGI Altix 3000	10	Itanium 2, 1.6	Fork	Linz
4	gescher	COW, Gigabit Ethernet	10	Pentium 4, 3	PBS	Vienna
2	altix1.uibk	NUMA, SGI Altix 350	10	Itanium 2, 1.6	Fork	Innsbruck
3	schafberg	NUMA, SGI Altix 350	10	Itanium 2, 1.6	Fork	Salzburg
5	agrid1	NOW, Ethernet	10	Pentium 4, 1.8	PBS	Innsbruck
6	arch19	NOW, Ethernet	10	Pentium 4, 1.8	PBS	Innsbruck

Grid site in Salzburg. As expected, the overheads increase with the number of aggregated Grid sites, as shown in Figures 7(c) (5.669%) and 7(d) (25.933%).

We can rank the importance of the measured overheads as follows: data transfer overhead, load imbalance overhead, workflow preparation, middleware overhead, job preparation overhead.

We configured DEE to perform a checkpoint after each main phase of the WIEN2k execution: `lapw0`, `lapw1`, and `lapw2`. Moreover, we configured the master enactment engine to perform input data checkpointing and slave engines to do output data checkpointing. Fig 8(a) compares the overheads of the light-weight workflow checkpointing and the workflow-level checkpointing for a centralized and a distributed enactment engine. The overhead of the light-weight workflow checkpointing is very low and relatively constant, since it only stores the workflow state and URLs to the intermediate data. The overhead of workflow-level checkpointing for a centralized enactment engine increases with

DEE: A Distributed Fault Tolerant Workflow Enactment Engine

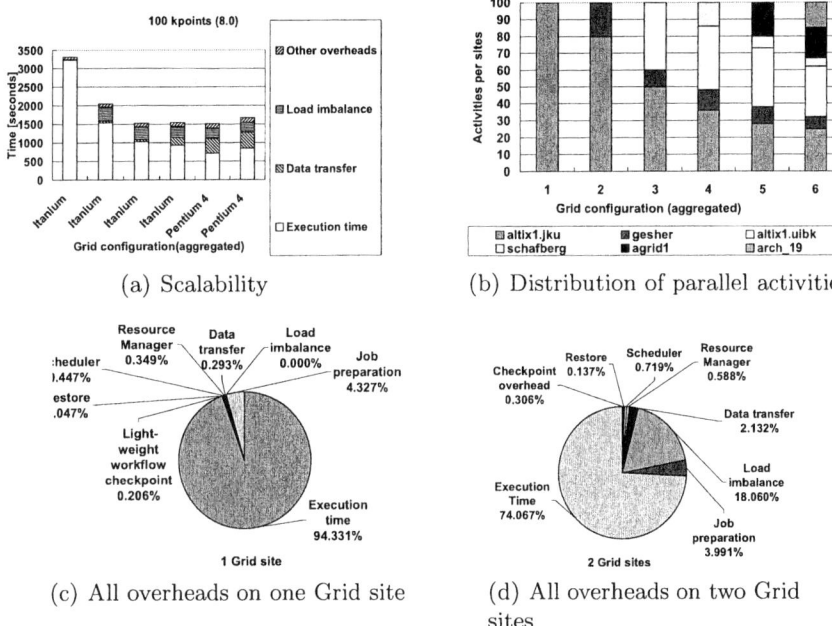

Fig. 7. Enactment Engine overheads

the number of Grid sites because there are more intermediate data to be transferred to the checkpoint database. For a distributed enactment engine, the workflow-level checkpointing overhead is much lower since every slave enactment engine uses a local checkpoint database to store the intermediate data files, which eliminates the wide area network file transfers. Fig 8(b) presents the gains obtained in the single site workflow execution because of checkpointing. We define the *gain* as the difference between the time stamp when the last checkpoint is performed T_{ckpt} minus the time stamp of the previous checkpoint T_{ckpt}^{prev}:

$$Gain = T_{ckpt} - T_{ckpt}^{prev}.$$

The biggest gains are obtained after checkpointing the parallel regions lapw1 and lapw2. The gain for the workflow level checkpointing is lower, since it subtracts the time required to copy the intermediate data to the checkpoint database. On two sites using two enactment engines (one master and one slave) the gain obtained will be only half, if we assume that only one of the two enactment engines crashes. Therefore, a distributed enactment engine provides lower losses upon failures, which are isolated in separate workflow partitions.

Fig 8(c) shows that the size of the data checkpointed at the workflow level is bigger than the overall size of intermediate data transferred for a small number of Grid sites (up to three, when scalability is achieved). Beyond four sites, the size of the intermediate data exceeds the workflow-level checkpointing data size. The

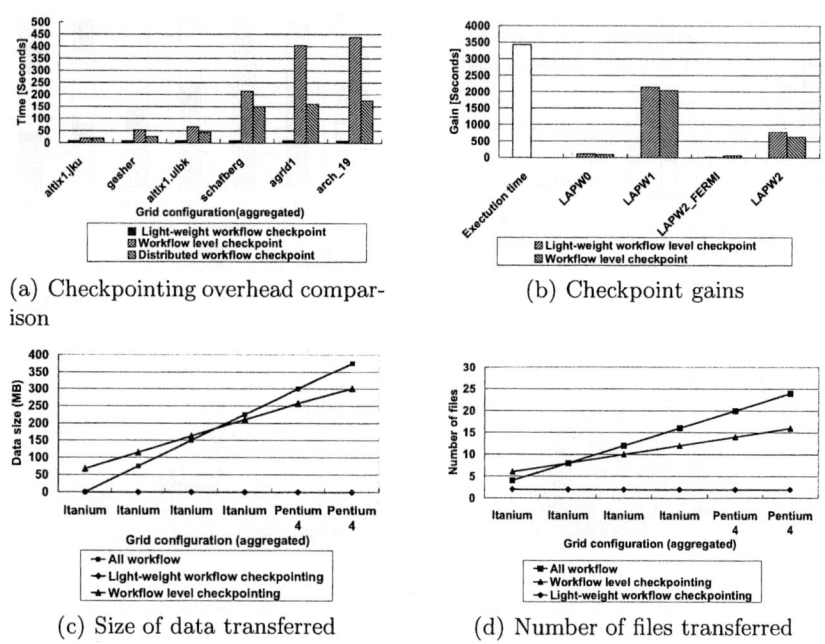

Fig. 8. Enactment Engine checkpointing results

data size of the light-weight workflow checkpointing is, of course, negligible. The number of files transferred preserves, more or less, this behavior (see Fig 8(d)).

5 Related Work

Checkpointing is one of the most important tasks required by fault tolerance. DAGMan [9] and the Pegasus [5] workflow management systems support activity-level checkpointing and restart techniques. However, they do not support checkpointing at the workflow level. Other Grid workflow management projects [3,11] also do not consider workflow-level checkpointing, because their intermediate data management is independent from the control flow management system.

GrADS [1] only supports rescheduling at the activity level, but not at the workflow level. Rescheduling at the workflow level, like is done in ASKALON with support from DEE, has the potential of producing better schedules since it reconsiders the entire remaining sub-workflow for optimized mapping onto the Grid resources.

None of the existing works [3,8,11] is based on a distributed architecture, like the one proposed in this paper. A distributed architecture improves the overall fault tolerance, decreases the overhead of the enactment engine and of the checkpointing, reduces the losses upon enactment engine failures, improves the data locality, and reduces the complexity of the typically large-scale scientific workflows through partitioning.

Finally, we introduced a systematic approach to understanding the overheads produced by the enactment engine and the fault tolerance techniques to the distributed Grid execution. We did not see a similar systematic approach in any of the related works.

6 Conclusion

This paper we presented the motivation, features, design, and implementation of a distributed workflow enactment engine, which is part of the ASKALON application development environment for Grid computing. A distributed architecture has several advantages over a centralized approach. It improves the overall fault tolerance of the enactment engine itself, it improves the data locality by appropriate workflow partitioning, reduces the enactment engine overhead due to simplified control flow and data flow structures, and reduces the losses upon enactment engine crashes. We have defined and implemented two approaches to workflow-level checkpointing that improves the robustness of workflow executions in unstable Grid environments. We have demonstrated our techniques for two real-world workflow applications from the theoretical chemistry and the hydrological fields.

Future work aims at improving the enactment engine in various aspects, including scalability, workflow optimization, and workflow partitioning. We will also study new real-world applications from the meteorological and astrophysics domains.

References

1. Francine Berman and et al. The GrADS Project: Software support for high-level Grid application development. *The International Journal of High Performance Computing Applications*, 15(4):327–344, 2001.
2. P. Blaha, K. Schwarz, G. Madsen, D. Kvasnicka, and J. Luitz. *WIEN2k: An Augmented Plane Wave plus Local Orbitals Program for Calculating Crystal Properties*. Institute of Physical and Theoretical Chemistry, Vienna University of Technology, 2001.
3. Junwei Cao and et al. GridFlow: Workflow Management for Grid Computing. In *Proceedings of the 3rd International Symposium on Cluster Computing and the Grid (CCGrid 2003)*, Tokyo, Japan, May 2003. IEEE Computer Sociery Press.
4. The Austrian Grid Consortium. http://www.austriangrid.at.
5. Ewa Deelman and et al. Mapping abstract complex workflows onto grid environments. *Journal of Grid Computing*, 1(1):9–23, 2003.
6. Dietmar W. Erwin and David F. Snelling. UNICORE: A Grid computing environment. *Lecture Notes in Computer Scicncc*, 2150, 2001
7. T. Fahringer. ASKALON - A Programming Environment and Tool Set for Cluster and Grid Computing. http://dps.uibk.ac.at/askalon, Institute for Computer Science, University of Innsbruck.

8. IT Innovation. Workflow enactment engine, October 2002. http://www.it-innovation.soton.ac.uk/mygrid/workflow/.
9. The Condor Team. Dagman (directed acyclic graph manager). http://www.cs.wisc.edu/condor/dagman/.
10. Jun Qin Thomas Fahringer and Stefan Hainzer. Specification of Grid Workflow Applications with AGWL: An Abstract Grid Workflow Language. In *Proceedings of IEEE International Symposium on Cluster Computing and the Grid 2005 (CCGrid 2005)*, Cardiff, UK, May 2005. IEEE Computer Society Press.
11. Gregor von Laszewski and et al. GridAnt-Client-side Workflow Management with Ant. Whitepaper, Argonne National Laboratory, 9700 S. Cass Avenue, Argonne, IL 60439, U.S.A., July 2002.

An OpenMP Skeleton for the A* Heuristic Search*

G. Miranda and C. León

Dpto. de Estadística, I.O. y Computación, Universidad de La Laguna,
E-38271 La Laguna, Canary Islands, Spain
{gmiranda, cleon}@ull.es

Abstract. This work presents a skeleton for the A* heuristic search. This skeleton provides to the user a sequential solver and a parallel solver based in the shared memory paradigm. The user interface for the specification of the necessary classes and methods to solve an specific problem is described. Also the internal operation mode of the two solvers is showed in detail. Computational results obtained on a multiprocessor machine for the Two-Dimensional Cutting Stock Problem are presented.

1 Introduction

An *Algorithmic Skeleton* must be understand as a set of procedures that compose the structure to use in the development of programs for the resolution of a given problem using a particular algorithmic technique. They provide an important advantage in comparison to a direct implementation of the algorithm from the beginning, not only in terms of code reuse but also in methodology and concept clarity. Skeletons introduce modularity in the design of algorithms. In general, the software that supply skeletons presents declarations of empty classes. The user must fill these empty classes to adapt the given scheme for the resolution of a particular problem.

In the resolution of combinatorial optimization problems, the search of a solution (for example, maximize or minimize a certain objective function) consists in finding a path in the state space or search tree that began in the root node and finished in a leaf node. One possibility to find the best solution to a problem would consist in exploring all its search space, analysing all the possible solutions for, finally, select the best one. But in most complex problems, the search tree size gets too big. By this reason, many techniques to avoid exploring all the search space are frequently used.

An heuristic exploration, contrarily to an algorithmic one, uses some information or specific knowledge about the problem to guide the search into the state space. This kind of information is determined by an *heuristic evaluation function, h'(n)*. This function represents the estimation cost to get the objective node from the node n. The function $h(n)$ represents the real cost to get to the solution node. The A* search belongs to this type of strategies. It uses an *estimated total cost function, f'(n)* for each node n. This

---

* This work has been supported by the EC (FEDER) and by the Spanish Ministry of Science and Technology with contract number TIC2002-04498-C05-05. Also by the Canary Government Project COF2003/022. The work of G. Miranda has been developed under the grant FPU-AP2004-2290.

function is determined by: $f'(n) = g(n) + h'(n)$. Where, $g(n)$ represents the accumulated cost from the root node to the node n and $h'(n)$ represents the estimated cost from the node n to the objective node (that is, the estimated remaining cost). For each problem, the evaluation function to apply must be defined. Depending on the defined function, the optimal solution will be found or not and it will be reached after exploring more or less nodes. An algorithmic skeleton for the A* strategy is presented in this work. The implementation has been done in C++ and it provides to the user two resolution patterns. One pattern works sequentially and the other gives a parallel operation based in the shared memory paradigm. The OpenMP [6] tool has been used to develop the parallel scheme. The parallel skeleton proposed is the main contribution of this article. In the literature, can be found message passing proposals for different heuristics searches [1] but not schemes based on shared memory. The skeleton has been tested with many academic and complex problems [5]. Here are presented the computational results obtained for the Two-Dimensional Cutting Stock Problem on a multiprocessor machine.

The remaining content is structured in the following way: In section 2 is presented the user interface and the operation mode of the A* search sequential scheme. The parallel skeleton is described in detail in section 3. Computational results are presented in section 4. Finally, the conclusions and future works are given.

2 The Sequential Skeleton

When using a skeleton, the user must define some classes and implement a set of necessary methods. In this case, the required classes are: `Problem` (stores the characteristics of the problem to solve), `Solution` (defines how to represent the solutions) and `SubProblem` (represents a node in the tree or search space). The user can also modify some properties of the search to do by using a configuration class [5]. The `SubProblem` class defines the search for a particular problem and it must contain a field of type `Solution` in which store the (partial) solution. The methods to define are: `initSubProblem(pbm, subpbms)` creates the initial subproblem or subproblems from the original problem. `lower_bound(pbm)` calculates the subproblem accumulated cost $g(n)$. `upper_bound(pbm, sol)` calculates the estimated total cost $f'(n)$ of the subproblem. `branch(pbm, subpbms)` generates a set of new subproblems from the current subproblem.

The code for an A* sequential search uses two linked list: `open` and `best`. The `open` list contains all the nodes generated but not expanded yet. These nodes are sorted by their $f'(n)$ value. When a minimal solution is being searched, the first subproblems in the list are the ones with the lower value of $f'(n)$. In maximization problems, the subproblems with higher value of $f'(n)$ are inserted at the head of the list. The subproblems that have been analysed are inserted in the `best` list. Depending on the specifications made by the user, this list must control or not the analysis of similar subproblems.

The skeleton allows different types of searches, the simplest one works as follows: The first node in the `open` list is removed. By the estimations done, this node is the best of all nodes waiting to be explored. If it is not a solution, the node is inserted in the list `best`. In case that the node can not be inserted, because the list has a similar and

```
1    ... Code to run by the master
2    while (!(open.empty()) && (!finished)) {
3      omp_set_lock(open.headLock);
4      node = open.remove();              // Extract next subproblem
5      omp_unset_lock(open.headLock);
6      if (node.lower != node.upper) {    // If it is not the solution
7        if (best.insert(node) == true) { // Subproblem not analysed before
8          #pragma omp flush
9          omp_set_lock(node.lock);
10         if (!(node.assigned) && !(node.done)) {
11           node.assigned = true;        // Node is now assigned
12           omp_unset_lock(node.lock);
13           genSpbms(pbm, node);         // Generate new subproblems
14           genNodes++;
15           omp_set_lock(node.lock);     // Node is not assigned now
16           node.assigned = false;
17         }
18         omp_unset_lock(node.lock);
19         while (node.assigned) {        // Wait if someone is working on the node
20           #pragma omp flush
21         }
22         insertSpbms(node);             // Insert node subpbms into the Open list
23         if (this->setup.getMemoryHard())
24           delete (node);
25      } } else {                        // If the solution has been found
26        bestSolution = node.lower;
27        solution = node.sp.solution;
28        finished = true;
29        delete (node);
30    } }
31    finished = true;
32    ...
```

Fig. 1. Code for the Master

better one, the solver selects next node in open to explore it. In other case, the node subproblem is branched using the function branch(pbm, subpbms) implemented by the user. All new generated subproblems are inserted by order in the list of open subproblems. If there is an identical subproblem in this list, the new subproblem is not inserted, in order to avoid exploring duplicated subproblems. At this point, it returns to the first step and continues like that until the selected node from the open list verifies that lower is equal to upper. In this case, the search finishes because the solution to the problem has been found. If the open list gets empty before finding the solution, the instance of the problem has no solution. It has to be taken into account that the skeleton will give the optimal solution only if the function upper_bound is always and upper estimation (or sub-estimation in case of a minimization problem) of the real value.

3 The Parallel Skeleton

This solver is based in a shared memory scheme to store the subproblem lists (open and best) used during the search process. Both lists have the same functionality than in the sequential skeleton. The difference is that lists are now stored in shared memory and it makes possible that several threads can work simultaneously in the generation of subproblems from different nodes. One of the main problems is to hold the open list sorted. The order in the list is necessary to get always the optimal solution. Moreover, if

```
1     ... // Code to run by the slaves
2     while (!finished)) {
3       #pragma omp flush
4       omp_set_lock(open.headLock);
5       node = open.head;                    // First unassigned subpbm with work to do
6       if (node != NULL) omp_set_lock(node.lock);
7       omp_unset_lock(open.headLock);
8       while ((node != NULL) && ((node.assigned) || (node.done))) {
9         lastnode = node;
10        node = node.next;
11        omp_unset_lock(lastnode.lock);
12        if (node != NULL) omp_set_lock(node.lock);
13        #pragma omp flush
14      }
15      if (node != NULL) {            // Work on this node or subproblem
16        node.assigned = true;        // Mark the node as assigned
17        omp_unset_lock(node.lock);
18        genSpbms(pbm, node);
19        omp_set_lock(node.lock);     // Now the node is unassigned
20        node.assigned = false;
21        omp_unset_lock(node.lock);
22      }
23      #pragma omp flush (finish)
24    } ...
```

Fig. 2. Code for the slaves

worse subproblems are first branched, the two lists would grow unnecesarily and useless work would be done. By this reason, standard mechanisms for the work distribution could not be used.

The technique applied is based on a *master-slave* model. Before the threads began to work together, the master generates the initial subproblems and inserts them into the open list. At that moment, the master and the slaves begin their work to solve the problem. The *Master* code is shown in Figure 1. At each step, the master extracts the first node of the open list (line 4). If its subproblem is a solution, the search finishes (lines 26-29). In other case, the current node is inserted into the list of best subproblems, only if no other similar and better subproblem has been analysed before (line 7). Assuming that the node is inserted in best, next step consists in verifying if there is some one doing its branch or if it had been done before. If the node is still unbranched and nobody is working on it, the master does this work (line 13). If the node is assigned, the master must wait until the thread which works on it finishes to generate its subproblems (lines 19-21). Once all the node subproblems have been generated, the master inserts them by order in the open list (line 22).

Figure 2 shows a pseudo-code of the slaves. Until the master does not notify the end of the search, each slave works generating subproblems from the unexplored nodes of the open list. Each slave explores the open list to find a node not assigned to any thread and with work to do (lines 3-14). When a node like that is found, the slave marks it as assigned, generates all its subproblems, indicating that the work is done and finally leaves it unassigned (lines 15-22). Then the slave begins to find another node to work on, and continues like that until the master notifies the end of the search.

The code in the figures represents the case where the subproblems generation is of independent type. So, when a slave or the master calls the method genSpbms, it can generate in one call all the node subproblems. The list of generated subproblems keeps

Table 1. Sequential and Parallel executions - Cutting Stock Problem

P.	sequential			1 thread			2 threads			3 threads			4 threads		
	Comp.	Gen.	Time	Comp.	Gen.	Time	Comp.	Gen.	Time	Comp.	Gen.	Time	Comp.	Gen.	Time
1_	3502	8236	2.8	4105	12436	7.7	4135	12165	4.9	4220	12058	3.8	4044	11368	3.4
A4	864	38458	75.9	868	40120	93.9	854	17377	9.5	848	16172	9.0	860	14798	7.4
A5	1674	13279	6.2	1524	13332	12.7	1510	6282	6.6	1527	6268	3.9	1526	5875	3.1

referenced from the spbms field in the node and the done field is set to true, because no more subproblems can be generated by the combination of the current one. However in case of dependent generation of subproblems, the thread will be able to generate more or less subproblems, depending on the number of elements in best that are still pending to combinate with.

Problems in the model appear when different slaves find an unassigned node and both decide to branch it at the same time. There are also problems if the master tests if a node is assigned at the same moment that a slave is going to mark it. To avoid these kind of conflicts, the skeleton adds to each node a lock variable. These variables are of omp_lock_t type and they are used to have write or read exclusively access to the assigned field of the node. It is also necessary to add a lock variable to the head of the open list, because the master is eliminating nodes from that list while the slaves are exploring it. This variable must be taken before modifying (line 3 of Figure 1) or reading (line 4 of Figure 2) the open list head. On the other hand, it is very important that all threads had a consistent view of all shared variables. Many flush primitives are used for this proposal.

4 Computational Results

The problem selected to do the skeleton efficiency analysis is the Two-Dimensional Cutting Stock Problem [2,7]. The implementation given to the skeleton is partially based in the Viswanathan and Bagchi's algorithm [4,8]. For the computational study, we have selected some instances from the ones exposed in [3]. The selected problem instances are: 1_, A4 and A5. The experiments have been run in a multiprocessor machine with 4 processors Intel Xeon 1400 MHz.

The number of computed nodes, generated nodes and average execution time of the sequential and parallel solver using 1 to 4 threads are shown in Table 1. Execution times of the sequential solver and the parallel solver that uses only one thread can be compared in order to study the overhead introduced by the parallel scheme. Note that the number of generated nodes is quite lower in the parallel schemes. That involves doing less insertions by order into the open list. So the speedup of this scheme is gotten because the slaves contribute to reach better bounds faster, making possible to discard subproblems than in the sequential case must to be inserted. This is the reason of the superlineal speedup in the A4 problem. Computed nodes (branched or explored nodes) are similar in both schemes. The number of generated nodes in the case of the parallel solver differs very little from the ones generated in the sequential case because of the implementation done. In the sequential execution, all created subproblems are inserted into the open list. In the parallel scheme the subproblems are stored in the spbms list of the node and they are inserted into the open list only when the node is analysed. The last generated subproblem is the first to be inserted. That is, just contrary to the sequential scheme. This can cause that two subproblems with the same *upper_bound* value interchange their order (in relation to the sequential scheme) in the open list.

5 Conclusions

In this work, we have presented a sequential and a parallel solver for the A* heuristic search. The parallel solver implemented using a shared memory tool is the main contribution. We have presented computational results obtained for the implementation of a complex problem. As we have shown, the necessary synchronization to allow having the list in shared memory, introduces an important overhead to the parallel version. The improvements introduced by the parallel solver are obtained because the number of generated nodes decreases when more threads collaborate in the problem resolution. That is due to the fact that the update of the best current bound is done simultaneous by all threads, allowing to discard subproblems that in the sequential case have to be inserted and perhaps also explored. Other advantage of the parallel scheme is that the work distribution between the slaves is balanced.

Actually, we are working to obtain more results with other problems in order to study the behaviour and efficiency of the skeleton in the implementation of different cases. Also, improvements on the insertion scheme are been studied. Finally, some work is done in the implementation of a parallel version based on the message passing paradigm to compare with the one presented here.

References

1. M. Aguilar, B. Hirsbrunner, *Pact: An Environment for Parallel Heuristic Programming*, International Conference Massivelly Parallel Processing: Application and Development, Delft, The Netherlands, June 21-23, 1994.
2. H. Dyckhoff, *A Typology of Cutting and Packing Problems*, European Journal of Operational Research, Volume 44, 145–159, 1990.
3. M. Hifi, *Improvement of Viswanathan and Bagchi's Exact Algorithm for Constrained Two-Dimensional Cutting Stock*, Computer Operations Research, Volume 24, 727–736, 1997.
4. G. Miranda, C. Len and C. Rodrguez, *Implementacin del Problema de Corte Bidimensional con la Librera MaLLBa*, Actas de las XV Jornadas de Paralelismo, Almera, Spain, 96–101, 2004.
5. G. Miranda and C. Len, *Esqueletos para Algoritmos A**, Dpto. de Estadstica, I.O. y Computacin, Universidad de La Laguna, DT-DEIOC-04-06, 2004.
6. OpenMP Architecture Review Board, *OpenMP C and C++ Application Program Interface*, Version 1.0, http://www.openmp.org, 1998.
7. P.E. Sweeney and E.R. Paternoster, *Cutting and Packing Problems: A Categorized, Application-Orientated Research Bibliography*, Journal of the Operational Research Society, Volume 43, 691–706, 1992.
8. K.V. Viswanathan and A. Bagchi, *Best-First Search Methods for Constrained Two-Dimensional Cutting Stock Problems*, Operations Research, Volume 41, 768–776, 1993.

A Proposal and Evaluation of Multi-class Optical Path Protection Scheme for Reliable Computing

Shoichiro Seno[1], Teruko Fujii[1], Motofumi Tanabe[1], Eiichi Horiuchi[1], Yoshimasa Baba[1], and Tetsuo Ideguchi[2]

[1] Information Technology R&D Center, Mitsubishi Electric Corporation,
Kamakura, Kanagawa 247-8501, Japan
{senos, teruko, motofumi, eiichi, y_baba}@isl.melco.co.jp
[2] Faculty of Information Science and Technology, Aichi Prefectural University,
Nagakute-cho, Aichi-gun, Aichi 480-1198, Japan
ideguchi@ist.aichi-pu.ac.jp

Abstract. Emerging GMPLS (Generalized Multi-Protocol Label Switching)-based photonic networks are expected to realize dynamic allocation of network resources for high-performance computing. To address diverse reliability requirements of computing, various optical path recovery schemes, some of them under GMPLS standardization, should be supported on a photonic network. This paper proposes multi-class optical path protection scheme, supporting 1+1 protection, M:N protection, and extra path reusing back-up resources, for a unified approach to build such a network. The proposed scheme was implemented on an OXC prototype and its path establishment time and failure recovery time were evaluated. It was confirmed that a node supporting a wide range of recovery capabilities could realize small or relatively small recovery time to various applications according to their diverse reliability requirements.

1 Introduction

Emergence of GMPLS (Generalized Multi-Protocol Label Switching) and next-generation photonic network technologies is seen as a great opportunity for high-performance computing, because it can provide means to automatically allocate dedicated optical paths to computers attached to a photonic network. For this reason, research testbeds have been built [1], and the Grid community is studying networking issues related to realization of a Grid infrastructure over a photonic network [2][3].

Reliability is a major requirement in high-performance computing. Here *reliability* means not only a short recovery time to switch to a back-up path, but also includes various capabilities to deal with failures, e.g., re-establishment of an optical path to route around the failure, survivability to multiple failures, optimal use of back-up resources when there is no failure, and fair sharing of remained bandwidth when a path's bandwidth is partially out of service. Also, applications can differ with respect to reliability requirements: some are sensitive to data loss and some require real-time transmission [2].

In the GMPLS-controlled networks, distributed and automatic control of various equipment, i.e., OXCs (Optical Cross Connects) and routers, provides of dynamic establishment and release of broadband optical paths [4]-[7]. To support reliability of

GMPLS-controlled optical paths, failure management by LMP (Link Management Protocol) [8] was defined and various optical path recovery schemes have been discussed and standardized [9]-[12].

GMPLS's optical path recovery schemes can be divided into two categories: protection and restoration. With protection, a primary path is protected by a corresponding back-up path whose resources (links, switch ports, etc.) are pre-assigned at the time of path establishment. Protection enables fast recovery of traffic transported over the primary path by switching it over the back-up path upon its failure, because there is no need of signaling for back-up path establishment. With restoration, a back-up path will be newly established upon a primary path's failure. Compared to protection, restoration is slower and less likely to secure a back-up path because it is not pre-assigned. But restoration has an advantage of higher resource utilization than protection since resources pre-assigned for back-up paths in the case of protection are available for other paths. To improve resource utilization of protection, transport of extra traffic by a back-up path and shared use of back-up paths have been proposed [10]. Here extra traffic means a class of low priority traffic subject to preemption when a failure occurs on the primary path of the back-up path it is using.

As seen in the above, optical path recovery schemes have a trade-off between recovery time/availability and resource utilization efficiency. Because reliable computing applications will have diverse requirements, and some services will demand short recovery time while other services can tolerate longer recovery time while enjoying lower cost, a photonic network supporting a wide range of recovery capabilities is desirable.

This paper proposes a multi-class optical path protection scheme as a unified approach to support a wide range of recovery capabilities. The multi-class optical path protection scheme includes 1+1 protection, M:N protection (inclusive of 1:1 protection), and *extra paths* to carry low-priority extra traffic. An *extra path* is an extension of an extra traffic path occupying a back-up path of 1:1 protection [10]. Whereas previous protection schemes only allow reuse of an entire back-up path for extra traffic, an *extra path* may reuse a part of a back-up path.

To realize the multi-class optical path protection scheme, the authors implemented it over an OXC prototype with extensions to the GMPLS protocols, and evaluated its path establishment time and failure recovery time. It was confirmed that a node supporting a wide range of recovery capabilities could provide small or relatively small recovery time to various applications according to their diverse reliability requirements.

In the following, chapter 2 proposes the multi-class protection scheme and Chapter 3 describes its implementation. Chapter 4 presents evaluation of the implementation. Chapter 5 is the conclusion.

2 The Multi-class Optical Path Protection Scheme

In GMPLS path recovery works, there are two general optical path protection classes: 1+1 optical path protection and 1:1 optical path protection. Both classes use a pair of paths called a primary path and a back-up path; traffic will be transported over the former, and it will be switched to the latter upon a failure on the former. A primary path and a back-up path should be independent from each other's failure to assure recovery. This is usually achieved by assigning path resources, i.e., nodes, links, etc., to each path from different pools so that a single failure may not affect both of them. If one resource is assigned to both primary and back-up paths, it is called a *shared risk*. Path selection algorithms must minimize *shared risks*.

A 1+1 Optical Path Protection

1+1 optical path protection transports the same traffic over both the primary path and the back-up path at the path's ingress node, and selects traffic from one of the paths at its egress node. Fig. 1 shows transport mechanism of 1+1 optical path protection.

Because the egress node can select traffic from the back-up path when it detects a failure on the primary path, 1+1 optical path protection can provide the fastest recovery time.

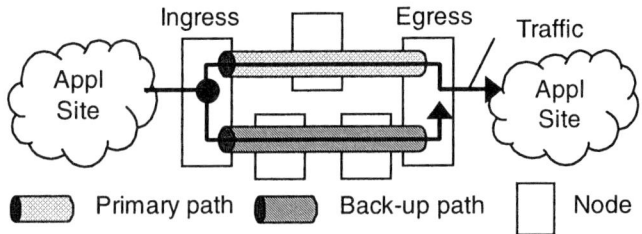

Fig. 1. 1+1 Optical Path Protection

B 1:1 Optical Path Protection

1:1 optical path protection uses a primary path and a back-up path like 1+1 optical path protection, but transports traffic only to one of them on which no failure is detected. Because the path's ingress node and egress node must agree with which path to transport the traffic by signaling, its recovery time is longer than that of 1+1 optical path protection, but it is still shorter than that of restoration, as this signaling procedure does not require hop-by-hop signaling.

In 1:1 optical path protection, a back-up path's resources may be reused to transport extra traffic which can be pre-empted upon detection of a failure on the primary path, as shown in Fig. 2. Thus, overall resource utilization will be improved by 1:1 optical path protection.

To further improve resource utilization, the authors proposed an extension of resource usage of extra traffic, in which a path carrying extra traffic can consist of the whole or a part of a back-up path and additional unassigned resources, as shown in Fig. 3 [13]. This class of paths is called *extra path*.

Extra path enables efficient utilization of back-up resources as well as flexible resource assignment, e.g., an extra path's link which is not a part of a back-up path may be assigned to a new back-up path.

C M:N Optical Path Protection

The above described optical path protection classes protect traffic from a single point of failure. If a network has to recover from more than one simultaneous failures, another protection class will be necessary. If we can choose a set of primary optical paths from the same ingress node to the egress node without *shared risks* and a back-up optical path also without *shared risks*, we can expect essentially the same level of survivability from a single failure as 1:1 optical path protection. On the other hand, if we can select two back-up paths for a single primary path, where each pair of them has no shared risks, the primary path will be protected from double failures.

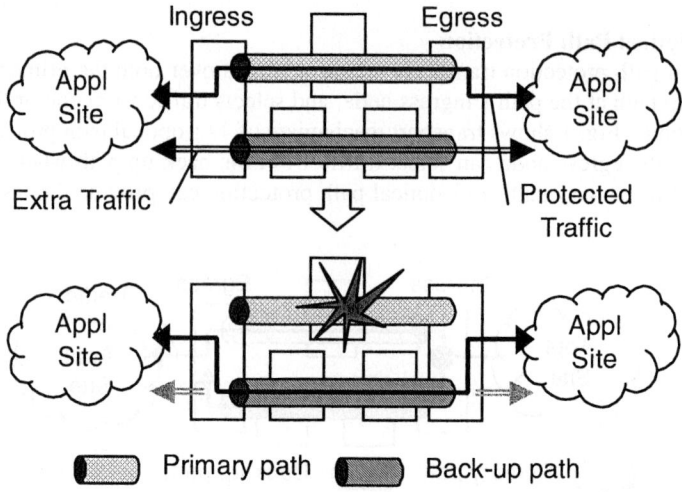

Fig. 2. 1:1 Optical Path Protection and Extra Traffic

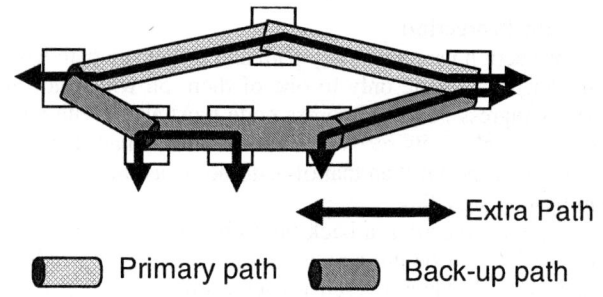

Fig. 3. Extra Path

Based upon the above observation, this paper proposes to add M:N optical path protection, an extension of 1:1 optical path protection, to protection capabilities of a photonic network.

In M:N optical path protection, N primary paths are protected by M back-up paths, where M+N paths do not have *shared risks*. When a failure is detected on a primary path, one of M back-up paths is chosen to transport the affected traffic. Thus, M:N optical path protection provides survivability against M simultaneous failures.

Another feature of M:N optical path protection is grouping of multiple optical paths including primary paths and back-up paths. This enables assignment of recovery priority to primary paths and choice of a back-up path not used for extra traffic.

Therefore, M:N optical path protection can support more diverse quality of reliability compared with 1:1 optical path protection.

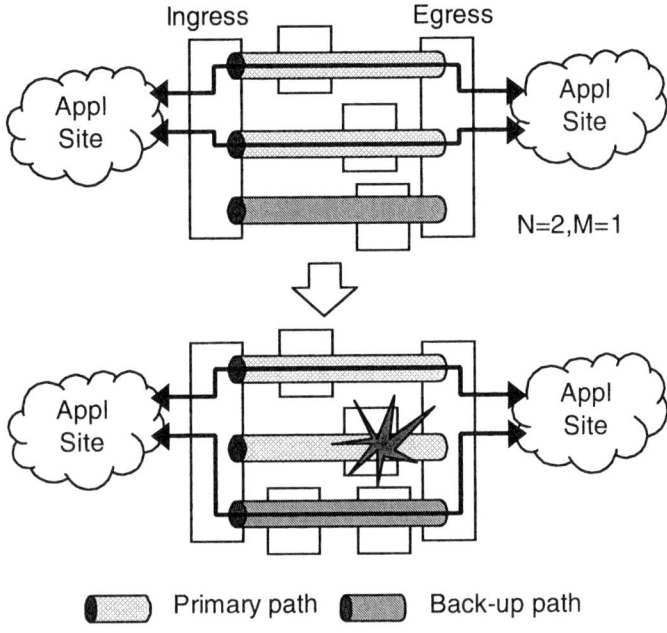

Fig. 4. An Example of M:N Optical Path Protection

D Multi-class Optical Path Protection

Table 1 shows comparison of the above described protection classes with respect to availability, recovery time, and cost. Please note that each protection class provides different trade-off among these features.

Table 1. Multi-class Protection Trade-off

Protection Class	Availability	Recovery time	Cost
1+1 Protection	High	Short	High
M:N Protection	⇩	⇩	⇩
Unprotected	Low	Long	Low
Extra Path			

This paper proposes provision of all these protection classes by a photonic network where a single protection scheme, *the Multi-class Optical Path Protection Scheme*, supports multiple classes by a common GMPLS-based routing and signaling procedure. Thus, in a network implementing this scheme, an application can choose a most suitable class according to its requirements. For example, a real-time application can choose 1+1 protection to minimize disruption by switch-over. An application requiring large bandwidth but tolerable to data loss can use extra paths in addition to usual paths.

3 Implementation of the Multi-class Optical Path Protection Scheme

The authors have implemented and evaluated the GMPLS protocol on an OXC prototype [14][15]. The proposed *Multi-class Optical Path Protection Scheme* is also implemented on the same OXC prototype and evaluated.

Table 2 shows the OXC prototype's specifications and Fig. 5 shows its architecture. The prototype consists of CPU Module and Optical Switch, and the former processes the GMPLS protocols and controls the latter to provide optical paths. The proposed *Multi-class Optical Path Protection Scheme* is implemented in the Protection Manager.

Table 2. OXC Prototype Specification

ITEM		DESCRIPTION
Optical Switch	Architecture	2-D MEMS, Bascule
	Capacity	8 x 8 or 16 x 16
	Insertion Loss	Less than 8 dB
	Wavelength Range	1,290 ~ 1,330 nm, 1,530 ~ 1,570 nm
GMPLS	Signaling	RSVP-TE
	Routing	OSPF-TE
	Link Management	LMP
	Switching Capability	Fiber, Lambda
	Path Computation	CSPF
Dimensions		700 x 600 x 1,800 mm
Power Supply		AC 100 V☐50/60 Hz

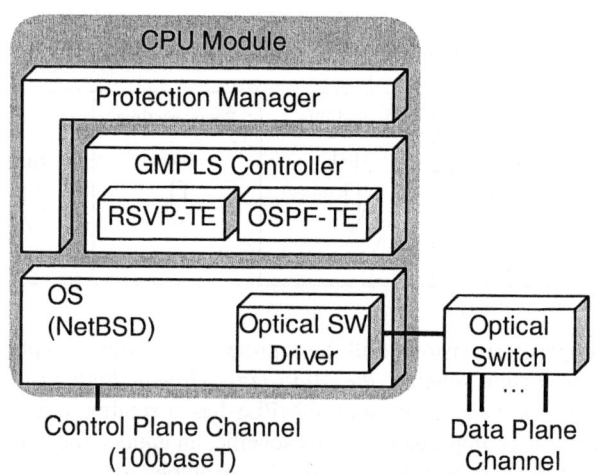

Fig. 5. OXC Prototype's Architecture

A Path Group

GMPLS's protection protocol [12] associates a back-up path and its one or more primary path(s). Although this can be applied to M:N protection where N primary paths are associated with M back-up paths, establishment of association may not be straight-forward.

In the proposed scheme's implementation, the authors introduced *path group* to clearly identify a group consisting of primary paths sharing the same set of back-up paths and these back-up paths. For example, 1+1 protection's path group includes one primary path and one back-up path, M:N protection's path group consists of N primary paths and M back-up paths, and unprotected path's path group or extra path's path group contains only the path itself. A path group represents a unit of recovery processing unit of the Protection Manager.

B Extension to RSVP-TE

1. Support of Path Group

The implementation uses a new opaque object to convey protection type and path group identifier. This object is carried by a PATH message to signal establishment of a primary or back-up path.

2. Support of Extra Path

Because an extra path's terminating nodes may not be its primary path's ingress or egress node, it is necessary to define a procedure to signal preemption of the extra path from the ingress or egress node, which triggers the protection switch-over. Also it is necessary to resolve a back-up path's resource request conflicts caused by simultaneous extra path requests. To achieve these goals, the implementation uses the following procedure:

(1) When a back-up path's resource is requested by an extra path at one of the path's transient nodes or egress node during an extra path establishment procedure, the request is reported to the path's ingress node and the ingress node acknowledges the request if there are no resource conflicts. The implemented procedure shown in Fig. 6, a transient node (Node 2) requests use of a part of the back-up path (between Node 2 and Node 4 via Node 3) to the ingress node (Node 1) by the Extra Request message, and it is acknowledged by the Extra Response message.

(2) When a failure of the primary path is detected by or signaled to the path's ingress node, the ingress node signals preemption of extra paths to the nodes requesting back-up resources in (1). When the signaled nodes tear down extra paths and report back, the primary path's traffic is switched-over to the back-up path. The implemented procedure shown in Fig. 7, the ingress node detects a failure and signals the Extra Path Preemption Request messages to transient nodes (Node 2 and Node 4). When the extra path is preempted, the Extra Path Preemption Response messages are signaled to the ingress node and the primary path's traffic is switched-over to the back-up path by exchanging Switch-over Request/Response messages.

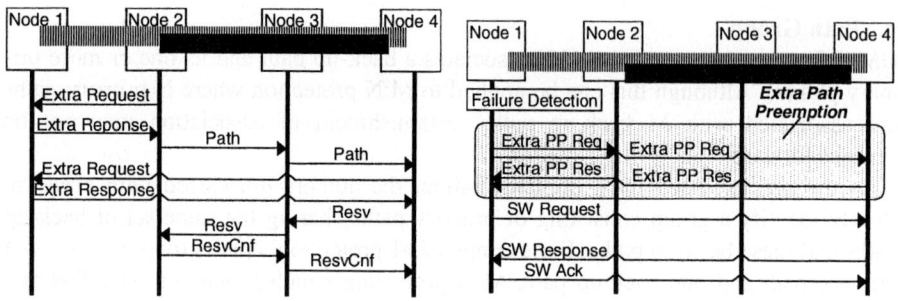

Fig. 6. Extra Path Establishment

Fig. 7. Failure Recovery

4 Evaluation of the Implementation

The authors evaluated the implementation's path establishment time and failure recovery time for multiple protection classes.

Fig. 8 shows the evaluated network configuration, where three OXCs were connected in tandem to allow establishment of a three-hop path. Path establishment time was measured by a path's ingress node by software timer. Failure recovery time was measured by an SDH analyzer connected to both ends of an optical path.

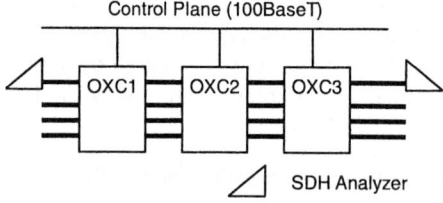

Fig. 8. Evaluated Network Configuration

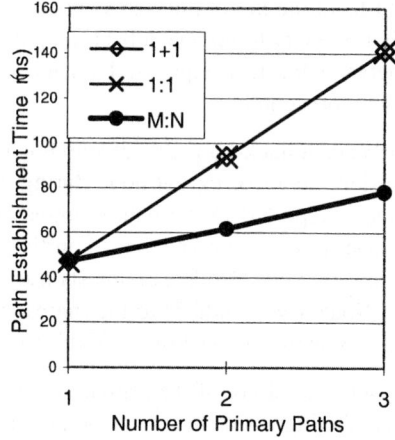

Fig. 9. Path Establishment Time

Table 3. Measured Recovery Time

Protection Class	Recovery Time (ms)
1+1 Protection	4
1:1 Protection	15
1:3 Protection	15
1:1 Protection + Extra Path[*]	24

[*] An extra path is set between OXC2 and OXC3.

Fig. 9 compares path establishment time for 1+1, 1:1, and M:N protection with increasing number of primary paths. Path establishment time of M:N protection was measured for 1:1, 1:2, and 1:3, and that of other protection was multiple of the 1+1

(1:1) case. When a back-up path is shared by more than one primary paths, path establishment time per primary/back-up pair decreases with increase of back-up path sharing by M:N protection. Thus M:N protection requires less time for path establishment than shared mesh protection of 1:1 protection applied to the same path configuration.

Table 3 shows measured recovery time of implemented protection classes where the paths were established between OXC1 and OXC3. 1+1 protection achieved the shortest recovery time while 1:1 protection and 1:3 protection were almost the same. Switching time of the optical switch occupied more than half of 1+1 protection's recovery time, and the rest was processing delay. The difference between 1+1 protection and 1:1 (M:N) protection, caused by switch-over signaling, was not significantly large, and preemption of an extra path did not add a large overhead to 1:1 protection.

5 Conclusion

GMPLS-based photonic network technologies will emerge as a preferred environment for high-performance computing. To address diverse reliability requirements of computing, various optical path recovery schemes, some of them under GMPLS standardization, should be supported on a photonic network. This paper proposes multi-class optical path protection scheme, supporting 1+1 protection, M:N protection, and extra path reusing back-up resources, for a unified approach to realize such a network. The proposed scheme was implemented on an OXC prototype and its path establishment time and failure recovery time were evaluated. It was confirmed that a node supporting a wide range of recovery capabilities like these could provide small or relatively small recovery time to various applications with diverse reliability requirements.

Further study items include precise characterization of reliability requirements of high-performance computing, and more elaborate evaluation of the proposed scheme in the light of such requirements.

References

1. L. Smarr et al, "The OptIPuter, Quartzite, and Starlight Projects: A Campus to Global-Scale Testbed for Optical Technologies Enabling LambdaGrid Computing," OWG7, OFC 2005, March, 2005.
2. D. Simeonidou et al, ed., "Optical Network Infrastructure for Grid," GFD-I.036, Global Grid Forum, August, 2004.
3. V. Sander, ed., "Networking Issues for Grid Infrastructure," GFD-I.037, Global Grid Forum, November, 2004.
4. E. Mannie, ed., "Generalized Multi-Protocol Label Switching Architecture," RFC 3945, IETF, October 2004.
5. L. Berger, ed., "Generalized Multi-Protocol Label Switching (GMPLS) Signaling Functional Description," RFC 3471, IETF, January 2003.
6. L. Berger, ed., "Generalized Multi-Protocol Label Switching (GMPLS) Signaling Resource ReserVation Protocol-Traffic Engineering (RSVP-TE) Extensions," RFC 3473, IETF, January 2003.

7. K. Kompella et al, ed., "OSPF Extensions in Support of Generalized Multi-Protocol Label Switching," draft-ietf-ccamp-ospf-gmpls-extensions-12.txt (work in progress), IETF, October 2003.
8. J. Lang, ed., "Link Management Protocol (LMP)," draft-ietf-ccamp-lmp-10.txt (work in progress), IETF, October 2003.
9. J. Lang et al, ed., "RSVP-TE Extensions in support of End-to-End GMPLS-based Recovery," draft-ietf-ccamp-gmpls-recovery-e2e-signaling-03.txt (work in progress), IETF, April 2005.
10. J. Lang et al, ed., "Generalized MPLS Recovery Functional Specification," draft-ietf-ccamp-gmpls-recovery-functional-01.txt (work in progress), IETF, September 2003.
11. E. Mannie et al, ed., "Recovery (Protection and Restoration) Terminology for GMPLS," draft-ietf-ccamp-gmpls-recovery-terminology-06.txt (work in progress), IETF, April 2005.
12. D. Papadimitriou et al, ed., "Analysis of Generalized Multi-Protocol Label Switching (GMPLS)-based Recovery Mechanisms (including Protection and Restoration)," draft-ietf-ccamp-ccamp-gmpls-recovery-analysis-05.txt (work in progress), IETF, April 2005.
13. S. Seno et al, "An Efficient Back-up Resource Management Scheme for GMPLS-based Recovery," OECC/COIN 2004, Yokohama, July, 2004.
14. M. Akita et al, "A Study on Technology to Realize Optical Cross Connects for All Optical Networks," OECC 2002, Yokohama, July, 2002.
15. S. Seno et al, "Optical Cross-Connect Technologies," Mitsubishi Electric ADVANCE, Vol.101, March 2003.

Lazy Home-Based Protocol: Combining Homeless and Home-Based Distributed Shared Memory Protocols

Byung-Hyun Yu, Paul Werstein, Martin Purvis, and Stephen Cranefield

University of Otago, Dunedin 9001, New Zealand
{byu, werstein}@cs.otago.ac.nz,
{mpurvis, scranefield}@infoscience.otago.ac.nz

Abstract. This paper presents our novel protocol design and implementation of an all-software page-based DSM system. The protocol combines the advantages of homeless and home-based protocols. During lock synchronization, it uses a homeless diff-based memory update using the update coherence protocol. The diff-based update during lock synchronization can reduce the time in a critical section since it reduces page faults and costly data fetching inside the critical section. Other than the update in lock synchronization, it uses a home-based page-based memory update using the invalidation coherence protocol. The protocol is called *"lazy home-based"* since the home update is delayed until the next barrier time. The lazy home update has many advantages such as less interruption in home nodes as well as less data traffic and a smaller number of messages. We present an in-depth analysis of the effects of the protocol on DSM applications.

1 Introduction

Parallel programming by means of distributed shared memory (DSM) has many advantages since it can hide data communication between nodes. The ease of programming is in contrast to message passing in which a programmer controls data communication between nodes explicitly. Parallel programming with message passing can be very cumbersome and complicated when a programmer deals with fine-grained data structures. However, the programming convenience in DSM comes with the extra cost of achieving memory consistency over all nodes. In a page-based shared virtual memory system, the extra data traffic for memory consistency becomes worse due to the large granularity of the memory unit, which can cause false sharing. Therefore, it is more challenging to implement an efficient page-based DSM system.

Weak memory consistency models can reduce the data traffic and the number of messages required for memory consistency by relaxing the memory consistency conditions. For example, Entry consistency (EC) [1] and Scope consistency (ScC) [2] models provide the most relaxed memory consistency conditions by taking advantage of the relationship between a synchronization object and data that are protected by the synchronization object. With more relaxed models, more optimized implementations of DSM are possible in terms of less data traffic, fewer messages transferred, less false sharing and fewer page faults, though they create more restrictions on the conditions for a correctly executing program.

2 Homeless Versus Home-Based DSM Protocol

Apart from the relaxed memory models, it is also important to implement the models in an efficient way in order to take advantage of the relaxed constraints that the models can provide. For example, a homeless [3] or a home-based protocol [4] can be used to implement the memory models. In the homeless protocol, the required up-to-date memory is constructed by distributed diffs which contain all the write history of the shared pages. A diff can be identified by its base page number, creating node and vector time-stamp. The vector time-stamps are used in order to apply diffs in the manner of the happens-before partial order. In the home-based protocol, each shared page is assigned a home node. The home node takes responsibility for keeping the most up-to-date copy of the assigned pages. Non-home nodes should send diffs of the page to the home node in order to keep the home pages up to date.

The two protocols have their strengths and weaknesses. In page-based DSM systems, the fine-grained diff-based update used in the homeless protocol has an advantage since unnecessary data in a page are less often transferred. Although it is dependent on an application's memory access patterns, by the nature of the protocol, the homeless protocol has less data traffic since it is not necessary to update a home node at synchronization points. However, a diff in the homeless protocol cannot be removed until all nodes have the diff. This will require more memory space to store unnotified diffs. Ultimately, garbage collection is required when the memory size for the diffs exceeds the predetermined threshold, which is very costly due to global diff exchange. Also diffs created from the same page could be accumulated in a migratory memory access pattern, which makes the homeless protocol less scalable. In terms of coherence-related load balance, the homeless protocol is more susceptible to having one node service requests from many other nodes [5], which is known as a *hot spot*. A hot spot in the homeless protocol makes it more difficult to be scalable.

The home-based protocol can solve much of the scalability problem of the homeless protocol. A diff created can be removed immediately after it is known to the home node, so garbage collection is not needed. There is no diff accumulation since each diff is incorporated into the page of the home node immediately at the synchronization time. Moreover, an efficient implementation is possible with the home-based protocol at synchronization points, with each node updating home nodes with one message containing aggregated diffs created from different pages but belonging to one home node. This optimization reduces the number of messages thus avoiding many send and receive network operations which would be very inefficient. With the same scenario in the homeless protocol, each node just waits until other nodes ask for the diff of a faulting page. In case of multiple writer applications, these diff requests are sent to many nodes which have a diff of the page. This is inefficient compared to one round trip page request in the home-based protocol.

The weaknesses of the home-based protocol are also well known. First, without good home assignment the home-based protocol may suffer. In particular, DSM applications that have regular and exclusive memory access patterns are very sensitive to good home assignment. Second, upon a page fault, even if one byte of data is needed, the whole virtual page, which is 4096 bytes in Linux must be transferred. Therefore, a dynamic home assignment scheme and more fine-grained diff-based update are needed to avoid these weaknesses.

3 Lazy Home-Based Protocol

3.1 Protocol Design

To combine the advantages of the homeless and home-based protocols, we adopt ScC as the memory consistency model and use a hybrid protocol. During lock synchronization, the update protocol is applied but during barrier synchronization, the invalidation protocol is applied. The reason for the hybrid coherence protocol is that during lock synchronization a data access pattern is relatively predictable and fine-grained, but during barrier synchronization, it is more unpredictable and large. Therefore the update protocol can selectively send the data updated inside a critical section. More efficiently, the data are piggybacked with the lock ownership transfer. Also, page faults inside the next critical section of the next lock owner are significantly reduced. During barrier synchronization, stale data are invalidated so that the most up-to-date data can be obtained from a home node.

Compared with previous implementations of the home-based protocol, our protocol is more "lazy" because it does not send diffs at a lock release time in order to update home nodes but delays home update until the next barrier time. To illustrate this laziness, Figure 1 compares previous home-based protocol implementations with our lazy home-based implementation. As seen in Figure 1, our lazy home-based protocol implementation eliminates two home update diff messages sent by N0 and N1 and two page faults in N1 as compared with previous home-based ScC protocol implementations.

Our implementation takes advantage of ScC more aggressively than previous ones in a sense of "laziness". The *lazy* home update is still correct under the ScC model since other nodes should not read diffs made in non-critical sections before the next barrier and sufficient diffs created in critical sections are transferred to the next lock owner by the update protocol.

3.2 Implementation

To implement the LHScC protocol, we distinguished a non-critical section (NCS) diff and a critical section (CS) diff. A NCS diff is a diff created in a non-critical section, and a CS diff is a diff created in a critical section. Intuitively, according to

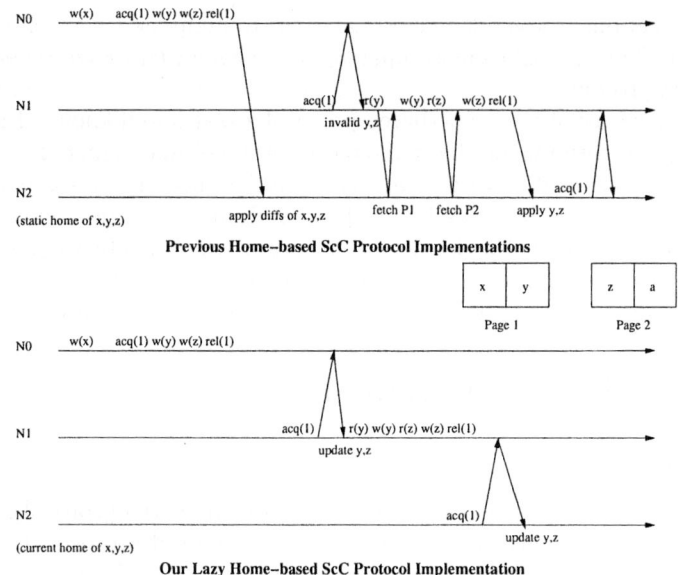

Fig. 1. Difference between Previous Home-based and Our Lazy Home-based Implementations

our memory model, NCS diffs made between two consecutive barriers should be mutually exclusive to one another. NCS diffs are kept until the next barrier when non-home nodes send their non-home NCS diffs to corresponding home nodes. In this way, all NCS diffs are safely preserved at the corresponding home nodes. As for CS diffs, they are sent to the next lock owner during a lock ownership change. Intuitively, the last lock owner before a barrier should have the most up-to-date data that the lock protects. Therefore, the CS diffs from the last lock owners are sufficient to construct the most up-to-date CS data. Upon arrival at a barrier, the last owner of each lock sends its CS diffs to corresponding home nodes unless a node is the home of the CS diffs. In the case that a node owns the same lock consecutively, diffs created from a same page are numbered so that they are applied at the next lock owner in the happens-before partial order.

We developed a *diff integration* technique to reduce data traffic [6]. Since we used fine-grained diff update during lock synchronization, a diff accumulation problem can arise as in the homeless protocol. The diff accumulation problem can be found in a migratory application in which each node writes on the same page in sequence. The diff integration technique incorporates multiple diffs created from the same page during a critical section into one diff, which solves the diff accumulation problem.

To add a dynamic home assignment feature, we also developed a dynamic home assignment scheme [6]. Our scheme is different from others [7,8,9]. Our protocol updates home nodes after all nodes have a knowledge of optimum home locations. This guarantees minimum data traffic related to home page updates by non-home nodes. On the other hand, other dynamic home assignment schemes predict optimum home locations based on previous memory access patterns. This prediction would work well for applications showing regular and coarse-grained memory access patterns without a migratory pattern. However, for applications showing irregular, fine-grained or migratory memory access pattern these schemes do not work since a future memory access pattern would be different from the previous patterns.

4 Performance Evaluation

Our purpose for the performance measurements is to measure the benefits and side effects of LHScC. We chose seven applications to evaluate our LHScC protocol. The applications are obtained from the TreadMarks application suite except PNN which was implemented by us. The problem sizes and sequential execution times of the applications can be found in Table 1. Note that each application tested over the different protocols is identical. We briefly describe the applications.

- **Parallel Neural Network (PNN)** is a parallel implementation of two neural network algorithms: forward and backward propagations. The data set we trained is the shuttle set obtained from the University of California, Irvine machine learning repository. The data set is divided and allocated to each node to be trained in parallel. In the main loop, each node trains part of the data in parallel. Then the local weight changes calculated in parallel previously are summed sequentially through lock synchronization in order to create the global new weight changes. Since each node's local weight changes are transferred and summed, the memory access pattern is migratory during the lock synchronization.
- **Barnes-Hut** is a simulation of gravitational forces using the Barnes-Hut N-Body algorithm. Our Barnes-Hut application uses only barrier synchronization. The memory access pattern of Barnes-Hut is known to be irregular and fine-grained.
- **Integer Sort (IS)** ranks numbers represented as an array of keys by using a bucket sort. Two different implementations of IS were tested— one has

Table 1. Problem Sizes, Iterations and Sequential Execution Times (secs.)

Application	Problem Size	Iterations	Execution Time
PNN	44,000	235	613.56
Barnes-Hut	64k Bodies	3	79.58
IS-B	$2^{24} \times 2^{15}$	20	71.61
IS-L	$2^{22} \times 2^{13}$	30	16.39
3D-FFT	64x64x64	50	45.10
SOR	4000x4000	50	49.58
Gauss	1024x1024	1023	15.26

only barrier synchronization (**IS-B**) in the main loop and the other has lock and barrier synchronizations (**IS-L**) in the main loop.
- **3D-Fast Fourier Transform (3D-FFT)** solves a partial differential equation using forward and inverse FFTs. In the main loop only one barrier synchronization happens at the end of each loop. The memory access pattern of each loop is regular.
- **Successive Over-Relaxation (SOR)** calculates the average of neighbouring cells' four values (up, down, left and right). The shared matrix is divided into N blocks of rows on which N processors work. Only the values in boundary rows that two nodes share are sent to each other. Therefore the memory access pattern is very regular, coarse-grained and exclusive.
- **Gauss** solves a matrix equation of the form $Ax = b$. In the main loop, only one node finds a pivot element. After finding the pivot element, all the nodes run Gaussian elimination in parallel. The memory access pattern is very regular and exclusive.

To evaluate our protocol efficiency, we compared our lazy home-based ScC DSM implementation (LHScC) with TreadMarks, which is regarded as the state of the art homeless LRC implementation, and our home-based LRC implementation (HLRC). Note that performance comparisons between LRC and EC [10] or LRC and ScC [2] have been presented previously. Adve et al. [10] concluded that there is no clear winner between EC and LRC. Rather, performance difference between them is not because of the model adopted but because of the unique memory access pattern of each application and coherence unit size. However, Iftode et al. [2] concluded that a ScC-adopted DSM implementation showed better performance than an LRC-adopted DSM implementation in applications that have a false sharing memory access pattern inside a critical section. The applications we chose have no false sharing memory access pattern inside a critical section. Therefore, performance improvements under LHScC have nothing to do with reduction of false sharing due to ScC.

Our dedicated cluster network consists of 32 nodes, each one having a 350 MHz Pentium II CPU and running Gentoo Linux with gcc 3.3.2. All nodes are connected with a 100 Mbit switched Ethernet. Each node is equipped with a 100 Mbit network interface card (NIC) and 192 MB of RAM except Node 0 which has a 1 Gbit NIC and 318 MB of RAM. In previous experiments [5], N0 had a 100 Mbit NIC — the same as the rest of the nodes. After we found out that N0 can cause a hot spot due to the nature of the homeless protocol, we replaced the 100 Mbit NIC with a 1 Gbit NIC. The replacement will benefit the homeless protocol most since the home-based protocol is less susceptible to a hot spot.

4.1 Overall Speedups

As can be seen in Table 2, LHScC retains the scalability of the home-based protocol and avoids most of the poorer performances of HLRC for SOR and Gauss, even though Gauss over LHScC with 32 nodes is slightly worse than over HLRC. LHScC also avoids the poorer performances of TreadMarks in PNN,

Table 2. Comparison of Speed-ups between TreadMarks (TM), Home-based LRC (HLRC) and Lazy Home-based ScC (LHScC)

Apps	4 nodes			8 nodes			16 nodes			32 nodes		
	TM	HLRC	LHScC	TM	HLRC	LHScC	TM	HLRC	LHScC	TM	HLRC	LHScC
PNN	3.9	3.9	3.9	7.1	6.9	7.5	8.8	9.5	13.0	4.8	8.2	17.2
B-H	2.1	2.0	2.1	2.4	2.4	2.7	1.8	2.7	3.1	1.5	3.0	3.4
IS-B	3.4	3.6	3.6	4.1	6.0	5.6	2.5	8.3	7.3	0.9	7.7	6.9
IS-L	2.3	2.9	2.7	1.4	2.6	2.9	0.5	1.6	2.0	0.1	0.9	1.1
FFT	1.4	0.9	1.3	2.1	1.2	2.0	2.9	1.9	2.9	2.9	3.2	3.2
SOR	3.4	1.9	3.2	6.0	2.4	5.4	11.3	3.4	9.4	15.5	3.9	13.4
Gauss	2.1	0.2	1.3	1.6	0.3	1.0	0.9	0.5	0.7	0.4	0.5	0.4

Barnes-Hut, IS-B and IS-L over more than 16 nodes. In particular, LHScC has shown significantly better performance with applications showing coarse-grained migratory memory access patterns in a critical section such as PNN and IS-L.

LHScC showed better scalability compared with the homeless protocol in TreadMarks. For example, over 4 nodes LHScC has no clear performance superiority over the other two protocols. However, over 32 nodes, the LHScC implementation was 3.6 times, 1.7 times, 7.7 times, and 11 times faster than TreadMarks in PNN, Barnes-Hut, IS-B and IS-L, respectively, and 3.4 times faster than HLRC in SOR.

The better performance of SOR and Gauss in TreadMarks can be explained by the lazy diffing technique [3]. The lazy diffing technique, which does not create a diff unless it is requested, is very effective for single writer applications without migratory memory access patterns. But this low protocol overhead is only applied to applications such as SOR or Gauss which have a very regular and exclusive memory access pattern. Migratory applications such as PNN and IS showed much better performance under the home-based protocol. In particular, significant improvements can be achieved in PNN and IS-L under LHScC by diff integration and efficient implementation of the update protocol during lock synchronisation.

The super slow-down shown in PNN, Barnes-Hut and the two IS over a large number of nodes in TreadMarks proves that the homeless protocol is vulnerable to the scalability problem. Up to 8 nodes, the homeless protocol showed relatively comparable performance with the two home-based protocols. However over 32 nodes, TreadMarks showed rapid slow-down except for FFT and SOR. A hot spot, garbage collection and diff accumulation are three major hindrances to scalability in the homeless protocol [5]. SOR and Gauss have the most benefit from the dynamic home migration technique, even though Gauss over 32 nodes showed the adverse effect of the technique due to its frequent barrier synchronisation use.

Generally, Table 3 indicates that the larger the data traffic, the poorer the performance, as strongly suggested in the two IS applications over TreadMarks, and SOR over HLRC. The exceptions are the difference between PNN over TreadMarks and HLRC, and Barnes-Hut over TreadMarks and the two home-based protocols. The reason for the exception of PNN is a frequent occurrence of a hot spot. The reason for Barnes-Hut is that even though it produced less data

Table 3. Comparison of Data Communication Traffic between TreadMarks (TM) and Two Home-based Protocols — HLRC and LHScC

Apps.	16 Processors						32 Processors					
	Number of Messages(K)			Amount of Traffic(MB)			Number of Messages(K)			Amount of Traffic(MB)		
	TM	HLRC	LHScC	TM	HLRC	LHScC	TM	HLRC	LHScC	TM	HLRC	LHScC
PNN	114.2	159.4	102.4	277.6	191.2	97.2	228.1	325.8	204.7	1011.0	403.9	190.3
B-H	2809.5	413.4	594.7	622.9	891.9	858.5	8696.5	792.2	1142.3	1211.0	1635.8	1595.4
IS-B	62.0	65.5	66.8	604.3	118.7	79.7	200.5	158.9	159.1	2327.6	285.6	166.0
IS-L	23.9	26.3	26.3	247.4	44.8	44.2	68.7	54.3	54.2	987.6	93.6	91.4
FFT	134.5	152.2	159.5	223.1	612.5	219.4	320.7	408.4	323.8	453.9	838.6	441.9
SOR	45.5	57.0	69.1	65.6	221.8	87.5	63.1	148.2	96.6	76.0	297.2	118.5
Gauss	120.7	321.5	123.7	124.4	1502.0	130.1	250.4	950.3	253.9	259.2	1756.0	263.7

traffic in the homeless protocol than in the home-based protocol, the number of messages produced was much more, for example nearly ten times more for 32 nodes compared with the home-based protocol. This shows that write-write false sharing applications such as Barnes-Hut over the homeless protocol will produce many diff requests sent to many nodes in order to construct the up-to-date copy of an invalid page. On the contrary, with the home-based protocol, only one page request sent to a home node of the page is sufficient to have the up-to-date copy of the page, which is much more efficient.

In the case of IS-B over 16 nodes, even though TreadMarks produces fewer messages compared to the other two systems, it produces much more data traffic—more than 7.6 times compared with LHScC. The data traffic statistics of IS-B over TreadMarks means that the average size of a message is quite large due to diff accumulation. For example, the average size of a packet in TreadMarks over 32 nodes is 11605 bytes, compared to 1043 bytes in LHScC. This data shows that in IS-B over TreadMarks, when a node requests the required diffs of a stale page from the last writers, the diffs received can be large due to diff accumulation.

To better identify the benefits of LHScC, we measured the times taken during lock synchronization (lock acquire and release) and critical sections in PNN over the three different protocols. As shown in Figure 2, Node 0 (N0) over TreadMarks suffers the most from lock contention due to a hot spot in N0. The hot spot in N0 occurs since only N0 writes on the shared pages at the end of barrier in each loop and those pages are accessed after the barrier by all nodes. Therefore all nodes request the pages from the last writer (N0) at the same time, which makes a hot spot. The hot spot becomes worse due to a migratory access pattern in PNN causing diff accumulation. Meanwhile, a hot spot is removed in HLRC since the shared pages are assigned evenly to the nodes. In PNN pages 1 to 4 are most written by all nodes. Since home assignment in HLRC is statically performed, nodes 1 to 4 share the responsibility of sending the most up-to-date four pages. However, eager home update after lock synchronization interrupts the main computation in nodes 1 to 4. Also nodes 1 and 2 which are two lock manager nodes have another interruption due to lock requests from other nodes. That is why nodes 1 to 4, and in particular, nodes 1 and 2, have relatively long lock synchronization times as shown in Figure 3 (note that the vertical scale in

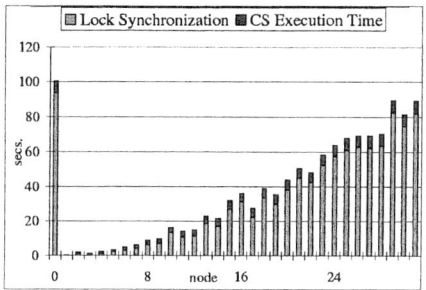 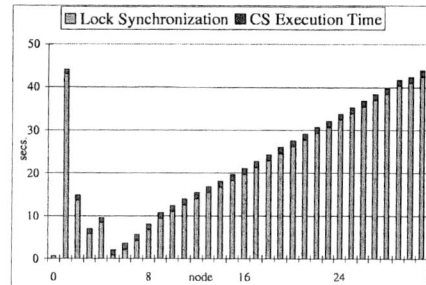

Fig. 2. Lock Synchronization and CS Times in PNN over TreadMarks

Fig. 3. Lock Synchronization and CS Times in PNN over HLRC

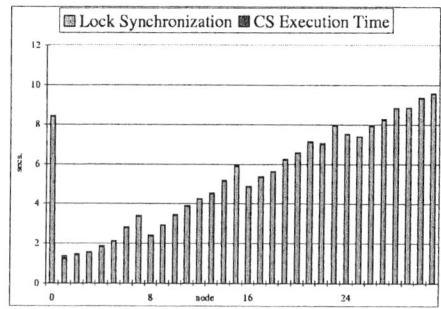

Fig. 4. Lock Synchronization and CS Times in PNN over LHScC

Figure 3 is different from that in Figure 2). Finally, LHScC greatly reduces the time taken for lock synchronization and critical section execution compared with the two other protocols as shown in Figure 4 (again, the vertical scale is reduced). The critical section execution time in LHScC is negligible, just 0.04 seconds on average, compared with 4.46 seconds and 1.29 seconds in TreadMarks and HLRC, respectively. In PNN over LHScC, N0 is a hot spot node as in TreadMarks, but this time the hot spot effect is weakened thanks to our diff integration technique.

5 Related Work

As far as we know, there are two all-software ScC DSM implementations similar to LHScC: Brazos and JiaJia. JiaJia [11] employs a home-based protocol. On the other hand, Brazos [12] is essentially a homeless DSM system. There are also several LRC DSM implementations that have similar ideas to LHScC. Below, we present the comparisons between those systems and LHScC.

Brazos is a homeless page-based all-software ScC DSM system. In Brazos, stale data are made up-to-date by receiving diffs from other nodes efficiently by exploiting multicast communication, compared to LHScC that uses both diffs and pages in order to update stale data without multicast support. Since Brazos uses multicasting for memory coherence during lock and barrier synchronisation, it reduces many complexities of implementing a ScC homeless DSM system. Even though Brazos claims that it uses an adaptation technique to update stale data between homeless and home-based protocols, it is dependent on a page's memory access pattern, and it still has to pay the adaptation and page ownership finding overheads since it uses essentially a homeless protocol. On the contrary, LHScC is much more efficient in combining the two protocols, as it is essentially a home-based protocol with the lazy home update and there is no overhead of combining the two protocols.

JiaJia is a home-based all-software ScC DSM system. However, it has no concept of the lazy home update and only uses the write-invalidate coherence protocol. Also, the implementation of ScC differs between JiaJia and LHScC. In JiaJia, a lock manager manages ScC coherence information so that the manager determines which pages should be invalidated for the next lock owner, whereas in LHScC each local node determines the required diffs for the next lock owner. In this way, LHScC can prevent an added burden on a lock manager and is a more fine-grained implementation of ScC. The implementation of ScC in JiaJia is not only inefficient compared to LHScC but also cannot prevent write-write false sharing inside a critical section due to the use of the write-invalidate protocol and the large page granularity.

ADSM [13] is a homeless all-software LRC DSM system in which two adaptations between single and multiple writer protocols, and write-update and write-invalidate protocols, are selectively used based on a page's memory access pattern. Basically it uses the invalidation protocol. However the update protocol is used for migratory pages inside a critical section and producer/multiple consumer pages during barrier synchronisation. Compared to the update protocol used in LHScC, CS data transferred by the update protocol in ADSM are limited to migratory pages only. Also the granularity of the update is the size of a page, whereas there is a more fine-grained diff size in LHScC.

There have been similar ideas of using the two coherence protocols selectively in order to implement a more efficient coherence protocol in a software DSM system. In KDSM [14], instead of using only the invalidation coherence protocol as in most home-based systems, the update coherence protocol is also used only at lock synchronisation times to solve an inefficient page fetch process occurring in a critical section. However, the efficiency obtained by their implementation is still limited due to the LRC model implementation. For example, at the time of release a node has to send modified data to the corresponding homes, which is unnecessary in LHScC. Also, in the case of diff accumulation, the efficiency of their protocol can be severely diminished.

Another similar use of a hybrid protocol is found in the Affinity Entry Consistency (AEC) system [15] even though the AEC system employs the homeless protocol only. In their system, a Lock Acquirer Prediction (LAP) technique is used to predict the next lock owner in order to prefetch required CS data to the next lock owner. We believe that the LAP technique is not required for most lock-based DSM applications since the next lock owner is already determined many times before the release time. When the next lock owner is not determined at the release time, employing the LAP technique leads to unnecessary updating if the prediction is wrong. Rather than updating eagerly based on prediction, it would be better to wait until the next lock owner is determined. In a similar scenario, at the release time, our implementation first creates diffs modified inside a critical section but waits until the next lock owner is determined. Upon receiving a lock request, the diffs previously created and stored are sent to the lock requester with the lock ownership. That is, LHScC eagerly creates and stores required CS diffs, but can lazily transfer those diffs when the next lock request is received after the diff creation.

Finally, Orion [16] is a home-based LRC DSM system. It has a different approach to other adaptation techniques. It exploits a home node which collects all data access information from other non-home nodes. When other non-home nodes are detected as the frequent readers of its home pages, the home node notifies all non-home nodes about the frequent reader nodes. Next, when a non-home node sends diffs to the home node, it also updates the frequent reader nodes, hoping that the diffs are accessed by them.

6 Conclusions

In this paper, we presented the design and implementation of our novel lazy home-based ScC protocol (LHScC). LHScC is different from a conventional home-based protocol in that home update time is delayed until the next barrier time. LHScC combines diff-based update in the homeless protocol and page-based update in the home-based protocol. Our implementation of LHScC includes a dynamic home assignment scheme and a diff integration technique in order to solve wrong home assignment and diff accumulation problems, respectively. Our performance evaluation shows that LHScC retains good scalability of the home-based protocol and removes a static home assignment problem.

References

1. Bershad, B., Zekauskas, M., Sawdon, W.: The Midway distributed shared memory system. In: Proc. of the IEEE Computer Conference (Compcon). (1993) 528–537
2. Iftode, L., Singh, J.P., Li, K.: Scope consistency: A bridge between release consistency and entry consistency. In: Proc. of the 8th ACM Annual Symp. on Parallel Algorithms and Architectures (SPAA'96). (1996) 277–287
3. Keleher, P., Dwarkadas, S., Cox, A.L., Zwaenepoel, W.: Treadmarks: Distributed shared memory on standard workstations and operating systems. In: Proc. of the Winter 1994 USENIX Conference. (1994) 115–131
4. Zhou, Y., Iftode, L., Li, K.: Performance evaluation of two home-based lazy release consistency protocols for shared memory virtual memory systems. In: Proc. of the 2nd Symp. on Operating Systems Design and Implementation (OSDI'96). (1996) 75–88

5. Yu, B.H., Huang, Z., Cranefield, S., Purvis, M.: Homeless and home-based lazy release consistency protocols on distributed shared memory. In: Proceedings of the 27th Conference on Australasian Computer Science, Australian Computer Society, Inc. (2004) 117–123
6. Yu, B.H., Werstein, P., Cranefield, S., Purvis, M.: Performance improvement techniques for software distributed shared memory. In: Proceedings of the 11th International Conference on Parallel and Distributed Systems (ICPADS 2005), IEEE Computer Society Press (2005)
7. Hu, W., Shi, W., Tang, Z.: Home migration in home-based software DSMs. In: Proc. of the 1st Workshop on Software Distributed Shared Memory (WSDSM'99). (1999)
8. Fang, W., Wang, C.L., Zhu, W., Lau, F.C.: A novel adaptive home migration protocol in home-based DSM. In: Proc. of the 2004 IEEE International Conference on Cluster Computing (Cluster2004). (2004) 215–224
9. Cheung, B., Wang, C., Hwang, K.: A migrating-home protocol for implementing scope consistency model on a cluster of workstations. In: The 1999 International Conference on Parallel and Distributed Processing Techniques and Applications (PDPTA'99), Las Vegas, Nevada, USA. (1999) 821–827
10. Adve, S.V., Cox, A.L., Dwarkadas, S., Rajamony, R., Zwaenepoel, W.: A comparison of entry consistency and lazy release consistency implementations. In: Proc. of the 2nd IEEE Symp. on High-Performance Computer Architecture (HPCA-2). (1996) 26–37
11. Hu, W., Shi, W., Tang, Z.: JIAJIA: An SVM system based on a new cache coherence protocol. In: Proc. of the High-Performance Computing and Networking Europe 1999 (HPCN'99). (1999) 463–472
12. Speight, W.E., Bennett, J.K.: Brazos: A third generation DSM system. In: Proc. of the USENIX Windows NT Workshop. (1997) 95–106
13. Monnerat, L., Bianchini, R.: Efficiently adapting to sharing patterns in software DSMs. In: HPCA '98: Proceedings of the The Fourth International Symposium on High-Performance Computer Architecture, Washington, DC, USA, IEEE Computer Society (1998) 289–299
14. Yun, H.C., Lee, S.K., Lee, J., Maeng, S.: An Efficient Lock Protocol for Home-Based Lazy Release Consistency. In: Proceedings of the 3rd International Workshop on Software Distributed Shared Memory System, Brisbane, Australia, IEEE Computer Society (2001) 527–532
15. Seidel, C.B., Bianchini, R., de Amorim, C.L.: The affinity entry consistency protocol. In: ICPP '97: Proceedings of the international Conference on Parallel Processing, IEEE Computer Society (1997) 208–217
16. Ng, M.C., Wong, W.F.: Orion: An adaptive home-based software distributed shared memory system. In: 7th International Conference of Parallel And Distributed System (ICPADS 2000), Iwate, Japan, IEEE Computer Society (2000) 187–194

Grid Enablement of the Danish Eulerian Air Pollution Model

Cihan Sahin, Ashish Thandavan, and Vassil N. Alexandrov

Centre for Advanced Computing and Emerging Technologies,
University of Reading, Reading RG2 7HA
{c.sahin, a.thandavan, v.n.alexandrov}@rdg.ac.uk

Abstract. The Danish Eulerian Air Pollution Model (DEM model), developed by the Danish National Environmental Research Institute, simulates the transport and distribution of air pollutants in the atmosphere. This paper describes the deployment of the DEM model as a grid service by using Globus Toolkit version 3.2 (GT3). The project starts by porting the DEM model onto the IBM pSeries machine at the University of Reading and investigates a coarse-grain grid implementation. Components including resource management, grid service development and client application development are designed and implemented within a working grid infrastructure.

Keywords: Grid, Globus Toolkit 3, OGSA, Large Scale Air Pollution Model, MPI.

1 Introduction

Large-scale air pollution models are used to simulate the variation and the concentration of emissions into the atmosphere and can be used to develop reliable policies and strategies to control the emissions [1][2]. These models are successfully described mathematically and lead to large systems of ordinary differential equations of orders up to 80 millions [2].

Operational use of an air pollution model requires the model to run within a limited time period, hence the increased need for peak computational power and throughput. Beside the operational use, continuous improvement of the model needs to be done by running experiments because of reasons like improvements to the physical and chemical representations of the model, applying different numerical methods, testing different algorithms and parallelization methods and the re-analysis of the improved model.

Operational air pollution models and their experimental runs usually exceed the computational capabilities of an average-sized organization and the computing resources may act as a limiting factor for further development. This limitation may be overcome either by purchasing more powerful computers or using computing resources offered by other organizations. Using resources from other organizations is not an easy task as the terms should be clearly defined and questions like what to use, when to use, and how to use should be identified and completely addressed by a structured framework. From such a framework emerges a new computing technology – *Grid Computing*[4].

The demand for resources by computationally intensive applications like a large-scale air pollution model can be addressed by grid technologies, which provide a pool of heterogeneous computing resources for scientists to carry out their experiments. The Globus Toolkit [9], developed by Globus Alliance [8], offers middleware to build grids and grid services. This paper investigates the porting of the DEM air pollution model onto the Open Grid Services Infrastructure (OGSI) [6], using the Globus Toolkit version 3.2.1, which was the latest stable release at the time this work was undertaken.

The work described in this paper arises from the work done for the OGSA Testbed project, which was a one year UK E-Science project funded by EPSRC. The aims were to install and evaluate the middleware based on the Open Grid Services Architecture (OGSA) on resources owned by the partners and deploy various applications on this testbed.

2 Danish Eulerian Air Pollution Model

The large scale air pollution model used in this project is the Danish Eulerian Model (DEM model) developed by National Environmental Research Institute (NERI) in Denmark. The different contituent physical and chemical processes [2][3] are illustrated by Figure 1.

These processes lead to a system of partial differential equations. Partial differential equations are difficult to solve computationally so splitting procedures are applied, which produces five sub-models where different numerical methods can be applied to exploit available computational resources more efficiently [3].

The space domain of the DEM air pollution model is 4800 km x 4800 km, with Europe at the centre. This domain is divided horizontally into 96 x 96

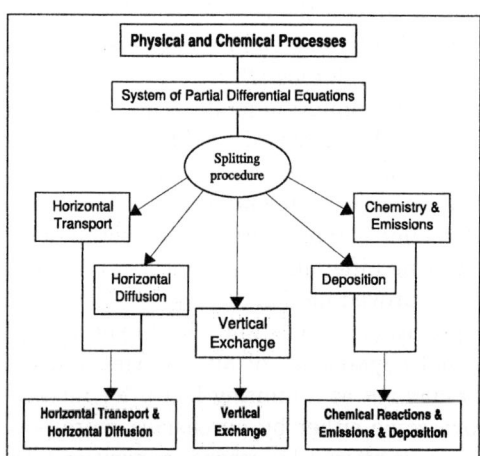

Fig. 1. Physical and chemical processes of the DEM model

squares and 480x480 squares in coarse and fine resolutions respectively. Two different vertical layers are used - 1-layer and 10-layer - which corresponds to 2-D and 3-D versions respectively. The DEM model is implemented as a parallel application by using OpenMP and MPI in NERI, Denmark. This project uses the MPI implementation.

3 Porting of the Code and Results from Runs

The ACET centre at the University of Reading has an IBM pSeries cluster comprising of four p655 nodes, with eight 1.5 GHz POWER4 processors and 16GB memory each and running AIX 5.2 which uses Loadleveler as batch scheduler. In the remainder of this paper, the IBM parallel computer is referred to as an 'HPC resource'. As the DEM model was originally developed and tested on a grid of Sun computers [3], it was necessary to recompile it for the HPC resource at the University of Reading.

Initially, during execution, the model used to hang just after starting. As illustrated by Figure 2, each process sends out and receives two sets of data, except the processes with the lowest and highest ranks. In the code, the above communication is provided with several MPI_send and MPI_receive functions [5]. At the end of each communication, MPI_Waitall is also set to maintain synchronization.

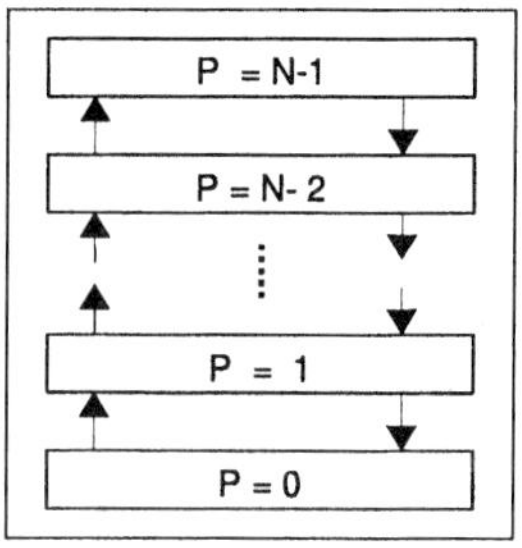

The original version of the model waits for four communication operations to be completed per process before it carries on with the execution – which does not apply for the zeroth and the N-1th process. In order for the code to successfully complete execution, it was modified accordingly. The results from runs with different number of processors listed below demonstrate the scalability of the code on the HPC resource.

Fig. 2. Communication between the processes

Table 1 presents the DEM run results for the two dimensional version with the low resolution mode over a year (1997). It demonstrates that the DEM code

Results from runs of the DEM (in seconds)

Table 1. Low Resolution

Sub-model	N = 1	N = 2	N = 4	N = 8	N = 16
Wind + Sinks	120.6	61.5	34.4	18.1	19.8
Advection	1362.5	654.1	317.8	165.7	84.9
Chemistry	5884	2971.2	1483.7	778.1	411.4
Outso2	375.9	305.2	280.2	319.3	343.2
Comm	0.08	51.9	45.1	105.1	283.9
Total	7749.4	4051	2168.2	1388.5	1143.2

Table 2. High Resolution

Sub-Model	N=8
Wind + Sinks	632.2
Advection	5253.5
Chemistry	21439.4
Outso2	9398.2
Comm	3367.7
Total	42082.4

scales well on the HPC resource - the advection and chemistry sub-models scale almost linearly. Table 2 shows how the DEM code can be very computationally demanding when it runs in high resolution mode. The communication in the tables refers to the communication along the inner boundaries. The number of inner boundaries increase with increasing number of processors.

4 Globus Toolkit v3 on AIX

From the beginning, we encountered problems with compiling and installing the Globus Toolkit v3 [7] on our resource. As GT3 packages are needed on the resource for grid enablement of the application, an alternate method had to be designed and implemented. A third machine with the necessary services running was introduced between the clients and the HPC resource.

As illustrated in Figure 3, the Globus machine, on which the Globus Toolkit v3.2 is installed, handles all the grid-based services including the security, lifecycle management and data transfers between clients and the HPC resource. This design further needs the development of a resource management strategy between the Globus machine and the HPC resource, which includes job control and monitoring mechanisms. The communication between the globus machine and the HPC resource are provided by two pairs of client/server applications. One pair deals with job control, forwarding the job requirements to the HPC resource and returning the relevant job information back to the DEM grid service. An XML type resource specification language is used to communicate job requirements. Also, a second pair of client/server applications keeps track of the jobs running on the HPC resource. This monitoring service provides the status of the running jobs and the system status of the HPC resource to the grid service. This design acts like a simple replacement of the resource management (WS-GRAM [10]) service of GT3 on the HPC resource.

The overall system consists of four main parts as shown in Figure 3. Component 1 is the actual DEM grid service implementation through which clients can interact with the HPC resource. Component 2 represents the communication

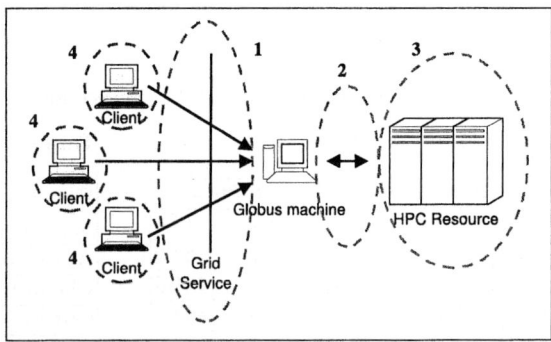

Fig. 3. Proposed solution and its main components

between the Globus machine and the HPC resource. Component 3 represents the interface to the local scheduler on the resource, where actual job control and monitoring is carried out. Component 4 represents the client applications.

5 DEM Grid Service

The DEM grid service component of the design enables the communication between clients and the Globus machine and facilitates remote method invocation by the clients. The service interface specifies what functionality the service offers to the outer world. These include the job control functions (submission, cancellation, hold, release), data transfers (in both directions) and monitoring functions (job and system status). The high-level class diagram of the service implementation is shown in Figure 4.

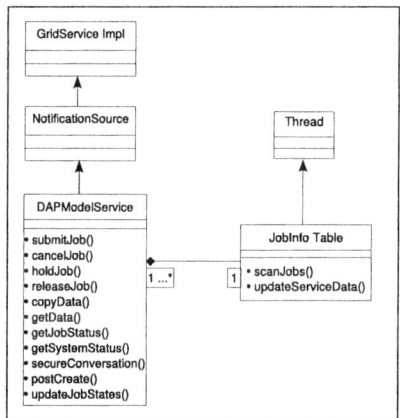

Fig. 4. Service implementation

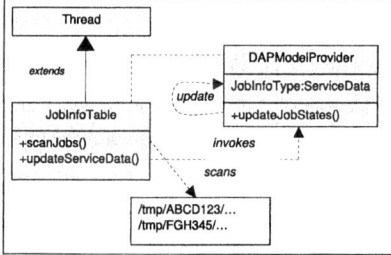

Fig. 5. Notification mechanism

5.1 Notifications and Server Side Security

Status information about a running job is made available by the use of service data elements (SDEs) and the notification service that keeps the client application informed. A threaded object, which is fired up by the grid service regularly checks the available information about a running job (status, current output status, job description file and job identification number) and updates the corresponding service data element whenever a change occurs. This functionality is offered by a GT3 service and is enough to keep any client who listens to the service data element updated. All that a client has to do is to implement the deliverNotification method of grid notifications to receive any change made on the service data element.

The grid service uses the authentication and authorization services provided by GT3 to authenticate and authorize clients as well as to encrypt messages.

The Globus Toolkit's GSI security is added to the grid service by configuring the deployment description file accordingly. The service uses GT3's default security configuration file (securityConfig) as it requires a secure conversation for all the service. The service uses the `gridmap` file for authorizing users. As notification messages are not direct method invocations to operations determined in the service interface, they have to be encrypted (on the message level) as well. This is provided in the service implementation code.

6 Resource Management

Due to the unavailability of the necessary GT3 services on the HPC resource, the resource management functionality - job control (submission, cancellation, hold, release) and job monitoring and data transfer (between Globus machine and HPC resource) - has to be replaced by a suitable alternative (Figure 6).

6.1 Job Control and Data Transfer

The arrows numbered 1 to 8 in Figure 6 show the communication steps involved in the job control mechanism and data transfer. **A** denotes the client program invoked by the grid service and **B** denotes the multi-threaded server listening for the clients on the HPC resource. The steps are listed as below:

1. Client application submits a job request to the DEM model grid service
2. DEM grid service fires up a client application (communication client **A**)
3. Client **A** talks to the server **B** running on the HPC resource via TCP
4. Server **B** passes on the client requirements to the local scheduler
5. Local scheduler returns a job identification number (job ID) and a message
6. The Job ID and any relevant messages are passed back to client **A**
7. Client **A** passes on the information to DEM grid service and quits
8. DEM grid service passes on the information back to the client application

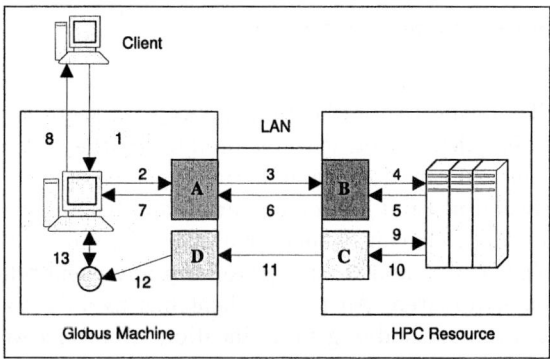

Fig. 6. Job control, monitor and data transfer mechanism between Globus machine and the HPC resource

At this point, the client application has the all information required to identify its job(s) on the HPC resource. The server **B** is multi-threaded and must be running all the time as a daemon on the HPC resource. It receives the messages from the client **A** and invokes a perl script, which maps the request to Loadleveler. The client/server pair (**A** and **B**) is also capable of transferring text and binary files between the Globus machine and the HPC resource.

6.2 Job Monitoring

An important aspect of the design is to provide job and system monitoring information to the grid service to be passed on to clients that may request it. One could argue that the client/server pair **A** and **B** could also be used for providing the status information, but for robustness reasons explained shortly, another pair of client/server applications are developed.

A typical run of the DEM application varies from a few hours to days. The DEM grid service, which keeps the communication client alive, may not be up all through the execution time. As every communication pair would need to be kept alive, there may be hundreds of connections between the globus machine and the HPC resource at any given time and cause the server to slow down or disconnect frequently. For each job, a different call to Loadleveler is needed. This can cause Loadleveler's status table to crash and its scheduler to stop scheduling new jobs. In Figure 6 the arrows numbered 9 to 13 shows the communication steps for the job and system monitoring. These are :

1. Client **C** on the HPC resource periodically requests job and system status from the resource's local scheduler, Loadleveler
2. LoadLeveler returns the job status to the client **C**
3. **C** sends the job and system status information through a TCP network connection to the Globus machine
4. Server **D** on the Globus machine receives job and system status as a serialised object and writes to the local disk
5. The DEM grid service's relevant methods can look up the job status or system status by reading the serialised object

The server **D** is designed as a multi-threaded application to be able to handle more then one resource. It thus provides a very flexible and easily extendable way to handle different schedulers running on different resources.

7 Client Application

The login interface of the Client Application is used to generate a proxy certificate. A passphrase must be issued at this stage to obtain a valid proxy. The user can select the resource and the application at this stage as well. A threaded object, `ClientListener`, is fired up to listen for the service data elements reporting information and state of the jobs currently running under the client's username. It implements a standard grid service notification interface called

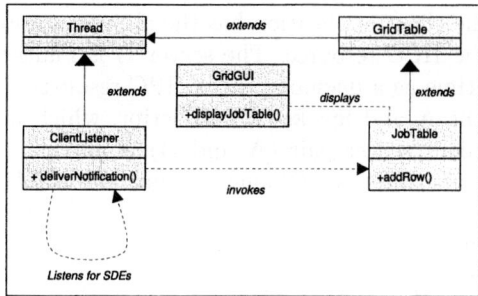

Fig. 7. Conceptual level diagram to update job states on GUI

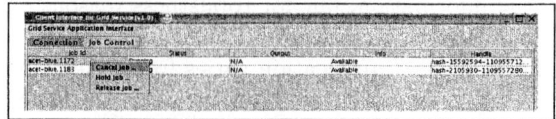

Fig. 8. The list of executing jobs

`DeliverNotification`. Whenever a service data element's value changes on the service side, this interface delivers it to the clients subscribed to it and displayed by a threaded object (`JobTable`) (Figure 7).

On the client application's interface, each job has job ID, job status, output status and information cells. They are updated seamlessly as and when a change occurs. `JobTable` provides the control of the jobs as well as collects output information or job description information from the globus machine. Control of the currently running jobs is provided by interacting with listed running jobs as shown in Figure 8.

7.1 Client Side Security

The contects of the SOAP messages sent to the grid service have to be encrypted. Another security issue is the notification messages received from the server. The client application must ensure that the notification messages it is listening for are from the service it subscribes to. This mechanism is provided during the subscription of the client to the service. Prior to subscription, the client authenticates the service using the `gridmap` file.

7.2 Job Configuration and Submission Interface

The most important functionality of the application is to be able to submit jobs to the HPC resource. The client can enter the specifications of the job submission like the input data, execution environment (serial or parallel job, number of processors required, scheduler type, etc.), package information or output data as shown in Figure 9.

The submission process involves first copying the relevant files followed by the invocation of the remote service method `submitJob` (Figure 10). The `submitJob`

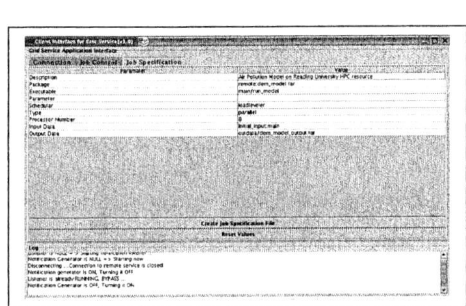 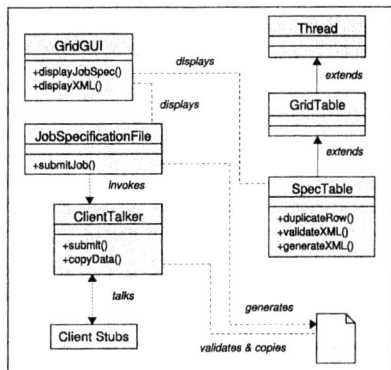

Fig. 9. Job submission interface of the application

Fig. 10. Job submission stages

method returns the job ID if submission is successful. The submitted job information immediately appears on the incomplete-job table for future interaction.

8 Conclusions

In this project, the Danish Eulerian Air Pollution Model was ported to Open Grid Service Infrastructure (OGSI). Porting the DEM application to OGSI required the deployment of the DEM application as a grid service by using an OGSI-compliant software toolkit. Globus Toolkit version 3.2 was used as it implements the OGSI specifications. The project was carried out in three main stages.

The first stage was porting the DEM application to the HPC resource at Reading University which included the compilation and the successful running of the DEM model code on the HPC resource. The second stage was enabling the DEM application as a grid service using the Globus Toolkit version 3.2 on the HPC resource. Due to problems with the installation of GT3 on the resource [7], an alternative design was developed and implemented. The proposed design had another machine acting as a gatekeeper and facilitating communication between the clients and the HPC resource. Furthermore, a resource management system was needed to provide the communication and data transfers between the Globus machine and the HPC resource. This solution allows wider grid implementations by plugging in different compute resources. As the client application was designed and implemented to handle any application, it can easily be used as a generic grid client application. Our HPC resource is the only computing resource used in the project at Reading. However, the overall design of the system is flexible enough for it to include new resources with a minimum effort. Hence, a recource discovery mechanism, such as GT3's Monitoring and Discovery System (MDS), could be implemented into the solution.

There were major problems and difficulties throughout the the project. It was not possible to install GT3 on our HPC resource leading to development of the new design. OGSI, and thus GT3, is complicated to work with due to limited documentation, examples and tutorials. This led to rather steep learning and development curves.

References

1. Zlatev Z.: *Computer treatment of large air pollution models* Kluwer Academic Publishers, Dorsrecht-Boston-London (1995)
2. Zlatev, Z., Dimov, I. and Georgiev, K.: *Three-dimensional version of the Danish Eulerian Model* Zeitschrift für Angewandte Mathematik und Mechanik, Vol. 76 **473-476** (1996) S4
3. Alexandrov, V.N., Owczarz W., Thomsen, P.G., Zlatev, Z.: *Parallel runs of a large air pollution model on a grid of sun computers* Mathematics and Computers in Simulation, **65** (2004).
4. Foster, I., Kesselman, C., Tuecke, S.: *The Anatomy of the Grid: Enabling Scalable Virtual Organizations* Intl J. Supercomputer Applications, **15(3)** 2001
5. Pacheco P.S.: *Parallel Programming with MPI* San Fransisco (1997)
6. Tuecke S., Czajkowski K., Foster I., Frey J., Graham S., Kesselman C., Maquire T., Sandholm T., Snelling D., Vanderbilt T.: *Open Grid Services Infrastructure (OGSI) version 1.0* Global Grid Forum Draft Recommendation (2003)
7. Thandavan A., Sahin C., Alexandrov V.N.: *Experiences with the Globus Toolkit on AIX and deploying the Large Scale Air Pollution Modelling application as a grid service* Cracow Grid Workshop, Cracow, Poland (2004)
8. Foster, I., Kesselman, C.: *The Globus Project: A Status Report* Proc. IPPS/SPDP '98 Heterogeneous Computing Workshop, **4-18** (1998)
9. Foster, I., Kesselman, C.: *Globus: A Metacomputing Infrastructure Toolkit* Intl J. Supercomputer Applications, **11(2):115-128**, (1997)
10. WS-GRAM website *The Globus Alliance* http://www-unix.globus.org/toolkit/docs/3.2/gram/ws/index.html

Towards a Bayesian Statistical Model for the Classification of the Causes of Data Loss

Phillip M. Dickens[1] and Jeffery Peden[2]

[1] Department of Computer Science, University of Maine, Orono Maine, 04429,
dickens@umcs.maine.edu
[2] Department of Mathematics and Computer Science, Longwood University,
Farmville, Virginia
jpeden@mcs.longwood.edu

Abstract. Given the critical nature of communications in computational Grids it is important to develop efficient, intelligent, and adaptive communication mechanisms. An important milestone on this path is the development of classification mechanisms that can distinguish between the various causes of data loss in cluster and Grid environments. The idea is to use the classification mechanism to determine if data loss is caused by contention within the network or if the cause lies outside of the network domain. If it is outside of the network domain, then it is not necessary to trigger aggressive congestion-control mechanisms. Thus the goal is to operate the data transfer at the highest possible rate by only backing off aggressively when the data loss is classified as being network related. In this paper, we investigate one promising approach to developing such classification mechanisms based on the analysis of the patterns of packet loss and the application of Bayesian statistics.

1 Introduction

Computational Grids create large-scale distributed systems by connecting geographically distributed computational and data-storage facilities via high-performance networks. Such systems, which can harness and bring to bear tremendous computational resources on a single large-scale problem, are becoming an increasingly important component of the national computational infrastructure. At the heart of such systems is the high-performance communication infrastructure that allows the geographically distributed computational elements to function as a single (and tightly-coupled) computational platform. Given the importance of Grid technologies to the scientific community, research projects aimed at making the communication system more efficient, intelligent, and adaptive are both timely and critical.

An important milestone on the path to such next-generation communication systems is the development of a classification mechanism that can distinguish between causes of data loss in cluster/Grid environments. The idea is to use the classification mechanism to respond aggressively only when the loss is classified as being network related. Thus the communication system can take full advantage of the underlying bandwidth when system conditions permit, can back off in response to observed (or predicted) contention *within the network*, and can accurately distinguish between these two situations. This research is developing such a mechanism.

The approach we are pursuing is to analyze what may be termed *packet-loss signatures*, which show the distribution (or pattern) of those packets that successfully traversed the end-to-end transmission path and those that did not. These signatures are essentially large selective-acknowledgment packets that are collected by the receiver and delivered to the sender upon request. We chose the name "packet-loss signatures" because of the growing set of experimental results [15, 16] showing that different causes of data loss have different "signatures".

The question then is how to quantify the differences between packet-loss signatures in a way that can be used by a classification mechanism. The approach we have chosen is to differentiate based on the underlying structure of the packet-loss signatures. The metric we are using to quantify this structure is the *complexity* of the packet-loss signatures.

In this paper, we show how complexity measures are computed and demonstrate how such measures can be mapped to the dominant cause of packet loss. We then show that the statistical properties of complexity measures are different enough to be used as the basis for a sophisticated probability model for the causes of packet loss based on the application of Bayesian statistics. In this paper we show how such a model can be developed and used; its implementation and (real-time) evaluation will be the focus of forthcoming papers.

The major contribution of this paper is the development of an approach by which the control mechanisms of a data transfer system can determine the dominate cause of data loss in real time. This opens the door to communication systems that are made significantly more intelligent by their ability to respond to data loss in a way that is most appropriate for the particular cause of such loss. While the longer-term goal of this research is to develop and analyze a wide range of responses based on the causes of data loss, the focus of this paper is on the development of the classification mechanism itself. This paper should be of interest to a large segment of the Grid community given the interest in and importance of exploring new approaches by which data transfers can be made more intelligent and efficient.

The rest of the paper is organized as follows. In Section 2, we provide background information. In Section 3, we discuss related work. In Section 4, we describe the computation of complexity measures. In Section 5, we describe the experimental design. In Section 6, we present our experimental results. In Section 7, we discuss the application of Bayesian statistics to the problem of determining the dominant cause of packet loss. In Section 8, we provide our conclusions and current research directions.

2 Background

The test-bed for this research is FOBS[1]: a high-performance data transfer system for computational Grids [11-14]. FOBS is a UDP-based data transfer system that provides reliability through a selective-acknowledgment and retransmission mechanism. It is precisely the information contained within the selective-acknowledgment packets that is collected and analyzed by the classification mechanism. FOBS uses a rate-based congestion control mechanism that is similar to the TCP friendly rate control

---

[1] Fast Object-Based data transfer System.

protocol specification (TFRC) [3]. The primary difference is that FOBS waits until the end of the current transmission window before reacting to data loss while the TFRC protocol responds immediately to a single loss event.

There are three advantages of using FOBS as the test-bed for this research. First, FOBS is an application-level protocol. Thus the control mechanisms can collect, synthesize, and leverage information from a wide variety of sources. Currently, values from the hardware counters (the /proc pseudo-file system) and the complexity measures are utilized in the classification mechanism. However, we expect that other information sources will be developed as the research progresses. Second, the complexity measures are obtained as a function of a constant sending rate. Thus the values of the variables collected are (largely) unaffected by the behavior of the algorithm itself. Third, FOBS is structured as a feedback-control system. Thus the external data (e.g., the complexity measures) can be analyzed at each control point, and this data can be used to determine the duration and sending rate of the next control interval.

In this paper, we focus on distinguishing between contention for network resources and contention for CPU resources. This distinction is important for two reasons. First, contention for CPU cycles can be a major contributor to packet loss in UDP-based protocols. This happens, for example, when the receiver's socket-buffer becomes full, additional data bound for the receiver arrives at the host, and the receiver is switched out and thus unavailable to pull such packets off of the network. The receiver could be switched out for a number of reasons including preemption by a higher priority system process, interrupt processing, paging activity, and multi-tasking. To illustrate the importance of this issue, consider a data transfer with a sending rate of one gigabit per second and a packet size of 1024 bytes. Given these parameters, a packet will arrive at the receiving host around every 7.9 micro-seconds, which is approximately the amount of time required to perform a context switch on the TeraGrid systems [1] used in this research (as measured by Lmbench [20]). Thus even minor CPU/OS activity can cause packets to be lost.

The second reason this distinction is important is that data loss resulting from CPU contention is outside of the network domain and should not be interpreted and treated as growing network contention. This opens the door for new and significantly less aggressive responses for this class of data loss, and also explains the reason for reacting to data loss on a longer timescale that that suggested by the TFRC protocol: We want to look at the patterns of packet loss to determine the cause of such loss. While this approach is not appropriate for Internet-1 networks, we believe that it is quite reasonable for very high-performance networks such as the TeraGrid. We have found that 16 round-trip times (RTTs, approximately 10,000 packets at a sending rate of one gigabit per second) are sufficient for the complexity analysis. In the results presented here, the control interval was on the order of 120 RTTs. As noted above, however, the control interval and sending rate can be modified dynamically.

It is important to point out that we used contention at the NIC as a proxy for network contention for two reasons: First, while it was relatively easy to create contention at the NIC, it was extremely difficult to create contention within the 40 gigabit per second networks that connect the facilities on the TeraGrid. Second, we have collected empirical evidence supporting the idea that, at least in some cases, the complexity measures taken from intermediate gigabit routers of other networks are quite similar to those measured at the NIC [16]. This issue is discussed in more detail in Section 4.1.

3 Related Work

The issue of distinguishing between causes of data loss has received significant attention within the context of TCP for hybrid wired/wireless networks (e.g., [4, 5, 7, 8, 19]). The idea is to distinguish between losses caused by network congestion and losses caused by errors in the wireless link, and to trigger TCP's aggressive congestion control mechanisms only in the case of congestion-induced losses. This ability to classify the root cause of data loss, and to respond accordingly, has been shown to improve the performance of TCP in this network environment [4, 19]. These classification schemes are based largely on simple statistics on observed round-trip times, observed throughput, or the inter-arrival time between ACK packets [7, 9, 19]. Debate remains, however as to how well techniques based on such simple statistics can classify loss [19]. Another approach being pursued is the use of Hidden Markov Models where the states are characterized by the mean and standard deviation of the distribution of round-trip times [19]. Hidden Markov Models have also been used to model network channel losses and to make inferences about the state of the channel [21].

Our research has similar goals, although we are developing a finer-grained classification system to distinguish between contention at the NIC, contention in the network, and contention for CPU resources. Also, complexity measures of packet-loss signatures appear to be a more robust classifier than (for example) statistics on round-trip times, and could be substituted for such statistics within the mathematical frameworks established in these related works. Similar to the projects discussed above, we separate the issue of classification of root cause(s) of data loss from the issue of implementing responses based on such knowledge.

4 Diagnostic Methodology

The packet-loss signatures can be analyzed as time series data with the objective of identifying diagnostics that may be used to characterize causes of packet loss. A desirable attribute of a diagnostic is that it can describe the dynamical structure of the time series. The approach we are taking is the application of *symbolic dynamics* techniques, which have been developed by the nonlinear dynamics community and are highly appropriate for time series of discrete data. As discussed below, this approach to classifying causes of packet loss works because of the differing timescales over which such losses occur.

In symbolic dynamics [18], the packet-loss signature is a sequence of symbols drawn from a finite discrete set, which in our case is two symbols: 1 and 0. One diagnostic that quantifies the amount of structure in the sequence is *complexity*. There are numerous ways to quantify complexity. In this discussion, we have chosen the approach of d'Alessandro and Politi [10], which has been applied with success to quantify the complexity and predictability of time series of hourly precipitation data [17].

The approach of d'Alessandro and Politi is to view the stream of 1s and 0s as a language and focus on subsequences (or *words*) of length **n** in the limit of increasing values of **n** (i.e., increasing word length). First-order complexity, denoted by C_1, is a measure of the richness of the language's vocabulary and represents the asymptotic growth rate of the number of *admissible words* of fixed length **n** occurring within the string as **n** becomes large. The number of admissible words of length **n**, denoted by **Na(n)**, is simply a count of the number of distinct words of length **n** found in the given sequence. For example, the string **0010100** has **Na(1)** = 2 (0,1), **Na(2)** = 3 (00,01,10), **Na(3)** = 4 (001, 010, 101, 100). The *first-order complexity* (C_1) is defined as

$$C_1 = \lim_{n \to \infty} \log_2 (Na(n) / n). \qquad (1)$$

The first-order complexity metric characterizes the level of randomness or periodicity in a string of symbols. A string consisting of only one symbol will have one admissible word for each value of **n**, and will thus have a value of $C_1=0$. A purely random string will, in the limit, have a value of $C_1=1$. A string that is comprised of a periodic sequence, or one comprising only a few periodic sequences, will tend to have low values of C_1.

4.1 Rational for Complexity Measures

Complexity measures work because of the different timescales at which loss events occur. In the case of contention for CPU cycles due to multi-tasking, packets will start being dropped when the data receiver is switched out and its data buffer becomes full. It will continue to drop packets until the receiver regains control of the CPU. The amount of time a receiver is switched out will be on the order of, for example, the time-slice of a higher-priority process for which it has been preempted or the aggregate time-slice of the processes ahead of it in the run queue. Such events are measured in milliseconds, where, as discussed above, the packet-arrival rate is on the order of microseconds. Therefore if contention for CPU cycles is the predominant cause of data loss, the packet-loss signatures will consist of strings of 0's created when the receiver is switched out, followed by strings of 1's when it is executing and able to pull packets off of the network. Thus the packet-loss signature will be periodic and will thus have a low complexity measure. Different operating systems will have different scheduling policies and time-slices, but the basic periodic nature of the events will not change.

Data loss caused by contention at the NIC, however, operates on a much smaller timescale, that being the precise order in which the bit-streams of competing packets arrive at and are serviced by the hardware. The packet-loss signatures represent those particular packets that successfully competed for NIC resources and those that did not. This is an inherently random process, and this fact is reflected in the packet-loss signatures.

An important question is whether contention at the NIC sheds any light on the packet-loss signatures that are generated by collisions at intermediate routers in the transmission path. This is obviously dependent on the particular queuing discipline of each router. However, routers that employ a Random Early Detection (RED) policy can be expected to produce packet-loss signatures very similar to those produced by contention at the NIC. This is because once the average queue length surpasses a given threshold, the router begins to drop (or mark) packets, and the particular packets to be dropped are chosen randomly. Also, we have developed a small set of experimental results showing that packet-loss signatures caused by contention at intermediate routers are largely equivalent to those observed with contention at the NIC [15, 16]. This is *not* to say however that the aggregate policies of a large set of intermediate routers will not produce packet-loss signatures that will be misinterpreted by the classification mechanism. This is an important issue and is the focus of current research.

5 Complexity Analysis

The classification mechanism uses a sliding window of length **n** to search for the admissible words of length **n** in the signature. If, for example, it is searching for words of length **n = 3**, then the first window would cover symbols 0-3, the second window would cover symbols 1-4, and so forth. Recall that the symbols are either 1 or 0, and represent the received/not-received status of each packet in the transmission window. As each word is examined, an attempt is made to insert it into a binary tree whose branches are either 1 or 0. Inserting the word into the tree consists of following the path of the symbol string through the tree until either (1) a branch in the path is not present or (2), the end of the symbol string is reached. If a branch is not present, it is added to the tree and the traversal continues. In such a case, the word has not been previously encountered and the number of admissible words (for the current word size) is incremented. If the complete path from the root of the tree to the end of the symbol string already exists, then the word has been previously encountered and the count of admissible words is unchanged. Given the count of admissible words of length **n**, it is straight-forward to calculate the complexity value for that word length. The calculation of the complexity measures is performed in real-time with a computational overhead of approximately 15%.

6 Experimental Design

All experiments were conducted on the TeraGrid [1]: a high-performance computational Grid that connects various supercomputing facilities via networks operating at speeds up to 40 gigabits per second. The two facilities used in these experiments were the Center for Advanced Computing Research (CACR, located at the California Institute of Technology), and the National Center for Supercomputing Applications (NCSA, located at the University of Illinois, Urbana). The host platform at each facility was an IA-64 Linux cluster where each compute node consisted of dual 1.5 GHz Intel Itanium2 processors. The operating system at both facilities was Linux 2.4.21-SMP. Each compute node had a gigabit Ethernet connection to the TeraGrid network.

The experiments were designed to capture a large set of complexity measures under known conditions. In one set of experiments, the data receiver executed on a dedicated processor within CACR, and additional compute-bound processes were spawned on this same processor to create CPU contention. As the number of additional processes was increased, the amount of time the data receiver was switched out similarly increased. Thus there was a direct relationship between CPU load and the resulting packet loss rate. To investigate loss patterns caused by contention for NIC resources, we initiated a second (background) data transfer. The data sender of the background transfer executed on a different node within the NCSA cluster, and sent UDP packets to the node on which the data receiver was executing (there was not a receiver for the background transfer). Initially, the combined sending rate was set to the maximum speed of the NIC (one gigabit per second), and contention for NIC resources was increased by increasing the sending rate of the background transfer. The packet loss experienced by both data transfers was a function of the combined sending rate, and this rate was also set to provide a wide range of loss rates. In all cases the primary data transfer had a sending rate of 1 gigabit per second. The complexity values were computed at each control point. All experiments were performed late at night when there was observed to be little (if any) network contention. These experiments were conducted with the congestion-control mechanisms *disabled*.

We attempted to keep the loss rate within a range of 0 – 10% and to gather a large number of samples within that range. In theecase of contention at the NIC, we collected approximately 3500 samples in this range (each of which represented a data transfer of 100 Megabytes). The scheduler made it more difficult to keep the loss rate within the desired range when looking at CPU contention. This was because we had no control over how it switched processes between the two CPUs after they were spawned. We were able to collect approximately 1500 samples within the 0 to 10% loss range.

For reasons related to the development of the Bayesian analysis (discussed below), we needed to have a large sample of complexity values at discrete (and small) intervals within the loss range. To accomplish this, we sorted the data by loss rate, and binned the data with 30 complexity measures per bin. The bins were labeled with the mean loss rate of the 30 samples.

6.1 Experimental Results

Figure 1shows the complexity values associated with contention for NIC resources at each data bin. As can be seen, there is a very strong relationship between complexity measures and loss rates. This figure also shows that complexity measures, and the inherent randomness in the system they represent, increases very quickly with increasing loss rate. It can also be observed that the distribution of complexity measures at each data bin appear to be tightly packed around the mean.

As shown in Figure 2the behavior of complexity measures associated with contention for CPU resources show little change with increasing loss rates. While the distribution of complexity measures in each bin is not as tightly packed as that of NIC contention, the measures do appear to be spread out evenly around the mean. Figure 3 shows the mean complexity measures for both causes of data loss, at each data bin, with 95% confidence intervals around the mean. As can be seen, there is a very clear separation of complexity measures even at very low loss rates. In fact, the confidence intervals do not overlap for any loss rate greater than 0.3%. These are very encouraging results, and provide strong support for the claim that complexity measures are excellent metrics upon which a sophisticated classification mechanism can be built. The development of such a mechanism is considered in the next section.

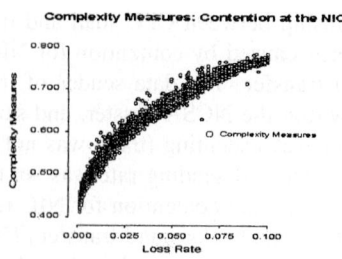
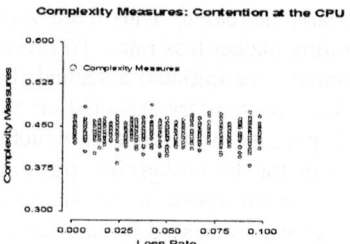

Fig. 1. This figure shows the complexity measures associated with contention for NIC resources as a function of the loss rate

Fig. 2. This figure shows the complexity measures associated with contention for CPU resources as a function of the loss rate

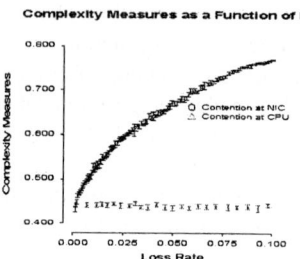

Fig. 3. The figure shows the mean complexity measures and 95% confidence intervals around the mean as a function of the cause of data loss and the loss rate

7 Statistical Analysis

In the previous section it was shown that the statistical properties of the complexity measures associated with NIC and CPU contention diverge very quickly with increasing loss rate. The next step is to leverage this difference in statistical properties to develop classifiers for the causes of data loss. Similar to the approach taken by Liu, *etal* [19], the classification mechanism will be based on the application of Bayesian statistics. This approach centers on how values of a certain metric – for example, first-order complexity values – can identify a cause of packet loss. We denote the cause of packet loss by K, and consider two cases: the cause of loss is related to contention at the NIC ($K = K_{NIC}$) or contention at the CPU ($K = K_{CPU}$). Denoting the complexity metric by C, Bayes rule states

$$P(K \mid c) = \frac{P(c \mid K) \cdot P(K)}{P(c)} \qquad (2)$$

Here $P(K \mid c)$ is the *posterior probability of K given complexity value* $C = c$. That is, it represents the probability of a given cause of data loss in light of the current complexity measure. The expression $P(c \mid K)$ is the conditional probability that the value c for metric C is observed when the cause K is dominant. It represents the *like-*

lihood of observing complexity measure $C = c$ assuming the model ($K = K_{NIC}$) or ($K = K_{CPU}$). $P(K)$ is the probability of cause K occurring, and $P(c)$ is the probability that value c of metric C is observed. In the Bayesian approach, the experimenter's prior knowledge of the phenomenon being studied is included in the *prior distributions*. In this case, we are able to develop models for $P(c \mid K)$ (that is, the likelihood of observing metric $C = c$ given cause K), and the distribution for $P(c)$. Similarly, $P(K)$ can be measured empirically. We begin by looking at the densities of the complexity measures for $P(c \mid K = K_{NIC})$ and $P(c \mid K = K_{CPU})$.

7.1 Data Models

The complexity measures appear to be well behaved suggesting that it may be possible to develop simple empirical models to describe the distributions for
$P(c \mid K = K_{NIC})$ and $P(c \mid K = K_{CPU})$. The approach we used was to feed the complexity measures through a sophisticated statistical program for fitting models to observed data (Simfit [6]). Given the shape of the curve for the complexity measures associated with NIC contention, it seemed reasonable to investigate models involving exponentials. We tested many models (including polynomial), and the best fit for the data was obtained using the sum of two exponentials. The observed complexity measures and the derived data model are shown in Figure 4. Figure 5 shows the mean value of the observed complexity measures within each bin, 95% confidence intervals around the means, and how they lay on the fitted data model. Visually, it appears to be an excellent fit.

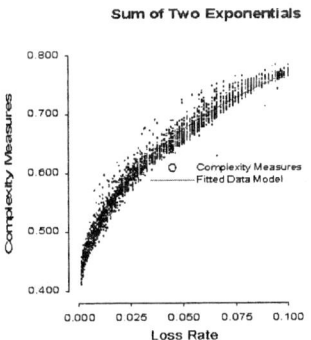
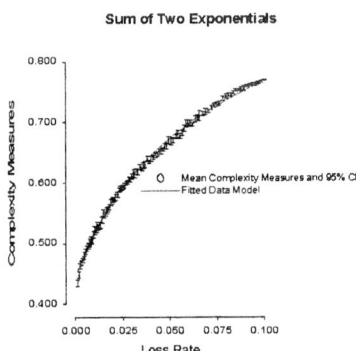

Fig. 4. This figure shows the fitted data model when the cause of loss is contention for NIC resources

Fig. 5. This figure shows the mean complexity measure within each bin, 95% confidence intervals around the mean, and how the observed data lays along the fitted model

Based on the behavior of complexity measures associated with contention for CPU resources, it appeared that a straight line would provide the best fit for the data. Extensive testing showed this to be true, and Figure 6 shows the fitted data model and the distribution of the observed complexity measures around the fitted model. Figure 7 shows the mean complexity measure for each data bin, 95% confidence intervals around the mean, and how the observed data fits the data model.

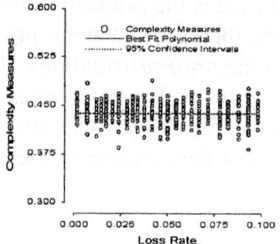
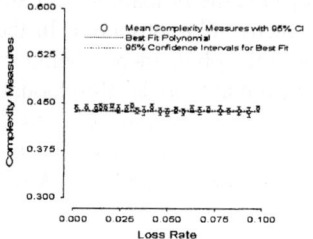

Fig. 6. This figure shows the fitted data model when the loss is caused by contention for CPU resources

Fig. 7. This figure shows the mean complexity measure within each bin, 95% confidence intervals around the mean, and how the observed data lays across the fitted data model

7.2 Goodness of Fit

Visually, it appears that the models describe the empirical data quite well, but more than a visual "match" is required. To construct a model to fit data, there are several criteria that need to be satisfied. Factors such as largest relative residual, the mean relative residual, the sum of the squared residuals, the percentage of residuals greater than a chosen value, and the correlation coefficient should also be considered.

The analysis of residuals for the two data models is shown in Table 1. As can be seen, the largest relative residual for both models is approximately 14%, while the mean relative residual is 1.37% for the NIC model and 2.58% for the CPU model. The sum of squared relative residuals is particularly good for both models, especially considering the number of data points. The number of data values above and below the theoretical prediction is very close for both data models. The correlation coefficient for the NIC model is excellent. These results indicate a very good fit for both models.

Table 1. This table shows the metrics used to determine goodness of fit for both fitted data models

1. LRR:	13.32%	13.92%
2. MRR:	1.37%	2.58%
3. SSQ:	0.447	0.156
4. CC:	0.987	0
5. NRA:	1615	772
6. NRB:	1782	709
7. RR 10-20%	0.12%	0.81%
8. RR 20-80%	0%	0%

1. Largest relative residual
2. Mean relative residual
3. Sum of the squared relative residuals
4. Correlation coefficient
5. Total number of residuals above the theoretical prediction
6. Total number of residuals below the theoretical prediction
7. Relative Residuals in range of 10-20%.
8. Relative Residuals in range of 20-80%.

7.3 Distribution of Complexity Measures

The empirically derived models fit the data quite well and we will use these as our starting point for the analysis. The approach is to use the fitted data model as the "theoretical" mean complexity measure at each loss rate. Our working hypothesis is that the data values are normally distributed around the mean of the empirical data. We go about justifying this working conclusion as follows. The obvious first step is to try and rule out a Normal distribution by checking to see if the distribution is clearly *non*-Normal, and this is clearly not the case. The next step is to try to rule out a Normal distribution by testing the numbers of data points contained within the various standard deviations away from the mean. This does not allow us to rule out a Normal distribution, either. The final step is to perform a hypothesis test using a statistical test for Normality. The one we have chosen is the Shapiro-Wilks test statistic. We applied this test to the data in each data bin, and when outliers were removed from the data, we were, without exception, not required to reject our hypothesis. Based on this, we accept our working hypothesis that this distribution is Normal.

7.4 Use of the Theoretical Model

We now have models for $P(c \mid K = K_{NIC})$, $P(c \mid K = K_{CPU})$ and, by extension, $P(c)$.[2] Also, we have a working hypothesis that the complexity measures are normally distributed around the mean, and we have a good estimate of the standard deviation around the mean based on the empirical data. We use these models as follows.

First, the classification mechanism computes the complexity measure for the current transmission window and computes the loss rate for the window. Next, the values associated with each fitted data model, at the given loss rate, are determined. These values are used as the "theoretical" mean complexity measure, for each cause of data loss, at that given loss rate. The standard deviation around both means is that derived from the empirical data. Next, and given the working hypothesis that the complexity measures are normally distributed about the mean, we are able to compute the likelihood of the complexity measure assuming each model. We do this as follows.

For each complexity value we compute the distance from the "theoretical" mean. We then standardize this distance by using the standard deviation so that we have an equivalent distance in the standard normal distribution equal to

$$\frac{x_C - \mu_C}{\sigma_C} \quad (3)$$

where the denominator is the empirically derived standard deviation of our hypothesized Normal distribution. The final step is to calculate the area under the curve between the observed complexity measure and the means of the two distributions. However, since this area approaches zero as the observed value approaches the theoretical mean, we must subtract this value from 1 to arrive at the desired probability value. It

---

[2] $P(C) = P(C \mid K = K_{NIC} * P(K = K_{NIC}) + P(C \mid K = K_{CPU} * P(K = K_{CPU})$.

is this probability value that will be used by the control mechanisms to determine the appropriate response to observed data loss.

8 Conclusions and Future Research

In this paper, we have shown that complexity measures of packet-loss signatures can be highly effective as a classifier for causes of packet loss over a wide range of loss rates. It was shown that the divergence of complexity measures, and thus the ability to discriminate between causes of packet loss, increases rapidly with increasing loss rates. Also, it was shown that complexity measures can be used as the basis for a sophisticated Bayesian statistical model for the causes of data loss. These are very encouraging results, and suggest that the control mechanisms of a high-performance data transfer system can determine the dominant cause of data loss, and, by extension, can respond to data loss in a way that is appropriate for the given cause of such loss. However, it was also shown that complexity measures in and of themselves are not powerful enough to discriminate between causes of data for loss rates less than approximately 0.003. Thus one focus of current research is the identification of other metrics or statistical models that can be effective at very low loss rates.

What this research does not answer is the generality of the empirical models. Based on empirical results obtained for other architectures however [15, 16], we believe that the statistical characteristics associated with the different causes of data loss will be consistent across architectures. This research also does not answer the question of whether random network events can create complexity measures for which our interpretation is incorrect. Both of these issues are the focus of current research.

References

[1] The TeraGrid Homepage. http://www.teragrid.org
[2] Allcock, W., Bester, J., Breshahan, J., Chervenak, A., Foster, I., Kesselman, C., Meder, S., Nefedova, V., Quesnel, D. and Tuecke, S., Secure, Efficient Data Transport and Replica Management for High-Performance Data_Intensive Computing. In the Proceedings of *IEEE Mass Storage Conference*, (2001).
[3] Allman, M., Paxson, V. and W.Stevens. *TCP Congestion Control, RFC 2581*, April, 1999.
[4] Balakrishnan, S., Padmanabhan, V., Seshan, S. and Katz, R. A Comparison of Mechanisms for Improving TCP Performance Over Wireless Links. *IEEE/ACM Transactions of Networking, 5(6).* 756-769.
[5] Balakrishnan, S., Seshan, S., Amir, E. and Katz, R., Improving TCP/IP performance over wireless networks. In the Proceedings of *ACM MOBICON*, (November 1995).
[6] Bardsley, W. SimFit: A Package for Simulation, Curve Fitting, Graph Plotting and Statistical Analysis. http://www.simfit.man.ac.uk
[7] Barman, D. and Matta, I., Effectiveness of Loss Labeling in Improving TCP Performance in Wired/Wireless Networks. In the Proceedings of *ICNP'2002: The 10th IEEE International Conference on Network Protocols*, (Paris, France, November 2002).
[8] Biaz, S. and Vaidya, N., Discriminating Congestion Losses from Wireless Losses using Inter-Arrival Times at the Receiver. In the Proceedings of *IEEE Symposium ASSET'99*, (Richardson, TX, March, 1999).

[9] Biaz, S. and Vaidya, N., Performance of TCP Congestion Predictors as Loss Predictors. Texas A&M University, Department of Computer Science Technical Report 98-007 February, 1998.

[10] D'Alessandro, G. and Politi, A. Hierarchical Approach to Complexity with Applications to Dynamical Systems. *Physical Review Letters*, *64* (14). 1609-1612.

[11] Dickens, P., FOBS: A Lightweight Communication Protocol for Grid Computing. In the Proceedings of *Europar 2003*, (2003).

[12] Dickens, P., A High Performance File Transfer Mechanism for Grid Computing. In the Proceedings of *The 2002 Conference on Parallel and Distributed Processing Techniques and Applications (PDPTA)*. (Las Vegas, Nevada, 2002).

[13] Dickens, P. and Gropp, B., An Evaluation of Object-Based Data Transfers Across High Performance High Delay Networks. In the Proceedings of *the 11th Conference on High Performance Distributed Computing*, (Edinburgh, Scotland, 2002).

[14] Dickens, P., Gropp, B. and Woodward, P., High Performance Wide Area Data Transfers Over High Performance Networks. In the Proceedings of *The 2002 International Workshop on Performance Modeling, Evaluation, and Optimization of Parallel and Distributed Systems.*, (2002).

[15] Dickens, P. and Larson, J., Classifiers for Causes of Data Loss Using Packet-Loss Signatures. In the Proceedings of *IEEE Symposium on Cluster Computing and the Grid(ccGrid04)*, (2004).

[16] Dickens, P., Larson, J. and Nicol, D., Diagnostics for Causes of Packet Loss in a High Performance Data Transfer System. In the Proceedings of *Proceedings of 2004 IPDPS Conference: the 18th International Parallel and Distributed Processing Symposium*, (Santa Fe, New Mexico, 2004).

[17] Elsner, J. and Tsonis, A. Complexity and Predictability of Hourly Precipitation. *Journal of the Atmospheric Sciences*, *50* ((3)). 400-405.

[18] Hao, B.-l. *Elermentary Symbolic Dynamics and Chaos in Dissipative Systems*. World Scientific, 1989.

[19] Liu, J., Matta, I. and Crovella, M., End-to-End Inference of Loss Nature in a Hybrid Wired/Wireless Environment. In the Proceedings of *Modeling and Optimization in Mobile, Ad Hoc, and Wireless Networks (WiOpt'03)*, (Sophia-Antipolis, France, 2003).

[20] LMbench. http://www.bitmover.com/lmbench/

[21] Salamatian, K. and Vaton, S., Hidden Markov Modeling for Network Communication Channels. In the Proceedings of *ACM SIGMETRICS 2001 / Performance 2001*, (Cambridge, Ma, June 2001).

Parallel Branch and Bound Algorithms on Internet Connected Workstations

Randi Moe[1] and Tor Sørevik[2]

[1] Dept. of Informatics, University of Bergen, Norway
[2] Dept. of Mathematics, University of Bergen, Norway

Abstract. By the use of the GRIBB software for distributed computing across the Internet, we are investigating the obstacles and the potential for efficient parallelization of Branch and Bound algorithms. Experiments have been carried out using two different applications, i.e. the Quadratic Assignment Problem (QAP) and the Traveling Salesman Problem (TSP). The results confirm the potential of the approach, and underline the requirements of the problem, algorithm and architecture for the approach to be successful.

1 Introduction

Harvesting idle cycles on Internet connected workstations has been used for solving a number of large scale computational problems, such as such as the factorization of RSA 129 [AGLL94] and finding Mersenne prime numbers [GIM]. Ongoing computations include SETI@home [SET], Folding@home [Fol], FightAIDS@home [Fig] and Compute Against Cancer [Com]. The record breaking Traveling salesman computation by Bixby et. al. [ABCC98] and the Nugent30 problem for the Quadratic Assignment Problem [ABGL02] has also been solved by distributed Internet computing.

Obviously not all large scale problems can easily be solved this way. First of all the Internet has orders of magnitude higher latency and lower bandwidth than the internal network of a tightly coupled parallel computer, thus communication intensive parallel applications don't scale at all on WAN (Wide Area Network) connected computers. Secondly, the computational nodes in a NOW (Network of Workstations) different and fluctuating power. Consequently one need a dynamical load balancing scheme which can adapt to available resources of the NOW and the demand of the application. Most existing parallel programs do apply static load balancing, and for that reason can not be expected to run well on NOWs. These two architectural characteristics set restriction on the problems applicable to scalable parallel GRID-computing. There is however still important subclasses of large scale problems which are sufficient coarse grain and do have dynamic scheduling of subtasks to run efficiently.

One class of such problems is large scale Branch and Bound (B&B) problems. Here a dynamic scheduling is always called for in any parallelization, as it is impossible apriori to know the computational effort needed to examine the different subtrees. Whether or not the problem is "communication thin" or not will depend on the kind of problem at hand. In this paper we investigate the obstacles and the potential for efficient parallelization of B&B-algorithms. We do this experimentally by two different applications, i.e. the Quadratic Assignment Problem (QAP) and the Traveling Salesman Problem (TSP). In particular we try to find out which obstacles are problem dependent, algorithm dependent, or architectural dependent.

In Section 2 we give a brief overview of our parallel B&B algorithm, and it is implementation using GRIBB. For details on GRIBB see [Moe02, Moe04b, Moe04a]. In Section 3 we identify and discuss four potential obstacles to good parallel performance. In Section 4 we report on some numerical experiments and discuss when the different obstacles come into play, and whether the problem can be related to the algorithm, the particular test case or the computer architecture.

2 Parallel B&B in GRIBB

The GRIBB system set up a server process running on a local computer and a number of client processes, possible spread all over the world, communicating with the server. Clients have no knowledge of other clients - communication occurs exclusively between the server and the individual clients.

As we need a coarse grained parallelism, we parallelize the B&B search by processing independent subtrees. In order to rapidly create a sufficient number of subtrees, we start out doing a width-first search. The computation is then parallelized in a server-client style, where the server keeps a pool of tasks. Tasks are distributed to the clients on their demand. The clients report back to the server when their tasks are done, the server record results, update bounds and provide a new task for the client. This process is repeated until convergence. As soon as the server is up and running clients may be added as they become available, and deleted if needed for other purposes. A more detailed description of the algorithms is found in [Moe04a].

An overall design criterion for GRIBB has been to provide a general framework, easy to adapt to new problems. This calls for object oriented programming, and we have used Java. The fundamental objects of GRIBB are `tasks`. These are created by the server program and solved by the client program. Thus they have to be communicated between the two programs. This is done using RMI. The RMI-methods are implemented in the `Server` class and invoked remotely by the client program.

The more of the manipulation of `tasks` which can be done on the generic `task` class, the more could be kept in the generic GRIBB-framework. In particular we don't want the top-level `Server` class and the `Client` class to act directly on derived objects, but on objects defined in the high level generic classes.

3 Obstacles to Perfect Scaling

In this section we discuss some potential obstacles to the scaling of our parallel B&B algorithm. Some of these obstacles applies to any parallel implementation regardless of chosen programming model or architecture while others appear in this specific server-client model.

Parallel Inefficiency. The performance of B&B routines crucially depends on the quality of the available bounds. The bounds typically improves as the computation proceeds. The rate of the improvements depends on the order in which the subtasks are done, and might have huge effects on the overall computation time. When apriori knowledge on where there is high possibilities for finding good solutions, and hence good bounds, is available, a best-first selection strategy can be used. A strict application of this task selection criterion is however purely sequential and must be abandoned in a parallel execution, where many tasks are done simultaneously. Thus tasks may be distributed with poorer bounds as they can not wait for for all results that would have be available in the sequential case. Consequently the total work of doing B&B search in parallel is likely to be more than for the corresponding sequential case.

This potential problem is inherit in the parallelization strategy of our B&B algorithm, and do exist independent of whether we run our algorithm on a tightly coupled system or on a set of distributed workstations. The importance of it is clearly problem dependent.

Load Balancing. Our algorithm distributes the tasks to the clients on a first-come-first-served basis. This simple strategy for distribution of workloads is applied as we assume no apriori knowledge of the computational cost of the different subtasks. Nor do we assume any knowledge about the computational capacity of the different compute nodes. As long as the number of tasks is large compared to the number of clients, the "law-of-the-large-numbers" suggest that we will have a pretty good load balance this way, while we may see severe load imbalance when the tasks are few compared to the number of participating clients.

The Server-Bottleneck. The server-client model for parallel computing has several advantages. It provides straightforward ways for dynamic work distribution, simple termination criterion, and simple communication patterns. This communication pattern does however create a potential bottleneck at the server. If the server becomes saturated no further scaling of the algorithm is possible. Again, this is not a feature of distributed computing, but of the the server-client model used. How severe this problem is, and when it eventually kicks in, depends on the system as well as the problem.

Communication Bounded Computation. For parallel computation to scale we always have to have the communication to computation ratio low. This of course becomes more difficult as the bandwidth decreases and the latency increases, when moving from departmental network to Internet connected workstations. For our problems the amount of data to communicate increase only slowly with the problem size, while the computation costs increase much more rapidly. Thus we need the individual tasks to be large. With the number of tasks large for good load balancing and scaling, we conclude that only large problems are applicable to scalable, parallel solution on Internet connected computers.

For our problems and the parallelization strategy we have chosen, we usually have a choice on how many tasks to generate. For good load balancing we like the number of tasks to be large. This of course makes each task smaller and the likelihood of communication domination or server-bottleneck increases. Efficient execution depends on finding the right balance. This balance-point depends on the problem size as well as the architecture. For the architecture of our special interest, Internet connected workstations, a good balance-point only exists when the number of tasks as well as their size is large. Thus only huge problems can be expected to run well. However, having a huge problem which splits into sufficiently many well-sized subtask is only a necessary requirement for successful Internet computing, we may still be hit hard by "parallel inefficiency".

4 Numerical Experiments

Through some experiments we have tried to illustrate the effects of the different obstacles to scaling. Our first experiment is a single client configuration, designed to show the effect of the network architecture, and the granularity of the problem needed for efficient parallelization on Internet connected computers. The experiment also provide some insight into the load balancing problem, and demonstrate a simple upper bound to scaling. Our second test case is a multi client configuration designed to study the effect of "parallel inefficiency" and server bottleneck.

First we run 3 different problems with 2 different configurations i.e. a) over the Internet (WAN) and b) on a local area network (NOW), here a 100 Mbit Ethernet. In each case we used only one client. The server had thus a very low load and all timings are as seen on the client. The server lived in Norway at the University of Bergen. For the Internet based solver the client was situated in Greece, at the Technical University of Crete. All systems used are Pentium based, running Linux. A 300 Mhz Celeron in Crete and a 2.6 Ghz Pentium 4 in Bergen.

As our test cases we have used QAP with two standard test problems from QAPLIB [QAP], QAP-Nug15 and QAP-Nug17. Our QAP-implementation is based on the method described by Hahn et al. [HGH98, HG98] using a Dual Procedure similar to the Hungarian method for solving the Linear Assignment Problem. Further details are given in [Moe04a]. We have also implemented a one-tree based solver for the TSP-problem [HK70, HK71]. The reported results are based on a case taken from TSPlib [TSP].

For each task we have recorded the time for fetching a task (*fetch*), the computation time (*comp*) and the time taken to return the task (*return*). In Table 1 we summarize the results. Each 3×3 box in the table represent one particular problem (described in the left column) executed on a NOW or across the Internet. Each of the 3 measurements is summarized by 3 numbers: minimum, average and maximum over the n tasks.

The differences in the computing times within a specific problem and architecture can be contributed to the differences in the computational demand of the different tasks. The TSP problem has the largest differences in the size of the tasks. Here 1/6 of the computing time is spent on the largest task. Limiting the speed-up to 6. To increase efficiency of our TSP-implementation we have sorted the tasks, such that those most likely to produce good bounds come first. These are also the ones most expensive to compute. The others become much cheaper to compute as they received good bounds from the first one. In parallel this need not happen. While the first client works on the difficult first task, the other $p-1$ clients struggle with the next $p-1$ tasks with far worse bounds than in the sequential case. This is exactly the scenario where we expect parallel inefficiency to occur, and numerical experiments confirm that this happens here. To obtain the maximal speed-up of 6, limited by load imbalance, more than 20 workstations were needed.

The measured computing time reflects the real cost of the computing well as we had single user access to the client. However, for the communication the situation is different. The networks where available to public use and probably quite busy. For the different tasks within one problem each of the "fetching" or "returning" operation is exactly the same. We can see no other reason for the huge differences in the measured communication time for different instances of the same operations on a given network than the ever changing load on them. As illustrated by the columns of the single client case in Figure 2 the majority of the tasks (77.5%) complete their communication work at less than 25 milliseconds, which should be sufficient for the problem at hand. However, 5% need more than 4 times this amount. On the Internet life is different. Here timing seems to be not only larger, but truly random as well. This unpredictability of the communication time is a strong argument for the necessity of dynamic scheduling.

For the problem size represented by QAP-Nug15 and TSP-gr21, the computation to communication ratio is small, and there is essentially no chance for an effective parallelization on Internet connected systems. As we move to QAP-Nug17 this changes. The computational work of a task increases by a factor of approximately 40, while the communication increased by a factor of roughly 2, and the computing part is still 57% of the total consumed time on the clients as seen in Table 1.

Table 1. Single client experiment: Three different problems are run on NOW and on Internet. All timings are in millisecond.

Problem	#tasks n		Local network			Internet		
			min	aver	max	min	aver	max
TSP gr21	2697	fetch	3	6	130	1,756	8,971	56,197
		comp	1	532	256,032	1	3,722	1,913,412
		return	3	10	109	1,071	9,943	89,774
QAP Nug15	2104	fetch	5	8	308	2,640	23,075	311,116
		comp	5	221	8,092	41	1,583	67,307
		return	2	6	95	1,075	4,654	403,539
QAP Nug17	3366	fetch	11	13	306	4,669	34,967	774,060
		comp	219	8,402	640,432	1,593	63,859	4,793,558
		return	7	24	142	2,980	12,558	1,208,140

Our second test case is designed to study the effect of "parallel inefficiency" and server bottleneck to scalability. To have a (relatively) controlled experimental setting for this test, we have run it on a NOW. As discussed in Section 3, the two obstacles under scrutiny, are not an effect of the architecture, but primarily an effect of the number of tasks and the task size. Only the two QAP problems have potential for scaling. Here the largest individual task contributes to less than 1/40 of the total computation cost. Consequently work load imbalance should not be a serious problem for less than 30 clients.

The smallest QAP-problem is the one most likely to stress the network/server, thus this has been chosen as our test problem. Our testbed is a student laboratory at the University of Bergen.

The average communication time as well as the average compute time for different tasks increases as the number of clients increase as shown in Figure 1. Again huge differences are hidden behind the average numbers. On one client 77.5% of the communication tasks were completed in less than 25 milliseconds. For higher number of clients, $p=4,16$, or 30, the corresponding numbers where 70.5%, 33.5%, and 19.8%, respectively. The histogram in Figure 2 shows how communication time of the different tasks shifts towards high-valued intervals with increasing number of clients, This must be a server bottleneck as the network becomes saturated. The increase in average compute time is different. A large part of this is due to the increased in the time for computing task $2, \ldots, p$, which now are computed with the poor initial bound. For $p > 8$ the increased compute time on these initial tasks contribute to more than half the total increase shown in Figure 1. For this particular case scaling appears to be limited by poor initial bounds for the computation part and by a server bottleneck on the communication side. The latter would be avoided with an increasing size of the computation for each task (i.e. the QAP-Nug17 case). Any cure for parallel

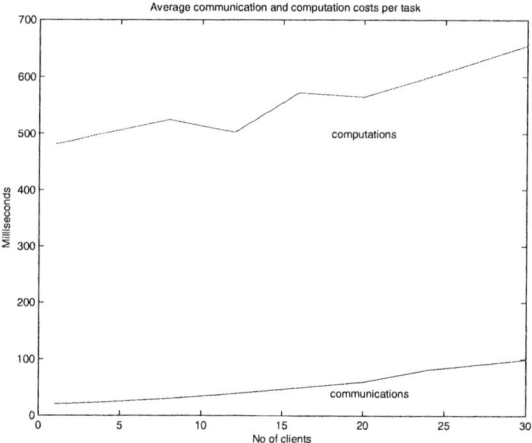

Fig. 1. The average communication and computation costs in milliseconds are reported for various number of clients

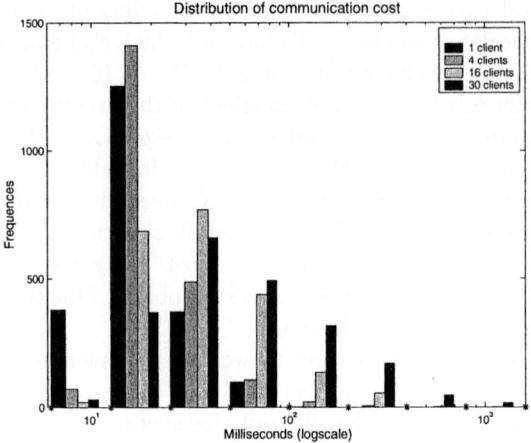

Fig. 2. A histogram for the cost distribution of the communication of tasks on NOW. The different tasks are collected in separate cost-intervals (in milliseconds).

inefficiency have to be problem dependent. Speed-up numbers, show that in the QAP-Nug15 case, the speed-up is almost linear for up to 12 clients. Then it tails off, and for 30 clients it is 19.5

5 Conclusions

We find the results of our experiments quite encouraging. The requirement that the problem to solve being huge with many large subtasks is no real limitation. This is what this kind of computation is for. More experiments are needed to test the scalability on real huge problems, but the limited experiments we have done indicates that the worst hurdle to scalability is the "parallel inefficiency". In cases as for the QAP-Nug15 used here, initial bounds were bad, and all clients were forced to struggle hard with their first tasks. Little speed-up would be achieved by adding an extra 1000 clients to the resource pool. Another limitation to scalability demonstrated by the TSP problem is when one single subtask becomes a significant part of the total work. However, we believe this to be a non-typical case.

An obstacle to distributed computing, not discussed here, is the difficulty of programming. In our case that was made easy, as we built it on the GRIBB software. GRIBB took care of the management of the search tree, and all issues related to parallelization, communication, distribution of work, etc. That reduced the programming effort to essentially programming of efficient QAP- or TSP-kernels.

Acknowledgment. It is with great pleasure we acknowledge the kind hospitality of the Technical University of Crete where most of this research were conducted. We also like to acknowledge the assistance of Brage Breivik in programming of the TSP-code.

References

[ABCC98] D. Applegate, R. E. Bixby, V. Chvatal, and W. Cook. On the solution of traveling salesman problems. *Documenta Mathematica*, ICM(III):645–656, 1998.

[ABGL02] K. M. Anstreicher, N. W. Brixius, J.-P. Goux, and J. T. Linderoth. Solving Large Quadratic Assignment Problems on Computational Grids. *Mathematical Programming, Series B*, 91:563–588, 2002.

[AGLL94] Derek Atkins, Michael Graff, Arjen K. Lenstra, and Paul C. Leyland. The magic words are squeamish ossifrage. In Josef Pieprzyk and Reihanah Safavi-Naini, editors, *Advances in Cryptology – ASIACRYPT '94*. Springer, 1994.

[Com] Compute Against Cancer. *http://www.parabon.com/cac.jsp/*.

[Fig] FightAIDS@home. *http://www.fightaidsathome.org/*.

[Fol] Folding@home. *http://folding.stanford.edu/*.

[GIM] GIMPS-Great Internet Mersenne Prime Search. *http://www.mersenne.org/*.

[HG98] Peter Hahn and Thomas Grant. Lower Bounds for the Quadratic Assignment Problem Based Upon a Dual Formulation. *Operations Research*, 46(6), Nov-Dec 1998.

[HGH98] Peter Hahn, Thomas Grant, and Nat Hall. A Branch-and-Bound Algorithm for the Quadratic Assignment Problem Based on the Hungarian Method. *European Journal of Operational Research*, August 1998.

[HK70] M. Held and R.M. Karp. The traveling-salesman problem and minimum spanning trees. *Operations Research*, 18:1138–1162, 1970.

[HK71] M. Held and R.M. Karp. The traveling-salesman problem and minimum spanning trees: Part ii. *Math. Programming*, 1:6–26, 1971.

[Moe02] Randi Moe. GRIBB - An Infrastructure for Internet Computing. In Dag Haugland and Sin C. Ho, editors, *Proceedings of Nordic MPS'02*. Norway, 2002. The Eight Meeting of the Nordic Section of the Math. Prog. Society.

[Moe04a] Randi Moe. GRIBB - A User Guide. Technical report, Dep. of Informatics, Univ. of Bergen, Norway, 2004.

[Moe04b] Randi Moe. GRIBB - Branch-and-Bound Methods on the Internet. In *LNCS 3019*, pages 1020–1027. Springer Verlag, 2004. Fifth Int. Conf. on Parallel Processing and Appl. Math., Sep 2003.

[QAP] QAPLIB - A Quadratic Assignment Problem Library. R.E. Burkard, E. Çela, S.E. Karisch and F. Rendl, Eds. *http://www.seas.upenn.edu/qaplib/*.

[SET] SETI@home. *http://www.setiathome.ssl.berkeley.edu/*.

[TSP] TSPLIB - A library of sample instances for the TSP (and related problems) from various sources and of various types. *http://www.iwr.uni-heidelberg.de/groups/comopt/software/TSPLIB95/*.

Parallel Divide-and-Conquer Phylogeny Reconstruction by Maximum Likelihood*

Z. Du[1], A. Stamatakis[2], F. Lin[1], U. Roshan[3], and L. Nakhleh[4]

[1] Bioinformatics Research Center, Nanyang Technological University,
Nanyang Avenue, Singapore 639798
[2] Institute of Computer Science, Foundation for Research and Technology-Hellas,
P.O. Box 1385, Heraklion, Crete, GR-71110 Greece
[3] College of Computing Sciences, Computer Sciences Department, New Jersey,
Institute of Technology, University Heights, Newark, NJ 07102
[4] Department of Computer Science, Rice University,
6100 Main St. MS 132, Houston, TX 77005

Abstract. Phylogenetic trees are important in biology since their applications range from determining protein function to understanding the evolution of species. Maximum Likelihood (ML) is a popular optimization criterion in phylogenetics. However, inference of phylogenies with ML is NP-hard. Recursive-Iterative-DCM3 (Rec-I-DCM3) is a divide-and-conquer framework that divides a dataset into smaller subsets (subproblems), applies an external *base method* to infer subtrees, merges the subtrees into a comprehensive tree, and then refines the global tree with an external *global method*. In this study we present a novel parallel implementation of Rec-I-DCM3 for inference of large trees with ML. Parallel-Rec-I-DCM3 uses RAxML as external base *and* global search method. We evaluate program performance on 6 large real-data alignments containing 500 up to 7.769 sequences. Our experiments show that P-Rec-I-DCM3 reduces inference times and improves final tree quality over sequential Rec-I-DCM3 and stand-alone RAxML.

1 Introduction

Phylogenetic trees describe the evolutionary relationship among a group of organisms and are important for answering many biological questions. Some applications of phylogenetic trees are multiple sequence alignment [5], protein structure prediction [19], and protein function prediction [12]. Several research groups such as CIPRES (www.phylo.org) and ATOL (tolweb.org) are working on novel heuristics for the reconstruction of very large phylogenies.

There exist two basic models for calculating phylogenetic trees based on DNA or Protein sequence data: distance-based and character-based methods. *Distance-based* methods construct a tree from a matrix of pairwise distances between the input sequences. Current popular distance-based methods such as

---

* Part of this work is funded by a Postdoc-fellowship granted by the German Academic Exchange Service (DAAD).

Neighbor Joining [18] (NJ) tolerate only a limited amount of error in the input; thus, they do not perform well under challenging conditions [15]. The two most popular character based optimization criteria are Maximum Parsimony (MP) [9] and Maximum Likelihood (ML) [6]. The optimal ML-tree is the tree topology which is most likely to have given rise to the observed data under a given statistical model of sequence evolution. This problem has always been known to be very hard in practice—the number of different tree topologies grows exponentially with the number of sequences n, e.g. for $n = 50$ organisms there already exist $2.84 * 10^{76}$ alternative topologies; a number almost as large as the number of atoms in the universe ($\approx 10^{80}$)—and was recently proven to be NP-hard [3].

For constructing the Tree of Life, i.e., the evolutionary tree on all species on Earth, efficient heuristics are required which allow for analysis of large and complex datasets in reasonable time, i.e. in the order of weeks instead of years. Heuristics that can rapidly solve ML are of enormous benefit to the biological community. In order to accelerate the analysis of large datasets with thousands of sequences via a divide and conquer approach the family of Disk Covering Methods (DCMs) [8,17] has been introduced. DCMs are divide and conquer methods which decompose the dataset, compute subtrees on smaller subproblems using a *base method* and then merge the subtrees to yield a tree on the full dataset. To the best of our knowledge Recursive-Iterative-DCM3 (Rec-I-DCM3) [17] is the best-known DCM in conjunction with the MP and ML criteria. In combination with Rec-I-DCM3 we use RAxML [20] as base method and a re-implementation of parallel RAxML [21] as global method. RAxML is among the currently fastest, most accurate, as well as most memory-efficient ML heuristics on real biological datasets. The goal of the parallel MPI implementation is to exploit the computational power of PC clusters, i.e. to calculate better trees within the same amount of total execution time. P-Rec-I-DCM3 is available as open source software from the authors. It can be compiled and executed on any Linux PC cluster.

2 Related Work

The survey of related work is restrained to parallel/distributed implementations of maximum likelihood programs and to divide and conquer approaches. The phylogenetic navigator (PHYNAV [13]) which is based on a zoom-in/zoom-out approach represents an interesting alternative to Rec-I-DCM3. However, the program has a relatively high memory consumption (crashed on a 1.000-taxon tree with 1GB RAM) compared to RAxML. Furthermore, the global optimization method (fast Nearest Neighbor Interchange adapted from PHYML [7]) is not as efficient on real alignment data as RAxML [20]. Thus, it is not suited to handle large real-data alignments of more than 1.000 sequences.

Despite the fact that parallel implementations of ML programs are technically very solid, they significantly drag behind algorithmic development. This means that programs are parallelized that mostly do not represent the state-of-the-art algorithms any more. Therefore, they are likely to be out-competed by the most recent sequential algorithms. For example the largest tree computed with parallel fastDNAml [22] which is based on the fastDNAml algorithm (1994) contains 150 taxa. The same holds for a technically very interesting JAVA-based distributed implementation of fastDNAml: DPRml [11]. Performance penalties are also caused by using JAVA due to unfavorable memory efficiency and speed of numerical calculations. Those language-dependent limitations will become more significant when trees comprising more than 417 taxa (currently largest tree with DPRml, personal communication) are computed with DPRml. M.J. Brauer et al. [2] have implemented a parallel genetic tree-search algorithm which has been used to compute trees of up to approximately 3.000 taxa with the main limitation being also memory consumption (personal communication).

Finally, there exists the previous parallel implementation of RAxML which has been used to compute one of the largest ML-based phylogenies to date containing 10.000 organisms [21]. Due to the high memory efficiency of RAxML (which the program inherited from fastDNAml) and the good performance on large real-world data it appears to be best-suited for use with Rec-I-DCM3. The goal of P-Rec-I-DCM3(RAxML) consists in a further acceleration of the algorithm.

3 Our New Method: Parallel Recursive-Iterative-DCM3

3.1 Parallelizing Recursive-Iterative-DCM3

Rec-I-DCM3 is the latest in the family of Disk Covering Methods (DCMs) which were originally introduced by Warnow *et. al.* [8]. Rec-I-DCM3 [17] was designed to improve the performance of MP and ML heuristics. The method applies an external *base method* (any MP or ML heuristic), to smaller subsets of the full datasets (which we call subproblems). The division into subproblems is executed recursively until all subproblems contain less taxa than the user-specified maximum subproblem size m. The subproblems which are smaller in size and evolutionary diameter [17], are easier and faster to analyze. Rec-I-DCM3 has been shown to significantly improve upon heuristics for MP [17]. In this paper we initially show that sequential Rec-I-DCM3 with RAxML as external method can improve upon stand-alone RAxML for solving ML. Parallel Rec-I-DCM3 is a modification of the original Rec-I-DCM3 method. The P-Rec-I-DCM3 algorithm is outlined below (differences from the sequential version are highlighted in bold letters).

Outline of Parallel Rec-I-DCM3.

- Input:
  - Input alignment S, #iterations n, base heuristic b, parallel global search method g, starting tree T, maximum subproblem size m
- Output: Phylogenetic tree leaf-labeled by S.
- Algorithm: For each iteration do
  - Set $T' =$ **Parallel-Recursive-DCM3**(S, m, b, T).
  - **Apply the parallel global search method g starting from T' until we reach a local optimum.** Let T'' be the resulting local optimum.
  - Set $T = T''$.

The Parallel-Recursive-DCM3 routine performs the work of dividing the dataset into smaller subsets, solving the subproblems in parallel (using the base method) with a master-worker scheme, and then merging the subtrees into the full tree. However, the sizes of individual subproblems vary significantly and the inference time per subproblem is not known a priori and difficult to estimate. This can lead to significant load imbalance (see Section 4.3). A distinct method of subtree-decomposition which yields subproblems of equal size for better load-balance does not appear promising (unpublished implementation in RAxML). This is due to the fact that Rec-I-DCM3 constructs subproblems intelligently with regard to closely-related taxa based on the information of the guide tree.

The global search method further improves the accuracy of the Recursive-DCM3 tree and can also find optimal global configurations that were not found by Recursive-DCM3, which only operates on smaller—local—subsets. To effectively parallelize Rec-I-DCM3 one has to parallelize the Recursive-DCM3 *and* the global search method. A previous implementation of parallel Rec-I-DCM3 for Maximum Parsimony [4] only parallelizes Recursive-DCM3 and not the global method. Our unpublished studies show that only parallelizing Recursive-DCM3 does not significantly improve performance over the serial Rec-I-DCM3. Our Parallel Rec-I-DCM3, however, also parallelizes the global search method, a key and computationally expensive component of Rec-I-DCM3. Note, that due to the complexity of the problem and the ML criterion it is not possible to avoid global optimizations of the tree altogether. All divide and conquer approaches for ML to date execute global optimizations at some point (see Section 2). For this study we use sequential RAxML as the base method and a re-implementation of parallel RAxML (developed in this paper) as the global search method.

3.2 Parallelizing RAxML

In this Section we provide a brief outline of the RAxML algorithm, which is required to understand the difficulties which arise with the parallelization.

Provided a comprehensive starting tree (for details see [20]), the likelihood of the topology is improved by subsequent application of topological alterations. To evaluate and select alternative topologies RAxML uses a mechanism called *lazy subtree rearrangements* [20]. This mechanism initially performs a rapid pre-scoring of a large number of topologies. After the pre-scoring step a few (20) of the best pre-scored topologies are analyzed more thoroughly. To the best of our knowledge, RAxML is currently among the fastest *and* most accurate programs on real alignment data due to this ability to quickly pre-score a large number of alternative tree topologies.

As outlined in the Figure available on-line at [1] the optimization process can be classified into two main computational phases:

Difficult Parallelization: The *Initial Optimization Phase (IOP)* where the likelihood increases steeply and many improved pre-scored topologies are found during a single iteration of RAxML.

Straight-Forward Parallelization: The *Final Optimization Phase (FOP)* where the likelihood improves asymptotically and practically all improved topologies are obtained by thoroughly optimizing the 20 best pre-scored trees.

The difficulties regarding the parallelization of RAxML are mainly due to hard-to-resolve dependencies caused by the detection of many improved trees during the IOP. Moreover, the *fast* version of the hill-climbing algorithm of RAxML which is used for global optimization with Rec-I-DCM3 further intensifies this problem, since it terminates after the IOP. During one iteration of RAxML all n subtrees of the candidate topology are subsequently removed and re-inserted into neighboring branches. The dependency occurs when the lazy rearrangement of a subtree i yields a topology with a better likelihood than the candidate topology even though it is only pre-scored. In this case the improved topology is kept and rearrangement of subtree $i+1$ is performed on the new topology. Especially, during the IOP improved pre-scored topologies are frequently encountered in the course of one iteration, i.e. n lazy subtree rearrangements. Since the lazy rearrangement of *one* single subtree is fast, a coarse-grained MPI-parallelization can only be based on assigning the rearrangement of distinct subtrees within the current candidate tree simultaneously to the workers. This represents a non-deterministic solution to the potential dependencies between rearrangements of subtrees i and $i+1$. This means that when two workers w_0 and w_1 simultaneously rearrange subtrees i and $i+1$ within the currently best tree and the rearrangement of subtree i yields a better tree, worker w_1 will miss this improvement since it is still working on the old tree. It is this frequently occurring dependency during the IOP between steps $i \rightarrow i+1$ ($i = 1...n$, $n =$ number of subtrees) which leads to parallel performance penalties. Moreover, this causes a non-deterministic behavior since the parallel program traverses another path in search space each time and might yield better or worse final tree topologies compared to the sequential program. The scalability for smaller number of processors is better since every worker misses less improved trees.

The aforementioned problems have a significant impact on the IOP only, since the FOP can be parallelized more efficiently. Furthermore, due to the significantly larger proportion of computational time required by the FOP the parallel performance of the slow hill-climbing version of RAxML is substantially better (see [21] for details). The necessity to parallelize and improve performance of RAxML fast hill-climbing has only been recognized within the context of using RAxML in conjunction with Rec-I-DCM3 and is an issue of future work.

3.3 Parallel Recursive-Iterative-DCM3

In this section we summarize how Rec-I-DCM3 and RAxML are integrated into one single parallel program. A figure with the overall program flow is available on-line at [1]. The parallelization is based on a simple master-worker architecture. The master initially reads the alignment file and initial guide tree. Thereafter, it performs the DCM-based division of the problem into smaller subproblems and stores the merging order which is required to correctly execute the merging step. All individual subproblems are then distributed to the workers which locally solve them with sequential RAxML and then return the respective subtrees to the master. Once all subproblems have been solved the master merges them according to the stored merging order into the *new guide tree*. This guide tree is then further optimized by parallel RAxML. In this case the master distributes the IDs of the subtrees (simple integers) which have to be rearranged along with the currently best tree (only if it has improved) to the worker processes. The worker rearranges the specified subtree within the currently best tree and returns the tree topology (only if it has a better likelihood) along with a new work request. This process continues until no subtree rearrangement can further improve upon the tree. Finally, the master verifies if the specified amount of P-Rec-I-DCM3 iterations has already been executed and terminates the program. Otherwise, it will initiate a new round of subproblem decomposition, subproblem inference, subtree merging, and global optimization. The time required for subproblem decomposition and subtree merging is negligible compared to ML inference times.

4 Results

Test Datasets: We used a large variety of biological datasets ranging in size and type (DNA or RNA). The datasets have been downloaded from public databases or have been obtained from researchers who have manually inspected and verified the alignments: **Dataset1:** 500 *rbcL* DNA sequences (1398 sites) [16], **Dataset2:** 2,560 *rbcL* DNA sequences (1,232 sites) [10], **Dataset3:** 4,114 16s ribosomal Actinobacteria RNA sequences (1,263 sites) [14], **Dataset4:** 6,281 small subunit ribosomal Eukaryotes RNA sequences (1,661 sites) [24], **Dataset5:** 6,458 16s ribosomal Firmicutes (bacteria) RNA sequences (1,352 sites) [14], **Dataset6:** 7,769 ribosomal RNA sequences (851 sites) from three phylogenetic domains, plus organelles (mitochondria and chloroplast), obtained from the Gutell Lab at the Institute for Cellular and Molecular Biology, The University of Texas at Austin.

Test Platform: P-Rec-I-DCM3 and RAxML are implemented in C and use MPI. As test platform we used a cluster with 16 customized Alpha Server ES45 compute nodes; each node is equipped with 4 Alpha-EV68 1GHz processors and has 8 GB/s memory bandwidth and an interconnect PCI adapter with over 280 MB/s of sustained bandwidth. The central component of the cluster consists of a Quadrics 128-port interconnect switch chassis which delivers up to 500 MB/s per node, with 32 GB/s of cross-section bandwidth and MPI application latencies of less than 5 microseconds.

4.1 Sequential Performance

In the first set of experiments we examine the sequential performance of standalone RAxML over Rec-I-DCM3(RAxML). The respective maximum subset sizes of Rec-I-DCM3 are adapted to the size of each dataset: Dataset1 (max. subset size: 100 taxa), Dataset2 (125 taxa), Dataset3 to Dataset6 (500 taxa). In our experiments both methods start optimizations on the *same* starting tree. Due

Table 1. Rec-I-DCM3 versus RAxML log likelihood (LLH) values after the same amount of time

Dataset	Rec-I-DCM3 LLH	RAxML LLH	Dataset	Rec-I-DCM3 LLH	RAxML LLH
Dataset1	-99967	-99982	Dataset4	-1270920	-1271756
Dataset2	-355071	-355342	Dataset5	-901904	-902458
Dataset3	-383578	-383988	Dataset6	-541255	-541438

to the relatively long execution times we only executed one Rec-I-DCM3 iteration per dataset. The run time of one Rec-I-DCM3 iteration was then used as inference time limit for RAxML. Table 1 provides the log likelihood values for RAxML and Rec-I-DCM3 after the same amount of execution time. Note that, the apparently small differences in final likelihood values *are significant* because those are logarithmic values and due to the requirements for high score accuracy in phylogenetics [23]. The experiments clearly show that Rec-I-DCM3 improves over stand-alone RAxML on *all* datasets (a more thorough performance study is in preparation).

4.2 Parallel Performance

In our second set of experiments we assess the performance gains of P-Rec-I-DCM3 over the original sequential version. For each dataset we executed three individual runs with Rec-I-DCM3 and P-Rec-I-DCM3 on 4, 8, and 16 processors respectively. Each individual sequential and parallel run was executed using the *same* starting tree and the previously indicated subset sizes. In order to determine the speedup we measured the execution time of one sequential and parallel Rec-I-DCM3 iteration for each dataset/number of processors combination. The average sequential and parallel execution times per dataset and number of processors over three individual runs are available on-line at [1]. Due to the dependencies in parallel RAxML and the load imbalance the overall speedup and scalability of P-Rec-I-DCM3 are moderate. However, we consider the present work as proof-of-concept implementation to demonstrate the benefits from using P-Rec-I-DCM3 with RAxML and to analyze the technical and algorithmic challenges which arise with the parallelization. A separate analysis of the speedup values in Table 2 for the *base, global,* and *whole method* shows that the performance penalties originate mainly from parallel RAxML. Note however, that in order to improve the unsatisfying performance of parallel RAxML it executes a slightly more thorough search than the sequential global optimization with RAxML. Therefore, the comparison can not only be based on speedup values but must also consider the significantly better final likelihood values of P-Rec-I-DCM3. To demonstrate this P-Rec-I-DCM3 is granted the overall execution time of one sequential Rec-I-DCM3 iteration (same response time). The final log likelihood values of Rec-I-DCM3 and P-Rec-I-DCM3 (on 16 processors) after the same amount of total execution time are listed in Table 3. Note that, the apparently small differences in final likelihood values *are significant*. Furthermore, the computational effort to attain those improvements is not negligible due to the asymptotic increase of the log likelihood in the *FOP* (see Section 3.2).

Table 2. Average base method, global method, and overall speedup values (over three runs) for P-Rec-I-DCM3 over Rec-I-DCM3 for one iteration of each method

Procs	base	global	overall	Procs	base	global	overall	Procs	base	global	overall
Dataset1				Dataset2				Dataset3			
4	4	2.4	2.6	4	3	2.68	2.7	4	1.95	2.6	2.2
8	4.7	2.8	3.6	8	5.3	3.2	3.45	8	5.5	5	5.3
16	4.85	2.78	3.5	16	7	4.2	4.6	16	6.7	5.7	6.2
Dataset4				Dataset5				Dataset6			
4	2.9	2.3	2.6	4	2.3	2.7	2.5	4	3.2	1.95	2.2
8	4.2	4.9	4.6	8	4.8	4.4	4.7	8	4.8	2.5	3
16	8.3	5.3	6.3	16	7.6	5.1	5.8	16	5.4	2.8	3.3

Table 3. Average Log likelihood (LLH) scores of Rec-I-DCM3 (Sequential) and P-Rec-I-DCM3 (Parallel, on 16 processors) per dataset after the same amount of total execution time over three individual runs

Dataset	Sequential LLH	Parallel LLH	Dataset	Sequential LLH	Parallel LLH
Dataset1	-99967	-99945	Dataset4	-1270785	-1270379
Dataset2	-355088	-354944	Dataset5	-902077	-900875
Dataset3	-383524	-383108	Dataset6	-541019	-540334

4.3 Parallel Performance Limits

The general parallel performance limits of RAxML have already been outlined in Section 3.2. At this point we discuss the parallel performance limits of the base method by example of Dataset3 and Dataset6 which initially appear to yield suboptimal speedups and show that those values are near-optimal. We measured the number of subproblems as well as the inference time per subproblem for one Rec-I-DCM3 iteration on those Datasets. The main problem consists in a significant load imbalance caused by subproblem sizes. The computations for Dataset3 comprise 19 subproblems which are dominated by 3 inferences that require more than 5.000 seconds (maximum 5.569 seconds). We determined the optimal schedule of those 19 subproblems on 15 processors (since 1 processor serves as master) and found that the maximum inference time of 5.569 seconds is the limiting factor, i.e. the minimum execution time for those 19 jobs on 15 processors is 5.569 seconds. With this data at hand we can easily calculate the maximum attainable speedup by dividing the sum of all subproblem inference times through the minimum execution time ($37353secs/5569secs = 6.71$) which corresponds to our experimental results. There is no one-to-one correspondence since the values in Table 2 are average values over several iterations and three runs per dataset with different guide trees and decompositions.

The analysis of Dataset6 shows a similar image: there is a total of 43 subproblems which are dominated by 1 long subtree computation of 12.164 seconds and three smaller ones ranging from 5.232 to 6.235 seconds. An optimal schedule for those 43 subproblems on 15 processors shows that the large subproblem which requires 12.164 is the lower bound on the parallel solution of subproblems. The optimal speedup for the specific decomposition on this dataset is therefore $63620 secs/12164 secs = 5.23$.

5 Conclusion and Future Work

In this paper we have introduced P-Rec-I-DCM3(RAxML) for inference of large phylogenetic trees with ML. Initially, we have shown that Rec-I-DCM3(RAxML) finds better trees than stand-alone RAxML. The parallel implementation of P-Rec-I-DCM3 significantly reduces response times for large trees and significantly improves final tree quality. However, the scalability and efficiency of P-Rec-I-DCM3 still need to be improved. We have discussed the technical as well as algorithmic problems and limitations concerning the parallelization of the *global method* and the load imbalance within the *base method*. Thus, the development of a more scalable parallel algorithm for *global* optimization with RAxML and a more thorough investigation of subproblem load imbalance constitute main issues of future work. Nonetheless, Rec-I-DCM3(RAxML) currently represents one of the fastest and most accurate approaches for ML-based inference of large phylogenetic trees.

References

1. Additional on-line material:. www.ics.forth.gr/~stamatak.
2. M.J. Brauer, Holder M.T., Dries L.A., Zwickl D.J., Lewis P.O., and Hillis D.M. Genetic algorithms and parallel processing in maximum-likelihood phylogeny inference. *Molecular Biology and Evolution*, 19:1717–1726, 2002.
3. B. Chor and T. Tuller. Maximum likelihood of evolutionary trees is hard. In *Proc. of RECOMB05*, 2005.
4. C. Coarfa, Y. Dotsenko, J. Mellor-Crummey, L. Nakhleh, and U. Roshan. Prec-i-dcm3: A parallel framework for fast and accurate large scale phylogeny reconstruction. Proc. of HiPCoMP 2005, to be published.
5. Robert C. Edgar. Muscle: multiple sequence alignment with high accuracy and high throughput. *Nucleic Acids Research*, 32(5):1792–1797, 2004.
6. J. Felsenstein. Evolutionary trees from DNA sequences: A maximum likelihood approach. *Journal of Molecular Evolution*, 17:368–376, 1981.
7. S. Guindon and Gascuel O. A simple, fast, and accurate algorithm to estimate large phylogenies by maximum likelihood. *Syst. Biol.*, 52(5):696–704, 2003.
8. D. Huson, S. Nettles, and T. Warnow. Disk-covering, a fast-converging method for phylogenetic tree reconstruction. *J. Comp. Biol.*, 6:369–386, 1999.
9. Camin J. and Sokal R. A method for deducing branching sequences in phylogeny. *Evolution*, 19:311–326, 1965.
10. M. Kallerjo, J. S. Farris, M. W. Chase, B. Bremer, and M. F. Fay. Simultaneous parsimony jackknife analysis of 2538 *rbcL* DNA sequences reveals support for major clades of green plants, land plants, seed plants, and flowering plants. *Plant. Syst. Evol.*, 213:259–287, 1998.
11. T.M. Keane, Naughton T.J., Travers S.A.A., McInerney J.O., and McCormack G.P. Dprml: Distributed phylogeny reconstruction by maximum likelihood. *Bioinformatics*, 21(7):969–974, 2005.

12. D. La, B. Sutch, and D. R. Livesay. Predicting protein functional sites with phylogenetic motifs. *Prot.: Struct., Funct., and Bioinf.*, 58(2):309–320, 2005.
13. S.V. Le, Schmidt H.A., and Haeseler A.v. Phynav: A novel approach to reconstruct large phylogenies. In *Proc. of GfKl conference*, 2004.
14. B. Maidak. The RDP (ribosomal database project) continues. *Nucleic Acids Research*, 28:173–174, 2000.
15. B.M.E. Moret, U. Roshan, and T. Warnow. Sequence length requirements for phylogenetic methods. In *Proc. of WABI'02*, pages 343–356, 2002.
16. K. Rice, M. Donoghue, and R. Olmstead. Analyzing large datasets: *rbcL* 500 revisited. *Systematic Biology*, 46(3):554–563, 1997.
17. U. Roshan, B. M. E. Moret, T. Warnow, and T. L. Williams. Rec-i-dcm3: a fast algorithmic technique for reconstructing large phylogenetic trees. In *Proc. of CSB04*, Stanford, California, USA, 2004.
18. N. Saitou and M. Nei. The neighbor-joining method: A new method for reconstructing phylogenetic trees. *Molecular Biology and Evolution*, 4:406–425, 1987.
19. I.N. Shindyalov, Kolchanov N.A., and Sander C. Can three-dimensional contacts in protein structures be predicted by analysis of correlated mutations? *Prot. Engng.*, 7:349–358, 1994.
20. A. Stamatakis, Ludwig, T., Meier, and H. Raxml-iii: A fast program for maximum likelihood-based inference of large phylogenetic trees. *Bioinformatics*, 21(4):456–463, 2005.
21. A. Stamatakis, Ludwig T., and Meier H. Parallel inference of a 10.000-taxon phylogeny with maximum likelihood. In *Proc. of Euro-Par2004*, pages 997–1004, 2004.
22. C. Stewart, Hart D., Berry D., Olsen G., Wernert E., and Fischer W. Parallel implementation and performance of fastdnaml - a program for maximum likelihood phylogenetic inference. In *Proc. of SC2001*, 2001.
23. T.L. Williams. The relationship between maximum parsimony scores and phylogenetic tree topologies. In *Tech. Report, TR-CS-2004-04,*. Department of Computer Science, The University of New Mexico, 2004.
24. J. Wuyts, Y. Van de Peer, T. Winkelmans, and R. De Wachter. The European database on small subunit ribosomal RNA. *Nucl. Acids Res.*, 30:183–185, 2002.

Parallel Blocked Algorithm for Solving the Algebraic Path Problem on a Matrix Processor

Akihito Takahashi and Stanislav Sedukhin

Graduate School of Computer Science and Engineering, University of Aizu,
Tsuruga, Ikki-machi, Aizuwakamatsu City, Fukushima 965-8580, Japan
{m5081123, sedukhin}@u-aizu.ac.jp

Abstract. This paper presents a parallel blocked algorithm for the algebraic path problem (APP). It is known that the complexity of the APP is the same as that of the classical matrix-matrix multiplication; however, solving the APP takes much more running time because of its unique data dependencies that limits data reuse drastically. We examine a parallel implementation of a blocked algorithm for the APP on the one-chip Intrinsity FastMATH adaptive processor, which consists of a scalar MIPS processor extended with a SIMD matrix coprocessor. The matrix coprocessor supports native matrix instructions on an array of 4 × 4 processing elements. Implementing with matrix instructions requires us to transform algorithms in terms of matrix-matrix operations. Conventional vectorization for SIMD vector processing deals with only the innermost loop; however, on the FastMATH processor, we need to vectorize two or three nested loops in order to convert the loops to equivalent one matrix operation. Our experimental results show a peak performance of 9.27 GOPS and high usage rates of matrix instructions for solving the APP. Findings from our experimental results indicate that the SIMD matrix extension to (super)scalar processor would be very useful for fast solution of many matrix-formulated problems.

1 Introduction

Matrix computations and problems of finding paths in networks are among the most important and challenging computational problems [12]. The algebraic path problem (APP) generalizes a number of well-known problems such as transitive closure, all-pairs shortest path problem, minimum-cost spanning tree, tunnel problem, and matrix inversion. It is known that although the complexity of the APP is the same as that of the classical matrix-matrix multiplication, the standard cache-friendly optimizations such as blocking or tiling used in dense linear algebra problem cannot directly be applied to the APP because of its unique data dependencies.

The issue has been addressed in [1], as a loop ordering issue in designing a blocked Floyd's algorithm for the all-pairs shortest path problem. Although the Floyd's algorithm [5] has the same three nested loops as matrix-matrix multiplication, the order of the three loops cannot be changed freely like matrix-matrix multiplication. We can change only the order of the two inner loops. The outermost loop must remain outermost, that limits data reuse drastically. A blocked Floyd's algorithm and its unique data dependencies have been also discussed in [2,3,4].

Extending the results in the blocked Floyd's algorithm, this paper describes a universal blocked algorithm applicable to the all APP applications. Blocked algorithms for the APP have been studied in [8] for a systolic array and in [13] for linear SIMD/SPMD arrays. In our work, we consider a parallel implementation of a blocked algorithm on a matrix processor, in particular, the FastMATH processor [14,15,16] which provides SIMD matrix instructions that operate on matrix data in parallel.

Implementing on the FastMATH processor requires us to transform algorithms in terms of matrix-matrix operations and to arrange sequential data in memory into blocks of matrix data in order to efficiently utilize the SIMD matrix instructions. Conventional vectorization for SIMD vector processing deals with only the innermost loop; however, on the FastMATH processor, we need to vectorize two or even three nested loops in order to convert the loops to equivalent one matrix operation. We state the vectorization of two or three nested loops as *2-D vectorization* or *3-D vectorization* respectively to differentiate them from the conventional vectorization which deals with only the innermost loop.

The remainder of this paper is organized as follows: In Section 2, theoretical background of the APP and its applications are given. In Section 3, we describe a universal blocked algorithm applicable to the all APP applications. In Section 4, the FastMATH processor is described. In Section 5, we discuss a parallel implementation of the blocked Floyd's algorithm on the FastMATH. In Section 6, experimental results are presented. Finally, the paper is summarized in Section 7.

2 The Algebraic Path Problem

The algebraic path problem (APP) unifies a number of well-known problems into a single algorithmic scheme. The APP is defined as follows [6,7,11,12]. Let $G = (V, E, w)$ be a weighted graph, where $V = \{0, 1, ..., n-1\}$ is a set of vertices, $E \in V \times V$ is a set of edges, and $w : E \to S$ is an edge weighting function whose values are taken from the set S. This function belongs to a path algebra $(S, \oplus, \otimes, *, \bar{0}, \bar{1})$ together with two binary operations, *addition* $\oplus : S \times S \to S$ and *multiplication* $\otimes : S \times S \to S$, and a unary operation called *closure* $* : S \to S$. Constants $\bar{0}$ and $\bar{1}$ belong to S. Hereby, $\oplus$ and $\otimes$ will be used in infix notation, whereas the closure of an element $s \in S$ will be noted as s^*. Additionally, $*$ will have precedence over $\otimes$ and this over $\oplus$. This algebra is a closed semiring, i.e. an algebra fulfilling the following axioms: $\oplus$ is associative and commutative with $\bar{0}$ as neutral; $\otimes$ is associative with $\bar{1}$ as neutral, and distributive over $\oplus$. Element $\bar{0}$ is absorptive with respect to $\otimes$. For $s \in S$, the equality $s^* = \bar{1} \oplus s \otimes s^* = \bar{1} \oplus s^* \otimes s$ must be hold.

A path p is a sequence of vertices $(v_0, v_1, ..., v_{t-1}, v_t)$, where $0 \leq t$ and $(v_{i-1}, v_i) \in E$. The weight of the path p is defined as follows:

$$w(p) = w_1 \otimes w_2 \otimes ... \otimes w_t,$$

where w_i is the weight of the edge (v_{i-1}, v_i). The APP is the determination of the sum of weights of all possible paths between each pair of vertices (i, j). If $P(i, j)$ is the set of all the possible paths from i to j, the APP is to find the values:

$$d_{ij} = \bigoplus_{p \in P(i,j)} w(p).$$

The APP can be formulated in matrix form. In this case, we associate an $n \times n$ matrix $A = [a_{ij}]$ with the weighted graph $G = (V, E, w)$, where $a_{ij} = w((i,j))$ if $(i,j) \in E$, and $a_{ij} = \bar{0}$ otherwise. This matrix representation permits us to formulate the APP as the computation of a sequence of matrices $A^{(k)} = [a_{ij}^{(k)}]$, $1 \leq k \leq n$. The value $a_{ij}^{(k)}$ is the weight of all the possible paths from vertex i to j with intermediate vertices v, $1 \leq v \leq k$. Initially, $A^{(0)} = A$, then $A^* = A^{(n)}$, where $A^* = [a_{ij}^*]$ is an $n \times n$ matrix satisfying $a_{ij}^* = d_{ij}$ if $(i,j) \in P(i,j)$, and $a_{ij}^* = \bar{1}$ otherwise. $\bar{1}$ is the weight of the empty path [10].

Instances of the APP are defined by specializing the semiring $(S, \oplus, \otimes, *, \bar{0}, \bar{1})$. Some of them are detailed below:

- All-pairs shortest distances in a weighted graph: the weights a_{ij} are taken from $S = R_+ \cup +\infty$ (it is assumed that $a_{ij} = +\infty$ if the vertices i and j are not connected.). R_+ is the set of non-negative real numbers. $\oplus$ = min, $\otimes$ = add, and $*$ operation is $a^* = 0$ for all $a \in S$. This specialization leads to the Floyd's algorithm for the all-pairs shortest path problem.
- Transitive and reflexive closure of a binary relation: the weights a_{ij} are taken from $S = \{false, true\}$. $\oplus$ and $\otimes$ are the Boolean or and and operations respectively, and $*$ operation is $a^* = true$ for all $a \in S$. This specialization produces the Warshall's algorithm for transitive closure of a Boolean matrix.
- Minimum-cost spanning tree: the weights a_{ij} are taken from $S = R_+ \cup +\infty$ (it is assumed that $a_{ij} = +\infty$ if the vertices i and j are not connected.). R_+ is the set of non-negative real numbers. $\oplus$ = min, $\otimes$ = max, and $*$ operation is $a^* = 0$ for all $a \in S$. This specialization yields the Maggs-Plotkin algorithm [9].
- Tunnel problem (also called the network capacity problem): the weights a_{ij} are taken from $S = R_+ \cup +\infty$ (it is assumed that $a_{ij} = 0$ if the vertices i and j are not connected.). R_+ is the set of non-negative real numbers. $\oplus$ = max, $\otimes$ = min, and $*$ operation is $a^* = 0$ for all $a \in S$.
- Inversion of a real non-singular $n \times n$ matrix $A = [a_{ij}]$: the weights a_{ij} are taken from $S = R$. $\oplus$ and $\otimes$ are the conventional arithmetic add and multiply operations on R respectively, and $*$ operation is $a^* = 1/(1-a)$ for all $a \in R$ with $a \neq 1$, and a^* is undefined for $a = 1$. Solving the APP yields $(I_n - A)^{-1}$, where I_n is the $n \times n$ identity matrix. Minor changes in the algorithm allow A^{-1} to be computed directly [6]. This specialization generates the Gauss-Jordan algorithm.

3 Blocked Algorithm for Solving the APP

Fig. 1 shows a scalar algorithm for solving the APP. By specializing the semiring, this basic algorithm can be either the Floyd's algorithm, Warshall's algorithm, Maggs-Plotkin algorithm, algorithm for the tunnel problem, or Gauss-Jordan algorithm. The scalar algorithm is computationally intensive, which involves n^3 "multiply-add" operations for a given $n \times n$ matrix. Note that a "multiply-add" operation is defined by specializing the semiring for each APP application.

To achieve high performance, reusing data stored in cache memory and registers is crucial because of the processor-memory gap on modern microprocessors. Blocking or tiling is a common optimization to increase a degree of data reuse. Fig. 2 shows a universal blocked algorithm for the APP.

The blocked algorithm solves the APP as computations of a sequence of $B \times B$ matrices, where B is a block size. For the sake of simplicity, we assume that a problem size n is in multiples of B. In the blocked algorithm, $\oplus$ and $\otimes$ are extended to $\oplus_b$ and $\otimes_b$ as matrix operations. $\oplus_b$ is element-wise matrix-add operation: $A \oplus_b B = [a_{ij} \oplus b_{ij}]$. $\otimes_b$ is matrix-multiply operation: $A \otimes_b B = [c_{ij}]$, where $c_{ij} = \bigoplus_{k=1}^{n} a_{ik} \otimes b_{kj}$.

For each k-th iteration, a diagonal block $A_{K,K}$ (*black* block) is computed as solving the APP with the scalar algorithm; secondly, column blocks (*red* blocks) are updated with the block $A_{K,K}$; thirdly, *white* blocks are updated with the row/column blocks; finally, row blocks (*blue* blocks) are updated with the block $A_{K,K}$. When $B = 1$ or $B = n$, the blocked algorithm is transformed to the scalar algorithm with no cache and n^2 cache size, respectively.

The code of lines 9-12 in Fig. 2 is a 3-D vectorized form of

for all $I = 1 : B : n$ $(I \neq K)$ % White block update
   for all $J = 1 : B : n$ $(J \neq K)$
      for $i = 1 : B$
         for $j = 1 : B$
            for $k = 1 : B$
               $A_{I,J}(i,j) = A_{I,J}(i,j) \oplus A_{I,K}(i,k) \otimes A_{K,J}(k,j)$
            end k **end** j **end** i **end** J **end** I

for $k = 1 : n$
   $a_{kk} = a_{kk}^*$ % Black element update
   for all $i = 1 : n$ $(i \neq k)$ % Red element update
      $a_{ik} = a_{ik} \otimes a_{kk}$
   end i
   for all $i = 1 : n$ $(i \neq k)$ % White element update
      for all $j = 1 : n$ $(j \neq k)$
         $a_{ij} = a_{ij} \oplus a_{ik} \otimes a_{kj}$
      end j **end** i
   for all $j = 1 : n$ $(j \neq k)$ % Blue element update
      $a_{kj} = a_{kk} \otimes a_{kj}$
   end j **end** k

Fig. 1. Scalar algorithm for solving the APP

```
1  for K = 1 : B : n
2    A_{K,K} = A*_{K,K}   % Black block update (computed as shown in Fig. 1)
3    if B == n
4      break
5    end
6    for all I = 1 : B : n (I ≠ K) % Red block update
7      A_{I,K} = A_{I,K} ⊗_b A_{K,K}
8    end I
9    for all I = 1 : B : n (I ≠ K) % White block update
10     for all J = 1 : B : n (J ≠ K)
11       A_{I,J} = A_{I,J} ⊕_b A_{I,K} ⊗_b A_{K,J}
12     end J end I
13   for all J = 1 : B : n (J ≠ K) % Blue block update
14     A_{K,J} = A_{K,K} ⊗_b A_{K,J}
15 end J end K
```

Fig. 2. 3-D vectorized blocked algorithm for solving the APP

As described, the series of scalar operations can be replaced with one matrix operation by vectorizing three nested loops (i-loop, j-loop, and k-loop). The 3-D vectorization reduces the loop overhead and presents an appropriate algorithmic scheme suitable for exploiting SIMD matrix instructions.

The blocked algorithm is applicable to the all APP applications; for example, the blocked Floyd's algorithm studied in [1,2,3,4] can be derived by replacing $\oplus_b$ with element-wise matrix-min operations and $\otimes_b$ with matrix-multiply operations of the underlying algebra, i.e., $c_{ij} = \min_{k=1}^{n}(a_{ik} + b_{kj})$.

The *black block update* requires B^3 "multiply-add" operations and $2B^2$ load/store operations (B^2 load operations and B^2 store operations). The *blue block update* and *red block update* requires $B^3 \times n/B$ "multiply-add" operations and $3B^2 \times n/B$ load/store operations ($2B^2$ load operations and B^2 store operations), respectively. The *white block update* requires $B^3 \times n^2/B^2$ "multiply-add" operations and $4B^2 \times n^2/B^2$ load/store operations ($3B^2$ load operations and B^2 store operations). The total number of "multiply-add" operations is

$$n/B(B^3 + 2 \cdot B^3 \cdot n/B + B^3 \cdot n^2/B^2) = n^3 + 2n^2 B + nB^2, 1 < B < n.$$

The total number of load/store operations is

$$n/B(2B^2 + 2 \cdot 3B^2 \cdot n/B + 4B^2 \cdot n^2/B^2) = 4n^3/B + 6n^2 + 2nB, 1 < B < n.$$

In the blocked algorithm, a degree of data reuse, i.e., ratio of "multiply-add" operations to load/store operations, is roughly $B/4$. In the scalar algorithm, n^3 "multiply-add" operations and $4n^3$ load/store operations are required, i.e., a degree of data reuse is minimum, $O(1)$. When $B = n$, the algorithm involves n^3 "multiply-add" operations and $2n^2$ load/store operations since an $n \times n$ matrix A is initially loaded into cache from main memory and finally stored in main memory from cache after all computations are done. In this case, a degree of data reuse is maximum, $O(n)$.

4 FastMATH Processor

The Intrinsity FastMATH processor is a fixed-point embedded processor optimized for matrix and vector computations. The processor integrates a MIPS32 core with a SIMD matrix coprocessor capable of 64 giga operations per second (GOPS) peak performance at 2 GHz.

The MIPS32 core is a dual-issue core feasible one MIPS32 instruction and one matrix instruction in one clock cycle. The MIPS 32 core contains 16 KB instruction cache, 16 KB data cache, and 16-entry fully associative translation look-aside buffer.

The matrix coprocessor consists of a 4×4 array of 32-bit processing elements. Each processing element has sixteen 32-bit general-purpose registers, two 40-bit accumulators, and four 3-bit condition code registers. From a programmer's perspective for the coprocessor, there are sixteen matrix registers that each can hold a 4×4 matrix of 32-bit elements, two matrix accumulators that each can hold a 4×4 matrix of 40-bit elements, and four matrix condition control registers that each can hold a 4×4 matrix of 3-bit condition code elements. The coprocessor is directly coupled to the 1 MB on-chip L2 cache (64 GB/s) for feeding data fast to its SIMD execution unit.

The coprocessor provides the following matrix operations: element-wise operations (addition, subtraction, Boolean, compare, pop count, shift, rotate, saturate, multiply, multiply-add, multiply-subtract), data movement and rearrangement operations (select row, select column, transpose, shift row, shift column, table lookup, block rearrangement), computation operations (matrix-multiply, sum row, sum column), and matrix load/store operations.

Fig. 3 shows performance of the matrix-matrix multiplication of the vendor provided BLAS. As shown in Fig. 3, although the performance is not well sustained for large matrices, the matrix-matrix multiplication shows remarkably high performance, 48.3 GOPS as peak performance for a matrix size is 352×352 on the one-chip processor.

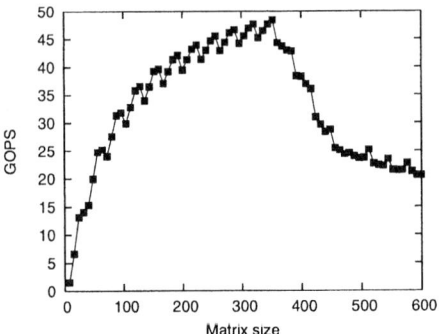

Fig. 3. Performance of the matrix-matrix multiplication on the FastMATH processor

5 FastMATH Implementation

The FastMATH processor supports matrix instructions that operate on 4×4 matrix data (32-bit integer) in parallel. In this section, we present an implementation of a 2-D vectorized blocked Floyd's algorithm with 4×4 matrix operations and the optimal block size that maximizes data reuse of matrix registers.

The blocked Floyd's algorithm has four different types of data dependencies. Because of the data dependencies, the number of matrices needed for each of block updates is different. The *black block update* involves one $B \times B$ matrix. The *red/blue block update* needs two $B \times B$ matrices. The *white block update* requires three $B \times B$ matrices. Obviously, the *white block update* is the most compute and memory intensive part of the algorithm.

In order to minimize processor-memory traffic as much as possible for the *white block update*, we consider data reuse of the matrix registers and chose the optimal block size of B. On the FastMATH processor, there are sixteen general-purpose matrix registers that each can hold a 4×4 matrix. Since two matrix registers are needed to store intermediate results, remaining fourteen matrix registers can be used for storing three $B \times B$ matrices. Therefore, the optimal size of B can be chosen by using the following equation: $3B^2 \leq \lfloor 14 \cdot 4^2 \rfloor$; i.e., $B \leq 8$. Accordingly, we employ the optimal block size of $B = 8$. Matrix data is loaded into a matrix register in row major order. Thus, it is necessary to arrange sequential data into blocks of matrix data. An 8×8 matrix is stored with four matrix registers by dividing it into four blocks of a 4×4 matrix.

Fig. 4 shows a 2-D vectorized 8×8 *white block update* with the 4×4 matrix operations. Since the FastMATH processor does not support the matrix-multiply instruction of the underlying algebra for the all-pairs shortest path problem, we vectorized two loops (i-loop and j-loop) and replaced series of scalar operations in those loops with 4×4 matrix-min or 4×4 matrix-add FastMATH instructions. The Y = repmat(X,m,n) function creates a large matrix Y consisting of an $m \times n$ tiling of copies of X [17]; for example, suppose $X = \begin{bmatrix} 3 & 4 \end{bmatrix}$, the repmat(X,2,1) creates a 2×2 matrix $Y = \begin{bmatrix} 3 & 4 \\ 3 & 4 \end{bmatrix}$.

for $I = 1 : 8 : n$ $(I \neq K)$
  for $J = 1 : 8 : n$ $(J \neq K)$
    for $k = 1 : 4$
      $A_{I,J}$=min($A_{I,J}$,repmat($A_{I,K}(:,k),1,4$)+repmat($A_{K,J}(k,:),4,1$))
      $A_{I,J+4}$=min($A_{I,J+4}$,repmat($A_{I,K}(:,k),1,4$)+repmat($A_{K,J+4}(k,:),4,1$))
      $A_{I+4,J}$=min($A_{I+4,J}$,repmat($A_{I+4,K}(:,k),1,4$)+repmat($A_{K,J}(k,:),4,1$))
      $A_{I+4,J+4}$=min($A_{I+4,J+4}$,repmat($A_{I+4,K}(:,k),1,4$)+repmat($A_{K,J+4}(k,:),4,1$))
      $A_{I,J}$=min($A_{I,J}$,repmat($A_{I,K+4}(:,k),1,4$)+repmat($A_{K+4,J}(k,:),4,1$))
      $A_{I,J+4}$=min($A_{I,J+4}$,repmat($A_{I,K+4}(:,k),1,4$)+repmat($A_{K+4,J+4}(k,:),4,1$))
      $A_{I+4,J}$=min($A_{I+4,J}$,repmat($A_{I+4,K+4}(:,k),1,4$)+repmat($A_{K+4,J}(k,:),4,1$))
      $A_{I+4,J+4}=$
        min($A_{I+4,J+4}$,repmat($A_{I+4,K+4}(:,k),1,4$)+repmat($A_{K+4,J+4}(k,:),4,1$))
    end k **end** J **end** I

Fig. 4. 2-D vectorized 8×8 *white block update* with 4×4 matrix operations

The code in Fig. 4 is based on a 2-level blocked algorithm in the case where the block size of 1-level tiling is 8×8 and that of 2-level tiling is 4×4. The other block updates also can be represented in the same way, based on the 2-level blocked algorithm.

6 Experimental Results

To evaluate performance, we measured CPU cycles and calculated performance in GOPS. In addition, we investigated a percentage of matrix instructions by counting matrix and MIPS32 instructions in a program on the FastMATH processor.

Fig. 5 shows the performance of the blocked Floyd's algorithm described in Section 5 on the FastMATH processor. Experiments were carried out for problem sizes ranging from $n = 8$ to $n = 600$ in multiples of 8. The peak performance is 9.27 GOPS for a matrix size of 360×360. For matrix sizes larger than 360×360, the performance degrades dramatically. This is because the block size is too small for L2 cache blocking, results in the increased L2 cache miss ratios; however, note that the optimal block size for this parallel implementation was chosen taking into account the maximum data reuse of matrix registers, not L2 cache.

In order to reuse data in the L2 cache and matrix registers, we implemented a 3-level blocked Floyd's algorithm. In this implementation, the problem is tiled into blocks of a 128×128 matrix. Each 128×128 block is tiled further into sub-blocks of an 8×8 matrix, and each of 8×8 blocks is updated with the parallel 4×4 matrix operations.

The result of the 3-level blocked Floyd's algorithm is shown in Fig. 6. Experiments were performed for matrix sizes varying from 128×128 to 1024×1024 in multiples of 128. The peak performance is 7.59 GOPS for a matrix size of 256×256. Although the cache-friendly implementation shows slightly slower performance when matrix sizes are from 128×128 to 384×384, for larger matrices, it shows better performance of more than 4 GOPS, which is roughly $4\times$ improvement compared to the performance of the 2-level blocked algorithm shown in Fig. 5.

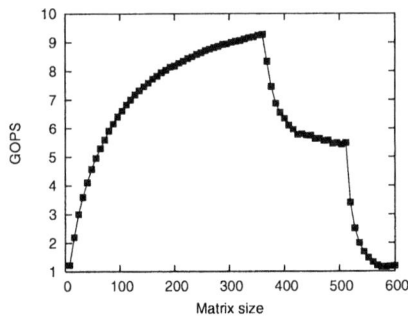

Fig. 5. Performance of the 2-level blocked Floyd's algorithm

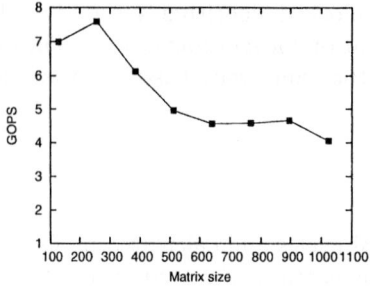
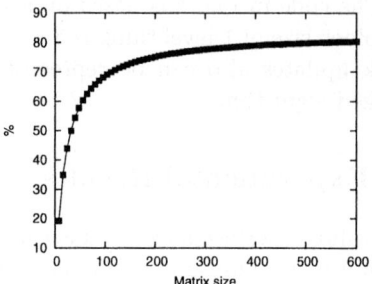

Fig. 6. Performance of the 3-level blocked Floyd's algorithm

Fig. 7. Percentage of matrix instructions

Fig. 7 shows the percentage of matrix instructions over the total number of instructions issued during execution of the 2-level blocked Floyd's algorithm. The percentage of matrix instructions is relatively high. It is saturated very quickly and reached to 80% for large matrices. We can conclude that according to the high percentage of matrix instructions, the blocked Floyd's algorithm is well-suited for implementing with matrix instructions, as well as the other APP applications.

7 Conclusion

This paper has presented a universal blocked algorithm applicable to all APP applications and a parallel implementation of the blocked Floyd's algorithm with SIMD matrix instructions as an instance of the APP. The blocked implementation with SIMD matrix instructions showed high performance and high usage rates of matrix instructions on the FastMATH processor. According to findings from our experimental results, the SIMD matrix extension to (super)scalar processor would be very useful for fast solution of many matrix-formulated problems.

References

1. Venkataraman, G., Sahni, S., Mukhopadhyaya, S.: A blocked all-pairs shortest-paths algorithm, in: Proceedings of the the 7th Scandinavian Workshop on Algorithm Theory (SWAT2000), Bergen, Norway, July 2000
2. Penner, M., Prasanna, V.K.: Cache-friendly implementations of transitive closure, in: Proceedings of the 2001 International Conference on Parallel Architectures and Compilation Techniques (PACT'01), Barcelona, Spain, September 2001.
3. Griem, G., Oliker, G.: Transitive closure on the Imagine stream processor, in: the 5th Workshop on Media and Stream Processors (MSP-5), San Diego, CA, December 2003
4. Park, J.-S., Penner, M., Prasanna, V.K.: Optimizing graph algorithms for improved cache performance, IEEE Transactions on Parallel and Distributed Systems 15 (9) (2004) pp. 769–782.

5. Floyd, R.W.: Algorithm 97: Shortest path, Communications ACM 5 (6) (1962) pp. 345.
6. Rote, G.: A systolic array algorithm for the algebraic path problem (shortest paths; matrix inversion), Computing 34 (1985) pp. 191–219.
7. Robert, Y., Trystram, D.: Parallel implementation of the algebraic path problem, in: Proceedings of the Conference on Algorithms and Hardware for Parallel Processing (CONPAR 86), 1986, pp. 149–156.
8. Nunez, F.J., Valero, M.: A block algorithm for the algebraic path problem and its execution on a systolic array, in: Proceedings of the International Conference on Systolic Arrays, 1988, pp. 265–274.
9. Maggs, B.M., Plotkin, S.A.: Minimum-cost spanning tree as a path-finding problem, Information Processing Letters 26 (1988) pp. 291–293.
10. Rote, G.: Path problems in graphs, Computing Supplementum 7 (1990) pp. 155–189.
11. Sedukhin, S.: Design and analysis of systolic algorithms for the algebraic path problem, Computers and Artificial Intelligence 11 (3) (1992) pp. 269–292.
12. Fink, E.: A survey of sequential and systolic algorithms for the algebraic path problem, Technical report CS-92-37, Department of Computer Science, University of Waterloo, 1992.
13. Cachera, D., Rajopadhye, S., Risset, T., Tadonki, C.: Parallelization of the algebraic path problem on linear SIMD/SPMD arrays, Technical report 1346, Irisa, 2001.
14. Olson, T.: Advanced processing techniques using the Intrinsity FastMATH processor, in: Embedded Processor Forum, California, USA, May 2002.
15. Anantha, V., Harle, C., Olson, T., Yost, G.: An innovative high-performance architecture for vector and matrix math algorithms, in: Proceedings of the 6th Annual Workshop on High Performance Embedded Computing (HPEC 2002), Massachusetts, USA, September 2002
16. Intrinsity Software Application Writer's Manual, ver. 0.3, Intrinsity, Inc., 2003.
17. Using MATLAB Version 6, The Math Works, Inc., 2002.

A Parallel Distance-2 Graph Coloring Algorithm for Distributed Memory Computers[*]

Doruk Bozdağ[1], Umit Catalyurek[1], Assefaw H. Gebremedhin[2], Fredrik Manne[3], Erik G. Boman[4], and Füsun Özgüner[1]

[1] Ohio State University, USA
{bozdagd, ozguner}@ece.osu.edu, umit@bmi.osu.edu
[2] Old Dominion University, USA
assefaw@cs.odu.edu
[3] University of Bergen, Norway
Fredrik.Manne@ii.uib.no
[4] Sandia National Laboratories, USA
egboman@sandia.gov

Abstract. The distance-2 graph coloring problem aims at partitioning the vertex set of a graph into the fewest sets consisting of vertices pairwise at distance greater than two from each other. Application examples include numerical optimization and channel assignment. We present the first distributed-memory heuristic algorithm for this NP-hard problem. Parallel speedup is achieved through graph partitioning, speculative (iterative) coloring, and a BSP-like organization of computation. Experimental results show that the algorithm is scalable, and compares favorably with an alternative approach—solving the problem on a graph G by first constructing the square graph G^2 and then applying a parallel distance-1 coloring algorithm on G^2.

1 Introduction

An archetypal problem in the efficient computation of sparse Jacobian and Hessian matrices is the distance-2 (D2) vertex coloring problem in an appropriate graph [1]. D2 coloring also finds applications in channel assignment [2] and facility location problems [3]. It is closely related to a strong coloring of a hypergraph which in turn models problems that arise in the design of multifiber WDM networks [4]. The D2 coloring problem is known to be NP-hard [5].

In many parallel applications where a graph coloring is required, the graph is already distributed among processors. Under such circumstances, gathering the graph on one processor to perform the coloring may not be feasible due to memory constraints. Moreover, in some parallel applications the coloring needs to be performed repeatedly

---

[*] This work was supported in part by NSF grants ACI-0203722, ACI-0203846, ANI-0330612, CCF-0342615, CNS-0426241, NIH NIBIB BISTI P20EB000591, Ohio Board of Regents BRTTC BRTT02-0003, Ohio Supercomputing Center PAS0052, and SNL Doc.No: 283793. Sandia is a multiprogram laboratory operated by Sandia Corporation, a Lockheed Martin company, for the U.S. DOE's National Nuclear Security Administration under contract DE-AC04-94AL85000.

due to changes in the structure of the graph. Here, the coloring may take up a substantial amount of overall computation time unless a scalable algorithm is used.

A number of papers dealing with the design of efficient parallel distance-1 (D1) coloring algorithms have appeared [6,7,8,9]. For D2 coloring we are not aware of any work other than [10] where an algorithm for shared memory computers was presented.

In this paper, we present an efficient parallel D2 coloring algorithm suitable for distributed memory computers. The algorithm is an extension of the parallel D1 coloring algorithm presented in [6]. The latter is an iterative data parallel algorithm that proceeds in two-phased rounds. In the first phase, processors concurrently color the vertices assigned to them. Adjacent vertices colored in the same parallel step of this phase may result in inconsistencies. In the second phase, processors concurrently check the validity of the colors assigned to their respective vertices and identify a set of vertices that needs to be re-colored in the next round to resolve the detected inconsistencies. The algorithm terminates when every vertex has been colored correctly. To reduce communication frequency, the coloring phase is further decomposed into computation and communication sub-phases. During a computation sub-phase, a group of vertices, rather than a single vertex, is colored based on currently available color information. In a communication sub-phase processors exchange recent color information.

The key issue in extending this approach to the D2 coloring case is devising an efficient means of information exchange between processors hosting a pair of vertices that are two edges away from each other. We use a scheme in which the host processor of a vertex v is responsible for (i) coloring v, (ii) relaying color information to processors that store the D1 neighbors of v, and (iii) detecting inconsistencies that involve the D1 neighbors of v.

Our parallel D2 coloring algorithm has been implemented using MPI. Results from experiments performed on a 32-node PC cluster using a number of real-world as well as random graphs show that the algorithm is efficient and scalable. We have also compared our D2 coloring algorithm on a given graph G with the parallel D1 coloring algorithm from [6] applied to the square graph G^2. These results in general show that our algorithm scales better and uses less memory and storage.

In the sequel, we discuss preliminary concepts in Section 2; present our algorithm in Section 3; report experimental results in Section 4 and conclude in Section 5.

2 Distance-2 Graph and Hypergraph Coloring

Two distinct vertices in a graph $\mathcal{G} = (\mathcal{V}, \mathcal{E})$ are *distance-k* neighbors if the shortest path connecting them consists of at most k edges. A *distance-k coloring* of $\mathcal{G} = (\mathcal{V}, \mathcal{E})$ is a mapping $C : \mathcal{V} \to \{1, 2, \ldots, q\}$ such that $C(v) \neq C(w)$ whenever vertices v and w are distance-k neighbors. The associated optimization problem aims at minimizing q. A distance-k coloring of a graph $\mathcal{G} = (\mathcal{V}, \mathcal{E})$ is equivalent to a D1 coloring of the kth *power* graph $\mathcal{G}^k = (\mathcal{V}, \mathcal{F})$ where $(v, w) \in \mathcal{F}$ whenever vertices v and w are distance-k neighbors in $\mathcal{G}$. We denote the set of distance-k neighbors of vertex v by $N_k(v)$, and the set $N_k(v) \cup \{v\}$ by $N_k[v]$. For simplicity, we drop the subscript in the case where $k = 1$.

Let A be a symmetric matrix with nonzero diagonal elements and $\mathcal{G}_a(A) = (\mathcal{V}, \mathcal{E})$ be the *adjacency* graph of A, where $\mathcal{V}$ corresponds to the columns of A. As illustrated

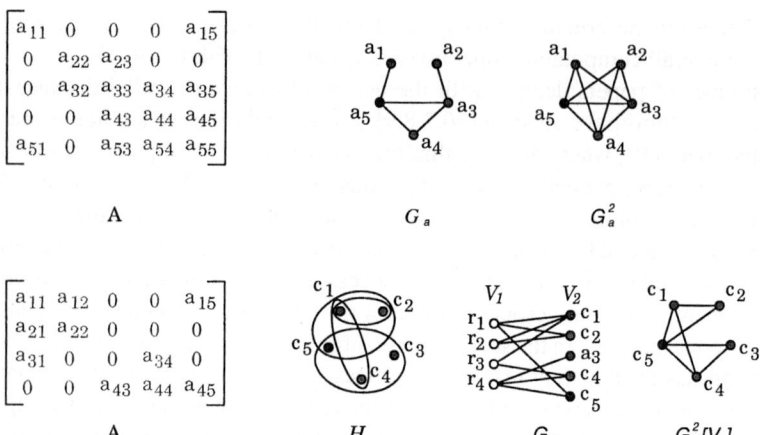

Fig. 1. Equivalence among structurally orthogonal column partition of A, D2 coloring of $\mathcal{G}(A)$ and D1 coloring of $\mathcal{G}^2(A)$. Top: symmetric case. Bottom: non-symmetric case (also shows equivalence with strong coloring of hypergraph H).

in the upper row of Figure 1, a partitioning of the columns of A into groups of structurally orthogonal columns is equivalent to a D2 coloring of $\mathcal{G}_a(A)$. (Two columns are structurally orthogonal if they do not have nonzero entries in the same row.) The right most subfigure in Figure 1 shows the equivalent D1 coloring in the square graph $\mathcal{G}_a^2$.

Now let A be non-symmetric. The *bipartite* graph of A is the graph $\mathcal{G}_b = (\mathcal{V}_1, \mathcal{V}_2, \mathcal{E})$ where $\mathcal{V}_1$ is the row vertex set, $\mathcal{V}_2$ is the column vertex set, and there exits an edge between row vertex r_i and column vertex c_j whenever $a_{ij} \neq 0$. As the lower row of Figure 1 illustrates, a partitioning of the columns of A into groups of structurally orthogonal columns is equivalent to a partial D2 coloring of $\mathcal{G}_b(A)$ on $\mathcal{V}_2$. The right most subfigure shows the equivalent D1 coloring of $\mathcal{G}_b^2[\mathcal{V}_2]$, the subgraph of the square graph $\mathcal{G}_b^2$ induced by $\mathcal{V}_2$.

D2 coloring of a bipartite graph is also related to a variant of hypergraph coloring. A hypergraph $\mathcal{H} = (\mathcal{V}, \mathcal{E})$ consists of a vertex set $\mathcal{V}$ and a collection $\mathcal{E}$ of subsets of $\mathcal{V}$ called *hyperedges*. A *strong hypergraph coloring* is a mapping $C : \mathcal{V} \rightarrow \{1, 2, \ldots, q\}$ such that $C(v) \neq C(v)$ whenever $\{v, w\} \subseteq e \in \mathcal{E}$. As Figure 1 illustrates, a strong coloring of a hypergraph is equivalent to a partial D2 coloring of its hyperedge-vertex incidence bipartite graph. For further discussion on the equivalence among matrix partitioning, D2 graph coloring and hypergraph coloring as well as their relationships to computation of Jacobians and Hessians, see [1].

3 Parallel Distance-2 Coloring

In this section we describe our new parallel D2 coloring algorithm for a general graph $\mathcal{G} = (\mathcal{V}, \mathcal{E})$. Initially, the input graph is assumed to be distributed among p processors. The set V_i of vertices in the partition $\{V_1, \ldots, V_p\}$ of $\mathcal{V}$ is assigned to and colored by processor P_i; we say that P_i *owns* V_i. P_i also stores the adjacency list of its vertices and the IDs of the processors owning them. This classifies $\mathcal{V}$ into *interior* and *boundary*

Algorithm 1 An iterative parallel distance-2 coloring algorithm
---
procedure PARALLELCOLORING($\mathcal{G} = (\mathcal{V}, \mathcal{E}), s$)
  Initial data distribution: $\mathcal{G}$ is divided into p subgraphs $G_1 = (V_1, E_1), \ldots, G_p = (V_p, E_p)$ where $V_1, \ldots, V_p$ is a partition of the set $\mathcal{V}$ and $E_i = \{(v, w) : v \in V_i, (v, w) \in \mathcal{E}\}$. Processor P_i owns the vertex set V_i, and stores the edge set E_i and the ID's of the processors owning the other endpoints of E_i.
  on each processor P_i, $i \in P = \{1, \ldots, p\}$
    Color interior vertices in V_i
    $U_i \leftarrow$ boundary vertices in V_i ▷ U_i is to be iteratively colored by P_i
    while $\exists j \in P, U_j \neq \emptyset$ **do**
      $W_i \leftarrow$ COLOR(G_i, U_i, s)   ▷ W_i is examined for conflicts by P_i
      $U_i \leftarrow$ DETECTCONFLICTS(G_i, W_i)
---

vertices. All D1 neighbors of an interior vertex are owned by the same processor as itself. A boundary vertex has at least one D1 neighbor owned by a different processor.

Clearly, any pair of interior vertices, that are assigned to different processors, can safely be colored concurrently. This is not true for a pair containing a boundary vertex. In particular, if such a pair is colored at the *same* parallel superstep, then the partners may receive the same color and result in a *conflict*. However, if we enforce that interior vertices be colored before or after boundary vertices, then a conflict can only occur for pairs of boundary vertices. Thus, the presented algorithm is concerned with parallel coloring of boundary vertices.

The main idea in our algorithm is to color boundary vertices concurrently in a speculative manner and then detect and rectify conflicts that may have arisen. The algorithm is iterative—it proceeds in *rounds*. Each round consists of a *tentative coloring* and a *conflict detection* phase. Both of these phases are performed in parallel. The latter phase detects conflicts in a current coloring and accumulates a list of vertices to be recolored in the next round. Given a pair of vertices involved in a conflict, only one of them needs to be recolored to resolve the conflict; the choice is done randomly. The algorithm terminates when there are no more vertices to be colored. The high-level structure of the algorithm is outlined in Algorithm 1.

Notice the termination condition of the while-loop. Even if a processor P_i currently has no vertices to color ($U_i = \emptyset$), it could still be active since other processors may require color information from P_i. Furthermore, P_i may participate in detecting conflicts on other processors.

For every path v, w, x, the host processor for vertex w is responsible for detecting D2 conflicts that involve vertices v and x as well as D1 conflicts involving w and its adjacent vertices $N(w)$. The set W_i in Algorithm 1 contains the current set of vertices residing on processor P_i that will be examined for detecting these conflicts. W_i includes both 'middle' vertices, and vertices from U_i. The discussion of the routines COLOR and DETECTCONFLICTS in Sections 3.1 and 3.2 will clarify these points.

3.1 The Tentative Coloring Phase

The tentative coloring phase is organized as a sequence of *supersteps*. In each superstep, each processor colors s vertices sequentially, and sends the colors of these vertices to

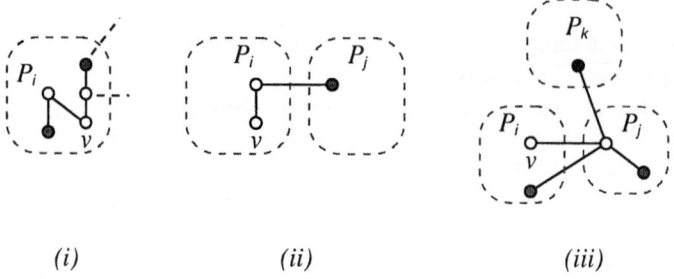

Fig. 2. Distribution scenarios of the distance-2 neighbors of vertex v across processors

processors owning their D1 neighbors. To perform the coloring, a processor first gathers information from other processors to build a (partial) list of forbidden colors for each of its boundary vertices scheduled to be colored in the current superstep. Such a list for a vertex v consists of the colors used by its already colored D2 neighbors. The colors of the off-processor vertices in $N(v)$ colored in previous supersteps are easily available since the host processor of v has already received and stored them. We refer to such and on-processor colors as *local*. However, the colors used by vertices exactly two edges away from v may have to be obtained from another processor.

Figure 2 illustrates the three scenarios in which the vertices on a path v, w, x may be distributed among processors. Case *(i)* corresponds to the situation where both w and x are owned by P_i. Case *(ii)* shows the situation where w is owned by P_i and x is owned by P_j, $j \neq i$. In these two cases, the color of w is local to P_i. Case *(iii)* shows the situation where w is owned by P_j, and vertices v and x do not have a common D1 neighbor owned by P_i. Vertex x may be owned by any one of the three processors P_i, P_j, or P_k, $i \neq j \neq k$. In case *(iii)*, the color of x is not local to P_i and needs to be relayed through P_j which is capable of detecting the situation. In particular, P_j builds and sends a list of forbidden colors for each vertex owned by P_i that P_i cannot access directly. Since P_j does not know the internal structure of the vertices in P_i, it includes the color of every x in the list of forbidden colors for v for each path v, w, x where w is owned by P_j.

At the beginning of the algorithm, each processor sends a coloring-schedule of its boundary vertices to neighboring processors. In this way, each processor will know the D2 color information it needs to send in each superstep. Note that it is only necessary to send information regarding D1 neighbors of a vertex owned by another processor. Each processor then computes a list X_i of vertices on neighboring processors for which it must supply color information. With the knowledge of X_i, processor P_i can now be "pro-active" in building and sending lists of relevant color information. When a processor receives the partial lists of forbidden colors from all of its neighboring processors, it merges these lists with local color information to determine a complete list of forbidden colors for its vertices scheduled to be colored in the current superstep. Using this information, a processor then speculatively colors these vertices and sends the new color information to processors owning D1 neighbors.

In addition to coloring vertices in the current set U_i, a processor also computes a list W_i of vertices that it needs to examine in the conflict detection phase. Two vertices are involved in a conflict only if they are colored in the same superstep. Thus W_i consists of (i) every vertex that has at least two neighbors on different processors that are colored in the same superstep, and (ii) every vertex v in U_i that has at least one neighbor on a processor P_j, $j \neq i$, colored in the same superstep as v. The tentative coloring routine sketched so far is outlined with more details in Algorithm 2.

The set W_i is efficiently determined in the following manner. The vertices in $X_i \cup U_i$ are traversed a superstep at a time. For each superstep, first, each vertex in U_i and its neighboring boundary vertices are *marked*. Then for each vertex $v \in X_i$ the vertices in $N(v)$ owned by processor P_i are marked. If this causes some vertex to be marked twice in the same superstep, then the vertex is added to W_i. The combined sequential work carried out by P_i and its neighboring processors to perform the coloring of U_i is $O(\Sigma_{v \in U_i} |h(v)|)$ where $h(v)$ is the graph induced by the edges incident on $N[v]$. Summing over all processors, the total work involved in coloring the vertices in $U = \cup U_i$ is $O(\Sigma_{v \in U} |h(v)|)$ which is equivalent to the complexity of a sequential algorithm.

Algorithm 2 Speculative coloring

1: **function** COLOR(G_i, U_i, s)
2:     Partition U_i into ℓ_i subsets $U_{i,1}, U_{i,2}, \ldots, U_{i,\ell_i}$, each of size s, and send the schedule to relevant processors
3:     $X_i \leftarrow \bigcup_{j,k} U_{j,k}^i$ ▷ $U_{j,k}^i$: vertices received by P_i to be colored by P_j in step k
4:     **for each** $v \in U_i \cup X_i$ **do**
5:         $C(v) \leftarrow 0$ ▷ (re)initialize colors
6:     $W_i \leftarrow \emptyset$ ▷ W_i is used for detecting conflicts
7:     **for each** $v \in V_i$ s.t v has at least two neighbors in $X_i \cup U_i$ on different processors, both colored in the same superstep **do**
8:         $W_i \leftarrow W_i \cup \{v\}$
9:     **for each** $v \in U_i$ s.t v has at least one neighbor in X_i that is colored in the same superstep as v **do**
10:        $W_i \leftarrow W_i \cup \{v\}$
11:     **for** $k_i \leftarrow 1$ **to** ℓ_i **do**   ▷ each k_i corresponds to a superstep
12:         **for each** neighboring P_j where $k_j < \ell_j$ **do**   ▷ P_j is not in its last superstep
13:             Build and send lists of forbidden colors to P_j for relevant vertices in U_{j,k_j+1}
14:         Receive and merge lists of forbidden colors for relevant vertices in U_{i,k_i}
15:         Update lists of forbidden colors with local color information
16:         **for each** $v \in U_{i,k_i}$ **do**
17:             $C(v) \leftarrow c$ s.t. $c \neq 0$ is the smallest permissible color for v
18:         Send colors of relevant vertices in U_{i,k_i} to neighboring processors
19:     **while** $\exists j \in P$, s.t. P_j is a neighbor of P_i and $k_j \leq l_j$ **do**
20:         Receive color information for superstep k_j from P_j
21:         **if** $k_j < l_j$ **then**
22:             Build and send list of forbidden colors to P_j for relevant vertices in U_{j,k_j+1}
23:     **return** W_i

For each $v \in U_i$, processor P_i receives the colors of vertices exactly two edges from v from processors hosting vertices in $N(v)$. Meanwhile, such processors receive the color of v from P_i. The only time a color might be sent to P_i more than once is when there exists a triangle v, w, x where w is owned by P_j, x is owned by P_k, and

i, j and k are all different. In such a case, both P_j and P_k would send the color of x to P_i. In any case, the overall size of communicated data is bounded by $\Sigma_{v \in U U_i} |h(v)|$.

The discussion above implies that a partitioning of $\mathcal{G}$ among processors where the number of boundary vertices is small relative to the number of interior vertices on each processor is highly desirable as it reduces communication cost.

3.2 The Conflict Detection Phase

A conflict involving a pair of adjacent vertices is detected by both processors owning these vertices. A conflict involving a pair of vertices exactly two edges apart is detected by the processor owning the middle vertex. To resolve a conflict, one of the involved vertices is randomly chosen to be recolored in the next round. Algorithm 3 outlines the parallel conflict detection phase DETECTCONFLICTS executed on each processor P_i. This routine returns a set of vertices to be colored in the next round by P_i.

Each processor P_i accumulates and sends a list $R_{i,j}$ of vertices to be recolored by each P_j in the next round. P_i is responsible for recoloring vertices in $R_{i,i}$ and therefore adds received notifications $R_{j,i}$ from each neighboring processor P_j to $R_{i,i}$.

To efficiently determine the subset of W_i that needs to be recolored, we use two color-indexed tables $seen[]$ and $where[]$. The assignment seen[c] = w for a vertex $w \in W_i$ is effected if at least one vertex in N[w] of color c has already been encountered. The entry $where[c]$ stores the vertex with the lowest random value among these. Initially both $seen[C(w)]$ and $where[C(w)]$ are set to w. This ensures that any conflict involving w and a vertex in $N(w)$ will be discovered. For each neighbor of w, a

---

Algorithm 3 Conflict Detection

1: **function** DETECTCONFLICTS(G_i, W_i)
2:     $R_{i,j} \leftarrow \emptyset$ for each $j \in P$ ▷ $R_{i,j}$ is a set of vertices P_i notifies P_j to recolor
3:     **for each** vertex $w \in W_i$ **do**
4:        $seen[C(w)] \leftarrow w$
5:        $where[C(w)] \leftarrow w$
6:        **for each** $x \in N(w)$ **do**
7:           **if** $seen[C(x)] = w$ **then**
8:              $v \leftarrow where[C(x)]$
9:              **if** $r(v) \leq r(x)$ **then**   ▷ $r(x)$ is a random number associated with x
10:                 $R_{i,I(x)} \leftarrow R_{i,I(x)} \cup \{x\}$ ▷ $I(u)$ is ID of processor owning u
11:              **else**
12:                 $R_{i,I(v)} \leftarrow R_{i,I(v)} \cup \{v\}$
13:                 $where[C(x)] \leftarrow x$
14:           **else**
15:              $seen[C(x)] \leftarrow w$
16:              $where[C(x)] \leftarrow w$
17:     **for each** $j \neq i \in P$ **do**
18:        send $R_{i,j}$ to processor P_j
19:     **for each** $j \neq i \in P$ **do**
20:        receive $R_{j,i}$ from processor P_j
21:        $R_{i,i} \leftarrow R_{i,i} \cup R_{j,i}$
22:     **return** $R_{i,i}$

check on whether its color has already been seen is done. If the check turns positive, the vertex that needs to be recolored is determined based on a comparison of random values and the table *where* is updated accordingly (Lines 3–16).

Note that in Line 6, it is sufficient to only check for conflicts using vertices that are both in $N(w)$ and in either U_i or X_i. However, determining which vertices in $N(w)$ this applies to takes more time than testing for a conflict. Also, it is not necessary to notify a neighboring processor on the detection of a conflict involving adjacent vertices as the conflict will also be discovered by the other processor.

4 Experimental Results

We carried out experiments on a 32-node PC cluster equipped with dual 2.4 GHz Intel P4 Xeon CPUs with 4 GB of memory. The nodes are interconnected via a switched 10Gbps Infiniband network. Our test set consists of 21 graphs from molecular dynamics and finite element applications [11,7,12,13]. We report average results for each class of graphs, instead of individual graphs. Each result is in turn an average of 5 runs.

The left half of Table 4 displays the structural properties of the test graphs. The first part of the right half lists the number of colors and the runtime in milliseconds used by a

Table 1. Structural properties of the test graphs classified according to application area (left). Performance results (right). Sources: MD [13]; FE [7]; CA, SH [11]; ST, AU, CE [12].

app	name	$|V|$	$|E|$	Degree max	Degree avg	D1 time	D1 colors	D2 time (norm.)	D2 colors	D1 on $\mathcal{G}^2$ (norm.) $\times |E|$	conv. time	color. time
MD	popc-br-4	24,916	255,047	43	20	3.3	21	20.6	75	4.7	33.0	4.8
MD	er-gre-4	36,573	451,355	42	25	5.5	19	23.2	66	5.0	34.6	5.1
MD	apoa1-4	92,224	1,131,436	43	25	16.5	20	18.1	73	5.0	28.5	4.3
FE	144	144,649	1,074,393	26	15	44.3	12	20.5	41	4.8	25.8	4.0
FE	598a	110,971	741,934	26	13	35.0	11	20.9	38	4.7	28.3	4.0
FE	auto	448,695	3,314,611	37	15	248.9	13	16.1	42	4.9	16.6	4.5
CA	bmw3_2	227,362	5,530,634	335	49	53.3	48	42.7	336	3.2	35.7	3.0
CA	bmw7st1	141,347	3,599,160	434	51	34.4	54	43.8	435	3.3	35.6	3.0
CA	inline1	503,712	18,156,315	842	72	179.5	51	70.5	843	7.0	63.8	6.2
ST	pwtk	217,918	5,708,253	179	52	50.7	48	45.5	180	2.9	34.8	2.7
ST	nasasrb	54,870	1,311,227	275	48	11.7	41	42.8	276	3.2	35.5	3.2
ST	ct20stif	52,329	1,323,067	206	51	12.5	49	46.0	210	3.8	37.7	3.5
AU	hood	220,542	5,273,947	76	48	58.5	42	35.8	103	3.2	29.2	2.7
AU	ldoor	952,203	22,785,136	76	48	249.7	42	35.8	112	3.2	29.0	2.7
AU	msdoor	415,863	9,912,536	76	48	106.2	42	36.9	105	3.2	29.8	2.7
CE	pkustk10	80,676	2,114,154	89	52	20.1	42	43.6	126	2.9	33.0	2.7
CE	pkustk11	87,804	2,565,054	131	58	23.7	66	57.6	198	4.2	45.6	3.8
CE	pkustk13	94,893	3,260,967	299	69	29.2	57	72.5	303	6.0	62.2	6.0
SH	shipsec1	140,874	3,836,265	101	54	34.9	48	48.5	126	3.1	37.7	8.2
SH	shipsec5	179,860	4,966,618	125	55	46.4	50	48.5	140	3.2	37.2	3.0
SH	shipsec8	114,919	3,269,240	131	57	29.1	54	52.9	150	3.5	41.2	3.4

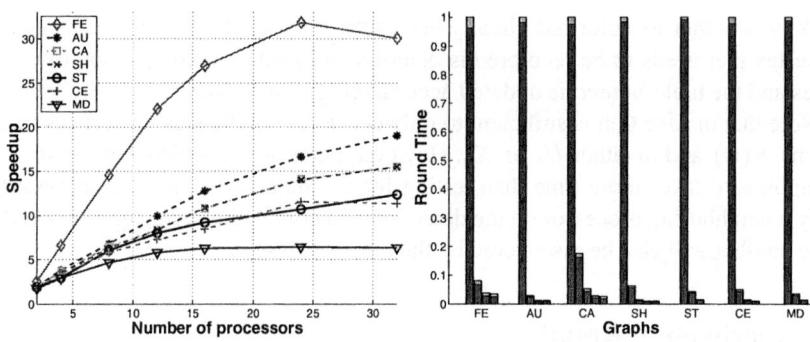

Fig. 3. (a) Speedup while varying number of processors for $s = 100$. (b) Breakdown of execution time into rounds and coloring and conflict detection phases for $p = 16$. Each bar is divided into time spent on coloring (bottom) and time spent on conflict detection (top).

sequential D1 coloring algorithm. The second part shows timings for two different ways of sequentially achieving a D2 coloring: a direct D2 coloring on G and a D1 coloring on G^2. The time spent on constructing $\mathcal{G}^2$ from $\mathcal{G}$ is given under column *conv. time*. The reported times have been normalized with respect to the corresponding time required for performing a D1 coloring. We also list the ratio of the number of edges in $\mathcal{G}^2$ to that in $\mathcal{G}$, to show the relative increase in storage requirement.

As one can see, $\mathcal{G}^2$ requires a factor of nearly 3 to 7 more storage than $\mathcal{G}$. D2 coloring on $\mathcal{G}$ is in most cases slightly slower than constructing and then D1 coloring $\mathcal{G}^2$. We believe this is due to the fact that a D1 coloring on $\mathcal{G}^2$ accesses memory more sequentially in comparison with a D2 coloring on $\mathcal{G}$.

Figure 3(a) shows the speedup obtained in using our D2 coloring algorithm on G while keeping the superstep size fixed at 100. For most graph classes, reasonable speedup is obtained as the number of processors is increased. We have also conducted experiments to investigate the impact of superstep size. We found that with the exception of extreme values, superstep size does not significantly influence speedup.

We observed that the number of conflicts increases with increasing number of processors and superstep size. Still, it stays fairly low and does not exceed 10% of the number of vertices with the exception of MD graphs with up to 32 processors for $s = 100$. The number of rounds the algorithm had to iterate was observed to be consistently low, increasing only slowly with superstep size and number of processors. This is due to the fact that the number of initial conflicts drops rapidly between successive rounds. To further show how the time within each round is spent we present Figure 3(b). The figure shows the time spent on coloring boundary vertices and conflict detection in each round for 16 processors. All timings are normalized with respect to the time spent in the first round, excluding the time spent on coloring interior vertices.

Figure 4(a) shows how the total time is divided into time spent on coloring interior vertices, coloring boundary vertices, and conflict detection. All timings are normalized with respect to the sequential coloring time. As the number of processors increases, the time spent on coloring boundary vertices does not change much while the time spent on coloring interior vertices decreases almost linearly. This should be seen in light of

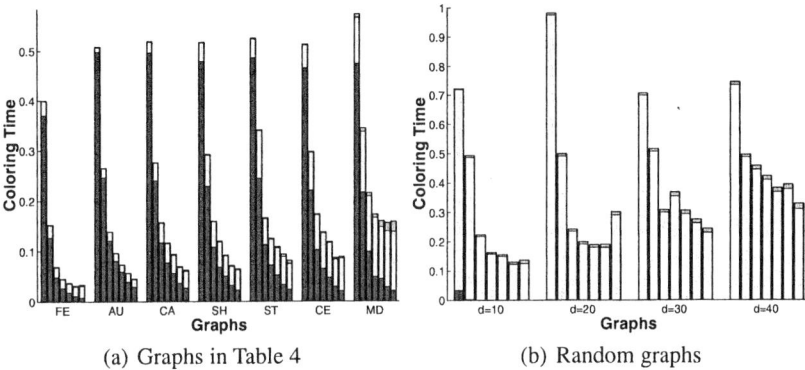

(a) Graphs in Table 4   (b) Random graphs

Fig. 4. Breakdown of execution time into time spent on coloring internal vertices (bottom), coloring boundary vertices (middle), and conflict detection (top). For each graph class, timings for $p = 2, 4, 8, 12, 16, 24, 32$ are reported.

the fact that the number of boundary vertices increases as more processors are applied whereas the coloring of interior vertices does not involve any communication.

To investigate scalability on boundary vertices, we performed experiments on random graphs. For a random graph almost every vertex becomes a boundary vertex regardless of how the graph is partitioned. We generated random graphs with 100,000 vertices and with average degrees of 10, 20, 30, and 40. Figure 4(b) shows that the algorithm scales fairly well and almost all the time is spent on coloring boundary vertices.

We have also evaluated D1 coloring on $\mathcal{G}^2$ and experimental results (omitted due to space constraints) indicate that this approach is less scalable and requires more storage than the D2 coloring on $\mathcal{G}$ approach due to large density of $\mathcal{G}^2$. We intend to implement parallel construction of $\mathcal{G}^2$ and investigate trade-offs between these two approaches in more detail in a future work.

5 Conclusion

We have presented an efficient parallel distance-2 coloring algorithm suitable for distributed memory computers and experimentally demonstrated its scalability. In a future work we plan to adapt the presented algorithm to solve the closely related strong hypergraph coloring problem. This brings up the open problem of finding a suitable partition of the vertices and edges of a hypergraph.

References

1. Gebremedhin, A.H., Manne, F., Pothen, A.: What color is your jacobian? Graph coloring for computing derivatives. SIAM Rev. (2005) To appear.
2. Krumke, S., Marathe, M., Ravi, S.: Models and approximation algorithms for channel assignment in radio networks. Wireless Networks **7** (2001) 575 – 584
3. Vazirani, V.V.: Approximation Algorithms. Springer (2001)

4. Ferreira, A., Pérennes, S., Richa, A.W., Rivano, H., Stier, N.: Models, complexity and algorithms for the design of multi-fiber wdm networks. Telecommunication Systems **24** (2003) 123–138
5. McCormick, S.T.: Optimal approximation of sparse hessians and its equivalence to a graph coloring problem. Math. Programming **26** (1983) 153–171
6. Boman, E.G., Bozdağ, D., Catalyurek, U., Gebremedhin, A.H., Manne, F.: A scalable parallel graph coloring algorithm for distributed memory computers. (EuroPar 2005, to appear)
7. Gebremedhin, A.H., Manne, F.: Scalable parallel graph coloring algorithms. Concurrency: Practice and Experience **12** (2000) 1131–1146
8. Gebremedhin, A.H., Manne, F., Woods, T.: Speeding up parallel graph coloring. In: proceedings of Para 2004, Lecture Notes in Computer Science, Springer (2004)
9. Jones, M.T., Plassmann, P.: A parallel graph coloring heuristic. SIAM J. Sci. Comput. **14** (1993) 654–669
10. Gebremedhin, A.H., Manne, F., Pothen, A.: Parallel distance-k coloring algorithms for numerical optimization. In: proceedings of Euro-Par 2002. Volume 2400., Lecture Notes in Computer Science, Springer (2002) 912–921
11. : (Test data from the parasol project) http://www.parallab.uib.no/projects/parasol/data/.
12. : (University of florida matrix collection) http://www.cise.ufl.edu/research/sparse/matrices/.
13. Strout, M.M., Hovland, P.D.: Metrics and models for reordering transformations. In: proceedings of MSP 2004. (2004) 23–34

Fast Sparse Matrix-Vector Multiplication by Exploiting Variable Block Structure

Richard W. Vuduc[1] and Hyun-Jin Moon[2]

[1] Lawrence Livermore National Laboratory
richie@llnl.gov
[2] University of California, Los Angeles
hjmoon@cs.ucla.edu

Abstract. We improve the performance of sparse matrix-vector multiplication (SpMV) on modern cache-based superscalar machines when the matrix structure consists of multiple, irregularly aligned rectangular blocks. Matrices from finite element modeling applications often have this structure. We split the matrix, A, into a sum, $A_1 + A_2 + \ldots + A_s$, where each term is stored in a new data structure we refer to as *unaligned block compressed sparse row (UBCSR) format* . A classical approach which stores A in a block compressed sparse row (BCSR) format can also reduce execution time, but the improvements may be limited because BCSR imposes an alignment of the matrix non-zeros that leads to extra work from filled-in zeros. Combining splitting with UBCSR reduces this extra work while retaining the generally lower memory bandwidth requirements and register-level tiling opportunities of BCSR. We show speedups can be as high as 2.1× over no blocking, and as high as 1.8× over BCSR as used in prior work on a set of application matrices. Even when performance does not improve significantly, split UBCSR usually reduces matrix storage.

1 Introduction

The performance of diverse applications in scientific computing, economic modeling, and information retrieval, among others, is dominated by sparse matrix-vector multiplication (SpMV), $y \leftarrow y + A \cdot x$, where A is a sparse matrix and x, y are dense vectors. Conventional implementations using compressed sparse row (CSR) format storage usually run at 10% of machine peak or less on uniprocessors [16]. Higher performance requires a compact data structure and appropriate code transformations that best exploit properties of both the sparse matrix—which may be known only at run-time—and the underlying machine architecture. Compared to dense kernels, sparse kernels incur more overhead per non-zero matrix entry due to extra instructions and indirect, irregular memory accesses. We and others have studied techniques for selecting good data structures and for automatically tuning the resulting implementations (Section 6). Tuned SpMV can in the best cases achieve 31% of peak and 4× speedups over CSR [6,16,18].

The best performance occurs for the class of applications based on finite element method (FEM) modeling, but within this class there is a performance gap between matrices consisting primarily of dense blocks of a single size, uniformly aligned, and matrices whose structure consists of multiple block sizes with irregular alignment. For the former class, users often rely on so-called block compressed sparse row (BCSR) format, BCSR, which stores A as a sequence of fixed-size $r \times c$ dense blocks. BCSR uses one integer index of storage per block instead of one per non-zero as in CSR, reducing the index storage by $\frac{1}{rc}$. Moreover, fixed-sized blocks enable unrolling and register-level tiling of each block-multiply.

However, two difficulties arise in practice. First, *the best $r \times c$ varies both by matrix and by machine* [16], motivating automatic tuning. Secondly, *speedup is mitigated by fill-in of explicit zeros*. We observed cases in which BCSR reduces the execution time of SpMV to $\frac{2}{3}$ that of CSR (1.5× speedup) while also requiring storage of 50% additional explicit zero entries [16]. To reduce this work and achieve still better speedups, this paper considers simultaneously (a) *splitting A* into the sum $A = A_1 + \cdots + A_s$, where each A_i may be stored with a different block size, and (b) storing each A_i in a *unaligned block compressed sparse row (UBCSR) format* that relaxes *both* row and column alignments of BCSR, at the cost of indirection to x and y instead of just x as in BCSR and CSR.

We recently compared BCSR-based SpMV to CSR on 8 platforms and 44 matrices, and identified 3 classes of matrices [16, Chap. 4]: FEM matrices 2–9, whose structure consists essentially of a single block size uniformly aligned, FEM matrices 10–17, whose structure contains mixed block structure, and matrices from other applications (*e.g.*, economic modeling, linear programming).[1] The median speedups for FEM 2–9 were between 1.1× and 1.54× higher than the median speedups for FEM 10–17. Split UBCSR reduces this gap, running in as little as half the time of CSR (2.1× speedup) and $\frac{5}{9}$ the time of BCSR (1.8× speedup). Moreover, splitting can significantly reduce matrix storage. We are making our techniques available in OSKI [17], a library of automatically tuned sparse matrix kernels that builds on the SPARSITY framework [6,5].

This paper summarizes our recent technical report [19]. We will refer the reader there for more detail when appropriate.

2 Characterizing Variable Block Structure in Practice

We use variable block row (VBR) format [12,11], which logically partitions rows and columns into block rows and columns, to distinguish the dense block substructure of FEM Matrices 10–17 from 2–9 in two ways:[2]

- **Unaligned blocks**: Consider any $r \times c$ dense block starting at position (i, j) in an $m \times n$ matrix, where $0 \leq i < m$ and $0 \leq j < n$. BCSR typically assumes a *uniform alignment constraint*, $i \bmod r = j \bmod c = 0$. Relaxing the column constraint so that $j \bmod c$ is any value less than c has yielded some improvements in practice [2]. However, most non-zeros of Matrices 12 and 13 lie in blocks of the same size, with $i \bmod r$ and $j \bmod c$ distributed

---

[1] We omit test matrix 1, a dense synthetic matrix stored in sparse format.
[2] We treat Matrix 11, which contains a mix of blocks and diagonals, using other techniques [16, Chap. 5]; Matrices 14 and 16 are eliminated due to their small size.

uniformly over values up to $r-1$ and $c-1$, respectively. One goal of our UBCSR is to relax the row alignment as well.
– **Mixtures of "natural" block sizes:** Matrices 10, 15, and 17 possess a mix of block sizes when viewed in VBR format. This motivates the decomposition, $A = A_1 + A_2 + \cdots + A_s$, where each term A_i consists of the subset of blocks of a particular size. Each term can then be tuned separately.

These observations apply to the matrices listed in Table 1, which include a subset of the matrices referred to previously as Matrices 10–17, and 5 additional matrices (labeled Matrices A–E) from other FEM applications.

VBR can reveal dense block substructure. Consider storage of Matrix 12 in VBR format, using a CSR-to-VBR conversion routine available in the SPARSKIT library [12]. This routine partitions the rows by looping over rows in order, starting at the first row, and placing rows with identical non-zero structure in the same block. The same procedure is used to partition the columns, with the result shown in Figure 1 (*top*). The maximum block size in VBR turns out to be 3×3, with 96% of non-zeros stored in such blocks (Figure 1 (*bottom-left*), where a label of '0' indicates that the fraction is zero when rounded to two digits but there is at least 1 block at the given size). Moreover, these blocks are not uniformly aligned on row boundaries as assumed by BCSR. Figure 1 (*bottom-right*) shows the distributions of $i \mod r$ and $j \mod c$, where (i,j) is the starting position in A

Table 1. Variable block test matrices. Problem sources are summarized elsewhere [16, Appendix B].

#	Matrix	Dimension	No. of Non-zeros	Dominant block sizes (% of non-zeros)
10	ct20stif	52329	2698463	6×6 (39%)
	Engine block			3×3 (15%)
12	raefsky4	19779	1328611	3×3 (96%)
	Buckling problem			
13	ex11	16614	1096948	1×1 (38%)
	3D flow			3×3 (23%)
15	vavasis3	41092	1683902	2×1 (81%)
	2D partial differential equation			2×2 (19%)
17	rim	22560	1014951	1×1 (75%)
	Fluid mechanics problem			3×1 (12%)
A	bmw7st_1	141347	7339667	6×6 (82%)
	Car body analysis			
B	cop20k_m	121192	4826864	2×1 (26%), 1×2 (26%)
	Accelerator cavity design			1×1 (26%), 2×2 (22%)
C	pwtk	217918	11634424	6×6 (94%)
	Pressurized wind tunnel			
D	rma10	46835	2374001	2×2 (17%)
	Charleston Harbor model			3×2 (15%), 2×3 (15%)
				4×2 (9%), 2×4 (9%)
E	s3dkq4m2	90449	4820891	6×6 (99%)
	Cylindrical shell			

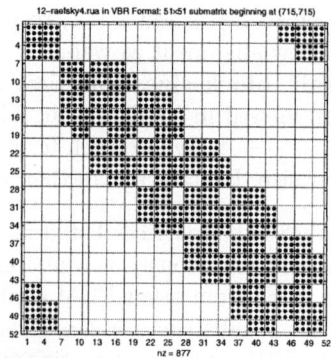

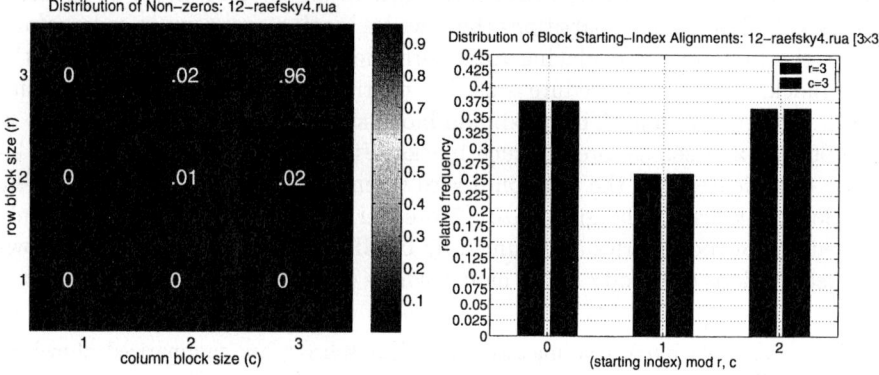

Fig. 1. Logical grid (block partitioning) after greedy conversion to variable block row (VBR) format: Matrix 12-**raefsky4**. (*Top*) Partitioning after conversion to VBR format. (*Bottom-left*) 96% of the non-zero blocks are in 3×3 blocks. (*Bottom-right*) The 3×3 blocks have varying alignments.

of each 3×3 block, and the top-leftmost entry of A is $A(0,0)$. The first row index of a given block row can start on any alignment, with 26% of block rows having $i \mod r = 1$, and the remainder split equally between 0 and 2. This observation motivates UBCSR which allows flexible alignments.

We summarize the variable block structure of the matrix test set used in this paper in the rightmost column of Table 1. This table includes a short list of dominant block sizes after conversion to VBR format, along with the fraction of non-zeros for which those block sizes account. The reader may assume that the dominant block size is also irregularly aligned except in the case of Matrix 15. More information on the distribution of non-zeros and block size alignments appears elsewhere [16, Appendix F].

3 A Split Unaligned Block Matrix Representation

We consider an SpMV implementation which

1. Converts (or stores) the matrix A first in VBR format, allowing for padding of explicit zeros to affect the block size distribution.

2. Splits A into a sum of s terms, $A = A_1 + \cdots + A_s$, according to the distribution of block sizes observed when A is in VBR.
3. Stores each term A_i in UBCSR format, a modification of BCSR which relaxes row and column alignment.

The use of VBR is a *heuristic* for identifying block structure. Finding the maximum number of non-overlapping dense blocks in a matrix is NP-Complete [15], so there is considerable additional scope for analyzing dense block structure.

Though useful for characterizing the structure (Section 2), VBR yields poor SpMV performance. The innermost loops, which carry out multiplication by an $r \times c$ block, cannot be unrolled in the same way as BCSR because c may change from block to block within a block row. VBR performance falls well below that of alternative formats on uniprocessors [16, Chap. 2].

This section very briefly summarizes a recent technical report [19, Sec. 3].

3.1 Converting to Variable Block Row Format with Fill

The default SPARSKIT CSR-to-VBR conversion routine only groups rows (or columns) when the non-zero patterns between rows (columns) match exactly. However, this convention can be too strict on some matrices in which it would be profitable to fill in zeros, just as with BCSR. We instead use the following measure of similarity between columns (or equivalently, rows). Let u^T and v^T be two row (or u, v for column) vector patterns, *i.e.*, whose non-zero elements are equal to 1. Let k_u and k_v be the number of non-zeros in u and v, respectively. Then we use $S(u^T, v^T)$ to measure the similarity between u^T and v^T:

$$S(u^T, v^T) = \frac{u^T \cdot v}{\max(k_u, k_v)} \qquad (1)$$

This function is symmetric, equals 0 when u^T, v^T have no non-zeros in common, and equals 1 when u^T and v^T have identical patterns. Our partitioning procedure greedily examines rows sequentially, starting at row 0, computing S between the first row of the current block row and the candidate row. If the similarity exceeds a specified threshold, θ, the row is added to the current block row. Otherwise, the procedure starts a new block row. We partition columns similarly if the matrix pattern is non-symmetric, and otherwise use the same partition for rows and columns. We fill in explicit zeros to make the blocks conform to the partitions.

At $\theta = 0.7$, Matrix 13 has 81% of non-zeros in 3×3 blocks, instead of just 23% when $\theta = 1$ (Table 1). Moreover, the fill ratio is just 1.01: a large blocks become available at the cost of only a 1% increase in flops.

3.2 Splitting the Non-zero Pattern

Given the distribution of work (*i.e.*, non-zero elements) over block sizes for a matrix in VBR at a given threshold θ, given the desired number of splittings s, and given a list of block sizes $\{r_1 \times c_1, \ldots, r_{s-1} \times c_{s-1}\}$, we greedily extract blocks from the VBR representation of A to form a splitting $A = A_1 + \cdots + A_s$ where the terms are structurally disjoint and each A_i is stored in $r_i \times c_i$ UBCSR format (see Section 3.3) and A_s is stored in CSR format. We use a greedy splitting procedure (see the full report [19] in which the order of the block sizes specified matters.

3.3 An Unaligned Block Compressed Sparse Row Format

We handle unaligned block rows in UBCSR by simply augmenting the usual BCSR data structure with an additional array of row indices Arowind such that Arowind[I] contains the starting index of block row I. Each block-multiply is fully unrolled as it would be for BCSR.

4 Experimental Methods

The split UBCSR implementation of SpMV has the following parameters: the similarity threshold θ which controls fill, the number of splitting terms s, and the block sizes for all terms, $r_1 \times c_1, \ldots, r_s \times c_s$. Given a matrix and machine, we select these parameters by a constrained empirical search procedure, described precisely in the technical report [19]. This procedure is not completely exhaustive, but could examine up to roughly 250,000 implementations for a given matrix depending on the block size distribution. However, fewer than 10,000 were examined in all the cases in Table 1.

Nevertheless, any such an exhaustive search is generally not practical at runtime, owing to the cost of conversion (between 5–40 SpMVs [16]). While automated methods exist for selecting a block size in the BCSR case [2,6,16,18], none exist for the split UBCSR case. Thus, our results (Section 5) are an empirical upper-bound on how well we expect to be able to do.

5 Results and Discussion

The split UBCSR implementation of Section 3 often improves performance relative to a traditional register-tiled BCSR implementation, as we show for the matrices in Table 1 and on four hardware platforms. When performance does not improve significantly, we may still reduce the overall storage.

We are mainly interested in comparing the best split UBCSR implementation (see Section 4) against the best BCSR implementation chsen by by exhaustively searching all block sizes up to 12×12, as done in prior work [6,16,18]. We also sometimes refer to the BCSR implementation as the *register blocking* implementation following the convention of earlier work. We summarize the 3 main conclusions of our technical report as follows. "Speedups" compare actual execution time, and summary statistics (minimum, maximum, and median speedups) are taken with respect to the matrices and shown in figures by end- and mid-points of an arrow. The full report shows data for each matrix and platform [19].

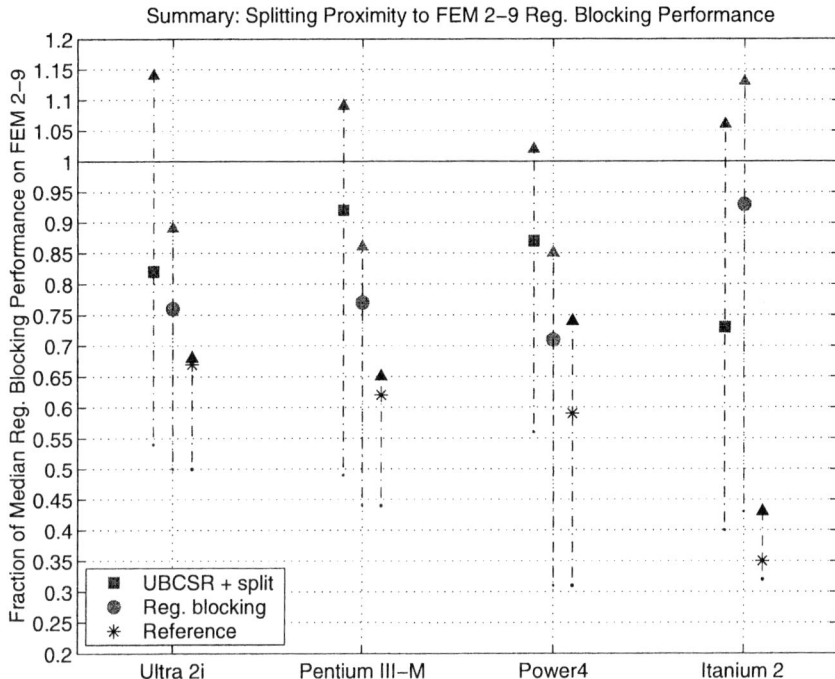

Fig. 2. Fraction of median BCSR performance over Matrices 2–9

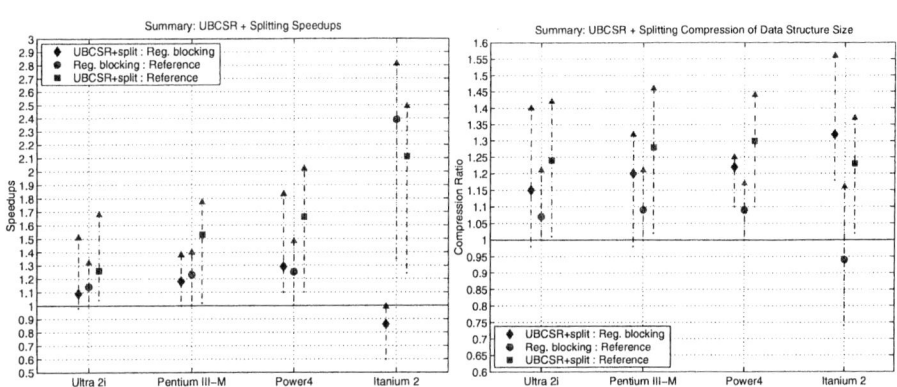

Fig. 3. Speedups and compression ratios after splitting + UBCSR storage, compared to BCSR

Finding 1: *By relaxing block row alignment using UBCSR storage, it is possible to approach the performance seen on Matrices 2–9.* Figure 2 shows split UBCSR performance as a fraction of median BCSR performance taken over Matrices 2–9. We also show statistics for BCSR only and reference implementations. The median fraction achieved by splitting exceeds the median fraction achieved by BCSR on all but the Itanium 2. On the Pentium III-M and Power4, the median fraction of splitting exceeds the maximum of BCSR only, demonstrating the potential utility of splitting and the UBCSR format. The maximum fraction due to splitting slightly exceeds 1 on all platforms.

Finding 2: *Median speedups relative to the reference performance range from $1.26\times$ (Ultra 2i) up to $2.1\times$ (Itanium 2).* Figure 3 (*left*) compares the speedups of (a) splitting over BCSR (blue solid diamonds), (b) BCSR over the reference (green solid circles), and (c) splitting over the reference (red solid squares). Splitting is at least as fast as BCSR on all but the the Itanium 2 platform. Median speedups, taken over the matrices in Table 1 and measured relative to the reference performance, range from $1.26\times$ (Ultra 2i) up to $2.1\times$ (Itanium 2). Relative to BCSR, median speedups are relatively modest, ranging from 1.1–$1.3\times$. However, these speedups can be as much as $1.8\times$ faster.

Finding 3: *Splitting can lead to a significant reduction in total matrix storage.* The compression ratio of splitting over the reference is the size of the reference (CSR) data structure divided by the size of the split+UBCSR data structure. We summarize the compression ratios in Figure 3 (*right*). The median compression ratios compared to the reference are between 1.26–$1.3\times$. Compared to BCSR, the compression ratios of splitting can be as high as $1.56\times$.

The asymptotic storage for CSR is roughly 1.5 doubles per non-zero when the number of integers per double is 2 [16, Chap. 3]. When abundant dense blocks exist, the storage decreases toward a lower limit of 1 double per non-zero. Relative to the reference, the median compression ratio for splitting ranges from 1.24 to 1.3, but can be as high as 1.45, which is close to this limit.

6 Related Work

The inspiration for this study is recent work on splitting by Geus and Röllin [4], Pinar and Heath [10], and Toledo [14], and the performance gap observed in prior work [6,18,5]. Geus and Röllin explore up to 3-way splittings based on row-aligned BCSR format. (The last splitting term in their implementations is also fixed to be CSR, as in our work.) Pinar and Heath examine 2-way splittings where the first term is $1\times c$ format and the second in 1×1. Toledo also considers 2-way splittings and block sizes up to 2×2, as well as low-level tuning techniques (*e.g.*, prefetching) to improve memory bandwidth. The main distinction of this paper is the relaxed row-alignment of UBCSR.

Split UBCSR can be combined with other techniques that improve register-level reuse and reuse of the matrix entries, including multiplication by multiple vectors where $7\times$ speedups over CSR are possible [5,1]. Exploiting numerical symmetry with BCSR storage yields speedups as high as $2.8\times$ over non-symmetric CSR, $2.1\times$ over non-symmetric BCSR, and reduces storage [7].

Matrices 18–44 of the SPARSITY benchmark suite largely remain difficult, though they should be amenable to cache-level blocking in which the matrix is stored as a collection of smaller, disjoint rectangular [9,5] (or even diagonal [13]) blocks to improve temporal access to elements of x.

Better low-level tuning of the CSR SpMV implementation may also be possible. Recent work on low-level tuning of SpMV by unroll-and-jam (Mellor-Crummey, et al. [8]), software pipelining (Geus and Röllin [4]), and prefetching (Toledo [14]) are promising starting points. On the vector Cray X1, just one additional permutation of rows in CSR, with no other data reorganization, yields order of magnitude improvements [3].

For additional related references, see our full report [19].

7 Conclusions and Future Work

This paper shows that it is possible to extend the classical BCSR format to handle matrices with irregularly aligned and mixed dense block substructure, thereby reducing the gap between various classes of FEM matrices. We are making this new split UBCSR data structure available in the Optimized Sparse Kernel Interface (OSKI), a library of automatically tuned sparse matrix kernels that builds on SPARSITY, an earlier prototype [17,6].

However, our results are really only empirical bounds on what may be possible since they are based on exhaustive search over split UBCSR's tuning parameters (Section 4). We are pursuing effective and cheap heuristics for selecting these parameters. Our data already suggest the form this heuristic might take. One key component is a cheap estimator of the non-zero distributions over block sizes, which we measured in this paper exactly using VBR. This estimator would be similar to those proposed in prior work for estimating fill in the BCSR case [6,16], and would suggest the number of splittings and candidate block sizes. Earlier heuristic models for the BCSR case, which use benchmarking data to characterize the machine-specific performance at each block size, could be extended to the UBCSR case and combined with the estimation data [2,6,16].

We are also pursuing combining split UBCSR with other SpMV optimizations surveyed in Section 6, including symmetry, multiple vectors, and cache blocking. In the case of cache blocking, CSR is often used as an auxiliary data structure; replacing the use of CSR with the 1×1 UBCSR data structure itself could reduce some of the row pointer overhead when the matrix is very sparse.

References

1. A. H. Baker, E. R. Jessup, and T. Manteuffel. A technique for accelerating the convergence of restarted GMRES. Technical Report CU-CS-045-03, University of Colorado, Dept. of Computer Science, January 2003.
2. A. Buttari, V. Eijkhout, J. Langou, and S. Filippone. Performance optimization and modeling of blocked sparse kernels. Technical Report ICL-UT-04-05, Innovative Computing Laboratory, University of Tennessee, Knoxville, 2005.
3. E. D'Azevedo, M. R. Fahey, and R. T. Mills. Vectorized sparse matrix multiply for compressed sparse row storage. In *Proceedings of the International Conference on Computational Science*, LNCS 3514, pages 99–106, Atlanta, GA, USA, May 2005. Springer.

4. R. Geus and S. Röllin. Towards a fast parallel sparse matrix-vector multiplication. In E. H. D'Hollander, J. R. Joubert, F. J. Peters, and H. Sips, editors, *Proceedings of the International Conference on Parallel Computing (ParCo)*, pages 308–315. Imperial College Press, 1999.
5. E.-J. Im. *Optimizing the performance of sparse matrix-vector multiplication*. PhD thesis, University of California, Berkeley, May 2000.
6. E.-J. Im, K. Yelick, and R. Vuduc. Sparsity: Optimization framework for sparse matrix kernels. *International Journal of High Performance Computing Applications*, 18(1):135–158, 2004.
7. B. C. Lee, R. Vuduc, J. Demmel, and K. Yelick. Performance models for evaluation and automatic tuning of symmetric sparse matrix-vector multiply. In *Proceedings of the International Conference on Parallel Processing*, Montreal, Canada, August 2004.
8. J. Mellor-Crummey and J. Garvin. Optimizing sparse matrix vector multiply using unroll-and-jam. In *Proceedings of the Los Alamos Computer Science Institute Third Annual Symposium*, Santa Fe, NM, USA, October 2002.
9. R. Nishtala, R. Vuduc, J. Demmel, and K. Yelick. When cache blocking sparse matrix vector multiply works and why. In *Proceedings of the PARA'04 Workshop on the State-of-the-art in Scientific Computing*, Copenhagen, Denmark, June 2004.
10. A. Pinar and M. Heath. Improving performance of sparse matrix-vector multiplication. In *Proceedings of Supercomputing*, 1999.
11. K. Remington and R. Pozo. NIST Sparse BLAS: User's Guide. Technical report, NIST, 1996. gams.nist.gov/spblas.
12. Y. Saad. SPARSKIT: A basic toolkit for sparse matrix computations, 1994. www.cs.umn.edu/Research/arpa/SPARSKIT/sparskit.html.
13. O. Temam and W. Jalby. Characterizing the behavior of sparse algorithms on caches. In *Proceedings of Supercomputing '92*, 1992.
14. S. Toledo. Improving memory-system performance of sparse matrix-vector multiplication. In *Proceedings of the 8th SIAM Conference on Parallel Processing for Scientific Computing*, March 1997.
15. V. Vassilevska and A. Pinar. Finding nonoverlapping dense blocks of a sparse matrix. Technical Report LBNL-54498, Lawrence Berkeley National Laboratory, Berkeley, CA, USA, 2004.
16. R. Vuduc. *Automatic performance tuning of sparse matrix kernels*. PhD thesis, University of California, Berkeley, Berkeley, CA, USA, December 2003.
17. R. Vuduc, J. Demmel, and K. Yelick. OSKI: An interface for a self-optimizing library of sparse matrix kernels, 2005. bebop.cs.berkeley.edu/oski.
18. R. Vuduc, J. W. Demmel, K. A. Yelick, S. Kamil, R. Nishtala, and B. Lee. Performance optimizations and bounds for sparse matrix-vector multiply. In *Proceedings of Supercomputing*, Baltimore, MD, USA, November 2002.
19. R. Vuduc and H.-J. Moon. Fast sparse matrix-vector multiplication by exploiting variable blocks structure. Technical Report UCRL-TR-213454, Center for Applied Scientific Computing, Lawrence Livermore National Laboratory, Livermore, CA, USA, July 2005.

Parallel Transferable Uniform Multi-round Algorithm for Achieving Minimum Application Turnaround Times for Divisible Workload

Hiroshi Yamamoto, Masato Tsuru, and Yuji Oie

Department of Computer Science and Electronics,
Kyushu Institute of Technology, Kawazu 680-4, Iizuka, 820-8502, Japan
yamamoto@infonet.cse.kyutech.ac.jp, {tsuru, oie}@cse.kyutech.ac.jp

Abstract. A parallel transferable uniform multi-round (PTUMR) scheduling algorithm is proposed for mitigating the adverse effect of the data transmission time by dividing workloads and allowing their parallel transmissions to distributed clients from the master in a network. The performance of parallel computing using the master/worker model for distributed grid computing tends to degrade when handling large data sets due to the impact of data transmission time. Multiple-round scheduling algorithms have therefore been proposed to mitigate the effects of the data transmission time by dividing the data into chunks that are sent in multiple rounds so as to overlap the time required for computation and communication. However, standard multiple-round algorithms assume a homogeneous network environment with uniform link transmission capacity, and as such cannot minimize the turnaround time effectively in real heterogeneous network environments. The proposed PTUMR algorithm optimizes the size of chunks, the number of rounds, and the number of workers to which data is to be transmitted in parallel, and is shown through performance evaluations to mitigate the adverse effects of data transmission time between the master and workers significantly, achieving turnaround times close to the theoretical lower limits.

Keywords: Master/Worker model, Divisible Workload, UMR, Parallel Data Transmission, Constraint Minimization Problem.

1 Introduction

As the performance of wide-area networks and off-the-shelf computers increases, large-scale distributed computing, namely grid computing, has become feasible, realizing a virtual high-performance computing environment by connecting geographically distributed computers via the Internet [1,2]. The master/worker model is suitable for loosely coupled applications and particularly suited to grid computing environments involving a large number of computers. In this model, a master with application tasks (and data to be processed) dispatches subtasks to several workers, which process the data allocated by the master. A typical instance of applications based on such a model is a divisible workload application [3,4]. The master divides the data required for the application into an arbitrary number of chunks, and then dispatches the chunks to multiple workers. For a given application, all chunks require identical processing. This divisibility is encountered in various applications such as image feature extraction [5].

Recent applications have involved large amount of data, and the transmission time of data sent from the master to workers is no longer negligible compared to the computation time [6]. Therefore, the master is required to schedule the processing workload of workers effectively by considering the impact of communication time and computation by each worker on the application turnaround time. A number of scheduling algorithms have been proposed in which the master dispatches workload to workers in a 'multiple-round' manner to overlap the time required for communication with that for computation [3,7,8,9]. These methods can mitigate the adverse effects of time spent to transmit large amounts of data, and have made it possible to achieve good application turnaround times. However, these methods assume a homogeneous environment, in which both the networks associated with the master and workers have the same transmission capacity, adopting a sequential transmission model. In this model, the master uses its network connection in a sequential manner where the master transmits data to one worker at a time [10,11]. In actual networks, where the master and workers are connected via a heterogeneous network, the sequential transmission model is unable to minimize the adverse effects of data transmission time on the application turnaround time. Therefore, in this study, a new scheduling algorithm called parallel transferable uniform multi-round (PTUMR) is proposed based on the UMR algorithm [7,8]. The PTUMR scheme determines how the application data should be divided and when data should be sent to workers by allowing parallel data transmission to multiple workers. The proposed algorithm is examined analytically, and a solution is derived for the optimization problem in terms of the size of chunks, the number of rounds, and the number of parallel transmissions employed in the algorithm. Performance evaluations indicate that the algorithm can reduce the adverse effects of data transmission time on application turnaround time, allowing turnaround times close to the lower limits to be achieved.

This paper is organized as follows. In Section 2, the conceptual basis for multiple-round scheduling and the conventional UMR algorithm are introduced. The proposed PTUMR algorithm is presented in Section 3, and its performance is investigated in Section 4. The paper is finally concluded in Section 5.

2 Conventional UMR Scheduling Algorithm

The concept behind the multiple-round scheduling method is introduced briefly followed along with a description of the standard UMR algorithm as an example of existing multiple-round scheduling algorithms.

2.1 Multiple-Round Scheduling Method

Recently, a number of scheduling methods have been proposed in which the master dispatches data to workers in a multiple-round manner in order to overlap communication with computation and thereby reduce the application turnaround time. Figure 1 shows an example of this scenario, where the master dispatches a workload of the application to one worker. In single-round scheduling, the master transmits the entire set of application data W_{total} [units] to the worker at one time. In multiple-round scheduling, on the other hand, the data W_{total} is divided into multiple chunks of arbitrary size and processed in M rounds. Here, in Round $j(= 0, \cdots, M-1)$, the master allocates one chunk of $chunk_j$ [units] in length to the worker. In data transmission, the master transmits data to the worker at a rate of B [units/s], and the worker computes the data allocated by the master at a rate of S [units/s]. In addition, it is assumed that the time required for the workers to return computation results to the master is small enough to be neglected. Prior to the start of data transmission, a fixed time interval $nLat$ [s] independent of the size of the chunk is added due to some initial interaction between the master and workers such as TCP (Transmission Control Protocol) connection establishment. Similarly, a part of the computation time is independent of the chunk size is defined as $cLat$ [s].

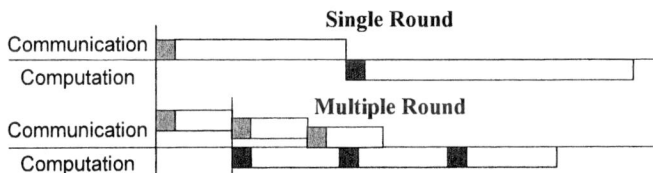

Fig. 1. Multiple-round scheduling

Multiple-round scheduling is briefly compared below with single-round scheduling in terms of the application turnaround time T_{ideal} under the ideal assumption that each worker never enters the idle computation state once the first chunk has been received. In single-round scheduling, the worker starts computation after receiving the entire set of data W_{total} from the master, as shown in Fig. 1. Therefore, the application turnaround time in single-round scheduling is represented as follows.

$$T_{ideal} = nLat + \frac{W_{total}}{B} + cLat + \frac{W_{total}}{S}. \quad (1)$$

In multiple-round scheduling (over M rounds), the worker can start computation after receiving only $chunk_0$ units of the data in the first round (Round 0). The application turnaround time is therefore given by

$$T_{ideal} = nLat + \frac{chunk_0}{B} + \sum_{j=0}^{M-1}\left(cLat + \frac{chunk_j}{S}\right),$$
$$= nLat + \frac{chunk_0}{B} + M \times cLat + \frac{W_{total}}{S}. \quad (2)$$

In multiple-round scheduling, the adverse effects of data transmission time on the application turnaround time can be reduced because the data transmission time in Round 2 and later can be overlapped with the computation time. However, the use of a large number of rounds would lead to an increase in the total overhead, and eventually result in a degradation of application turnaround time. Thus, optimizing both the number of rounds and the size of chunks so as to minimize the application turnaround time is a key issue in multiple-round scheduling.

2.2 UMR

UMR is an example of multiple-round scheduling algorithm [7,8]. The network model for UMR is shown in Fig. 2. The master and workers are connected to a high-speed backbone network that is free of bottlenecks. It is assumed that the master allocates chunks of identical size to all workers in each round. UMR adopts the sequential transmission model, by which the master transmits a chunk to one worker at a time. Figure 3 illustrates how the data is transmitted to workers and then processed under UMR. To reduce the data transmission time in the first round, small chunks are transmitted to workers in the first round, and the size of chunks then grows exponentially in subsequent rounds.

In UMR, the number of rounds and the chunk size in the first round have been approximately optimized under the sequential transmission model [8]. However, the sequential transmission model adopted in UMR prevents the minimization of application turnaround time in real network environments, particularly when the transmission capacity of the master-side link is larger than that of the worker-side links. Under such conditions, the UMR algorithm cannot utilize the network resource to their maximum.

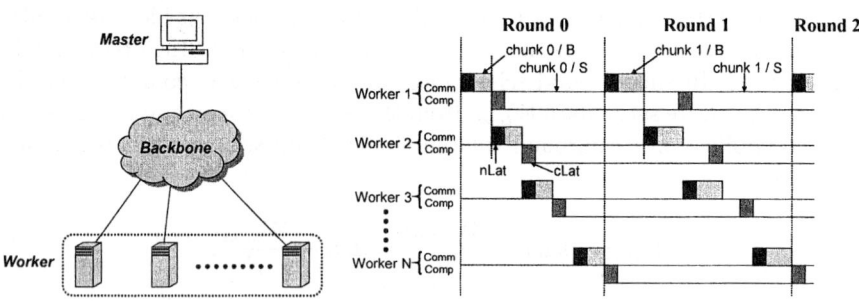

Fig. 2. Distributed computing model

Fig. 3. Timing chart of data transmission and worker processing under UMR

3 PTUMR Scheduling Algorithm

In this section, the new PTUMR scheduling algorithm is presented. The PTUMR algorithm determines how the application data should be divided and when the data should be transmitted to workers in network environments that allow the master to transmit data to multiple workers in parallel by establishing multiple TCP connections. The PTUMR is based on the same distributed computing model as standard UMR, as shown in Fig. 2. The master and N workers are connected to a backbone network that is assumed to be free of bottlenecks as in UMR. The set of workers is assumed to be homogeneous in terms of the computational power of each worker and the transmission capacity of the link attached to the worker. For this assumption, the master dispatches identically sized chunks to all workers in each round except for the last round (See Subsection 3.1).

In contrast to UMR, the PTUMR algorithm allows the master to transmit chunks to m workers simultaneously (Fig. 4) assuming that $nLat$ can be overlapped among concurrent transmissions. This extension is applicable and suitable especially for asymmetric networks, where the transmission capacity (B_{master}) of the master-side link is greater than that (B_{worker}) of the worker-side links, as

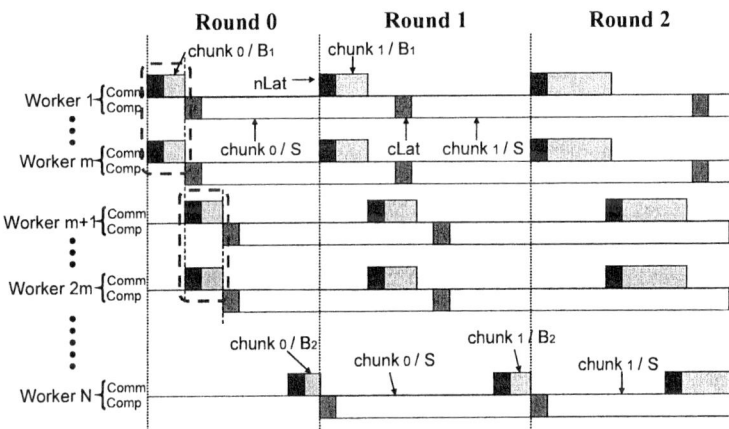

Fig. 4. Timing chart of data transmission and worker processing under PTUMR

Table 1. Model parameters and their values examined in performance evaluation

W_{total}	$100, 200, \cdots, 2000$ [units]
N	$1, 2, \cdots, 100$
S	1 [units/s]
B_{master}	$120, 240, \cdots, 1200$ [units/s]
B_{worker}	120 [units/s]
$nLat$	$0.001, 0.005, 0.01, 0.05, 0.1$ [s]
$cLat$	0.5 [s]
m	$1, 2, \cdots, N$

is often the case in actual network environments. More precisely, the master simultaneously transmits chunks to the first m workers in parallel at transmission rate B_1 in the first stage, and then performs the action repeatedly in subsequent stages for the next m workers. If $N \bmod m$ is not equal to 0, as the final stage at the end of each round, the master transmits chunks to the last $N \bmod m$ workers in parallel at transmission rate B_2. The transmission rate B_1 and B_2 are expressed as follows.

$$B_1 = \min\left\{B_{worker}, \frac{B_{master}}{m}\right\}, \quad B_2 = \min\left\{B_{worker}, \frac{B_{master}}{N \bmod m}\right\}. \quad (3)$$

The parameters for this model are listed in Tab. 1. From the assumption of a homogeneous worker set, the computation rate S, worker-side link capacity B_{worker}, and latency parameters $cLat$ and $nLat$ corresponding to related overhead, are identical for all workers.

The goal is then to find the number M^+ of rounds, the size $chunk_0^+$ of the first chunk (initial chunk size), and the number m^* of workers to which data will be transmitted in parallel such that the application turnaround time is minimized given a total amount W_{total} of application data and N workers under an environment characterized by S, B_{master}, B_{worker}, $cLat$, and $nLat$.

By solving an appropriate constraint minimization problem, the PTUMR algorithm obtains the number $M^+(m)$ of rounds and the size $chunk_0^+(m)$ of the first chunk that minimizes the application turnaround time for any given number m of workers to which data is to be transmitted in parallel ($1 \leq m \leq N$). An appropriate parallelism m^* is then determined in a greedy manner. The algorithm is explained in further detail below.

3.1 Chunk Size and Number of Rounds

The algorithm determines the number $M^+(m)$ of rounds and initial chunk size $chunk_0^+(m)$ that are nearly optimal in terms of minimizing the application turnaround time for any given parallelism m. In general, since workers may enter a computationally idle state while waiting for transmission of the next chunk to complete, T_{real} is difficult to express analytically. Therefore, instead of T_{real}, the ideal application turnaround time T_{ideal}, which can be readily represented in analytical form, is derived under the ideal assumption that no worker ever enters the idle computation state once it has received its first chunk of data.

From this assumption, the master should complete the transmission of chunks to all workers for computation in Round $j + 1$, just before worker N completes the processing of the chunk received in Round j. This situation is formulated as follows for given number m of workers to which data is to be transmitted in parallel.

$$\lfloor \frac{N}{m} \rfloor \left(nLat + \frac{chunk_{j+1}}{B_1}\right) + e\left(nLat + \frac{chunk_{j+1}}{B_2}\right) = cLat + \frac{chunk_j}{S}, \quad (4)$$

$$e = \begin{cases} 0 & \text{if } N \bmod m = 0, \\ 1 & \text{otherwise.} \end{cases}$$

From Eq. (4), the desired chunk size $chunk_j$ for Round j can be determined by the initial chunk size $chunk_0$ as follows.

$$chunk_j = \beta^j(chunk_0 - \alpha) + \alpha, \qquad (5)$$

$$\alpha = \frac{B_1 B_2 S}{B_1 B_2 - (eB_1 + \lfloor N/m \rfloor B_2)S}(\lceil N/m \rceil nLat - cLat),$$

$$\beta = \frac{B_1 B_2}{(eB_1 + \lfloor N/m \rfloor B_2)S}.$$

On the other hand, since the sum of chunks allocated to all workers should be equal to the total workload (application data size), the relation between the chunk size $chunk_0$ in the first round and the number M of rounds is given by

$$G = N \times \sum_{j=0}^{M-1} chunk_j - W_{total}$$

$$= N \times \left\{ \frac{1-\beta^M}{1-\beta}(chunk_0 - \alpha) + M\alpha \right\} - W_{total} = 0. \qquad (6)$$

Using this relation, the application turnaround time T_{ideal} under this ideal assumption can be derived for a given parallelism m, which can be regarded as a function of the initial chunk size $chunk_0$ and the number M of rounds, as follows.

$$T_{ideal} = \frac{W_{total}}{NS} + M \times cLat + \gamma \times chunk_0 + \delta \times nLat, \qquad (7)$$

$$\gamma = \frac{1}{2B_1 B_2 N}\{2eB_1(N \bmod m) + B_2\lfloor N/m \rfloor(N+m+N \bmod m)\},$$

$$\delta = \lceil N/m \rceil - \frac{m\lfloor N/m \rfloor}{2N}\{\lfloor N/m \rfloor + 2e - 1\}.$$

where the size of chunk transmitted to each worker in the last round is chosen in a way that every worker completes the processing of the last chunk with one accord.

Consider the constraint minimization problem with real-number parameters. $T_{ideal}(M, chunk_0)$ in Eq. (7) is the objective function that should be minimized subject to the constraint of Eq. (6), where M and $chunk_0$ are treated as real numbers. Let $(M^*, chunk_0^*)$ denote the optimal solution of this minimization problem, which can be readily solved by numerical solvers using the Lagrange multipliers associated with the constraints. Then, since the number of rounds used in PTUMR should be an integer, it is necessary to determine an appropriate number of rounds M^+ as an integer and the corresponding initial chunk size $chunk_0^+$ satisfying Eq. (6), if M^* is not an integer. Note that, supposing Eq. (5) holds, the application turnaround time $T_{real}(M, chunk_0)$ can be calculated as a function of the given initial chunk size $chunk_0$ and the number M of rounds. Therefore, the better M^+ of two integers near M^*, $\lfloor M^* \rfloor$ and $\lceil M^* \rceil$, is chosen by comparing $T_{real}(\lfloor M^* \rfloor, chunk_0(\lfloor M^* \rfloor))$ and $T_{real}(\lceil M^* \rceil, chunk_0(\lceil M^* \rceil))$. Here, $chunk_0(M)$ denotes the initial chunk size corresponding to the number M of rounds so as to satisfy the Eq. (6).

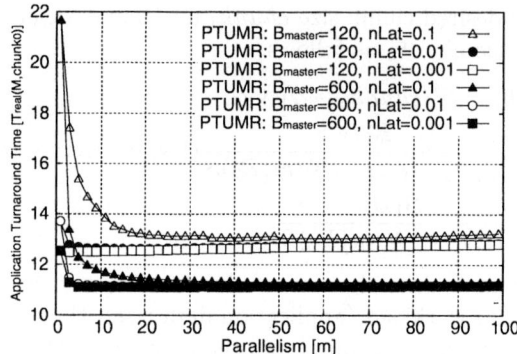

Fig. 5. Impact of parallelism on application turnaround time

3.2 Number of Parallel Transmissions

For any given number m of workers to which data is to be transmitted in parallel, let $M^+(m)$ be the nearly-optimal number of rounds and $chunk_0^+(m)$ be the initial chunk size, as obtained above. To find the optimal parallelism m^* achieving the application turnaround time $T_{real}(M^+(m^*), chunk_0^+(m^*))$ close to the minimum value, the impact of parallelism m on the nearly-optimal application turnaround time $T_{real}(M^+(m), chunk_0^+(m))$ is examined in Fig. 5. Note that the results for the sequential transmission model (i.e., $m = 1$) corresponds to those for the standard UMR algorithm. Here, the parameters $W_{total} = 1000$ [units], $N = 100$, $B_{master} = 120, 600$ [units/s], and $nLat = 0.1, 0.01, 0.001$ [s] are set. All other parameters are set according to Tab. 1.

The case of $B_{master} = 120$ represents a symmetric network, in which the transmission capacity B_{master} of the master-side link is equal to that B_{worker} of the worker-side, which is the case considered in UMR. Figure 5 indicates, however, that even in a symmetric network, the transmission to multiple workers in parallel ($m > 1$) yields better performance than UMR ($m = 1$). For a wide range of $nLat$, increased parallelism m can reduce the adverse effects of overhead on the application turnaround time by overlapping $nLat$ for multiple workers. On the other hand, increasing m also reduces the transmission speeds B_1 and B_2 for each worker due to sharing of the transmission capacity of the master-side link by multiple data transmissions as shown in Eq. (3). Therefore, there exists an optimal number m^* of parallel transmissions that minimizes the application turnaround time T_{real}. For example, $m^* = 54, 17$ and 3 for $nLat = 0.1, 0.01$, and 0.001, respectively.

The case of $B_{master} = 600$ represents an asymmetric network in which the master-side link has a larger capacity than the worker-side links, as often occurs in real network environments. In such an asymmetric network, as also indicated in Fig. 5, increasing m dramatically reduces the application turnaround time T_{real} regardless of $nLat$. This occurs because an increase in m (up to $m = 5$) does not reduce the transmission speed for each worker, and the optimal degree of parallelism m^* is therefore larger than in the case of a symmetric network. For example, $m^* = 100$, 28 and 10 for $nLat = 0.1, 0.01$ and 0.001, respectively.

4 Performance Evaluation

The performance of the PTUMR algorithm is evaluated in terms of application turnaround time $T_{real}(M^+(m^*), chunk_0^+(m^*))$ for the number M^+ of rounds, the initial chunk size $chunk_0^+$, and the number m^* of parallel transmissions. The effectiveness of PTUMR is evaluated by comparing the achievable turnaround time T_{real} with the lower bound T_{bound}, which corresponds to the best possible application turnaround time in an environment where the network resources are sufficient to render the data transmission time negligible. T_{bound} is obtained as follows.

$$T_{bound} = cLat + \frac{W_{total}}{NS}. \qquad (8)$$

The impact of computational and network resources on the application turnaround time T_{real} is investigated below, and the performance of PTUMR is compared to that of UMR using the parameters listed in Tab. 1. The appropriate scale of applications for the proposed algorithm is then determined based on the total amount of application data W_{total}.

4.1 Impact of Computational and Network Resources on Application Turnaround Time

The impact of the computational and network resources on the application turnaround time is examined here by assuming a total size W_{total} of 1000. Figure 6 shows the application turnaround time T_{real} as a function of the number N of workers. As N increases, the application turnaround times under both the PTUMR and UMR algorithms remarkably decrease, and PTUMR furthermore

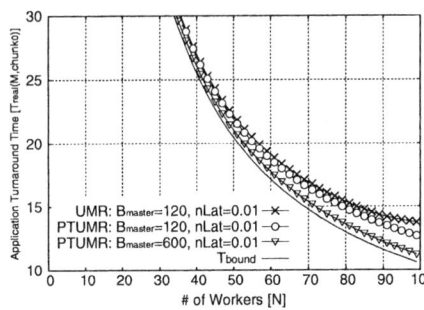
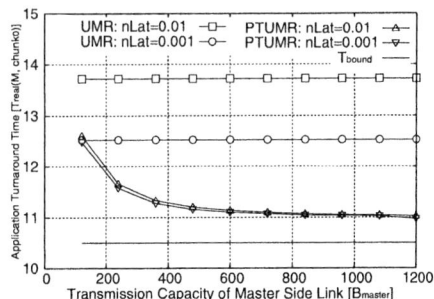

Fig. 6. Impact of number of workers on application turnaround time

Fig. 7. Impact of transmission capacity of master-side link on application turnaround time

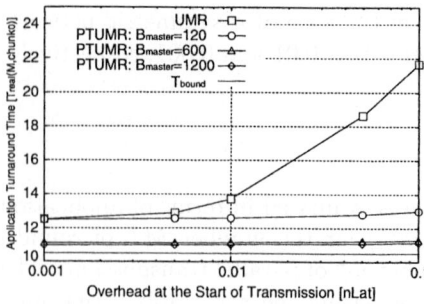 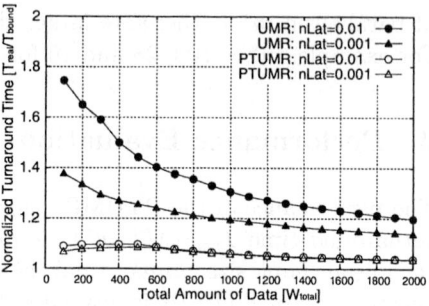

Fig. 8. Impact of overhead at start of data transmission on application turnaround time

Fig. 9. Impact of total workload (data size) on normalized application turnaround time

achieves T_{real} close to the lower bound T_{bound} for any N by increasing the parallelism m to utilize the network resources at the master-side to their maximum, while the difference between T_{real} of UMR and T_{bound} increases with N.

Furthermore, the performance of the PTUMR algorithm is evaluated in more detail below in reference to Fig. 7 and 8, which show the effect of B_{master} and $nLat$ on performance for $N = 100$. Figure 7 shows the application turnaround time T_{real} as a function of the transmission capacity B_{master} on the master-side link. The sequential transmission model in UMR limits the transmission speed between the master and each worker to the capacity B_{worker} of worker-side link. Therefore, even if B_{master} increases, the UMR algorithm cannot effectively utilize the network capacity, and thus cannot improve the application turnaround time. By contrast, the application turnaround time under PTUMR decreases with increasing B_{master} because the algorithm can utilize the full capacity by transmitting chunks to multiple workers in parallel.

Figure 8 shows the relationship between the overhead $nLat$ at the start of data transmission and the application turnaround time T_{real}. Under the PTUMR, T_{real} close to the lower bound T_{bound} can be achieved for a wide range of overhead $nLat$, because PTUMR reduces the overhead in the application turnaround time by aggressively overlapping the overhead $nLat$ for multiple workers.

This evaluation demonstrates that the PTUMR algorithm can achieve application turnaround time close to the lower bound through effective utilization of the transmission capacity on the master-side and overlapping of the overhead for multiple workers.

4.2 Impact of Total Workload on Application Turnaround Time

Finally, the effect of the total amount W_{total} of application data on the application turnaround time T_{real} is evaluated assuming a master-side transmission capacity B_{master} of 600, and 100 workers (N). Figure 9 shows the application turnaround time T_{real} normalized by the lower bound T_{bound} as a function of the application data size W_{total}. PTUMR provides excellent performance very close to the lower bound for any W_{total} and any $nLat$, that is, the PTUMR algorithm effectively eliminates the performance degradation associated with these factors. Under UMR, however, the normalized turnaround time becomes quite poor as the total data size W_{total} decreases, although good performance is achieved at large W_{total}. The degradation of performance at low W_{total} under UMR can be attributed to the increased proportion of overhead with small data sets: this problem can be overcome by PTUMR. These results therefore show that the PTUMR algorithm can effectively schedule applications of any size by minimizing the adverse effects of overhead on the application turnaround time.

5 Conclusion

The amount of data handled by distributed applications has recently increased, with the result that the time required for data transmission from the master to the workers has begun to degrade the application turnaround time. The adverse effects of data transmission time on the application turnaround time can be mitigated to a certain extent by employing a multiple-round scheduling algorithm such as UMR to overlap the data transmission time with the computation time. However, as UMR adopts the sequential transmission model, it cannot minimize the application turnaround time effectively especially in asymmetric networks where the master-side link capacity is greater than the worker side link capacity.

In this study, a new multiple-round scheduling algorithm adopting a multiple transmission model was introduced as an extension of UMR that allows for application data to be transmitted to multiple workers in parallel by establishing multiple TCP connections simultaneously. The proposed PTUMR algorithm determines appropriate parameters of chunk size, number of rounds, and number of parallel transmissions, by solving a constraint minimization problem. The performance of PTUMR was evaluated in various environments, and it was found that the PTUMR algorithm mitigates the adverse effects of data transmission time significantly, achieving turnaround times close to the lower bound over a wide range of application data size and network condition.

Acknowledgments

This work was supported in part by the Ministry of Education, Culture, Sports, Science and Technology, Japan, under the Project National Research GRID Initiative (NAREGI) and in part by the Ministry of Education, Culture, Sports, Science and Technology, Japan, Grant-in-Aid for Scientific Research on Priority Areas (16016271).

References

1. I. Foster and C. Kesselman, *The GRID Blueprint for a New Computing Infrastructure,* Morgan Kaufmann Publishers, 1998.
2. I. Foster, C. Kesselman, and S. Tuecke, "The Anatomy of the Grid," *International Journal of Supercomputer Applications,* Vol. 15, No. 3, pp. 200–222, 2001.

3. V. Bharadwaj, D. Ghose, V. Mani, and T. G. Robertazzi, "Scheduling Divisible Loads in Parallel and Distributed Systems," *IEEE Computer Society Press*, 1996.
4. T. G. Robertazzi, "Ten Reasons to Use Divisible Load Theory," *Jounal of IEEE Computer*, Vol. 36, No. 5, pp. 63–68, May 2003.
5. D. Gerogiannis and S. C. Orphanoudakis, "Load Balancing Requirements in Parallel Implementations of Image Feature Extraction Tasks," *IEEE Trans. Parallel and Distributed Systems*, Vol. 4, No. 9, pp. 994–1013, 1993.
6. A. Chervenak, I. Foster, C. Kesselman, C. Salisbury, and S. Tuecke, "The Data Grid: Towards an Architecture for the Distributed Management and Analysis of Large Scientific Datasets," *Journal of Network and Computer Applications*, 23:187–200, 2001.
7. Y. Yang and H. Casanova, "UMR: A Multi-Round Algorithm for Scheduling Divisible Workloads," *Proc. of International Parallel and Distributed Processing Symposium (IPDPS'03)*, Nice, France, April 2003.
8. Y. Yang and H. Casanova, "A Multi-Round Algorithm for Scheduling Divisible Workload Applications: Analysis and Experimental Evaluation," *Technical Report of Dept. of Computer Science and Engineering, University of California CS20020721*, 2002.
9. O. Beaumont, A. Legrand, and Y. Robert, "Optimal Algorithms for Scheduling Divisible Workloads on Heterogeneous Systems," *Proc. of International Parallel and Distributed Processing Symposium (IPDPS'03)*, Nice, France, April 2003.
10. C. Cyril, O. Beaumont, A. Legrand, and Y. Robert, "Scheduling Strategies for Master-Slave Tasking on Heterogeneous Processor Grids," *Technical Report 2002-12*, LIP, March 2002.
11. A. L. Rosenberg, "Sharing Partitionable Workloads in Heterogeneous NOWs: Greedier Is Not Better," *Proc. of the 3rd IEEE International Conference on Cluster Computing (Cluster 2001)*, pp. 124–131, California, USA, October 2001.

Application of Parallel Adaptive Computing Technique to Polysilicon Thin-Film Transistor Simulation

Yiming Li

Department of Communication Engineering,
Microelectronics and Information Systems Research Center,
National Chiao Tung University, 1001 Ta-Hsueh Rd., Hsinchu, 300, Taiwan
ymli@faculty.nctu.edu.tw

Abstract. In this paper, parallel adaptive finite volume simulation of polysilicon thin-film transistor (TFT) is developed using dynamic domain partition algorithm on a PC-based Linux cluster with message passing interface libraries. A set of coupled semiconductor device equations together with a two-dimensional model of grain boundary for polysilicon TFT is formulated. For the numerical simulation of polysilicon TFT, our computational technique consists of the Gummel's decoupling method, an adaptive 1-irregular meshing technique, a finite volume approximation, a monotone iterative method, and an a posteriori error estimation method. Parallel dynamic domain decomposition of adaptive computing technique provides scalable flexibility to simulate polysilicon TFT devices with highly complicated geometry. This parallel approach fully exploits the inherent parallelism of the monotone iterative method. Implementation shows that a well-designed load balancing simulation significantly reduces the execution time up to an order of magnitude. Numerical results are presented to show good efficiency of the parallelization technique in terms of different computational benchmarks.

1 Introduction

Development of thin-film transistors (TFTs) has recently been of great interest in modern optical and electronic industries [1,2,3]. Mathematical modeling and numerical simulation of TFT devices theoretically provide an efficient way to interpret the experimental results on the material and the device structure. Drift-diffusion (DD) model consisting of the Poisson equation, the electron current continuity equation, and the hole current continuity equation has successfully been applied to explore transport phenomena of electron and hole in very large scale integration (VLSI) semiconductor devices [4,5]. Compared with conventional metal-oxide-semiconductor field effect transistors (MOSFETs), polysilicon TFTs possess significant grain structures. Therefore, it leads to significant grain boundaries in the substrate of TFT. Different physical effect of impurity traps in the interface of grains affects the charge distribution and terminal transport properties [6,7]. Effective technology computer-aided design (TCAD) tools for

TFT simulation have to incorporate effect of grain structures in the DD model. Unfortunately, it complicates formulation of mathematical model and encounters numerical convergence issues due to highly strong nonlinear property of grain boundary with respect to the potential energy.

In this paper, parallel adaptive simulation of polysilicon TFT devices, shown in Fig. 1, with dynamic domain partition algorithm is presented. A set of two-dimensional (2D) drift diffusion equations together with modeling of nonlinear grain boundary effects is formulated and numerically solved using adaptive computing technique. Our adaptive computing technique consists of the Gummel's decoupling method, the adaptive 1-irregular meshing technique, the finite volume approximation, the monotone iterative method, and a posteriori error estimation method. This adaptive computing technique for VLSI device simulation has been developed in our recent work [8,9]. According to an estimation of the computed solutions, mesh is adaptively refined with respect to each iteration loop which produces load-unbalancing among processors. Therefore, dynamic domain partition algorithm is applied to re-calculate the number of jobs to be solved and equally re-distributed them to each processor. This approach enables adaptive simulation of the polysilicon TFTs with significant grain boundary effect and highly complicate geometry. Numerical results are presented to show the efficiency of the method in terms of different computational benchmarks.

This paper is organized as follows. In Sec. 2, we state a formulated model for the simulation of polysilicon TFTs. In Sec. 3, we show the adaptive computing algorithms. In Sec. 4, we state the parallel dynamic partition algorithm. In Sec. 5, simulation results are presented and discussed. Finally, we draw the conclusions.

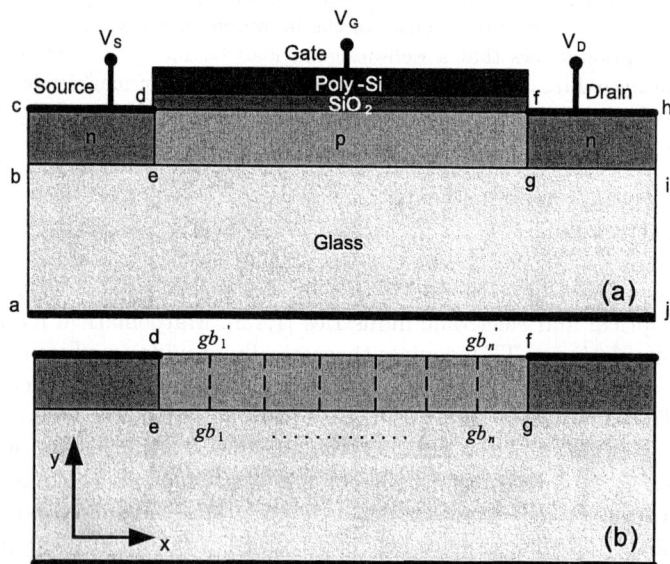

Fig. 1. (a) A cross-section view of the simulated polysilicon TFT. (b) An illustration of the interfaces among grains. Grain boundaries are in the interfaces of grains.

2 Mathematical Model of Thin-Film Transistor

Derived from the Maxwell's equation as well as the charges' conservation law, the fundamental DD equations assuming local isothermal conditions have received much attention in the past years [5]. Consisting of the Poisson equation, the electron current continuity equation, and the hole current continuity equation, a set of DD equations has successfully been applied to simulate intrinsic property of MOSFETs and calculate their terminal transport characteristics [4,5]. Taking the effect of grain boundary into consideration, the 2D DD model for the polysilicon TFT simulation is given by [6,7,8,9]

$$\Delta\phi = \frac{q}{\varepsilon_S}(n - p + D + BT(\phi)), \quad (1)$$

$$BT(\phi) = N_{At}f_{po}e^{-|(\phi-\phi_B)/V_T|}, \quad (2)$$

$$\frac{1}{q}\nabla \cdot (-q\mu_n n \nabla\phi + qD_n\phi_n) = R(n,p), \quad (3)$$

and

$$\frac{1}{q}\nabla \cdot (-q\mu_p p \nabla\phi - qD_p\phi_p) = -R(n,p). \quad (4)$$

The equation (1) is so-called the Poisson equation, where the unknown ϕ to be solved in the domain, shown in Fig. 1a, is the electrostatic potential. Eqs. (3) and (4) are the current continuity equations of electron and hole, respectively, where the unknowns n and p to be solved are the densities of electron and hole. Shown in Eq. (1), q is the elementary charge, ε_S is silicon permittivity, and D is the spatial-dependent doping profile [4]. Equation (2) is the distribution function of the grain boundary concentration which occurs in the interface of grains $\overline{g_{bi}g_{bi'}}$, $i = 1...l$, where l is the number of grain boundaries, shown in Fig. 1b. Equation (2) is a nonlinear equation of electrostatic potential ϕ and has to be solved together with the Poisson equation in Eq. (1). N_{At} is the concentration of acceptor trap and $f_{po}e^{-|(\phi-\phi_B/V_T)|}$ is the occupation probabilities of hole. ϕ_E is assumed to be the periodical energy band structure along the conducting channel in the neighborhood of the grain boundaries. f_{po} is an initial probability of hole. μ_n and μ_p in Eqs. (3) and (4) are the mobility of electron and hole [10]. D_p and D_p are the diffusion coefficients of electron and hole, and $R(n,p)$ is the term of generation-recombination of electron and hole [4,5].

Equations (1)-(3) are subject to proper boundary conditions for (ϕ, n, p), shown in Fig. 1a. Boundaries $\overline{cd}$, $\overline{fh}$ and $\overline{aj}$ are specified by the type of Dirichlet boundary condition. The boundaries $\overline{ac}$ and $\overline{hj}$ are assumed to be the Neumann boundary condition. The boundaries and $\overline{df}$ are $\overline{eg}$ with the Robin boundary conditions [4,5,8,9].

3 Adaptive Finite Volume Simulation Technique

We state the adaptive computing technique in the numerical simulation of polysilicon TFT devices. The Gummel's decoupling algorithm, the adaptive finite volume method, the monotone iterative method, and a posterior error estimation method are sequentially discussed in this section. One of efficient decoupled solution methods in semiconductor device simulation is with the Gummel's decoupling procedure, where the three coupled partial differential equations (PDEs) are decoupled and solved sequentially. The basic idea of the Gummel's decoupled method for the DD equations is that the Poisson equation is solved for $\phi^{(g+1)}$ given the previous states $n^{(g)}$ and $p^{(g)}$. The electron current continuity equation is solved for $n^{(g+1)}$ given $\phi^{(g)}$ and $p^{(g)}$. The hole current continuity equation is solved for $p^{(g+1)}$ given $\phi^{(g)}$ and $n^{(g)}$. The superscript index g denotes the Gummel's iteration loops. We solve each decoupled PDE with the adaptive finite volume computing technique [8].

Theoretical concept of the adaptive finite volume method relies on an estimations of the gradients of computed solutions, for example the gradient of potential, electron concentration, and hole concentration. In general, variations of carrier lateral current density along the device channel surface physically produces a good estimation and is also taken into the consideration. A posteriori error estimation provides a global assessment of the quality of numerical solutions, where a set of local error indicators are incorporated for the refinement strategy. This physical-based error estimation and error indicators applied here are not restricted to any particular types of mesh structures. In this work, we use 1-irregular mesh refinement in the numerical simulation of polysilicon TFT. This approach will enable us to simulate device characteristics including effects of grain boundary. The data structure of the 1-irregular mesh is designed with hierarchical and is suitable for the implementation of the adaptive algorithm by using object-oriented programming concepts.

We firstly partition the 2D solution domain of polysilicon TFT into a set of finite volumes and each decoupled PDE is approximated by the finite volume method. For the electron and hole current continuity equations, we also apply the Scharfetter-Gummel exponential fitting scheme to locate the sharp variation of the solutions. The monotone iterative solver [9] is directly applied to solve the system of nonlinear algebraic equations. We note that the monotone iterative method applies here for TFT simulation is a global method in the sense that it does not involve any Jacobian matrix. However, the Newton's iterative method not only has Jacobian matrix but also inherently requires a sufficiently accurate initial guess to begin with the solutions. The monotone iterative method is highly parallel; consequently, it is cost-effective in terms of both computational time and storage memory [9].

Once an approximated solution is computed, an a posteriori error analysis is performed to assess the quality of approximated solution. The error analysis will produce error indicators and an error estimator. If the estimator is less than a specified error tolerance, the adaptive process will be terminated and the approximated solution can be post-processed for further physical analysis.

Otherwise, a refinement scheme will be employed to refine each element. A finer partition of the domain is thus created and a new solution procedure is repeated. Mesh discretization is adaptively generated by using the maximum gradient of electrostatic potential ϕ in Eq. (1) and the variation of current density J_n in Eq. (3) as error estimation. The refinement scheme applied to refine an element depends on the magnitude of the error indicator for that element.

4 Parallel Dynamic Domain Partition Algorithm

After the calculation of error estimation and error indicators, and checking the stages of adaptation, the next step is to determine whether workload balance still exists or not for all processors. When a refined tree structure is created, the number of processors for the next computing should be dynamically assigned and allocated according to the total number of nodes firstly. Then a geometric dynamic graph partition method in the x- and y-directions, shown in Fig. 2, is applied to partition the number of nodes with respect to each processor.

A computational procedure for domain decomposition is as follows: (1) initialize the message passing interface (MPI) environment and configuration parameters; (2) based on the 1-irregular unstructured meshing rule, a tree structure and the corresponding mesh are created; (3) count the number of nodes and apply dynamic partition algorithm to determinate the number of processors in the simulation. All nodes are numbered, besides that the boundary and critical points are identified; (4) all assigned jobs are solved with the monotone iterative algorithm. The computed data communicates by the MPI protocol; (5) perform convergence test for all elements and run the adaptive refinement for those needed elements; (6) repeats the steps (2)-(5) until the error of all elements is less than a specified error bound; (7) and the host processor collects data and stops the MPI environment.

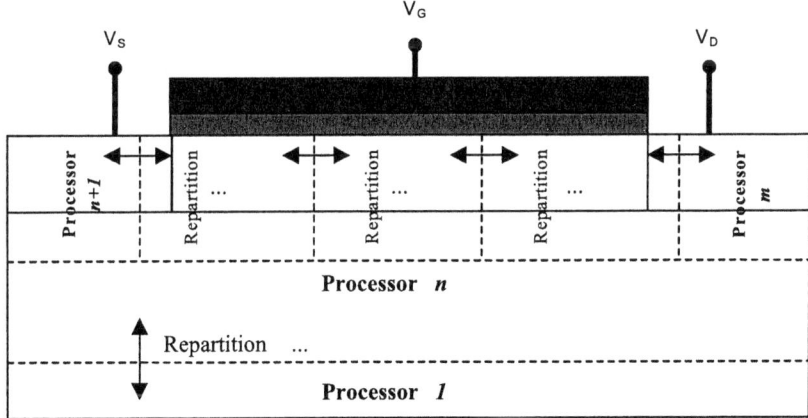

Fig. 2. An illustration of the parallel dynamic partition of the domain decomposition for a simulated 2D polysilicon TFT

The dynamic partition algorithm implemented in the step (3) above is outlined as follows: (3.1) count the number of total nodes; (3.2) find out the optimal number of processors; (3.3) calculate how many nodes should be assigned to each processor; (3.4) along the x- or y-direction in device domain, search (from left to right and bottom to top) and assign nodes to these processors sequentially. Repeats this step until all nodes have been assigned; (3.5) and in the neighborhood of the p-n junction, one may have to change search path for obtaining a better load-balancing.

5 Results and Discussion

We present in this section several numerical simulations to demonstrate the performance of the adaptive computing technique. As shown in Fig. 1a, the simulated polysilicon TFT is with $\overline{aj} = 4\mu m$, $\overline{be} = \overline{gi} = 1\mu m$, and $\overline{ac} = \overline{hj} = 0.5\mu m$. The gate oxide thickness of the SiO_2 layer is equal to $0.1\mu m$. The junction depth is with $\overline{bc} = \overline{hi} = 0.05\mu m$. The polysilicon TFT is assumed to have an elliptical-shaped Gaussian doping profile, where the peak concentration is equal to $2*10^{20} cm^{-3}$.

Figure 3 shows the process of mesh refinements for a simulation case, where the gate and drain biases are fixed at 1.0 V and 13 V. The mechanism of 1-irregular mesh refinement is based on the estimation of solution error element by element. Figure 3a is the initial mesh which contains 52 nodes, Fig. 3b is the 3rd refined mesh containing 1086 nodes, and Fig. 3c is the 5th mesh containing 3555 nodes. We note that the process of mesh refinement is guided by the result of error estimation automatically. As shown in Fig. 3c, at the 5th refined level we find that the refined meshes are intensively located near the surface of channel and the junction of the drain side due to large variation of the solution gradient. The preliminary result, shown in Fig. 3c, indicates the boundaries of grain. If the refinement process goes continuously, the grain boundaries will be located automatically in the adaptive simulation.

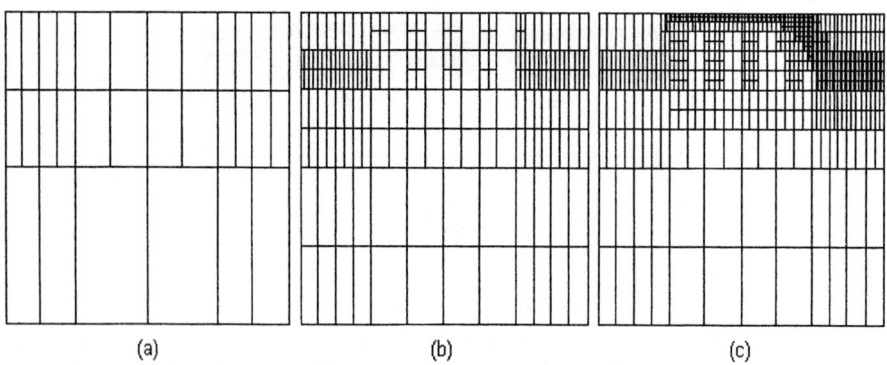

Fig. 3. (a) The initial mesh contains 52 nodes, (b) the 3rd refined mesh contains 1086 nodes, (c) and the 5th refined mesh contains 3555 nodes

The number of nodes (and elements) versus the refinement levels with and without grain boundary traps cases are shown in Fig. 4a. The number of refined elements and nodes is increased as the level of refinements is increased. We note that the increase is saturated when the levels of refinement are increased. It results from the small variation of solutions as well as the error of the solutions. This result shows that our error estimation and mesh refinement have good computational efficiency in the numerical simulation of the polysilicon TFT. Our next example is designed to demonstrate the robustness of the simulation algorithm for the same test device. As shown in Fig. 4b, the global convergence behavior of the Gummel's iteration is confirmed, and the maximum norm error for the error of the Gummel's iteration is less than 10^{-3} within 20 iterations for all conditions. For a specified Gummel's iteration loop, the convergence property within each monotone iteration loop is monotonically for all simulation cases [9].

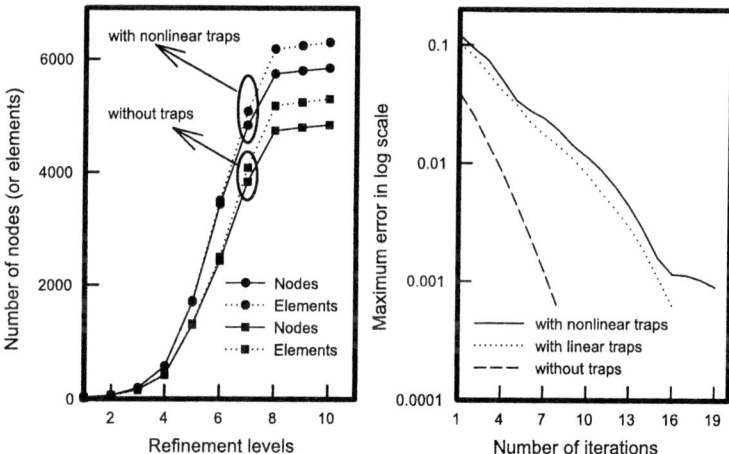

Fig. 4. (a) The number of nodes and elements versus the level of refinements with and without considering the trap model of the grain boundary. (b) A convergence plot of the Gummel's iteration with and without including the trap model of grain boundary, where the simulated polysilicon TFT is with $V_D = V_G = 1.0V$.

Table 1. A list of benchmark for the device simulated with 2000 nodes, $V_D = 13V$, and $V_G = 1V$

Number of Processors	Time(sec.)	Speedup	Efficiency
1	438	1.00	–
2	248	1.76	88.00%
4	132	3.31	82.75%
8	74	5.91	73.87%
16	51	8.58	53.62%

Furthermore, performance of the parallel adaptive computation of the polysilicon TFT device is explored. The structure of polysilicon TFT is the same with the first example and the refinement criterion for each element is setting to be $1V_T$ ($V_T = 0.0259V$ under room temperature) in this simulation. Tables 1 and 2 show the parallel performance benchmarks of the simulated TFT device with 2000 and 16000 nodes for various cluster configurations, and good speedup

Table 2. Achieved benchmark of the polysilicon TFT device simulation with 16000 nodes, where the device is with $V_D = 13V$ and $V_G = 1V$ for various parallel conditions

Number of Processors	Time(sec.)	Speedup	Efficiency
1	16251	1.00	–
2	8814	1.84	92.00%
4	4685	3.46	86.50%
8	2449	6.63	82.87%
16	1317	12.33	77.06%

Table 3. Achieved parallel performance with respect to the number of nodes for the simulated TFT device on a 16-processor PC cluster

Number of nodes	Sequential time (sec.)	Parallel time (sec.)	Speedup	Efficiency
2000	438	51	8.58	53.62%
4000	1425	154	9.24	57.81%
8000	6102	594	10.26	64.18%
16000	16251	1317	12.33	77.06%

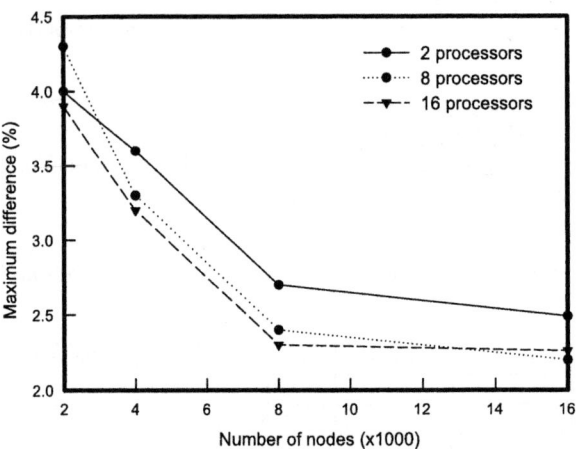

Fig. 5. Load-balancing of the domain decomposition for the polysilicon TFT simulation on a PC-based Linux cluster

in parallel time is observed. The efficiency of parallel configuration with 16 processors for 2000 nodes and 16,000 nodes are over 53.62% and 77.06%. For the same number of processors, as shown in the table 2, the parallel efficiency of the 16-processors is higher than that of the result shown in the table 1. It suggests that the parallel efficiency is proportional to the workload of the simulated task. The superior scalability of the parallel processing is mainly due to the nature of the monotone iterative method. Table 3 shows the achieved parallel performance with different number of nodes for the simulated polysilicon TFT device on a 16-processor PC cluster. The parallel efficiency increases when the number of nodes increases according to the well-designed parallel dynamic domain partition algorithm. We note that a good load-balancing is also obtained.

The performance of the dynamic load-balancing of the domain decomposition for the polysilicon TFT simulation is shown in Fig. 5. The maximum difference is defined as the maximum difference of the code execution time divided by the maximum execution time. The variation of the maximum difference for TFT device is less than 4.3% and the difference decreases when the number of nodes increases. It shows a good load-balancing of the proposed parallelization scheme when the number of nodes is larger than 8000.

6 Conclusions

In this paper, we have successfully implemented the dynamic domain partition technique to perform parallel adaptive simulation of polysilicon TFTs. This solution scheme mainly relies on the adaptive finite volume method on the 1-irregular mesh and the monotone iterative method. The grain boundary effect has been considered and successfully modeled when solving the TFT's DD model. Numerical results and benchmarks for a typical polysilicon TFT are presented to show the robustness and efficiency of the method. We have computationally found significant difference on the electrostatic potential, the electron density, and the I-V curves for the polysilicon TFT with and without including grain traps. Comparison between the refined mesh and the computed potential has demonstrated very good consistency of the adaptive simulation methodology for the testing cases. Good performance of the parallelization has been obtained in terms of the speed-up, the efficiency, and the load-balancing. We believe this simulation approach will benefit the development of TCAD tools for polysilicon TFT devices in display industry.

Acknowledgments

This work was supported in part by Taiwan National Science Council under contracts NSC-93-215-E-429-008, NSC-94-2752-E-009-003-PAE, the Ministry of Economic Affairs, Taiwan under contract 93-EC-17-A-07-S1-0011, and research grants from the Toppoly optoelectronics Corp. in Miao-Li, Taiwan in 2003-2005.

References

1. Singh, J., Zhang, S., Wang, H., Chan, M.: A unified predictive TFT model with capability for statistical simulation. Proc. Int. Semicond. Dev. Research Symp. (2001) 657-660.
2. Lee, M.-C., Han, M.-K.: Poly-Si TFTs with asymmetric dual-gate for kink current reduction. IEEE Elec. Dev. Lett. **25** (2004) 25-27.
3. Armstrong, G. A., Ayres, J. R., Brotherton, S. D.: Numerical simulation of transient emission from deep level traps in polysilicon thin film transistors. Solid-State Elec. **41** (1997) 835-84.
4. Sze, S. M.: Physics of Semiconductor Devices. Wiley-Interscience. New York (1981).
5. Selberherr, S.: Analysis and Simulation of Semiconductor Devices. Springer-Verlag. Wein-New York (1984).
6. Bolognesi, A., Berliocchi, M., Manenti, M., Carlo, A. Di, Lugli, P., Lmimouni, K., Dufour, C.: Effects of grain boundaries, field-dependent mobility, and interface trap states on the electrical characteristics of pentacene TFT. IEEE Trans. Elec. Dev. **51** (2004) 1997-2003.
7. Chan, V. W., Chan, P.C.H., Yin, C.: The effects of grain boundaries in the electrical characteristics of large grain polycrystalline thin-film transistors. IEEE Trans. Elec. Dev. **49** (2002) 1384-1391.
8. Li, Y., Yu, S.-M.: A Parallel Adaptive Finite Volume Method for Nanoscale Double-gate MOSFETs Simulation. J. Comput. Appl. Math. **175** (2005) 87-99.
9. Li, Y.: A Parallel Monotone Iterative Method for the Numerical Solution of Multidimensional Semiconductor Poisson Equation. Comput. Phys. Commun. **153** (2003) 359-372.
10. Lin, H.-Y., Li, Y., Lee, J.-W., Chiu, C.-M., Sze, S. M.:A Unified Mobility Model for Excimer Laser Annealed Complementary Thin Film Transistors Simulation. Tech. Proc. Nanotech. Conf. **2** (2004) 13-16.

A Scalable Parallel Algorithm for Global Optimization Based on Seed-Growth Techniques

Weitao Sun

ZHOU PEI-YUAN Center for Applied Mathematics, Tsinghua University,
Beijing 100084, China
sunwt@tsinghua.edu.cn

Abstract. Global optimization requires huge computations for complex objective functions. Conventional global optimization based on stochastic and probability algorithms can not guarantee an actual global optimum with finite searching iteration. A numerical implementation of the scalable parallel Seed-Growth (SG) algorithm is introduced for global optimization of two-dimensional multi-extremal functions. The proposed parallel SG algorithm is characterized by a parallel phase that exploits the local optimum neighborhood features of the objective function assigned to each processor. The seeds are located at the optimum and inner neighborhood points. Seeds grow towards nearby grids and attach flags to them until reaching the boundary points in each dimension. When all grids in the subspace assigned to each CPU have been searched, the local optimum neighborhood boundaries are determined. As the definition domain is completely divided into different subdomains, the global optimal solution of each CPU is found. A coordination phase follows which, by a synchronous interaction scheme, optimizes the partial results obtained by the parallel phase. The actual global optimum in the total definition space can be determined. Numerical examples demonstrate the high efficiency, global searching ability, robustness and stability of this method.

Keywords: Global optimization; Parallel Seed-Growth; Space partition

1 Introduction

Nonlinear objective function optimization has many difficulties, such as having multi-parameter and non-linearity. Since Backus and Gilbert[1] proposed the BG theory in 1968, research on inversion theory and application has begun to develop quickly. At present, optimization methods for inversion problems can be classified into two categories. The first kind is the deterministic local optimization methods based on objective function gradient information, which includes Steepest Descent techniques, Newton's method and the Conjugate Gradient method. This kind of method has very fast computation and convergence speed. However, the quality of an optimal solution depends closely on the selection of initial models. The other is stochastic searching and heuristic methods, which includes the Simulated Annealing method, the Genetic Algorithm, the Neural Network and the Chaos Optimization method, etc. Global convergence property and independent final optimal solution are achieved. But the computation costs are very high and the convergence is slow. Research on the generation of the stochastic searching and heuristic methods is becoming the leading edge in geophysics inversion problems [2]-[4]. Hibbert[5] solved a wave equation problem by combining the Genetic Algorithm and linear inversion method. Chunduru[6] and Zhang[7] provided a hybrid optimization method based on the Simulated Annealing algorithm and Conjugate Gradient. Liu et al.[2], Ji and Yao[3] and Macias et al.[8] developed optimization methods by a combination of the Simulated Annealing, Neural Network, Simplex Search and Uniform Design. These improved methods are more efficient than previous ones, although the convergence is still very slow.

Based on the scalable parallel Seed-Growth and neighborhood determination scheme, we developed a definition space division and reduction algorithm. The total definition space is partitioned into subdomains each containing one local optimum (set) and its neighborhood. Optimizations beginning at the points that belong to the neighborhood will converge to the same local optimum. Thus further optimizations are only needed in the undivided domain.

2 Deterministic Global Optimization

Global optimization algorithms have been proven to be a powerful tool for escaping from the local minima. Unfortunately, it is still too slow to be convenient in many applications. The global convergence of the SA algorithm requires the generation of an infinite number of random trial points at each value of the temperature. Current attempts to practically overcome these limitations have been presented by Dowsland[9]. The main efforts involve reduction of computation and prevention from falling into the local optimum.

A complex objective function usually has multiple local optima. Local optimization within the neighborhood of each local optimum will converge to this local optimum. This paper introduced a deterministic non-linear optimization method based on the neighborhood determination algorithm. First, the local optima are found by the conventional local optimization algorithm. The corresponding neighborhoods are determined and eliminated from the total parameter space by a scalable parallel Seed-Growth algorithm. Then, new starting points are distributed in the remaining parameter space to search for new local optima and their neighborhood. As the definition space is totally divided, the actual global optimum can be found. The mathematical definition of Seed-Growth algorithm can be found in Sun[10].

3 Algorithm Description

The SG method is carried out in a recursive manner to find the local optimum neighborhood boundary. The pseudo program of the SG method for a 2D problem is illustrated as follows.

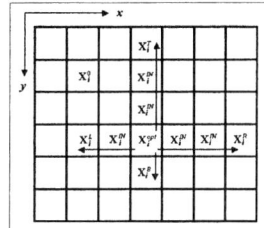

Fig. 1. The Seed-Growth algorithm for local optimum neighborhood determination

Step 0: Assign the total definition space to different CPUs. Given an assigned definition subspace, divide it into a $m \times n$ grid system (Fig. 1).

Step 1: Having any randomly distributed starting point $X_i^0 \in D$ in an unsearched domain, compute the local optimum X_i^{opt} by conventional local optimization algorithm.

Step 2: Taking X_i^{opt} as the starting seed-growth point, boundary points are searched for in x and y directions.
Grids at the left side of X_i^{opt} are taken as the starting points one by one for local optimization. If the local optimization result converges to X_i^{opt}, the starting point is an inner point (X_i^{IN}) of X_i^{opt} neighborhood and is pushed into the stack as a new seed for later searching. Otherwise, the starting point is the left boundary (X_i^L) of X_i^{opt} neighborhood. The right boundary point X_i^R is found in the same way. The boundary points in y direction are found in the same way.

Step 3: Pop a new seed from the stack and repeat steps 2 and 3 until there are no more seeds in the stack.

Step 4: Divided the definition space into different subspaces, each containing a local optimum (set) and corresponding neighborhood.

Step 5: Compare the optima from different CPUs, and the global optimum can be deterministically found.

After an iteration of local optimizing and neighborhood boundary searching from a certain starting point, the determined neighborhood N_i is found to be an irregular subdomain in the solution space. Elimination of N_i from the total solution space prevents unnecessary local optimization and cuts down the computation time.

4 Numerical Example

Although a much larger real data example can be used to test the limits of the parallel Seed-Growth algorithm, it is more important to illustrate code accuracy and efficiency by a relatively small typical example. A non-linear function with a complex surface was used to test the new algorithm. The example was run on 9 PCs connected by Myrinet networks. Each grid computer had two 700 Hz Xeon CPUs and 1Gbyte memory. The program was implemented in C language.

Test function:

$$F(x, y) = 0.5 - \frac{(\sin^2(\sqrt{x^2 + y^2}) - 0.5)}{[1 + 0.001(x^2 + y^2)]^2}. \tag{1}$$

The definition domain was $x \in [-100, 100]$, $y \in [-100, 100]$. The global maximum value 1 was located at (0,0). There were infinite secondary global maxima around the actual one (Fig. 2a). Because of the oscillation property, it is very difficult to find the global maximum by conventional optimization methods. In this example, a subdomain $x \in [-9, 9]$, $y \in [-9, 9]$ containing the global maximum was chosen as the searching space. The grid side length was 3.0. First, optimization was carried out on one CPU using 23.6 seconds (Fig. 3a). Among the 332 local optima (sets), the global optimum 1 was found at (-9.862863e-015, -8.496365e-015) (Fig. 4a). Then the definition was divided into 3×3 subdomains (Fig. 2b) and each subspace was assigned to one CPU. The global optimum 1 was found at (-8.691454e-9, 0.000000e+000) by the parallel Seed-Growth method (Fig. 4b). The maximal computation time of the 9 CPUs was 3.522659 seconds (Fig. 3b), which means a speedup of 6.70 was achieved.

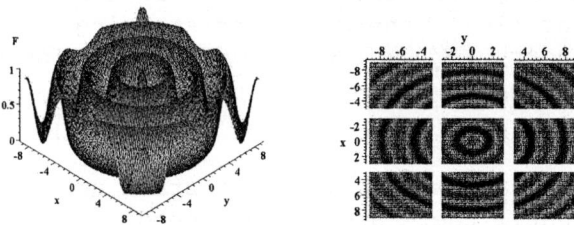

Fig. 2. (a) The objective function surface (left); (b) The 3×3 definition space division. The subspaces from left to right and top to bottom are respectively assigned to 9 CPUs (right).

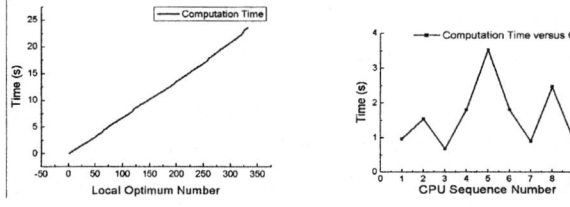

Fig. 3. (a) The computation time for one CPU Seed-Growth optimization (left); (b) The computation time for the 9-CPU parallel Seed-Growth optimization (right)

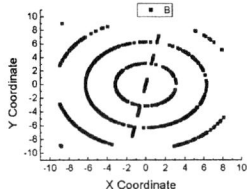 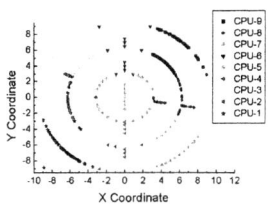

Fig. 4. (a) The local optima distribution of one CPU Seed-Growth optimization (left); (b) The local optima found by 9-CPU parallel Seed-Growth optimization (right)

In Fig. 3 the computation costs of the serial and parallel Seed-Growth optimization method are compared. When the optimization problem was assigned to multiple CPUs, the neighborhood determination time decreased dramatically. The scalable parallel Seed-Growth algorithm provides a fast definition space partition scheme, which saves a great amount of computation costs for the neighborhood determination. The global optimum was found in only 3.52 seconds.

Table 1. The SG algorithm testing results and comparison with Simulated Annealing algorithm

AT	GM	FV	XGS	YGS	EN	TGM	TFV
SG algorithm	[-9.86e-15, -8.50e-15]	1.0	3	3	6629	[0,0]	1
Parallel SG algorithm	[-8.69e-9, 0.0]	1.0	3	3	1261	[0,0]	1
SA algorithm	[-7.91e-6, 5.78e-5]	-0.9999	2.30e-3	1.56e-3	27001	[0,0]	1

AT: Algorithm type; GM: Global minima; FV: Function value of global minima; XGS: Grid size in x-direction; YGS: Grid size in y-direction; EN: Function evaluation number, average evaluation number for parallel SG algorithm; TGM: True global minima; TFV: True function value of global minima

A Simulated Annealing (SA) algorithm was also used to optimize the test functions for comparison (Table 1). The initial annealing temperature was 5 and the cooling factor was 0.7. The maximal iteration number at each annealing temperature was 5. The maximal inner loop bound at each temperature was 20. The maximal function evaluation number for SA was 1e5. The SA final temperature was 7.64e-7. The experiment results and comparison show that the parallel SG algorithm is more efficient than the serial SG algorithm and the SA algorithm.

5 Conclusions

A parallel Seed-Growth algorithm was proposed for neighborhood determination. In the proposed algorithm, the definition space is partitioned into subdomains and assigned to different CPUs. In each parallel task, seeds grow towards nearby unsearched grids until they reach the neighborhood boundary in each dimension. The definition space assigned to each CPU is divided into different neighborhoods. By determining a new local optimum and its neighborhood, unnecessary optimizations in these subdomains can be avoided high efficiency can be achieved. A comparison of various methods using a numerical example showed that an obviously improvement in global optimization by using the parallel Seed-Growth method.

Acknowledgement

The work described in this paper was supported by the National Natural Science Foundation of China (No. 10402015) and the China Postdoctoral Foundation (No. 2004035309).

References

1. Backus, G. E., Gilbert, F. The resolving power of gross earth data. Geophysical Journal of the Royal Astronomical Society, Vol. 16, (1968) 169-205.
2. Liu, P., Ji, C., et al. An improved simulated annealing-downhill simplex hybrid global inverse algorithm. Chinese Journal of Geophysics, Vol. 38, No. 2, (1995) 199-205.
3. Ji, C. and Yao, Z. The uniform design optimized method for geophysics inversion problem. Chinese Journal of Geophysics, Vol.39, No. 2, (1996) 233-242.
4. Ai, Y., Liu, P. and Zheng, T. Adaptive global hybrid inversion. Science in China (series D) Vol.28, No.2, (1991) 105-110.
5. Hibbert, D. B. A hybrid genetic algorithm for the estimation of kinetic parameters. Chemometrics and Intelligent laboratory systems, Vol.19, (1993) 319-329.
6. Chunduru, R., Sen, M. K., et al. Hybrid optimization methods for geophysical inversion. Geophysics, Vol. 62, (1997) 1196-1207.
7. Zhang, L., Yao, Z. Hybrid optimization method for inversion of the parameters. Progress in Geophysics, Vol.15, No.1, (2000) 46-53.
8. Macias, C. C., Sen, M. K. et al. Artificial neural networks for parameter estimation in geophysics. Geophysical Prospecting, Vol.48, (2000) 21-47.
9. Dowsland. Simulated annealing. In Reeves, C. R.(ed.), Modern heuristic techniques for combinatorial optimization problems. McGraw Hill Publisher, (1995) 20-69.
10. Sun, WT. Global Optimization with Multi-grid Seed-Growth Parameter Space Division Algorithm. Proceedings of the 2005 International Conference on Scientific Computing, Las Vegas, USA, 2005,32-38

Exploiting Efficient Parallelism for Mining Rules in Time Series Data

Biplab Kumer Sarker[1], Kuniaki Uehara[2], and Laurence T. Yang[3]

[1] Faculty of Computer Science, University of New Brunswick, Fredericton, Canada
sarker@unb.ca
[2] Department of Computer and Systems Engineering, Kobe University, Japan
uehara@kobe-u.ac.jp
[3] Department of Computer Science, St. Francis Xavier University, Antigonish, Canada
lyang@stfx.ca

Abstract. Mining interesting rules from time series data has earned a lot of attention to the data mining community recently. It is quite useful to extract important patterns from time series data to understand how the current and the past values of patterns in the multivariate time series data are related to the future. These relations can basically be expressed as rules. Mining these interesting rules among patterns is time consuming and expensive in multi-stream data. Incorporating parallel processing techniques is helpful to solve the problem. In this paper, we present a parallel algorithm based on a lattice theoretic approach to find out the rules among patterns that sustain sequential nature in the multi-stream data of time series. The human motion data considered as multi-stream multidimensional data used as data set for this purpose is transformed into sequences of symbols of lower dimension due to its complex nature. Then the proposed algorithm is implemented on a Distributed Shared Memory (DSM) multiprocessors system. The experimental results justify the efficiency of finding rules from the sequences of the patterns for time series data by achieving significant speed up comparing with the previous reported algorithm.

Keywords: Data Mining, Time series data, Parallel algorithm, Association rule, Multiprocessor system.

1 Introduction

Time series data mining has recently become an important research topic and is earning substantial interest from both academia and industry. It sustains the nature of multi-stream data due to its characteristics. An example of such type could be *"if company A's stock goes up on day 1, B's stock will go down on day 2 and C's stock will goes up in day 3*, i.e. some of the events of company A influences to occur some of the events of company B and C. If we analyze the multi-stream of time series for some stock prices and can discover correlations among all the streams, then the correlations can help us to decide better time to buy stocks. So, the correlations can be expressed as rules. It is quite effective and useful to analyze the underlying behavior of time series data to investigate the correlations among its multi-stream. Strong dependencies capture structures in the streams because it indicates that there exists relationship between their constituent patterns, that occurrences of those patterns are not independent. These correlations depicted as rules are useful to describe the sequential nature of time series data. Therefore, the dependencies can be termed as association rules in the multi-stream.

Researchers have been concentrating to find out these dependency rules from the large and complex data sets such as prices of stocks, intensive weather patterns, astrophysics databases, human motion data etc. Basically, except human motion data, these are one dimensional data in nature. In this paper, we focus on human motion data deemed as high dimensional multi-stream due to its features [4]. These correlations that are discovered from multi-stream of human motions data characterize a specific motion data. Furthermore, those correlations become basic elements that can be used to construct motion with combinations of themselves, just as phonemes of human voice do. These basic elements are called primitive motions. As a result, we can use these primitive motions as indices to retrieve and recognize motion for creating SFX movies, computer graphics, animations etc.

The following section discusses the problem statement and various research findings related to the topics. Section 2 introduces the structure of human body as multi-stream and briefly discusses the way of converting high dimensional motion data into sequence of symbols of multi-stream representing as lower dimensional data. The next section discusses the discovery process of association rules from the symbols of multi-stream. It is very expensive, time consuming and computational intensive task to discover association rules from these kinds of huge amount of time series data represented as symbols of multi-streams. Hence, section 3 illustrates the advantage of using lattice theoretic based approach over Distributed Shared Memory (DSM) Multiprocessor system. A parallel algorithm using the lattice theoretic approach is also presented here. The algorithm can efficiently find the sequence of correlations among the body parts that perform various motions in a small amount of time. The results are also discussed elaborately in the section and compared with one of our previous reported algorithm. Finally section 4 concludes the paper by presenting the direction of future research.

1.1 Problem Statement

Let A be a set of distinct attributes. With no loss of generality, any subset of A can be represented as a sequence that is sorted according to the lexicographic order of attribute names. For instance, $\{a, c\}$ and $\{c, a\}$ represent the same subset of $\{a, b, c\}$ that is identified by the sequence ac. We term such a sequence as a *canonical attribute sequence* (*cas*) [12]. There exists a one-to-one mapping between the set of all *cas*s and the power set, denoted 2^A, so that the set of *cas*s can be identified with to 2^A. 2^A can be treated as a Boolean lattice where $\emptyset$ (i.e. empty *cas*) and A (i.e. the complete *cas*) are, respectively, the bottom and top elements. The order in 2^A is denoted as $\leq$, coinciding with set inclusion; where $s \leq u$ reads as s is a *partial cas* or a *subcas* of *u*.

Given a threshold ratios σ and γ such that $0 \leq \sigma \leq 1$ and $0 \leq \gamma \leq 1$, mining a database D for association rules consists of extracting all pairs of *cas*s s and u for which the relation $s \Rightarrow u$ is a $\sigma\gamma$-valid association rule; i.e. $support(s \Rightarrow u) \geq \sigma$ and $confidence(s \Rightarrow u) \geq \gamma$. The problem can be solved with the two-step procedure below.

1) Find all σ-frequent *cas v*, that is to say all v such that
   $support(v) \geq \sigma$.
2) For all *cas v* found in step 1, generate all association rules $s \Rightarrow u$ such that
   1. $s < v$ and $u = v\text{-}s$ and
   2. $confidence(s \Rightarrow u) \geq \gamma$.

1.2 Related Works

A good number of serial and parallel algorithms have been developed for mining association rules for basket data (supermarket transaction data) [6-8]. Several researchers have applied data mining concepts on time series to find patterns including [1, 3, 5, 13, 16-19]. However, these algorithms are sequential in nature except [13, 16].

The problem of mining sequential patterns was introduced in [14]. Three algorithms have been proposed in this work for the purpose. However, in subsequent work [15], the same authors proposed *GSP* algorithm that outperformed *AprioriAll* by up to 20 times which is considered as the best in their earlier works. In [9], the algorithm SPADE was shown to outperform *GSP* by a factor of two by using the advantage of lattice based approach. A parallel version of SPADE, was introduced in [10] for Shared Memory Multi-processor (SMP) systems. Sequence discovery can be essentially thought of as association discovery over a temporal database. While association rules discovery only applies to intra-relationship between patterns and sequence mining refers to inter-transactional patterns [14]. Due to this similarity, sequence mining algorithms like *AprioriAll*, *GSP*, etc., utilize some of the ideas initially proposed for the discovery of association rules [9]. Our algorithm proposed in section 3 is basically based on [10]. But we use prefix based classes rather than suffix based classes [10] and implemented the algorithm to discover the sequence of correlations over the multi-stream data. To the best of our knowledge there is no suitable algorithm proposed in the literatures for discovering rules from multidimensional multi-stream time series data. Hence, our approach, discovering association rules from a large amount of time series data differs from the above algorithms in following ways:

1) We transform the large amount of multi-dimensional time series data into symbols of multi streams to make the data into lower dimensions. Because, it is very expensive, time consuming and computational intensive task to discover association rules from these kinds of huge amount of time series data represented as symbols of multi-streams.

2) We use lattice-theoretic based approach to decompose the original search space into smaller pieces i.e. into the prefix based classes which can be processed independently in main-memory of DSM multi-processors systems.

2 Transformation of Multi-dimensional Multi-stream Data

The motion data captured by motion capturing system consists of various information of the body parts. This means that, motion captured data can be represented by the 3 dimensional time series stream considering various positions of various body joints [4]. The following body parts data can be obtained as: lower torso, upper torso, the root of neck, head, root of collar bones, shoulder, elbows, wrists, hips, knees and ankles. Moreover, body parts can be represented as the tree structure that is shown in Figure.1.

To get time series of 3-D motion data, we use an optical motion capture system. The system consists of six infrared cameras, 18 infrared ray reflectors, and a data processor. To get time series of motion data an actor puts 18 markers on the selected body joints and performs an action in the field surrounded by the installed cameras.

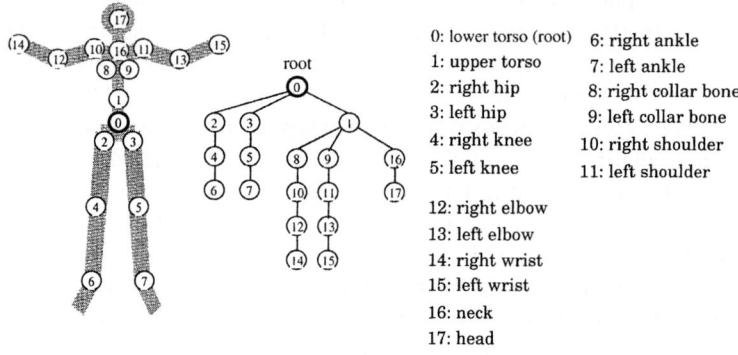

0: lower torso (root) 6: right ankle
1: upper torso   7: left ankle
2: right hip   8: right collar bone
3: left hip   9: left collar bone
4: right knee   10: right shoulder
5: left knee   11: left shoulder
12: right elbow
13: left elbow
14: right wrist
15: left wrist
16: neck
17: head

Fig. 1. Human body parts used for the experiments and considered as tree structure

The recorded actions are processed in order to calculate the 3-D locations of the reflectors. Figure 2 shows an example of the motion that represents the correlation such as "after one finished raising one's right hand, one starts lowering the left hand".

The example shows that the motion data has features as multi-stream, such as: "unfixed temporal interval between events on each stream: consistent contents on each stream do not always occur in fixed temporal interval. For instance, "raising the right hand" and "lowering the left hand" occur twice in each stream, however, temporal interval of occurrence between "raising the right" and "lowering the left hand" are different ($T1 \neq T2$).

In order to find motion association rules with easy analysis with considerations of various occurrences and reduce the cost of the task, we convert the high dimension multi-stream motion into sequence of symbols of lower dimension. That is, motion data can be converted into a small number of symbol sequences. Each symbol represents a basic content and motion data that can be expressed as the set of the primitive motions. We call this process content-based automatic symbolization [4]. For the content-based automatic symbolization, we focused on each change in the velocity of a body part that can be considered as a break point. We divide motion data into segments at those breakpoints where velocity changes. However the variation of the curve for motion data between these breakpoints are small and include noise occurred by unconscious movements, which are independent from occurrence of the changes of contents. The unconscious movements are mainly caused by the vibrations of the body parts. These are too short in time and tiny movements and contain no significant content of motion. Thus they are discarded by evaluating time scale and displacement

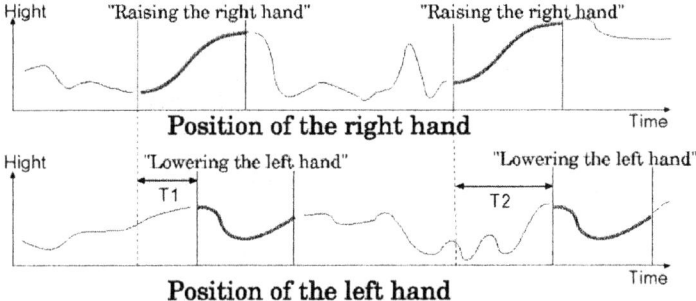

Fig. 2. An example of corelation of multi-stream of human motion

of positions considering the 3D distances between points. Segmented data is clustered into groups according to their similarity. However, even segmented data with same content have different time lengths, because nobody can perform exactly in the same manner as past. In order to find out the similarity between time series data with different lengths, we employ Dynamic Time Warping (DTW) [1], which was developed in the speech recognition domain. By applying DTW on our motion data, the best correspondence between two time series was found. Human voice has fixed number of consistent contents, phonemes, but human motion does not have pre-defined patterns. So it is unknown that how many consistent contents exist on our motion time series data. For this reason a simple and powerful unsupervised clustering algorithm, Nearest Neighbor (NN) algorithm [2] is used with DTW to classify an unknown pattern (content) and to choose the class of the nearest cluster by measuring the distance.

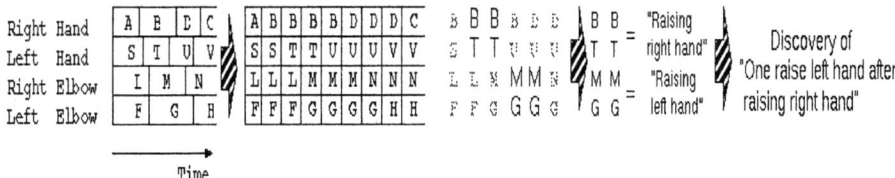

Fig. 3. Symbols of multi-stream

Thus, motion data is converted into symbol streams based on its content by using symbol that are given to clusters (Fig. 3). For more details of these processes please refer to [4]. After segmenting and clustering processes, a multi-stream that represents human motion is expressed as a multiple sequence of symbols that we call the sequence of symbols of multi-stream.

3 Mining Sequential Association Rules

Recently, lattice based sequence mining has been successfully employed by the researchers using supermarket dataset [5]. The advantage lies in this approach that it helps to decompose original search space into smaller pieces termed as the prefix based equivalence classes in our case, which can be processed independently in main-memory with the advantages of the DSM multi-processor systems. The decomposition is recursively applied within each parent class to produce even smaller classes in the next level. For example, in figure 4, the parent classes can be denoted as a, b, c, d, e and $\theta_1, \theta_2, \theta_3, \theta_4, \theta_5$ represent the level of the search space.

3.1 Search Space Partition

As a first step to find the association rules, the first step essentially consists of enumerating the frequent item sets. The database is very large, leading to a large search space. In order to lower the main memory requirements and perform enumeration in parallel, the search space is splitted into several parts that can be processed independently in parallel. This can be accomplished via prefix-based equivalence classes. It is possible that each class is a sub-lattice of the original sequences lattice and can be processed independently. For example, in the figure 4, it is shown that the effect of decomposing the frequent sequence lattice for the sample database, by collapsing all sequences with the same 1-length prefix into a single class. For details about lattice theory and structure please refer to [9].

The following way describes how the database can be partitioned. For example, given the number m in which search space is to be partitioned by satisfying the condition that k is the smallest integer such that $m \leq 2^k$. Here, $A = \{a, b, c, d, e\}$ and k is an integer. For example, for $m = 4$, $k = 2$ which satisfies the above condition. For this value, it is possible to generate 4 sets of classes for splitting (abc), (abd), (abe), (ac), (ad), (ae), (bc), (bd), (be), (c), (d), (e). The splitted parts are shown in the Fig. 4 by dotted lines. For $m = 8$, the value of k will be 3, then the search space will be splitted in 8 parts considering their corresponding prefixes. This is how the search space can be splitted by prefix based approach among the multiple processors.

3.2 Lattice Decomposition-Prefix Based Classes

In figure 4, there are five resulting prefix classes, namely, [RightLeg], [LeftLeg], [RightArm], [LeftArm], [Trunk] in our case, which are referred to parent classes. It is to mention that for simplicity purpose to show the figure, the example using only 5 parts of the body is presented here. They are denoted in the figure as a, b, c, d and e respectively. However, the figure for 17 parts also can be generated in the same way. Each class is independent in the sense that it has complete information for generating all frequent sequences that share the same prefix.

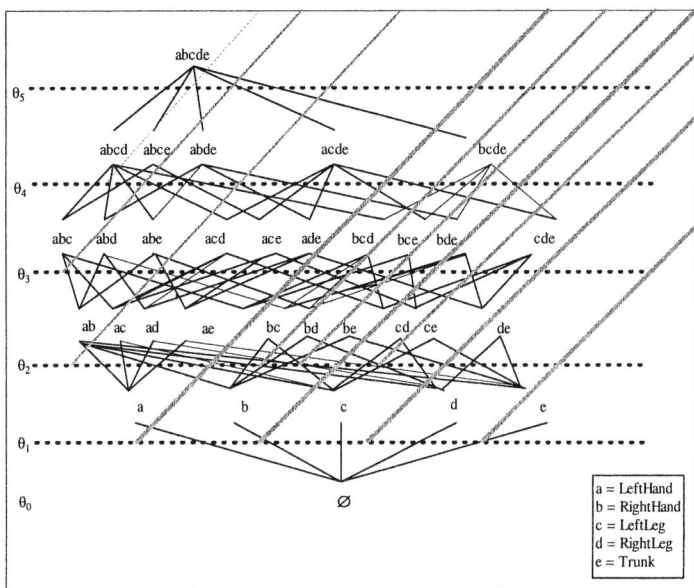

Fig. 4. Partitioning search space with the prefix based equivalence classes

For example, if a class [a] has the elements [b] → [a], and [c] → [a]. The only possible frequent sequences at the next step can be [b] → [c] → [a] or [c] → [b] → [a] and [bc] → [a]. It should be obvious that no other parts such as y can lead to a frequent sequence with the prefix [a], unless (ya) or y → x is also in [x]. The method decomposes the sequences at each new level into smaller independent classes. The figure shows the effect of using 2-level prefixes i.e. ab, ac, bd, bc etc. The figure also shows the 3-level, 4-level and 5-level prefixes. For all the levels, it can be obtained as a tree like structure of independent classes. This composition tree is to be processed in a breath-first manner, within each parent classes. In other words, parent classes are processed one-by-one, but within a parent class we process the new classes in a breath-first search. Frequent sequences are generated for the databases list by considering their support threshold *min_sup*. The sequences are being found to be frequent at the current level from the classes for the next *New_Level*. The level-wise process will be repeated until all frequent sequences have been enumerated.

In terms of memory management, it is easy to see that we need memory to store intermediate lists for the most 5 consecutive levels within a parent class (Fig. 4). Once all the frequent sequences or the next level have been generated, the sequences at the current level can be deleted.

3.3 Algorithmic Implementation

As the data set, each motion consists of repetition of 2 or 3 times for one kind. Test data are about 5 to 20 seconds long and the sampling frequency is 120 times/second. The motions are performed 23 times in 6 different types each of which lasts for about 6-12 sec. The database consists of 50 different types of motion such as walking, running, dancing, pitching etc. All of the experiments were performed on a 64-node SGI Origin 3400 DSM multi-processor system. The database is stored on an attached 6 GB local disk. The system runs IRIX 7.3.

There are two main paradigms that may be utilized in the implementation of parallel data mining: a data parallel approach or a task parallel approach. In data parallelism, P processors work on distinct portions of the database, but synchronously process the global computation tree. The parallelism is available within class. In task parallelism, the processors share the database, but work on different classes in parallel, asynchronously processing the computation tree. We need to keep the temporary list of all newly generated candidates both infrequent and frequent since we cannot say if a candidate is frequent until all processors have finished the current level. In the task parallelism all processors have access to one copy of the database, but they work on separate classes.

```
Algorithm Mining_Sequencial_Parts
Input: P, min_sup
Output: Plist
begin
1.  P = {Total body parts representing the parent
        classes P_i, where i=1,2,...,17}
2.  for each P_i ∈ P
3.  {
4.  while (Previous_level ≠ ∅);
5.  do in parallel for all processors p
6.        {
7.           New_Level=New_Level ∪ FindNewClasses
             (Previous_Level.bodyparts());
8.           Previous_Level = Previous_Level.next()
9.        }
10. barrier;
11. New_Level= ∪_{p∈P} New_Level_p
12. Repeat steps 5-7 if (New_Level≠ ∅)
13. }
end

Procedure FindNewClasses( P, min_sup)
begin
1.  for all the sequences in P_i∈ P
2.  {
3.         for all the sequences P_j∈ P, with j ≤ i
4.         {
5.            L= P_i∪P_j;
6.  do in parallel for all processors P
7.         {
8.            X= P_i ∩ P_j
9.         }
10. if (σ(L) ≥  min_sup) then P_i=P_i∪{L});
11.        }
12. }
13. return Plist=Plist ∪ P_i
end
```

We use the task parallel approach to solve our problem. The pseudo code of the algorithm is given below. This is done within the level of the computation tree (see fig. 4). In other words, at each new level of the computation tree (within a parent class like *a, b, c*, etc.), each processor processes all the new classes at that level, performing intersections for each candidate, but only over its local block (steps 7-11 in algorithm `Mining_Sequencial_Parts`). The local supports are sorted in a local array to prevent false sharing among processors (step 10 in procedure *FindNewClasses*). After barrier synchronization signals that all processors have finished processing the current level, a sub-reduction is performed in parallel to determine the global support of the each candidate. The frequent sequences are then retained for the next level, and the same process is repeated for other levels until no more frequent sequence are found (steps 10-12 in algorithm `Mining_Sequencial_Parts`). The level-wise task parallelism requires modifications by performing local intersection for all classes at the current level, followed by a barrier before the next level can begin.

In our case, with regards to 17 parts of the body, we used 17 processors to process the each prefix based class independently in the memory. It is to mention that as reported earlier, the fig.4 represents only prefix based search space with 5 parts. However, in case of 17 parts the structure will be of same type using 17 individual (parent) classes. After implementing the proposed algorithm for 17 body parts as 17 classes using 17 processors, we found efficiently many sequential associations rules for the body parts that took part in performing a motion. As an example of such discovered rules for performing *"walking"* is "raising the *RightHand* forward, lowering the *LeftHand* back, raising the *RightLeg* forward, directing the *LeftLeg* backward and move forward the *trunk*". We can find such kind of sequential rules for other sets of motion data like running, pitching, dancing etc. The running time presented in figure 6 indicated by "time using lattice based approach" justifies that good speed up is obtained by our approach.

3.4 Comparison of the Results

To justify the effectiveness of our algorithm we compared with our proposed algorithm presented in [13, 16]. For ready reference here, we present a brief overview of our previous approach. In our previous papers [13, 16], we introduced two parallel algorithms from the same data sets that we used here for mining association rules based on well known *apriori* algorithm [6] in data mining for super market set data; one of them for mining association rules from 17 body parts and the other one from 5 parts of the body by reducing the search space due to the large number of combinations that used to occur during the search process and the complexity of search process itself. Note that the algorithm using the 5 parts of the body regarded as the better solution among the proposed two algorithms [13, 16]. But it was considered as a limitation to our goal as our motivation was to discover rules using all of the parts of the body i.e.from 17 parts. So, as a new approach (in this paper), we propose the lattice based parallel algorithm for this purpose. Hence, the approach that we use in this paper is prevailed over with our previous problem. As described in the above section, we can discover rules from our datasets consist of 17 parts and thus fulfill our objective using the algorithm presented here. So, we present the comparison between the results using our previous approach and the present approach, to figure out the efficiency of our present proposed algorithm. The results shown in Fig. 6 indicate that our present lattice based parallel algorithm can effectively reduce the time required for discovering rules in terms of scalability whereas the time taken by the parallel algorithm of our previously reported algorithm varies with the number of processors.

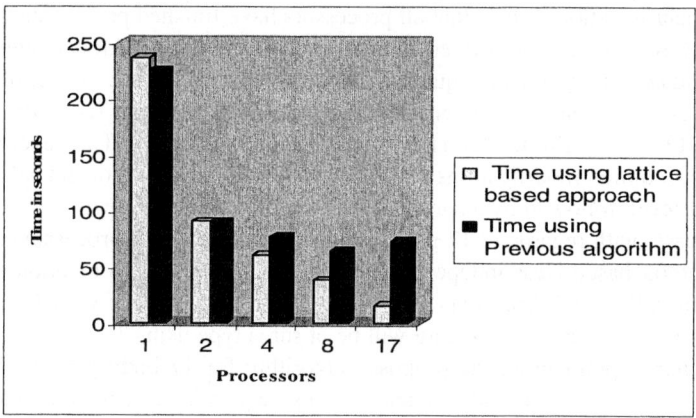

Fig. 6. The comparison results of the time required for discovering the rules between two algorithms for the same data set

4 Conclusion and Future Work

In this paper, we have presented a parallel algorithm for finding association rules from sequences of the body parts that performs different kinds of motion from multi-stream time series data such as human motion data. The algorithm has considered a large number of combinations and the depth of the sequence of parts that perform motions (such as walking, dancing, pitching etc.). The extraction technique of motion data into symbols of multi-stream has also been discussed briefly. The experimental results have demonstrated that the algorithm can efficiently determine rules of sequence of body parts that performs motion in our case by using the advantage of lattice based structure. It also has outperformed the result of our previously reported algorithm.

It is to mention that the structure of the algorithm and the way of implementation is not specific to the problem of searching the sequence of patterns (in our case body parts). It infers that our motivation was to reduce the time and find the sequence of the body parts using various combinations efficiently, and hence we achieved that for multidimensional data. This technique can be implemented with other data sets in the multidimensional multi-stream time series domain for finding interesting sequential rules for patterns in the domain of medicine and business. As a future work we aim to use the techniques for such data sets.

References

1. Berndt D.J. and Clifford J., Finding Patterns in Time Series: A Dynamic Programming Approach, in Proc. of Advances in Knowledge Discovery and Data Mining, pp. 229- 248, 1996.
2. Bay S.D., Combining Nearest Neighbor Classifiers Through Multiple Feature Subsets, in Proc. of 15th International Conference on Machine Learning, pp. 37-45, 1998.
3. Oates T. and Cohen P.R., Searching for Structure in Multiple Stream of Data, in Proc. of 13th International Conference on Machine Learning, pp. 346-354, 1996.
4. Shimada M. and Uehara K., Discovery of Correlation from Multi-stream of Human Motion, Lecture Notes in Artificial Intelligence, Vol. 1967, pp. 290-294, 2000.
5. Das G., Lin K., Mannila H., Renganathan G. and Smyth P., Rule Discovery from Time Series, in Proc. of Fourth International Conference on Knowledge Discovery and Data mining, AAAI Press, pp. 16-22, 1998.
6. Agrawal R.and Shafer J., Parallel Mining of Association Rules, IEEE Transactions on Knowledge and Data Engineering, Vol. 6, No.8, pp. 962-969, 1996.
7. Agrawal R. and. Srikant R., Fast Algorithms for Mining Associations Rules, in Proc. of 20th VLDB Conference, pp. 487-499, 1994.
8. Park, J.S. Chen M.S. and Yu P.S., Efficient Parallel Data mining for Association Rules, in Proc. of CIKM, pp. 31-36, 1995.
9. Zaki M.J., Efficient Enumeration of Frequent Sequences, in Proc. of Intl. Conf. on Information and Knowledge Management, pp. 68-75, 1998.
10. Zaki M.J., Parallel Sequence Mining of Shared Memory Machine, in Proc. of Workshop on Large-Scale Parallel KDD Systems, SIGKDD, pp. 161-189, 1999.
11. Rosenstein, M.T. and Cohen, P.R., Continuous Categories for a Mobile Robot, in Proc. of 16th National Conference on Artificial Intelligence, pp. 634-640, 1999.
12. Jean-Marc Adamo: Data Mining for Association Rules and Sequential Patterns- Sequential and Parallel Algorithms, Springer-Verlag, Berlin Heidelberg New York (2001).
13. Sarker, B. K., Mori, T., Hirata, T. and Uehara, K. Parallel Algorithms for Mining Association Rules in Time Series Data, Lecture Notes in Computer Science, LNCS-2745, pp. 273-284, 2003.
14. Agrawal R. and Srikant R., Mining Sequential Patterns, in Proc. of 11th ICDE Conf., pp. 3-14, 1995.
15. Agrawal R. and Srikant R., Mining Sequential Patterns: Generalization and Performance Improvements, in Proc. of 5th Intl. Conf. on Extending Database Technology, pp. 3-14, 1996.
16. Sarker B.K., Hirata T., and Uehara K., Parallel Mining of Associations Rules in Time Series Multi-stream Data, Special issue in the Journal of Information, to be appeared in the issue 1 of 2006.
17. Zhu Y. and Shasha D, StatStream: Statistical Monitoring of Thousands of Data Streams in Real Time, Proc. of the 28th VLDB Conf., (2002), pp. 358-369.
18. Roddick J.F. and Spiliopoulou M., A Survey of Temporal Knowledge Discovery Paradigms and Methods, IEEE Transactions on Knowledge and Data Engineering, 14(4) (2002), pp.750-767.
19. Honda R. and Konishi O.,Temporal Rule Discovery for Time Series Satellite Images and Integration with RDB, Proc. 5th European Conf. PKDD, (2001), pp. 204-215.

A Coarse Grained Parallel Algorithm for Closest Larger Ancestors in Trees with Applications to Single Link Clustering

Albert Chan[1], Chunmei Gao[2], and Andrew Rau-Chaplin[3]

[1] Research partially supported by the Natural Sciences and Engineering Research Council of Canada
[2] Department of Mathematics and Computer Science, Fayetteville State University, Fayetteville NC 28301, USA
achan@uncfsu.edu
http://faculty.uncfsu.edu/achan
[3] Faculty of Computer Science, Dalhousie University, Halifax, NS B3J 2X4, Canada
cgao@cs.dal.ca
[4] Faculty of Computer Science, Dalhousie University, Halifax, NS B3J 2X4, Canada
arc@cs.dal.ca
http://www.cs.dal.ca/~arc

Abstract. Hierarchical clustering methods are important in many data mining and pattern recognition tasks. In this paper we present an efficient coarse grained parallel algorithm for Single Link Clustering; a standard inter-cluster linkage metric. Our approach is to first describe algorithms for the Prefix Larger Integer Set and the Closest Larger Ancestor problems and then to show how these can be applied to solve the Single Link Clustering problem. In an extensive performance analysis an implementation of these algorithms on a Linux-based cluster has shown to scale well, exhibiting near linear relative speedup.

Keywords: Single Link Clustering, Closest Larger Ancestor, Parallel Graph Algorithms, Coarse Grained Multicomputer, Hierarchical Agglomerative Clustering.

1 Introduction

Clustering is one of the key processes in data mining. *Clustering* is the process of grouping data points into classes or *clusters* so that objects within a cluster have high similarity in comparison to one another, but are very dissimilar to objects in other clusters [12].

Hierarchical agglomerative clustering methods are important in many data mining and pattern recognition tasks. They typically start by creating a set of singleton clusters, one for each data point in the input and proceeds iteratively by merging the most appropriate cluster(s) until the stopping criterion is achieved. The appropriateness of a cluster(s) for merging depends on the (dis)similarity of cluster elements. An important example of dissimilarity between two points is the Euclidian distance between them. To merge subsets of points, the distance between individual points has to be generalized to the distance between subsets. Such a derived proximity measure is called a *linkage metric*. The type of the linkage metric used significantly affects hierarchical algorithms, since it reflects a particular concept of *closeness* and *connectivity*.

Major inter-cluster linkage metrics [17,18] include single link, average link, and complete link. The pair-wise dissimilarity measures can be described as: $d(C_1, C_2) = \oplus\{d(x,y) | x \in C_1, y \in C_2\}$, where $\oplus$ is minimum (single link), average (average link), or maximum (complete link), C_1, C_2 represent two clusters, and d is the distance function. The output of these hierarchical agglomerative clustering methods is a cluster hierarchy or, a tree of clusters, also known as a *dendrogram* (see Figure 1). A dendrogram can easily be broken at selected links to obtain clusters of desired cardinality or radius.

In single link clustering, we consider the distance between clusters to be equal to the shortest distance from any member of one cluster to any member of the other cluster. Many sequential single link clustering (SLC) algorithms are known [10,11,16,19,20,21]. However, since both the data size and computational costs are large, parallel algorithms are also of great interest. Parallel SLC algorithms have been described for a number of SIMD architectures including hypercubes [14,15], shuffle-exchange networks [13], and linear arrays [1]. For the CREW-PRAM, Dahlhaus [5] described an algorithm to compute a single link clustering from a minimum spanning tree. Given a minimum spanning tree with n data points, this algorithm takes $O(\log n)$ time using $O(n)$ processors to compute the single link dendrogram, i.e. a cluster tree for its single link clustering.

In this paper we present an efficient coarse grained parallel algorithm for the Single Link Clustering (SLC) problem. Our parallel computational model is the Coarse Grained Multicomputer (CGM) model [6]. It is comprised of a set of p processors $P_0, P_1, P_2, ...P_{p-1}$ with $O(\frac{N}{p})$ local memory per processor and an arbitrary communication network, where N refers to the problem size. All algorithms consists of alternating local computation and global communication rounds.

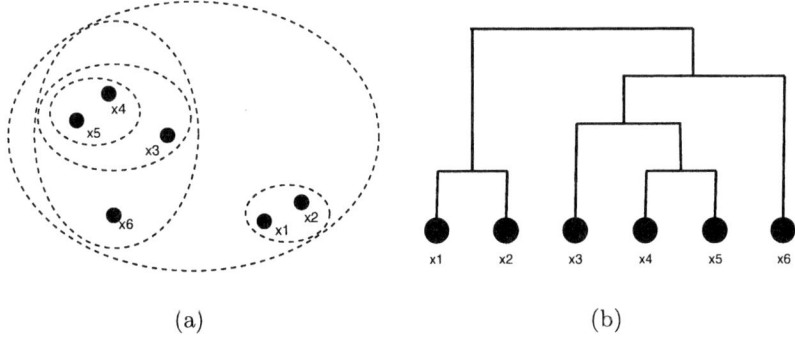

Fig. 1. (a) Single link clustering of a set of points (b) The corresponding dendrogram

Our algorithm follows the basic approach described by Dahlhaus [5] in his PRAM algorithm in which the proximity matrix representing the Euclidean distances between points in the input data set is transformed into a weighted complete graph from which a Minimum Spanning Tree (MST) is constructed. This MST is then transformed into a Modified Pseudo Minimum Spanning Tree (MPMST) from which the dendrogram hierarchy can easily be constructed. The key step is the construction of the MPMST, which is based on solving the Closest Larger Ancestor (CLA) problem.

In the Closest Larger Ancestor problem we have a tree T of n vertices. Each vertex v in T is associated with an integer weight w_v. We want to find out for each vertex v in T, the ancestor u of v that is closest to v and $w_u > w_v$.

In this paper we first describe an algorithm for solving the Closest Larger Ancestor problem in $O(\frac{n^2}{p})$ time, using $\log^2 p$ communication rounds, and $O(\frac{n^2}{p})$ storage space per processor. We then show how this algorithm can be used as a key component in a Single Link Clustering algorithm that runs in time $O(\frac{n^2}{p})$, using $\log^2 p$ communication rounds, and $O(\frac{n^2}{p})$ storage space per processor. Although the parallel Closest Larger Ancestor algorithm is not optimal, it is practical to implement and fits well in the Single Link Cluster algorithm without increasing the overall complexity.

In the final section of this paper we describe a systematic evaluated our parallel single link clustering algorithm on a CGM cluster with 24 nodes. We investigate the performance of our algorithms in terms of running time, relative speedup, efficiency and scalability. Our single link clustering algorithm exhibits near linear speedup when given at least 250,000 data points per processor. For example, it scales near perfectly for large data sets of 64 million points on up to 24 processors. The parallel Closest Larger Ancestor and Single Link Clustering algorithms described in this paper are, to our knowledge, the first efficient coarse-grained parallel algorithms given for these problems.

2 Closest Larger Ancestors

Before tackling on the Closest Larger Ancestors problem we first study a simpler, but related problem.

2.1 Computing Closest Larger Predecessors (CLP)

The Closest larger predecessor problem is defined as follows: Given a list of n integers $x_1, x_2, \cdots, x_n$, find for each integer x_i, another integer x_j such that $j < i$, $x_j > x_i$ and j is as large as possible.

For example, if the integer list is $\{19, 25, 17, 6, 9\}$, then 19 and 25 do not have CLPs; the CLP for 17 is 25, and the CLPs for both 6 and 9 is 17.

We need to define some operations to help us solving the CLP problem. Given an integer set S and an integer m, we define the integer set $S_{>m}$ to be $S_{>m} = \{x | x \in S \ \&\&\ x > m\}$.

We can now define a new operation, called the "Larger Integer Set" (LIS) operation over an integer set and an integer. We use the "$\otimes$" symbol to represent the LIS operation.

Definition 1. *Given a set of integers S and an integer m. The LIS operation is defined as $S \otimes m = S_{>m} \cup \{m\}$.*

Definition 2. *Given a set of integer S, define S_{max} to be the maximum element in S; also define S_{min} to be the minimum element in S.*

Theorem 1. *Given a list of n integers $x_1, \cdots, x_n$, let R be the integer set $R = ((\cdots(\{x_1\} \otimes x_2) \otimes \cdots) \otimes x_n)$. If the "closest larger predecessor" of x_n is x_i then $x_i \in R_{>x_n}$ and $x_i = (R_{>x_n})_{min}$.*

Proof. Assume $x_i \notin R_{>x_n}$, then there must be another integer x_j (can be x_n) after x_i such that $x_j > x_i$ (this is the only reason why x_i will disappear from $R_{>x_n}$). However, this implies that x_i cannot be the "closest larger predecessor" of x_n and contradict to our assumption. Therefore, $x_i \in R_{>x_n}$.

Now assume $x_i \neq (R_{>x_n})_{min}$, then there must be another integer $x_k \in R_{>x_n}$ such that $x_k < x_i$. Observe that $\forall x_m \in R_{>x_n}, x_m > x_n$. Therefore we know that x_k should come *before* x_i, since otherwise the "closest larger predecessor" of x_n would be x_k instead of x_i. However x_i comes after x_k implies that x_k will be excluded from $R_{>x_n}$. This is a contradiction, and therefore, $x_i = (R_{>x_n})_{min}$. □

We now extend the "LIS" operation to two integer sets:

Definition 3. *Given two integer sets U and V, we define the integer set $U_{>V}$ to be $U_{>V} = U_{>V_{max}}$.*

Definition 4. *Given two integer sets U and V, we define the LIS operation over U and V to be $U \otimes V = U_{>V} \cup V$*

This is a "proper" extension since we have $S \otimes \{m\} = S \otimes m$ and now the "LIS" operation is associative.

If a and b are integers, and S is a set of integers. We also define:

- $a \otimes b = \{a\} \otimes \{b\}$; and
- $a \otimes S = \{a\} \otimes S$.

To prove that the LIS operation is associative, we need the following definition:

Definition 5. *Let U, V, and W be three integer sets, define $U_{>VW} = \{x | x \in U\ \&\&\ x > V_{max}\ \&\&\ x > W_{max}\}$, i.e. $U_{>VW} = U_{>V} \cap U_{>W}$.*

Lemma 1. *The LIS operation is associative.*

Proof. We have $(U \otimes V) \otimes W = (U_{>V} \cup V) \otimes W = U_{>VW} \cup V_{>W} \cup W$, and $U \otimes (V \otimes W) = U \otimes (V_{>W} \cup W) = U_{>VW} \cup V_{>W} \cup W$. Therefore the $\otimes$ operation is associative. □

Theorem 2. *Given n integers $x_1, x_2, \cdots, x_n$, let $R = x_1 \otimes x_2 \otimes \cdots \otimes x_n$. If $x_i, x_j \in R$ and $i < j$ then $x_i > x_j$.*

Proof. Let $R_j = x_1 \otimes x_2 \otimes \cdots \otimes x_j$. From the definition of the "LIS" operation, if $x_i, x_j \in R$, it must be $x_i, x_j \in R_j$ and $x_i \in R_{j-1}$. If $x_i <= x_j$, then $x_i \notin (R_{j-1})_{>x_j}$ and therefore, $x_i \notin R_j$. This is a contradiction, so $x_i > x_j$. □

Lemma 2. *The "LIS" operation over two integer sets U and V can be completed in $O(\log |U| + |V|)$ time.*

Proof. If we implement the integer sets as ordered lists using arrays of enough capacity and enforce the union $(U \cup V)$ operation to append all the integers from V to U without disturbing the original orders, then by Theorem 2 the integers in the ordered lists will appear in reversely sorted order. That means V_{max} will be the first integer in the ordered list in V, and this can be obtained in $O(1)$ time. In $O(\log |U|)$ time, we can determine the first integer in U that is smaller than V_{max}. It then takes $O(|V|)$ time to copy all integers over. □

2.2 CGM Computing Prefix LIS in Trees

Definition 6. *For each node u in a tree, let n_u be the set of the nodes in the path from u to the root, inclusive. Also assume that each node is associated with an integer weight. The prefix LIS for each node u is defined as the LIS for all the weights in p_u.*

We now describe a CGM algorithm to find the prefix LIS in a tree. Figure 2(a) shows an example of LIS in a tree.

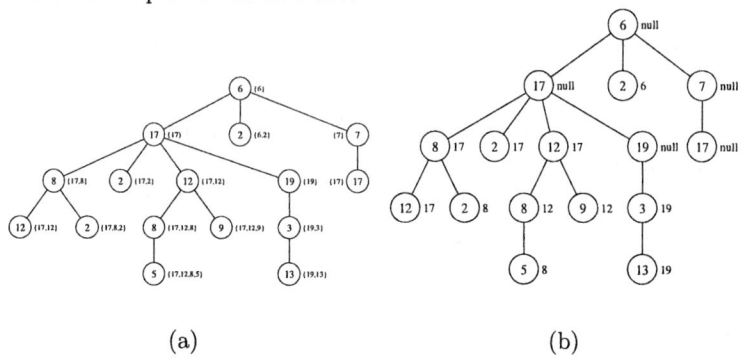

(a)   (b)

Fig. 2. (a) LIS in a Tree and (b) CLA values in a Tree

Algorithm 1 CGM prefix LIS in Tree.
  Input: A tree T of n vertices, evenly distributed over p processors. Each vertex
        v in T is associated with an integer weight w_v. The root of T is r.
  Output: For each vertex v in T, the ordered set $S_v = r \otimes \ldots \otimes v$.
  (1) If $p = 1$, solve the problem sequentially.

(2) Find the centroid c^1 of T [2].
(3) Broadcast c to every processor.
(4) Each vertex checks its parent, if the parent is c, temporarily set the vertex's parent to itself. This effectively partitions T into a set of subtrees $T_0, T_1, \ldots, T_k$ where k is the number of children c has.
(5) Group all the vertices of the same subtree into adjacent processors. This can be done by finding the connected components of the (partitioned) tree, and sort the vertices by component ID [8].
(6) Partition the CGM into a set of sub-CGM, according to the partitioning of T. Recursively solve the CGM prefix LIS in Tree problem for each sub-tree using the sub-CGM. Let the result for each vertex v be S_v.
(7) The processor containing c broadcasts S_c and the component ID of c.
(8) Each vertex v that is not in the same sub-tree as c update S_v to $S_c \otimes S_v$.
— End of Algorithm —

Theorem 3. *Algorithm 1 solves the CGM prefix LIS in Tree problem using $O(\frac{n^2}{p})$ local computation, $\log^2 p$ communication rounds and $O(\frac{n^2}{p})$ storage space per processor.*

Proof. The correctness of the algorithm comes immediately from the fact that the LIS operation is associative (Lemma 1).

Let $T(n,p)$, $C(n,p)$, and $S(n,p)$ be the running time, number of communication rounds, and storage requirement of Algorithm 1, respectively. Step 1 is the terminating condition of the recursive calls. Here we have $T(\frac{n}{p}, 1) = O(\frac{n}{p} \log n)$, $C(\frac{n}{p}, 1) = 0$, and $S(\frac{n}{p}, 1) = O(\frac{n^2}{p^2})$, respectively. Step 6 is the recursive calls, and we have $T(n,p) = O(T(\frac{n}{2}, \frac{p}{2}))$, $C(n,p) = O(C(\frac{n}{2}, \frac{p}{2}))$, and $S(n,p) = O(S(\frac{n}{2}, \frac{p}{2}))$. All other steps have upper bounds of $T(n,p) = O(\frac{n^2}{p})$, $C(n,p) = O(\log p)$, and $S(n,p) = O(\frac{n^2}{p})$. (Note that the actual bounds for each individual step are different, but the above is the maximum of them). Solving the recurrence, we have $T(n,p) = O(\frac{n^2}{p})$, $C(n,p) = O(\log^2 p)$, and $S(n,p) = O(\frac{n^2}{p})$, respectively. □

2.3 Computing Closest Larger Ancestors

We are now finally ready to describe a CGM algorithm for calculating the "closest larger ancestor" values in trees. Figure 2(b) shows the "CLA" values for each vertex in the tree shown in Figure 2(a).

Algorithm 2 CGM Closest Larger Ancestors.
Input: A tree T of n vertices, evenly distributed over p processors. Each vertex v in T is associated with an integer weight w_v.
Output: For each vertex v in T, the ancestor u of v closest to v and $w_u > w_v$.

---
[1] Recall that the centroid of a tree T is the vertex c such that removing c resulting T to be partitioned into subtrees of size no more than $\frac{|T|}{2}$. Do not confuse the tree centroid with the cluster centroid.

(1) Apply Algorithm 1 to find out S_v.
(2) Once S_v is computed, each vertex can compute the closest larger ancestor by calculating $((S_v)_{>w_v})_{min}$ (or root if null).
— End of Algorithm —

Theorem 4. *Algorithm 2 solves the Closest Larger Ancestor problem in $O(\frac{n^2}{p})$ time, using $\log^2 p$ communication rounds, and $O(\frac{n^2}{p})$ storage space per processor.*

Proof. The correctness of the algorithm is obvious from Theorem 1 and 3. Step 1 takes $O(\frac{n^2}{p})$ time, using $\log^2 p$ communication rounds and $O(\frac{n^2}{p})$ storage space per processor. Step 2 takes $O(\frac{n^2}{p^2})$ time, with no communication and $O(\frac{n^2}{p^2})$ storage space per processor. Adding the values leads to the declared bounds. □

3 Single Link Clustering

In this section we sketch a CGM single link clustering algorithm. Single link clustering is one of the most widely studied hierarchical clustering techniques and it is closely related to finding the Euclidean Minimum Spanning Tree (MST) of a set of points [18]. Based on solving the Closest Larger Ancestor (CLA) problem, we are able to transform an MST to an Modified Pseudo Minimum Spanning Tree (MPMST) in parallel on CGM. Following the basic approach described by Dahlhaus [5], the MPMST can easily be transformed to a single link clustering dendrogram. For a detailed description of this algorithm suitable for implementation see [9].

Algorithm 3 CGM Single Link Clustering.
Input: Each processor stores a copy of the set S of n input data points in d dimensional Euclidean space and the distance function $D(i,j)$ which defines the distance between all points i and $j \in S$.
Output: A tree H which is the SLC dendrogram corresponding to S.
(1) From S and $D(i,j)$ construct the proximity matrix A and the corresponding complete weighted graph G.
(2) Compute the Minimum Spanning Tree T of G using the MST algorithm given in [8].
(3) Transform the MST T into a MPMST T' by redirecting each vertices parent link to its closest larger ancestor as computed by Algorithm 2.
(4) From the MPMST T' compute H, the single link clustering dendrogram, following the algorithm given in [5].
— End of Algorithm —

Theorem 5. *Algorithm 3 solves the CGM Single Link Cluster problem using $O(\frac{n^2}{p})$ local computation, $\log^2 p$ communication rounds and $O(\frac{n^2}{p})$ storage space per processor.*

Proof. The correctness of the algorithm follows immediately from Theorem T-CLA and Theorem 1 and 2 of [5]. Step 1 takes $O(\frac{n^2}{p})$ time, using $O(1)$ communication rounds and $O(\frac{n^2}{p})$ storage space per processor. Step 2 takes $O(\frac{n^2}{p})$

time, using $\log^2 p$ communication rounds and $O(\frac{n^2}{p})$ storage space per processor. Step 3 takes $O(\frac{n^2}{p})$ time, using $\log^2 p$ communication rounds and $O(\frac{n^2}{p})$ storage space per processor. Step 4 takes $O(\frac{n^2}{p})$ time, using $O(1)$ communication rounds and $O(\frac{n^2}{p})$ storage space per processor. Summing up the values and we have the declared bounds. □

4 Experimental Evaluation

In this section we discuss the performance of our CGM single link clustering algorithm under a variety of test scenarios. We use CGM*graph*/CGM*lib* [3,4] to provide the infrastructure for the CGM model. We report results for our experiments on a variety of synthetic data sets and show the parallel performance in relative speedup as the number of processors is increased, as well as the scalability of the algorithms in terms of the size of the input data sets. All parallel times are measured as the wall clock time between the start of the first

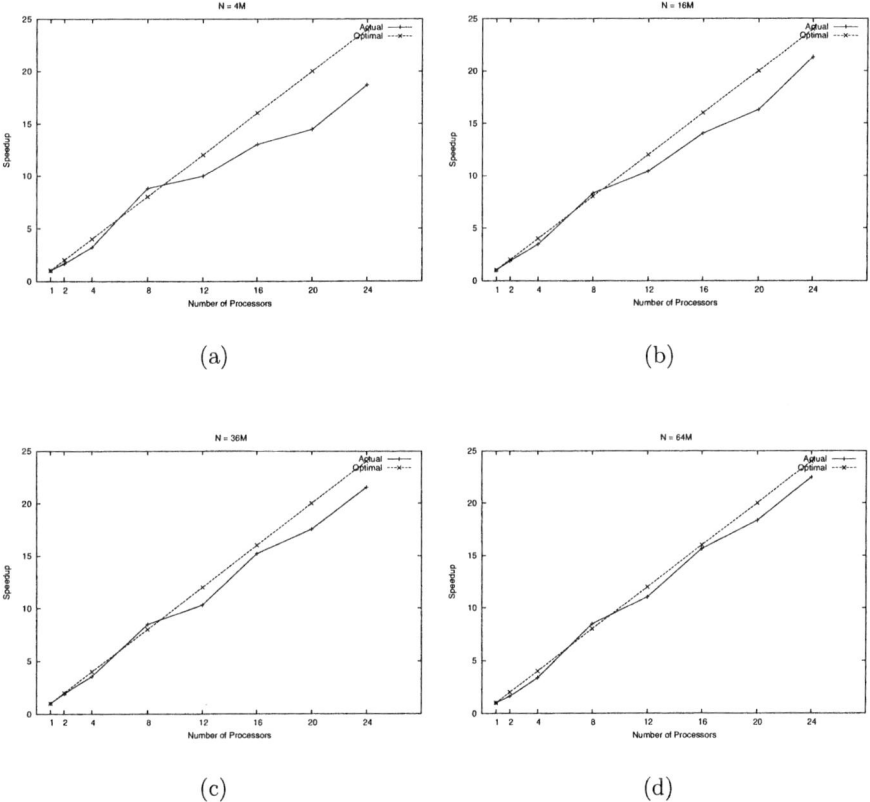

Fig. 3. Relative Speedup for (a) N = 4M, (b) N = 16M, (c) N = 32M, and (d) N=64M.

process and the termination of the last process. The times also include data distribution and routing.

Throughout these experiments, as we increased the number of processors we observed two countervailing trends. Increasing processors while holding total data size constant, leads to less data per processor and therefore better relative speedup because of reduced local computational time and cache effects. On the other hand, increasing the number of processors increases the total time for CGM barrier synchronization and communication time, even when total data size communicated is held constant, and therefore tends to reduce relative speedup.

Speedup is one of the key metrics for evaluation of parallel database systems [7] as it indicates the degree to which adding processors decreases the running time. The relative speedup for p processors is defined as $S_p = \frac{T_1}{T_p}$, where T_1 is the running time of the parallel program using one processor where all communication overhead have been removed from the program, and T_p is the running time using p processors. Figures 3 shows the relative speedup observed for data sets of $N = 4M$, $16M$, $36M$ and $64M$, as a function of the number of processors used, where $N = n^2$ and n is the number of data points.

Relative speedup improves as we increase the size of the input since better data and task parallelism could be achieved. As we can see, when we increase the data size from $N = 4M$ to $N = 64M$, the relative speedup curve tracks more closely to the linear optimal speedup curve. For 24 processors at $N = 64M$, our method achieves a speedup of 22.45.

5 Conclusion

In this paper we have investigated the problem of computing the Prefix Larger Integer Set (LIS), the Closest Larger Ancestor (CLA), and Single Linkage Clustering (SLC) on coarse grained distributed memory parallel machines. An implementation of these algorithms on a distributed memory cluster has demonstrated that they perform well in practice.

References

1. S. Arumugavelu and N. Ranganathan. SIMD Algorithms for Single Link and Complete Link pattern clustering. In *Proc. of Intl. Conf. on Pattern Recognition*, 1996.
2. A. Chan and F. Dehne. A coarse grained parallel algorithm for maximum weight matching in trees. In *Proceedings of 12th IASTED International Conference Parallel and Distributed Computing and Systems (PCDS 2000)*, pages 134–138, 2000.
3. A. Chan and F. Dehne. CGMlib/CGMgraph: Implementing and testing CGM graph algorithms on PC clusters. In *Proceedings of 10th European PVM/MPI User's Group Meeting (Euro PVM/MPI.03)*, pages 117–125, 2003.
4. A. Chan, F. Dehne, and R. Taylor. Cgmgraph/cgmlib: Implementing and testing cgm graph algorithms on pc clusters and shared memory machines. *The international Journal of High Performance Computing Applications*, 19(1):81–97, 2005.
5. E. Dahlhaus. Fast parallel algorithm for the single link heuristics of hierarchical clustering. In *Proceedings of the Fourth IEEE Symposium on Parallel and Distributed Processing*, pages 184–187, 1992.

6. F. Dehne, A. Fabri, and A. Rau-Chaplin. Scalable parallel geometric algorithms for coarse grained multicomputers. *Proc.ACM Symposium on Computational Geometry*, pages 298–307, 1993.
7. D. DeWitt and J. Gray. Parallel database systems: the future of high performance database systems. *Communication of the ACM*, 35(6):85–98, 1992.
8. A. Ferreira P. Flocchini I. Rieping A. Roncato N. Santoro E. Cáceres, F. Dehne and S. W. Song. Efficient parallel graph algorithms for coarse grained multicomputers and bsp.
9. C. Gao. Parallel single link clustering on coarse-grained multicomputers. Master's thesis, Faculty of Computer Sceince, Dalhousie University, April 2004.
10. S. Guha, R. Rastogi, and K. Shim. CURE: an efficient clustering algorithm for large databases. In *ACM SIGMOD International Conference on Management of Data*, pages 73–84, 1998.
11. S. Guha, R. Rastogi, and K. Shim. ROCK: A robust clustering algorithm for categorical attributes. In *International Conference on Data Engineering*, volume 25, pages 345–366, 1999.
12. J. Han and M. Kamber. *Data Mining: Concepts and Techniques*. Morgan Kaufmann Publishers, 2000.
13. X. Li. Parallel algorithms for hierarchical clustering and cluster validity. *IEEE Transactions on Pattern Analysis and Machine Intelligence*, 12(11):1088–1092, 1990.
14. X. Li and Z. Fang. Parallel algorithms for clustering on Hypercube SIMD computers. In *Proceedings of 1986 Conference on Computer Vission and Pattern Recognition*, pages 130–133, 1986.
15. X. Li and Z. Fang. Parallel clustering algorithms. *Parallel Computing*, 11(3):275–290, 1989.
16. Ankerst M., Breunig M. M., Kriegel H. P., and Sander J. Optics: Ordering points to identify the clustering structure. *ACMSIGMOD Int. Conf. on Management of Data*, 1999.
17. F. Murtagh. Multidimensional clustering algorithms. *Physica-Verlag, Vienna*, 1985.
18. C. Olson. Parallel algorithms for hierarchical clustering. *Parallel Computing*, 21:1313–1325, 1995.
19. Sibson R. Slink: an optimally efficient algorithm for the single-link cluster method. *The Computer Journal*, 16-1:30–34, 1973.
20. T. Zhang, R. Ramakrishnan, and M. Livny. BIRCH: an efficient data clustering method for very large databases. In *ACM SIGMOD International Conference on Management of Data*, pages 103–114, 1996.
21. T. Zhang, R. Ramakrishnan, and M. Livny. Birch: A new data clustering algorithm and its applications. *Data Mining and Knowledge Discovery*, 1(2):141–182, 1997.

High Performance Subgraph Mining in Molecular Compounds

Giuseppe Di Fatta[1,2] and Michael R. Berthold[1]

[1] University of Konstanz, Dept. of Computer and Information Science,
78457 Konstanz, Germany
berthold@uni-konstanz.de
[2] ICAR, Institute for High Performance Computing and Networking,
CNR, Italian National Research Council, 90018 Palermo, Italy
difatta@pa.icar.cnr.it

Abstract. Structured data represented in the form of graphs arises in several fields of the science and the growing amount of available data makes distributed graph mining techniques particularly relevant. In this paper, we present a distributed approach to the frequent subgraph mining problem to discover interesting patterns in molecular compounds. The problem is characterized by a highly irregular search tree, whereby no reliable workload prediction is available. We describe the three main aspects of the proposed distributed algorithm, namely a dynamic partitioning of the search space, a distribution process based on a peer-to-peer communication framework, and a novel receiver-initiated, load balancing algorithm. The effectiveness of the distributed method has been evaluated on the well-known National Cancer Institute's HIV-screening dataset, where the approach attains close-to linear speedup in a network of workstations.

1 Introduction

A crucial step in the drug discovery process is the so-called High Throughput Screening and the subsequent analysis of the generated data. During this process, hundreds of thousands of potential drug candidates are automatically tested for a desired activity, such as blocking a specific binding site or attachment to a particular protein. This activity is believed to be connected to, for example, the inhibition of a specific disease. Once all these candidates have been automatically screened it is necessary to select few promising candidates for further, more careful and cost-intensive analysis. A promising approach focuses on the analysis of the molecular structure and the extraction of relevant molecular fragments that may be correlated with activity. Relevant molecular fragment discovery can be formulated as a frequent subgraph mining (FSM) problem [1] in analogy to the association rule mining (ARM) problem [2,3]. While in ARM the main structure of the data is a list of items (itemset) and the basic operation is the subset test, FSM is based on graph and subgraph isomorphism.

In this paper we present a high performance application of the frequent subgraph mining problem applied to the analysis of molecular compounds.

The rest of the paper is structured as follows. In the next section we introduce the molecular fragment mining problem and discuss related approaches. In Sect. 3 we discuss alternative definitions of discriminative molecular fragments and briefly describe the sequential algorithm on which the distributed approach is based. In Sect. 4 and 5 we present, respectively, a high performance distributed computing approach for subgraph mining and the adopted dynamic load balancing policy. Section 6 describes the experiments we conducted to verify the performance of the distributed approach. Finally, we provide conclusive remarks.

2 Problem Definition and Related Works

The problem of selecting discriminative molecular fragments in a set of molecules can be formulated in terms of frequent subgraph mining in a set of graphs. Molecules are represented by attributed graphs, in which each vertex represents an atom and each edge a bond between atoms. Each vertex carries attributes that indicate the atom type (i.e., the chemical element), a possible charge, and whether it is part of a ring. Each edge carries an attribute that indicates the bond type (single, double, triple, or aromatic). Frequent molecular fragments are subgraphs that have a certain minimum support in a given set of graphs, i.e., are part of at least a certain percentage of the molecules. Discriminative molecular fragments are contrast substructures, which are frequent in a predefined set of molecules and infrequent in the complement of this subset. In this case two parameters are required: a minimum support ($minSupp$) for the focus subset and a maximum support ($maxSupp$) for the complement.

These topological fragments carry important information and may be representative of those components in the compounds that are responsible for a positive behavior. Such discriminate fragments can be used to predict activity in other compounds [4] and to guide the synthesis of new ones.

A number of approaches to find frequent molecular fragments have recently been published [5,6,7,8] but they are all limited by the complexity of graph and subgraph isomorphism tests and by the combinatorial nature of the problem. Some of these algorithms can therefore operate on very large molecular databases but only find small fragments [5,6], whereas others can find larger fragments but are limited by the maximum number of molecules they can analyse [7,8].

Finding frequent fragments in a set of molecules can be seen as analysing the space of all possible fragments that can be found in the entire molecular database. Obviously, this set of all existing fragments is enormous even for relatively small datasets: a single molecule of average size can already contain in the order of hundreds of thousands of different fragments. Existing methods usually organize the space of all possible fragments in a lattice, which models subgraph relationships, that is, edges connect fragments that differ by exactly one atom and/or bond. The search then reduces to traversing this lattice and reporting all fragments that fulfill the desired criteria. Based on existing data mining algorithms for market basket analysis [2,3] these methods conduct depth-first [7] or breadth-first searches [6,5]. An example of a search tree is depicted in Fig. 1, which also shows the region of discriminative fragments.

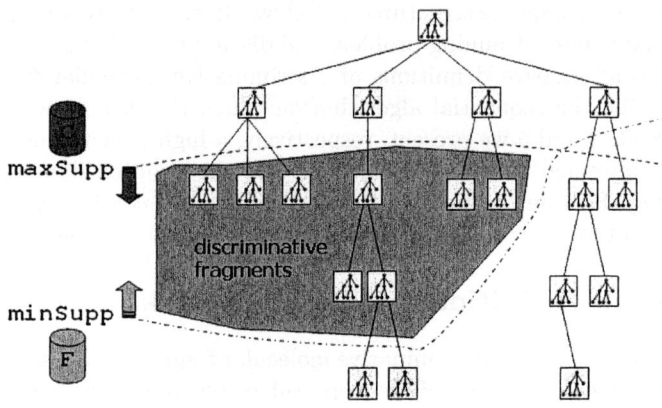

Fig. 1. Discriminative molecular fragment search tree

However, none of the current sequential algorithms in a single processor can be used for extremely large datasets (millions of molecules) and unlimited size of the fragments that can be discovered. Quite obviously, parallel approaches to this type of problem are a promising alternative. Although, in recent years, several parallel and distributed algorithms have been proposed for the association rule mining problem (D-ARM) [9], very few have addressed the FSM problem [10,11]. The approach in [10] achieved a relatively good performance in a small-scale computational environment, but its scalability and efficiency are limited by two main factors. First, the approach is based on a master-slave communication model, which clearly cannot scale well to a large number of computing nodes. Secondly, the communication overhead due to the large number of frequent fragments limits the efficiency of the overall process. In this paper, we overcome these two limitations by adopting a better definition of discriminative fragments and by providing a more efficient and scalable distributed computing framework.

3 Efficient Frequent Subgraph Mining

We can assume that the molecular compounds in the dataset can be classified in two groups, the focus set F (active molecules) and its complement C (non-active molecules). For example, during the High Throughput analysis, compounds are tested for a certain active behaviour and a score associated to their activity level is determined. In this case, a threshold (*thres*) on the activity value allows the classification of the molecules in the two groups.

The aim of the data mining process is to provide a list of molecular fragments that are frequent in the focus dataset and infrequent in the complement dataset. However, the high number of frequent fragments that can be found in a large dataset suggests the adoption of the closed frequent subgraphs (CFS). A closed frequent subgraph is a frequent subgraph whose support is higher than the support of all its proper supergraphs. Given the CFS set, it is possible to directly generate all frequent subgraphs without any further access to the dataset. Moreover, the support of all frequent subgraphs is implicitly defined by the support of the closed ones. For this reason we adopt the CFS in more efficient definitions of discriminative molecular fragments.

Given a dataset D and a frequency threshold $minSupp$, the sets of frequent and closed frequent subgraphs are defined, respectively, as

$FS_D = \{s \mid supp(s,D) \geq minSupp\}$ and
$CFS_D = \{s \mid supp(s,D) \geq minSupp$ and $\nexists\, x \in FS_D, x \supset s$ and $supp(x,D) = supp(s,D)\}$,

where s is a graph, $supp(s,D)$ is the number of graphs in D, which are supersets of s, i.e. the support of s in D.

In our context, we have to extend the concept of the closure to the duality of active and non-active compounds. The following alternative definitions can be adopted for the discriminative fragments (DF).

Definition 1 (Constrained FS). DF_{all} *is the set of frequent subgraphs in the focus dataset constrained to infrequency in the complement dataset, according to*

$$DF_{all} = \{s \in FS_F \mid supp(s,C) \leq maxSupp\}.$$

Definition 2 (Constrained Focus-closed FS). DF_F *is the set of closed frequent subgraphs in the focus dataset constrained to infrequency in the complement dataset, according to*

$$DF_F = \{s \in CFS_F \mid supp(s,C) \leq maxSupp\}.$$

Definition 3 (Constrained Closed FS). DF_{FC} *is the set of frequent subgraphs in the focus dataset constrained to infrequency in the complement dataset, which are closed w.r.t. both sets of graphs, according to*

$$DF_{FC} = \{s \in FS_F \mid supp(s,C) \leq maxSupp$$
$$\text{and } \nexists\, x \in FS_F, x \supset s, supp(x,F) = supp(s,F) \text{ and } supp(x,C) = supp(s,C)\}.$$

The first definition considers the subgraphs that are frequent in the focus dataset and are constrained to a maximum support in the complement dataset. In the other two definitions, the constrained frequent subgraphs are restricted by the closure, respectively, only in the focus and in both datasets.

Closed frequent substructures can be considered a compact representation of the complete set of frequent substructures and lead to a significant improvement of the efficiency of the mining process.

Table 1 provides an example of the number of discriminant fragments for the NCI HIV dataset (cf. Sect. 6) when the different definitions are adopted. It should be pointed out that the alternative definitions do not reduce the number of nodes in the search tree, but only the number of stored and reported molecular fragments.

Table 1. Discriminative molecular fragments in 37171 NCI compounds ($thres = 0.5$)

minSupp(%)	DF_{all}	DF_F	DF_{FC}	minSupp(%)	DF_{all}	DF_F	DF_{FC}
20	1091	35	827	20	0	0	0
15	9890	53	2158	15	5279	5	248
10	17688	120	4874	10	9403	24	977
8	59270	241	6044	8	50222	98	1728
6	OutOfMem	441	9522	6	127773	155	2629

(a) $maxSupp = 1\%$ (b) $maxSupp = 0.1\%$

3.1 Sequential Subgraph Mining

The distributed approach presented in this paper is based on the sequential algorithm described in [7]. The algorithm organizes the space of all possible fragments in an efficient depth-first search tree. Each possible subgraph of the molecular structures is evaluated in terms of the number of embeddings that are present in the molecular database. Each node of the search tree represents a candidate frequent fragment. A search tree node evaluation comprises the generation of all the embeddings of the fragment in the molecules. When a fragment meets the minimum support criterion, it is extended by one bond to generate new search tree nodes. When the fragment meets both criteria of minimum support in active molecules and maximum support in the inactive molecules, it is then reported as a discriminative frequent fragment. The algorithm prunes the DFS tree according to three criteria. The support-based pruning exploits the anti-monotone property of the fragment support. The size-based pruning exploits the anti-monotone property of the fragment size. And, finally, a partial structural pruning is based on a local order of atoms and bonds. For further details on the algorithm we refer to [7].

The analysis of the sequential algorithm pointed out the irregular nature of the search tree. An irregular problem is characterized by a highly dynamic or unpredictable domain. In this application the complexity and the exploration time of the search tree, and even of a single search tree node cannot be estimated, nor bounded.

4 Distributed Subgraph Mining

The distributed approach we propose is based on a search space partitioning strategy, distributed task queues with dynamic load balancing and a peer-to-peer communication framework.

A static load balancing policy cannot be adopted as the work load is not known in advance and cannot be estimated. We adopted a receiver-initiated DLB approach based on two components, a quasi-random polling for the donor selection and a work splitting technique for the subtask generation. Both components contribute to the overall DLB efficiency and to its suitability to heterogeneous computing resources.

It is worth mentioning that all the algorithms, which have been proposed for D-ARM in the past years, assume a static, homogeneous and dedicated computation environment and do not provide dynamic load balancing [9].

In the next section, we discuss some details of the distributed application related to the search space partitioning.

4.1 Search Space Partitioning

Partitioning a Depth First Search (DFS) tree, i.e. parallel backtracking [12], has been widely and successfully adopted in many applications. In general, it is quite straightforward to partition the search tree to generate new independent jobs, which can be assigned to idle processors. In this case, no synchronization is required among remote jobs.

A job assignment contains the description of a search node of the donor worker, which becomes the initial fragment from which to start a new search at the receiving worker. The job assignment must contain all the information needed to continue the search from exactly the same point in the search space. In our case, this is essential in order to exploit the efficient search strategy provided by the sequential algorithm and based on advanced pruning techniques. Thus, a job description includes the search node state to rebuild the same local order necessary to prune the search tree as in the sequential algorithm (cf. structural pruning in [7]). This requires an explicit representation of the state of the donated search node. For this aim, we adopted the Simplified Molecular Input Line Entry Specification (SMILES) [13], a notation for organic structure description, which we enhanced with numerical tags for atoms. These tags are used to represent the subscripts of the atoms in a fragment according to the local order, i.e. the order in which atoms have been added to the fragment.

The enhanced-SMILES representation of the fragment plus the last extension performed (last extended atom subscript, last extended bond type and last added atom type) are sufficient to re-establish the same local order at a remote process. The receiving worker has to re-compute all the embeddings of the core fragment into all molecular compounds in order to re-start the search. This extra computation is necessary and is by far preferred over the expensive communication cost of an explicit representation of the embeddings. The number of embeddings of a fragment in the molecules can be very large, especially in the lower part of the search tree. Moreover, the donor worker can also perform a selection and a projection of the dataset based on the donated search node.

Each worker maintains only a local and partial list of substructures found during the execution of subtasks. Therefore, at the end of the search process we perform a reduction operation. Workers are organized in a communication tree and the number of communication steps required is in the order of $O(\log N)$, where N is the number of processes. However, the determination of the closed fragments includes expensive graph and subgraph isomorphism tests and may represent a non-trivial computational cost. Therefore, the selection of the closed fragments has to be distributed as well. This is performed during the reduction operation in parallel by several concurrent processes.

A static partition of the search space can be adopted when job running times can be estimated, which is not our case. We adopted a dynamic search tree partitioning with a self-adaptive job-granularity based on a quasi-randomized dynamic load balancing, which is discussed in the next section.

5 Dynamic Load Balancing (DLB)

Many DLB algorithms for irregular problems have been proposed in the literature and their properties have been studied. Most of them rely on uniform [14] or bounded [15] task times or the availability of workload estimates [16]. However, none of these assumptions holds in our case; we cannot guarantee that the computation cost of a job is greater than the relative transmitting time, nor provide minimum or maximum bounds for the running time of subtasks. It is quite challenging to efficiently parallelize irregular problems with such an unpredictable workload.

In general, the DLB policy has to provide a mechanism to fairly distribute the load among the processors using a small number of generated subtasks to reduce the communication cost and the computational overhead. In particular, the quality of both the selection of donors and the generation of new subtasks is fundamental for an effective and efficient computational load distribution. These two tasks are carried out, respectively, by the DLB algorithm and the work splitting-mechanism discussed in the next two sections.

5.1 Quasi-Random Polling

When a worker completes its task, it has to select a donor among the other workers to get a new subtask. In general, not all workers are equally suitable as donor. Workers that are running a mining task for a longer time, have to be preferred. This choice can be motivated by two reasons. The longest running jobs are likely to be among the most complex ones. And this probability increases over time. Secondly, a long job-execution time may also depend on the heterogeneity of the processing nodes and their loads. With such a choice we provide support to the nodes that are likely overloaded either by their current mining task assignment or by other unrelated processes.

The DLB approach we adopted is a receiver-initiated algorithm based on a distributed quasi-random polling. Each worker keeps an ordered list of potential donors and performs a random polling over them to get a new task. The probability of selecting a donor from the list is not uniform. In particular, we adopt a simple linearly decreasing probability, where the donor list is ordered according to the starting time of the latest job assignment. This way, long running jobs have a high probability of being further partitioned, while most recently assigned tasks do not.

In order to maintain statistics of job executions, we adopted a centralized approach. At the starting and at the completion of a job execution, workers notify the bootstrap node, which collects global job statistics. Workers keep the local donor list updated by an explicit query to the bootstrap node.

Approaches based on global statistics are known to provide optimal load balancing performance, while randomized techniques provide high scalability.

In order to reduce latency, each worker also keeps a local pool of unprocessed jobs. This way at the completion of a job, the request and reception of a new one can be overlapped to the execution of a job from the local pool.

Furthermore, each worker keeps a list of donated and not completed jobs in order to support mechanisms for fault tolerance and termination detection.

It should be noticed that the server for job statistics plays the same role as the centralized directory of the first-generation P2P systems. The current implementation of our P2P computing framework allows the dynamic joining of peers and a basic fault-tolerance mechanism in case of abrupt peer disconnection.

5.2 Work Splitting

In problems with uniform or bounded subtask times the generation of either too small or too big jobs is not an issue. In our case, wrong job granularity may decrease the efficiency and limit the maximum speedup tremendously. While a coarse job granularity may induce load imbalance and bounds on the maximum speedup, a fine granularity may decrease the distributed system efficiency and more processing nodes will be required to reach the maximum speedup. Thus, it is important to provide an adaptive mechanism to find a good trade-off between load balancing and job granularity.

In order to accomplish this aim we introduce a mechanism at the donor to reduce the probability of generating trivial tasks and of inducing idling periods at the donor processor itself. Search nodes from the stack can only be donated (a) if they have sufficient support in the active compounds and (b) if they do not have a very restrictive local order. A node with a restrictive local order is likely to generate a small subtree even in the case of high support.

A worker follows three rules to donate a search node from its local stack. A search tree node n can only be donated if

1. $stackSize() \geq minStackSize$,
2. $support(n) \geq (1 + \alpha) * minSupp$,
3. $lxa(n) \leq \beta * atomCount(n)$,

where α and β are tolerance factors, $lxa()$ is the subscript of the last extended atom in the fragment (see below), $atomCount()$ provides the number of atoms in a fragment and $minStackSize$ specifies a minimum number of search nodes in the stack to avoid starvation of the donor. The values of these parameters are not critical and in our experiments we adopted $minStackSize = 4$, $\alpha = 0.1$ and $\beta = 0.5$. These rules for selecting nodes of the local search tree guarantee that the worker does not run out of work while donating non-trivial parts of its search tree.

While rules 1 and 2 are quite straightforward, in order to explain rule 3, we have to refer to the structural pruning technique adopted in the sequential algorithm (cf. [7]). An atom subscript indicates the order in which the atom has been added to the fragment. All the atoms of the fragment with a subscript less than lxa cannot be further extended according to the sequential algorithm. As a consequence, subtrees rooted at a node with a high lxa value (close to the number of atoms in the fragment) are expected to have a low branching factor.

6 Experimental Results

The distributed algorithm has been tested for the analysis of a set of real molecular compounds - a well-known, publicly available dataset from the National Cancer Institute, the DTP AIDS Antiviral Screen dataset. This screen utilized a soluble formazan assay to measure protection of human CEM cells from HIV-1 infection [17]. Compounds able to provide at least 50% protection to the CEM cells were retested. Compounds that provided at least 50% protection on retest were listed as moderately active (CM). Compounds that reproducibly provided 100% protection were listed as confirmed active (CA). Compounds not meeting these criteria were listed as confirmed inactive (CI). We used a total of 37169 total compounds, of which 325 belong to class CA, 875 are of class CM and the remaining 35969 are of class CI. In order to carry out tests on different sizes of the focus dataset we combined these compounds as follows. We joined the CA set with a different number of CM compounds to form four focus datasets with, respectively, 325, 650, 975 and 1200 compounds.

Experimental tests have been carried out on a network of workstations[1]. The software has been developed in Java; the communication among processes has been implemented using TCP socket API and XML data format.

In our tests, we introduced a synchronization barrier to wait for a number of processors to join the P2P system before starting the mining task only in order to collect performance results. In general, this is not necessary, but in the following results we did not want to consider the latency that is required to simply start up the remote peers.

In general, the mining task becomes more difficult when the absolute value of the minimum support decreases. In this case, a bigger and deeper part of the fragment lattice has to be explored. We fixed $minSupp = 6\%$ and varied the number of molecules in the focus dataset in order to show the influence of the different definitions of Sect. 3 on the running time. For the different focus datasets that have been defined above, this corresponds to an absolute minimum support, respectively, of 20, 39, 59, and 72 molecules.

A comparison of running times of the serial and distributed (over 8 processors) algorithms is shown in Fig. 2. The serial algorithm (serial DF_{FC}) and one parallel version (parallel DF_{FC}) search for the closed frequent fragments according to definition 3. The other two parallel versions search for all frequent fragments (parallel DF_{all}) and for the discriminative fragment of definition 2 (parallel DF_F). It is evident that mining the dataset for all frequent fragments (DF_{all}) can become quite an expensive task. The running time of parallel DF_{all}

---

[1] Nodes have different hardware and software configurations. The group of the eight highest performing machines is equipped with a CPU Intel Xeon 2.40GHz, 3GB RAM and run Linux 2.6.5-7.151 as well as Java SE 1.4.2_06.

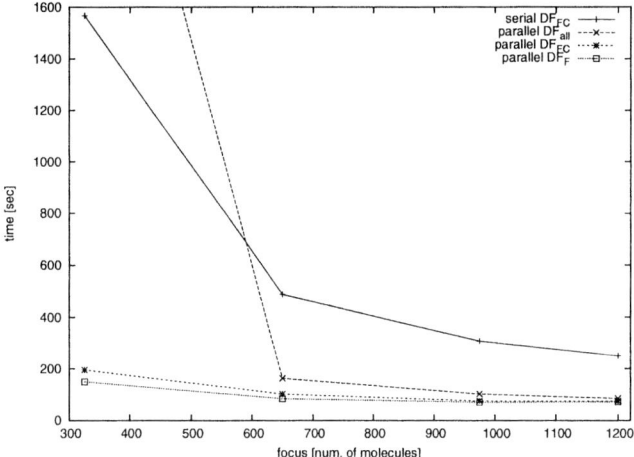

Fig. 2. Running time comparison (minSupp=6% maxSupp=1%)

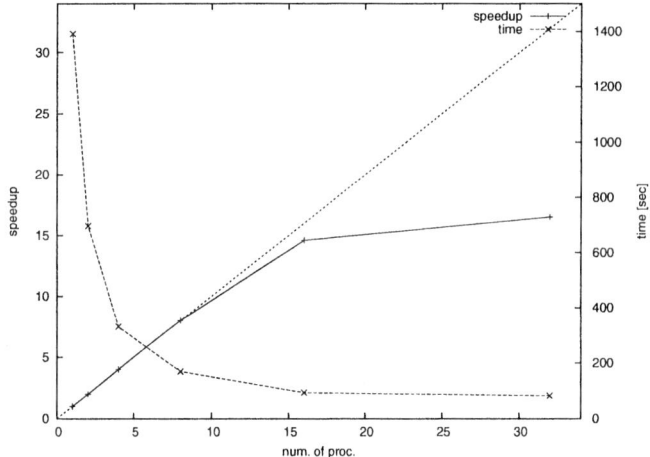

Fig. 3. Speedup and running time (minSupp=6%, maxSupp=1%)

for 325 active molecules was above 3000 seconds. This is due to the combinatorial explosion of the number of frequent fragments. It should be mentioned that, in this case (DF_{all}), the sequential algorithm cannot even complete the mining task due to the single-system memory limitations.

Mining the dataset for the closed fragments (DF_F and DF_{FC}) is feasible for the serial algorithm and is significantly sped up by the parallel execution.

We complete the analysis of the distributed approach by showing the speedup curve (Fig. 3) of the parallel over the serial algorithm, when they search for the discriminative fragments of definition 3 (DF_{FC}). The speedup is linear in the first part of the chart. Then, it is evident that more resources cannot further decrease the running time because the amount of work is not significant enough and additional computational resources cannot be effectively exploited. Nevertheless, it is positive that the running time does not increase when unnecessary resources are used as one might expect because of the additional communication and computation overheads. This provides evidence of the good scalability properties of the system.

7 Conclusions

In this paper we presented a high performance computing approach to the frequent subgraph mining problem for the discovery of discriminative molecular fragments. The adopted approach is based on three components, which are a dynamic partitioning of the search space, a novel dynamic load balancing policy and a peer-to-peer communication framework. Very low communication and synchronization requirements, quasi-randomized receiver-initiated load balancing and high scalability of the communication framework make this distributed data mining application suitable for large-scale, non-dedicated, heterogeneous computational environments like Grids. Furthermore, the proposed approach naturally tolerates node failures and communication latency and supports dynamic resource aggregation. Experimental tests on real molecular compounds confirmed its effectiveness.

Future research effort will focus on very large-scale systems, where the centralized server for collecting job statistics could potentially become a bottleneck.

Acknowledgements

This work was supported by the Italian National Research Council (CNR) and the DFG Research Training Group GK-1042 "Explorative Analysis and Visualization of large Information Spaces".

References

1. Washio, T., Motoda, H.: State of the art of graph-based data mining. ACM SIGKDD Explorations Newsletter **5** (2003) 59–68
2. Agrawal, R., Imielinski, T., Swami, A.N.: Mining association rules between sets of items in large databases. In: Proceedings of the 1993 ACM SIGMOD International Conference on Management of Data, (Washington, D.C.)
3. Zaki, M., Parthasarathy, S., Ogihara, M., Li, W.: New algorithms for fast discovery of association rules. In: Proceedings of 3rd Int. Conf. on Knowledge Discovery and Data Mining (KDD'97). (1997) 283–296
4. Deshpande, M., Kuramochi, M., Karypis, G.: Frequent sub-structure-based approaches for classifying chemical compounds. In: Proceedings of IEEE International Conference on Data Mining (ICDM'03), Melbourne, Florida, USA (2003)
5. Deshpande, M., Kuramochi, M., Karypis, G.: Automated approaches for classifying structures. In: Proceedings of Workshop on Data Mining in Bioinformatics (BioKDD). (2002) 11–18

6. Yan, X., Han, J.: gSpan: Graph-Based Substructure Pattern Mining. In: Proc. of the IEEE Int. Conference on Data Mining, Maebashi City, Japan (2002)
7. Borgelt, C., Berthold, M.R.: Mining molecular fragments: Finding relevant substructures of molecules. In: IEEE International Conference on Data Mining (ICDM 2002), Maebashi, Japan (2002) 51–58
8. Kramer, S., de Raedt, L., Helma, C.: Molecular feature mining in hiv data. In: Proceedings of 7th Int. Conf. on Knowledge Discovery and Data Mining, (KDD'01), San Francisco, CA (2001) 136–143
9. Zaki, M.J.: Parallel and distributed association mining: A survey. IEEE Concurrency **7** (1999) 14–25
10. Di Fatta, G., Berthold, M.R.: Distributed mining of molecular fragments. In: IEEE DM-Grid Workshop of the Int. Conf. on Data Mining, Brighton, UK (2004)
11. Wang, C., Parthasarathy, S.: Parallel algorithms for mining frequent structural motifs in scientific data. In: Proceedings of the 18th Annual International Conference on Supercomputing (ICS'04), Saint Malo, France (2004)
12. Finkel, R., Manber, U.: DIB - a distributed implementation of backtracking. ACM Transactions on Programming Languages and Systems **9 (2)** (1987) 235–256
13. Daylight Chemical Information Systems, Inc.: SMILES - Simplified Molecular Input Line Entry Specification. (In: http://www.daylight.com/smiles)
14. Karp, R., Zhang, Y.: A randomized parallel branch-and-bound procedure. In: Proceedings of the 20 Annual ACM Symposium on Theory of Computing (STOC 1988). (1988) 290–300
15. Chakrabarti, S., Ranade, A., Yelick, K.: Randomized load-balancing for tree-structured computation. In: Proceedings of the Scalable High Performance Computing Conference (SHPCC '94), Knoxville, TN (1994) 666–673
16. Chung, Y., Park, J., Yoon, S.: An asynchronous algorithm for balancing unpredictable workload on distributed-memory machines. ETRI Journal **20** (1998) 346–360
17. Weislow, O., Kiser, R., Fine, D., Bader, J., Shoemaker, R., Boyd, M.: New soluble formazan assay for hiv-1 cytopathic effects: Application to high flux screening of synthetic and natural products for aids antiviral activity. Journal of the National Cancer Institute, University Press, Oxford, UK, **81** (1989) 577–586

Exploring Regression for Mining User Moving Patterns in a Mobile Computing System

Chih-Chieh Hung, Wen-Chih Peng*, and Jiun-Long Huang

Department of Computer Science, National Chiao Tung University,
Hsinchu, Taiwan, ROC
{hungcc, wcpeng, jlhuang}@csie.nctu.edu.tw

Abstract. In this paper, by exploiting the log of call detail records, we present a solution procedure of mining user moving patterns in a mobile computing system. Specifically, we propose algorithm LS to accurately determine similar moving sequences from the log of call detail records so as to obtain moving behaviors of users. By exploring the feature of spatial-temporal locality, we develop algorithm TC to group call detail records into clusters. In light of the concept of regression, we devise algorithm MF to derive moving functions of moving behaviors. Performance of the proposed solution procedure is analyzed and sensitivity analysis on several design parameters is conducted. It is shown by our simulation results that user moving patterns obtained by our solution procedure are of very high quality and in fact very close to real user moving behaviors.

1 Introduction

User moving patterns refer to the areas where users frequently travel in a mobile computing environment. It is worth mentioning that user moving patterns are particularly important and are able to provide many benefits in mobile applications. A significant amount of research efforts has been elaborated upon issues of utilizing user moving patterns in developing location tracking schemes and data allocation methods [3][5][6]. Clearly, it has been recognized as an important issue to develop algorithms to mine user moving patterns so as to improve the performance of mobile computing systems.

The study in [5] explored the problem of mining user moving patterns with the moving log of mobile users given. Specifically, in order to capture user moving patterns, a moving log recording each movement of mobile users is needed. In practice, generating the moving log of all mobile users unavoidably leads to the increased storage cost and degraded performance of mobile computing systems. Consequently, in this paper, we address the problem of mining user moving patterns from the existing log of call detail records (referred to as CDR) of mobile computing systems. Generally, mobile computing systems generate one call detail record when a mobile user makes or receives a phone call. Table 1 shows an example of selected real call detail records where Uid is the identification of an individual user that makes or receives a phone call and Cellid indicates the corresponding base station that serves that mobile user. Thus, a mobile computing system produces daily a large amount of call detail records which contain hidden valuable information about the moving behaviors of mobile users. Unlike the

---

* The corresponding author of this paper.

Table 1. An example of selected call detail records

Uid	Date	Time	Cellid
1	01/03/2004	03:30:21	A
1	01/03/2004	09:12:02	D
1	01/03/2004	20:30:21	G
1	01/03/2004	21:50:31	I

moving log keeping track of the entire moving paths, the log of call detail records only reflects the fragmented moving behaviors of mobile users. However, such fragmented moving behaviors are of little interest in a mobile computing environment where one would naturally like to know the complete moving behaviors of users. Thus, in this paper, with these fragmented moving behaviors hidden in the log of call detail records, we devise a solution procedure to mine user moving patterns. The problem we shall study can be best understood by the illustrative example in Fig. 1 where the log of call detail records is given in Table 1. The dotted line in Fig. 1 represents the real moving path of the mobile user and the cells with the symbol of a mobile phone are the areas where the mobile user made or received phone calls. Explicitly, there are four call detail records generated in the log of CDRs while the mobile user travels. Given these fragmented moving behaviors, we explore the technique of regression analysis to generate user moving patterns (i.e., the solid line in Fig. 1). If user moving patterns devised are close to the real moving paths, one can utilize user moving patterns to predict the real moving behaviors of mobile users.

In this paper, we propose a solution procedure to mine user moving patterns from call detail records. Specifically, we shall first determine similar moving sequences from the log of call detail records and then these similar moving sequences are merged into one moving sequence (referred to as *aggregate moving sequence*). It is worth mentioning that to fully explore the feature of periodicity and utilize the limited amount of call detail records, algorithm LS (standing for Large Sequence) devised is able to accurately extract those similar moving sequences in the sense that those similar moving sequences are determined by two adjustable threshold values when deriving the aggregate moving sequence. By exploring the feature of spatial-temporal locality, which refers to the feature that if the time interval between the two consecutive calls of a mobile user is

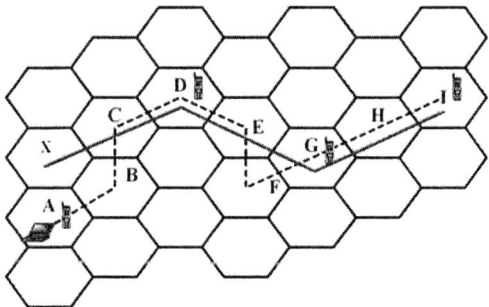

Fig. 1. A moving path and an approximate user moving pattern of a mobile user

small, the mobile user is likely to move nearby, algorithm TC (standing for Time Clustering) developed should group those call detail records into a cluster. For each cluster of call detail records, algorithm MF (standing for Moving Function), a regression-based method, devised is employed to derive moving functions of users so as to generate approximate user moving patterns. Performance of the proposed solution procedure is analyzed and sensitivity analysis on several design parameters is conducted. It is shown by our simulation results that approximate user moving patterns obtained by our proposed algorithms are of very high quality and in fact very close to real moving behaviors of users.

The rest of the paper is organized as follows. Some preliminaries and definitions are presented in Section 2. Algorithms for mining moving patterns are devised in Section 3. Performance results are presented in Section 4. This paper concludes with Section 5.

2 Preliminary

In this paper, assume that the moving behavior of mobile users has periodicity and the consecutive movement of the mobile user is not too far. Therefore, if the time interval of two consecutive CDRs is not too large, the mobile user is likely to move nearby. To facilitate the presentation of this paper, a *moving section* is defined as a basic time unit. A *moving record* is a data structure that is able to accumulate the numbers of base station identifications (henceforth referred to as *item*) appearing in call detail records whose occurring times are within the same moving section. Given a log of call detail records, we will first convert these CDR data into multiple moving sequences where a moving sequence is an ordered list of moving records and the length of the moving sequence is ε. The value of ε depends on the periodicity of mobile users and is able to obtain by the method proposed in [1]. As a result, a moving sequence i is denoted by $<MR_i^1, MR_i^2, MR_i^3, ..., MR_i^\varepsilon>$, where MR_i^j is the jth moving record of moving sequence i. We consider four hours as one basic unit of a moving section and the value of ε is six. Given the log data in Table 1, we have the moving sequence $MS_1 = < \{A : 1\}, \{\}, \{D : 1\}, \{\}, \{\}, \{G : 1, I : 1\} >$. *Time projection sequence* of moving sequence MS_i is denoted as TP_{MS_i}, which is formulated as $TP_{MS_i} = < \alpha_1, ..., \alpha_n >$, where $MR_i^{\alpha_j} \neq \{\}$ and $\alpha_1 < ... < \alpha_n$. Explicitly, TP_{MS_i} is a sequence of numbers that are the identifications of moving sections in which the corresponding moving records are not empty. Given $MS_1 =< \{A : 1\}, \{\}, \{D : 1\}, \{\}, \{\}, \{G : 1, I : 1\} >$, one can verify that $TP_{MS_1} =< 1, 3, 6 >$. By utilizing the technique of sequential clustering, a time projection sequence TP_{MS_i} is divided into several groups in which time intervals among moving sections are close. For the brevity purpose, we define a clustered time projection sequence of TP_{MS_i}, denoted by $CTP(TP_{MS_i})$, which is represented as $< CL_1, CL_2, ..., CL_x >$ where CL_i is the ith group and $i = [1, x]$. Note that the value of x is determined by our proposed method.

3 Mining User Moving Patterns

The overall procedure for mining moving patterns comprises three phases, i.e., data collection phase, time clustering phase and regression phase. The details of algorithms in each phases are described in the following subsections.

3.1 Data Collection Phase

As mentioned early, in this phase, we shall identify similar moving sequences from a set of w moving sequences obtained and then merge these similar moving sequences into one *aggregate moving sequence* (to be referred to as AMS). Algorithm LS is applied to moving sequences of each mobile user to determine the aggregate moving sequence that comprises a sequence of large moving records denoted as LMR^i, where $i = [1, \varepsilon]$. Specifically, large moving record LMR^j is a set of items with their corresponding counting values if there are a sufficient number of MR_i^j of moving sequences containing these items. Such a threshold number is called *vertical_min_sup* in this paper. Once the aggregate moving sequence is generated from these recent w moving sequences, we will then compare this aggregate moving sequence with these w moving sequences so as to further accumulate the occurring counts of items appearing in each large moving record. The threshold to identify the similarity between moving sequences and the aggregate moving sequence is named by *match_min_sup*. The algorithmic form is given below.

Algorithm LS

input: w moving sequences with their lengths being ε, two threshold:*vertical_min_sup* and *match_min_sup*
output: Aggregate moving sequence AMS
1 **begin**
2　**for** $j =1$ to ε
3　　**for** i=1 to w
4　　　LMR^j =large 1-itemset of MR_i^j;
　　　　(by *vertical_min_sup*)
5　**for** $i = 1$ to w
6　　**begin**
7　　　$match = 0$;
8　　　**for** $j = 1$ to ε
9　　　　**begin**
10　　　　$C(MR_i^j, LMR^j) =$
　　　　　　$|x \in MR_i^j \cap LMR^j| / |y \in MR_i^j \cup LMR^j|$;
11　　　　$match = match+|MR_i^j|*C(MR_i^j, LMR^j)$;
12　　　**end**
13　　**if** $match \geq match\_min\_sup$ **then**
14　　　accumulate the occurring counts of
　　　　items in the aggregate moving sequence;
15　**end**
16 **end**

In algorithm LS (from line 2 to line 4), we first calculate the appearing counts of items in each moving sections of w moving sequences. If the count of an item among w moving sequences is larger than the value of *vertical_min_sup*, this item will be weaved into the corresponding large moving record. After obtaining all large moving records, AMS is then generated and is represented as $< LMR^1, LMR^2, ...,$

$LMR^\varepsilon >$, where the length of the aggregate moving sequence is ε. As mentioned before, large moving records contain frequent items with their corresponding counts. Once obtaining the aggregate moving sequence, we should in algorithm LS (from line 5 to line 12) compare this aggregate moving sequence with w moving sequences in order to identify those similar moving sequences and then calculate the count of each item in each large moving record. Note that a moving sequence (respectively, AMS) consists of a sequence of moving records (respectively, large moving records). Thus, in order to quantity how similar between a moving sequence (e.g., MS_i) and AMS, we shall first measure the closeness between moving record MR_i^j and LMR^j, denoted by $C(MR_i^j, LMR^j)$. $C(MR_i^j, LMR^j)$ is formulated as $\frac{|\{x \in MR_i^j \cap LMR^j\}|}{|\{y \in MR_i^j \cup LMR^j\}|}$ that returns the normalized value in $[0, 1]$. The larger the value of $C(MR_i^j, LMR^j)$ is, the more closely MR_i^j resembles LMR^j. Accordingly, the similarity measure of moving sequence MS_i and AMS is thus formulated as $sim(MS_i, AMS) = \sum_{i=1}^{\varepsilon} |MR_i^j| * C(MR_i^j, LMR^j)$. Given a threshold value $match\_min\_sup$, for each moving sequence MS_i, if $sim(MS_i, AMS) \geq match\_min\_sup$, moving sequence MS_i is identified as a similar moving sequence containing sufficient moving behaviors of mobile users. In algorithm LS (from line 13 to line 14), for each item in large moving records, the occurring count is accumulated from the corresponding moving records of those similar moving sequences.

3.2 Time Clustering Phase

In this phase, two threshold values (i.e., δ and σ^2) are given in clustering a time projection sequence. Explicitly, the value of δ is used to determine the density of clusters and σ^2 is utilized to make sure that the spread of the time is bounded within σ^2. Algorithm TC is able to dynamically determine the number of groups in a time projection sequence.

Algorithm TC (from line 2 to line 3) first starts clustering coarsely TP_{AMS} into several marked clusters if the difference between two successive numbers is smaller than the threshold value δ. As pointed out before, CL_i denotes the ith marked cluster. In order to guarantee the quality of clusters, a spread degree of CL_i, denoted as $Sd(CL_i)$,

Algorithm TC

input: Time projection sequence TP_{AMS}, threshold δ and σ^2
output: Clustered time projection sequence $CTP(TP_{AMS})$
1 **begin**
2   group the numbers whose differences are within δ;
3   mark all clusters;
4   **while** there exist marked clusters and $\delta \geq 1$
5   **for** each marked clusters CL_i
6   **if** $Sd(CL_i) \leq \sigma^2$
7   unmark CL_i;
8        $\delta = \delta - 1$;
9   **for** all marked clusters CL_i

10	group the numbers whose differences are within δ in CL_i;
11	**end while**
12	**if** there exist marked clusters
13	**for** each marked cluster CL_i
14	$k = 1$;
15	**repeat**
16	k++;
17	divide evenly CL_i into k groups ;
18	**until** the spread degree of each group$\leq \sigma^2$;
19	**end**

is defined to measure the distribution of numbers in cluster CL_i. Specifically, $Sd(CL_i)$ is modelled by the variance of a sequence of numbers. Hence, $Sd(CL_i)$ is formulated as $\frac{1}{m}\sum_{k=1}^{m}(n_k - \frac{1}{m}\sum_{j=1}^{m}n_j)^2$, where n_k is the kth number in CL_i and m is the number of elements in CL_i. As can be seen from line 5 to line 7 in algorithm TC, for each cluster CL_i, if $Sd(CL_i)$ is smaller than σ^2, we unmark the cluster CL_i. Otherwise, we will decrease δ by 1 and with given the value of δ, algorithm TC (from line 8 to line 10) will re-cluster those numbers in unmark clusters. Algorithm TC partitions the numbers of TP_{AMS} iteratively with the objective of satisfying two threshold values, i.e., δ and σ^2, until there is no marked cluster or $\delta = 0$. If there is no marked clusters, $CTP(TP_{AMS})$ is thus generated. Note that, however, if there are still marked clusters with their spread degree values larger than σ^2, algorithm TC (from line 12 to line 18) will further finely partition these marked clusters so that the spread degree for each marked cluster is constrained by the threshold value of σ^2. If the threshold value of δ is 1, a marked cluster is usually a sequence of continuos numbers in which the spread degree of this marked cluster is still larger than σ^2. Given marked cluster CL_i, algorithm TC initially sets k to be 1. Then, marked cluster CL_i is evenly divided into k groups with each group size $\lceil \frac{n}{k} \rceil$. By increasing the value of k each run, algorithm TC is able to partition the marked cluster until the spread degree of each partition in the marked cluster CL_i satisfies σ^2.

3.3 Regression Phase

Assume that AMS is $< LMR^1, LMR^2, ..., LMR^\varepsilon >$ with its clustered time projection sequence $CTP(TP_{AMS}) = CL_1, CL_2, ..., CL_k$, where CL_i represents the ith cluster. For each cluster CL_i of $CTP(TP_{AMS})$, we will derive the estimated moving function of mobile users, expressed as $E_i(t) = (\hat{x}_i(t), \hat{y}_i(t), valid\_time\_interval)$, where $\hat{x}_i(t)$ (respectively, $\hat{y}_i(t)$) is a moving function in x-coordinate axis (respectively, in y-coordinate axis)) and $valid\_time\_interval$ indicates the time interval when the moving function is valid.

Without loss of generality, let CL_i be $\{t_1, t_2, ..., t_n\}$ where t_i is one of the moving section in CL_i. As described before, a moving record has the set of the items with their corresponding counts. Therefore, we could extract those large moving records from AMS to derive the estimated moving function for each cluster. In order to derive moving functions, the location of base stations should be represented in geometry model through a map table provided by tele-companies. Hence, given AMS and a cluster of $CTP(TP_{AMS})$, for each cluster of $CTP(TP_{AMS})$, we could have geometric coordinates of frequent items with their corresponding counts, which are able to represent as $(t_1,x_1,y_1,w_1), (t_2,x_2,y_2,w_2), ...(t_n,x_n,y_n,w_n)$. Accordingly, for each cluster of $CTP(TP_{AMS})$, regression analysis is able to derive the corresponding estimated moving function.

Given a cluster of data points (e.g., $(t_1, x_1, y_1, w_1), (t_2, x_2, y_2, w_2), ..., (t_n,x_n,y_n,w_n)$), we first consider the derivation of $\hat{x}(t)$. If the number of distinct time points in a given cluster is $m+1$, a m-degree polynomial function $\hat{x}(t) = a_0 + a_1 t + ... + a_m t^m$ will be derived to approximate moving behaviors in x-coordinate axis. Specifically, the regression coefficients $\{a_0, a_1, ...a_m\}$ are chosen to make the residual sum of squares $\epsilon_x = \sum_{i=1}^{n} w_i e_i^2$ minimal, where w_i is the weight of the data point (x_i, y_i) and $e_i = (x_i - (a_0 + a_1 t_i + a_2(t_i)^2 ... + a_m(t_i)^m))$. To facilitate the presentation of our paper, we define the following terms:

$$T = \begin{bmatrix} 1 & t_1 & (t_1)^2 & ... & (t_1)^m \\ 1 & t_2 & (t_2)^2 & ... & (t_2)^m \\ ... & ... & ... & ... & ... \\ 1 & t_n & (t_n)^2 & ... & (t_n)^m \end{bmatrix}, a^* = \begin{bmatrix} a_0 \\ a_1 \\ ... \\ a_m \end{bmatrix}, b_x = \begin{bmatrix} x_1 \\ x_2 \\ ... \\ x_n \end{bmatrix}, e = \begin{bmatrix} e_1 \\ e_2 \\ ... \\ e_n \end{bmatrix}^T, W = \begin{bmatrix} w_1 & & & \\ & w_2 & & \\ & & ... & \\ & & & w_n \end{bmatrix}.$$

The residual sum of squares can be expressed as $\epsilon_x = e^T W e$. Since w_i are positive for all i, W is written as: $W = \sqrt{W}\sqrt{W}$, where $\sqrt{W}$ is a diagonal matrix with its diagonal entries to be $[\sqrt{w_1}, \sqrt{w_2}, ..., \sqrt{w_n}]$. Thus, $\epsilon_x = e^T W e = (b_x - Ta^*)^T \sqrt{W}\sqrt{W}(b_x - Ta^*) = (\sqrt{W}b_x - \sqrt{W}Ta^*)^T(\sqrt{W}b_x - \sqrt{W}Ta^*)$. Clearly, ϵ_x is minimized w.r.t. a^* by the normal equation $(\sqrt{W}T)^T(\sqrt{W}T)a^* = (\sqrt{W}T)^T \sqrt{W}b_x$ [2]. The coefficients $\{a_0, a_1, ...a_m\}$ can hence be obtained by solving the normal equation: $a^* = [(\sqrt{W}T)^T(\sqrt{W}T)]^{-1}(\sqrt{W}T)^T \sqrt{W}b_x$. Therefore, $\hat{x}(t) = a_0 + a_1 t + ... + a_m t^m$ is obtained. Following the same procedure, we could derive $\hat{y}(t)$. As a result, for each cluster of $CTP(TP_{AMS})$, the estimated moving function $E_i(t) = (\hat{x}(t), \hat{y}(t), [t_1, t_n])$ of a mobile user is devised.

Algorithm MF

input: AMS and clustered time projection sequence $CTP(TP_{AMS})$
output: A set of moving functions $F(t) = \{E_1(t), U_1(t), E_2(t), ..., E_k(t), U_k(t)\}$
1 begin
2   initialize $F(t)$=empty;
3   for i= 1 to k-1
4   begin
5   doing regression on CL_i to generate $E_i(t)$;
6   doing regression on CL_{i+1} to generate $E_{i+1}(t)$;
7   t_1 =the last number in CL_i;

8   t_2 =the first number in CL_{i+1};
9   using inner interpolation to generate
    $U_i(t) = (\hat{x}_i(t), \hat{y}_i(t), (t_1, t_2))$;
10   insert $E_i(t), U_i(t)$ and $E_{i+1}(t)$ in $F(t)$;
11 end
12 if($1 \notin CL_1$)
13   generate $U_0(t)$ and Insert $U_0(t)$ into the head of $F(t)$;
14 if($\varepsilon \notin CL_k$)
15   generate $U_k(t)$ and Insert $U_k(t)$ into the tail of $F(t)$;
16 return $F(t)$;
17 end

4 Performance Study

In this section, the effectiveness of mining user moving patterns by call detail records is evaluated empirically. The simulation model for the mobile system considered is described in Section 4.1. Section 4.2 is devoted to experimental results and comparison with the original algorithm of mining moving patterns [5].

4.1 Simulation Model for a Mobile System

To simulate the base stations in a mobile computing system, we use a eight by eight mesh network, where each node represents one base station and there are hence 64 base stations in this model [4][5]. A moving path is a sequence of base stations travelled by a mobile user. The number of movements made by a mobile user during one moving section is modeled as a uniform distribution between mf-2 and mf+2. Explicitly, the larger the value of mf is, the more frequently a mobile user moves. To model user calling behavior, the calling frequency is employed to determine the number of calls during one moving section. If the value of cf is large, the number of calls for a mobile user will increase. Similar to [5], the mobile user moves to one of its neighboring base stations depending on a probabilistic model. To make sure the periodicity of moving behaviors, the probability that a mobile user moves to the base station where this user came from is modeled by P_{back} and the probability that the mobile user routes to the other base stations is determined by $(1-P_{back})/(n-1)$ where n is the number of possible base stations this mobile user can move to. The method of mining moving patterns in [5], denoted as UMP, is implemented for the comparison purposes. For interest of brevity, our proposed solution procedure of mining user moving patterns is expressed by $AUMP$ (standing for approximate user moving patterns). The location is represented as the identification of a base station. To measure the accuracy of user moving patterns, we use the hop count (denoted as hn), which is measured by the number of base stations, to represent the distance from the location predicted by moving functions derived to the actual location of the mobile user. Intuitively, a smaller value of hn implies that the more accurate prediction is achieved.

4.2 Experiments of UMP and AUMP

To conduct the experiments to evaluate UMP and $AUMP$, we set the value of w to be 10, the value of cf to be 3 and the value of ε to be 12. In order to reduce the

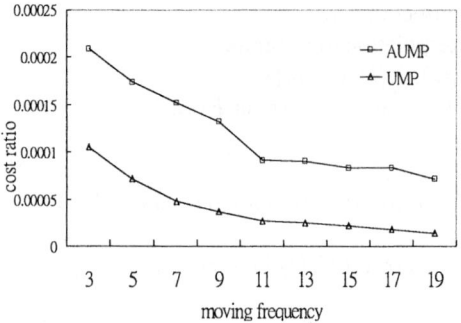

Fig. 2. The cost ratios of AUMP and UMP with the moving frequency varied

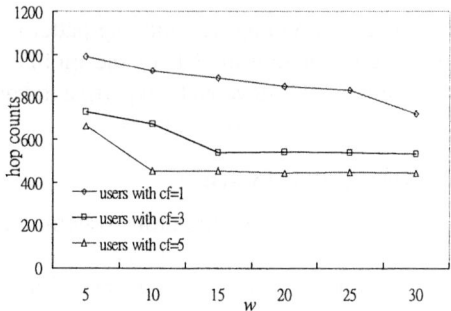

Fig. 3. The performance of AUMP with the value of w varied

amount of data used in mining user moving patterns, $AUMP$ explores the log of call detail records. The cost ratio for a user, i.e., $\frac{\frac{1}{hn}}{amount\ of\ log\ data}$, means the prediction accuracy gained by having the additional amount of log data. Fig. 2 shows the cost ratios of UMP and $AUMP$. Notice that $AUMP$ has larger cost ratios than UMP, showing that $AUMP$ employs the amount of log data more cost-efficiently to increase the prediction accuracy.

The impact of varying the values of w for mining moving patterns is next investigated. Without loss of generality, we set the value of ε to be 12, that of mf to be 3 and the values of cf to be 1, 3 and 5. Both $vertical\_min\_sup$ and $match\_min\_support$ are set to 20%, the value of δ is set to be 3, and σ^2 is set to be 0.25. With this setting, the experimental results are shown in Fig. 3.

As can be seen from Fig. 3, the hop count of AUMP decreases as the value of w increases. This is due to that as the value of w increases, meaning that the number of moving sequences considered in AUMP increases, AUMP is able to effectively extract more information from the log of call detail records. Note that with a given the value of w, the hop count of AUMP with a larger value of cf is smaller, showing that the log of data has more information when the value of cf increases. Clearly, for mobile users having high call frequencies, the value of w is able to set smaller in order to quickly obtain moving patterns. However, for mobile users having low call frequencies, the value of w should be set larger so as to increase the accuracy of moving patterns mined by $AUMP$.

5 Conclusions

In this paper, without increasing the overhead of generating the moving log, we presented a new mining method to mine user moving patterns from the existing log of call detail records of mobile computing systems. Specifically, we proposed algorithm LS to capture similar moving sequences from the log of call detail records and then these similar moving sequences are merged into the aggregate moving sequence. By exploring the feature of spatial-temporal locality, algorithm TC proposed is able to group call detail records into several clusters. For each cluster of the aggregate moving sequence, algorithm MF devised is employed to derive the estimated moving function, which is able to generate user moving patterns. It is shown by our simulation results that user moving patterns achieved by our proposed algorithms are of very high quality and in fact very close to real user moving behaviors.

Acknowledgment

The authors are supported in part by the National Science Council, Project No. NSC 92-2211-E-009-068 and NSC 93-2213-E-009-121, Taiwan, Republic of China.

References

1. J. Han, G. Dong, and Y. Yin. Efficient Mining of Partial Periodic Patterns in Time Series Database. In *Proceeding of the 15th International Conference on Data Engineering*, March 1999.
2. R. V. Hogg and E. A. Tanis. *Probability and Statistical Inference*. Prentice-Hall International Inc., 1997.
3. D. L. Lee, J. Xu, B. Zheng, and W.-C. Lee. Data Management in Location-Dependent Information Services. In *Proceeding of IEEE Pervasive Computing*, pages 65–72, 2002.
4. Y.-B. Lin. Modeling Techniques for Large-Scale PCS Networks. *IEEE Communications Magazine*, 35(2):102–107, February 1997.
5. W.-C. Peng and M.-S. Chen. Developing Data Allocation Schemes by Incremental Mining of User Moving Patterns in a Mobile Computing System. In *Proceeding of IEEE Transactions on Knowledge and Data Engineering, Volume 15*, pages 70–85, 2003.
6. H.-K. Wu, M.-H. Jin, J.-T. Horng, and C.-Y. Ke. Personal Paging Area Design Based On Mobile's Moving Behaviors. In *Proceeding of IEEE INFOCOM*, 2001.

A System Supporting Nested Transactions in DRTDBSs*

Majed Abdouli[1], Bruno Sadeg[1],
Laurent Amanton[1], and Adel Alimi[2]

[1] Laboratoire d'Informatique du Havre, BP 540, 25 Rue P. Lebon,
76600 Le Havre, France
{Majed.Abdouli, Bruno.Sadeg, Laurent.Amanton}@univ-lehavre.fr
[2] Research Group on Intelligent Machines,
BP 5403038 Sfax, Tunisia
Adel.Alimi@ieee.org

Abstract. Extended transaction models in databases were motivated by the need of complex applications such as CAD/CAM and software engineering. Nested transaction models have so far been shown to play an important role in such applications. However, these models are not yet fully studied. In this paper, we focus on the applicability of such models to real-time database systems, particularly to issues related to the global serializability of distributed real-time nested transactions. Our contribution in this field is twofold: we propose (i) a real-time concurrency control, called **2PL-NT-HP**, to solve data conflicts problem between nested transactions and (ii) a real-time commit protocol to guarantee the uniform commitment of distributed nested transactions. To this purpose, we have adapted the *PROMPT* real-time commit protocol which is designed specifically for the real-time flat transactions. This protocol causes intra-aborts cascade in nested environment and hence decreases its real-time performances. To alleviates this drawback, the borrowing subtransaction carries out a speculative execution by accessing both before and after-image[1] of the lending subtransaction. Simulations we have carried out show that **S-PROMPT**[2] approach is very useful in DRT-DBSs compared to the classical approaches.

1 Introduction

The main protocol for ensuring atomicity of multi-site transactions in a distributed environment is the **2PC** (2-Phase Commit) protocol. Typically, the 2PC protocol is combined with **2PL** (2-Phase Locking) protocol, as the means of ensuring the global serializability of transaction in a distributed database. The implications of this combination on the delay a transaction may hold locks on various data-items might be severe. At each site, and for each transaction,

---

* Distributed Real-Time Database Systems.
[1] Prepared data-item.
[2] Speculative-PROMPT.

locks must be held until either a commit or an abort message will be received by the coordinator of the 2PC protocol. Since the 2PC protocol is a blocking[3] protocol if the failures occur, the locks delays can be unbounded. Moreover, even if no failure occurs, since the protocol involves three rounds of messages (see Section 3), e.g., request of vote, vote and decision, the delay can be intolerable. The impact of long delays greatly decreases the real-time database performances. By using PROMPT[1] (Permits Reading of MOdified Prepared data for Timeliness) real-time commit protocol, that is specifically designed for the real-time domain, the prepared data can be used by a conflicting transaction, so the delay of waiting greatly decreases and the real-time performances are greatly increased. PROMPT allows transaction to *optimistically* borrow, in a controlled manner to avoid the inherent problem of cascading abort due to use of dirty data[4], the update data of transaction currently in their commit phase. The controlled borrowing reduces the data inaccessibility and the priority inversion that is inherent in distributed real-time commit processing. In [1], the simulation-based evaluation shows PROMPT to be highly successful, as compared to the classical commit protocols, in minimizing the number of missed transaction deadlines. In this paper we use this protocol to an extended transaction model, called *nested transaction models* motivated by the need of complex application such as CAD and software engineering. Nested transaction models have so far been shown to play an important role in such applications.

A nested transaction is considered as a hierarchy of subtransactions, and each subtransaction may contain other subtransactions, or contain the atomic database operations (read or write), on one hand. On the other hand, a nested transaction is a collection of subtransactions that compose the whole atomic execution unit [2]. A nested transaction is represented by a transaction tree [2,3]. A nested transaction offers more decomposable execution units and finer grained control over concurrency and recovery than flat transaction. Nested transactions provide intra-parallelism, (subtransactions running in parallel) as well as better failure recovery options, i.e., when a subtransaction fails and aborts, there is a chance to restart it by its parent instead of restarting the whole transaction (which is the case of flat transactions). The major driving force in using nested transaction is the need to model long-lived applications, nested transactions may also be efficiently used to model short transactions (like in the context of real-time database applications).

In this paper, our contribution in the field of distributed real-time nested transaction is twofold: (i) firstly, we have proposed a real-time concurrency control protocol for ensuring the consistency of database and to finish the transactions within their timing quota. And (ii) for ensuring the atomicity of multi-site transactions in real-time nested environment, we have adapted and extended the *PROMPT* real-time commit protocol. In this protocol, the major drawback is the potential problem due to abort that occur if the transaction which

---

[3] This problem is alleviated in the sense that transaction deadlines resolve naturally this problem: a deadline plays the timeout role.

[4] Uncommitted data is generally not recommended in traditional database systems.

has lent uncommitted data finally abort. This phenomenon greatly affect the real-time performance of nested transactions, since a subtransaction can only access data-item on its resident site and need to invoke subtransactions in order to access a data-item on the other sites. If the borrowing subtransaction finally aborts, then all its descending subtransaction have to be aborted causing then *intra-aborts cascade*. To alleviate this problem, the borrowing subtransaction carries out a speculative execution by accessing both before and after-image of the lending subtransaction. This **S-PROMPT** approach is very useful in real-time nested applications in general. Deadlock detection and resolution are also investigated.

The remainder of this paper is organized as follows: In the next Section we introduce the real-time concurrency control for nested transactions called **2PL-NT-HP**. In Section 3, we present PROMPT in more details and we introduce the **S-PROMPT** protocol. Deadlock detection and resolution is given in Section 4, system model and the results of simulation experiments are given in Section 5. The last Section consists of conclusion and some future research directions.

2 Real-Time Concurrency Control

In the following, we present a real-time concurrency control for nested transaction models called **2PL-NT-HP**. It combines two-phase locking protocol [2,3] and high priority based scheme [4,5]. It solves the conflicts between (sub)-transactions by allowing (sub)transaction with higher priority to access data-item firstly and blocks or aborts (sub)transactions with lower priority. The $2PL\text{-}NT\text{-}HP$ is based on the standard $2PL\text{-}NT$ [2], where are added some real-time characteristics[5]. These characteristics are a combination of *priority inheritance* [4,5], *priority abort* [6,5], *conditional restart* [7,5] and a *controller of wasted resources*[8]. When (sub)transactions having access conflicts to different trees of transaction, the EDF^5 policy is used to block or abort the lower priority (sub)transactions. Conditional restart scheme is employed to avoid the *starvation* problem occurred when using the classical scheduler EFD, the mechanism is as follow: If the slack time of (sub)transaction(s) with a higher priority is enough to be executed after finishing all lower priority transactions, and then allowing transactions with lower priority to access data first instead of aborting lower priority transaction. Otherwise, lower priority transactions must be aborted and restarted only if the controller of wasted resources authorizes it. This controller of wasted resources checks if the restarted transaction can achieve its work before its deadline, using this mechanism, we reduce the wasting of resources as well as the wasting of time. Recall that in nested transaction, subtransactions are executed on behalf of their TL-transaction, then subtransactions of the same tree of transactions should not abort each other when the conflicts occur. When intra-conflict happens, the priority inheritance scheme [4,5] is used to avoid priority inversion problem. The problem of priority inversion occurs where a highpriority

---

[5] Earliest Deadline First.

Algorithm 1 2PL-NT-HP : Real-Time Concurrency Controller

begin
AcquisitionLock
case 1 : Specific case (transaction (T_i) requests a lock)
if (no transaction holds a lock) **then**
　the lock is granted to (T_i)
end if
case 2 : Conflict intra-transactions (T_i) (i = 1) requests a lock on data-item which is locked by another transactions in the same transaction tree (T_j) (j =2,3,...,n) with an incompatible mode
while $(j \leq n)$ **do**
　if $(P(Tj) < P(T_i))$ **then**
　　(T_j) inherits $P(T_i)$
　else
　　put (T_i) in the waiting list of a data-item
　end if
　j:= j + 1
end while
case 3 : Conflict inter-Transactions: (conflict between transaction trees)
if $(P(T_i) > P(T_j))$ **then**
　if $(SlackTime(T_i) >= \sum_{j=1}^{n} RemExTime(T_j))$ **then**
　　put T_i in the waiting list of a data-item
　　(T_j) inherits $P(T_i)$
　else
　　Abort and put T_j in the Controller of Wasted Resources
　　if (it is possible to run T_j after restarting it) **then**
　　　restart T_j
　　end if
　end if
else
　put (T_i) in the waiting list of a data-item
end if
LockInheritance: the inheritance can be separated into two-types : LT^6 to PT and PT to PT
case 1 : **LT to PT**
whenever a LT holds a lock in r- or w-mode, its parent PT inherits and retains the lock in the corresponding mode
case 2 : **PT to PT, recall that a PT include the super PT (TLT)**
whenever a sub-transaction PT retains or holds a lock in r- or w-mode , its ancestor PT retains a lock in the corresponding mode
Commit Phase
when TLT retains all locks, it is so far ready to commit : the 2PC is performed.
end

transaction waits for a low priority transaction to commit, violating the notion of priority in the priority-based scheduling. This happens when high priority transaction requests a lock on a data-item which is already locked by a low priority transaction, and the lock modes are in conflict.

3 Real-Time Commit Protocols

3.1 PROMPT Commit Protocol

The **PROMPT** [1] protocol is integrated with $2PC^7$ protocol by allowing transactions to access uncommitted data-item held by prepared transactions in the optimistic belief that this data-item will eventually be committed. The main feature of the *PROMPT* commit protocol is that transactions requesting data-items held by other transactions in the prepared state are allowed to access this data-item. Prepared subtransaction lend their uncommitted data-item to concurrently executing subtransaction (without releasing the update locks). The mechanics of the interactions between such lenders and their associated borrowers are captured in the following three scenarios, only one of witch will occur for each lending.

- Lender receives decision before borrower completes data processing: Here, the lending subtransaction receives its global decision before the borrowing subtransaction has completed its local data processing. If the global decision is to commit, then the lending subtransaction completes in the normal fashion. Otherwise, the lender is aborted. In addition, the borrower is also aborted since it has utilized dirty data.
- Borrower completes data processing before lending receives decision: In this case, the borrower is made to wait and not allowed to send a $WORKDONE$ message to its coordinator. It has to wait until either the lender receives its global decision or its own deadline expires, whichever occurs earlier.
- Borrower aborts during data processing before lender receives decision: Here, the lending is nullified.

To further improve its real-time performance, three additional features are included in the *PROMPT* commit protocol:

- Active Abort: The subtransactions that abort locally[8] must inform their coordinator as soon as they decide to abort. This active abort is beneficial in two ways: (i) it provides more time for the restarted transaction to complete before its deadline, and (ii) it minimize the wastage of both logical and physical system resources.
- Silent Kill: For a transaction that is killed before the coordinator enters its commit phase, there is no need for the coordinator to invoke the abort protocol.
- Healthy Lending: A committing transaction that is close to its deadline may be killed due to deadline expiry before its commit processing is finished. Lending by such transaction must be avoided, since they are likely to result in the aborts of all the associated borrowers. To address this issue, the authors of [1] have added a feature to *PROMPT* whereby only healthy transactions,

---

[7] A particularly attractive feature of PROMPT is that it can be integrated with many of other optimizations suggested for 2PC.
[8] Due to conflicts with higher priority transaction.

that is, transactions whose deadlines are not very closed, are allowed to lend their prepared data. A health factor, HF_T, is associated with each transaction T and a transaction is allowed to lend its data-item only if its HF is greater than a (system-specified) minimum value $MinHF$.

Aborts in *PROMPT* do not arbitrarily cascade[9] because the borrowing subtransaction cannot simultaneously be a lender, e.g., the borrower is not allowed to enter in its prepared state. Therefore, if the lending subtransaction abort for some reasons, the abort chain is bounded and is of length one. This situation is accepted in flat transaction models but not in nested models, the next subsection alleviates the problem.

3.2 S-PROMPT Commit Protocol

The **S-PROMPT** commit protocol is motivated by the following two reasons: Firstly, in flat transaction models, if the lending subtransaction[10] finally aborts, all its associated borrowing subtransactions have to be restarted, and secondary, specific to nested transactions models, a subtransaction T_i can only access data-item on its resident site and has to invoke subtransactions in order to access data-item on the other site. If T_i has lent a prepared data and after the lending subtransaction aborts, T_i must abort as well all its descendants, so PROMPT causes *intra-aborts cascade* in nested transaction models. To alleviate these problems, in this paper we propose a Speculative PROMPT commit protocol to resolve this intra-aborts cascade in nested transactions. In S-PROMPT protocol, whenever a subtransaction enters in its prepared state, it produces after-image (prepared data). The conflicting subtransactions access both before and after-images of lending subtransaction and then carries out speculative execution. The borrowing subtransaction selects appropriate execution after the termination of the lending subtransaction. If the lending subtransaction commit successfully, the borrowing subtransaction selects to continue its work with the after-image, otherwise, it work with the before image. The principle of S-PROMPT is as follows: when an arriving subtransaction T_i has a conflict with an active subtransaction T_j, then T_i is duplicated such as one copy executes using before-image (non-updated data-item) and the second copy executes uses after-image of the lending subtransaction T_j. This scheme is repeated whatever a new arriving subtransaction enters the system and has a conflict with prepared subtransaction.

4 Deadlock Detection and Resolution

Deadlock is the situation where two or more transactions wait for each other to release locks. In nested transaction models, two types of deadlocks may occur: *intra-transaction deadlocks* and *inter-transaction deadlocks*. Inter-transaction

---

[9] In flat transaction models.
[10] In flat transaction models, transaction is divided into subtransactions according to the localization of the data-item they have to access.

deadlocks arise between nested transactions and intra-transaction deadlocks arise between subtransactions in the same nested transactions. Priority abort based concurrency control algorithms avoid inter-transaction deadlocks, since a higher priority transaction is never blocked by a lower priority transaction. Therefore, this priority abort protocol is deadlock-free among transactions of different transaction trees provided that no transactions have the same priority. However, in the same transaction tree, there are subtransactions running with the same priorities. So, locking protocol may lead to intra-transaction deadlocks. The basic idea of intra-transaction deadlocks detection is to construct a **WFG**[11] [9,10] for every non-leaf subtransactions containing wait-for information of the subtransaction's descendants to detect internal deadlocks. The resolution of intra-transaction deadlocks must chose the victim subtransaction(s) according to several criteria, since abort or restart subtransaction causes the aborting or restarting of all its descendants. In our real-time application, we have chosen the candidate subtransaction(s) that has the latest execution start time that has spent less time in execution than the subtransaction started previously[10], e.g., its ancestor, this choice makes safe that the victim subtransaction is a subtransaction child since its parent has the greater latest execution start. Using this policy, we minimize the wasted time due to aborts.

5 Simulation Results

5.1 System Model

To evaluate the real-time performance of the various protocols described in the previous section, we have used a detailed simulation. The model consists of a database that is distributed, in a non-replicated manner, over N sites connected by a network. Each site in the model has five components : a *Transaction Generator* **GT** which generates transactions, a *Transaction Manager*, **TM** which models the execution behavior of a transaction, a *Concurrency Control Manager* **CCM** which implements the concurrency control algorithm, a *Message Server* **MS** which handles message exchanges between TM of its own site and those of the other sites, and a *Commit Manager* **CM** which implements the details of commit protocols. In addition to these per-site components, the model has also a *Network Manager* **NM** which models the behavior of the communication network.

5.2 Performance Results

The performance metric in our experiments is the *SuccessRatio*, which is the percent of number of nested transactions completed before their firm deadline over the total number of nested transactions submitted to the system: $\frac{Number\ of\ Completed\ transactions\ before\ its\ deadline}{Number\ of\ Transactions\ submitted\ to\ the\ system}$. For each of experiment, the results were calculated as a average of 10 independents executions. Each execution

---

[11] Wait-for-graph.

ends after 500 transaction arrivals exponentially distributed. The $MaxActTr$ parameter specifies the maximum number of active nested transactions. In our model, we suppose that the system works without crashes and the database is memory-resident.

We have begun to simulate regular flat transactions, e.g., nested transaction with LevelSize = 1, to see if the nested transaction models are more advantageous or not for DRTDBS. To simulate regular flat transactions, the system parameters are the same as in (table 1), except NumberOps, SubProb and LevelSize. The number of operations generated for each flat transaction is uniformly distributed from the interval [50-100]. This creates transactions in average the same size as nested transactions. Then, LevelSize = 1, e.g., only a TLT, e.g., TLT is the flat transaction is this case, is generated, and SubProb is set as $\{0.0, 0.0, 0.0\}$ so that no child subtransactions are created. $2PL$-HP and $2PC$ are performed for these experiments, simulation results show that the nested transaction models seen more advantageous for real-time application than flat transaction models, this is due to the previously noted reason that for nested transactions, when subtransaction fails and aborts, there is a chance to restart it by its parent instead of restarting the whole transaction which is the case of flat transactions, (Fig.1.).

Our second experiment was conducted the default setting for all model parameters (Table 1). Simulation results show that at normal loads and under the proposed concurrency controller, the success ratio is better when using the new S-$PROMPT$ commit protocol than when using the classical $PROMPT$, the enhancement is about 10% (Fig.2.(a)). Under heavy loads, the success ratio is

Table 1. System Parameters of the Performance Model

Parameter	Default	Parameter	Default
NumCPU	1	NumSites	8
DBSize	500	MaxActTr	50
DistDegree	5	SlakFactor	3
InterArTime	10-100	CPUtime	1ms
MsgProcTime	1ms	SubProb	Level 1: 0.3 Level 2: 0.2 Level 3: 0.0
NumberOps	5-10	LevelSize	3
UpdateProb	0.5	ConflictLevel	x=80, y=20
LockProc	1ms	DecesionCoorTime	1ms
DeadDetRes	1ms	MinHF	1ms

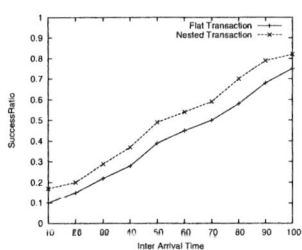

Fig. 1. Flat vs Nested Transactions (under 2PC Protocol)

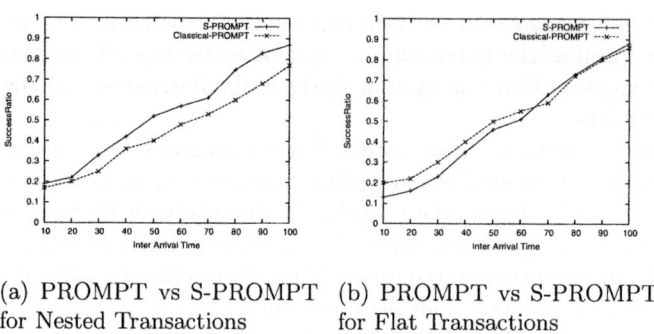

(a) PROMPT vs S-PROMPT for Nested Transactions (b) PROMPT vs S-PROMPT for Flat Transactions

Fig. 2. PROMPT vs S-PROMPT

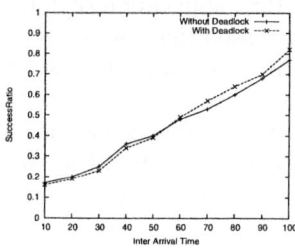

Fig. 3. Deadlock Detection and Resolution Results(under S-PROMPT)

virtually identical. For flat transactions, simulation result show that the success ratio is virtually identical and no significant performance improvement is observed because the abort chain is bounded and is of length one (Fig.2.(b)). However, this S-PROMPT is more advantageous for real-time nested applications . The policy proposed for deadlock detection and resolution is very expensive, especially in distributed setting, on one hand. On the other hand, very few deadlocks occur among subtransaction of individual nested transactions under priority inheritance scheduling. So, no performance improvement is observed (Fig.3.).

6 Conclusion and Future Works

In this paper, we have proposed, to ensure the global serializability of distributed firm real-time nested transactions, a real-time concurrency control approach and a speculative PROMPT commit protocol. The proposed concurrency controller protocol employs a combination scheme of priority inheritance, priority abort, a conditional restart and a controller of wasted resources. S-PROMPT is designed specially for distributed nested transactions that play an important role in real-time applications, simulation results show that nested transactions out perform flat transactions and S-PROMPT is very useful in nested environment as compared to the classical PROMPT. S-PROMPT resolve the problem of intra-aborts cascade of the borrowing subtransaction by carrying out a speculative execution by accessing both before and after-image of the lending subtransaction. As the part of future work, we will incorporate $S\text{-}PROMPT$ with many of the optimization suggested for 2PC.

References

1. Haritsa, J., Ramamritham, K.: The prompt real-time commit protocol. IEEE Transactions on Parallel and Distributed Systems Journal **11** (2000)
2. Moss, J.: Nested Transactions: an Approach to Reliable Distributed Computing. PhD thesis, University of Massachusetts (1986)
3. Reddy, P.K., Kitsuregawa, M.: Speculation based nested locking protocol to increase the concurrency of nested transactions. In: International Database Engineering and Application symposium, Yokohama, Japan, Addison (2000) 18–28
4. Huang, J.: On using priority inheritance in real-time databases. Real-Time Systems Journal **4** (1992)
5. Chen, H.R., Chin, Y.: An efficient real-time scheduler for nested transaction models. In: Ninth International Conference on Parallel and Distributed Systems(ICPADS'02). (2002)
6. Ramamritham, K.: Real-time databases. Distributed and Parallel Databases Journal **1** (1993)
7. John A. Stankovic, K.R., Towsley, D.: Scheduling in real-time transaction systems. In: 8th Real-Time Systems Symposium, Adition (1991)
8. Majed Abdouli, B.S., Amanton, L.: Scheduling distributed real-time nested transactions. In: IEEE ISORC, IEEE Computer Society (2005)
9. Chen, Y., Gruenwald, L.: Research issues for a real-time nested transaction. In: 2nd Workshop on Real-Time Applications, Addison (1994) 130–135
10. Dogdu, E.: Real-Time Databases: Extended Transactions and the utilization of Execution Histories. PhD thesis, Western Reserve University (1998)

Efficient Cluster Management Software Supporting High-Availability for Cluster DBMS

Jae-Woo Chang and Young-Chang Kim

Dept. of Computer Engineering, Chonbuk National University,
Chonju, Chonbuk 561-756, South Korea
{yckim, jwchang}@dblab.chonbuk.ac.kr

Abstract. A cluster management software is needed for monitoring and managing cluster systems or cluster DBMSs. The cluster system management software have been extensively studied by a couple of research groups, whereas the design of cluster DBMS management software is still in the early stages. In this paper, we design and implement a high-availability cluster management software that monitors the status of the nodes in a cluster DBMS, as well as the status of the various DBMS instances at a given node. This software enables users to visualize a single virtual system image and provides them with the status of all of the nodes and resources in a system. In addition, when combined with a load balancer, our cluster management software can improve the performance of a cluster DBMS and can overcome the limitations associated with the existing parallel DBMSs.

1 Introduction

Cluster systems, which are formed by connecting PCs and workstations using a high-speed network [1,2], are required to support 24 hours nonstop service for the Internet. Therefore, based on these cluster systems, much research is being conducted regarding the use of cluster DBMS that offer a mechanism to support high performance, high availability and high scalability [3,4,5]. To manage the cluster DBMS efficiently, a cluster DBMS management software is needed. First, such a software should enable users to visualize a cluster system consisting of multiple nodes as a single virtual system image. Secondly, by using a graphical user interface (GUI), the cluster DBMS management software should provide users with the status of all of the nodes in the system and all of the resources (i.e. CPU, memory, disk) at a given node. Thirdly, a load balancing function is needed to make all the nodes perform effectively by evenly distributing the users' requests. Finally, a fail-over technique is needed to support high availability when node failure occurs [6,7].

In this paper, we design and implement of a high-availability cluster management software, which monitors the status of all of the nodes in a cluster DBMS, as well as the status of the DBMS instances at a given node. This software enables users to recognize cluster system as a single virtual system image and provides them with the status of all of the nodes and resources in a system. In addition, when combined with a load balancer, our cluster management software can increase the performance of the cluster DBMS, as well as overcoming the limitations associated with existing parallel DBMSs.

The remainder of this paper is organized as follows. The next section introduces the OCMS as a related work on existing cluster DBMS management software. In section 3, we describe the design of our cluster DBMS management software. In section 4, we discuss performance analysis of our cluster DBMS management software. Finally, we draw our conclusions in section 5.

2 OCMS (Oracle Cluster Management System)

In this section, we introduce the existing management software; including OCMS(Oracle Cluster Management System) [8], which is a well known cluster DBMS management software. OCMS is included as a part of the Oracle8i Parallel Server product on Linux and provides cluster membership services, a global view of clusters, node monitoring and cluster reconfiguration. It consists of a watchdog daemon, node monitor and cluster manager. First, the watchdog daemon offers services to the cluster manager and to the node monitor. It makes use of the standard Linux watchdog timer to monitor selected system resources for the purpose of preventing database corruption. Secondly, the node monitor passes node-level cluster information to the cluster manager. It maintains a consistent view of the cluster and informs the cluster manager of the status of each of its local nodes. The node monitors also cooperates with the watchdog daemon to stop nodes which have an abnormally heavy load. Finally, the cluster manager passes instance-level cluster information to the Oracle instance. It maintains process-level status information for a cluster system. The cluster manager accepts the registration of Oracle instances to the cluster system and provides a consistent view of the Oracle instances.

3 Design of Cluster DBMS Management Software

A cluster-based DBMS consists of multiple server nodes and uses a shared disk. All of the server nodes are connected to each other by means of a high-speed gigabit Ethernet. The master node acts as both a gateway to the cluster system and a server node. The master node manages all of the information gathered from all of the server nodes and makes use of a Linux virtual server for its scheduling algorithms.

3.1 Cluster DBMS Management Software

To design a good monitoring software, we first minimize the objects to be monitored, so that we may avoid the additional load of monitoring itself. Secondly, we change the frequency of the dynamic monitoring, so that we may control the amount of network traffic which occurs whenever a node transmits its status information to the master node. The cluster DBMS management software consists of four components; the probe, handler, CM(Cluster Manager) and NM(Node Manager). In addition, the probes (or handlers) can be classified into the system, DB and LB(Load Balancer) ones. Figure 1 shows the components of the cluster DBMS management software.

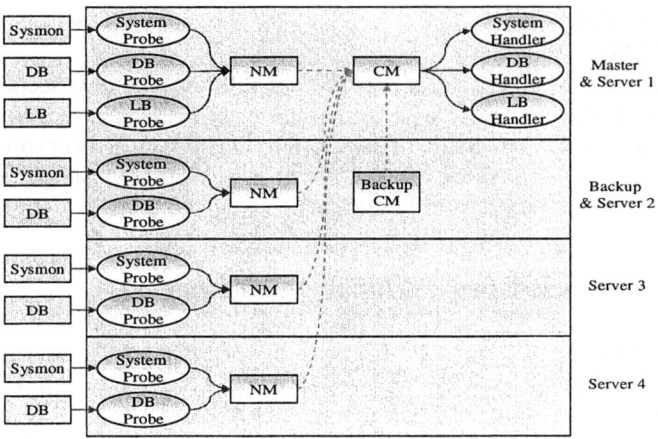

Fig. 1. Components of the cluster DBMS management software

Probes & Handlers. There are three kinds of probes and handlers: system, DB, LB. System, DB, LB probe monitor the status of system resources, DBMS instances, and load balancer, respectively. The probes also generate events according to the status of the system, DBMS instances, and load balancer. Then they send the events and the monitored information to the each handler. The handlers then store them into the service status table and perform a specific procedure for each type of event.

NM(Node Manager). The NM is a component which manages network communication with the CM at the master node. It transmits both the events generated by each probe and the monitored information to the CM at the master node. It also perceives the status of the CM. If the NM detects that the master node has encountered an error during the communication with the CM, the NM makes a connection with the backup node and transmits all of the available information to it. The NM also generates an event and transmits it to the CM, when it detects an error from any of the probes. If the NM does not receive a response from the CM during a defined interval, the NM considers that the CM has failed and makes a connection with a backup node.

CM(Cluster Manager). The CM running on the master node manages a service status table and perceives the status of the networks by sending a ping message to each server node through both the cluster network and the service network. It also analyzes the events received from each server node and transmits them to the handler to perform the appropriate procedure. If the CM detects that the NM has encountered an error, it restarts the NM. If the CM detects an error within the network by means of the ping message, it generates an event and transmits it to the corresponding service handler, requesting it to perform its recovery procedure. Because the CM runs on the master node, a failure of the master node causes the failure of the whole cluster DBMS. To solve the problem, the CM selects a backup node that can assume the role of the master node when the master node fails. The backup CM running on the backup node communicates regularly with the CM of the master node and keeps track of all of the information which the master CM manages in its local disk. If the master

CM perceives that the backup node has failed, it selects one of the remaining server nodes as the new backup node.

3.2 Recovery Procedures in the Case of Failure

The server nodes comprising a cluster system can be classified into the master node, the backup node, and the database server nodes. Also the statuses of the nodes can be classified into four types according to the status of both the service and cluster networks. First, if there is no failure in either network, the cluster system runs with a normal situation. Secondly, if the service network fails, a server node can communicate with the other nodes, but it cannot receive user requests or return the results to the user. Thirdly, if the cluster network fails, a given server node cannot communicate with the others and so the master node cannot distribute user requests to any of the server nodes. Finally, if both networks fail, the situation is considered as node failure since the nodes cannot function anymore. We can classify the status of the nodes according to the type of network failure as shown in Table 1.

Table 1. Classification of network failures

Status of node \ Network	Cluster network	Service network
Service network failure	O	X
Cluster network failure	X	O
Node failure	X	X

Master node failure. The master node sends a ping message to each server node in the cluster system and detects the failure of a node by analyzing the response of each server node. If the master node fails to receive any response from any of the server nodes, it regards the situation as cluster network failure. Meanwhile, when the backup node cannot communicate with the master node using a ping message, it regards the situation as master node failure. To prevent this situation, the backup node regularly checks for the failure of the master node by sending it a ping message periodically and itself becomes the new master node if the master node fails.

Backup node failure. If the backup node fails, the cluster DBMS management software perceives the failure and terminates the backup node. In addition, the backup node terminates the active DB, Sysmon, NM and its backup CM. During this time, the master node performs its recovery procedure, in order to remove the failed backup node from the list of available server nodes and to select a new backup node from amongst these remaining nodes.

DB server node failure. In the cluster system, the failure of a server node can cause the failure of the entire cluster DBMS, because the cluster DBMS uses a shared disk. Therefore, the cluster DBMS management software should perform a recovery procedure to preserve the integrity of the data by preventing the failed server node from accessing any data. First, if the service network fails, all of the server nodes should stop their transactions and perform their recovery procedure, since they cannot return the results of any user requests to the users. Secondly, if the cluster network fails, all of the server nodes should react in the same way as that described above, because they can neither receive any user requests from the master node, nor communicate with any of the other server nodes. Finally, if a server node fails, the cluster DBMS management software should remove this server node from the list of available nodes of the Linux virtual server and inform the other server nodes of the failure of this server node. During this time, the failed server node performs its recovery procedure to recovers its transactions and terminate its database. The other server nodes should recover the data of the failed node.

4 Performance Analysis of Cluster DBMS Management Software

In this section, we describe performance analysis of our cluster management software using iBASE/Cluster. For our implementation, we make use of the Redhat Linux 7.1 operating system and iBASE/Cluster DBMS [10]. Also, we make use of Linux virtual server as the load balancer [9].

We conduct the performance analysis of our cluster DBMS management software using iBASE/Cluster. We cannot directly compare the performance of our cluster DBMS management software with that of OCMS, because the latter is a cluster DBMS management software for Oracle, which does not support the recovery procedure for the iBASE/Cluster. According to [11], the sensing and recovery time of OCMS for Oracle is more than 20 seconds. Table 2 shows both the sensing time for the three types of node failures and the time required to perform the recovery procedure for each of them. It is shown from the result that OCMS is much worse than our cluster DBMS management software because OCMS has to wait until Oracle instance's reconfiguration.

Table 2. Sensing time and recovery time for the three types of node failure

	Sensing time	Recovery time
Master node failure	0.91	0.78
Backup node failure	0.89	0.51
Database server node failure	0.81	0.71

First, when the master node fails, the time required for sensing the master node failure is 0.91 seconds and the time required for performing the recovery procedure is 0.78 seconds. Since the backup node is assumed to have the role of the master node, it resets its virtual IP and creates a thread to monitor the network status of the nodes in the cluster system. Secondly, when the backup node fails, the time required for sensing the backup node failure is 0.89 seconds and the time required for performing the recovery procedure is 0.51 seconds. The new backup node creates a thread to monitor the master node and periodically receives the information of the service status table of the master node. Finally, when the database server node fails, the time required for sensing the database server node failure is 0.87 seconds and the time required for performing the recovery procedure is 0.71 seconds. If the database server node fails, the master node removes the server node from the available server list, in order that user requests are no longer transmitted to the failed server node.

5 Conclusions

In this paper, we described the design and implementation of cluster DBMS management software, which provides efficient cluster DBMS management. Our cluster DBMS management software monitors the system resources of all of the server nodes and detects the failure of any of these nodes. When such a failure occurs, our cluster DBMS management software performs its recovery procedure in order to provide normal service, regardless of the failure. Our cluster DBMS management software enables users to visualize a single virtual system image and provides them with the status of all of the nodes and. Finally, we analyzed performance of our cluster management software using iBASE/Cluster DBMS. The result shows that our cluster DBMS management software can support nonstop service by performing its recovery procedure even in the case where a node has failed.

References

1. J. M. Kim, K. W. On, D. H. Jee, "Clustering Computing Technique", 1999
2. Rajkumar Buyya , High Performance Cluster Computing Vol 1,2, Prentice Hall PTR, 1999.
3. High Performance Communication, http://www-csag.cs.uiuc.edu/projects/communication.html.
4. Michalewicz, Z.: Genetic Algorithms + Data Structures = Evolution Programs. 3rd edn. Springer-Verlag, Berlin Heidelberg New York (1996)
5. J. Y. Choi, S. C. Whang, "Software Tool for Cluster", Korea Information Science Society, Vol 18, No 3. pp40~47.
6. Gregory, F.Pfister, In Search of Clusters 2nd Edition, Prentice-Hall, 1998
7. Linux Clustering, http://dpnm.postech.ac.kr/cluster/index.htm
8. Oracle Corporation, "Oracle 8i Administrator's Reference Release3(8.1.7) for Linux Intel", chapter 7, Oracle Cluster Management Software, 2000
9. Linux Virtual Server, http://www.linuxvirtualserver.org
10. Hong-Yeon Kim, Ki-Sung Jin, June Kim, and Myung-Joon Kim, "iBASE/Cluster: Extending the BADA-IV for a Cluster Environment", Proceeding of the 18th Korea Information Processing Society Fall Conference, Vol. 9, No. 2, pp. 1769-1772, Nov. 2002.
11. Oracle Corporation, "Oracle 9i Administrator's Reference Release2(9.2.0.1.0)", chapter F, Oracle Cluster Management Software, 2002

Distributed Query Optimization in the Stack-Based Approach

Hanna Kozankiewicz[1], Krzysztof Stencel[2], and Kazimierz Subieta[1,3]

[1] Institute of Computer Sciences of the Polish Academy of Sciences, Warsaw, Poland
hanka@ipipan.waw.pl
[2] Institute of Informatics Warsaw University, Warsaw, Poland
stencel@mimuw.edu.pl
[3] Polish-Japanese Institute of Information Technology, Warsaw, Poland
subieta@pjwstk.edu.pl

Abstract. We consider query execution strategies for object-oriented distributed databases. There are several scenarios of query decomposition, assuming that the corresponding query language is fully compositional, i.e. allows decomposing queries into subqueries addressing local servers. Compositionality is a hard issue for known OO query languages such as OQL. Thus we use the Stack-Based Approach (SBA) and its query language SBQL, which is fully compositional and adequate for distributed query optimization. We show flexible methods based on decomposition of SBQL queries in a distributed environments. Decomposition can be static or dynamic, depending on the structure of the query and distribution of data. The paper presents only the main assumptions, which are now the subject of our study and implementation.

1 Introduction

Distributed query optimization has been thoroughly investigated for relational database systems (see surveys, e.g. [1,2,3,4,5]), but the methods are hard to generalize for object-oriented or XML-oriented databases. In order to execute a query in a distributed environment, one has to decompose it into queries which are to be run by a number of participating servers. Unfortunately, queries in the most widely known query languages OQL and XQuery are hard to decompose due to irregular, non-orthogonal syntax and imprecise semantics. Therefore, optimization methods may work for simple queries when data is horizontally fragmented and centralized or distributed indices are present. More advanced methods, however, present a hard issue.

A distributed query processor should minimize: (1) communications costs, (2) CPU costs and I/O costs at the client (the global application) and (3) CPU costs and I/O costs at servers. Furthermore, parallel computations at many servers should be exploited. If all computations are performed by clients (clients are *fat*), communication lines will be heavily loaded. If all computations are performed by servers (clients are *thin*), the most heavily loaded server can become the bottleneck.

Execution strategies can be global and local. A global strategy concerns a global application and a coordinating server, while local strategies are implemented by all participating sites. Global strategies should be supplemented by appropriate local strategies. For example, before application of any global strategy, local optimization e.g. using indices and rewriting (factoring out independent subqueries, pushing expensive operators down a syntactic tree, etc.) should be applied.

In this paper we consider query execution strategies for object-oriented distributed databases. We use the Stack-Based Approach (SBA) [6,7] as the framework, since it provides the fully compositional query language called SBQL (Stack-Based Query Language). Its flexibility allows designing many decomposition/optimization strategies which can also exercise opportunities of parallel computation. The paper deals with global optimization strategies. We discuss several optimization scenarios. In our discussion and simple examples we assume the most common case of horizontal data fragmentation and distributive queries.

The paper is organized as follows. In Section 2 we consider various execution strategies which can be applied in distributed environments. In Section 3 we sketch the Stack-Based Approach. In Section 4 we present optimization strategies in case of horizontally fragmented data. Section 5 concludes.

2 Query Execution Strategies

Popular techniques of distributed query processing are based on decomposition [8,9,10,11,12]. A global application (client) usually processes queries in the following way: (1) the query is parsed and (2) decomposed into subqueries to be sent to particular sites, (3) the order of execution (including parallelism) of these subqueries is established, (4) the subqueries are sent to servers, (5) results of these subqueries are collected and (6) combined by the client. There can be several strategies which can be used in this framework, as follows:

Total data shipping. The client requests all the data required by a query from servers and processes the query itself. Obviously, the strategy is conceptually simple and easy to implement, but causes a high or unacceptable communication overhead.

Static decomposition. The query is decomposed into a number of subqueries to be sent to servers simultaneously. Servers work in parallel. The client collects the results and combines them into a global result.

Dynamic decomposition. The client analyses a query and then generates a subquery to be sent to one of servers. Then collects the result from this server and uses this result to generate another subquery to be sent to next server. The result returned by the second server is collected, and so on. Servers do not work in parallel.

Dynamic decomposition with data shipped to servers. Subqueries sent to servers are accompanied with a data collection, which facilitates better optimization at these servers. Semi-joins [13] are a well-know application of this strategy.

Hybrid decomposition exploits the advantages of both dynamic and static decomposition as well as the possibility to ship data which facilitates local optimizations. At the beginning several subqueries may be generated and executed simultaneously by a number servers. The collected results are used to generate subsequent batch of subqueries and data fragments. The process is repeated until the final result of the query is computed.

3 Stack-Based Approach (SBA)

Query optimization is efficient only if it is possible to verify that the query before optimization is equivalent to the query after optimization for all database states. Therefore, we must have a precise definition of the query language semantics. Specification of semantics is a weak point of current query languages such as OQL and XQuery. They present semantic through rough explanations and examples, which leave a lot of room for different interpretations. In contrast, SBA and its query language SBQL have fully formal, precise and complete semantics. In SBA a query language is treated as a kind of a programming language. Thus evaluation of queries is based on mechanisms well known from programming languages. The approach precisely determines the semantics of all query operators (including hard non-algebraic ones, such as dependent joins, quantifiers or transitive closures), their relationships with object-oriented concepts, constructs of imperative programming, and programming abstractions, including procedures, functional procedures, methods, views, etc. The stack-based semantics causes that all the abstractions can be recursive.

In the Stack Based Approach four data store models are defined, with increasing functionality and complexity. The M0 model described in [6] is the simplest data store model. In M0 objects can be nested (with no limitations on nesting levels) and can be connected with other objects by links. M0 covers relational and XML-oriented structures. It can be easily extended [7] to comply with more complex models which include classes and static inheritance (M1), dynamic object roles and dynamic inheritance (M2), encapsulation (M3) and other features of object-oriented databases.

The basis of SBA is the environment stack. It is the most basic auxiliary data structure in programming languages. It supports the abstraction principle, which allows the programmer to consider the currently written piece of code to be independent of the context of its possible uses. SBA respects the naming-scoping-binding discipline, which means that each name occurring in a query is bound to a run-time entity (an object, an attribute, a method, a parameter, etc.) according to the scope of its name.

4 Distributed Queries to Horizontally Fragmented Data

Due to compositionality of SBQL queries can be easily decomposed into subqueries, up to atomic ones (single names or values). This property can be used to decompose queries into subqueries addressing particular servers. For instance, if name N occurs in a query and we know that data named N is horizontally fragmented among servers A and B, then we can substitute name N by $(N_A \cup N_B)$, where N_A is to be bound on server A and N_B on the server B. Such decomposition works for a general case, but it is easier for horizontally fragmented data (most usual case) and queries distributive with respect to set/bag union. If a query is not distributive, a dynamic schema (similar to semi-joins) can be used. If data is horizontally fragmented, a query can be executed in a simplest variant of the static decomposition scenario, which assumes sending the query in parallel to all servers and then summing their results. This requires, however, checking if the query is distributive, i.e. the union operator, as shown above, must be recursively pushed to the root of the query syntactic tree.

A distributed optimizer must have access to a global schema which describes structure and location of distributed objects. The global schema models and reflects the data store and itself it looks like a data store. Some nodes of the global schema have the attribute *loc*. Its value indicates names of servers which store data described by a node. If the value of a node's attribute *loc* contains a number of servers, the collection of objects represented by this node is fragmented among the indicated servers. Only root object nodes have the attribute *loc*, since we assume that an entire object is always stored on the same server (no vertical fragmentation). Pointers from objects to objects may cross the boundaries of servers. A thorough description of database schemata for SBQL can be found in [14]. Due to space limit in this paper we use a very simple schema with a collection of objects *Emp*, which are fragmented and stored in Milano and Napoli.

4.1 Analyzing Queries

1. A query is statically analyzed against the global schema in order to determine its nodes where particular names occurring in the query are to be bound. The method of the analysis involves static environment and result stacks, see e.g. [7, 15]. During this process, if a name is to be bound to a node of the global schema which contains the attribute *loc*, the name is marked with the value of this attribute. In this way we involve names of local servers into the query syntactic tree.
2. The distributiveness of query operators is employed. All nodes of the syntactic tree associated with servers are marked with the flag "*distributive*". Next, if a node marked *distributive* is the left argument of a distributive operator, the whole subquery rooted at this operator is also marked distributive. This process is repeated up the syntax tree of the query until no more markings are possible. If this process meets a non-distributive operator (e.g. an aggregate function), it is terminated.
3. Each node of the syntax tree marked *distributive* and having associated names of servers is split by the union operator into as many nodes as servers. Then, the splitting is continued on bigger and bigger query syntax sub-trees, till its root or a non-distributive node.
4. In this way the query is decomposed into subqueries addressing single servers. Depending on interdependencies of the subqueries they can be executed sequentially or simultaneously. If a subquery q_1 is parameterized by subquery q_2, q_2 must be executed before q_1. If they are independent, they can be executed in parallel.

4.2 Example of Static Decomposition

Majority of queries are simply distributive with respect to horizontal fragmentation. They can be statically decomposed into subqueries addressed to a number of servers. The results collected from the servers allow merging the final result out of them. Consider the following query:

$$Emp \, . \, name \quad (1)$$

The name *Emp* is marked distributive and is assigned to servers Milano, Napoli:

$$Emp_{\{Milano, Napoli\}} \, . \, name \quad (2)$$

and equivalently

$$(Emp_{Milano} \cup Emp_{Napoli}) . name \tag{3}$$

Since the dot operator is distributive with respect to horizontal fragmentation, we can push $\cup$ up the syntax tree of the query and eventually receive:

$$(Emp.name)_{Milano} \cup (Emp.name)_{Napoli} \tag{4}$$

This query can be executed in parallel on different servers and their results are eventually merged by the client.

4.3 Example of Dynamic Decomposition

Sometimes a query contains a subquery being its parameter. Consider the following query (find employees who earn more than Blake):

Emp **where** *sal* > ((*Emp* **where** *name* = "Blake") . *sal*) (5)

In this query the right argument of the outer selection contains a so-called independent subquery that can be exevuted independently of the outer one. Each name *Emp* can be decorated by the names of servers:

$Emp_{\{Milano,Napoli\}}$ **where** *sal* > (($Emp_{\{Milano,Napoli\}}$ **where** *name* = "Blake") . *sal*) (6)

After this operation we can split the query syntax tree, as shown in the previous example. Because we expect that the Blake's salary is a value that can be found on one of the servers, the final result of the decomposition can be presented as follows:

$P := ((Emp$ **where** $name = $ "Blake" $) . sal)_{Milano}$

if P **is null then** $P := ((Emp$ **where** $name = $ "Blake" $) . sal)_{Napoli}$ (7)

$(Emp$ **where** $sal > P)_{Milano} \cup (Emp$ **where** $sal > P)_{Napoli}$

The decomposition employs distributivity of the **where** operator. In effect, the inner subquery is executed in some order and then the outer query is executed in parallel. The decomposition (the order of execution of subqueries) may involve the cost model estimating the execution cost of subqueries on particular servers, e.g. we can check whether to look for the Blake's salary first in Milano and then in Napoli, or v/v.

5 Conclusions

In this paper we have proposed optimization techniques for distributed object databases. We have shown that if the query language is fully compositional, it facilitates many decomposition methods (either static or dynamic). The Stack-Based Approach (SBA) and the query language SBQL have the unique query decomposition potential. Unlike OQL and XQuery, each fragment of an SBQL, even an atomic name or value, is also a query. Therefore, the optimizer has full freedom in partitioning a query into subqueries sent to remote servers. The decomposition can support various scenarios of distributed query processing, including static and dynamic decomposition.

We have discussed optimization techniques for the data grid architecture originally proposed in [16]. This architecture is currently being developed on top of our object oriented DBMS ODRA (Object Database for Rapid Application development) devoted to grid applications, composing web services and distributed web content management.

References

1. C.T.Yu, C.C.Chang: Distributed Query Processing. ACM Comput. Surv. 16(4): 399-433 (1984)
2. S.Ceri, G.Pelagatti: Distributed Databases: Principles and Systems McGraw-Hill Book Company 1984
3. M.T.Özsu, P.Valduriez: Principles of Distributed Database Systems, Second Edition, Prentice-Hall 1999
4. D.Kossmann: The State of the Art in Distributed Query Processing. ACM Comput. Surv. 32(4): 422-469 (2000)
5. C.T.Yu, W.Meng. Principles of Database Query Processing for Advanced Applications, Morgan Kaufmann Publishers, 1998
6. K.Subieta, Y.Kambayashi, and J.Leszczyłowski. Procedures in Object-Oriented Query Languages. Proc. VLDB Conf., Morgan Kaufmann, 182-193, 1995
7. K.Subieta. Theory and Construction of Object-Oriented Query Languages. Polish-Japanese Institute of Information Technology Editors, Warsaw 2004, 522 pages
8. V.Josifovski, T.Risch: Query Decomposition for a Distributed Object-Oriented Mediator System. Distributed and Parallel Databases 11(3): 307-336 (2002)
9. D.Suciu: Query Decomposition and View Maintenance for Query Languages for Unstructured Data. VLDB 1996: 227-238
10. K.Evrendilek, A.Dogac: Query Decomposition, Optimization and Processing in Multidatabase Systems. NGITS 1995
11. E.Leclercq, M.Savonnet, M.-N.Terrasse, K.Yétongnon: Objekt Clustering Methods and a Query Decomposition Strategy for Distributed Objekt-Based Information Systems. DEXA 1999: 781-790
12. E.Bertino: Query Decomposition in an Object-Oriented Database System Distributed on a Local Area Network. RIDE-DOM 1995: 2-9
13. P.A.Bernstein, N.Goodman, E.Wong, C.L.Reeve, J.B.Rothnie Jr.: Query Processing in a System for Distributed Databases (SDD-1). ACM Trans. Database Syst. 6(4): 602-625 (1981)
14. R.Hryniów, M.Lentner, K.Stencel, K.Subieta: Types and Type Checking in Stack-Based Query Languages, Institute of Computer Science, Polish Academy of Sciences, Report 984, March 2005
15. J.Płodzień, A.Kraken: Object Query Optimization through Detecting Independent Subqueries. Inf. Syst. 25(8): 467-490 (2000)
16. H.Kozankiewicz, K.Stencel, K.Subieta. Implementation of Federated Databases through Updatable Views. Proc. of the European Grid Conference, Amsterdam, The Netherlands, Springer LNCS 3470: 610-620 (2005)

Parallelization of Multiple Genome Alignment

Yiming Li[1,2] and Cheng-Kai Chen[2,3]

[1] Department of Communication Engineering
[2] Microelectronics and Information Systems Research Center
[3] Department of Computer and Information Science,
National Chiao Tung University, Hsinchu City, Hsinchu 300, Taiwan
ymli@faculty.nctu.edu.tw

Abstract. In this work, we implement a genome alignment system which applies parallelization schemes to the ClustalW algorithm and the interface of database querying. Parallel construction of the distance matrices and parallelization of progressive alignment in the ClustalW algorithm are performed on PC-based Linux cluster with message-passing interface libraries. Achieved experiments show good speedup and significant parallel performance.

1 Introduction

Multiple sequence alignment searches the best configuration of similar subsequences between two or more sequences. This NP-hard problem has been of great interests in these years [1,2,3,4]. However, they still suffer the issue of massive computation. Clustal is currently one of the most popular multiple alignment algorithms [5]. In the alignments, all pairwise similarity scores are calculated firstly, and the distance matrix is constructed. According to the distance matrix, the guide tree is built in the second step. Progressive multiple alignments are performed by sequentially aligning groups of sequences considering their branching order in the guide tree. There are two variants, one is a local alignment algorithm, ClustalW, and the other is a global alignment method, ClustalV.

PC- or workstation-based computing technique provides a computationally cost-effective way to genome alignment [4,6,7]. In this work, we implement a parallel computing method to accelerate the alignment process, where the message passing interface (MPI) libraries are used to perform the data communication on a PC-based Linux cluster. PC-based Linux cluster for parallel semiconductor device simulation has been developed in our earlier work [8]. Alignment process consists of three parallelization steps. First, parallel construction of the distance matrices in the ClustalW algorithm is performed on each processor. Parallelization of the progressive alignment is then carried out on each processor. The interface of database querying is also parallelized. The parallelization of construction of distance matrices has the highest parallel performance. Achieved examples show good speedup and parallel performance.

The paper is organized as follows. In Sec. 2, we state the algorithm of multiple genome alignment. In Sec. 3, we describe the parallelization methods. Achieved results are discussed in the section 4. Finally, we draw the conclusions.

2 Multiple Sequence Alignment

Sequence alignments can be categorized into two parts, one is the so-called global alignment and the other is the local alignment. When aligning sequences, the global alignment adds gaps in all sequences to maintain the same length, while the local alignment produces only the similar parts. In this section, the adopted multiple sequence alignment algorithm and its basis, the pairwise sequence alignment algorithm, are introduced.

2.1 The Pairwise Sequence Alignment Algorithm

The pairwise sequence alignment contributes the optimal alignment for two sequences $S1$ and $S2$. To perform the pairwise alignment, the score function $S(x_i, y_j)$ describing the similarity between the individuals of $S1$ and $S2$ must be determined firstly, where the x_i and y_i indicate the i-th character of the sequences x and y, respectively. If $x_i = y_j$, the score function $S(x_i, y_j) = 1$, else $S(x_i, y_j) = 0$.

In addition, we define the gap penalty g as a negative value. When procedures above are settled, the matrix M for alignment is built with the height and width according to the length of two sequences plus one, i.e., the width equals the length of $S2 + 1$ and the height equals the length of $S1 + 1$. The distance matrix M is given by

Begin Construct distance matrix M
  If $i = j = 0$, then $M(i, j) = 1$
  Else if $j = 0$, then $M(i, j) = i * g$
  Else if $i = 0$, then $M(i, j) = j * g$
  Else Max$\{M(i, j-1) + g, M(i-1, j) + g, M(i-1, j-1) + S(x_i, y_j)\}$
End Construct distance matrix M

The procedure is started with the top-left element, and each element references one of the neighbors, the upper-left, the left, or the upper element. Once the matrix M is built, the alignment is retrieved by back-tracking from the bottom right element along with the reference path. With this algorithm, the optimal alignment of two sequences can be retrieved shown below.

Begin Back-tracking algorithm
  $i =$ length of $S1$ and $j =$ length of $S2$
  while($i \neq 0$ or $j \neq 0$)
    If $M(i, j)$ references $M(i, j-1)$
      Keep y_i in $S2$ and add gap in $S1$; $j = j + 1$
    Else if $M(i, j)$ references $M(i-1, j)$
      Keep x_j in $S1$ and add gap in $S2$; $i = i + 1$
    Else
      Keep both x_i and y_i; $j - j + 1$; $i - i + 1$
  End while
End Back-tracking algorithm

2.2 The Multiple Sequence Alignment Algorithm

As shown in Fig. 1, the ClustalW method includes building distance matrix, constructing guide tree, and performing progressive alignment [9]. The distance matrix is built by series of pairwise sequence alignments shown in Figs. 1b and 1c. It requires $n(n-1)/2$ pairwise alignments to align n sequences. For each pairwise alignment, the distance $D(i,j)$ between sequences can be obtained by the percentage of the equivalent non-gap characters. Guide tree indicating order of alignment in the next step is then constructed after obtaining the distance matrix. The guide tree traverses the distance matrix from the highest similarity to lower ones until all sequences are included in the guide tree. Finally, the sequences are aligned by the order of guide tree instructs shown in Fig. 1d.

Progressive alignment is then performed according to the order indicated by the guide tree. Shown in Fig. 1, $S2$ and $S4$ aligned and generate the aligned $S2'$ and $S4'$. $S1$ is then aligned with $S4'$ and so on. Shown in Fig. 1e, the sequences are aligned. In the step 3, once certain gaps are generated during the alignment of aligned sequence and unaligned sequence, these gaps are added to all other aligned sequences.

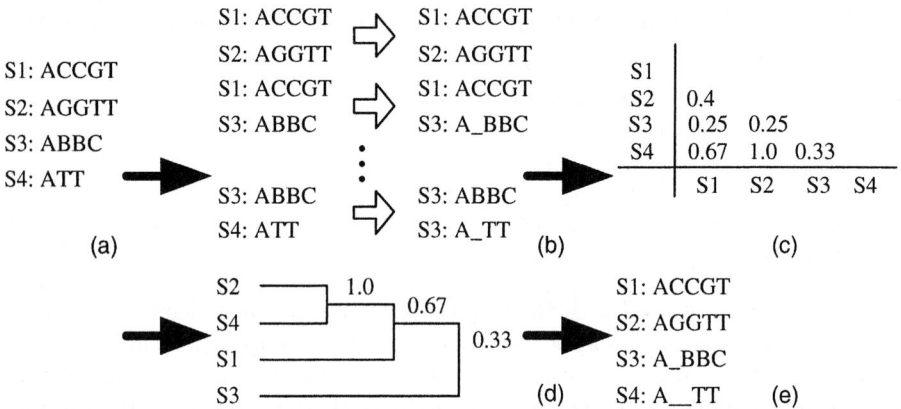

Fig. 1. Steps in the ClustalW algorithm are (a) sequences to be aligned; (b) perform pairwise alignment for each combination; (c) build the distance matrix; (d) construct the guide tree; and (e) obtain the aligned result

3 Parallelization Methodology

The parallelization technique includes two parts. One is to decompose the inner multiple alignment algorithm and the other is to keep the load balancing among the distributed processors. Parallelization is running on a PC-based Linux cluster system [8]. To show the performance of the multiple sequence alignment, the inner decomposition of the algorithm is necessary. The ClustalW has three main procedures: construct the distance matrix, construct the guide tree, and perform the progressive alignment. According to our experience, the first procedure dominates the whole computational time (more than 90%). Therefore, in order to perform the pairwise alignment for obtaining the distance matrix, our parallelization scheme divides the whole workload into many pieces. We note that the pairwise alignment is highly computational independent with respect to other alignments, so a high parallelization benchmark can be achieved in this work. If there are n sequences to be aligned, each P processor performs $n(n-1)/2P$ aligned operations. In the second procedure, we adopt the neighbor-joining method [10] to construct the guide tree effectively. Compared with the total computing time, the computing time in this procedure is insignificant(about 2% of the total computing time), where the computing cost may come from the communication of the parallelization scheme. Whether perform the parallelization of this procedure or not, it should not significantly affect the overall speedup. We do parallelize the final procedure of the progressive alignment. The parallelization efficiency is highly dependent on the constructed guide tree. The quality of the guide tree depends upon the sequences to be aligned, but an ideal speedup approaching to $N/logN$ can be achieved through an optimal well-balanced guide tree.

We further present a task-distributed parallelization scheme for the genome alignment database. The scheme provides a flexible environment to keep the load-balancing for each processor in the database of huge genome. When user submits a sequence alignment request, the system has to perform large-scale comparison of entire organisms. The proposed system will divide the entire organism database into many equal-sized pieces and dispatch the alignment job to each processor, thus the load-balancing is maintained.

4 Results and Discussion

Three experiments are designed and performed to show the parallel performance of the established parallel system. In the experiment 1, we estimate speedup versus the number of processors for different configurations of multiple sequences alignment. Figure 2a shows the average benchmark results, where the numbers of aligned sequences are fixed. The speedup is proportional to the sequences length in the best and worst cases. Running on 16 processors, for the best case (100 seqs. with 1600 chars.), the speedup is 14.95. For the worst case (100 seqs. with 200 chars.), the speedup is 7.27 running. It states the overhead cost of the message passing communication becomes an encumbrance when performing short sequences alignment in parallel. The benchmark results for the similar case with fixing the numbers of characters in each sequence are shown in Fig. 2b. It is found that the excellent results are still confirmed.

In the experiment 2, the parallel performance by exploring realistic genome data sets is presented. Two genome data sets, the "ecoli.nt" and the "drosoph.nt" from NCBI [11] are investigated. For both the testing cases, there are two randomly generated sequences; each sequence contains 100 characters for performing the multiple sequence alignment with the realistic genome sequences. Figure 3a shows the achieved speedup and efficiency verse the number of processors. For

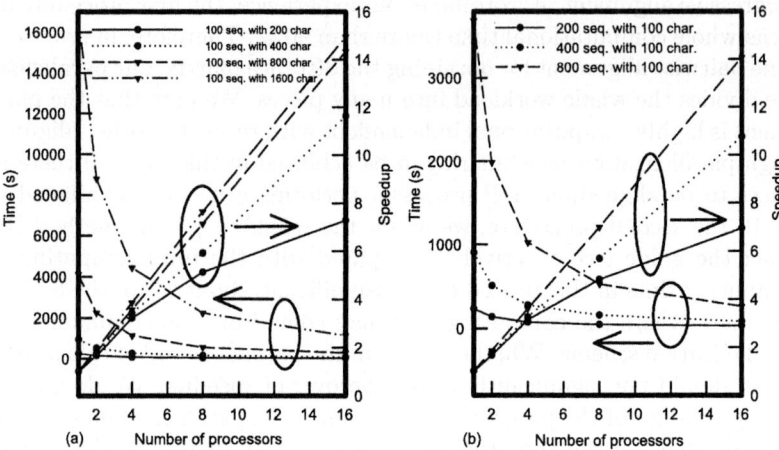

Fig. 2. CPU time vsersus the numbers of processors, where (a) the number of aligned sequences and (b) characters in each sequence are fixed

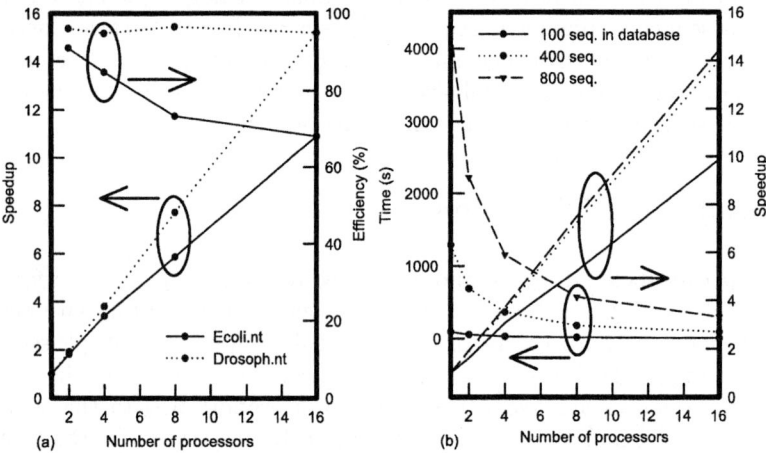

Fig. 3. Speedup and efficiency (a) of the parallel system with respect to different data sets and (b) for performing 10 sequences with 100 characters multiple sequences alignment with three databases

the genome data set "drosoph.nt", the speedup is larger than 15 and the efficiency is larger than 90% on 16-processor cluster. In the experiment 3, benefits of the task- distributed alignment system is examined. Figure 3b shows the comparison of the speedup and efficiency for performing 10 sequences with 100 characters multiple sequences alignment among three databases with different sizes 100, 400, and 800 sequences, respectively. To perform the multiple alignment procedure simultaneously, the proposed system divides whole database into several small pieces and dispatches them to each node. Results show that such domain decomposition schemes have good efficiency.

5 Conclusions

In this paper, a parallel computing system for multiple sequences alignment has been implemented on a 16-processor PC-based Linux cluster. Parallelizing the interface of database querying, the construction of the distance matrices, and progressive alignment in the ClustalW algorithm has been investigated. Results of different experiments have demonstrated good speedup and efficiency.

Acknowledgments

This work is supported in part by the National Science Council of Taiwan under contracts NSC-93-2215-E-429-008 and NSC-94-2752-E-009-003-PAE, and the Ministry of Economic Affairs, Taiwan under contract 93-EC-17-A-07-S1-0011.

References

1. Vingron, M., Argos, P.: A fast and sensitive multiple sequence alignment algorithm. Comput. Appl. Biosci. **5** (1989) 115-121.
2. Gupta, S. K., Kececioglu, J. D., Schäffer, A. A.: Improving the Practical Time and Space Efficiency of the Shortest-Paths Approach to Sum-of-Pairs Multiple Sequence Alignment. J. Comput. Bio. **2** (1995) 459-472.
3. Zhang, C., Wong, A.K.C. Toward efficient multiple molecular sequence alignment: a system of genetic algorithm and dynamic programming. IEEE Trans. Systems, Man and Cybernetics, Part B. **27** (1997) 918-932.
4. Luo, J., Ahmad, I., Ahmed, M., Paul, R.: Parallel Multiple Sequence Alignment with Dynamic Scheduling. IEEE Int. Conf. Info. Tech.: Coding and Computing. **1** (2005)8-13.
5. Higgins, D. G., Sharp, P. M.: CLUSTAL: a Package for Performing Multiple Sequence Alignment in a Microcomputer. Gene. **73** (1988) 237-244.
6. Kleinjung, J., Douglas, N., Herringa, J.: Parallelized multiple alignment. Bioinformatics. **18** (2002) 1270-1271.
7. Li, K.-B.: ClustalW-MPI: ClustalW analysis using distributed and parallel computing. Bioinformatics. **19** (2003) 1585-1586.
8. Li, Y. Sze, S. M., Chao, T.-S.: A Practical Implementation of Parallel Dynamic Load Balancing for Adaptive Computing in VLSI Device Simulation. Engineering with Computers **18** (2002) 124-137.
9. Feng, D. F., Doolittle, R. F.: Progressive sequence alignment as a prerequisite to correct phylogenetic trees. J. Mol. Evol. **25** (1987) 351-360.
10. Saitou, N. Nei, M., The neighbor-joining method: a new method for reconstructing phylogenetic trees. J. Mol. Evol. **4** (1987) 406-425.
11. National Center for Biotechnology Information. [Online]. Available: http://www.ncbi.nlm.nih.gov/

Detonation Structure Simulation with AMROC

Ralf Deiterding

California Institute of Technology,
1200 East California Blvd., Mail-Code 158-78, Pasadena, CA 91125
ralf@cacr.caltech.edu

Abstract. Numerical simulations can be the key to the thorough understanding of the multi-dimensional nature of transient detonation waves. But the accurate approximation of realistic detonations is extremely demanding, because a wide range of different scales needs to be resolved. In this paper, we summarize our successful efforts in simulating multi-dimensional detonations with detailed and highly stiff chemical kinetics on recent parallel machines with distributed memory, especially on clusters of standard personal computers. We explain the design of AMROC, a freely available dimension-independent mesh adaptation framework for time-explicit Cartesian finite volume methods on distributed memory machines, and discuss the locality-preserving rigorous domain decomposition technique it employs. The framework provides a generic implementation of the blockstructured adaptive mesh refinement algorithm after Berger and Collela designed especially for the solution of hyperbolic fluid flow problems on logically rectangular grids. The ghost fluid approach is integrated into the refinement algorithm to allow for embedded non-Cartesian boundaries represented implicitly by additional level-set variables. Two- and three-dimensional simulations of regular cellular detonation structure in purely Cartesian geometry and a two-dimensional detonation propagating through a smooth 60 degree pipe bend are presented. Briefly, the employed upwind scheme and the treatment of the non-equilibrium reaction terms are sketched.

1 Introduction

Reacting flows have been a topic of on-going research since more than hundred years. The interaction between hydrodynamic flow and chemical kinetics can be extremely complex and even today many phenomena are not very well understood. One of these phenomena is the propagation of detonation waves in gaseous media. Detonations are shock-induced combustion waves that internally consist of a discontinuous hydrodynamic shock wave followed by a smooth region of decaying combustion. In a self-sustaining detonation, shock and reaction zone propagate essentially with an identical supersonic speed between 1000 and 2000 m/s that is approximated to good accuracy by the classical Chapman-Jouguet (CJ) theory, cf. [26]. But up to now, no theory exists that describes the internal flow structure satisfactory. The Zel'dovich-von Neumann-Döring (ZND) theory is widely believed to reproduce the one-dimensional detonation structure correctly, but experiments [21] uncovered that the reduction to one space dimension is not even justified in long combustion devices. It was found that detonation waves usually exhibit non-neglectable instationary multi-dimensional sub-structures in the millimeter range and do not remain exactly planar. The multi-dimensional instability manifests itself in instationary shock waves propagating perpendicular to the detonation front. A complex flow pattern is formed around each *triple point*, where the detonation front is intersected by a transverse shock. Pressure and temperature are increased enormously leading to a drastic enhancement of the chemical reaction. Hence, the accurate representation of triple points is essential for safety analysis, but also in technical applications, e.g. in the pulse detonation engine. Some particular mixtures, e.g. low-pressure hydrogen-oxygen with high argon diluent, are known to produce very regular triple point movements. The triple point trajectories form regular "fish-scale" patterns, so called detonation cells, with a characteristic length L and width λ (compare left sketch of Fig. 1).

Figure 1 displays the hydrodynamic flow pattern of a detonation with regular cellular structure as it is known since the early 1970s, cf. [21]. The right sketch shows the periodic wave configuration around a triple point in detail. It consists of a Mach reflection, a flow pattern well-known from non-reactive supersonic hydrodynamics [3]. The undisturbed detonation front is called the incident shock, while the transverse wave takes the role of the reflected shock. The triple point is driven forward by a strong shock wave, called Mach stem. Mach stem and reflected shock enclose the slip line, the contact discontinuity. The shock front inside the detonation cell travels as two Mach stems from point A to the line BC, see left graphic of Fig. 1. In the points B and C the triple point configuration is inverted nearly instantaneously and the front in the cell becomes the incident shock. Along the symmetry line AD the change is smooth and the shock strength decreases continuously. In D the two triple points merge exactly in a single point. The incident shock vanishes completely and the slip line, which was necessary for a stable triple point configuration between Mach stem and incident shock, is torn off and remains behind. Two new triple points with two new slip lines develop immediately after.

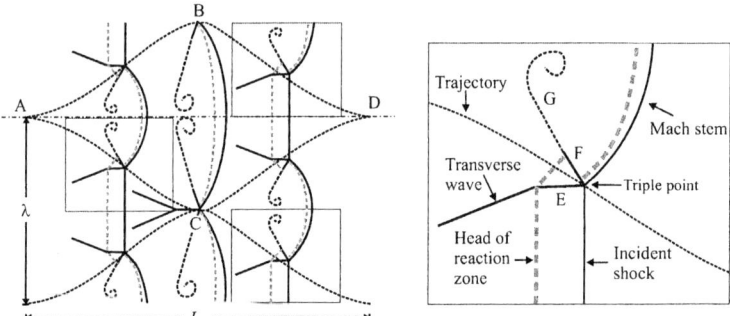

Fig. 1. Left: regular detonation structure at three different time steps on triple point trajectories, right: enlargement of a periodical triple point configuration. E: reflected shock, F: slip line, G: diffusive extension of slip line with flow vortex.

2 Governing Equations

The appropriate model for detonation propagation in premixed gases with realistic chemistry are the inviscid Euler equations for multiple thermally perfect species with reactive source terms [26]. These equations form a system of inhomogeneous hyperbolic conservation laws that reads

$$\begin{aligned}
\partial_t \rho_i + \nabla \cdot (\rho_i \boldsymbol{u}) &= W_i \dot{\omega}_i, \quad i = 1, \ldots, K, \\
\partial_t (\rho \boldsymbol{u}) + \nabla \cdot (\rho \boldsymbol{u} \otimes \boldsymbol{u}) + \nabla p &= 0, \\
\partial_t (\rho E) + \nabla \cdot ((\rho E + p) \boldsymbol{u}) &= 0.
\end{aligned} \quad (1)$$

Herein, ρ_i denotes the partial density of the ith species and $\rho = \sum_{i=1}^{K} \rho_i$ is the total density. The ratios $Y_i = \rho_i/\rho$ are called mass fractions. We denote the velocity vector by $\boldsymbol{u}$ and E is the specific total energy. We assume that all species are ideal gases in thermal equilibrium and the hydrostatic pressure p is given as the *sum* of the partial pressures $p_i = \mathcal{R} T \rho_i / W_i$ with $\mathcal{R}$ denoting the universal gas constant and W_i the molecular weight, respectively. The evaluation of the last equation necessitates the previous calculation of the temperature T. As detailed chemical kinetics typically require species with *temperature-dependent* material properties, each evaluation of T involves the approximative solution of an implicit equation by Newton iteration [4]. The chemical production rate for each species is derived from a reaction mechanism of J chemical reactions as

$$\dot{\omega}_i = \sum_{j=1}^{J} (\nu_{ji}^r - \nu_{ji}^f) \left[k_j^f \prod_{l=1}^{K} \left(\frac{\rho_l}{W_l} \right)^{\nu_{jl}^f} - k_j^r \prod_{l=1}^{K} \left(\frac{\rho_l}{W_l} \right)^{\nu_{jl}^r} \right], \quad i = 1, \ldots, K, \quad (2)$$

with $\nu_{ji}^{f/r}$ denoting the forward and backward stoichiometric coefficients of the ith species in the jth reaction. The rate expressions $k_j^{f/r}(T)$ are calculated by an Arrhenius law, cf. [26].

3 Numerical Methods

We use the time-operator splitting approach or method of fractional steps to decouple hydrodynamic transport and chemical reaction numerically. This technique is most frequently used for time-dependent reactive flow computations. The *homogeneous* Euler equations and the usually stiff system of ordinary differential equations

$$\partial_t \rho_i = W_i \dot{\omega}_i (\rho_1, \ldots, \rho_K, T), \quad i = 1, \ldots, K \quad (3)$$

are integrated successively with the data from the preceding step as initial condition. The advantage of this approach is that a globally coupled implicit problem is avoided and a time-implicit discretization, which accounts for the stiffness of the reaction terms, needs to be applied only *local* in each finite volume cell. We use a semi-implicit Rosenbrock-Wanner method [10] to integrate

(3). Temperature-dependent material properties are derived from look-up tables that are constructed during start-up of the computational code. The expensive reaction rate expressions (2) are evaluated by a mechanism-specific Fortran-77 function, which is produced by a source code generator on top of the Chemkin-II library [11] in advance. The code generator implements the reaction rate formulas without any loops and inserts constants like $\nu_{ji}^{f/r}$ directly into the code.

As detonations involve supersonic shock waves we use a finite volume discretization that achieves a proper upwinding in all characteristic fields. The scheme utilizes a quasi-one-dimensional approximate Riemann solver of Roe-type [8] and is extended to multiple space dimensions via the method of fractional steps, cf. [22]. To circumvent the intrinsic problem of unphysical total densities and internal energies near vacuum due to the Roe linearization, cf. [6], the scheme has the possibility to switch to the simple, but extremely robust Harten-Lax-Van Leer (HLL) Riemann solver. Negative mass fraction values are avoided by a numerical flux modification proposed by Larrouturou [12]. Finally, the occurrence of the disastrous carbuncle phenomena, a multi-dimensional numerical crossflow instability that destroys every simulation of strong grid-aligned shocks or detonation waves completely [18], is prevented by introducing a small amount of additional numerical viscosity in a multi-dimensional way [20]. A detailed derivation of the entire Roe-HLL scheme including all necessary modifications can be found in [4]. This hybrid Riemann solver is extended to a second-order accurate method with the MUSCL-Hancock variable extrapolation technique by Van Leer [22].

Higher order shock-capturing finite volume schemes are most efficient on rectangular Cartesian grids. In order to consider complex moving boundaries within the scheme outlined above we use some of the finite volume cells as ghost cells to enforce immersed boundary conditions [7]. Their values are set immediately before the original numerical update to model moving embedded walls. The boundary geometry is mapped onto the Cartesian mesh by employing a scalar level set function φ that stores the signed distance to the boundary surface and allows the efficient evaluation of the boundary outer normal in every mesh point as $n = \nabla\varphi/|\nabla\varphi|$ [15]. A cell is considered to be a valid fluid cell in the interior, if the distance in the cell *midpoint* is positive and is treated as exterior otherwise. The numerical stencil by itself is not modified, which causes a slight diffusion of the boundary location throughout the method and results in an overall non-conservative scheme. We alleviate such errors and the unavoidable staircase approximation of the boundary with this approach by using the dynamic mesh adaptation technique described in Sec. 4 to also refine the Cartesian mesh appropriately along the boundary.

For the inviscid Euler equations (1) the boundary condition at a rigid wall moving with velocity w is $u \cdot n = w \cdot n$. Enforcing the latter with ghost cells, in which the discrete values are located in the cell centers, involves the mirroring of the primitive values ρ_i, u, p across the embedded boundary. The normal velocity in the ghost cells is set to $(2w \cdot n - u \cdot n)n$, while the mirrored tangential velocity remains unmodified. The construction of the velocity vector within the ghost cells therefore reads $u' = (2w \cdot n - u \cdot n)n + (u \cdot t)t = 2((w - u) \cdot n) n + u$ with t denoting the boundary tangential. The utilization of mirrored cell values in a ghost cell center x requires the calculation of spatially interpolated values in the point

$$\tilde{x} = x + 2\varphi n \qquad (4)$$

from neighboring interior cells. For instance in two space dimensions, we employ a bilinear interpolation between usually four adjacent cell values, but directly near the boundary the number of interpolants needs to be decreased, cf. Fig. 2. It has to be underlined that an extrapolation in such situations is inappropriate for hyperbolic problems with discontinuities like detonation waves that necessarily require the monotonicity preservation of the numerical solution. Figure 2 highlights the reduction of the interpolation stencil for some exemplary cases close to the embedded boundary. The interpolation location according to (4) are indicated by the origins of the red arrows.

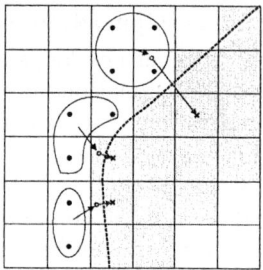

Fig. 2. Construction of values from interior cells used in internal ghost cells (gray)

4 An Adaptive Mesh Refinement Framework

Numerical simulations of detonation waves require computational meshes that are able to represent the strong local flow changes due to the reaction correctly. In particular, the induction zone between leading shock and head of reaction zone needs a high local resolution. The shock of a self-sustained detonation is very sensitive to changes in the energy release from the reaction behind and the inability to resolve all reaction details usually causes a considerable error in approximating the correct speed of propagation. In order to supply the necessary temporal and spatial resolution efficiently, we employ the blockstructured adaptive mesh refinement (AMR) method after Berger and Colella [2], which is tailored especially for hyperbolic conservation laws on logically rectangular finite volume grids. We have implemented the AMR method in a generic, dimension-independent object-oriented framework in C++. It is called AMROC (Adaptive Mesh Refinement in Object-oriented C++) and is free of charge for scientific use [5]. An effective parallelization strategy for distributed memory machines has been found and the codes can be executed on all systems that provide the MPI library.

Instead of replacing single cells by finer ones, as it is done in cell-oriented refinement techniques, the Berger-Collela AMR method follows a patch-oriented approach. Cells being flagged by various error indicators (shaded in Fig. 3) are clustered with a special algorithm [1] into non-overlapping rectangular grids. Refinement grids are derived recursively from coarser ones and a hierarchy of successively embedded levels is thereby constructed, cf. Fig. 3. All mesh widths on level l are r_l-times finer than on level $l - 1$, i.e. $\Delta t_l := \Delta t_{l-1}/r_l$ and $\Delta x_{n,l} := \Delta x_{n,l-1}/r_l$ with $r_l \geq 2$ for $l > 0$ and $r_0 = 1$, and a time-explicit finite volume scheme (in principle) remains stable on all levels of the hierarchy. The recursive temporal integration order is an important difference to usual unstructured adaptive strategies and is one of the main reasons for the high efficiency of the approach.

The numerical scheme is applied on level l by calling a single-grid routine in a loop over all subgrids. The subgrids are computationally decoupled by employing additional ghost cells around each computational grid. Three types of different ghost cells have to be considered in the sequential case: Cells outside of the root domain are used to implement physical boundary conditions. Ghost cells overlaid by a grid on level l have a unique interior cell analogue and are set by copying the data value from the grid, where the interior cell is contained (synchronization). On the root level no further boundary conditions need to be considered, but for $l > 0$ also internal boundaries can occur. They are set by a conservative time-space interpolation from two previously calculated time steps of level $l - 1$.

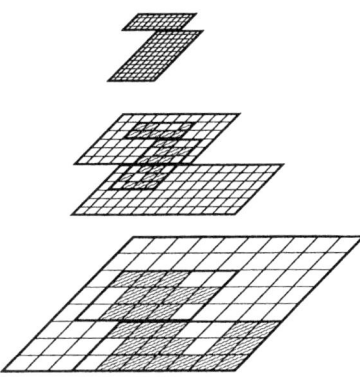

Fig. 3. AMR hierarchy

Beside a general data tree that stores the topology of the hierarchy, the AMR method utilizes at most two regular arrays assigned to each subgrid. They contain the discrete vector of state for the actual and updated time step. The regularity of the data allows high performance on vector and super-scalar processors and cache optimizations. Small data arrays are effectively avoided by leaving coarse level data structures untouched, when higher level grids are created. Values of cells covered by finer subgrids are overwritten by averaged fine grid values subsequently. This operation leads to a modification of the numerical stencil on the coarse mesh and requires a special flux correction in cells abutting a fine grid. The correction replaces the coarse grid flux along the fine grid boundary by a *sum* of fine fluxes and ensures the discrete conservation property of the hierarchical method at least for purely Cartesian problems without embedded boundaries. See [2] or [4] for details.

Up to now, various reliable implementations of the AMR method for single processor computers have been developed. Even the usage of parallel computers with shared memory is straight-forward, because a time-explicit scheme allows the parallel calculation of the grid-wise numerical update [1]. But the question for an efficient parallelization strategy becomes more delicate for distributed memory architectures, because on such machines the costs for communication can not be neglected. Due to the technical difficulties in implementing dynamical adaptive methods in distributed memory environments only few parallelization strategies have been considered in practice yet, cf. [19,17].

In the AMROC framework, we follow a rigorous domain decomposition approach and partition the AMR hierarchy from the root level on. The key idea is that all higher level domains are required to follow this "floor-plan". A careful analysis of the AMR algorithm uncovers that the only parallel operations under this paradigm are ghost cell synchronization, redistribution of the AMR hierarchy and the application of the previously mentioned flux correction terms. Interpolation and averaging, but in particular the calculation of the flux corrections remain strictly local [4]. Currently, we employ a generalization of Hilbert's space-filling curve [16] to derive load-balanced root level distributions at runtime. The entire AMR hierarchy is considered by projecting the accumulated work from higher levels onto the root level cells. Although rigorous domain decomposition does not lead to a perfect balance of workload on single levels, good scale-up is usually achieved for moderate CPU counts. Figure 4 shows a representative scalability test for a three-dimensional spherical shock wave problem for the computationally inexpensive Euler equations for a single polytropic gas without chemical reaction. Roe's approximate Riemann solver within the multi-dimensional Wave Propagation Method [13] is used as efficient single-grid scheme. The test was run on the ASC Linux cluster (ALC) at Lawrence Livermore National Laboratories that connects Pentium-4-2.4 GHz dual processor nodes with Quadrics Interconnect. The base grid has 32^3 cells and two additional levels with refinement factors 2 and 4. The adaptive calculation uses approx. 7.0 M cells in each time step instead of 16.8 M cells in the uniform case. The calculation on 256 CPUs employs between 1,500 and 1,700 subgrids on each level. Displayed are the average costs for each root level time step, which involve two time steps on the middle level and eight on the highest. All components of the dynamically adaptive algorithm, especially regridding and parallel redistribution are activated to obtain realistic results. Although we utilize a single-grid update routine in Fortran 77 in a C++ framework with full compiler optimization, the fraction of the time spent in this Fortran routine are 90.5 % on four and still 74.9 % on 16 CPUs. Hence, Fig. 4 shows a satisfying scale-up for at least up to 64 CPUs.

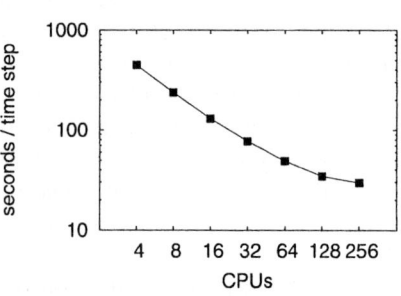

Fig. 4. Representative AMROC scale-up test for fixed problem size

5 Numerical Results

An ideal candidate for fundamental detonation structure simulations is the self-sustaining $H_2 : O_2 : Ar$ CJ detonation with molar ratios $2 : 1 : 7$ at $T_0 = 298$ K and $p_0 = 6.67$ kPa that is known to produce extremely regular detonation cell patterns [21]. The analytical solution according to the one-dimensional ZND theory is extended to multiple space dimensions and transverse disturbances are initiated by placing a small rectangular unreacted pocket behind the detonation front, cf. [14] or [4]. Throughout this paper, only the hydrogen-oxygen reac-

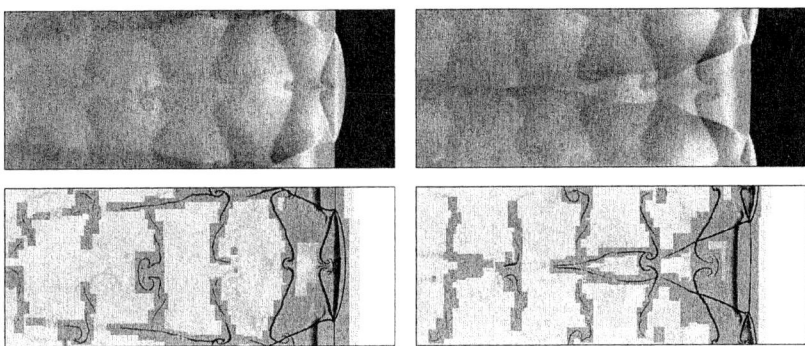

Fig. 5. Color plots of the temperature and schlieren plots of the density on refinement regions in the first (left) and second half (right) of a detonation cell

tion mechanism extracted from the larger hydrocarbon mechanism assembled by Westbrook is used [24]. The mechanism consists of 34 elementary reactions and considers the 9 species H, O, OH, H_2, O_2, H_2O, HO_2, H_2O_2 and Ar. According to the ZND theory, the induction length, the distance between leading shock and head of reaction zone in one space dimension, is $l_{ig} = 1.404$ mm for this mechanism in above configuration. The detonation velocity is 1626.9 m/s.

The application of the numerical methods of Sec. 3 within the parallel AMROC framework allowed a two-dimensional cellular structure simulation that is four-times higher resolved ($44.8\,\text{Pts}/l_{ig}$) than earlier calculations [14]. Only recently Hu et al. presented a similarly resolved calculation for the same CJ detonation on a uniform mesh [9]. Unfortunately, no technical details are reported for this simulation. In our case, the calculation was run on a small Beowulf-cluster of 7 Pentium 3-850 MHz-CPUs connected with a 1 Gb-Myrinet network and required 2150 h CPU-time. The calculation is in a frame of reference attached to the detonation. Because of the regularity of the oscillation only one cell is simulated. The adaptive run uses a root level grid of 200×40 cells and two refinement levels with $r_{1,2} = 4$. A physically motivated combination of scaled gradients and heuristically estimated relative errors is applied as adaptation criteria. See [4] for details. Two typical snapshots with the corresponding refinement are displayed in Fig. 5.

The high resolution of our simulation admits a remarkable refinement of the triple point pattern introduced in Sec. 1. Figure 6 displays the flow situation around the primary triple point A that is mostly preserved before the next collision. An analysis of the flow field uncovers the existence of two minor triple points B and C along the transverse wave downstream of A. While B can be clearly identified by a characteristic inflection, the triple point C is much weaker and very diffused. B is caused by the interaction of the strong shock wave BD with the transverse wave. The slip line emanating from B to K is clearly present. C seems to be caused by the reaction front and generates the very weak shock wave CI. A detailed discussion of the transient flow field is given in [4].

On 24 Athlon-1.4 GHz double-processor nodes (2 Gb-Myrinet interconnect) of the HEidelberg LInux Cluster System (Helics) our approach allowed a

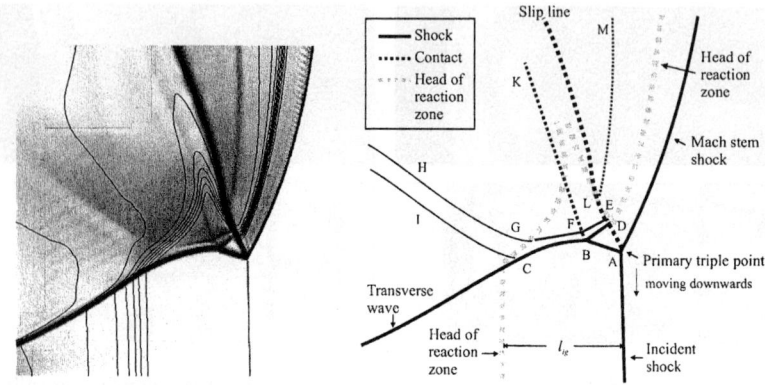

Fig. 6. Flow structure around a triple point before the next collision. Left: isolines of Y_{OH} (black) on schlieren plot of velocity component u_2 (gray).

sufficiently resolved computation of the three-dimensional cellular structure of a hydrogen-oxygen detonation. The maximal effective resolution of this calculation is $16.8 \, \text{Pts}/l_{ig}$ and the run required 3800 h CPU-time. Our adaptive results are in perfect agreement with the calculations by Tsuboi et al. for the same configuration obtained on a uniform mesh on a super-scalar vector machine [23]. A snapshot of the regular two-dimensional solution of the preceding section is used to initialize a three-dimensional oscillation in the x_2-direction and disturbed with an unreacted pocket in the orthogonal direction. We use a computational domain that exploits the symmetry of the initial data, but allows the development of a full detonation cell in the x_3-direction. The AMROC computation uses a two-level refinement with $r_1 = 2$ and $r_2 = 3$ on a base grid of $140 \times 12 \times 24$ cells and utilizes between 1.3 M and 1.5 M cells, instead of 8.7 M cells like a uniformly refined grid.

After a settling time of approx. 20 periods a regular cellular oscillation with identical strength in x_2- and x_3-direction can be observed. In both transverse directions the strong two-dimensional oscillations is present and forces the creation of rectangular detonation cells with the same width as in two dimensions, but the transverse waves now form triple point lines in three space dimensions. During a complete detonation cell the four lines remain mostly parallel to the boundary and hardly disturb each other. The characteristic triple point pattern can therefore be observed in Fig. 7 in all planes perpendicular to a triple point line. Unlike Williams et al. [25] who presented a similar calculation for an overdriven detonation with simplified one-step reaction model, we notice no phase-shift between both transverse directions. In all our computations for the hydrogen-oxygen CJ detonation only this regular three-dimensional mode, called "rectangular-mode-in-phase", or a purely two-dimensional mode with triple point lines just in x_2- or x_3-direction did occur.

In order to demonstrate the enormous potential of the entire approach even for non-Cartesian problems we finally show an example that combines highly efficient dynamic mesh adaptation with the embedded boundary method sketched

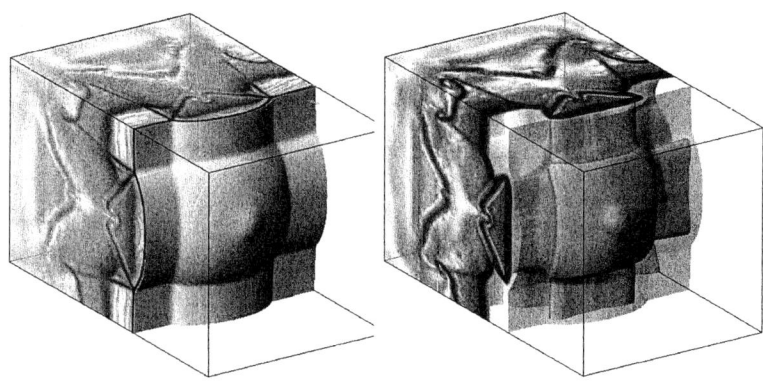

Fig. 7. Schlieren plots of ρ (left) and Y_{OH} (right) in the first half of a detonation cell (computational domain mirrored for visualization at lower boundary). The plot of Y_{OH} is overlaid by a translucent blue isosurface of ρ at the leading shock wave that visualizes the variation of the induction length l_{ig} in three space dimensions.

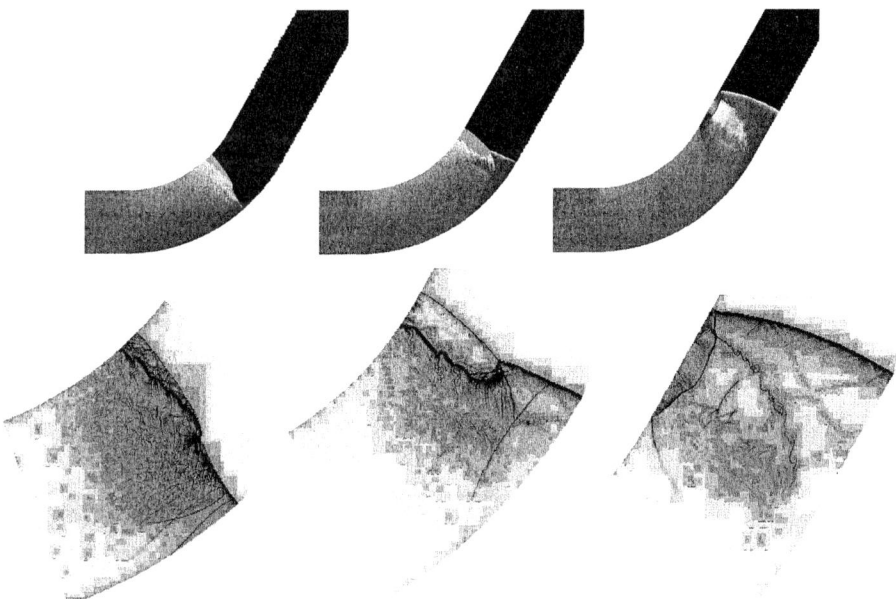

Fig. 8. Color plots of the temperature (upper row) and corresponding enlarged schlieren plots of the density on refinement regions (lower row) for a regular oscillating hydrogen-oxygen detonation propagating upwards a pipe bend.

in Sec. 3. A two-dimensional regular oscillating detonation is placed into a pipe of width 5λ. The pipe bends at an angle of 60 degree and with inner radius 9.375λ. When the detonation propagates through the bend it gets compressed and consequently overdriven near the outer wall, but a continuous shock wave diffraction occurs near the inner wall. This diffraction causes a pressure decrease below the limit of detonability that leads to a continuous decoupling of shock and reaction front. This effect is clearly visible in the earliest graphics of Fig. 8. The detonation exits the bend before the decay to a flame occurs across the entire tube width. A re-ignition wave arises from the successfully transmitted region and reinitiates the detonation in the decoupled area, cf. middle graphics of Fig. 8. It propagates in the direction normal to the pipe middle axis and causes a strong shock wave reflection as it hits the inner wall, compare last graphics of Fig. 8. This simulation uses a base grid of 300 × 248 cells, four levels of refinement with $r_{1,2,3} = 2$, $r_4 = 4$, and has an effective resolution of 16.9 Pts/l_{ig}. Approximately 1.0 M to 1.5 M cells are necessary on all levels instead $\approx$ 76 M in the uniform case. See lower row of Fig. 8 for some snapshots of the dynamic mesh evolution. The simulation used approximately 3000 CPU hours on 64 CPUs of the ASC Linux cluster.

6 Conclusions

We have described an efficient solution strategy for the numerical simulation of gaseous detonations with detailed chemical reaction. All temporal and spatial scales relevant for the complex process of detonation propagation were successfully resolved. Beside the application of the time-operator splitting technique and the construction of a robust high-resolution shock capturing scheme, the key to the high efficiency of the presented simulations is the generic implementation of the blockstructured AMR method after Berger and Collela [2] in our AMROC framework [5]. AMROC provides the required high local resolution dynamically and follows a parallelization strategy tailored especially for the emerging generation of distributed memory architectures. An embedded boundary method utilizing internal ghost cells extends the framework effectively to non-Cartesian problems. All presented results have been achieved on Linux-Beowulf-clusters of moderate size in a few days real time which confirms the practical relevancy of the approach.

References

1. Bell, J., Berger, M., Saltzman, J., Welcome, M.: Three-dimensional adaptive mesh refinement for hyp. conservation laws. SIAM J. Sci. Comp. **15** (1994) (1):127–138
2. Berger, M., Colella, P.: Local adaptive mesh refinement for shock hydrodynamics. J. Comput. Phys. **82** (1988) 64–84
3. Courant, R., Friedrichs, K. O.: *Supersonic flow and shock waves*. Applied mathematical sciences **21** (Springer, New York, Berlin, 1976)
4. Deiterding, R.: *Parallel adaptive simulation of multi-dimensional detonation structures* (PhD thesis, Brandenburgische Technische Universität Cottbus, 2003)
5. Deiterding, R.: AMROC - Blockstructured Adaptive Mesh Refinement in Object-oriented C++. Available at http://amroc.sourceforge.net (2005)
6. Einfeldt, B., Munz, C. D., Roe, P. L., Sjögreen, B.: On Godunov-type methods near low densities. J. Comput. Phys. **92** (1991) 273–295

7. Fedkiw, R. P., Aslam, T., Merriman, B., Osher, S.: A non-oscillatory Eulerian approach to interfaces in multimaterial flows (the ghost fluid method). J. Comput. Phys. **152** (1999) 457–492
8. Grossmann, B., Cinella, P.: Flux-split algorithms for flows with non-equilibrium chemistry and vibrational relaxation. J. Comput. Phys. **88** (1990) 131–168
9. Hu, X. Y., Khoo, B. C., Zhang, D. L., Jiang, Z. L.: The cellular structure of a two-dimensional $H_2/O_2/Ar$ detonation wave. Combustion Theory and Modelling **8** (2004) 339–359
10. Kaps, P., Rentrop, P.: Generalized Runge-Kutta methods of order four with stepsize control for stiff ordinary differential equations. Num. Math. **33** (1979) 55–68
11. Kee, R. J., Rupley, F. M., Miller, J. A.: *Chemkin-II: A Fortran chemical kinetics package for the analysis of gas-phase chemical kinetics*. (SAND89-8009, Sandia National Laboratories, Livermore, 1989)
12. Larrouturou, B.: How to preserve the mass fractions positivity when computing compressible multi-component flows. J. Comput. Phys. **95** (1991) 59–84
13. LeVeque, R. J.: Wave propagation algorithms for multidimensional hyperbolic systems. J. Comput. Phys. **131** (1997) (2):327–353
14. Oran, E. S., Weber, J. W., Stefaniw, E. I., Lefebvre, M. H., Anderson, J. D.: A numerical study of a two-dimensional H_2-O_2-Ar detonation using a detailed chemical reaction model. J. Combust. Flame **113** (1998) 147–163
15. Osher, S., Fedkiw, R.: *Level set methods and dynamic implicit surfaces*. Applied Mathematical Science **153** (Springer, New York, 2003)
16. Parashar, M., Browne, J. C.: On partitioning dynamic adaptive grid hierarchies. In Proc. of 29th Annual Hawaii Int. Conf. on System Sciences (1996)
17. Parashar, M., Browne, J. C.: System engineering for high performance computing software: The HDDA/DAGH infrastructure for implementation of parallel structured adaptive mesh refinement. In *Structured Adaptive Mesh Refinement Grid Methods*, IMA Volumes in Mathematics and its Applications (Springer, 1997)
18. Quirk, J. J.: Godunov-type schemes applied to detonation flows. In J. Buckmaster, editor, *Combustion in high-speed flows*, Proc. Workshop on Combustion, Oct 12-14, 1992, Hampton (Kluwer Acad. Publ., Dordrecht, 1994) 575–596
19. Rendleman, C. A., Beckner, V. E., Lijewski, M., Crutchfield, W., Bell, J. B.: Parallelization of structured, hierarchical adaptive mesh refinement algorithms. Computing and Visualization in Science **3** (2000)
20. Sanders, R., Morano, E., Druguett, M.-C.: Multidimensional dissipation for upwind schemes: Stability and applications to gas dynamics. J. Comput. Phys. **145** (1998) 511–537
21. Strehlow, R. A.: Gas phase detonations: Recent developments. J. Combust. Flame **12** (1968) (2):81–101
22. Toro, E. F.: *Riemann solvers and numerical methods for fluid dynamics* (Springer, Berlin, Heidelberg, 1999)
23. Tsuboi, N., Katoh, S., Hayashi, A. K.: Three-dimensional numerical simulation for hydrogen/air detonation: Rectangular and diagonal structures. Proc. Combustion Institute **29** (2003) 2783–2788
24. Westbrook, C. K.: Chemical kinetics of hydrocarbon oxidation in gaseous detonations. J. Combust. Flame **46** (1982) 191–210
25. Williams, D. N., Bauwens, L., Oran, E. S.: Detailed structure and propagation of three-dimensional detonations. Proc. Combustion Institute **26** (1997) 2991–2998
26. Williams, F. A.: *Combustion theory* (Addison-Wesley, Reading, 1985)

Scalable Photon Monte Carlo Algorithms and Software for the Solution of Radiative Heat Transfer Problems*

Ivana Veljkovic[1] and Paul E. Plassmann[2]

[1] Department of Computer Science and Engineering,
The Pennsylvania State University, University Park, PA 16802, USA
[2] The Bradley Department of Electrical and Computer Engineering,
Virginia Tech, Blacksburg, VA 24061, USA
veljkovi@cse.psu.edu, plassmann@vt.edu

Abstract. Radiative heat transfer plays a central role in many combustion and engineering applications. Because of its highly nonlinear and nonlocal nature, the computational cost can be extremely high to model radiative heat transfer effects accurately. In this paper, we present a parallel software framework for distributed memory architectures that implements the photon Monte Carlo method of ray tracing to simulate radiative effects. Our primary focus is on applications such as fluid flow problems in which radiation plays a significant role, such as in combustion. We demonstrate the scalability of the framework for two representative combustion test problems, and address the load balancing problem resulting from widely varying physical properties such as optical thickness. This framework allows for the incorporation of other, user-specified radiative properties, which should enable its use within a wide variety of other applications.

1 Introduction

Because Radiative Heat Transfer (RHT) rates are generally proportional to the fourth power of the temperature, applications that simulate combustion processes, nuclear reactions, solar energy collection and many other high-temperature physical phenomena, are highly influenced by the accuracy of the models used for radiative effects. RHT differs from other physical phenomena by its strong nonlocal effects. This nonlocality comes from the fact that the photons that carry radiation have variable mean free path (the path from the point where photon is released to the point where it is absorbed), as illustrated in the left part of Fig. 1. Because of these nonlocal effects, conservation laws cannot be applied over an infinitesimal volume (as is common for many fluid mechanics problems), but must be applied over the entire computational domain. The resulting formulation leads to complex models based on differo-integral equations. The additional issues that make radiative heat transfer calculations difficult include the complex geometry of the enclosure, non-uniform temperature field, scattering and nonuniform, nonlinear and nongray radiative properties.

Classical methods, such as the method of spherical harmonics [2] and the method of discrete ordinates [3], have been successfully implemented in numerous applications,

---

* This work was partially supported by NSF grants DGE-9987589, CTS-0121573, EIA-0202007, and ACI-0305743.

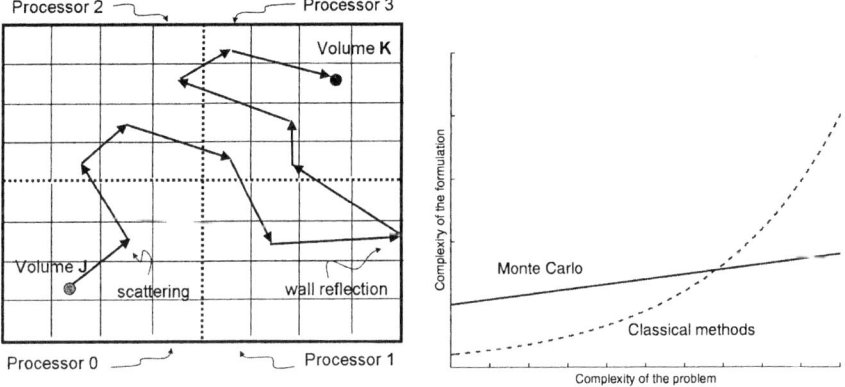

Fig. 1. Left: An example illustrating the nonlocal influence of radiation. Here we show the path of a photon in a participating medium with scattering and reflective walls. We observe that energy emitted from volume J can affect the absorbed energy of volume K, even though they are quite distant and do not belong to the same processor sub-domain. Right: A cartoon comparing the complexity of the Monte Carlo method and other conventional solution techniques, from [1].

but they have failed to address various important issues that can arise in the applications where more complex radiative properties have to be modelled. However, general thermal radiation problems can be effectively solved with sampling techniques by the Photon Monte Carlo (PMC) method. This method is based on a model of radiative energy travelling in discrete packets, like photons, and the computation of their effect while travelling as rays, scattering, and interacting with matter within the computational domain. As illustrated in the right part of Fig. 1, we note that the complexity of the formulation for Monte Carlo methods grows linearly with the complexity of the problem and the same relationship holds for CPU time. For other conventional methods, the complexity grows almost exponentially.

In this paper, we will consider the development of the parallel algorithms and software to implement a general version of the Photon Monte Carlo (PMC) method of ray tracing. This method can effectively deal with complex geometries, non-uniform temperature fields and scattering and it can employ a great variety of methods to calculate radiative properties of the enclosure. In previous work, software libraries that implement this method limit themselves to a particular data distribution, specific mesh structure and special functions to calculate radiative properties of the domain. While these choices may be suitable for a particular application, they greatly limit the usability of this method and its successful application in modelling simulations. Therefore, in this paper we will describe a robust and extensible framework for sequential and parallel implementation of PMC ray tracing method.

The paper is organized as follows. In section 2 we describe the formulation of Radiative Transfer Equation (RTE) suitable for the PMC method and we give an overview of the PMC algorithm. Section 3 offers a description of the framework for the PMC method and discussion about some of the most important properties and advantages of this framework. In section 4 we describe a parallel extension of the PMC software and discuss some of the issues of incorporating it within full-scale combustion simulations. Finally, in section 5 we present experimental results for two representative test problems and discuss the overall efficiency and observed load balancing for our parallel framework.

2 The Photon Monte Carlo Algorithm

In numerical simulations, the computational domain containing a participating medium is typically discretized into N sub-surfaces and M sub-volumes. The PMC method traces a statistically significant sample of photon bundles from their point of emission within surface/volume j to their point of absorption within surface/volume k. When the photon bundle is absorbed, its energy is added to the absorbed energy of the absorbing element. With this approach, the PMC method is able to calculate the energy gains/losses for every element (or fluid particle) in the enclosure.

Algorithm 1: The PMC ray tracing algorithm
Let N be the number of photon bundles to trace
Let M be the pointer to the mesh
$(x, \hat{d})$ are the starting position and unit direction of the photon
$(x', \hat{d}')$ are the new position and unit direction of the photon
$\mathcal{L}$ = generate_list_of_photons(M,N)
E = compute_emissive_energy(M)
while ($\mathcal{L} \neq \emptyset$)
  photon($x, \hat{d}$) = pop($\mathcal{L}$)
  i = photon.elementID
  if(i == surface)
    $[x', \hat{d}']$ = surfaceInteractions(i,M,photon)
  else //if volume element
    $[x', \hat{d}']$ = volumeInteractions(i,M,photon)
  endif
  if($[x', \hat{d}'] \neq$ NULL)
    push($[x', \hat{d}'], \mathcal{L}$)
  endif
endwhile

Algorithm 2: Determining volume interactions
i is mesh element ID, M is pointer to the mesh
j is the neighboring element
volumeInteractions(i, M, photon)
  $[j,x']$ = determine_intersection(i, M, photon)
  //extract coordinates, faces and other data
  E = get_element_data(i,M)
  $[Q,d']$ = radiative_properties(i,E)
  update_mesh_energy(Q,i,M)
  if($d' \neq$ NULL)
    $[x', d']$ = update_direction_position(i,j,M)
    photon.elementID = j
  endif
end

Algorithm 1 presents an outline of our implementation of the PMC ray tracing algorithm. Algorithm 2 describes the procedure of determining element-photon interactions. An illustration of possible element-photon interactions is shown in the Fig. 2.

3 A Software Framework for the PMC Method

The PMC library is developed to be only a part of large-scale software system that can include a number of different components, such as mesh generators, CFD solvers, and

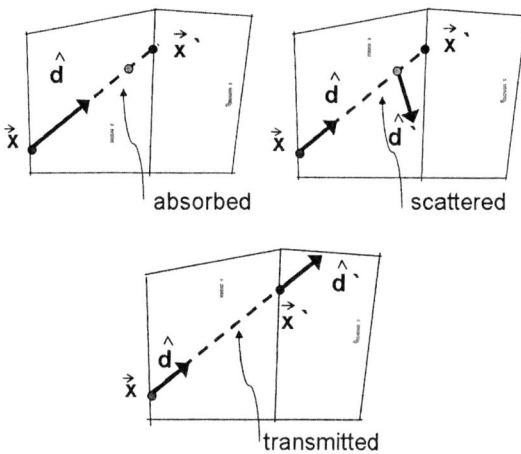

Fig. 2. Possible photon-volume interactions in a two-dimensional domain

numerous methods for modelling chemistry. Hence, it becomes crucial to develop a software framework that will enable the incorporation of our PMC implementation into a wide variety of potential applications. To calculate the radiation sources by tracing the history of a large number of photon bundles, we need to know how to pick statistically meaningful photon bundles. For every photon bundle, PMC algorithm has to determine point of emission within a surface or a volume, direction of emission, wavelength, point of absorption, and various other properties. These properties should be chosen at random, but they also have to reflect radiative properties of the medium. For example, after choosing a point of emission on a given surface element for the photon bundle P, the wavelength is calculated as a function of the radiative properties in a given location on the surface and of the random number R_λ generated in PMC. Similar relations can be derived for other properties of the photon bundles, and they are called *random number relations*.

There are many methods available for computing the random number relations, and their complexity usually depends on the complexity of the radiative properties and needs of the application. For example, for most current combustion applications that involve radiation-turbulence interactions, the use of a second-order polynomial approximation to compute the starting location of the photon bundle yields satisfactory accuracy. However, if the Direct Numerical Simulation (DNS) method is used to solve for the turbulent field, it may be necessary to use a higher-order polynomial approximation to get satisfactory accuracy. Therefore, a robust software that implements PMC ray tracing should be able to employ different methods for the calculations of the random number relations and enable the addition of the new ones, depending on the needs of the underlying application.

These issues lead to a conclusion that we need a software environment with 'plug and play' capabilities. Namely, the functions that determine random number relations are only given as placeholders. Depending on the nature of the problem, the user of the library can then 'plug' the functions that already exist in the library or she/he can

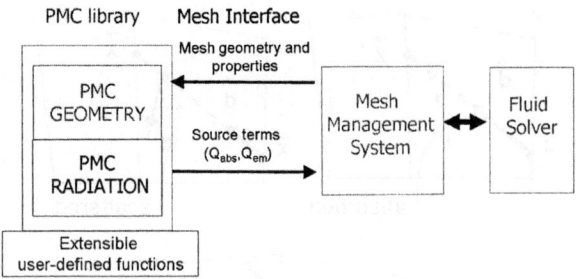

Fig. 3. A block diagram of the interactions within the PMC software framework

write new functions that comply with the certain established interface. For example, we often have to calculate some radiative property of the medium, such as absorption coefficient as a function of temperature, pressure, mass fraction of chemical species and wavelength. The collection of functions that calculate radiative properties are presented in Algorithm 2 as radiative_properties(). There are numerous models for calculating absorption coefficient and in our library only some of these models will be implemented. Therefore, if the user does not find a suitable function in the PMC library, he can write his own function to calculate absorption coefficient and 'plug' it into the software framework. Thus, our PMC framework consists of two types of functions— PMC Geometry, functions that perform the ray tracing, and PMC Radiation, functions that determine random number relations for PMC based on radiative properties of the medium and that can be extended to include various user-defined modules. The overall structure of the PMC software is illustrated in the Fig. 3.

4   Parallelization of the Monte Carlo Algorithm

The parallel software for Monte Carlo ray tracing method is usually designed to solve a particular type of radiation problem [4, 5], or it is tailored for specific parallel architectures, such as Cray X-MP/Y-MP architectures [6], that are not readily available to a wide scientific community. The efficiency of these codes varies and depends on the type of problems they are solving.

When developing the Monte Carlo algorithms for distributed memory environments, we have two broad choices of how to conduct parallelization of data —ray partitioning and domain partitioning. In the ray partitioning method, the total number of photon bundles is divided among processors. A processor tracks each of the assigned bundles through its entire path until it is absorbed or its energy is depleted. If the problem is such that the entire computational domain can fit the memory of one processor, than ray partitioning is by far the most efficient method to use. However, problems arise if a spatial domain decomposition must be performed as well.

The data for the combustion problems often cannot fit the memory of one processor. This practically means that domain decomposition (described bellow in subsection 4.1) must be used for computational load distribution for PMC as well, which can lead to significant load balancing problems and communication overhead issues.

4.1 Parallelization Through Domain Partitioning

In the domain partitioning method, the underlying mesh is divided among processors and every processor traces photon bundles through its portion of the mesh. When the bundle crosses the mesh boundary, it is sent to the neighboring processor. An advantage of this method is that the domain partitioning is inherited from the mesh generator and therefore, no additional partitioning methods are needed. However, the domain partitioning also introduces a problem with the load balancing. If one part of the enclosure is hot, it will emit more photon bundles than regions with lower temperature, since the emitted energy depends on the fourth power of temperature. In this case, the processor that owns the hot, emitting region will be overwhelmed with work while the processor which owns the cold region will have much fewer bundles to track. This problem can be solved with an adaptive domain decomposition strategy which ensures an even distribution of particles among the processors.

Unfortunately, this domain decomposition has to be balanced with the CFD program, and this balancing can be a complicated procedure. This issue can be mitigated by the use a partitioning algorithm that includes estimates for the computational load for PMC in the partitioning strategy. However, because the properties of the sub-domains may change during the execution of the application and the mesh repartitioning may be very costly, the parallel PMC software should be able to provide reasonable efficiency even if the partitioning is not well balanced with respect to the PMC computation.

Algorithm 3: A parallel PMC ray tracing algorithm
Let N be the number of bundles to trace
Let M be the pointer to the mesh
Let $\mathcal{N}_p$ be the list of photon bundles to send to neighboring processor p
$\mathcal{L}$ = generate_list_of_photons(M,N)
while ($\mathcal{L} \neq \emptyset$)
  photon($x, \hat{d}$) = pop($\mathcal{L}$)
  i = photon.elementID
  if(i == surface)
    [$x', \hat{d}'$] =
      surfaceInteractions(i,M,photon)
  else //if volume element
    [$x', \hat{d}'$] =
      volumeInteractions(i,M,photon)
  endif
  if([$x', \hat{d}'$] $\neq$ NULL)
    if(photon at the border with processor p)
      push([$x', \hat{d}'$],$\mathcal{N}_p$)
    else
      push([$x', \hat{d}'$],$\mathcal{L}$)
  endif
  exchange_photonbundles($\mathcal{L}$)
endwhile

Algorithm 4: Communication between processors
exchange_photonbundles(list of photons $\mathcal{L}$)
  foreach(neighbor p)
    if($\mathcal{N}_p$ has sufficient photon bundles)
      send_package($\mathcal{N}_p$,p)
    endif
    if(there is a package from p)
      $\mathcal{L}_p$ = receive_package(p)
      $\mathcal{L}$ = append($\mathcal{L}, \mathcal{L}_p$)
    endif
  foreach
end

Algorithm 3 presents an outline of our implementation of parallel Photon Monte-Carlo ray tracing algorithm. Algorithm 4 describes the communication procedure between processor.

5 Experimental Results

If in a given problem conduction and/or convection are important, the RTE equation must be solved simultaneously with the overall conservation of energy equation. The energy equation is usually solved by conventional numerical methods, such as finite volume or finite element methods where radiative heat flux is included as a source term. In that case, iteration of the temperature field is necessary; the temperature field is used by the PMC method to calculate radiative sources, which are used to update the temperature field.

To demonstrate the capabilities of our parallel PMC software framework, we completed two simple experiments, whose geometries are illustrated in Fig. 4. The first experiment models the radiative heat transfer between two infinite parallel plates, and the second example computes one step of PMC during a simulation of a combustion process inside a jet engine. The amount of work grows linearly with the overall number of photon bundles we track through the domain. Therefore, as we increase the number of processors, we proportionally increase the number of photon bundles and keep the number of cells constant. This approach enables us to examine the scalability of the PMC implementation.

We used the Parametric Binary Dissection algorithm [8] which tries to minimimize the difference for a chosen metric between the two subregions it generates at each step. The metric chosen for these experiments is the weighted sum of the computational and communication load of the sub-domain. The aspect ratio of the sub-domains was used as the measure of communication cost, and the volume of the sub-domains was taken as the computational load, which is a reasonable measure when we consider fluid flow calculations performed with a near-uniform discretization. We decided to minimize the computational load of fluid flow calculations, rather then minimizing the computational load of PMC algorithm, since we believe that this is a typical scenario for combustion

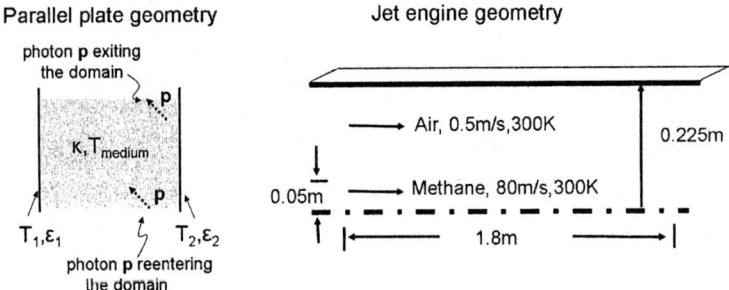

Fig. 4. Left: the geometry of the parallel plate problem. Right: The geometry of the jet engine, from [7].

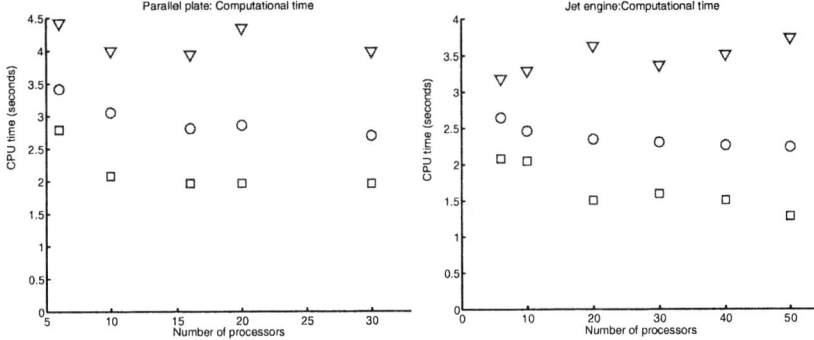

Fig. 5. The maximum, average and minimum computational load for individual processors on the parallel machine for the parallel plate problem (left) and the jet engine problem (right)

simulations. For both experiments we plotted the CPU time for computation, (the maximum, minimum and average CPU time for individual processors). To achieve good scalability for our experiments, it is desirable for these quantities to remain asymptotically constant as we vary the number of processors.

In both cases, the combustion simulation supplies the PMC functions with the temperature and absorption coefficient field, and the PMC library outputs the absorbed energy for each element in the domain. In the case of the jet engine simulation, the flame is optically thin, so we can expect that extensive communication between processors will take place. Also, the initial temperature profile is such that we have a region with very high temperature and a region that is relatively cold.

We tested our software on a Beowulf cluster with 94 dual processor AMD Opteron 250 nodes each with 4GB RAM with a fully connected 4-way Infiniband Topspin switch interconnection network. In Fig. 5 we present the range of individual CPU times needed for just the PMC computation to illustrate the load balancing for the parallel plate and jet engine problems. One can observe that the maximum and minimum computational load differs by a factor of two in both cases, but that this ratio stays constant as we increase the number of processors, illustrating the scalability of the implementation.

For the second set of experiments, we tested a different metric for the domain decomposition algorithm. The metric chosen is the computational load with respect to the PMC algorithm—no communication costs or the computational load of the fluid-flow calculations are considered. The number of rays emitted from a given sub-domain is proportional to the total emission of the sub-domain and this value can be calculated from the temperature and absorption coefficient distribution. Therefore, the natural choice for modelling computational load of PMC calculations is using total emission.

In the left part of Fig. 6 we plotted the ratio between the minimum and the maximum of the emitted rays over the computational environment. We can observe that the difference is negligible, but in the right part of Fig. 6 we can also note that this initial load balancing did not improve the overall load balancing since the efficiency has not increased compared to the original case when no radiation metrics was used. The overall computational load was, in this case, proportional to the total number of volumes that were traversed during the ray tracing, not to the number of emitted rays.

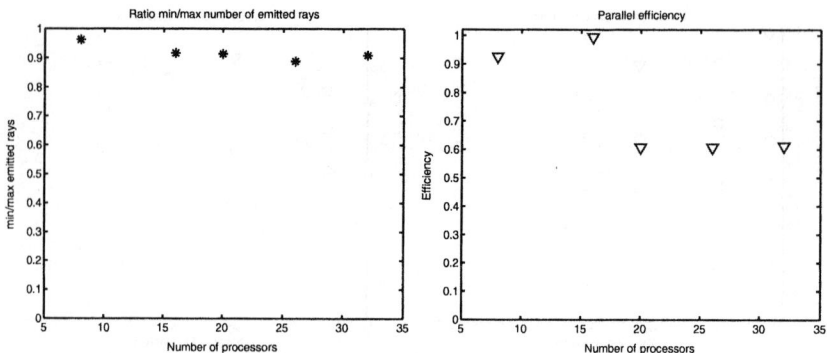

Fig. 6. The ratio between minimum and maximum number of emitted rays in the parallel environment (left) and the parallel efficiency (right) for the case when we only considered computational load with respect to the radiation calculation for the domain decomposition

We can conclude that modelling computational load for the PMC algorithm is a very complex problem. We plan to conduct further research to investigate various partitioning algorithms and find the suitable model for the PMC computational load that would yield maximum performance of the applications that use the PMC framework.

6 Conclusions and Future Work

The Photon Monte Carlo ray tracing method is a powerful tool for solving the radiative heat transfer equation in applications where complex radiative properties of the domain have to be modelled. In this paper we have described the advantages of designing a framework that is able to model many radiative properties that emphasize various aspects of radiative heat transfer. We have also presented a parallel implementation of this framework and demonstrated its scalability and efficiency. For future work, we plan to implement modules that deal with the cases when the spectral dependence of the radiative properties is large (non-gray radiation). We plan to employ full k-g distribution schemes that are extracted from a database for narrow-band k-g distributions developed by Wang and Modest (see [9]). To speed up the extraction of the full k-g distribution, we plan to incorporate a function approximation scheme such as one described in [10, 11].

References

[1] Modest, M.F.: Radiative Heat Transfer, second edition. Academic Press (2003)
[2] Jeans, J.: The equations of radiative transfer of energy. Monthly Notices Royal Astronomical Society (1917) 28–36
[3] Chandrasekhar, S.: Radiative Transfer. Dover Publications (1960)
[4] Sawetprawichkul, A., Hsu, P.F., Mitra, K.: Parallel computing of three-dimensional Monte Carlo simulation of transient radiative transfer in participating media. Proceedings of the 8th AIAA/ASME Joint Thermophysics and Heat Transfer Conf., St Louis, Missouri (2002)

[5] Dewaraja, Y., Ljungberg, M., A. Majumdar, A.B., Koral, K.: A parallel Monte Carlo code for planar and SPECT imaging: Implementation, verification and applications in I-131 SPECT. Computer Methods and Programs in Biomedicine (2002)
[6] Burns, P., Christon, M., Schweitzer, R., Lubeck, O., Wasserman, H., Simmons, M., Pryor, D.: Vectorization of Monte Carlo particle transport: An architectural study using the lanl benchmark GAMTEB. Proceedings, Supercomputing 1989 (1988) 10–20
[7] Li, G.: Investigation of Turbulence-Radiation Interactions by a Hybrid FV/PDF Monte Carlo Method. PhD thesis, The Pennsylvania State University (2002)
[8] Bokhari, S., Crockett, T., Nicol, D.: Binary dissection: Variants and applications. ICASE Report No. 93-39 (1993)
[9] Wang, A., Modest, M.F.: High-accuracy, compact database of narrow-band k-distributions for water vapor and carbon dioxide. In: Proceedings of the ICHMT 4th International Symposium on Radiative Transfer, Turkey. (2004)
[10] Veljkovic, I., Plassmann, P.E., Haworth, D.C.: A scientific on-line database for efficient function approximation. Computational Science and Its Applications—ICCSA 2003, The Springer Verlag Lecture Notes in Computer Science (LNCS 2667) series, part I (2003) 643–653
[11] Veljkovic, I., Plassmann, P., Haworth, D.: A parallel implementation of scientific on-line database for efficient function approximation. Post-Conference Proceedings, The 2004 International Conference on Parallel and Distributed Processing Techniques and Applications, Las Vegas, USA (2004) 24–29

A Multi-scale Computational Approach for Nanoparticle Growth in Combustion Environments

Angela Violi[1,2,3] and Gregory A. Voth[1]

[1] Department of Chemistry, University of Utah, 84112, Salt Lake, City (UT)
[2] Department of Chemical Engineering, University of Utah, 84112, Salt Lake, City (UT)
[3] Department of Mechanical Engineering, University of Michigan, Ann Arbor, 48109 (MI)

Abstract. In this paper a new and powerful computer simulation capability for the characterization of carbonaceous nanoparticle assemblies across multiple, connected scales, starting from the molecular scale is presented. The goal is to provide a computational infrastructure that can reveal through multi-scale computer simulation how chemistry can influence the structure and function of carbonaceous assemblies at significantly larger length and time scales. Atomistic simulation methodologies, such as Molecular Dynamics and Kinetic Monte Carlo, are used to describe the particle growth and the different spatial and temporal scales are connected in a multi-scale fashion so that key information is passed upward in scale. The modeling of the multiple scales are allowed to be dynamically coupled within a single computer simulation using the latest generation MPI protocol within a grid-based computing scheme.

1 Introduction

A detailed description of high molecular mass particle formation in combustion systems, such as a flame environment, can be viewed as comprised of two principal components: gas-phase chemistry and particle dynamics, which describes the evolution of the particle ensemble. Figure 1 [1] summarizes the different pathways describing the formation of nanoparticle agglomerates starting from simple fuels, going through the formation of polycyclic aromatic hydrocarbons (PAH) [2-8], particle inception, particle coagulation leading to primary particles (50 nm in diameter) [9-17], ending up with their particle agglomerates (~500 nm in diameter).

As reported in fig. 1, the processes involved in the particle formation exhibit a wide range of time scales, spanning pico- or nanoseconds for intramolecular processes to milliseconds for intermolecular reactions. At the same time, the length scale also undergoes significant changes going from a few angstroms for small PAH to hundreds of nanometers for particle aggregates.

The goal of this work is the development of multi-scale atomistic approaches to study the formation and fate of carbonaceous material, in a chemically specific way. The reason for seeking the structural information of the particles is due to the importance that particles emitted from combustion sources have in the environment. Two main problems are related to the presence of aerosols in the atmosphere: the health impact and the climate change. Once the chemical structures of these particles are known, challenging problems, such as their interaction with biological systems and the optical properties relevant to direct radiative forcing, can be addressed.

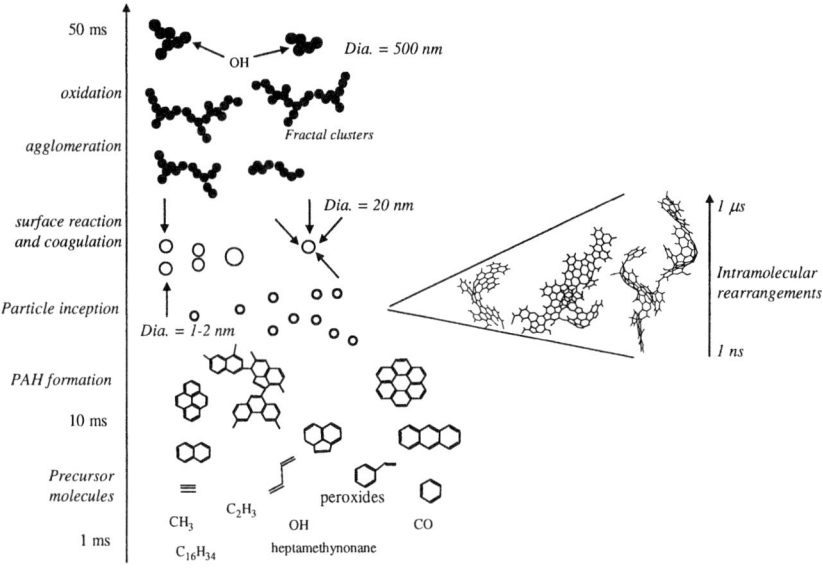

Fig. 1. Processes involved in the formation of particles in flames

There is no attempt in the literature to study the processes involved in the formation of particles using an atomistic representation, and this paper reports on the work we pioneered for nanoparticle growth in combustion.

In the following sections, it will be shown how a new multi-scale simulation can be created and applied to the formation of carbon nanoparticles. The methodology will allow us to follow the transformations that occur during nanoparticle formation in a chemically specific way, providing information on both the chemical structure and the configuration of the nanoparticles and their agglomeration.

The primary goal of this work is to provide the computational infrastructure to help connect molecular properties with the behavior of particle systems at much larger length and time scales. This new computational approach can reveal the role of chemical changes at the molecular level in defining the functional behavior of nanoparticle assembly. In particular, this approach will help us to understand how specific chemical structures or changes at the scale of the molecular building blocks will influence the emergent structure and function of particle assembly at larger length and time scales.

This effort will involve the development of underlying modeling concepts at the relevant scales, the multi-scale linking of these scales, and finally the computational execution of the multi-scale simulations using a novel MPI grid computing strategy.

2 Methodology

Keeping in mind that the overarching goal of this work is the development of a computational paradigm capable of coupling chemical scales to much longer length-scale assembly processes, the multi-scaling thus begins at an atomistic-level of

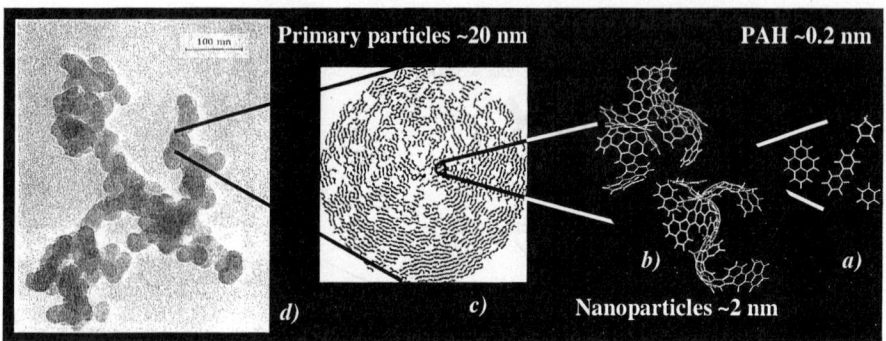

Fig. 2. Computational multi-scale coupling scheme for particle formation

resolution where gas-phase PAH monomers (*a* in Fig. 2) are used to build a larger system of nanoparticles having a diameter of 2-3 nm (*b*), which in turn represent the monomers for the following growth into primary particles with an average diameter of ~20 nm (*c*) and eventually particle agglomerates of ~500 nm (*d* in Fig. 2).

After a description of the methodologies used to go from simple gas-phase PAH monomers to the formation of nanoparticles (*a-b*) and then to particle agglomeration (*c-d*), section 2.3 will report on the computational modeling infrastructure that allows the atomistic methods to be dynamically coupled within a single computer simulation using the latest generation MPI protocol within a grid-based computing scheme.

It is important to keep in mind that although presented in separate sections, the phenomena of nanoparticle formation and their further agglomeration and growth into bigger units occur simultaneously. In other words, once nanoparticles are formed and start to agglomerate, PAH monomers from the gas phase can still contribute to the growth through surface reactions and the information is produced at the lower scale needs to be carried out to the largest scale.

2.1 From Gas-Phase Monomers to Nanoparticle Inception

The simulation methods for many-body systems can be divided into two classes of stochastic and deterministic simulations, which are largely covered by Monte Carlo method and Molecular Dynamics method, respectively. Kinetic Monte Carlo (KMC) simulations [18-19] solve the evolution of an ensemble of particles in configuration space, provided some initial condition, and a complete set of transition events (chemical reactions, diffusion events, etc.). By contrast Molecular Dynamics (MD) solves the trajectories of a collection of atoms in phase space (configuration + momentum) given the initial conditions and potential. This is an advantage of MD simulation with respect to KMC since not only is the configuration space probed but also the whole phase space, which can give more information about the dynamics of the system, subject to any additional constraints on the equations of motion. However, the limitation in the accessible simulation time represents a substantial obstacle in making useful predictions with MD. Resolving individual atomic vibrations requires a time step of approximately femtoseconds in the integration of the equations of motion, so that reaching times of microseconds is very difficult.

With the aim of keeping an atomistic description of the system and at the same time to be able to cover large time scales, such as milliseconds for the particle inception (see time scale in Figure 1), the MD and KMC methodologies have been coupled to extend the accessible time scale by orders of magnitude relative to direct MD, (Atomistic Model for particle Inception Code – AMPI) [20-21]. The time-step for a single KMC iteration is a "real time," determined by the kinetics system.

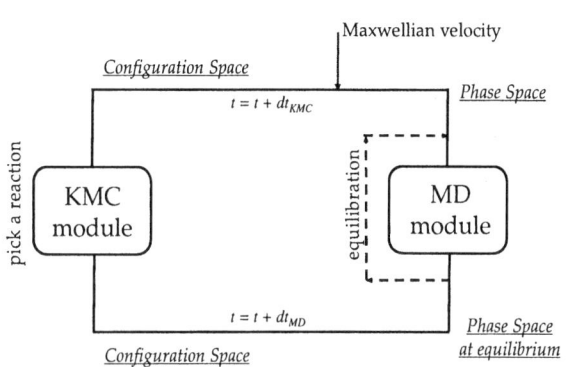

Fig. 3. Schematic of the AMPI code

The reaction rates among the compounds present in the system are specified as probabilities, and the surface configuration over time is then given by a master equation, describing the time evolution of the probability distribution of system configurations. The AMPI code couples the two techniques in a novel way, placing both of them on equal footing with a constant alternation between the two modules. Figure 3 shows a schematic representation of the AMPI code. Since the output of the KMC module is a configuration space, while the input to the MD part is represented by the phase space (configuration + momentum), velocities are assigned to the atoms of the growing seed according to a Maxwell–Boltzmann distribution function after exiting the KMC module. The KMC and MD methods therefore, differ in their effective ability to explore conformation space.

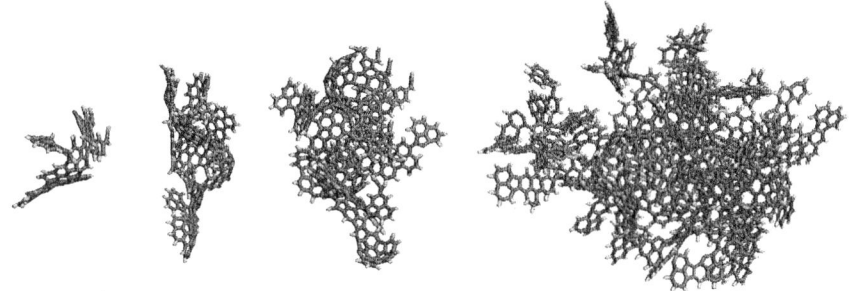

Fig 4. Carbonaceous nanoparticles

In a typical simulation, an aromatic seed molecule is placed in a reacting environment described by 1) temperature, 2) number and concentrations of the monomers in the gas-phase that can contribute to the growth process and 3) reaction rates between the monomers and the seed. The code alternates between the KMC and MD modules: during a KMC time step one reaction is executed at one site on the growing aromatic structure. The probability of choosing a particular reaction is equal to the rate at which the reaction occurs relative to the sum of the rates of all of the possible reactions for the current configuration. For each time step, one reaction is chosen from the probability-weighted list of all of the events that can possibly occur at that step. The molecular system is then altered, and the list of relative rates is updated to reflect the new configuration. The MD module, run after each KMC step, allows for relaxation of the molecules towards thermal equilibrium. Figure 4 shows the structures of carbonaceous nanoparticles produced with the AMPI code. For the specific example reported in Fig. 4 acetylene, naphthalene, acenaphthylene and their radicals, respectively are used as gas-phase monomers that can contribute to the growth of an initial seed represented by naphthalene in a specific environment of benzene-oxygen laminar premixed flame.

2.2 From Nanoparticle Inception to Agglomerates

As soon as the first nanoparticles are formed, they start reacting with each other. Carbonaceous nanoparticle agglomeration is influenced by large length and time scale motions that extend to mesoscopic scales, i.e., one micrometer or more in length and one microsecond or more in time. At the same time, the effects of various important intra-nanocluster processes, such as fragmentation, slow conformational rearrangements, ring closure and associated reactions need to be addressed to produce a realistic description of the particle growth phenomenon.

In order to increase the time and length scales accessible in simulations to be able to simulate nanoparticle assembly, a different methodology, via accelerated dynamics is proposed.

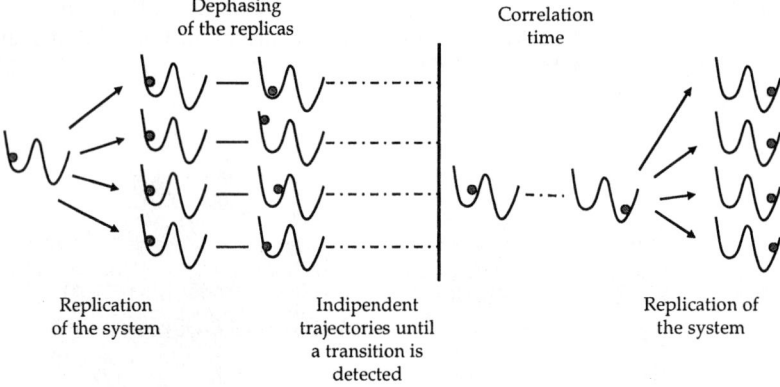

Fig. 5. Schematic of the PR MD architecture [23]

Given a resource of M processors in a parallel architecture, we employed a temporal parallelism to simulate different sections of the long-time dynamics on each of the M processors, rather than attempting to simulate each timestep M times faster. Among accelerated dynamic methods, the PRMD [22-23-24] is the simplest and most accurate method and it dramatically extends the effective time scale of an MD simulation. In the PR method, starting with an N atom system in a particular state, the system is replicated on many processors. After a dephasing stage, where momenta

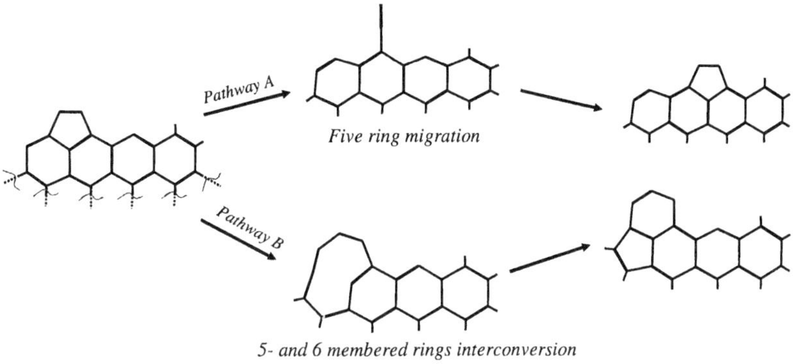

Five ring migration

5- and 6 membered rings interconversion

Fig. 6. Surface rearrangement reactions

are randomized, each processor carries out an independent MD trajectory. Each replica trajectory is monitored for a transition event by performing a quench after each ΔT_{block} of integration time. When one processor (i) detects an event, all processors are notified to stop. The simulation clock is incremented by the sum of the trajectory times accumulated by all M replicas since the beginning, and the PR simulation is restarted, continuing onward until the next transition occurs and the process is repeated. Figure 5 reports a schematic for the Parallel Replica architecture [24]. Within the PR method, the effective time of the MD simulation is "boosted" significantly: the overall rate of some event is thus enhanced by a factor of M so that the probability of observing an event for the first time at a simulation time in the interval (t_1, t_1+dt_1) changes from $p(t_1)dt_1 = ke^{-kt_1}dt_1$ for a single-processor simulation to $p(t_1)dt_1 = Mke^{-kt_1}dt_1$ when simulating on M processors.

The use of PR MD for an ensebles of nanoparticles produced with the AMPI code allows the identification of intramolecular as well as intermolecular reactions [1-2]. As an example, Fig. 6 shows possible rearrangement reactions that can occur on the particle surface. The particle considered has 289 C atoms but in Fig. 6, for clarity, only the region where the modification occurs is reported. Two mechanisms are highlighted: in pathway A the transformation is initiated by β-scission followed by the formation of a five membered ring similar to the initial one but shifted by one aromatic ring. The second pathway involves the transposition of 5- and 6-membered rings.

4 Multi-scale Coupling and Computational Grid Implementation

As stated before, nanoparticle inception from gas-phase species and particle agglomeration processes occur simultaneously and reconstituting the system in different spatial and temporal domains without bridging and linking the different scales, results in a sequence of independent spatial/temporal representations of the system. Without some means of coupling the scales each new representation of the overall multi-scale simulation methodology remains isolated from the other constituents. This limitation can restrict the ability of the multi-scale simulation methodology in directly complementing experimental work, as most of the experimental combustion spans a wide range of length and time-scales. For the formation of nanoparticles, the chemistry that occurs at the lower scale modifies the properties of particles and dictates the clustering behavior. The example reported in Fig. 6 for the particle surface can clarify this concept. The five membered ring is formed through the addition of acetylene from the gas-phase to the aromatic particle surface and the five-member-ring migration has important implications to surface growth of particles. For example, a "collision" of these propagating fronts may create a site that cannot be filled by cyclization and thus cannot support further growth. Formation of such surface defects may be responsible for the loss of reactivity of soot particle surface to growth. In essence, the newly nucleated particles grow by coagulation and coalescent collisions as well as surface growth from gas-phase species and the changes that can occur on single nanoparticles due to surface growth needs to propagate upward to the PR module.

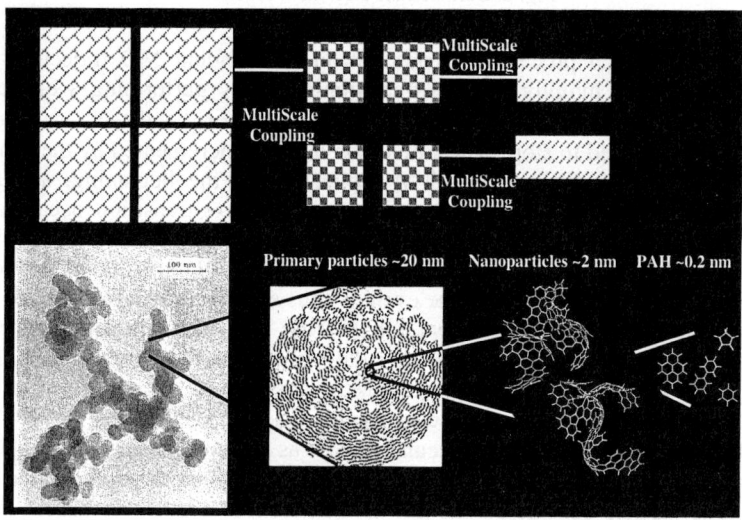

Fig. 7. Different grid schemes for each parallelization level

Each different process of the nanoparticle formation is modeled with a separate simulation suite. The parallel decomposition of the full multi-scale methodology is a critical component of the infrastructure. Each of the platforms (AMPI code and PR MD) has its own internal degree of parallelization. The novel infrastructure parallelization component consists in a new level of intra-parallelism: the aim is to construct a dynamic communication interface between the various simulation components of the multi-scale methodology, such that different codes can essentially speak to each other on the fly. The novelty of this approach is that a new level of communication is established between different programs with each program running on a separate computational platform. When combined with the intrinsic internal process communication, the result is a dual parallel communication scheme characterized by infrequent communication and the more familiar frequent large communication. Thanks to relatively recent adoption of the MPI-2 standard in major

MPI implementations, it is now feasible to implement communication schemes between separate programs. In particular, new programs (slave processes – PR MD in this case) can be launched via MPI-2 from an already running application (the master process – AMPI code) via an MPI-2 dynamic process creation. A series of interconnected communication channels can be established between the separate programs resulting in a communication interface that can be used to couple the slave and master processes, see Fig. 7. In other words, a set of distinct simulation methodologies, can be fused into one unified computational platform. The degree of fusion can be tuned to the specific problem. This extension leads to the opportunity of creating multi-level parallel applications where each of the sub-levels also operates under its own internal parallelization sub-grid. Both the fast internal parallelization as well as the mew multi-scale inter-process communication can be incorporated with each code using the MPI-2 standard. Figure 8 shows two examples of particle agglomeration obtained through the use of this new multi-scale infrastructure. The figure shows two snapshots from the multi-scale simulations of a system composed of 27 identical round particles (left panel) and a system of 27 flat particles produced in aliphatic flames (right panel). The simulations were performed at 1600K. The round particles tend to cluster and they show a preferred orientation that is back to back. The sheet-like particles show a different behavior: some of them drift away from the ensemble, some do cluster in small agglomerates. Therefore, the agglomeration behavior of nanoparticles is influenced by their morphology. This is an example of the kind of information necessary to build a realistic model for particle formation (for this example 164 Opteron nodes, 328 processors, 2 Gbytes memory per node were used).

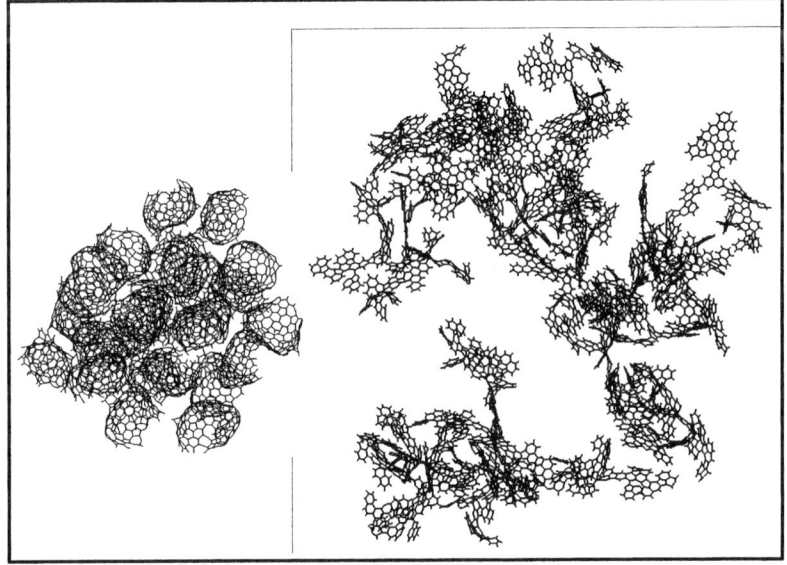

Fig. 8. Particle agglomerates

6 Conclusions

A new multi-scale computer simulation and modeling capability is presented to study the formation of carbonaceous nanoparticles in combustion systems. This methodology provides a framework that can better mimic the chemical and physical processes leading to nanoparticle soot particle inception and growth. The systematic multi-scaling begins at an atomistic-level of resolution where atomistic molecular dynamics simulations are employed to give quantitative predictions for the formation of the first nanoparticles and provide a computational template in which to base the next steps of the overall multi-scale computational paradigm.

This approach provides a connection between the various time scales in the nanoparticle self-assembly problem, together with an unprecedented opportunity for the understanding of the atomistic interactions underlying carbonaceous nanoparticle structures and growth. The scales are dynamically linked with each other and the execution of the full multi-scale simulation involves the use of a novel MPI and grid computing strategy.

Acknowledgments

This research is funded by the University of Utah Center for the Simulation of Accidental Fires and Explosions (C-SAFE), funded by the Department of Energy, Lawrence Livermore National Laboratory, under subcontract B341493 and by a National Science Foundation Nanoscale Interdisciplinary Research Team grant (EEC-0304433). The calculations presented in this paper were carried out at the Utah Center for High Performance Computing (CHPC), University of Utah that is acknowledged for computer time support. The authors thank Dr. A. Voter for providing us with the PRMD code.

References

1. Bockhorn, H. in Bockhorn (Ed.) Soot Formation in Combustion: Mechanisms and Models of Soot Formation, Springer Series in Chemical Physics, vol. 59, Springer-Verlag, Berlin (1994) 1-3.
2. Haynes B.S., Wagner H.Gg.: Soot formation. Prog. Energy Combust. Sci. 7(4) (1981) 229-273.
3. Calcote H.F. Combust. Flame 42 (1981) 215-242.
4. Homann K.H. Proc. Combust. Inst. 20 (1985) 857-870.
5. Glassman I.: Soot formation in combustion processes. Proc. Combust. Inst. 22 (1989) 295-311.
6. Bockhorn H., Schaefer T. in Soot Formation in Combustion: Mechanisms and Models, Springer Series in Chemical Physics 59 (1994) 253-274.
7. Richter H., Howard J.B.: Prog. Energy Combust. Sci. 26 (2000) 565-608.
8. Frenklach M. Phys. Chem. Chem. Phys. 4(11) (2002) 2028-2037.
9. Megaridis C.M., Dobbins R.A.: Combust. Sci. Tech. 66(1-3) (1989) 1-16.
10. Dobbins R.A., Subramaniasivam H. in Bockhorn (Ed.) Soot Formation in Combustion: Mechanisms and Models of Soot Formation, Springer Series in Chemical Physics, vol. 59, Springer-Verlag, Berlin, (1994) 290-301.

11. Vander Wal R.L., "A TEM methodology for the study of soot particle structure" Combust. Sci. Tech. 126(1-6): 333-357 (1997).
12. Köylü Ü.Ö., McEnally C.S., Rosner D.E., Pfefferle L.D. Combust. Flame 110(4): 494-507 (1997).
13. D'Alessio A., D'Anna A., D'Orsi A., Minutolo P., Barbella R., Ciajolo A. Proc. Combust. Inst. 24: 973-980 (1992).
14. Minutolo P., Gambi G., D'Alessio A., D'Anna A. Combust. Sci. Tech. 101(1-6): 311-325 (1994).
15. D'Alessio A., Gambi G., Minutolo P., Russo S., D'Anna A. Proc. Combust. Inst. 25: 645-651 (1994).
16. Dobbins, R. A. Combustion Science and Technology Book Series, 4(Physical and Chemical Aspects of Combustion), pp.107-133 (1997).
17. Dobbins, R. A. Combust. Flame 130(3) (2002) 204-214.
18. Bortz, A. B, Kalos, M.H., Lebowitz, J.L., J. Comp. Phys. 17 (1975) 10-18.
19. Fichthorn, K.A., Weinberg, W.H., J. Chem. Phys. 95 (1991) 1090-1096.
20. Violi A., Sarofim A.F., Voth G.A. Comb. Sci. Tech. 176(5-6) (2004) 991-997.
21. Violi A. Combust. Flame, 139 (2004) 279-287.
22. Voter, A.F., Phys. Rev. B 34 (1986) 6819-6839.
23. Voter, A.F., Phys. Rev. B 57 (1998) 13985-13990.
24. Voter, A.F., Montalenti, F., Germann, T.C., Annu. Rev. Mater. Res. 32 (2002) 321-326.
25. A. Violi, G.A. Voth, A.F. Sarofim Proceedings of the third International Mediterranean Combustion Symposium June 8-13 2003 Marrakech, Morocco, p. 831-836.
26. A. Violi, M. Cuma, G.A. Voth, A.F. Sarofim, Third Joint Meeting of the U.S. Sections of the Combustion Institute March 16-19 2003 Chicago, Illinois

A Scalable Scientific Database for Chemistry Calculations in Reacting Flow Simulations*

Ivana Veljkovic[1] and Paul E. Plassmann[2]

[1] Department of Computer Science and Engineering,
The Pennsylvania State University, University Park, PA 16802, USA
[2] The Bradley Department of Electrical and Computer Engineering,
Virginia Tech, Blacksburg, VA 24061, USA
veljkovi@cse.psu.edu, plassmann@vt.edu

Abstract. In many reacting flow simulations the dominant computational cost is the modelling of the underlying chemical processes. An effective approach for improving the efficiency of these simulations is the use of on-line, scientific databases. These databases allow for the approximation of the computationally expensive chemical process through interpolation from previously computed exact values. A sequential software implementation of these database algorithms has proven to be extremely effective in decreasing the running time of complex reacting flow simulations. To take advantage of high-performance, parallel computers, a parallel implementation has been designed and shown to be successful for steady-state problems. However, for transient problems this implementation suffers from poor load balancing. In this paper, we propose a modified algorithm for coordinating the building and the distributed management of the database. We present initial experimental results that illustrate improved load balancing algorithms for this new database design.

1 Introduction

The increasing availability of high-performance, parallel computers has made possible the simulation of extremely complex and sophisticated physical models. Nevertheless, the complete, multi-scale simulation of many complex phenomena can still be computationally prohibitive. For example, in reacting flow simulations, the inclusion of detailed chemistry with a Computational Fluid Dynamics (CFD) simulation that accurately models turbulence can be intractable.

This difficulty occurs because a detailed description of the combustion chemistry usually involves tens of chemical species and thousands of highly nonlinear chemical reactions. The numerical integration of these systems of differential equations is often stiff with time-scales that can vary from 10^{-9} seconds to 1 second. Solving the conservation equations for such chemically reacting flows, even for simple, two-dimensional laminar flows, with this kind of chemistry is computationally very expensive [1].

One approach to address this problem involves the implementation of a specialized database to archive and retrieve accurate approximations to repetitive, expensive

---
* This work was supported by NSF grants CTS-0121573, ACI-9908057, and DGE-9987589 and ACI-0305743.

calculations. That is, instead of solving the equations at every point required by the simulation, one can solve it for some significantly smaller number of points, archive these values, and interpolate from these values to obtain approximations at nearby points. The motivation for this approach is the idea that only a low dimensional manifold within the composite space is typically accessed [2, 3, 4] (the composite space is a normalized space of dimension the number of chemical species plus the temperature and pressure). This approach was originally proposed by Pope for combustion simulations [2]. In his paper, Pope proposed to tabulate previously computed function values; when a new function value is required, the set of previously computed values is searched and, if possible, the function is approximated based on these values.

In previous work we introduced a sequential algorithm that extended Pope's approach in several directions [5]. The new algorithms are implemented in the software system DOLFA and have been successfully used in combustion applications [6]. Because of the importance of the use of parallel computers to solve large-scale, combustion problems, it is desirable to develop a parallel version of this sequential algorithm. We proposed an approach for distributing the functional database and introduced three communication strategies for addressing the load balancing problems in the parallel implementation [7]. We have presented a detailed formulation of these three communication heuristics and demonstrated the advantage of a hybrid strategy [8].

This hybrid strategy is effective at redistributing the computational load if the global partitioning of the database search tree is reasonable. However, in some transient cases it can happen that, even though the initial distribution of the database is well balanced, after some number of time-steps the database distribution becomes progressively skewed. To attempt to alleviate this problem, in this paper we introduce a new heuristic for building and maintaining the global search tree. This heuristic ensures a good quality initial database distribution and, in addition, performs a reordering of the database after each time-step that redistributes the computational load based upon performance statistics obtained from the current time-step.

The remainder of this paper is structured as follows. In Section 2 we briefly review the sequential algorithm as implemented in the software package DOLFA [5]. In Section 3 we describe the parallel implementation of DOLFA with new algorithms for building the global BSP search tree and database redistribution heuristic. Finally, in Section 4 experimental results are presented demonstrating the parallel performance of the improved implementation.

2 The Sequential Scientific Database Algorithm

In reacting flow simulations the underlying chemistry calculations can be implemented as function evaluations. The function values to be computed, stored, and approximated by the database are n-dimensional functions of m parameters, where both n and m can be large (in the 10's or 100's for typical combustion problems). We denote this function as $F(\theta)$ which can be stored as an n-dimensional vector and where the parameters θ can be represented as an m-dimensional vector. The sequential database algorithm works by storing these function values when they are directly (exactly) computed. If possible, other values are approximated using points stored in the database. In addition to the function values, the database stores other information used to estimate the error based on an Ellipsoid Of Accuracy (EOA) as described in [5].

To illustrate how the sequential database works, when $F(\theta)$ is to be calculated for some new point, we first query the database to check whether this value can be approximated with some previously calculated value. For computational combustion, the parameter space is often known as the *composite space*. The composite space consists of variables that are relevant for the simulation of a chemical reaction, such as temperature, pressure and mass fractions of chemical species.

In DOLFA, the database is organized as a Binary Space Partition (BSP) tree to enable the efficient search of the composite space. Internal nodes in the BSP tree denote planes in the m-dimensional, composite space. A BSP tree is used rather than a generalization of a quadtree or octree because of the high dimensionality of the search space. In the search tree terminal nodes (leafs) represent convex regions determined by the cutting planes on the path from the given terminal node to the root of the BSP tree. This construction is illustrated in the left side of Fig. 1. For each region we associate a list of database points whose EOAs intersect the region. In addition, to avoid storing points that are unlikely to be used again for approximation, we employ techniques based on usage statistics that allow us to remove such points from the database.

The software system DOLFA has demonstrated significant speedups in combustion applications. For example, the software was tested with a combustion code that implements a detailed chemistry mechanism in an HCCI piston engine [9]. For this application, when compared to simulation results based only on direct integration, insignificant effects on the accuracy of the computed solution were observed.

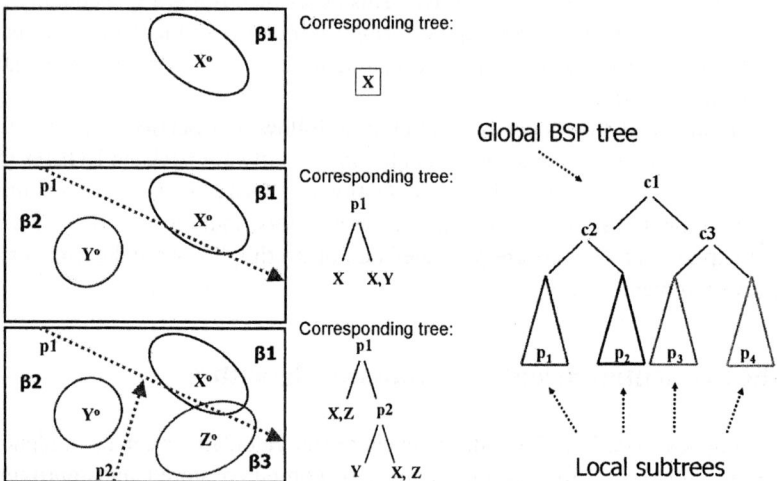

Fig. 1. Left: the generation of a BSP tree and the resulting spatial decomposition of a two-dimensional domain from the insertion of three data points. The arrows illustrate the orientation of the cutting planes and the ovals represent the EOAs. This figure is taken from [5]. Right: An illustration of how the global BSP tree is partitioned among processors in parallel implementation—each processor owns a unique subtree of the global tree.

3 The Parallel Database Algorithm

In developing a parallel version of the database, perhaps the most straightforward approach would be to use the sequential version on each processor, without any interprocessor collaboration in building a distributed database. This approach would work; however, it suffers from two significant drawbacks: a significant number of redundant function calculations and no memory scalability. In this section, we present parallel algorithms that address each of these issues.

The importance of addressing the first issue of redundant evaluations becomes obvious if we compare typical times for function evaluation and interprocessor communication. Namely, for the HCCI combustion simulation which was mentioned earlier in Section 2, one function evaluation takes 0.3 seconds (solving an ODE system with 40 chemical species). On the other hand, on the Beowulf cluster where our experiments were conducted the time to communicate the computed results (40 double precision values) is approximately 2.7 microseconds. Thus, our goal is to design a parallel implementation of DOLFA that minimizes the total number of direct calculations performed on all processors. However, this minimization must be obtained while maintaining good load balancing and while minimizing the total interprocessor communication.

Before each call to the database from the main driver application, all processors must synchronize to be able to participate in global database operations. This synchronization may be very costly if it is conducted often. For example, if the database is called with a single point at a time as is done with the sequential version of DOLFA. Modifying the DOLFA interface by allowing calls that are made with lists of query points, instead of one point at a time, minimizes the synchronization overhead and enables more efficient interprocessor communication.

Fortunately, in most combustion simulations, updates to the fluid mechanics variables and chemical species mass fractions happen independently of one another on the mesh subdomain assigned to each processor. More precisely, the chemical reaction at point P at time t does not affect the chemical reaction at point Q at time t. Therefore, we can pack the points for which we want to know function values into a list and submit it to the database with one global function call. The output to this call is a list of function values at the points given in the input list.

The original parallel algorithm proposed in [7] consisted of two main parts. First, the approach used for building the distributed (global) BSP tree and second, the heuristics developed for managing the interprocessor communication and the assignment of direct function evaluations to processors. In the next subsection we will offer a new algorithm for building the BSP tree based on all the points that are queried to the database in the first call. Subsection 3.2 will discuss the communication patterns and the assignment of direct function evaluations. Finally, in the subsection 3.3 we describe new heuristics for redistributing the database to ensure good load balancing in the case of transient simulations.

3.1 Building the Global BSP Tree

As with the sequential case, we typically have no *a priori* knowledge of the distribution of queries within the search space. However, since the database is called with the list of points for which we want to know the function values, we have an acceptable description of initial composite space we are accessing before we start evaluating the queries. Thus, the global BSP tree can be built before we start the database operations, based on the list of points that are queried. The BSP tree can then be partitioned into unique subtrees which are assigned to processors as illustrated in the right side of Fig. 1.

These subtrees are computed by recursively decomposing the global search space by introducing cutting planes that roughly divide the set of points equally among each subregion. The resulting subtrees (and their corresponding subregions) are then assigned to individual processors. The choice of cutting planes is limited to coordinate planes chosen by computing the dimension with maximum variance.

As we mentioned before, each processor owns its subtree but also has the information about the portion of the search space stored on every other processor in the environment. Therefore, the search for a suitable point ϕ to use for approximation for point θ is conducted in a coordinated manner between processors. More precisely, if the query point θ is submitted to processor i, but this point belongs to the subtree of processor j, processor i will be able to recognize this fact. The cost of building the global BSP tree is proportional to $\log_2(p)$ where p is the number of processors, under the assumption that the network latency is much bigger than both the time needed for one floating point operation and network bandwidth.

3.2 Managing the Interprocessor Communication

Based on our assumption that the function evaluation is expensive relative to communication, we want to design a parallel search scheme that will minimize the number of direct evaluations that must be performed. Whenever there is a possibility that another processor may be able to approximate the function value at a given point, we would like to communicate that query to that processor for processing. This means that, for a given query point X submitted to processor p_1, we have one of two possibilities. First, when point X belongs to a subtree T_1 of p_1, as in case of points $\{Y_2, Y_4, Y_5\}$ in the left side of Fig. 2, we simply proceed with the sequential algorithm. Second, the more challenging case is when the point X belongs to a subtree of some other processor, say p_2, such as point Y_1 in the left side of Fig. 2, $Y_1 \in T_2$. We call such points 'non-local'. As p_1 can store some number of points that belong to processor p_2, p_1 may still be able to approximate this query. If it cannot, the processor sends the point to processor p_2. If p_2 can approximate $F(Y_1)$, it sends back the result. But, if p_2 also cannot satisfy the query, the open question is to determine which processor should conduct the direct integration.

Two issues affect the choice of a strategy for answering this question. First, as we stated above, we want to minimize the number of redundant calculations. However, we also want to minimize the load imbalance between the processors. If the computational load is measured based on the number of direct calculations performed on each processor, we would like this number to be roughly equal for all the processors in the environment.

The first solution proposes that the owner of the subtree (p_2 in the above example) performs the direct integration. Because problems in combustion chemistry usually include turbulence and nonhomogeneous media, this approach can result in significant

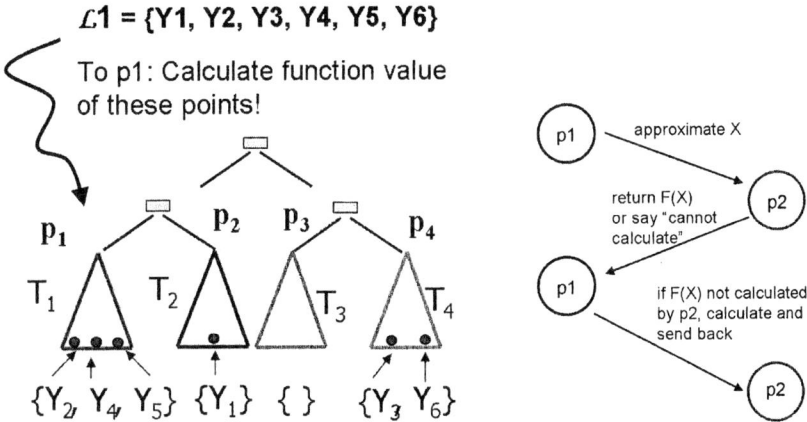

Fig. 2. Left: three different types of points that are submitted to a particular processor (p_1 in this case) for evaluation. Right: an illustration of the "ping-pong" communication between processors.

load imbalance between processors. If the BSP tree is not well distributed with respect to the current lists of points submitted to processors for evaluation, it can happen that majority of points belong to a small number of subtrees. However, this approach leads to the best possible retrieval rates. As such, it can be used as a benchmark for determining a lower bound for the overall number of direct calculations.

A second approach is to have the processor to which the query was submitted (p_1 in the above example) do the direct calculations. The communication between processors would consist of processor p_2 letting processor p_1 know that it cannot approximate $F(Y_1)$ and processor p_1 returning to p_2 the calculated function value $F(Y_1)$. We describe this resulting communication pattern as a "ping-pong," and it is illustrated in the right side of Fig. 2.

This second approach mitigates the load imbalance, but introduces additional communication overhead and increases the number of redundant calculations. For example, lists on processors p_3 and p_4 may contain points that are close to Y_1 and that could be approximated with $F(Y_1)$. In the first approach, we would have only one direct calculation for $F(Y_1)$. In this second approach, since p_2 will send off point Y_1 back to original processors to calculate it (in this case, p_1, p_3 and p_4), we would have these three processors redundantly calculate $F(Y_1)$.

Since each of these two approaches performs well for one aspect in the overall parallel implementation, a natural solution to this problem is to design a hybrid approach that will combine the best properties of the previous two solutions. With the hybrid approach, in the above example processor p_2 decides whether to calculate $F(Y_1)$ for some Y_1 from processor p_1 based on the average number of direct retrievals in the previous iteration and additional metric parameters.

In previous work [8], we demonstrated that the hybrid strategy jointly minimizes the number of redundant calculations and the load imbalance and, therefore, has a significant advantage over the other two approaches. However, if the BSP tree is extremely imbalanced with respect to the points that are submitted to the processors, the hybrid strategy will fail to ensure reasonable performance results. As these cases are not rare (in particular for non-steady state combustion calculations) we have to design strategies that will deal with this case. In the following subsection we propose such an approach.

3.3 Redistributing the BSP Tree

Combustion simulations are usually an iterative process that incorporate computational fluid dynamics (CFD), chemical reaction mechanisms, and potentially various radiation and turbulence models. We have observed that, even if the BSP tree is well balanced regarding the initial points queried to the database, it can occur that in some future calls, most of the points will belong to subtrees assigned to a small subset of processors. We have also noticed that this load shifts gradually, so between two iterations (i.e., two database calls) the difference between number of direct integrations on a particular processor is not large, but that this difference accumulates over time. Therefore, we propose to redistribute the BSP after each iteration to prevent this accumulation of the load imbalance.

We note that the internal BSP tree structure, without the leaves that represent convex regions, only takes approximately 0.1% of the memory for the entire tree for typical combustion simulations. Therefore, we first extend our global BSP tree distribution so that every processor keeps in its memory a local copy of most of the internal structure of the entire global BSP tree. Therefore, after each iteration, processors shell exchange the information about parts of the BSP tree (i.e. new partition hyperplanes) that were added in the previous iteration.

In our algorithm, we first sort the list of n processors $\{p_i\}$ according to their computational load C_i during the previous database call. As a measure for the computational load, we use the number of direct calculations performed on the processor during the last iteration. In addition, we maintain the number of direct calculations performed for each region in the subtree that belongs to this processor. Every processor p_i (in the sorted list) will have its partner p_{n-i} with which it will exchange some parts of the BSP tree. Namely, processor p_{n-i} will transfer some convex regions to processor p_i

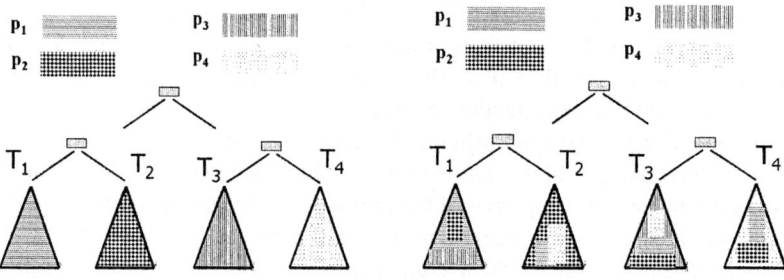

Fig. 3. An illustration of how the global BSP tree is distributed among the processors. The left tree shows the initial distribution after building the global BSP tree. The right tree shows the database organization after several calls to the database and the subsequent BSP tree redistributions.

such that the sum of computational loads for these regions (i.e., the number of direct calculations in the regions during the previous iteration) is roughly $\frac{C_{n-i}-C_i}{2}$. Processor p_{n-i} will mark those regions that it sent as if they belong to processor p_i and will not do any direct calculations for the points belonging to these regions. However, if possible, it will retrieve the points from those regions.

We observe that after a number of iterations our subtree does not look like the one in the left side of Fig. 3, because now processors do not own the entire subregions that were initially assigned to them. An illustration of the new BSP tree distribution is shown in the right side of Fig. 3. We can also observe that this algorithm will introduce additional overhead— both memory overhead for storing the structures that will describe the newly added nodes in the BSP tree and the communication overhead of transferring regions from one processor to another. However, this overhead is not be significant when compared to advantages of load balancing.

4 Experimental Results

We have designed an application testbed that mimics typical combustion simulations to test the performance of our parallel database implementation [8]. This application mimics a two-dimensional, laminar, reacting flow. The user-defined tolerance is 0.001 and it takes roughly 0.1 second to perform one direct evaluation. The user tolerance is set in such way that approximately 85% of queries are approximated using the database. We tested our software on a Beowulf cluster with 94 dual processor AMD Opteron 250 nodes each with 4GB RAM and an interconnection network consisting of a fully connected 4-way Infiniband Topspin switch.

We can use various metrics to illustrate the effectiveness of our new scheme. In this case, we used load balancing which was measured by computing the ratio between the minimum and average number of direct calculations across all processors. If this relative difference is near 1.00, this implies that roughly every processor conducted the same amount of work over the entire calculation; however, there may still be load imbalance at individual time-steps.

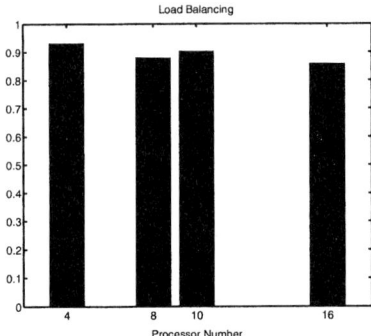

Fig. 4. The improved load balancing for a steady state simulation based on the new algorithm for BSP tree construction. The bar graph shows the improved load balancing between processor pairs resulting from the new BSP tree design strategy.

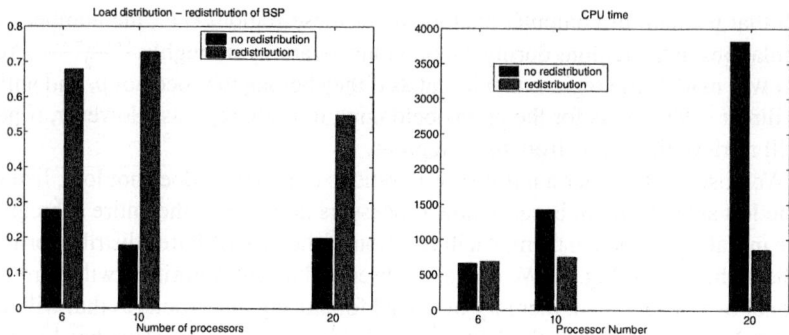

Fig. 5. Left: comparison in load balancing for a typical transient application using DOLFA with and without redistribution. Right: comparison in time for above mentioned cases.

In the previous algorithm we first accumulated some number of points into the database and then build the BSP tree. In the cases where these initial points did not represent the overall accessed space well, we observed this ratio to be smaller than 0.1 (indicating that the maximum load is 10 times larger than the minimum load). In this simulation we used the first approach to manage the direct calculation of the 'non-local' points.

In Fig. 4 we demonstrate the improved load balancing for the new method of building the BSP tree. With our new approach where the BSP tree is built based on all the points submitted to the database in the first call, we observe a significant improvement in the overall load balancing. This testbed application is not transient, so the initial load balancing is maintained until the end of the simulation. This result is extremely significant, since the first approach to manage the direct calculation of the 'non-local' points minimized the number of redundant calculations. Even though this approach can result in a large load imbalance, it can be successfully used in some steady-state simulations.

In the left part of Fig. 5, we compared the load balancing for a representative transient simulation for two cases - with and without redistribution of the BSP tree. Here we also used the first approach to manage the direct calculation of the 'non-local' points. We can observe a superior performance of the redistribution algorithm and its effect on the load balancing. In the right part of Fig. 5 we compared the average times to complete the same simulation for these two cases. We can observe the tremendous impact of load balancing on the CPU time and again, the advantage of the redistribution algorithm.

5 Conclusions and Future Work

We have demonstrated in previous work that sequential database system DOLFA can significantly reduce the execution time for large-scale, reacting flow simulations that involve frequent expensive chemistry calculations. In addition, we have introduced a parallel extension of this database system, based on maintaining a global BSP tree which can be searched on each processor in a manner analogous to the sequential database algorithm. In this paper, we have described new algorithms for the initial design of the global BSP tree and balancing the computational load based on iteratively redistributing the BSP tree among processors and demonstrated their effect on load balancing and overall execution time. In the future work, we plan to test our parallel implementation with an application that models reacting flow with complex chemistry and incorporates radiative heat transfer using a high-order, Direct Numerical Simulation (DNS) fluid simulation.

References

[1] Maas, U., Pope, S.B.: Laminar flame calculations using simplified chemical kinetics based on intrinsic low-dimensional manifolds. Twenty-Fifth Symposium (International) Combustion/The Combustion Institute (1994) 1349–1356
[2] Pope, S.: Computationally efficient implementation of combustion chemistry using in-situ adaptive tabulation. Combustion Theory Modelling 1 (1997) 41–63
[3] Maas, U., Schmidt, D.: Analysis of the intrinsic low-dimensional manifolds of strained and unstrained flames. In: 3rd Workshop on Modelling of Chemical Reaction Systems, Heidelberg. (1996)
[4] Rabitz, H., Alis, O.: General foundations of high dimensional model representations. Journal of Math. Chemistry 25 (1999) 197–233
[5] Veljkovic, I., Plassmann, P.E., Haworth, D.C.: A scientific on-line database for efficient function approximation. Computational Science and Its Applications—ICCSA 2003, The Springer Verlag Lecture Notes in Computer Science (LNCS 2667) series, part I (2003) 643–653
[6] Haworth, D., Wang, L., Kung, E., Veljkovic, I., Plassmann, P., Embouazza, M.: Detailed chemical kinetics in multidimensional CFD using storage/retrieval algorithms. In: 13th International Multidimensional Engine Modeling User's Group Meeting, Detroit. (2003)
[7] Veljkovic, I., Plassmann, P., Haworth, D.: A parallel implementation of scientific on-line database for efficient function approximation. Post-Conference Proceedings, The 2004 International Conference on Parallel and Distributed Processing Techniques and Applications, Las Vegas, USA (2004) 24–29
[8] Plassmann, P., Veljkovic, I.: Parallel heuristics for an on-line scientific database for efficient function approximation. Post-Conference Proceedings,Workshop on State-Of-The-Art in Scientific Computing PARA04,June 2004, Denmark (2004)
[9] Embouazza, M., Haworth, D., Darabiha, N.: Implementation of detailed chemical mechanisms into multidimensional CFD using in situ adaptive tabulation: Application to HCCI engines. Society of Automotive Engineers, Inc (2002)

The Impact of Different Stiff ODE Solvers in Parallel Simulation of Diesel Combustion*

Paola Belardini[1], Claudio Bertoli[1], Stefania Corsaro[2], and Pasqua D'Ambra[3]

[1] Istituto Motori (IM)-CNR,
Via Marconi, 8 I-80125 Naples, Italy
{p.belardini, c.bertoli}@im.cnr.it

[2] Department of Statistics and Mathematics for Economic Research,
University of Naples "Parthenope", Via Medina 40, I-80133 Naples, Italy
stefania.corsaro@uniparthenope.it

[3] Institute for High-Performance Computing and Networking (ICAR)-CNR,
Via Pietro Castellino 111, I-80131 Naples, Italy
pasqua.dambra@na.icar.cnr.it

Abstract. In this paper we analyze the behaviour of two stiff ODE solvers in the solution of chemical kinetics systems arising from detailed models of Diesel combustion. We consider general-purpose solvers, based on Backward Differentiation Formulas or Runge-Kutta methods and compare their impact, in terms of reliability and efficiency, on the solution of two different chemical kinetics systems, modeling combustion in the context of realistic simulations of Common Rail Diesel Engines. Numerical experiments have been carried out by using an improved version of KIVA3V, interfacing a parallel combustion solver. The parallel combustion solver is based on the CHEMKIN package for evaluating chemical reaction rates and on the general-purpose stiff ODE solvers for solving chemical kynetic equations.

1 Introduction

One of the main computational kernels in multidimensional modeling of Diesel engines is a system of ordinary differential equations, driving the time evolution of the mass density of M chemical species involved in R chemical reactions. The equations have the following form:

$$\dot{\rho}_m = W_m \sum_{r=1}^{R}(b_{mr} - a_{mr})\dot{\omega}_r(\rho_1, \ldots, \rho_m, T), \quad m = 1, \ldots, M. \quad (1)$$

In equation (1), ρ_m is the mass density of species m, W_m is its molecular weight, a_{mr} and b_{mr} are integral stoichiometric coefficients for reaction r, $\dot{\omega}_r$ is the kinetic reaction rate, whose expression, in Arrhenius form [9], is given by:

---

* This work has been supported by the Project: *High-Performance Algorithms and Software for HSDI Common Rail Diesel Engine Modeling*, funded by Campania Regional Board.

$$\dot{\omega}_r(\rho_1,\ldots,\rho_m,T) = K_{fr}(T) \prod_{m=1}^{M} (\rho_m/W_m)^{a'_{mr}} - K_{br}(T) \prod_{m=1}^{M} (\rho_m/W_m)^{b'_{mr}}, \quad (2)$$

where $K_{fr}(T)$ and $K_{br}(T)$ are the reaction coefficients depending on temperature T and a'_{mr} and b'_{mr} are the reaction orders. System (1), provided with suitable initial conditions, has to be solved at each time-step and at each grid cell of a 3d computational domain. Our aim is its solution for complex and detailed chemical reaction models, used for predictive simulations of exhaust emissions in Diesel engines, in order to match the stringent limits imposed by government laws (e.g EURO V in Europe, Their 2 Bin 5 in USA). In these models the concentrations of the involved chemical species change with very different time scales, varying in time and space, and introducing high degrees of stiffness. Therefore, the solution of system (1) requires reliable and efficient numerical software both in sequential and in parallel computing environments. In this work we consider two different off-the-shelf general-purpose stiff ODE solvers and analyze their performance for two different Diesel combustion models in the context of the simulations of a FIAT engine.

In Section 2 we briefly present the different models for Diesel combustion, in Section 3 we introduce the solvers. We test two existing stiff ODE solvers, based on adaptive implicit methods, named SDIRK4 and VODE. In Section 4 we describe an improved version of the KIVA3V-II code for engine applications, where the original code has been interfaced with a parallel combustion solver developed by the authors for effective solution of detailed chemistry models in the context of engine simulations. In Section 5 we describe the test cases and the setup of the experiments and we discuss some preliminary results. In Section 6 we report some conclusions.

2 Diesel Combustion Models

Diesel engine combustion is characterized by liquid fuel injection in a turbulent environment. The related physical model includes submodels driving the different phenomena involved in the liquid phase, (the *cold phase*), and in the *combustion phase*, that is characterized in turn by two kinetic mechanisms, one for the low temperature combustion, responsible for the *ignition delay*, one for the high temperature combustion, responsible for the *heat release*. Main goals of numerical simulations include the analysis of the complex chemistry of Diesel combustion, using detailed chemical kinetics schemes. These schemes involve a very high number of intermediate species whose concentrations are very low but dramatically affect reaction rates, leading to high physical stiffness. In our experiments, two different detailed kinetic schemes have been employed. In particular, we used a chemistry model developed by Gustavsson e Golovitchev [7], involving 57 chemical species and 290 kinetic equations and considering N-eptane (C_7H_{16}) as primary fuel. The chemistry model is conceptually reported in Fig. 1. It considers the H abstraction and the oxidation of the primary fuel, with production of alchil-peroxy-radicals, followed by the ketoydroperoxide branching.

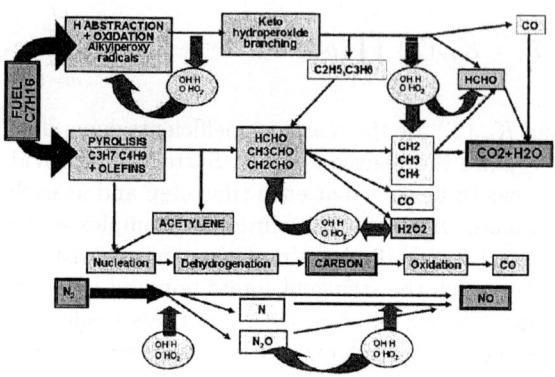

Fig. 1. Kinetic scheme for the N-eptane combustion

In the model the fuel pirolysis determines the chetons and olefins formation. Moreover, a scheme of soot formation and oxidation is provided, together with a classical scheme of NOx formation. The model is characterized by a high degree of stiffness, indeed during the simulation the ratio between minimum and maximum characteristic times of the species destruction rates[1] reaches the order of about 10^{18}. Since the liquid phase parameters of the N-eptane are significantly different from the diesel fuel, a different scheme is also considered, using N-dodecane ($C_{12}H_{26}$) as primary fuel. The two schemes are very similar, except that in the N-dodecane one a preliminary block of equations is responsible for the reduction of the primary fuel to eptil-radicals. Also in this model stiffness reaches the order of about 10^{18}.

3 The ODE Solvers

In this section we briefly describe main features of the general-purpose ODE solvers we tested, namely, SDIRK4 and VODE packages. The first one is based on a one-step integration formula, while VODE is based on linear multi-step formulas. In order to motivate our choices, we recall some concepts that play a special role in the framework of the solution of stiff systems. For a deep insight the matter we address to scientific literature about the topic.

In the following, we refer to a well-posed and stable (no eigenvalue of the Jacobian matrix has a positive real part) initial value ordinary differential problem $\dot{y} = f(t, y)$, with $t \in I \subseteq \Re$, $y \in \Re^N$ and initial condition $y(t_0) = y_0$. The problem is called *stiff* in the interval $I_s \subseteq I$ if at least one eigenvalue of the Jacobian matrix has large absolute value of its real part in I_s. Efficient and reliable integration of stiff equations requires implicit methods with suitable stability properties, in order to allow large step sizes and not be sensitive to initial transients. Let S be the region of linear stability and $R(z)$ the associate, complex,

---

[1] The characteristic times of the species destruction rates are an estimate of the eigenvalues of the Jacobian matrix of the right-hand side of system (1).

linear stability function of an ODE discretization method. We start recalling that a method is *absolutely stable (A-stable)* if $S \supseteq \{z \in C : Re(z) \leq 0\}$. It is well known [2,8] that A-stability does not guarantee efficient solution of stiff problems; ad hoc properties, such as, stiff accuracy, stiff decay, stiff and L-stability have been stated in order to characterize methods that are appropriate in this framework. We consider $R_1 = \{z \in C : Re(z) \leq -a\}$ and $R_2 = \{z \in C : -a \leq Re(z) < 0, -c \leq Im(z) \leq c\}$ with $a, c > 0$. If $S \supseteq R_1 \cup R_2$ the method is said *stiffly stable*. Stiff stability was introduced by Gear [6] in his analysis of linear multi-step methods. It is a weaker property than absolute stability, since, in practice, it requires that the method is stable in a subplane of the complex plane C when dealing with complex numbers with large absolute value of real part, while it relaxes the required conditions close to the origin. *L-stability* is instead stronger than A-stability: if the method is A-stable and $\lim_{z \to \infty} R(z) = 0$ the method is said L-stable. The latter property is motivated by the fact that A-stability can not preserve a method from producing values, in successive iterations, that have the same magnitudo order; this, obviously, is not desiderable in the solution of stiff problems. Stiff accuracy is a property of some Runge-Kutta (RK) methods that in some cases implies L-stability [8]. It is defined stating special relations among the parameters that characterize a RK method; in particular, if a RK method is stiffly accurate, then it satisfies a *stiff decay* property [2]. Let us consider the test equation $\dot{y} = \lambda(y - g(t))$, with g bounded. Let $t_n > 0$ and h_n be the step size at step n of a discretization method. We say that the method has stiff decay if $h_n Re(\lambda) \to -\infty \Rightarrow |y_n - g(t_n)| \to 0$; roughly speaking, rapidly varying solution components are skipped, the others are accurately described.

SDIRK4. This package [8] is based on a 5-stages Singly Diagonally Implicit Runge-Kutta (SDIRK) method of order four, with variable step size control. We recall that an s-stages RK method can be expressed in the general form:

$$\begin{cases} Y_i = y_{n-1} + h_n \sum_{j=1}^{s} a_{ij} f(t_{n-1} + c_j h_n, Y_j), & i = 1, s \\ y_n = y_{n-1} + h_n \sum_{i=1}^{s} b_i f(t_{n-1} + c_i h_n, Y_i), \end{cases} \quad (3)$$

that is, any particular RK method is characterized by a special choice of matrix $A = (a_{ij})$ and vector $b = (b_i)$, and $c_i = \sum_{j=1}^{s} a_{ij}$, $i = 1, \ldots, s$. A RK method is said singly diagonally implicit if A is lower triangular and $a_{ii} = a \ \forall \ i = 1, s$. SDIRK method implemented in SDIRK4 is L-stable and has stiff decay property [8]. Furthermore, one-step formulas such as RK methods are very attractive for applications in the context of time-splitting procedures, where a large number of solver restarts are required. Indeed, one-step formulas allow fast increase in step size after a restart. The main computational kernel of SDIRK4 is the solution of s non-linear systems of dimension N, at each time step. Applying a simplified Newton method for non-linear systems, at each non-linear iteration we need to solve linear systems of dimension $N \cdot s$, all having the same coefficient matrix $I - h_n a \partial f(t_{n-1}, y_{n-1})/\partial y$; thus, only one Jacobian evaluation and one LU factorization is required at each time step. In our experiments, we consider the default choice for parameter setting and the software option for a numerical internally computed full Jacobian. Linear algebra kernels are solved via routines from package DECSOL [8].

VODE. This package is designed for both stiff and non-stiff systems. It uses variable coefficient Adams-Moulton methods and variable coefficient Backward Differentiation Formulas (BDF) [8] in the non-stiff and stiff case, respectively. A basic linear multi-step formula has the general following expression:

$$\sum_{i=0}^{K_1} \alpha_{n,i} y_{n,i} + h_n \sum_{i=0}^{K_2} \beta_{n,i} \dot{y}_{n,i} = 0. \qquad (4)$$

BDF are obtained setting $K_1 = q$, $K_2 = 0$. BDF of order not exceeding six are stiffly stable [8] and satisfy stiff decay property [2]. VODE implements BDF of orders q from one through five, with an automatic, adaptive technique for the selection of order and step size. VODE provides several methods for the solution of the systems of algebraic equations; we used modified Newton method with numerically computed Jacobian. We set the VODE option for Jacobian reusing: in this case the software automatically detects the need to recompute Jacobian, thus preserving efficiency. At each step of the nonlinear solver, the linear systems are solved using routines from LINPACK [5]. Adaptivity properties of VODE, both in the formula order selection and in the Jacobian reusing, make this software very effective in the solution of stiff ODEs. However, in a parallel setting they motivate load imbalance.

4 Interfacing KIVA3V with Parallel Combustion for Realistic Simulations

In order to obtain realistic simulations of Diesel combustion, we developed a parallel combustion solver to be interfaced with KIVA3V-II [3,4]. KIVA3V-II is a software code, originally developed at Los Alamos National Laboratories, for numerical simulations of chemically reactive flows with spray in engine applications [1]. Numerical procedures implemented into KIVA3V-II solve, by a finite volume method, the complete system of unsteady Navier-Stokes equations for a turbulent multi-component mixture of ideal gases, coupled to equations for modeling vaporizing liquid fuel spray and combustion. Numerical solutions of the equations are obtained by applying a time-splitting technique that decouples fluid flow phenomena from spray and combustion, leading to a solution strategy for which, at each time step, a sequence of three different sub-models has to be solved. In order to improve capability of KIVA3V-II code while dealing with arbitrary large and complex reaction models, we integrated the original sequential code with a parallel combustion solver, based on new software for setting of chemical reaction systems and for solving the resulting ODEs. The parallel combustion solver is based on a SPMD parallel programming model; the processes solve concurrently the subset of ODE systems, modeling chemical reactions on subgrids of the overall computational grid. It is written in Fortran and uses standard MPI API for message passing. Our software is based on CHEMKIN-II [9], a software package for managing large models of chemical reactions in the context of simulation software. It provides a database and a software library for computing model parameters involved in system (1).

In the original KIVA3V-II code, in order to reduce computational costs for solving system (1), some equilibrium assumptions are applied and chemical reactions that proceed kinetically are separated from ones considered in equilibrium. In our approach, no equilibrium assumption has been made. The full set of differential equations in (1) is solved in each computational cell by means of the general-purpose ODE solvers. Furthermore, in the original KIVA3V-II solution procedure, at each simulation cycle an overall time step is chosen, according to accuracy criteria, to solve both the chemical kinetic equations and the transport phenomena. In the improved KIVA3V-II code, we choose the overall time step as the dimension of the time-splitting interval in which chemical reaction equations have to be solved, demanding to the ODE solver possible selection of time sub-step sizes for the ODE system integration in each grid cell. In this way, chemical reaction time sub-steps are related to the local stiffness conditions of every sub-domain in which we divide the overall 3d domain, rather than to the overall conditions of the reactive system, refining the quality of the computation coherently with the complexity of the physical phenomena. In a parallel setting, this approach leads to computational load imbalance and grid cells partitioning strategy becomes a critical issue for efficiency. Our code supports three different grid distribution strategies, namely pure-block, pure-cyclic and random. Performance analysis on the impact of the different data distributions in simulations involving the Gustavsson-Golovitchev chemistry scheme for N-eptane combustion are reported in [3,4].

5 Numerical Experiments

Simulations have been carried out for a prototype, single cylinder Diesel engine, having characteristics similar to the 16 valves EURO IV Fiat Multijet with two different compression ratios (CR), in order to analyze the effects of the CR on the emissions. The engine characteristics are reported in Table 1. In order to fit a wide range of operative conditions both in the experimental measurements and

Table 1. Multijet 16V Engine Characteristic

Bore[mm]	82.0
Stroke[mm]	90
Compression ratio	15.5:1/16.5:1
Displacement[cm^3]	475
Valves per cylinder	4
Injector	microsac 7 holes Φ 0.140 mm
Injection apparatus	Bosch Common Rail III generation

in the numerical simulations, the production engine has been modified for having a swirl variable head. The engine is equipped with an external supercharging system to simulate intake conditions deriving from turbo-charging application. In addition, the exhaust pipe is provided with a motored valve to simulate the backpressure due to the turbocharger operation. In the experimental tests, performed at 1500 rpm and with an injected fuel quantity corresponding to a Mean Effective Pressure of 2 bars on the 4 cylinder engine, different values of Exhaust Gas Recirculation (EGR), rail pressure, and injection timing have been used.

Numerical experiments have been carried out on a Beowulf-class Linux cluster, made of 16 PCs connected via a Fast Ethernet switch, available at IM-CNR. Eight PCs are equipped with a 2.8GHz Pentium IV processor, while the others have a 3.8GHz Pentium IV processor. All the processors have a RAM of 1 GB and an L2 cache of 256 KB. We used the GNU Fortran compiler (version 3.2.2) and the LAM implementation (version 7.0) of MPI. In the stopping criteria for the application of the ODE solvers, we defined vectors *rtol* and *atol* for the local relative and absolute error control respectively. In all the experiments here analyzed we used the same tolerances: *atol* values were fixed in dependence of the particular chemical species, while all components of *rtol* were set to 10^{-3}.

Figure 2 shows some comparisons between experimental (continuous line) and simulated (dotted line) values of combustion pressure all obtained with CR=16.5:1, but with different injection timings (15 and 5 CA BTDC), EGR (0 % and 40 %) and rail pressure values (500 and 900 bar); in the lower part of the diagram the measured electro injector energy current (indicating injection timing) is also reported. Numerical predictions of Fig. 2 are referred to the results obtained using the parallel combustion solver based on the VODE package for the ODEs arising from the N-dodecane combustion model. Similar results could be shown for the N−eptane combustion model and also with the highest value of CR. We note that numerical results are in good agreement with experimental ones both in the low and in the high temperature combustion phase. This

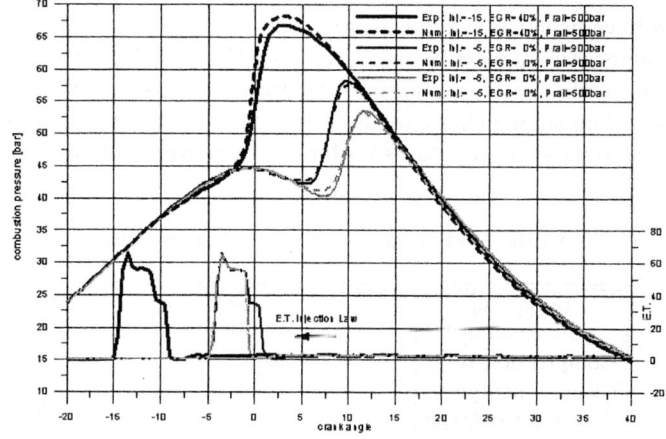

Fig. 2. Numerical and experimental data of combustion pressure

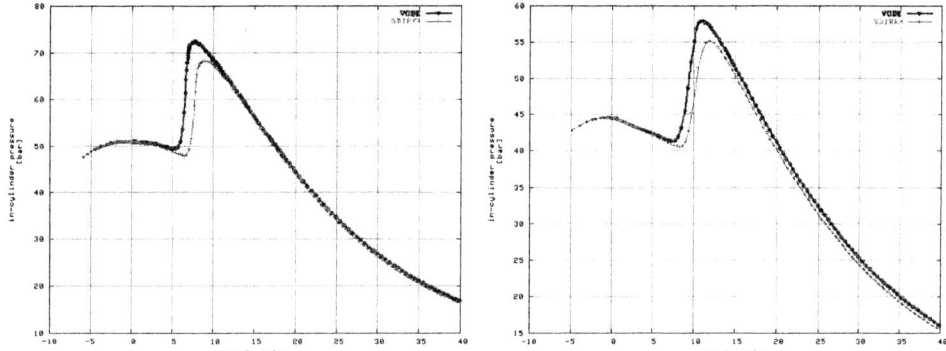

Fig. 3. Comparison between VODE and SDIRK4 simulated in-cylinder pressure. Left: N-eptane combustion model with CR=16.5:1; Right: N-dodecane combustion model with CR=17.5:1.

reveals that the physical submodels of the combustion process and the employed numerical algorithms can be reliably used to predict the in-cylinder pressure and the pollutant emissions. The same deep analysis on the impact of SDIRK4 on the reliability of the simulation results, varying operative conditions of the engine, is work in progress. However, we show in Figure 3 some preliminary results concerning simulated in-cylinder pressure, obtained with VODE and SDIRK4, both for N-eptane and N-dodecane models. They indicate that SDIRK4 has the same qualitative behaviour of VODE even though it seems to predict a higher ignition delay and, as a consequence, a lower peak of pressure. This aspect is at the moment under investigation.

In the following we show some performance results in using SDIRK4 and VODE packages, in conjunction with pure-block and random distribution of the computational grid. We show here results obtained for simulations involving the N-eptane combustion model and the test case with CR=16.5:1. The same analysis is work in progress for the N-dodecane model. Details on the partitioning strategies we adopted can be found in [3,4]. Our typical grid is a 3d cylindrical sector representing a sector of the engine cylinder and piston-bowl. It is formed by about 3000 cells. In our experiments we focus on the starting phase of the combustion process within each simulation; this phase essentially involves the part of the grid representing the piston bowl. The active grid involved in this phase, corresponding to the crank angle interval $[-5°, 40°]$, includes a number of cells that ranges between 800 and 1030.

In Fig. 4 the overall simulation time expressed in hours, for each process configuration, is shown. We observe that the random distribution, as discussed in [4], gives the best performance due to the best load balancing. In particular, SDIRK4 with random distribution allows to obtain the minimum simulation time, that is about fifty minutes on 16 processes. In any case SDIRK4 outperforms VODE. We now focus on the number of function evaluations performed within each simulation, that is a measure of the computational complexity of the

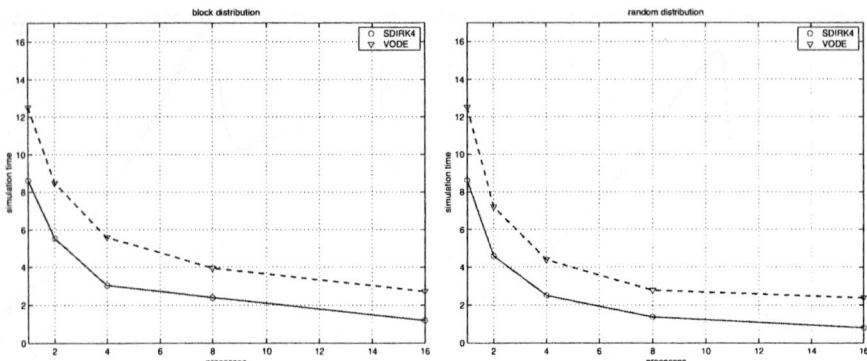

Fig. 4. Simulation time expressed in hours. Left: pure block distribution; Right: random distribution

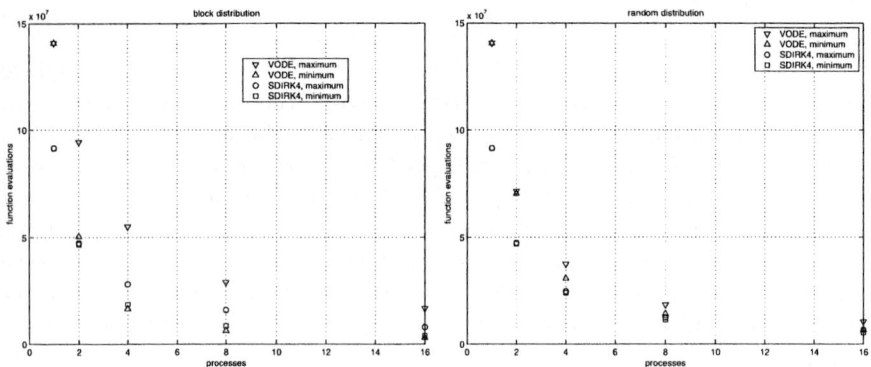

Fig. 5. Maximum and minimum number of performed function evaluations versus number of processes. Left: pure block distribution; Right: random distribution.

combustion solver. In Fig. 5 we report the minimum and the maximum number of performed function evaluations for each process configuration and for each solver, for both partitioning strategies. More precisely, for a fixed number of processes, we computed the total number of function evaluations performed by each process and reported the maximum and the minimum among such values. We note that, in agreement with results shown in Fig. 4, the maximum value of function evaluations performed by VODE is always larger than the one required by SDIRK4, leading to larger simulation times. Moreover, the distance between the maximum and the minimum is always lower for SDIRK4, for both distribution strategies. As we expected, this behaviour reveals that SDIRK4 produces better load balancing than VODE in a parallel setting. In Fig. 6 we report the speed-up lines for both the solvers and the partitioning strategies. Note that random partitioning produces the best speed-ups. In any case, SDIRK4 shows larger speed-up than VODE and better scalability properties. The previous analysis on the comparison between VODE and SDIRK4 on the N-eptane model indicates

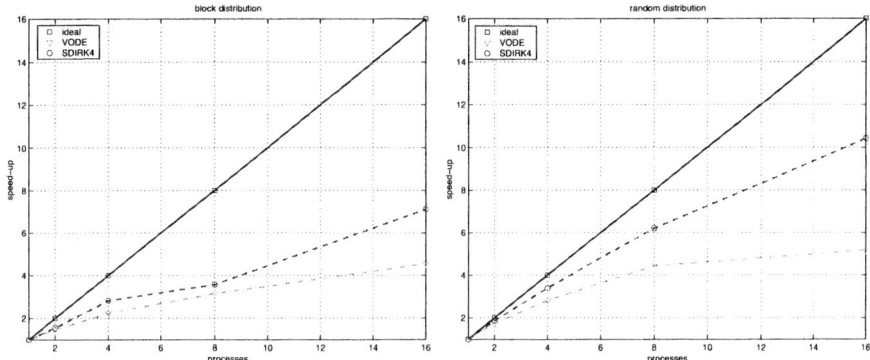

Fig. 6. Speed-up. Left: pure block distribution; Right: random distribution.

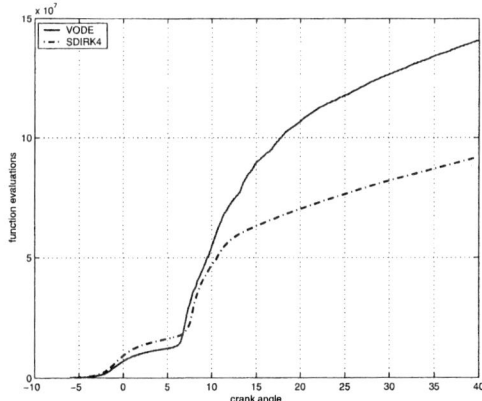

Fig. 7. Number of function evaluations versus crank angles in a single process simulation

that SDIRK4 outperforms VODE for our combustion simulations. However, in the following we show that VODE is more effective of SDIRK4 in the low temperature combustion phase, corresponding to the crank angle interval in which the combustion model produces highly stiff ODE systems. In Fig. 7 the number of function evaluations versus crank angles is shown. The reported values refer to a single process numerical simulation, the same analysis has to be carried on while increasing the number of processors. VODE is more efficient than SDIRK4 in the very first part of the engine cycle simulation phase we focus on, while, for values of crank angles in the interval $[8^o, 40^o]$, corresponding to the high temperature combustion, the computational complexity of the simulation employing VODE strongly increases with respect to the one performed with SDIRK4. The previous consideration leads to the idea that a multi-method simulation software, based on different ODE solvers, automatically chosen in dependence on the feature of the simulation phase, could be taken into account.

6 Conclusions

In this work we reported preliminary results on an analysis of the impact of different stiff ODE solvers in parallel simulations of combustion in Diesel engines. We compared a one-step ODE solver, based on a Single Diagonally Implicit Runge-Kutta formula of order four, with an adaptive multi-step solver, based on BDF formulas with order from one through five. Performance results seem to indicate that the one-step solver is more effective for our purpose, since it produces better load balancing and hence larger speed-up and good scalability properties. However, the multi-step solver seems to be better in the simulation phase in which stiffness is high. This aspect and some other aspects related to the accuracy and reliability of the one-step solver in realistic simulations are under investigation.

References

1. A. A. Amsden, KIVA-3V: A Block-Structured KIVA Program for Engines with Vertical or Canted Valves, *Los Alamos National Laboratory Report No. LA-13313-MS*, (1997).
2. U. M. Ascher, L. R. Petzold, Computer Methods for Ordinary Differential Equations and Differential-Algebraic Equations, *SIAM*, (1998).
3. P. Belardini,C. Bertoli,S. Corsaro,P. D'Ambra, Parallel Simulation of Combustion in Common Rail Diesel Engines by Advanced Numerical Solution of Detailed Chemistry, in , *Applied and Industrial Mathematics in Italy*, Proc. of the 7th Conference of SIMAI, M. Primicerio, R. Spigler, V. Valente eds., World Scientific Pub., (2005).
4. P. Belardini,C. Bertoli,S. Corsaro,P. D'Ambra, Multidimensional Modeling of Advanced Diesel Combustion System by Parallel Chemistry, *Society for Automotive Engineers (SAE) Paper*, 2005-01-0201, (2005).
5. Dongarra, J.J., Moler, C. B., Bunch J. R., Stewart, G. W., LINPACK Users' Guide, *SIAM*, (1979).
6. C. W. Gear, Numerical Initial Value Problems in Ordinary Differential Equations, *Prentice-Hall*, Englewood Cliffs, NY, (1973).
7. Gustavsson, J., Golovitchev, V.I., Spray Combustion Simulation Based on Detailed Chemistry Approach for Diesel Fuel Surrogate Model, *Society for Automotive Engineers (SAE) Paper*, 2003-0137, (2003).
8. E. Hairer, G. Wanner, Solving Ordinary Differential Equations II. Stiff and Differential-Algebraic Problems, second edition, *Springer Series in Comput. Mathematics*, Vol. 14, Springer-Verlag, (1996).
9. R. J. Kee, F. M. Rupley, J. A. Miller, Chemkin-II: A Fortran chemical kinetics package for the analysis of gas-phase chemical kinetics, *SAND89-8009, Sandia National Laboratories*, (1989).

FAST-EVP: An Engine Simulation Tool

Gino Bella[1], Alfredo Buttari[1], Alessandro De Maio[2],
Francesco Del Citto[1], Salvatore Filippone[1], and Fabiano Gasperini[1]

[1] Faculty of Engineering, Università di Roma "Tor Vergata",
Viale del Politecnico, I-00133, Rome, Italy
[2] N.U.M.I.D.I.A s.r.l. Rome, Italy

Abstract. FAST-EVP is a simulation tool for internal combustion engines running on cluster platforms; it has evolved from the KIVA-3V code base, but has been extensively rewritten making use of modern linear solvers, parallel programming techniques and advanced physical models.

The software is currently in use at the consulting firm NUMIDIA, and has been applied to a diverse range of test cases from industry, obtaining simulation results for complex geometries in short time frames.

1 Introduction

The growing concern with the environmental issues and the request of reduced specific fuel consumption and increased specific power output are playing a substantial role on the development of automotive engines. Therefore, in the last decade, the automotive industries have undergone a period of great changes regarding the engines design and development methods with an increased commitment to research by the industry. In particular, the use of CFD have revealed to be of great support for both the design and the experimental work in order to quickly achieve the projects targets while reducing the product development costs. However, considerable work is still needed since CFD simulation of realistic industrial applications may take many hours or even weeks that not always agrees with the very short development times that are required to a new project in order to be competitive in the market.

The KIVA code [10] solves the complete system of general time-dependent Navier-Stokes equations and it is probably the most widely used code for internal combustion engines modeling. Its success mainly depends on its open source nature, which means having access to the source code. KIVA has been significantly modified and improved by researchers worldwide, especially in the development of sub-models to simulate the important physical processes that occurs in an internal combustion engine (i.e. fuel-air mixture preparation and combustion).

In the following we will review the basic mathematical model of the Navier-Stokes equations as discretized in our application; we will describe the approach to the linear system solution based on the library routines from [8,7]. In particular we outline the new developments in the spray dynamics and explicit flux phases that resulted in the code being used today; finally, we show some experimental results.

2 Mathematical Model

The mathematical model of KIVA-3 is the complete system of general *unsteady Navier-Stokes equations*, coupled with chemical kinetic and spray droplet dynamic models. In the following the equations for fluid motion are reported.

- Species continuity:

$$\frac{\partial \rho_m}{\partial t} + \nabla \cdot (\rho_m \mathbf{u}) = \nabla \cdot [\rho D \nabla(\frac{\rho_m}{\rho})] + \dot{\rho}_m^c + \dot{\rho}_m^s \delta_{ml} \tag{1}$$

where ρ_m is the mass density of species m, ρ is the total mass density, $\mathbf{u}$ is the fluid velocity, $\dot{\rho}_m^c$ is the source term due to chemistry, $\dot{\rho}_m^s$ is the source term due to spray and δ is the Dirac delta function.

- Total mass conservation:

$$\frac{\partial \rho}{\partial t} + \nabla \cdot (\rho \mathbf{u}) = \dot{\rho}^s \tag{2}$$

- Momentum conservation:

$$\frac{\partial (\rho \mathbf{u})}{\partial t} + \nabla \cdot (\rho \mathbf{u} \mathbf{u}) = -\frac{1}{\alpha^2}\nabla p - A_0 \nabla(\frac{2}{3}\rho k) + \nabla \cdot \overline{\sigma} + \mathbf{F}^s + \rho \mathbf{g} \tag{3}$$

where $\overline{\sigma}$ is the viscous stress tensor, $\mathbf{F}^s$ is the rate of momentum gain per unit volume due to spray and $\mathbf{g}$ is the constant specific body force. The quantity A_0 is zero in laminar calculations and unity when turbulence is considered.

- Internal energy conservation:

$$\frac{\partial (\rho I)}{\partial t} + \nabla \cdot (\rho I \mathbf{u}) = -p\nabla \cdot \mathbf{u} + (1 - A_0)\overline{\sigma} : \nabla \mathbf{u} - \nabla \cdot \mathbf{J} + A_0 \rho \epsilon + \dot{Q}^c + \dot{Q}^s \tag{4}$$

where I is the specific internal energy, the symbol : indicates the matrix product, $\mathbf{J}$ is the heat flux vector, $\dot{Q}^c$ and $\dot{Q}^s$ are the source terms due to chemical heat release and spray interactions.

Furthermore the standard $K - \epsilon$ equations for the turbulence are considered, including terms due to interaction with spray.

2.1 Numerical Method

The numerical method employed in KIVA-3 is based on a variable step *implicit Euler* temporal finite difference scheme, where the time steps are chosen using accuracy criteria. Each time step defines a cycle divided in three phases, corresponding to a physical splitting approach. In the first phase, spray dynamic and chemical kinetic equations are solved, providing most of the source terms; the other two phases are devoted to the solution of fluid motion equations [1]. The spatial discretization of the equations is based on a *finite volume method*, called the *Arbitrary Lagrangian-Eulerian method* [16], using a mesh in which positions of the vertices of the cells may be arbitrarily specified functions of time. This approach allows a mixed Lagrangian-Eulerian flow description. In the Lagrangian phase, the vertices of the cells move with the fluid velocity and there is no convection across cell boundaries; the diffusion terms and the terms associated with pressure wave propagation are implicitly solved by a modified version of the SIMPLE (Semi Implicit Method for Pressure-Linked Equations) algorithm [15]. Upon convergence on pressure values, implicit solution of the diffusion terms in the turbulence equations is approached. Finally, explicits methods, using integral sub-multiple time-steps of the main computational time step, are applied to solve the convective flow in the Eulerian phase.

3 Algorithmic Issues

One of the main objectives in our work on the KIVA code [7] was to show that general purpose solvers, based on up-to-date numerical methods and developed by experts, can be used in specific application codes, improving the quality of numerical results and the flexibility of the codes as well as their efficiency.

The original KIVA code employs the Conjugate Residual method, one member of the Krylov subspace projection family of methods [3,12,13,17]. Krylov subspace methods originate from the Conjugate Gradient algorithm published in 1952, but they became widely used only in the early 80s. Since then this field has witnessed many advances, and many new methods have been developed especially for non-symmetric linear systems. The rate of convergence of any given iterative method depends critically on the eigenvalue spectrum of the linear system coefficient matrix. To improve the rate of convergence it is often necessary to *precondition* it, i.e. to transform the system into an equivalent one having better spectral properties. Thus our work started with the idea of introducing new linear system solvers and more sophisticated preconditioners in the KIVA code; to do this we had to tackle a number of issues related to the code design and implementation.

3.1 Code Design Issues

KIVA-3 is a finite-volume code in which the simulation domain is partitioned into hexahedral cells, and the differential equations are integrated over the cell to obtain the discretized equation, by assuming that the relevant field quantities are constant over the volume. The scalar quantities (such as temperature, pressure and turbulence parameters), are evaluated at the centers of the cells, whereas the velocity is evaluated at the vertices of the cells. The cells are represented through the coordinates of their vertices; the vertex connectivity is stored explicitly in a set of three connectivity arrays, from which it is possible by repeated lookup to identify all neighbours, as shown in figure 1. The original implementation of the CR solver for linear employs a *matrix-free* approach, i.e. the coefficient matrix is not formed explicitly, but its action is computed in an equivalent way whenever needed; this is done by applying the same physical considerations that would be needed in computing the coefficients: there is a main loop over all cells, and for each cell the code computes the contribution from the given cell into the

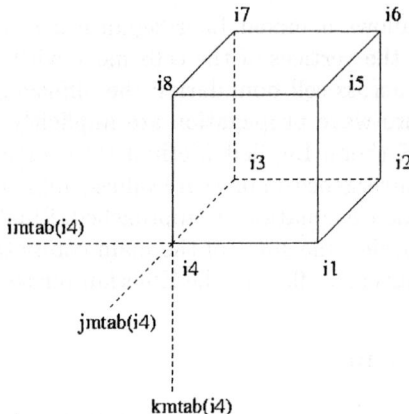

Fig. 1. Vertex numbering for a generic control volume

components corresponding to all adjacent cells (including itself), from $i1$ through $i8$. This is a "scatter" approach, quite different from the usual "gather" way of computing a matrix-vector product.

The major advantage of a matrix-free implementation is in terms of memory occupancy. However it constraints the kind of preconditioners that can be applied to the linear system; in particular, it is difficult to apply preconditioners based on incomplete factorizations. Moreover, from an implementation point of view, data structures strictly based on the modeling aspects of the code do not lend themselves readily to transformations aimed at achieving good performance levels on different architectures.

The above preliminary analysis has influenced the design of the interface to the sparse linear solvers and support routines in [8] with the following characteristics:

1. The solver routines are well separated into different phases: matrix generation, matrix assembly, preconditioner computation and actual system solution;
2. The matrix generation phase requires an user supplied routine that generates (pieces of) the rows of the matrix in the global numbering scheme, according to a simple storage scheme, i.e. coordinate format;
3. The data structures used in the solvers are parametric, and well separated from those used in the rest of the application.

4 Integration of the Numerical Library

The basic groundwork for the parallelization and integration of the numerical library has been laid out at the time of [7], which we briefly review below.

The code to build the matrices coefficients has been developed starting from the original solver code: the solvers in the original KIVA code are built around routines that compute the residual $r = b - Ax$, and we started from these to build the right hand side b and the matrix A. Since the solution of equations for thermodynamic quantities (such as temperature, pressure and turbulence) requires cell center and cell face values, the non-symmetric linear systems arising from temperature, pressure and turbulence equations have coefficient matrices with the same symmetric sparsity pattern, having no more than 19 nonzero entries per row. In the case of the linear systems arising from the velocity equation, following the vectorial solution approach used in the original code, the unknowns are ordered first with respect to the three Cartesian components and then with respect to the grid points. The discretization scheme leads to non-symmetric coefficient matrices with no more than 27 entries per (block) row, where each entry is a 3×3 block.

4.1 Algorithmic Improvements

The original KIVA-3 code solution method, the Conjugate Residual method, is derived under the hypothesis of a symmetric coefficient matrix; thus, there is no guarantee that the method should converge on non-symmetric matrices such as the ones we encounter in KIVA. Therefore we went to search for alternative solution and preconditioning methods. Since the convergence properties of any given iterative method depend on the eigenvalue spectrum of the coefficient matrices arising in the problem domain, there no reason to expect that a single method should perform optimally under all circumstances [11,14,17]. Thus we were led to an experimental approach, in searching for the best compromise between preconditioning and solution methods. We settled on the Bi-CGSTAB method for all of the linear systems in the SIMPLE loop; the critical solver is that for the pressure correction equation, where we employed a block ILU preconditioner, i.e. an incomplete factorization based on the local part of A. The BiCGSTAB method always converged, usually in less than 10 iterations, and practically never in more than 30 iterations, whereas the original solver quite often would not converge at all.

Further research work on other preconditioning schemes is currently ongoing, and we plan to include its results in future versions of the code [4,5].

5 Parallelization Issues and New Developments

Since the time of [7] the code has undergone a major restructuring: FAST-EVP code is now based on the KIVA-3V version, and thus it is able to model valves. This new modeling feature has no direct impact on the SIMPLE solvers interface, but it is important in handling mesh movement changes. While working on the new KIVA-3V code base, we also reviewed all of the space allocation requirements, cleaning up a lot of duplications; in short we have now an application fully parallelized, even in its darkest parts.

All computations in the code are parallelized with a domain decomposition strategy: the computational mesh is partitioned among the processors participating in the computation. This partitioning is induced by the underlying assumptions in the linear system solvers; however it is equally applicable to the rezoning phase. The support library routines allow us to manage the necessary data exchanges throughout the code based on the same data structures employed for the linear system solvers; thus, we have a unifying framework for all computational phases.

The rezoning phase is devoted to adjusting the grid points following the application of the fluid motion field; the algorithm is an explicit calculation that is based on the same "gather" and "scatter" stencils found in the matrix-vector products for the linear systems phase in Fig. 1. It is thus possible to implement in parallel the explicit algorithm by making use of the data movement operations defined in the support library [8].

The chemical reaction dynamics is embarassingly parallel, because it treats the chemical compounds of each cell independently.

For the spray dynamics model we had to implement specific operators that follow the spray droplets in their movement, transferring the necessary information about the droplets whenever their simulated movement brings them across the domain partition boundaries.

5.1 Mesh Movement

The simulation process modifies the finite volume mesh to follow the (imposed) piston and valve movement during the engine working cycle (see also Fig. 2. The computational mesh is first deformed by reassigning the positions of the finite volume surfaces in the direction of the piston movement, until a critical value for the cell aspect ratio is reached; at this point a layer is cut out (or added into) the mesh to keep the aspect ratio within reasonable limits. When this "snapper" event takes place it is necessary to repartition the mesh and to recompute the patterns of the linear system matrices. The algorithm for matrix assembly takes into account the above considerations by preserving the matrix structure between two consecutive "snapper" points, and recomputing only the

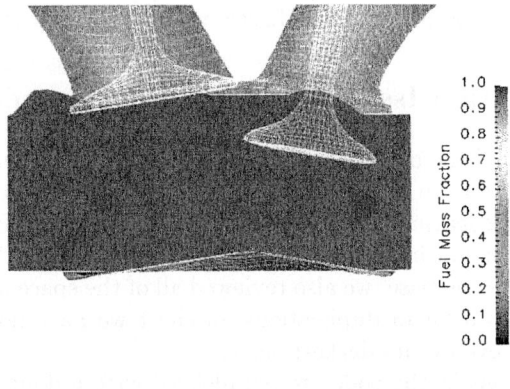

Fig. 2. Competition engine simulation

values of the non-zero entries at each invocation of the linear system solver; this is essential to the overall performance, since the computation of the structure is expensive.

Similarly, the movement of valves is monitored and additional "snapper" events are generated accordingly to their opening or closing; the treatment is completely analogous to that for the piston movement.

6 Experimental Results

The first major test case that we discuss is based on a high performance competition engine that was used to calibrate our software. The choice of this engine was due to the availability of measurements to compare against, so as to make sure not to introduce any modifications in the physical results. Moreover it is a very demanding and somewhat extreme test case, because of the high rotation regime, high pressure injection conditions, and relatively small mesh size.

A section of the mesh, immediately prior to the injection phase, is shown in Fig. 2; the overall mesh is composed of approximately 200K control volumes. The simulated comprises 720 degrees of crank angle at 16000 rpm, and the

Table 1. Competition engine timings

Processes	Time steps	Total time (min)
1	2513	542
2	2515	314
4	2518	236
5	2515	186
6	2518	175
7	2518	149

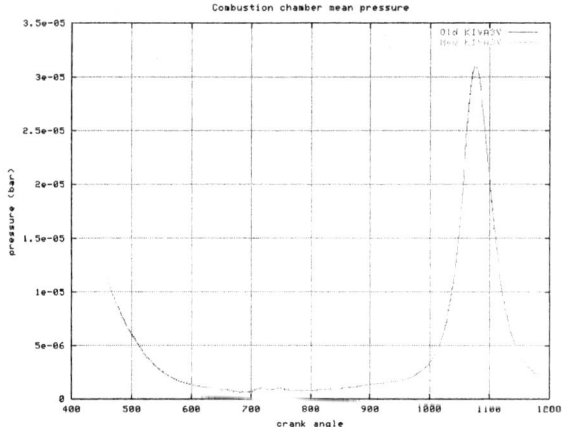

Fig. 3. Competition engine average pressure

overall timings are shown in Table 1. The computation has been carried out at NUMIDIA on a cluster based on Intel Xeon processors running at 3.0 GHz, equipped with Myrinet M3F-PCIXD-2 network connections. The physical results were confirmed to be in line with those obtained by the original code, as shown in fig. 3 and 4.

Another interesting test case is shown in Fig. 5; here we have a complete test of a commercial engine cylinder coupled with an air box, running at 8000 rpm, of which we detail the resulting airflow. The discretization mesh is composed of 483554 control volumes; the simulation comprises the crank angle range from 188

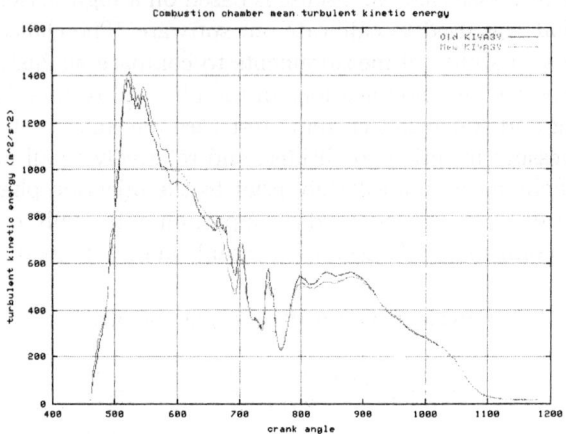

Fig. 4. Competition engine average turbulent energy

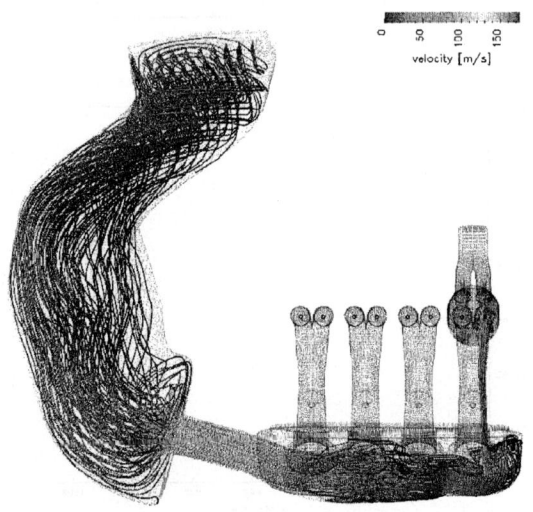

Fig. 5. Commercial engine air flow results

to 720 degrees, with 4950 time steps, and it takes 703 minutes of computation on 9 nodes of the Xeon cluster. In this particular case the grid had never been tested on a serial machine, or on smaller cluster configurations, because of the excessive computational requirements; thus the parallelization strategy has enabled us to obtain results that would have been otherwise unreachable.

7 Conclusion

We have discussed a new engine design application, based on an extensive revision and rewrite of the KIVA-3V application code, and parallelized by use of the PSBLAS library. The application has been tested on industrial test cases, and has proven to be robust and scalable, enabling access to results that were previously impossible.

Future development work includes further refinement of the explicit rezoning and spray dynamics phases, and experimentation with new preconditioners and linear system solvers.

Acknowledgments

The authors wish to thank the anonymous reviewers for their comments on the paper draft.

References

1. A.A. Amsden, P.J. O'Rourke, T.D. Butler, *KIVA-II: A Computer Program for Chemically Reactive Flows with Sprays*, Los Alamos National Lab., Tech. Rep. LA-11560-MS, 1989.
2. A.A. Amsden, *KIVA 3: A KIVA Program with Block Structured Mesh for Complex Geometries*, Los Alamos National Lab., Tech. Rep. LA-12503-MS, 1993.
3. R. Barrett, M. Berry, T. Chan, J. Demmel, J. Donat, J. Dongarra, V. Eijkhout, R. Pozo, C. Romine, and H. van der Vorst. *Templates for the solution of linear systems*. SIAM, 1993.
4. A. Buttari, P. D'Ambra, D. di Serafino and S. Filippone: *Extending PSBLAS to Build Parallel Schwarz Preconditioners*, in Proceedings of PARA'04, to appear.
5. P. D'Ambra, D. Di Serafino, and S. Filippone: *On the Development of PSBLAS-based Parallel Two-level Schwarz Preconditioners*, Preprint n. 1/2005, Dipartimento di Matematica, Seconda Università di Napoli, 2005.
6. P. D'Ambra, S. Filippone, P. Nobile, *The Use of a Parallel Numerical Library in Industrial Simulations: The Case Study of the Design of a Two-Stroke Engine*, Proc. of Annual Meeting of ISCS'97, Istituto Motori-CNR ed., 1998.
7. S. Filippone, P. D'Ambra, M. Colajanni: *Using a Parallel Library of Sparse Linear Algebra in a Fluid Dynamics Applications Code on Linux Clusters*. Parallel Computing - Advances & Current Issues, G. Joubert, A. Murli, F. Peters, M. Vanneschi eds., Imperial College Press Pub. (2002), pp. 441 448.
8. S. Filippone and M. Colajanni. PSBLAS: A library for parallel linear algebra computation on sparse matrices. *ACM Trans. Math. Softw.*, 26(4):527–550, December 2000.

9. J. Dongarra and R. Whaley, *A user's guide to the BLACS v1.0.* LAPACK working note #94 (June), University of Tennessee, 1995. http://www.netlib.org/lapack/lawns.
10. P.J. O'Rourke, A.A. Amsden, *Implementation of a Conjugate Residual Iteration in the KIVA Computer Program*, Los Alamos National Lab., Tech. Rep. LA-10849-MS, 1986.
11. N. M. Nachtigal, S. C. Reddy, L. N. Threfethen: How fast are nonsymmetric matrix iterations?, SIAM J. Matrix Anal. Applic., 13(1992), 778-795.
12. C. T. Kelley: Iterative Methods for Linear and Nonlinear Equations, SIAM, 1995.
13. A. Greenbaum: Iterative Methods for Solving Linear Systems, SIAM, 1997.
14. L. N. Threfethen: Pseudospectra of linear operators, SIAM Review, 39(1997), 383-406.
15. S.V. Patankar, *Numerical Heat Transfer and Fluid Flow*, Hemisphere Publ. Corp.
16. W.E. Pracht, *Calculating Three-Dimensional Fluid Flows at All Speeds with an Eulerian-Lagrangian Computing Mesh*, J. of Comp. Physics, Vol.17, 1975.
17. Y. Saad, Iterative Methods for Sparse Linear Systems, PWS Pub., Boston, 1996.

A Mobile Communication Simulation System for Urban Space with User Behavior Scenarios

Takako Yamada, Masashi Kaneko, and Ken'ichi Katou

The University of Electro-Communications, Graduate School of Information Systems,
1-5-1 Choufugaoka, Choufu-shi, Tokyo, 182-8585, Japan
{takako, masashi, kkatou}@ymd.is.uec.ac.jp

Abstract. We present herein a simulation system to model mobile user behavior in urban spaces covering roads, railways, stations and traffic signals. The proposed simulation system was developed in order to provide analysis in situations that cannot be dealt with analytically, such as mobile users moving along roads and the arrival or departure on commuter trains of group users at stations. In the present paper we observe mobile traffic, which changes with the user distribution, in order to discuss base station location policies. The proposed simulation system can deal with these user behaviors in actual urban space with roads, crossings and stations. The movement of mobile users and the effects thereof on the performance of the mobile system are examined with respect to service quality, such as call loss probability (C.L.P). Using the proposed simulation system, we observe and compare several cell location patterns and discuss the influence of retrial calls.

1 Introduction

The number of mobile phones in Japan is approximately 80 million and continues to increase. Mobile phone networks are considered to be an important component of the infrastructure of Japan. In addition, the demand for data packet service in mobile communication networks is increasing. New communication services are emerging, including high-speed data packet transfer in 3rd generation mobile services and wireless LAN and IP phone services using PHS networks. Thus the importance of reliability and service quality for these services has become a very important issue. The utilization of channels in wireless communication services is quite important because radio frequency is a finite resource.

In cellular mobile services, service areas are divided into small areas called cells, each of which has a base station at the cell center. Recently, hot spots served by public area wireless networks (PAWNs) have been expected to become a cost-effective complementary infrastructure for third-generation cellular systems. PAWNs are base-station-oriented wireless LANs that offer tens of megabits per second to public or private hot spot users. One of the important problems for mobile communication services is to provide homogeneous service. In cellular service, the resource provided by a single base station is limited. Channel exhaustion in crowded cells is more frequent than in cells with fewer users. Thus, accurate estimation of traffic demands in the service area is important with respect to base station allocation. Moreover, the traffic due to mobile users who move from one cell to another requires handover procedures between related base stations, and such handover calls are lost when there are no free channels in the following cells.

Performance evaluation has been discussed analytically in previous studies[1,2,3,4]. In most of these studies, the queueing models, such as M/M/k/k models (in which channels are regarded as servers in the system)[5], are used to evaluate the stochastic characteristics, such as channel occupancy time. Typical analytical models assume an exponential distribution for call arrival intervals, channel occupation time, and user sojourn time in each cell. Thus, in analytical models, the following situations are difficult to consider.

- Arrival of groups of users at stations and crosswalks.
- Uneven distribution of users, moving directions caused by roads and buildings and fluctuations of these distributions which may change from time to time.
- Retrial calls that independently arrive after failed calls.

The distribution by computer simulation in two-dimensional space of actual population distribution data and road traffic data were used by Tutschku[6] and Matsushita[7]. Nousiainen, Kordybach, and Kemppi[8] investigated user movement using a city map, although the movement behavior was very simple. However, few models have discussed user mobility along roads and user call retrial behavior. We have been investigating walking pedestrian models[9] based on rogit model[10] and mobile user models[11]. In the present paper, we propose a moving pedestrians (users) model in an urban space in order to consider urban congestion in areas that include crosswalks, signals along sidewalks, stations and buildings. The walking mobile users in the proposed model call and retry their calls when the call fails. We develop a simulation system based on the proposed moving user model. The variations of users having different call behavior features and movement characteristics are generated by scenarios written in custom macros. Using these scenarios, the simulation has sufficient flexibility to deal with various situations easily.

The proposed simulation system can evaluate traffic for a cellular telecommunication system using user distribution data together with actual city maps, as shown in Figure 1. We can confirm and clarify the simulation process and the traffic situation in each cell through graphical user interfaces and visualization of user behavior on the map. In the present paper, we selected Choufu city, a typical suburb of Tokyo. Based on statistics for pedestrians, population distribution for daytime and nighttime and on the railroad traffic data for Choufu station, we present the user distribution, movement, and call behavior in order to evaluate the call loss probability (C.L.P.). Furthermore, we examine the effect of retrial calls and base station location policies.

2 Mobile Communication Simulation System

A schematic diagram of the simulation system is shown in Figure 1. Input data of the simulation system are initial user location distribution, road network data, route connection data, cell data and simulation scenarios of user behavior, traffic signals and train schedules. The proposed simulation system can deal with service areas of several square kilometers, through which users move on foot along roads, or remain in residential areas or offices. With a simple editor developed for this simulation system, we can edit road networks that are converted into non-directional graph networks data. Furthermore, we can edit traffic signals, crossings, and other factors in order to organize the urban space and influence user movements. A macro language developed for this simulation is used to describe the simulation scenario. Scenarios are prepared for each simulation execution. The scenarios control the traffic equipment and define files for the initial user location distribution and the user movement patterns with origin and destination map data and route selection settings. Other distribution functions for call arrival rate, holding time and call retrial intervals can be included in the scenarios. By applying scenario files using the macro language, we can provide high flexibility to the proposed simulation system. Figure 2 is a map of the road network for pedestrians around Choufu station.

In this simulation, mobile phone users are connected to the nearest base station, and thus the service area is divided by a Voronoi diagram[12] of individual cells. When a mobile user moves from one cell to another cell while the user is talking, the base station should transfer the call to next base station. This procedure is called the handover process. The mobile user has a single chance to find a free channel in the approaching cell. If there is a free channel, then the base station transfers the call to the base station in the approaching cell. If there is no empty channel, then the call fails. We count the number of call failures as the number of terminated calls.

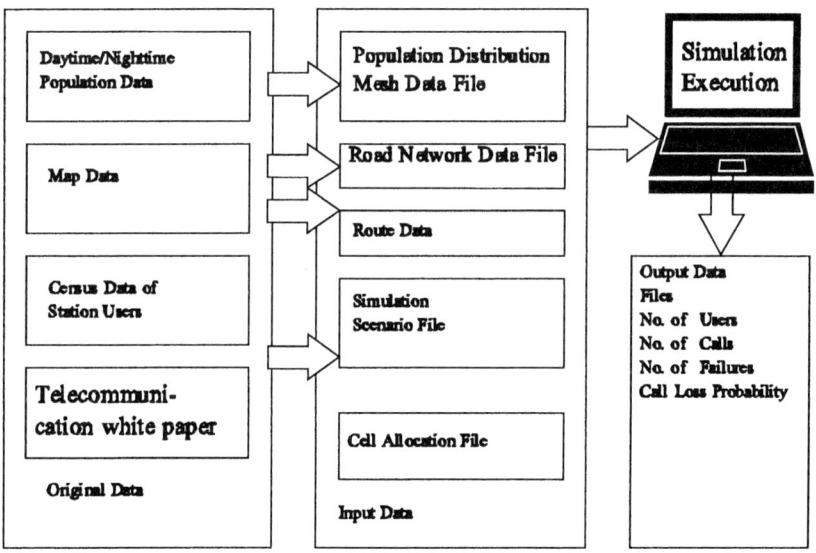

Fig. 1. Schematic diagram of the simulation system

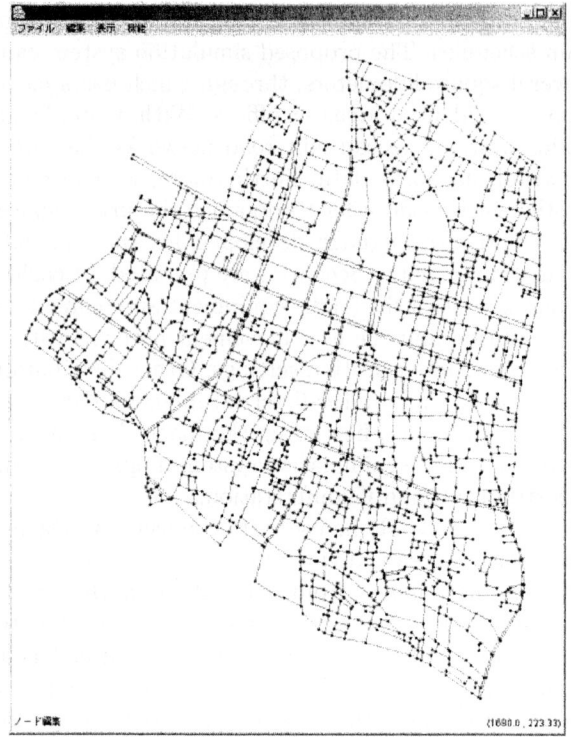

Fig. 2. Pedestrian road network around Choufu Station (graph diagram)

2.1 Simulation Setting by Scenarios

In the simulation scenarios used herein, we can define various simulation parameters, such as simulation clock, execution time, and file path, to indicate data file locations. These parameters and files define user behaviors according to type, as well as urban spaces in the service area. Table 1 shows an example of the description of a signal, "TESTSIGNALB", which controls user arrival. When the system receives the signal, one user is generated. By repeating this procedure, the users are generated according to an exponential distribution with average of 1.5-second time interval. Similarly a control signal, "TRAIN" generates commuter trains that arrive periodically at Choufu station in 150-second intervals.

2.2 Presentation of User Behavior

When we see pedestrians in a downtown area, each pedestrian has a unique destination and selects a unique route. The compilation of the walking patterns of each pedestrian consists of the flow of pedestrians on roads. In the proposed simulation system, we define typical movement patterns of pedestrians, who may or may not be mobile users. The pedestrian type defines the arrival and

Table 1. Example of signal definition

```
@SIGNAL TESTSIGNALB
@@LABEL
@@WAITE@1.5
@@ACTIVE
@@GOTO
@SIGNAL TRAIN
@@LABEL
@@WAITD@150
@@ACTIVE
@@GOTO
```

Table 2. Example of a scenario for the "USERA" pedestrian type

```
@USER USERA TESTSIGNALB 1 PDMESH
@@200000/30.0 300.0 GM
@@MOVE STATIONMESH
@@WAITSIGNAL TRAIN
@@END
```

departure of the pedestrian from the system by scenarios in which the probability distribution is written. The pedestrians walk from each origin to each destination according to the origin-destination distribution file. The walking patterns and average speeds are also defined in the scenarios.

The example shown in Table 2 is for a user type. The first line of the scenario defines the user generation procedure. In this scenario, a user is generated according to an exponential distribution given by "TESTSIGNALB" in Table 1 and the initial population density is defined by a data file called "PDMESH". The second line defines user behavior in using mobile phone. In this scenario, the average call interval has an exponential distribution with an average of 200,000 seconds. The average time interval for retrial calls is 30 seconds, and the average holding time for each call has an exponential distribution with an average of 300 seconds. The third line defines user walking behavior with regard to traffic signals and commuter trains.

The initial user population distribution file "PDMESH" is based on national census data. The file "STATION MESH" defines the movement pattern as origin-to-destination (OD pairs). This file is generated according to the difference between the daytime and nighttime populations. The number of pedestrians who walk from residential areas to Choufu Station in the morning is calculated by subtracting the nighttime population from the daytime population, and these pedestrians leave the service area by commuter trains. On the other hand, the pedestrians who arrive at the station move from the station to the business area. Their destinations are selected randomly according to the population density in the service area. The number of pedestrians arriving at the station is determined according to statistical census data on station users. In the present simulation, the traffic signals change every 60 seconds. Commuter trains arrive periodically every 150 seconds, based on the arrival schedule of Choufu station during the morning rush hour.

The pedestrians walk according to type, for example, moving straight toward his or her destination by the shortest route or taking detours along the way. The degree of detour from the shortest route is also defined in the scenarios. The basic route of each pedestrian is the shortest route, and alternative routes that involve random detours are given according to the logit model[10]. The speed of each pedestrian is calculated periodically by considering individual ideal walking speeds and the effects of road congestion caused by other pedestrians, traffic signals or the widths of roads.

2.3 User Parameters for Simulation

From the 2000th telecommunications white paper, which reports the current statistics of mobile phone usage in Japan, we set the average user holding time as 300 seconds (2.5 minutes) and the average time interval of mobile phone calls as 21,221 seconds (approximately 6 hours). The other parameters used in the simulation are shown in Table 3.

Table 3. Fixed simulation parameter

Simulation time	13 hours
Data Extraction	from 1 to 13 hours
Type1 (Pedestrians in residential area)	Poisson arrival with 4.27/sec.
Type2 (Pedestrians arrive at Choufu Station)	426 pedestrians arrive per 150 sec.

3 Simulation Results and Discussion

3.1 Regular Cell Allocation and Simulation Results

Cells having a 100-m radius and five channels cover the service area of this report. The cells are located regularly and are called Pattern "A", as shown in Figure 3. An example of the statistics of the cell that covers Choufu Station and its adjacent five cells are shown in Table 4. These cell traffic statistics show the number of new calls in each cell, the number of call failures, the number of handover failures and the rate of call failure for all calls in the cells under consideration. When the cells are located regularly, the cell for Choufu station has a high call loss probability of 0.66. However, the call loss probabilities of the other cells, which are not included in Table 4, are less than 0.1, even though some of these cells covering crossings with signals. In the following tables, we denote the number of new calls as "N.C.", the number of handover calls as "H.C.", the number of failed new calls as "F.N.C.", the number of failed handover calls as "F.H.C.", and the call loss probability as "C.L.P.".

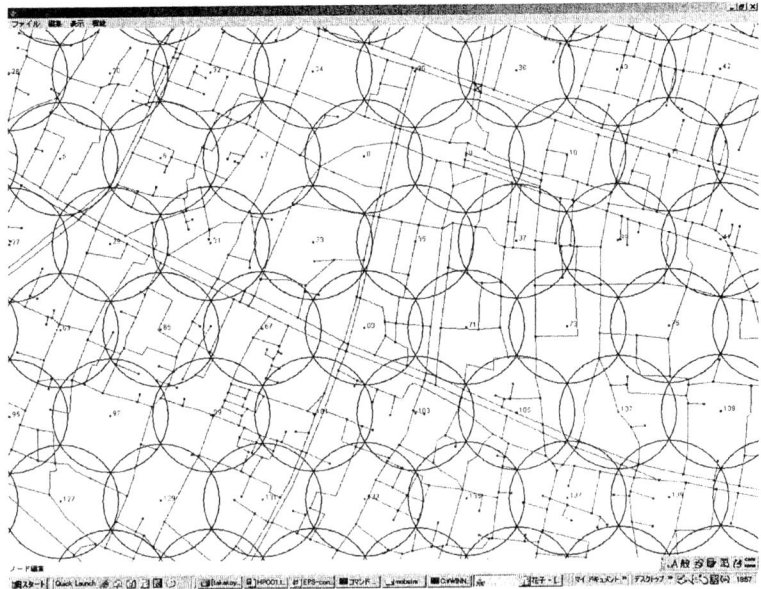

Fig. 3. Pattern A cell allocation

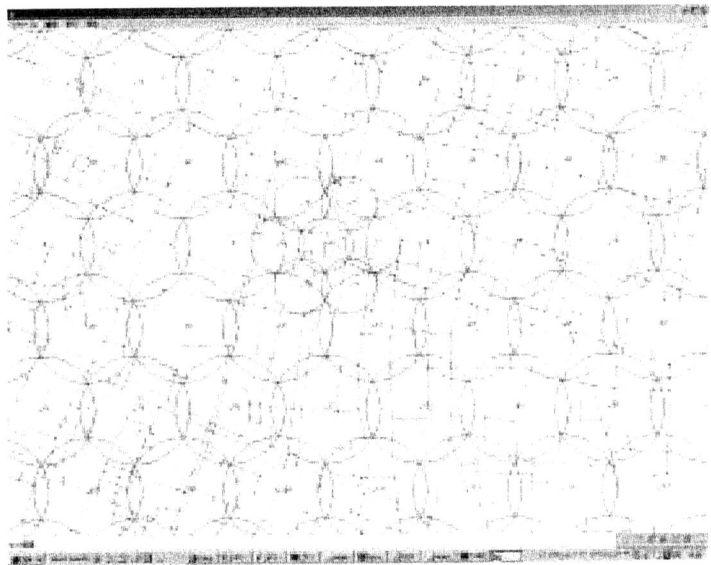

Fig. 4. Pattern B cell allocation

3.2 Variations of Cell Allocations

Figure 4 shows another cell allocation pattern, in which the cell that covers Choufu Station is divided and replaced by six smaller cells. This allocation pattern is called "Pattern B". Figure 5 and 6 shows the distribution of call loss probability for the "Pattern A" allocation pattern and "Pattern B" allocation pattern respectively. The replacement of a congested cell with smaller cells decreases call loss probability by half. In Tables 4 and 5, the numbers of new calls, handover calls, and failure calls are shown for the seven center cells of the "Pattern A" cell allocation and for 13 cells for the "Pattern B" cell allocation. When "Pattern A" allocation is applied, the call loss probability within the cells marked by the bold line in Figure 5 is 0.344, whereas the corresponding call loss probability is 0.147 for the "Pattern B" allocation in Figure 6. The number of handover calls increases when "Pattern B" allocation is applied because the length of the cell border increases. In contrast, the performances of the individual cells are approximately the same, even if the total traffic in the congested cells increases.

Let us consider two allocation Patterns A and B. First, we compare the number of pedestrians in the cells covering the station and the surrounding cells.

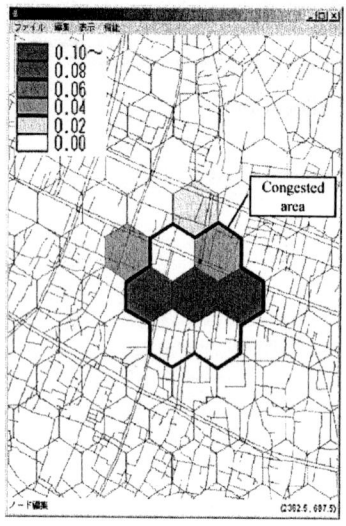 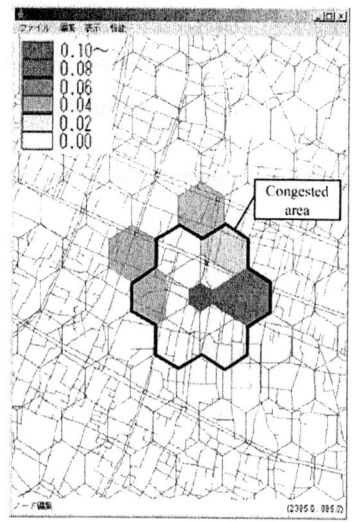

Fig. 5. Distribution of C.L.P. for Pattern A allocation

Fig. 6. Distribution of C.L.P. for Pattern B allocation

Table 4. Performance values for Pattern A allocation(without retrial calls)

	Cell 9	Cell 10	Cell 8	Cell 38	Cell 37	Cell 35	Cell 36	Total
N.C.	3010	888	569	539	301	207	136	5650
H.C.	1860	947	905	595	376	302	157	5142
F.N.C.	2059	114	49	15	2	0	0	2239
F.H.C.	1238	133	73	24	3	0	0	1471
C.L.P.	0.677	0.135	0.083	0.034	0.007	0.000	0.000	
C.L.P. of bordered area								0.344

Table 5. Performance values for Pattern B allocation (without retrial calls)

	Cell 10	Cell 11	Cell 16	Cell 8	Cell 44	Cell 9	Cell 15
N.C.	1869	799	672	398	391	284	357
H.C.	1761	1068	1012	1022	1003	1018	752
F.N.C.	956	89	59	13	17	2	3
F.H.C.	881	119	61	34	21	3	3
C.L.P.	0.593	0.188	0.074	0.043	0.025	0.011	0.011
	Cell 43	Cell 42	Cell 41	Cell 12	Cell 13	Cell 14	Total
N.C.	322	170	149	92	84	24	5611
H.C.	429	172	355	226	906	101	9825
F.N.C.	4	0	0	0	0	0	1143
F.H.C.	2	0	0	0	1	0	1125
C.L.P.	0.020	0.000	0.002	0.000	0.000	0.000	
C.L.P. of bordered area							0.147

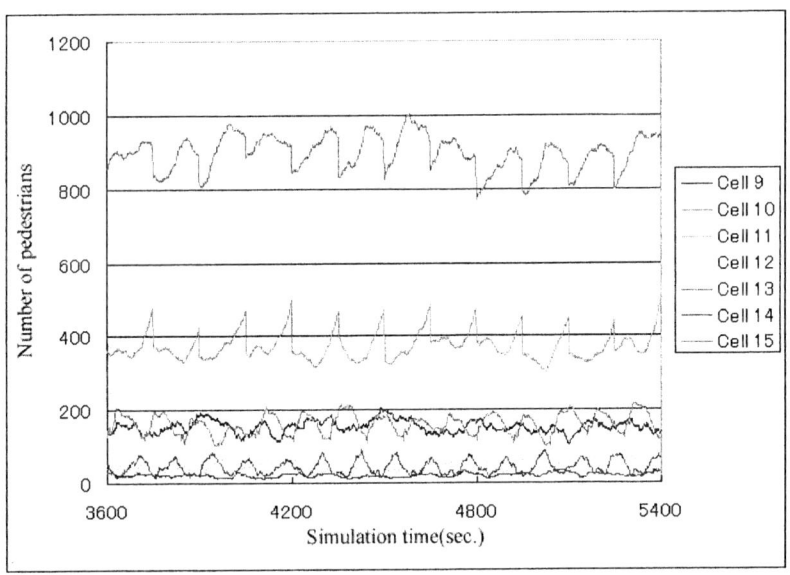

Fig. 7. Number of pedestrians in congested cells

The results are shown in Figure 5 and Figure 6. The observed area corresponds to the congested cells in the service area. The number of pedestrians in these cells periodically changes as commuter trains arrive at the station and groups of passengers arrive. Figure 7 shows the number of pedestrians in the congested cells around the station. This figure shows the transition of the number of pedestrians in the stable state. The number of pedestrians in the cells around the station changes periodically as commuter train passengers arrive at the station and walk through the cells. In Figure 8, we can see a transition of the number of

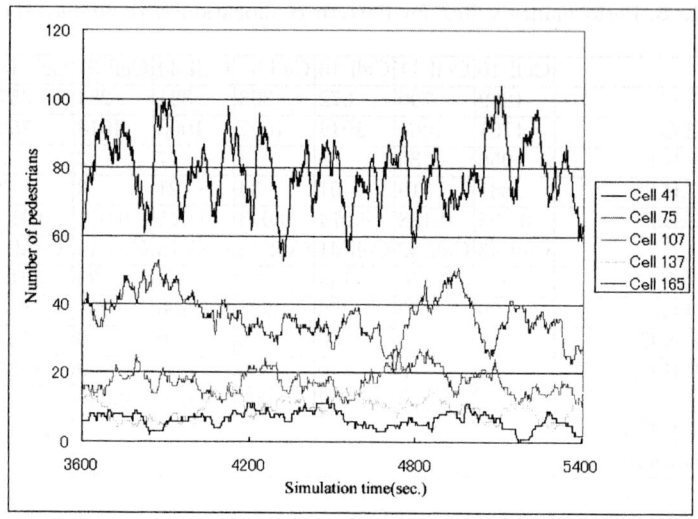

Fig. 8. Number of pedestrians in cells to southwest direction

pedestrians in Cells 41,75,107,137 and 165. As the cell number becomes larger, the locations of the cells separate from the station to the southwest direction. We can see that the periodic change subsides as the distance from the station becomes larger.

3.3 Considerations for Retrial Calls and Channel Allocation

In this section, we examine the effect of retrial calls at first. Here, we assume that the interval time of each retrial call has an exponential distribution with an average of 30 seconds. By comparing Tables 6, the call loss probability increases in the congested Cells 8 and 9 when we consider retrial calls. In such cells, the number of users experiencing call failure and retry increases. These users move from one cell to another while they continue call retrial. Therefore, the traffic of related cells increases. As a result, the call loss probability of these related cells increases. This result implies that the error due to the retrial calls appears in neighboring cells around congested cells. A number of studies on traffic evaluation of cell phone systems have assumed independent call arrival. However, the present result indicates that the interactions of related calls should be considered for more precise traffic evaluation.

When we consider retrial calls, cells with shorter call arrival intervals and higher traffic by group users, such as Cell 9, which covers Choufu Station, increase the call loss probability. In Cell 9, the users who continue to retry calls increase, and these users move to adjacent cells as they continue to dial. Thus, the traffic of adjacent cells around Cell 9 become heavier, and this leads to higher call loss probability. This observation indicates that the interaction of Cell 9 and its adjacent cells caused by retrial calls of moving users amplifies the estimation gap of traffic.

Table 6. Performance values for Pattern B allocation(without retrial calls)

	Cell 10	Cell 11	Cell 16	Cell 8	Cell 44	Cell 9	Cell 15
N.C.	1869	799	672	398	391	284	357
H.C.	1761	1068	1012	1022	1003	1018	752
F.N.C.	956	89	59	13	17	2	3
F.H.C.	881	119	61	34	21	3	3
C.L.P.	0.593	0.188	0.074	0.043	0.025	0.011	0.011
	Cell 43	Cell 42	Cell 41	Cell 12	Cell 13	Cell 14	Total
N.C.	322	170	149	92	84	24	5611
H.C.	429	172	355	226	906	101	9825
F.N.C.	4	0	0	0	0	0	1143
F.H.C.	2	0	0	0	1	0	1125
C.L.P.	0.020	0.000	0.002	0.000	0.000	0.000	
C.L.P. of bordered area							0.147

Table 7. Effect of channel increment (without retrial calls)

Cell Number	8	9	10	11	38	40
C.L.P.						
no retrial calls	0.080	0.665	0.129	0.000	0.033	0.000
retrial calls	0.178	0.838	0.254	0.005	0.112	0.000

In many studies on performance, the evaluation of cellular mobile communication systems assume independent call arrivals, although our observations assert the estimation for the traffic interactions among related cells.

4 Future Works

In the present paper, we propose a model with moving mobile users for a service area of several square kilometers. This model deals with mobile users who move in an urban space with road networks, traffic signals and commuter trains. We developed a computer simulation system using map data, population data and road data. In this simulation system, we examined various user behaviors, including retrial calls, by applying various scenarios and discussing the base station allocation policy effect of additional channels and the effect of user retrial calls. At the same time, this simulation dealt with group arrival by commuter trains. We determined that group arrival by commuter train and the deviation caused by this arrival process causes an estimation error in call loss probability and other estimations of traffic in cells.

References

1. T.Takahashi,T.Ozawa,Y.Takahashi, "Bounds of Performance Measures in Large-Scale Mobile Communication Networks",*Performance Evaluation*,Vol.54,pp.263-283.2003.

2. M. Kanno, M. Murata, H. Miyahara, "Performance Evaluation of CDMA Cellular Network with Consideration of Softhandoff", *IEICE Technical Report(SSE99-37)*, RCS99-61, pp.43-48, July 1999 (in Japanese).
3. Takashi Okuda, Yohei Ando, Tetsuo Ideguchi, Xejun Tian "A Simplified Performance Evaluation for Delay of Voice End-User in TDMA Integrated Voice/Data Cellular Mobile Communications Systems" C*IEEE Globecom2002*, Taipei, 2002.
4. Kenneth Mitchell, Khosrow Sohraby , "An Analysis of the Effects of Mobility on Bandwidth Allocation Strategies in Multi-Class Cellular Wireless Networks", *IEEE Infocom 2001 Technical Program.*, pp.1005-1011, 2001.
5. L. Kleinrock, "Queueing Systems Volume I:Theory", Wiley&Sons, 1975.
6. Kurt Tutschku, "Models and Algorithms for Demand-oriented Planning of Telecommunication Systems", Thesis, Wurzburg University, pp.37-47, 1999.
7. Satoshi Matsushita, Shigeyuki Okazaki, "A study of simulation model for pedestrian movement with evacuation and queuing", *Proc. of the International Conference on Engineering for crowd safety*, pp.17-18 march, 1993.
8. VTT Information Technology(Finland), Sami Nousiainen, Krzysztof Kordybach, Paul Kemppi, "User Distribution and Mobility Model Framework for Cellular Network Simulations" , *IST Mobile & Wireless Telecommunications Summit 2002*, 2002.
9. M. Kaneko, T. Yamada, K. Katou, "A Pedestrian Model and Simulation System for Mobile Communication with Urban Space", Proc. Queue Symposium, pp.218-227, Jan 2004 (in Japanese).
10. E. Teramoto CH. Kitaoka CM. Baba CI. Tanahashi, "Broadarea Traffic Simulator NETSTREAM *Proc. 22th Japan Society of Traffic Engineers* Cpp.133- 136 C2002 (in Japanese).
11. T.Yamada,Y.Takahashi,and L.Barolli,"A Simulation System for Allocation of Base Stations in MobileCommunication Networks: A Case Study", *IPSJ Journal*,Vol.42,No.2,pp.276-285,2001.
12. A. Okabe, A. Suzuki, "Mathematics for Optimal Allocation", Asakura, 1992(in Japanese).

Distributing Multiple Home Agents in MIPv6 Networks

Jong-Hyouk Lee and Tai-Myung Chung

Internet Management Technology Laboratory,
Dept. of Computer Engineering, Sungkyunkwan University,
300 Cheoncheon-dong, Jangan-gu, Suwon-si, Gyeonggi-do, 440-746, Korea
{jhlee, tmchung}@imtl.skku.ac.kr

Abstract. Mobile IPv6 is a protocol that guarantees mobility of mobile node within the IPv6 environment. However current Mobile IPv6 supports simple the mobility of the mobile nodes. If a mobile node is moved away from the home link, it takes time for mobile node to make a registration and binding update at the home agent. In this paper, we propose Distributing Multiple Home Agents (DMHA) scheme. Using this scheme, a mobile node performs binding update with the nearest home agent, so that the delay of binding update process can be reduced. It also reduces the fault rate of handoff up to about 38.5% based on the same network parameters. Experimental results presented in this paper shows that our proposal has superior performance comparing to standard scheme.

1 Introduction

The user demand for mobile service has increased today. However, current IP protocol has a fatal problem that the mobile node (MN) can not receive IP packets on its new point when the MN moves to another network without changing its IP address. To solve this problem, Internet Engineering Task Force (IETF) has proposed a new protocol entitled Mobile IP. Mobile IPv6 (MIPv6) is a protocol that guarantees mobility of MN within the IPv6 environment. Especially, it basically provides the route optimization and doesn't need Foreign Agent (FA), which is used for Mobile IPv4 (MIPv4). In MIPv6 world, a MN is distinguished by its home address. When it moves to another network, it gets a care of address (CoA), which provides information about MN's current location. The MN registers its newly given CoA with its home agent (HA). After that, it can directly communicate with correspondent node (CN) [1], [2]. The basic operation of MIPv6 is shown in Fig. 1.

The binding information is valid only for the life time (420 sec) [3]. It should be updated after the life time is expired. Several packets should be sent and received between the MN and the HA for updating. These packets starts with ICMP Home Agent Address Discovery Request Message, ICMP Home Agent Address Discovery Reply Message, ICMP Mobile Prefix Solicitation Message, ICMP Mobile Prefix Advertisement Message, Authentication and Registration, and ends with Binding Update Acknowledgment packet [1].

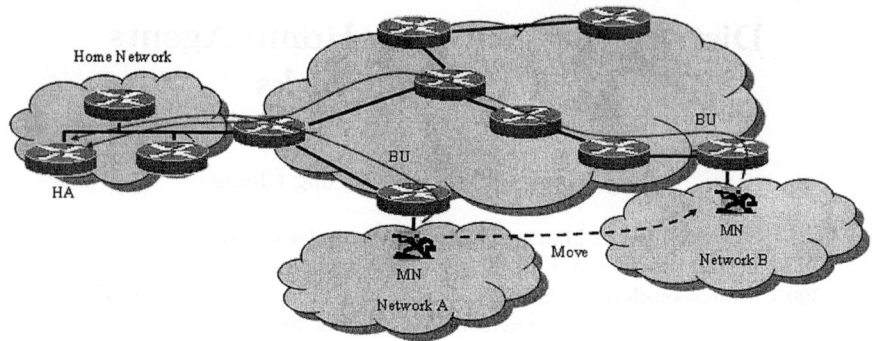

Fig. 1. The basic operation of MIPv6

There is a problem in the current Mobile IP standard. If the MN is located at a foreign network such as network A, or network B shown in Fig. 1. The MN does not care about the distance to the HA. It may be far away from the home network, or it may be close to the HA. MN uses the same method for Binding Update (BU). This is not efficient because when the MN is far away such as in network B, the BU takes much time. It also generates extra traffic over the network. Therefore, handoff failure rate is increased.

In order to resolve this problem, this paper proposes Distributing Multiple Home Agents (DMHA) scheme. When MN moves from a network to another network, it performs BU with the nearest HA, so that the delay of BU process can be reduced. It also reduces the handoff delay. Therefore, handoff failure rate is decreased. The selection of HAs is dynamic. Since those HAs have the same anycast Address, the nearest HA can be chosen easily. It is based on Dynamic Home Agent Address Discovery (DHAAD) [1] and uses IPv6 anycast [4].

The rest of this paper is organized as follows. In section 2, we will see how the handoff process takes place section by section when a MN moves to a new network. Then, we will analyze the delay caused during BU. Also, we will see about the anycast, which is used for BU between multiple HAs and MN. In section 3, we define the DMHA scheme and describe how it works and how it is organized. In Section 4, we evaluate the performance of the DMHA scheme. The final section gives our conclusions.

2 Background and Motivation

2.1 The Handoff Process

When a MN detects an L3 handoff, it performs Duplicated Address Detection (DAD) on its link-local address and selects a new default router in result of Router Discovery, and then performs Prefix Discovery with that new router to form new CoA. It registers its new primary CoA with its HA. After updating its home registration, the MN updates associated mobility bindings in CNs which are performing route optimization.

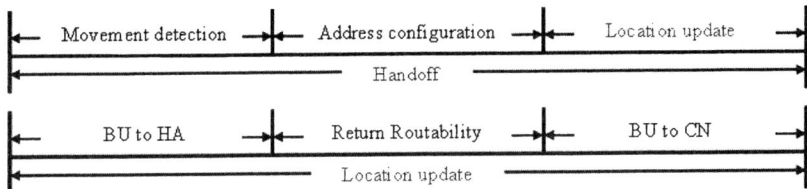

Fig. 2. The handoff process

The time is needed to detect MN's movement detection, to configure a new CoA in the visiting network, and to resister with the HA or CN together make up the overall handoff delay.

Movement detection can be achieved by receiving Router Advertisement (RA) message from the new access router. The network prefix information in the RA message can be used to determine L3 handoff by comparing with the one received previously. After movement detection, the MN starts address configuration which makes topologically correct addresses in a visited network and verifies that these addresses are not already in use by another node on the link by performing DAD. Then the MN can assign this new CoA to its interface. This CoA must be registered with the HA for mobility management and CN for route optimization [5], [6]. Fig. 2 illustrates the handoff process occurring as the MN is transferred to a new network.

Fig. 2 illustrates the MN carrying out BU to HA of the new network during the location update process. Such a process increases the delay required within the BU process as the MN moves further away from HA and incurs the final result of increasing the location update delay.

2.2 The IPv6 Anycast

The anycast routing arrives onto the node thats nearest to the anycasts destination address. The nodes with same address can be made to function as web mirrors and can gain the function of dispersing the node traffic function and can be chosen and utilized as the nearest server by the user. For example, this process can be considered when a user wishes to download a file from a web page. In order for the user to download a file from the nearest file server, the user doesnt choose a file server but it can be automatically directed to the nearest file server to download the file via the operation of anycast [7]. The file download speed via anycast not only has an improved speed but can reduce the overall network traffic.

3 Distributing Multiple Home Agents Scheme

We proposed DMHA scheme uses anycast to find the nearest HA among several HAs. The HAs are geographically distributed. And these are belong to the same anycast group. A packet sent to an anycast address is delivered to the closest member in the group. The closeness is measured by the distance in routing protocol. All the HAs have the same anycast address. A mobile node can find the nearest HA by the Dynamic Home Agent Address Discovery (DHAAD) mechanism. According to the IPv6 anycast protocol, the DHAAD packet sent to an anycast address is routed to the nearest HA. The actual route is determined by the distance in the network [8].

When a MN detects that it has moved from a network to another network, it receives a new CoA. In the MIPv6, the MN will send the BU to the HA if it keeps the address of HA in the binding cache. In our proposal, MN deletes HA record in the binding cache when it moves from a network to another network. MN should find a new HA. It finds the nearest HA by using anycast. This scheme has no single point of failure. If one of the HAs is down, the MN can find the second nearest one.

Each message in Fig. 3 is described as follows. (1) MN arrives at another network. (2) MN takes the message of new access router (advertisement), and finds out new CoA (nCoA). (3) MN deletes home agent record in the binding cache. (4) MN sends ICMP Home Agent Address Discovery Request message to the anycast address. (5) The nearest HA receives the packet and replies ICMP Home Agent Address Discovery Reply message. (6) MN updates home agent record in the binding cache. (7) MN sends BU request message to nearest HA. (8) The nearest HA replies the BU reply message. (9) If the nearest HA to a MN is not the original HA in the home network, the nearest HA sends the BU request message to the original HA. (10) The original HA replies the BU reply message. (11) Also, CN sends the packet to the home address at the original HA. (12) The packet will be forwarded to the nCoA by the original HA.

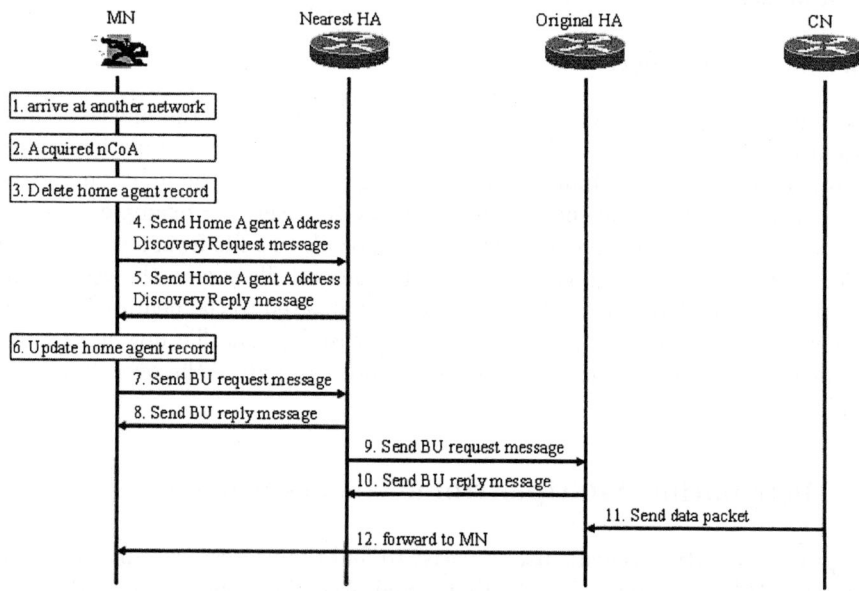

Fig. 3. Flow of total message

4 Performance Evaluation

4.1 Basic Idea on Modeling

At each step, the time required to send a message is composed of the transmission time, the propagation time and the processing time, i.e. $M^i = \alpha^i + \beta^i + \gamma^i$, where i represents the step i. The transmission time α^i is computed by the size of the control message in bits over the bit rate of the link on which the message is sent, i.e. $\alpha^i = \frac{M_b}{B}$, where M_b is a control message assuming the fixed size. The B is the bandwidth of the link, B_l for the wired line, and B_w for the wireless case. The propagation time β^i varies depending on the transmission medium, i.e. β_l is the time for the wired line, and β_w is for the wireless one. The processing time γ^i has the same value at intermediate routers, MN, HA, and CN. The wired medium is more stable than the wireless one, so the retransmission is not needed. Therefore, the physical transmission time T^i is represented by $M^i (= M_l)$. Later in each step the message processing time on the wired and wireless cases are represented as M_l and M_w, respectively. At the wireless link, the message retransmission is necessary because a message can be lost in any moment. MN retransmits the message when lost in the air transmission. By considering the number of link failures (N_{fail}) and the probability of link failure, we obtain the additional signal processing time at these steps in the wireless case, i.e. $T_w^i = \sum_{N_{fail}=0}^{\infty} \{\tilde{T}_w^i(N_{fail}) \cdot Prob(N_{fail}\ failure\ and\ 1\ success)\}$. Whenever ACK signal may not be received for T_{out} after the request signal is sent, we assume that the message is lost and the control signal is retransmitted. If there are N_{fail} failures, then T_{out} and the message retransmission occur N_{fail} times. So $\tilde{T}_w^i(N_{fail})$ is induced as $\tilde{T}_w^i(N_{fail}) = M_w + N_{fail} \cdot (T_{out} + M_w)$. Therefore signal processing time for retransmission steps becomes

$$T_w^i = \sum_{N_{fail}=0}^{\infty} \{M_w + N_{fail} \cdot (T_{out} + M_w)\} \cdot Prob(N_{fail}\ failure\ and\ 1\ success)$$

$$= M_w + (T_{out} + M_w) \sum_{N_{fail}=0}^{\infty} N_{fail} \cdot Prob(N_{fail}\ failure\ and\ 1\ success) \quad (1)$$

Here, $\sum_{N_{fail}=0}^{\infty}\{N_{fail} \cdot Prob(N_{fail}\ failure\ and\ 1\ success)\}$ is obtained by the infinite geometric progression. Usually, link failure probability q is smaller than 1. Therefore, $\sum_{N_{fail}=0}^{\infty}\{N_{fail} \cdot Prob(N_{fail}\ failure\ and\ 1\ success)\} = \frac{q}{1-q}$ Generally, q has the value of 0.5. So (1) becomes

$$T_w^i = M_w + (T_{out} + M_w) \cdot \frac{0.5}{0.5}$$
$$= 2M_w + T_{out} \quad (2)$$

And the additional message processing time else M_i may be required. It is assumed to be the message processing time T_{proc}.

4.2 Total Handoff Time

The system parameters used to analyze the system are listed in Table 1. Each value is defined based on [9], [10]. Fig. 3 represents the flows of messages. Based on the scheme in this figure, we compute the total handoff time. Handoff completion time is acquired by summing I, II, and III below.

Table 1. System parameters

Variables	Definitions	Values
B_l	bit rate of wired link	155 Mbps
B_w	bit rate of wireless link	144 Kbps
β_l	wired link message propagation time	0.5 msec
β_w	wireless link message propagation time	2 msec
γ	message processing time	0.5 msec
T_{proc}	extra processing time	0.5 msec
T_{out}	message loss judgment time	2 msec
q	probability of message loss	0.5
$H_{a,b}$	number of hops between a and b	

I. Sum of the processing time (SPT)
The processing time is required in steps 2, 3, and 6. So, $SPT = T^2_{proc} + T^3_{proc} + T^6_{proc}$. In the above case T^i_{proc} has a fixed value (T_{proc}) at each step which is just as much as the processing time. Therefore,

$$SPT = 3T_{proc} \qquad (3)$$

II. Sum of the message transmission time in wired links (SMT_l)
Message transmission in the wired line states are in steps 9 and 10. In these cases, the total time for message transmission is $SMT_l = T^9_l \cdot H_{nearestHA, originalHA} + T^{10}_l \cdot H_{originalHA, nearestHA}$. We assume that each $T^i_l (i = 9, 10)$ has a fixed value (T_l), and thus

$$SMT_l = T_l \cdot \{2H_{nearestHA, originalHA}\} \qquad (4)$$

III. Sum of the message transmission time in wireless links (SMT_w)
Message transmissions are in steps 4, 5, 7, and 8 in wireless links. Here message transmission time is $SMT_w = T^4_w + T^5_w + T^7_w + T^8_w$. From the equation T^i_w, becomes

$$SMT_w = 4T_w(T_w = 2M_w + T_{out}) = 4(2M_w + T_{out}) \qquad (5)$$

Therefore, we obtain the total required time for the handoff completion by the summation from step I to step III.

$$\begin{aligned} T_{req} &= SPT + SMT_l + SMT_w \\ &= 3T_{proc} + T_l \cdot \{2H_{nearestHA, originalHA}\} + 4(2M_w + T_{out}) \end{aligned} \qquad (6)$$

4.3 Handoff Failure Rate

The T is a random variable of the time for MN staying in the overlapped area and the T_{req} is the time required for the handoff completion. Hence, the handoff failure rate is represented as $P = Prob(T < T_{req})$, where we assume that T is exponentially distributed. Thus, $P = Prob(T < T_{req}) = 1 - exp(-\lambda T_{req}) < P_f$. Here is λ the arrival rate of MN into the boundary cell and its movement direction is uniformly distributed on the interval $[0, 2\pi)$. So λ is calculated by the equation $\lambda = \frac{VL}{\pi S}$ [11]. Here V is the expected velocity for MN that varies in the given environment and L is the length of the boundary at the overlapped area assuming a circle with radius l, i.e. $L = \frac{1}{6} 2\pi l 2 = \frac{2}{3} \pi l$. The area of the overlapped space S is $S = 2(\frac{1}{6}(\pi l^2 - \frac{\sqrt{3}}{4} l^2))$. Therefore, λ is induced by equations regarding as l. Hence we get the handoff failure rate by T_{req}.

$$l > \frac{4VT_{req}}{(2\pi - 3\sqrt{3})log(1/(1-P_f))} \quad (7)$$

Similarly, when the desired handoff failure rate is given, then V is calculated. Thus,

$$V < \frac{l(2\pi - 3\sqrt{3})log(l/(1-P_f))}{4T_{req}} \quad (8)$$

Consequently, the desired maximum velocity is acquired.

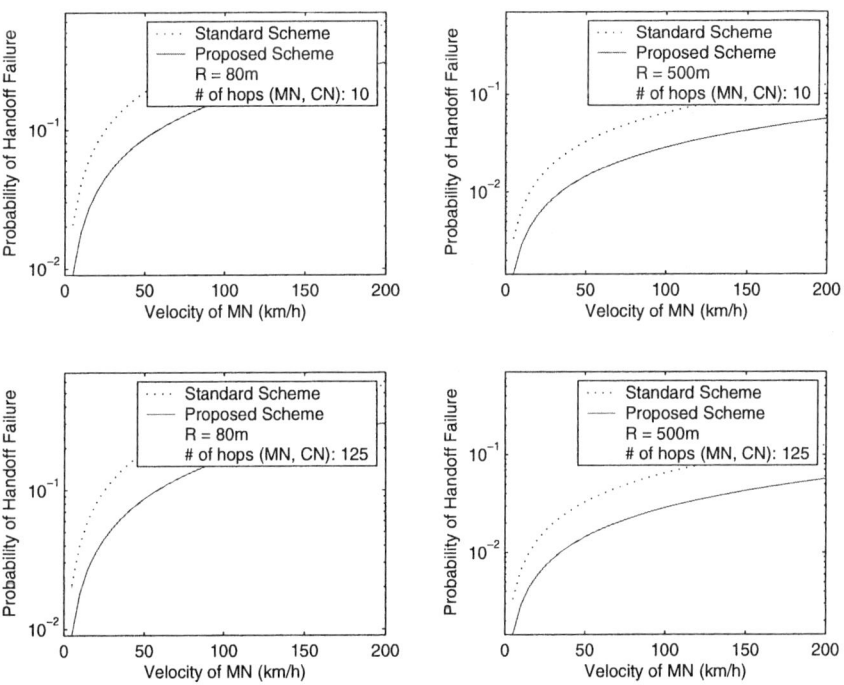

Fig. 4. Handoff failure rate by the velocity of MN

4.4 Experimental Results

Fig. 4 compares results for the probability of handoff failure between the existing scheme and the proposed one. It is obtained by using system parameters given in Table 1. The probability of handoff failure is influenced by few factors that are the velocity of MN, hop count ($H_{MN,CN}$) from MAR to CN and the radius of a cell(l). The increase of MN velocity V means the handoff should be completed within relatively short period of time. If MN moves faster than the regular speed, handoff time may not be sufficient and consequently the handoff is failed. In Fig. 4 for the various V, The proposed scheme shows the relatively low handoff failure rate comparing to the previous one so it provides the more stable performance.

The graph in Fig. 5 shows the handoff failure rate as a function of the radius of the cell. As the radius of the cell increases, the overlapped area becomes larger, so that it is easy to complete the handoff. As a result, the handoff failure rate decreases. When handoff occurs with an MN which has a high velocity, it is difficult to complete the handoff, because the MN moves out of the overlapped area quickly. We calculate the cell size when ($H_{MN,CN}$) is 10, the probability of handoff failure is 2%, and the velocity of MN is 5km/h. By the equation (7), the minimum cell radius of the existing scheme is 53.4m, meanwhile the proposed one is 32.8m. Therefore, the proposed scheme is more efficient than the existing one in terms of cell radius and the efficiency is improved up to about 38.5%.

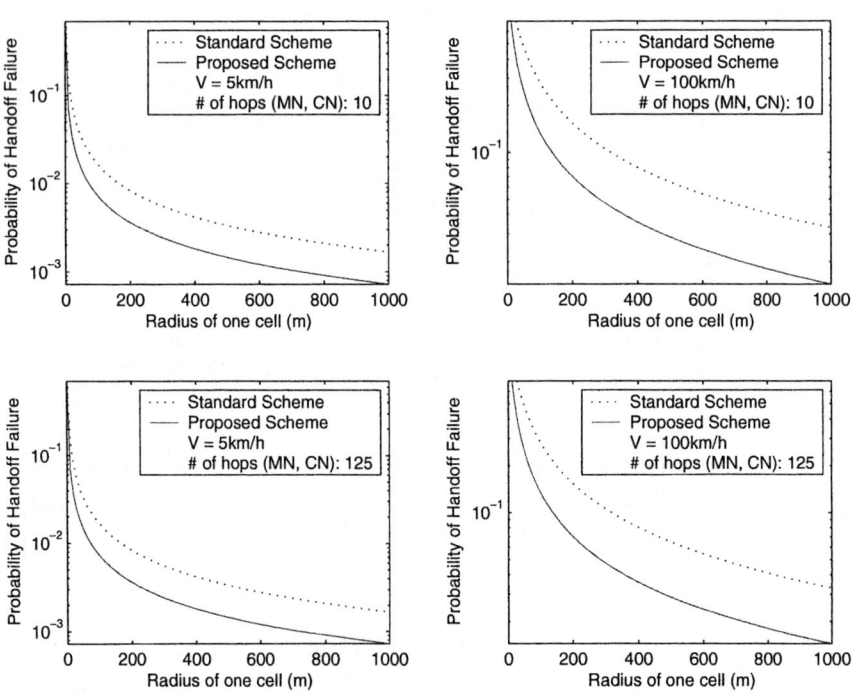

Fig. 5. Handoff failure rate by the radius of a cell

Similarly, by the equation (8), we are able to calculate the speed of MN. In case that ($H_{MN,CN}$) is 10, the probability of handoff failure is 2% and the radius of l is 500m. Then the maximum velocity of MN in the existing scheme is 25.5km/h while the proposed one is 35.3km/h. As a result, the proposed scheme is overall superior to the existing one and even it supports more stable environments.

5 Conclusions

In this paper, we propose Distributing Multiple Home Agents (DMHA) scheme. When MN moves from a network to another network, it performs BU with the nearest HA, so that the delay of BU process can be reduced. It also reduces the handoff delay. Therefore, handoff failure rate is decreased. According to the analytical model and the comprehensive experimental results, the handoff execution time on MN is relatively longer comparing to the previous approach and the handoff failure rate is significantly decreased up to about 38.5% with almost negligible signaling overhead. Moreover, it would be a good example of and function well in anycast mobile environment, one of IPv6 address modes. In the future, we will describe the requirements for the efficient network management policy, and the access control.

References

1. D. Johnson, C. Perkins, J. Arkko, "Mobility Support in IPv6", RFC 3775, June 2004.
2. C. Perkins, "Mobility Support in IPv4", RFC 3344, August 2002.
3. Kame Project, http://www.kame.net, Accessed on March 18 2005.
4. D. Johnson, S. Deering, "Reserved IPv6 Subnet Anycast Addresses", RFC 2526, March 1999.
5. R. Hsieh, Z.G. Zhou, A. Seneviratne, "S-MIP: a seamless handoff architecture for mobile IP", INFOCOM 2003, Vol. 3, pp. 1173 - 1185, April 2003.
6. S. Sharma, Ningning Zhu, Tzi-cker Chiueh, "Low-latency mobile IP handoff for infrastructure-mode wireless LANs", Selected Areas in Communications, IEEE Journal, Vol. 22, Issue 4, pp. 643 - 652, May 2004.
7. S. Weber, Liang Cheng, "A survey of anycast in IPv6 networks", IEEE Communications Magazine, Vol. 42, Issue 1, pp. 127 - 132, January. 2004.
8. R. Hinden, S. Deering, "IP Version 6 Addressing Architecture", RFC 2373, July 1998.
9. J. McNair, I. F. Akyildiz, M. D. Bender, "Handoffs for real-time traffic in mobile IP version 6 networks", IEEE GLOBECOM, vol.6, pp. 3463 - 3467, 2001.
10. J. Xie, I. F. Akyildiz, "An optimal location management scheme for minimizing signaling cost in mobile IP", IEEE ICC, vol.5 , pp. 3313 - 3317, 2002.
11. R. Thomas, H. Gilbert, G. Mazziotto, "Influence of the moving of the mobile stations on the performance of a radio mobile cellular network", in Proceedings of the 3rd Nordic Seminar, 1988.

Dynamically Adaptable User Interface Generation for Heterogeneous Computing Devices

Mario Bisignano, Giuseppe Di Modica, and Orazio Tomarchio

Dipartimento di Ingegneria Informatica e delle Telecomunicazioni,
Università di Catania, Viale A. Doria 6, 95125 Catania, Italy
{Mario.Bisignano, Giuseppe.DiModica,
Orazio.Tomarchio}@diit.unict.it

Abstract. The increasing number of personal computing devices today available for accessing online services and information is making more difficult and time-consuming to develop and maintain several versions of user interfaces for a single application. Moreover, users want to access services they have subscribed, no matter the device they are using, always maintaining their preferences. These issues demand for new software development models, able to easily adapt the application to the client's execution context, while keeping the application logic separated from its presentation. In this work we present a framework that allows to specify the user's interaction with the application, in an independent manner with respect to the specific execution's context, by using an XML-based language. Starting from such a specification, the system will subsequently "render" the actual user's application interface on a specific execution environment, adapting it to the end user's device characteristics.

1 Introduction

Many different personal computing devices are available today to users: they range from full powered notebooks, to portable devices such as PDAs, and to smartphones. All of these devices allow users to access and execute different kind of applications and services, even when they are outside of their office. Moreover, the heterogeneity of the available wireless access technologies gives rise to new network scenarios, characterized by frequent and dynamic topology changes and by joining/leaving of network nodes. Due to the continuous mobility of users' devices and, as a consequence, to network dynamics, applications cannot rely on reliable and stable users' execution contexts. Context-aware mechanisms are needed in order to build adaptive applications, location-based services, able to dynamically react to changes in the surrounding environment[13,3]. Current approaches, both as traditional models for distributed computing and related middleware [4], do not fully satisfy all the requirements imposed by these environments.

One of the main requirements to be able to build ubiquitous and pervasive computing scenarios is the possibility for the user to access the same service from heterogeneous terminals, through different network technologies using different access techniques [12]. In order to accomplish this, while avoiding at the same time the burden of reimplementing from scratch the same application for a different device and/or for a different network technology, new programming paradigms along with the related middleware are needed, providing the adequate level of transparency to the application developer. In particular, from the developer's perspective, developing a new user interface and new content types each time a new device penetrates the market is not a feasible solution.

In this paper we present the architecture of a framework whose goal is to make the presentation layer of the application, i.e. the user's interaction level, adaptive and (as much as possible) independent from the specific execution context. The actual user interface will be dynamically generated at runtime according to context information. The presentation level (user interface), together with the user's interaction model and the associated task model is described at an high and abstract level, in a device independent way. Moreover, the system is supported by a set of adaptation components (and/or renderer), each one specific for the current execution context at the client side (user terminal characteristics, current network features, user preferences, etc). A given render is in charge of adapting the application's user interface for a specific execution environment, according to the actual end user device's features, which will be represented using the CC/PP standard for profile information representation. Depending on the user's needs (user preferences) and the application characteristics, this step can be done either off-line [1], thus distributing the result to the client later on, or dynamically on-line [16]. As further described in this paper, from an architectural point of view, the system has been structured in such a way to promote the dynamic insertion of new adaptation modules (even at runtime), specific for some functionality not foreseen in advance. A prototype of the framework has already been implemented, together with two "renderers": one for standard PCs, equipped with a complete J2SE environment (Java Standard Edition) and the other one for mobile phones equipped with the Java Micro Edition (J2ME) environment.

The rest of the paper is organized in the following way. Section 2 presents a review of related work in this area, trying to outline merits and limits of existing approaches. Then, in Section 3, the system architecture is presented, together with the description of its components and their behavior. Finally we draw the conclusions in Section 4.

2 Related Work

The development of services that all kind of end user's devices should be able to access, despite their different features, had given rise to several different approaches to the problem, both in academic and in commercial environment. Nevertheless, today many of existing approaches often address Web content fruition [14,6,7], having as target devices well defined categories of terminals (typically PDAs like Pocket-PC and cellular phones WAP-enabled). All of these approaches can be generally classified into some categories: scaling, manual authoring, transducing [14,6] and transforming [7]. Among these approaches, the one based on the description of the user interface by means of an XML-based vocabulary seems the most promising. Some of the system/languages aiming at this purpose, adopting an "intent oriented" scheme are: UIML [1], AUIML [2], XForms [17], Dygimes [9], Teresa [10]. In these systems only the interaction of the application with the user is described, but not its graphic components, whose rendering is committed to a specific module. UIML [1] (User Interface Markup Language) is one of the first attempt to designing user interfaces adaptable to the device: even though it was proposed as an open specification, the renderer implementation is carried out by means of proprietary tools developed by Harmonia Inc. (the company that cared about its development). AUIML [2] (Abstract User Interface Markup Language) is an XML-based markup language created by IBM. At the beginning it was designed as a remote administration's tool to be run on devices with different features, ranging from cellular phones and PDA to desktop computers. Notwithstanding, the original developing process of AUIML was stopped and the concepts of "intent oriented" language have been integrated in XForms [17]. The latter is an XML-based markup language developed by the W3C consortium [17]. Recently it has been endorsed as a "W3C Recommendation", and represents the evolution of the web forms. Nevertheless, its strength is proportional to its complexity: this is not an problem for desktop PCs, but for sure it is for all the devices with limited computing power like PDAs and smartphones. However, we considered the definition of the interaction elements of the user interface in XForms as the base for the creation of the XML vocabulary presented in this work. Dygimes [9] is a complete environment for building multi-device interfaces: it uses XML-based user interface descriptions that, combined with task specifications, enable to generate actual user interfaces for different kinds of devices at runtime. The specification of graphical layout constraints has inspired our work during the design of the renderers. Finally Teresa [10], which is a tool for designing multi-device applications by using a model-based approach. It is mainly targeted towards application developers, which could easily generate different user interfaces for different kind of devices in a semi-automatic way by starting from a device independent task model specification of the application. So it is not a framework for the dynamic generation of user interface at runtime: however, the CTT notation for its task model specification has been adopted in our work. Apart the strength and the weakness of each of the previous approaches, all of them do not (explicitly) deal with the multimedia content fruition, which is becoming in the latest years a very common feature of many applications/services. When considering multimedia contents, content adaptation is also needed, a factor which can also be strongly affected by network parameters (e.g. bandwidth availability, packet delay). Although this paper is not specifically focused on this aspect, in this area different approaches can be found in literature [8,15].

3 System Architecture

We consider the overall scenario, actually very common nowadays, where a service provider intends to offer a set of information services to a wide variety of potential users; on his turn, each user is willing to access the set of the offered services disregarding either the device he is using or the location he is visiting. Services include in general both some kind of information (accessible through a Web or a Wap browser) and some application/scripts to run onto the user's devices in order to correctly query the business logic executing at the server side. To date, software developers have satisfied these user's needs by building up different versions of a service's user interface according to the user's device features and the overall execution environment. Our purpose is to design a software framework through which an application developer can easily implement and distribute his application to different kind of end user's devices. The framework being introduced will relieve application developers from rewriting the software code related to an application's user interface for each device's execution environment. The core of the framework includes an abstract XML based specification language (*XML vocabulary*), as a means to describe the user interaction with the application. The adoption of an XML based vocabulary enables the developer to create just abstract descriptions (*Application Interaction Specification*) of the basic elements that populates a user interface (UI), and of a task model (*XML Task Model*) that describes the dynamic transitions among the different frames of the application. All this abstract specification will be later transformed (*"rendered"*) into the interaction components of the actual user interface for the specific device being used by the user. Device features have been managed by a *Profile Management* module, that is in charge of handling information by using the CC/PP standard [5]. At design time, we decided to make the choice of where to perform the rendering process be flexible: according to the user's device capabilities, the rendering process can take place either at the client side (i.e., on the user's device itself) or at the server side. An event handling system is also present, which is able to manage both *local* interaction events (events that can be locally processed), and *remote* interaction events (events that should be remotely processed by the application business logic). The *Task Model Manager* will keep trace of the dynamic behavior of the designed application. Whenever required, the framework can also perform the adaptation of media contents according to the user's device profile. Figure 1 depicts the overall architecture of the designed framework. The bottom component referred to as Runtime coordination and execution is in charge of coordinating the following tasks: execution of the operations needed at the application start-up; recovering the information on the devices from the profile repository; activation of the task model manager for the extrapolation of the application's dynamics; activation of the correct renderer for the generation of the actual user interface. A complete prototype has been implemented in Java, together with two renderers, one for the J2SE environment and the other one for J2ME enabled devices. In the following the features of the modules that the architecture is composed of, together with their mutual interactions, will be further described.

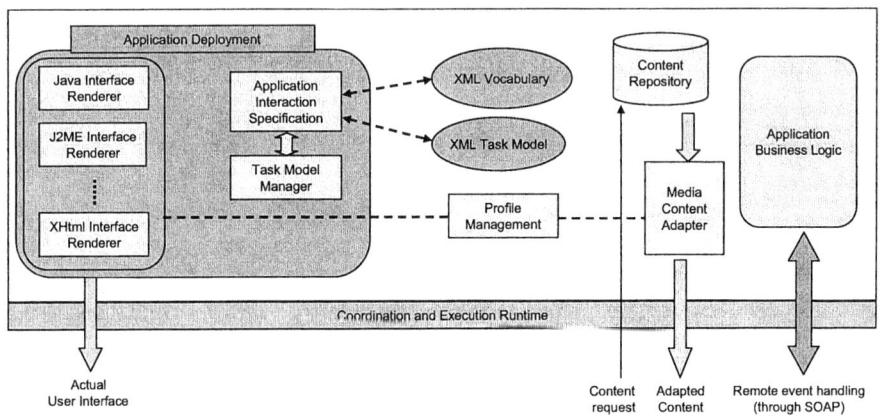

Fig. 1. Framework architecture

3.1 XML Vocabulary

A model for the user interface, the single interaction elements, their state and associated events have been defined (*XML vocabulary*). To define this model, technologies based on standard markup languages such as XML have been adopted, so as to guarantee an high level of flexibility and the possibility to customize the model itself. In defining this interaction model, an *intent-oriented* approach has been adopted, where each element that the user interacts with is modeled at an abstract level, independently of what its actual rendering on the user's device would be. This is the main reason why we prefer to talk about "user interaction specification" rather than "user interface". When defining the items of the vocabulary we strove to consider all the possible combinations of objects representing the interactions with the interface. Unfortunately, in this paper there is no room to provide a detailed description of the interaction elements populating the vocabulary. However all of them could be classified in the following categories:

- *Data Input*: The input object enables the user-application interaction through the insertion of data that, furthermore, undergo validation and restriction procedures (type, length, etc..).
- *Data Output*: this object describes a simple transfer of information from the user interface to the user.
- *Choices and multiple selections*: These elements define the interactions enabling the user to choose among a list of items.
- *Information uploading*: these elements deals with the transfer of information (in the form of a particular file and of simple textual data) to a remote server.
- *Triggers*: This tag triggers the generation of events that can be locally or remotely handled. Every application provides a mechanism for the activation of some available functions: the framework permits to specify whether the event associated to a given interaction object must be either locally or remotely handled on a server.
- *Content displaying*: Besides the common **input** and **output** element, new elements have been added to our vocabulary in order to manage multimedia contents (images, audio and video). However, since only a few devices are able to display graphic and multimedia elements, some attributes allow the developer to specify whether the graphic and multimedia content are required by the semantic of the application or not.

In order to give the reader an idea of the different graphical output generated by the same XML description, we provide in Fig. 2 an example of a description of an interaction based on a choice tag, namely the `select1` tag. In the figure, in addition to the XML code fragment, two possible renderings (the first one for a desktop device and the second one for a J2ME devices) are also depicted.

3.2 Task Model Management

The *XML task model* describes the dynamics of the application: it is specified by using the well known ConcurTaskTree (CTT) notation [11]. This notation supports a hierarchical description of task models with the possibility of specifying a number of temporal relations among them (such as enabling, disabling, concurrency, order independence,

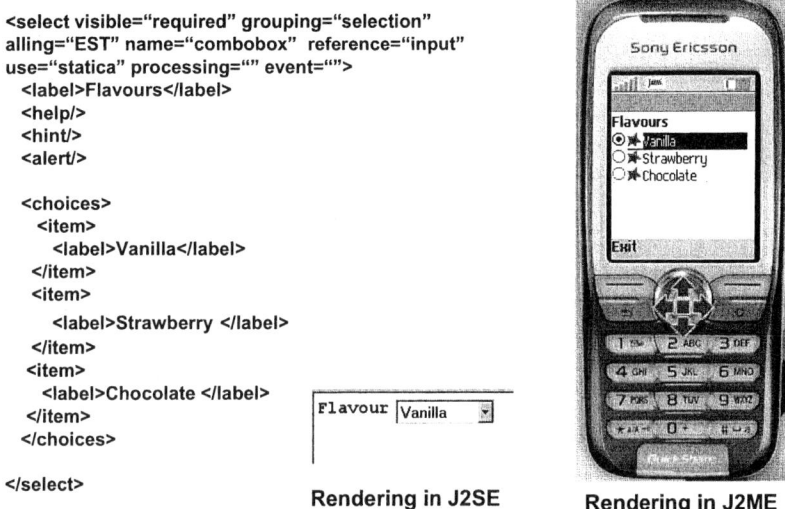

Fig. 2. Example of **select1** element usage

and suspend-resume). The *Task Model Manager* is in charge of retrieving and subsequently processing the XML specification associated to each task. The application logic of this component can be executed either on the server or on the client itself: both the approaches are possible (see Section 4 for details).

3.3 Profile Management and Media Content Adaptation

The Profile Management module deals with device features and user preferences management. Whenever a service is to be delivered to a given device, the information regarding the profile of such device are retrieved and used to optimally tailor the service's user interface to the device's capabilities.

Although it has not been fully implemented, the mechanism allows to take into account three different kind of profiles: device profile, network profile, user profile. The W3C recommendation on CC/PP (Composite Capability/Preference Profile)[5], which is the basis for the UAProf (User Agent Profile) specifications, defined by the Open Mobile Alliance for mobile telephony, has been taken as a reference for the definition of the profile concept within our work. These profiles are very complex, including many information both related to the hardware features of a device and to the software environment available on it. In our implementation work we considered only a limited number of profiles parameters, the ones we believe are fundamental to check the viability of the approach, e.g. screen size, number of supported colors, supported image type, etc.

When also considering services delivering multimedia contents, it becomes likewise important to deal with content adaptation. In this case, apart from considering the user device's features, network parameters should be taken into account, since, for example, the bandwidth can be a constraint for streaming multimedia contents. The system includes the possibility to integrate a "media content adapter", capable of performing the real work of adapting a specific content according to the global context of the end user. Techniques that could be adopted will depend on the actual content format: they could include image resizing, transcoding from a video/audio format to another one, color numbers scaling, resolution changing, etc. However, technological solutions addressing this issue are for some aspects already available in literature or in the commercial arena, and are thus beyond the scope of this paper.

Finally, user preferences have been taken into account by our framework: the rendering of the actual user interface will depend also on the user preferences in addition to the device constraints. As a proof of concept we allow the user to specify if image, audio and video have to displayed when accessing services: this way, for example, even if the device he is using is capable to display images, the user may choose to not display them if he is connecting through a slow and/or expensive network connection.

3.4 Application Business Logic

This issue is strictly correlated to the adopted application's deployment model. The framework does not impose any fixed deployment model: this means that the interface generation can be performed both on-line, when the application is requested, and off-line, by delivering the application together with the rendered interface for the device on which it is supposed to run. If the application is a stand-alone one, and the user device has enough computing power, both the user interface and the application business logic can run on the user device. In this case handling the events triggered by the user's interaction is a straightforward task, since all of them will be locally handled. But, if the user device is not able to execute the business application logic or, as it often happens, the application is not stand-alone and needs to access to remote resources, then the problem to manage events (that will be called *remote events*) generated by the interface arises.

The developed framework deals with this issue by distinguishing between *local events* and *remote events*. The formers are locally handled by the user device, while the latters are handled by a suitable business logic available on a remote server. This distinction can be directly made at application's design time: when describing the user interaction specification, it can be specified if a given interaction will generate a local or a remote event. A remote interaction, though, requires a communication protocol supporting the information exchanging between the application server and the user interface. To this end, we decided to adopt the SOAP protocol, allowing services developed on top of our framework to be automatically accessed if structured as Web services. Trigger elements of the XML vocabulary allow to specify, using the *event* attribute, the procedure to call on the server.

3.5 Renderers

The Renderer component is the main component of the architecture. Its task is to carry out the entire rendering process that will produce the device-adapted user's interface. Once the application developer has specified the user-application interaction using the previous XML vocabulary, a specific "interface renderer" will produce the actual user

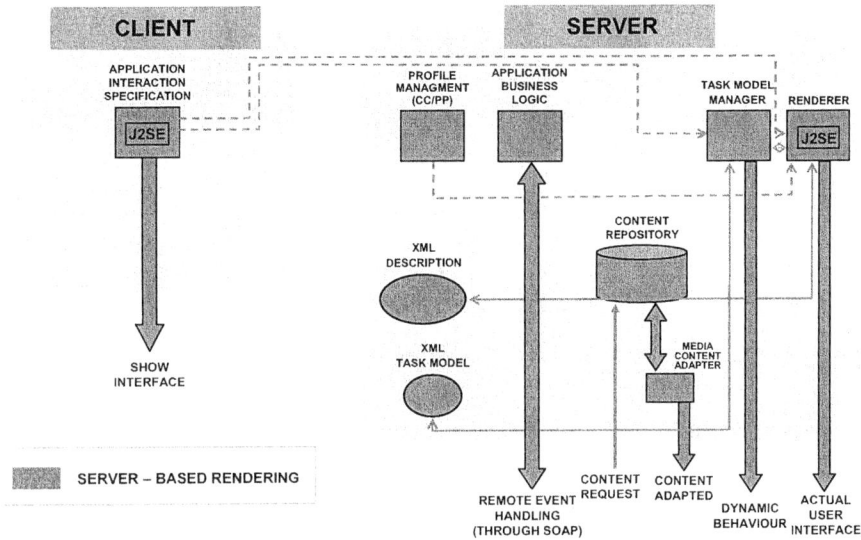

Fig. 3. Server based architecture

interface to be displayed on the user terminal. Each interaction scenario at the applicative level (for example, "selection of an element from a list", "insertion of a numeric value", "pushing a button", etc) is realized by means of an ad-hoc "renderer" (specific for each type of client's execution environment), that produces a user interface that satisfies functional requirements of the application to the best allowed by the client capabilities.

The choice of the interface renderer is made in a dynamic fashion, according to the client profile. The profile management module has just the role to manage different client profiles, choosing the suitable interface renderer for that device. Based on the user's device profile such module will be able to:

- send the rendered interface to the user's device
- send both the renderer and the XML interface description to the user's device
- send the XML interface description only to the user's device

This means that the architecture allows for a client-based or for a server-based rendering approach. As far as the current implementation of the system, three different scenarios involving two environments are available:

− Server-based rendering for a J2SE client
− Client-based rendering on a J2SE client
− Client-based rendering on a J2ME client

Server-based rendering for a J2SE client. The components of the framework involved in this scenario are those depicted in the figure 3. The functionality of each component is the same as the one that has been described in the overall architecture. In this case, only the Application Interaction Specification is located at client side, while both the

Task Model Manager and the renderer are located at server side. It is therefore necessary to transfer on the client's device the data structure concerning both the dynamics of the tasks and the actual interfaces (frames) generated at server side. In the figure 4 an example of the temporal sequence of the actions triggered by all the involved components is depicted.

In order for a server based rendering to be feasible, there are some requirements that the user's device must meet. It must be provided with a Web browser, enabled to download and execute Java applets, and there must be a support for a serialization mechanism. As a first step, the client access from its browser the URL corresponding to the requested service: typically an HTML page will be shown that triggers the downloading of a Java applet containing a jar archive including all the Java classes required for the execution of the application at client side. After the application is started, it connects to a remote Servlet on the server and asks for the profile information of the device it is running onto. Based on the received information, the device will be enabled to display certain types of contents, such as images, sounds, videos.

Then, the logic on the device contacts another Servlet in order to get the data structure containing all the information regarding the dynamics of the specific application generated by the task model manager on the server. Finally, the client sends to a third Servlet (the actual implementation of the server-side renderer) the name of the form to be rendered, together with the preferences expressed by the user. Based on those data, the Servlet will retrieve from the repository the proper XML form representing the abstract description of the required interface and will suitably render it (creating the frame with the appropriate graphic interactors). This mechanism is repeated every time a new window interface has to be displayed.

As far as the execution of the application logic is concerned, we have depicted only the phase related to the invocation of remote events, since this is the only need for

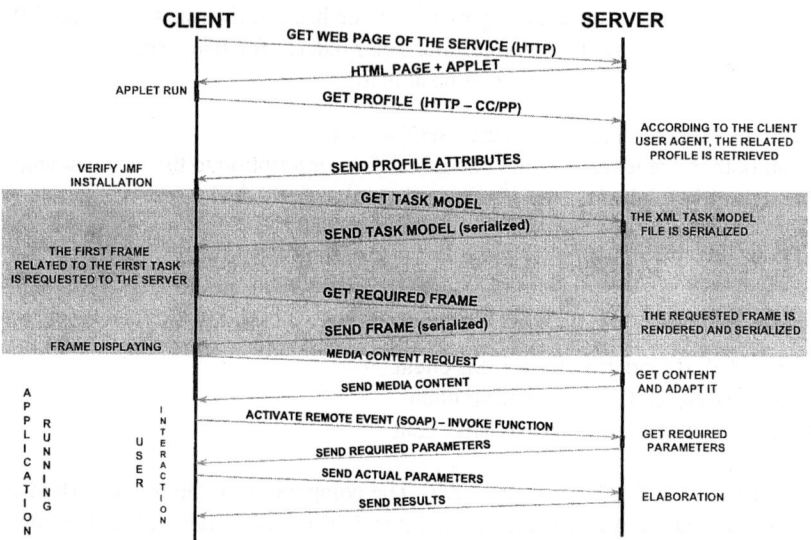

Fig. 4. Server based rendering: sequence diagram

a remote communication. The exchange mechanism involves four steps: 1) the client communicates to the server the name of the remote function to be invoked; 2) the client gets the needed parameters as a response; 3) the parameters are packed into a request that is sent to the remote server, who receives it and execute the required function; 4) the server sends the client a response containing the results of the just executed function.

Client-based rendering on a J2SE client. In this scenario the involved functional blocks are the same as those of the previous one. What changes is the location of the task model manager's logic and the renderer's logic, which are now located at client side. As a direct consequence, the interaction between the client and the server components differs only for the actions highlighted in figure 4: the client only contacts the remote server to just get the XML interface's abstract description and any needed multimedia elements, as indicated in the XML document. Then, the entire rendering process is locally executed.

Client-based rendering on a J2ME client. The components of the framework involved in this scenario are the same as those of the previous one, with the only difference that the client execution environment is J2ME instead of J2SE. As far as the sequence diagram, the only difference concerns the deployment mechanism which is based on the "on the air" (OTA) provisioning.

4 Conclusion

The wide plethora of personal computing devices today available, requires new models of user interface generation. Implementing dynamic user interfaces (UIs) with traditional models, by hard-coding all the requirements for heterogeneous computing devices adaptation, is an error prone and expensive task. The approach followed in our work relies on the separation between the presentation of the application (user interaction) from its business logic. The presentation layer has been structured in such a way to promote the automatic generation of the actual user interface starting from an abstract specification of the user interaction. For this purpose, an XML based language has been defined, giving the application developer the means to represent the user interaction with the application in an abstract way. This "intent-oriented" language can be interpreted by apposite renderer modules which in turn, according to the client device's profile, will generate the user interface fitting to its features. A standard and interoperable mechanism based on SOAP allowed us to also easily manage remote events generated by the interface, by allowing applications to easily access remote services. Finally, two prototypes of renderer have been implemented: one for standard PCs equipped with full Java environment, and the other one for mobile phones equipped with the J2ME environment (CLDC/MIDP profile).

References

1. M. Abrams, C. Phanouriou, A.L. Batongbacal, S.M. Williams, and J.E. Shuster. UIML: an appliance-independent XML user interface language. *Computer Networks*, 11-16(31):1695–1708, May 1999.

2. P. Azevedo, R. Merrick, and D. Roberts. OVID to AUIML - User-Oriented Interface Modelling. In *Proc. of TUPIS'2000*, York, UK, October 2000.
3. S. Banerjee, M. Youssef, R. Larsen, A.U. Shankar, and A. Agrawala et al. Rover: Scalable Location-Aware Computing. *IEEE Computer*, 35(10):46–53, October 2002.
4. L. Capra, W. Emmerich, and C. Mascolo. Middleware for mobile computing. In *Tutorial Proc. of the International Conf. on Networking 2002*, Pisa, Italy, May 2001. LNCS 2497, Springer Verlag.
5. CC/PP Specifications. Available at http://www.w3.org/Mobile/CCPP/.
6. A. Fox et al. Experience with Top Gun Wingman, a proxy-based Graphical Web browser for the 3Com PalmPilot. In *Proc. IFIP Int. Conf. On Distributed Systems Platforms and Open Distributed Processing (Middleware'98)*, Lake District, England, September 1998.
7. O. Buyukkokten et al. Power Browser:Efficient Web Browsing for PDAs. In *Proc. Conf Human Factors in Computing Systems (CHI'00)*, The Hague, Netherlands, April 2000. ACM Press.
8. W.Y. Lum and F.C.M. Lau. A Context-Aware Decision Engine for Content Adaptation. *IEEE Pervasive Computing*, 3(1):41–49, July-September 2002.
9. K. Luyten, K. Coninx, C. Vandervelpen, J.V.den bergh, and B. Creemers. Dygimes: Dynamically Generating Interfaces for Mobile Computing Devices and Embedded Systems. In *Proc. of MobileHCI 2003*, Udine (IT), September 2003.
10. G. Mori, F. Patern, and C. Santoro. Design and Development of Multi-Device User Interfaces through Multiple Logical Descriptions. *IEEE Transactions on Software Engineering*, 30(8):507–520, August 2004.
11. G. Mori, F. Paterno', and C. Santoro. CTTE: Support for Developing and Analysing Task Models for Interactive System Design. *IEEE Transactions on Software Engineering*, 28(8):797–813, August 2002.
12. M. Satyanarayanan. Pervasive computing: vision and challenges. *IEEE Personal Communications*, 4(8):10–17, April 2001.
13. B.N. Schilit, D.M. Hilbert, and J. Trevor. Context-aware communication. *IEEE Wireless Communications*, 9(5):46–54, October 2002.
14. B.N. Schilit, J. Trevor, D. M. Hilbert, and T. K. Koh. Web interaction using very small internet devices. *IEEE Computer*, 10(35):37–45, October 2002.
15. O. Tomarchio, A. Calvagna, and G. Di Modica. Virtual Home Environment for multimedia services in 3rd generation networks. In *Networking 2002*, Pisa (Italy), May 2002.
16. O. Tomarchio, G. Di Modica, D. Vecchio, D. Hovanyi, E. Postmann, and H. Portschy. Code mobility for adaptation of multimedia services in a VHE environment. In *IEEE Symposium on Computer Communications (ISCC2002)*, Taormina (Italy), July 2002.
17. XForms 1.0 . W3C Recommendation. Available at http://www.w3.org/TR/2003/REC-xforms-20031014/, October 2003.

A Communication Broker for Nomadic Computing Systems

Domenico Cotroneo, Armando Migliaccio, and Stefano Russo

Dipartimento di Informatica e Sistemistica,
Universitá di Napoli Federico II, Via Claudio, 21,80125 Napoli, Italy
{cotroneo, armiglia, sterusso}@unina.it

Abstract. This paper presents the Esperanto Broker, a communication platform for nomadic computing applications. By using this broker, developers can model application components as a set of objects that are distributed over wireless devices and interact via remote method invocations. The Esperanto Broker is able to guarantee remote object interactions despite device movements and/or disconnections. We describe the conceptual model behind the architecture, discuss implementation issues, and present preliminary experimental results.

Keywords: Communication Paradigm, Distributed Object Model, Mobility Management, Nomadic Computing.

1 Introduction

Recent advantages achieved in wireless and in mobile terminals technologies are leading to new computing paradigms, which are generally described as mobile computing. Mobile Computing encompasses a variety of models which have been proposed in literature [11], depending on the specific class of mobile terminals (such as PDAs, laptops or sensors), and on the heterogeneity level of the network infrastructure. In this work, focus is on Nomadic Computing (NC), which is a form of mobile computing where communication takes place over strongly heterogeneous network infrastructure, composed of several wireless domains, which are glued together by a fixed infrastructure (the core network) [9]. A wireless domain may represent a building floor, a building room, or a certain zone of a university campus. Several kinds of devices, with different characteristics in terms of resources, capabilities, and dimensions, may be dynamically connected to the core network from different domains, and may use provided services.

It is widely recognized that traditional middleware platforms appear inadequate to be used for the development of nomadic computing applications [7,11], since they operate under a broad range of networking conditions (including rapid QoS fluctuations) and scenarios (including terminals and users movements).

During past years, a great deal of research has been conducted on middleware for mobile settings: solutions proposed in [6,5,14] provide innovative mechanisms, such as context awareness, reconfiguration, dynamic and spontaneous discovery, and adaptation. Solutions proposed in [8,10,12] deal with Quality of Service aspects of mobile computing infrastructures, such as security, network resources allocation, and fault-tolerant strategies. These issues become more challenging due to mobile computing features, in terms of both network characteristics (i.e. bandwidth, latency, topology) and mobile devices characteristics (i.e. hardware and software capabilities).

While we recognize that these studies have represented fundamental milestones for the pursuit of new middleware solutions for mobile computing, most of them do not focus on mobile-enabled interaction mechanisms. Wireless networks have an unpredictable behavior, and users are able to move among different service domains. Therefore, the definition of a mobile-enabled interaction platform is a crucial step of the entire mobile computing middleware design process. Most of cited works delegate the mobility support to the underlying (network and transport) layers, by using protocols such as Mobile IP, Session Initiation Protocol (SIP), or a TCP tailor-made for mobile computing environments [13,17,4]. However, as stated in [15], mobility support is needed on each layer of the ISO/OSI stack, especially on the middleware layer. As for the interaction paradigms, it is recognized that traditional remote procedure call does not accommodate itself to mobile/nomadic computing environments. However, although many different solutions have been proposed for adapting interaction paradigms to mobile settings (such as tuple space, and publish/subscribe [7]), such paradigms result in a programming model which is untyped, unstructured, and thus complex to program. Moreover, the W3C has recently carried out a standardized set of interaction paradigms in order to enable the interoperability among service providers and requesters in the Web [18]. Following such a specification appears a crucial step to enable interoperability among different middleware solutions in the promising Wireless Internet.

The above motivations conduct us to claim that a communication broker, which acts as a building block for the realization of enhanced services, is needed. This paper proposes a new communication platform which implements a Broker for nomadic computing environments, named the Esperanto Broker (EB), which has the following properties: i) it provides a flexible interaction paradigm: a crucial requirement for such a paradigm is the decoupling of the interacting entities. EB provides an object oriented abstraction of paradigms proposed in WSDL specification [18]. It thus reduces the development effort, by preserving a structured and typed programming abstraction ii) it supports terminal mobility by providing mechanisms to handle handoff occurrences: by this way, interactions between involved parts are preserved despite terminal movements among different domains and/or terminal disconnections. By using the proposed platform, developers can model application components as a set of objects that are distributed over wireless domains and interact via remote method invocations.

The rest of paper is organized as follows. Section 2 describes the conceptual architecture of the overall system, including the adopted interaction model, the programming abstraction, and mobility support. Section 3 deals with the implementation issues. Section 4 presents preliminary experimental results. Section 5 concludes the paper with some final remarks about results achieved and direction of future work.

2 Overall Architecture

The figure 1 shows the Esperanto layered architecture. The architecture is composed of two platforms. The Mediator-side platform runs on each domain of the NC infrastructure, and it is in charge of handling connections with Device-side platform running of mobile terminals which are located in its domain. Mediators are connected each other via the core network. The Device-side platform is in charge of connecting devices with the domain's Mediator, and of providing applications with distributed object view of services (i.e., services are requested as remote invocations on a target server object). The Mediator-side decouples service interactions in order to cope with terminal disconnections and/or terminal handoff. To this aim, among the mechanisms aiming to achieve flexibility and decoupling of interactions, we adopt the tuple space model thanks to its simple characterization. In our approach, the shared memory is conceived as a distributed space among mediators, and the Device-side platform provides functionalities to access to it. In the following, we describe both Mediator and Device side platforms, in order to highlight design issues we have addressed.

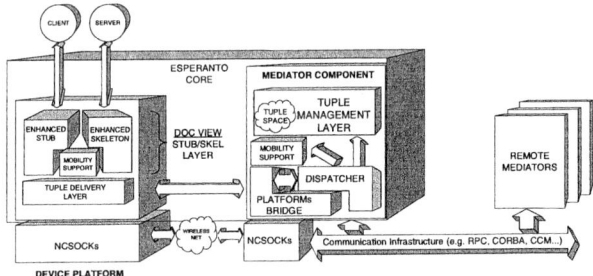

Fig. 1. The Esperanto Broker layered architecture

2.1 Device-Side Platform

The Device-side platform addresses the following issues: i) the NCSOCKS layer provides mobile-enabled communication facilities, which allow applications to interact regardless of wireless technologies being used (e.g., WiFi, Bluetooth, and IrDA). Nomadic Computing SOCKetS (NCSOCKS) library is an API, which provides applications with location and connection awareness support; ii) The TDL layer encapsulates service method invocations into a list of tuples which enables the Mediator to keep the interaction decoupled. The provided primitives are the common *write*, *read*, and *take*. All the provided primitives are non-blocking, but it is possible to specify a timeout value for the *read/take* primitive, in order to indicate the maximum blocking time (so that implementing partial blocking strategies). This layer provides also an asynchronous notification primitive. Once a tuple is written into the space, it is forwarded to the subscribed applications, directly. Similarly to JavaSpaces [2], a lease is associated to each tuple: a tuple will stay in the space until either it is taken, or its lease expires; and iii) the Stub/Skeleton provides applications with a distributed object interface where objects can interact via the WSDL communication paradigms (*request/response, oneway, solicity/response, notify*) [18]. To support the defined invocation strategies, the standard OMG IDL is enriched. As an example, the following interface

```
interface MyService {
    oneway void fooA(in string name);
    reqres bool fooB(in long op, out long result);
    solres void fooC(in string info, out string position);
    notify void fooD(in string weather);
};
```

defines an Esperanto IDL (E-IDL) interface `MyService` which includes the defined invocation strategies. Parameter-passing direction emphasizes respectively a server-side view, for *oneway* and *reqres*, and a client-side view for *solres* and *notify*. More precisely, `name` and `op` are input parameters for server objects, but `info` and `weather` are input parameters for client objects. Stubs and Skeletons translate and marshal/unmarshal remote invocations in terms of tuple-based operations.

2.2 The Mediator Component

The Mediator component is in charge of i) managing connections with terminals by keeping transparent communication technologies (since a domain may be composed of heterogeneous wireless networks); and ii) providing the tuple space service. Mediators communicate to one another via a CORBA ORB. We thus realize the distributed tuple space, as set of CORBA Objects. The Mediator itself is a distributed component, implemented as a set of distributed objects:

Bridge and Dispatcher: since the Device-side is not implemented via CORBA, the Mediator-side needs a Bridge. Tuple-oriented operations are implemented as method invocations on Dispatcher object. The Dispatcher is in charge of providing tuple location transparency, allowing Device-side platform to interact only with the Mediator being currently connected. The Dispatcher copes with the mobility of terminals, by implementing the protocols TDPAM and HOPAM, as described later.

Tuple Management: they are in charge of providing tuple space primitives, namely *read, write, take, subscribe/unsubscribe*. This layer implements the access to a local shared memory. It is worth noting that the tuple space is conceived as a distributed space where each Dispatcher handles a part of the space.

Mobility support: the proposed solution gives to each Mediator two responsibilities: i) as *Home* Mediator, it keeps track of domain handoff of devices registered to it; and ii) as *Host* Mediator, it notifies *Home* Mediators when a roamer device is currently connected to it. To this aim, each Mediator has a Device Manager which stores the list of terminals being connected to NC infrastructure. Furthermore, the Mediator-side platform implements two protocols in order to guarantee the correctness of a tuple delivery process despite device movements and/or disconnections: the Tuple Delivery Protocol Among Mediators (TDPAM) and the HandOver Protocol Among Mediators (HOPAM). TDPAM defines how a tuple is delivered among Mediators ensuring "seamless" device migrations and works as follows: if the tuple receiver (i.e. the object which receives either a service request or a service reply) is running on a mobile device located in the Domain where the request is coming from, the Mediator writes the tuple locally; if the tuple receiver is running on a mobile device located in a different Domain, it forwards the tuple to the Mediator where the device is currently located; if the tuple receiver refers to a group of Esperanto objects, the Mediator forwards the tuple to each remote Mediator available. As for the read/take operation, thanks to the above mentioned *write* strategy, read requests are always processed as tuple retrieval on the local Mediator. HOPAM is in charge of managing domain changes as mobile device moves, preserving the correctness of the TDPAM. It works as follows: once the handoff has been triggered, the new Mediator i) notifies the *Home* Mediator that the *Host* Mediator ha been changed; and ii) transfers all the tuples concerning objects running on the mobile device, from the old shared memory to the local shared space.

3 Implementation Issues

3.1 Device-Side Platform

NCSOCKS Layer. The NCSOCKS provides a C++ socket-like API to communicate on an IP-based channel. *DatagramPacket*, and *DatagramSocket* are classes provided to send and to receive datagrams using UDP: DatagramPacket represents a packet used for connectionless payload delivery, whereas, DatagramSocket is a socket used for sending and receiving datagram packets over a network. We implemented the UDP abstraction since TCP is unsuitable for mobile computing [3]. The implemented API has a *mobility-aware* behavior. In this way applications can be able to perform actions depending on their specific needs and on the current wireless network condition. By example, the send() primitive sends data only if the connection is established and retries to send data if an handoff is occurring. The number of data delivery tries is programmable according to a timeout specified by the programmer. If the process is not successful, an exception will be raised.

Tuple Delivery Layer. In order to provide primitives to access to tuple space (i.e., *write, read, take, subscribe and unsubscribe*), the TDL is in charge of exchanging PDU between the Device-side and Mediator-side platforms using API provided by NCSOCKS layer. Moreover, the TDL provides a primitive to detect domain changes and/or long device disconnections. Using such a primitive, the upper Stub/Skeleton layer can prevent the tuple delivery process from failures due to handoff and device disconnections.

Enhanced Stub/Skeleton layer. The enhanced Stub/Skeleton layer is in charge of providing the DOC view to the application developers. The compilation process of an interface `Interface` produces the following stub/skeleton classes: `InterfaceStubC` and `InterfaceSkelC`, for the client-side and `InterfaceStubS` and `InterfaceSkelS` for the server side. `InterfaceStubC` and `InterfaceSkelS` classes implement the stub/skeleton pattern for the *request/response* and *oneway* paradigms; `InterfaceSkelC` and `InterfaceStubS` classes implement the pattern for the *solicit/response* and *notify* paradigms. To exemplify, let us consider the case of a simple `reqres` paradigm: On the stub-side (skeleton-side), the mapping of *request/response* into tuple-oriented primitives consists of the following operations: i) callback interface subscription in order to retrieve responses (requests) (we used the subscription to avoid busy wait); ii) marshaling of the request parameters into a tuple to write to the server (notification from the client and unmarshaling of parameters from the tuple taken from the space); iii) wait for notification by the server (marshaling of the response parameters into a tuple to write for the client) iv) unmarshaling of parameters from the tuple taken from the space (notification to the client). If the tuple can not be notified to the interested entities, due to temporarily disconnections, it is written in the tuple space. When the network is available again, interested stubs/skeletons can retrieve the tuple via the *take* operation.

3.2 Mediator-Side Platform

Figure 2 depicts a detailed UML CORBA diagram of the Mediator-side platform. As far as the implementation is concerned, we used TAO [16] as CORBA platform.

Tuple. A tuple consists of a XML descriptor containing the following tags: i) *sender*, which is the Source Endpoint of the remote interaction, i.e. the object that sends messages; ii) *receiver*, which is the Destination Endpoint of the remote interaction, i.e. the object that receives messages; iii) *paramList*, which is a list

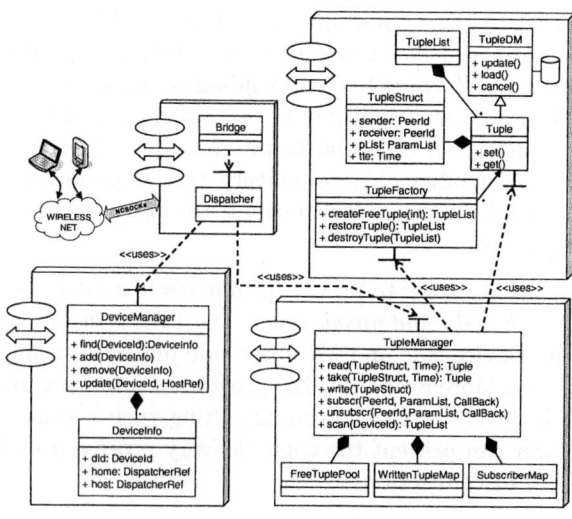

Fig. 2. Detailed Component diagram of Mediator-Side platform

of parameters, where each one has a name, an attribute, and a value; and iv) *tte* (Time To Expiration), which indicates the lease time of the tuple. Each communication endpoint is represented by a *PeerId*: a *PeerId* allows EB to locate an Esperanto object, despite its terminal movements. It is composed of a *PeerIndex*, which allows EB to dispatch requests to several objects that are running on the same device, and a *DeviceId*, which consists of a device identifier, and of terminal's home domain identifier. In this way, the EB is able to locate a terminal in the entire NC environment. In order to achieve persistence of XML descriptors, our solution mandates such responsibility to the Tuple object. Such an approach allows to preserve the transparency of a specific database technology, encapsulating the persistence strategy in the CORBA servant implementation. In the current prototype, persistence is achieved by the `TupleDM` class, which implements a serialization of the *TupleStruct* attribute into a XML file.

Tuple Factory. It is a factory of *Tuple* servants. Servants can be created in two ways: i) factory creates ready-to run *Tuple* servants; ii) factory creates recovered instances of *Tuple* servants, i.e., it creates tuples by XML tuple representations from the mass storage. As far as the allocation strategy is concerned, since it is expensive to create a new tuple for each incoming request, a pool of tuple servants is created at initialization time.

Tuple Manager. It provides the tuple space primitives. The *read* and the *take* methods implement the template matching algorithm. The algorithm performs a template matching between a tuple template (i.e. a tuple with both formal and actual parameters) provided as input, and all the tuples residing on the Mediator's space. A template matches a tuple if the following conditions hold: i) the tuple and the template have the same XML schema or the template has a `null` schema (i.e. the parameters list is empty); ii) the tuple and the template have the same DE; iii) the tuple and the template have the same SE, if specified in the template, otherwise this condition will be ignored; and iv) the list of parameters matches, i.e. each parameter has the same name, type, and (if specified) the same value. The *write* algorithm works as follow: i) it extracts the tuple structure provided as input parameter, and points out to the *receiver* field; ii) it checks if the provided Destination Endpoint is already subscribed to the write event; iii) it forwards tuple to all subscribed Esperanto objects, and if there are some unreachable destinations (i.e., they may be temporarily disconnected) or not subscribed, it stores the tuple into the space. The *scan* method is in charge of retrieving all tuples addressed to a specific mobile device. It processes the whole shared space, returning to the caller the list of tuples whose *receiver* field contains the specified *DeviceId*. This method is crucial for HOPAM implementation because it allows tuples belonging to a roamer device to be moved from old *Host Mediator* to new *Host Mediator*.

Device Manager. It is in charge of keeping the information about devices being connected to the NC infrastructure. More precisely, for each connected device identified by its *DeviceId*, `DeviceManager` holds IOR to its current *Host Mediator* and its *Home Mediator*.

Bridge and Dispatcher. The Bridge interprets (creates) PDUs coming from (directed to) devices, and the Dispatcher is charge of performing operations described in the PDU. From this point on, i) the Bridge extracts the PDU from the tuple and invokes the *write* operation on the Dispatcher object; ii) the Dispatcher extracts *DeviceId* from the tuple and retrieves device's state from the `DeviceManager` (namely, its current host domain and its home domain); then iii) it checks if the tuple is addressed to a device residing in the same domain; iv) if it is, it invokes the *write* operation on the `TupleManager` object, otherwise, it invokes a remote *write* on the Dispatcher which is currently connected to the destination device. When a `GEETINGS PDU` is delivered to the current Mediator, it potentially triggers an handoff procedure. Once the handoff has been triggered, the new Mediator's Dispatcher i) notifies the current device location to device's *Home Mediator* (by invoking the `notify()` method); and ii) transfers all the tuples concerning the terminal, from the old domain space to its local space (by invoking the `moveTuples()` on the old Dispatcher.

4 Preliminary Experimental Results

In this section we discuss the results obtained from performance experiments we have conducted, as compared with MIwCO, an open-source implementation of OMG Wireless CORBA [1]. The Wireless CORBA reference architecture is similar to the Esperanto Broker, however, the OMG Wireless CORBA does not offer all the paradigms standardized by the W3C, and method invocations are implemented by means of synchronous invocations. Moreover, network monitoring algorithms needed to trigger handover procedures are not specified in the Wireless CORBA, and they are not implemented in the current version of MIwCO. The Esperanto Broker addresses such issues in its architecture and implementation. The main objective of our experiments is to evaluate the performance penalty due to the adoption of a CORBA implementation for connecting Mediators; and the decoupled interactions. All experiments were performed with following testbed: Mediators and AB+HLA are distributed on two 1.8Ghz CPU PIV computer with 1GB of RAM running on Linux 2.4.19. As far as mobile devices are concerned, we use Compaq IPAQ 3970, equipped with Linux familiar v0.7.1 and Bluetooth modules. As far as the core network is concerned, we used a Fast Ethernet switch which links all the servers where Mediators (ABs) are running. In order to interconnect Mediators, we used the CORBA TAO version 1.4. As for performance measurements, we evaluate round-trip latency of a method invocations. More precisely, in order to compare ESPERANTO measurements with MIwCO ones, we evaluate the two-way *request/response* interaction paradigm. As for MIwCO, we used both GTP over TCP and GTP over Bluetooth L2CAP protocols. For this reason we used GTP (for MIwCO) and NCSOCKS (for ESPERANTO) over Bluetooth. As for the comparison, we referred to a simple interface named Measure furnished with the following method signature `long foo(in string op)`. The generated stub and skeleton contains the same signature of the `foo` method, both for MIwCO and for ESPERANTO.

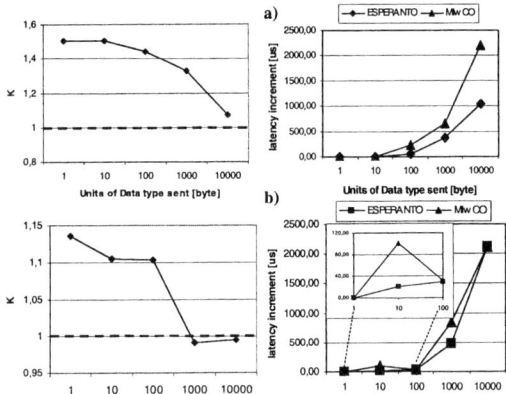

Fig. 3. a) Client/Server interaction by means of same Mediator/Access Bridge; b) Client/Server interaction by means of different Mediators/Access Bridges

The performance penalty was measured as the the ratio of the ESPERANTO to the MIWCO round-trip latency time, i.e. $K = t_{ESPERANTO}/t_{MiWCO}$. We also measured the latency increment (calculated as $t_{i+1} - t_i$) for two subsequent invocations, as function of the dimension of the input parameter, namely op. Measurements are performed under two scenarios. In the first scenario, objects are connected to the same domain (the same Mediator, for ESPERANTO, and the same AB, for MIwCO). Results are depicted in figure 3.a, which shows that the obtained performance are quite similar. In the second scenario, objects are located in two distinct domains. Results are depicted in figure 3.a. As figure shows, ESPERANTO has a cost in terms of performance, due to the fact that it uses more complex interactions among Mediators.

5 Conclusions and Future Work

This paper has described the Esperanto Broker for Nomadic Computing, which provides application developers with an enhanced Distributed Object programming model, integrating mobility management support and adopting mobile-enabled interaction mechanisms. A prototype was developed and tested over distributed heterogeneous platform. From a methodological point of view, this experience has shown that Esperanto provides an effective means for supporting applications running over a nomadic environment. From an experimental point of view, result demonstrated that the proposed architecture has a cost in term of performance. Our future work aims to integrate QoS-based mechanisms, and to support applications with requirements in terms of desired QoS.

Acknowledgements

This work has been partially supported by the Italian Ministry for Education, University, and Research (MIUR) in the framework of the FIRB Project "Middleware for advanced services over large-scale, wired-wireless distributed systems (WEB-MINDS)", and by Regione Campania in the framework of "Centro di Competenza Regionale ICT".

References

1. Mico project: MIWCO and Wireless CORBA home page, 2004. http://www.cs.helsinki.fi/u/jkangash/miwco/.
2. Sun Microsystems: Sun Microsystem Home Page, 2004. http://www.sun.com.
3. A. Bakre and B. R. Badrinath. M-RPC: a remote procedure call service for mobile clients. *in Proc. of 1st Int. Conf. on Mobile Computing and Networking (MobiCom)*, pages 97–110, 1995.
4. A. Bakre and B. Bradinah. I-TCP: Indirect TCP for mobile hosts. *in Proc. of 15th Int. Conf. on Distributed Computing Systems*, 1995.
5. L. Capra, W. Emmerich, and C. Mascolo. CARISMA: Context-Aware Reflective mIddleware System for Mobile Applications. *IEEE Transactions on Software Engineering*, 29(10):929–945, 2003.
6. A. Chan and S. Chuang. MobiPADS: a reflective middleware for context-aware mobile computing. *IEEE Transactions on Software Engineering*, 29(12):1072–1085, 2003.
7. A. Gaddah and T. Kunz. A Survey of Middleware Paradigms for Mobile Computing. Techical Report, July 2003. Carleton University and Computing Engineering.
8. J. He, M. Hiltunen, M. Rajagopalan, and R. Schlichting. Providing QoS Customization in Distributed Object Systems. *Proceedings of the IFIP/ACM International Conference on Distributed Systems Platforms*, LNCS 2218:351–372, 2001.
9. L. Kleinrock. Nomadicity: Anytime, Anywhere in a disconnected world. *Mobile Networks and Applications*, 1(1):pages 351 – 357, December 1996.
10. J. Luo, P. Eugster, and J. Hubaux. Pilot: Probabilistic Lightweight Group Communication System for Ad Hoc Networks. *ACM transaction on mobile computing*, 3(2):164–179, April-June 2004.
11. C. Mascolo, L. Capra, and W. Emmerich. Mobile Computing Middleware. *Lecture Notes In Computer Science, advanced lectures on networking*, pages 20 – 58, 2002.
12. A. Montresor. The Jgroup Reliable Distributed Object Model. *In Proc. of the 2th IFIP WG 6.1 Int. Conf. on Distributed Applications and Interoperable Systems*, 1999.
13. Network Working Group, IETF. *IP mobility support, RFC 2002*, 1996.
14. A. Popovici, A. Frei, and G.Alonso. A proactive middleware platform for mobile computing. *Proc. of the 4th ACM/IFIP/USENIX International Middleware Conference*, 2003.
15. K. Raatikainen. Wireless Access and Terminal Mobility in CORBA. Technical Report, December 1997. OMG Technical Meeting, University of Helsinky.
16. D. C. Schmidt, D. L. Levine, and S. Mungee. The Design of the TAO Real-Time Object Request Broker. *Computer Communications*, 21(4):pages 294–324, 1998.
17. H. Schulzrinne and E. Wedlund. Application-Layer Mobility Using SIP. *Mobile Computing and Communications Reviews*, 4(3):47–57, 1999.
18. W3C. The World Wide Web Consortium. Web Services Description Language (WSDL) 1.1, 2004. http://www.w3.org/TR/wsdl.html.

Adaptive Buffering-Based on Handoff Prediction for Wireless Internet Continuous Services

Paolo Bellavista, Antonio Corradi, and Carlo Giannelli

Dip. Elettronica, Informatica e Sistemistica - Università di Bologna,
Viale Risorgimento, 2 - 40136 Bologna, Italy
Phone: +39-051-2093001; Fax: +39-051-2093073
{pbellavista, acorradi, cgiannelli}@deis.unibo.itl

Abstract. New challenging deployment scenarios are accommodating portable devices with limited and heterogeneous capabilities that roam among wireless access localities during service provisioning. That calls for novel middlewares to support different forms of mobility and connectivity in wired-wireless integrated networks, to provide runtime service personalization based on client characteristics, preferences, and location, and to maintain service continuity notwithstanding temporary disconnections due to handoff. The paper focuses on how to predict client horizontal handoff between IEEE 802.11 cells in a portable way, only by exploiting RSSI monitoring and with no need of external global positioning, and exploits mobility prediction to preserve audio/video streaming continuity. In particular, handoff prediction permits to dynamically and proactively adapt the size of client-side buffers to avoid streaming interruptions with minimum usage of portable device memory. Experimental results show that our prediction-based adaptive buffering outperforms traditional static solutions by significantly reducing the buffer size required for streaming continuity and by imposing a very limited overhead.

1 Introduction

The increasing availability of wireless Internet Access Points (APs) and the popularity of wireless-enabled portable devices stimulate the provisioning of distributed services to a wide variety of mobile terminals, with very heterogeneous and often limited resources. Even if device/network capabilities are always increasing, the development of wireless applications is going to remain a very challenging task. Their design should take into account several factors, from limited client-side memory to limited display size/resolution, from temporary loss of connectivity to frequent bandwidth fluctuations and high connectivity costs, from extreme client heterogeneity to local resource availability that may abruptly change in the case of client roaming. [1]

Let us consider the common deployment scenario where wireless solutions extend the accessibility to the traditional Internet via APs working as bridges between fixed hosts and wireless devices [2]. The most notable example is the case of IEEE 802.11 APs that support connectivity of Wi-Fi terminals to a wired local area network [3]. In the following, we will indicate these integrated networks with fixed Internet hosts, wireless terminals, and wireless APs in between, as the *Wireless Internet* (WI).

WI service provisioning must consider the specific characteristics of client devices, primarily their limits on local resources and their high heterogeneity. Limited processing power, memory, and file system make portable devices unsuitable for traditional services designed for fixed networks. These constraints call for both assisting wireless terminals in service access and downscaling service contents depending on terminal resource constraints. In addition, portable devices currently exhibit extreme heterogeneity of hardware capabilities, operating systems, installed software, and network technologies. This heterogeneity makes hard to provide all needed service versions with statically tailored contents and calls for on-the-fly adaptation. Client limits and heterogeneity are particularly crucial when providing *continuous services*, i.e., applications that distribute time-continuous flows of information to their requesting clients, such as audio and video streaming [4]. WI continuous services should address the very challenging issue of avoiding temporary flow interruptions when clients roam from one wireless locality to one another, also by considering client memory limitations, which do not allow traditional buffering solutions based on proactive client caching of large chunks of multimedia flows.

We claim the need of middleware solutions to support WI service provisioning to portable devices, by locally mediating their service access and by dynamically adapting results to client terminal properties, location, and runtime resource availability [3-6]. Middleware components should follow client roaming in different WI localities and assist them locally during their service sessions. Moreover, client memory limitations suggest deploying middleware components over the fixed network, where and when needed, while portable devices should only host thin clients. By following the above guidelines, we have developed an application-level middleware, based on Secure and Open Mobile Agent (SOMA) proxies, to support the distribution of location-dependent continuous services to wireless devices with limited on-board resources [7, 8]. The primary design idea is to dynamically deploy mobile proxies acting on behalf of wireless clients over fixed hosts in the network localities that currently offer client connectivity.

The paper focuses on an essential aspect of our middleware: how to avoid interruptions of continuous service provisioning when a client roams from one WI locality to one another (*wireless cell handoff*) at runtime. To achieve this goal, handoff prediction is crucial. On the one hand, it permits to migrate mobile proxies in advance to the wireless cells where mobile clients are going to reconnect, so to have enough time to proactively reorganize user sessions in newly visited localities, as detailed in a previous work [9]. On the other hand, service continuity requires maintaining client-side buffers of proper size with flow contents to play during the handoff process and to reconnect to mobile proxies in the new WI localities. Handoff prediction can enable the adaptive management of client buffers, by increasing buffer size (of the amount expectedly needed) only in anticipation of client handoffs, thus improving the efficiency of memory utilization, which is essential for portable devices. Let us observe that the proposed adaptive buffering, specifically developed for our mobile proxy-based middleware to avoid streaming interruptions, can help any class of WI applications that benefit from content pre-fetching in the client locality.

In particular, the paper presents how to exploit handoff prediction to optimize the utilization of client-side pre-fetching buffers for streaming data. The primary guideline is to provide handoff prediction-based adaptive management of client buffers with the twofold goal of minimizing buffer size and triggering pre-fetch operations only when needed. Our adaptive buffering exploits an original solution to predict handoffs between IEEE 802.11 cells, by using only client-side Received Signal Strength Indication (RSSI) from the wireless APs in visibility. Our prediction mechanisms originally adopt a first-order Grey Model (GM) to reduce RSSI fluctuations due to signal noise; they are portable, lightweight, and completely decentralized, and do not require any external global positioning system [8,10].

The paper also reports a thorough performance evaluation of our handoff prediction-based adaptive buffering in a wide-scale simulated environment, which can model nodes randomly roaming among IEEE 802.11 APs. In addition, we have collected in-the-field experimental results by deploying our system prototype over an actual set of Wi-Fi laptops. Experimental results show that our adaptive buffering outperforms not only traditional solutions based on statically pre-determined buffer size, but also adaptive dimensioning solutions based on non-GM-filtered RSSI values, also when considering different implementations of communication-level handoff by different Wi-Fi client card vendors. The proposed solution has shown to reduce the buffer size needed to maintain streaming continuity and to impose a very limited overhead, by only exploiting local RSSI data already available at clients.

2 Related Work

Few researches have addressed position prediction in wireless networks, most of them by proposing solutions based on either the estimate of current position/speed or usual movement patterns. [11] predicts future location/speed by exploiting a dynamic Gauss-Markov model applied to the current and historical movement data. [12] bases its trajectory prediction on paths followed in the recent past and on the spatial knowledge of the deployment environment, e.g., by considering admissible path databases. Note that exploiting these position prediction solutions as the basis for handoff prediction requires coupling them with the knowledge of AP coverage area maps. In addition, in open and extremely dynamic scenarios, with medium/short-range wireless connectivity, user mobility behaviors change very frequently and irregularly, thus making almost inapplicable handoff predictions based on past user mobility habits.

There are a first few approaches in the literature that have already investigated RSSI prediction. [13] predicts future RSSI values by using a retroactive adaptive filter to mitigate RSSI fluctuations; device handoff is commanded when the difference between the current and the predicted RSSI values is greater than a threshold. [14] exploits GM to decide when to trigger the communication handoff by comparing RSSI predictions with average and current RSSI values. However, both [13] and [14] apply RSSI prediction to improve communication-level handoff, e.g., to reduce unnecessary bouncing, and not to predict the movements of wireless clients so to adaptively manage their streaming buffers.

Adaptive buffer management for stream interruption avoidance is a well investigated area in traditional fixed distributed systems. [15] exploits predicted network delay to adapt buffer size to the expected packet arrival time; its goal is to avoid stream interruptions due to packet jitter variations. [16] compares different algorithms for adaptive buffer management to minimize the effect of delay jitter: packet delay and/or packet peak recognition are exploited to dynamically adapt the buffer size. [17] predicts buffer occupation with a Proportional Integral Derivative predictor; the goal is to minimize packet loss due to buffer overflow. [18] is the only proposal dealing with wireless clients: it suggests a proxy-based infrastructure exploiting neural networks to predict client connection state; proxies execute on the wired network to manage the state/connections of their associated clients. Its primary goal is to preserve session state and not to maintain service continuity.

To the best of our knowledge, our middleware is definitely original in integrating a lightweight, portable, completely decentralized, and modular handoff prediction solution, only based on RSSI data, and in exploiting it to increase the efficiency of adaptive buffer management for continuous services.

3 Client-Side Adaptive Buffering Based on Handoff Prediction

Given the crucial role of efficiently managing client buffers to prevent streaming interruptions during client roaming, we propose an innovative buffer management solution that tends to save limited client memory by increasing buffer size only when a client handoff is going to occur. To this purpose, it is first necessary to exactly clarify how communication-level handoff works. In fact, the IEEE 802.11 standard does not impose any specific handoff strategy and communication hardware manufacturers are free to implement their own strategies, as detailed in the following. The different communication-level handoff strategies motivate different variants of our prediction-based adaptive buffering: for this reason, the paper proposes and compares two alternative buffer implementations specifically designed for the two most relevant classes of possible handoff strategies, i.e., Hard Proactive (HP) and Soft Proactive (SP).

3.1 Reactive and Hard/Soft Proactive Communication-Level Handoff

Several communication-level handoff strategies are possible, which mainly differ in the event used to trigger handoff. We can identify two main categories: reactive and proactive. Reactive handoff strategies tend to delay handoff as much as possible: handoff starts only when wireless clients lose their current AP signal. Reactive strategies are effective in minimizing the number of handoffs, e.g., by avoiding to trigger a handoff process when a client approaches a new wireless cell without losing the origin signal and immediately returns back to the origin AP. However, reactive handoffs tend to be long because they include looking for new APs, choosing one of them, and asking for re-association. In addition, in reactive handoffs the RSSI of associated APs tends to be low in the last phases before de-association, thus producing time intervals before handoff with limited effective bandwidth. Proactive strategies, instead, tend to trigger handoff before the complete loss of the origin AP signal, e.g., when the new cell RSSI is greater than the origin one. These strategies are less effective in reducing the number of useless handoffs but are usually prompter, by performing search operations for new APs before starting the actual handoff procedure and by limiting time intervals with low RSSI before handoffs.

By concentrating on proactive strategies, two primary models can be identified. On the one hand, HP strategies trigger a handoff any time the RSSI of a visible AP overcomes the one of currently associated AP of more than an Hysteresis Handoff Threshold (HHT); HHT is usually introduced to prevent heavy bouncing effects. On the other hand, SP strategies are "less proactive" in the sense that they trigger handoff only if i) the HP condition applies (there is an AP with RSSI greater than current AP RSSI plus HHT), and ii) the current AP RSSI is lower than a Fixed Handoff Threshold (FHT). For instance, the handoff strategies implemented by Cisco Aironet 350 and Orinoco Gold Wi-Fi cards follow, respectively, the HP and SP models. More in detail, Cisco Aironet 350 permits to configure its handoff strategy with the "Scan for a Better AP" option: if the current AP RSSI is lower than a settable threshold, the Wi-Fi card monitors RSSI data of all visible APs, looking for a better AP; for sufficiently high threshold values, the Cisco cards behave according to the HP model. Orinoco Gold cards implements the SP strategy, applied to the Signal to Noise Ratio (SNR) in place of RSSI, without giving any possibility to configure the used thresholds.

3.2 HP- and SP-Handoff Predictors

The goal of our handoff prediction-based buffer management is to have client buffers of the maximum size and full exactly when actual re-associations to destination APs occur. Wrong handoff predictions produce incorrect dimensioning of client-side buffers; correct but late handoff predictions cause correctly-sized buffers that are not fulfilled with the needed pre-fetched streaming data.

Just to give a rough idea of the magnitude order of the advance time needed in handoff prediction, let us briefly consider the example of a client receiving a multimedia stream played at 1000Kbps constant bit-rate and a handoff procedure taking 1.5s to complete. That time interval includes the time for communication-level handoff and the time to locally reconnect to the migrated companion proxy, and largely overestimates the actual time measured in [8]. In this case, the client-side buffer size must be at least 187.5KB. If the client available bandwidth is 1500Kbps on average, the buffer fills with a speed of 500Kbps on average, by becoming full (from an empty state) in about 3s. Therefore, in the worst case, our handoff prediction should be capable of anticipating the actual handoff of 3s, to trigger buffer pre-fetching in time.

Our buffer management solution exploits a predictor structured in three pipelined modules. The first module (*Filter*) is in charge of filtering RSSI sequences to mitigate RSSI fluctuations due to signal noise. The second module tries to estimate handoff probability in the near future (*Prob*) based on RSSI values provided at its input. The last module (*Dim*) determines the correct buffer size to enforce, depending on handoff probability. More formally, our predictor consists of three modules, each of them implementing a function with the following domain and co-domain of definition:

- Filter: RSSI → FilteredRSSI
- Prob: FilteredRSSI → HandoverProbability
- Dim: HandoverProbability → BufferSize

The modular architecture of our predictor permits a completely separated implementation and deployment of Filter, Prob, and Dim modules, thus simplifying the exploitation and experimentation of different filtering, handoff prediction, and buffer size management mechanisms, even dynamically composed at provision time by downloading the needed module code [19]. For instance, the experimental results in Section 4 will show the performance of our middleware when feeding the Prob module with either actual RSSI sequences (Filter is the identity function) or GM-filtered ones (by pointing out the improvement due to the only GM-based RSSI filtering).

By delving into finer details about the already implemented predictor modules available in our middleware, Prob can assume three different states:

- LowProb, if handoff is considered highly improbable in the near future;
- HighProb, if handoff is considered almost certain in the near future;
- MedProb, otherwise.

Dim exploits the state delivered by Prob to dynamically modify the size of associated client-side buffers: in the current implementation, when in the HighProb state, Dim sets the buffer size at the maximum for that multimedia flow (flow bit-rate * 1.5s); when in LowProb, Dim sets the size at the minimum (maximum/10); and when in MedProb, it sets the size at (maximum+minimum)/2. We are currently evaluating if more complex processing functions for Prob and Dim modules (e.g., with finer granularity for the discrete states of handoff probability and buffer size) could improve our middleware performance; first results encourage to exploit simple and lightweight module functions, which can achieve the needed performance results with a limited computational overhead (see Section 4).

The Prob module runs at the client side, is completely decentralized, and only exploits local monitoring information about the RSSI of all IEEE 802.11 APs in current visibility. The awareness of RSSI monitoring data is achieved in a completely portable way over heterogeneous platforms, as detailed in Section 3.4. In particular, we have implemented two variants of our Prob module, one suitable for communication-level HP handoffs and the other for SP ones. We have decided not to develop Prob modules for reactive strategies because reactive handoffs are inherently unsuitable for continuous service provisioning environments, given their longer time needed to complete handoff. In addition, handoff prediction is less challenging in the case of reactive communication-level handoffs than when dealing with proactive ones: the triggering of a reactive handoff only depends on one AP RSSI data.

The HP-variant of our Prob module is in the state:

- LowProb, if the filtered value for the current AP RSSI is greater than the filtered RSSI values for any visible AP plus a Hysteresis Superior Threshold (HST);
- HighProb, if the filtered value for the current AP RSSI is lower than at least one filtered RSSI value for one visible AP plus a Hysteresis Inferior Threshold (HIT);
- MedProb, otherwise.

Figure 1 represents filtered RSSI values for current and next APs, in proximity of a HP handoff. A wireless client, moving from the origin AP locality to the destination AP one, is first associated with the origin AP (white background), then with the destination AP (grey background). The HP Prob module state changes from LowProb to MedProb and finally to HighProb. When the actual RSSI of destination AP overcomes the actual RSSI of origin AP plus HHT, the handoff is triggered (for the sake of simplicity, the figure considers filtered RSSI values equal to actual ones in HO).

Adaptive Buffering-Based on Handoff Prediction for Wireless Internet   1027

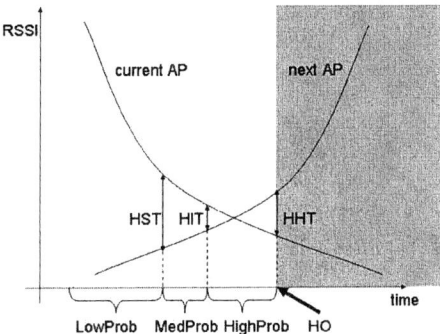

Fig. 1. The different states of the Prob HP-variant

The SP-variant of the Prob module can assume the following states:

- LowProb, if the filtered RSSI value for current AP is greater than either a Fixed Superior Threshold (FST) or the filtered RSSI value for any visible AP plus HST;
- HighProb, if the filtered RSSI value for current AP is lower than a Fixed Inferior Threshold (FIT) and than at least one filtered RSSI of a visible AP plus HIT;
- MedProb, otherwise.

Similarly to Figure 1, Figures 2a and 2b represent filtered RSSI values for the origin and destination APs in proximity of an SP handoff. Figure 2a depicts a case where filtered RSSI values for current and next APs change relatively slow: in this case, changes in Prob results and actual handoff are triggered by the overcoming of hysteresis thresholds. In Figure 2b, instead, filtered RSSI values rapidly evolve and it is the passing of fixed thresholds that triggers Prob state variations and handoff.

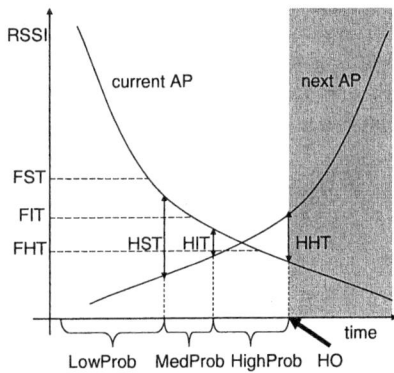 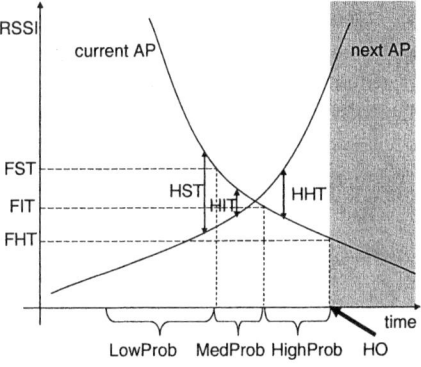

Fig. 2a. The different states of the Prob SP-variant for slow RSSI evolution

Fig. 2b. The different states of the Prob SP-variant for relatively fast RSSI evolution

Both HP and SP variants of the Prob module take advantage of our GM-based Filter module. Filter exploits a very simple and lightweight first-order GM [10] to obtain filtered RSSI values on the basis of RSSI values monitored in the recent past, as described in the following section. Let us stress again that our modular handoff predictor is completely local and decentralized: each wireless client hosts its handoff predictor, whose Prob results only depend on either monitored or filtered RSSI values for all APs in visibility, with no need of interacting with any other client or SOMA-based middleware component running in the wired infrastructure.

3.3 Grey Model-Based RSSI Prediction

Given one visible AP and the set of its actual RSSI values measured at the client side $R0 = \{r0(1), ..., r0(n)\}$, where $r0(i)$ is the RSSI value at the discrete time i, it is possible to calculate $R1 = \{r1(1), ..., r1(n)\}$, where:

$$r_1(i) = \sum_{j=1}^{i} r_0(j) \tag{1}$$

Then, from the GM(1,1) discrete differential equation of the first order [8]:

$$\frac{dr_1(i)}{di} + ar_1(i) = u \tag{2}$$

the wireless client can autonomously determine a and u, which are exploited to obtain the predicted RSSI value pr(i) at discrete time i according to the GM(1,1) prediction function [10]:

$$pr(i) = \left(r_1(1) - \frac{u}{a}\right)e^{-ak} + \frac{u}{a} \tag{3}$$

Let us observe that the average accuracy of the RSSI prediction may also depend on the number of actual RSSI values $r_0(i)$ employed by the adopted GM(1,1). In principle, longer the finite input series R_0, more regular the RSSI predicted values, and slower the speed with which the GM(1,1) prediction anticipates the actual RSSI sequence in the case of abrupt evolution [10]. We have evaluated the performance of our predictors, as extensively presented in Section 4, also while varying the number n of values in R_0. We did not experience any significant improvement in the predictor performance, on average, by using n values greater than 15. For this reason, all the experimental results reported in the following will refer to the usage of R_0 sets with 15 past RSSI values.

4 Experimental Results

To quantitatively evaluate the effectiveness of the proposed modular handoff predictor and of its application in our adaptive buffer management infrastructure, we have identified some performance indicators and measured them both in a simulated environment, with a large number of Wi-Fi clients roaming among a large number of wireless AP localities, and in our campus deployment scenario, where four laptops move among the different coverage areas of six APs. Two laptops are Linux-based, while the other two host Microsoft Windows.NET; they alternatively exploit Cisco Aironet 350 (HP handoff) and Orinoco Gold (SP handoff) IEEE802.11 cards. In addition, we have compared the performance of our HP/SP Prob modules when connected to either our GM(1,1) Filter function or an identity Filter function that provides output values identical to its input: the goal was of understanding the isolated contribution of the GM(1,1) Filter function to the overall performance of our adaptive buffer management infrastructure.

In particular, we have considered the following performance indicators:

- *Average Buffer Size (ABS)* = $\dfrac{1}{T}\int_0^T BS(t)dt$,

where $BS(t)$ is the time-varying buffer size. In other words, ABS is the time-weighted average of the buffer size;

- *Average Buffer Duration (ABD)* = $\dfrac{1}{T}\int_0^T BD(t)dt$

where $BD(t)$ is the time-varying validity of a chosen buffer size. In other words, *ABD* is the average time interval between two successive operations of buffer re-sizing;

- *Successful Handoff (SH%)* = $\left(1 - \dfrac{DH}{NH}\right)*100$

where *DH* is the number of actual client handoffs and *NH* is the number of handoffs predicted by the proposed HP/SP predictors.

In general, the goal of an optimal buffer management solution is to contemporarily achieve minimum values for *ABS* and sufficiently large *ABD* values, with maximum $SH_\%$. Obviously, *ABS* and $SH_\%$ are strongly correlated: on the one hand, very good values for $SH_\%$ can be easily achieved with large *ABS* values; on the other hand, it is possible to obtain very low *ABS* values by simply delaying as much as possible the buffer size enlargement, but with the risk of streaming interruptions (too low $SH_\%$ value). Moreover, it is necessary to maintain sufficiently large *ABD* values not to continuously perform useless and expensive buffer re-size operations.

We have measured the three indicators above in a challenging simulated environment where 17 APs are regularly placed in a 62m x 84m area and RSSI fluctuation has a 3dB standard deviation. Wireless clients follow trajectories with a randomly variable speed and with a randomly variable direction (with a Gaussian component for the standard deviation of Π/6). The speed is between 0.6m/s and 1.5m/s to mimic the behavior of walking mobile users; FST = 66dB; FIT = 70dB; FHT = 80dB; HST = 10dB; HIT = 6dB; HHT = 6dB. On the average, each wireless client has the visibility of ten APs at the same time, which represents a worst case scenario significantly more complex than the actually deployed Wi-Fi networks (where no more than five APs are usually visible at any time and from any client position).

Table 1 reports the average results for the three performance indicators over a large set of simulations, each one with about 500 handoffs. For the video streaming exploited in all the experiments, the buffer size required to avoid interruptions in the case of static fixed dimensioning is 200KB. The most important result is that any proposed Prob module, when provided with either GM-filtered RSSI values or actual RSSI values, significantly reduces *ABS* (between 27.5% and 33.5%), thus relevantly improving the client memory utilization. In addition, Prob modules fed with GM-filtered RSSI values largely outperform the cases with actual RSSI values, especially with regard to the *ABD* performance indicator. In fact, even if *ABS* has demonstrated to maintain good values in all cases, directly monitored non-filtered RSSI (with its more abrupt fluctuations) tends to trigger a higher number of useless handoff predictions and, consequently, more useless modifications in the enforced buffer size.

Table 1. Performance indicators for HP and SP predictors. In the case of static fixed buffer: ABS=200KB, SH%=100, and ABD=∞

Handover Strategy	Filter Function	ABS (KB)	$SH_\%$	ABD (s)
HP	Identity	140	92.1	2.80
	GM(1,1)	133	92.8	5.20
SP	Identity	145	91.6	2.79
	GM(1,1)	138	97.5	5.66

Figure 3 points out the correlation between GM-filtered RSSI and buffer size. In particular, it depicts the time evolution of buffer size (dotted line) depending on both GM-filtered RSSI of the currently associated AP (grey line) and the greatest RRSI among the non-associated visible APs (black line). Let us stress that when the currently associated AP RSSI is significantly greater than RSSI from other APs, our buffer management infrastructure maintains buffer size at its minimum (20KB in the example); when the RSSI of the currently associated AP, instead, is similar to the RSSI of another AP, buffer size increases at its maximum (200KB); otherwise, our infrastructure works to manage a medium-sized buffer (110KB).

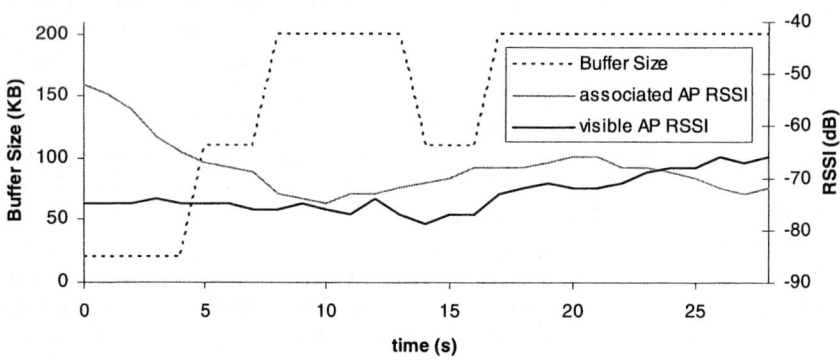

Fig. 3. Buffer size variations depending on time evolution of GM-filtered RSSI values for the currently associated AP and for another AP in visibility

In addition to simulations, we have evaluated the performance of HP/SP Prob modules also by using a service prototype, built on top of our middleware, and by moving four client laptops among the campus WI localities during streaming provisioning. Even if the number of considered in-the-field handoffs is largely lower than the simulated one (thus, less relevant from the statistical point of view), in-the-field performance results confirm the simulation-based ones. In particular, the prototype-based *ABS*, *SH%*, and *ABD* results have demonstrated to be better than simulation-based ones, on the average, also due to the lower number of considered APs, and the consequently simpler handoff prediction. However, we have experienced a significant degradation of prototype-based performance indicators in the case of extreme RSSI fluctuations, e.g., when a client follows a trajectory in strict proximity of relevant obstacles, such as the reinforced concrete walls of our campus buildings.

The code of the handoff prediction prototype, additional details about prototype implementation, and further simulation/prototype-based experimental results are available at http://lia.deis.unibo.it/Research/SOMA/SmartBuffer/

5 Conclusions and On-going Research

The exploitation of mobile middleware proxies that work over the fixed network on behalf of their resource-constrained clients is showing its suitability and effectiveness in the WI, especially when associated with handoff prediction. In particular, handoff prediction can help in realizing novel adaptive buffering solutions that optimize buffer size and pre-fetching. Our work of design, implementation, and evaluation of different buffering solutions has shown that our dynamic and simple GM-based proposal outperforms static buffering strategies, by preserving service continuity with limited requirements on client memory capabilities.

The promising performance results obtained are stimulating further related research activities. We are experimenting other handoff prediction techniques based on either higher-level GM models or the GM application to Ekahau-provided estimates of client positions (not directly to RSSI data) [20]. The goal is to evaluate whether a greater complexity of the Prob module of our predictor can significantly improve prediction quality, thus justifying the replacement of the currently adopted GM(1,1), which is extremely simple and lightweight.

Acknowledgements

Work supported by the MIUR FIRB WEB-MINDS and the CNR Strategic IS-MANET Projects.

References

1. Bacon, J., Bates, J., Halls, D.,: Location-oriented Multimedia. IEEE Personal Communications, Vol. 4, No. 5, Oct. 1997.
2. Corson, M.S., Macker, J.P., Park, V.D.,: Mobile and Wireless Internet Services: Putting the Pieces Together. IEEE Communications, Vol. 39, No. 6, June 2001.
3. Stallings, W.: Wireless Communications and Networks. Pearson Education, Aug. 2001.
4. Ramanathan, P., Sivalingam, K.M., Agrawal, P., Kishore, S.: Dynamic Resource Allocation Schemes during Handoff for Mobile Multimedia Wireless Networks. IEEE Journal on Selected Areas in Communications, Vol. 17, No. 7, July 1999.
5. Saha, S., Jamtgaard, M., Villasenor, J.: Bringing the Wireless Internet to Mobile Devices. IEEE Computer, Vol. 34, No. 6, June 2001.

6. Curran, K., Parr, G.: A Middleware Architecture for Streaming Media over IP Networks to Mobile Devices. IEEE Int. Conf. Wireless Communications and Networking (WCNC), Mar. 2003.
7. Bellavista, P., Corradi, A., Stefanelli, C.: Application-level QoS Control and Adaptation for Video on Demand. IEEE Internet Computing, Vol. 7, No. 6, Nov.-Dec. 2003.
8. Bellavista, P., Corradi, A.: A QoS Management Middleware based on Mobility Prediction for Multimedia Service Continuity in the Wireless Internet. IEEE Int. Symp. on Computers and Communications (ISCC), July 2004.
9. Bellavista, P., Corradi, A., Giannelli, C.: Mobile Proxies for Proactive Buffering in Wireless Internet Multimedia Streaming. Int. Workshop on Services and Infrastructure for the Ubiquitous and Mobile Internet (SIUMI), June 2005.
10. Deng, J.L.: Introduction to Grey Theory. The Journal of Grey System, Vol. 1, No. 1, 1989.
11. Liang, B., Haas, Z.J.: Predictive Distance-Based Mobility Management for Multidimensional PCS Network. IEEE/ACM Transactions on Networking, Vol. 11, No. 5, Oct. 2003.
12. Karimi, H.A., Liu, X.: A Predictive Location Model for Location-based Services. ACM Int. Workshop Advances in Geographic Information Systems (GIS), Nov. 2003.
13. Kapoor, V., Edwards, G., Sankar, R.: Handoff Criteria for Personal Communication Networks. IEEE Int. Conf. Communications (ICC), May 1994.
14. Sheu, S.T., Wu, C.C.: Using Grey Prediction Theory to Reduce Handoff Overhead in Cellular Communication Systems. IEEE Int. Symp. Personal, Indoor and Mobile Radio Communications (PIMRC), Sep. 2000.
15. DeLeon, P., Sreenan, C.J.: An Adaptive Predictor For Media Playout Buffering. IEEE Int. Conf. Acoustics, Speech, and Signal Processing (ICASSP), Mar. 1999.
16. Narbutt, M., Murphy, L.: Adaptive Playout Buffering for Audio/Video Transmission over the Internet. IEE UK Teletraffic Symposium, 2001.
17. Ip, M.T.W., Lin, W.W.K., Wong, A.K.Y., Dillon, T.S., Wang, D.: An Adaptive Buffer Management Algorithm for Enhancing Dependability and Performance in Mobile-Object-Based Real-time Computing. Int. Symp. Object-Oriented Real-Time Distributed Computing (ISORC), May 2001.
18. Kumar, B.P.V., Venkataram, P.: A Neural Network–based Connectivity Management for Mobile Computing Environment. Int. Journal of Wireless Information Networks, Apr. 2003.
19. Bellavista, P., Corradi, A., Montanari, R., Stefanelli, C.: Context-aware Middleware for Resource Management in the Wireless Internet. IEEE Trans. on Software Engineering, Vol. 29, No. 12, Dec. 2003.
20. Ekahau, Inc. – The Ekahau Positioning Engine v2.1. http://www.ekahau.com/products/positioningengine/

A Scalable Framework for the Support of Advanced Edge Services

Michele Colajanni[1], Raffaella Grieco[2], Delfina Malandrino[2], Francesca Mazzoni[1], and Vittorio Scarano[2]

[1] Dipartimento di Ingegneria dell'Informazione,
Università di Modena e Reggio Emilia, 41100 Modena, Italy
{colajanni, mazzoni.francesca}@unimore.it
[2] Dipartimento di Informatica ed Applicazioni "R.M. Capocelli",
Università di Salerno, 84081 Baronissi, Salerno, Italy
{rafgri, delmal, vitsca}@dia.unisa.it

Abstract. The Ubiquitous Web requires novel programming paradigms and distributed architectures for the support of advanced services to a multitude of user devices and profiles.

In this paper we describe a Scalable Intermediary Software Infrastructure (SISI) that aims at efficiently providing content adaptation and combinations of other complex functionalities at edge servers on the WWW. SISI adopts different user profiles to achieve automatic adaptation of the content according to the capabilities of the target devices and users.

We demonstrate SISI efficiency by comparing its performance against another framework for content adaptation at edge servers.

1 Introduction

The World Wide Web, thanks to its simplicity, visibility, scalability and ubiquity, is considered the inexhaustible universe of information available through any networked computer or device and is also considered as the ideal testbed to perform distributed computation and to develop complex distributed applications.

In the *Pervasive and Ubiquitous Computing* era the trend is to access Web content and multimedia applications by taking into account four types of important requirements: *anytime-anywhere* access to *any data* through *any device* and by using *any access network*. Nowadays, the existing and emerging wireless technologies of the 3rd generation (GSM, GPRS and UMTS), involve a growing proliferation of new rich, multimedia and interactive applications that are available on the Web. On the other hand, these appealing services are requested from a growing variety of terminal devices, such as Desktop PC, pagers, personal digital assistants (PDAs), hand-held computers, Webphones, TV browsers, laptops, set top boxes, smart watches, car navigation systems, etc. The capabilities of these devices widely range according to different hardware and software properties. In particular, for mobile devices relevant constraints concern storage, display capabilities (such as screen size and color depth), wireless network connections, limited bandwidth, processing power and power consumption. These constraints involve several challenges for the delivery and presentation of complex personalized applications towards these devices, especially if there is the necessity of guaranteeing specified levels of service.

The most simple solution (still used by many Web portals) is to follow the *"one-size-fits-all"* philosophy. The idea is to provide content that is specifically designed for Desktop PC, without taking care of the troubles that, for example, could affect mobile users.

The effective presentation of Web content requires additional efforts and in particular new computation patterns. Providing a tailored content in an efficient way for different client devices, by addressing the mismatch between rich multimedia content and limited client capabilities, is becoming increasingly important because of the rapid and continuous evolving of the pervasive and ubiquitous devices.

One of the current research trend in distributed systems is how to extend the traditional client/server computational paradigm in order to allow the provisioning of *intelligent* and *advanced* services. One of the most important motivation is to let heterogeneous devices access WWW information sources through a variety of emerging 3G wireless technologies. This computational paradigm introduces new actors within the WWW scene, the intermediaries [1,2] that is, software entities that act on the HTTP data flow exchanged between client and server by allowing content adaptation and other complex functionalities, such as geographical localization, group navigation and awareness for social navigation [3,4], translation services [5], adaptive compression and format transcoding [6,7,8], etc.

The major research in this area is how to extend the capabilities of the Web to provide content adaptation and other complex functionalities to support personalization, customization, mobility and ubiquity. The idea of using an intermediate server to deploy adapted contents is not novel: several popular proxy systems exist, which include RabbIT [9], Muffin [10], WebCleaner [11], FilterProxy [12] and Privoxy [13]. These systems provide functionalities such as compressing text, removing and/or reducing images, removing cookies, killing Gif animations, removing advertisement, java applets and javascripts code, banners, pop-ups, and finally, protecting privacy and controlling access.

The main objective of this paper is to present SISI, a flexible and distributed intermediary infrastructure that enables universal access to the Web content. This framework has been designed with the goal of guaranteeing an efficient and scalable delivery of personalized services. SISI adds to the existing frameworks two main novelties:

- *per user* profiles. Many existing proxies only allow just one system profile, which is applied to all requests, coming from any user. That is, all the requests involve the same adaptation services. SISI, instead, allows each user to define one (or more) personal profiles, in such a way that the requests of a user may involve the application of some services, and those of another user may involve the application of completely different services.
- high scalability, because these services are computationally onerous and the large majority of existing frameworks does not support more than few units of contemporary requests per second.

The rest of the paper is organized as following: in Section 2 we present SISI, an intermediary software infrastructure, whose main objective is to provide advanced edge services by efficiently tackling the dynamics nature of the Web. To this end, we present the integration of a per-user profile mechanism into the SISI framework to dynamically allow different Web content presentations according to users's preferences. In Section 3 we present SISI advanced functionalities. Section 4 describes the workload model that we used to benchmark the SISI framework, whose results are shown in Section 5; finally, some remarks and comments will conclude the paper in Section 6.

2 Scalable Intermediary Software Infrastructure (SISI)

In this section we provide a description of the SISI architecture. This framework is based on top of existing open-source, mainstream applications, such as Apache Web server [14] and mod_perl [15]. The motivations are threefold: first of all, it will make our work widely usable because of the popularity of these products. Then, our results will be released as open source (by using some widely accepted open-source license) so that it will be available for improvements and personalizations to the community. Last but not least, Apache is a high quality choice, since it represents one of the most successful open-source projects (if not the most successful) that delivers a stable, efficient and manageable software product to the community of Web users.

2.1 SISI Overview

The main guidelines and objectives of the SISI project are described in [16]. The idea is to create a new framework that aims at facilitating the deployment of efficient adaptation services running on intermediate edge servers.

SISI framework uses a simple approach to assemble and configure complex and distributed applications from simple basic components. The idea is to provide functionalities that allow programmers to implement services without taking care of the details of the infrastructure that will host these services.

SISI provides a modular architecture that allows an easy definition of new functionalities implemented as building blocks in Perl. These building blocks, packaged into *Plugins*, produce transformations on the information stream as it flows through them. Moreover, they can be combined in order to provide complex functionalities (i.e. a translation service followed by a compression service). Thus, multiple Plugins can be composed into SISI edge services, and their composition is based on preferences specified by end users. Technically, SISI services are implemented as Apache handlers by using mod_perl.

An handler is a subroutine, implemented by using mod_perl, whose goal is to manipulate HTTP requests/responses. Since Apache has twelve different phases in its HTTP Request Life-cycle, it provides different *hooks* to have the full control on each phase. mod_perl provides a Perl interface for these hooks. In such a way Perl modules will be able to modify the Apache behavior (for example, a PerlResponseHandler configures an Apache Response object). Handlers can be classified according to the offered functionality, in different categories that is, Server life cycle, Protocols, Filters and HTTP Protocol. We used the last two to implement our handlers under Apache.

A detailed description of the functionalities of SISI modules is presented in [16].

3 SISI Advanced Functionalities

SISI is a modular architecture composed of basic building blocks applications that can be easily assembled to cooperate and provide complex functionalities.

SISI programmability is a crucial characteristics since it allows an easy implementation and assembling of edge services that can enhance the quality and the user perception of the navigation. Often the introduction of new services into existing networks is an expensive and time-consuming process. To simplify this task, SISI provides mechanisms to obtain a general-purpose programmable environment. Programming under existing intermediaries could involve difficulties in term of efficiency, generality or integration. For this reason SISI offers an execution environment, in which a compositional framework provides the basic components for developing new services, and a programming model is used to make this execution environment highly *programmable*. The SISI programming model provides APIs and software libraries (Perl language) for programming and deploying new services into the intermediary infrastructure.

Another important aspect of the SISI framework is the user and device profiling management. In particular, the administrator manages users' accounts, by adding or removing users from the system, resolving incorrect situations (for example, forgotten passwords), providing information about the allowed services for a given user. When a new user is added to the system, a default profile is automatically generated and he/she can modify the profile the first time he/she enters the system.

SISI approach in user profiling management is to explicit ask the user what he/she needs and use this information with a rule-based approach to personalize the content. In particular, users have to fill-out forms to create new profiles and to modify or delete existing ones. For example, when a user connects with a PDA, he/she could want his/her device to display only black and white images or not to be given images at all to save bandwidth. Through this services configuration, SISI is able to affect the adaptation of a given delivery context, and to change the user experience accordingly. Moreover, SISI allows a simple configuration mechanism to add, remove, enable or disable functionalities and a more complete configuration mechanism where service parameters can be specified by asking the user to fill-out Web-based forms.

In this context another important characteristics is the *hot-swap* of services composition i.e. the capability to load and execute different services according to different profiles at run-time without recompiling and restarting the software infrastructure.

SISI supports the deployment and un-deployment of advanced services, by making these tasks automatic and accessible through local and remote locations. Application deployment is an important system functionality that provides clients with an anytime-anywhere access to services and applications. By making the deployment an automatic task (i.e. wizard) it is possible to add new functionalities into the intermediary system without taking care of the complexity of the system itself.

Finally, SISI supports logging and auditing functionalities. The intermediary entity provides mechanisms to record security-related events (logging) by producing an audit trail that allows the reconstruction and examination (auditing) of a sequence of events. The process of capturing user activities and other events on the system, storing this information and producing system reports is important to understand and recover from security attacks. Logging is also important to provide billing support, since services can be offered with different price models (flat-rate, per-request, per-byte billing options).

4 Workload Models and Testbed

Since studies on real traces show great differences, the difficulty of defining a "typical" workload model is a well known issue. Traces were collected form a real dynamic Web site, but they were modified in order to artificially stress the content adaptation process, so to emulate a worst case scenario. The number of HTML pages has been raised, so that adaptation occurs more frequently, many GIF images were substituted with animated GIFs in order to stress GIF de-animation process.

Requests are referred to a mix of content types consisting of images (49%), HTML documents (27%), others (24%). HTML resources typically contain embedded objects ranging from a minimum of 0 to a maximum of 25, with a mean value of 10. Embedded objects are images (GIFs and JPGs), CSS or SWF files. Animated GIF images are about 6% of the whole workload. These percentages tend, once again, to stress the content adaptation process. HTML pages also contain links to other pages, ranging from a minimum of 0 to a maximum of 25, with a mean value of 5.

The workload is characterized by small inter-arrival times for client requests.

To avoid possible non predictable network effects, the experiments were conducted in a LAN environment.

In order to test the SISI framework we set up a testbed composed of three nodes connected through a switched fast Ethernet LAN. One node, equipped with a Pentium 4 1.8GHz and 512MB RAM, running Gentoo Linux with kernel 2.6.11, ran an Apache Web server deploying the origin resources. Another node, equipped with a Pentium 4 2.4GHz and 1GB RAM, running Fedora Core 3 Linux with kernel 2.6.9, ran the application proxy, being it SISI rather than Muffin, deploying adapted contents. Finally a third node, equipped with a Pentium 4 1.8GHz and 1.5GB RAM, running Gentoo Linux with kernel 2.6.11, ran *httperf* [17] by D. Mosberger, which is is a tool for measuring Web server performance. It provides a flexible facility for generating various HTTP workloads and for measuring server performance.

5 Performance Evaluation

The SISI architecture is implemented on top of the Apache Web server software [14]. The prototype was extensively tested to verify its scalability and to compare its performance with Muffin [10], which is a Web HTTP proxy that provides content adaptation functionalities such as removing cookies, killing GIF animations, removing advertisements, adding, removing, or modifying any HTML tag, etc. It is a Java-based programmable and configurable proxy: new services or filters can be developed through a set of provided APIs and can be added at run time, using the provided graphical interface.

The first set of experiments focuses on the different response times the user gets when connecting to a proxy, instead of directly connecting to the origin Web Server. A second set of experiments aims at comparing SISI with another intermediary: Muffin.

It is worth noting that SISI supports user authentication, while Muffin does not. Thus, we can expect that Muffin is advantaged with respect to SISI because of this lack. A third set of experiments, finally, aims at verifying SISI scalability by applying different services at the same time.

In all the experiments, the response time is referred to a whole page, including its embedded objects. That is, the response time represents the necessary time to completely download both the (possibly adapted) page content and all of its (possibly adapted) embedded objects.

5.1 Intermediary Overhead

In this section we evaluate the impact on the user response time given by the overhead of contacting a proxy adaptation server instead of the origin Web server. We configured SISI and Muffin so that they only forward resources to users, without applying any content adaptation. This is done in order to understand how much the user response time is affected by the use of an intermediary. We configured *httperf* to request 3000 pages, with the respective embedded objects. At a slow rate of 5 pages per second, the 90 percentile of user response time is 44 ms, 87 ms and 120 ms for the origin Web server, Muffin and SISI, respectively. The user response time contacting an intermediary is slightly less than two to three times the one obtained by contacting the origin Web server. It is worth noting that contacting an intermediary implies an increase of one hop, thus there is a slight network delay due to the additional three way handshaking, necessary to open the connection between the intermediary and the origin Web server.

Furthermore, SISI authenticates the user, thus there is an additional computational step, in order to satisfy the request, which possibly justifies a further increase in the user response time, with respect to Muffin.

To sum up, with a very slow rate and without any adaptation process, Muffin outperforms SISI, but the ranking will definitely change, as soon as we apply some content adaptation and/or increase the request rate, as we will see in the next subsections.

5.2 Performance Comparison of Two Intermediaries

In this section we aim to compare Muffin and SISI performance. It is worth noting that we could not find two services exactly performing the same adaptation process, but we may assume that Muffin Painter service is very similar to SISI RemoveLink service, from a computational point of view. That is, both of them parse the HTML content, search for some tags and rewrite them. Actually, Muffin Painter service can remove and/or modify colors and background images found in HTML documents, while SISI RemoveLink searches for a `href=""` tags and replaces them with plain text. Our claim is that, even though they do not perform the same tasks, the two services are computationally comparable.

We set up *httperf* to request 3000 pages with the respective embedded objects, at varying rates. Figure 1 and Table 1 show the 90 percentile of user response time when Muffin or SISI, respectively, are contacted as intermediaries and apply the adaptation service. An X sign in Table 1 (as well as in the following) means the intermediary is overloaded.

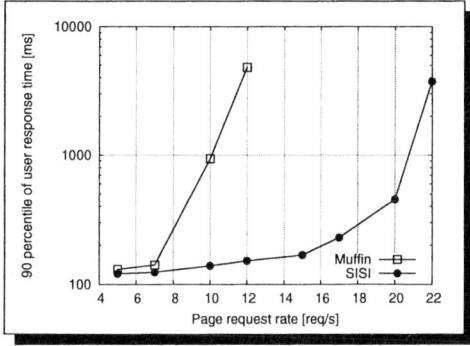

Fig. 1. 90 percentile of user response time as a function of the page request rate

Table 1. 90 percentile of user response time as a function of the page request rate

Page Request Rate [pages/s]	5	7	10	12	15	17	20	22
Muffin Painter service	131	141	938	X	X	X	X	X
SISI RemoveLink service	121	124	139	152	168	229	455	X

For Muffin we only report three page request rates because the system is overloaded with higher rates. SISI, instead, shows a better stability, with a 90 percentile of the user response time below half a second with a rate of 20 pages per second. Muffin shows a very poor stability: increasing the page rate from 5 to 7 brings to a 7.6% increment of the user response time, while passing from 7 to 10 pages per second, brings an increment of 565.25%. This is a clear sign that Muffin is overloaded. SISI performs much better: the increase is about 10% up to 15 pages per second, passing from 15 to 17 brings an increment of about 36%. When the page request rate further increases from 17 to 20 the user response time is nearly doubled, but still under half a second. Finally with a rate of 22 pages per second there is an increment of about 718%, which evidences an overloaded intermediary. To sum up, Muffin can sustain a rate of 7, while SISI up to 20 pages per seconds, nearly tripling the sustainable load.

5.3 SISI Scalability

GifDeanimate Service. In this section we aim to evaluate SISI scalability. To this purpose, we choose SISI GifDeanimate service, which parses a Graphics Interchange Format (GIF) image into its component parts. The GifDeanimate service produces a de-animated version of the original image, showing only its first frame. Our main objective is to save bandwidth in the delivery of Web pages with a lot of animated embedded images and to spare CPU cycles at the client device.[1]

In this case also, we set up *httperf* to request 3000 pages with the respective embedded objects, at varying rates. Figure 2 and the third row of Table 2 show the 90 percentile of user response time when SISI GifDeanimate service is applied to the requests.

---

[1] Muffin seems to have a similar service to SISI GifDeanimate, which is called AnimationKiller, but Muffin adaptation is far away from the adaptation performed by SISI, which keeps only the first frame of the animated GIF. This is why we do not compare SISI with Muffin results.

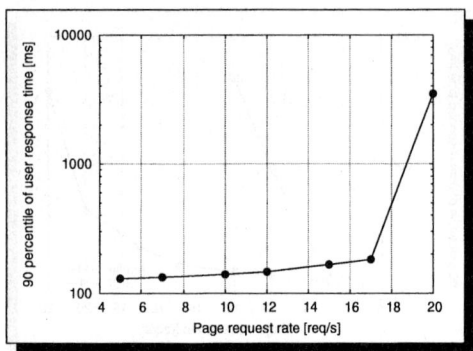

Fig. 2. SISI: 90 percentile of user response time with GifDeanimate service

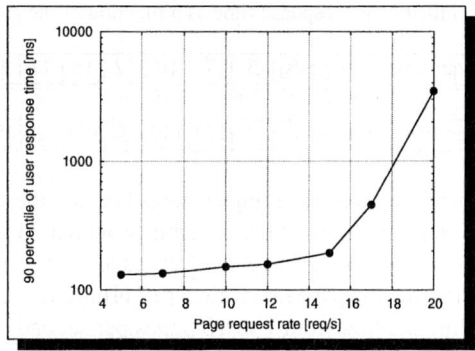

Fig. 3. SISI: 90 percentile of user response time with GifDeanimate and RemoveLink services

Table 2. SISI: 90 percentile of user response time with RemoveLink and GifDeanimate services

Page Request Rate [pages/s]	5	7	10	12	15	17	20
RemoveLink	121	124	139	152	168	229	455
GifDeanimate	130	133	140	147	167	183	X
Both services	131	134	151	158	193	457	X

We can roughly draw the same observations as for the previous RemoveLink service, with a limited increase in the user response time up to a rate of 17 pages per second. The fact that the system gets overloaded earlier (with a rate of 20 pages per second, instead of 22) is a sign that GifDeanimate is a heavier service than RemoveLink.

Composition of Services. Finally, we want to test SISI scalability when more services are applied one after the other to the same request. To this goal, we set up SISI to perform both RemoveLink and GifDeanimate services on each request. In this experiment we have nearly one third of all the requests being adapted.

In this case also, we set up *httperf* to request 3000 pages with the respective embedded objects, at varying rates. Figure 3 and Table 2 show the 90 percentile of user response time when SISI GifDeanimate and Remove Link services are applied to the requests.

From Table 2 we can notice that the influence of the RemoveLink service is very limited on the whole user response time. If we compare the third and the fourth row of Table2 for the first two rates the difference is very low. Increasing the rate brings to slightly bigger differences up to 17 pages per second, when the system is getting overloaded.

6 Conclusions and Future Work

In this paper we presented a Scalable Intermediary Software Infrastructure (SISI), whose main goal is to create a framework that aims at facilitating the deployment of adaptation services running on intermediate edge server on the WWW. It provides a modular architecture that allows an easy definition of new functionalities implemented as building blocks in Perl.

Users can define one or more personal profiles, in such a way that the requests of a user may involve the application of some services, and those of another user may involve the application of completely different services.

The prototype was extensively tested. In particular our experiments demonstrate that the response times the user gets when connecting to a proxy, instead of directly connecting to the origin Web Server are increased from two to three times by proxies which do not or do use user authentication, respectively. When comparing SISI with another intermediary (Muffin), we found that SISI provides much better performance. In particular, it can nearly triple the sustainable load. Finally, SISI is able to efficiently deliver adapted content on a per-user basis in a scalable way.

Our goal in the very next future is to deploy SISI in a distributed network environment and to port it to different operating systems.

Acknowledgments

This research work was financially supported by the Italian FIRB 2001 project number RBNE01WEJT "WEB–MiNDS" (Wide-scalE, Broadband MIddleware for Network Distributed Services). http://web-minds.consorzio-cini.it/

References

1. Barrett, R., Maglio, P.P.: Intermediaries: An approach to manipulating information streams. IBM Systems Journal **38** (1999) 629–641
2. Luotonen, A., Altis, K.: World-Wide Web proxies. Computer Networks and ISDN Systems **27** (1994) 147–154
3. Barrett, R., Maglio, P.P.: Adaptive Communities and Web Places. In: Proceedings of 2th Workshop on Adaptive Hypertext and Hypermedia, HYPERTEXT 98., Pittsburgh (USA), ACM Press (1998)
4. Calabrò, M.G., Malandrino, D., Scarano, V.: Group Recording of Web Navigation. In: Proceedings of the HYPERTEXT'03, ACM Press (2003)

5. Almaden Research Center, I.: Web Based Intermediaries (WBI) (2004) http://www.almaden.ibm.com/cs/wbi/.
6. Ardon, S., Gunningberg, P., LandFeldt, B., Y. Ismailov, M.P., Seneviratne, A.: MARCH: a distributed content adaptation architecture. Intl. Jour. of Comm. Systems, Special Issue: Wireless Access to the Global Internet: Mobile Radio Networks and Satellite Systems. **16** (2003)
7. Hori, M., Kondoh, G., Ono, K., Hirose, S., Singhal, S.: Annotation-Based Web Content Transcoding. In: Proceedings of the 9th International World Wide Web Conference, Amsterdam (The Netherland), ACM Press (2000)
8. WebSphere: IBM Websphere Transcoding Publisher (2005) http://www-3.ibm.com/software/webservers/transcoding.
9. Olofsson, R.: RabbIT proxy (2005) http://rabbit-proxy.sourceforge.net/.
10. Boyns, M.: "Muffin - a filtering proxy server for the World Wide Web". (2000) http://muffin.doit.org.
11. WebCleaner: A filtering HTTP proxy. (2005) http://webcleaner.sourceforge.net.
12. McElrath, B.: FilterProxy in perl (2002) http://filterproxy.sourceforge.net/.
13. Burgiss, H., Oesterhelt, A., Schmidt, D.: Privoxy Web Proxy. (2004) http://www.privoxy.org/.
14. The Apache Software Foundation: Apache (2005) http://www.apache.org.
15. The Apache Software Foundation: mod_perl (2005) http://perl.apache.org.
16. Grieco, R., Malandrino, D., Mazzoni, F., Scarano, V., Varriale, F.: An intermediary software infrastructure for edge services. In: Proceedings of the 1st Int. Workshop on Services and Infrastructure for the Ubiquitous and Mobile Internet (SIUMI'05) in conjunction with the 25th International Conference on Distributed Computing Systems (ICDCS'05). (2005)
17. Mosberger, D., Jin, T.: httperf - a tool for measuring web server performance. SIGMETRICS Performance Evaluation Review **26** (1998) 31–37 http://www.hpl.hp.com/research/linux/httperf.

A Novel Resource Dissemination and Discovery Model for Pervasive Environments Using Mobile Agents

Ebrahim Bagheri, Mahmood Naghibzadeh, and Mohsen Kahani

Department of Computing, Ferdowsi University of Mashhad,
Mashhad, Iran
Eb_ba63@stu-mail.um.ac.ir,naghib@um.ac.ir,
kahani@um.ac.ir

Abstract. Pervasive computing has been the new buzz word in the distributed computing field. Vast range of heterogeneous resources including different devices, computational resources and services world wide are virtually summoned under agreed conditions to form a single computational system image [1].In this approach a central controller is needed to coordinate the resources which is called the resource management system (RMS).A typical RMS is responsible for three main jobs including matching, scheduling and executing received requests based on suitable and available resources. Regarding the matching module, resource dissemination and discovery forms the main part. In this article we introduce two new approaches to tackle this problem using the mobile agents. The different strategies are dividend into hierarchical and non-hierarchical models. The performed simulations show a better performance for the hierarchical approach compared with the other model.

Keywords: Pervasive Environments, Resource Discovery, Resource Dissemination, Virtual Organization, Mobile Agents.

1 Introduction

Ubiquitous computing environment is a virtual unified networked set of heterogeneous computers that share their local resources with one another in order to serve the users needs in transparent way. They can encompass many different administrative domains and include lots of different organizations. Resources are usually categorized into independently managed groups based on some criteria and are called virtual organizations [2]. Theses criteria are usually based on the ownership model.

With the emergence of the mobile computing infrastructure the demand for omnipresent services and information has been more than ever before. The devices used in this model are very resource limited and thus should be able to exploit the outer available resources for their needs and also be able to discover these resources in a very cost inexpensive manner. For this reason many different resource management models have been proposed and implemented. Different models serve different intentions and requirements of a specific environment and based upon the resources available, but what is clear is that the main functionality of the RMS is to accept incoming distributed requests, find the matching machines containing the required resources and allocate the resources to the requests and schedule them. These operations are

allocate the resources to the requests and schedule them. These operations are mainly executed with the optimized QoS offered [3].

Resource dissemination and discovery form a very important part of the resource management service. Although they can be seen as two separate operations but they are mainly integrated to create much more effective results. Resource dissemination mainly focuses on introducing new resources added to the environment to the other machines available. The current dissemination models can be categorized into a periodic pull or push model and on the other hand an on demand request approach.

Resource discovery aims at finding suitable resources residing on disperse machines to serve the incoming requests. This function can be performed using distributed or centralized queries. Different agent models can also be applied. In this article we introduce two new models of resource dissemination and discovery using mobile agents. The models have been completely explained through the rest of the article and simulation results have been shown. The article is conducted in 5 sections. Section 2 elaborates on the architecture of the environment while section 3 will introduce the algorithms and explain the different parts in an in-depth manner. The simulation results will be shown in section 4 and then the article will be concluded in section 5.

2 Environment Architecture

There are an enormous number of resources lying dispersedly on different machines in a pervasive computing environment. The architecture in which the resources are categorized is of much importance due to scheduling decisions, communication structure, etc that need to be considered [4].

In our approach we assume independent administrative domains to be in charge of a set of resources owned by a specific authority. These domains are called Virtual Organizations (VO). Resources are hence owned by exclusive virtual organizations. In our model, resources belonging to a specific virtual organization can be scattered throughout the environment and so geographical placement of resources doesn't shape the administrative domain but is the ownership model which is used [5]. This model was selected so that service level agreements and the economical models could be effectively created.

The machines form the different parts of a massive network but in the conceptual layer the virtual organizations play the main role by establishing agreements to share their resources, and cooperating in creating a rich computational environment.

3 Module Structure

The module created for the algorithm consists of two main parts: Resource Dissemination, and Resource Discovery.

3.1 Resource Dissemination

The resource dissemination algorithm is used to introduce new resources added to different machines in the environment to the other machines and virtual organizations [6]. In our resource dissemination approach, every time that a resource is added to a machine inside a specific virtual organization, the resource dissemination algorithm broadcasts the static attributes of the added resource to all of the existing machines inside the same virtual organization. In this phase, every machine residing on the same virtual organization as the machine owning the new resource will be aware of the new resource. To inform other virtual organizations of the existence of such resource, a message is sent to every virtual organization, declaring the existence of the new resource in the source virtual organization. The message passing theme used here is different from the one used inside the same virtual organization in the way that the machine inside the same virtual organization can exactly spot the resource but the other virtual organizations only have a vague awareness of the resource placement inside the corresponding virtual organization.

3.2 Resource Discovery

One of the most important parts of the resource management system is the resource discovery algorithm [7]. We propose a new resource discovery model which utilizes mobile agents for resource discovery. In this regard, two hierarchical and non-hierarchical models have been created.

In the non-hierarchical model, when a resource request arrives at a specific machine the environment knowledge based inside the machine is consulted for matching results. As the machine has only information on the static attributes of the available resources and on the other hand can not exactly locate the position of resources on the other virtual organizations, decision on a suitable resource could not be taken. For this reason a number of mobile agents, according to the number of available resources, are created. The agents are then sent to the attorney of the corresponding virtual organizations. The attorney of a virtual organization is not statically designated but as we have defined is randomly selected every time from the available machines in the destination virtual organization based on network load and originating virtual organization closeness and will avoid bottleneck creation which is anticipated in the case of a static attorney. The attorney will then again consult its knowledge base to exactly locate the resource(s) available in that virtual organization. It will hence create a number of mobile agents which are again as the same number as the available resources, to move to the corresponding machines and assess the dynamic attributes and also evaluate connection parameters. The mobile agents will then return to the attorney and pass back the results. The attorney will select the best resource and introduce it to the waiting agent. The agent will, on its turn, return to the originating machine on the source virtual organization. The mobile agents returned from many different virtual organizations will feedback the information collected and the best resource is then selected. Using the mobile agents for the discovery purpose not only allow us to precisely accumulate the dynamic parameters of a resource such as the cpu load, the length of the printer queue and etc. but it will also enable us to roughly estimate the network conditions and estimate the time to execution of the job. As the resources may suddenly leave or stop functioning, the agents will also update the knowledge base of every machine in the visiting virtual organization resulting in up-to-date machines.

The hierarchical algorithm on the hand uses a cascading model. The source virtual organization is primarily queried for suitable resources to match the request and if suitable results are not found, agents are sent to external virtual organizations to discover available resources and gather required information. The decision to which machine has the suitable resource is then taken.

4 Simulation and Evaluation

The model was simulated and implemented and fifteen different experiments were conducted. The simulations' test data were chosen from different natures to test algorithm behavior under different conditions.

The first five experiments studied the effect of variance of the number of resources (SVSNDR) while the second five set of simulations focused on the change in the number of nodes (SVDNSR) while the number of resources and virtual organization were left untouched. The last set of conducted simulations focused on the outcome of the virtual organization quantity change (DVSNSR).The resource distributions in the 15 undertaken experiments are shown in figures 1 to 3.For comparison purposes different performance evaluation criteria where introduced which are defined as 1.(Selected/Maximum) Path Ratio: This criterion divides the length of the discovered path to the length of the farthest suitable resource path available in the environment which shows the extent of path length optimization. The path length is calculated based on the hops to reach the destination. (Figure 4). 2. Mean Agent Life Time (MALT): This factor shows on an average basis how long an agent created in the algorithm lives in the environment. 3. Number of Active Agents multiplied by their Average Life Span (SL): This factor introduces the exact amount of environment resource consumption by the agents created in the algorithm since the number of active agents multiplied by their life time will allow us to compute the amount of algorithm activity. 4. Maximum Discovery Time (MXDT): This criterion is equal to the effective resource discovery time. 5. Minimum Discovery Time (MNDT): The exploration time of the fastest agent to return the expected results (Figure 5). The minimum discovery time in both approaches belong to the resource discovery process inside the same virtual organization; therefore the diagram for this criterion in both approaches should be similar. The simulation results show the same fact and hence

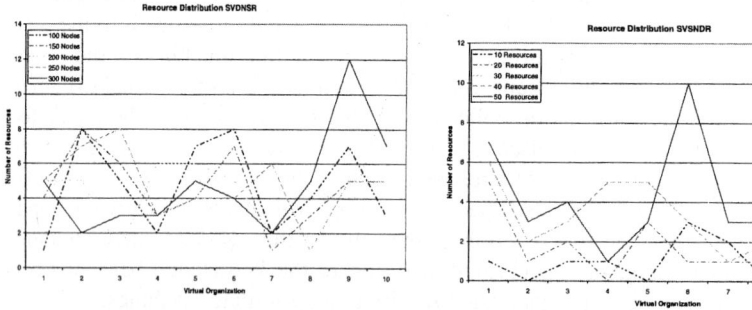

Fig. 1. Dynamic change in the number of resources

Fig. 2. Dynamic change in the number of nodes

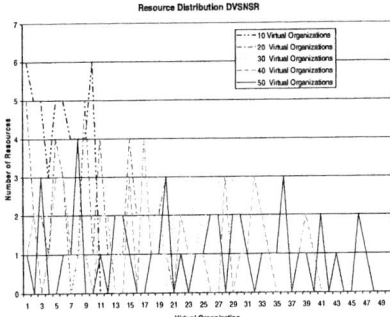

Fig. 3. Dynamic change in the number of Resources

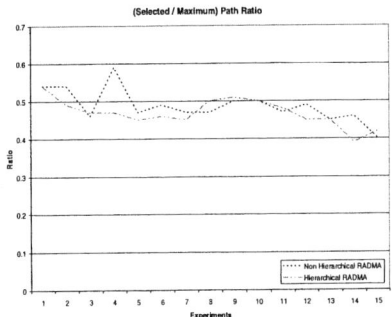

Fig. 4. (Selected / Maximum) Path Ratio

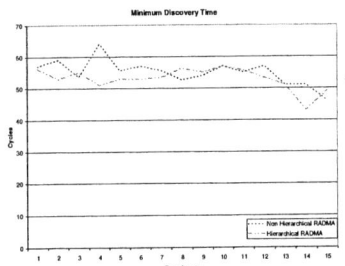

Fig. 5. -Minimum Discovery Time (MNDT)

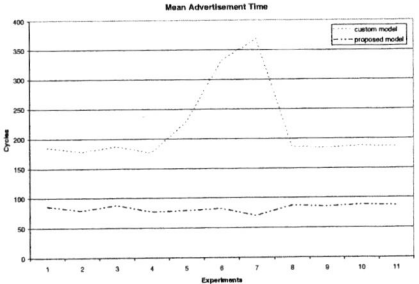

Fig. 6. –Mean Resource Dissemination Time

validate the simulation. On the other hand The MNDT and MXDT diagrams for the hierarchical model were predicted to have similarities caused by the initial originating virtual organization search which is also proved by the simulation outcome.

Although the hierarchical algorithm creates agents with longer life span but the overall environment resource consumption is balanced by creating less mobile agents. The experiments show that the hierarchical algorithm performs much better with higher speed in resource discovery, lower overall system burden on the network traffic and lower environment resource consumption.

The resource dissemination algorithm also shows much better performance compared with the normal resource broadcast model (Figure 6). It is worth noting that the efficiency of this resource dissemination algorithm is enhanced while it is accompanied with resource discovery models explained in this article.

5 Conclusion

In this article we have proposed two new models (hierarchical and non-hierarchical) for resource discovery and a complementing resource dissemination algorithm to complete them and provide an efficient module to use inside the resource management system of the pervasive environment management package. Fifteen different experiments were conducted which show a better performance for the hierarchical algorithm. This algorithm provides better results along with higher speed in resource discovery, lower overall system burden on the network traffic, and lower environment resource consumption.

References

1. Kong, Q., Berry, A., "A General Resource Discovery System for Open Distributed Processing", ICODP'95, 1995
2. Foster, I., Kesselman, C., "The Grid: Blueprint for a New Computing Infrastructure " 2nd ed. Morgan Kaufmann, 2004.
3. Krauter, K., Buyya, R., and Maheswaran, M., "A Taxonomy and Survey of Grid Resource Management Systems", Technical Report: University of Manitoba (TR-2000/18) and Monash University (TR-2000/80)
4. Czerwinski, S., Zhao, B., Hodes, T., Joseph, A., Katz, R., "An architecture for a secure service discovery service", Fifth Annual International Conference on Mobile Computing and Networks (MobiCom '99)
5. Foster, I., Kesselman, C., Tuecke, S., "The Anatomy of the Grid Enabling Scalable Virtual Organizations", Intl J. Supercomputer Applications, 2001.
6. Rakotonirainy, A., Groves, G., "Resource Discovery for Pervasive Environments", 4th International Symposium on Distributed Objects and Applications (DOA2002), 2000.
7. McGrath, R., "Discovery and Its Discontents: Discovery Protocols for Ubiquitous computing". Technical Report UIUCDCS-R-99-2132, 2000

A SMS Based Ubiquitous Home Care System

Tae-seok Lee, Yuan Yang, and Myong-Soon Park

Internet Computing Lab. Dept. of Computer Science and Engineering,
Korea Univ., Seoul, 136-701 Korea
{tsyi, yy, myongsp}@ilab.korea.ac.kr

Abstract. In this study, we defined requirements of ubiquitous environment, which users can monitor and control situations of home anytime, anywhere. In addition we developed and built a model that can be combined with latest home network technologies. Various middleware technologies such as UPnP, JINI and LonWork, were being studied and developed to build home network. However, they still remains in wired environment and can not call or respond service request whenever user wants. But thanks to advancement of mobile technologies, propagation of CDMA mobile devices keep growing and its cost keeps lowering. Therefore, new services which utilize CDMA can be materialized. It has symbolic importance that ubiquitous concept, which can monitor and control home environment by converging home network technologies and mobile communication technologies, can be used to develop a new model. In this paper, we developed control message protocol for SMS(Short Message Service)-HCS(Home Care System) and CPE (Customer Premises Equipment) interface, and embodied with Embedded Linux board which has PCMCIA CDMA modem. SMS-HCS can make a family member to monitor inside-home environment, control home through home network, and it provides security and safety service which can alarm emergency situation immediately.

1 Introduction

Home network uses various technologies from physical network equipment at lower layer to numerous application programs. Among them, middleware technology provides general control method which application program at higher layer can deal with physical equipments [1, 6]. Various technologies including Universal Plug and Play (UPnP) [9], Jini [7, 13], Open Service Gateway Initiative (OSGi) [10], LonWork and HAVi etc. are suggested and evolved rapidly. Ubiquitous computing concept, which shifts the paradigm that physical space can communicate with electronic space through computer and network anytime and anywhere, has been integrated different areas together and is being spread rapidly to diverse domains. Therefore, home network technologies will be evolves as one of ubiquitous elements in near future [2, 8, 11, 14, 15]. Some home appliances that apply to this trend are being introduced in home networking area.

While radical advancement of mobile network and proliferation of mobile device, a system that controls home appliances using mobile devices remotely can be new way to control CPEs within ubiquitous home environment anytime & anywhere [3, 4, 12]. Connecting wireless internet using mobile device can be good way to connect existing wired network which is connected to home network directly. But wireless internet ready areas are limited and its cost is still high. Therefore CDMA SMS which send short message characters can be a good alternative [5]. In addition CDMA SMS has wide potential for adapting to other applications because of transition to MMS (Multi-media Message Services) in near future. Existing sensors had been developed for a long time and these sensors are used to monitor conflict region, disaster and environment. You may leave home without worry if you have Home Care System which can monitor inside situations by installing these kinds of sensor at home.

In this paper, we define requirements of SMS-HCS, develop a model, which can be grafted with the latest home network technologies, build the system using Embedded Linux board and evaluate how this system can correspond with ubiquitous computing concept, the new paradigm of home network. SMS-HCS can control home appliances or equipment remotely whenever and wherever it is needed and can report automatically when a problem occurs. In chapter 2, we will discuss requirements, considerations and problems for SMS remote control service. We will describe architecture and operating method of this system in chapter 3 and show design and implementation in chapter 4. Conclusion will be provided in chapter 5.

2 Requirements SMS-HCS

SMS-HCS services for user (home owner) include home appliances remote controlling service through SMS of mobile device, home remote monitoring service, home security & disaster prevention service. Figure 1 shows four services which will be materialized in this paper.

A user (home owner) can monitor status of home appliances, temperature, gas leak or intrusion from Home Care System through mobile phone. A user usually pays bills by periodic meter reading of gas, electricity and water supply. In order to read meter, meterman should visit every single house and check meters. This practice generates too much costs and time. SMS-HCS which can provide handy monitoring practice is needed to reduce time and cost. Figure 2 illustrates two service scenarios which are developed for this paper.

Assumptions and corresponded logical reason for implementing each scenario of Figure 2 are as follows.

- A home appliance is connected through home network and is controlled by SMS-HCS.
- SMS-HCS has CDMA modem card which can transmit and receive SMS.
- A user (home owner) knows telephone number and authentication number of SMS-HCS.

There is temperature, gas leak and intrusion detection sensors in a user's home and detected date will be transmitted to SMS-HCS.

Fig. 1. SMS Home Care System

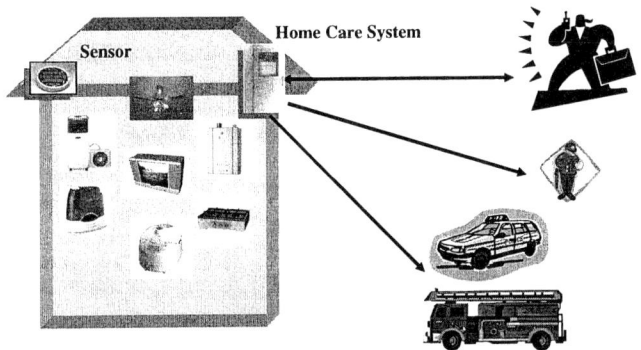

Fig. 2. Environment for SMS-HCS service scenario

3 Architecture of SMS-HCS

SMS-HCS services includes user authentication function for home appliances working management, a user's action control command analysis function, home appliance preset management and preset message transmission function, appliance status SMS message sending function and appliance status display function using Qt/Embedded. Home Remote Monitoring Service consists of remote monitoring command analysis function and home status message sending function using SMS.

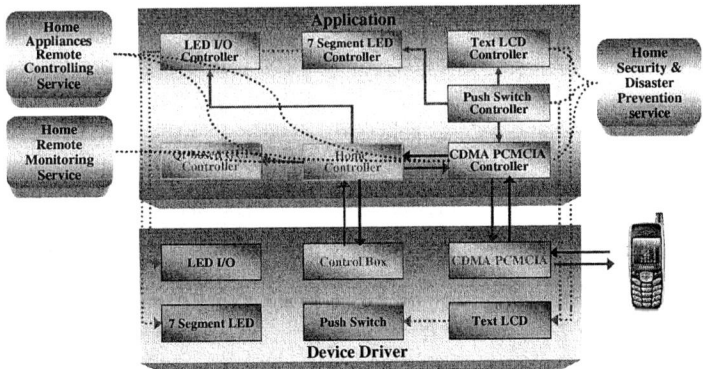

Fig. 3. SMS Home Care System Structure

Figure 3 show total architecture of SMS Home Care system. In Application Area, Seven controller such as Text LCD Controller, Push Switch Controller, 7 Segment LED Controller, LED I/O Controller, CDMA PCMCIA Controller, Home Controller, Qt/Embedded GUI Controller) are materialized. In Kernel Area, Control Box which is device drive is used.

Home appliance action control service uses LED I/O Controller, Qt/Embedded GUI Controller, CDMA PCMCIA Controller and Home Controller, and Remote monitoring service uses CDMA PCMCIA Controller and Home Controller. Home security and disaster prevention service uses Push Switch Controller, TextLCD Controller and 7 Segment LED Controller.

3.1 Action Preset: Home Control Service

When a user send signal to control specific device, Home Care System make this device take action or preset for action with wanted value immediately and then send conform message to the user. Inside of Home Care System consists of CDMA controller, Qt/Embedded GUI and control box. In particular, control box plays a role which manages each device's status and acts as preset timer. Figure 4 shows signal flow of each module.

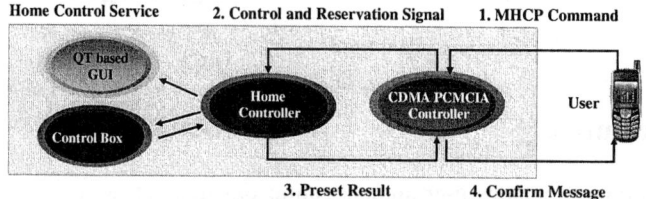

Fig. 4. Action preset and control diagram

3.2 Remote Monitoring: Home Control Service

If a user sent SMS in order to capture current status of a device at home, Home Care System checks target device's status through control box and report the result to the user. Figure 5 shows working flow of each module of remote monitoring service among Home Control Service.

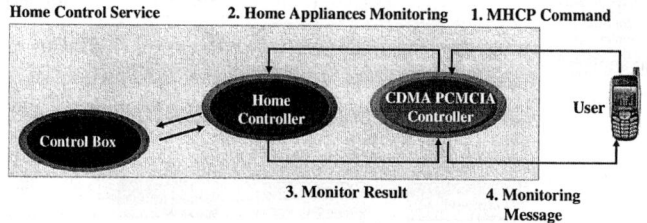

Fig. 5. Diagram of Remote Monitorin

3.3 Disaster Alarm System: Security/Disaster Prevention Service

When a problem such as gas leak, fire, water pipe freeze and intrusion at a user's home, Home Care System generate warning message automatically and send it to the user. Push button controller acts as a substitute of abnormal condition sensing part. And status of abnormal condition can be checked from Text LCD. Figure 6 shows action flows of each module of security/disaster prevention system.

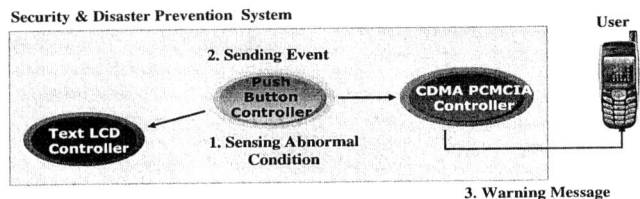

Fig. 6. Diagram of Disaster Alarm

4 Design and Implementation

Hardware consists of Linux Embedded Board, Development host for PCMCIA CDMA modem and mobile phone. And software includes Embedded Linux, Qt/Embedded Library and ARM Linux cross-compiler was used.

First, we have developed SMS Home Control Protocol in order to manage home appliances using CDMA SMS (Short Message Service), and then materialized SMS Home Control Service. We designed MHCP (Mobile Home Control Protocol) for Home Control Service between a home owner and Home Care System and developed detailed modules which are applied to SMS Home Control Service. This system provides service for busy moderners which can control and monitor home appliances easily with SMS message anytime and anywhere. Services can be divided into home appliance preset service, Status and preset details monitoring service and emergence security/disaster prevention alarm service.

4.1 SMS MHCP(Mobile Home Control Protocol) Definition

Due to using short SMS message, fixed-length command frame was designed as follows.

It is not easy to input MHCP commands, which are illustrated in Table 1, to mobile phone directly. However, if menu type mobile application program which is shown in Figure 8 is offered, it is ease to use even though a user does not have knowledge of MHCP commands by utilizing menu.

Protocol Name	Protocol Type	Security Code	Device ID	Action Type	Preset Time	User Data

Fig. 7. Structure of MHCP Control Command Frame

Table 1. MHCP Fields Description

Name	Bytes	Description
Protocol Name	4	MHCP(Mobile Home Control Protocol) V1.0
Protocol Type	1	0: Request, 1: Response
Security Code	4	User Authentication Code
Appliance	2	1: Air Conditioner, 2: Boiler, 3: Humidifier, 4: TV, 5: Electric Rice Pot, 6: Gas Range, 7: Light(1), 8: Light(2)
Operation	1	1: On, 2: Off, 3: Monitor
Preset Time	3	Reseve time to operate appliance (Zero: operate without delay)
Operation Value	3	Temperature, Humidity, Luminosity, Channel, etc (1-999)

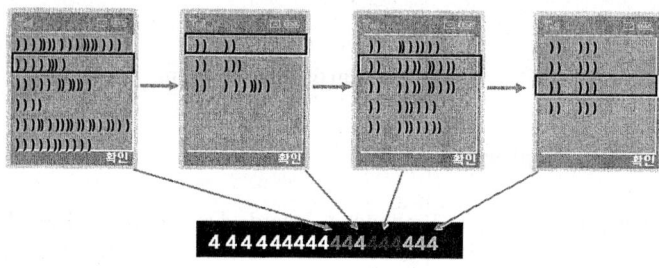

Fig. 8. Example of MHCP Menu Input Method

As shown in Figure 8, if let security code be generated automatically by presetting and select type of appliance, setting status, preset time and value from menu, MHCP message(MHCP028140210110030) that a boiler can be turn on ten minutes later to raise temperature up to 30 ℃ is generated automatically and is transmitted. Consequently, structured MHCP message can send a message efficiently as well as being used easily.

4.2 Implementation Details

(1) Action Display: Qt/Embedded GUI application program

Table 2. Examples of Home Appliances Setting

No	Appliance	Setting Value	Time	Status
1	Air Conditioner	Temperature Setting (0-30)	Preset Time	On/Off
2	Boiler	Temperature Setting (0-70)	Preset Time	On/Off
3	Humidifier	Humidity Setting (0-100)	Preset Time	On/Off
4	Television	Channel Setting (0-999)	Preset Time	On/Off
5	Electric Rice Pot	No Setting(00)	Preset Time	On/Off
6	Gas Range	No Setting(00)	Preset Time	On/Off
7	Light(1)	Luminosity Setting (0-100)	Preset Time	On/Off
8	Light(2)	Luminosity Setting (0-100)	Preset Time	On/Off

In order to show home control state just as what really happens, image graphic which is processed visually will be shown on board's LCD Display.

(2) Communication between home owner and home control system

Although setting values of controlled appliances, which use MHCP control message and are transmitted to Home Control system, are different device by device, they are within 0~999 and defined as following table.

4.3 Home Appliance Action Management Service

It is service that a home owner can preset and control his or her appliances from outside. That is, a home owner can send or receive a message about presetting or confirming. And by receiving preset-action execution message, owner's request can be processed based on situation. Detailed description of Home appliances action management service set-up is as follows.

(1) Preset Action Scenario

Once SMS Home Control System starts, all devices' images are displayed using Qt/Embedded GUI as turn-off status and eight LEDs are turned off. Also, system remains on wait mode until MHCP SMS message arrives. After receiving MHCP SMS message, system passes this message and sends response message. During these processes, every processed log will be saved in console.

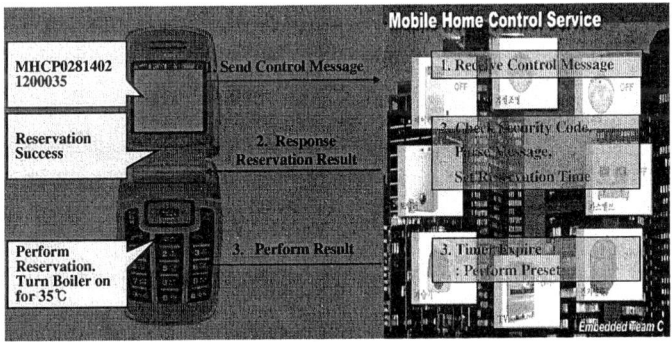

Fig. 9. Preset Action

In the scenario which is shown on Figure 9, a MHCP message that instructs system to turn boiler on 200 minutes later for 35℃ of setting temperature is sent then preset message submitted message will be received. After preset 200 minutes passed, Home Control System turns boiler on automatically for 35℃ of setting temperature then sends command complete message to Call-Back number.

(2) Remote Monitoring

Next, we introduced Home Control Monitoring service. Figure 10 shows monitoring flow for preset status or current status of home appliances.

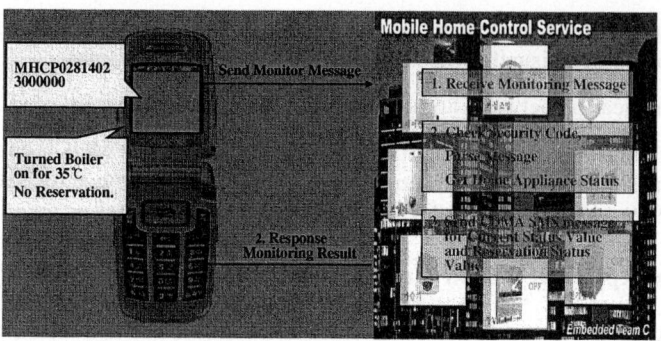

Fig. 10. Remote Monitoring

MHCP message is made by input which contains desired home appliance's number and a number as defied for monitoring action. While it is request message, the name of protocol is MHCP and the type of protocol is '0'. Therefore, all input value is '000000' because preset time and value has no meaning. Although any preset value or setting value is ignored when it is passing, chipper should be kept. As we described before, monitoring command will not be accepted if authentication number of security card is matched. If exact home control remote monitoring message is transmitted, Home Control System reads current status value and preset status value of designated appliance, and then sends these values by generating response message.

5 Conclusion

In this paper, our objective was development of Home Control system and related three services that can be enabled as follows through SMS-HCS which use SMS in home network environment.

- Home Appliance Action Management Service
- Home Remote Monitoring Service
- Home Security and Disaster Prevention Service

SMS-HCS, which is based on SMS service, can partially materialize ubiquitous computing environment. SMS Home Care System was defined by installing CDMA PCMCIA modem on Embedded Linux system. From this successful result, new method which can control home appliances anytime, anywhere was introduced and its potential was proved. SMS-based system can give cost and performance benefits compare to existing wireless internet-based system. Also SMS-based system guarantees high fidelity by utilizing control and contributes that a home owner can get real-time situation of his or her home not only by existing sound-based alarm but also SMS-based information from ubiquitous environment in home. From now on it will be needed to study for more handy message interface using MMS (Multimedia Message Service) which is the next generation messaging service.

References

1. A. Crabtree, T. Rodden and T. Hemmings, Supporting communication in domestic settings, Proceedings of the 2003 Home Oriented Informatics and Telematics Conference, Irvine, California: International Federation for Information Processing, 2002.
2. A. Crabtree, T. Rodden, T. Hemmings and S. Benford, Finding a place for UbiComp in the home, Proceedings of the 5th International Conference on Ubiquitous Computing, pp. 208-226, Seattle: Springer, 2003.
3. A. Schmidt, A. Takaluoma et al., Context-Aware Telephony over WAP, personal Technologies, vol 4, no. 4, Sept. 2000.
4. J. Tang, N. Yankelovich et al., ConNexus to Awarenex: Extending Awareness to Mobile Users, Proc. SIGCHI Conf. Human Factors in Comp. Sys., 2001, Seattle, WA, Mar. 31-Apr. 5, 2001.
5. C. Schmandt et al., Everywhere Messaging, IBM Sys J. 2000.
6. D. Hindus, S.D. Mainwaring, N. Leduc, A.E. Hagström and O. Bayley, Casablanca: designing social communication devices for the home, Proceedings of the 2001 CHI Conference on Human Factors in Computing Systems, pp. 325-332, Seattle: ACM Press, 2001.
7. J. Waldo, The Jini architecture for networkcentric computing, Communications of the ACM, pp. 76-82, vol. 42 (7), pp. 76-82, 1999
8. K. Edwards and R. Grinter, At home with ubiquitous computing: seven challenges, Proceedings of the 3rd International Conference on Ubiquitous Computing, pp. 256-272, Atlanta, Georgia: Springer, 2001.
9. Micorosoft Coporation, Universal Plug and Play Device Architecture, Jun. 2000, Available to: http://www.upnp.org
10. P. Dobrev, D. Famolari, C. Kurzke, B. A. Miller, Device and Service Discovery in Home Networks with OSGi, IEEE Communications Magazine, Vol. 40, Issue 8, Aug. 2002.
11. P. Tandler, Software Infrastructure for Ubiquitous Computing Environments: Supporting Synchronous Collaboration with Heterogeneous Devices, presented at Ubicomp 2001: Ubiquitous Computing, Atlanta, Georgia, 2001.
12. R. Wakikawa et al., Roomotes: Ubiquitous Room based Remote Control from Cell phones, Extended Abstracts of the SIGCHI Conf. Human Factors in Comp. Sys., Seattle, WA, Mar. 31-Apr. 5, 2001.
13. Sun Microsystems, Jini 1.1, 2000, Available to: http://developer.java.sun.com
14. T. Rodden and S. Benford, The evolution of buildings and implications for the design of ubiquitous domestic environments, Proceedings of the 2003 CHI Conference on Human Factors in Computing Systems, pp. 9-16, Florida: ACM Press, 2003.
15. Tom Rodden, Andy Crabtree, Between the Dazzle of a New Building and its Eventual Corpse: Assembling the Ubiquitous Home, DIS2004: ACM, 2004.

A Lightweight Platform for Integration of Mobile Devices into Pervasive Grids[*]

Stavros Isaiadis and Vladimir Getov

Harrow School of Computer Science,
University of Westminster, London, U.K
{S.Isaiadis, V.S.Getov}@westminster.ac.uk

Abstract. For future generation Grids to be truly pervasive we need to allow for the integration of mobile devices, in order to leverage available resources and broaden the range of supplied services. For this integration to be realized, we must reconsider the design aspects of Grid systems that currently assume a relatively stable and resourceful environment. We propose the use of a lightweight Grid platform suitable for resource limited devices, coupled with a proxy-based architecture to allow the utilization of various mobile devices in the form of a single virtual "cluster". This virtualization will hide the heterogeneity and dynamicity, mask the failures and quietly recover from them, provide centralized management and monitoring and allow for the federation of similar services or resources towards advanced functionality, quality of service, and enhanced performance. In this paper we are presenting the major functional components of the platform.

1 Introduction

Consumer mobile electronic devices such as personal digital assistants (PDA), mobile phones and digital cameras currently hold a very big share of the computer market pie. The trends are very likely to increase in the future, resulting in a very big mobile community. Such devices increasingly provide support for integrated multimedia equipment, intelligent positioning systems, and a diverse range of sensors.

The emergence of the Grid [17, 18] as the new distributed computing infrastructure has accelerated many changes in this field. Distributed systems can now cross several different organizational boundaries and different administrative and security domains, creating what has been termed "a virtual organization". However, most of these systems do not take into consideration mobile and/or resource limited devices and their integration into the Grid as resource providers and not just consumers is very difficult.

For the Grid community such an integration is an opportunity to utilize available resources in the mobile community, increase its performance and capacity and broaden the range of available services. Current Grid system designers consider this unnecessary and bring in the argument that a couple of extra servers will provide

---

[*] This research work is carried out partly under the FP6 Network of Excellence CoreGRID funded by the European Commission (Contract IST-2002-004265).

much better performance. But for future generation Grids to be truly ubiquitous we must have the option of integrating mobile devices into Grid systems. In addition, currently, research and industry are focusing on mobile computing and hence mobile devices are destined to become more and more powerful in the near future.

There are a number of current efforts focusing on the integration of mobile devices into the Grid but they do not provide implementation considerations [2, 8], they deal only with infrastructure-less ad hoc networks [1] or they impose restrictions to the programming model or underlying platform [9] that can be used. The potential benefit and challenges of this integration, has been the theme in many other papers and research projects lately [3, 4, 5], but none of these provide any implementation methodology or propose an architecture to support this integration.

The aim of this paper is to present the status of our work in progress on a lightweight Grid platform, suitable for resource limited devices, coupled with a proxy-based architecture and a set of middleware components that will provide the foundations for enhanced functionality, increased reliability (even in unreliable wireless infrastructures) and higher service availability. The lightweight components based Grid platform has been introduced in [10] and is only briefly described here.

The rest of this paper is organized as follows: section 2 presents the lightweight Grid platform and the accompanying architecture. Section 3 briefly describes the most important functional components. Finally, section 4 concludes the paper and lists our future plans.

2 Lightweight Platform

2.1 Description

Contemporary Grid implementations/platforms have a very rich set of features – they were designed with built-in exhaustive set of functions. Current standards, software and toolkits for implementing and deploying Grids [11, 12] are also motivated to provide a generic computational Grid with all possible features built-in. The Open Grid Services Architecture [13], on which most of the current Grid platforms are based, is built as a feature rich platform. This approach ensures that service requests from applications are included in a complete set of features offered by the platform.

However, complexity (in terms of interactions, manageability and maintainability) of the implementation of any Grid platform based on this philosophy will be very significant. Additionally, deployment of these Grid systems demand considerable computing resources.

In our approach, instead of building the underlying platform with an exhaustive rich set of features, we create a lightweight core platform, built only with the minimal essential features. The authors of this paper have tried to identify this set of core components that are absolutely necessary for a functional lightweight platform in [12] were more details of the platform can also be found. The resultant platform is generic and will be used as the foundation in the architectural context presented below, to provide an efficient and highly available yet lightweight infrastructure in unreliable and resource-limited environments.

Figure 1 illustrates the interaction of some of the major components of the platform. More details about some of these components are provided in section 3.

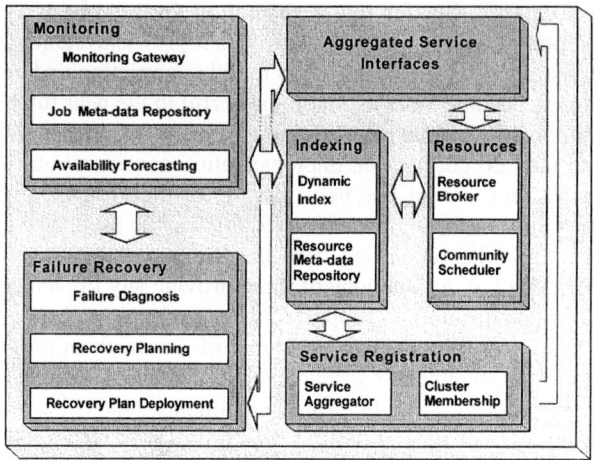

Fig. 1. Lightweight platform components

2.2 Architecture

A lightweight Grid platform is the necessary first step that will allow resource limited devices to contribute services and/or resources to the Grid. But when trying to merge an unreliable and very dynamic environment –as in mobile devices making use of wireless communications links, with a relatively reliable and static one –the Grid in its more traditional form, inevitably this will affect the overall reliability, and therefore performance, towards lower levels.

The fragile nature of such devices –due to inherent limitations and unreliable communication links, increases the failure rate. Whenever a failure occurred, the Grid components would have to reschedule and reallocate resources for the active application, possibly migrating data around the Grid thus reducing response and performance times. Considering that in the mobile end of the Grid, the failure rate is increased, this is not something we would like in busy, heavily loaded and complex Grid environments.

To overcome these problems, we are following a different approach: instead of directly presenting each mobile device to the Grid, we are grouping all mobile devices that fall into the same subnet (whether this is defined by physical or logical boundaries) and creating a virtual "cluster" that will be presented to the Grid. This cluster-based design requires a number of proxies that provide the interface point between the Grid and the lightweight cluster of devices. More details about this architectural design can be found on [19].

This architecture will provide a virtualization/abstraction layer in order to present a single interface to all similar resources in the cluster. For example, all storage resources will be presented to the Grid as one big storage pool available through a single interface at the proxy. This "aggregator" service as we call it, will provide access to the aggregated resources in the cluster in a uniform, location and device independent way. Underlying dynamicity and heterogeneity are hidden from the Grid.

This design has a number of significant advantages when built on top of the lightweight Grid platform: provides the foundations for failure recovery and high service availability, hides device and location details from the Grid, masks the dynamicity and unreliability of the environment, provides a uniform interface to the lightweight cluster making job submission and also programming easier, and can provide enhanced functionality and advanced services transparently to the Grid clients.

3 Platform Characteristics

3.1 Service Availability and Failure Recovery

For the purpose of failure detection and failure recovery, we are suggesting a monitoring framework based on tested schemes like heartbeats adapted for our centralized cluster based environment, and a well defined set of intelligent agents (or intelliagents as mentioned in [14]) installed in the mobile devices that will gather necessary information. This information may include dynamic resource state information like usage statistics, failure ratios, downtime/uptime ratio or static device meta-data like hardware and software information, bandwidth capacity and more. Failure recovery based on the approach of intelliagents, makes use of predefined scenarios and trigger lists. If an agent predicts a possible failure (like e.g. severe signal degradation that might lead to an offline status, or low battery levels) data and/or task migration may be required to pre-emptively deal with the failure and save as much computation effort as possible. Human administrator intervention is kept to a minimum.

Collected information can be fed to an analyzing/forecasting component that in cooperation with the cluster's indexing components will provide recommendations as to what is the best possible resource usage plan. A dynamic index classifies available resources/services according to information supplied by the forecasting facilities e.g. from a set of identical resources, the classification could be based on the downtime-to-uptime ratio of the hosting devices.

3.2 Job Submission

Service aggregators, present a uniform interface to the underlying aggregated resources available in the cluster. The Grid only sees a single virtual resource/service. This makes job submission easier and minimizes the workload on the Grid brokers and schedulers that would otherwise be much higher due to a much bigger number of resources in the Grid. The job of scheduling and brokering resources is now delegated to a cluster community scheduling system, deployed at the proxies. The aggregator services make use of a delegation mechanism that forwards invocations to service and resource providers in the "cluster". This abstraction layer provides a homogeneous interface to the Grid for all underlying aggregated resources and services independent of the specific service architecture used.

3.3 Enhanced Functionality

Aggregator services and the virtualization of resources, allow us to transparently provide enhanced functionality, like collective operations on similar services available in the "cluster". Examples of such collective operations on available "cluster" services are:

- Mirrored execution of a job in many "cluster" hosts to provide increased reliability and the best possible response time for highly prioritized or critical applications.
- Mathematical and statistical collective operations on groups of resources or sets of sensors. For example, we could perform merging or reporting operations on independent data sets from a distributed database or on a set of sensors that collect and store formatted data, in order to present a complete and comprehensive data set to the client. This would be done transparently and the client that requested the data set may have no idea of the distributed nature of the database.
- Automatic distribution of load, whenever this is feasible and adequate resources are available in the "cluster".

4 Conclusion and Future Plans

This project is a work in progress that has only recently started. We are developing an implementation prototype using Java technologies to ensure a certain degree of interoperability. So far, we have implemented components for service aggregation, forwarding of invocation requests and indexing of the available services/resource in the "cluster". We have tested our platform's aggregation and indexing functions against a wide variety of services and it has been proved to perform reasonably well –even for a very preliminary prototype. Nevertheless, there is still plenty of work to do and our plan for the near future center around dynamic discovery of services, implementation of the monitoring framework, the forecasting components and finally support for mobility and roaming between "clusters".

The participation of the authors of this paper in the CoreGRID Network of Excellence [16] and the collaboration with partners dealing with many different aspects of Grid technologies, makes us confident that the outcome of this project will be more than just successful and will lay the roadmap for further development in the area of integrating mobile and resource limited devices into the Grid. Our vision of the Grid is to become a truly ubiquitous and transparent virtual computing infrastructure available from all types of end-user devices, whether that is a powerful server, a mid-range desktop or laptop, or a lower-end limited smart phone.

References

1. L. Cheng, A. Wanchoo, I. Marsic, "Hybrid Cluster Computing with Mobile Objects", Proc. of Fourth International Conference on High-Performance Computing in the Asia-Pacific Region, pp. 909-914, 2000.
2. T.Phan, L. Huang, C. Dulan, "Challenge: Integrating Mobile Wireless Devices into the Computational Grid", Proc. of the 8th Annual International Conference on Mobile Computing and Networking, pp 271-278, 2002
3. Chlamtac, J. Redi, "Mobile Computing: Challenges and Potential", Encyclopedia of Computer Science, 4th edition, International Thomson Publishing, 1998
4. B. Chen, C. H. Chang, "Building Low Power Wireless Grids", TUwww.ee.tufts.edu/~brchen/pub/LowPower_WirelessGrids_1201.pdfUT

5. D. Bruneo, M. Scarpa, A. Zaia, A. Puliafito, "Communication Paradigms for Mobile Grid Users", IEEE/ACM International Symposium on Cluster Computing and the Grid, p. 669, 2003
6. K. Czajkowski, D. Ferguson, I. Foster, J. Frey, S. Graham, T. Maguire, D.Snelling, S. Tuecke, "From Open Grid Services Infrastructure to WS-Resource Framework: Refactoring & Evolution", TU www.ibm.com/developerworks/library/ws-resource/ogsi_to_wsrf_1.0.pdfUT, 2004
7. The *AKOGRIMO* project: TUhttp://www.akogrimo.org
8. J. Hwang, P. Aravamudham, "Proxy-based Middleware Services for Peer-to-Peer Computing in Virtually Clustered Wireless Grid Networks", International Conference on Computer, Communication and Control Technologies, 2003
9. D. Chu, M. Humphrey, "Mobile OGSI.NET: Grid Computing on Mobile Devices", TUwww.cs.virginia.edu/~humphrey/papers/MobileOGSI.pdfUT , 2004
10. J. Thiyagalingam, S. Isaiadis, V. Getov, "Towards Building a Generic Grid Services Platform: A Component-Oriented Approach", in V. Getov and T. Kielmann (Eds), "Component Models and Systems for Grid Applications", 39-56, Springer, 2005
11. Grimshaw, A. Ferrari, G. Lindahl, K. Holcomb, "Metasystems", in Communications of the ACM, vol. 41, n. 11, 1998.
12. The Globus Project, TUwww.globus.orgUT
13. Foster, D. Gannon, H. Kishimoto, "The open grid services architecture", GGF-WG Draft on OGSA Specification, 2004
14. S. Corsava, V. Getov, "Self-Healing Intelligent Infrastructure for computational clusters", SHAMAN workshop proceedings, ACM ISC conference, New York, 2002.
15. R. M. Badia, J. Labarta, R. Sirvent, J. M. Pérez, J. M. Cela , R. Grima, "Programming Grid Applications with GRID Superscalar", Journal of Grid Computing, Volume 1, Issue 2, 2003
16. CoreGRID Network of Excellence, TUwww.coregrid.netUT
17. Foster, C. Kesselman, J. M. Nick, S. Tuecke, "The Physiology of the Grid", http://www.globus.org/research/papers/ogsa.pdf, 2002
18. Foster, C. Kesselman, S. Tuecke, "The Anatomy of the Grid: Enabling scalable Virtual Organizations", International Journal Supercomputer Applications, 2001
19. S. Isaiadis, V. Getov, "Integrating Mobile Devices into the Grid: Design Considerations and Evaluation", to appear in the EuroPar proceedings, 2005

High-Performance and Interoperable Security Services for Mobile Environments

Alessandro Cilardo[1], Luigi Coppolino[1], Antonino Mazzeo[1], and Luigi Romano[2]

[1] Universita' degli Studi di Napoli Federico II Via Claudio 21 – 80125 Napoli, Italy
{acilardo, lcoppoli, mazzeo}@unina.it
[2] Universita' degli Studi di Napoli Parthenope Via Acton 38 - 80133 Napoli, Italy
lrom@uniparthenope.it

Abstract. This paper presents a multi-tier architecture, which combines a hardware-accelerated back-end and a Web Services based web tier to deliver high-performance, fully interoperable Public Key Infrastructure (PKI) and digital time stamping services to mobile devices. The paper describes the organization of the multi-tier architecture and provides a thorough description of individual components.[1]

1 Introduction

Recent advances in wireless technologies have enabled pervasive connectivity to Internet scale systems which include heterogeneous mobile devices (such as mobile phones and Personal Digital Assistants), and a variety of embedded computer systems (such as smart cards and vehicle on-board computers). This trend, which is generally referred to as ubiquitous computing [1], leads to the need of providing security functions to applications which are partially deployed over wireless devices, and poses a number of new challenges, since security-related functions – such as digital signature, time stamping, and secure connection – are mostly based on cryptographic algorithms which are typically computational-intensive.

Digital Signature and Time Stamping, in particular, are the building blocks for innumerable security applications. A digital signature is a numeric string or value that is unique to the document to which is associated and has the property that only the signatory can create it [2]. Digital time stamping, on the other hand, is a set of techniques enabling one to ascertain whether an electronic document was created or signed at a certain time [3]. Due to the ever growing importance of distributed, wireless, pervasive computing, it is mandatory that such cryptographic primitives be provided to the emerging class of low-end mobile devices. However, achieving this goal raises a number of challenging issues,

---

[1] This work was supported in part by the Italian National Research Council (CNR), by Ministero dell'Istruzione, dell'Universita' e della Ricerca (MIUR), by the Consorzio Interuniversitario Nazionale per l'Informatica (CINI), and by Regione Campania, within the framework of following projects: Centri Regionali di Competenza ICT, and Telemedicina.

which arise from specific characteristics and intrinsic limitations of the deployment environment. First, servers have to guarantee higher levels of performance and responsiveness with less predictable workloads. Second, mobile devices have intrinsic limitations in terms of computing power. Third, the heterogeneity of client systems exhacerbates interoperability problems.

This paper presents a multi-tier architecture, which combines a hardware-accelerated back-end and a Web Services based web tier to deliver high-performance, fully interoperable Public Key Infrastructure (PKI) [4] and digital time stamping services to mobile devices. The multi-tier architecture promotes a development approach which exploits a clean separation of responsibilities. In particular, the back-end includes specialized hardware accelerators which can be used to improve the security and the performance of digital signature and digital time stamping services. The accelerator integrates three crucial security- and performance-enhancing facilities: a time-stamping hash processor, an RSA crypto-processor, and an RSA key-store. The web-tier, on the other hand, addresses client-side interoperability issues. A major advantage of our solution is that the web-tier exposes the service interface using Web Services technologies. This allows software entities to directly access (i.e. without any human interaction) the service. This makes it possible to use the service in novel contexts, such as transactional environments where transactions are executed by automated software entities, without human supervision. Using Web Services also favors independence from specific characteristics of the client side technology, thus allowing deployment of the services over a variety of heterogeneous platforms.

2 System Architecture

We propose an architecture with a multi-tier organization. The overall structure of the system is depicted in Figure 1. The Client Tier, the Web Tier, the Server Tier and the Data Tier are located in the first, second, third and fourth tier of the architecture, respectively. This favors a development approach which exploits clean separation of responsibilities. We use the emerging Web Services technology to facilitate service access from/to any platform. Unlike traditional client/server models, such as a Web server/Web page system, Web services do not provide the user with a GUI. Web services instead share business logic, data, and processes through a programmatic interface across a network. The main objective is to allow applications to interoperate without any human intervention. The *back-end* is in charge of actually providing the specific security services. In other words, the back-end satisfies the application functional requirements. Since these services are based on cryptographic algorithms which are usually computationally intensive, this tier has to fully exploit the potential of server platform to boost performance. To this end, and also to achieve better availability and security, we have provided the server with a hardware engine that will be better illustrated in the following section. The conceptual model of the back-end is comprised of three components:

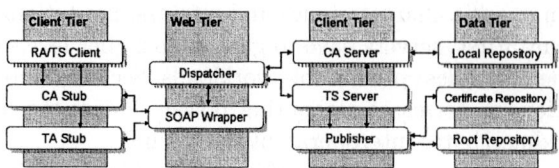

Fig. 1. System architecture

CA Server: it is the system that provides the digital-signature services. It is responsible for creating and signing X.509 certificates [4] and contacting the Publisher Component for the publication of signed certificates. It is also responsible for the processing of requests for certificate revocation.

TS Server: it is the system that provides the digital-time-stamp services. In particular, this component is responsible for creating and retrieving the digital time stamp and the hash chain. It allows to verify the digital time stamping too.

Publisher: this component is responsible for the publication of any information which needs to be made public. Specifically it receives X.509 certificates from the CA Server and publishes them on the Certificate Repository. It also receives revocation requests, and accordingly updates the Certificate Revocation List(CRL) [4]. The Publisher is also responsible for receiving and publishing the hash tree root from the TS Server.

The CA Server uses a local Data Base (localDB) to store the data that it needs for the certificates management. The localDB and the Certificate Repository are logically deployed in the Data Tier.

The *web-tier* wraps the services provided by the back-end server, and addresses interoperability issues, which are due to the heterogeneity of the clients. It also decouples services implementation (i.e., the back-end) from their interfaces. It consists of the following components:

SOAP wrapper: it is in charge of i) processing requestor's messages, and ii) coordinating the flow of SOAP messages through the subsequent components. It is also responsible for ensuring that the SOAP semantics are followed;

Dispatcher: it is responsible for acting as a bridge between the SOAP wrapper and the functional components; it i) converts the SOAP messages from XML into the specific protocol for communicating with the functional back-end, and ii) converts any possible response data back into XML and places it in the response SOAP message.

As for the *Client* side, the client tier consists of the *CA Stub* and the *TS Stub*, both composed of a simple interface to the middle tier . Clients access the service by using the Simple Object Access Protocol (SOAP), and are not aware of the implementation details of the services provided by the middle-tier. The actual client is composed by a Registration Authority (RA) and a Time Stamping Client. The Registration Authority entity allows: i)to assemble the data needed for building the X.509 certificate, ii)to request the certificate through the CA

Stub, *iii*)to prepare and store on the local machine the PKCS12 format [5] of the X.509 certificate, and *iv*)to revoke a certificate. As far as the TS client is concerned, it allows to submit a document for receiving the related digital time stamping and the hash chain needed to verify such stamping. It also allows to submit a request of verification for a digital time stamping.

As far as security is concerned, every Client holds a private key in a security token and an X.509 certificate. The Client signs each request to the Service Manager with its own private key. The Service Manager holds a list of trusted certificates, which it uses to verify the authenticity of incoming requests. After having verified the authenticity of a request, the Service Manager builds a new request to the CAServer – which includes the data received from the client – and signs the request with its private key. The activated Server verifies the source of the incoming request and discards requests from unauthorized sources. If the CAServer receives a valid request for a new certificate, it enrolls an X.509 certificate, stores it in the Certificates Repository and sends it back to the RA. If the request concerns a certificate revocation, the CAServer updates the CRL including a reference to the specified certificate. If the TSServer receives a valid request for a time stamp, it evaluates the digital time stamping and the related hash chain, and sends them, signed with its private key, to the TS client. If the request concerns a digital time stamping verification, the Time Stamping Authority retrieves from the request the document and the hash chain, verifies the digital time stamping and replies to the request with the result of the verification. This security chain requires that every entity holds a private key and an associated certificate. The CAServer's certificate can be either self-signed (the CAServer generates its own certificate) or emitted by a separate CA.

3 Implementation

This section provides some details about the actual implementation of the proposed architecture. The user who needs a certificate applies for it at the RA (to this end he provides the personal data which is to be inserted in the certificate). The RA enrolls a pkcs10 request, which contains the user's personal data, signs such request and than sends it to the Dispatcher using a SOAP message. The Dispatcher verifies the signature of the pkcs10 to be sure that it comes from a trusted RA, selects a CA Server instance (from the cluster of active servers), and forwards the certificate request to it. The CA Server verifies that the request is coming from a trusted party and enrolls an X509 compliant certificate. The CA Server retrieves the unique number of the certificate from the "local database" shared by all the CA in the cluster and stores a new updated value for that number. The CA also signs the X509 certificate and publishes it in an LDAP repository. Finally it encodes the certificate in DER format and sends it back to the RA through the Dispatcher. The RA enrolls the certificate in a pkcs12 safe bag and sends it back to the client.

Activities conducted at the RA's side entail a procedure for enrolling the pkcs10 request. The RA acquires client's data and generates a key pair for the client. It then enrolls the pkcs10, as described in the previous section, while loading its own private key from a local keystore. Then the pkcs10 – encoded in der format – is signed and used to obtain a certificate. The request of the certificate is delivered by means of a web service, whose stub builds the SOAP request. Finally the RA stores the certificate in the pkcs12, securing it with a passphrase obtained by the client.

Server's activities, as far as the CA Server is concerned, entail loading its private key and waiting for a request. When the server receives a request, it gets a der encoded pkcs10 signed by the Dispatcher. The Server verifies that the Dispatcher is a trusted one and that the entity's signature matches the entity's public key. If both verifications are satisfied, it prepares the certificate (otherwise it logs an error event and returns an error message to the Dispatcher). Before enrolling the certificate, the Server increments the unique number to be assigned to the certificate, it stores it in the data base. In parallel, it prepares the X.509 certificate and sends it back (der encoded).

The back-end implementation is based on hardware acceleration, in order to improve performance and security of digital signature and time stamping operations. The aims of the hardware architecture are:

(a) to migrate all security cricital operations and sensitive data to a separate hardware device and make them inaccessible to software programs. In particular, this is a concern for signature keys;

(b) to provide hardware acceleration for hash processing;

(c) to provide hardware acceleration for digital signature.

In order to address the above-mentioned objectives we developed the architecture depicted in Fig. 2 The architeture is plugged into the host system via the standard PCI interface. On board input and output memory banks serve as buffers for handling input and output hash value streams which are written to and read by the host systems with Direct Memory Access (DMA) operations. The architecture include three main security components [6,7,8]:

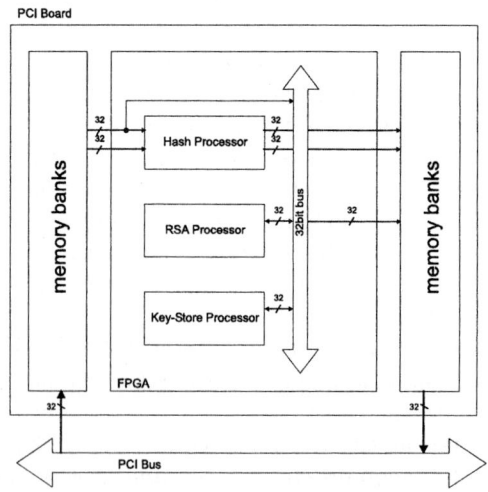

Fig. 2. Overall Architecture of the crypto-processor

(1) a hash processor which is in charge of building the hash trees ([8,9]),
(2) a Key-Store which is in charge of handling secret RSA signature keys,
(3) an RSA processor carrying out the security and performance critical operation for digital signature, i.e. modular exponentiation.

All non-critical pre-processing operations on data to be signed, including padding, are performed by the host system. Fixed-length root hash values are merged with such pre-built data by the hardware accelerator before performing modular exponentiation. A prototype of the crypto-accelerator was implemented in Field-Programmable Gate Array (FPGA) technology [6,7,8,9].

4 Conclusions

This paper presented a multi-tier architecture, combining a hardware-accelerated back-end and a Web Services based web tier to deliver high-performance, fully interoperable Public Key Infrastructure (PKI) and digital time stamping services to mobile devices. The presented work provided an approach and proof-of-concept implementation for provisioning of security services in mobile environments, effectively addressing the problems of server-side performance and client limitations in term of computing power and interoperability.

References

1. Weiser, M.: Some computer science issues in ubiquitous computing. Commm ACM, **36**(7) (2003) 72–84.
2. Diffie, W., Hellman, M. E.: New Directions in Cryptography. IEEE Trans. Info. Theory, (1976), 644-654.
3. Lipmaa, H.: Secure and Efficient Time-Stamping Systems. PhD Thesis (1999), available at http://www.tcs.hut.fi/ helger/papers/thesis/
4. Housley, R., Polk, W., Ford, W., Solo, D.: Internet X.509 Public Key Infrastructure Certificate and Certificate Revocation List (CRL) Profile. Internet Engineering Task Force (IETF) – RFC 3280, (2002), available at http://www.ietf.org/rfc/rfc3280.txt?number=3280
5. PKCS 12: Personal Information Exchange Standard. Version 1.0 Draft, (1997).
6. Cilardo, A., Saggese, G. P., Mazzeo, A., Romano, L.: Exploring the Design-Space for FPGA-based Implementation of RSA. Elsevier Microprocessors and Microsystems, **28** (2004) 183–191.
7. Saggese, G. P., Mazzeo, A., Mazzocca, N., Romano, L.: A tamper resistant hardware accelerator for RSA cryptographic applications. Journal of Systems Architecture, **50** (2004) 711–727.
8. Cilardo, A., Cotroneo, D., di Flora, C., Mazzeo, A., Romano, L., Russo, S.: A High Performance Solution for Providing Digital Time Stamping Services to Mobile Devices. Computer Systems Science and Engineering Journal, (to appear).
9. Cilardo, A., Saggese, G. P., Mazzeo, A., Romano, L.: An FPGA-based Key-Store for Improving the Dependability of Security Services. Procs. of the Tenth IEEE International Workshop on Object-oriented Real-time Dependable Systems, (2005).

Distributed Systems to Support Efficient Adaptation for Ubiquitous Web

Claudia Canali[1], Sara Casolari[2], and Riccardo Lancellotti[2]

[1] Department of Information Engineering, University of Parma, 43100 Parma, Italy
claudia@weblab.ing.unimo.it
[2] Department of Information Engineering,
University of Modena and Reggio Emilia, 41100 Modena, Italy
{casolari.sara, lancellotti.riccardo}@unimore.it

Abstract. The introduction of the Ubiquitous Web requires many adaptation and personalization services to tailor Web content to different client devices and user preferences. Such services are computationally expensive and should be deployed on distributed infrastructures. Multiple topologies are available for this purpose: highly replicated topologies mainly address performance of content adaptation and delivery, while more centralized approaches are suitable for addressing security and consistency issues related to the management of user profiles used for the adaptation process.

In this paper we consider a two-level hybrid topology that can address the trade-off between performance and security/consistency concerns We propose two distributed architectures based on the two-level topology, namely core-oriented and edge-oriented. We discuss how these architectures address data consistency and security and we evaluate their performance for different adaptation services and network conditons.

1 Introduction

The current trend in the evolution of the Web is towards an ever increasing complexity and heterogeneity. The growth in Web popularity is related to the diffusion of heterogeneous client devices, such as handheld computers, mobile phones, and other pervasive computing devices that enable ubiquitous Web access. Furthermore, we observe an increasing complexity of the Web services aiming to match the user preferences. The need for personalization lead to the introduction of content adaptation services that tailor Web resources to user preferences and to device capabilities.

Content adaptation is a computationally intensive task that places a significant overhead on CPU and/or disk if compared to the delivery of traditional Web resources. As a consequence, much interest is focused towards high performance architectures capable of providing efficient and scalable adaptation services. The use of an intermediary infrastructure of multiple server nodes that are interposed between the clients and the traditional Web servers seems the most suitable approach to build an efficient system for Web content adaptation and delivery. Different distributed topologies of collaborative nodes have been evaluated ranging from flat to hierarchical schemes [1]. However, most of these studies only focus on simple adaptation services, which do not introduce significant security and consistency problems. On the other hand, more complex adaptation tasks usually rely on stored user profiles that possibly contain sensitive or critical information about the users. In this case consistency and security issues [2] suggest to avoid replication of such information over a large number of nodes.

We have a clear trade-off between increasing system replication by moving services towards the network edge for performance reasons and reducing replication to preserve security and consistency. The first is a common trend in the development of distributed Web systems, as testified by edge-based technologies such as the ESI system or by the ACDN architecture [3], while the latter issue has been recently pointed out in studies (such as [4]) on dynamic Web-based services.

This paper explores the trade-off and evaluates the impact on performance of different placements of adaptation services in a distributed infrastructure. This study is particularly interesting in the case of content adaptation because the assumption that the most significant contribution to the system performance is related to the network delays (which is one of the key reasons behind edge-based technologies) may be no longer true.

We present two architectures for efficient content adaptation and delivery, namely *edge-oriented* and *core-oriented*, based on a two-level topology. The two architectures differ on the placement of adaptation services on the levels of the infrastructure, close or far from the network edge. We carry out an experimental performance evaluation to outline which choice can provide better performance for different adaptation services and for different network conditions.

The remainder of this paper is organized as follows. Section 2 describes the adaptation services that can be provided and presents the two-level architectures for efficient content adaptation and delivery. Section 3 provides the results of the performance evaluation of the edge-oriented and the core-oriented architectures. Finally, Section 4 provides some concluding remarks.

2 Content Adaptation in a Two-Level Topology

In this section we discuss the content adaptation services that can be provided by a distributed infrastructure and their deployment on a two-level topology. In particular, we present two architectures that provide a different mapping between adaptation services and the nodes of the distributed infrastructure.

Content adaptation services are extremely heterogeneous [5]. However, for the goal of our study we can classify adaptation services in two main categories, each of them introducing different constraints on the infrastructure, especially depending on the type of information needed to carry out the content adaptation. State-less adaptation includes transcoding, which is a content adaptation to the capabilities of the client device, and state-less personalization which aims to adapt Web resources to the user preferences. This class of service extracts from the user request every information needed for the adaptation; for example, information on the client device or on the preferred language(s) can be extracted from the HTTP headers of each request. Hence, transcoding and state-less personalization require no special handling of (possibly) sensitive information and, consequently, do not place consistency nor security constraints on the topology. On the other hand, state-aware adaptation usually involves personalization which is carried out on the base of a stored user profile. In this case we possibly have to handle sensitive information. The choices available for the deployment of the content adaptation infrastructure depend on the nature of the adaptation service to be provided. Indeed, the type of service affects the balance of the trade-off between the need for highly distributed systems related to performance issues and the need for few controlled data replicas related to guaranteeing data consistency and security.

To address the above-mentioned trade-off we focus on a two-level topology that allows to replicate or centralize services and information depending on the performance, security and consistency constraints of each provided adaptation service. The topology is based on a subdivision of nodes in two levels that we call *edge* and *internal*.

The edge level is characterized by a large amount of nodes, each located close to the network edge. Such nodes are usually placed in the points of presence of ISPs to be as close as possible to clients.

The internal level of the content adaptation system is composed by a smaller number of powerful nodes. Such nodes can be placed in *well connected* locations, which means in Autonomous Systems with a high peering degree, to reduce communication costs especially with the edge nodes. Internal nodes are suitable for handling data characterized by consistency requirements: the most common approach in this case is to reduce data replication. Literature shows that data consistency can hardly be provided over more than 10 nodes and the bound of synchronous updates is to be relaxed if more replication is needed [2]. The use of internal nodes can reduce data replication based on hashing mechanisms that *partition* the space of user profiles. The user profile is replicated only on one internal node (or few nodes, depending on the hash algorithm used and on whether backups are kept) to avoid replica consistency issues. The same scheme would be unfeasible on edge nodes which are less reliable and not so well connected. Due to the limited replication of the internal nodes, we can also guarantee higher security standards for them. Hence, these nodes are also suitable for adaptation tasks requiring sensitive information stored in the user profile.

The two-level topology allows the deployment of different architectures depending on the mapping between the adaptation services and the nodes of the two-level structure. In particular, we describe two different solutions, namely *core-oriented* and *edge-oriented*, that represent two opposite approaches in addressing the trade-off between performance and consistency/security concerns.

The *core-oriented* architecture is more conservative in addressing consistency and security issues as it forces all adaptation services to be deployed on internal nodes, while the edge nodes are extremely simple gateways which only have to forward client requests to the appropriate internal node. Even if most operations in the core-oriented architecture are carried out by the internal nodes, computational power is not an issue. Indeed, sophisticated local replication strategies (e.g. clustering) can provide the amount of computation power required without introducing management problems thanks to the reduced degree of replication of the internal nodes.

The *edge-oriented* architecture relaxes consistency and security constraints related to user information and allows content adaptation to be carried out at the edge level. To this aim, the portion of the user profile not containing critical data can be replicated on the edge nodes, while the internal nodes are only used for the most critical personalization. It is not possible to identify a clear distinction between critical and non-critical

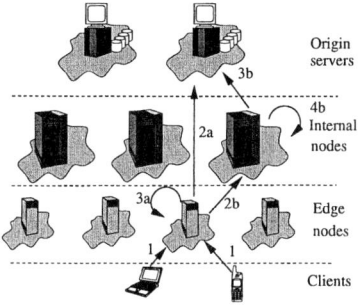
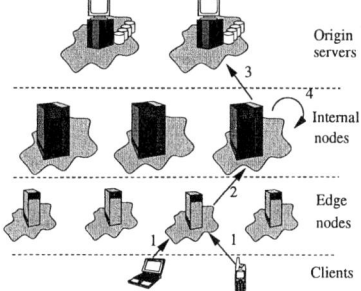

Fig. 1. Edge-oriented architecture **Fig. 2.** Core-oriented architecture

user profile information due to the high heterogeneity in adaptation services and in the user preferences. For example, the geographic location of the user can be considered critical or not, while numbers of credit cards always require high security guarantees.

Figure 1 and 2 describe how client requests are served in the edge-oriented and core-oriented architecture, respectively. These figures show the two levels of nodes and the origin servers (belonging to the content provider) that host the repository of the Web resources.

In the edge-oriented architecture the request is received by an edge node (step 1 in Figure 1). The node decides, based on the requested service, if adaptation is to be carried out locally (path 1-2a-3a in the figure) or if the request is to be forwarded (path 1-2b-3b-4b) to an internal node for processing (the actual content adaptation is to be carried out in step 3a or 4b, depending on the level in which it takes place). A fetch operation from the origin server (steps 2a or 3b) has to be carried out prior of the content adaptation because we assume that no caching is adopted. Resource caching has been used to improve the performance of content adaptation systems [1]. However, caching does not change the conclusions of the performance comparison of the proposed architectures because both variants would be affected in the same way by caching.

The core-oriented architecture shown in Figure 2 carries out content adaptation on the internal nodes in every case. Hence, two steps are always required for the content adaptation because the request flow passes always from the edge to the internal nodes.

3 Performance Evaluation

In this section we present an experimental evaluation of the core-oriented and edge-oriented architectures that aims to analyze the impact on performance of the different approaches to the trade-off of advanced content adaptation architectures.

For our experiments we consider three adaptation services with different computational requirements. Even if they do not involve complex personalization tasks, these services are of state-aware type because they are based on a stored user profile. At the best of our knowledge, there are no studies that point out which user profile information is "critical" in a content adaptation context. Our experience suggests that nearly 20% of user requests requires access to information that can be considered critical, but different experiments carried out with percentages of critical requests ranging from 10% to

30% do not show significant differences. In the case of the edge-oriented architecture, we configured our experimental testbed in such a way that only critical client requests require adaptation to be carried out on the internal nodes, while the rest of the requests are handled by edge nodes.

For our experiments, we set up a Web delivery infrastructure with a client node, a Web server and a two-level content adaptation infrastructure composed of five edge and three inner nodes. Each node is connected to the others through a Fast Ethernet LAN, however, we use WAN emulation to take into account the effect of geographic links among the nodes of the content adaptation infrastructure and in the connection to the origin server. Through a WAN emulator, that is able to simulate packet delay and bandwidth limitation in the links connecting the adaptation system to the origin servers (delay=100ms, bandwidth=2Mbit/s). We also create two different network scenarios for the links among the nodes of the content adaptation architecture. The first scenario, namely *good connection*, represents a best case for the network and is characterized by network delays of 10 ms and bandwidth of 50 Mbit/s, while for the second scenario, namely *poor connection*, we consider high latency (100 ms) and low bandwidth (1 Mbit/s).

We define three workloads with different computational requirements. In all three cases the client request are based on synthetic traces with a frequency of 5 requests per second. The first workload, namely *banners*, is rich of small GIF images that are processed through aggressive compression algorithms to reduce image size and color depth in order to save bandwidth. The second workload, namely *photo album*, contains several photos in the form of large, high-resolution JPG files and aims to test a content adaptation system that enables ubiquitous access to a Web album application. To enable ubiquitous access, photographic images must be processed by reducing image size to adapt the photo to the small displays typical of mobile clients. Furthermore, to reduce download time, we can reduce the JPEG quality factor. The third workload focuses on *multimedia* resources. In this case, content adaptation aims to enable ubiquitous access to audio clips stored as MP3 files and is mostly related to save bandwidth by recoding the audio streams at lower bit rates. The core of such adaptation (the MP3 recoding

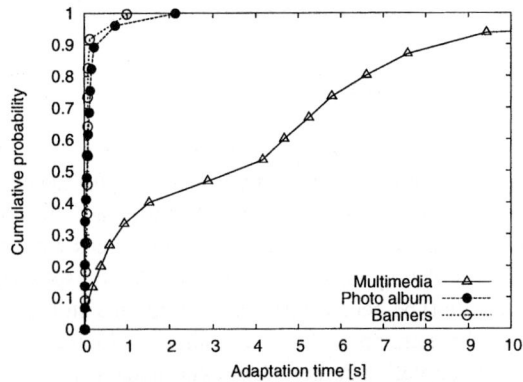

Fig. 3. Content adaptation time

algorithm) is an example that suits other more complex systems, as in the case of the text-to-speech application proposed in [6].

To provide a deeper insight on the performance evaluation of the two architectures, we carried out a preliminary experiment to measure the content adaptation time for the three workloads. Figure 3 shows the cumulative distribution of adaptation time for the different workloads. We can see from the figure that the banner scenario is the lightest adaptation service with a 90-percentile of adaptation time of 0.13 seconds, while for the photo album workload, that requires processing of larger files (pictures), the 90-percentile is almost three times higher that is, 0.32 seconds. Figure 3 shows that adaptation is even more expensive in the multimedia workload, with a 90-percentile of 8.30 seconds, that is one order of magnitude higher than that of the other two workloads.

The results of our experiments are outlined in Table 1 for the good and poor network scenarios. Columns 2 and 3 show the 90-percentile of response time for the edge-oriented and core-oriented architectures, respectively, as a function of the workload.

The edge-oriented solution always provides a performance gain, that is reported in the last column of Table 1, over the core-oriented architecture. In the case of poor network connection the impact of the two steps required by the core-oriented architecture is more evident and the performance gain of the edge-oriented architecture is up to nearly 8 times higher with respect to the case of good connection. Besides this consideration, it is interesting to note that for both network scenarios the performance gain of the edge-oriented architecture is highly variable and presents a strong dependence on the workload. Indeed, if we consider the corresponding column of Table 1, we see that for each network scenario the performance gain decreases as we pass from the banners to the photo album to the multimedia workloads. This can be explained by considering the content adaptation cost. The performance gain of the edge-oriented architecture is reduced as the computational requirements of the workload grow. When adaptation time is low, the edge-oriented architecture outperforms the core-oriented approach, as shown for the banner workload. In this case, even for the best network scenario (good connection) the performance gain of the edge-oriented architecture is higher than 10%.

On the other hand, most computationally-expensive adaptations require a time that far overweights the cost of the additional step, as shown for the multimedia workload, where adaptation time is in the order of seconds. In this latter case, both the core-

Table 1. Core-oriented and edge-oriented performance for good and poor connection

Good connection			
Workload	90-percentile of response time [s]		Performance
	Edge-oriented	Core-oriented	gain [%]
Banners	0.29	0.34	12%
Photo album	0.60	0.64	6.3%
Multimedia	9.12	9.29	1.8%
Poor connection			
Workload	90-percentile of response time [s]		Performance
	Edge-oriented	Core-oriented	gain [%]
Banners	0.46	0.80	42%
Photo album	1.47	2.17	32%
Multimedia	17.05	19.15	10%

oriented and the edge-oriented cases are a viable solution, even in the case of poor network connectivity.

4 Conclusions

In this paper we investigated the issues introduced by sophisticated adaptation services that enable the ubiquitous Web access, which are characterized by two main issues: high computational requirements and management of user profiles containing potentially sensitive information. We proposed two architectures (edge-oriented and core-oriented) based on a two-level topology for the deployment of content adaptation services. We evaluate how these architectures address the trade-off between replication for performance reasons and centralization for security and consistency concerns.

Our experiments show that the edge-oriented architecture provides always better performance than the core-oriented solution. However, we found that the performance gain derived from locating adaptation functions on the network edge is highly dependent on both the network status and the computational cost of the provided adaptation services, depending on wheter the network-related time is comparable with adaptation time or not.

References

1. Cardellini, V., Colajanni, M., Lancellotti, R., Yu, P.S.: A distributed architecture of edge proxy servers for cooperative transcoding. In: Proc. of 3rd IEEE Workshop on Internet Apps. (2003)
2. Gray, J., Helland, P., O'Neil, P.E., Shasha, D.: The dangers of replication and a solution. In: Proc. of the 1996 ACM SIGMOD International Conference on Management of Data. (1996)
3. Rabinovich, M., Xiao, Z., Aggarwal, A.: Computing on the edge: A platform for replicating internet applications. In: Proc. of 8th Int'l Workshop on Web Content and Distribution, Hawthorne, NY (2003)
4. Gao, L., Dahlin, M., Nayate, A., Zheng, J., Iyengar, A.: Application specific data replication for edge services. In: Proc. of 12th WWW Conference, Budapest, HU (2003)
5. Colajanni, M., Lancellotti, R.: System architectures for web content adaptation services. IEEE Distributed Systems online (2004)
6. Barra, M., Grieco, R., Malandrino, D., Negro, A., Scarano, V.: Texttospeech: a heavy-weight edge service. In: Poster Proc. of 12th WWW Conference, Budapest, HU (2003)

Call Tracking Management Using Caching Scheme in IMT-2000 Networks

Dong Chun Lee

Dept. of Computer Science Howon Univ., South Korea
ldch@sunny.howon.ac.kr

Abstract. To support call locality in IMT-2000 networks, we locate the cache in Local Signaling Transfer Point (LSTP) level. The hosts Registration Area (RA) crossings within the LSTP area do not generate the Home Location Register (HLR) traffic. The RAs in LSTP area are grouped statically, which removes the signaling overhead and mitigates the Regional STP (RSTP) bottleneck in dynamic grouping. Grouping RAs statically, we lessen the Inconsistency Ratio (IR). The idea behind LSTP level caching is to decrease the IR with static RA grouping. The proposed scheme solves the HLR bottleneck problem and reduces to the mobile traffic cost.

1 Introduction

The mobility management schemes are based on Interim Standard-95 (IS-95) and Global System for Mobile Communication (GSM) standard. Those schemes use the two level hierarchies composed of HLR and Visitor Location Register (VLR) [3, 12]. Whenever a terminal crosses a RA or a call originate, HLR should be updated or queried. Frequent DB accesses and message transfers may cause the HLR bottleneck and degrade the system performance. To access the DBs and transmit the signaling messages frequently cause the HLR bottleneck and load the wireless network [2, 4].

Many papers in the literature have demonstrated that the IS-95 standard strategy does not perform well. This is mainly because whenever a mobile host moves. The VLR of a registration area which detected the arrival of the client always reports to the HLR about the host's new location. While a call is placed, the callee is also located by going to the HLR's database to find the callee's new location. As the HLR could be far away from a VLR, communication to the HLR is costly. The Forwarding Point (FP) strategy and the Local Anchor (LA) strategy [3, 7, 10] are representatives of the old VLR to the new VLR. Update of the client's location to the HLR's database is not always needed to minimize communications to the HLR. To locate a mobile host (callee) however, some extra time is required to follow the forwarding link(s) to locate the host. When the number of the forwarding links is high, the locating cost becomes significant.

The LA strategy is somewhat similar to the FP strategy. A LA is itself a VLR. It records the location information of mobile host. When a mobile host moves out of the LA into an adjacent VLR, the VLR report the new location of the mobile host to this LA rather than to the HLR. For this mobile host, the LA builds a link from the LA to the VLR. While the mobile client moves to another VLR, the LA's link (a record) is updated and redirected to the new VLR. The HLR's record about this mobile host always points to the LA during this time. To reduce the communication cost between the VLR and the LA the system will reassign the VLR as the new LA. Although the registration cost is reduced, the call tracking cost is increased because an intermediate LA has to be traversed.

In IMT-2000 networks environments, one of the distinct characteristics of call patterns is the call locality. The call locality is related to the regional distribution with which the callers request calls to a given callee. We propose call tracking method to decrease the signaling traffic load greatly applying the call locality to tracking a call in an efficient manner. This implies that the HLR query traffic caused whenever a call originates is reduced to some extent by applying the cache. The key idea behind applying the cache is to make the cached information be referred frequently and maintain the consistency to some extent.

2 IS-95 Standard

According to the IS-95 standard strategy, the HLR always knows exactly the ID of the serving VLR of a mobile terminal. We outline the major steps of the IS-95 call tracking is outlined as follows [3]:

1. The VLR of caller is queried for the information of callee. If the callee is registered to the VLR, the call tracking process is over and the call is established. If not, the VLR sends a Location Request (LOCREQ) message to the HLR.
2. The HLR finds out to which VLR the callee is registered, and sends a Routing Request (ROUTREQ) message to the VLR serving the callee. The VLR finds out the location information of the callee.
3. The serving MSC assigns a Temporary Local Directory Numbers (TLDN) and returns the digits to the VLR which sends it to the HLR.
4. The HLR sends the TLDN to the MSC of the caller.
5. The MSC of the caller establishes a call by using the TLDN to the MSC of the callee.

Among the above 5 steps, the Call Tracking process is composed of step1 and step2.

3 Proposed Scheme

We define the LSTP area as the one composed of the RAs which the corresponding MSCs serve. Then we statically group the VLRs in LSTP area in order to maintain the consistency rate high. Even though it is impossible to maintain the consistency at all times as the computer system does, we can effectively keep it high to some extents by grouping method. The cache is used to reduce the call tracking load and make the fast call setup possible. It is also possible to group RAs dynamically regardless of the LSTP area. Suppose that the post VLR and the VLR which serves the callee's RA are related to the physically different LSTPs. In this case, we should tolerate the signaling overhead even though the caller and callee belong to same dynamic group. A lot of signaling messages for tracking a call are transmitted via RSTP instead of LSTP. If the cost of transmitting the signaling messages via RSTP is large enough compared to LSTP, dynamic grouping method may degrade the performance although it solves the Ping-Pong effect. If a call originates and the entry of the callee's location exists in LSTP level cache, it is delivered to the post VLR. We note that we don't have to consider where the callee is currently. It is because the P_{VLR} keeps the callee's current location as long as the callee moves within its LSTP area. Without the terminals movements into a new LSTP area, there is no HLR registration.

The location information of terminals is stored in the cache in LSTP level, where the number of call requests to those terminals is over threshold K. The number of call requests is counted in the post VLR or the VLR which serves the callee's RA. If we take the small K, there are a lot of the entries in cache but the possibility that the cached information and the callee's current location are inconsistent is increased. To the contrary, if we take the large K, the number of terminals of which location information is referred is decreased but the consistency is maintained high.

As for the hierarchical level in which the cache is located, the MSC level is not desirous considering the general call patterns. It is effective only for the users of which working and resident areas are regionally very limited, e.g., one or two RAs.

When a call is tracked using the cached information, the most important thing is how to maintain the ICR low to some extent that the tracking using the cache is cost effective compared to otherwise. In the proposed scheme, the terminals with high rate of call request regardless of the RA crossings within LSTP area are stored in cache. As for the cached terminal, it is questionable to separate the terminals by CMR. Suppose that the terminal A receive only one call and cross a RA only one time and the terminal B receive a lot of calls and cross the RAs n times. Then the CMR of two terminals is same but it is desirous that the location information of terminal B is stored in the cache. As for caching effect, the improvement of the performance depends on the reference rate of the user location information more than the number of entries in cache, i.e., the cache size. The object of locating cache in LSTP level is to extend the domain of the region in which the callee's location information is referred for a lot of calls.

Algorithm *LSTP_Cache_Tracking( )*
{Call to mobile user is detected at local switch;
If callee is in same RA then
return;
Redial_ Cache ( )   */* optional*/*
If there is an entry for callee in L_Cache then   */* Cache hit*/*
                                                 / L_Cache: Cache in LSTP level*/*
  {Query corresponding P_{VLR} specified in cache entries;
  If there exists the entry for callee in PVLR then
      Callee VLR, V return callee's location inf. to calling switch;

  else   */* Inconsistent state */*
  Switch which serves P_{VLR} invokes Basic_Tracking();
  / skip the first and second statements in Basic_ Tracking ( )*
  }
  else   */* Cache miss*/*

continue Basic_Tracking ( )
                           /* Basic_Tracking( ) is used in IS-95standard */
{ If callee is in same RA then
return;
Query Callee's HLR;
Callee's HLR queries callee's current VLR, V;
V returns callee's location info. to HLR;
HLR returns the location info. to calling switch;
} }

Using the above algorithms, we support the locality shown in user's general call behavior. The tracking steps are depicted in Fig. 1.

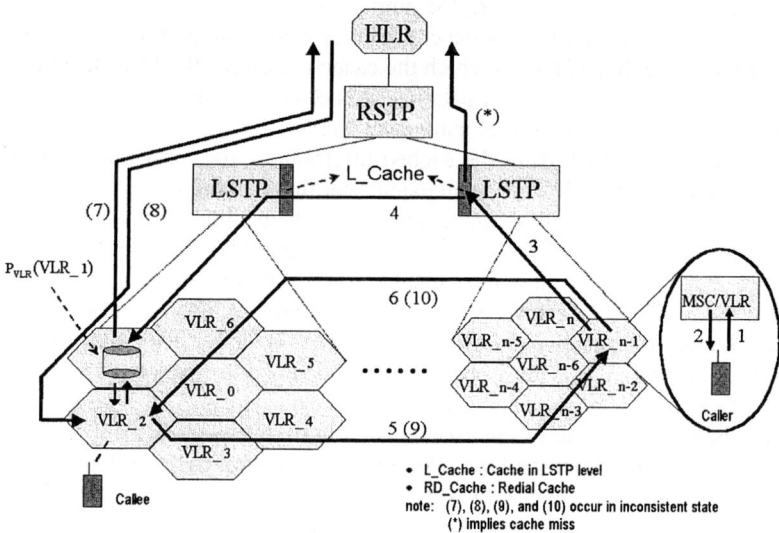

Fig. 1. Call tracking in LSTP level caching scheme

4 Performance Analysis

For numerical analysis, mobile hosts moving probability should be computed. We adopt square model as geometrical RA model. Generally, it is assumed that a VLR serves one RA.

As shown in Fig. 2, RAs in LSTP area can be grouped. The terminals in RAs inside circle n area still exist in circle n area after their first RA crossings. That is, the terminals inside circle area cannot cross their LSTP area when they cross the RA one time. While the terminals in RAs which meet the line of the circle can move out from their LSTP area. We can simply calculate the number of moving out terminals. Intuitively, the terminals in arrow marked areas move out in Fig. 2. Using the number of outside edges in arrow marked polygons (i.e., squares), we can compute the number of terminals which move out from the LSTP area as follows.

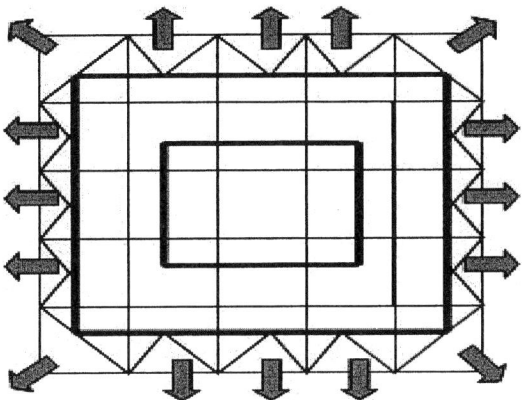

Fig. 2. Square RA models

$$\left(\frac{Total\ No.\ of\ outside\ edges\ in\ arrow\ marked\ polygon\ s}{No.\ of\ edges\ of\ hexagon \times No.\ of\ RAs\ in\ LSTP\ area} \right)$$
$$\times\ No.\ of\ terminals\ in\ LSTP\ area$$

In Fig. 2, there are 25 RAs in square 2 area. The number of terminals which move out from LSTP area is

$$the\ terminals\ in\ LSTP\ area \times \frac{5}{25}.$$

The number of VLRs in the LSTP area represented as square n can be generalized as

$$No.\ of\ VLRs\ in\ LSTP\ area = \frac{1}{(2n+1)^2} \qquad (where\ n = 1, 2, ...).$$

The rate of terminals which move out from LSTP area can be generalized as

$$R_{move\_out,\ No.\ of\ VLRs\ in\ LSTP\ area} = \frac{1}{(2n+1)} \qquad (where\ n = 1, 2, ...).$$

In this model, the rate of terminals movements within the LSTP area is

$$R_{move\_out,\ No.\ of\ VLRs\ in\ LSTP\ area} = \left(1 - R_{move\_out,\ No.\ of\ VLRs\ in\ LSTP\ area}\right)$$

The RAs is said to be locally related when they belong to the same LSTP area and remotely related when the one of them belongs to the different LSTP area. Because the relations are determined according to the terminals moving patterns, *P (local)* and *P (remote)* can be written as follows.

$$P(local) = R_{move\_in,\ No.\ of\ VLRs\ in\ LSTP\ area}$$
$$P(remote) = R_{move\_in,\ No.\ of\ VLRs\ in\ LSTP\ area}$$

P (local) and *P (remote)* are varied according to the number of VLRs in LSTP area. Consider the terminals n time RA changes, where the number of RAs in LSTP area is 7. If a terminal moves into a new LSTP area in n^{th} movement, the terminal is located in one of outside RAs - 6 RAs - of the new LSTP area. Therefore, when the terminal moves $(n + 1)^{th}$ times, two probabilities, $R_{move\_out, No.\ of\ VLRs\ in\ LSTP area}$ and $R_{move\_out, No.\ of\ VLR\ in\ LSTP area}$ are both 3/7. If the terminals $(n + 1)^{th}$ movement occurs within the LSTP area, the probabilities of terminals $(n + 2)^{th}$ movement are the two probabilities are 4/7, 3/7, respectively. Otherwise, they are both 3/7. We classify the terminals moving patterns and compute the probabilities to evaluate the traffic cost correctly.

4.1 Call Tracking Cost

We define the rate of the cached terminals of which the number of call_count is over threshold K in PVLR as $R_{No.\ of\ Cached\ TRs}$. If there is an entry in LSTP level cache for the requested call, pointed VLR, PVLR, is queried. If not, HLR is queried and the pointed VLR in HLR entry is queried subsequently. We define the Signaling Costs (SCs) as follows.

SC1: Cost of transmitting a message from one VLR to another VLR through HLR
SC2: Cost of transmitting a message from one VLR to another VLR through RSTP
SC3: Cost of transmitting a message from one VLR to another VLR through LSTP

We evaluate the performance according to the relative values of SC1, SC2, and SC3 which need for call tracking. Generally, we can assume SC3 ≤ SC2 < SC1 or SC3 ≤ SC2 << SC1. Even though the absolute values can not be determined, the difference of relative values can be computed.

Case 1: SC3 < SC2 < SC1, Case 2: SC3, SC2 << SC1, Case 3: SC3 << SC2, SC1
Case 4: SC3, SC2 < SC1

A call is said to originate locally if the caller and callee are located in same LSTP area. Otherwise, it is said to originate remotely. In call tracking, we consider those two call types and classify each call type into two cases of consistent call and inconsistent one. We define ICR, which is the ratio that the cached information and the entry in P_{VLR} are inconsistent. Therefore, ICR is the ratio of the number of terminals which move out from their LSTP area to the number of terminals in LSTP area to which calls are requested. Described simply, it is the rate of terminals which move out from their LSTP area if the calls are requested to all terminals in LSTP area.

The call tracking cost is influenced by the number of VLRs in LSTP area, the terminals moving speed, the call origination rate, and cache reference rate. In other words, the caching effect is dependent on HR in LSTP level cache and the ICR in callee's LSTP area. We mentioned above that the call relations between caller and callee should be considered. According to the relations, the message transmitting route via LSTP or RSTP is determined. The followings are the generalized tracking costs according to the call relations.

○ **Local Relation between Caller and Callee**

$C_{LSTP\ level caching, Call\ Tracking} = (1 - R_{No.\ of\ cached\ TRs}) C_{IS-95,\ CallTracking} + R_{No.\ of\ Cached\ TRs}$

$[\ HR \times \{(1\text{-}ICR) \times 2SC3 + ICR \times 2(SC1+SC3)\} + (1\text{-}HR) \times 2SC1]$

$+ C_{cache}$

○ **Remote Relation between Caller and Callee**

$$C_{LSTP\ level\ caching, Call\ Tracking} = (1 - R_{No.\ of\ cached\ TRs})C_{IS-95,\ CallTracking} + R_{No.\ of\ Cached\ TRs}$$

$$[\ HR\ \{(1\text{-}ICR) \times 2SC2 + ICR \times 2(SC1 + SC2)\} + (1\text{-}HR) \times 2SC1]$$
$$+ C_{cache}$$

1) IS-95 Standard
Whenever a call originates, the HLR is queried. The call tracking cost is computed as follows.

$$C_{Is\text{-}95,\ Call\ Tracking} = 2 \times SC1$$

2) LA Scheme
If a call originates, the LA and the serving VLR of the RA in which callee is located are queried subsequently after the HLR query. I should consider the local and remote relations between the LA serving RA and the RA in which the callee is located. The call tracking cost is written as follows.

$$C_{LA,\ Call\ Tracking} = 2SCI + 2SC2 \times P(remote) + 2SC3 \times P(local)$$

3) FP Scheme
If a call originates, it is delivered to the serving MSC of the RA in which the callee is located through the link pointer. In this scheme, the call tracking cost and setup delay is dependent on the threshold T. For example, if T is 3, T is reset after a terminal crosses RAs 3th time. In maximum, (T – 1) additional traverses are needed to track a call. In this case, the tracking cost is varied according to the terminal moving patterns. Therefore, we should consider the local and remote

Relations among the RA where T=0, the intermediate RAs in moving route, and the RA in which the callee is located. The followings are the call tracking costs according to the terminal RA crossing patterns when the number of VLRs is 7. When T is 3, I can consider 4 moving patterns, respectively before T is expired. Table 1 shows the probabilities and the required costs according to the moving patterns when T is 3. According to T, the call tracking cost can be generalized as follows.

$$C_{FP\ .Call\ Tracking} = \sum_{i=1}^{2t} \{P(X(i)) \times Cost\ (X(i))\}$$

Table 1. Example of call tracking cost in FP scheme (T = 3)

i	X(i)	P(Xi)	Cost(P(Xi))
1	move_in+move_in	(4/7) × (4/7)	2(SC1 + 2SC3)
2	move_in+move_out	(4/7) × (3/7)	2(SC1 + SC2 + SC3)
3	move_out+move_in	(3/7) × (1/2)	2(SC1 + SC2 + SC3)
4	move_out+move_out	(3/7) × (1/2)	2(SC1 + 2SC2)

In performance evaluation, the rate of the local calls is assumed as 0.7. As I simulated with various rates, there were slight changes in the results. In fact, the performance is mainly dependent on SC1 instead of SC2 or SC3. If the value difference between SC2 and SC3 is not great, I can regard them as the same cost terms. If the HR is lessen in local and remote calls, the relative portion of SC1 to SC2 and SC3 in tracking a call is increased. In case of cache miss, I submit to the call tracking mechanism in IS-95 standard.

The tracking cost is reduced according to the number of the VLRs in LSTP area. It implies that the more VLRs in LSTP area are, the lower $R_{move\_out No.\ of\ VLR\ in\ LSTP area}$ is subsequently.

Note that the call relation should be considered separately. The degree of the call locality is related to the size of area in which the calls generate to a terminal. Considering the geographical region to which caller and callee belong, the caller's moving speed heavily affects the call relation but little affects call locality when the cache is in LSTP level. With the LSTP level caching, we can get over the limit of the degree of call locality. If so, the reference rate of the callee's location information is increased.

4.2 Traffic Comparison

In performance evaluation, the rate of the local calls is assumed as 0.7. As we simulated with various rates, there was slight changes in the results. In fact, the performance is mainly dependent on SC1 instead of SC2 or SC3. If the value difference between SC2 and SC3 is not great, we can regard them as the same cost terms. If the HR is lessen in local and remote calls, the relative portion of SC1 to SC2 and SC3 in tracking a call is increased. In case of cache miss, we submit to the call tracking mechanism in IS-95 standard.

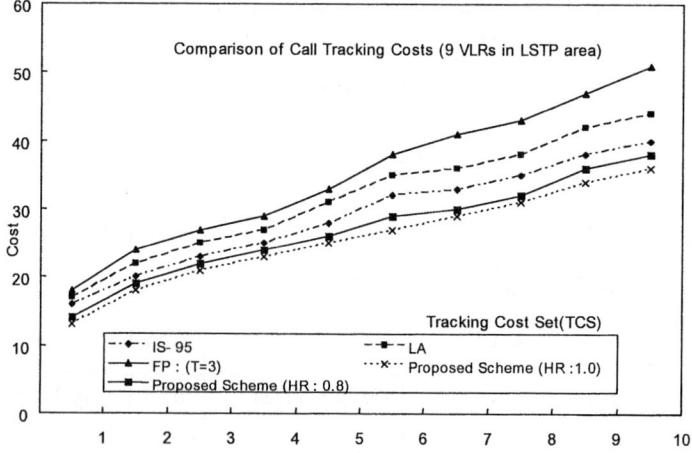

Fig. 3. Tracking cost comparison (where SC3, SC2 << SC1)

In Fig. 3, the costs are decreased in order as FP > LA > IS-95 > Proposed scheme. LA scheme has one more traverse to track a call compared to IS-95 standard. The difference between two costs in FP scheme is due to one more traverse via LSTP or RSTP. Additionally, the cost for one more traverse varies according to the local or remote relation between callee's RA and the previous RA in tracking route. T should be taken considering cost effectiveness in call tracking and the degeneration overhead together. If T is too large, we should tolerate the setup delay too. As for tracking cost, those two schemes degrade the performance due to the inherent additional traverses and degeneration overhead. The larger the relative value difference of SC2 and SC3 to SC1 is, the performance in proposed scheme is improved greater compared to those in other schemes.

5 Conclusions

The proposed scheme is mainly focused on supporting the call locality efficiently. The object that we locate the cache in LSTP level is to reuse the location information of the callee to which there are lots of call requests and furthermore to increase the reference rate. It generalizes the degree of temporal and spatial localities shown in the call patterns of mobile users. In addition, the RAs in LSTP area are grouped statically to decrease the ICR and mitigate the traffic bottleneck in HLR and RSTP. The grouping method also decreases the signaling traffic caused by terminals frequent RA crossings. With the static RA grouping, the LSTP level caching scheme supports the call locality in efficient manner. In supporting the call locality, the proposed scheme is relatively insensitive to the CMR, compared to the other schemes.

As a result of cost evaluation, the more the VLRs in LSTP area are and the higher the call origination rate is, the performance is improved greater. The proposed scheme is much less affected by the hosts frequent RA crossings compared to the previous schemes including IS-95 standard. Consideration of the call locality is attractive in mobility management scheme. The ratio of the number of cached hosts to the number of total ones in LSTP area is considered in applying the LSTP level caching scheme.

Acknowledgements

This work was supported by the Howon University Fund, 2005

References

1. A. Bar-Noy, I.Kessler and M. Sidi, Mobile Users: To Update or not to Update?, Proc. of IEEE INFOCOM'94, 1994.
2. A.D. Malyan, L.J. Ng, Victor C.M. Leung, and Robert W. Donaldson, Network Architecture and Signaling for Wireless Personal Communications, IEEE JSAC., Vol. 11, No. 6, 1997.
3. EIA/TIA, Cellular Radio telecommunications Intersystem Operations: Automatic Roaming, Technical Report IS-95 (Revision C), EIA/TIA, July 1998.

4. G.P. Pollini, Signaling Traffic Volume Generated by Mobile and Personal Communications, IEEE Comm. Mag., Vol.33 No. 6, pp. 60-65, June 1995.
5. G.P. Pollini and D.J.Goodman, Signaling System Performance Evaluation for Personal Communications, IEEE Trans. on Veh. Tech., 1994.
6. Ian F. Aky ld z, Janise McNair, Joseph Ho, and Wenye Wang, Mobility Management in Current and Future Communications Networks, IEEE Network Mag., Vol. 12, No. 4, 1998.
7. J.S.M. Ho and I.F.Akyildiz, Local Anchor scheme for Reducing Location Tracking Cost in PCNs, Proc. of ACM MOBICOM'95, 1995
8. J.Z. Wang, A Fully Distributed Location Registration Strategy for Universal Personal Communication Systems, IEEE JSAC., Vol. 11, No. 6, pp. 850-860, 1998.
9. R. Jain and Y.B Lin, A Caching Strategy to Reduce Network Impacts of PCS, IEEE JSAC., Vol. 12, No. 8, pp. 1434-1444, Oct. 1996.
10. R. Jain and Y.B.Lin, An Auxiliary User Location Strategy Employing Forwarding Pointers to Reduce Network Impacts of PCS, Proc. Of IEEE ICC'95, 1995
11. Russell. T, Signaling System #7, McGraw-Hill, 1995.
12. S. Mohan and R. Jain, Two User Location Strategies for Personal Communications Services, IEEE Personal Comm. Mag., Vol. 1, No.1, pp. 42-50, First Quarter, 1994.
13. Seung Joon Park, Dong Chun Lee, and Joo Seok Song, Querying User Location Using Call Locality Relation in Hierarchically Distributed Structure, Proc. of IEEE GLOBECOM'98, 1998.
14. Y.B. Lin, Determining the User Locations for Personal Communications Networks, IEEE Trans. on Veh. Tech., Vol. 43, pp. 466-473, 1996.

Correction of Building Height Effect Using LIDAR and GPS

Hong-Gyoo Sohn[1], Kong-Hyun Yun[1], Gi-Hong Kim[2], and Hyo Sun Park[3]

[1] School of Civil and Environmental Engineering, Yonsei University, 134 Shinchon-Dong,
Seodaemoon-Gu, Seoul, 120-749, Korea
{sohn1, ykh1207}@yonsei.ac.kr
[2] Department of Civil Engineering, Kangnung University,
Kangnung, 210-702, Korea
ghkim@kangnung.ac.kr
[3] Department of Architectural Engineering, Yonsei University, 134 Shinchon-Dong,
Seodaemoon-Gu, Seoul, 120-749, Korea
hspark@yonsei.ac.kr

Abstract. Correction of building height effects is a critical step in image interpretation from aerial imagery in urban area. In this paper, an efficient scheme to correct building height effects from aerial color imagery using multi source data sets is presented. The following steps have been performed to remove the shadow effect from aerial color imagery. First, the shadow regions of the orthorectified aerial color image are precisely located using the solar position and the heights of ground objects derived from LIDAR (Light Detection and Ranging) data. The shadow area is composed of many different ground features. To accurately recover the original spectral information the step for accurate segmentation of shadow regions needed. This step has been performed by utilizing the existing digital map which contains surface cover information. To assign correction information corresponding to the surface features on the shadowed area, correction factor needed to be modeled. For this three basic assumptions are proposed and comparison between the context region and the same non-shadowed context region is made. Finally, the shadow-effect-corrected image is generated by using the correction factor obtained from the previous step.

1 Introduction

In the analysis of urban areas images may be generated from optic sensors and radars. Especially black-and-white aerial images have especially been widely used to extract geometric and semantic information for spatial data generation and updates. In urban analysis, however, the quality of images can be degraded due to the presence of shadows and occlusions especially in built-up urban areas. The compensation of hidden area can be treated by using multi-view images [1]. However, due to the characteristic of cast shadow occurrence regardless of sensor position, total correction of shadowed area is not easy. The problem of shadow effects can be solved by correcting shadow effects using multi-source data sets such as LIDAR and an existing digital map. In other words, the correction of shadow effects is feasible by combing height information obtained from LIDAR data with high height accuracy and feature polygons obtained from an existing digital map which contains positional and attribute data of diverse ground features.

To deal with shadow effects in remotely sensed images, attempts have been made to correct gray values that are dimmed by cast shadows [2-4]. Most researches, however, are focused on mountainous terrains in satellite imagery. Rau and others [1] have observed that in urban area imagery, shadow effects could be corrected with the histogram-matching method, which is applied to minimize the gray-value differences between a shadowed area and its surroundings. The histogram-matching method basically manipulates the pixel values of one image to match these with the pixel distribution of another image. The pixel values of shadowed regions are calculated by their surrounding images in terms of histogram distribution, although the results are not visually satisfactory.

This paper aims to correct shadow effects from aerial color imagery using multi-source data sets. For accurate shadow modeling, LIDAR data were selected for the generation of DSM, since it has been reported that the elevation accuracy of LIDAR data is approximately ±15 cm [5]. Moreover, to accurately correct shadow effects, shadow regions were segmented using an existing digital map showing surface cover types. The shadowed regions were accurately modeled using the solar position obtained from two data sets. Based on our three basic assumptions, the identified shadowed regions were then given new color values. Finally, the evaluation of our scheme was made using the region-growing method by comparing the color imagery before and after the correction of the shadow effect.

2 Details of Study Area and Data Used

To test our proposed scheme, the study area and the data sets for this research were carefully selected. The data sets are composed of LIDAR data, a digital map with a scale of 1:1000, and an aerial color image showing some parts of Sungnam City (127.01°E, 37.36°N), Korea. The aerial color image was acquired on 10 December 2002, at around 11:19 a.m., which corresponds to early winter and casts long shadows on the ground surfaces.

As Fig. 1 shows, the image shows high buildings, trees, roads, streets, grass, and other elements. The aerial color photos were scanned at 50 μm, making the pixel resolution approximately 25 cm on the ground. To combine the image with other data sources, it was orthorectified using a digital map. The overall accuracy of the orthorectified image was approximately 0.75 m. The DEM was created through the triangulated irregular network (TIN) method using spot heights and elevation contours from 1:1000-scale digital maps, and was converted into a grid format at a 25 cm spacing. The overall accuracy of the digital maps with a 1:1000 scale was about 0.7 m horizontally and 0.5 m vertically. The ground height was approximately 30-55 meters and the height of the buildings was about 45-150 meters from the mean sea level.

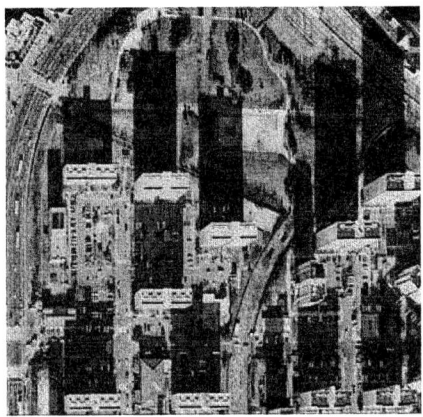

Fig. 1. Aerial color image of study area

2.1 The Proposed Scheme

Fig. 2 shows the schematic diagram of the proposed method. First, the aerial color image, the LIDAR data and the digital map should be co-registered. The LIDAR data (WGS84 coordinate) must be transformed into the Transverse Mercator coordinate being used in the national coordinate system in Korea, since the three data sets must have the same coordinate system. Due to the horizontal error of the LIDAR data, they

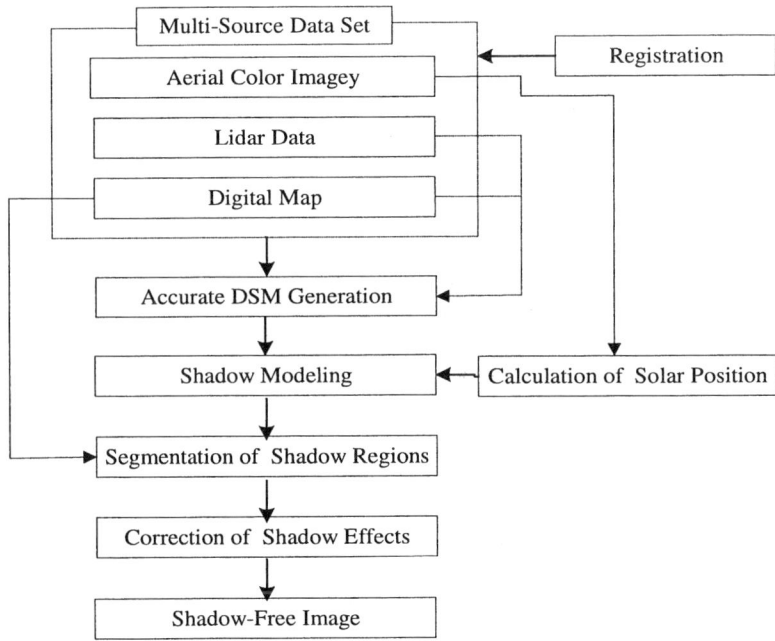

Fig. 2. Schematic diagram of the proposed method

should be precisely co-registered with the aerial image or the digital map. In this research, the peak points of the LIDAR data between the building boundaries and the ground were used for the registration. The difference between the coordinates of the peak points and the building boundaries of the image was calculated and used for the horizontal shift of all the LIDAR data.

A grid DEM was then produced using the elevation contours and spot heights from the digital map with a 1:1000 scale. Next, the building layer was extracted from the digital map—in particular, the building polygon, out of the map's many urban-area-related features. The building was thus represented as a single polygon in the digital map. As shown in Fig. 1, however, the upper part of the building was not flat but undulating. For accurate extraction of shadow regions, a polygon representing the rising part of the building was added in the building layer. Therefore, the building heights that were acquired as mean values of the LIDAR data within each polygon were entered. Subsequently, the shadowed regions were delineated by using the solar position, after which the DSM was produced. In the next stage, the shadowed regions were segmented for accurate shadow effect correction using the digital map. Finally, the shadow effects in the segmented shadowed regions were corrected.

3 Extraction of Shadow Regions

The building heights were determined directly using LIDAR data. The LIDAR data within a building polygon had two homogeneous values. One represented the height of the bulk part of the building, while the other represented the building height except for its bulk part. However, the building polygon was shown as only one polygon in the digital map. To solve this problem, a polygon representing the bulk part of the building was added. The boundary between the bulk part and the other part was then delineated. In other words, one polygon was created by grouping peak points above 1.0 m in comparison with the average of the LIDAR data within the polygon.

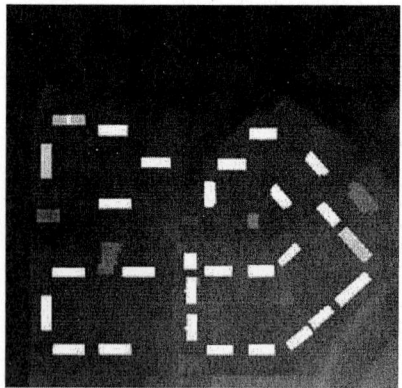 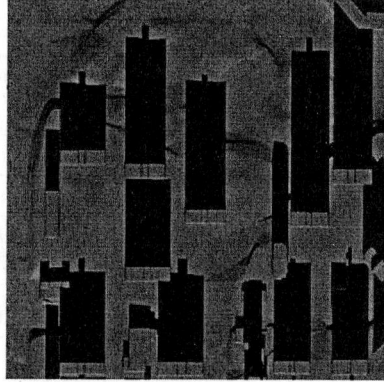

Fig. 3. Generated DSM and shadow modeling result

For the extraction of shadowed regions, in addition to building heights, the solar position at the time of image acquisition is required. To calculate the solar position (the solar elevation angle and the azimuth), the accurate time and date of the image acquisition and the latitude and longitude of the study area are needed. . The image was acquired on 10 November 2002, at around 11:19 a.m. The longitude of the study area was E 127° 6′ 58″, while the latitude was N 37° 22′ 01″. Fig. 3 shows the results of the DSM, which is composed of the DEM and the building height and final result of shadow modelling.

3.1 Assessment of Shadow Modeling Result

The proportion of the shadowed regions in an aerial image depends mainly on two factors: the height of the buildings on the terrain and the solar elevation angle. Higher buildings and a lower solar elevation angle result in larger shadowed regions. Therefore, the accuracy of shadow modeling can vary according to the proportion of the shadowed region.

To evaluate the accuracy of the shadow-extraction algorithm, the cast shadows in our test aerial color image were compared to the test output. For this purpose, a volunteer was recruited to independently interpret the shadowed regions in the aerial image based on the patterns of the reduced gray values. Since this volunteer was not only experienced in interpreting aerial images but was also given instructions for this exercise, his interpretation of the shadowed regions was assumed to be accurate. A comparison of the regions marked by the volunteer to those delineated by the proposed shadow modeling showed an overall accuracy of 97.62%. Table 1 illustrates the results of the shadow modeling. The numbers in Table 1 represent the pixels that were extracted, and the reference data refer to the results of the human interpretation.

Table 1. Error matrix resulting from shadow modeling (unit: pixel)

Category		Reference data		
		shadow	Non-shadow	Total
Modeling data	Shadow	571963	15144	587107
	Non-shadow	13925	·	13925
	total	585888	15144	·

4 Segmentation of Shadow Regions

Segmenting shadows is very important since the results of the segmentation have a direct effect on the correction of shadow effects. Shadowed regions have many different surface cover types, each of which has a spectral reflectance different from the others, and the degree of influence of a shadow also varies with the surface type. Polidorio and others [6] have suggested a very effective technique for segmenting shaded areas in aerial color images. This method is based on the physical phenomenon of the atmospheric dispersion of sunlight, most popularly known as the Rayleigh scattering effect. Because the results of our test showed that the shadowed regions were not clearly segmented, however, this method was not relevant. In this study, an existing digital map was utilized for segmentation.

5 Algorithms for Shadow Effect Correction

Stockham [7] has recognized that an image consists of two components: the amount of source light incident on the scene being viewed and the amount of image points that reflect light. These are called illumination and reflectance components. The image could be modeled as a product of two components, $\alpha_{i,j}$ $I_{i,j} = r_{i,j} \times L_{i,j}$. The parameter $L_{i,j}$ represents the illumination source, while $r_{i,j}$ is the reflectivity function of the image. The model of the image that is occluded by shadows could be expressed as $I'_{i,j} = \alpha_{i,j} \times r_{i,j} \times L_{i,j}$ To generate a shadow-free image, the attenuation factor of the shadow, , must be calculated and removed. For this purpose, the Eq., $I'_{i,j} = \alpha_{i,j} r_{i,j} L_{i,j}$, was modified as follows

$$\frac{I'}{\alpha_{i,j}} = r_{i,j} L_{i,j} \tag{1}$$

$$\frac{I'}{\alpha_{i,j}} = I' - I'(1 - \frac{1}{\alpha_{i,j}}) \tag{2}$$

$$I' - I'(1 - \frac{1}{\alpha_{i,j}}) = r_{i,j} L_{i,j} \tag{3}$$

$$I' - k = r_{i,j} L_{i,j} \tag{4}$$

$$I' - k = I \tag{5}$$

Eq. 5 means that the original pixel that is not occluded by a shadow could be produced by correcting a factor in the pixel that is occluded by a shadow. In this study, the correction factor was defined as k. Thus, if the correction factors for each segmented shadowed region could be calculated, then the correction of shadow effects would be possible. The correction factor could be calculated by obtaining the difference between the average of the shadowed region and the average of the region with the same surface type, which is the so-called *shadow-free region*.

Before applying Eq. 5 to the correction of shadow effects, three basic assumptions in treating shadows in aerial images were considered. These are:

- Radiometric information on the region occluded by a shadow is available.
- The influence of the cast shadow caused by each object is uniform.
- The DN value of a similar surface cover characteristic is uniform.

Eqs. 6 and 7 below were applied in this study:

$$O^t_{R,G,B}(x, y) = I^t_{R,G,B}(x, y) + k \qquad (6)$$

$$k = I^r_{R,G,B}(m) - I^t_{R,G,B}(m) \qquad (7)$$

where $O^t_{R,G,B}(x, y)$: the DN value of output image at each band.
$I^t_{R,G,B}(x, y)$: the DN value of original image at each band.
$I^r_{R,G,B}(m)$: the mean value of reference area at each band.
$I^t_{R,G,B}(m)$: the mean value of target area at each band.

In Eq. 7, the target area refers to the shadowed region that will be corrected, while the reference area represents the corresponding non-shadowed region.

5.1 Results of Shadow Effect Correction

Fig. 4 shows the corrected image using the proposed algorithm. The image is much better than the original image in visual attraction. The radiometric correction of the asphalt road on the image was particularly very successful. The regions with diverse and complex ground features showed relatively low correction effects. For example, a parking lot made of paved asphalt showed randomly distributed cars, but these cars were not represented as unique single polygons in the digital map. While the cars could be ignored from the viewpoint of the interpretation of urban facilities, these neglect in our paper resulted in the wrong correction of the shadow effect. Table 2 gives the average and standard deviation of main surface cover types before and after shadow effect correction.

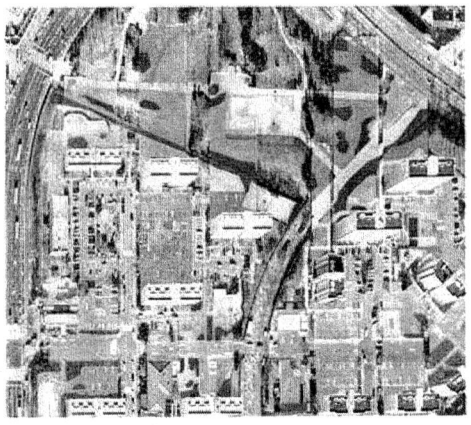

Fig. 4. Shadow effect corrected image

Table 2. Average & Standard Deviation of each feature

(Blue band)

Categoty	Shadow		Non-shadow	
	Average	Std.	Average	Std.
Asphatl road	118.55	20.27	196.50	19.65
Unpaved road	121.31	24.92	215.97	24.02
Soil	98.11	27.15	181.19	24.79
Building	172.86	29.11	242.30	11.81
Trees	72.34	19.60	122.00	22.42

(Green band)

Categoty	Shadow		Non-shadow	
	Average	Std.	Average	Std.
Asphatl road	97.14	19.18	214.75	15.24
Unpaved road	95.53	15.15	244.88	9.39
Soil	80.05	19.26	223.16	17.03
Building	152.20	23.66	248.41	7.49
Trees	50.84	16.57	140.61	23.71

(Red band)

Categoty	Shadow		Non-shadow	
	Average	Std.	Average	Std.
Asphatl road	67.94	16.54	199.14	13.97
Unpaved road	67.00	15.70	238.70	11.69
Soil	66.13	20.25	236.44	11.78
Building	121.08	21.73	244.50	10.71
Trees	35.89	5.57	128.20	19.05

6 Conclusions

This paper described the experiment that was conducted on the effectiveness of the proposed algorithm for correcting shadow effects to enhance the interpretability of aerial color images, by combining a digital map and LIDAR data in the test image. To delineate the shadowed regions in the image, shadow modeling was implemented using LIDAR data and the solar position. Compared to manual extraction, the use of the LIDAR data to delineate the shadowed regions in the test image resulted in the detection of 97.62% of the cast shadow pixels. This could be attributed to the high vertical accuracy of the LIDAR data. This paper also shows how to make a wise choice among possible available data for urban areas, and illustrates how to fuse multi-source data sets to achieve a given purpose using extracted information from individual data.

Acknowledgement

This material is based on work sponsored by Korea Research Foundation under grant KRF-2001-042-E00137, which is gratefully acknowledged.

References

1. Rau, J.Y., N.Y., Chen, and L.C. Chen. 2002. True orthophoto generation of built-up areas using multi-view images. *Photogrammetric Engineering and Remote Sensing*, 68(6). 581-588.
2. Crippen, R.E., R.G. Blom, and J.R. Heyada. 1988. Directed band ratioing for the retention of perceptually-independent topographic expression in chromaticity-enhanced imagery. *International Journal of Remote Sensing*, 9(4). 767-776.
3. Civco, D.L. 1989. Topographic normalization of Landsat thematic mapper digital imagery. *Photogrammetric Engineering and Remote Sensing*, 55(9). 1303-1309.
4. Pouch, G.W. and D.J. Campagna. 1990. Hyperspherical direction cosine transformation for separation of spectral and illumination information in digital scanner data. *Photogrammetric Engineering and Remote Sensing*, 56(4). 475-479.
5. Cowen, D.J., J. R. Jensen, C. Hendrix, M.E. Hodgson, and S.R. Schill. 2000. A GIS-Assisted rail construction econometric model that incorporates LIDAR data. *Photogrammetric Engineering and Remote Sensing*, 66(11). 1323-1328.
6. Polidorio, A.M, F. C. Flores, N. N. Imai, A .M. G. Tommaselli, and C. Fransco. 2003. Automatic shadow segmentation in aerial color images. *Proceeding of the XVI Brazilian symposium on computer graphics and image processing*, 270-277.
7. Stockham, J. T.G. 1972. Image processing in the context of a visual model. Proceeding of the IEEE, 60, 828-842.

Minimum Interference Path Selection Based on Maximum Degree of Sharing

Hyuncheol Kim[1,*] and Seongjin Ahn[2,**]

[1] Dept. of Electrical and Computer Engineering, Sungkyunkwan University,
300 Chunchun-Dong, Jangan-Gu, Suwon, Korea, 440-746
hckim@songgang.skku.ac.kr

[2] Dept. of Computer Education, Sungkyunkwan University,
53 Myungryun-Dong, Jongro-Gu, Seoul, Korea, 110-745
sjahn@comedu.skku.ac.kr

Abstract. The explosive growth of Internet traffic has led to a dramatic increase in demand for data transmission capacity, which requires high transmission rates beyond the conventional transmission capability. This demand has spurred tremendous research activities in new high-speed transmission and switching technologies. The present paper focuses on the point of minimizing recovery blocking probability and contention probability to ensure fast and seamless recovery. For this purpose, first the recovery characteristics and grooming architectures of broadband convergence networks is analyzed. Based on the analyzed information, this paper proposes a new dynamic path selection scheme. The simulation results verify that the proposed revised SRLG schemes achieve good network performance and provides good network connectivity as well.

1 Introduction

To cope with the continuous growth of Internet Protocol (IP) traffic, backbone connections must have very high bandwidth capability, including optical fibers with Wavelength Division Multiplexing (WDM). The Internet transport infrastructure is moving toward a model of high-speed routers interconnected by intelligent optical core networks [1][2][3].

WDM divides the bandwidth of a fiber into many non-overlapping wavelengths and provides enormous bandwidth on the physical layer. With recent advancements in WDM technology, the amount of raw bandwidth available in fiber links has increased in orders of magnitude. Using WDM, the bandwidth of a fiber link can be divided into tens (or hundreds) of non-overlapping wavelength channels, each of which can operate at a peak electronic processing rate, i.e., over a few gigabits per second (Gbps).

However, in order to minimize the failure interference, recent efforts in the research have focused on sharing links between backup lightpaths or between

---

* This work was supported by grant No. R01-2004-000-10618-0(2004) from the Basic Research Program of the Korea Science & Engineering Foundation.
** Corresponding Author.

working lightpaths (also called active lightpath or primary lightpath) and backup lightpaths (also called secondary lightpath), called the Shared Risk Link Group (SRLG). Several algorithms have been presented to find the maximum disjoint paths between source and destination nodes [6][7]. However, no efficient algorithm has addressed finding maximum disjoint lightpaths by taking into account the SRLG at the optical layer recovery.

This paper proposes a new hybrid recovery algorithm in optical broadband convergence networks that uses both protection and restoration schemes. The proposed schemes support protection lightpaths that can transform dynamically. It also provides efficient recovery functions for different types of service by supporting dynamic restoration lightpaths.

The core purpose of the proposed schemes is to use the SRLG concept with the extended method to provide as many working and backup lightpaths as possible. Moreover, the proposed schemes do not need an extension of optical network signaling (routing) protocols for support.

The structure of the paper is as follows. Chapter 2 clarifies some key terminologies with vague and complex meanings, and also examines some previous studies on optical network survivability and traffic grooming. Chapter 3 concentrates on the network model and node architecture of the proposed schemes and presents some features that provide useful information on the considered network in terms of performance improvement. Chapter 4 models and analyzes the proposed survivable schemes. Chapter 5 concludes this paper by summarizing some key points made throughout and assessing the representation of analyzed results. Chapter 5 also includes concluding remarks and future research topics.

2 Related Works

With optical fiber bandwidth and node capacity increasing at an explosive rate, breaks in fiber span or node failures can cause serious damage to customers and service providers. Therefore, it is imperative for network service providers to design a survivable network in order to minimize communication loss, and this need has led to research efforts in recovery blocking probability and resource utilization [3][4][7][8][9].

In optical broadband networks, a link between two nodes is composed of several optical fibers, each of which carries many wavelengths (or channels). The channels are divided into the Control Channel Group (CCG) for the transmission of control messages and the Data Channel Group (DCG) for the transmission of data. Thus, several types of optical network faults can exist depending on the location of the fault: data channel failure, control channel failure, fiber failure, link failure, span failure, and node failure. Fig. 1 shows the different types of failure in optical networks.

To minimize the failure interference and to capture diversity, recent efforts in the research have focused on sharing links between backup paths or between working paths and backup paths, called the SRLG. In more recent practice, where spare capacity is explicitly designed into the network for survivability,

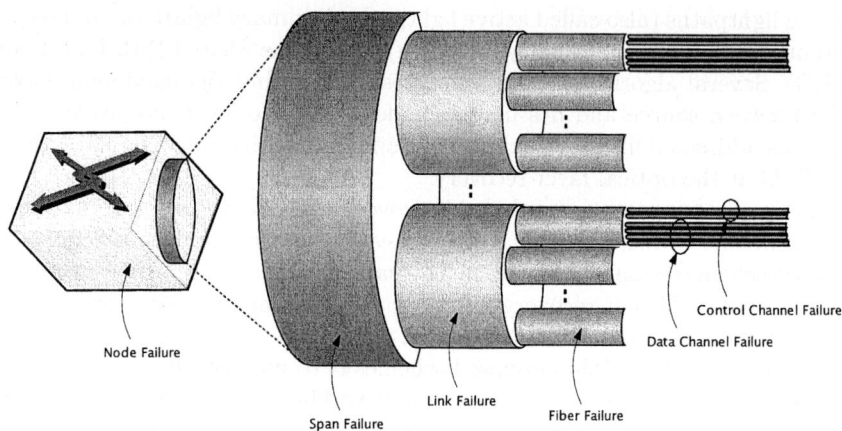

Fig. 1. Types of network failure

it is important to take the existence of any "shared risk entities" into account in the capacity design. A Shared Risk Group (SRG) or SRLG defines a set of transport channels or paths that have something in common: they all fail together if a certain single item of physical equipment fails. Fig. 2 illustrates a genuine SRLG with two nominally disjoint fiber cables that share the same duct or other common physical structures such as a bridge crossing.

At present, however, SRLG information cannot be self-discovered. Instead, the information must be manually configured for each link. Furthermore, in a large network, it is very difficult to maintain accurate SRLG information. The problem becomes particularly overwhelming when multiple administrative domains are involved, for instance, after the acquisition of one network operator by another [1].

In [6], the authors proposes a disjoint path selection scheme for Generalized Multi-protocol Label Switching (GMPLS) networks with SRLG constraints. It

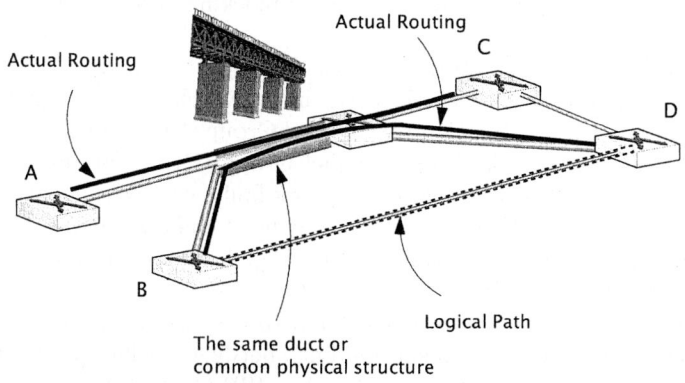

Fig. 2. Concept of Shared Risk Link Group (SRLG)

is called the weighted-SRLG (WSRLG) scheme. It treats the number of SRLG members related to a link as part of the link cost when the shortest path algorithm is executed. In WSRLG, a link that has many SRLG members is rarely selected as the shortest path.

The authors in [10] propose a mechanism for providing inter-region protection, which is used for highly reliable carrier services. The key concept of [10] lies in a multi-layer disjoint routing algorithm, called MLD, which finds multi-layer protection paths [10].

However, the SRLG-based survivability frameworks mentioned above only address the relationship between one originator (source) and one end (destination), and do not consider the multiple originators that can exist in the network [6][7]. Moreover, they assume that network level link-state information is accurate when they calculate the path. In practice, the network state information is not always accurate because it is not updated in time when the failure occurs [5][11].

3 Minimum Interference Path Selection Scheme

Failure recovery may be done at the electronic layers in order to recover from line card or electronic switch failures. Electronic recovery mechanisms, e.g., like those used in SONET, can also be used to protect against failures like fiber cuts or a malfunctioning optical switch, which occur at the optical layer. However, in many cases providing recovery at the optical layer can be more robust and efficient. For example, optical layer recovery can protect electronic services that do not have built-in recovery mechanisms or whose recovery mechanisms are slow (e.g., IP).

In a path-restorable network or path-protected network, the reaction to a failure takes place from an end-to-end viewpoint for each service path affected by the failure. For this reason, path-oriented survivability schemes have several advantages.

One such advantage is that they are more liable to customer-level control and visibility. Span restoration is rather inherently a function of the transport network itself, whereas path-level protection or path reestablishment can be a function that is put in the user's control or under service layer node control such as the router. In Shared Backup Path Protection (SBPP), it is also possible for users to recognize in advance and even control exactly where their service will be rerouted during the failure. This is sometimes said to be more important to customers.

Path-oriented schemes also provide an inherent form of response against the node failure that is not as readily available with span restoration. Path-oriented schemes only require fault detection at the path end-nodes, and this can be attractive for transparent or translucent optical networks where signal monitoring may not be available at intermediate nodes.

3.1 Algorithm

When the source nodes are different, which is the most typical post-failure scenario for path restoration, coordination among the source nodes (so they do not

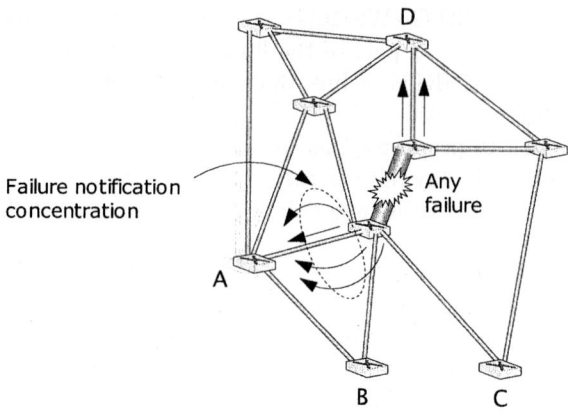

Fig. 3. Failure notification concentration

compete for capacity on the bottleneck links) is difficult, if not impossible. The most obvious solution is to alleviate the restoration contention probability at the connection setup stage and allow the source nodes to retry when the initial restoration effort fails.

To overcome these restrictions, as shown in Fig. 3, working and backup lightpaths in the extended SRLG scheme are selected so as many simultaneous setup requests as possible come near to a single node. In the extended SRLG, an originator takes into account the number of ultimate source-destination pairs of the pertaining link. An originator can increase its routing computation accuracy by incorporating information on the pending lightpaths.

Fig. 4 shows the concept of proposed scheme. Along with fast path restoration, the object of proposed scheme is to use the SRLG with the expanded method to satisfy as many demands as possible for a fixed amount of network resources.

The proposed scheme introduce a new parameter, f^j: the degree of sharing, which refers to the number of lightpaths that can be allocated between one ultimate source-destination node pair on a link. More specifically, $f^j = 1$ means that one lightpath that can be allocated between one ultimate source-destination node pair, and can be assigned on link j. Let f^{max} be the maximum degree of sharing on link j.

Fig. 5, 6, 7, 8, and 9, illustrate the operational procedures of the proposed scheme. Let us suppose that every edge in the figures corresponds to a bidirectional fiber and every node has grooming ports.

Consider the example of the network topology shown in Fig. 5. In the proposed scheme, a connection can traverse multiple spans before it reaches the destination. Thus, a connection can be groomed with different connections on different lightpaths.

Each ultimate source-end pair follows SRLG constraints until the entire fully disjointed path is used as shown in Fig. 6 and 7. As shown in Fig. 7, however, the diversity of connections can cause failure notification flooding at a failure.

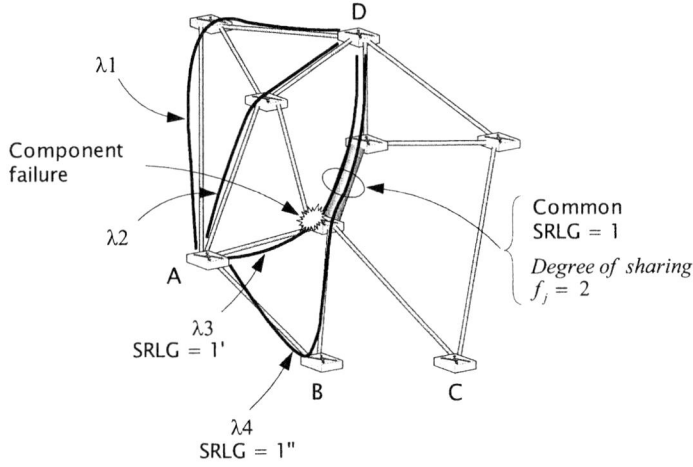

Fig. 4. The concept of extended SRLG

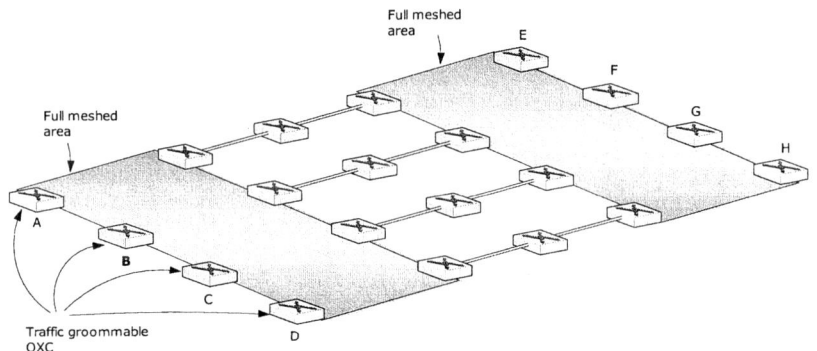

Fig. 5. Initial network configuration

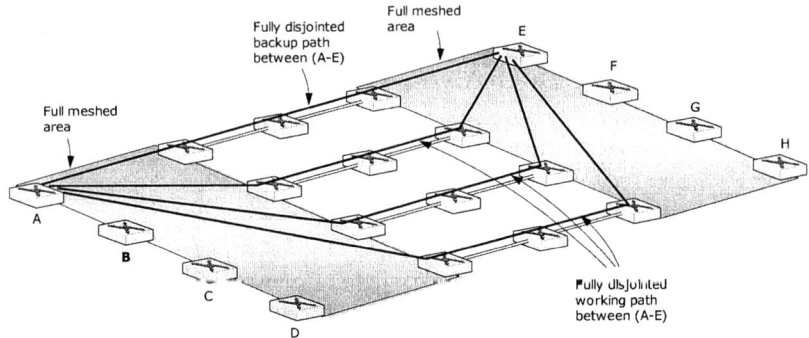

Fig. 6. Provisioning disjoint connections between nodes A-E

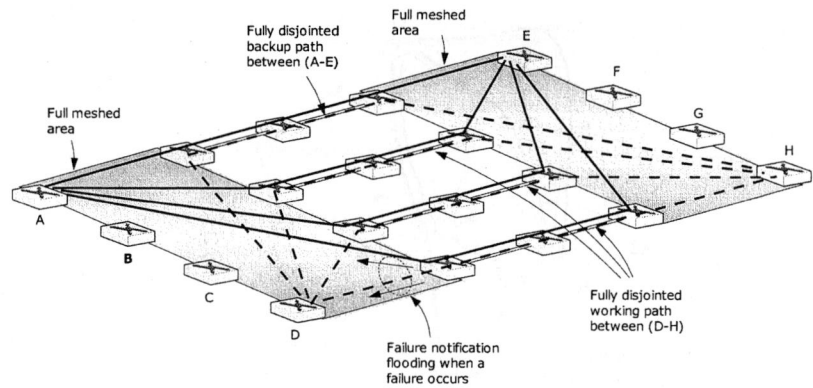

Fig. 7. Provisioning disjoint connections between nodes D-H

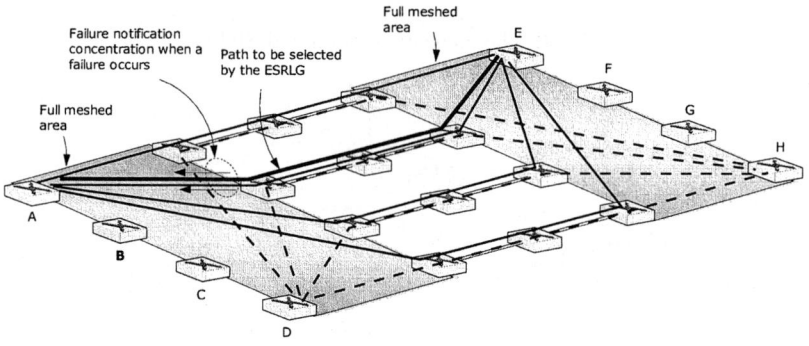

Fig. 8. Provisioning connection between nodes A-E by the proposed scheme

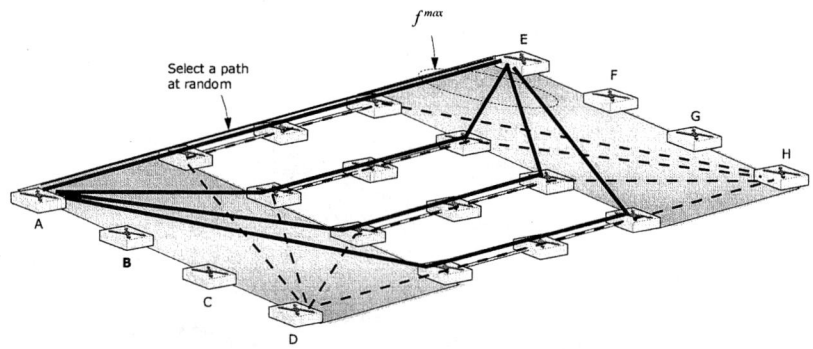

Fig. 9. Provisioning connection between nodes A-E after the proposed scheme

In Fig. 8, in order to alleviate the recovery contention probability of the case in which there is no more fully disjoint paths that satisfy the setup requirements, the proposed scheme selects a path that is most often used by the originator if it satisfies the QoS and other constraints.

4 Analysis

This section mainly focuses on the contention of path restoration request issues with inaccurate network information that affect the restoration performance.

The results based on a 23-node topology with 20 wavelengths per link indicate that the proposed algorithms reduce the overall restoration contention probability at each node by 75% compared to existing schemes based on k-shortest path routing and random routing.

As a concomitant phenomenon, the number of primary paths undergoing conversion is also significantly reduced while the overhead introduced is negligible. The overall path restoration blocking probability was reduced by an order of magnitude, with higher improvements seen for heavy loads. In this paper, distributed optical network architecture is considered.

The simulations use a dynamic traffic model and lightpath restoration (including new lightpath or recovery restoration) requests are uniformly distributed across all ingress-egress node pairs.

The lightpath restoration requests arrive at the network in an independent Poisson process and the connection holding time is exponentially distributed. The Inter Arrival Time (IAT) is Poisson-based, while the connection Holding Time (HT) has been modeled with exponential statistics.

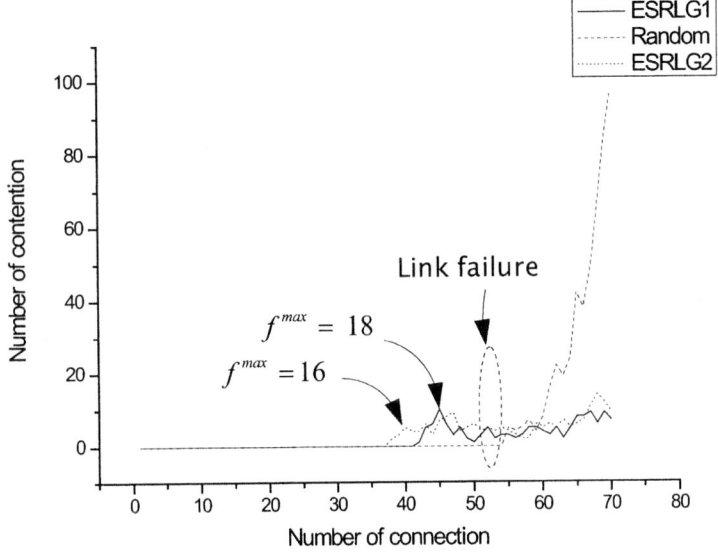

Fig. 10. Simulation result: Number of Contention

Three different models have been simulated and compared: Bandwidth First (BF) routing, randomized QoS routing, and the proposed routing scheme. In BF routing, the path from the source to the destination is determined by the residual bandwidth. In randomized QoS routing, randomized QoS routing algorithms bring in randomness and select paths with probability [5].

The simulation results, as shown in Fig. 10, show that when a fault occurs at connections which request different bandwidths, instead of using the method of requesting connection with just the BF or that of allocating channel randomly, it would be much better to use the proposed scheme since it can reduce connection setup contention and recovery channel setup time from an entire network standpoint.

The results also show, by means of path restoration blocking probability and setup delay supported by high-end and core routers, how the proposed strategy improves network performance compared with traditional BF or randomized optical routing.

5 Conclusion

The present paper focussed on the point of minimizing recovery blocking probability and contention probability to ensure fast and seamless recovery. For this purpose, first the recovery characteristics and grooming architectures of optical broadband networks was analyzed. Based on the analyzed information, this paper proposed two new dynamic traffic grooming schemes.

In this paper, we employed the OPNET simulation platform with the mathematical analysis to implement revised SRLG schemes in WDM-based optical broadband network. Then, we compared the performance of the schemes. The simulation results verify that the proposed revised SRLG schemes achieve good network performance and provides good network connectivity as well.

References

1. Greg Bernstein, Bala Rajagopalan, Debanjan Saha, "Optical Network Control: Architecture, Protocols, and Standards," Addison Wesley, 2003.
2. Raja Rajagopalan, Dimitrios Pendarakis, Debanjan Saha and Ramu S. Ramamoorthy, "IP over Optical Networks: Architectureal Aspects," *IEEE Communications Magazine*, Vol. 38, Issue 9, Sep. 2000, pp. 94-102.
3. Gisli Hjalmtysson, Jennifer Yates and Sid Chaudhuri, "Restoration Services for the Optical Internet," *Proceeding of SPIE, Terabit Optical Networking: Architecture, Control, and Management Issues*, Vol. 4213, Oct. 2000
4. Wayne D. Grover, "Mesh-Based Survivable Networks - Options and Strategies for Optical, MPLS, SONET, and ATM Networking," Prentice Hall, 2004.
5. Wang Jianxin, Wang Weiping, Chen Jianer, Chen Songqiao, "A Randomized Qos Routing Algorithm On Networks with Inaccurate Link-state Information," *International Conference on Communication Technology, WCC-ICCT 2000* vol. 2, Aug. 2000, pp. 1617-1622.

6. Eiji Oki, Nobuaki Matsuura, Kohei Shiomoto, Naoaki Yamanaka, "A Disjoint Path Selection Schemes With Shared Risk Link Groups in GMPLS Networks," *IEEE Communications Letters*, Vol. 6, No. 9, Sep. 2002, pp. 406-408
7. S. Ramamurthy, Biswanath Mukherjee, "Survivable WDM Mesh Networks, Part II-Restoration," *ICC '99*, Vol. 3. Jun. 1999, pp. 2023-2030.
8. Yinghua Ye, Sudhir Dixit, Mohamed Ali, "On Joint Protection/Restoration in IP-Centric DWDM-Based Optical Transport Networks," *IEEE Communications Magazine*, Vol. 38, Issue 6, Jun. 2000, pp. 174-183.
9. Jian Wang, Laxman Sahasrabuddhe, Biswanath Mukherjee, "Path vs. Subpath vs. Link Restoration for Fault Management in IP over WDM Networks: Performance Comparisions Using GMPLS Control Signaling," *IEEE Communications Magazine*, Vol. 40, Issue 11, Nov. 2002, pp. 80-87.
10. Takashi Miyamura, Takashi Kurimoto, Michihiro Aoki, Shigeo Urushidani, "A Multi-layer Disjoint Path Selection Algorithm for Highly Reliable Carrier Services," *IEEE GLOBECOM'04*, Vol. 3, Nov. 2004, pp. 1974-1978.
11. Koushik Kar, Murali Kodialam, T.V. Lakshman, "Minimum Interference Routing of Bandwidth Guaranteed Tunnels with MPLS Traffic Engineering Applications," *IEEE J. Sel. Areas in Communications*, Vol. 18, No. 12, DEC. 2000, pp. 2566-2579.

The Effect of the QoS Satisfaction on the Handoff for Real–Time Traffic in Cellular Network

Dong Hoi Kim[1] and Kyungkoo Jun[2]

[1] Mobile Telecommunication Research Laboratory,
Electronics and Telecommunications Research Institute, Korea
donghk@etri.re.kr
[2] Department of Multimedia System Engineering,
University of Incheon, Korea
kjun@incheon.ac.kr

Abstract. In this paper, we study the relationship between the QoS satisfaction and the handoff performance in cellular networks. The handoff drop rate is observed when changing allocated effective bandwidth and required packet loss rate, respectively. By proposing to use the token bucket model to calculate the effective bandwidth, a table of effective bandwidth and corresponding packet loss rate of VoIP and video streaming traffic is constructed for use in the simulation. Results show that the increase in the factor benefiting the QoS leads to the degradation of the handoff performance, i.e., the decrease in the packet loss rate increases the handoff drop rate.

1 Introduction

QoS-guaranteed service characteristic of next generation cellular networks brings challenges and complexities which have not been experienced in best–effort based networks in the past. It has been a well known dilemma that the bandwidth allocations in favor of QoS guarantee sacrifices the network utilization, and vice versa.

In addition, in next generation cellular systems, it is anticipated that the demand for real–time service will grow very rapidly. Thus, the search for the effective bandwidth satisfying QoS requirement, e.g., packet loss rate, for real-time traffic is required.

The problem of satisfying QoS requirements while maximizing resource utilization has been addressed in quite a lot of works. Among them, in [6], they introduce the packet level quality constraints expressed as rate or delay to tackle the problem of the QoS provision. However, how to search the proper level of rate or delay satisfying the required QoS constraints are not discussed. In [7], adaptive bandwidth allocation scheme is proposed under the assumption that users have a list of candidate QoS constraints, which vary in terms of required amount of resource. However, such bandwidth-adaptable services are not easy to be adopted in full scale. In [8], starting with the assumption of the existence of QoS-adaptable services, it introduces the QoS degradation ratio and degradation frequency as another category of QoS constraints.

In this paper, we study the change of the handoff drop rate against varying effective bandwidth and corresponding packet loss rate of VoIP and video streaming which are used as simulated traffic in our experiments. To compute the effective bandwidth and the packet loss rate, we propose another way different from established methods, which applied statistical analysis on the traffic traces captured in read-time in order to draw the effective bandwidth satisfying certain level of QoS condition.

This paper is organized as follows. Section 2 introduces the calculation of the effective bandwidth by the use of the token bucket model. Section 3 presents the details of the simulation environment and also discusses the simulation results with analysis. Section 4 concludes this paper.

2 Approximated Effective Bandwidth

In this section, we propose a procedure of calculating *effective bandwidth* [5]of simulated traffic by the use of the token bucket model, and present calculated effective bandwidth of VoIP and real-time video streaming, which are intended for use in the next generation cellular network simulation of the following section. We use the term *approximated effective bandwidth* for the outcomes of our suggested process to differentiate them from the effective bandwidth of original literature.

Our proposed scheme for the effective bandwidth is different from existing methods in that, by the use of token bucket model, we do not calculate the effective bandwidth from traffic traces, but set a value as the effective bandwidth and then compute an associated QoS parameter, e.g., packet loss rate.

The token bucket model was developed with the intention of shaping, policing and marking of broadband traffic. The fundamental idea of the token bucket is the use of a buffer (token buffer) which is filled at the rate of λ and can hold maximum B bytes at any moment. Whenever traffic packets are processed, i.e., either transmitted or received, tokens are consumed; if no token is available in the buffer during the packet processing, the corresponding packet is considered as *out-of-profile* and treated according to predefined rules, e.g., drop when policing traffic, delay when shaping traffic, or mark and drop it in time of congestion when marking traffic. Thus, the token buffer can control the amount of traffic passing over any time interval of length t less than $B + \lambda t$.

When utilizing the token bucket model for the calculation of the approximated effective bandwidth, λ can be thought as the effective bandwidth. Therefore, for a single traffic, we are able to compute a set of effective bandwidths each of which has a corresponding QoS constraint (packet loss rate). These effective bandwidths can be selectively used in the simulation according to required QoS level, i.e., the maximum allowable packet loss rate.

The characteristics of VoIP and video streaming traffic used in this paper are as follows.

VoIP traffic. VoIP traffic shows a typical on–off pattern stemming from the behavior of a voice source; it is *on* if a talker is speaking, otherwise remains at *off* state [3][4]. The distribution of the lengths of on and off periods is assumed exponential with mean values equal respectively to 350 ms and 650 ms. In addition, during on period, the voice source is presumed to generate packet at regular intervals, corresponding to a bit rate of 32 Kbps.

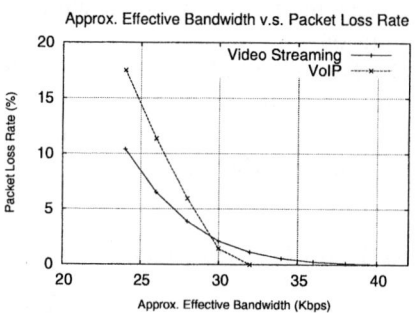

Fig. 1. Approximated Effective Bandwidth of VoIP and video streaming traffic

Video Streaming. Video streaming consists of a sequence of frames which are emitted at regular interval T. Every frame of the stream is identically composed of a same number of slices, and each slice corresponds to a single packet. The number of slices in every frame is same though, the sizes of the slices even belonging to a same frame are different. Moreover, the distribution of the slice sizes (packet sizes) as well as that of the inter–slice time in a frame follows a truncated Pareto. In this paper, we assume a video streaming with a rate of 32 Kbps with frame interval 100 ms ($T = 100$), eight slices per frame.

With the above described VoIP and video streaming traffic, we generate a set of the approximated effective bandwidths along with associated packet loss rates as shown in Figure 1. For both cases of VoIP and video streaming, as the effective bandwidth assigned to traffic increases, the corresponding packet loss rate obviously decreases.

3 Simulation and Performance Analysis

3.1 Simulator

Through simulation, we try to analyze the effect of assigned effective bandwidth on QoS satisfaction of users as well as on the handoff drop rate of the network. For this purpose, we develop a simulator, which consists largely of three parts: simulation map, Mobile Host(MH) and its mobility model, and Base Station (BS) algorithms. Among the BS algorithms, we mainly focus on the Call Admission Control (CAC) algorithm in this paper and will present the details of the algorithm shortly.

Simulation Map. We consider a cellular network of 19 cells each of which cell radius is set to be 1 Km. By using wrap–around technique, we are able to avoid border effect, and at the same time, simulate the performance of a large cellular network.

Mobile Host Modeling. The MH creation interval in each cell follows Poisson process with rate λ. Created MHs have life–time during which the MHs are able to exist in the simulation. The life–time of each MH is uniformly distributed within a range of $[0, t_{life}]$. To simulate the movement of MHs, we configure the simulator to have an option to select one of the random models: Random Direction Model (RDM) or Random Waypoint Model (RWM) [2].

Call Admission Control of the Base Station. Among a set of algorithms running on base stations, in this section, we discuss Call Admission Control (CAC) algorithm used in our simulation. The purpose of the CAC is that, upon arrival of new call or handoff requests, base stations determine whether to accept the requests under the condition that the acceptance does not harm the QoS level of existing calls while maximizing resource utilization.

Since the purpose of our simulation is not in the suggestion of new CAC algorithm, but in the analysis of the effect of the effective bandwidth on the handoff drop rate of the networks, we use a simplified version of the CAC algorithm proposed in [1] in which each MH is classified either into class 1 or class 2 based on the priority of their traffic.

3.2 Performance Analysis

In the following, we present the experiment results produced by running the simulator described in previous section. In the simulation, MHs are assumed to generate VoIP and video streaming traffic, and the handoff drop rate is measured at every 19 cells in the simulation map. We perform the simulation respectively for VoIP and video streaming traffic, and for each case, several experiments each with different effective bandwidth are executed to measure the handoff drop rate. The average rate λ of Poisson process of MH creation is set to 0.10 and every experiment is executed twice respectively with RDM and RWM.

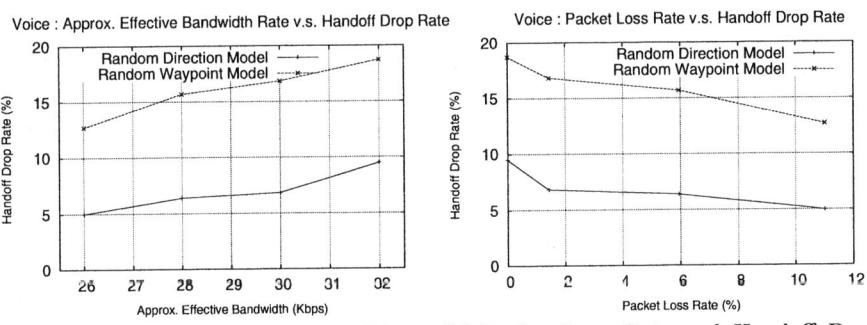

(a) Effective Bandwidth and Handoff Drop Rate
(b) Packet Loss Rate and Handoff Drop Rate

Fig. 2. Handoff drop rate of VoIP

Figure 2(a) shows the handoff drop rate in the case of VoIP as the required effective bandwidth changes. Figure 2(b) shows the handoff drop rate as the packet loss rate, which represents the QoS satisfaction of MHs, changes. The handoff drop rate in the figure is the average of the measurements of 19 cells in the simulation map. The measurement results of the cases of both RDM and RWM are presented together.

In Figure 2(a), it is observed that, as the effective bandwidth allocated to VoIP traffic increases, the handoff drop rate also increases, in other words, the more effective bandwidth is allocated, the more often the handoff fails. It is because the increase in required bandwidth for VoIP causes MHs to request increased bandwidth from the cells to which they are handed off. The linearly increasing relationship between the effective bandwidth and handoff drop rate is observed in both cases of mobility models. Note that the handoff drop rate of RWM case is two times higher than that of RDM. It is discussed shortly.

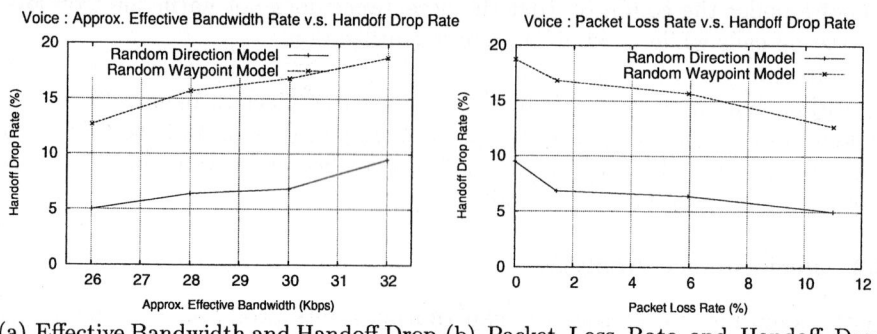

(a) Effective Bandwidth and Handoff Drop Rate   (b) Packet Loss Rate and Handoff Drop Rate

Fig. 3. Handoff drop rate of Video Streaming

Figure 2(b) shows the effect of the packet loss rate on the handoff drop rate in the case of VoIP traffic. As explained in Figure 1, the packet loss rate is another representation of the effective bandwidth. Therefore, it is obvious that, as the packet loss rate increases, the handoff drop rate decreases. Note also that that handoff drop rate of RWM case is two times higher than that of RDM.

Figure 3 shows the handoff drop rate in the case of video streaming. The results are similar to those of Figure 2; the handoff drop rate increases in proportion to the effective bandwidth and decreases as the packet loss rate increases.

4 Conclusions and Future Work

In this paper, we studied the effect of the user-level QoS satisfaction on the handoff of next generation cellular networks through the simulation, which experimented the change of handoff drop rate against varying assigned effective bandwidths and corresponding packet loss rates. We used the token bucket model to compute the effective bandwidth of simulated VoIP and video streaming.

Throughout the simulation results, it was observed that the increase in the factors benefiting the QoS satisfaction leads to the degradation of the overall performance of the cellular network; the decrease in the packet loss rate (i.e., the increase of the effective bandwidth) brings the degradation of the handoff drop rate. It is because the raise of the effective bandwidth reduces the number of users that a single cell can host.

References

1. Ye, J., Hou, J., Papavassiliou, S. : A Comprehensive Resource Management Framework for Next Generation Wireless Networks : IEEE Transactions on Mobile Computing, vol. 1, No. 4, 2002.
2. Camp, T., Boleng, J., Davies, V. : A Survey of Mobility Models for Ad Hoc Network Research : Wireless Communication & Mobile Computing, vol. 2, no. 5, 2002.
3. Bruno, R., Garroppo, R., Giordano, S. : Estimation of Token Bucket Parameters of VoIP Traffic : Proceedings of IEEE ATM workshop, 2000.
4. Jiang, W., Schulzrinne, H. : Analysis of On-Off Patterns in VoIP and Their Effect on Voice Traffic Aggregation : Proceedings Ninth International Conference on Computer Communications and Networks, 2000.
5. Kelly, F. : Notes on effective bandwidths : Stochastic Networks: Theory and Applications, Oxford University Press, 1996.
6. Solana, A., Bardaji, A., Palacio, F. : Capacity Analysis and Performance Evaluation of Call Admission Control for Multimedia Packet Transmission in UMTS WCDMA System : Proceedings of 5th IEE European Personal Mobile Communications Conference (EPMCC'03), 2003.
7. Chou, C., Shin, K. : Analysis of Adaptive Bandwidth Allocation in Wireless Networks with Multilevel Degradable Quality of Service : IEEE Transactions on Mobile Computing, vol. 3, no. 1, 2004.
8. Xiao, Y., Chen, C. : QoS for Adaptive Multimedia in Wireless/Mobile Networks : Proceedings of 9th International Symposium in Modeling, Analysis, and Simulation of Computer and Telecommunication Systems, 2001.

Throughout the simulation results, it was observed that the increase in the factors tending the QoS satisfaction leads to the degradation of the overall performance of the cellular networks: the decrease in the packet loss rate (i.e. the increase of the effective bandwidth) brings the degradation of the handoff drop rate. It is to reduce the rate of this effect in hand with reduces the number of users that a single cell can host.

References

1. Naghshineh, M., Schwartz, M.: Distributed Call Admission Control in Mobile/Wireless Networks. IEEE Transactions on Mobile Computing, vol. 14, no. 4 (1996)

2. Zhang, T., et al.: Local Predictive Resource Reservation for Handoff in Multimedia Wireless IP Networks. IEEE Journal on Selected Areas in Communications, vol. 19, no. 10 (2001)

3. Chuoh, Y., Gunupudi, R., Gokhale, S.: Estimation of Loss Rates Parameters of VoIP Traffic. Proceedings of IEEE ATM workshop 2000

4. Chuoh, Y., et al.: Voice Traffic Aggregation. Proceedings Ninth International Conference on Computer Communications and Networks, 2000

5. Beigi, B., et al.: An effective bandwidth-based admission theory and simulation, 1998.

6. Adam, J., Gaudino, E., et al.: Capacity Analysis of WCDMA Systems by means of Simulation Method. Tutorials on CDMA WCDMA Mobile Technologies for Third Generation Mobile communications (IEWCCDMA), 2003.

7. Chen, L., Shu, et al.: Analysis of Adaptive Bandwidth Allocation in Wireless Networks with Multimedia Support. Journal of Computing and Information Computing, vol. 3, no. 3, 2003

8. Xiao, Y., Chen, C.L.: QoS for Adaptive Multimedia in Wireless/Mobile Networks. The Handbook of Information in Wireless Networks: Algorithms and Protocols, Chapman and Hall International Series, 2007

Author Index

Abdouli, Majed 888
Acacio, Manuel E. 213
Ahmadi, Hossein 156
Ahn, Kyeongrim 660
Ahn, Sanghyun 255
Ahn, Seongjin 1096
Ahn, Youngjin 277
Alexandrov, Vassil N. 745
Alimi, Adel 888
Amanton, Laurent 888
An, Hui-Yao 321
Antoniu, Gabriel 429
Antonopoulos, Christos D. 223
Aung, Khin Mi Mi 567
Awan, Irfan 67
Ayguadé, Eduard 366
Azim, Mostafa Mohamed A. 196

Baba, Yoshimasa 723
Bagheri, Ebrahim 1043
Bahig, Hazem M. 271
Bang, Young-Cheol 4, 117, 277
Belardini, Paola 958
Bella, Gino 969
Bellavista, Paolo 1021
Beltran, Vicenç 366
Berthold, Michael R. 866
Bertoli, Claudio 958
Bisignano, Mario 1000
Boman, Erik G. 796
Bononi, Luciano 640
Bozdağ, Doruk 796
Bracuto, Michele 640
Buttari, Alfredo 969

Canali, Claudia 1070
Carrera, David 366
Casolari, Sara 1070
Catalyurek, Umit 796
Chan, Albert 856
Chang, Jae-Woo 245, 898
Chen, Cheng-Kai 910
Chen, Daoxu 455
Chen, Haitao 265
Choi, Jun Kyun 24

Choo, Hyunseung 117, 277
Chung, Jinwook 660
Chung, Kwang Sik 123
Chung, SungTaek 4
Chung, Tai-Myung 991
Chung, Yon Dohn 190
Chu, Xiaowen 388
Cilardo, Alessandro 1064
Colajanni, Michele 1033
Collins, Eli D. 573
Comito, Carmela 672
Coppolino, Luigi 1064
Corradi, Antonio 1021
Corsaro, Stefania 958
Cotroneo, Domenico 180, 1011
Cranefield, Stephen 733
Cuenca, Pedro 605

D'Ambra, Pasqua 958
D'Angelo, Gabriele 640
Daoud, Sameh S. 271
de Amorim, Claudio L. 629
Deiterding, Ralf 916
Del Citto, Francesco 969
De Maio, Alessandro 969
de Mello, Rodrigo Fernandes 487
Dickens, Phillip M. 755
Di Fatta, Giuseppe 866
Di Lorenzo, Giusy 395
Di Modica, Giuseppe 1000
Donatiello, Lorenzo 640
D'Onofrio, Salvatore 401
Duan, Rubing 704
Du, Z. 776
Dziok, Dominik 666

Fahringer, Thomas 704
Fardis, Masoum 87
Filippone, Salvatore 969
Fiore, Ugo 511
Frattolillo, Franco 401
Frenz, Stefan 465
Fujii, Teruko 723
Funika, Wlodzimierz 666
Fürlinger, Karl 595

Galily, Mehdi   87
Gao, Chunmei   856
García, José M.   213
Gasperini, Fabiano   969
Gebremedhin, Assefaw H.   796
Ge, JiDong   378
Gemelli, Riccardo   652
Gerndt, Michael   595
Getov, Vladimir   1058
Giannelli, Carlo   1021
Giudici, Francesco   77
Goeckelmann, Ralph   465
Gong, Zheng-hu   265, 321
Grieco, Raffaella   1033
Guitart, Jordi   366
Guo, Minyi   409, 455

Habibipour, Farzad   87
Ha, JaeCheol   541, 549
Han, Zongfen   419
Hong, Chun Pyo   524, 560
Hong, Jinkeun   555
Ho, Pin-Han   196
Horiguchi, Susumu   196
Horiuchi, Eiichi   723
Huang, Ching-Wei   440
Huang, Jiun-Long   878
Huang, Min   111
Huang, Zunguo   265
Huedo, Eduardo   499
Hu, HaiYang   378
Hung, Chih-Chieh   878

Ideguchi, Tetsuo   723
Isaiadis, Stavros   1058
Izaiku, Takato   409

Jang, Sang Hoon   348
Jang, Yeong Min   348
Jang, Yoonchul   340
Jan, Mathieu   429
Jiang, Xiaohong   196
Jin, Hai   419
Joo, Seong-Soon   4
Jung, Ilhyung   24
Jun, Kyungkoo   315, 360, 1106

Kacsuk, Péter   684
Kahani, Mohsen   1043
Kaneko, Masashi   979

Kang, Chung Gu   305
Kao, Odej   505
Karl, Wolfgang   694
Katou, Ken'ichi   979
Kim, Backhyun   354
Kim, Byunggi   340
Kim, Chang Hoon   524, 560
Kim, ChangKyun   541, 549
Kim, Dong Hoi   305, 315, 360, 1106
Kim, Gi-Hong   1087
Kim, HaeSook   93
Kim, Howon   567
Kim, Hyuncheol   660, 1096
Kim, Iksoo   354
Kim, Junghwan   477
Kim, KapDong   93
Kim, Kihong   555
Kim, Kwanjoong   340
Kim, Moonseong   4, 117, 277
Kim, Myoung Ho   190
Kim, Pyung Soo   14, 535
Kim, Sung-Hyun   549
Kim, Sun Ja   190
Kim, Young-Chang   245, 898
Kim, Young Man   283
Koch, Marcin   666
Ko, Myeong-Cheol   477
Kozankiewicz, Hanna   904
Kranzlmüller, Dieter   2
Kwon, Soonhak   524, 560

Lancellotti, Riccardo   1070
Landfeldt, Björn   33
Lauria, M.   180
Lee, Byunggil   567
Lee, Byung kwan   331
Lee, Dong Chun   1077
Lee, Eung Hyuk   14, 535
Lee, Hak Goo   14
Lee, Hee-Bong   348
Lee, Hyo Keun   24
Lee, Jaehwoon   255
Lee, Jeong-Ook   477
Lee, Jong-Hyouk   991
Lee, Sang-Il   57
Lee, Tae-Seok   293, 1049
Lee, Tai-Chi   331
Lee, Yong-Jin   99
Lee, Young Choon   203
León, C.   717

Liao, Xiaofei   419
Li, Jia   111
Lim, Yujin   255
Lin, Chuang   388
Lin, F.   776
Lin, Woei   166
Li, Yiming   829, 910
Llorente, Ignacio M.   499
Lobosco, Marcelo   629
Loques, Orlando   629
Lu, Jian   378
Lu, Xi-Cheng   321

Magaldi, Massimo   395
Malandrino, Delfina   1033
Malony, Allen D.   617
Mancini, Emilio P.   143
Manne, Fredrik   796
Mastroianni, Carlo   672
Mazzeo, Antonino   1064
Mazzocca, Nicola   395
Mazzoni, Francesca   1033
Migliaccio, Armando   1011
Miller, Barton P.   573
Min, Geyong   67
Miranda, G.   717
Moe, Randi   768
Montagna, Sergio   652
Montero, Rubén S.   499
Moon, Hyun-Jin   807
Moon, SangJae   541, 549
Morris, Alan   617
Mosca, Paola   395
Moscato, Francesco   395

Naeini, Maryam Moslemi   156
Naghibzadeh, Mahmood   1043
Nakhleh, L.   776
Nam, In Gil   560
Nikolopoulos, Dimitrios S.   223
Noblet, David A.   429

Oie, Yuji   817
Oono, Takamasa   409
Orozco-Barbosa, Luis   605
Özgüner, Füsun   796

Pagani, Elena   77
Palmieri, Francesco   511
Paolillo, G.   180

Park, HeaSook   93
Park, Hyo Sun   1087
Park, IlHwan   541
Park, Jong Sou   567
Park, Kiejin   567
Park, Myong-Soon   293, 1049
Park, Sangjoon   340
Park, Seongbin   123
Park, Wongil   340
Peden, Jeffery   755
Pei, Changxing   105
Peng, Wei   321
Peng, Wen-Chih   878
Pignolo, Maurizio   652
Piquer, José M.   133
Plassmann, Paul E.   928, 948
Premadasa, Kaushalya   33
Prodan, Radu   704
Purvis, Martin   733

Quaing, Boris   694
Quan, Dang Minh   505
Quan, Dongxiao   105

Rak, Massimiliano   143
Rau-Chaplin, Andrew   856
Ren, Fengyuan   388
Rho, Jungkyu   477
Ripke, Andreas   45
Romano, Luigi   1064
Roshan, U.   776
Russo, Stefano   180, 1011
Ruz, Cristian   133

Sadeg, Bruno   888
Sahin, Cihan   745
Saitoh, Yoshihiro   409
Santana, Marcos José   487
Santana, Regina Helena Carlucci   487
Sarbazi-Azad, Hamid   156
Sarker, Biplab Kumer   845
Scarano, Vittorio   1033
Schneider, Scott   223
Schoettner, Michael   465
Schulthess, Peter   465
Sedukhin, Stanislav   786
Senger, Luciano José   487
Seno, Shoichiro   723
Shan, Zhiguang   388
Shende, Sameer   617

Shin, Ki-Jeong 293
Sipos, Gergely 684
Smetek, Marcin 666
Sohn, Hong-Gyoo 1087
Song, Hyewon 57
Sørevik, Tor 768
Stamatakis, A. 776
Stencel, Krzysztof 904
Subieta, Kazimierz 904
Sumitomo, Kenichi 409
Sunderam, Vaidy 1
Sun, Weitao 839

Tabirca, Sabin 233
Tabirca, Tatiana 233
Takahashi, Akihito 786
Tak, Byung Chul 190
Talia, Domenico 672
Tanabe, Motofumi 723
Tang, Chi-Feng 166
Tao, Jie 694
Tao, XianPing 378
Thandavan, Ashish 745
Tomarchio, Orazio 1000
Torella, Roberto 143
Torres, Jordi 366
Träff, Jesper Larsson 45
Tsaur, Ding-Jyh 166
Tsujita, Yuichi 585
Tsuru, Masato 817
Tu, Xuping 419

Uehara, Kuniaki 845

Veljkovic, Ivana 928, 948
Villa, Francisco J. 213

Villalón, José 605
Villano, Umberto 143
Violi, Angela 938
Vittorini, Valeria 395
Voth, Gregory A. 938
Vuduc, Richard W. 807

Wang, Hui 409
Wang, Lan 67
Wang, Xingwei 111
Wei, Yaya 388
Werstein, Paul 733
Wismüller, Roland 666
Wolf, Felix 617
Wu, Chin-Chi 166

Yagyu, Kazuhiko 409
Yamada, Takako 979
Yamamoto, Hiroshi 817
Yang, Laurence Tianruo 233, 487, 845
Yang, Seung Hae 331
Yang, Wuu 440
Yang, Yuan 293, 1049
Yazdian, Ali 87
Ye, Baoliu 455
Yen, Sung-Ming 549
Yi, Yunhui 105
Yoo, HyungSo 541
Youn, Chan-Hyun 57
Youn, Cheong 93
Yu, Byung-Hyun 733
Yu, Heon-Chang 123
Yun, Kong-Hyun 1087

Zhu, Changhua 105
Zomaya, Albert Y. 3, 203

Lecture Notes in Computer Science

For information about Vols. 1–3618

please contact your bookseller or Springer

Vol. 3728: V. Paliouras, J. Vounckx, D. Verkest (Eds.), Integrated Circuit and System Design. XV, 753 pages. 2005.

Vol. 3726: L.T. Yang, O.F. Rana, B. Di Martino, J. Dongarra (Eds.), High Performance Computing and Communcations. XXVI, 1116 pages. 2005.

Vol. 3718: V.G. Ganzha, E.W. Mayr, E.V. Vorozhtsov (Eds.), Computer Algebra in Scientific Computing. XII, 502 pages. 2005.

Vol. 3715: E. Dawson, S. Vaudenay (Eds.), Progress in Cryptology – Mycrypt 2005. XI, 329 pages. 2005.

Vol. 3714: H. Obbink, K. Pohl (Eds.), Software Product Lines. XIII, 235 pages. 2005.

Vol. 3713: L. Briand, C. Williams (Eds.), Model Driven Engineering Languages and Systems. XV, 722 pages. 2005.

Vol. 3712: R. Reussner, J. Mayer, J.A. Stafford, S. Overhage, S. Becker, P.J. Schroeder (Eds.), Quality of Software Architectures and Software Quality. XIII, 289 pages. 2005.

Vol. 3711: F. Kishino, Y. Kitamura, H. Kato, N. Nagata (Eds.), Entertainment Computing - ICEC 2005. XXIV, 540 pages. 2005.

Vol. 3710: M. Barni, I. Cox, T. Kalker, H.J. Kim (Eds.), Digital Watermarking. XII, 485 pages. 2005.

Vol. 3708: J. Blanc-Talon, W. Philips, D. Popescu, P. Scheunders (Eds.), Advanced Concepts for Intelligent Vision Systems. XXII, 725 pages. 2005.

Vol. 3706: H. Fuks, S. Lukosch, A.C. Salgado (Eds.), Groupware: Design, Implementation, and Use. XII, 378 pages. 2005.

Vol. 3703: F. Fages, S. Soliman (Eds.), Principles and Practice of Semantic Web Reasoning. VIII, 163 pages. 2005.

Vol. 3702: B. Beckert (Ed.), Automated Reasoning with Analytic Tableaux and Related Methods. XIII, 343 pages. 2005. (Subseries LNAI).

Vol. 3699: C.S. Calude, M.J. Dinneen, G. Paun, M.J. Pérez-Jiménez, G. Rozenberg (Eds.), Unconventional Computation. XI, 267 pages. 2005.

Vol. 3698: U. Furbach (Ed.), KI 2005: Advances in Artificial Intelligence. XIII, 409 pages. 2005. (Subseries LNAI).

Vol. 3697: W. Duch, J. Kacprzyk, E. Oja, S. Zadrożny (Eds.), Artificial Neural Networks: Formal Models and Their Applications – ICANN 2005, Part II. XXXII, 1045 pages. 2005.

Vol. 3696: W. Duch, J. Kacprzyk, E. Oja, S. Zadrożny (Eds.), Artificial Neural Networks: Biological Inspirations – ICANN 2005, Part I. XXXI, 703 pages. 2005.

Vol. 3695: M.R. Berthold, R. Glen, K. Diederichs, O. Kohlbacher, I. Fischer (Eds.), Computational Life Sciences. XI, 277 pages. 2005. (Subseries LNBI).

Vol. 3694: M. Malek, E. Nett, N. Suri (Eds.), Service Availability. VIII, 213 pages. 2005.

Vol. 3693: A.G. Cohn, D.M. Mark (Eds.), Spatial Information Theory. XII, 493 pages. 2005.

Vol. 3691: A. Gagalowicz, W. Philips (Eds.), Computer Analysis of Images and Patterns. XIX, 865 pages. 2005.

Vol. 3690: M. Pěchouček, P. Petta, L.Z. Varga (Eds.), Multi-Agent Systems and Applications IV. XVII, 667 pages. 2005. (Subseries LNAI).

Vol. 3687: S. Singh, M. Singh, C. Apte, P. Perner (Eds.), Pattern Recognition and Image Analysis, Part II. XXV, 809 pages. 2005.

Vol. 3686: S. Singh, M. Singh, C. Apte, P. Perner (Eds.), Pattern Recognition and Data Mining, Part I. XXVI, 689 pages. 2005.

Vol. 3685: V. Gorodetsky, I. Kotenko, V. Skormin (Eds.), Computer Network Security. XIV, 480 pages. 2005.

Vol. 3684: R. Khosla, R.J. Howlett, L.C. Jain (Eds.), Knowledge-Based Intelligent Information and Engineering Systems, Part IV. LXXIX, 933 pages. 2005. (Subseries LNAI).

Vol. 3683: R. Khosla, R.J. Howlett, L.C. Jain (Eds.), Knowledge-Based Intelligent Information and Engineering Systems, Part III. LXXX, 1397 pages. 2005. (Subseries LNAI).

Vol. 3682: R. Khosla, R.J. Howlett, L.C. Jain (Eds.), Knowledge-Based Intelligent Information and Engineering Systems, Part II. LXXIX, 1371 pages. 2005. (Subseries LNAI).

Vol. 3681: R. Khosla, R.J. Howlett, L.C. Jain (Eds.), Knowledge-Based Intelligent Information and Engineering Systems, Part I. LXXX, 1319 pages. 2005. (Subseries LNAI).

Vol. 3679: S.d.C. di Vimercati, P. Syverson, D. Gollmann (Eds.), Computer Security – ESORICS 2005. XI, 509 pages. 2005.

Vol. 3678: A. McLysaght, D.H. Huson (Eds.), Comparative Genomics. VIII, 167 pages. 2005. (Subseries LNBI).

Vol. 3677: J. Dittmann, S. Katzenbeisser, A. Uhl (Eds.), Communications and Multimedia Security. XIII, 360 pages. 2005.

Vol. 3675: Y. Luo (Ed.), Cooperative Design, Visualization, and Engineering. XI, 264 pages. 2005.

Vol. 3674: W. Jonker, M. Petković (Eds.), Secure Data Management. X, 241 pages. 2005.

Vol. 3672: C. Hankin, I. Siveroni (Eds.), Static Analysis. X, 369 pages. 2005.

Vol. 3671: S. Bressan, S. Ceri, E. Hunt, Z.G. Ives, Z. Bellahsène, M. Rys, R. Unland (Eds.), Database and XML Technologies. X, 239 pages. 2005.

Vol. 3670: M. Bravetti, L. Kloul, G. Zavattaro (Eds.), Formal Techniques for Computer Systems and Business Processes. XIII, 349 pages. 2005.

Vol. 3666: B.D. Martino, D. Kranzlmüller, J. Dongarra (Eds.), Recent Advances in Parallel Virtual Machine and Message Passing Interface. XVII, 546 pages. 2005.

Vol. 3665: K. S. Candan, A. Celentano (Eds.), Advances in Multimedia Information Systems. X, 221 pages. 2005.

Vol. 3664: C. Türker, M. Agosti, H.-J. Schek (Eds.), Peer-to-Peer, Grid, and Service-Orientation in Digital Library Architectures. X, 261 pages. 2005.

Vol. 3663: W.G. Kropatsch, R. Sablatnig, A. Hanbury (Eds.), Pattern Recognition. XIV, 512 pages. 2005.

Vol. 3662: C. Baral, G. Greco, N. Leone, G. Terracina (Eds.), Logic Programming and Nonmonotonic Reasoning. XIII, 454 pages. 2005. (Subseries LNAI).

Vol. 3661: T. Panayiotopoulos, J. Gratch, R. Aylett, D. Ballin, P. Olivier, T. Rist (Eds.), Intelligent Virtual Agents. XIII, 506 pages. 2005. (Subseries LNAI).

Vol. 3660: M. Beigl, S. Intille, J. Rekimoto, H. Tokuda (Eds.), UbiComp 2005: Ubiquitous Computing. XVII, 394 pages. 2005.

Vol. 3659: J.R. Rao, B. Sunar (Eds.), Cryptographic Hardware and Embedded Systems – CHES 2005. XIV, 458 pages. 2005.

Vol. 3658: V. Matoušek, P. Mautner, T. Pavelka (Eds.), Text, Speech and Dialogue. XV, 460 pages. 2005. (Subseries LNAI).

Vol. 3657: F.S. de Boer, M.M. Bonsangue, S. Graf, W.-P. de Roever (Eds.), Formal Methods for Components and Objects. VIII, 325 pages. 2005.

Vol. 3656: M. Kamel, A. Campilho (Eds.), Image Analysis and Recognition. XXIV, 1279 pages. 2005.

Vol. 3655: A. Aldini, R. Gorrieri, F. Martinelli (Eds.), Foundations of Security Analysis and Design III. VII, 273 pages. 2005.

Vol. 3654: S. Jajodia, D. Wijesekera (Eds.), Data and Applications Security XIX. X, 353 pages. 2005.

Vol. 3653: M. Abadi, L. de Alfaro (Eds.), CONCUR 2005 – Concurrency Theory. XIV, 578 pages. 2005.

Vol. 3652: A. Rauber, S. Christodoulakis, A M. Tjoa (Eds.), Research and Advanced Technology for Digital Libraries. XVIII, 545 pages. 2005.

Vol. 3650: J. Zhou, J. Lopez, R.H. Deng, F. Bao (Eds.), Information Security. XII, 516 pages. 2005.

Vol. 3649: W.M. P. van der Aalst, B. Benatallah, F. Casati, F. Curbera (Eds.), Business Process Management. XII, 472 pages. 2005.

Vol. 3648: J.C. Cunha, P.D. Medeiros (Eds.), Euro-Par 2005 Parallel Processing. XXXVI, 1299 pages. 2005.

Vol. 3646: A. F. Famili, J.N. Kok, J.M. Peña, A. Siebes, A. Feelders (Eds.), Advances in Intelligent Data Analysis VI. XIV, 522 pages. 2005.

Vol. 3645: D.-S. Huang, X.-P. Zhang, G.-B. Huang (Eds.), Advances in Intelligent Computing, Part II. XIII, 1010 pages. 2005.

Vol. 3644: D.-S. Huang, X.-P. Zhang, G.-B. Huang (Eds.), Advances in Intelligent Computing, Part I. XXVII, 1101 pages. 2005.

Vol. 3642: D. Ślezak, J. Yao, J.F. Peters, W. Ziarko, X. Hu (Eds.), Rough Sets, Fuzzy Sets, Data Mining, and Granular Computing, Part II. XXIII, 738 pages. 2005. (Subseries LNAI).

Vol. 3641: D. Ślezak, G. Wang, M. Szczuka, I. Düntsch, Y. Yao (Eds.), Rough Sets, Fuzzy Sets, Data Mining, and Granular Computing, Part I. XXIV, 742 pages. 2005. (Subseries LNAI).

Vol. 3639: P. Godefroid (Ed.), Model Checking Software. XI, 289 pages. 2005.

Vol. 3638: A. Butz, B. Fisher, A. Krüger, P. Olivier (Eds.), Smart Graphics. XI, 269 pages. 2005.

Vol. 3637: J. M. Moreno, J. Madrenas, J. Cosp (Eds.), Evolvable Systems: From Biology to Hardware. XI, 227 pages. 2005.

Vol. 3636: M.J. Blesa, C. Blum, A. Roli, M. Sampels (Eds.), Hybrid Metaheuristics. XII, 155 pages. 2005.

Vol. 3634: L. Ong (Ed.), Computer Science Logic. XI, 567 pages. 2005.

Vol. 3633: C. Bauzer Medeiros, M. Egenhofer, E. Bertino (Eds.), Advances in Spatial and Temporal Databases. XIII, 433 pages. 2005.

Vol. 3632: R. Nieuwenhuis (Ed.), Automated Deduction – CADE-20. XIII, 459 pages. 2005. (Subseries LNAI).

Vol. 3631: J. Eder, H.-M. Haav, A. Kalja, J. Penjam (Eds.), Advances in Databases and Information Systems. XIII, 393 pages. 2005.

Vol. 3630: M.S. Capcarrere, A.A. Freitas, P.J. Bentley, C.G. Johnson, J. Timmis (Eds.), Advances in Artificial Life. XIX, 949 pages. 2005. (Subseries LNAI).

Vol. 3629: J.L. Fiadeiro, N. Harman, M. Roggenbach, J. Rutten (Eds.), Algebra and Coalgebra in Computer Science. XI, 457 pages. 2005.

Vol. 3628: T. Gschwind, U. Aßmann, O. Nierstrasz (Eds.), Software Composition. X, 199 pages. 2005.

Vol. 3627: C. Jacob, M.L. Pilat, P.J. Bentley, J. Timmis (Eds.), Artificial Immune Systems. XII, 500 pages. 2005.

Vol. 3626: B. Ganter, G. Stumme, R. Wille (Eds.), Formal Concept Analysis. X, 349 pages. 2005. (Subseries LNAI).

Vol. 3625: S. Kramer, B. Pfahringer (Eds.), Inductive Logic Programming. XIII, 427 pages. 2005. (Subseries LNAI).

Vol. 3624: C. Chekuri, K. Jansen, J.D. P. Rolim, L. Trevisan (Eds.), Approximation, Randomization and Combinatorial Optimization. XI, 495 pages. 2005.

Vol. 3623: M. Liśkiewicz, R. Reischuk (Eds.), Fundamentals of Computation Theory. XV, 576 pages. 2005.

Vol. 3622: V. Vene, T. Uustalu (Eds.), Advanced Functional Programming. IX, 359 pages. 2005.

Vol. 3621: V. Shoup (Ed.), Advances in Cryptology – CRYPTO 2005. XI, 568 pages. 2005.

Vol. 3620: H. Muñoz-Avila, F. Ricci (Eds.), Case-Based Reasoning Research and Development. XV, 654 pages. 2005. (Subseries LNAI).

Vol. 3619: X. Lu, W. Zhao (Eds.), Networking and Mobile Computing. XXIV, 1299 pages. 2005.